ANIMMA 2025 - Advancements in Nuclear Instrumentation Measurement Methods and their Applications

EPJ Web of Conferences Volume 338 (2025)

Valencia, Spain
9-13 June 2025

ISBN: 979-8-3313-3019-4

Printed from e-media with permission by:

Curran Associates, Inc.
57 Morehouse Lane
Red Hook, NY 12571

CURRAN ASSOCIATES INC. proceedings.com

Some format issues inherent in the e-media version may also appear in this print version.

For additional information, please contact EDP Sciences – Web of Conferences at the address below.

EDP Sciences – Web of Conferences
17, Avenue du Hoggar
Parc d'Activité de Courtabœuf
BP 112
F-91944 Les Ulis Cedex A
France

Phone: +33 (0) 1 69 18 75 75
Fax: +33 (0) 1 69 28 84 91

contact-edps@webofconferences.org

Additional copies of this publication are available from:

Curran Associates, Inc.
57 Morehouse Lane
Red Hook, NY 12571 USA
Phone: 845-758-0400
Fax: 845-758-2633
Email: curran@proceedings.com
Web: www.proceedings.com

TABLE OF CONTENTS

FUNDAMENTAL PHYSICS

SPACE SCIENCES AND TECHNOLOGY

FUSION DIAGNOSTICS AND TECHNOLOGY

RESEARCH REACTORS AND PARTICLE ACCELERATORS

NUCLEAR POWER REACTORS AND NUCLEAR FUEL CYCLE

NUCLEAR SAFEGUARDS, HOMELAND SECURITY AND CBRN

DECOMMISSIONING, DISMANTLING AND REMOTE HANDLING

ENVIRONMENTAL AND MEDICAL SCIENCES

CURRENT TRENDS IN DEVELOPMENT RADIATION DETECTORS

EDUCATION, TRAINING AND OUTREACH

EPJ Web of Conferences **338**, 00001 (2025) https://doi.org/10.1051/epjconf/202533800001
ANIMMA 2025

Dear Colleagues and Friends, dear ANIMMA attendees,

Nowadays nuclear measurement techniques and associated instrumentation are still in progress, and, within high-energy physics experiments, still bring a key contribution to fundamental research and to our knowledge of the Universe and its fundamental laws, the "physics of the two infinites". They also enable us to control nuclear reactors, to monitor the dosimetry, to contribute in public health in the field of diagnosis and therapy, to ensure some controls and characterization related to transport safety, radioactive wastes and dismantling operations, to check the compliance of components and industrial processes, to monitor pollution, to cope with nuclear proliferation… Not less than that, so to say!

I think we are all well aware of the fact that, over the past few decades, we have made tremendous progress in the development of measuring devices, data processing and analysis tools in many areas of interest such as computing (data mining, Artificial Intelligence, Algorithmic, Machine Learning and Deep Learning), robotics, space, health technologies, energy, environment, and transport. These advances unavoidably affect the requested performances of predictive modelling tools as well as experimental capabilities linked to our fields of activities. In this respect, ANIMMA scientific community has both the legitimacy and the leadership to face these challenges, and continuously develop a relevant international network.

As such, ANIMMA conference and the associated network is a seedbed of new scientific international collaborations dealing with experimental sciences and technologies.

The ANIMMA Conference has always strived to create a special meeting place shared by all those working in nuclear instrumentation and measurement and their applications, as we strongly believe that cross-border exchanges between scientists, engineers and industrialists can only lead to the most developed ideas, the best solutions and the most efficient collaborations and partnerships. Therefore, ANIMMA continues to maintain a high level of scientific and technical quality by presenting not only the latest advances but also the state of the art in each field through the participation of international specialists and experts.

ANIMMA-2025 is the ninth edition of a series of conferences with a scientific program dealing with:

- 3 full days dedicated to ANIMMA International Summer School (AISS) from June 7 to 9
- A full-day dedicated to 4 international workshops on Monday June 9
- 3 prestigious keynote lectures on Tuesday June 10 during the opening ceremony
- 3 relevant plenary talks on Wednesday 11th, Thursday 12th and Friday 13th at the start of each day followed by parallel oral sessions
- Intensive oral sessions to promote posters on Tuesday 10th and Wednesday 11th afternoon followed by plenary poster session
- Permanent poster and industrial exhibition from Tuesday 10th to Friday 13th

More than 280 scientific presentations were given during the conference.

ANIMMA-2025 welcomed 356 attendees from 27 countries including 80 graduate students. We awarded 2 best student papers and 3 best posters.

The attendees deserve our warmest thanks and consideration for making this eighth edition a great success in the wonderful city of Valencia, Spain.

All the committee members are looking forward to welcoming everyone again at ANIMMA2027 in Prague, Czech Republic.

Prof. Dr. Abdallah LYOUSSI

ANIMMA General Chairman and Program Committee Chair

edp sciences

EPJ Web of Conferences

Statement of Peer review

In submitting conference proceedings to EPJ Web of Conferences, I certify to the Publisher that I adhere to the **Policy on Publishing Integrity** of the journal in order to safeguard good scientific practice in publishing.

1. All articles have been subjected to peer review administered by the proceedings editors.
2. Reviews have been conducted by expert referees, who have been requested to provide unbiased and constructive comments aimed, whenever possible, at improving the work.
3. Proceedings editors have taken all reasonable steps to ensure the quality of the materials they publish and their decision to accept or reject a paper for publication has been based only on the merits of the work and the relevance to the journal.

Title, date and place of the conference

ANIMMA : Advancement in Nuclear Instrumentation Measurement Methods and their Applications
Valencia, Spain, June 9-13 2025

Proceedings editor(s):

Abdallah Lyoussi, Michel Carette, Erica Fanchini, Malcolm Joyce, Rastislav Hodak, Lory Laera, Patrick Ledu, Carlos Marinas, Massimo Morichi, Gabriela Llosa, Stanislav Pospisil, Christelle Reynard-Carette, Luka Snoj, Ludo Vermeeren

Date and editor's signature

October, 14th 2025

Upgrade of the ATLAS Tile Calorimeter for the High Luminosity-LHC

Antonio Cervello Duato[1*] on behalf of the ATLAS Tile Calorimeter System
[1]Institut de Física Corpuscular (CSIC-UV), València, Spain
(*) antonio.cervello@ific.uv.es

***Abstract*— The Tile Calorimeter (TileCal) is a sampling hadronic calorimeter covering the central region of the ATLAS experiment with steel as the absorber and plastic scintillators as the active medium. Scintillators are read out by wavelength shifting fibers coupled to photomultiplier tubes, positioned at the outermost part of the modules. TileCal is used for detecting hadronic particles, discriminating muons, and measuring hadronic particle energy and transverse missing momentum.**

The upcoming High-Luminosity phase of the LHC, starting in 2030, will increase the nominal instantaneous luminosity by a factor of at least five, alongside an upgraded ATLAS Trigger and Data Acquisition architecture. This upgrade necessitates a complete redesign of the readout electronics and power systems of TileCal. It improves calorimeter signal precision for the trigger system, enabling the development of more effective trigger algorithms. The TileCal upgrade program has involved extensive research and development, including test beam studies and the construction of a Demonstrator module. These proceedings outline recent progress in the development of the on- and off-detector systems, expected performance characteristics, and results from test beam campaigns featuring the latest electronics designs.

***Keywords* — ATLAS, Tile Calorimeter, HL-LHC, Readout Electronics, Test Beam, Detector Upgrade**

I. Introduction

The ATLAS Tile Calorimeter (TileCal) [1] serves as the central hadronic calorimeter of the ATLAS experiment [2], playing a key role in the reconstruction of hadronic jets, missing transverse energy, and tau leptons decaying hadronically. It also contributes with signals to the Level-1 calorimeter trigger system and assists in the identification of muons. TileCal is mechanically organized into three large cylindrical structures centered along the LHC beam axis: a central Long Barrel (LB) and two Extended Barrels (EBs), one on each side. The LB is itself split into two halves along the vertical plane to allow access and installation, effectively making the system composed of four cylindrical segments. In total, TileCal consists of 256 modules, with 64 modules in each of the four segments.

Each module is a wedge-shaped structure (see Fig. 1) composed of alternating layers of steel absorbers and plastic scintillating tiles. The tiles are oriented perpendicular to the beam axis, and scintillation light is collected via WaveLength Shifting (WLS) fibers routed radially to the PhotoMultiplier Tubes (PMTs) mounted at the module's outer radius.

Readout is performed through Super-Drawers (SDs), mechanical enclosures that house both the PMTs and the front-end electronics. In the LB modules, each module is equipped with one SD, each containing up to 45 PMTs. In the EB modules, each SD host up to 32 PMTs. The full detector contains 9852 PMTs. Most calorimeter cells are read out by two PMTs to ensure redundancy, while a small number of special cells are read out using a single PMT.

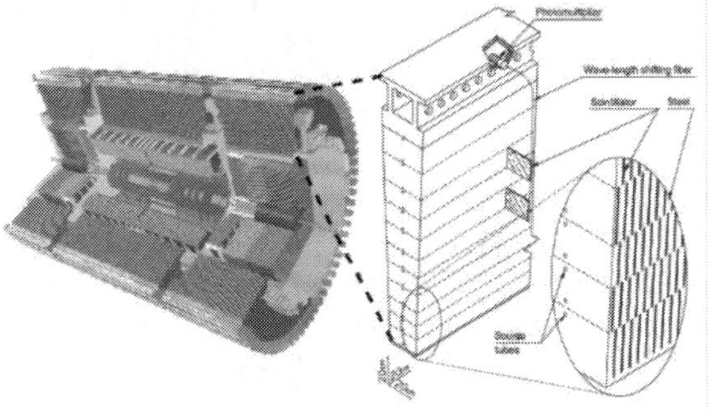

Fig. 1 TileCal sub-detector with a slice of the mechanical structure with its optical readout, showing tiles, WLS fibers, and PMTs.

With the forthcoming High Luminosity-LHC (HL-LHC) phase [3], the instantaneous luminosity is expected to increase by a factor of at least five compared to Run 2, leading to up to 200 simultaneous proton-proton collisions per bunch crossing. The resulting increase in radiation levels, trigger rates, and pile-up conditions necessitates a comprehensive upgrade of TileCal [4]. The upgraded system is designed to support a fully digital trigger architecture with enhanced granularity and improved signal precision.

II. Mechanics

Each SD is composed of several Mini-Drawers (MDs), which are mechanically and electrically independent units, as can be seen in Fig. 2. These MDs support the PMTs and associated electronics. Four MDs are combined to form a LB SD, while three MDs are used for an EB SD. The EB SDs also includes two micro-Drawers (microD) that hold PMTs in the correct position. These microDs do not contain electronics.

The MDs are constructed using aluminum and high-density fire-retardant polyethylene, materials selected for mechanical integrity and radiation resistance. Thermal management is ensured via a cooling bridge and integrated water channels

embedded in the aluminum frame, providing efficient heat dissipation for the mounted electronics.
This modular mechanical design simplifies handling, installation, and maintenance procedures.

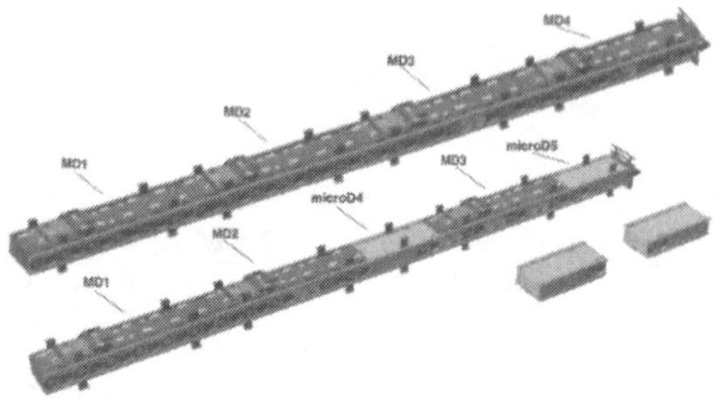

Fig. 2 SD architecture for a LB module (top) and an EB module (bottom).

III. Power distribution

A. Low-Voltage system

The Low Voltage (LV) power distribution system in the TileCal HL-LHC upgrade is designed to ensure robust and radiation tolerant delivery of power to the front-end electronics [5]. It follows a three-stage architecture.

In the first stage, 200 VDC bulk power supplies the voltage, monitoring and control of the system from the counting room.

In the second stage, low-voltage power supplies boxes (LVBox) are located on-detector and perform the conversion from the standard 200 VDC input provided from the counting room to 10 VDC. These LVBox units (Fig. 3) are based on commercial off-the-shelf components, but due to their placement inside the MDs, they are among the most radiation-exposed components. Extensive irradiation campaigns were conducted to qualify suitable radiation-hard components for use in the final design. Each LVBox hosts up to eight LVBricks, which are custom DC-DC converters responsible for performing the 200 VDC to 10 VDC conversion. Each half of the MD's electronics is powered by a dedicated LVBrick to ensure system resiliency in the event of component failure.

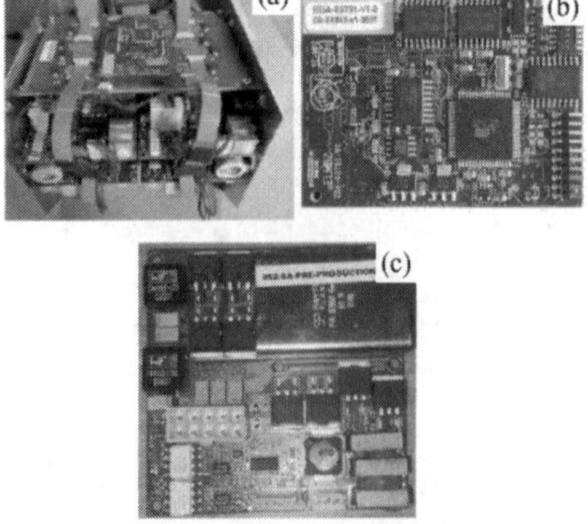

Fig. 3 LVBox (a) fully assembled with an ELMBv2 (b) and the LVBrick (c).

The monitoring of the LVBricks is handled by the Embedded Local Monitoring Board version 2 (ELMBv2) , which is housed inside the LVBox, connected to the ELMB-motherboard. The ELMBv2 communicates with the Detector Control System (DCS) via a CANbus interface, enabling real-time monitoring of voltage, current, and temperature per LVbricks. The ELMBv2 is a radiation-hard redesign of the legacy ELMB and retains functional equivalence.

To increase system robustness, monitoring and control tasks are separated. The system responsible for LVBrick—ON/OFF operations—is achieved using a tri-state voltage scheme on the ELMB-motherboard. The tristate signal transitions are provided by the Auxilary Board (located off-detector in the counting room) upon request from DCS. This system allows individual control of each MD's LVBricks.

The third and final stage of the LV distribution consists of point-of-load regulators situated directly on the on-detector electronics, installed in the MDs. These regulators provide final voltage conditioning close to the electronics, improving stability and minimizing noise.

B. High-Voltage system

The TileCal High Voltage (HV) distribution system for the HL-LHC upgrade is designed to reliably supply HV to all PMTs.

The system is composed of HVRemote (Fig. 4a) and HVSupply boards (Fig. 4b), which are installed in custom HV-crates located in the counting room. These crates are connected to the on-detector components through approximately 100 m long cables. Inside the modules, the HVBus board (Fig. 4 (c))—a passive component—is used to distribute the HV individually to each PMT located in the MDs [6]. The HVBus boards are built with four layers to protect high-voltage tracks by embedding them in the inner Printed Circuit Board (PCB) layers.

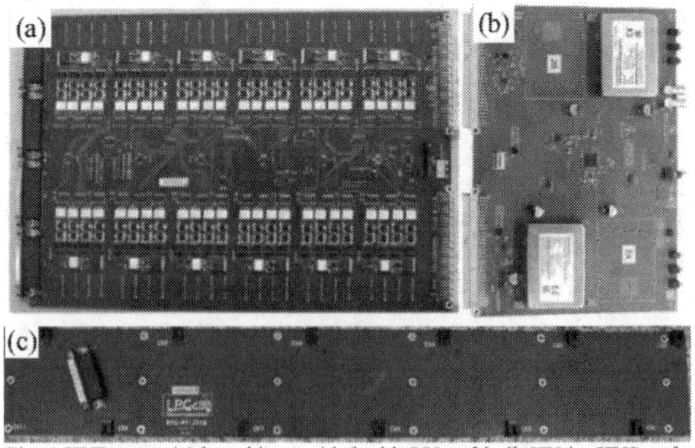

Fig. 4 HVRemote (a) board is provided with LV and bulk HV by HVSupply (b). The HVBus board is installed in the MD and distributes the individual HV for each one of the 12 PMTs.

Each HVRemote board includes independent regulation loops for 48 HV channels, with the total number of boards in the system reaching 256. The regulation loop for each channel maintains the output voltage within a precision of ±0.5 V. Compared to the legacy system, the HL-LHC HV system introduces ON/OFF control for each group of 4 channels and jumpers for enabling or disabling individual channels, thus enhancing operational flexibility.

A single input HV, either −830 V or −950 V, is provided to each group of 24 channels. The HV is generated by Hamamatsu C12446-12 modules mounted on the HVSupply boards. Two such HV modules are used to supply power up to 48 channels in a LB module or 32 channels in an EB module.

Importantly, the regulation electronics are located off-detector, where they are not exposed to radiation. This has allowed the use of a simplified and more robust regulation loop compared to the legacy system, while still meeting stringent precision requirements.

IV. Readout Electronics

The upgraded data acquisition system for the TileCal is designed to meet the high-performance demands of the new architecture for the Trigger and Data AcQuisition (TDAQ) in ATLAS for the HL-LHC. It features a fully digital trigger architecture with on-detector electronics integrated within the module mechanics, responsible for signal amplification, digitization, and data transmission. For the off-detector side, the system reconstructs energy and timing per calorimeter cell and produces trigger primitives for the trigger system using digital filtering algorithms. Fig. 5 shows the schema of the signal and data path for the TileCal system.

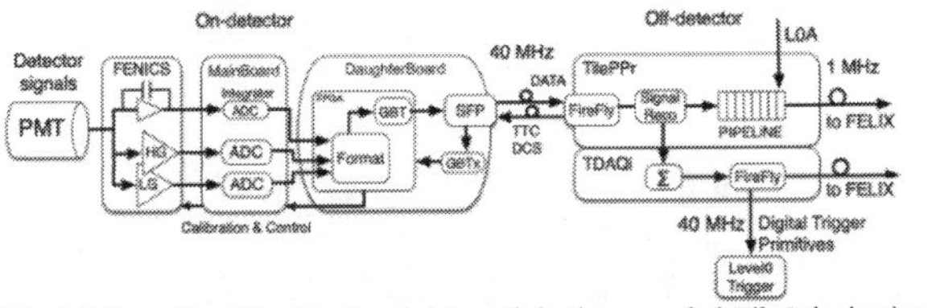

Fig. 5 Schematic of the signal and data path in the upgraded TileCal: shaping with the FENICS cards, digitization on the MainBoard, high-speed transmission via the DaughterBoard, and data processing and forwarding by the TilePPr through the FELIX and TDAQi to the DAQ and trigger systems.

A. On-detector electronics

The on-detector electronics are fully integrated within the MD mechanics. These electronics are responsible for signal amplification, digitization, control, and data transmission to the off-detector readout systems. The system is designed to meet the stringent requirements of radiation tolerance, signal integrity, synchronization with the LHC clock, and high-speed data transmission.

1) PMT Block

Each MD contains up to twelve PMT blocks, which house the critical components of the on-detector electronics chain. A PMT block (Fig. 6) consists of a light mixer, a PMT, an active high-voltage divider, and an analog front-end board known as Front-End board for the New Infrastructure with Calibration and signal Shaping (FENICS).

Maintaining energy scale precision at the level of 1% in the calorimeter requires the high voltage delivered to the PMTs to remain stable within ±0.5 V, over an operational range between 600 V and 900 V. Under HL-LHC conditions, the increased luminosity results in PMT currents reaching up to 40 μA in the most exposed regions. To maintain response linearity under these conditions, High-Voltage Active Dividers (HVAD) are employed. The function of HVAD is to divide HV to 8 PMT dynodes.

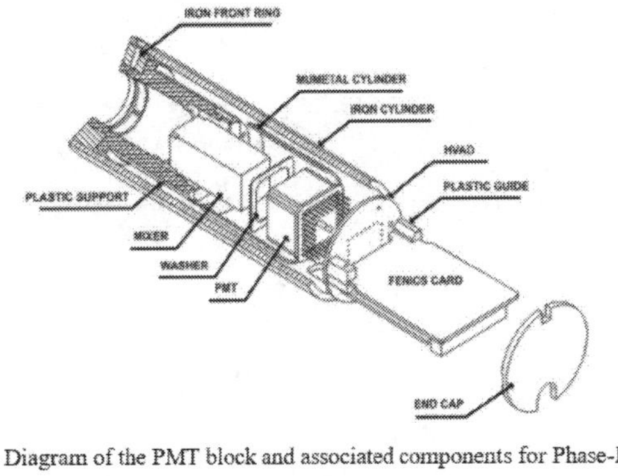

Fig. 6 Diagram of the PMT block and associated components for Phase-II.

The FENICS board (Fig. 7) implements two parallel signal paths. The first is the fast readout path, which digitizes physics signals at a frequency of 40 MHz. It operates with two gain settings in parallel to cover a dynamic range from 200 fC to 1000 pC [7]. This allows precise energy measurements for signals ranging from low-energy muons to high-energy hadronic jets reaching multi-TeV scales. The second is the integrator readout path, which continuously integrates the PMT current. This mode is used during calibration runs with a ^{137}Cs radioactive source and for monitoring the instantaneous luminosity at the ATLAS interaction point.

Fig. 7 FENICS card for signal shaping, amplification, and calibration.

In addition to signal processing, the FENICS board includes a precise Charge Injection System (CIS) that allows electronic calibration by injecting a known charge and measuring it in ADC counts.

In preparation for the HL-LHC upgrade, approximately 10% of the TileCal PMTs will be replaced. This replacement targets units that show degraded performance or non-conformities and aims to ensure that the full detector meets the stringent requirements on energy resolution, linearity, and stability over the extended HL-LHC operational period.

2) MainBoard

Each MD includes a MainBoard (Fig. 8) responsible for digitizing the signals from up to 12 PMTs. For the fast readout path, the board uses 12 dual-channel 12-bit ADCs operating at 40 MHz, providing high-resolution digitization of the physics signals [8]. For the integrator readout, it employs 12 16-bit

Successive Approximation Register ADCs to ensure precise measurement of the integrated PMT current. In addition to data acquisition, the MainBoard also manages the configuration of the FENICS boards for calibration and physics runs.

Fig. 8 Picture of the MainBoard.

Since each calorimeter cell is read by two separate PMTs, these are connected to opposite sides of the MainBoard to minimize the impact of single-point failures.

3) DaughterBoard

Each MD is equipped with a single DaughterBoard (Fig. 9), which manages high-speed communication between the on-detector and off-detector electronics [9]. Operating at data rates of 9.6 Gbps (redundant uplinks) and 4.8 Gbps (Trigger, Timing and Control (TTC) commands and the synchronized clock), the DaughterBoard transmits both precision physics data and slow control and monitoring information from the front-end electronics.

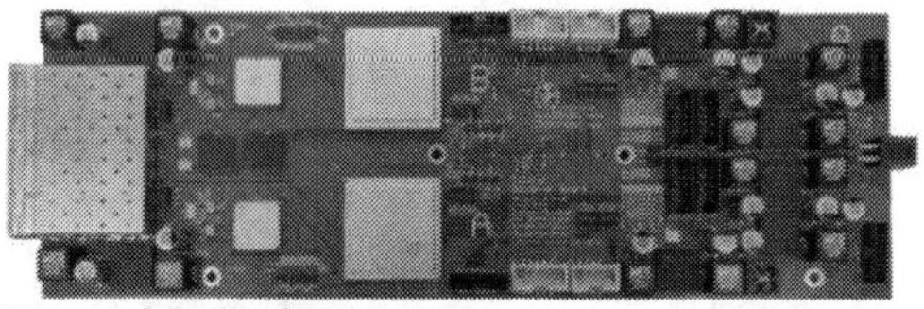

Fig. 9 Picture of the DaughterBoard.

It also receives the LHC clock and distributes it across the on-detector system, while handling configuration exchanges and control commands directed to the FENICS and Main Boards. The DaughterBoard is built around two Kintex UltraScale FPGAs, which perform the required data handling, synchronization, and control functions.

To enhance robustness and redundancy, the DaughterBoard—and its companion MainBoard—are designed with two electrically independent sides.

B. Off-detector electronics

The off-detector system performs energy and time reconstruction per PMT channel and generates trigger primitives for the ATLAS Level-0 trigger system. These primitives are computed using digital algorithms and are then sent to the trigger system, while the full data stream is temporarily stored in pipeline buffers awaiting a trigger decision.

The off-detector electronics are implemented in a modular architecture hosted within Advanced Telecommunications Computing Architecture (ATCA) crates. Each crate contains up to eight Tile PreProcessor (TilePPr) blades, each of which is subdivided into three main subsystems: the ATCA Carrier BaseBoard (ACBB), the Compact Processing Modules (CPMs), and the TDAQ interface (TDAQi) board.

1) ATCA Carrier BaseBoard

The ACBB (Fig. 10) is the core hardware platform of the TilePPr, designed to host up to four CPMs, it also integrates three additional key mezzanines: Intelligent Platform Management Controller (IPMC), Tile Computer-on-Module (TileCoM) and Tile Gigabit Ethernet (TileGbE). It provides power distribution, clock management, and high-speed communication infrastructure. The ACBB ensures reliable operation through redundant power supplies and monitoring systems, delivers synchronized timing to all modules via a centralized clock system, and supports fast data exchange over the ATCA backplane.

The TileCoM is a custom DDR4 SO-DIMM format mezzanine card designed to configure, control, and monitor both on-detector and off-detector electronics. It integrates a Zynq UltraScale+ SoC, which runs a Linux OS to manage complex tasks. TileCoM supports remote FPGA configuration over TCP/IP, interfaces with the ATLAS TDAQ system via IPbus for control and provides monitoring of around 2000 sensors, by means of OPC UA servers, to DCS.

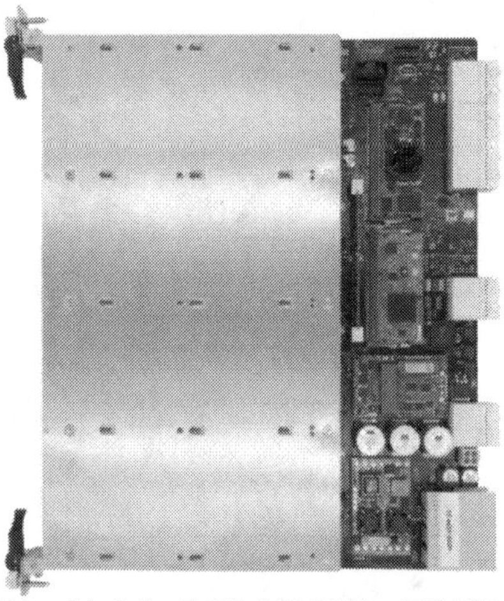

Fig. 10 ACBB board, including the TileCoM, IPMC and TileGbE switch mezzanine boards.

The TileGbE switch is a custom board based on the Broadcom BCM5396 chip, which functions as an unmanaged 16-port switch. It provides Ethernet connectivity for the TileCal off-detector electronics, including three ports per CPM (12 in total), one port for the TileCoM, one port for the TDAQi, and two ports connected to Zone 2, ensuring internal communication and system control within the ATCA crate.

2) Compact Processing Modules

The CPMs (Fig. 11) are the main processing and high-speed interface units of the off-detector electronics at the HL-LHC. A total of 128 CPMs, distributed in 32 ACBBs, are needed to handle the full calorimeter readout. Each CPM manages the data from two modules—one LB and one EB—processing up to 77 PMT channels.

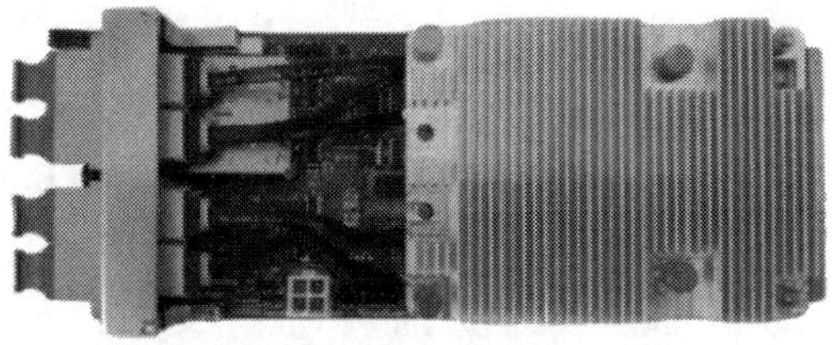

Fig. 11 Compact Processing Module board.

The CPMs receive digitized PMT signals via high-speed GigaBit Transceiver (GBT) optical links (four uplinks at 9.6 Gbps and two downlinks at 4.8 Gbps per MD) from the DaughterBoard. They perform real-time energy reconstruction and calibration per cell, providing preprocessed trigger data to the ATLAS trigger system via the TDAQi with a maximum latency of 1.7 us. In parallel, all input and processed data are buffered in on-board memory until a Level-0 Accept signal is received [10]. The selected data are then sent to the Front-End LInk eXchange (FELIX) system using the FULL Mode protocol at trigger rates up to 1 MHz. FELIX is the new common readout interface developed for the ATLAS Phase-II upgrade to meet the data throughput demands of the HL-LHC. It functions as a data router, receiving packets from front-end electronics and forwarding them over a high-bandwidth commodity network to programmable processing nodes.

Additionally, TTC information is received from FELIX over a deterministic 4.8 Gbps GBT link, ensuring synchronization for accurate processing and trigger decisions within the ATLAS TDAQ system, as well as to provide the clock distribution to the front-end electronics.

3) Trigger and Data AcQuisition interface

The TDAQi module (Fig. 12) connects the TilePPr to the ATLAS TDAQ system. Implemented as an ATCA Rear Transition Module it receives real-time calorimeter cell energy data from the CPMs and generates trigger primitives with varying granularity and precision. These primitives are sent to trigger processors, L0Calo, L0Muon, and Global systems [11]. The TDAQi features multiple high-speed optical transceivers for communication with CPMs, the FELIX system, and trigger processors.

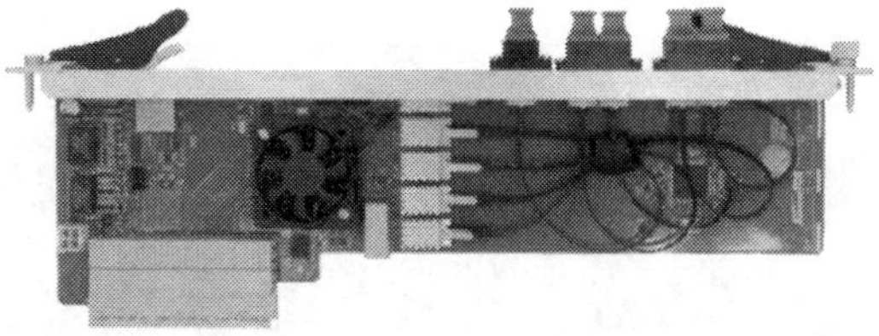

Fig. 12 TDAQi board.

V. Calibration System Upgrades

To ensure accurate energy measurements and maintain the performance of the TileCal, a comprehensive calibration system is employed, shown in Fig. 13. This system corrects for variations in the detector response by applying channel-specific calibration constants [12]. These constants account for the behavior of the front-end electronics, PMTs, and optical components, which may drift over time due to radiation and aging effects.

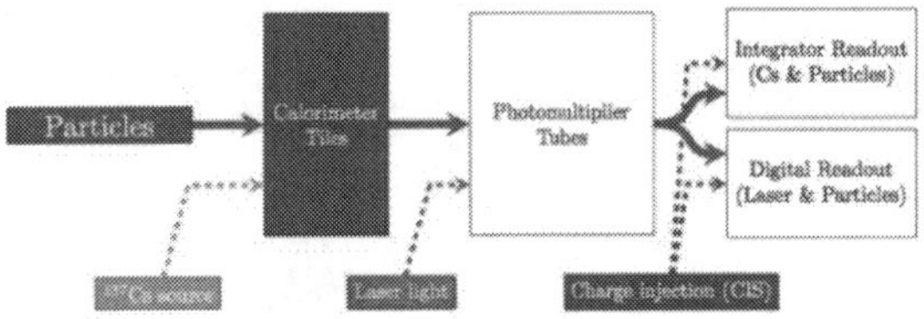

Fig. 13 Scheme of the TileCal calibration chain.

The calibration relies on multiple subsystems: the CIS calibrates the readout electronics, as explained in the PMT Block section; the Laser system monitors the PMT gain stability and the Cesium system, using a movable gamma source, provides absolute calibration of the optical and PMT response.

A. Laser system

The Interface for the Laser system to the New Acquisition infrastructure (ILANA), shown in Fig. 14, is the upgraded interface for the TileCal Laser calibration system, developed to be compatible with the legacy LASer Calibration Rod (LASCAR) system used in Run 2 (2015—2018) and Run 3 (2022—2026) [13].

Fig. 14 Crate with the ILANA board installed during tests.

ILANA manages key operations such as controlling the laser pump, filter wheel, and shutter, injecting calibration charges for the photodiodes, measuring signals from PMTs and photodiodes, and recording PMT signal timing. It communicates with the FELIX system, receiving TTC and control signals, and transmitting data at a calibration request rate.

B. Cesium system

The Cesium system of the TileCal will retain its current architectural and mechanical framework but will undergo significant enhancements to meet the demands of HL-LHC operation. The upgraded system will feature 402 Sensor based in INductance (SIN) detectors for capsule tracking, 9 Geiger counters for source detection, and 56 pressure sensors. It will operate through three independent monitoring setups, one for each calorimeter barrel, with up to 11 Cesium boxes per detector side [14].

New control hardware will be introduced, including SIN+ Embedded Monitoring and Control Interface (EMCI) cards for capsule monitoring, Garage+EMCI cards for source lock control, and ADC cards for pressure readout. Communication

 https://doi.org/10.1051/epjconf/202533801001

will transition from electrical to optical data transfer, improving efficiency and reliability. The system will be fully integrated into the DCS through the OPC UA in the Embedded Monitoring Processor, ensuring robust, centralized control and continuous calibration under HL-LHC conditions.

VI. Verification Test Benches

The TileCal upgrade involves a full replacement of both the on-detector and off-detector electronics, as well as the substitution of the most radiation-affected 10% of the PMTs. This large-scale upgrade presents a significant challenge in terms of testing, validation, and certification of the new components, as well as the refurbishment of the 9,852 PMT blocks with the new electronics.

To meet this challenge, the Portable ReadOut ModulE for Tile Electronics (PROMETEO) system, shown in Fig. 15, has been developed as a compact, versatile test bench. It is used for the certification of fully assembled PMT blocks in conjunction with the PMT Block test bench, as well as for the standalone testing of the front-end and back-end electronics.

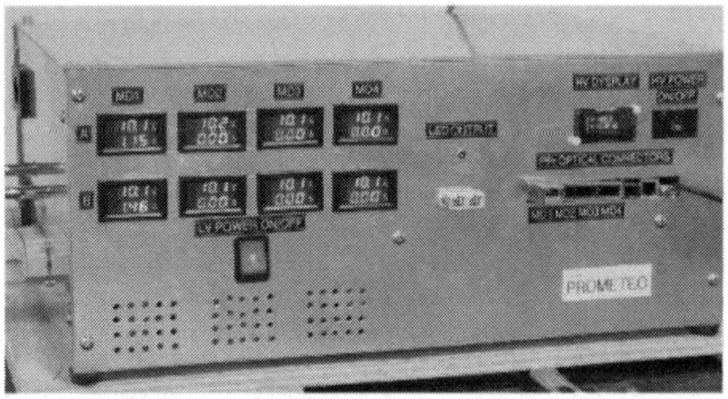

Fig. 15 PROMETEO box with all the connections for module certification.

VII. Test beam Campaigns

Several test beam campaigns were carried out at the H8 beam line of the CERN SPS North Area from 2015 to 2025. These campaigns allowed for extensive testing of TileCal modules using particle beams of various types and energies, both aligned with and perpendicular to the beam axis. Fig. 16 shows a picture of the TileCal test beam modules on the moving table.

Fig. 16 TileCal test beam modules at the H8 beam line of the CERN SPS North Area.

The main objectives were to validate the Phase-II readout and trigger architectures, assess the stability and performance of the upgraded electronics, and optimize the measurement of the electromagnetic energy scale [15]. It enabled the first complete tests of the full data acquisition chain, including high-speed data transmission to FELIX and integration with the ATLAS TDAQ system. These test beam campaigns were essential in verifying the readiness and robustness of the upgraded electronics for the HL-LHC era. Fig. 17 shows the comparison of dE/dx distributions in one of the TileCal cells using 160 GeV/c muons from data and simulation with the upgrade electronics and the FELIX readout system during a test beam campaign.

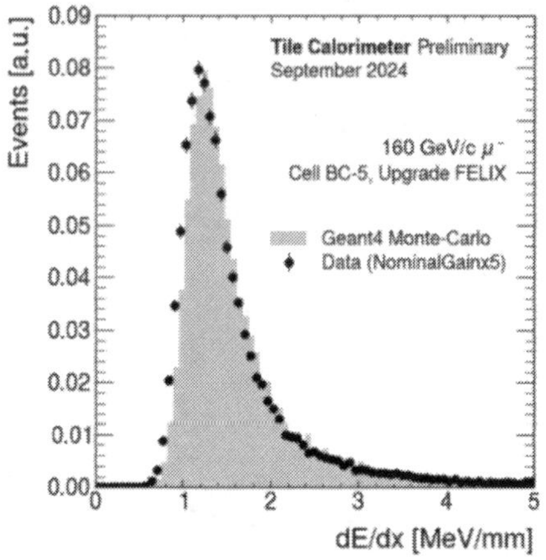

Fig. 17 Comparison of dE/dx distributions from data and simulation with upgrade electronics and the FELIX readout system.

VIII. TileCal Demonstrator Module

The TileCal Demonstrator was developed to validate the upgraded readout architecture for the HL-LHC. It integrates all major Phase-II components—MainBoard, DaughterBoard, HVbus, and upgraded 3-in-1 cards (not FENICS cards, to provide analog trigger signals)—across four MDs, holding up to 48 PMTs. Its off-detector readout is managed by an ACBB and a CPM, which distributes configuration and clock signals while receiving LHC-synchronized digitized data.

Fig. 18 Demonstrator electronics SD in theTileCal.

Inserted in the LBA14 module of ATLAS in 2019 and fully operational by 2020, the Demonstrator (Fig. 17) has continuously taken data throughout Run 3. While using updated electronics, it maintains compatibility with the legacy system by providing analog trigger signals and adapting data formats, ensuring seamless operation within the existing ATLAS infrastructure [16].

Fig. 19 shows the average CIS calibration constants for all high-gain ADCs in LBA14 across validated runs from April to November 2024 (black circles). As an example, the channel 7 is shown for comparison (blue triangles), with a ±0.7% systematic uncertainty (yellow band).

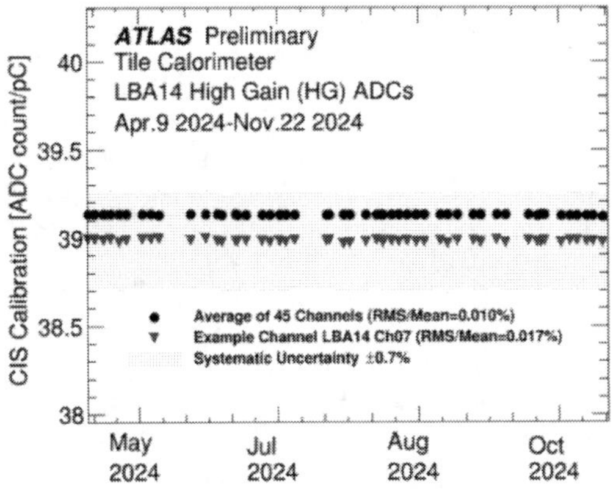

Fig. 19 Average CIS calibration constants for all high-gain ADCs in LBA14 across validated runs from April to November 2024.

IX. Conclusions

The upgrade of the TileCal for the HL-LHC represents a comprehensive redesign of both on-detector and off-detector systems to meet the demands of increased luminosity, radiation and data throughput. The new architecture enhances granularity, improves signal precision, and enables a fully digital trigger system, all while ensuring reliability and maintainability over the long-term HL-LHC operation.

The on-detector electronics have been fully integrated into a modular mechanical design, featuring radiation-tolerant components, precise calibration capabilities, and redundant readout paths to ensure robustness. The off-detector systems, implemented in ATCA architecture provides trigger primitives to the trigger systems for trigger decision, deliver real-time energy reconstruction and high-speed triggered data transmission to the ATLAS TDAQ system through the DAQ system for ATLAS, the FELIX interface.

Upgraded power distribution systems, both low- and high-voltage, now provide finer control, enhanced safety, and radiation resilience. The calibration infrastructure has been significantly enhanced with new Laser and Cesium system interfaces.

Dedicated test benches such as PROMETEO and extensive test beam campaigns have played an important role in validating the functionality and integration of the upgraded components, while the Demonstrator module has provided invaluable operational experience within the ATLAS detector during the Run 3 data taking period.

Altogether, the results obtained have demonstrated readiness for HL-LHC operation, combining technological innovation with operational reliability to maintain the calorimeter's essential role in ATLAS physics program in the coming decade.

Acknowledgment

This work was supported by the Spanish Ministry of Science and Innovation (MCIN) and the State Research Agency (AEI) under project PID2021-15069OB-100. This contribution would not have been possible without the grant for the hiring of pre-doctoral research staff (ACIF) 2021 funded by Generalitat Valenciana ACIF/2021/098.

References

[1] ATLAS Collaboration, "Operation and performance of the ATLAS Tile Calorimeter in LHC Run 2," *EPJ Conf.*, vol. 84, no. 12, Dec. 2024, doi: 10.1140/epjc/s10052-024-13151-4.

[2] ATLAS Collaboration, "The ATLAS experiment at the CERN Large Hadron Collider," *JINST*, vol. 3, S08003, 2008, doi: 10.1088/1748-0221/3/08/S08003.

[3] O. Brüning *et al.*, "High-Luminosity Large Hadron Collider (HL-LHC): Preliminary Design Report," CERN, Geneva, Switzerland. CERN-2015-005, 2015, doi: 10.5170/CERN-2015-005.

[4] ATLAS Collaboration, "Technical Design Report for the Phase-II Upgrade of the ATLAS Tile Calorimeter," CERN, Geneva, Switzerland, Tech. Rep. CERN-LHCC-2017-019, https://cds.cern.ch/record/2285583

[5] P. Starovoitov *et al.*, "Upgrade of ATLAS hadronic Tile Calorimeter for the High Luminosity LHC," *Instruments*, vol. 6, no. 4, p. 54, Aug. 2022, doi: 10.3390/instruments6040054.

[6] A. Gomes *et al.*, "Upgrade of the ATLAS Tile Calorimeter for the High Luminosity LHC," *PoS*, vol. ICHEP2024, p. 885, 2025, doi: 10.22323/1.476.0885.

[7] D. Calvet *et al.*, "Upgrade of ATLAS hadronic Tile Calorimeter for the High Luminosity LHC," *Nucl. Instrum. Methods Phys. Res., Sect. A*, vol. 1066, p. 169614, Sep. 2024, doi: 10.1016/j.nima.2024.169614.

[8] F. Tang *et al.*, "Design of Main Board for ATLAS TileCal Demonstrator," in *Proc. 19th IEEE-NPSS Real Time Conf.*, Nara, Japan, 2014, doi: 10.1109/RTC.2014.7097541.

[9] E. Valdes *et al.*, "Readiness of the ATLAS Tile Calorimeter link daughterboard for the High-Luminosity LHC era," *PoS TWEPP2019*, Apr. 2020, doi: 10.22323/1.370.0087

[10] F. Carrió *et al.*, "The PreProcessor module for the ATLAS Tile Calorimeter at the HL-LHC," *Nucl. Instrum. Methods Phys. Res., Sect. A*, vol. 958, Apr. 2020, doi: 10.1016/j.nima.2019.162487.

[11] X. Yue *et al.*, "Tile TDAQ interface module for the Phase-II Upgrade of the ATLAS Tile Calorimeter," in *Proc. IEEE Nucl. Sci. Symp. Med. Imaging Conf.*, Manchester, U.K., Oct. 2019, doi: 10.1109/NSS/MIC42101.2019.9059861.

[12] A. Cortes-Gonzalez *et al.*, "ATLAS Tile Calorimeter calibration and monitoring systems," *EPJ Web Conf.*, vol. 170, p. 01003, May 2017, doi: 10.1051/epjconf/201817001003.

[13] M. Agaras *et al.*, "Laser calibration of the ATLAS Tile Calorimeter during LHC Run 2," *J. Instrum.*, vol. 18, no. 6, P06023, Jun. 2023, doi: 10.1088/1748-0221/18/06/P06023.

[14] G. Blanchot *et al.*, "The Cesium Source Calibration and Monitoring System of the ATLAS Tile Calorimeter: Design, Construction and Results," *J. Instrum.*, vol. 15, no. 3, P03017, Feb. 2020, doi: 10.1088/1748-0221/15/03/P03017.

[15] J. Abdallah *et al.*, "Study of energy response and resolution of the ATLAS Tile Calorimeter to hadrons of energies from 16 to 30 GeV," *Eur. Phys. J. C*, vol. 81, no. 6, p. 549, Jun. 2021, doi: 10.1140/epjc/s10052-021-09292-5.

[16] F. Carrió *et al.*, "The data acquisition system for the ATLAS Tile Calorimeter Phase-II upgrade demonstrator," *IEEE Trans. Nucl. Sci.*, vol. 69, no. 4, pp. 727–734, Apr. 2022, doi: 10.1109/TNS.2022.3143233.

Results of the LEGEND-200 experiment in the search for $0\nu\beta\beta$ decay

C. Romo-Luque* on behalf of the LEGEND Collaboration
Los Alamos National Laboratory, Los Alamos, NM 87545, USA
(*) carmenromoluque@gmail.com

***Abstract*—The LEGEND experiment is looking for the extremely rare neutrinoless double beta ($0\nu\beta\beta$) decay of ^{76}Ge using isotopically-enriched high-purity germanium (HPGe) detectors. The detection of this process would imply that the neutrino is a Majorana particle and the total lepton number would not be conserved, which could be related to the cosmological asymmetry between matter and antimatter through leptogenesis. The long-term goal of the collaboration is LEGEND-1000: a 1-ton detector array planned to run for 10 years, with a projected half-life sensitivity exceeding 10^{28} years, fully covering the inverted neutrino mass hierarchy. A first search for the $0\nu\beta\beta$ decay has been carried out by LEGEND-200 building on the experience gained from GERDA and the MAJORANA DEMONSTRATOR. The experiment has been collecting physics data for a year at the Gran Sasso National Laboratory in Italy with 140 kg of HPGe detectors. With a total exposure of 61 kg yr, LEGEND-200 has achieved a background index of $5^{+3}_{-2} \times 10^{-4}$ counts/(keV kg yr) in the $0\nu\beta\beta$ decay signal region from the highest performing detectors. After combining the results from GERDA, the MAJORANA Demonstrator and LEGEND-200, an exclusion sensitivity $> 2.8 \times 10^{26}$ yr has been obtained at 90% confidence level for the $0\nu\beta\beta$ decay half-life, with no evidence for a signal. A new observed lower limit of $T^{0\nu}_{1/2} > 1.9 \times 10^{26}$ yr at 90% confidence level has been established.**

***Keywords* —** *Neutrinoless double beta decay, Majorana neutrinos, Germanium detectors, Liquid argon, Low-background experiment.*

I. INTRODUCTION

Neutrino oscillation experiments have established that neutrinos possess mass and that lepton flavor is not conserved—findings that imply physics beyond the Standard Model. At the same time, the observed imbalance between matter and antimatter in the Universe remains one of the most profound unsolved problems in cosmology. A process that could provide insight into these mysteries is neutrinoless double beta decay ($0\nu\beta\beta$). First proposed by Furry in 1939 [1], this hypothetical nuclear transition would involve the emission of two electrons without accompanying neutrinos, and could occur if neutrinos are Majorana particles—that is, if they are their own antiparticles [2]. The process would signify violation of lepton number conservation and offer a potential window into the origin of the matter-antimatter asymmetry, as well as unresolved questions about neutrino properties, such as their absolute mass and mass hierarchy.

In $0\nu\beta\beta$, the decay proceeds via the exchange of light Majorana neutrinos. Unlike standard two-neutrino double beta decay ($2\nu\beta\beta$), in which the emitted electrons share energy with neutrinos, the neutrinoless mode would result in the electrons carrying the full decay energy. This would manifest as a narrow peak in the summed electron energy spectrum, just at the endpoint of the broad spectrum produced by the standard process with neutrinos.

Current experimental limits place the half-life of $0\nu\beta\beta$ decay at greater than 10^{26} years. As a result, any experiment searching for this rare process must fulfill several stringent criteria: an extremely low background rate to maximize discovery sensitivity, excellent energy resolution to clearly identify the expected peak at the decay energy, large exposure (i.e., a high product of detector mass and measurement time) to increase the probability of observing a decay, and a high $Q_{\beta\beta}$ value to minimize interference from low-energy backgrounds.

II. GERMANIUM FOR $0\nu\beta\beta$ SEARCHES

There are 35 known isotopes that can undergo 2-neutrino double beta decay, of which 11 have been experimentally observed. Among them, ^{76}Ge—decaying to ^{76}Se—is one of the most well-suited candidates for $0\nu\beta\beta$ searches. High-purity germanium (HPGe) detectors can be fabricated using enriched ^{76}Ge, offering several key advantages for $0\nu\beta\beta$ experiments. Germanium detectors achieve the best energy resolution of any $0\nu\beta\beta$ experiment, reaching approximately 0.1% FWHM at $Q_{\beta\beta}$, which allows precise discrimination between signal and background. Additionally, they offer detailed information on the event topology, enabling improved identification of true decay events and further suppression of background noise. These detectors can be isotopically enriched beyond 90% in ^{76}Ge resulting in a very low intrinsic background. In addition, in HPGe detectors the source of the decay and the active volume are the same, which leads to high detection efficiency with minimal signal loss and do not require self-shielding. Germanium detectors have undetectably low ^{232}Th- and ^{238}U-chain internal contamination and no known background source produces a peak near $Q_{\beta\beta}$. Fig. 1 shows an HPGe inverted coaxial detector with the electric field lines.

Germanium has been studied for over 50 years, with notable efforts in the last decade led by the GERDA and MAJORANA DEMONSTRATOR (MJD) collaborations. Thanks to the complementarity of the LAr scintillation light detection and the event topology reconstruction of Ge detectors, GERDA achieved the world-leading background index of $5.2^{+1.6}_{-1.3} \times 10^{-4}$ cts/(keV kg yr) around $Q_{\beta\beta}$ and set one of the most stringent limits on the $0\nu\beta\beta$ half-life ($> 1.8 \times 10^{26}$ yr at 90% CL) [4]. MJD developed low noise electronics allowing to achieve a record energy resolution for large-scale $0\nu\beta\beta$ decay experiments of 2.52 ± 0.08 keV (0.124%) FWHM [3].

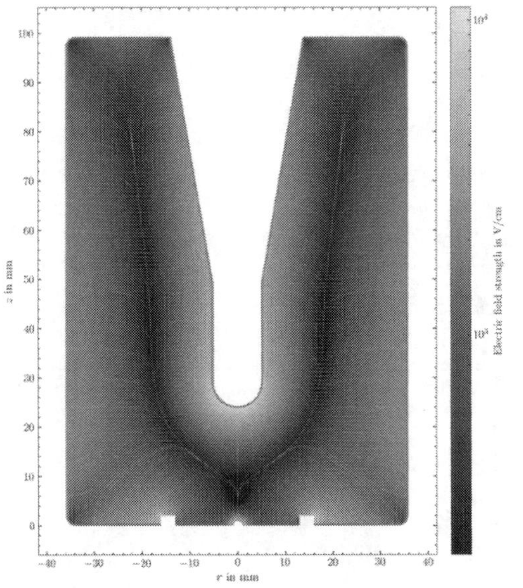

Fig. 1. Electric field lines of an inverted coaxial point contact (ICPC) detector. Contacts are shown in gray.

In HPGe detectors, particle interactions within the detector generate electron-hole pairs. The applied electric field drives electrons toward the n-type contact and holes toward the p-type contact, inducing signal in the p-type contact. The total charge is proportional to the energy of the event and the resulting charge and current drift patterns differ between signal and background events, enabling more effective background discrimination.

III. The LEGEND project

The Large Enriched Germanium Experiment for Neutrinoless $\beta\beta$ Decay (LEGEND) [5] is an international collaboration established in 2016 to unite the technical expertise and leadership from both the MJD and GERDA collaborations, while also incorporating new members to strengthen core capabilities. Today LEGEND comprises nearly 300 collaborators from 60 institutions across 12 countries.

The LEGEND project is designed to carry out a two-phase experimental campaign with the ultimate goal of $0\nu\beta\beta$ decay discovery potential for half-lives exceeding 10^{28} years. The first phase, LEGEND-200, is currently running in an upgrade of the GERDA infrastructure at the Laboratori Nazionali del Gran Sasso (LNGS) in Italy. LEGEND-200 is expected to operate with 200 kg of enriched germanium detectors immersed in liquid argon (LAr) over a period of 5 years, achieving a total exposure of 1 ton yr. The target background index is 2×10^{-4} counts/(keV kg yr) and the projected discovery sensitivity for the $0\nu\beta\beta$ half-life is around 10^{27} yr at 3σ confidence level. The second phase, LEGEND-1000, will use 1 ton of enriched HPGe detectors and will operate for 10 years at Hall C at LNGS. By conducting a search with near-zero background, it aims to cover the inverted-ordering neutrino mass scale region, with a projected $0\nu\beta\beta$ half-life discovery sensitivity $> 1.3 \times 10^{28}$ yr.

IV. LEGEND-200

LEGEND-200 [6], shown in Fig. 2, consists of a core of individual HPGe detectors enriched in ^{76}Ge up to $86\% - 92\%$ and arranged in vertical strings. The detectors are mounted on scintillating polyethylene naphthalate plates using copper rods that are made from underground-electroformed Cu, with electrical isolation provided via plastic insulators. The detector arrays are immersed in a 64 m^3 LAr bath, where the scintillation light from interacting particles in the argon is detected via two barrels of wavelength-shifting polystyrene fibers. One fiber barrel is positioned inside the circle formed by the detector strings and comprises 18 readout channels, while the outer barrel surrounds the array and contains 40 channels. The fibers are coated with TPB and coupled to silicon photomultiplier (SiPM) photodetectors, which provide single photo-electron detection capabilities. The granular nature of the Ge detector array and its immersion within an active LAr scintillating volume offers additional strong discrimination between $0\nu\beta\beta$ decay signal events, which are isolated to a single Ge detector, and background-generating events in multiple coincident detectors.

An ultra-pure water tank instrumented with 63 photomultiplier tubes (PMTs) surrounds the argon cryostat and serves as a passive shielding from muon-induced neutrons preventing them from reaching sensitive detectors and as veto for through-going muons based on Cherenkov light. The inner sufaces are covered with a reflective, wavelength-shifting film that enhances the light yield and converts the scintillation light to 400 nm, matching the sensitivity range of the PMTs.

In October 2022, 142 kg of enriched germanium detectors were installed at LNGS. The underground location of the laboratory, with approximately 3500 m water equivalent of overburden, provides effective shielding from external radiation and cosmic rays, significantly reducing backgrounds. A total of 101 Ge detectors were deployed, with four different geometries: P-type Point Contact (PPC) from MJD, Broad Energy Germanium (BEGe) plus semicoaxial (Coax) from GERDA and the novel Inverted-Coaxial Point-Contact (ICPC) specifically designed for LEGEND. Detector masses ranged from 0.7 kg in the BEGe units to up to 4 kg in some of the larger ICPC detectors.

V. LEGEND-200 results

The first physics data-taking period of LEGEND-200 started in March 2023 and concluded in February 2024, with a total exposure of 61 kg yr selected for the final $0\nu\beta\beta$ decay analysis. After a commissioning phase, a blinding strategy was implemented: all events with an energy between $\pm$ 25 keV around $Q_{\beta\beta}$ were withheld to prevent bias during analysis.

Quality cuts are used in the analysis to exclude events that do not match typical energy depositions in the HPGe array and events with energy depositions in multiple HPGe detectors (multiplicity cut) are discarded. Calibrations are performed weekly using twelve ^{228}Th sources of around 5 kBq each

Fig. 2. Cross-sectional rendering of the LEGEND-200 experiment, showing (a) the arrangement of HPGe detectors, (b) the wavelength-shifting fibers coupled to SiPMs, (c) the liquid argon cryostat, and (d) the surrounding ultra-pure water tank.

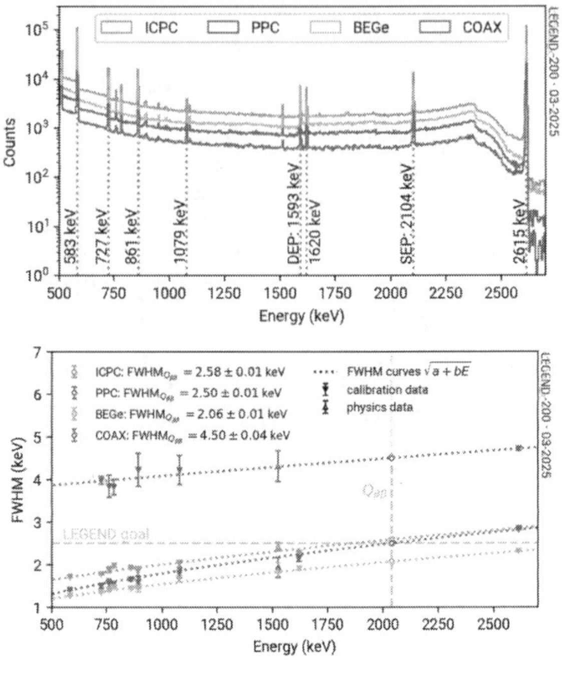

Fig. 3. Top: energy spectrum from a ^{228}Th calibration run for the four different detector types. Bottom: FWHM energy resolution curves (dotted line) derived from calibration data as a function of energy for the four different detector types. For completeness, energy resolutions from physics data at the ^{42}K line (1525 keV) are included. The empty circle indicates the extrapolated resolution at $Q_{\beta\beta}$. The LEGEND-200 energy resolution goal of 2.5 keV at $Q_{\beta\beta}$ is shown in the dashed gray line.

[7] to assess the energy scale, resolution and pulse shape discrimination (PSD) performance of the HPGe detectors. Between calibrations, physics runs are taken. Data are split into partitions, that is, different stable periods for each HPGe. Within each partition, the energy resolution, the energy bias and the PSD parameters remain without significant variations for a certain number of runs. The sensitivity to $0\nu\beta\beta$ decay critically depends on both energy scale stability and superior energy resolution. Fig. 3 shows the energy spectrum from a calibration run (top) and the energy resolution curves (bottom) for the different detector types. All detectors except the Coax (which are not planned for future configurations) meet the LEGEND requirements for energy resolution, with stable values of 0.1% FWHM at $Q_{\beta\beta}$ throughout the entire data taking.

Muon signals recorded by the PMTs in the water tank are processed offline, and events coincident with HPGe triggers within a ± 1.75 μs window are excluded from the dataset.

The gray line in Fig. 4 shows the energy spectrum of the acquired data after quality, HPGe multiplicity and muon veto cuts. Below $Q_{\beta\beta}$ the continuous distribution of $2\nu\beta\beta$ decay events dominates the spectrum. Gamma rays from the decays of ^{40}Ar, ^{238}U and ^{232}Th in structural materials, as well as from the decay of ^{42}K (progeny of ^{42}Ar) in liquid argon, contribute to the gamma lines and continuum observed below 3 MeV. At higher energies, the spectrum is dominated by energy-degraded α particles from the ^{238}U chain. In addition, β particles from the decay of ^{42}K, which has a Q_β value of 3.53 keV, can reach the detector and penetrate into the active volume. While the spectrum shows all anticipated features, comparison with radioassay-based background predictions indicates an elevated background level, including near $Q_{\beta\beta}$.

Two PSD techniques are employed to identify signal-like events based on the rising edge of the HPGe waveforms. The first is the *A/E* classifier, which measures the ratio of the maximum current amplitude (*A*) to the total energy (*E*) of the event. This method is effective for rejecting multi-site events, which typically exhibit lower *A/E* values compared to single-site events, and α-particle interactions, characterized by higher *A/E* values. The second PSD classifier, late charge (LQ), is defined as the area above the last 20% of the charge signal normalized by the energy. Events near the electrodes have high LQ values.

Scintillation light registered in the liquid argon veto in coincidence with signals from the HPGe detectors is used to suppress background. The coincident window is defined as [-1, 5] μs relative to the onset of the HPGe rising edge. Events are excluded from the analysis if the total amplitude summed across all SiPM channels exceeds four photoelectrons, or if more than four channels register signals above threshold.

The subsets of events passing the PSD cut and the combined PSD plus LAr veto cuts are shown in Fig. 4 as the gray-filled and red histograms, respectively. After PSD, there is strong suppression of events originating in the surfaces, including α events from the ^{238}U decay chain and β events from ^{42}K. Compton scatters from the 2615 keV gamma ray of ^{208}Tl remaining after PSD are effectively removed by the LAr veto filtering selection. The resulting energy spectrum after all

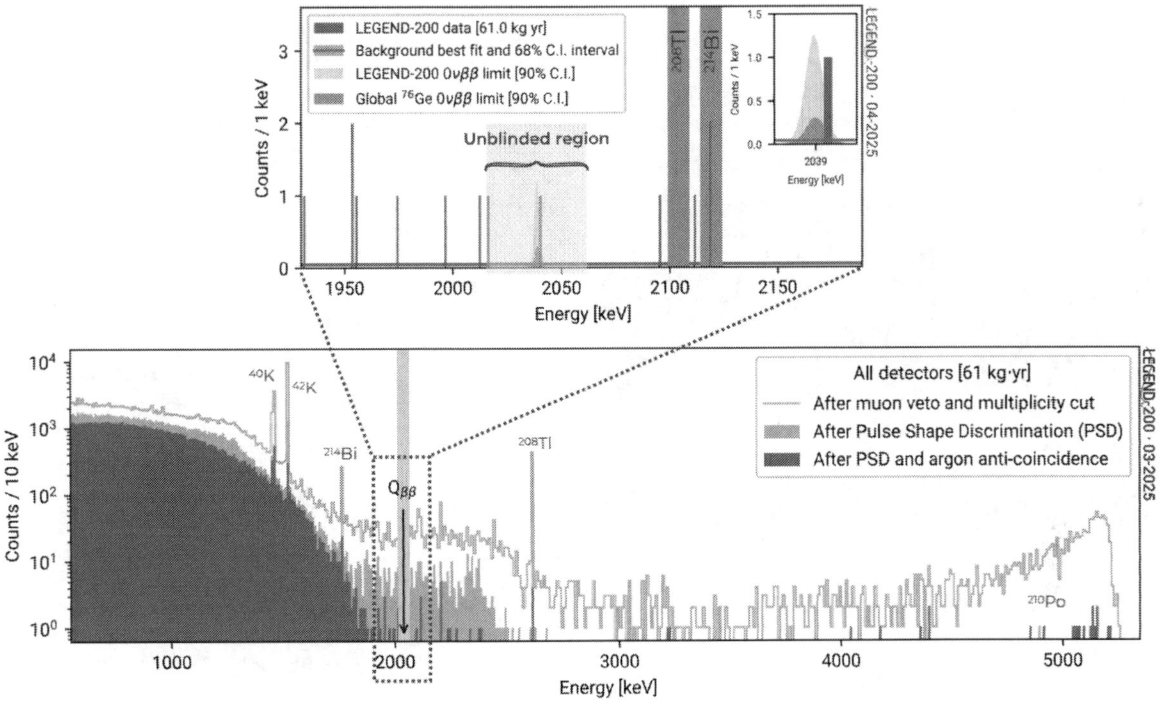

Fig. 4. Energy spectrum of LEGEND-200 from its first year of physics data collection, with a total exposure of 61 kg yr. An energy window of ± 25 keV around $Q_{\beta\beta}$ was blinded. The gray histogram represents events that passed the quality, muon veto and multiplicity cuts, with the major background sources labeled. The filled gray histogram includes events that also passed pulse shape discrimination, resulting in a significant suppression of surface-origin α and β events. The subset of events that additionally passed the anti-coincidence liquid argon (LAr) cut is shown in red, where remaining Compton events and those from the ^{232}Th decay chain are further strongly suppressed. The inset above shows the unblinded analysis window with the events surviving all cuts. The known gamma lines (± 5 keV) are indicated by gray bands, which are excluded from the statistical analysis.

analysis cuts consists of a $2\nu\beta\beta$ distribution at lower energies, with only a few events surviving near $Q_{\beta\beta}$.

The inset of Fig. 4 shows the energy spectrum after all analysis cuts in the range [1930-2190] keV, used for the statistical analysis and calculation of the background index. The gamma lines from ^{208}Tl (2104 keV) and ^{214}Bi (2119 keV) are excluded from the analysis with a window of ± 5 keV around each line. Two separate unblindings were performed for distinct subsets of data. The first, with an exposure of 48.3 kg yr, corresponds to the highest-performing HPGe detectors: BEGe, PPC and ICPC produced by Mirion. The second includes data from the Coax and ICPC detectors with abnormally wide p+ pads, with an exposure of 12.7 kg yr. As shown in the inset of Fig. 4, a total of eleven events survive all cuts—seven from the first subset and four from the second.

The results obtained with the frequentist analysis show no evidence of signal, setting a lower limit on the $0\nu\beta\beta$ decay half-life $> 0.5 \times 10^{26}$ yr at 90% confidence level (C.L.). The background index is determined to be $5^{+3}_{-2} \times 10^{-4}$ counts/(keV kg yr) and $13^{+8}_{-5} \times 10^{-4}$ counts/(keV kg yr) for the first and second datasets, respectively. The elevated background index observed in the second dataset is attributed to the Coax detectors, which are not included in the plans for future phases of the experiment. The frequentist fit joining the results with GERDA (127.2 kg yr) and MJD (64.5 kg yr) data yields a combined lower limit on the $0\nu\beta\beta$ decay half-life of 1.9×10^{26} yr (90% C.L.), compatible with the Bayesian result [6]. The combined exclusion sensitivity obtained is $> 2.8 \times 10^{26}$ yr, the best achieved among $0\nu\beta\beta$ decay searches to date.

After the initial year of physics data taking, a dedicated background characterization phase was conducted to investigate the sources of the observed background, which exceeded expectations. With the planned deployment of additional large-mass ICPC detectors, data acquisition will resume using an enhanced detector array.

VI. LEGEND-1000

LEGEND-1000 is planned to begin operation in the early 2030s, with 1 tonne of high-purity p-type, ICPC Ge semiconductor detectors, enriched above 90% in ^{76}Ge. These large detectors have demonstrated low background levels, excellent pulse shape discrimination and energy performance, reaching energy resolutions of 0.12% FWHM (0.05% σ) at $Q_{\beta\beta}$.

An additional 20-fold reduction in background is anticipated for LEGEND-1000 relative to LEGEND-200, achieved through the use of underground-sourced argon—reduced in the ^{42}Ar isotope— newly developed low-radioactivity electronics and cabling, and the exclusive deployment of ICPC detectors.

The experiment, shown in Fig. 5, will be installed in Hall C at LNGS. It will re-deploy 130 kg of the ICPC detectors

Fig. 5. Schematic design of the LEGEND-1000 experiment, featuring arrays of HPGe detectors at the center, immmersed in underground-sourced liquid argon. The neutron moderator surrounds the cryostat and is housed within atmospheric-sourced liquid argon, with the entire system enclosed in an ultra-pure water tank.

currently used in LEGEND-200, along with 870 kg of newly produced detectors. Approximately 340 ICPC detectors each with an average mass of 3.0 kg will be arranged in 42 strings to enable independent operation and phased commissioning. The strings will be supported by ultra-clean materials and read out using ultra-low-background application-specific itegrated circuit electronics. The detector arrays will be immersed in underground LAr and enclosed within an electroformed copper reentrant tube. The reentrant tube will be surrounded by conventional atmospheric-sourced LAr, contained within a vacuum-insulated cryostat. The argon volumes will be instrumented with a curtain of wavelength-shifting fibers and read out by SiPMs. The cryostat will be housed in a water tank providing additional shielding and instrumented with PMTs. A neutron moderator made of a hydrogen-rich polymer will be installed outside the underground argon volume to effectively thermalize cosmogenic neutrons, thereby reducing the probability of neutron capture on ^{76}Ge. It will be instrumented with light guides read out by SiPMs to tag neutron captures on ^{40}Ar, which produce a γ-ray cascade with a total energy of up to 6 MeV.

The Collaboration aims to construct the LEGEND-1000 experiment with the objective of probing ^{76}Ge $0\nu\beta\beta$ decay half-lives beyond 10^{28} yr with a 3σ discovery sensitivity, covering the full inverted neutrino mass hierarchy. LEGEND-1000 is optimized to operate in a quasi-background-free regime and is expected to enable an unambiguous discovery of $0\nu\beta\beta$ decay with a few detected counts at $Q_{\beta\beta}$.

VII. Conclusions

The results from the first year of LEGEND-200 physics data taking, corresponding to a total exposure of 61 kg yr, have been presented. The achieved background levels, $5^{+3}_{-2} \times 10^{-4}$ counts/(keV kg yr), while higher than predicted, are the lowest reported by any $0\nu\beta\beta$ experiment to date. No evidence of a $0\nu\beta\beta$ signal has been observed, and a lower limit on the half-life of $T^{0\nu}_{1/2} \geq 0.5 \times 10^{26}$ yr at 90% C.L. has been set based on a frequestist analysis. A combined fit of data from GERDA, the MAJORANA Demonstrator and LEGEND sets a lower limit on the ^{76}Ge $0\nu\beta\beta$ half-life at $T^{0\nu}_{1/2} \geq 1.9 \times 10^{26}$ yr (90% C.L.). Data acquisition will resume with an enhanced detector array after deploying additional large-mass ICPC detectors and further background reduction by refining surface treatments of nearby components. LEGEND ultimately aims to operate up to 1 tonne of ^{76}Ge-enriched HPGe detectors within the future LEGEND-1000. The program is designed to reach discovery sensitivities to $0\nu\beta\beta$ decay half-lives beyond 10^{28} years, thereby probing the inverted neutrino mass ordering.

Acknowledgment

This work is supported by the U.S. DOE and the NSF, the LANL, ORNL and LBNL LDRD programs; the European ERC and Horizon programs; the German DFG, BMBF, and MPG; the Italian INFN; the Polish NCN and MNiSW; the Czech MEYS; the Slovak RDA; the Swiss SNF; the UK STFC; the Canadian NSERC and CFI; the LNGS and SURF facilities.

References

[1] W. H. Furry, "On transition probabilities in double beta-disintegration," Phys. Rev. **56**, 1184-1193 (1939).
[2] E. Majorana, "Teoria simmetrica dell'elettrone e del positrone," Nuovo Cim. **14**, 171-184 (1937).
[3] J. Arnquist *et al.* (Majorana), "Final Result of the Majorana Demonstrator's Search for Neutrinoless Double-β Decay in ^{76}Ge", Phys. Rev. Lett. **130**, 062501 (2023).
[4] M. Agostini *et al.* (Gerda), "Final Results of GERDA on the Search for Neutrinoless Double-β Decay," Phys. Rev. Lett. **125**, no.25, 252502 (2020).
[5] N. Abgrall *et al.* [LEGEND], "The Large Enriched Germanium Experiment for Neutrinoless $\beta\beta$ Decay: LEGEND-1000 Preconceptual Design Report," [arXiv:2107.11462 [physics.ins-det]].
[6] H. Acharya *et al.* [LEGEND], "First Results on the Search for Lepton Number Violating Neutrinoless Double Beta Decay with the LEGEND-200 Experiment," [arXiv:2505.10440 [hep-ex]].
[7] L. Baudis, G. Benato, E. M. Bond, P. J. Chiu, S. R. Elliott, R. Massarczyk, S. J. Meijer and Y. Müller, "Calibration sources for the LEGEND-200 experiment," JINST **18**, no.02, P02001 (2023).

Upgrade of the Belle II Vertex Detector with Depleted Monolithic Active Pixel Sensors

A. Gabrielli[5,*], M. Babeluk[1], T. Bergauer[1], C. Irmler[1], A. B. Kaliyar[1], B. Pilsl[1], C. Schwanda[1], L. Gaioni[2], V. Re[2], E. Riceputi[2], G. Traversi[2], S. Giroletti[3], L. Ratti[3], S. Bettarini[4,5], F. Bosi[5], G. Casarosa[4,5], L. Corona[5], F. Forti[4,5], M. Massa[5], L. Massaccesi[4,5], M. Minuti[5], A. Moggi[5], S. Mondal[4,5], G. Rizzo[4,5], V. Vobbilisetti[5], M. Barbero[6], P. Barrillon[6], R. Boudagga[6], P. Breugnon[6], D. Fougeron[6], P. Pangaud[6], J. Serrano[6], D. Xu[6], D. Auguste[7], J. Bonis[7], Y. Peinaud[7], M. Winter[7], J. Baudot[8], G. Bertolone[8], A. Dorokhov[8], G. Dujany[8], L. Federici[8], C. Finck[8], A. Himmi[8], C. Hu-Guo[8], A. Kumar[8], M. Maushart[8], F. Morel[8], H. Pham[8], I. Ripp-Baudot[8], R. Sefri[8], P. Stavroulakis[8], I. Valin[8], F. Bernlochner[9], C. Bespin[9], J. Dingfelder[9], H. Krüger[9], L. Schall[9], M. Vogt[9], M. Karagounis[10], Y. Buch[11], A. Frey[11], B. Schwenker[11], M. Schwickardi[11], T. Hara[12,13], D. Jeans[12,13], T. Kishishita[12], K. R. Nakamura[12,13], Y. Okazaki[12,13], T. Higuchi[14], Y. Onuki[15], S. Wang[15], C. Lacasta[16], C. Marinas[16], J. Mazorra de Cos[16], L. Molina-Bueno[16], A. Bevan[17], M. Bona[17], D. Howgill[17], W. Song[18], J. Gong[18], X. Gao[18], A. Fernandez Prieto[19], and A. Gallas Torreira[19]

[1]Institute of High Energy Physics, Austrian Academy of Sciences, Nikolsdorfer Gasse 18, 1050 Vienna, Austria
[2]Department of Engineering and Applied Sciences, University of Bergamo, Viale Marconi 5, I-24044 Dalmine (BG), Italy
[3]Department of Electrical, Computer and Biomedical Engineering, University of Pavia, Via Ferrata 5, I-27100 Pavia, Italy
[4]Dipartimento di Fisica "E. Fermi", Universita' di Pisa, L.go B. Pontecorvo 3, I-56127 Pisa, Italy
[5]INFN Sezione di Pisa, L.go B. Pontecorvo 3, I-56127 Pisa, Italy
[6]Aix Marseille Univ, CNRS/IN2P3, CPPM, Marseille, France
[7]Laboratoire de Physique des 2 infinis Irène Joliot-Curie – IJCLab, Université Paris-Saclay, CNRS/IN2P3, IJCLab, 91405 Orsay, France
[8]Université de Strasbourg, CNRS, IPHC UMR 7178, F-67000 Strasbourg, France
[9]Physikalisches Institut, Rheinische Friedrich-Wilhelms-Universität Universität Bonn, Nussallee 12, 53115 Bonn, Germany
[10]University of Applied Sciences and Arts Dortmund, Sonnenstraße 96-100, 44139 Dortmund, Germany
[11]II. Physikalisches Institut, Georg-August-Universität Göttingen, Friedrich-Hund-Platz 1, 37077 Göttingen, Germany
[12]High Energy Accelerator Research Organization (KEK), Tsukuba 305-0801, Japan
[13]The Graduate University for Advanced Studies (SOKENDAI), Hayama 240-0193, Japan
[14]Kavli Institute for the Physics and Mathematics of the Universe (WPI), University of Tokyo, Kashiwa-no-ha 5-1-5, Kashiwa 277-8583, Japan
[15]Department of Physics, University of Tokyo, Hongo 7-3-1, Tokyo 113-0033, Japan
[16]Instituto de Fisica Corpuscular (IFIC), CSIC-UV, Catedratico Jose Beltran, 2. E-46980 Paterna, Spain
[17]School of Physical and Chemical Sciences, Department of Physics and Astronomy, Queen Mary University of London, 327 Mile End Road, London, E1 4NS, United Kingdom
[18]College of Physics, Jilin University , 2699 Qianjin Street, Changchun, Jilin, China
[19]Universidade de Santiago de Compostela, 2010 Instituto Galego de Física de Altas Enerxías (IGFAE), Colexio de San Xerome, PZ Obradoiro, S/N. E-15782 Santiago de Compostela, Spain
(*) alice.gabrielli@pi.infn.it

Abstract - **The Belle II experiment at the SuperKEKB e^+e^- collider is preparing for an upgrade of its vertex detector to cope with an increased luminosity of up to $6 \times 10^{35}\,\mathrm{cm^{-2}s^{-1}}$. The new vertex detector (VTX) will consist of six layers of depleted monolithic active pixel sensors (DMAPS), with a total material budget of about 3% of X_0. The OBELIX chip, developed for this upgrade, is derived from the TJ-Monopix2 sensor and manufactured using the Tower Semiconductor 180 nm CMOS technology. It features a $33\,\mu$m pixel pitch, time-stamping capability with 50 ns resolution, and a dedicated digital periphery compatible with the Belle II trigger system, supporting rates up to 30 kHz. The sensor is designed to operate under the expected background hit rate at the target luminosity, with high radiation tolerance, up to $5 \times 10^{14}\,\mathrm{n_{eq}/cm^2}$ and 1 MGy, while maintaining a power density in the range of 200-300 mW/cm^2, corresponding to hit rates from a few MHz/cm^2 up to 120 MHz/cm^2. This paper presents results from laboratory measurements and beam tests performed on TJ-Monopix2 chips, including both not-irradiated and irradiated devices. Particular focus is given to the performance of irradiated sensors as a function of temperature, a key aspect for defining the maximum allowable operating temperature for OBELIX. These studies provide essential input for the thermal design of the VTX cooling system, especially for the innermost layers where power density and hit rates are highest.**

Keywords — **Belle II, Vertex detector, VTX, Upgrade, CMOS Pixel Sensor, Depleted Monolithic Ac-**

tive Pixel Sensor DMAPS, Particle tracking detectors.

I. INTRODUCTION

The Belle II experiment [1] is a high-precision B-factory designed to probe potential new physics beyond the Standard Model by operating at the intensity frontier. The experiment is hosted at the SuperKEKB collider [2] in Japan, which provides asymmetric collisions between 7 GeV electrons and 4 GeV positrons tuned to the $\Upsilon(4S)$ resonance. To achieve its goals, SuperKEKB uses a nano-beam scheme in combination with extremely high beam currents, targeting a peak luminosity of $6 \times 10^{35}\,\mathrm{cm^{-2}s^{-1}}$ and aiming for an integrated luminosity of $50\,\mathrm{ab^{-1}}$. These demanding conditions place stringent requirements on the detector, especially on the vertex detector (VXD) located closest to the interaction point.

The existing VXD is composed of six layers of silicon-based sensors for precise track reconstruction near the interaction point. Its inner layers employ DEPFET pixel sensors (PXD) [3], which deliver high spatial resolution thanks to their 50–70 μm pixel pitch, albeit with a relatively long integration time of 20 μs. The outer layers use double-sided silicon strip detectors (SVD) [4], which offer excellent time resolution of 3 ns and extended spatial coverage. This configuration allows to achieve spatial resolutions between 10 and 25 μm, with polar angle coverage spanning from 17° to 150°. The material budget per layer varies between 0.25% and 0.75% of a radiation length, leading to an overall material budget of approximately $3.5\% X/X_0$.

Although the VXD has proven effective in environments with moderate background levels, its performance could be compromised under the intense background conditions expected at the SuperKEKB target luminosity. Increased background rates from the machine and uncertainties in their projections, especially in light of possible modifications to the interaction region, could lead to performance degradation due to high occupancy.

To overcome the aforementioned limitations and to enhance both tracking and vertexing capabilities, a new pixel vertex detector (VTX) [5] has been proposed. This upgraded system is based on CMOS monolithic active pixel sensors (MAPS), which offer superior spatial resolution, increased tolerance to background radiation, and improved tracking performance (see Section II). Moreover, the VTX will be engineered with sufficient flexibility to accommodate potential changes to the interaction region (IR), thereby ensuring that Belle II can maintain its precision measurement capabilities well into the future.

II. UPGRADE REQUIREMENTS AND VTX PROPOSAL

The upgrade of the Belle II vertex detector is motivated by the need to ensure reliable performance under the challenging background conditions expected at the target luminosity, particularly in view of a potential redesign of the interaction region [5]. Background extrapolations indicate that the current detector operates with limited safety margins, with occupancies projected to reach up to 2% in the innermost pixel layer (PXD layer 1) and 9% in the third strip layer (SVD layer 3) in high-background scenarios, approaching the limits of acceptable performance. To preserve tracking and vertexing capabilities under these conditions, the new vertex detector must offer higher granularity and improved time resolution.

The Conceptual Design Report [5] defines a set of key performance requirements for the VTX to meet the physics goals of Belle II at full luminosity. These include a spatial resolution better than 15 μm, with pixel pitches in the range of 30–40 μm, and a material budget of 0.2%–$0.8\% X/X_0$ per layer for the inner and outer sections, respectively. The detector must handle hit rates up to $120\,\mathrm{MHz/cm^2}$ with triggered readout and provide fast timestamping capabilities in the range of 50–100 ns. Radiation hardness is also essential, with tolerance to a total ionizing dose of up to 1 MGy and a Non-Ionizing Energy Loss (NIEL) fluence of $5 \times 10^{14}\,\mathrm{n_{eq}/cm^2}$ in the innermost layers over ten years of operation. To minimize the impact of cooling infrastructure on the material budget, the power dissipation must be in the range of $200\,\mathrm{mW/cm^2}$.

The technology chosen to meet these requirements is based on Depleted Monolithic Active Pixel Sensors (DMAPS), in particular the TJ-Monopix2 chip [6], developed in the Tower Semiconductor 180 nm CMOS process and originally designed for the ATLAS ITk outer tracker [7].

The baseline design of the new VTX features six layers of planar sensors, all using the same OBELIX (Optimised BELle II pIXel) chip, derived from the TJ-Monopix2 architecture. Compared to the current VXD, this configuration prioritizes finer spatial and timing granularity while keeping the material budget low. The integration of DMAPS technology is expected to significantly improve tracking efficiency, especially for low-momentum particles and in high hit-rate conditions.

An overview of the VTX layout and its transverse geometry is shown in Figure 1.

The VTX is structured in two main regions: an inner section (iVTX) and an outer section (oVTX). The iVTX consists of two layers located at radii of 14 mm and 22 mm, employing an all-silicon ladder concept. Each ladder is composed of four OBELIX chips from the same wafer, connected via a post-processed redistribution layer (RDL) and potentially thinned to approximately 50 μm. These layers achieve a material

 https://doi.org/10.1051/epjconf/202533801003

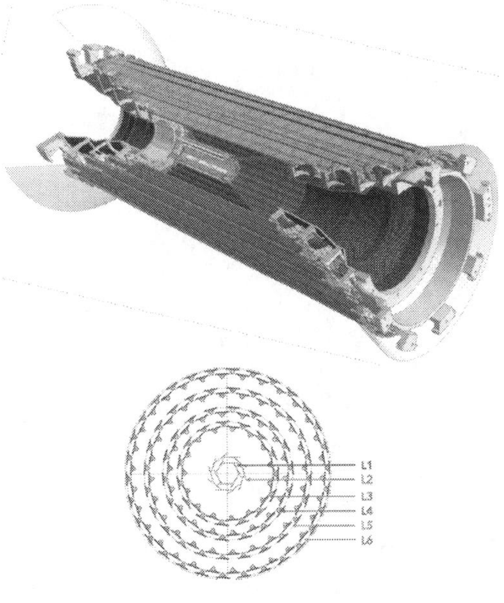

Figure 1: Schematic view of the baseline VTX layout, showing the six detection layers and a transverse cross-section of the detector. The cross-section illustrates the concentric arrangement of the ladders, highlighting the windmill configuration of the innermost layers (iVTX) and the staggered design of the outer layers (oVTX) at two distinct radii.

budget of about 0.3% X/X_0 per layer. Power dissipation in the iVTX is expected to range between 200 and 300 mW/cm^2, depending on the local hit rate, as estimated from chip simulations.

Given the relatively high power density and the degradation of chip performance with increasing temperature after irradiation, several cooling strategies are currently under investigation. To enhance thermal performance alternative approaches to air cooling, which was the first choice, are being explored, including thermal conduction via layers of thermal pyrolytic graphite (TPG) and liquid cooling using thin aluminum pipes positioned along the ladder edges.

Studies of the TJ-Monopix2 chip, described in Section III, play a crucial role in defining the operational temperature limits of the OBELIX sensor. In particular, measurements on irradiated TJ-Monopix2 chips at elevated temperatures provide essential input for evaluating performance degradation and determining safe operating conditions. These results directly inform the thermal design requirements and the cooling strategy for the iVTX.

The oVTX extends the detector volume up to a radius of 140 mm and comprises four additional layers. These are mounted on a lightweight support structure inspired by the ALICE ITS2 [8] and based on an omega-shaped mechanical design. The structure features two carbon fiber skins sandwiching a Rohacell foam core, with a cold plate and embedded liquid cooling pipes. Sensors are attached to the cold plate, and flexible circuits board provide the necessary read-out and power interconnections. This layout maintains a material budget of up to 0.8% X/X_0 per layer for perpendicular tracks while providing robust mechanical support and effective thermal management, enabling the integration of four distinct detection layers within the oVTX region.

III. TJ-MONOPIX2 AS A FORERUNNER FOR OBELIX

Manufactured using the Tower Semiconductor 180 nm CMOS process, modified to enhance radiation tolerance [6], the TJ-Monopix2 sensor acts as the predecessor of the OBELIX chip. Fig. 2 shows the cross-section of the in-pixel DMAPS structure used in both sensors, with the small collecting electrode DMAPS pixel structure.

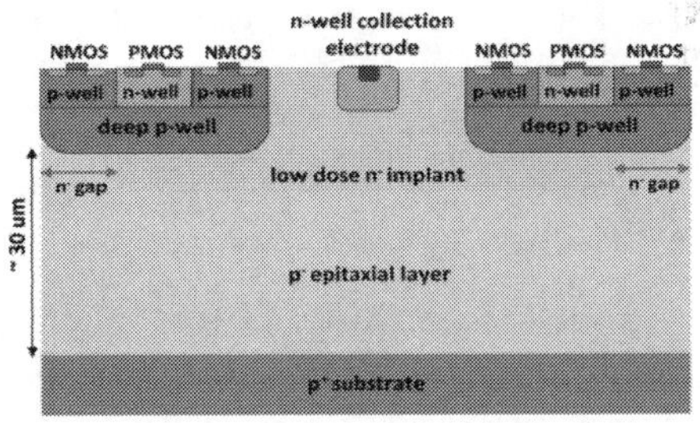

Figure 2: Cross section of a DMAPS pixel manufactured in the modified TowerJazz 180 nm CMOS process, designed for improved radiation hardness.

TJ-Monopix2 comprises a pixel matrix with 512 × 512 pixels, with a pitch of $33 \times 33\,\mu$m. It provides 25 ns hit timestamping and delivers analog Time-over-Threshold (ToT) information with 7-bit precision. The chip includes four different front-end implementations, including both DC- and AC-coupled variants; two of these incorporate a cascode transistor within the voltage amplifier to improve performance. The readout relies on a column-drain architecture, which allows the device to sustain hit rates exceeding 100 MHz/cm^2 [6].

TJ-Monopix2 operates with a triggerless readout mode and lacks peripheral memory and logic to manage trigger signals. OBELIX integrates a new digital periphery specifically designed to meet the requirements of Belle II, enabling triggered operation, latency buffering, and additional functionalities such as track trigger information and high-resolution timing reconstruction.

Extensive laboratory and test beam studies were conducted to assess the performance of TJ-Monopix2 to validate key performance parameters critical for the OBELIX and the VTX design. In the following, particular emphasis is placed on characterisation and testing of irradiated samples, as this is crucial for assessing the degradation expected in the innermost layer during operation at the target luminosity.

Several test campaigns with not-irradiated chips have confirmed stable and reliable performance: operation at thresholds of approximately 250 e^-, a threshold dispersion of 17 e^-, and noise levels around 8 e^- [9], [10]. Beam tests using 3–5 GeV electrons at DESY demonstrated a hit efficiency above 99% and a spatial resolution of about 9 μm at normal incidence. These results validate the expected baseline performance of TJ-Monopix2 for its use in OBELIX.

The latest test beam campaign, carried out in March 2025, focused on investigating the behaviour of irradiated sensors across a wide temperature range, enabling a detailed evaluation of their performance under realistic operating conditions. Studying performance degradation as a function of temperature after irradiation is crucial to define a safe and efficient operating temperature range for OBELIX.

To enable these studies, a dedicated temperature control setup was developed using a Peltier cell for active cooling and heating of the chip. The setup was designed for use both in the laboratory and on the beam line, while minimizing additional material in the beam path between the chip and the telescope. Although the chip itself lacks integrated temperature sensors, the setup was calibrated using the NTC sensor on the PCB, placed next to the chips edge, to provide a reliable estimate of the chip temperature. The system demonstrated stable and precise temperature control throughout the campaign. The temperatures quoted in this paper refer to the NTC sensor readings; the actual chip temperature is estimated to be approximately 7°C higher. Four irradiated sensors were tested during the campaign: three devices were irradiated with 24 MeV proton to 1 MeV neutron equivalent fluences of 1×10^{14}, 2.5×10^{14}, and 5×10^{14} n_{eq}/cm^2 respectively and one sensor was irradiated with 90 MeV electrons to an equivalent neutron fluence of 5×10^{14} n_{eq}/cm^2. Prior to the test beam, all sensors were characterised in the laboratory using the same heating/cooling setup. As expected from bulk damage, the leakage current increased significantly after irradiation and scaled strongly with temperature. An increase in threshold with temperature was also observed, due to the higher leakage current accelerating the discharge of the collecting electrode.

Efficiency measurements at the beam test confirmed a dependence on the irradiation level (Fig. 3). Less irradiated sensors showed higher efficiency, even when operated at higher thresholds. For the most irradiated proton-irradiated chip (5×10^{14} n_{eq}/cm^2), the detection efficiency dropped more rapidly with increasing threshold, as expected. Notably, the electron-irradiated sensor exhibited higher efficiency than the proton-irradiated one at the same fluence of 5×10^{14} n_{eq}/cm^2, suggesting that electron irradiation induces less bulk damage than proton irradiation at the same 1 MeV equivalent fluence. The comparison between DC-coupled (filled dots) and AC-coupled (empty dots) front-ends shows that, across all irradiation levels, the DC-coupled configuration generally achieves slightly higher efficiency for a given threshold.

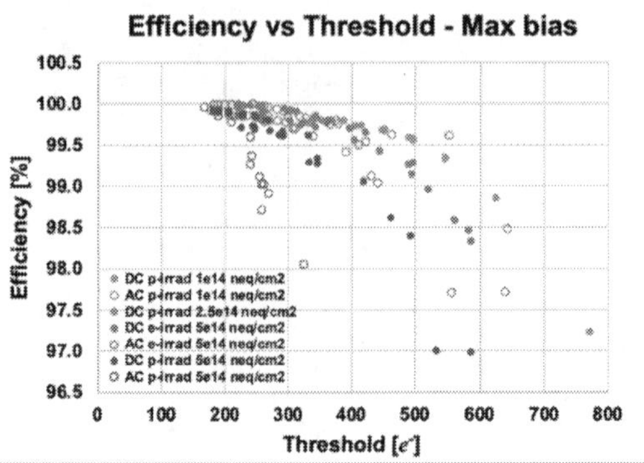

Figure 3: Efficiency vs threshold for the different irradiated sensors and both DC-coupled (filled dots) and AC-coupled (empty dots) frontends.

Efficiency degradation after irradiation was also found to be temperature dependent, particularly for the AC-coupled front-end (Fig. 4). The DC-coupled cascode front-end maintained efficiency above 99% up to approximately 40°C when operated at a threshold of around 350 e^-. At lower temperatures, stable operation with thresholds down to 200 e^- was achieved, while AC-coupled sensors exhibited a more pronounced efficiency drop with increasing temperature.

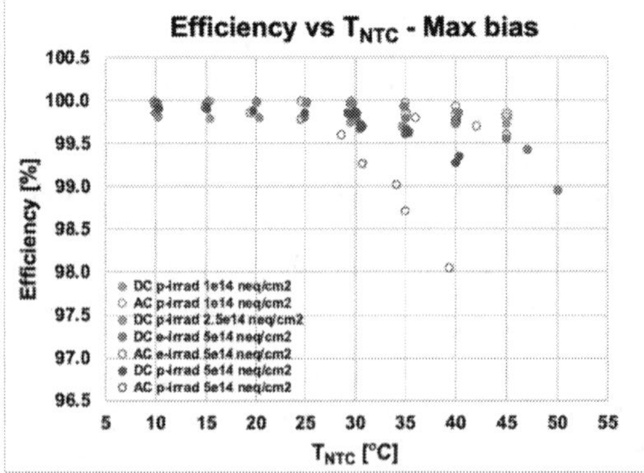

Figure 4: Efficiency vs temperature for the different irradiated sensors and both DC-coupled (filled dots) and AC-coupled (empty dots) frontends.

At the same threshold, the AC-coupled front-end showed stronger degradation compared to the DC-coupled one on the proton-irradiated chip at 5×10^{14} n_{eq}/cm^2. In-pixel efficiency maps (Fig. 5) illustrate that the AC front-end suffers a visible drop in efficiency at the pixel corners, consistent with reduced charge collection performance in those regions. This effect is not

https://doi.org/10.1051/epjconf/202533801003

observed in the corresponding DC-coupled front-end maps, highlighting the superior radiation tolerance of the DC design.

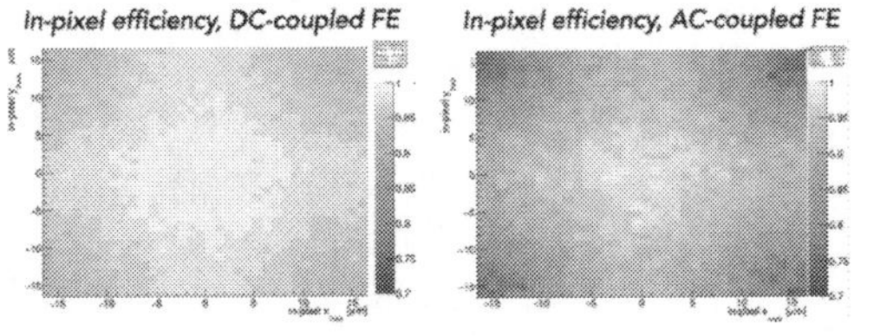

Figure 5: In-pixel efficiency maps at a threshold of about 550 e^- for the DC-coupled (left) and AC-coupled (right) front-ends on a proton-irradiated sensor at a NIEL fluence of $5 \times 10^{14}\,\mathrm{n_{eq}/cm^2}$. The measurement was performed at maximum bias voltage and a NTC temperature of approximately 40°C. The AC-coupled front-end shows localized efficiency degradation at the pixel corners, attributed to reduced charge collection.

These results provide critical input for defining the safe and efficient operating temperature range of the OBELIX sensor in the iVTX. In particular, they confirm the need for a carefully engineered thermal management solution to limit temperature-induced degradation and ensure robust performance under the high hit-rate and radiation environment expected for the innermost layers of VTX at target luminosity.

IV. THE OBELIX CHIP FOR BELLE II

To fulfill the stringent demands of the upgraded vertex detector, the OBELIX chip [11] has been developed as a DMAPS solution based on the TJ-Monopix2 design. It features an expanded matrix composed of 464 rows and 896 columns, offering hit timestamping with 47 ns resolution, a 7-bit time-over-threshold (ToT) readout, and 3-bit local threshold adjustment to mitigate threshold dispersion across the matrix. Figure 6 displays the chip's floorplan and overall dimensions. To ensure stable operation over the enlarged area, on-chip power regulation is implemented using low-dropout regulators (LDOs) along with additional power conditioning circuits.

Compared to its predecessor TJ-Monopix2, OBELIX introduces a new custom digital periphery adapted to the specific requirements of the Belle II environment. One of the central innovations is the implementation of a triggered readout scheme managed by the Trigger Readout Unit (TRU). This system supports average hit rates up to $120\,\mathrm{MHz/cm^2}$, with peak tolerance reaching $600\,\mathrm{MHz/cm^2}$, and operates efficiently at trigger rates of 30 kHz with latencies up to $10\,\mu$s. According to simulation studies, the architecture sustains a data bandwidth of 4.8 MHits/s while keeping hit loss below 0.02%.

Two additional functional blocks have been integrated into the digital periphery. The first is the Track

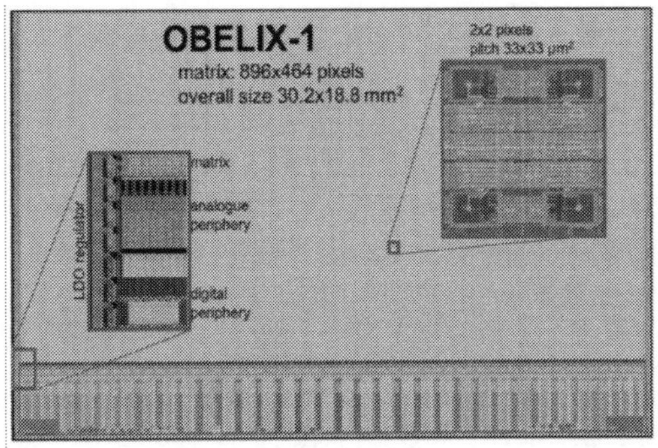

Figure 6: Floorplan and dimensions of the OBELIX chip. Matrix inherit from TJ-Monopix2, size adjusted

Trigger Transmission (TTT) module, which provides low-latency hit data to the Belle II trigger system. This allows the implementation of a fast track-finding algorithm in an FPGA, comparing detector data to predefined patterns in real time. Only hits from the outer VTX layers contribute to the TTT, as inner layers face much higher occupancies that, combined with strict power constraints, would otherwise lead to pile-up and degraded trigger performance.

The second module, known as the Periphery Time-to-Digital (PTD), is optimized for precision timing in the outer VTX layers, where the expected hit rate does not exceed $10\,\mathrm{MHz/cm^2}$. Although the internal timestamp granularity in-pixel is limited to 47 ns, the chip exploits the column-wise HitOr signal, *i.e.* a logical OR of all pixel discriminators, sampled at a clock running 16 times faster. Timing measurements performed with TJ-Monopix2 indicate that, with appropriate corrections for time walk and signal propagation delays, a timing resolution as fine as 3 ns is achievable [12]. This refined reconstruction is constrained to low-occupancy regions, as multiple simultaneous hits within a column can introduce ambiguities, making it suitable only for the outer detector layers.

By enabling the addition of high-precision time information to a subset of hits, OBELIX enhances the suppression of off-time background tracks without impacting the primary timestamp resolution or increasing power consumption. Simulation results and preliminary measurements support the expected performance, and the first full-scale prototype, OBELIX-1, is scheduled for submission in winter 2025.

V. CONCLUSIONS

The upgrade of the Belle II Vertex Detector with OBELIX DMAPS sensors represents a major advancement for high-luminosity operations at SuperKEKB. The extensive laboratory and test beam studies of TJ-Monopix2, the forerunner of OBELIX, confirmed good performance for non-irradiated sensors with high detection efficiency ($> 99\%$), excellent spatial resolution ($\sim 9\,\mu$m), and stable operation with low noise and threshold dispersion.

Tests on irradiated sensors at NIEL fluences up to $5 \times 10^{14}\,n_{eq}/cm^2$, investigating performance degradation as a function of temperature, are crucial to define a safe and efficient operating range for OBELIX and to guide the VTX cooling system design, especially for the innermost layers. After irradiation, efficiency degradation was found to be temperature dependent, with the AC-coupled front-end showing stronger performance loss. In contrast, the DC-coupled front-end maintained efficiency above 99% up to 40°C, demonstrating its superior radiation tolerance.

OBELIX builds upon the TJ-Monopix2 architecture with significant enhancements, including triggered readout, latency buffering, and improved time resolution and data handling, enabling efficient operation in the high-background environment of Belle II. Its ability to maintain high performance under high hit rates and radiation will be essential for meeting the experiment's physics goals and sets a technological benchmark for future vertex detectors in high-energy physics.

ACKNOWLEDGMENT

This work has received the support from the European Union's Horizon 2020 Research and Innovation programme under Grant Agreements no 101004761 (AIDAinnova), Multilateral Scientific and Technological Cooperation in the Danube Region (MULT 03/23), Horizon 2020 ERC-Consolidator Grant No. 819127, TY-FJPPN (Toshiko Yuasa-France Japan Particle Physics Network), the MCIU with funding from the European Union NextGenerationEU (PRTR-C17.I01) and Generalitat Valenciana (GVANEXT), Project ASFAE/2022/016.

The measurements leading to these results have been performed at the Test Beam Facility at DESY Hamburg (Germany), a member of the Helmholtz Association (HGF).

REFERENCES

[1] T. Abe *et al.*, "Belle II Technical Design Report," Nov. 2010. arXiv: 1011.0352 [physics.ins-det].

[2] Y. Ohnishi *et al.*, "Accelerator design at SuperKEKB," *PTEP*, vol. 2013, 03A011, 2013. DOI: 10.1093/ptep/pts083.

[3] P. Avella, "DEPFET sensors development for the Pixel Detector (PXD) of Belle II," *JINST*, vol. 9, no. 01, P. Govoni, P. S. Marrocchesi, F.-L. Navarria, M. Paganoni, A. Perrotta, and T. Rovelli, Eds., p. C01057, 2014. DOI: 10.1088/1748-0221/9/01/C01057.

[4] K. Adamczyk *et al.*, "The design, construction, operation and performance of the Belle II silicon vertex detector," *JINST*, vol. 17, no. 11, P11042, 2022. DOI: 10.1088/1748-0221/17/11/P11042. arXiv: 2201.09824 [physics.ins-det].

[5] H. Aihara *et al.*, "The Belle II Detector Upgrades Framework Conceptual Design Report," Jun. 2024. arXiv: 2406.19421 [hep-ex].

[6] K. Moustakas, "Design and Development of Depleted Monolithic Active Pixel Sensors with Small Collection Electrode for High-Radiation Applications," Ph.D. dissertation, U. Bonn, 2021.

[7] C. Bespin *et al.*, "DMAPS Monopix developments in large and small electrode designs," *Nucl. Instrum. Meth. A*, vol. 978, p. 164460, 2020. DOI: 10.1016/j.nima.2020.164460. arXiv: 2006.02297 [physics.ins-det].

[8] A. Fantoni, "Upgrade of the ALICE inner tracking system: Construction and commissioning," *Phys. Scripta*, vol. 95, no. 8, L. Bravina, S. Kabana, V. Volkova, and V. I. Zakharov, Eds., p. 084011, 2020. DOI: 10.1088/1402-4896/aba0f7.

[9] M. Schwickardi, "Characterisation of CMOS Sensors for the Belle II Vertex Detector Upgrade," Ph.D. dissertation, Gottingen U., 2024. DOI: 10.53846/goediss-10611.

[10] C. Bespin Valero, "Characterization of the TJ-Monopix2 Depleted Monolithic Active Pixel Sensor for High-Energy Physics Experiments," Ph.D. dissertation, Bonn U., 2024. DOI: https://hdl.handle.net/20.500.11811/12562.

[11] M. Babeluk *et al.*, "The OBELIX chip for the Belle II VTX upgrade," *Nucl. Instrum. Meth. A*, vol. 1067, p. 169659, 2024. DOI: 10.1016/j.nima.2024.169659.

[12] M. Babeluk *et al.*, "The DMAPS upgrade of the Belle II vertex detector," *Nucl. Instrum. Meth. A*, vol. 1064, p. 169428, 2024. DOI: 10.1016/j.nima.2024.169428.

EPJ Web of Conferences **338**, 01004 (2025)
ANIMMA 2025
https://doi.org/10.1051/epjconf/202533801004

FPGA performance for signal reconstruction of the ATLAS Tile Calorimeter in the HL - LHC environment

Sonakshi Ahuja[1*], Fernando Carrió[1], Antonio Cervelló[1], Francesco Curcio[1], Luca Fiorini[1], Antonio Gómez[1], David Hernandez[1], Francisco Hervás[1], Arantxa Ruiz[1], Alberto Valero[1]
[1] Instituto de Física Corpuscular, Spain
(*) sonakshi.ahuja@cern.ch

***Abstract*—The ATLAS Tile Calorimeter (TileCal) will face considerable challenges from increasing radiation and high data throughput in the High Luminosity Large Hadron Collider (HL - LHC) era. This study aims to find a balance between FPGA resource usage and latency with efficient signal reconstruction algorithms of the calorimeter pulses. The precise reconstruction of signal pulse from calorimeter pulses is crucial as it corresponds to the energy of the particle deposited in the calorimeter. Optimal Filtering (OF) is currently used for reconstructing the pulse amplitude from digitised samples. While efficient, the performance degrades considerably with increasing pile up. A solution to combat this challenge is to work with Neural Networks. This study shows the performance of OF contrasted with a Single Layer Perceptron (SLP) with data trained on simulated ATLAS TileCal pulses. The results show that the SLP outperforms the OF algorithm in the pile-up case, which is expected in the HL - LHC era.**

***Keywords* — HL - LHC, TileCal, Signal Reconstruction**

I. Introduction

THE Tile Calorimeter (TileCal) [1] is the central hadronic calorimeter of the ATLAS [2] detector at the Large Hadron Collider (LHC). It is a sampling calorimeter that uses steel plates as the absorber material and plastic scintillators as the active medium. Fig 1. shows the TileCal barrels and electronics drawer unit.

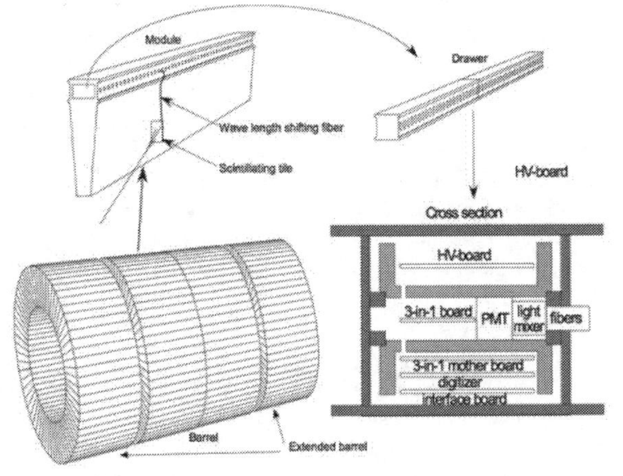

Fig. 1 The Tile Calorimeter extended and long barrel sections with the drawers containing the electronics [3]

The calorimeter is partitioned in three main sections: the central barrel consists of the the Long Barrel ($\eta < 1.0$), followed by two extended barrels covering a pseudorapidity (η) range of 0.8 - 1.7. Each barrel is segmented into 64 azimuthal modules covering 2π in ϕ. Each cell in the calorimeter corresponds to a combination of radial and azimuthal segmentation and is read out by two PMT's for redundancy.

When particles traverse the detector, they deposit energy in the scintillators, producing light that is collected by wavelength-shifting fibres and read out by PMTs. These analog signals are shaped, amplified, and digitised by the front-end electronics.

Accurate reconstruction of the particle energy from these digitised samples is critical, especially under the high pile-up conditions anticipated at the HL - LHC [4]. Currently, TileCal employs the OF algorithm for signal reconstruction, a method used extensively during Run 2 and Run 3. OF estimates the signal amplitude as a weighted sum of the digitised samples, using precomputed weights optimised for the nominal pulse shape and expected noise. While this algorithm is robust and effective under stable conditions, its performance degrades when the signal is distorted by pile-up or is misaligned with the sampling window — conditions that are expected to become more common at the HL - LHC.

To address these limitations, this study investigates the use of neural networks for signal reconstruction in TileCal. Neural Networks can learn complex, non-linear correlations from data and adapt to varying pulse conditions.

This work focuses on a SLP, a simple feed-forward model that is both efficient and suitable for implementation in firmware on FPGAs. This makes it a promising candidate for real-time processing in the upgraded TileCal readout system. The SLP, trained on pseudo TileCal pulses, is evaluated in pile-up conditions and compared with the performance of the OF algorithm under the same conditions.

II. Readout Architecture and Signal Reconstruction

This section begins with the setup and implementation of the two signal reconstruction algorithms, Optimal Filtering and a Neural Network based Single Layer Perceptron. We also detail the data preparation, implementation on FPGA firmware and reconstruction algorithm design.

The upgraded TileCal [4] readout system introduced in the Phase-II upgrade is designed to operate under the demanding conditions of the HL-LHC, including higher luminosity, increased radiation, and substantial pile-up. A key feature of this architecture is the full digitisation of signals from each PMT individually, ensuring that the granularity of the detector

 https://doi.org/10.1051/epjconf/202533801004

is fully preserved. Importantly, the data from every bunch crossing is transmitted off-detector, for the reconstruction of the signals.

The analog signals generated by the PMTs are processed by the front-end electronics, which perform signal shaping, amplification, and digitisation. These include the upgraded 3-in-1 cards, responsible for shaping the signal and providing both high-gain (HG) and low-gain (LG) amplification paths

Digitisation is performed with a 10-bit ADC operating at 40 MHz, corresponding to one sample per LHC bunch crossing, The upgraded front-end also changes from the 3-in-1 to new boards such as FENICS, while the digitisation uses a 12-bit width which manages digitised signal routing to the off-detector electronics [5].

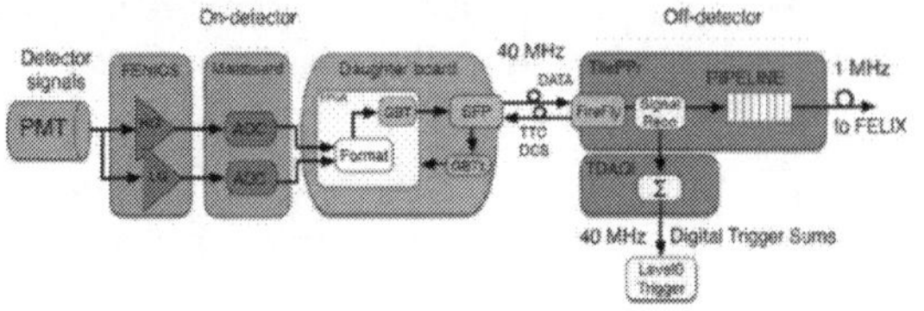

Fig. 2 The Tile Calorimeter upgraded readout architecture for the HL - LHC era.

Once off-detector, the digitised samples are passed to the PreProcessor module where energy reconstruction algorithms such as Optimal Filtering or Neural Network models are implemented in either software or firmware. The consistent digitisation and full granularity provided by the upgraded readout system are crucial for the performance of these reconstruction methods, especially under the high pile-up conditions expected at HL-LHC.

III. Signal reconstruction algorithms

To extract the signal amplitude from digitised samples, we explored different reconstruction methods. This study compares the well-established Optimal OF algorithm with a neural network-based approach, namely the SLP.

A. Optimal Filtering Algorithm

The OF algorithm is a lightweight and robust technique that has been used extensively in the ATLAS Tile Calorimeter readout during Run 2 and Run 3. Its primary objective is to reconstruct the amplitude of a pulse — which is directly related to the energy deposited in the calorimeter — by computing a weighted sum of the input samples, as shown in figure 2. The reconstructed amplitude A is given by:

$$A = \sum_{i=0}^{n-1} a_i \cdot y_i,$$

where y_i are the digitised samples and a_i are optimal weights precomputed based on expected conditions.

These weights are calculated using the known pulse shape, the autocorrelation matrix of the electronic noise, and assumptions on the signal timing and phase. While the OF method is efficient and works well under ideal conditions, it is known to be sensitive to pulse shape distortions and out-of-time signals (pile-up), which can lead to degraded performance in high-luminosity environments. In the TileCal readout, OF typically uses 7 input samples, centred around the expected signal peak. The signal itself is amplified in two gain stages — high gain and low gain — with a ratio of 1:40, to cover a dynamic range from approximately 200 fC to 1000 pC

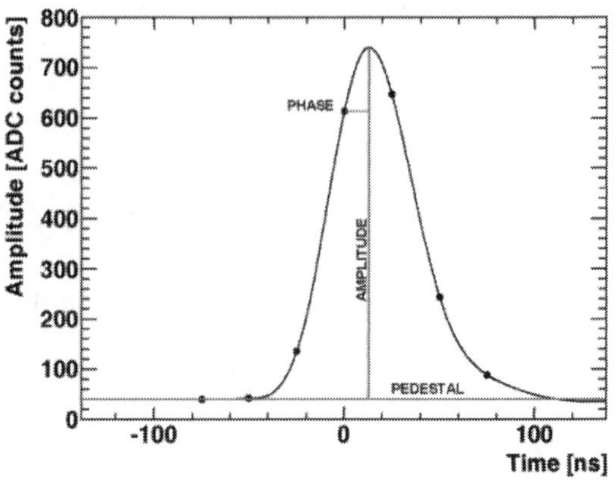

Fig. 3 Optimal Filtering *signal* reconstruction in the Tile Calorimeter using 7 input samples, and a non - centred pulse on the reconstructed amplitude.

B. Neural Network Model

The study and implementation of a neural network, specifically a SLP, was motivated by the challenges faced by the OF algorithm in conditions of high pile-up and increased data throughput. Unlike OF, which relies on fixed, pre-determined weights, the SLP offers a more flexible, data-driven approach that can learn complex, non-linear relationships from the data.

The model takes as input a sliding window of 9 samples, which correspond to the digitised signal samples from the detector. This input is processed through a single weight layer followed by a hyperbolic tangent (tanh) activation function.

The goal is to reconstruct the true amplitude of the signal for each event. A simple Mean Squared Error (MSE) loss is used to evaluate the difference between the predicted and actual amplitudes during training.

To maintain a fair comparison, the SLP is trained and tested on the same dataset used for OF. Although training is performed only on High Gain (HG) samples, a straightforward conversion factor allows switching to Low Gain (LG) whenever HG saturates. One of the main strengths of using a neural network lies in its ability to adapt to variations in the pulse shape — including those that are distorted — through learning from realistic or simulated data.

Finally, the model output, which is a floating-point amplitude, is converted into a fixed-point format suitable for efficient implementation on the FPGA, keeping in mind the constraints on the resources.

IV. Testbench development

The main goal of developing the firmware testbench is to evaluate the performance and efficiency of signal reconstruction algorithms before deploying them in the ATLAS cavern for the HL-LHC era. The testbench environment replicates the actual hardware configuration, using the same version of the FPGA that will be installed in the detector. This enables a realistic validation of the firmware's behaviour and timing in a controlled setting.

The SLP model is implemented in VHDL, structured into components representing the input, output, and activation logic. The activation function, in this case a piecewise-linear approximation of the tanh function, is implemented separately and integrated into the overall architecture. Computations within the model are performed using fixed-point arithmetic, and normalisation and de-normalisation stages are included to properly scale input and output values to match the training regime and restore physical meaning to the reconstructed amplitude.

Since the model processes a sliding window of 9 samples, a first-in-first-out (FIFO) buffer is used to maintain the sliding sample window. The design is tested for different fixed-point precisions (bit widths) to evaluate the trade-off between resource usage and reconstruction accuracy. Latency is also studied and adjusted to meet clock timing constraints and ensure deterministic behaviour.

The full firmware compilation proceeds in three stages: synthesis, placement and routing, and finally bitstream generation. The bitstream corresponds to the binary configuration loaded onto the FPGA.

Firmware simulation is carried out using a cocotb-based testbench[6], which enables co-simulation of the VHDL implementation and Python control code. The simulation flow is tightly synchronised with the firmware clock, where the testbench waits for valid signals from the firmware before supplying new data or proceeding to the next step. Inputs to the device under test (DUT) include the 9-sample high-gain window and the true amplitude, simulating conditions like those expected in real detector data.

To ensure reproducibility and avoid version mismatches between tools such as Vivado (a hardware design suite by Xilinx for synthesis and implementation of FPGA designs), cocotb (a coroutine-based Python framework for testing VHDL), Python, and ROOT, the entire setup is encapsulated within a Docker container. This provides a consistent and isolated environment for firmware development and testing.

The DUT outputs the reconstructed amplitude, which is written to a ROOT ntuple file. PyROOT is used for further analysis and visualisation, including histogramming and performance comparison. The reconstructed output from the FPGA is compared against the reference software model, allowing a direct evaluation of how closely the firmware reproduces the trained behaviour of the SLP.

V. PERFORMANCE EVALUATION AND RESULTS

The performance evaluation focuses on the firmware implementation in the high-gain range. A total of 1 million events were processed through the testbench to assess and compare the OF algorithm and the SLP under realistic conditions. We begin by contrasting their performance in scenarios with and without pile-up. In the absence of pile-up, the OF algorithm demonstrates excellent linearity in its reconstruction, clearly reflecting its efficiency in low-noise environments. However, under pile-up conditions, its reconstruction quality degrades significantly, failing to maintain a reliable amplitude response.

Given the expected increase in pile-up and data throughout the HL-LHC era, the limitations of OF in such scenarios highlight the necessity for more adaptable models.

The SLP, trained on pseudo TileCal pulses, outperforms OF in the presence of pile-up, showing a stronger linear correlation between the predicted and true amplitudes as shown in figure 4. When comparing the FPGA output of the SLP model with its software counterpart, a small discrepancy is observed — the software model offers slightly higher precision.

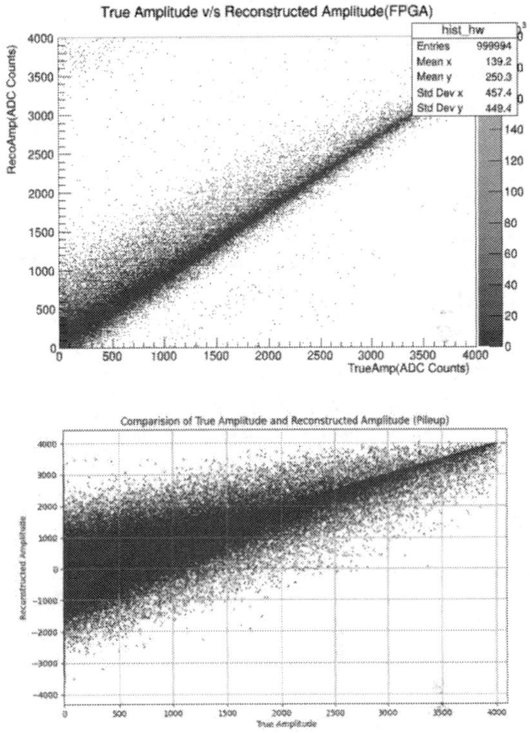

Fig. 4. Comparison between the NN (top) and OF (below) algorithm for True Amplitude and Reconstructed Amplitude in the FPGA.

This difference arises from the use of fixed-point arithmetic in the FPGA, as opposed to floating-point operations in software. Nevertheless, this minor trade-off is acceptable when considering the resource constraints and latency requirements of the firmware environment.

The firmware operates using two distinct clock domains: a 40 MHz clock aligned with the LHC bunch crossing, and a faster 400 MHz processing clock used internally for algorithm execution. In terms of latency, the OF implementation requires approximately 7 bunch crossings to produce an output, while the SLP introduces a slightly higher latency of about 12 bunch crossings, still well within acceptable bounds for real-time reconstruction.

VI. CONCLUSIONS AND RESULTS

The aim of this work was to assess the performance of signal reconstruction algorithms implemented in firmware, under the conditions expected at the HL-LHC—namely, increased pile-up, radiation, and data rates. We focused on two approaches: the Optimal Filtering algorithm, known for its simplicity and reliability, and a more flexible Single Layer Perceptron neural network model. Our results show that while OF performs well under low pile-up conditions, the neural network method provides improved reconstruction accuracy in high pile-up scenarios.

This study implemented both reconstruction algorithms directly in the FPGA firmware. Their performance was evaluated using a custom testbench that simulates realistic TileCal conditions, allowing us to validate the firmware before deployment. The study demonstrates that both OF and neural networks can be deployed and evaluated in a controlled environment before integration in the ATLAS detector.

The work also underscores the importance of the upgraded readout architecture, which—by providing full-granularity digitised data at 40 MHz—enables robust real-time reconstruction and opens the door to more sophisticated algorithmic strategies like machine learning. Overall, this study serves as a proof of concept for implementing neural network-based reconstruction in firmware and contributes to the broader effort of preparing the TileCal readout for HL - LHC operation.The document must be in this two-column format.

VII. Future Work

The next steps involve exploring more complex neural network architectures such as Multi-Layer Perceptrons (MLPs) and Convolutional Neural Networks (CNNs) to improve reconstruction accuracy and robustness. We also plan to investigate different configurations of these models, including the use of various activation functions like ReLU and PReLU, optimisers such as ADAM, and alternative loss functions. Additionally, we will scan over different window sizes to understand how temporal context affects reconstruction. These studies aim to assess the trade-offs between reconstruction performance and FPGA resource usage.

Acknowledgment

This research was funded by Ministerio de Ciencia e Innovación with project reference PID2021-125069OB-I00.

References

[1] ATLAS Collaboration, "Operation and performance of the ATLAS tile calorimeter in LHC Run 2", Eur. Phys. J. C 84, 1313 (2024), https://doi.org/10.1140/epjc/s10052-024-13151-4
[2] ATLAS Collaboration, "The ATLAS Experiment at the CERN Large Hadron Collider", JINST 3 (2008) S08003
[3] High-Luminosity Large Hadron Collider (HL-LHC) : Preliminary design 304 report, CERN (2015), http://dx.doi.org/10.5170/CERN-2015-005
[4] ATLAS Collaboration, "Technical Design Report for the Phase-II Upgrade of the ATLAS Tile Calorimeter", CERN-LHCC-2017-019, https://cds.cern.ch/record/2285583
[5] E. Valdes, et al., Readiness of the ATLAS Tile Calorimeter link 309 daughterboard for the High-Luminosity LHC era, PoS(TWEPP2019)087 310 (2019). http://refhub.elsevier.com/S0168-9002(22)01152-4/sb6
[6] T. Martin and F. Schmidt, "Cocotb: Coroutine-based co-simulation testbench for verifying RTL using Python," GitHub Repository, https://github.com/cocotb/cocotb.

https://doi.org/10.1051/epjconf/202533801005

The Compact Processing Module for the Upgrade of the ATLAS Tile Calorimeter Towards the High-Luminosity LHC

F. Carrió[1,*], A. Cervelló[1], L. Fiorini[1], D. Hernández[1], A. Ruiz[1], and A. Valero[1]
[1]Instituto de Física Corpuscular (CSIC-UV), Valencia, Spain
(*) fernando.carrio@cern.ch

Abstract— **The Tile Calorimeter (TileCal) is the central hadronic calorimeter of the ATLAS experiment at the Large Hadron Collider (LHC). In 2026, the LHC will undergo a series of upgrades leading to the High-Luminosity LHC (HL-LHC), which will provide an instantaneous luminosity 5 to 7 times larger than the nominal LHC design value. The ATLAS Tile Calorimeter Phase-II Upgrade will completely replace the readout electronics with a new clock distribution and readout architecture with a fully digital trigger system to process more complex physics events while maintaining the trigger selection efficiency.**

The upgraded on-detector electronics will transmit detector data every 25 ns to the Compact Processing Modules (CPMs) in the counting rooms, which handle real-time energy reconstruction, trigger data transmission, and clock distribution. A total of 128 CPMs hosted in 32 ATCA carriers will read out the entire detector with a total data bandwidth of 40 Tbps. Each CPM, equipped with a Xilinx Kintex UltraScale KU115 FPGA and six FireFly modules, processes data from two TileCal modules, transmitting both reconstructed and raw data to TDAQi and FELIX systems. In addition, the CPM distributes the accelerator clock and configuration commands to the DaughterBoards in the on-detector. This work presents the current status of the CPM developments for the HL-LHC and the roadmap towards its installation in the ATLAS experiment.

Keywords — **ATLAS Tile Calorimeter (TileCal), Data Acquisition (DAQ) systems, Field-Programmable Gate Array (FPGA), High Energy Physics, High-Speed Electronics.**

I. INTRODUCTION

THE Tile Calorimeter (TileCal) [1] is the central hadronic calorimeter of the ATLAS experiment [2], covering a pseudorapidity of $|\eta| < 1.7$. TileCal is a sampling calorimeter made of steel plates and scintillator material. This calorimeter contributes to the measurement of jets, tau leptons, and missing transverse energy. TileCal is composed of three cylindrical barrels along the beam axis: the central Long Barrel and two Extended Barrels. Each barrel comprises 64 modules, each one hosting the photomultiplier tubes (PMTs) and readout electronics in the outermost part of the module. Fig. 1 shows in detail the structure of a TileCal module. The complete readout of the detector is done using approximately 10,000 PMTs.

In the on-detector electronics, the analog PMT signals are shaped and amplified with two different gains (1:64 ratio), digitized with 10-bit Analog-to-Digital Converters (ADCs) and then stored in pipeline memories. The low-gain PMT signals are summed forming pseudoprojective trigger towers and transmitted to the ATLAS trigger system through electrical cables. Upon the reception of a first-level trigger signal, the TileCal modules send the selected events to the Read-Out Drivers (ROD) [3] which reconstruct the energy and time per channel and transmit all the processed data to the next elements in the ATLAS DAQ chain after the first-level trigger.

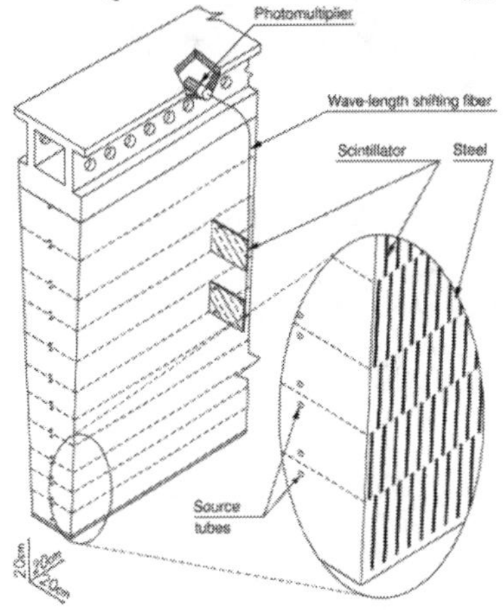

Fig. 1: Detail of a Tile module showing the 11 radial layers of scintillator tiles interleaved with the steel absorber, and the position of the PMTs and readout electronics [1].

II. TILECAL AT THE HL-LHC

During the Long Shutdown 3 (LS3) (2026–2030) the Large Hadron Collider (LHC) will be upgraded towards the High-Luminosity LHC (HL-LHC). The HL-LHC will deliver an instantaneous peak luminosity up to 7.5×10^{-34} cm^{-2}s^{-1}, and a total integrated luminosity of 4000 fb^{-1} by the end of its data-taking (2030–2041).

TileCal will undergo a series of upgrades as part of the ATLAS Phase-II Upgrade [4] during LS3, to meet the stringent requirements of the HL-LHC and the new ATLAS Trigger and Data Acquisition (TDAQ) architecture. These upgrades are necessary to cope with increased data rates, higher radiation levels, and low latency constraints. Both on-detector and off-detector readout electronics will be replaced introducing a new readout strategy to enable the transmission of full-digital data with fine granularity information to the ATLAS trigger system.

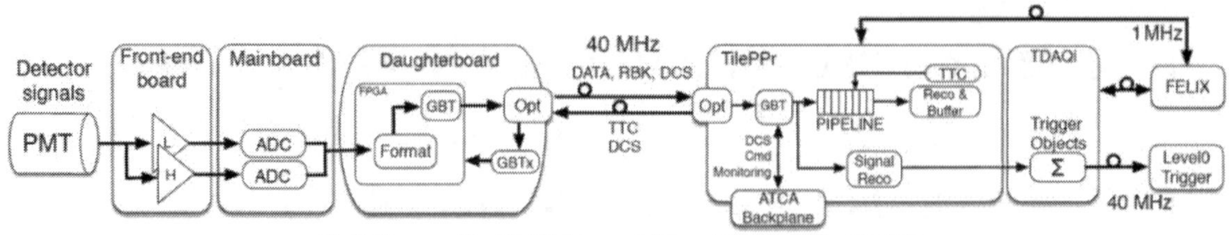

Fig. 2. Diagram of the TileCal data acquisition architecture for the HL-LHC.

In the new data acquisition architecture (Fig. 2), the on-detector readout electronics will transmit detector data every bunch crossing (25 ns) to the off-detector electronics. The Tile PreProcessor (TilePPr) [5] modules, in the off-detector electronics, will reconstruct cell energies per bunch crossing transferring the results to the Trigger and Data Acquisition Interface (TDAQi) [6] boards. The TDAQi boards will use the reconstructed cell energies to build trigger primitives for the ATLAS trigger system. The TilePPr modules will store the samples and reconstruct energy per channel in pipeline memories, until the reception of a trigger acceptance signal when the event is transmitted to the ATLAS Front End Link eXchange (FELIX) [7] system.

A. The upgraded on-detector electronics

The upgraded TileCal modules at the HL-LHC will be divided into independent and identical sub-modules called mini-drawers to enhance the reliability of the detector electronics and facilitate their maintenance. Long Barrel modules will contain four mini-drawers, and Extended Barrel modules three mini-drawers. Each mini-drawer constitutes an independent readout subsystem containing all the necessary services for operating 12 PMTs, including redundant readout electronics and power supplies.

As shown in Fig. 3, a mini-drawer hosts one MainBoard [8], one DaughterBoard [9], one high-voltage distribution board, and 12 PMTs equipped with FENICS [10] cards (Front-End for the New Infrastructure with Calibration and Signal Shaping).

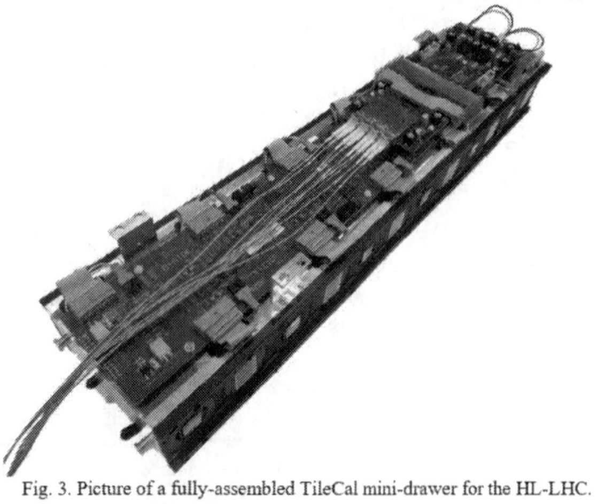
Fig. 3. Picture of a fully-assembled TileCal mini-drawer for the HL-LHC.

Each PMT block is equipped with one FENICS card which shapes and amplifies the PMT signals with two gains (1:40 ratio). The FENICS cards provide additional capabilities as a calibration circuit to calibrate the electronics over the full dynamic range, and an integrator circuit for luminosity measurements and Cesium calibration.

The MainBoards are responsible for the digitization of the PMT signals, and the configuration of the FENICS cards. Each MainBoard is equipped with twelve dual 12-bits ADCs to digitize the two gain signals coming from 12 PMTs, twelve 16-bit ADCs to read out the integrator signal, and four Intel Cyclone IV FPGAs for configuration and control of the FENICS cards. The analog PMT signals are digitized at 40 MHz (one sample per bunch crossing) and transferred to the DaughterBoards.

The DaughterBoards are responsible for transmitting the detector data to the off-detector electronics, as well as for the reception and distribution of the accelerator clock, timing control and configuration commands for the on-detector electronics. Each DaughterBoard is equipped with two Xilinx Kintex UltraScale XCKU035 FPGAs for the data acquisition and processing tasks, two GBTx [11] ASICs for the recovery of the accelerator clock, and four SFP modules for the communication with the off-detector electronics.

B. The upgraded off-detector electronics

The off-detector readout electronics of TileCal for the HL-LHC has been designed as a modular system composed of 32 TilePPr modules and 32 TDAQi boards. Each TilePPr consists of an Advanced Telecommunications Computing Architecture (ATCA) Carrier Blade Board (ACBB) and four Compact Processing Modules (CPMs), which implement the core data acquisition and processing functionalities.

The ACBB is a cutaway ATCA carrier specifically designed to provide power distribution, mechanical support, and connectivity to the TDAQi via the Zone 3 connector. Communication and control functions are distributed over three auxiliary mezzanine cards:

- TileCoM: A System-on-Chip board that manages the configuration on-detector and off-detector electronics, as well as the transmission of sensor monitoring data to the Detector Control System (DCS).
- GbE Switch: Provides high-speed Ethernet communication between the CPMs, TileCoM, and TDAQi, ensuring communication with the ATLAS network.
- Intelligent Platform Management Controller (IPMC): Implements ATCA-compliant monitoring and control features, such as hot-swap handling, temperature and power monitoring.

The TilePPr is responsible for receiving the digitized data from all 256 TileCal modules, buffering up to 10 µs of detector data per PMT channel, and reconstructing the energy deposited in each cell for every bunch crossing. The reconstructed energy is transmitted every 25 ns to the TDAQi modules for the computation of trigger primitives for the ATLAS Level-0 trigger system. Upon receiving a Level-0 accept signal, the selected events are extracted from the pipelines, and transmitted to the ATLAS FELIX system for further processing and event building.

III. THE COMPACT PROCESSING MODULE

The CPM is the central element of the Tile PreProcessor system, responsible for handling all functions related to detector data reception, energy and time reconstruction, data buffering, and interface with the ATLAS TDAQ system, more specifically with the L0-trigger and the FELIX systems.

A. Hardware design

Each CPM is implemented as a full-size Advanced Mezzanine Card, designed to read out and operate up to two TileCal modules. It is equipped with six Samtec FireFly optical transceivers to establish high-speed, low-latency optical links with the on-detector electronics using the GigaBit Transceiver (GBT) protocol [12]. Fig. 4 presents a picture of the CPM.

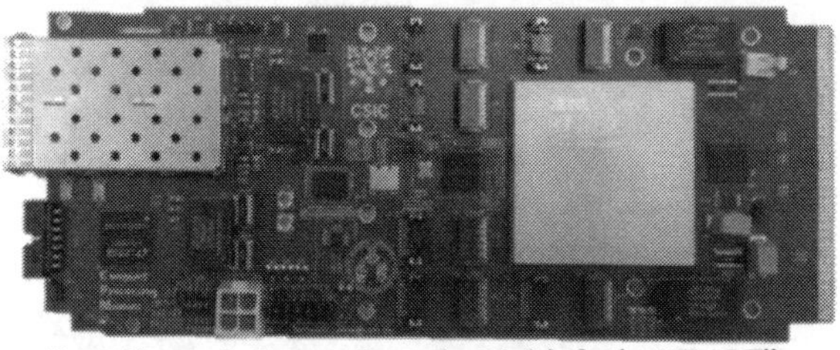

Fig. 4. Picture of a Compact Processing Module for the ATLAS Tile Calorimeter at the HL-LHC.

The PCB consists of 14 layers, out of which 6 layers are dedicated to power distribution and 8 to high-speed and control signals. The board has a total thickness of 1.6 mm ±10%, compliant with the ATCA standard. Panasonic Megtron 6 dielectric material was selected for its excellent electrical performance at high frequencies, being critical to keep a good signal integrity for the high-speed communication lines.

Four Samtec FireFly modules receive detector data via uplinks at 9.6 Gbps, while two Samtec FireFly modules transmit configuration commands and the accelerator bunch-crossing clock via downlinks at 4.8 Gbps. The clock distribution system is based on a low-jitter Skyworks Si5345 jitter cleaner, which guarantees low-jitter distribution of the accelerator clock with fixed and deterministic latency. This ensures synchronous sampling of the PMT signals by the on-detector electronics.

B. System functionalities

The core processing is implemented in a Xilinx Kintex UltraScale XCKU115 FPGA, which handles all the detector data acquisition and bunch cross alignment, real-time energy reconstruction, formatting, and transmission to the TDAQ subsystems.

The FPGA implements all the high-speed interfaces with the on-detector electronics and TDAQ subsystems using the embedded GTH MultiGigabit Transceivers [13]. A total of 32 GBT links are used to receive detector data and configure the on-detector electronics via the Samtec FireFly modules. Four FULL-Mode (9.6 Gbps) links are used for the transmission of trigger primitives to the TDAQi through Zone 3 connector of the ACBB, while two additional FULL-Mode links are used to transmit triggered data to FELIX. An additional high-speed interface is used for communication with the ATLAS Technical and Control Network via the GbE switch hosted by the ACBB.

Once the detector data is received, the FPGA performs real-time energy reconstruction for up to 77 PMT channels for every bunch crossing. In parallel with the reconstruction path, full-resolution detector data and their corresponding reconstructed values are stored in pipeline memories with a depth of 10 μs, awaiting a Level-0 trigger accept signal. Upon acceptance, the relevant event fragments are extracted from the pipelines and transmitted to the FELIX system.

Several techniques for energy reconstruction are currently under evaluation, including machine learning techniques such as Multi-Layer Perceptrons (MLPs), and Convolutional Neural Networks (CNNs), as well as other traditional techniques as Wiener and Kalman filters. All these techniques are being benchmarked in terms of latency, FPGA resource usage, and robustness to pile-up.

To ensure precise clock distribution to the on-detector electronics for the sampling of the PMT signals, the FPGA implements several Digital Dual Mixer Time Difference (DDMTD) circuits [14]. These blocks measure the phase drift between the recovered clock from the GBT links and the reference LHC clock extracted from the FELIX system with a resolution of approximately 6.1 ps with an uncertainty below 2 ps-RMS. The resulting phase measurement is used to dynamically correct the phase of the distributed clock if necessary. Fig 5 shows the results of the validation tests of the DDMTD circuit implemented in the CPM using the Phase Interpolator of the GTH transceiver to increment the phase of the recovered 40 MHz from FELIX in steps of 1.22 ps.

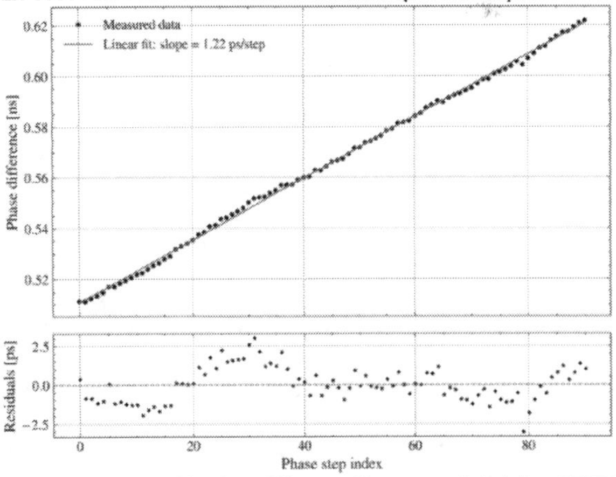

Fig 5. Measurement of the phase difference of the recovered clock from FELIX (40 MHz) and the same clock after being delayed in steps of 1.22 ps with the Phase Interpolator of a GTH Transceiver.

This phase monitoring system is crucial to ensure stable and synchronous sampling with the accelerator clock, correcting the phase variations in the distributed clock caused by temperature fluctuations or by timing uncertainties during the reinitialization of the communication blocks.

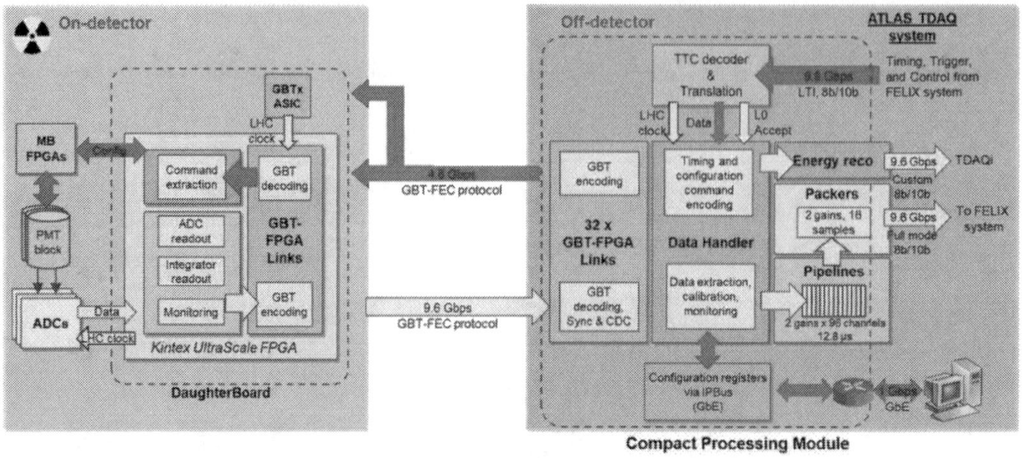

Fig 6. Block diagram of the firmware for the data acquisition system of the Demonstrator, including the CPM and DaughterBoard.

Finally, the Xilinx XCKU115 FPGA also receives monitoring data from the on-detector electronics, including temperature, voltage and current readings from the DaughterBoards and MainBoards. These data are transmitted to DCS via the TileCoM for real-time health monitoring.

C. Integration with the Tile Demonstrator module

The Compact Processing Module has undergone extensive validation through the operation of the TileCal Demonstrator module in the ATLAS experiment. As part of the ATLAS Tile Calorimeter R&D program for the HL-LHC, a Demonstrator module equipped with the upgraded on-detector electronics was built in 2014. This module is a hybrid module capable of transmitting selected data to both the current TDAQ system, and the FELIX system for the HL-LHC. Additionally, the Demonstrator module was designed to provide analog sums of the PMT signals to the ATLAS trigger system, maintaining backward compatibility with the existing TDAQ infrastructure.

The Demonstrator module was installed and commissioned in ATLAS in 2019 during the Long Shutdown 2. It has been operating since the start of Run 3 (2022–2026) to evaluate the performance of the new readout electronics under real operation conditions, as well as to identify early integration issues before the full-scale deployment of the new architecture.

The upgraded clock distribution and readout strategy for the HL-LHC was implemented in the operation of the Demonstrator module. Fig. 6 shows a complete block diagram of the main firmware blocks and interconnections of both on-detector and off-detector electronics of the Demonstrator module. The on-detector electronics transmit detector data to the CPM in the counting rooms every 25 ns. The LHC clock is recovered from the FELIX input link with fixed and deterministic latency and it is distributed to the on-detector electronics through the downlinks embedded within the control and timing commands.

The CPM receives detector samples at 40 MHz, with an input bandwidth of approximately 160 Gbps. The detector data are unpacked and stored in pipeline memories with a depth of 12.8 μs. Upon reception of a first-level accept signal, the corresponding data are extracted from the pipeline memories, formatted and transmitted to the RODs and to the FELIX system.

Extensive tests have been carried out during the installation and maintenance periods, in which the Demonstrator module was upgraded with newer prototypes of the readout electronics and additional functionalities. The Demonstrator module has demonstrated good performance in terms of noise during data taking, showing comparable results to the legacy modules.

IV. CONCLUSIONS

The upgrade of the TileCal readout electronics for the HL-LHC requires a complete redesign of both the on-detector and off-detector systems to cope with the increased data throughput, radiation levels, and trigger latency constraints imposed by the HL-LHC and the upgraded TDAQ architecture. The Compact Processing Module is the core component of the off-detector readout architecture, implementing data acquisition, real-time energy reconstruction, clock distribution, and communication with the ATLAS Trigger and FELIX systems.

The development of the CPM including both hardware and firmware has reached a mature state following extensive validation through integration tests and sustained operation within the ATLAS experiment during Run 3, as part of the preparation for the HL-LHC upgrade. The CPM ensures backward compatibility of the Demonstrator module with the legacy DAQ system while validating the new readout strategy in a real detector environment.

A pre-production of 14 CPMs is currently ongoing, to be followed by the full production of 128 CPMs by the end of the year. Future efforts will focus on completing energy reconstruction algorithms and firmware development, performing large-scale system integration, and finalizing the full commissioning in preparation for HL-LHC operations starting in 2030.

REFERENCES

[1] ATLAS Collaboration, "Operation and performance of the ATLAS Tile Calorimeter in LHC Run 2", EPJ Conf., vol. 84, no. 12, Dec. 2024. DOI: https://doi.org/10.1140/epjc/s10052-024-13151-4

[2] ATLAS Collaboration, "The ATLAS experiment at the CERN Large Hadron Collider", Journal of Instrumentation 3, p. S08003 (2008). DOI: https://doi.org/10.1088/1748-0221/3/08/S08003
[3] A. Valero et al., "ATLAS TileCal Read-Out Driver production", Journal of Instrumentation, 2: P05003, 2007. DOI: https://doi.org/10.1088/1748-0221/2/05/P05003
[4] ATLAS Collaboration, "Technical Design Report for the Phase-II Upgrade of the ATLAS Tile Calorimeter", CERN-LHCC-2017-019, https://cds.cern.ch/record/2285583
[5] F. Carrió and A. Valero, "The PreProcessor module for the ATLAS Tile Calorimeter at the HL-LHC", Nucl. Instrum. Methods Phys. Res., A, 958:162487, 2020. DOI: https://doi.org/10.1016/j.nima.2019.162487
[6] X. Yue et al., "Tile TDAQ interface module for the Phase-II Upgrade of the ATLAS Tile Calorimeter", in Proc. of the 2019 IEEE Nuclear Science Symposium and Medical Imaging Conference (NSS/MIC), 2019, United Kingdom, pp. 1-4. DOI: https://doi.org/10.1109/NSS/MIC42101.2019.9059861
[7] J. Anderson et al., "FELIX: a High-Throughput Network Approach for Interfacing to Front End Electronics for ATLAS Upgrades". In Proc. of the 21st International Conf. on Computing in High Energy and Nuclear Physics (CHEP2015), J. Phys. Conf. Ser., 664(8):082050, 2015. DOI: https://doi.org/10.1088/1742-6596/664/8/082050
[8] F. Tang et al., "Design of Main board for ATLAS TileCal Demonstrator", In Proc. of the 19th IEEE-NPSS Real Time Conference, 2014, Japan, pages 1–3. DOI: https://doi.org/10.1109/RTC.2014.7097541
[9] E. Valdes Santurio et al., "A Revised Version of the ATLAS Tile Calorimeter link DaughterBoard for the HL-LHC", IEEE Transactions on Nuclear Science, 68(9):2414–2420, 2021. DOI: https://doi.org/10.1109/TNS.2021.3103408
[10] D. Calvet et al., "Upgrade of ATLAS Hadronic Tile Calorimeter for the High Luminosity LHC", Nucl. Instrum. Methods Phys. Res., A, 958:169614, 2024. DOI: https://doi.org/10.1016/j.nima.2024.169614
[11] P. Leitao et al., "Test bench development for the radiation hard GBTX ASIC", Journal of Instrumentation, 10(01):C01038–C01038, 2015. DOI: https://doi.org/10.1088/1748-0221/10/01/C01038
[12] M. Barros Marin et al., "The GBT-FPGA core: features and challenges", Journal of Instrumentation, 10(03):C03021–C03021, 2015. DOI: https://doi.org/10.1088/1748-0221/10/03/C03021
[13] Xilinx Inc, "UltraScale Architecture GTH Transceivers User Guide - UG576 (v1.7.1)", August 2021. Online: https://docs.amd.com/v/u/en-US/ug576-ultrascale-gth-transceivers
[14] P. Moreira et al., "Digital dual mixer time difference for sub-nanosecond time synchronization in ethernet", In Proc. of the 2010 IEEE International Frequency Control Symposium, USA, 2010 pages 449–453. DOI: https://doi.org/10.1109/FREQ.2010.5556289

ACKNOWLEDGMENTS

This work was supported by the Spanish Ministry of Science and Innovation (MCIN) and the State Research Agency (AEI) under project PID2021-15069OB-100.

Charge signal formation in the TPC Vertical Drift design of DUNE

Joshua Pinchault[1,2*], Laura Zambelli[1], Dominique Duchesneau[1] and Johann Collot[2]
[1]LAPP, France
[2]LPSC, France
(*) pinchault@lpsc.in2p3.fr

***Abstract*—Giant Liquid Argon Time-Projection Chambers (LArTPC) will be used for the far detectors of the next generation of neutrino oscillation experiment: Deep Underground Neutrino Experiment (DUNE). This work focuses on the design of the Vertical Drift module (DUNE-VD), where the chamber consists of two drift volumes separated by a central cathode. This novel anode enables the detection of the ionization electrons through two inductions views and a charge collection view. Extensive R&D has been carried out at CERN in the past years to optimize the VD anode design. In this paper, a numerical simulation of the signal formation based on the Shockley Ramo's theorem and the physics of charge carriers' motion in liquid argon will be presented. Comparisons with radioactive sources data taken at CERN with prototypes will be shown. A good understanding of the collected signal shape and strength is essential to achieve the track reconstruction efficiency and the calorimetric measurement resolution needed by the DUNE experiment to study neutrino oscillations.**

***Keywords* — Simulation, Charge Signal and LArTPC**

I. Introduction

THE Vertical Drift (DUNE-VD) [1] configuration is based on the use of a novel anode design. The electrons produced by the ionization of argon atoms by charged particles passing through the chamber will drift to two independent anode planes located at the top and at the bottom of the detector.

This modular anode technology, developed for large-scale applications, is composed of two perforated Printed Circuit Boards (PCBs) as shown in Fig.1. On each side, copper strips for charge readout are etched in different orientations to enable track reconstruction, with each plane biased at a different voltage. The first plane acts as a shielding layer to protect the DAQ system against potential discharges due to the accumulation of charge carriers on the PCB surface. Moreover, this plane screens the signal that would otherwise be induced on the electrodes by charge moving in the drift volume. Thus, without this plane, the amplitude of the induced signals would be reduced. Then, two induction views produce bipolar signals from the motion of charges in their proximity. These planes are called induction 1 (view0) and induction 2 (view 1). Since the electrons are not collected, the integrated current must be zero due to charge conservation. Finally, the electrons reach the last plane called the collection (view 2), where calorimetric measurements are performed. The nominal bias voltage applied to each layer of the anode followed the configuration:

$$U_{shield} = -1\,500\text{ V};\ U_{view0} = -500\text{ V};\ U_{view1} = 0\text{ V};\ U_{view2} = +1000\text{ V}$$

All strips are connected to the grounded adapter board, and the output signals are convoluted with the impulse response of the front-end electronics. The electronics response differs between the top and bottom volumes.

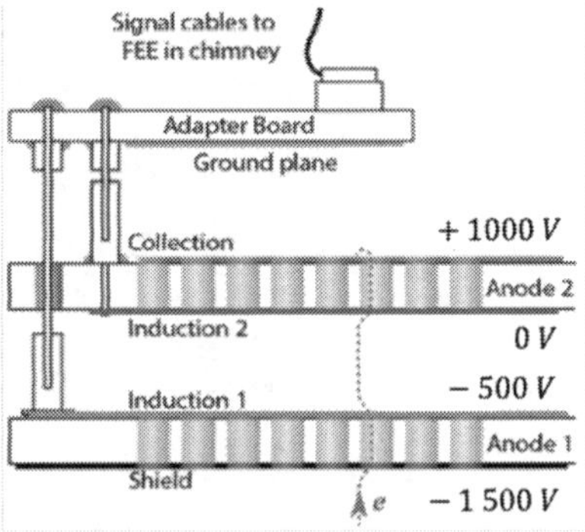

Fig. 1. Cross-sectional representation of the VD anode. The structure consists of a shielding plane (in black), followed by the induction 1 (view 0) in green, induction 2 (view 1) in blue, and finally the collection (view 2) in red. All views are connected to the ground (adapter board) to perform the charge readout. An example of a schematic electron trajectory under nominal bias conditions is also shown.

Each plane features its own hexagonal hole pattern, covering 60% of the total surface area and allowing electrons to drift through the anode. This value defines the minimal geometric transparency of the anode at zero electric field. The applied drift field focuses the electrons into the holes, thereby increasing the overall transparency. Unlike traditional wire planes commonly used in this type of detectors, drilled PCB are easier to manufacture and less expensive.

This paper describes the development of a numerical simulation of the induction signal formation. The objective of this work is to gain a deeper understanding of the waveform of the induced signals on each view and to characterize the systematic effects that could lead to charge loss within the anode. In this context, it is essential to account for the anode geometry and the layout of the various charge readout planes.

II. Simulation

The simulation relies on the Shockley–Ramo's theorem [2] to compute the electric currents induced on the readout electrodes by the motion of ionization electrons. This approach enables a precise modeling of the signal formation process in segmented detectors, by separating the influence of detector geometry (through the weighting fields) from the actual charge dynamics. The induced current $i_n(t)$ associated with the nth electrode is calculated as the projection of the electron drift velocity $\overrightarrow{v_d}$ onto the weighting field $\overrightarrow{E_w}$. The Shockley-Ramo's theorem is given by:

$$i_n(t) = q\,\overrightarrow{v_d}\cdot\overrightarrow{E_w} \qquad (1)$$

This formalism is particularly well suited for complex anode designs, such as the modular perforated PCB structure. The simulation is divided into two parts: First, field calculations, aimed at determining both the drift field $\overrightarrow{E_d}$ inside the anode and the weighting field associated with each readout electrode is computed. The second part describes the charge dynamics in liquid argon as a function of $\overrightarrow{E_d}$, including drift and signal induction processes.

A. Finite Difference Method

The electric potential is obtained by solving the generalized Poisson equation:

$$\vec{\nabla}\cdot\left(\varepsilon_r\,\vec{\nabla}\,\varphi(x,y,z)\right) = 0\,, \qquad (2)$$

where φ is the electric potential and ε_r is the relative dielectric constant.

This relationship takes into accounts for the dielectric interfaces between the PCB and the liquid argon. This equation is solved using the finite difference method in a Cartesian coordinate system. The use of Cartesian coordinates is motivated by the hexagonal geometry of the anode. The iterative schema of the finite difference method (FDM) including the dielectric constant is obtained from (2) and is given by:

$$V_{i,j,k}^{(n+1)} = \frac{M_{i+1,j,k}^{(n)}+M_{i-1,j,k}^{(n)}+M_{i,j+1,k}^{(n)}+M_{i,j-1,k}^{(n)}+M_{i,j,k+1}^{(n)}+M_{i,j,k-1}^{(n)}}{\epsilon_{i+1,j,k}+\epsilon_{i-1,j,k}+\epsilon_{i,j+1,k}+\epsilon_{i,j-1,k}+\epsilon_{i,j,k+1}+\epsilon_{i,j,k-1}}\,, \qquad (3)$$

where $V_{i,j,k}^{(n+1)}$ is the potential value at $(n+1)^{th}$-iteration, $\epsilon_{i,j,k}$is the dielectric tensor and $M_{i,j,k}^{(n)} \equiv \epsilon_{i,j,k}V_{i,j,k}^{(n)}$.

In order to simulate an infinite plane, mirror boundary conditions were applied. The copper readout strips, held at fixed potential, are one node thick, corresponding to 50 micrometers. The entire mesh is defined with a uniform step value h. The electric field components are then computed from central differences of the potential obtained via the FDM using (3). The results of the 3D simulation are presented in Fig. 2, where the electric field lines are shown to emanate from the collection plane and extend into the drift volume, accompanied by the corresponding PCBs contours.

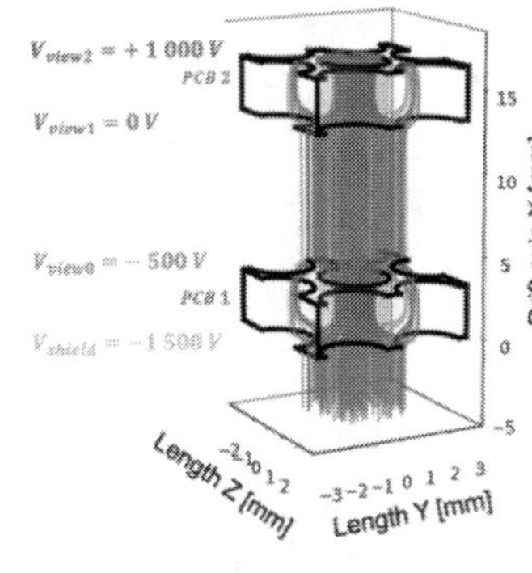

Fig. 2. Results of the drift field simulation inside the anode. In red, the electric field lines diverging from the collection electrode into the drift volume. The electric field is computed by FDM. In black, the representation of both perforated PCBs.

The electric field strength ranges from 0.5 kV/cm to 3 kV/cm inside the holes.

B. Weighting Potential

The weighting field is a virtual field that reflects the spatial distribution of the electrodes. In an isotropic medium, it is entirely determined by the geometry of the detector. To perform this calculation, the numerical method developed in the previous section is applied. Specifically, a unit potential (1 V) is set to a single readout strip, while all other strips are held at 0 V. The induction strips are 7.65 mm wide, whereas the collection strips have a width of 5.1 mm. Since the objective is to compute the average waveform for each readout view, it is not necessary to account for the relative orientation of the strips at this stage.

Instead, the weighting potential Φ_w is defined with respect to a specific electrode n and S_N the surfaces of the electrodes held at fixed potential as:

$$\vec{\nabla}\cdot\left(\varepsilon_r\,\vec{\nabla}\,\Phi_{w,n}(x,y,z)\right) = 0\,, \qquad (4)$$

with $\Phi_{w,n}|_{S_n} = 1\,;\ \Phi_{w,n}|_{S_{N\neq n}} = 0$

The boundary conditions on the planes colinear with the readout strips n are not known, which requires a larger calculation domain than that used for the drift field calculation. $\Phi_{w,n}(x,y,z)$ is screened by neighboring electrodes until it approaches zero. It has been found that the weighting field becomes almost zero beyond three adjacent strips, effectively defining the extend of the calculation domain. A 2D projection of the weighting field equipotentials in the plane orthogonal to the readout strips are shown in Fig. 3. This result includes the effect of the dielectric interface between the PCB and the liquid argon, as well as the drift field lines, represented in green.

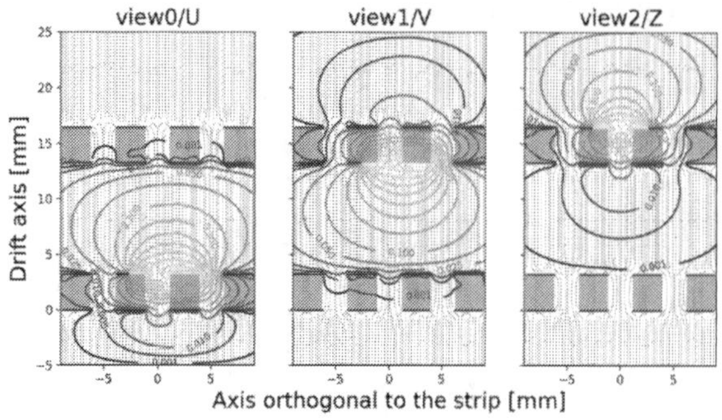

Fig. 3. Weighting potential for each view. The green lines represent the drift field lines, whereas the grey boxes correspond to the PCB regions. The colored contours show the equipotential lines of the weighting field.

C. *Motion of Charge Carriers in Liquid Argon*

The dynamics of charge carriers within the anode region are governed by the electric field and the interactions between electrons and argon atoms. The energy of ionization electrons is at sub-eV [3] : the mean free path is on the order of a nanometer, small compared to the dimensions of the system under study (on the order of a millimeter for the anode). As a result, electron dynamics are described using averaged quantities derived from models fitted to experimental measurements. The electron drift velocity is given by:

$$\vec{v_d} = \mu(\vec{E_d}, T) \times \vec{E_d} \ , \tag{5}$$

where $\mu(\vec{E_d}, T)$ is the electron mobility and T the liquid argon temperature.

For low drift fields, the mobility remains constant, and the velocity is directly proportional to the electric field. However, above approximately 100 V/cm [4] , drifting electrons exit the thermal regime, and their spatial distribution can no longer be described by a classical Maxwellian. In this hot electron regime, the mobility becomes field-dependent, and a portion of the electron energy is transferred to the argon atoms via elastic scattering processes. Fig. 4. shows the drift velocity of electrons as a function of the electric field for different temperatures based on the global data fit presented in [1] .

At the nominal drift field of 500 V/cm, the electron drift velocity in the active volume is approximately 1.6 mm/µs. This velocity is largely insensitive to small temperature variations inside the cryostat, which are estimated to be on the order of a few mK. One of the key features of the simulation is the inclusion of diffusion effects of the electron cloud in liquid argon. The spatial distribution of drifting electrons as a function of time is described by Fick's law [6] . The electrons follow a Gaussian distribution, whose standard deviation corresponds to the diffusion length $\sigma_{L,T}$, given by:

$$\sigma_{L,T} = \sqrt{2D_{L,T}t} \ , \tag{6}$$

where $D_{L,T}$ is the diffusion coefficient, t the drift time and the subscripts L and T refer to longitudinal and transverse components of the trajectory, respectively.

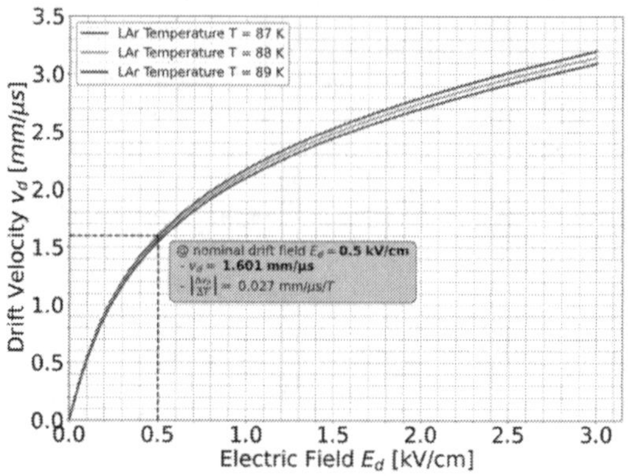

Fig. 4. Drift velocity profile in liquid argon as a function an external electric field from global data fit according to [5].

Measurements of electron diffusion in liquid argon are challenging due to its relatively low magnitude, which requires high-precision experimental setups. The work of Li et al. [1] measured longitudinal diffusion using a gridded drift cell technique. The simulation implements model for the longitudinal diffusion coefficient $D_L(\vec{E_d}, T)$ and uses the relation provided in [7] to obtain the corresponding transverse diffusion coefficient $D_L(\vec{E_d}, T)$:

$$D_T = D_L \left(1 + \frac{E_d}{\mu}\frac{\partial \mu}{\partial E_d}\right)^{-1} \ , \tag{7}$$

Fig. 5. shows the profile of the diffusion coefficients as a function of the drift electric field and the effective electron energies associated with each direction are shown, as obtained using the Einstein-Smoluchowski relation [8] such that:

$$D_{L,T} = \frac{\mu \, \epsilon_{L,T}}{e} \ , \tag{8}$$

where $\epsilon_{L,T}$ is the electron effective energy in eV.

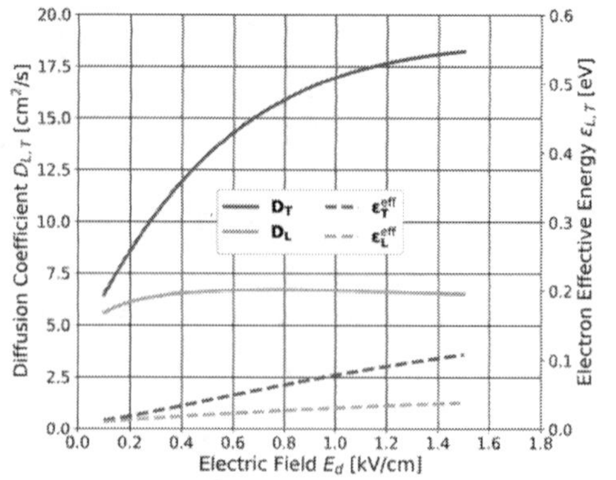

Fig. 5. Effective energy and diffusion coefficients as a function of an external electric field according to [5].

The electron diffusion is dominated by the transverse component. The model is extrapolated up to 3 kV/cm, which corresponds to the maximum electric field value obtained inside the anode.

D. *Induced signal by an electron cloud*

The physical quantities describing electron motion in liquid argon are used to propagate electrons through the anode structure and to directly apply the Shockley–Ramo's theorem as expressed in (1). Fig. 6. provides a qualitative representation of the product $< F_r > = < \overrightarrow{v_d} \cdot \overrightarrow{E_w} >$, known as the detector field response. This field identifies the regions within the detector where a drifting electron cloud would induce a signal on a readout charge strip. The field response is shown separately for each view. Its average value accounts for the periodic hole pattern along the charge readout strips. As the electron cloud crosses of positive and negative regions for the induction views, the resulting signals are bipolar. In contrast, the collection view corresponds to a region with only positive response field, leading to a unipolar signal whose integral equals the total collected charge.

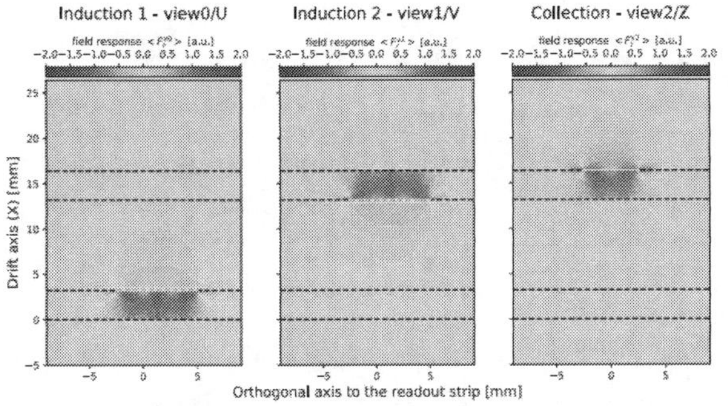

Fig. 6. The averaged field response of the detector for each view as a function of spatial coordinates. The third dimension represents the polarity and intensity of the field for a charge readout strip.

The field response is screened by the other electrodes, but not entirely suppressed. Consequently, electrons collected by adjacent strips induce a bipolar signal on the readout strip n. This effect contributes to the signal shape, affecting the overall waveform and requiring careful consideration in the simulation. The simulated signals based on the Shockley–Ramo's theorem for an electron cloud drifting through the anode are shown in Fig.7. (a). Bipolar waveforms for induction views and a unipolar waveform for collection view are obtained. For the latter, integral yields to the total collected charge. The signals $S_n(t)$ refer to the induced signal from electrons collected by electrode n. The signals $S_{n\pm1}(t)$ and $S_{n\pm2}(t)$ correspond to the contributions induced by electrons collected on neighboring strips. It is observed that the total charge collected on view 2 is slightly less than one. A systematic estimated at 7% of the charge on view 2 has been observed because a fraction of the electron being collected on view 0 as a result of diffusion effects.

The peak observed in collection view is caused by point-like effects near the collection electrode. The signal shape will be smoothed out by the output electronics response. In practice, the measured output signal $M(t)$ is the convolution of the induced signal with the electronics response, and can be written as:

$$M(t) = \int S(t') \, R(t - t') dt', \quad (9)$$

where $S(t)$ is the induced signal and $R(t)$ is the impulse response of the front-end electronics.

The Fig.7. (b) shows the convolved signals $M(t)$ for each view.

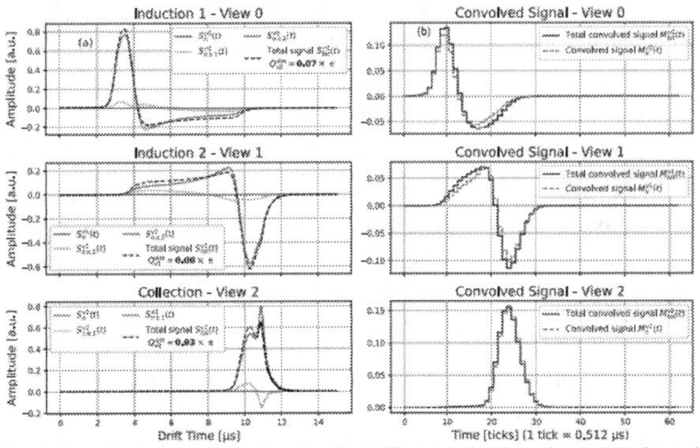

Fig. 7. Result of signal formation for the different induction views. (a) The red curve shows the induced signal considering only the readout strip n; the yellow curve corresponds to the contribution from neighboring strips at $n \pm 1$ and the red curve to those at $n \pm 2$. The black dash curve represents the total expected signal, obtained by summing all contributions. (b)

It was observed that accounting for the contribution from all adjacent strips modifies the shape of the convolved output waveform $M_{tot}(t)$ for the induction views. The simulation was then compared to data in order to validate the model.

III. Comparison with data

To validate the simulation, data from a 50 L R&D Time Projection Chamber (TPC) has been used. The drift length is 52 cm and the active area is 32 cm × 32 cm. The TPC is equipped with a perforated anode featuring three charge readout views as described previously and is surrounded by a field cage to maintain a uniform electric field. Fig.8. shows the experimental setup installed at CERN. The data were acquired under nominal operating conditions, with a drift electric field set to 500 V/cm. The bias voltages applied to each layer of the anode followed the configuration (−1500 V, −500 V, 0 V, +1000 V) as seen before.

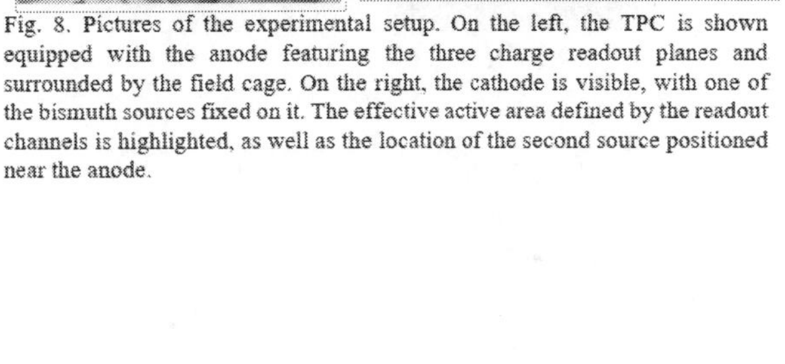

Fig. 8. Pictures of the experimental setup. On the left, the TPC is shown equipped with the anode featuring the three charge readout planes and surrounded by the field cage. On the right, the cathode is visible, with one of the bismuth sources fixed on it. The effective active area defined by the readout channels is highlighted, as well as the location of the second source positioned near the anode.

Inside the detector, two ^{207}Bi sources were installed, one close to the anode side and the other near the cathode. These sources emit two conversion electron rays around 1 MeV used for detector calibration studies. The range of these electrons in liquid argon is shorter than the strip width, allowing a direct comparison between the simulated waveforms and data. To reconstruct the data, we used the LARDON software [9] and tagged single hits. These are isolated hits, not associated with any reconstructed track, but correlated in space and time. The reconstructed single hits are shown on Fig. 9. The left-hand peak corresponds to the ^{207}Bi source located near the cathode, while the right-hand peak originates from the source placed close to the anode. The peak near the anode appears significantly more intense. To minimize the impact of argon purity and diffusion, we selected hits located within 5 mm of the source positioned near the anode. This selection corresponds to the expected range of 1 MeV electrons in liquid argon.

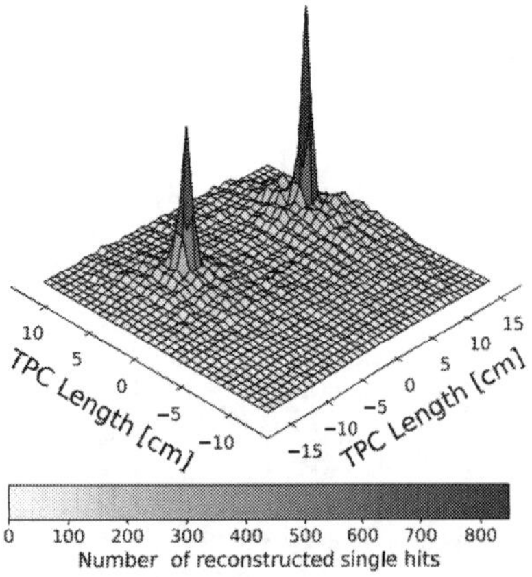

Fig. 9. All reconstructed single hits from a run under nominal operating conditions. The positions of the two ^{207}Bi sources inside the TPC are clearly identifiable.

The front-end electronics is cryogenic and immersed directly in the liquid argon to minimize noise, which is estimated to be below 1000 e^- equivalent noise charge (ENC). The system operates at a sampling frequency of 1.95 MHz and includes a low-pass filter to suppress high-frequency noise fluctuations.

All signals from each reconstructed hit are time-aligned. Induction signals are aligned based on their polarity reversal point, while collection signals are aligned at their maximum amplitude. A mean waveform is then constructed by performing a Gaussian fit to the ADC distribution in each time bin (1 bin $= 0.512\,\mu$s). Each waveform is normalized with respect to the integrated charge.

Fig. 10. presents the results of the study, showing a comparison between the simulated waveform to the average waveform extracted from data.

To quantify the agreement between simulation and data, we define the χ^2 value as:

$$\chi^2 = \sum_i \frac{\left(S_i^{sim} - D_i\right)^2}{\left(S_i^{sim}\right)^2}, \qquad (10)$$

where S_i^{sim} and D_i denote the simulated and average data waveform values at time bin i, respectively.

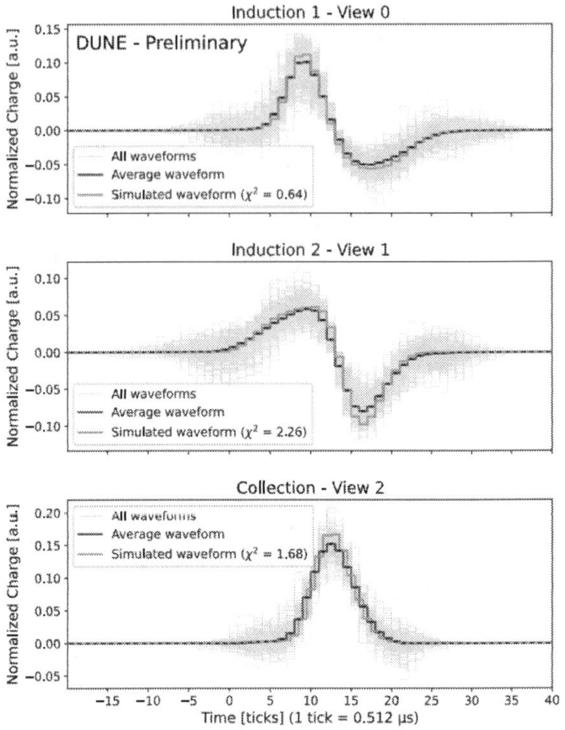

Fig. 10. Comparison between simulation and data. The selected individual waveforms are shown in grey, the averaged waveform extracted from data using a Gaussian fit is shown in purple, and the simulated waveform is displayed in orange.

This metric evaluates the similarity in waveform shape and duration across all views. The comparison shows good agreement, thereby validating the simulation for single hits.

IV. Conclusions

This work led to the development of a signal formation simulation, tailored specifically to the geometry of the perforated anodes used in the DUNE Vertical Drift TPC. The electron drift simulation, which incorporates charge transport models in liquid argon, revealed a systematic 6% charge loss on the collection plane under nominal operating conditions. Furthermore, data acquired with a small TPC at CERN, dedicated to R&D studies, were used to validate the simulation by comparing waveforms across the three charge readout views. A good agreement was found between the simulated signals accounting for the influence of adjacent strips and the waveforms extracted from reconstructed single hits originating from the source near the anode.

The next step involves investigating the information encoded in the waveforms shape. As shown in Fig. 11, the waveform varies with the incidence angle of the track, assuming a constant energy deposition. For example, for a zenith angle of $\theta = 30°$ and an azimuthal angle of $\varphi = 60°$, the waveform is modified on each view. This opens a promising direction to improve angular track reconstruction by studying the waveform shape.

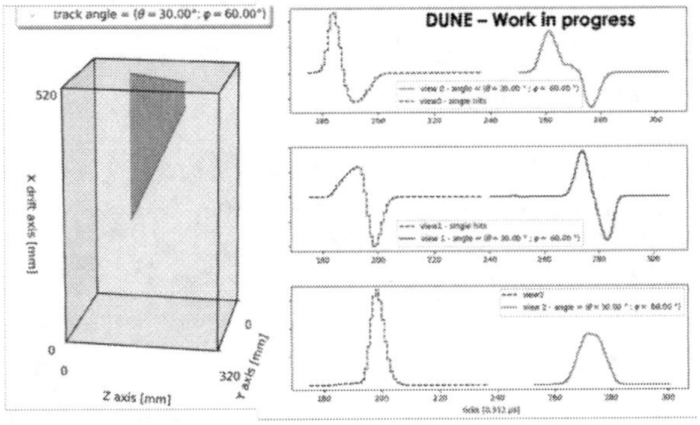

Fig. 11. Extended simulation at large scale. On the left, ionization electrons (in red) are shown drifting through the TPC volume, originating from an idealized track with a constant energy deposition of 2 MeV/cm. In green, the track projection on the anode plane. On the right, the resulting waveforms are compared: the dashed black line represents the single-hit response, while the solid lines show the modified waveforms on a given readout channel for each view.

In conclusion, the simulation results are consistent with data and enable detailed studies of waveform features as functions of geometry, bias configurations, and track angles. This provides a solid basis for improving signal reconstruction and understanding systematic effects in the detector.

References

[1] DUNE COLLABORATION, *et al.* The DUNE far detector vertical drift technology Technical design report. *Journal of Instrumentation*, 2024, vol. 19, no 8, p. T08004.

[2] RAMO, Simon. Currents induced by electron motion. *Proceedings of the IRE*, 2006, vol. 27, no 9, p. 584-585.

[3] ATRAZHEV, V. M. et IAKUBOV, I. T. Hot electrons in non-polar liquids. *Journal of Physics C: Solid State Physics*, 1981, vol. 14, no 33, p. 5139.

[4] SCHMIDT, Werner F. et ILLENBERGER, Eugen (ed.). *Electronic excitations in liquefied rare gases*. American Scientific Publishers, 2005.

[5] Y. Li, T. Tsang, C. Thorn, X. Qian, M. Diwan, J. Joshi, S. Kettell, W. Morse, T. Rao, J. Stewart et al., "Measurement of longitudinal electron diffusion in liquid argon," Nuclear Instruments and Methods in Physics Research Section A: Accelerators, Spectrometers, Detectors and Associated Equipment, vol. 816, pp. 160–170, 2016.

[6] Fick, A. (1855). Ueber diffusion. *Annalen der physik, 170*(1), 59-86.

[7] Robson, R. E. (1972). A thermodynamic treatment of anisotropic diffusion in an electric field. *Australian Journal of Physics, 25*(6), 685-694.

[8] Einstein, A. (1905). Über die von der molekularkinetischen Theorie der Wärme geforderte Bewegung von in ruhenden Flüssigkeiten suspendierten Teilchen. *Annalen der physik, 4*.

[9] Zambelli Laura, dune-lardon [Online]. Available: https://github.com/dune-lardon, Accessed on: Jul. 14, 2025

MC^2-1K, a Novel 250 μm CdTe based Imaging Spectrometer for Space Science

H. Allaire[1,2,*], O. Limousin[1], D. Baudin[3], A. Meuris[1], H. Bervas[3], F. Bouyjou[3], T. Chaminade[3], D. Chesnais[4], P. Couderc[4], O. Gevin[3], M. Onishi[5], M. Prieur[1], D. Renaud[1], F. Soufflet[4], S. Takeda[2,5], A. Toshihiko[5], F. Visticot[1]

[1]Université Paris-Saclay, Université Paris Cité, CEA, CNRS, AIM, 91191 Gif-sur-Yvette, France
[2]Kavli IPMU (WPI), UTIAS, The University of Tokyo, Kashiwa, Chiba 277-8583, Japan
[3]CEA, IRFU, Université Paris-Saclay, F-91191 Gif-sur-Yvette, France
[4] 3D PLUS SA, 78530 Buc, France
[5]iMAGINE-X Inc., 1-12-8 Shibuya, Tokyo 150-0002, Japan
(*) hugo.allaire@ipmu.jp

***Abstract* — Hard X-ray focusing optics technologies for solar physics space missions aiming at direct imaging of solar flares or hard X-ray space astronomy are challenging the development of time-resolved, high spectral and spatial resolution CdTe detectors. CEA and 3D PLUS, in the framework of the ALB3DO laboratory and with the support of the French space agency CNES, have initiated the development of an entirely new detector named MC^2. In the midterm, the detector will be a scalable, 4-side buttable 3D module including a 64 × 64-pixel CdTe sensor with a pitch of 250 μm. The patterned pixel sensor is flip-chipped on top of a mosaic of four full-custom readout ASICs named IDeF-X D^{2}R$_2$. In the development path, a preliminary prototype has been constructed and operated successfully. The device, named MC^2-1K, consists of a quarter of the MC^2 module, i.e. a 32 × 32-pixel CdTe detector with a 250 μm pitch flip chipped on top of a single IDeF-X D^{2}R$_2$ ASIC circuit. The hybrid detector is mounted onto a printed circuit board and installed into a vacuum chamber, associated with a dedicated readout bench comprised of the custom-made 32-channel Wilkinson ADC OWB-1 for system studies and performance assessment in a frame readout mode. An energy-calibrated spectrum comprising 958 channels has been successfully recorded under illumination with a ^{241}Am source. An energy resolution of 910 eV FWHM at 59.54 keV has been measured. This paper describes the detector prototype and demonstrates its first performance results. Performance limitations will be discussed and future work towards an entire module, which will be integrated in 3D, will be described.**

***Keywords* — CdTe, Hard X-ray, Imaging Spectroscopy, High Energy Astrophysics.**

I. Introduction

Hard X-ray focusing optics for future solar and astrophysical missions demand detector systems with high spatial and spectral resolution, and excellent photometry over a wide dynamic range (six orders of magnitude in counting). For astrophysics, the challenge is to reveal faint sources such as Compton-thick active galactic nuclei while, for solar physics, the challenge is to provide high-contrast images to capture flux and spectral information of the hot plasma in solar flares and in the scene context with the same quality. Our approach is to develop fine-pitch pixel arrays with independent spectroscopic channels by combining the sensitive material and the front-end electronics in a single hybrid detector system. Such capabilities are also relevant for adjacent applications in medical imaging, industrial monitoring, and non-destructive analysis.

We have been developing the Caliste hard X-ray detectors, which are 4-side buttable high performance, low power, pixelated Cadmium Telluride (CdTe)-based imaging spectrometers dedicated to space applications [1], [2]. The latest generation, Caliste-HD [3], features a 625 μm-pitch pixelated anode capable of single-photon counting with 100 ns time resolution, an energy resolution of 700 eV Full Width at Half Maxima (FWHM) at 60 keV and an energy range extending up to 1 MeV. The Solar Orbiter ESA mission, launched in February 2020, carries 32 Caliste-SO detector modules [4] in the STIX instrument [5] to perform indirect imaging of solar flares thermal and non-thermal X-ray emissions using a Fourier imaging technique combined with a Moiré pattern reconstruction method [6].

To answer future high energy observatory requirements aiming for angular resolutions of a few arcseconds FWHM [7], [8], [9], the imaging resolution of the focal plane detectors must lie in a 200 μm to 300 μm range. Due to technological constraints related to detector hybridization and module packaging, the Caliste architecture is unsuitable for such fine pitches. In the framework of ALB3DO, a joint laboratory between CEA-Irfu and 3D PLUS, we have initiated the Mini-CdTe on Chip (MC^2) project, a new concept of hard X-ray imaging spectrometer with a pixel pitch of 250 μm. The device's target performance comprises an energy resolution of 500 eV FWHM at 60 keV. The detector must cover an energy range extending from 2 keV to 250 keV and features a self-triggering architecture for handling high count rates. It must also have a radiation-tolerant design, together with a power consumption below 200 mW/cm^2.

This paper presents the first prototype of the MC^2 concept, along with data processing methods, including baseline correction and event pattern reconstruction. Performance metrics such as energy resolution and spatial uniformity are reported, current limitations discussed and the path toward the full 3D-integrated module outlined.

https://doi.org/10.1051/epjconf/202533802001

II. Instrumental developments and implementation

A. MC^2 component description

MC^2 comprises three main parts: a segmented-anode CdTe detector, hybridized with a mosaic of four Application-Specific Integrated Circuits (ASICs) held together in a resin panel after compression molding and read out by a custom Analog-to-Digital Converter (ADC).

CdTe detector: The sensor is a 1-mm-thick, high purity, high resistivity p-type CdTe monolithic crystal produced by Acrorad (Japan). The anode is segmented into a 32 × 32 array with a 250 μm pitch and a 50 μm gap, and faces a planar cathode for a total sensitive area of 8.05 mm × 8.05 mm. The Pt cathode contact is quasi-ohmic, while the anode is made of an Al/Ti/Au stack that creates a Schottky barrier contact, limiting the dark current on the collecting electrode when the detector is reversed-biased [10].

IDeF-X D^2R_2: The ASIC dedicated to MC^2 is named IDeF-X D^2R_2 (Imaging Detector Front-end for X-rays Dimension 2 Revision 2; D^2R_2 hereafter) and follows IDeF-X D^2R_1 [11]. It retains the X-FAB CMOS 0.18 μm technology from its predecessor but reduces the area of each spectroscopic channel (pixel) from 300 μm × 300 μm to 250 μm × 250 μm to match the CdTe segmented anode layout. This 32 × 32-pixel matrix ASIC is designed for low capacitance, low leakage current detectors and point-to-point DC hybridization. Each pixel is based on the continuous reset architecture [12]. It includes a Charge Sensitive Amplifier (CSA) that converts incoming charges into voltages, a pole-zero cancellation circuitry equivalent to a CR-RC band-pass filter, a peak detector holding the maximum voltage of the signal and a discriminator stage which triggers if the peak detector level crosses its channel-specific threshold. The integrated circuit self-triggered design autonomously reacts to a signal crossing the threshold value of at least one channel. Pixels are read out on demand externally, with up to 32 rows signals of a column multiplexed simultaneously to 32 output buffers.

Two bare ASICs have been extensively characterized, yielding a median Equivalent Noise Charge (ENC) of 54 electrons root mean square (el.rms hereafter) [13]. This corresponds to an expected energy resolution with a CdTe sensor of 734 eV FWHM at 60 keV. The best channel reaches an ENC of 27.3 el.rms, yielding an expected energy resolution of 548 eV FWHM at 60 keV. The average integral linearity measured between 2 keV and 250 keV is 2.5%. The channel-to-channel gain homogeneity has a standard deviation below 1% across all channels of a single chip.

OWB-1: The Optimized Wilkinson on Board version 1 (OWB-1) chip is a Wilkinson-architecture ADC built in CMOS 0.35 μm technology [14]. It has been optimized for direct interfacing with the 32 analog outputs of D^2R_2, providing 32 conversion channels. Each channel comprises an 8-bit gray-code counter operating at the ADC clock frequency (100 MHz), combined with a 5-bit sampling stage driven by a high frequency delay-locked loop, yielding a total resolution of 13 bits over a 2 V input range. The post-calibration noise level is measured below 1 least significant bit. The sampling rate reaches 370 kHz in the 32-parallel channel operational mode. The chip has been developed for space applications, with a radiation-hardened design providing full immunity against single-event latch-up and a temperature compensation system ensuring gain stability over a –40 °C to +40 °C temperature range.

B. Prototypes fabrication

Several technological challenges must be addressed before the realization of the full multi-ASIC MC^2 system. As an intermediate step toward this goal, a single ASIC demonstrator called MC^2-1K, with 1024 pixels, has been designed and produced. The point-to-point flip-chip hybridization of a monolithic Schottky-diode type, pixelated CdTe sensor on a D^2R_2 chip is based on the heritage of the Hitomi space mission [15] and is performed by Imagine-X (Japan). The resulting hybrid detector is shown in Fig. 1.

Fig. 1. View of the shifted configuration hybrid detector prototype MC^2-1K.

Due to wire bonding constraints between D^2R_2 and its Printed Circuit Board (PCB), and to preserve the sensor's integrity, we decided to shift the CdTe detector by one row and one column with respect to the ASIC pads, reducing the number of interconnected channels to a 31 × 31-pixel array. Future prototypes will benefit from wire-free resin encapsulation of the ASIC, allowing a fully hybridized matrix.

C. Experimental setup

The experimental setup used for detector operation and performance characterization is displayed in Fig. 2.

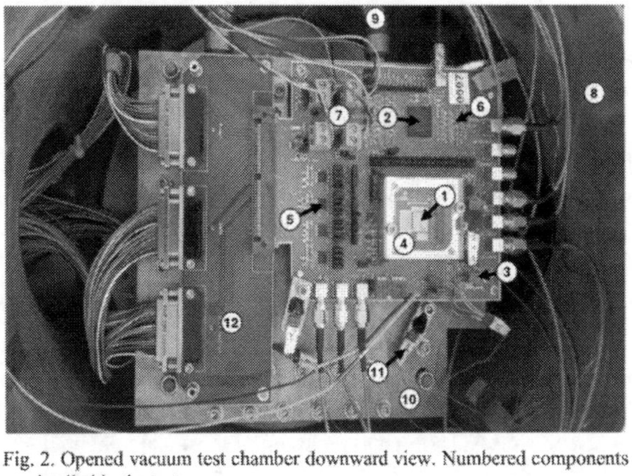

Fig. 2. Opened vacuum test chamber downward view. Numbered components are detailed in the text.

Three MC^2-1K prototypes have been produced (1 in Fig. 2), each comprising 961 connected pixels. They are mounted on dedicated PCBs, together with an interconnected OWB-1 ADC (2) and a high voltage power supply filter (3). The detectors are protected during transport and setup by an aluminum frame (4) holding a 5-mm-thick PMMA cover which is removed before experiment. The boards are equipped with digital-to-analog converters (5) for configurable references, passive components for fixed references (6) and low voltage power supplies (7).

The detectors are placed in a vacuum chamber (8) flushed with dried nitrogen. Detector temperature control and stability throughout operations are ensured by an external cooling system circulating refrigerant fluid (9) through a copper plate (10) to which the PCB is thermally coupled, while five temperature probes (11) are placed in critical locations of the setup. A cold finger also extends from the thermal plate to a conducting surface directly beneath D^2R_2. Sealed radioisotopes are placed outside the vacuum chamber, on a 200 µm-thick beryllium window partly transparent to X-ray photons and facing the top of the detector (i.e. the cathode side).

The PCB is connected through an adaptation board (12) to a custom testing board located outside the chamber. It is further plugged into a Xilinx ZedBoard equipped with a Zynq Field-Programmable Gate Array (FPGA) flashed with embedded firmware for bench control and data transfer to the user computer. In this setup, the detection matrix can be read out in *frame mode*. The FPGA holds the peak detectors of all pixels and proceeds to encode all data, column by column. Thanks to the 32 channels of OWB-1, the 32 rows are converted in parallel within 2.56 µs and serially output. The current setup allows a frame readout time of 1.32 ms.

III. Data treatment pipeline

The FPGA internal clock periodically triggers the frame readout sequence, synchronized with the OWB-1 conversion of the input signal. Sampled data are transferred to the FPGA memory, which is then read by an external computer running a Python acquisition script. Each frame is decoded from a binary format into a list of 1024 Analog-to-Digital Unit (ADU) integer values, accompanied by its acquisition time.

The challenges of the data treatment process are to (a) correct systematic dispersion between channels, (b) identify the relevant photon hit signals among the sampled signals of all channels, (c) group hits into events and (d) calibrate the data into physical units (i.e. from ADU to keV). The effects of the operating conditions on the detector are investigated for three cold finger temperatures T_{det} (–15.3 °C, –5.7 °C, +4.0 °C), three bias voltage V_{det} (–200 V, –300 V, –400 V) and a single D^2R_2 peaking time $\tau_{peaking}$ (3.6 µs).

A. *Channel dispersion characterization and correction*

Charge sharing is expected to have a significant effect in our detector's fine-pitch geometry. As the signal spreads across adjacent channels, it is critical to correct for the inter-pixel average rest level – the level at which the peak detector settles when no signal is present at the input of the analog channel – to ensure accurate energy reconstruction during event identification. Two scenarios are studied here: one without a source and one with a source illuminating the detector.

Without source: For any new acquisition, a set of 500 dark frames is measured to characterize the rest level behavior of each channel's peak detector output. For every pixel, the distribution of measured values is fitted with a Gaussian function, yielding a channel-specific mean rest level value μ_{rest} and standard deviation of the rest level value σ_{rest}. Fig. 3 displays the distribution of μ_{rest} values across the matrix.

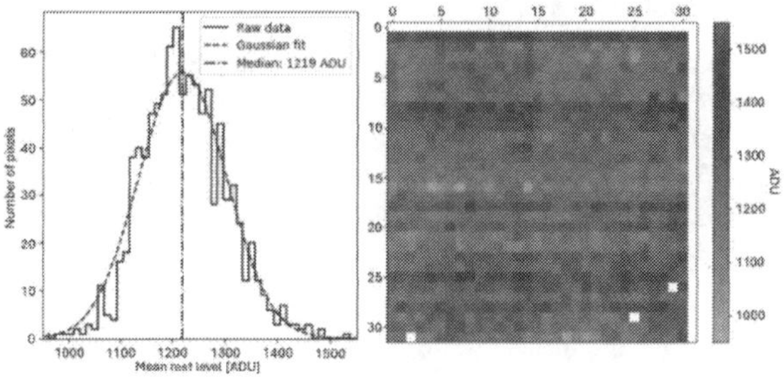

Fig. 3. Mean rest level values distribution (left) and map (right) of the MC^2-1K detector operating at T_{det} = +4.0 °C, V_{det} = –400 V and $\tau_{peaking}$ = 3.6 µs.

A symmetric distribution is observed in the histogram, with a fitted mean μ_{rest} value of approximately 1224 ± 84 ADU. The dispersion, corresponding to an energy of about 3 keV, is explained by the combination of technological variations between pixels during D^2R_2 fabrication and a common calibration for the 32 Wilkinson ADCs of OWB-1 leading to visible row-dependent features. A channel-specific rest level correction is performed on individual frames by subtracting the corresponding μ_{rest} value from each pixel's measurement. The mean μ_{rest} value after correction is centered around 0 ± 2 ADU.

With source: An additional effect is observed on the columns when charges are measured, as seen in a detector illuminated by an ^{241}Am source in the left map of Fig. 4.

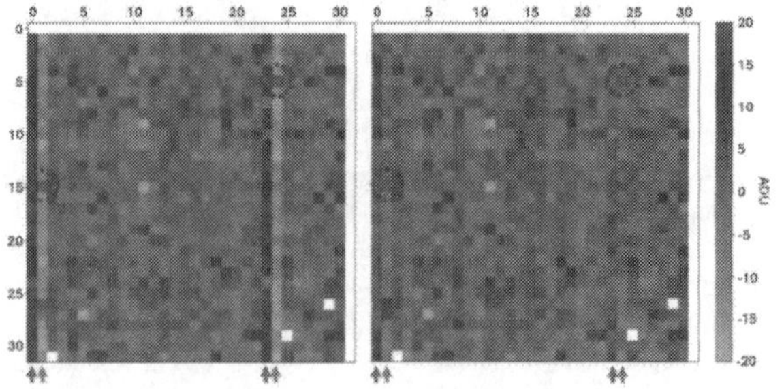

Fig. 4. On the left: sampled values cartography of the MC^2-1K detector illuminated by an ^{241}Am source operating at T_{det} = +4.0 °C, V_{det} = –400 V and $\tau_{peaking}$ = 3.6 µs. On the right: similar cartography for corrected data. Green arrows indicate columns comprising a hit; red arrows indicate the following column in a frame mode readout sequence. Red circles indicate hits.

The mean level of a column hit by a photon decreases, while the mean level of the following column in the full-frame readout sequence (i.e. the column to its left) increases. These effects are referred to as the *direct* and *indirect* column effects, respectively. Their amplitudes are found to be proportional to the total energy measured in each column, or sum value, as shown in Fig. 5.

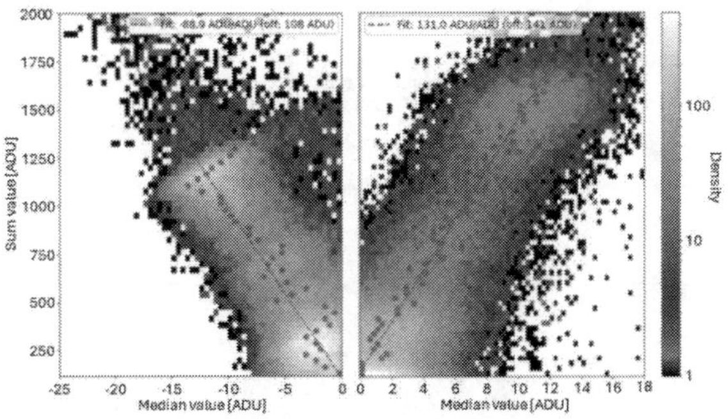

Fig. 5. Column effects observation and characterization for a MC^2-1K detector illuminated by an ^{241}Am source, operating at T_{det} = +4.0 °C, V_{det} = –400 V and $\tau_{peaking}$ = 3.6 µs. Each 2D distribution of sum values against median values is fitted by a linear equation. On the left: direct column effect for a given column. On the right: indirect column effect for the same column.

For a given column and acquisition, the 2D distribution of sum values versus median values can be approximated by a linear relationship, yielding a factor η_{direct} corresponding to the slope of the linear regression as shown on Fig. 5. A similar method is applied to the indirect column effect, for which the median value of a given column is related to the sum value of the one previously read out, yielding the $\eta_{indirect}$ factor. For an acquisition with $\tau_{peaking}$ = 3.6 µs:

- η_{direct} ranges across columns from –70 ADU/ADU to -90 ADU/ADU (-1.1% to -1.5%), with its amplitude increasing with V_{det} and decreasing with T_{det}.
- $\eta_{indirect}$ ranges across columns from 120 ADU/ADU to 140 ADU/ADU (0.7% to 0.8%), with no significant dependence on the operating conditions.

When considering for instance an energy deposit of 122 keV from a ^{57}Co source on a single column, the combined effects can cause a deviation of up to 2.8 keV, drastically increasing the low energy threshold of the affected channels and impacting the capability to reconstruct the photon energy deposited across multiple pixels. A linear correction is applied to each frame using the η_{direct} and $\eta_{indirect}$ column-dependent values, restoring the rest levels as seen in the right map of Fig. 4.

Our group has already observed these effects on a previously designed circuit. When one or several channels in a column exhibit a high output level, they draw current (primarily at the output buffer), thus lowering the level of the other connected channels. As the circuit subsequently multiplexes the next column, this causes an increase in the output level of the associated pixels.

B. *Event identification and clustering*

Threshold definition: After corrections have been applied, the 3σ dispersion of the μ_{rest} values across the matrix lies around 2 ADU. Moreover, σ_{rest} values of individual channels range between 2.5 and 10 ADU, as displayed in Fig. 6, physically corresponding to values from 90 eV to 350 eV.

σ_{rest} is lower than σ_{filter}, the standard deviation of the filter output which serves as the basis for defining the channel-dependent Low-Level Threshold (LLT). The relationship between σ_{filter} and σ_{rest} depends on the readout timing: $\sigma_{filter} = 3 \times \sigma_{rest}$ was found to be appropriate for our setup. Consequently, we chose LLT = 3 × σ_{filter} = 9 × σ_{rest} to discriminate between rest level fluctuations and signal. For a given frame, pixel values below the corresponding LLT values are discarded. For noisy channels, an increased LLT value is assigned by comparing the number of low energy counts to the estimated photopeak counts from a known source.

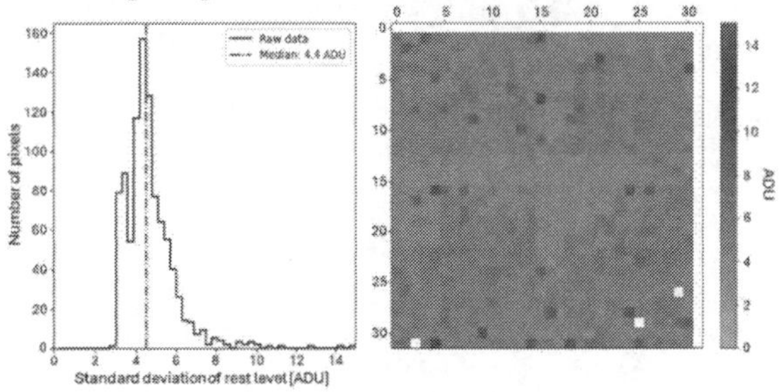

Fig. 6. σ_{rest} values distribution (left) and map (right) for a MC^2-1K detector operating at T_{det} = +4.0 °C, V_{det} = –400 V and $\tau_{peaking}$ = 3.6 µs.

A map of LLT values is displayed in Fig. 7 together with a filtered frame where counts below the LLT values have been removed, retaining only values considered as signal. For measurements at T_{det} = +4.0 °C, V_{det} = –400 V and $\tau_{peaking}$ = 3.6 µs, LLT values range from 20 ADU (~0.7 keV) up to 364 ADU (~12.9 keV), with a median value of 67 ADU (~2.4 keV) and a standard deviation of 39 ADU (~1.4 keV).

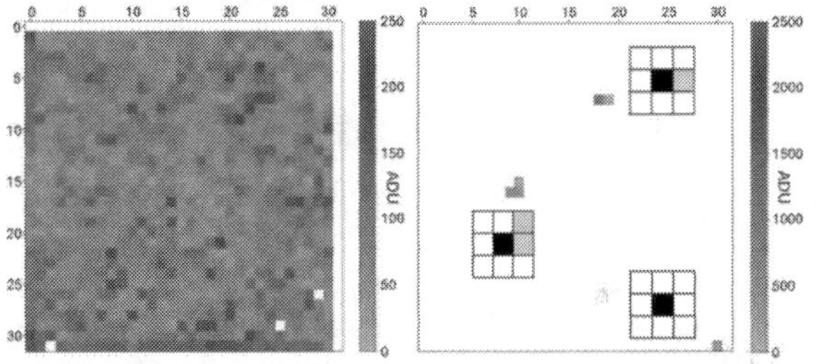

Fig. 7. On the left: LLT values in ADU for the MC^2-1K detector operating at T_{det} = +4.0 °C, V_{det} = –400 V and $\tau_{peaking}$ = 3.6 µs. On the right: example of extracted hits from a single frame, with pattern recognition results related to the library illustrated in Fig. 8.

Pattern identification: The event multiplicity is defined as the number of hits comprising an event. This results from diffusion in the sensor during charge transport towards the electrodes, as well as the non-uniqueness of charge generation due to Compton scattering and X-ray Fluorescence (XRF) emissions. Once the hit pixels have been identified, a pattern recognition algorithm groups so-called hits into events and classifies them according to the valid pattern library of Fig. 8.

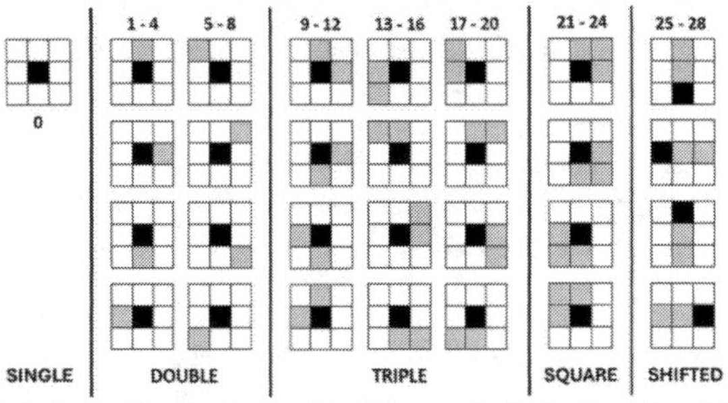

Fig. 8. Pattern identification algorithm library, originally developed for soft X-ray Si-detectors and adapted to the CdTe MC^2-1K hard X-ray detector. The main pixel, with highest measured signal, is shown in black.

This algorithm was originally developed for the data treatment of space missions dedicated to soft X-ray observations of GRBs employing Si detectors [16], [17]. We adapted the library to CdTe-specific XRF cases by adding patterns 5 to 8, 13 to 20, and 25 to 28. Indeed, for Cd and Te K-shell XRF photons, the mean free paths in the sensor are 115 μm and 59 μm, respectively — distances comparable to the interpixel gap of 50 μm, 70 μm for diagonal pixels, crossed by the XRF photons to generate the associated patterns.

C. *Individual channel energy calibration*

For the MC^2-1K acquisition system, the relationship between the measured Peak Height Amplitude (PHA) value, in ADU, and the deposited energy value E is pixel-dependent and linear over the dynamic range. It follows the equation

$$E = \frac{PHA}{G} + E_0 , \qquad (1)$$

with the two so-called calibration parameters *gain* G, in ADU/keV, and *offset* E_0, in keV. To determine the set of parameters of each channel, we use a calibration algorithm developed at CEA called Energy Calibration via Correlation (ECC) [18]. It employs a correlation method to globally fit a synthetic spectrum generated from a known source to the same source single-channel measurement. This method ensures the convergence of calibration parameters even for low statistics.

The calibration is performed on single events (i.e. measured by a single pixel), as multiple events (i.e. measured by two or more pixels) yield partial energy deposits from a given pixel's perspective. The distribution of calibration parameters is displayed in Fig. 9 for a given set of operating conditions.

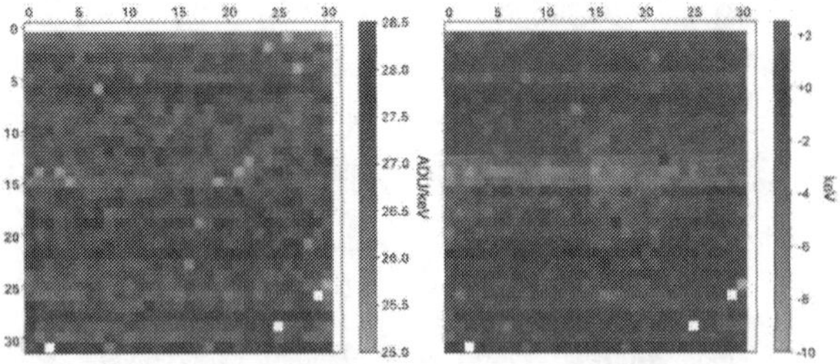

Fig. 9. Gain (left) and offset (right) calibration parameters provided by the ECC algorithm for an ^{241}Am measurement by the MC^2-1K detector operating at T_{det} = +4.0 °C, V_{det} = −400 V and $\tau_{peaking}$ = 3.6 μs.

The gain distribution is Gaussian and does not show any geometrical features under any operating conditions. The offset distribution shows lower values for center rows, down to −10 keV. This value decreases further as the temperature decreases. The offset distribution of the other channels can be described as Gaussian. The means and standard deviations of both calibration parameter distributions are extracted from Gaussian fits and are indicated in Table I.

The mean value and dispersion of the offset across the pixels decrease with temperature, the latter due to an associated decrease in the leakage current and thermal noise. The mean offset value also decreases with increasing bias voltage. LLT values are converted into physically defined Low energy Threshold (LT) values using the calibration parameters and are indicated in Table I. The LT distribution for a given set of operating conditions is shown on Fig. 10.

TABLE I
CALIBRATION PARAMETERS OBTAINED BY ECC FROM AN ^{241}AM SOURCE: AVERAGE VALUES AND DISPERSIONS OF PIXEL PARAMETERS.

T_{det} [°C]	V_{det} [V]	Gain [ADU/keV]	Offset [keV]	LT [keV]
+4.0	−200	27.1 (± 0.4)	−5.7 (± 2.0)	5.3 (± 1.8)
	−300	27.1 (± 0.4)	−5.6 (± 1.9)	5.3 (± 1.9)
	−400	27.2 (± 0.4)	−5.5 (± 2.0)	5.2 (± 1.8)
−5.7	−200	27.0 (± 0.4)	−3.1 (± 1.5)	4.3 (± 1.3)
	−300	27.1 (± 0.4)	−3.1 (± 1.5)	4.2 (± 1.3)
	−400	27.1 (± 0.4)	−3.2 (± 1.5)	4.4 (± 1.3)
−15.3	−200	27.1 (± 0.4)	−2.5 (± 0.9)	4.0 (± 1.2)
	−300	27.1 (± 0.4)	−2.3 (± 0.8)	3.7 (± 1.0)
	−400	27.1 (± 0.4)	−2.2 (± 0.8)	3.6 (± 0.9)

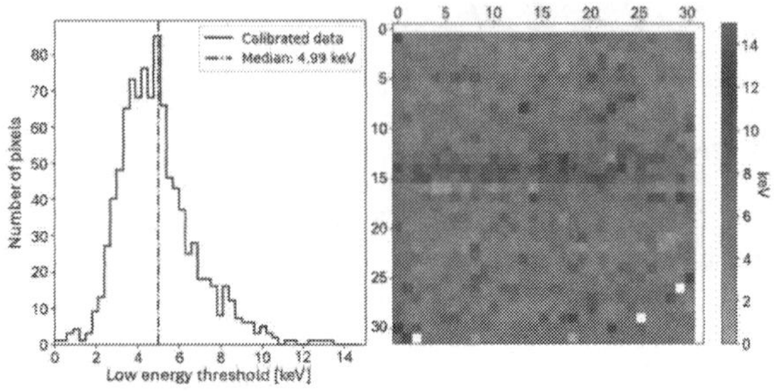

Fig. 10. Histogram (left) and map (right) of LT values obtained for the MC^2-1K detector operating at T_{det} = +4.0 °C, V_{det} = -400 V and $\tau_{peaking}$ = 3.6 μs.

The central row feature is clearly visible, with increased threshold values observed in this area. This region had already exhibited enhanced non-linear behavior during the bare ASIC characterization campaign. The distribution is asymmetrical, like the ENC distribution obtained during the same campaign. Both the mean and standard deviation of the LT distribution decrease with temperature, due to reduced leakage current and the resulting decrease in electronic noise of the analog channel.

IV. MEASURED PERFORMANCE

The MC^2-1K characterization includes measurement of an ^{241}Am source for calibration and performance evaluation. The reference operating conditions are T_{det} = +4.0 °C and V_{det} = −400 V, which provides the best spectral resolution as shown below, used with the experimentally determined optimum peaking time $\tau_{peaking}$ of 3.6 μs.

A. *Count rate spatial uniformity*

The count rate spatial distribution as measured by MC^2-1K in the reference operating conditions from an ^{241}Am illumination is displayed in Fig. 11. The bulk area is defined as channels more than one pixel away from the detector and ASIC borders, to mitigate the influence of the detector's shifted geometry and sensor borders.

The uniformity of the count rate across the 841 pixels of the bulk area has been validated, with an average count rate above 40 keV of 470 counts (± 6%), which is close to the 4.7% deviation expected from Poisson counting statistics. Three unconnected pixels were identified as measuring no counts above the threshold, validating CdTe-to-ASIC hybridization for 958/961 channels (99.7%).

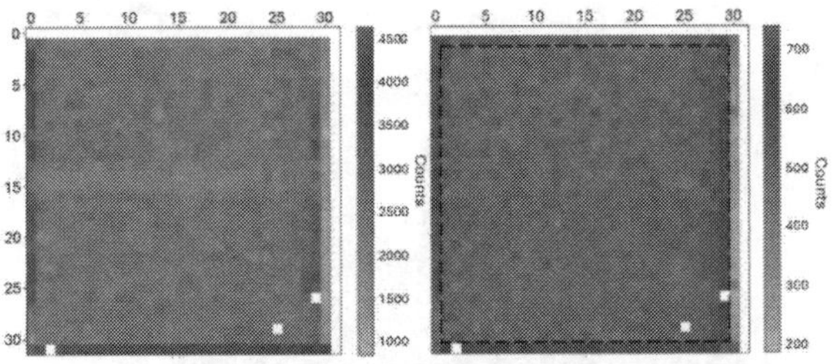

Fig. 11. Total count rate (left) and counts above 40 keV (right) measured by the MC^2-1K detector for an ^{241}Am source at T_{det} = +4.0 °C, V_{det} = –400 V and $\tau_{peaking}$ = 3.6 µs. Dashed lines delimit the bulk area.

B. *Energy resolution*

The individually calibrated spectra from connected pixels are combined to produce the ^{241}Am sum spectrum under the reference operating conditions shown on Fig. 12. Two sum spectra are displayed: one corresponds to the individual hits recorded by each of the 958 ASIC-connected pixels of the matrix, and the other corresponds to the reconstructed energy – summed over associated pixels – of each event.

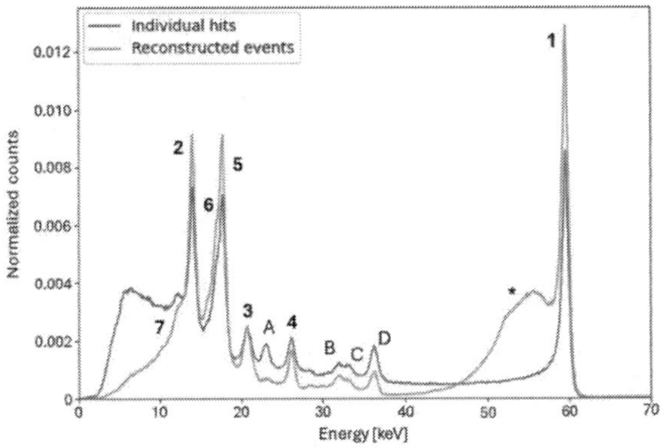

Fig. 12. Spectrum of hits and events summed over 958 pixels for an MC^2-1K ^{241}Am source measurements at T_{det} = +4.0 °C, V_{det} = –400 V and $\tau_{peaking}$ = 3.6 µs. Each spectrum is normalized to the total number of counts for comparison.

The source emission lines are clearly visible in both spectra, with the 59.54 keV (1) photopeak exhibiting a low energy tail mainly due to charge trapping in the sensor, manifesting as a charge loss in the measurement. Lower energy emission lines are observed at 13.85 keV (2), 21.16 keV (3) and at 26.34 keV (4). The line at 16.96 keV (5) displays a shoulder on its low energy side, associated with the merged 15.88 keV (6) emission line. The low intensity line at 11.89 keV (7) is barely distinguishable, buried within the low energy spectral component arising from the LT spread distribution spread.

XRF peaks are observed in the individual hits spectrum, partially reduced in the reconstructed events as the incident photon's initial energy deposit and the XRF photon deposited energy are combined – thus validating the event identification. Measured fluorescence lines are respectively Cd K_α and K_β at 23.1 keV (A) and 26.0 keV (4), and Te K_α and K_β at 27.3 keV and 31.0 keV (B). Each fluorescence line is associated with an escape line (when XRF photons leave the sensor before absorption), whose energy is found by subtracting the XRF energy from the incident photon energy. For a 59.54 keV photon, the respective escape line energies are 36.4 keV (D), 33.5 keV (C), 32.2 keV (B) and 28.5 keV.

The energy resolution is evaluated from a global fit on the single event sum spectrum using the *lmfit* Python library [19]. Each identified peak is fitted with a Crystal Ball function to account for the charge loss-related low energy tail [20]. The model is constrained by a unique value of ENC defined for the entirety of the fitting range. This so-called *estimated ENC* is derived from the measured energy resolution, after removing the Fano noise contribution. A background component is added in the form of a smoothed step function. Fig. 13 displays the fitted single event sum spectrum.

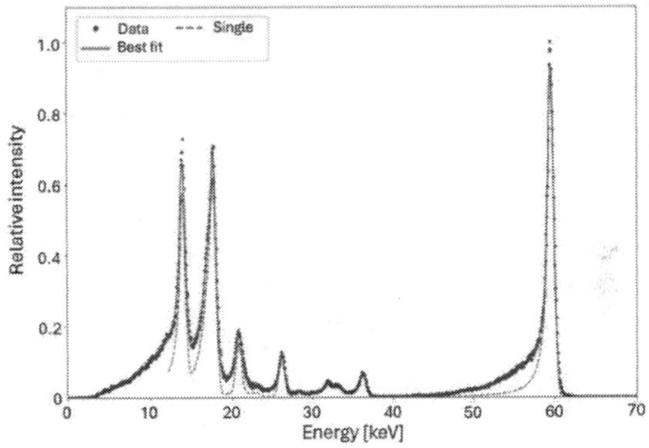

Fig. 13. Single events sum spectrum of an ^{241}Am source measured by the 958 connected pixels of MC^2-1K operating at T_{det} = +4.0 °C, V_{det} = –400 V and $\tau_{peaking}$ = 3.6 µs, normalized. A global fit is applied to the spectrum.

An energy resolution of 910 eV at 59.54 keV is obtained for a 958-pixel single events sum spectrum measured under the reference operating conditions of T_{det} = +4.0 °C, V_{det} = –400 V and $\tau_{peaking}$ = 3.6 µs. The estimated ENC is fitted at 75 el.rms. Other fitted energy resolutions at 17 keV and 60 keV are indicated together with the estimated ENC values in Table II for the explored operating conditions.

TABLE II
SPECTRAL RESOLUTION AND ELECTRONIC NOISE ESTIMATION OBTAINED FROM GLOBALLY FITTED SINGLE EVENTS SUM SPECTRA.

T_{det} [°C]	V_{det} [V]	FWHM at 17 keV [eV]	FWHM at 60 keV [eV]	Estimated ENC [el.rms]
+4.0	–200	972	1042	90
	–300	874	951	80
	–400	**833**	**910**	**75**
–5.7	–200	904	979	83
	–300	864	950	80
	–400	891	973	82
–15.3	–200	998	1066	92
	–300	956	1027	88
	–400	914	988	84

Increasing the bias voltage reduces the charge drift duration, consequently lowering the charge loss probability. Moreover, the calibration's accuracy for low statistics, single pixel datasets improves as the spectral charge loss-related tail weakens. These effects explain the observed improvements in energy resolution and estimated ENC at higher bias voltages. We further perform a Crystal Ball fit on the 59.54 keV peak, from 40 keV to 70 keV, as illustrated in Fig. 14.

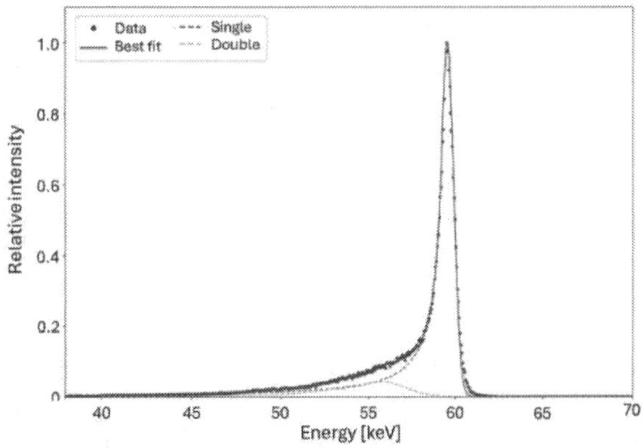

Fig. 14. Single events sum spectrum of an ^{241}Am source measured by the 958 connected pixels of MC^2-1K operating at T_{det} = +4.0 °C, V_{det} = –400 V and $\tau_{peaking}$ = 3.6 μs. A global fit is applied to the spectrum between 40 keV and 70 keV. Single events are indicated by a green dashed line. Double events incorrectly identified as single are indicated by a yellow dashed line.

Under the reference operating conditions, the energy resolution was fitted exclusively to the 59.54 keV peak, yielding a value of 855 ± 7 eV FWHM. This corresponds to an ENC of 69.0 el.rms. To properly fit the photopeak, a second crystal ball curve is fitted to the low energy tail. This feature accounts for double events incorrectly identified as single events due to the relatively high LT value of some channels. The shape of this adjustment is directly related to the LT low energy continuum observed in the individual hits spectrum of Fig. 12.

C. Multiplicities and charge loss

We investigate charge-sharing features as they directly impact the sensitivity of our relatively small-pitch geometry. Proper understanding of these features is required to apply spectral corrections. Table III summarizes the ratio of single and multiple events under different operating conditions, along with the estimated standard deviation across the matrix.

TABLE III
EVENT DISTRIBUTION BY MULTIPLICITY.

T_{det} [°C]	V_{det} [V]	Single events [%]	Double events [%]	Higher multiplicities [%]
+4.0	–200	62.8 (± 5.9)	28.9 (± 3.7)	8.3 (± 2.5)
	–300	64.1 (± 5.7)	27.2 (± 3.4)	8.7 (± 2.5)
	–400	65.2 (± 5.7)	27.3 (± 3.3)	7.5 (± 2.5)
–5.7	–200	57.8 (± 6.2)	32.2 (± 3.9)	10.0 (± 2.4)
	–300	59.6 (± 5.6)	30.5 (± 3.3)	9.9 (± 2.7)
	–400	59.9 (± 5.2)	30.9 (± 3.0)	9.2 (± 2.3)
–15.3	–200	61.3 (± 9.0)	29.9 (± 5.4)	8.8 (± 3.5)
	–300	57.6 (± 4.8)	31.4 (± 3.0)	11.0 (± 2.4)
	–400	57.2 (± 4.3)	31.2 (± 2.6)	11.6 (± 2.2)

The single event fraction decreases with temperature and slightly increases with high voltage. A high voltage trend is expected as it reduces the collection time, and thus the effect of diffusion. In contrast, the temperature dependency is opposite to expectation since diffusion increases with temperature. The explanation is that decreasing the temperature lowers the low energy threshold, as shown in Table III, due to decreased peak detector noise. Events previously perceived as single are then identified as double, causing the fraction of single events to decrease with temperature.

The correlation diagram presented in Fig. 15 shows the energy share of double events between two neighboring channels.

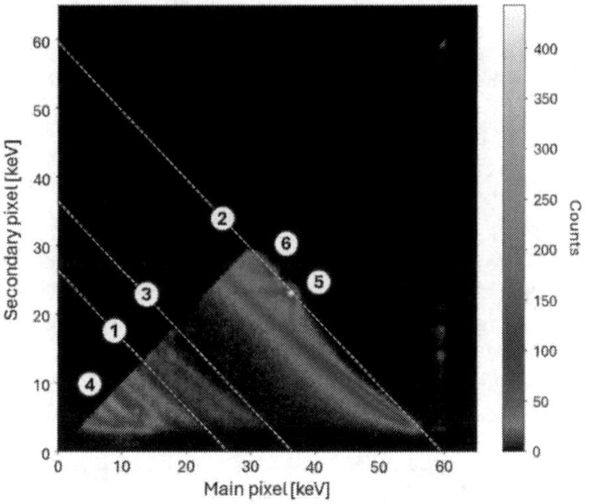

Fig. 15. Charge sharing between the main pixel and the secondary pixel for double events for an ^{241}Am source illumination, measured over the 958 channels of MC^2-1K operating at T_{det} = +4.0 °C, V_{det} = –400 V and $\tau_{peaking}$ = 3.6 μs. The dashed lines are emissions from the ^{241}Am source at 26.3 keV and 59.5 keV, and the Cd fluorescence escape line at 36.4 keV.

This distribution highlights the main photopeaks of the ^{241}Am source as traces. Ideally, these lines should be straight and follow the dashed white lines shown here. Instead, a curvature is observed, notably for the 26.3 keV (1) and 59.54 keV (2) lines, as well as for the Cd XRF escape line at 36.4 keV (3). This feature is also observed to a lesser extent in the X-ray lines between 14 keV and 21 keV (4). These curves are caused by charge loss in the inter-pixel area, a phenomenon previously observed in CdZnTe detectors with low pixel pitch [21].

Bright spots (5 and 6) are observed, corresponding to fluorescence events where the initial 59.54 keV photon and the XRF photon are simultaneously measured in two different pixels. Although flagged as double events, they do not suffer charge loss in the inter-pixel area and thus lie on the dashed white line at 59.54 keV. The interpixel charge loss phenomenon explains the "bump" on the low energy side of the 59.54 keV ^{241}Am photopeak (*) in the reconstructed events spectrum of Fig. 12. This bump is peaking at ~56 keV (-6% of the true photon energy), consistent with a charge loss comprised between 0% and 12%.

Maximum charge loss occurs at the center of the interpixel and varies with operating conditions. At +4.0 °C, it is estimated to be ~12.3%, ~11.5% and ~10.1% at -400 V, -300 V, -200 V, respectively. As the bias voltage decreases, the charge cloud widens due to longer drift time. A wider cloud causes charge carriers to be more likely positioned away from the interpixel center, where charge loss is strongest, thus reducing its impact. Charge loss is unaffected by temperature in the studied range, as diffusion changes negligibly.

V. Conclusions and perspectives

We presented the performance of the first MC^2-1K hybrid detector, based on CdTe pixel detectors with the D^{2}R$_2$ front-end and the radiation hard OWB-1 ADC. The hybridization connecting the CdTe anodes to the ASIC input pads over a matrix of 31 × 31 pixels with a 250 μm pitch was successful, yielding only three bad pixels and a uniform counting map under flat illumination. Fitting the single event sum spectrum from all good pixels (99.7%) yields an energy resolution of 910 eV FWHM at 59.54 keV, corresponding to an ENC of 75 el.rms at +4 °C.

This work is ongoing, with data acquisition system upgrades underway to increase frame data transfer rates. This is a key step toward enhancing photon statistics and improving the quality of spectral calibration under any test conditions. For future use in space instruments for hard X-ray astrophysics or solar physics, we aim to improve noise performance of our system to reach the CdTe Fano limit. This is especially important for resolving ~6 keV solar flare lines complex and for achieving a low energy threshold near 3 keV or lower. We have begun redesigning the charge-sensitive amplifier in the ANAQIN-X test chip, yielding promising results [22]. The ASIC redesign also offers an opportunity to implement self-triggering function and improve timing performance. The next step is to design a large matrix of spectroscopic channels, with the additional challenge of noise performance uniformity over 1024 pixels.

Acknowledgment

We thank Imagine-X for the fruitful collaboration in producing and mounting the hybrid detectors on printed circuit boards. We thank the French space agency (CNES) for the technical and financial support in the framework of a Research and Technology program.

References

[1] A. Meuris *et al.*, "Caliste 64, an innovative CdTe hard X-Ray micro-camera," in *2007 IEEE Nucl. Sci. Symp. Conf. Rec.*, Honolulu, HI, USA, 2007, pp. 2551-2557, doi: 10.1109/NSSMIC.2007.4436671

[2] O. Limousin *et al.*, "Caliste 256: A CdTe imaging spectrometer for space science with a 580μm pixel pitch," NIM-A, vol. 647, no. 1, pp. 46–54, Aug. 2011, doi: 10.1016/j.nima.2011.05.011.

[3] A. Meuris *et al.*, "Caliste HD: A new fine pitch Cd(Zn)Te imaging spectrometer from 2 keV up to 1 MeV," in *2011 IEEE Nucl. Sci. Symp. Conf. Rec.* Valencia, Spain, 2011, pp. 4485-4488, doi: 10.1109/NSSMIC.2011.6154695.

[4] A. Meuris *et al.*, "Caliste-SO, a CdTe based spectrometer for bright solar event observations in hard X-rays," NIM-A, vol. 787, pp. 72–77, Jul. 2015, doi: 10.1016/j.nima.2014.11.021.

[5] S. Krucker *et al.*, "The Spectrometer/Telescope for Imaging X-rays (STIX)," A&A, vol. 642, p. A15, Oct. 2020, doi: 10.1051/0004-6361/201937362.

[6] P. Massa *et al.*, "Imaging from STIX visibility amplitudes," A&A, vol. 656, p. A25, Dec. 2021doi: 10.1051/0004-6361/202140946.

[7] H. A. S. Reid *et al.*, "The Solar Particle Acceleration Radiation and Kinetics (SPARK) Mission Concept," Aerospace, vol. 10, no. 12, p. 1034, Dec. 2023, doi: 10.3390/aerospace10121034.

[8] S. D. Bongiorno *et al.*, "Assembly of the FOXSI-4 mirror modules," in *Proc. SPIE 12679*, San Diego, CA, USA, 2023, pp. 81–92. doi: 10.1117/12.2676955.

[9] N. E. Thomas, "SuperHERO (Super-High Energy Replicated Optics) balloon mission concept," in *Proc. SPIE 13093*, Yokohama, Japan, 2024, doi: 10.1117/12.3020830.

[10] T. Takahashi *et al.*, "High-resolution Schottky CdTe diode for hard X-ray and gamma-ray astronomy," NIM-A, vol. 436, no. 1, pp. 111–119, Oct. 1999, doi: 10.1016/S0168-9002(99)00606-3.

[11] D. Baudin *et al.*, "D^{2}R$_1$: A 2-D X-Ray Detector for CdTe-Based Fine Pitch and High-Energy Resolution Imaging Spectroscopy," IEEE Trans. Nucl. Sci., vol. 65, no. 7, pp. 1408–1415, Jul. 2018, doi: 10.1109/TNS.2018.2845461.

[12] G. De Geronimo and P. O'Connor, "A CMOS fully compensated continuous reset system," IEEE Trans. Nucl. Sci., vol. 47, no. 4, pp. 1458–1462, Aug. 2000, doi: 10.1109/23.872996.

[13] H. Allaire *et al.*, "First light of MC^2 -1K, a 250 μm pixel pitch CdTe imaging spectrometer for hard X-ray astronomy," NIM-A, vol. 1046, p. 167663, Jan. 2023, doi: 10.1016/j.nima.2022.167663.

[14] F. Bouyjou, O. Gevin, O. Limousin and E. Delagnes, "A 32-Channel 13-b ADC for Space Applications,", IEEE Trans. Nucl. Sci., vol. 64, no. 4, pp. 1071-1079, April 2017, doi: 10.1109/TNS.2017.2672863.

[15] T. Takahashi *et al.*, "The ASTRO-H (Hitomi) x-ray astronomy satellite," in *Proc. SPIE 9905*, Edinburgh, United Kingdom, 2016, pp. 212–228. doi: 10.1117/12.2232379.

[16] N. Gehrels *et al.*, "The Swift Gamma-Ray Burst Mission," ApJ, vol. 611, pp. 1005–1020, Aug. 2004, doi: 10.1086/422091.

[17] D. Götz *et al.*, "The scientific performance of the microchannel X-ray telescope on board the SVOM mission," Exp. Astron., vol. 55, pp. 487–519, 2023. doi: 10.1007/s10686-022-09881-6.

[18] D. Maier, O. Limousin, and G. Daniel, "Energy calibration via correlation using an adaptive mesh refinement," EPJ Web Conf., vol. 225, p. 01003, 2020, doi: 10.1051/epjconf/202022501003.

[19] M. Newville *et al.*, "LMFIT: Non-Linear Least-Squares Minimization and Curve-Fitting for Python", Zenodo, 2025, doi: 10.5281/zenodo.15014437.

[20] D. M. Santos and F. Dupertuis, "Mass distributions marginalized over per-event errors," NIM-A, vol. 764, pp. 150–155, 2014, doi: 10.1016/j.nima.2014.06.081.

[21] A. E. Bolotnikov, W. R. Cook, F. A. Harrison, A.-S. Wong, S. M. Schindler, and A. C. Eichelberger, "Charge loss between contacts of CdZnTe pixel detectors," NIM-A, vol. 432, no. 2, pp. 326–331, 1999, doi: 10.1016/S0168-9002(99)00388-5.

[22] H. Boulfani, A. Meuris, D. Baudin, O. Gevin, and O. Limousin, "ANAQIN-X: A prototype low noise ASIC for the readout of highly dense semiconductor X-ray detectors", submitted in these proceedings.

 https://doi.org/10.1051/epjconf/202533802002

ANAQIN-X: A prototype low noise ASIC for the readout of highly dense semiconductor X-ray detectors

H. Boulfani[1*], D. Baudin[1], O. Gevin[1], O. Limousin[2], A. Meuris[2], and H. Allaire [2]
[1] Université Paris-Saclay, IRFU, CEA, F-91191 GifsurYvette, France
[2] Université Paris-Saclay, Université Paris Cité, CEA, CNRS, AIM, 91191, Gif-sur-Yvette, France
(*) boulfanihani@gmail.com

***Abstract*—This article presents ANAQIN-X (Advanced Noise-optimized ASIC for Quantifying and Imaging in X-rays) a pixelated prototype readout chip dedicated to testing different architectures of Charge Sensitive Amplifiers (CSA), the first stage of the spectroscopic signal processing analog circuit. The chip is designed to read out low leakage current (1 pA), low capacitance (300 fF), hard X-rays detectors (Cd(Zn)Te/Si) in order to perform high-resolution (500 eV FWHM at 60 keV) hard X-ray (1-220 keV) spectroscopy. Its main block is a mini matrix of pixels of 250x250 μm^2 designed in X-FAB's 0.18 μm CMOS technology. Each channel consumes 250 µW in nominal mode, and contains an optimized Charge Sensitive Amplifier (CSA) followed by a CR-RC shaper with tunable peaking time (from 144 ns up to 3.6 μs) and a peak detector. The different CSA architectures integrated on this chip are based on continuous reset topology with nMOS or pMOS input transistors architectures with various W/L ratios. The chip includes also, an AC-coupled CSA with a tunable input capacitance (from 100 fF to 30 pF) and a Semi-AC coupling architecture allowing a reduction in the thermal noise generated by the reset transistor and the usage of small value coupling capacitance. The best performing pixel can reach a readout noise 20 el.rms without detector. A performance of 25 el.rms with a 300 fF detector can be considered. According to simulation, a noise reduction of 10 % is expected at 15 pA of leakage current, and a coupling capacitance of 1 pF is enough to get 98 % of gain. With an 300 fF input capacitor, the best performing pixel should be able to reach an Equivalent Noise Charge (ENC) of 21 el.rms.**

***Index Terms*—Application-specific integrated circuits (ASICs), CdTe, energy resolution, imaging spectroscopy, low-noise electronics, readout electronics, hard X-ray spectroscopy.**

I. Introduction

IDeF-X (Imaging Detector Front-end in X-rays) is an integrated circuit family devoted to read semiconductor detectors for space science applications. They are designed to be connected to multichannel hard X-ray detectors in order to form detection modules capable of performing high energy resolution imaging spectroscopy. The first space project embedding an IDeF-X ASIC was ESA's Solar Orbiter developed to perform close-up, high-resolution studies of our Sun and launched the 10th,February 2020. Solar Orbiter embeds STIX [1] (Spectrometer/telescope for Imaging X-rays), whose detection plane is made up of Caliste-SO [2], which integrates IDeF-X-HD [3] (High Dynamic IDeF-X), and EPD/STEP (Energetic Particle Detector/Suprathermal Electrons Protons) [4] containing IDeF-X-BD [5](Bi-Directional IDeF-X). The second project embedding IDeF-X ASICs was SVOM satellite [6] (Space Variable Objects Monitor) launched the 22th, June 2024, a French-Chinese mission born to observe gamma-ray bursts (GRBs). IDeF-X-ECLAIRs [7] has been integrated to ECLAIRs Telescope, the prime instrument of SVOM able to detect and localize GRBs.

Most IDef-X ASICs are based on one-dimensional analogue channels architecture. A new IDeF-X branch with 2-dimensional pixel arrays has been designed to enable fine imaging and ultra-low noise levels. We previously designed two pixelated integrated circuits, D^2R_1 and D^2R_2 , with promising results (see [8] and [9]). However, a high-level of noise has been spotted and yet to be explained. This article presents ANAQIN-X (Advanced Noise-optimized ASIC for Quantifying and Imaging in X-rays) the latest two-dimensional integrated circuit specifically designed to investigate the origin of that noise. ANAQIN-X is chip-test which main bloc is a small array of 12 pixels of 250x250 μm^2 designed in X-FAB 0.18 µm CMOS technology. The objective is to perform a readout of CdTe detectors with less than 20 el.rms of noise. The best performing pixel architecture will be selected to build a full-scale 32x32 or 48x48 pixel matrix.

This article will begin with a description of the main blocks of ANAQIN-X, followed by the measurements results. It will conclude with a recommendation for the pixel architecture that will satisfy the specifications of our imaging spectrometer.

II. ANAQIN-X Design Description

A. Design Overview

ANAQIN-X test chip embeds an array of twelve pixels, each measuring 250 x 250 μm^2. Each pixel contains a complete analog circuit for processing spectroscopic signal, as shown in figure 1. The ASIC is designed to test different Charge-Sensitive Amplifier (CSA) architectures in order to select the best-performing one. Figure 1 shows a photograph of the ANAQIN-X mini-matrix.

Fig. 1: ANAQIN-X Mini-matrix photography

B. Analog Chain Description

Figure 2 shows a block diagram of the pixel analog chain. The architecture is based on a continuous reset CSA [10] that converts the incoming charge into a voltage step with a conversion gain of 6.4 $\mu V/el$, followed by a Pole-Zero Cancelation (PZC) circuit operating as a CR-RC filter with a tunable peaking time (from 144 ns up to 3,6 μs). A discriminator compares the output of the filter with a threshold voltage to trigger the readout. Finally, a Peak Stretcher/Detector is used to store the maximum value of the energy signal until it is read.

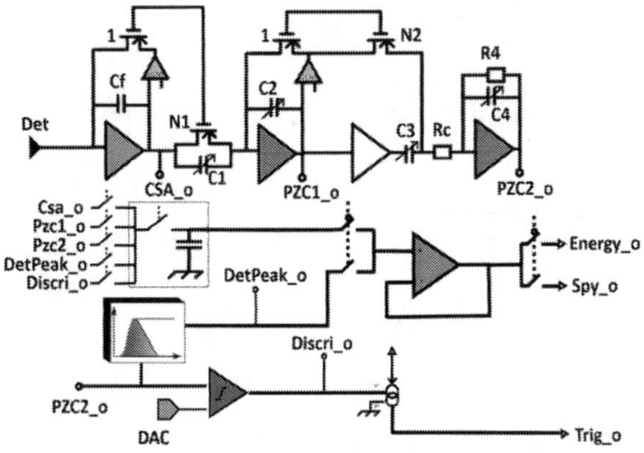

Fig. 2: ANAQIN-X analog chain schematic

C. Charge Sensitive Amplifier

1) DC Coupled CSA

Fig 3 and fig 4 show the schematics of nMOS and pMOS-based CSAs, respectively. In both cases, the system uses a folded cascode architecture with capacitive feedback and a continuous reset transistor to remove the accumulated charge [10]. The feedback capacitance defines the conversion gain, while the reset transistor ensures proper baseline restoration between events.

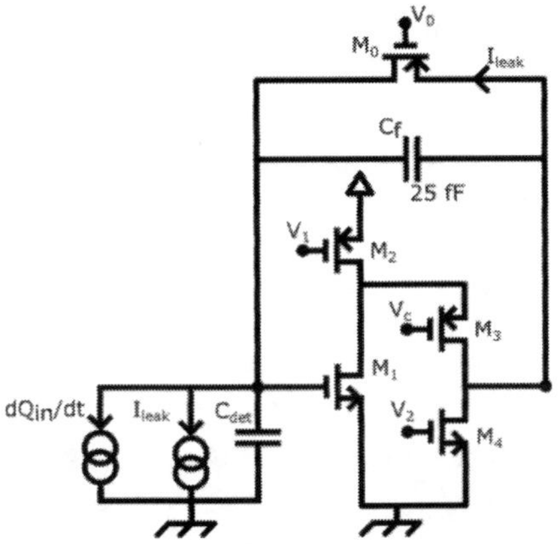

Fig. 3: nMOS Charge Sensitive Amplifier schematic

In the XH018 technology used to design D^2R_2, the nMOS transistor known as "Nelna" (NMOS Enclosed Low Noise Amplifier) is advertised as having low flicker noise. It was therefore selected to be the input transistor for the charge preamplifier in D^2R_2. The input transistor plays the most important role in determining the output noise of the analog chain. The total equivalent noise charge (ENC) is the square root of the sum of all noise contributions.

$$ENC^2 = ENC_{th}^2 + ENC_{1/f}^2 + ENC_{//}^2 \tag{1}$$

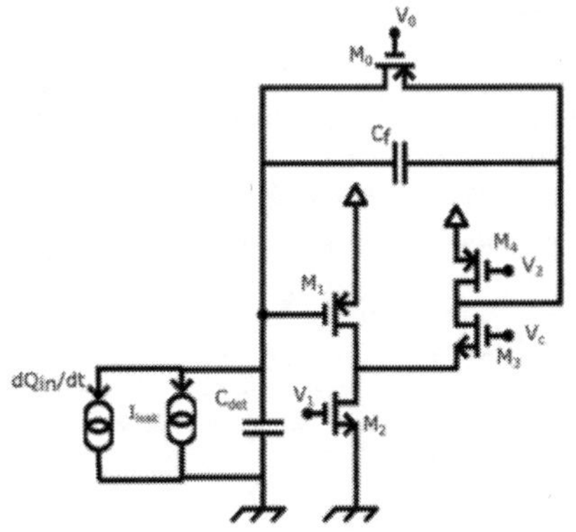

Fig. 4: pMOS Charge Sensitive Amplifier schematic

where ENC_{th}, $ENC_{1/f}$, and $ENC_{//}$ denote the equivalent thermal, flicker, and parallel noise charges, respectively.

As shown in system of equations 2, the contribution (or "weight") of each noise source to the total Equivalent Noise Charge (ENC) strongly depends on both the front-end transistor characteristics and shaping parameters. More specifically, the thermal noise decreases with the peaking time T_{peak} and increases with the total capacitance at the input; the flicker noise is independent of T_{peak} but also scales with the input capacitance; while the parallel noise increases linearly with T_{peak}. This means that, depending on the peaking time and the detector+ASIC input configuration, one noise component may become dominant over the others.

$$\begin{cases} ENC_{th}^2 &= \dfrac{A_{th}\alpha_{th}^2}{T_{peak}}(C_f + C_{det} + C_{in})^2 \\ ENC_{1/f}^2 &= A_{1/f}\alpha_{1/f}^2(C_f + C_{det} + C_{in})^2 \\ ENC_{//}^2 &= A_{//}\alpha_{//}^2 T_{peak} \end{cases} \tag{2}$$

- $\alpha_{1/f}^2, \alpha_{//}^2, \alpha_{th}^2$ are the coefficients of the different noise sources, corresponding respectively to flicker, parallel, and thermal noise. These coefficients depend on the design quality of the CSA and the shaper.
- $A_f, A_{th}, A_{//}$ are the filtering coefficients.
- C_f is the feedback capacitance of the CSA.
- C_{det} is the detector capacitance.
- C_{in} is the capacitance of the input transistor of the CSA.

In our architecture, the main noise sources are the thermal and flicker noises. For a semi-Gaussian shaping, the thermal noise ENC scales with the square of the input capacitance and inversely with the transconductance:

$$ENC_{th}^2 \propto \frac{C_{in}^2}{g_m T_{peak}} \tag{3}$$

While increasing the bias current or W/L ratio improves g_m, it also raises the gate capacitance and thus C_{in}. An optimal trade-off is typically achieved when:

$$C_{in} = \frac{C_{det}}{3} \tag{4}$$

In contrast, flicker noise decreases with the channel area $W \cdot L$, reaching its minimum when:

$$C_{in} = C_{det} \tag{5}$$

ANAQIN-X's optimization strategy

The results of that optimization are shown in Table II. In ANAQIN-X's mini-matrix 8 pixels were dedicated to series noise, with 4 nMOS and 4 pMOS based CSAs with different W/L sizing.

- 1 control pixel identical to that of D^2R_2: It is present in order to rule out possible degradation of the input transistor gate caused by matrix formation (a large-scale manufacturing step that could damage the gate oxide).
- 1 Pixel similar to D^2R_2, using the same Nelna transistor but with a resized channel to avoid a short channel effect.
- 2 Pixels based on another input transistor proposed by XH018 technology but of the same type (NMOS) to compare flicker noise reduction model with "Nelna" transistors.
- 4 Pixels with the same high-level structure as D^2R_2 but based on a P-type amplifier (different transistors offered by the technology).

Once optimized and with detector interconnection, parallel noise becomes non-negligible. In our topology, it consists of a combination of the shot noise generated by the detector leakage current and the parallel thermal noise generated by the reset transistor. Both are expressed by the following equation 6.

$$i^2_{PMO} = \frac{8qI_{leak}}{3} \quad ; \quad i^2_{Shot} = 2qI_{leak} \tag{6}$$

2) Special AC-Coupling CSA architecture

A CSA is reset by a high-value MOS transistor, which supplies the leakage current for the detector when DC-connected. Leakage current has a dual effect on noise:
a) It increases shot noise
b) It increases the thermal noise of the reset transistors by decreasing their transconductance. Here we proposed to take benefit of the

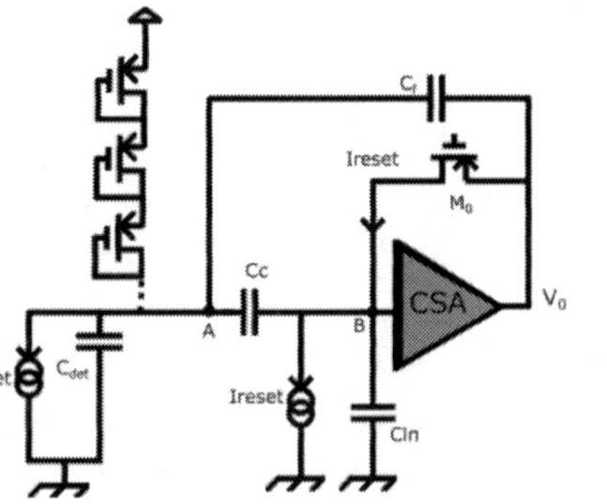

Fig. 5: Special AC-coupling CSA with a tunable I_{leak} generator

architecture depicted in figure 5 to reduce the thermal noise of the reset transistor of standard DC-coupled architecture. In this architecture, the CSA reset transistor that enables C_F discharge, and the detector leakage current generation are now separated by a coupling capacitor C_c. Unlike traditional AC-coupling, this capacitor is located inside the feedback loop.This configuration allows us, in theory:

- To have a much smaller coupling capacitor that would need to be installed to preserve the conversion gain of the CSA (this will be discussed below). Since this capacitor can be reduced in size, it becomes possible to integrate it in the pixel of the chip.
- A reduction in bias noise compared to classical AC-coupled architectures is possible since the coupling capacitor is in the chip and the biasing circuit can also be integrated. Thus,transistors in subthreshold regime can be used to bias the detector and achieve a high equivalent resistor. Additionally, since the transistor is not in the loop, several transistors can be stacked to make the equivalent resistor much higher.
- A reduction in the reset noise by making the reset current I_{reset} and I_{leak} independents, so that $i_{P_{M0}}$ does not scale with the detector's leakage current (will be discussed below).

The transfer function of the circuit proposed in figure 5 can be computed be the equation 7:

$$H_{csa} = \frac{A}{\frac{C_{in}}{C_{ac}} \cdot (C_{det} + C_f) + C_{in} + C_{det} + C_f + A \cdot C_f} \tag{7}$$

To get the closest possible to the ideal gain $1/C_f$, we need:

$$\begin{aligned} &A \cdot C_F >> C_{in} + C_{det} + C_f \\ &and \\ &A \cdot C_F >> \frac{C_{in}}{C_{ac}} \cdot (C_{det} + C_f) \end{aligned} \tag{8}$$

with $C_F = 25\,fF$ et $C_{Det} = 300\,fF$

- In order for $A \cdot C_F \gg C_{in} + C_{Det} + C_F$, we need $A > 1200$, assuming the same conditions as for standard AC-coupling.
- And for $A \cdot C_F \gg (C_{Det} + C_F) \cdot \frac{C_{in}}{C_{ac}}$, we need:

$$A \gg \frac{C_{in}}{C_{ac}} \cdot \left(1 + \frac{C_{det}}{C_{ac}}\right) \quad \Rightarrow A \gg 13 \cdot \frac{C_{in}}{C_{ac}} \tag{9}$$

Equation 12 implies that $C_C >> 13 \cdot \frac{C_{in}}{A}$, whereas for a standard AC-CSA, the condition on C_C is more restrictive ($C_C \gg C_{det}$). The figure 6 shows two simulation results for the conversion gain of the special AC-CSA compared to Standard AC-CSA function of the coupling capacitor C_C. It shows that the special-AC-CSA requires much less capacity to maintain most of its gain ($1pF$ is enough to have 98 % of the gain).

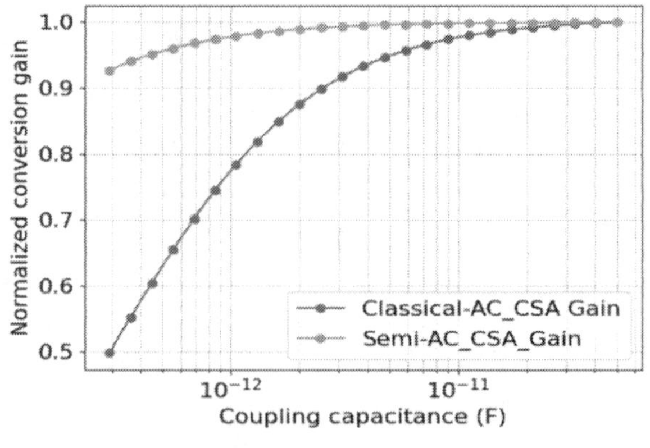

Fig. 6: Comparison of the normalized conversion gain of Semi-AC CSA with classical AC-CSA function coupling capacitor (C_C)

Noise's analytical description

The figure 7 displays our Semi-AC CSA with two input noise sources vn^2 and in^2. vn^2 represents the series noise voltage, and in^2 represents the parallel noise current. We will consider an equivalent noise source ieq^2 composed of vn^2 and in^2, which is expressed by equation 10:

$$i_{eq}^2 = i_n^2 + \frac{V_n^2}{Z_{ib}^2} \tag{10}$$

Adding C_C alters the noise behavior of the CSA in two ways. First,

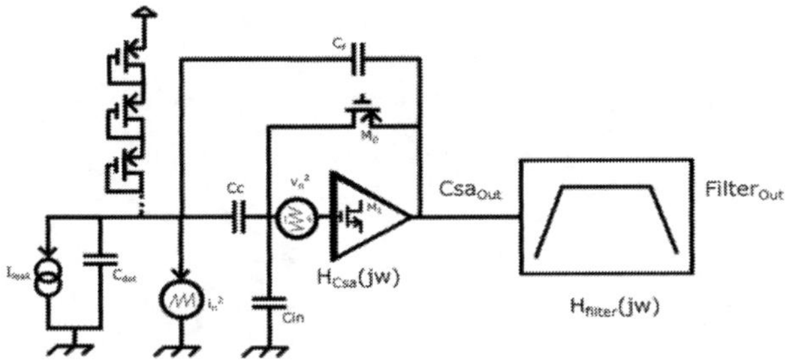

Fig. 7: Semi-AC CSA schematic with noise sources

it reduces reset noise, by de-correlating the reset and leakage currents. Secondly, it causes a slight increase in thermal and flicker noise, as C_C increases the input impedance seen by the series noise Z_{ib}. The expression of Z_{ib} in represented by the equation 11.

$$Z_{ib} = \frac{C_C}{(C_f + C_{det} + \frac{C_C * C_{in}}{C_C + C_{in}}) * (C_{in} + C_C)} \tag{11}$$

The different noise contributions equations becomes as expression in the system of equations 12.

$$\begin{cases} ENC_{th}^2 = \frac{A_{th}\alpha_{th}^2}{T_{peak}}\left(\frac{C_{ac}+C_{in}}{C_{ac}} \cdot (C_F + C_{Det}) + C_{in}\right)^2 \\ ENC_{1/f}^2 = A_{1/f}\alpha_{1/f}^2\left(\frac{C_{ac}+C_{in}}{C_{ac}} \cdot (C_F + C_{Det}) + C_{in}\right)^2 \\ ENC_{//}^2 = A_{//}\alpha_{Semi-AC//}^2 T_{peak} \end{cases} \tag{12}$$

with : $\alpha_{Semi-AC//}^2 = 2qI_{leak} + \frac{8KT}{3.N}$ and N is the number of diodes in the leakage current generator.

The overall noise performance of this architecture depends on the value of the coupling capacitor, the leakage current of the detector and the length N of leakage current generator .

Subtracting the ENC equations of a typical DC-CSA from this Semi-AC coupled one establishes the balance of benefits and risks, yielding the following equations 13, 14, and 15.

$$\Delta ENC_{//} = \frac{1}{q^2} * \left[A_{//} \cdot 2qI_{fuite}\frac{8kT}{3}\left(\frac{1}{N} - 1\right)T_{peak}\right] \tag{13}$$

$$\Delta ENC_{th} = \frac{A_{th}\alpha_{th}^2}{q^2 T_{peak}}\left(\frac{C_F + C_{det}}{\alpha}\left[2(C_f + C_{det} + C_{in}) + \frac{C_f + C_{det}}{\alpha}\right]\right) \tag{14}$$

$$\Delta ENC_f = \frac{A_f\alpha_{1/f}^2}{q^2}\left(\frac{C_F + C_{det}}{\alpha}\left[2(C_F + C_{det} + C_{in}) + \frac{C_F + C_{det}}{\alpha}\right]\right) \tag{15}$$

The figure 8 shows a comparison between those architectures. And

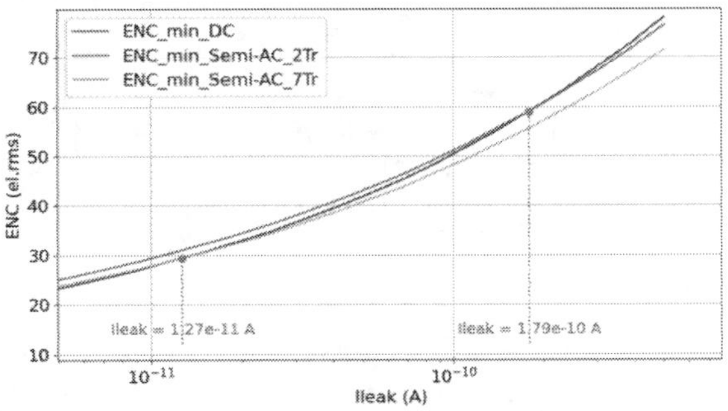

Fig. 8: Comparison of the ENC of Semi-AC CSA with DC CSA function leakage current (I_{leak})

the figure 9 shows that this "leakage current of profitability" gets smaller with the number of transistors in the biasing circuit. Since

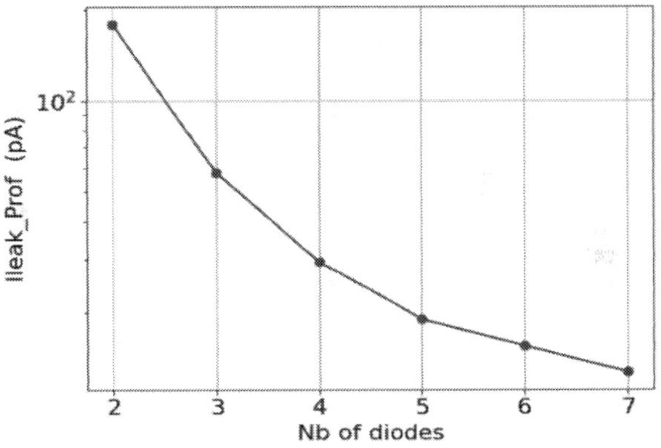

Fig. 9: Simulation results of the "I_{leak} of profitability" of Semi-AC CSA against DC CSA function of the number of diodes presents in the $Ileak$ generator

the special AC-Coupling CSA is designed to optimize parallel thermal reset noise, we can observe in the figure 8 that the noise is slightly worse for low leakage current values until 179 pA (with 2 diodes at C_C =1 pF), and 12.7 pA (with 7 diodes at C_C =1 pF).Thus, the profitability threshold decreases as Cc decreases. This continues until the point where the noise reduction provided by adding an additional diode to the leakage current generator becomes negligible. This statement is only correct if we had the possibility of adding as many diodes as we wanted. This is not the case, as we are limited by the supply voltage and the breakdown voltages of the transistors.

3) CSA architectures for validation

The two circuits shown in figure 10 have been implemented for the validation of the Semi-AC CSA architecture. To compare the performance of the proposed architecture with that of a "Classical" AC-Coupling CSA, a pixel with tunable input capacitance (from 100 fF to 30 pF) has been implemented. The purpose if the Multi Transistor Reset DC-Coupled CSA's purpose is to strengthen the simulation model of the noise reduction given by the leakage current generation system.

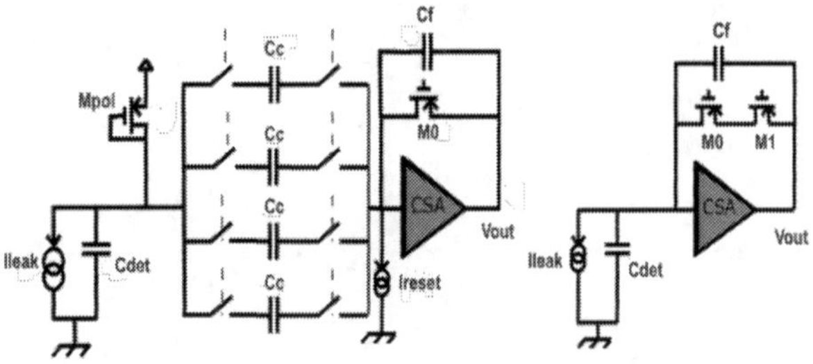

Fig. 10: Classical AC-Coupled CSA with tunable coupling capacitance (left) Double Transistor Reset DC-Coupled CSA (right)

TABLE I: Simulated ENC performances with $C_{det} = 300\ fF$ and $C_{det} = 0\ fF$ with $I_{leak} = 1\ pA$ at $T_{peak} = 3.6s$ for each DC-coupled pixel of ANAQIN-X.

Pixel	ENC expected (e.rms)	
	$C_{det} = 0\ fF$	$C_{det} = 300\ fF$
Reference D^2R_2	13.95	19.2
Nelna 50 µm / 300 nm	13.92	20.6
Ne 80 µm/ 1.1 µm	24.19	27.3
Ne 100 µm/ 1.3 µm	24.55	28.1
Pe3 100 µm/ 450 nm	13.17	21.03
Pe3 150 µm/ 350 nm	13.36	20.83
Pe3lna 50 µm/ 300 nm	12.98	22.41
Pe 70 µm/ 390 nm	15	24.23

In order to be able to compare the Semi-AC CSA model with that of a CSA-DC, the 4 pixels Semi-AC, Semi-AC with tunable number of diodes, Classic AC, and Double Transistor DC are based on the same voltage amplifier as one of the DC pixels. Namely, a pixel based on a pMOS input, the Pe3 100 µm/450 nm.

	nMOS	pMOS	Test
0	Low-Noise 1.8V 90 µm/180 nm	Low-power 3.3V 150 um/350 nm	Low-power 3.3V Special-AC [11]
1	Low-Noise 1.8V 50 um/300 nm	Low-power 3.3V 100 µm/450 nm	Low-power 3.3V DC-Multi-Tr
2	Low-Power 1.8V 80 um/1.1 um	low-power 1.8V 70 µm/390 nm	Low-power 3.3V Classic-AC
3	Low-Power 1.8V 100 µm/1.3 µm	Low-Noise 3.3V 50 µm/300 nm	Low-power 3.3V Semi-AC Comm Multi-Tr

TABLE II: Mini-Matrix pixels specificities

with:

- **Nelna** :1.8V low noise NMOS
- **Pe/Ne** :1.8 V low power PMOS/NMOS
- **Pe3/Ne3** :3.3V low power PMOS/NMOS
- **Pe3lna** :3.3V low noise PMOS

III. Tests Results

A. Gain and linearity

ANAQIN-X is tested by injecting a voltage pulses through on-chip capacitors implemented inside each pixel. No detector has been attached to the channel yet. Figure 12 shows the amplitude of the peak at the output of shaper as a function of the injected charge. The slope of the response of each pixel type is extracted in the linear region to determine the gain of each pixel. The mean conversion gain of the channel is found to be 25 µV/el, which matches the simulation within 10 %. Since the pMOS-based CSA were designed to read out charges equivalent to photons up to 120 keV, we can see that they loose linearity around 30 kel, as predicted.

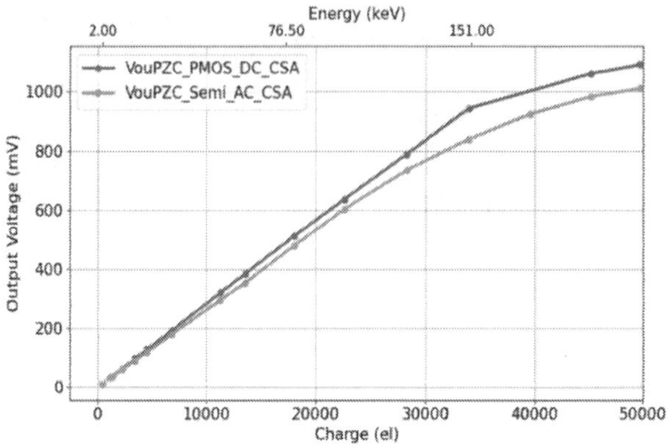

Fig. 11: PZC output voltage variation function of injected charge @ $T_{peak} = 3.6\ \mu s$ for pMOS DC and Semi-AC Pixels

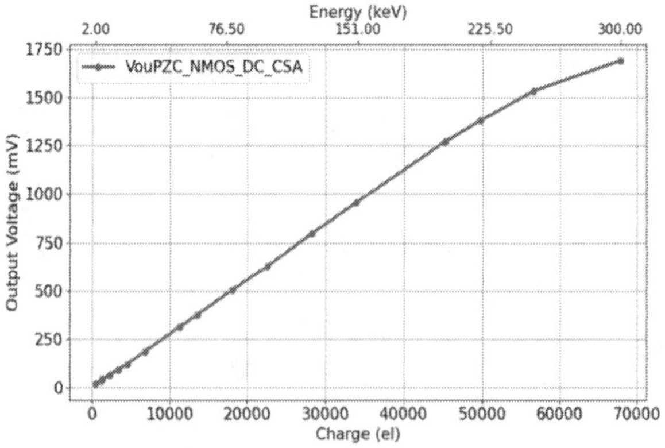

Fig. 12: PZC output voltage variation function of injected charge @ $T_{peak} = 3.6\mu s$ for nMOS DC Pixels

B. Equivalent Noise Charge (ENC)

The curves shown in Figure 13 represent the ENC performance of the 8 pixels with DC coupling as a function of peaking time. In this figure, we can see that the ENC decreases as the peaking time rises, due to a reduction in the contribution of thermal noise, as expressed in the equation 2.

As these measurements were taken without a detector attached to the channel, the ENC performance should be treated with caution.

 https://doi.org/10.1051/epjconf/202533802002

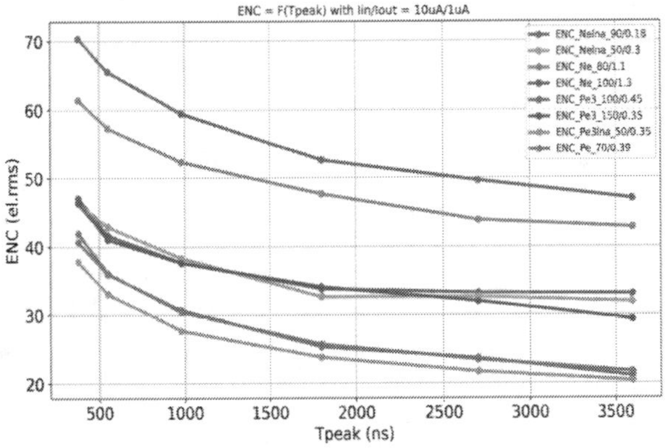

Fig. 13: Measurements of ENC for different DC CSAs function of peaking time T_{peak}

The measured noise performance is encouraging but below what was simulated. These are measurements without a detector. To compare them with the simulation, refer to the table I under the column $C_{det} = 0\ fF$. For example, the pixel "Pe3 150 µm/ 350 nm" was expected to be 13.36 el.rms and was measured at 20.33 el.rms at $T_{peak} = 3.6\ s$.

This discrepancy can be explained, in part, by the inability to perfectly filter the signals arriving at the ASIC. (power supply, reference voltages and currents, etc.) as well as by process variations, as the tests were only performed on a single chip. Measurements on a sample of several ANAQIN-X chips are planned and may shed light on the true nature of the discrepancy.

Inside ANAQIN-X, a leakage current generator (Icomp) is integrated

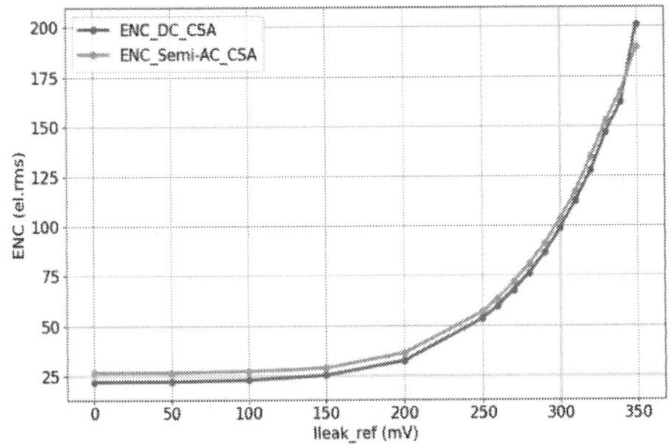

Fig. 14: Measurements of ENC of Semi-AC CSA VS DC CSA function of leakage current i_{leak} with 5 diodes in I_{leak} generator

to enable ASIC testing in the absence of a detector. At the output of Icomp, an N-fold current mirror distributes this current to each pixel, which can be activated individually via a switch. The current injected into the CSA is controlled by a voltage applied to the gate of the mirror transistor (Ileak-ref). The leakage current is estimated from this gate voltage by measuring the CSA fall time (T_{fall}), knowing C_f. This method assumes that the value of C_f is perfectly known. However, for our application, process variations on C_f are non-negligible. On top of this, the fall time measured at the CSA output is strongly affected by output noise, which makes its precise estimation challenging and potentially unreliable, especially in the absence of signal averaging or noise filtering.

Figure 14 presents a comparison of the ENC performance of a DC-CSA and the Semi-AC-CSA proposed in this article, based on Ileak-ref (image of the leakage current), in order to estimate the leakage current at which the Semi-AC architecture becomes less noisy than the DC-CSA. As I_{leak} increases, we can see that the Semi-AC-CSA does indeed become more efficient, eventually surpassing the standard topology starting at $Ileak - ref = 350mV$. There is a discrepancy with the simulation here as well, which can be explained by various factors. Firstly, by the Measurement/Simulation discrepancy already explained for DC-CSAs, and secondly by the uncertainty in measuring the injected leakage current accurately, due to T_{fall} uncertainties.

Although a direct measurement of the leakage current using a picoammeter is planned, it was not performed initially due to the stringent requirements of the setup. The high sensitivity of the instrument demands careful shielding, thermal stability, and mitigation of parasitic leakage paths, which made the indirect estimation based on CSA fall time a more practical first step despite its lower accuracy.

IV. Conclusion and Perspectives

ANAQIN-X shows promising results for the development of a new generation of 2-dimensional ASIC dedicated to read fine pitch CdTe detectors in the MC^2 project. Its noise performance is interesting in terms of achieving high energy resolution imaging spectrometers down to the 500 eV level at 60 keV. Measurements taken on the N pixels showed that the Nelna transistor did indeed suffer from a short channel effect that increased its noise (2 el.rms according to the measurement). However, this increase was not sufficient to explain the excess ENC compared to the simulation (22 el.rms). The measurement on the control pixel D^2R_2 shows an ENC of 33 el.rms, which is consistent with the performance of the best pixel in the ASIC D^2R_2 matrix. In order to rule out grid degradation due to matrix implementation, a test on several ANAQIN-X chips is underway to estimate the statistical distribution of the actual ENC of the pixel. The other two N pixels show the same type of simulation/measurement offset, demonstrating that the Nelna transistor does indeed reduce flicker noise, but to a lesser extent than predicted by the simulation model.

The most promising pixel chain showing the lowest ENC results is the "Pe3lna". Its noise was found to be 20.3 el.rms (with measurement 0.5 el.rms uncertainty). On the other hand, a Special AC-CSA topology has been implemented and tested, offering an attractive solution for readout CdTe/CZT detectors with parallel noise dominance. It is an interesting solution for applications where parallel noise would dominate, offering a 15 % improvement on ENC at Ileak = 100 pA compared with the DC-CSA pixel with identical input transistor and amplifier design.

References

[1] Säm Krucker, G. J. Hurford, O. Grimm, S. Kögl, H.-P. Gröbelbauer, L. Etesi, D. Casadei, A. Csillaghy, A. O. Benz, N. G. Arnold, F. Molendini, P. Orleanski, D. Schori, H. Xiao, M. Kuhar, S. Kobler, L. Iseli, M. Dreier, H. J. Wiehl, L. Kleint, and S. Bloomfield. The spectrometer/telescope for imaging x-rays (stix). *Astronomy & Astrophysics*, 642:A15, 2020.

Published online 30 September 2020; Section: Astronomical instrumentation.

[2] A. Meuris, G. Hurford, M. Bednarzik, O. Limousin, O. Gevin, I. Le Mer, J. Martignac, B. Horeau, O. Grimm, R. Resanovic, S. Krucker, and P. Orleański. Caliste-so x-ray micro-camera for the stix instrument on-board solar orbiter space mission. *Nuclear Instruments and Methods in Physics Research Section A: Accelerators, Spectrometers, Detectors and Associated Equipment*, 695:288–292, 2012. New Developments in Photodetection NDIP11.

[3] O. Gevin, O. Limousin, F. Lugiez, A. Michalowska, A. Meuris, D. Baudin, E. Delagnes, O. Lemaire, and F. Pinsard. IDeF-x HD: A CMOS ASIC for the readout of cd(zn)te detectors for space-borne applications. 10(2):2150009. Publisher: World Scientific Publishing Co.

[4] J. Rodríguez-Pacheco, R. F. Wimmer-Schweingruber, G. M. Mason, R. Gómez-Herrero, et al. The energetic particle detector: Energetic particle instrument suite for the solar orbiter mission. *Astronomy & Astrophysics*, 642:A7, 2020. Published online 30 September 2020.

[5] O. Gevin, E. Delagnes, D. Huynh, O. Limousin, and F. Lugiez. Idef-x bd: A low noise dual polarity asic for the readout of silicon and cdte detectors. In *2015 IEEE Nuclear Science Symposium and Medical Imaging Conference (NSS/MIC)*, pages 1–4, San Diego, CA, USA, Oct./Nov. 2015. IEEE. Conference held 31 Oct.–7 Nov. 2015.

[6] J.-L. Atteia, B. Cordier, J. Wei, and et al. The svom mission. *International Journal of Modern Physics D*, 31(05):2230008, 2022.

[7] O. Gevin, P. Baron, X. Coppolani, E. Delagnes, F. Daly, O. Limousin, F. Lugiez, A. Meuris, and F. Pinsard. Idef-x eclairs: An ultra low noise cmos asic for the readout of cd(zn)te detectors. In *2007 IEEE Nuclear Science Symposium Conference Record*, volume 1, pages 326–332, Honolulu, HI, USA, Oct. 2007. IEEE. Conference held 26 Oct.–3 Nov. 2007; Date added to IEEE Xplore 22 Jan. 2008.

[8] Alicja Michalowska. Studies and development of a readout asic for pixelated cdte detectors for space applications.

[9] B. David. Development of a cdte spectro-imaging for space application. https://theses.hal.science/tel-02901985v1, 2020. [Accessed 04-04-2025].

[10] G De Geronimo and P O'connor. A cmos fully compensated continuous reset system', 2000.

[11] Ivan Perić, Peter Fischer, Christian Kreidl, Hong Hanh Nguyen, Heiko Augustin, Niklaus Berger, Moritz Kiehn, Ann Kathrin Perrevoort, André Schöning, Dirk Wiedner, Simon Feigl, Timon Heim, Lingxin Meng, Daniel Münstermann, Mathieu Benoit, Dominik Dannheim, Frederic Bompard, Patrick Breugnon, Jean Claude Clemens, Denis Fougeron, Jian Liu, Patrick Pangaud, Alexandre Rozanov, Marlon Barbero, Malte Backhaus, Fabian Hügging, Hans Krüger, Florian Lütticke, Carlos Mariñas, Theresa Obermann, Maurice Garcia-Sciveres, Benjamin Schwenker, Alexander Dierlamm, Alessandro La Rosa, and Antonio Miucci. High-voltage pixel detectors in commercial cmos technologies for atlas, clic and mu3e experiments. *Nuclear Instruments and Methods in Physics Research, Section A: Accelerators, Spectrometers, Detectors and Associated Equipment*, 731:131–136, 2013.

Characterization of the KATANA water activation loop in the JSI TRIGA reactor by reactor pulse operation using neutron detectors

Julijan Peric[1,2,*], Domen Kotnik[1,2], Domen Govekar[1,2], Luka Snoj[1,2], and Vladimir Radulović[1,2]
[1] Jožef Stefan Institute, Reactor Physics Division, Slovenia
[2] Faculty of mathematics and physics, University of Ljubljana, Slovenia
(*) Julijan.peric@ijs.si

***Abstract*— This paper presents the characterization of the KATANA water activation facility in the TRIGA research reactor at the Jožef Stefan Institute (JSI), operating the reactor in pulse mode and using neutron detection techniques. The facility plays a crucial role in the benchmarking and validation of computational models relevant to nuclear fusion applications, particularly in the context of ITER. A dual approach was used to determine the system volume of the KATANA loop: a direct volumetric method involving a physical measurement and a novel neutron-based method using the circulation of neutron-activated water. The latter uses a ^{3}He proportional counter to track the flow and decay of the short-lived ^{17}N nuclides that emit neutrons with energies in the MeV range. Experimental results from both methods provided consistent volume estimates and confirmed the reliability of the neutron-based dynamic characterization technique. These results support future efforts in the development and validation of neutron and activation models for fusion-related systems.**

Keywords **— Water activation, KATANA, neutron, JSI TRIGA.**

I. Introduction

KATANA is a water activation facility located at the Jožef Stefan Institute TRIGA Mark II reactor in Ljubljana, Slovenia [1]. Since its commissioning at the end of 2023, it has completed three experimental campaigns that have contributed to a better understanding of the water activation processes and their modeling, which are important in the context of research in nuclear fusion, for the future operation of the ITER machine. The facility is designed to provide a flexible benchmark-quality environment for the validation of computational tools for the activation of water. It will also support shielding experiments, for ITER-relevant materials, and serve as a stable source of high-energy gamma rays and neutrons [2-5].
The computational fluid activation tools developed for ITER cooling loop characterization [6,7] consist of two methods: computational fluid dynamics (CFD) in conjunction with neutronic simulations. These methods are essential for the accurate modeling of spatially distributed radioactive sources, such as the activated cooling water in future fusion devices. CFD focuses on detailed thermal hydraulic modeling, while neutronic simulations are used to predict the behavior of neutrons and their interactions.

By integrating these two approaches, a more accurate and comprehensive understanding of water activation under different operating conditions can be achieved. Usually, these calculation tools are validated by dose rate measurements at different positions around the radiation source to ensure the accuracy and reliability of the simulations [7]. This approach is holistic, as the measurements combine neutron and flow information, which cannot be separated from each other. In the present study, a novel method was developed to characterize the system volume of the water activation loop using neutron detectors during operation of the JSI TRIGA reactor in pulse mode. Pulse operation allows the irradiation of a finite amount of water in the irradiation part of the KATANA water activation loop in a short length of time, typically below 1 s. As the highly activated package of water circulates through the loop, it emits neutrons along the way; by measurements using neutron detector it is possible to track the flow of the activated water and infer information on the system.

II. Water activation

The radioactivity of activated water consists mainly of high-energy gamma rays and neutrons, the most important isotopes being ^{16}N, ^{17}N and ^{19}O. Consequently, activated water acts as a moving radiation source, emitting radiation both during reactor operation and for a short time afterwards. Of the water activation reactions, the neutron interaction with ^{16}O has been studied most intensively. This is due to the high natural abundance of ^{16}O and its relatively large neutron activation cross-section.

TABLE I
WATER ACTIVATION REACTIONS [2]

Reaction	$t_{1/2}$	Major decay products	Threshold energy	Natural abundance
^{16}O (n, p) ^{16}N	7.13s	γ: 6.13 MeV γ: 7.12 MeV	≈ 10 MeV	99.76%
^{17}O (n, p)^{17}N	**4.17s**	**n: 0.38 MeV n: 1.17 MeV**	**≈ 9 MeV**	**0.04%**
^{18}O (n, γ) ^{19}O	26.9s	γ: 0.20 MeV γ: 1.36 MeV	< 1 eV	0.2%

The radioisotope ^{17}N is being studied in detail due to its ability to emit high-energy neutrons [4].

III. Reactor pulse operation

The JSI TRIGA research reactor is equipped with a pneumatic system that enables operation in pulse mode by facilitating the rapid ejection of the transient control rod. This rapid ejection renders the reactor prompt supercritical, initiating an exponential increase in power. However, as the power and therefore the temperature of the fuel increases, the negative temperature coefficient of reactivity of the TRIGA fuel leads to a rapid decrease in reactivity. This feedback mechanism quickly stabilizes the reactor and leads to a sharp drop in power and the establishment of a new equilibrium state. The peak power achieved during a pulse can be between 1 MW and several hundred megawatts, depending on the inserted reactivity and core configuration. Despite this high peak power, the total energy released remains relatively low (typically a few megajoules) due to the short duration of the pulse, which is on the order of 1 ms [8].

The dominant feedback effect in the TRIGA reactor is its large prompt negative temperature reactivity coefficient, a feature made possible by the unique composition of its fuel - a homogeneous mixture of 20 % enriched uranium and zirconium hydride with a Zr-H ratio of 1.6. The most important feedback mechanism is neutron spectrum hardening. s rather than in the surrounding water. As a result, changes in fuel temperature are immediately reflected in the moderator, enabling fast and intrinsic reactivity control that increases the safety and stability of pulse operation [9].

IV. KATANA water activation facility

The KATANA water activation facility is in the hall of the JSI TRIGA reactor. It was designed and built to provide detailed insights and the possibility to validate calculation methods for the water activation process under fusion-relevant conditions and much more, which is directly relevant for the preparations for the operation of the large fusion research reactor ITER.

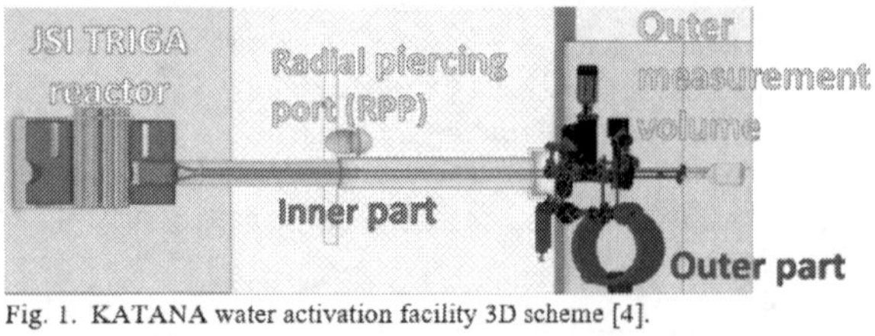

Fig. 1. KATANA water activation facility 3D scheme [4].

The design of the water activation facility consists of a closed loop partially inserted into the Radial Piercing Port (RPP). It consists of three main components: the inner activation part, which is located within the RPP in close proximity to the reactor core and where most of the neutron-induced water activation takes place; the transport piping connecting the outer and inner parts of the loop; and the outer observation part, which is located outside the RPP and where a number of gamma and neutron detectors are positioned to measure the emitted radiation. All auxiliary systems such as the pump, flow meters, pressure and temperature sensors are located outside the RPP to ensure safe and efficient operation of the system.

From a fluid dynamics perspective, the most important operating parameters that determine the behavior of the water activation circuit are the system volume, flow rate, temperature and pressure. Real-time monitoring of flow rate, temperature and pressure is achieved via a network of precisely positioned active sensors distributed throughout the system, allowing comprehensive characterization of the dynamic and thermodynamic states of the fluid. However, uncertainties in the simulated response of the water activation loop were observed due to the lack of geometric/volumetric specifications of some mechanical components, particularly the pump. Therefore, the system volume of the water activation loop was determined experimentally using two complementary methods.

The first approach consisted of a direct volumetric measurement in which the uncharacterized segments of the loop were filled with water and the volume of water introduced was quantified. However, this process was complicated by the presence of a multi-stage pump, which prevented easy filling due to its complex internal structure. For the remaining part of the circuit, the volumetric data was extracted from a detailed 3D CAD model.

The second method used a neutron-based technique. In this approach, the water was activated by a short neutron irradiation and then circulated through the loop. By analyzing the time-dependent neutron signal measured downstream, the circulation time of the activated water is determined. This temporal information was then used to infer the total volume of the system based on the known flow rate, providing a non-invasive, dynamic alternative to direct measurement.

A. Direct approach

For the direct volumetric measurement, the KATANA water activation loop was first partially dismantled to gain access to the multi-stage centrifugal pump. The pump was carefully removed from the system and underwent a decontamination process to remove residual radioactivity, mainly from activated corrosion products. After radiological clearance, the pump was thoroughly cleaned of residual water.

The pump was then gradually filled with deionized water in controlled steps of around 50 ml. Due to the complex internal structure of the multi-stage design, complete filling required manual rotation of the pump rotor to displace trapped air and ensure thorough filling of all internal cavities. This process enabled an accurate estimate of the internal volume of the pump and other previously uncharacterized components within the loop.

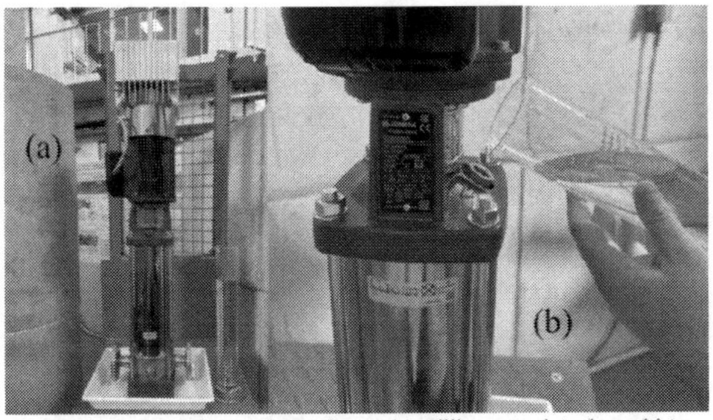

Fig. 2. Direct approach experimental setup (a). Filling water into the multistage Lowara pump (b). The experiment was carried out on February 6, 2025.

The internal volume of the multi-stage pump was successfully quantified using the direct measurement method. For the other components of the KATANA water activation loop, the corresponding volumes were determined from a detailed as-built model. The volume data from both the experimental measurement and the CAD analysis are listed in the table below.

TABLE II
DIRECT APPROACH RESULTS

Symbol	Quantity	
V_{PUMP}	Pump volume	5.2 l ± 0.1 l
V_{CAD}	CAD volume	7.6 l ± 0.05 l
$\mathbf{V_D}$	**KATANA volume (direct)**	**12.8 l ± 0.15 l**

B. Neutron approach

The KATANA facility was operated in stationary conditions with a flow rate of 0.66 l/s. The circulating water was subsequently exposed to a reactor pulse presented in the graph in Fig. 3.

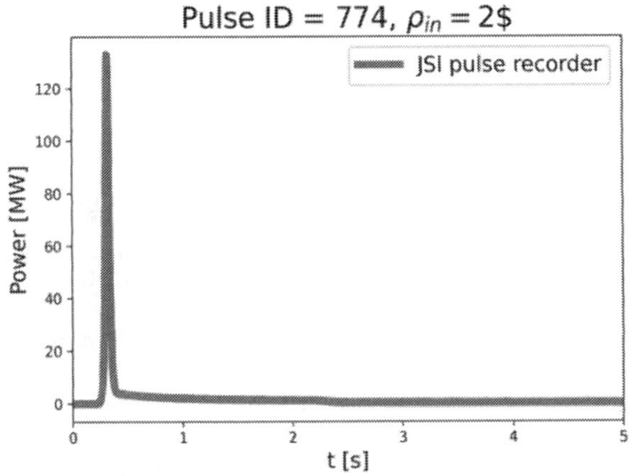

Fig. 3. Reactor pulse signal measured with the JSI pulse recorder during pulse 774. The experiment was carried out on October 24, 2024.

The graph illustrates the temporal development of the reactor power during pulse no. 774. A sharp power peak reaching more than 120 MW is observed shortly after the pulse is triggered, which activates the water in the irradiation volume of the KATANA water activation loop.

TABLE III
REACTOR PULSE PARAMETERS

Parameter	
Pulse ID	774
Inserted reactivity	2 $
Peak power	133 MW
Energy released	5.8 MWs
Ful width at half maximum	0.039s

The activated water package circulated through the closed loop system. A neutron detector positioned in the outer measurement volume recorded the time-dependent neutron signal. The neutron signal peaks were then used to determine the time delay between the first and second transition of the activated water.

Using the independently measured flow rate under these experimental conditions, the total volume of the system was calculated using the following relationship:

$$V = FLW1 \cdot \Delta t, \quad (1)$$

where V is the system volume, FLW1 is the volumetric flow rate and Δt is the measured circulation time. This method provided a reliable dynamic estimate of the system volume.

Fig. 4. Experimental setup of the KATANA volumetric characterizations with neutron approach. The experiment was carried out on October 24, 2024.

The neutron detection system used in the experiment consists of an LND ^{3}He proportional counter, paired with high-performance readout electronics for signal processing and data acquisition. The ^{3}He detector works based on the $^{3}He(n,p)^{3}H$ reaction, which has a very high reaction cross-section and enables high efficiency in the detection of thermal neutrons. The detector is housed in a moderating polyethylene holder to optimize the efficiency of neutron capture. The output signal from the detector is first fed into a CAEN A1422 preamplifier, which amplifies the weak charge pulses generated by neutron interactions in the proportional counter. After amplification, the signal is digitized using a CAEN V1782 Multi-Channel Analyzer (MCA). The MCA records the pulse height spectrum, allowing the differentiation of neutron-induced events from background and gamma radiation. The integrated system offers high temporal and energetic resolution, making it well suited for time-dependent neutron flux measurements, as shown in Fig. 5.

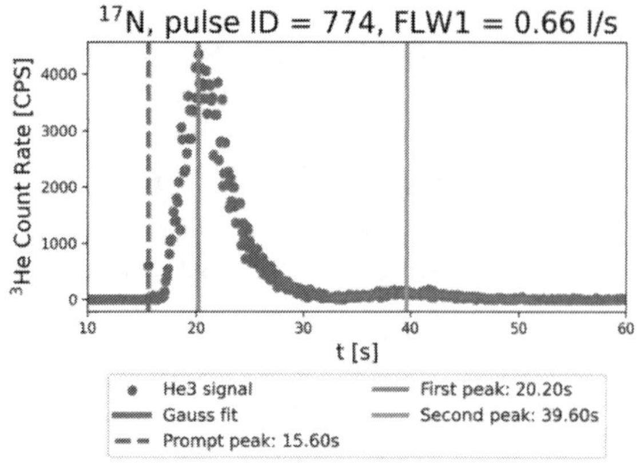

Fig. 5. Time-resolved ^{3}He detector response following activation of ^{17}O during pulse 774 at a KATANA flow rate of 0.66 l/s. Vertical lines denote peak positions: red dashed (prompt), green (first water activation peak), and orange (second water activation peak). The experiment was carried out on October 24, 2024.

The graph shows the measured signal of the ^{3}He neutron detector after the activation of water during pulse 774 with a KATANA flow rate of 0.66 l/s. The signal reflects the decay of the short-lived isotope ^{17}N, which emits high-energy neutrons. Three distinct features can be observed in the detector response: a prompt neutron peak at 15.60 s, due to the immediate neutron emission streaming out of the reactor and scattered on the reactor hall; a first water activation peak at 20.20 s, corresponding to the initial circulation of activated water to the detector position; and a second water activation peak at 39.60 s, indicating the second arrival of the activated water. A Gaussian fit was applied to the second peak to better determine the position of the maximum. The circulation time of the activated water was determined by calculating the time difference between the arrival of the second and the first activation peaks in the signal of the neutron detector.

TABLE IV
NEUTRON APPROACH RESULTS

Symbol	Quantity	
FLW1	Flow rate	0.660 l/s ± 0.005 l/s
Δt	Circulation time	19.40 s ± 0.05s
V_N	**KATANA volume (neutron)**	**12.80 l ± 0.10 l**

C. *Results*

The system volume of the KATANA water activation loop was determined using two independent methods. The direct volumetric measurement gave a volume of 12.80 l ± 0.15 l, while the neutron-based approach yielded a value of 12.80 l ± 0.10 l. The excellent agreement between the two methods confirms the reliability of the experimental procedures and validates the neutron-based technique as a robust, non-invasive alternative to volumetric characterization.

D. *Origins of uncertainty*

This study was motivated by the uncertainty of the system volume, as no detailed technical specifications for the internal volume of the pump were available. Uncertainties in direct volume determination arise primarily when checking whether the pump is filled with water and in the accuracy of the volume measurements during the filling process. With the neutron-based method, uncertainties arise from the temporal resolution of the detection system, small fluctuations in the water flow and the resolution of the water flow measurement system, which affect the accuracy of the circulation time measurement. Together, these factors limit the accuracy of the final volume estimate.

V. Conclusions

This paper successfully demonstrates the use of reactor pulse operation and time-resolved neutron detection for the characterization of the KATANA water activation loop in the JSI TRIGA reactor. Two complementary methods direct volumetric measurement and a dynamic neutron-based approach were used to determine the system volume. The results of both methods were in excellent agreement and provided a solid validation of the experimental methodology. In particular, the neutron-based technique proved to be a non-invasive and reliable alternative, especially suitable for the characterization of systems with limited accessibility or geometric complexity. In addition, the use of ^{3}He neutron detectors enabled high-resolution tracking of activated water, providing new insights into isotopic behavior and system dynamics under fusion-relevant conditions. These results underline the value of pulse operation for advanced diagnostic capabilities in nuclear research facilities and strengthen the basis for future benchmarking and validation of computational tools to support the development of fusion reactors.

Acknowledgment

The authors acknowledge the support of the Slovenian Ministry of Education, Science and Sport (project codes P2-0405 Fusion technologies; P2–0073 Reactor physics; PR-09769/PR-12326/PR-12842 Training of young researchers, NC-0022 Water activation in nuclear reactors). This work has been carried out within the framework of the EUROfusion Consortium, funded by the European Union via the Euratom Research and Training Programme (Grant Agreement No 101052200 — EUROfusion). Views and opinions expressed are, however, those of the author(s) only and do not necessarily reflect those of the European Union or the European Commission. Neither the European Union nor the European Commission can be held responsible for them.

References

[1] L. Snoj, et. al., A half-century of nuclear research, education and training: Story of the JSI TRIGA reactor, Annals of Nuclear Energy, vol. 214, 2025, DOI. 10.1016/j.anucene.2024.111122.

[2] D. Kotnik, J. Peric, D. Govekar, L. Snoj and I. Lengar, "KATANA - water activation facility at JSI TRIGA, Part I: Final design and activity calculations," Nuclear Engineering and Technology, vol. 57, 2025, DOI. 10.1016.j.net.2024.09.036.
[3] D. Kotnik, J. Peric, D. Govekar, L. Snoj and I. Lengar, "KATANA - water activation facility at JSI TRIGA, Part II: First experiments," Nuclear Engineering and Technology, vol. 57, 2025, DOI. 10.1016.j.net.2024.10.052.
[4] J. Peric, D. Kotnik, L. Snoj and V. Radulović, "Neutron Emission from Water Activation: Experiments and Modeling Under Fusion-Relevant Conditions at the KATANA Facility," Fusion Engineering and Design, vol. 216, 2025, DOI. 10.1016/j.fusengdes.2025.115052.
[5] D. Kotnik, A. Basavaraj, L. Snoj and I. Lengar, "Design optimization of the closed-water activation loop at the JSI irradiation facility", Fusion Engineering and Design, vol. 193, 2023, DOI. 10.1016/j.fusengdes.2023.113632.
[6] R. Pampin, E. Masia, F. Cau, D. Kotnik, R. Villari and A. Portone, Insights from applications of the RSTM tool for coupled CFD-activation fluid simulation, Fusion Engineering and Design vol. 203, 2024, DOI. 10.1016/j.fusengdes.2024.114466.
[7] M. De Pietri, J. Alguacil, E. Rodr.guez and R. Ju.rez, Development and validation in water of FLUNED, an open-source tool for fluid activation calculations, Computer Physics Communications, vol. 291, 2023, DOI. 10.1016/j.cpc.2023.108807.
[8] I. Švajger, D. Čalič, A. Pungerčič, A. Trkov and L. Snoj, "Evaluation of reactor pulse experiments," Nuclear Engineering and Technology, vol. 56, pp.1165-1203, 2024, DOI. 10.1016.j.net.2023.11.021.
[9] I.Vavtar and L. Snoj. "Evaluation of approximations in reactor pulse," EPJ Web of Conferences, Vol. 247, EDP Sciences, 2021.

Activation of the inner irradiation Snail after long-term operation of the KATANA irradiation facility

Ylenia Kogovšek Žiber[1,2*], Benjamin Barbarič[1], Domen Kotnik[1], Julijan Peric[1,2], Luka Snoj[1,2], and Igor Lengar[1]
[1]Jožef Stefan Institute, Slovenia
[2]Faculty of Mathematics and Physics, University of Ljubljana, Slovenia
(*) ylenia.ziber@ijs.si

***Abstract*—The primary coolant in fusion reactors such as ITER is water. In recent years,** research has increasingly focused **on the irradiation and activation of water, which leads to the formation of ionizing radiation fields. To better understand these phenomena, the KATANA irradiation facility for water activation was commissioned in early 2024 at the Jožef Stefan Institute. Since its initial testing, several upgrades and developments have been implemented. In 2026, the inner irradiation head where the water is irradiated in the reactor, is planned to be removed and replaced with an ITER relevant head design that more accurately replicates ITER's cooling components. Ahead of this replacement, a comprehensive neutron activation analysis of the current head, has been conducted to support the licensing process and ensure the safe removal. This paper presents the results of activation calculations for the inner head after long-term operation of KATANA, utilising the neutron irradiation field of the JSI TRIGA reactor. These preliminary calculations will aid in ensuring the safe maintenance of the KATANA facility and provide valuable insights for predicting the activation of its internal components.**

***Keywords* — Long-term operation, aluminium Snail activation, KATANA, JSI TRIGA, Fusion, ITER.**

I. Introduction

WATER as a primary coolant is used extensively in fission power plants and is expected to be equally important in future fusion reactors such as ITER [1] and DEMO [2]. When used in nuclear environments, water is exposed to intense neutron fluxes, leading to its activation and the subsequent emission of high-energy gamma rays and neutrons through radioactive decay. Oxygen activation is the dominant short-term mechanism; however, longer-term radiation exposure is mainly attributed to activated corrosion products (ACPs) [3–5], which can contribute significantly to occupational doses. In fusion reactors, the predominantly high-energy neutron environment, in combination with threshold reactions for high-energy neutrons, particularly the $^{16}O(n,p)^{16}N$ reaction above 10.5 MeV, leads to much greater activation of the cooling water, with activity levels up to five orders of magnitude higher than in comparable fission reactors [6].

Despite several computational studies focusing on water activation in fusion reactors [7], there remains an urgent need for experimental validation. This need arises from the scarcity of nuclear data, inconsistencies between libraries [8], and the challenges of accurately modelling time- and space-dependent radiation sources. Existing facilities, such as the Frascati Neutron Generator (FNG) in Italy [9] and the Fusion Neutronic Source (FNS) in Japan [10], have provided valuable data but are limited in water volume capacity and availability. With the recent shutdown of the JET tokamak, which concentrated mainly on water activation experiments during the last DTE3 campaign [11], the gap in experimental capabilities has widened, particularly in the production and study of high-energy gamma-ray sources (6–7 MeV).

To overcome the lack of experimental facilities, a dedicated closed-loop water activation facility, KATANA [12-15], was developed and commissioned in early 2024 at the Jožef Stefan Institute (JSI) to support benchmark experiments for validating fluid activation codes, with a focus on fusion applications. The inner irradiation component, referred to as the inner Snail due to its shape, is inserted into the radial piercing port (RPP) of the TRIGA MARK II research reactor [16, 17], placing it directly in a high-flux neutron field with neutron flux 6×10^{12} $cm^{-2}s^{-1}$ [18]. Aluminum was chosen as the structural material to mitigate long-term activation while maintaining experimental functionality. The facility will enable a number of different experiments, from shielding tests with ITER-relevant materials to assessing the behavior of radiation detectors under high-energy gamma fields.

Due to the high neutron flux in the inner Snail's location and thus the highest activation, a thorough activation analysis is required to ensure safe handling, facilitate maintenance, and support future upgrades. The current research is focused on quantifying the long-term activation (20 days after the end of irradiation) of the inner Snail after two years of reactor operation. This analysis will guide decisions regarding safe maintenance intervals and support the licensing process for replacing the current Snail with a more ITER-relevant version, scheduled for 2026. The calculations, will also be critical for decommissioning, planning and minimizing radiation damage to components. Section II of this paper describes the KATANA water activation facility in more detail. Section III presents the method and the framework for the activation calculations, while the results are presented in Section IV.

https://doi.org/10.1051/epjconf/202533803002

II. KATANA WATER ACTIVATION FACILITY AT JSI TRIGA REACTOR

The KATANA water activation facility at the TRIGA Mark II research reactor of the Jožef Stefan Institute [17] in Slovenia is a newly commissioned closed-loop irradiation system designed for advanced water activation experiments in support of fusion research [12,13]. The system is installed in the Radial Piercing Port (RPP), which provides high neutron flux and favorable shielding. It consists of an inner irradiation section near the reactor core, transport pipes, and an outer observation area with gamma and neutron detectors. A three-way valve allows control of the water distribution between the primary and secondary circuits. Shielding include special plugs and a thick concrete wall, minimize radiation exposure in the reactor hall. The system also allows for simulation of complex scenarios such as water mixing and leakage events. The final layout of KATANA was built and commissioned in early 2024 and the entire system is controlled via an external control panel for safe and flexible operation [13, 14]. A schematic of the KATANA irradiation facility is shown in Fig. 1.

The inner part of the KATANA loop, located inside the RPP of the JSI TRIGA reactor, includes the inner Snail, narrow transport pipes, and gamma/neutron shielding plugs. The inner Snail, a complex, four-block cylindrical assembly with a central vertical wall and instrumentation channel (1.2 cm inner diameter), maximizes water volume (2.95 L) and was 3D printed from aluminium alloy for minimal activation. To shield components and personnel, a specially designed plug filled with Boral plates, concrete, and lead shot was developed.

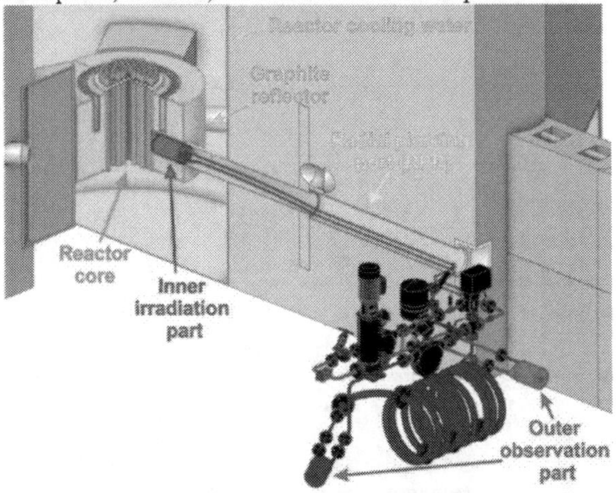

Fig. 1. Schematic of the KATANA irradiation facility. The primary loop (red) is located in the Radial Piercing Port (yellow) of the TRIGA research reactor and the inner irradiation part – Snail, is located near the reactor core (orange) in the graphite reflector (green). The reactor core is surrounded with the reactor cooling water (light blue). The secondary/delay loop is depicted in blue [13].

The outer part of the KATANA system is located outside the RPP and houses key instrumentation such as sensors, detectors, pumps, valves and the outer observation Snails. These Snails mirror the inner design but include a larger central cavity for the neutron detectors, while the water volume (2.94 l) remains almost identical to optimize activity and consistency. The KATANA irradiation facility is depicted in Fig. 2.

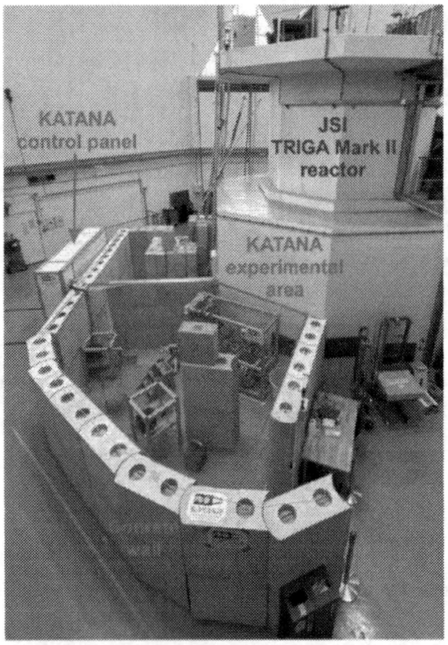

Fig. 2. The KATANA irradiation facility, fully constructed and operational with a closed-water activation loop at the JSI TRIGA research reactor, is remotely controlled and monitored via a dedicated control panel positioned outside the experimental zone, beyond the concrete shielding wall [13].

This paper is focused mainly on the inner Snail. The material used for 3D- printing is an aluminium alloy called EOS Aluminium (Al-Si10-Mg) with a density of ρ=2.67 g/cm^3 [19]. The manufactured complex inner Snail is shown in Fig. 3a and a schematic representation of the detailed model of the inner Snail is shown in Fig. 3b. The detailed material composition with the maximum amount of impurities specified by the manufacturer is listed in Table I.

TABLE I
MATERIAL COMPOSITION OF THE ALUMINIUM ALLOY USED FOR THE 3D-PRINTING OF THE INNER SNAIL [18]

Material	Atomic abundance [/]
^{27}Al	8.3938E-01
^{28}Si	9.56E-02
^{56}Fe	2.65E-03
^{63}Cu	2.1E-04
^{55}Mn	2.2E-03
^{24}Mg	3.59E-03
^{58}Ni	2.3E-04
^{64}Yn	4.1E-04
^{208}Pb	3.9E-05
^{120}Sn	1.1E-04
^{48}Tn	8.4E-04

To further support the development of fusion reactors, a new ITER relevant Snail head is being designed that takes into account as many relevant characteristics of the ITER first wall cooling panel as possible, such as the water resident time and the water flow [14]. The ITER first wall cooling panel and the new inner Snail with improved design can be seen in Fig. 3c and Fig. 3d respectively. The removal and replacement of the current inner Snail is planned for early 2026.

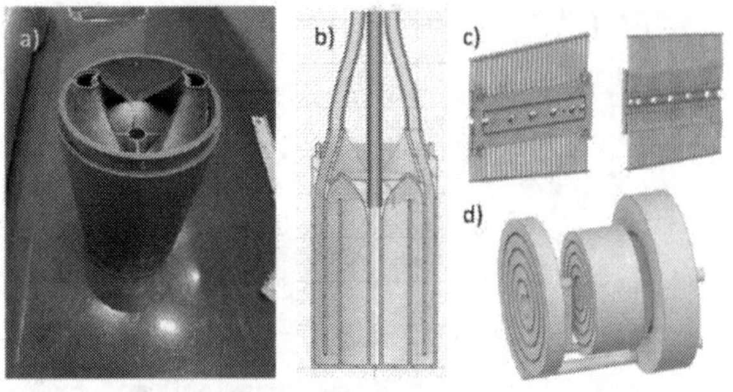

Fig. 3. The manufactured complex irradiation part, i.e. the inner Snail (completely 3D-printed from aluminium) (a), a schematic of the detailed model of the inner Snail (b), with the central empty pipe shown in red, that serves for instrumentation, a schematic of the ITER first wall cooling panel (c) and a schematic of the new inner Snail with the current design (d) [13].

III. ACTIVATION CALCULATION OF THE INNER SNAIL

Two main assumptions of the problem were applied for the activation calculations, namely negligible water flow and negligible water activation for long cooling times. This is possible due to the stagnation of the water after the shutdown of the facility and the short half-life time of the activated isotopes in water, such as ^{16}N with half-life time of 7.13 s [20] and due to the stagnation of the water after the shutdown of the facility and.

The neutron transport was calculated with the Monte Carlo Nuclear Particle transport code MCNP v6.2 [21] and the Isotopic inventory calculations were performed with the FISPACT-II 5.0 code [22].

A. Neutron transport

The neutron transport calculations with MCNP were performed with 2×10^8 number of source particles, with a mesh size of 2 mm × 2 mm × 2 mm over the area of the inner Snail. The nuclear data library used was ENDF/B-VIII.0 [19]. Due to the small inner Snail size in comparison to the TRIGA research reactor weight windows were calculated with the Automated Variance Reduction Generator ADVANTG [23] code using a 160 × 239 × 142 mesh to increase sampling in important regions. The mesh is not homogeneous and the voxels are significantly smaller near the inner Snail.

The neutron flux Φ_{MCNP} and neutron flux spectra were calculated from the neutron transport. Since the spectra were calculated in really small voxels and in 709 energy groups to match the flux definition in the FISPACT-II code, the component averaged neutron flux spectrum was used for all voxel calculations due to its lower statistical error and is shown in Fig. 4. The component averaged flux spectrum was calculated as the average flux spectrum from the flux spectra in each voxel of the inner Snail.

B. Important voxel determination

The voxels where the neutron flux was calculated using the cartesian mesh, do not by default match the complicated geometry of the inner Snail. The used cartesian mesh did include the Snail, parts of the RPP tube and some surrounding concrete. To speed up calculations and to really calculate only the activation of the inner Snail and not the other nearby components, the voxels of interest were determined with superimposing the CAD model of the inner Snail over the neutron flux mesh. This was performed using a JSI in-house workflow [24] for CAD over mesh voxel determination, however this can be also achieved with other tools.

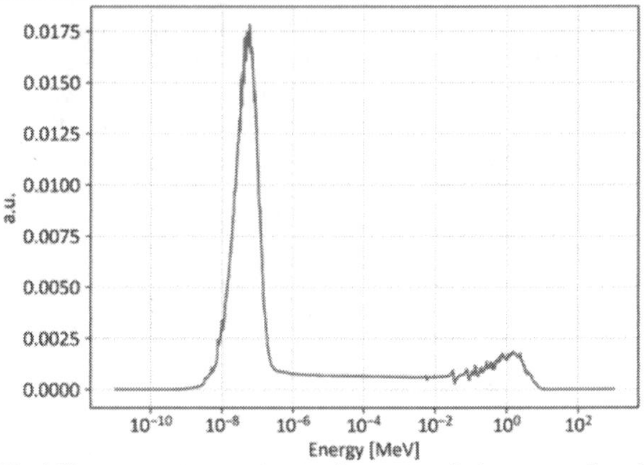

Fig. 4. The component averaged neutron flux spectrum in the energy scale used for the activation calculations.

C. Irradiation scenario

To perform the activation calculation of the inner Snail, an irradiation scenario is required. For this purpose, the irradiation history was derived from the TRIGA Mark II reactor. The irradiation timeline begins in November 2023, when the KATANA facility was initially tested, and extends through early March 2025 the starting point of these calculations. For the period from November 2023 to February 2025, an average monthly irradiation history used and is shown in Fig. 5 a. A more detailed irradiation profile was applied for February 2025, and is presented in Fig. 5 b.

D. Exploring the saturation model and activation calculations

To avoid performing activation calculations for each of the many small voxels of the inner Snail, the saturation model was explored [25]. In the saturation model, a simplification is made by assuming that already activated nuclei do not affect the reduction in the number of non-activated nuclei, or at least that this reduction is negligible. Similarly, it is assumed that the number of atoms in the target remains constant. It is also assumed that already activated nuclei cannot be reactivated.

Activated nuclei are produced at a reaction rate of $\Phi\sigma_* N_T(t)$, where Φ is the neutron flux, σ_* is the microscopic activation cross-section, and $N_T(t)$ is the number of non-activated nuclei in the target.

At the same time, activated nuclei (denoted by $*$) decay at a rate of $\lambda_* N_*(t)$, where λ_* is the decay constant and $N_*(t)$ is the number of activated nuclei.

The rate of change of activated nuclei, dN_*/dt, depends on the difference between the production $\Phi\sigma_* N_T(t)$ and the decay $\lambda_* N_*(t)$ of activated nuclei:

$$\frac{dN_T(t)}{dt} = \Phi\sigma_* N_T(t) - \lambda_* N_*(t), \quad (1)$$

Here, $N_T(t)$ represents the number of initial, non-activated nuclei. We define the initial conditions as $N_T(t = 0) = N_T(0)$

and $N_*(t=0) = N_*(0) = 0$. In the saturation model, the number of initial nuclei (atoms in the target) is constant, so $N_T(t) = N_T(0) =$ const. The differential equation for the activation dynamics becomes:

$$\frac{dN_*(t)}{dt} = \Phi\sigma_* N_{T;\,t=0}(t) - \lambda_* N_*(t), \quad (2)$$

This equation can be integrated:

$$\int_0^{N_*(t)} \frac{d\{\Phi\sigma_* N_{T;\,t=0} - \lambda_* N_*\}}{\Phi\sigma_* N_{T;\,t=0} - \lambda_* N_*} = -\lambda_* \int_0^t dt, \quad (3)$$

The solution to (3) is:

$$N_*(t) = \frac{\Phi\sigma_* N_{T;\,t=0}}{\lambda_*}\{1 - \exp(-\lambda_* t)\}, \quad (4)$$

Using the definition of the activity of a nuclide A_*:

$$A_*(t) = \frac{dN_*}{dt} = \lambda_* N_*(t), \quad (5)$$

the time-dependent activity can be expressed as:

$$A_*(t) = \Phi\sigma_* N_{T;\,t=0}\{1 - \exp(-\lambda_* t)\} = \\ = A_{*,sat}\{1 - \exp(-\lambda_* t)\}, \quad (6)$$

The activity approaches its saturated value $A_{*,sat}$ exponentially over time during the irradiation. After a long period, the asymptotic value of the activity represents the balance between the production and decay rates of the activated nuclide [26].

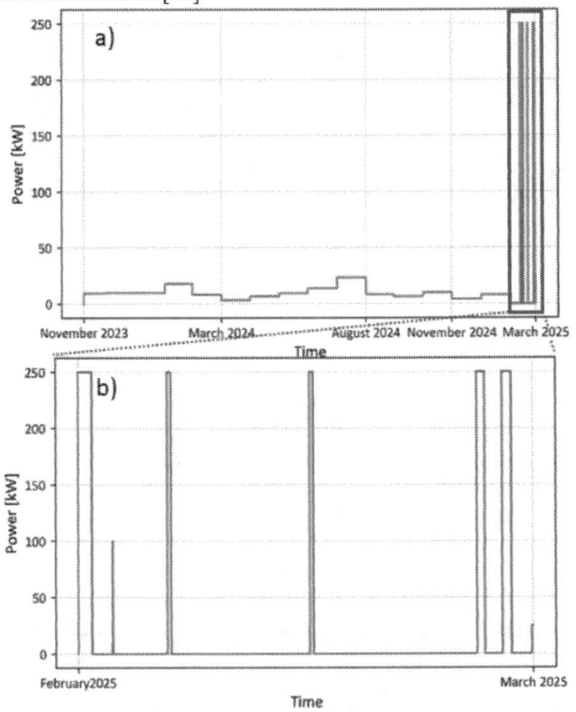

Fig. 5. The TRIGA Mark II operational history from the first testing of the KATANA facility in November 2023 to March 2025 (average values until February 2025) (a) and more detailed history from February 2025 to March 2025 (b). The maximum power is P=250 kW.

After the end of the irradiation, at any specific time t_i, the activity A_* depends solely on the neutron flux Φ, while the term $N_{T;t=0}$*exp($-\lambda_* t_i$) remains constant. This relationship holds even in the presence of multiple nuclides and varying fluxes, in which case the total activity is expressed as the sum over all individual activation processes:

$$\sum A_*(t=t_i) = \Phi \sum \sigma_{*,j} N_{Tj;\,t=0}(t_i)\{1 - \sum \exp(-\lambda_{*,j} t_i)\}, \quad (7)$$

This relation can be rewritten as:

$$A_*(t=t_i) = \Phi K(t=t_i) + C(t=t_i), \quad (8)$$

where $K(t=t_i)$ and $C(t=t_i)$ are the parameters of the linear function at time t_i. Theoretically, a zero intercept is expected however, an additional constant C is introduced to allow for numerical errors that may occur in later calculations.

The saturation model simplification can be considered applicable if at any given time t_i the relation between the activity and the flux is linear, as in (8), and the relation between the logarithm of the activity and the logarithm of the flux is also linear.

Some nuclides may be overlooked if their contribution to the total activity is very small, as their non-saturation behaviour has little impact on the overall activity. Therefore, a saturation model cannot be assumed for every isotope, even if the total activity appears to follow such a model. However, if the linear relationship in (8) holds for the total activity, and the material composition and neutron spectrum remain constant, a separate evaluation of each nuclide may not be necessary.

To simplify subsequent calculations, the relationship between specific activity (activity per unit mass) and neutron flux was examined instead of total activity. For each selected time step after irradiation ended, both the direct relationship between specific activity and flux, and the logarithmic relationship between these quantities, were evaluated. The flux range spanned from 1×10^9 to 5×10^{13} cm^{-2}s^{-1}, with the maximum value corresponding to the TRIGA reactor operating at full power (P=250 kW). Other flux values in the irradiation scenario (Fig. 5) were scaled proportionally to this maximum. These fluxes were chosen arbitrarily to cover the full range of possible fluxes in the TRIGA reactor.

For each maximum flux and its corresponding irradiation scenario, an input file was prepared, and activation calculations were performed using FISPACT-II with the component-averaged flux spectrum shown in Fig. 4. The specific activity was calculated for each maximum flux, and the parameters of the linear function in (8) were determined. These calculations were carried out for different, arbitrarily chosen time steps. The results, summarised in Table II, show a correlation coefficient R of 0.99 for all fits [27]. From these parameters the specific activation of the inner Snail can be calculated using the absolute flux Φ obtained from the neutron flux calculated in MCNP Φ_{MCNP}.

TABLE II
PARAMETERS OF THE LINEAR FIT IN THE SATURATION MODEL FOR DIFFERENT TIMES AFTER THE END OF IRRADIATION

Time	K	C
0 s	3.18E-04 ± 0.0001E-04	0
1 day	6.29E-07 ± 0.0007E-07	-1269
3 days	1.67E-07 ± 0.0006E-07	920
7 days	1.26E-07 ± 0.0002E-07	-87
20 days	1.21E-07 ± 0.0001E-07	117

The neutron flux calculated in MCNP Φ_{MCNP} is normalized per source particle, therefore a normalization to obtain the absolute flux Φ, was needed. The absolute normalization of the flux is derived from [28]:

$$\Phi = \frac{P\Phi_{MCNP}\nu}{W_f k}, \quad (9)$$

where, P is the reactor power, ν is the average number of fission neutrons, W_f is the average energy released as heat per fission event and k_{eff} is the effective multiplication factor.

For these calculations the full power of the TRIGA reactor, P = 250 kW, was used, since all linear relations calculations were scaled to this value. The average number of fission neutrons is ν = 2.43. The effective multiplication factor k_{eff} represents the ratio of neutrons produced by fission in one generation to those lost through absorption and leakage in the previous generation; k_{eff} =1 indicates a critical state; k_{eff}> 1 indicates supercritical; and k_{eff}< 1 indicates subcritical operation. The effective multiplication factor of the TRIGA reactor at full power (250 kW) is k_{eff}= 1.05387 and the average energy released is $W_f = 3.20 \times 10^{-11}$ J.**

IV. ACTIVATION RESULTS

The specific activation of the Snail was as shown in the previous chapter calculated from the relation between the absolute neutron flux and the activation. The specific activation of the Snail 3 days, 7 days and 20 days after the end of irradiation is shown in Fig. 6. The highest activation occurred in the Snail part nearest the reactor core and it is 2 orders of magnitude higher than the one furthest from the core.

The total activity of the inner Snail is calculated by adding the specific activity of all the voxels and multiplying the total with the mass of one voxel (V_{voxel}=8 mm^3, m_{voxel}=0.021 g) and it is presented in Table III. From Fig. 6 and Table III there can be seen that there is not much difference in the activity from 7 days to 20 days after the end of the irradiation. The uncertainty is calculated using the FISPACT-II code. This uncertainty does not take into account the statistical uncertainty in the MCNP calculations or the nuclear data uncertainty from the ENDF/B-VIII.0 data library.

V. CONCLUSIONS

The KATANA irradiation facility at the Jožef Stefan Institute is a state-of-the-art experimental facility for studying neutron activation in fusion-relevant environments by using the well characterized neutron flux in the JSI TRIGA research reactor. It consists of a modular, closed-loop water activation system with an inner irradiation part and an outer observation part. The inner part is a 3D-printed aluminium component, the Snail, positioned in a high-flux neutron field near the core of the TRIGA research reactor. In 2026, the current inner Snail will be replaced with an ITER-relevant version that better replicates the geometric and hydrodynamic properties of the ITER cooling at the first wall panel.

The activation calculations were performed over a simulated two-year operation period, utilising the applicability of the saturation model, which enables prediction of specific activity as a linear function of absolute neutron flux.

TABLE III
ACTIVITY OF THE INNER SNAIL

Time	Activity [Bq]
0 s	3.94E+12 ± 0.08E+12
1 day	7.78E+09 ± 0.16E+09
3 days	2.08E+09 ± 0.04E+09
7 days	1.56E+09 ± 0.03E+09
20 days	1.51E+09 ± 0.03E+09

The results show that the activity decreases rapidly in the first days after shutdown and the value is (1.56E+09 ± 0.03E+09) Bq at day 7 after the end of the irradiation, which is crucial for planning safe maintenance and decommissioning activities. These results not only support the licensing process for future upgrades, but also provide a solid framework for further experimental work at KATANA.

Future efforts will focus on the further uncertainty calculations, performing activation and shutdown dose rate calculations with the JSIR2S code [29-31], developed at the JSI. These efforts position KATANA as a vital experimental benchmark for ITER-oriented research, enabling accurate validation of models and aiding the refinement of future fusion reactor designs.

ACKNOWLEDGMENT

The authors acknowledge the support of the Slovenian Ministry of Education, Science and Sport (project codes P2-0405 Fusion technologies; P2–0073 Reactor physics; PR-12873/PR-12326 Training of young researchers, NC-0022 Water activation in nuclear reactors). This work has been carried out within the framework of the EUROfusion Consortium, funded by the European Union via the Euratom Research and Training Programme (Grant Agreement No. 101052200 — EUROfusion). Views and opinions expressed are, however, those of the author(s) only and do not necessarily reflect those of the European Union or the European Commission. Neither the European Union nor the European Commission can be held responsible for them.

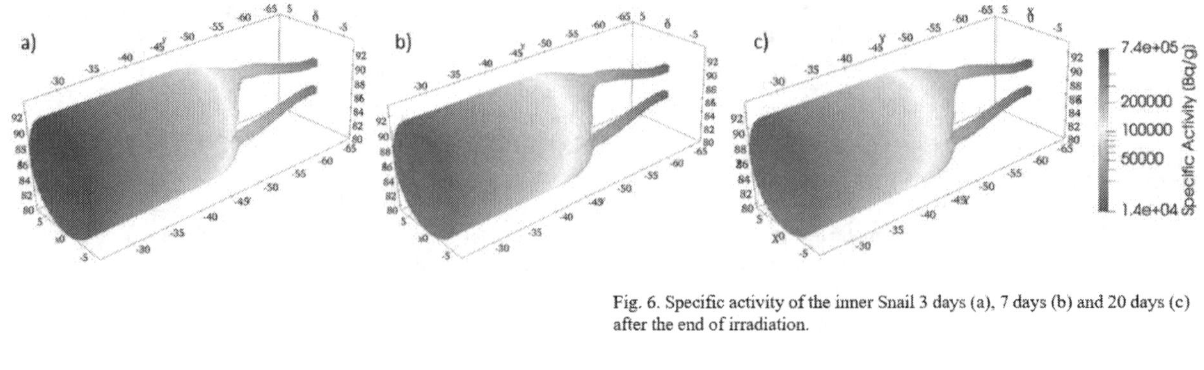

Fig. 6. Specific activity of the inner Snail 3 days (a), 7 days (b) and 20 days (c) after the end of irradiation.

REFERENCES

[1] N. ITER organization ITER, homepage (2024). Available: http://www.iter.org. Accessed on: 1.7.2025
[2] N. EUROfusion, "The demonstration power plant: DEMO O(n,p)N," 2024. Available: https://euro-fusion.org/programme/demo/. Accessed on: 1.7.2025
[3] D. Carloni, "Roadmap for activated corrosion products (ACPs) assessment in fusion reactors: A global collaborative approach," presented at 15th ITER Neutronics Meeting and Fusion Neutronics Workshop, 2024. [Online]. Available: https://indico.iter.org/e/117. Accessed on: 1.7.2025.
[4] S. Noce, "Overview of the ongoing EUROfusion R&D activities on the ITER activated corrosion products," presented at 15th ITER Neutronics Meeting and Fusion Neutronics Workshop, 2024. [Online]. Available: https://indico.iter.org/e/117. Accessed on: 1.72025.
[5] N. Terranova, S. Breidokaitė, G.M. Contessa, L. Di Pace, C. Gasparrini, T. Kaliatka, G. Mariano, "Activated corrosion products evaluations for occupational dose mitigation in nuclear fusion facilities," Environments, vol. 9, no. 7, p. 76, 2022.
[6] A. Žohar, L. Snoj, "On the dose fields due to activated cooling water in nuclear facilities," Prog. Nucl. Energy, vol. 117, 2019, Art. no. 103042. [Online]. Available: https://www.sciencedirect.com/science/article/pii/S0149197019301374
[7] P. Chiovaro, S. Ciattaglia, F. Cismondi, A. Del Nevo, P.A. Di Maio, G. Federici, C. Frittitta, I. Moscato, G. Spagnuolo, E. Vallone, "Investigation of the DEMO WCLL breeding blanket cooling water activation," Fusion Eng. Des., vol. 157, 2020, Art. no. 111697.
[8] A. Žohar, I. Lengar, L. Snoj, "Analysis of water activation in fusion and fission nuclear facilities," Fusion Eng. Des., vol. 160, 2020, Art. no. 111828.
[9] M. Martone, M. Angelone, M. Pillon, "The 14 MeV Frascati neutron generator," J. Nucl. Mater., vols. 212–215, pp. 1661–1664, 1994. [Online]. Available: https://www.sciencedirect.com/science/article/pii/0022311594911096
[10] C. Konno, I. Murata, "Fusion neutronics source (FNS) facility at Japan atomic energy agency," 2014. [Online]. Available: https://www.osti.gov/etdeweb/biblio/22318475
[11] V. Radulović, S. Rupnik, J. Naish, S. Bradnam, Z. Ghani, S. Popovichev, V. Kiptily, P. Batistoni, R. Villari, L. Snoj, "Preparation of a water activation experiment at JET to support ITER," Fusion Eng. Des., vol. 169, 2021, Art. no. 112410. [Online]. Available: https://www.sciencedirect.com/science/article/pii/S0920379621001861
[12] D. Kotnik, A.K. Basavaraj, L. Snoj, I. Lengar, "Design optimization of the closed-water activation loop at the JSI irradiation facility," Fusion Eng. Des., vol. 193, 2023, Art. no. 113632. [Online]. Available: https://www.sciencedirect.com/science/article/pii/S0920379623002168
[13] Domen Kotnik, Julijan Peric, Domen Govekar, Luka Snoj, Igor Lengar, "KATANA water activation facility at JSI TRIGA, Part I: Final design and activity calculations," Nucl. Eng. Technol., 2024. [Online]. Available: https://doi.org/10.1016/j.net.2024.09.036
[14] Domen Kotnik, Julijan Peric, Domen Govekar, Luka Snoj, Igor Lengar "KATANA water activation facility at JSI TRIGA, Part II: First experiments," Nucl. Eng. Technol., 2024. [Online]. Available: https://doi.org/10.1016/j.net.2024.10.052
[15] Julijan Peric, Domen Kotnik, Luka Snoj, Vladimir Radulović, "Neutron emission from water activation: Experiments and modeling under fusion relevant conditions at the KATANA facility," Fusion Eng. Des., 2025. [Online]. Available: https://doi.org/10.1016/j.fusengdes.2025.115052
[16] I. Mele, M. Ravnik, A. Trkov, "TRIGA Mark II benchmark experiment, part I: Steady-state operation," Nucl. Technol., vol. 105, no. 1, pp. 37–51, 1994.
[17] Ž. Štancar, L. Snoj, L. Barbot, "Reaction rate distribution experiments at the Slovenian JSI TRIGA Mark II research reactor," in Int. Handbook of Evaluated Reactor Physics Benchmark Experiments, TRIGA-FUND-RESR-002, Paris, France, 2017.
[18] L. Snoj, G. Žerovnik, A. Trkov, "Computational analysis of irradiation facilities at the JSI TRIGA reactor", Applied Radiation and Isotopes, Volume 70, Issue 3,2012,
[19] EOS GmbH Electro Optical Systems, "EOS Aluminium AlSi10Mg Process Data Sheet." [Online]. Available: https://www.eos.info/metal-solutions/metal-materials/data-sheets/mds-eos-aluminium-alsi10mg
[20] D. Brown, M. Chadwick, R. Capote, A. Kahler, A. Trkov, G. Žerovnik, *et al.*, "ENDF/B-VIII.0: The 8th major release of the nuclear reaction data library with CIELO-project cross sections, new standards and thermal scattering data," Nucl. Data Sheets, vol. 148, pp. 1–142, 2018.
[21] C. J. Werner, *et al.*, "MCNP version 6.2 release notes," Los Alamos Nat. Lab., Los Alamos, NM, USA, Tech. Rep., Feb. 2018.
[22] J.-C. Sublet, *et al.*, "FISPACT-II: An advanced simulation system for activation, trans-mutation and material modelling," Nuclear Data Sheets, vol. 139, pp. 77–137, 2017
[23] S. W. Mosher, *et al.*, "ADVANTG: An Automated Variance Reduction Parameter Generator, Rev. 1," Oak Ridge Nat. Lab., Tech. Rep., Aug. 2015.
[24] B. Barbarič, "Razvoj programske kode za prikaz rezultatov MCNP mreže cenilk v posameznih 3-D komponentah" 12th Conference of Young Nuclear Professionals, Podgorica Reactor Centre, 29 May 2025
[25] W. Abdel-Rahman and E. B. Podgoršak, "Neutron-Activation Revisited: The Depletion and Depletion-Activation Models", Medical Physics, 32, 2005.
[26] B. Barbarič, "Assessment of Core Component Activation in Krško Nuclear Power Plant," M.S. thesis, Dept. Nucl. Eng., Faculty of Mathematics and Physics, University of Ljubliana, Ljubiana, Slovenia, 2023.
[27] N. R. Draper, H. Smith, "Applied Regression Analysis", Wiley, 1998, Volume 3, Ch. 1,
[28] G. Žerovnik, M. Podvratnik, L. Snoj, "On normalization of fluxes and reaction rates in MCNP criticality calculations," Ann. Nucl. Energy, vol. 63, pp. 126–128, 2014. [Online]. Available: https://doi.org/10.1016/j.anucene.2013.07.045
[29] K. Ambrožič, L. Snoj, "Modeling of γ field around irradiated TRIGA fuel elements by R2S method", 13th International Conference on Radiation Shielding & 19th Topical Meeting of the Radiation Protection and Shielding Division of the American Nuclear Society, Volume 153, 2017
[30] K. Ambrožič, B. Kos, L. Snoj, G. Kramberger, "Application of JSIR2S for Dosimetry Calibration of Nurfet Semiconductor Dosimeters at the JSI TRIGA Reactor In-between Reactor Shutdowns", Proc. Int. Conf. Nuclear Energy for New Europe (NENE), 2019, Portorož, Slovenia,September 9–12
[31] K. Ambrožič, L. Snoj, "JSIR2S code for delayed radiation simulations: Validation against measurements at the JSI TRIGA reactor", Progress in Nuclear Energy, Volume 129, 2020

Development of an Event Tracking Feature in OpenMC for Neutron Noise Analysis

Michel Saliba[1,*], John Tramm[2], Paul Romano[2], Erik B Knudsen[3], Andreas Pautz[1,4], Oskari Pakari[1,4]
[1]École Polytechnique Fédérale de Lausanne (EPFL), 1015 Lausanne, Switzerland
[2]Argonne National Laboratory, 9700 S. Cass Ave, Lemont, IL 60439, United States
[3]United Neux, Copenhagen, Denmark
[4]Paul Scherrer Institut (PSI), 5232 Villigen, Switzerland
(*)michel.saliba@epfl.ch

***Abstract*—We present the development and implementation of a neutron event-tracking capability in OpenMC, an open-source, community-driven Monte Carlo radiation transport code. This feature can be utilized for explicit neutron noise simulations, enabling the estimation of integral kinetic parameters such as the prompt decay constant, which are important for accurate predictions of reactor transient behaviors and validation of computational tools. Experimentally, the prompt neutron decay constant (α) can be measured non-invasively during steady-state reactor operation via neutron noise analysis techniques such as the Rossi-alpha method. Accurately simulating these neutron noise measurements requires explicit tracking of individual neutron interactions, preserving detailed timing information essential for subsequent statistical analysis. Previously, such capabilities were primarily restricted to export-controlled software like MCNPX PoliMi or MCNP 6.3. Our open-source implementation in OpenMC offers customizable filters that discriminate and record specific neutron interactions, based on parameters such as reaction type, energy, material, and geometry. Recorded events are stored externally in either OpenMC's native HDF5 format or the MCPL format for versatility. We demonstrate the application of this feature by employing the Rossi-alpha method for noise analysis on simulated neutron event data. Benchmarking simulations using the Godiva I geometry and a model of the CROCUS zero-power reactor, employing ENDF/B.VII.0 nuclear data, confirm the reliability of our approach through comparisons with Serpent 2 results. This development significantly expands the accessibility of neutron noise simulation capabilities. The event-tracking feature could also see use in other detection applications, such as time-of-flight experiments or radiation imaging.**

***Index Terms*—Neutron Noise, Monte Carlo Neutron Transport, Prompt Decay Constant.**

I. Introduction

UNDERSTANDING reactor kinetics is important for ensuring the safety and efficiency of nuclear reactors. One key parameter in reactor transient analysis is the prompt neutron decay constant (α), which characterizes neutron population fluctuations [1]. Measuring α experimentally provides insight into reactor behavior, especially during transients, and serves as a reference integral quantity for validating computational tools [2], [3]. Moreover, noise techniques are also employed for reactor monitoring and fissile material inventory verification, particularly in the field of safeguards [4], [5].

Noise techniques in fissile systems have been investigated since the inception of nuclear engineering, as first demonstrated by Feynman [6], who proposed a method to extract integral parameters from the fluctuations of a detector signal in a steady-state fissile assembly. Recent examples of neutron noise experiments include programs at the MINERVE reactor [7]–[10], VENUS at SCKCEN [11], IPEN Brasil [12], and experiments on critical assemblies performed at the Los Alamos National Laboratory's NCERC facility [13].

Recent trends involve new experiments, methods, and instrumentation that attempt to advance noise analysis beyond traditional approaches, such as using distributed detector arrays to accurately measure the prompt decay constant (α) [14], or gamma-ray-based noise methods for ex-core reactor monitoring [5]. These developments are particularly interesting for emerging Generation IV reactor designs [15], whose novel materials and geometries introduce new modeling and nuclear data challenges.

One such challenge arises with the simulation of Tristructural Isotropic (TRISO)-coated particle fuel. Recent studies have highlighted discrepancies in Monte Carlo transport simulations involving TRISO fuel, particularly due to limitations in how current codes treat neutron scattering and resonance behavior [16]. To address these issues, a recent experiment at Los Alamos National Laboratory, called DEIMOS, investigated high-assay low-enriched uranium (HALEU) fuel with TRISO particles [17]. Noise experiments can here be part of zero-power reactor physics tests and characterization, and later provide kinetic parameter datasets for HALEU TRISO systems.

On the simulation side, deterministic codes have traditionally been employed to estimate the prompt neutron decay constant [18], [19]. Monte Carlo codes are also commonly used to estimate the prompt decay constant; however, many of these codes are export-controlled or rely on sophisticated algorithms, such as the iterative fission probability (IFP) method, to provide predictions of α [20]. Recent advances include explicit Monte Carlo simulations of the entire neutron population in a reactor using extensive computational resources [21]. Other methods include the estimation of α using explicit noise simulation (ENS) methods [22], in which every individual fission event is tracked. Usually, such analyses are conducted using MCNPX-PoliMi [23], and MCNP6.3 with the PTRAC feature [24], all of which are subject to export control restrictions.

In this study, we introduce a new feature for OpenMC, which is an open-source, community-driven Monte Carlo code [25]. The feature enables the explicit tracking of individual neutron interactions suitable for noise analysis. Our motivation aligns with the principles promoted by the International Atomic Energy Agency (IAEA) in their Open-source Nuclear Codes for Reactor Analysis (ONCORE) initiative [26], advocating for open-source, open-data, and open-science practices. The feature allows for user-defined custom filters to track arbitrary neutron interactions in arbitrary volumes or materials to track time, energy deposited and other relevant information of events. To verify and demonstrate the reliability of our implementation, we conducted comparisons with Serpent 2 [27], a well-established code already licensed for use within our laboratory. Our work aims to broaden access to advanced neutron noise simulation capabilities, promoting wider community engagement and supporting transparent and reproducible scientific research.

The outline of the paper is as follows. In the Benchmarks section, we present the geometries used in the simulations: the CROCUS zero-power reactor and the Godiva I critical assembly. Then in the Methods section, we describe the noise analysis techniques employed and the specific Monte Carlo codes used in the study. In the Results section, we present and analyze the simulation outcomes, discussing the key findings. Finally, the Conclusion summarizes the main results of the study and offers suggestions for future research directions.

II. Benchmarks

In this section, we describe the geometries used for the comparison between the different Monte Carlo transport codes.

A. The CROCUS zero-power reactor

The EPFL CROCUS reactor is a zero-power research facility [28] that operates as a critical assembly using uranium fuel and light water as a moderator. Primarily dedicated to education and research purposes [29], [30], CROCUS operates at a maximum power of 100 W. A distinctive feature of the reactor is its dual-zone fuel configuration, consisting of an inner zone with uranium oxide fuel and an outer zone with metallic uranium fuel, arranged concentrically as illustrated in Figure 1. The reactor's geometry is well-defined and included in the Nuclear Energy Agency's International Handbook of Evaluated Reactor Physics Benchmark Experiments [31].

CROCUS is equipped with four primary monitoring detectors: two compensated ionization chambers located to the north and south of the core for basic monitoring, and two fission chambers (FCs) positioned on the east and west sides, calibrated as power monitors [32]. Beyond these fixed systems, the reactor supports various neutron detection instruments, including different types of proportional counters (e.g., ^{3}He) and scintillation-based detectors.

CROCUS has been extensively studied via noise experiments [5], [22], [33] and provides therefore a good experimental basis for code validation.

Fig. 1. CROCUS zero-power research reactor, uranium fueled and water moderated with a maximum power of 100 W [28]

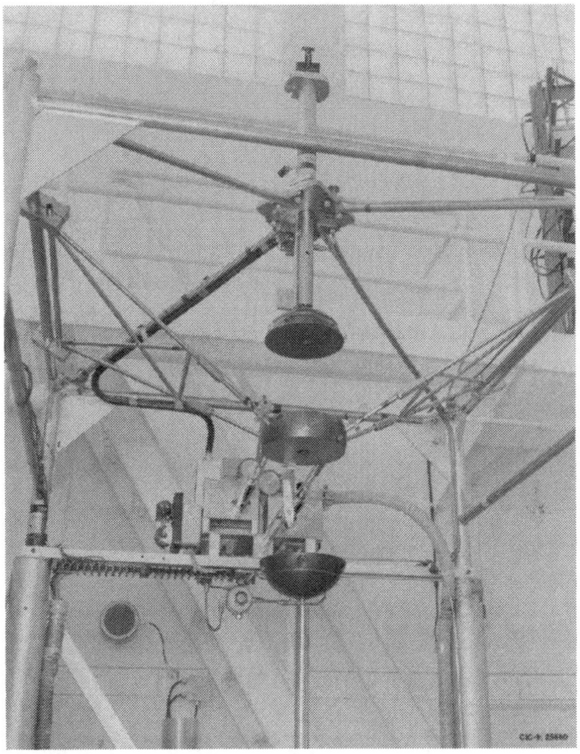

Fig. 2. GODIVA I (aka, Lady Godiva device) - the critical assembly made of highly enriched uranium (93.7% U^{235}) [35]

B. GODIVA I

Godiva I was a critical assembly composed of 93.7% highly enriched uranium, shown in Figure 2, with a radius of 8.737 cm, operated without a neutron reflector. It was operated at Los Alamos National Laboratory. Full details of the assembly are documented in the International Criticality Safety Benchmark Evaluation Project (ICSBEP) Handbook [34].

III. Methods

In this section, we describe the Monte Carlo codes employed and the noise analysis techniques used to estimate the prompt neutron decay constant (α).

A. OpenMC

OpenMC is an open-source, community-driven Monte Carlo particle transport code developed for simulating neutron and photon interactions [25]. It is widely used across various fields, including nuclear engineering, fusion energy research, medical physics, and radiation shielding. Designed with high-performance computing and scalability in mind, OpenMC is well-suited for modern, large-scale simulations. Since its initial release in 2012, it has evolved significantly through contributions from an active global user base. The code supports advanced geometry modeling, continuous-energy cross sections, and integration with multiphysics frameworks. Its open-source license and comprehensive Python API promote flexibility, making it easy to extend and adapt for specialized research and engineering applications. For the purposes of this work, a model of the CROCUS reactor and Godiva I was developed within OpenMC.

B. Serpent 2

To verify the OpenMC results, we employed Serpent 2 [27], a Monte Carlo neutron transport code extensively used in reactor physics and radiation shielding applications. While Serpent 2 is commonly utilized for generating multigroup cross-section libraries for deterministic reactor simulations, it is also capable of estimating reactor kinetic parameters with high fidelity [36]. For the purposes of this work, a model of the CROCUS reactor and Godiva I was developed within Serpent 2.

C. Event Tracking in Monte Carlo

The event tracking feature in Serpent 2 was custom-developed internally, recording only the incident neutron responsible for fission events, and is currently not part of the official Serpent release [14]. Although this feature is still under internal testing, it has been validated against experimental data from the CROCUS reactor [22]. Motivated by open-source and open-science principles, we opted to implement similar functionality in OpenMC.

The new event tracking feature in OpenMC, named *Collision Track*, enables recording of neutron interactions and is currently under development [37]. Its progress is publicly available via the OpenMC Git repository [38]. Integrated into OpenMC's Python API, Collision Track allows users to activate and configure tracking through specifying event types and recording conditions. While OpenMC can output full particle tracks, Collision Track offers a more efficient alternative by recording only user-defined events, significantly reducing memory and storage demands. Tracked events are stored externally in either MCPL [39] or HDF5 [40] formats. This feature, applicable to fixed-source simulations, offers fine-grained control over event data collection. The user can specify event filtering using IDs for cells, universes, materials, and nuclides. Reaction types follow the ENDF-6 MT format. Additional options include setting a minimum deposited energy threshold (E_threshold), limiting the number of stored particles (max_particles), and choosing the output file format (.mcpl or .h5).

Note that currently in OpenMC (version 0.15.2) the emission time of delayed neutrons is not sampled (i.e., set to zero). For this work, we therefore also modified the code to correctly sample delayed neutron emission times based on the sampled precursor group. This feature is not yet included in the official OpenMC release and is awaiting integration into future versions.

D. Simulation Routine

For both the Godiva and CROCUS configurations, simulations were performed under various subcritical and supercritical conditions. In the case of CROCUS, the water level was varied to adjust reactivity, while for Godiva, the fuel density was modified.

To efficiently run multiple simulations, we utilized the Python API in OpenMC, enabling automated execution of subsequent runs. For Serpent 2, the WATTS [41] Python library was employed to streamline the simulation workflow and parameter variation.

The simulation routine followed a two-step approach. First, a k-static calculation was performed to obtain the neutron multiplication factor (k_{eff}) and the spatial fission source distribution. These results were then used to initialize an external source simulation, during which individual fission event were recorded in an external file. The collected data were subsequently analyzed using the Rossi-alpha method, as described in Section III-E. In the case of Serpent 2, we also extracted the prompt neutron decay constant directly using the Iterated Fission Probability (IFP) method, detailed in Section III-F, for comparison.

E. The Rossi-α method - Noise Analysis

The Rossi-α method [42] is a neutron noise technique based on the statistical analysis of prompt fission chains as they evolve in time within a reactor. It relies on recording the timestamps of individual neutron detection events and calculating the time differences between event pairs, as expressed in Equation 1:

$$\tau_{i,j} = \text{TimeStamp}(j) - \text{TimeStamp}(i), \quad \text{with } i < j \tag{1}$$

For thermal reactors like CROCUS, these time differences are typically binned over a time window ranging from 0 to 0.1 s. For fast systems such as Godiva I, the range is in the order of few μs. An exponential function is then fitted to the binned distribution, as shown in Equation 2, to extract the prompt neutron decay constant α:

$$f(\tau) = Ae^{-\alpha|\tau|} + B \tag{2}$$

The value of α is determined by applying a non-linear least-squares fitting algorithm to the data.

A more comprehensive formulation of the Rossi-α function is given in Equation 3, which incorporates reactor-specific and detector-specific parameters. These include the detector efficiency ε (expressed as counts per fission), the fission rate F, the Diven factor D_ν, and the prompt neutron generation time Λ:

$$P(\tau) = \frac{\varepsilon D_\nu}{2\alpha\Lambda^2} e^{-\alpha|\tau|} + \varepsilon F \tag{3}$$

This extended is a more complete representation of the autocorrelation signal observed in neutron noise experiments.

In Monte Carlo neutron transport simulations, all recorded fission chains originate from a source neutron emitted at time zero. Therefore, a histogram of the recorded fission event timestamps can be constructed without further manipulation of the data, and the Rossi-α function (Equation 3) can be fitted. In this work we refer to this method as explicit noise simulation (ENS).

F. Iterated Fission Probability (IFP) Method

The Iterated Fission Probability (IFP) method is a Monte Carlo technique used to estimate adjoint-weighted kinetic parameters [20], such as the effective prompt neutron generation time and the effective delayed neutron fraction. It is based on tracking fission chains over multiple generations to approximate the adjoint neutron flux distribution.

In this method, a fixed-source simulation is performed starting from a single neutron with initial phase-space coordinates $(\mathbf{r}_0, \mathbf{v}_0)$. The simulation proceeds through $M + 1$ fission generations. If M is sufficiently large, the descendant neutron population p_i resulting from each initial neutron (ancestor i) reaches an asymptotic distribution. The importance I_k of the ancestor can then be estimated by summing the contributions from all its fission descendants.

To control population growth or extinction across the M latent generations, a renormalization factor of $1/k^{(g)}$ is applied at each generation g, where $k^{(g)}$ is the estimated multiplication factor at that generation. As $M \to \infty$, $k^{(g)}$ converges to the fundamental eigenvalue k, and the resulting importance I_k approximates the fundamental adjoint flux $\psi_k^\dagger(\mathbf{r}_0, \mathbf{v}_0)$, up to a normalization constant.

This method enables the computation of adjoint-weighted bilinear forms of the type:

$$\langle \psi_k^\dagger, A\psi_k \rangle, \tag{4}$$

where A is an operator of interest. In particular, the following expressions are used to compute the effective kinetic parameters:

$$\Lambda_{\text{eff}} = \frac{\langle \psi_k^\dagger, \frac{1}{v}\psi_k \rangle}{\langle \psi_k^\dagger, F\psi_k \rangle}, \tag{5}$$

$$\beta_{\text{eff}} = \frac{\langle \psi_k^\dagger, F_d\psi_k \rangle}{\langle \psi_k^\dagger, F\psi_k \rangle}, \tag{6}$$

where:

- Λ_{eff} is the effective prompt neutron generation time,
- β_{eff} is the effective delayed neutron fraction ,
- $F = F_p + F_d$ is the total fission operator (prompt plus delayed),
- F_j^d is the delayed fission operator for family j,
- v is the neutron speed.

At each generation g, a neutron is designated as an ancestor. Its importance is estimated at generation $g+M+1$. The scalar products above are computed by multiplying the score of the ancestor (from generation g) by its estimated importance.

For example, the adjoint-weighted total and delayed fission contributions are given by:

$$\langle \psi_k^\dagger, F\psi_k \rangle \propto k \sum_i p_i, \tag{7}$$

$$\langle \psi_k^\dagger, F_d\psi_k \rangle \propto k \sum_i (p_d)_i, \tag{8}$$

where $(p_d)_i$ is the total delayed neutron contribution associated with ancestor i.

Similarly, the adjoint-weighted total neutron lifetime is approximated as:

$$\langle \psi_k^\dagger, \frac{1}{v}\psi_k \rangle \propto \sum_i p_i t_i, \tag{9}$$

where t_i is the lifetime of the ancestor neutron that contributes to p_i.

In practical implementations, such as in SERPENT 2 and TRIPOLI-4, an overlapping-cycles version of the IFP method is used, where a new IFP cycle is initiated at each generation. This improves the statistical efficiency by approximately a factor of $\sqrt{M}$, although it may introduce inter-cycle correlations.

The prompt neutron decay constant coming from the IFP method will be defined as in equation:

$$\alpha_{IFP} = \frac{\rho - \beta_{eff,IFP}}{\Lambda_{eff,IFP}} \tag{10}$$

ρ is the reactivity of the system, which is calculated from the k-static estimation of k_{eff}.

IV. Results

We present the Rossi-α ENS results for various reactivity states of the CROCUS and Godiva assemblies. These results are then compared to the prompt decay constants obtained using the IFP method. We also extract the values of the effective delayed neutron fraction (β_{eff}) and the prompt neutron generation time (Λ_{eff}), and compare them with those derived from the IFP method.

A. Rossi-α Curves

For each simulation — for both Serpent 2 and OpenMC — 10^7 particles were banked, filtering specifically for fission events (MT = 18). The prompt neutron decay constant α was then extracted using the ENS approach by applying the Rossi-α method (see Section III-E) to the timestamps of the banked particles. This process was repeated for multiple subcritical and supercritical states of the selected assemblies, producing the results shown in Figures 3 and 4.

CROCUS is a light water reactor and therefore a thermal system. For its various critical states, the found α values are on the order of a few hundred per second. In contrast, Godiva

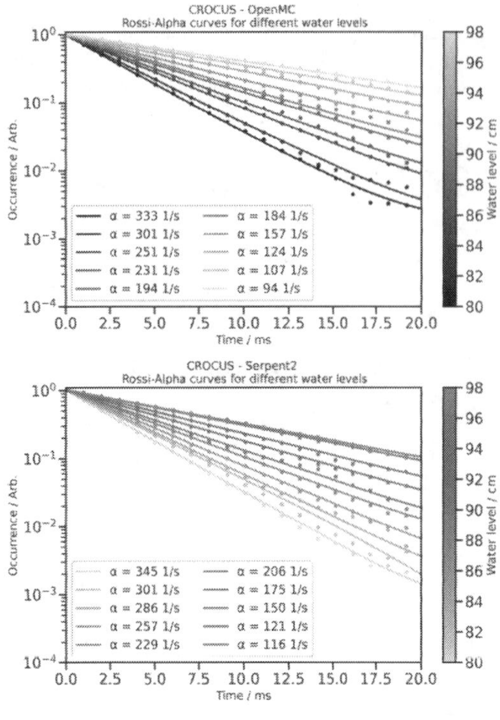

Fig. 3. (top) Rossi-Alpha curves for different states of the CROCUS geometry, simulated using OpenMC with the Collision Track feature. (bottom) Rossi-Alpha curves for different states of the CROCUS geometry, using a modified version of Serpent 2.

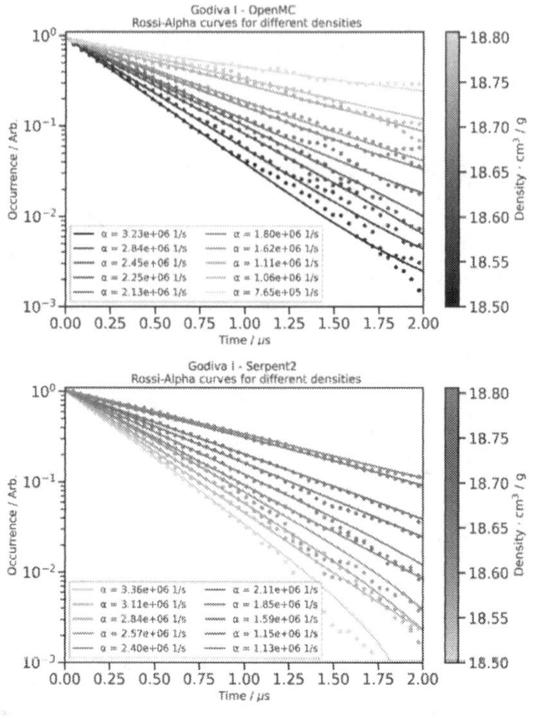

Fig. 4. (top) Rossi-Alpha curves for different states of the Godiva I geometry, simulated using OpenMC with the Collision Track feature. (bottom) Rossi-Alpha curves for different states of the Godiva I geometry, using a modified version of Serpent 2.

I is a fast system, where we find α values are on the order of millions per second. This expected difference serves as a first sanity check of the consistency of the feature across different system response time scales.

B. Verification and Comparison

To verify the functionality of the implemented feature in OpenMC, we compared the values of the α values obtained in Section IV-A with the values of α obtained with the Serpent 2 IFP method, detailed in III-F. We also added the results of the α results from an internal development of Serpent 2 for ENS to check how Collision Track in OpenMC perform against it [14].

As seen in Figure 5 (top), which shows the relative error for the CROCUS geometry (ENS versus IFP), the OpenMC results generally lie within 5% of the IFP estimates, with uncertainties overlapping or touching the zero line. In contrast, the Serpent 2 ENS results show an upward trend, with α values overestimated by up to 10–15% with higher reactivity. In Figure 5 (bottom), representing the Godiva I geometry, the OpenMC results fluctuate within a 10% range, and the error bars intersect the zero line. Conversely, the Serpent 2 ENS results tend to overestimate $|\alpha|$ more consistently.

The results obtained in Section IV-A were plotted against the reactivity values computed for each simulation. This analysis aims to extract the effective delayed neutron fraction β_{eff} and the effective prompt neutron generation time Λ_{eff} by fitting a linear model to the data. Starting from the known relation:

$$\alpha = \frac{\rho}{\Lambda_{\text{eff}}} - \frac{\beta_{\text{eff}}}{\Lambda_{\text{eff}}}, \tag{11}$$

we fit a linear function of the form:

$$\alpha = a \cdot \rho + b, \tag{12}$$

where the slope a and intercept b are determined from the fit. From these, the kinetic parameters can be extracted as:

$$\Lambda_{\text{eff}} = \frac{1}{a} \quad \text{and} \quad \beta_{\text{eff}} = -\frac{b}{a}. \tag{13}$$

The kinetic parameters for the CROCUS geometry are displayed for different methods in Table I, and the one of Godiva I are displayed in Table II.

TABLE I
COMPARISON OF β_{EFF} AND Λ_{EFF} FOR CROCUS USING DIFFERENT CODES AND METHODS. THE RELATIVE ERROR IS CALCULATED WITH RESPECT TO THE VALUE OBTAINED FROM THE SERPENT 2 IFP METHOD.

Code - Method	$\beta_{\text{eff}} - pcm$	Rel. Error	$\Lambda_{\text{eff}} - \mu s$	Rel. Error
Serpent 2 IFP	736	–	47.9	–
OpenMC ENS	745 ± 9	1%	48.6 ± 0.4	1%
Serpent 2 ENS	880 ± 20	20%	50.6 ± 1.3	6%

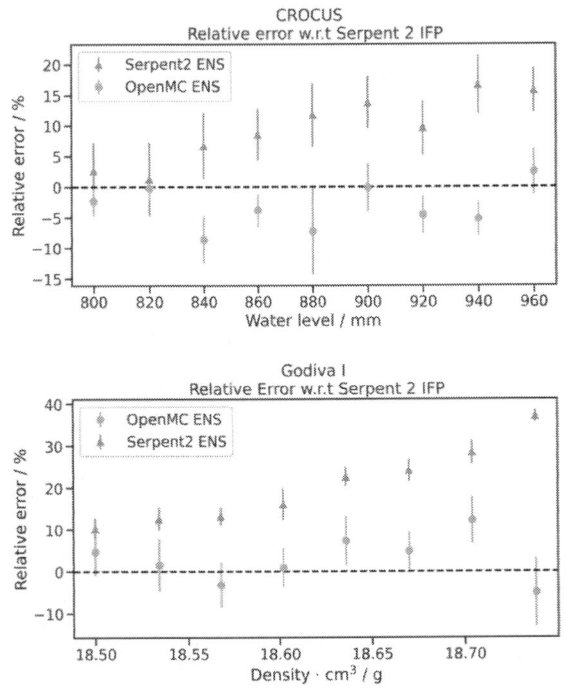

Fig. 5. The top plot represents the CROCUS geometry, while the bottom shows the Godiva I geometry. Both figures display the relative difference in the prompt neutron decay constant (α) obtained from OpenMC ENS and Serpent 2 ENS, relative to the reference values from the Serpent 2 IFP method. The dashed line indicates the zero reference, while the pink line represents Serpent 2 ENS and the yellow line corresponds to OpenMC ENS.

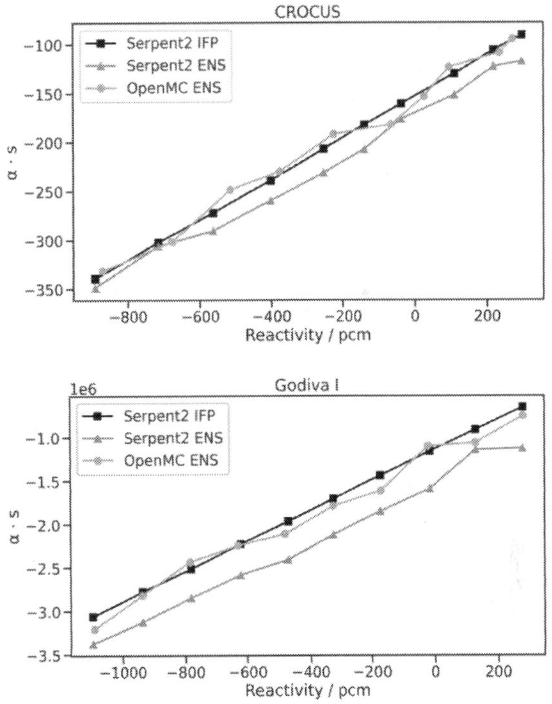

Fig. 6. The top plot represents the CROCUS geometry, while the bottom corresponds to the Godiva I geometry. Both figures illustrate the linear relationship between the prompt neutron decay constant (α) and the reactivity (ρ). The black line shows the Serpent 2 IFP results, the pink line corresponds to Serpent 2 ENS, and the yellow line represents OpenMC ENS.

TABLE II
COMPARISON OF β_{EFF} AND Λ_{EFF} FOR GODIVA I USING DIFFERENT CODES AND METHODS. THE RELATIVE ERROR IS CALCULATED WITH RESPECT TO THE VALUE OBTAINED FROM THE SERPENT 2 IFP METHOD.

Code - Method	$\beta_{\text{eff}} - pcm$	Rel. Error	$\Lambda_{\text{eff}} - ns$	Rel. Error
Serpent 2 IFP	648	–	5.69	–
OpenMC ENS	710 ± 30	9%	5.81 ± 0.17	2%
Serpent 2 ENS	902 ± 14	39%	5.87 ± 0.08	3%

The uncertainties were propagated using explicit sampling of the reported uncertainties and propagating them. From the tabulated values (Table I and Table II), it can be observed that the OpenMC ENS results are in good agreement with those obtained from the Serpent 2 IFP method. These findings provide preliminary verification of the event tracking feature implemented in OpenMC for neutron noise studies in both a thermal fission system (CROCUS), as well as a fast fission system (Godiva I).

V. CONCLUSION

In this work, we presented a newly developed event tracking feature implemented in the OpenMC Monte Carlo code. This feature enables explicit noise simulations and was used to estimate the prompt neutron decay constant through the Rossi-α method. To verify the functionality and accuracy of this implementation, we compared results obtained in OpenMC with those from established code, here the Serpent 2 Monte Carlo code using the IFP method.

To verify the approach across different system response times, two representative systems were studied: the CROCUS zero-power thermal reactor and the Godiva I fast critical assembly. These cases demonstrated the general capability of the event tracking feature in both thermal and fast neutron systems.

Our findings show that OpenMC, when equipped with the event tracking feature, produces results in close agreement with the IFP reference values. We demonstrated that other kinetic parameters — such as the effective delayed neutron fraction (β_{eff}) and the prompt neutron generation time (Λ_{eff}) — can be extracted from the same noise data using simple linear fitting techniques applied to the $\alpha - \rho$ data.

In future work, we aim to expand the range of benchmark cases and reactor types studied to further verify the implementation against established codes. Then, a set of experiments will be chosen to perform a validation of the tool. The goal is to produce a validated, reliable, open-source tool for reactor noise analysis to support the design of next generation of advanced reactor noise experiments and the analysis of legacy experiments.

Acknowledgment

We would like to thank the OpenMC community for their continuous support throughout the development of this project.

This work is supported by the U.S. Department of Energy, Office of Science, under contract DE-AC02-06CH11357.

References

[1] M. M. R. Williams, *Random processes in nuclear reactors*. Elsevier, 1974.

[2] R. D. Mosteller, "Validation suite for mcnp," Los Alamos National Laboratory (LANL), Los Alamos, NM (United States), Tech. Rep., 2002.

[3] B. C. Kiedrowski, F. B. Brown, N. A. Gibson, A. S. Bennett, and M. A. Gonzales, "Mcnp 6 shielding validation suite: Past, present, and future," *Transactions of the American Nuclear Society*, vol. 105, pp. 559–561, 2011.

[4] D. C. Henzlova, N. Ensslin, A. Favalli, W. H. Geist, L. Holzleitner, K. Koehler, M. S. Krick, M. M. Pickrell, T. D. Reilly, J. E. Stewart, and K. D. Veal, *Principles of Neutron Multiplicity*. Springer Nature Switzerland, 2024, pp. 439–482.

[5] O. Pakari, T. Mager, P. Frajtag, A. Pautz, and V. Lamirand, "Gamma noise to non-invasively monitor nuclear research reactors," *Scientific Reports*, vol. 14, no. 1, p. 8409, 2024.

[6] R. Feynman, F. D. Hoffmann, and R. Serber, "Dispersion of the neutron emission in u-235 fission," *Journal of Nuclear Energy (1954)*, vol. 3, no. 1-2, pp. 64–IN10, aug 1956.

[7] J.-P. Hudelot, R. Klann, B. Micklich, M. Antony, G. Perret, N. Thiollay, J.-M. Girard, V. Laval, and P. Leconte, "Minerve reactor characterization in support of the osmose program: Safety parameters," in *Proceedings of the PHYSOR-2004 Topical Meeting*, 2004.

[8] G. Perret, B. Geslot, A. Gruel, P. Blaise, J. Di-Salvo, G. De Izarra, C. Jammes, M. Hursin, and A. Pautz, "Kinetic parameter measurements in the minerve reactor," *IEEE Transactions on Nuclear Science*, vol. 64, no. 1, pp. 724–734, 2016.

[9] B. Geslot, A. Gruel, A. Pepino, J. Di Salvo, G. de Izarra, C. Jammes, C. Destouches, and P. Blaise, "Pile noise experiment in minerve reactor to estimate kinetic parameters using various data processing methods," in *2015 4th International Conference on Advancements in Nuclear Instrumentation Measurement Methods and their Applications (ANIMMA)*. IEEE, 2015, pp. 1–8.

[10] E. Gilad, B. Geslot, P. Blaise, and C. Dubi, "Sensitivity of power spectral density techniques to numerical parameters in analyzing neutron noise experiments," *Progress in Nuclear Energy*, vol. 101, pp. 288–298, 2017.

[11] X. Doligez, A. Billebaud, S. Chabod, T. Chevret, D. Fourmentel, A. Krasa, A. Kochetkov, F. Lecolley, J. Lecouey, G. Lehaut *et al.*, "Effective delayed neutron fraction measurement in the critical venus-f reactor using noise techniques," in *2015 4th International Conference on Advancements in Nuclear Instrumentation Measurement Methods and their Applications (ANIMMA)*. IEEE, 2015, pp. 1–6.

[12] A. dos Santos and R. Diniz, "A noise analysis approach for measuring effective delayed neutron parameters in the ipen/mb-01 reactor," in *AIP Conference Proceedings*, vol. 769, no. 1. American Institute of Physics, 2005, pp. 1397–1400.

[13] J. Goda, C. Bravo, T. Cutler, T. Grove, D. Hayes, J. Hutchinson, G. McKenzie, A. McSpaden, W. Myers, R. Sanchez *et al.*, "A new era of nuclear criticality experiments: The first 10 years of godiva iv operations at ncerc," *Nuclear Science and Engineering*, vol. 195, no. sup1, pp. S55–S79, 2021.

[14] M. Saliba, V. Lamirand, W. Monange, F. Vitullo, M. Hursin, A. Pautz, and O. Pakari, "Measurement of the prompt neutron decay constant in a zero-power reactor using a novel 3d detector system," *Nuclear Science and Engineering*, pp. 1–9, Mar. 2025.

[15] G. Locatelli, M. Mancini, and N. Todeschini, "Generation iv nuclear reactors: Current status and future prospects," *Energy Policy*, vol. 61, pp. 1503–1520, 2013.

[16] G. Lentchner, W. Fritsch, R. Crowder, N. Walton, A. Lewis, O. Chvala, K. Clarno, and V. Sobes, "Incorrect resonance escape probability in monte carlo codes due to the threshold approximation of temperature-dependent scattering," *Annals of Nuclear Energy*, vol. 207, p. 110717, Nov. 2024.

[17] T. Cutler, J. Lee, E. Luther, A. Maldonado, R. Sanchez, K. Stolte, H. Trellue, and N. Wynne, "The deimos experiment: Advanced reactor testbed," Los Alamos National Lab.(LANL), Los Alamos, NM (United States), Tech. Rep., 2023.

[18] J. M. Hykes and R. M. Ferrer, "Validation of casmo5 neutron kinetics parameters," *Annals of Nuclear Energy*, vol. 136, p. 107015, 2020.

[19] M. I. Radaideh, W. A. Wieselquist, and T. Kozlowski, "A new framework for sampling-based uncertainty quantification of the six-group reactor kinetic parameters," *Annals of Nuclear Energy*, vol. 127, pp. 1–11, 2019.

[20] N. Terranova and A. Zoia, "Generalized iterated fission probability for monte carlo eigenvalue calculations," *Annals of Nuclear Energy*, vol. 108, pp. 57–66, Oct. 2017.

[21] E. Dumonteil, R. Bahran, T. Cutler, B. Dechenaux, T. Grove, J. Hutchinson, G. McKenzie, A. McSpaden, W. Monange, M. Nelson *et al.*, "Patchy nuclear chain reactions," *Communications Physics*, vol. 4, no. 1, p. 151, 2021.

[22] O. V. PAKARI, "Experimental and numerical study of stochasticbranching noise in nuclear reactors," Ph.D. dissertation, École polytechnique fédérale de Lausanne, 2020.

[23] S. A. Pozzi, E. Padovani, and M. Marseguerra, "Mcnp-polimi: a monte-carlo code for correlation measurements," *Nuclear Instruments and Methods in Physics Research Section A: Accelerators, Spectrometers, Detectors and Associated Equipment*, vol. 513, no. 3, pp. 550–558, 2003.

[24] J. A. Kulesza, T. R. Adams, J. C. Armstrong, S. R. Bolding, F. B. Brown, J. S. Bull, T. P. Burke, A. R. Clark, R. A. A. Forster III, J. F. Giron *et al.*, "Mcnp® code version 6.3.0 theory & user manual," Los Alamos National Laboratory, Los Alamos, New Mexico, Tech. Rep. LA-UR-22-30006, Sep. 2022.

[25] P. K. Romano, N. E. Horelik, B. R. Herman, A. G. Nelson, B. Forget, and K. Smith, "Openmc: A state-of-the-art monte carlo code for research and development," *Annals of Nuclear Energy*, vol. 82, pp. 90–97, Aug. 2015.

[26] International Atomic Energy Agency (IAEA), "Open-source nuclear code for reactor analysis – oncore," 2024, accessed: Mar. 19, 2025. [Online]. Available: https://www.iaea.org/topics/nuclear-power-reactors/open-source-nuclear-code-for-reactor-analysis-oncore

[27] J. Leppänen, M. Pusa, T. Viitanen, V. Valtavirta, and T. Kaltiaisenaho, "The serpent monte carlo code: Status, development and applications in 2013," *Annals of Nuclear Energy*, vol. 82, pp. 142–150, aug 2015.

[28] J. Paratte, R. Früh, U. Kasemeyer, M. Kalugin, W. Timm, and R. Chawla, "A benchmark on the calculation of kinetic parameters based on reactivity effect experiments in the CROCUS reactor," *Annals of Nuclear Energy*, vol. 33, no. 8, pp. 739–748, may 2006.

[29] T. Ligonnet, A. Laureau, P. Simbsler, A. Pautz, and V. Lamirand, "Design of an open-loop pile-oscillation program in the crocus reactor," *IEEE Transactions on Nuclear Science*, vol. 71, no. 5, pp. 1001–1009, 2024.

[30] T. Mager, O. Pakari, M. Hursin, T. Ligonnet, A. Pautz, and V. Lamirand, "Intra-pin reaction rate measurements in the crocus experimental reactor," *IEEE Transactions on Nuclear Science*, vol. 71, no. 5, pp. 1047–1055, 2024.

[31] J. D. Bess, T. Ivanova, I. Hill, and L. Scott, "The 2019 edition of the IRPhEP handbook," *Trans. Am. Nucl. Soc.*, vol. 121, no. 1, pp. 1565–1568, Nov. 2019.

[32] Lamirand, Vincent, Laureau, Axel, Pakari, Oskari, Frajtag, Pravel, and Pautz, Andreas, "Power calibration methodology at the crocus reactor," *EPJ Web Conf.*, vol. 225, p. 04022, 2020. [Online]. Available: https://doi.org/10.1051/epjconf/202022504022

[33] O. Pakari, V. Lamirand, G. Perret, P. Frajtag, and A. Pautz, "Kinetic parameter measurements in the crocus reactor using current mode instrumentation," *IEEE Transactions on Nuclear Science*, vol. 65, no. 9, pp. 2456–2460, 2018.

[34] I. C. S. B. E. P. (ICSBEP), *International Handbook of Evaluated Criticality Safety Benchmark Experiments*. Paris, France: OECD Nuclear Energy Agency, 2021.

[35] C. Solomon, M. Krass, A. Garner, and M. Chichester, "Benchmark specifications for godiva i: A highly enriched uranium metal assembly," Los Alamos National Laboratory, Los Alamos, NM, USA, Tech. Rep. LA-14244-M, 2006, accessed: Mar. 19, 2025. [Online]. Available: https://ncsp.llnl.gov/sites/ncsp/files/2021-05/LA-14244-M.pdf

[36] J. Leppänen, M. Aufiero, E. Fridman, R. Rachamin, and S. van der Marck, "Calculation of effective point kinetics parameters in the serpent 2 monte carlo code," *Annals of Nuclear Energy*, vol. 65, pp. 272–279, mar 2014.

[37] M. Saliba, P. Romano, J. Tramm, E. B. Knudsen, D. Siefman, A. Pautz, and O. Pakari, "Development of an event tracking feature in openmc for neutron spectroscopy and scatter camera systems," in *Proceedings of the International Conference on Mathematics and Computational Methods Applied to Nuclear Science and Engineering (M&C 2025)*. American Nuclear Society, 2025, presented at the International Conference on

Mathematics and Computational Methods Applied to Nuclear Science and Engineering (M&C 2025), Denver, April 27–30, 2025. Paper #9271.

[38] M. Saliba, "Openmc retina feature branch (forked)," 2024, gitHub repository, forked from the OpenMC project. Available at https://github.com/saliba170499/openmc_retina/tree/retina_feature.

[39] T. Kittelmann, E. Klinkby, E. Knudsen, P. Willendrup, X. Cai, and K. Kanaki, "Monte carlo particle lists: Mcpl," *Computer Physics Communications*, vol. 218, pp. 17–42, Sep. 2017.

[40] M. Folk, G. Heber, Q. Koziol, E. Pourmal, and D. Robinson, "An overview of the hdf5 technology suite and its applications," in *Proceedings of the EDBT/ICDT 2011 Workshop on Array Databases*, ser. EDBT/ICDT '11. ACM, Mar. 2011.

[41] P. K. Romano, N. E. Stauff, Z. J. Ooi, Y. Miao, A. Lund, and L. Zou, "Watts: Workflow and template toolkit for simulation," *Journal of Open Source Software*, vol. 7, no. 79, p. 4735, Nov. 2022.

[42] B. Rossi and K. Greisen, "Cosmic-ray theory," *Reviews of Modern Physics*, vol. 13, no. 4, pp. 240–309, oct 1941.

HONEY: High-Resolution Reactor Kinetics experiments in the CROCUS reactor

Cecilia Montecchio[1,2*], Thomas Ligonnet[1], Davide Mancusi[2], Wilfried Monange[3], Oskari Pakari[1], Andrea Zoia[2], Vincent Lamirand[1]
[1] École polytechnique fédérale de Lausanne (EPFL), Switzerland
[2] Université Paris-Saclay, CEA Service d'Etudes des Réacteurs et de Mathématiques Appliquées, France
[3] Autorité de sûreté nucléaire et de radioprotection (ASNR), France
* cecilia.montecchio@epfl.ch

***Abstract*—This work presents high-resolution kinetic experiments in the CROCUS reactor, where control rod movements are monitored with unprecedented spatial detail using SAFFRON, a novel 3D detection system composed of 160 miniature detectors. These experiments are part of the HONEY (High-resolution Online Neutronics Experiments for dYnamics) program, a series of experimental campaigns developed through a collaboration between CEA (France), ASNR (France) and EPFL (Switzerland). The program is carried out in the CROCUS reactor, operated at EPFL, and aims to advance our understanding of reactor kinetics by investigating fission chain reactions, with particular emphasis on neutron clustering phenomena. As a stepping stone toward this goal, HONEY is devoted to improve the predictive capabilities of simulation tools under dynamic scenarios. The first campaign included global measurements of the reactor kinetic response using two reference fission chambers and, more importantly, it leveraged the high spatial resolution of SAFFRON. This array-type detection system, originally developed as part of a doctoral program at EPFL and supported by the H2020 CORTEX project, is specifically designed to investigate local neutron flux variations, which are of particular interest close to regions where perturbations are introduced. The results from the first campaign were used to validate the reactor kinetics capabilities of TRIPOLI-4, a high-fidelity Monte Carlo particle-transport code developed at CEA. HONEY offers a unique validation framework by combining global measurements with high-resolution local data, enabling detailed spatial assessments of dynamic reactor behavior and paving the way towards a thorough evaluation of how well computational models reproduce local phenomena.**

Keywords —Reactor kinetics, 3D high-resolution, code validation

I. Introduction

The HONEY experimental program is a collaborative effort involving EPFL in Switzerland, as well as CEA and ASNR in France, and it is conducted at the CROCUS zero-power reactor at EPFL. The first campaign targets comprehensive experimental measurements of the neutron flux behavior under reactivity insertions and withdrawals. More precisely, it aims at achieving high spatial resolution characterization of the reactor's kinetic response. Particular emphasis is given to capturing the kinetic response in the vicinity of the perturbation that drives the system away from its initial stable state. Although CROCUS is a tightly coupled system whose kinetic behavior is approximated effectively by point kinetics models, it still experiences the emergence of local effects. The SAFFRON system, featuring an array-like configuration of 160 detectors, enables the characterization of such localized perturbations. The first experiment of this campaign was conducted in April 2024, with the resulting data primarily intended to support the validation of reactor kinetics simulations performed with the Monte Carlo code TRIPOLI-4® [1]. The HONEY project enables unique insights that significantly enhance code validation, thanks to the simultaneous use of global and local detectors to collect data relative to the spatial kinetics of the reactor. Having access to precise local data, which is a challenge in itself, paves the way towards the validation of high-fidelity simulation tools capable of predicting the local behaviour of the neutron field as it evolves in space and time.

This contribution is organized as follows: in Sec.II we provide an overview of the CROCUS research reactor, followed by a description the SAFFRON detection system. Section III presents the experimental setup, while Section IV discusses the computational tool and model developed to fully leverage the experimental data for validation purposes. In Section V, we present the results of the comparisons between measurements and Monte Carlo simulations, with particular emphasis on the near-perturbation response. Finally, Section VI summarizes the conclusions and outlines future perspectives for the HONEY program.

II. Kinetics in the CROCUS reactor

A. CROCUS zero-power reactor

CROCUS is a zero-power light-water-moderated reactor licensed to operate up to a maximum thermal power of 100 W. The core consists of two types of fuel rods distributed in two zones: the inner one, which includes 336 uranium dioxide fuel rods enriched at 1.806 wt.% and the outer one, which includes a variable number of metallic uranium fuel rods enriched at 0.947 wt.%. The number of rods in the outer zone can be adjusted depending on the specific experimental configuration, and it currently includes 180 rods for hosting the SAFFRON detection system. In CROCUS, reactivity is primarily controlled by adjusting the vessel water level with a spillway. Water acts as a moderator: thus, by changing the level of immersion of the fuel rods it is possible to regulate

https://doi.org/10.1051/epjconf/202533804002

the neutron flux. The water level can be adjusted with an accuracy of ± 0.1 mm, corresponding to a reactivity worth of approximately ± 0.4 pcm near criticality. Reactivity can also be controlled via two optional rods made of B_4C, positioned in symmetrical channels in the outer zone of the core, with an accuracy of ± 0.5 mm, which is equivalent to a reactivity worth up to ± 0.2 pcm [2]. CROCUS is equipped with four monitors: two boron-coated compensated ionization chambers (north and south), and two fission chambers (east and west) calibrated in power and acting as safety monitors. Indeed, by operating in pulse mode with an adjustable dwell time, the fission chambers enable real-time flux measurements and accurate reconstruction of the power history.

B. SAFFRON detection system

SAFFRON is a three-dimensional (3D) full-core mapping array currently employed for thermal neutron detection in the core of the CROCUS reactor. The detector technology, originally developed in collaboration with the Paul Scherrer Institute in Switzerland [3]–[5], as well as the array were implemented and installed in the reactor as part of a doctoral thesis at EPFL [6] within the framework of the H2020 CORTEX project [7]. The system consists of 149 fixed and 11 mobile *miniature* and *minimalistic* (MiMi) neutron detectors, each with a sensitive area of approximately 1 mm^2 and a thickness of 0.2 mm. As shown in Fig. 1, the detectors are primarily arranged in three layers, with two supplementary layers containing fewer detectors, and are distributed across various radial inter-pin locations.

Fig. 1: Left: 3D CAD model of the CROCUS core with the SAFFRON detection system installed. The five axial layers are indicated at their precise locations. Image reproduced from Ref. [6]. Right: Photograph of CROCUS during the core loading procedure. The red support layers for the detector arrays and the black optical fibers are clearly visible. Image reproduced from Ref. [8].

Each MiMi consists of an inorganic ZnS(Ag):^{6}LiF scintillator screen coupled to a plastic optical fiber. Thermal neutrons are detected via the scintillation light emitted by the ZnS, excited by the particles released from the (n,α) interaction on ^{6}Li. This light is transmitted through the optical fiber to a silicon photomultiplier (SiPM), where, after appropriate signal processing, it conveys the count of thermal neutron interactions occurring within the active volume. The array configuration of these compact detectors enables real-time, high-resolution characterization of the thermal neutron flux across the entire core. This spatial coverage makes SAFFRON particularly well-suited for investigating space-dependent neutron field behaviour in CROCUS. The SAFFRON detection system currently operates effectively up to a maximum power of 1 W, with an optimal operating point reached around 600 mW [9]. Due to the non-negligible absorption cross section of ^{6}Li, the presence of SAFFRON in the core introduces a negative reactivity worth. This effect was estimated through Monte Carlo simulations using the ENDF/B-VII.1 nuclear data library [10]. To compensate for this effect, four metallic uranium rods were added to the periphery of the reactor core, increasing the total number of fuel rods from 176 rods in the standard configuration to 180.

SAFFRON has already demonstrated its versatility and reliability across a range of experimental conditions, such as variation of static flux maps, as thoroughly documented by Vitullo [6]. Significant efforts have been made to maximize the benefits of its installation in the core, including an inter-calibration campaign [11]. SAFFRON was recently employed to measure the prompt neutron decay constant [12]. It has also been used for absorber oscillation experiments conducted during the BLOOM program [13], as well as for the reproduction of neutron noise experiments initially performed during the CORTEX project [14]. Furthermore, SAFFRON is expected to play a central role in the HONEY project, contributing to advanced studies of reactor kinetics for code validation.

III. Experiments description

The first HONEY campaign included two separate experiments involving a positive reactivity insertion and a reactivity withdrawal. Both experiments began with the reactor in a critical configuration. As already mentioned, in CROCUS the reactivity can be controlled either by adjusting the water level or by moving the control rods. For these experiments, reactivity changes were introduced exclusively through control rod movements. This method was specifically chosen because it triggers the appearance of local effects.

The experimental setup was designed to slightly differ between the two cases of reactivity perturbation. For the negative reactivity insertion, the reactor was initially at its critical state without control rods, with the water level adjusted to only partially cover the fuel. To provide the reactivity perturbation, both control rods were simultaneously fully inserted into the reactor, resulting in a symmetrical flux response, consistent with the core's symmetrical configuration. In contrast, for the positive reactivity insertion experiment, the reactor was initially critical with only one control rod fully inserted and the other partially inserted. To introduce positive reactivity, one rod was fully withdrawn while the other, located in a symmetrical position in the reactor, remained fixed, creating an asymmetrical transient. This asymmetry is particularly useful for investigating local effects, as it accentuates their presence by offering a twin reference point unaffected by the perturbation.

To enable the comparison with the results of a computational simulation, it is necessary to define a fixed total duration for the experiment. The following assumptions were applied: the reactor remained critical for 30 seconds, after which the control rod(s) were moved, with a total motion time estimated to last 5 seconds until motion is complete. In this first campaign, the value was determined by analyzing video footage recorded by a camera pointed at the control rod lifting system. The rod(s) were then held in their transient position for 10 seconds, followed by an additional 5 seconds to return to the original configuration. At this point criticality was re-established, and the system was recorded for other 30 seconds. The fraction of the experimental time that will be reproduced by simulations is thus 1 minute and 20 seconds for both experiments. This duration was planned and selected according to the available computational resources dictated by the Monte Carlo code in order to achieve a target statistical accuracy on the observables of interest. The most important parameters related to the experiments are summarized in Tab. I.

TABLE I: EXPERIMENTAL PARAMETERS

	Water level [mm]	NW control rod position [mm]	SE control rod position [mm]	Total reactivity worth [pcm]
Negative $\Delta\rho$	952.1	$1000 \to 0 \to 1000$	$1000 \to 0 \to 1000$	~ 330
Positive $\Delta\rho$	1000	$0 \to 1000 \to 0$	671	~ 165

Summary of the key parameters related to the two experiments. The control rod position is expressed as the non-absorbing height (i.e 1000 mm corresponds to the fully withdrawn position, while 0 mm indicates the rod is fully inserted). The water level denotes the height from the bottom of the core. The estimated reactivity worth of the control rod is based on previously published values available in the literature [2].

To ensure statistically robust results and demonstrate the repeatability of the experiments, each transient was repeated nine consecutive times for the positive reactivity insertion. The initial power level was approximately 100 mW, increasing after each repetition. For the negative reactivity insertion, the transient was repeated six times, starting from approximately 40 W, in order to maintain a sufficiently high power level through the final repetition. A minimum waiting period of six minutes was observed between each repetition, primarily to restore the reactor to a critical state and allow the contribution of delayed neutrons to fade out. This procedure guarantees that each repetition can be treated as a replica of the same experiment. The uncertainties on the experimental data reflect the power history of the repeated experiments. The timing uncertainty associated with summing all the repetitions, primarily due to the control rod movement being triggered by the operator after a fixed 30 second delay and a subsequent fixed waiting time, has been neglected in this work.

IV. Code Validation

One of the primary purposes of the first HONEY campaign is to serve as a rigorous validation tool, with the ultimate ambitious goal of increasing the reliability of computational simulations as predictive instruments for kinetic scenarios. Traditionally, deterministic codes have been the standard tool for modeling the kinetic behavior of nuclear systems following a perturbation. However, recent advances in computational power, have increasingly enabled the use of Monte Carlo methods to reproduce dynamic scenarios. Pioneering work in this domain was conducted by Sjenitzer and Hoogenboom, who were the first to demonstrate the feasibility of simulating nuclear reactor kinetics by using Monte Carlo methods [15]–[17]. A particularly important advancement in this context concerns the development of advanced variance reduction techniques, which have greatly extended the applicability of Monte Carlo simulations to scenarios involving large flux excursions and long transients. These techniques include forced precursor decay [15] and population control strategies, encompassing combing [18] and the importance sampling scheme [9], [19]. The following section provides an overview of the kinetic capabilities of the Monte Carlo code TRIPOLI-4®, a high-fidelity particle-transport code developed at CEA. It also describes the computational model developed to reproduce the experimental results.

A. *The kinetic algorithm in* TRIPOLI-4®

The Monte Carlo code TRIPOLI-4® is equipped with a kinetic algorithm that enables the simulation of the time evolution of a nuclear system under a material or geometrical perturbation. The original algorithm has recently benefited from developments aimed at implementing a highly adaptive, adjoint-based importance sampling scheme [20], inspired from the zero-variance approach [21]. These additions have enhanced the kinetic capabilities of the code and facilitated the comparison between simulations and experimental results, thereby supporting the validation process. For a comprehensive description of the kinetic capabilities of TRIPOLI-4®, the reader is referred to the detailed work by Faucher et al. [22]. For the purpose of this contribution, it is sufficient to know that in kinetic simulations, the system is initially prepared in a critical state. Reactivity perturbations are assigned by setting user-defined changes in geometry and/or material compositions at given times. Neutrons are simulated from the initial time until some final time. To control the unbounded growth of particles in supercritical systems or the disappearance of neutrons in subcritical systems, population-control strategies are applied on a predefined time grid. These strategies aim at maintaining a convenient number of particles throughout the simulation, thereby optimizing computational cost in terms of both memory usage and simulation time. The quantities of interest (e.g., flux and reaction rates) are tallied over a time grid whose granularity can be adjusted as needed. For this study, the fiducial quantity estimated by the simulation and used for comparison with the experimental acquisition (i.e., counts over a dwell time) is the ensemble-averaged time integral, over a time step, of the energy and volume integrated neutron flux.

B. *Reproducing the experiments with the Monte Carlo code*

To validate the TRIPOLI-4® simulation capabilities, we developed a model corresponding to the CROCUS reactor configuration at the time of the experiment, including the 149 fixed MiMi detectors positioned at their exact locations within the core, as shown in Fig. 2. In order to simulate the system evolution, such as the control rod movements described in Sec. III, the code relies on a dynamic geometrical representation that reflects the insertion or withdrawal of the control rods over time. In TRIPOLI-4® this representation is achieved by defining a set of geometries, each corresponding to a distinct position of the rods along the transient. As long as all the geometries in this set have the same bounding box and there are no constraints on their internal structure (e.g., number of volumes, composition of volumes, etc.), it is possible to activate sequentially these geometries according to a predefined time grid. The number of geometrical representations required depends on the desired temporal resolution: the finer the time grid, the more accurately the rod motion is captured. However, increasing the number of geometries also increases the complexity of the simulation setup. Therefore, a trade-off must be considered between the level of detail and the practical feasibility of implementing a large number of geometry states. For this study, the control rod movement was discretized by defining a set of ten geometries for each experiment. To construct this set, a constant control rod velocity was assumed, neglecting both the initial acceleration and final deceleration phases. The time grid, including all the specific times at which geometry changes are imposed, was defined according to the timing constraints described in Section III.

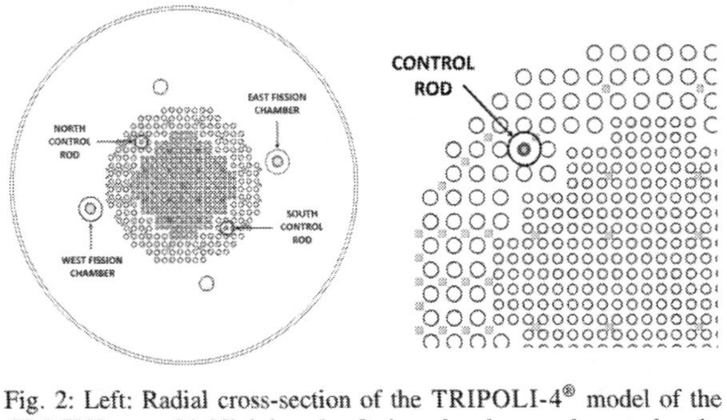

Fig. 2: Left: Radial cross-section of the TRIPOLI-4® model of the CROCUS core, highlighting the fission chambers and control rods. Right: Zoomed view of the north control rod at the mid-plane axial layer, showing the MiMi detectors (green squares).

V. Results of the experiments and Code Validation

This section presents the results of the code validation based on both experiments. We first compare what we refer to as the global response, which is the feedback of the entire reactor to the reactivity perturbation, as measured by the fission chambers. We then present the primary focus of this work, namely what we call the local response. This corresponds to the neutron flux behavior near the perturbation point, made accessible by the high-resolution measurements provided by the SAFFRON detection system.

A. *Global response*

The global response corresponds to the signal measured by the two fission chambers, symmetrically positioned outside the outermost fuel ring, as shown in Fig.2. This response is expected to reflect the overall power level of the reactor without being significantly influenced by the localized perturbations due to the control rod movement. For consistency in the direct comparison, the evolution of the fission chamber counts is compared to the simulated neutron flux scored directly within the physical volumes of the fission chambers. Separately, we have also verified that this local response is equivalent to the integral flux tally over the entire core. We therefore prefer using the integral response, which offers reduced statistical uncertainty. This choice is supported by experimental evidence showing that, in tightly coupled systems like CROCUS, the fission chamber signal reliably reflects global reactor behavior.

To assess the impact of nuclear data on the simulation results, we performed calculations relying on two libraries: `ENDF/B-VIII` [23] and `JEFF-3.3` [24]. The use of two libraries in this work allows illustrating the extent to which simulation results can vary, which is an essential step to probe the sensitivity of the code validation to this choice. In the subsequent analysis, the behavior predicted by solving the point kinetics equations, is represented as well, for comparison. The kinetic parameters used in the point kinetics equations are estimated with the IFP algorithm in TRIPOLI-4® [25], for both nuclear data libraries. Reactivity is computed from the effective multiplication factors of the initial and perturbed configurations, obtained using the power iteration method. The transient caused by the control rod movement is modeled with a linear reactivity evolution. Among all input parameters, reactivity uncertainty is identified as the main contributor to the overall uncertainty in the predicted flux. The kinetic simulation results in TRIPOLI-4® were obtained using 1000 neutrons per batch over 450 batches for the positive reactivity insertion, and 500 batches for the negative insertion. The result are shown in Fig. 3 and in Fig. 4. For the ease of comparison, the simulation and experimental results are independently normalized to their respective initial flux values. These values are calculated as the average over the first 30 seconds, which corresponds to the time window during which the reactor is maintained stable.

At first glance, the impact of nuclear data is significant, affecting both the trend predicted by the point kinetics equations and the behavior estimated by the kinetic simulations. For the negative reactivity insertion, the results of the kinetic simulations using both nuclear data libraries fall within the experimental range when considering one standard deviation. In the case of positive reactivity insertion, however, agreement within one standard deviation is observed only when using the `ENDF/B-VIII` nuclear data library, as the flux predicted by `JEFF-3.3` is systematically lower. Interestingly, for both perturbation cases, the experimental data fall entirely within the range bounded by the computational predictions of the two nuclear data libraries. In both experiments, kinetic simulations provide a more accurate reproduction of the experimental data than the point kinetics model. This is particularly evident in the case of positive reactivity insertion with JEFF-3.3, where the point kinetics model significantly underestimates the neutron flux, as illustrated in Fig. 3. As previously explained in Sec. IV, the transient is modeled in TRIPOLI-4® through geometry updates on a fixed time grid: despite these simplifications, good agreement is observed during the control rod movement phase.

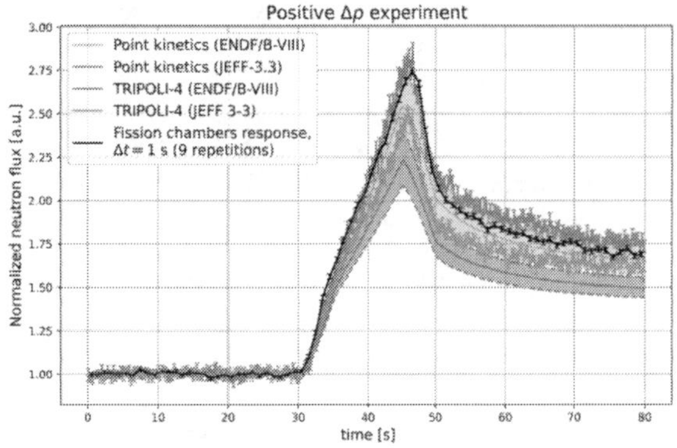

Fig. 3: Comparison between fission chambers acquisition, kinetic simulation in TRIPOLI-4® and point kinetics predicted behavior for the experiment featuring the positive reactivity insertion. Computational results are provided for two different data library, namely ENDF/B-VIII and JEFF-3.3, with uncertainties corresponding to one standard deviation. The experimental evolution of the system is well reproduced by the kinetic algorithm.

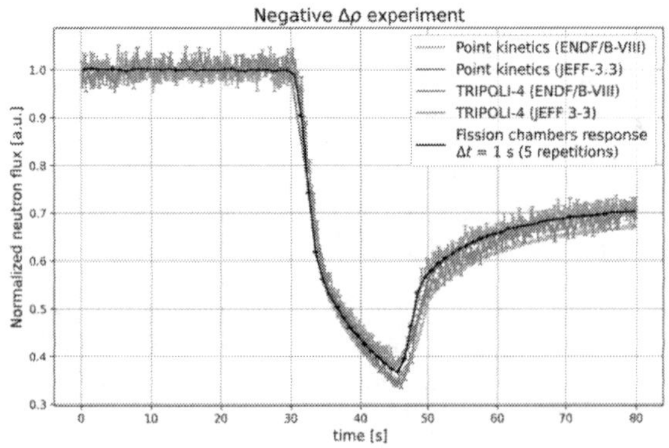

Fig. 4: Comparison between fission chambers acquisitions, kinetic simulation in TRIPOLI-4® and point kinetics predicted behavior for the experiment featuring the negative reactivity insertion. Computational results are provided for two different data library, namely ENDF/B-VIII and JEFF-3.3, with uncertainties corresponding to one standard deviation. The experimental evolution of the system is well reproduced by the kinetic algorithm.

B. Local response

The validation of TRIPOLI-4® ability to reproduce local effects relies on the comparison with measurements from the SAFFRON detection system. In principle, to fully exploit the high-spatial resolution offered by the array, one could analyze the response of each detector. However, for the sake of clarity and conciseness, this study focuses in detail on only three detectors, selected at increasing distances from the perturbation site, corresponding to the north control rod channel. These detectors belong to the mid-height axial layer, which is exposed to the highest flux and thus offers reduced experimental uncertainty and lower variance in simulation tallies. The first one, detector # 70, is the closest to the control rod channel (1.714 cm from center to center), followed by detector # 65 (4.778 cm), and finally detector # 84, positioned farther away, near the core center. This comparison is intended to showcase the fading of localized effects, which gradually vanish with increasing distance from the perturbation point, leading the detector response to relax toward the global behavior. Before discussing the results, it is important for our investigation to clarify the distinction between thermal and fast neutron fluxes. Indeed, the MiMi detection system is primarily sensitive to thermal neutrons, as it relies on neutron interactions with ^{6}Li, which has a significant absorption cross section in the thermal range. In our Monte Carlo simulations, the neutron flux was tallied using a standard two-group energy decomposition, with the customary cutoff at 0.625 eV to distinguish thermal from fast neutrons. Consequently, all comparisons between experimental and simulated results in this work are based on the thermal component of the flux.

The case of positive reactivity insertion is presented in Fig. 7. For clarity, only results based on the JEFF-3.3 nuclear data library are shown, although the same trend is observed with ENDF/B-VIII. From Fig.7b we can observe that, for the detector located very close to the perturbation, the measured flux significantly diverges from the behavior predicted by the point kinetics model. This is further confirmed by the difference plot, which shows absolute values exceeding one. The neutron flux scored using TRIPOLI-4® in kinetic mode, directly on the volume of the MiMi detector, appears instead to successfully captures this deviation. A good agreement within one standard deviation is observed with the measurements, as further confirmed by the difference plot. A larger difference, reaching an absolute value as high as 0.8, is observed during the transient phase associated with the control rod insertion. However, considering the approximations involved in this part of the simulation, as well as the unaccounted uncertainty in the experimental data regarding the triggering of the movement, this discrepancy does not affect the overall consistency of the results. For detector #65, shown in Fig.7d, a moderate deviation from the global response is observable, and the agreement between the experimental trend and the TRIPOLI-4® simulation remains good, falling within one standard deviation. Finally, for detector #84, located farther from the control rod channel, Fig.7f clearly shows a relaxation toward the global behavior: localized effects are no longer visible, and the point kinetics model begins to reproduce the flux evolution reasonably well, particularly in the initial phase of the transient, where agreement falls within one standard deviation. The TRIPOLI-4® simulation results also reflect this transition to the global response, maintaining agreement with the measurements within one standard deviation throughout the experiment, except during the control rod insertion phase, where the absolute value of the difference reaches a maximum of 0.33.

Additionally, we recall that only one control rod was

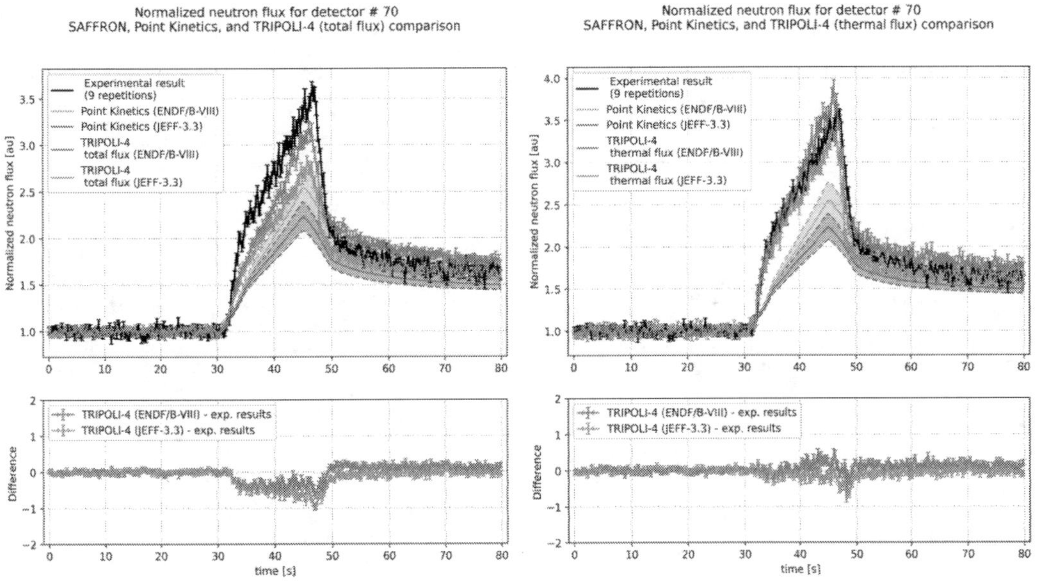

Fig. 5: Comparison between the experimental response to the positive reactivity insertion, the behavior simulated using the kinetic algorithm in TRIPOLI-4®, and the trend predicted by the point kinetics model. Computational results are shown for both the `JEFF-3.3` and `ENDF/B-VIII` libraries, with uncertainties corresponding to one standard deviation. On the left, the neutron flux returned by the kinetic simulation in TRIPOLI-4® is integrated over a broad energy range; on the right, it is restricted to the thermal range. Better agreement is observed in the thermal case, consistent with the fact that the SAFFRON detection system is primarily sensitive to thermal neutron interactions. Difference plots, shown below each case, further illustrate the deviations between the measurements and the computational results.

moved during the positive reactivity insertion experiment, while the other (located in a symmetric position within the code) remained at fixed height. This configuration allows for a meaningful comparison between the two MiMi detectors symmetrically positioned with respect to the control rods and placed as close as possible to them. As illustrated by the contrasting between Fig.7b and Fig.6, the detector near the perturbation (#70) captures a localized effect that is not visible in its symmetrical counterpart (#88), which only captures the global flux variation.

The impact of considering the thermal flux instead of the total flux in the score of a specific detector is illustrated in Fig. 5, which presents results for both nuclear data libraries for the detector closest to the perturbation point (i.e., # 70). This figure highlights the importance of aligning the simulation output with the physical sensitivity of the detector.

As a final observation, the maximum normalized neutron flux measured by the MiMi detector closest to the perturbation (i.e., with the detector #70) is approximately 30% times higher than the peak value predicted by the fission chambers. This difference is quite significant, and being capable of detecting it thanks to a high-resolution detection system is of particular interest, since localized effects can play a crucial role in reactor behavior. As a consequence, this also underscores the value of a simulation tool like TRIPOLI-4®, which, as demonstrated in this work, is capable of capturing through kinetic simulations such localized effects, undetectable by standard point kinetics.

VI. Conclusions

In this work, we presented the HONEY program, focusing on its first experimental campaign conducted at EPFL in April 2024. The campaign aimed to generate a comprehensive set of experimental data on reactor kinetics by measuring the system's response to reactivity perturbations, intended for validating the kinetic mode of the Monte Carlo code TRIPOLI-4®. We described the two experiments carried out, one involving a positive reactivity insertion and the other a reactivity withdrawal, both achieved throughout the control rod movement. We presented a comparison between the evolution of the neutron flux following the perturbation as measured by the two fission chambers and the corresponding behavior predicted by TRIPOLI-4® using a computational twin model of the CROCUS reactor. Good agreement was observed between the measured and the simulated trend. Simulations were carried out using two distinct nuclear data libraries, namely `JEFF-3.3` and `ENDF/B-VIII`: comparisons showed that the choice of the library may have a large impact on the obtained flux profiles. The second part of this work explored local effects near the perturbation point using the high-resolution SAFFRON system. TRIPOLI-4® simulations reproduced the spatial relaxation from local to global flux behavior, consistent with experimental observations. Future work will exploit the full 3D capabilities of SAFFRON with a new perturbation setup, originally developed for the BLOOM program, specifically designed for oscillation experiments [26]. This setup ensures more reproducible perturbations by eliminating un-

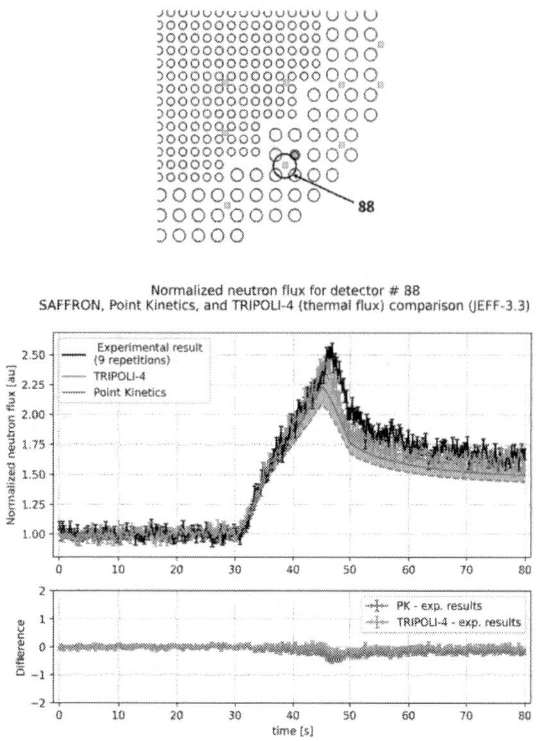

Fig. 6: Top: MiMi detector #88, positioned closest to the control rod that remains stationary, and symmetrically located with respect to the perturbation spot. Bottom: Kinetic simulation results relying on `JEFF-3.3` are compared with experimental data and point kinetics predictions, confirming that the trend observed by this detector is unaffected by localized effects and reflects the global reactor behavior.

certainties in the motion. In parallel, the HONEY program will carry a comprehensive investigation of neutrons spatial correlations, possibly leading to spatial clustering, which is critical for understanding localized neutron behavior. In this respect, the CROCUS reactor, equipped with the high-resolution SAFFRON detection system, offers a uniquely well-suited environment.

Acknowledgment

TRIPOLI-4® is a registered trademark of CEA. The authors thank EDF for partial financial support.

References

[1] F.-X. Hugot, A. Jinaphanh, C. Jouanne, C. Larmier, Y. K. Lee, D. Mancusi, O. Petit, T. Visonneau, and A. Zoia. Overview of the TRIPOLI-4® Monte Carlo code, version 12. *EPJ Nuclear Sci. Technol.*, 10:17, 2024.

[2] OECD/NEA. *Benchmark on the kinetics parameters of the CROCUS reactor*, volume IX of *Plutonium Recycling*. OECD Nuclear Energy Agency, 2007.

[3] J. B. Mosset et al. A 16-ch module for thermal neutron detection using ZnS:6LiF scintillator with embedded WLS fibers coupled to SiPMs and its dedicated readout electronics. *Nuclear Instruments and Methods in Physics Research Section A: Accelerators, Spectrometers, Detectors and Associated Equipment*, 845:494–498, 2016.

[4] A. Wolfertz, R. Adams, and G. Perret. First tests of a gamma-blind fast neutron detector using a ZnS:Ag-epoxy mixture cast around wavelength-shifting fibers. *Nuclear Instruments and Methods in Physics Research Section A: Accelerators, Spectrometers, Detectors and Associated Equipment*, 971(January):164003, 2020.

[5] F. Vitullo et al. A mm^3 Fiber-Coupled Scintillator for In-Core Thermal Neutron Detection in CROCUS. *IEEE Transactions on Nuclear Science*, 67(4):625–635, Apr. 2020.

[6] F. Vitullo. *Miniature and Minimalistic Neutron Detectors for Online High-Resolution Experiments in the Zero-Power Reactor CROCUS*. Ph.D. dissertation, EPFL, Lausanne, Switzerland, 2022.

[7] V. Lamirand, F. Vitullo, K. Ambrožič, O. Pakari, L. Braun, and D. Godat. Development of fibre-based neutron scintillators. Deliverable D2.3, CORTEX Project, 2021.

[8] V. Lamirand. Ten springs of experiments in CROCUS. *EPJ Web of Conferences*, 288:04026, 2023.

[9] M. Faucher. *Coupling between Monte Carlo Neutron Transport and Thermal-Hydraulics for the Simulation of Transients Due to Reactivity Insertions*. Ph.D. dissertation, Université Paris-Saclay, France, 2019.

[10] M. B. Chadwick et al. ENDF/B-VII.1 Nuclear Data for Science and technology: Cross Sections, Covariances, Fission Product Yields and Decay Data. *Nuclear Data Sheets*, 112(12):2887–2996, Dec. 2011.

[11] F. Vitullo, V. Lamirand, D. Godat, and A. Pautz. Simultaneous inter-calibration of 160 mimi neutron detectors. *EPJ Web of Conferences*, 288:04023, Nov. 2023.

[12] M. Saliba, V. Lamirand, W. Monange, F. Vitullo, M. Hursin, A. Pautz, and O. Pakari. Measurement of the Prompt Neutron Decay Constant in a Zero-Power Reactor Using a Novel 3D Detector System. *Nuclear Science and Engineering*, 0(0):1–9, 2025.

[13] T. Ligonnet, A. Pautz, D. Clément, N. Weiss, O. Pakari, V. Lamirand, and A. D. Bembinoff. 3D Flux Measurements of Pile-Oscillation: Initial Results of the BLOOM Program in CROCUS using the SAFFRON Detector Array. In *Proceedings of the 2025 International Conference on Nuclear Instrumentation and Methods (ANIMMA 2025)*, Valencia, Spain, June 2025.

[14] V. Lamirand, T. Ligonnet, A. Pautz, and O. Pakari. COLIBRI Phase 2 – Experimental Study of Fuel Rods Vibration in 3D Using the SAFFRON Core Mapping Array in CROCUS. In *Proceedings of the 2025 International Conference on Nuclear Instrumentation and Methods (ANIMMA 2025)*, Valencia, Spain, June 2025.

[15] D. Legrady and J. E. Hoogenboom. Scouting the Feasibility of Monte Carlo Reactor Dynamics Simulations. In *Proc. PHYSOR'08: International Conference on the Physics of Reactors*, Interlaken, Switzerland, 2008.

[16] B. L. Sjenitzer. *The Dynamic Monte Carlo Method for Transient Analysis of Nuclear Reactors*. Ph.D. dissertation, Delft University of Technology, Netherlands, 2013.

[17] B. L. Sjenitzer and J. E. Hoogenboom. Dynamic Monte Carlo method for nuclear reactor kinetics calculations. *Nuclear Science and Engineering*, 175:94–107, 2013.

[18] T. Booth. A weight (charge) conserving importance-weighted comb for Monte Carlo. In *Proc. RPSD Topical Meeting*, Falmouth, USA, 1996.

[19] D. Mancusi. *Reactors, reactions, and random numbers*. HDR thesis, Université Paris-Saclay, France, 2023.

[20] C. Montecchio, V. Lamirand, D. Mancusi, W. Monange, and A. Zoia. Towards a highly efficient and unbiased population-control algorithm for kinetic Monte Carlo simulations. *EPJ Web Conf.*, 302, 2024.

[21] D. Mancusi and A. Zoia. Zero-variance schemes for kinetic Monte Carlo simulations. *The European Physical Journal Plus*, 135:401, May 2020.

[22] M. Faucher, D. Mancusi, and A. Zoia. New kinetic simulation capabilities for TRIPOLI-4®: Methods and applications. *Ann. Nucl. Energy*, 120:74–88, 2018.

[23] D. A. Brown, M. B. Chadwick, R. Capote, et al. ENDF/B-VIII.0: The 8th Major Release of the Nuclear Reaction Data Library with CIELO-project Cross Sections, New Standards and Thermal Scattering Data. *Nuclear Data Sheets*, 148:1–142, 2018.

[24] A. J. M. Plompen et al. The Joint Evaluated Fission and Fusion Nuclear Data Library, JEFF-3.3. *European Physical Journal A*, 56:181, 2020.

[25] G. Truchet et al. Computing adjoint-weighted kinetics parameters in TRIPOLI-4® by the Iterated Fission Probability method. *Annals of Nuclear Energy*, 85:17–26, 2015.

[26] T. Ligonnet, A. Laureau, P. Simbsler, A.Pautz, and V. Lamirand. Design of an Open-Loop Pile-Oscillation Program in the CROCUS Reactor. *IEEE Transactions on Nuclear Science*, 71(5):1001–1009, 2024.

 https://doi.org/10.1051/epjconf/202533804002

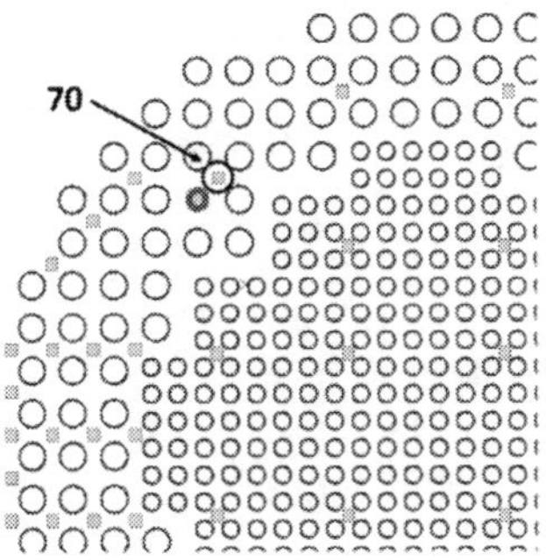

(a) MiMi detector # 70

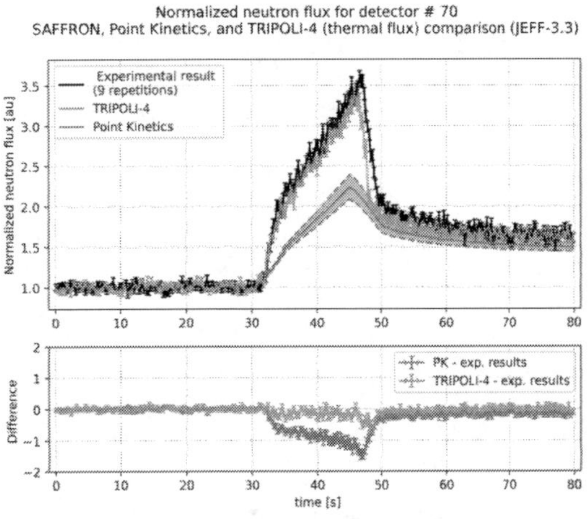

(b) Neutron flux at detector #70: measurement vs. kinetic simulation and point kinetics model

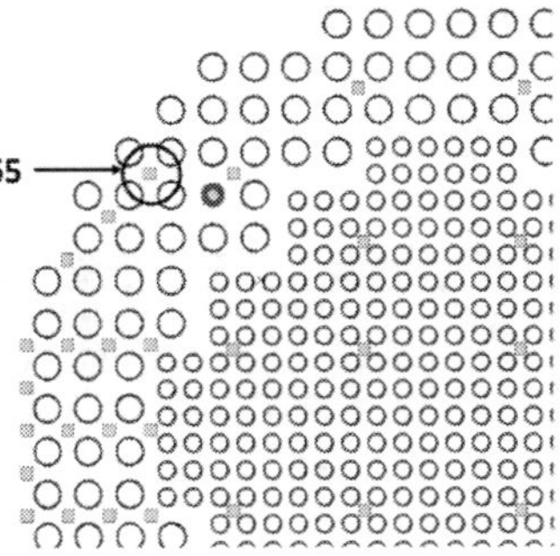

(c) MiMi detector # 65

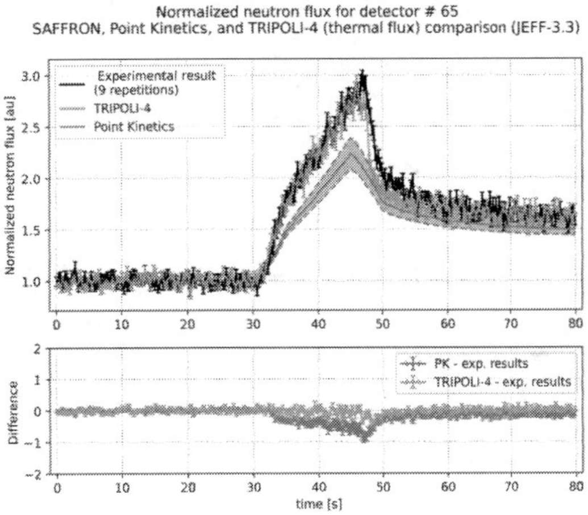

(d) Neutron flux at detector #65: measurement vs. kinetic simulation and point kinetics model

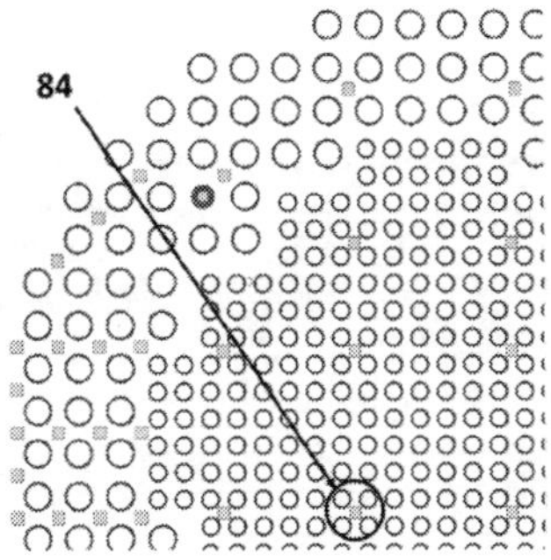

(e) MiMi detector # 84

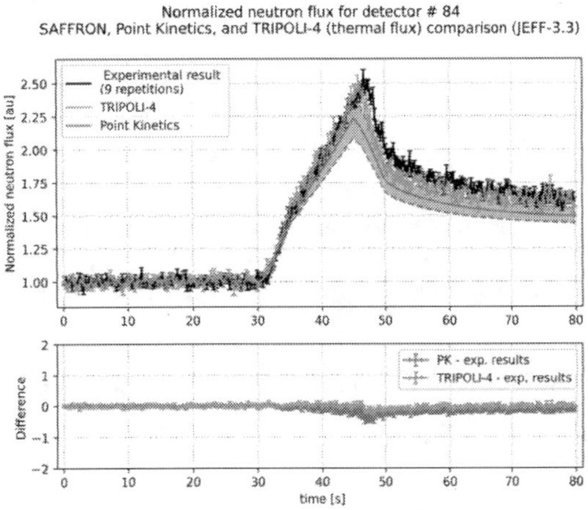

(f) Neutron flux at detector #84: measurement vs. kinetic simulation and point kinetics model

Fig. 7: Spatial analysis of the relaxation from local to global response. Kinetic simulation results using `JEFF-3.3` are compared with experimental data and point kinetics predictions, confirming the code's ability to accurately reproduce the deviation from the local response, in good agreement with SAFFRON measurements.

Design and testing of a fiber-coupled fast neutron scintillation detector for low-power research reactors

Alexis Dupont Bembinoff[1,*], Oskari Pakari[1,2], Daniel Clément[1], Thomas Ligonnet[1], Mathieu Hursin[1,2], Andreas Pautz[1,2], Vincent Lamirand[1,2]
[1]École Polytechnique Fédérale de Lausanne (EPFL), Switzerland
[2]Paul Scherrer Institut (PSI), Switzerland
(*)alexis.dupont-bembinoff@epfl.ch

***Abstract*—We present a new fiber-coupled organic scintillator fast neutron detector. Plastic scintillators are used in many applications to detect fast neutrons and gamma rays. The interest in small, fiber-coupled scintillation detectors is in their ability to perform high-resolution neutron flux measurements to ultimately provide multi-physics flux data in research nuclear reactors. Such measurements are planned in the framework of the EURATOM EVEREST project, specifically in the JSI-TRIGA and BME reactors to produce data for code validation.**

A 5x5x5 mm EJ-276D scintillator was tested with and without optical fiber coupling to a PMT with Pu-Be and ^{137}Cs sources. The Pulse Shape Discrimination histograms with coupling were still showing a region of interest for neutron detection. We produced a prototype reactor sensor with a surface of 1 mm^2 (V=0.3 mm^3) for high resolution experiments. The 0.3 mm^3 scintillators were coupled to a 20 m optical fiber for a set of experiments with a strong Co^{60} ($\approx$250 GBq) source and in the zero-power reactor CROCUS. The neutron detection capability was assessed in CROCUS for different power levels. The fiber-coupled small-sized scintillators have shown a linear response up to 25 W (total flux around 3.8x10^8 cm^{-2}.s^{-1}). Ongoing developments include testing alternative scintillation materials such as stilbene and organic glass. Future work will focus on experiments in the JSI-TRIGA reactor at higher neutron fluxes and smaller detector volumes.

***Index Terms*—Organic scintillator, fast neutron, detection, high-resolution, fiber-coupling, CROCUS, research reactor, photo-multiplier tube.**

I. Introduction

MODERN simulation codes for nuclear reactor neutronics can predict power with high spatial resolution, enabling the study of intra-fuel-pin neutron flux effects [1]. As part of the HORIZON-EURATOM-2023 initiative focused on advancing solutions for nuclear reactor pressure vessel aging and safety margin assessment, the EVEREST project (Experiments for Validation and Enhancement of the REactor preSsure vessel fluence assessmenT) is dedicated to the development of advanced multiphysics simulation tools for accurate prediction of neutron fluence in nuclear power plant reactor vessels. Tackling the multiphysics phenomena happening in a nuclear power plant could allow to optimize conventionally used conservative safety margins for the better, and therefore extend the lifetime of nuclear power plants. To validate these codes, we require measurements that are well and sufficiently localized (in between the fuel pins). The pursuit of cost-effective alternatives to miniature fission chambers is important to facilitate the deployment of a larger number of detectors. To observe multiphysics phenomena, powers on the order of 50 kW need to be reached to observe temperature feedback in the reactors of interest [2], [3], [4], [5]. A key challenge is thus the development of new high-resolution neutron detectors capable of handling the associated fluxes, whilst being exposed to comparatively high gamma ray fluxes, relatively increased temperatures (in the range of 50°C), and perturbations such as coolant movement.

Small detectors are of particular interest for performing highly localized measurements in confined geometries, such as inter-pin regions. In this context, IPEN conducted such measurements in 2024 using TLD-100 detectors within the IPEN/MB-01 core [6]. Furthermore, intra-pin dosimetry carried out in the NECTAR experiments [7], [8], [9] at the CROCUS reactor demonstrated the feasibility of achieving highly localized neutron flux measurements using gold foils. However, passive detection does not provide feedback on phenomena leading to a temporal variation of the measured quantity (occurring on the timescale of one second), whereas an active detection system can.

An important avenue of development is the use of optical fiber-coupled neutron scintillation detectors. The light collection electronics are often large (~cm), and placing scintillators into reactors or other spatially constrained environments is therefore challenging or altogether impossible. By fiber-coupling a <1 mm scintillator to the light collection electronics (usually a SiPM or PMT), new measurement locations - both in terms of space and electronics sensitivity - could be unlocked.

Scintillator-based fiber-coupled detectors were previously developed at Nagoya University in 1994 by Mori et al. [10] and at EPFL in 2020 [11], [12], [13], [14]. The work done at EPFL showed that small inorganic scintillators (ZnS:LiF based) can be used for thermal neutron detection in a zero power research reactor [15]. The detectors show linear responses up to 6.5W (total flux of 7.7x10^7 cm^{-2}.s^{-1} and thermal flux of 3.2x10^7 cm^{-2}.s^{-1}). They are unsuitable for higher power (limitation in the electronic) or even multi-physics relevant experiments in the current design [16]. Other researchers developed small fast neutron fiber-coupled detection system with inorganic scintillators [17], [18]. Our work explores a new angle on organic fast neutron scintillator development. Organic scintillators were previously tested in CROCUS [19], [20], but only at lower powers (up to 50mW), as the PMT was

 https://doi.org/10.1051/epjconf/202533804003

also in the reactor core.

In this article, we introduce a novel fast neutron detection system designed for high spatial resolution. The objective is to operate it in harsher environments than zero power reactors where multiphysics phenomena can be observed. The article is structured as follows: Section II introduce the different systems we use to develop the detector, section III presents the feasibility to perform Pulse Shape Discrimination (PSD). Section IV is dedicated to fiber-coupling neutron detection with an industrial sized scintillator (125 mm^3). Moreover, this section deals with the size reduction of the scintillator to decrease the gamma sensitivity and perform neutron counting in the zero power reactor CROCUS at different power states. Finally, section IV summarizes the different capacities of different scintillator sizes.

II. Design of the fiber-coupled organic fast neutron detection system

A. *Organic fast scintillators EJ-276D*

The detector material used in this development is EJ-276D, manufactured by Eljen Technologies [21]. It is a plastic scintillator designed for gamma-ray and fast neutron detection. It has been characterized for neutron monitoring [22], [23] , compared with other conventional detectors [24] or used for spectrometry purposes [25].

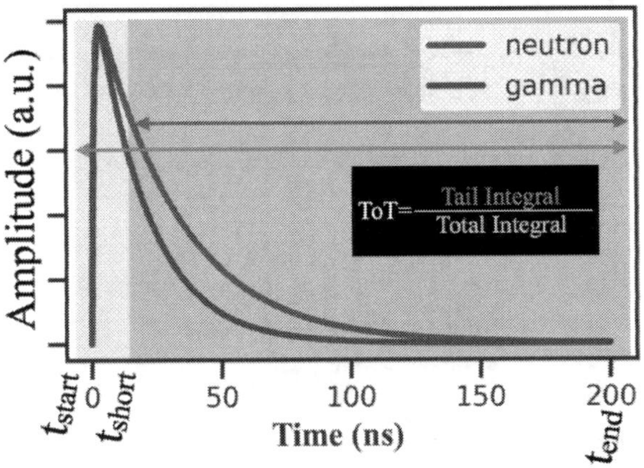

Fig. 1: Illustration of waveforms for neutron scintillation (in red) and gamma-ray scintillation (in blue) inside a PSD capable plastic scintillator. t_{short} (in green) is the starting time of the tail integration for calculating the ToT ratio

Gamma rays interact preferentially with the organic material via Compton scattering, leading to an electron causing scintillation. Neutron detection occurs through proton recoil interactions with hydrogen nuclei. Due to the much higher linear energy transfer (LET) of protons compared to electrons, the excited electronic states differ from those in gamma-ray induced scintillation. Higher energy states are reached, thus some electrons are jumping on triplet states back and forth inducing a longer decay shape as shown Figure 1. Neutron induced prompt fluorescence exhibits distinct *longer* decay times compared to gamma-ray interactions, enabling so-called Pulse Shape Discrimination (PSD) methods to classify pulses into gamma ray or neutron induced. Figure 1 illustrates the difference in expected waveforms, highlighting the small but significant difference in fluorescence decay of some 30 ns [21].

In order to use this difference, the principle of PSD is based on calculating the Tail-over-Total (ToT) ratio (Fig. 1) of the pulse, as shown in Eq. (1). The integral of the tail for neutron interactions is larger than that for gamma-ray interactions, so the ratio between the tail integral and the total integral allows for the discrimination of these two particle types in a 2D clustering histogram.

$$ToT = \frac{\int_{t_{short}}^{t_{end}} S(t)dt}{\int_{t_{start}}^{t_{end}} S(t)dt} \tag{1}$$

Where S(t) is the signal wave function, t_{start} is the starting time of the wave, t_{end} is the ending time of the wave, and t_{short} is the starting time of the tail integration. From the ToT we can plot a 2D histogram so called Pulse Shape Discrimination plot (PSD plot) with in y-axis the ToT and x-axis the deposited energy (or energy). The PSD plot would show two separated clusters for a mixed field measurement (n+γ): one for the neutron (higher ToT) and one for the gamma rays (lower ToT), and one peak for pure gamma-rays measurement corresponding to gamma scintillation detection. For the study, multiple scintillator sizes are investigated:

1) conventional 10 x 10 x 10 mm cube (1 cm^3)
2) standard 5 x 5 x 5 mm cube (125 mm^3)
3) custom cut 0.6 x 0.6 x 0.9 mm miniature scintillator (0.3 mm^3 mass based)

The latter 0.3 mm^3 scintillator is a piece manually cut at EPFL from a polished 5 × 5 × 5 mm^3 cube. To achieve this size, the cube was sawn while preserving a polished side. This piece was placed on modeling clay (Fig.2) for cutting and preventing the small parts to be projected somewhere else.

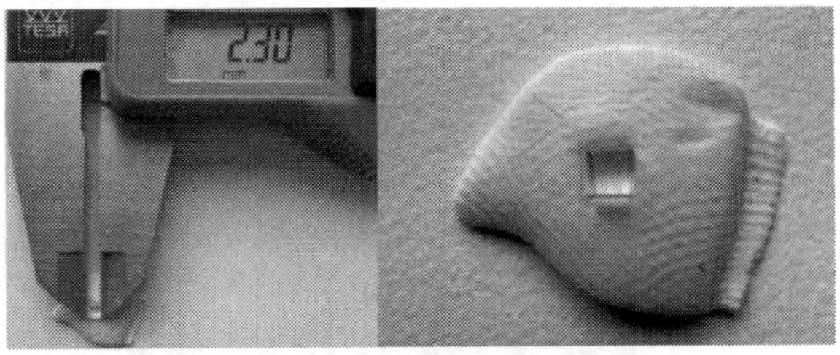

Fig. 2: Size of the first cut of EJ-276D scintillator (left), and picture of the modeling clay use to hold the piece in place for further cutting (right).

The small piece stuck in the modeling clay was then cut with a cutter multiple times to obtain a sub mm^3 scintillator with one side polished for the fiber coupling. The size reduction of the scintillator decreases the overall sensitivity of it. Previous reactor tests of organic scintillators revealed a volume-dependent effect on sensitivity [20].

B. *Optical fiber*

The scintillators were coupled to polished PMMA optical fibers by ESKA [26], [27] of two diameters. For the two largest scintillator cubes, the fiber core had a diameter of 2 mm and a 1 mm jacket. For the smallest scintillator, the fiber core had a diameter of 1 mm and a 1.2 mm jacket. The refractive index of the core is 1.49, and coupling was improved by using EJ-550 optical silicone grease (refractive index: 1.46) manufactured by Eljen Technologies [21] to reduce scintillation photon loss from the fiber/scintillator interface.

For the 1 cm^3 and 125 mm^3 scintillators, an aluminum fiber-holder was used to ensure the fiber is perpendicular to the contact surface (see Figure 3). The holder is a 5x5x5 mm cube with a 3.1 mm diameter hole to accommodate the fiber. Black electrical tape was placed around the three components (scintillator, holder, and fiber) to secure them together, ensuring a mechanically stable system during transport. The 0.3 mm^3 scintillator was positioned at the tip of a fiber (also using optical grease) and protected with an aluminum cap (see Figure 3). Black electrical tape was then used to wrap and secure the cap to the fiber, preventing bakcground light from entering.

Fig. 3: (a) difference between a ground fiber (left) and a polished fiber (right) for scintillator and PMT coupling; (b) 125 mm^3 scintillator with aluminum holder on top holding the 3 mm diameter fiber; (c) 0.3 mm^3 scintillator on the tip of the 2.2 mm diameter fiber (left) before putting the aluminum cap (on the right).

C. *Photo-Multiplier Tube (PMT)*

The used Photo-Multiplier Tube (PMT) is the H3178-51 model from Hamamatsu [28]. The PMT was powered with a bias voltage between 1.3 kV and 1.5 kV. The expected gain from the manufacturer datasheet is 7.9 × 10^5.

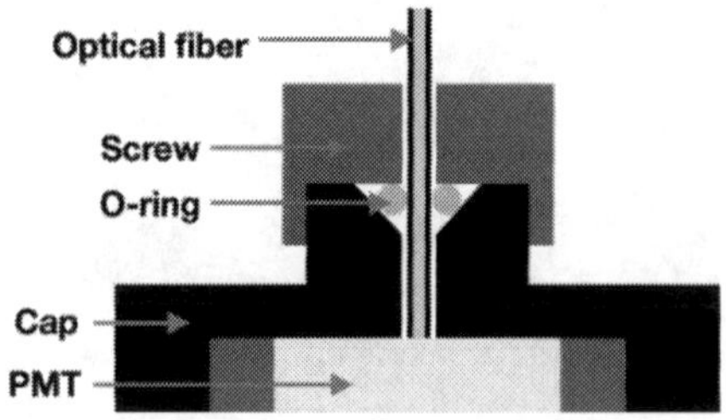

Fig. 4: Scheme of the fiber holder on top of the PMT with the screw and the O-ring for a sturdy and repeatable setup.

Depending on the experimental setup, either the scintillator or the fiber was coupled to the PMT using EJ-550 optical silicone grease to optimize scintillation photon transmission. For fiber coupling, the rear end of the fiber was attached to the PMT using an in-house developed screw holder (see Figure 4). The holder is essentially a black PVC cap with a thickness of 1 cm. It features a conical hole that houses an O-ring for gripping the fiber. When the screw is screwed to the cap, the O-ring is trapping the fiber. Thanks to this, the fiber is correctly held in the same position for each experiment and the background light is not interacting with the PMT photocathode.

D. *Data acquisition*

The PMT signal was read by a 500 MS/s CAEN DT5730S digitizer [29]. The DPP-PSD CAEN firmware [30] was used for signal acquisition, real-time pulse shape discrimination (PSD), energy spectrum generation, and waveform plotting. A waveform is recorded only if its rising edge exceeded the trigger threshold set for the experiment (0.12 mV in our case). Table I presents the fixed settings of the digitizer used in all experiments. These settings were arbitrary set during the experiments to obtain the best PSD plot configuration.

TABLE I: Digitizer settings for measurements

Discriminator mode	Leading edge
Trigger hold off	1024 ns
Pre-trigger	96 ns
Record length	992 ns
DC offset	20.0 %

To obtain the PSD plot from the digitizer, we defined what the digitzer internally calls a total gate and a short gate, as presented in Section II-A. The short gate is used for tail integration, while the total gate performs the full signal integration. From this point onward, the system—scintillator, fiber, PMT, and digitizer—is referred to as miniature organic scintillation system (MOSS).

III. Results: organic scintillator measurements without fiber coupling

A. *Gamma ray reference measurement with a Cs^{137} source*

The goal of this experiment was to observe the gamma-ray response of the scintillator coupled to the PMT and determine whether an energy calibration can be performed using a ^{137}Cs source, and to provide a baseline experiment to compare our later neutron experiments to.

Fig. 5: Experimental setup for gamma-ray detection with 5x5x5 mm EJ-276D scintillator coupled to a PMT (1.3 kV) in front of a 125 kBq ^{137}Cs source.

For these experiments, the 5x5x5 mm (125 mm^3) scintillator was directly coupled to the PMT without a fiber. The ^{137}Cs source was placed in front of the system (contact with the PMT cap). The bias voltage applied on the PMT was 1.3 kV, the energy coarse gain was 10fC/LSBxVpp and the short gate was set at 120 ns and the total gate at 600 ns. The waveform begins rising at 96 ns (pre-trigger), and 30 ns is sufficient to capture the peak height. The tail integral is calculated from 120 ns to 600 ns to ensure accurate horizontal clustering. The manner in which we herein quantify the detector response is the so-called PSD plot. The PSD plot consists of a 2D histogram where the deposited energy is on the x-axis and the ToT on the y-axis.

Figure 6 shows the PSD plot resulting from the ^{137}Cs experiment. The PSD plot shows a single band in the range ToT = [0.15, 0.21]. This band corresponds to gamma rays emitted and interacting within the scintillator. At lower energies (below

150 keV), pile-up effects are observed. In addition to the PSD plot, the pulse height integral spectrum (Fig. 7) was used to perform an energy calibration by using the Compton edge at 476 keV (with 661.7 keV being the full energy of emission of ^{137}Cs [31]).

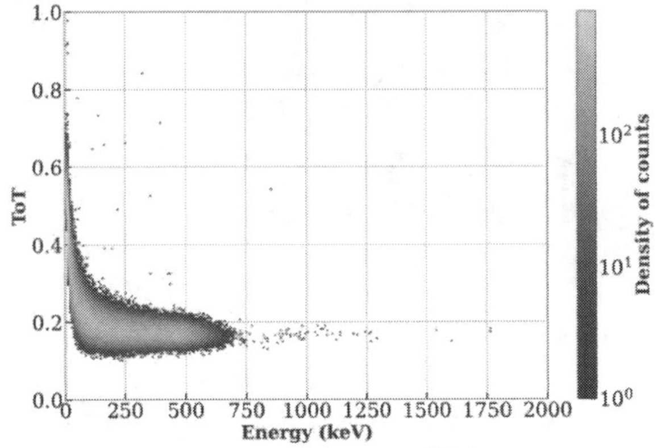

Fig. 6: Energy calibrated PSD plot of a 125 kBq ^{137}Cs source measurement with a 125 mm^3 EJ-276D scintillator coupled to a PMT.

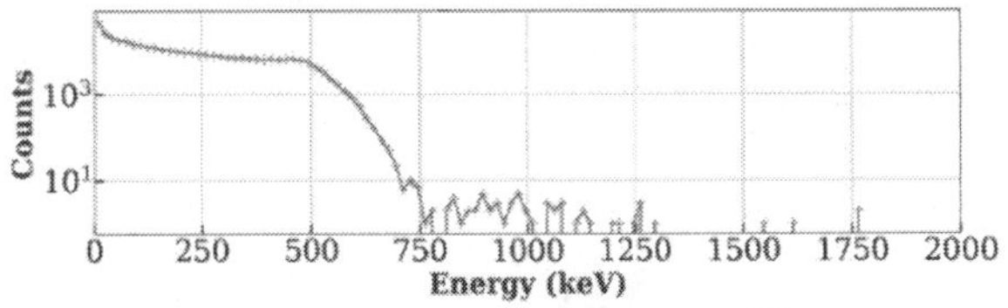

Fig. 7: Gamma ray energy spectrum for the 125 kBq ^{137}Cs source measurement with a 125 mm^3 EJ-276D scintillator coupled to a PMT. Energy calibration was performed with the 475 keV Compton edge.

B. Neutron and gamma-ray reference measurement with a PuBe source

The goal of this measurement was to test the detection of fast neutrons using Pulse Shape Discrimination (PSD), and establishing the optimal settings for distinguishing between the neutron and gamma contributions, i.e., the optimal choice for long and short gate settings.

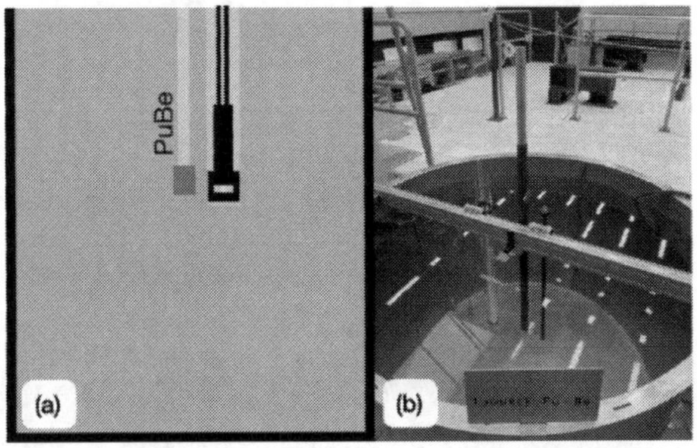

Fig. 8: (a) Setup of the mixed field (neutron & gamma-ray) irradiation in CARROUSEL with the 1 cm^3 EJ-276D scintillator coupled to a PMT with a bias voltage of 1.3 kV; (b) CARROUSEL facility with the PuBe source inserted in the central tube, and Plexiglass PMT holder tubes on its left and right sides.

We performed mixed field irradiation of the system with a Plutonium-Beryllium (PuBe) source in the CARROUSEL facility. CARROUSEL is a cylindrical water tank that can hold a PuBe in its center [32], [15]. The PuBe source has an intensity of 10^7 n.s^{-1}.

For this experiment, the scintillator used was the 10x10x10 mm one (1 cm^3). The PMT bias voltage was 1.3 kV and the optimized firmware settings are the same as section III-A: short integral 120 ns and total integral 600 ns. The PSD plot of the mixed-field measurement (Fig. 9) shows two distinct regions above few hundreds keV energy with different ToT values.

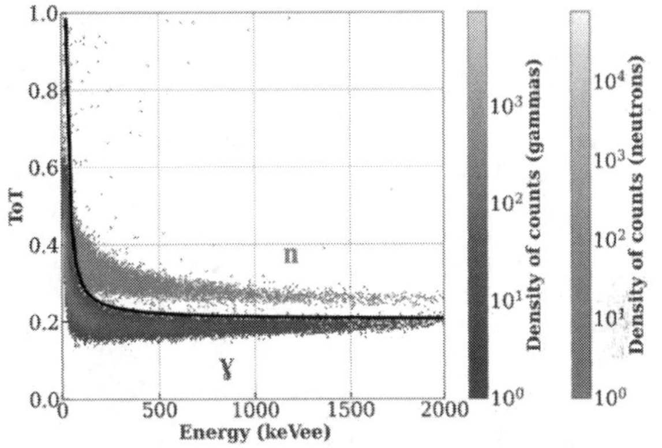

Fig. 9: Energy calibrated PSD plot with 1 cm^3 EJ-276D scintillator coupled to a PMT at 7 cm from a 185 GBq PuBe source with a PMT bias voltage of 1.3 kV. The detection events associated with neutrons are in red while the gamma ray detection events are in blue. The black line defines the discrimination limit.

According to the waveforms of the different scintillation processes, the red region of the plot corresponds to neutron detection, whereas the blue region corresponds to gamma-ray detection. For heavy charged particles (protons, in our case), the light yield per unit of energy is reduced compared to that for electrons [31], hence the y-axis in keVee (electron equivalent). The term keVee sets an absolute basis for the energy calibration.

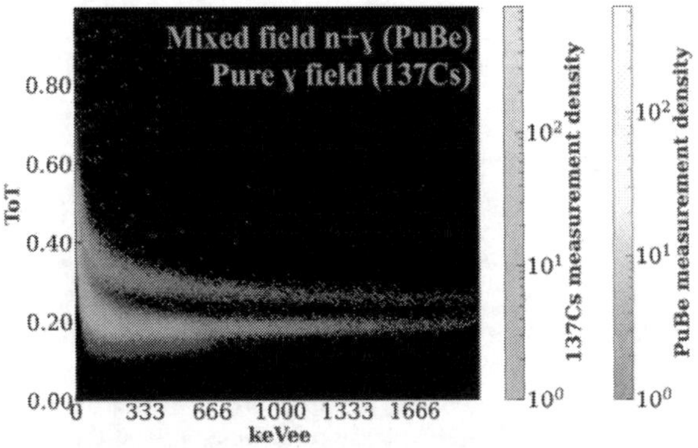

Fig. 10: PSD superposition between a mixed field neutrons & gamma ray (PuBe) PSD plot in orange-yellow and a pure gamma-ray field (Cs^{137}) PSD plot in blue-green.

The low energy region (below 150 keV) smearing is due to pile-up phenomena. The black curve in Figure 9 delimits conservatively the decision threshold for gamma ray and neutron detection, i.e. we accept to misclassify some neutrons and in turn reduce the odds of misclassifying gamma rays.

To further illustrate that the red region in Figure 9 corresponds to neutron events, we introduce a PSD plot superposition in Figure 10 with a black background. From now on, PSD plots with black backgrounds will be PSD superposition plots (one in orange, one in blue). We observe that the peak from the previous pure gamma-ray measurement overlaps with the presumed gamma-ray peak in the mixed-field measurement. It is to be noted that it overlaps in the lower ToT peak of the PuBe measurements, which is expected considering the larger energy range of PuBe gamma rays. The gamma-ray region for this scintillator, with these settings, is in ToT=[0.15;0.21].

IV. Results: fiber-coupled organic scintillator

In this Section, we present our results with fiber-coupled organic scintillators, starting with irradiation of the medium-size one (125 mm^3) in CARROUSEL, and finally the in-core test of the smaller one.

A. Mixed field and pure gamma-ray measurements with the 125 mm^3 fiber-coupled scintillator

The first test consisted in irradiating the 125 mm^3 medium-size scintillator (for detection efficiency reasons) with the PuBe neutron and Cs^{137} gamma source successively, in order to replicate the analysis of Section III.

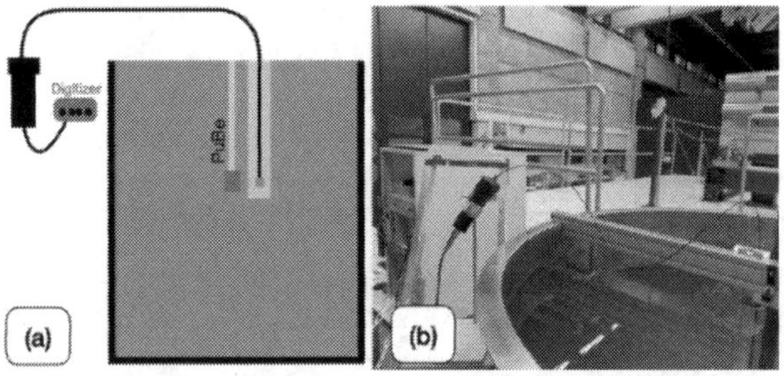

Fig. 11: (a) setup of the mixed field (neutron & gamma ray) detection in CARROUSEL with a 125 mm^{3}scintillator (fiber of 2m, 1.5 kV); (b) CARROUSEL facility with the PMT on a ladder and the optical fiber going into the experimental tube.

The scintillator was coupled to a 2m-long optical fiber (arbitrary length). As the coupling reduces the amount of scintillation light reaching the PMT, new digitizer settings are required for increasing the amplification. The PMT bias voltage was set to 1.5 kV, and the energy coarse gain was also adjusted to 2.5 fC/LSBxVpp. The short gate and total gate are unchanged. The setup in CARROUSEL is presented in Figure 11.

The PSD superposition plot in Figure 12 shows the PuBe measurement (orange) with the previous setup and the ^{137}Cs measurement (blue), similarly to Figure 10.

The coupling with the optical fiber changes the shape of peaks observed in the PSD plot. First, for the gamma irradiation, no Compton edge is observed, which is probably caused by the loss of scintillation photons at the different interfaces, and complicates energy calibration. From now on, the PSD plots with fiber coupling will be expressed in arbitrary deposited charges units.

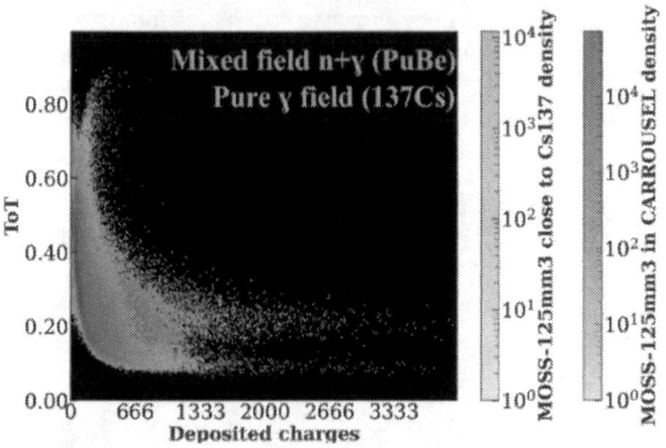

Fig. 12: Non energy calibrated PSD superposition plot between a 125 mm^3 scintillator (1.5 kV) with a 2m fiber in a neutron & gamma-ray mixed field in CARROUSEL (PuBe) and a pure gamma-ray field (^{137}Cs, 125 kBq) measurements

The two clusters present a globally larger mixed detection area. In the lower charge deposit region neutron detections, gamma-ray detection, and pile-up phenomena are overlapping. This is caused by the partial loss of photons in the fiber and the coupling, which broadens the PSD resolution due to pile-up effects. Nevertheless, two distinct regions corresponding to neutron and gamma-ray detections remain observable.

A major first conclusion is therefore that the fiber coupling still allows for discrimination between neutrons and gamma rays. The next step in our development is to reduce the size of the scintillator: first, to match the fiber size for a high-spatial-resolution system, and second, to reduce the number of overall detections per unit flux to increase operational range. Note that the smaller scintillator (0.3 mm^3) showed a count rate too low (below background) with the PuBe source to allow for further characterization.

B. Measurement in CROCUS with the 0.3 mm^3 fiber-coupled scintillator

The 0.3 mm^3 scintillator was tested under mixed field conditions in the zero-power reactor CROCUS. CROCUS is a light water moderated reactor, the criticality is controlled by the water level or with control rods. The neutron detection system, presented in the previous section, is inserted into the Southeast experimental air-channel of CROCUS, which consist of a guide tube with its control rod removed.

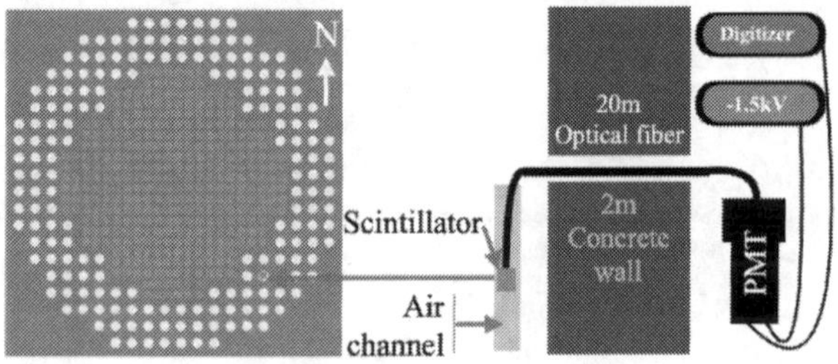

Fig. 13: 0.3 mm^3 scintillator measurement setup in CROCUS. The scintillator is placed in the middle of the South-East CROCUS control rod. A 20m fiber passes in the 2m concrete cavity wall and goes to the PMT and digitizer. This lattice was created with Honeycomb [33]

The goal of this experiment was to verify the PSD plot of a mixed field source with the 0.3 mm^3 scintillator as compared to the previous PuBe measurement with a larger volume detector. Additionally, we examined the scintillator neutron response with increasing power to assess linearity of counting.

The experiment involved data acquisition from reactor start-up up to 80W. Figure 15 shows the total PSD plot for the experiment in CROCUS. The PSD plot reveals a single cluster. The lower region of the plot is once again associated with gamma rays and pile-up phenomena.

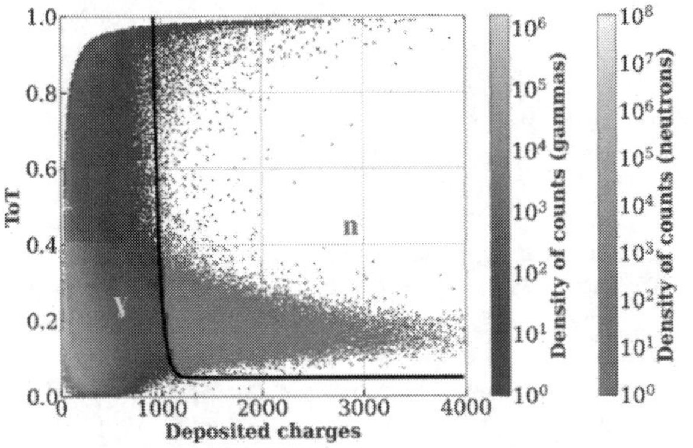

Fig. 14: Non-energy calibrated PSD plot for neutron & gamma ray mixed field in CROCUS with 0.3 mm^3 scintillator (bias voltage of 1.5 kV). The neutron events are in red while the gamma ray events are in blue.

We estimate, based on the mean free path of electrons in EJ276D, that indiviual gamma rays can only deposit up to 200 keV in the available volume of our smallest scintillator. Assuming that the neutrons transfer all their energy (average of 1-2 MeV in CROCUS [34]) to the proton, the energy deposited in the scintillator would be 2 MeV, based on proton energy loss mechanisms [35], [36], [37].

To further verify our hypothesis (neutron detection in the frame Figure 15), we performed an additional measurement conducted in the LOTUS cavity using the SILC Co^{60} irradiator (source activity: 107 GBq as of 04.2025). This irradiation setup was previously found to be comparable to the gamma component of CROCUS [13], and allows for testing the gamma-ray energy deposits. The results are presented in Fig. 15.

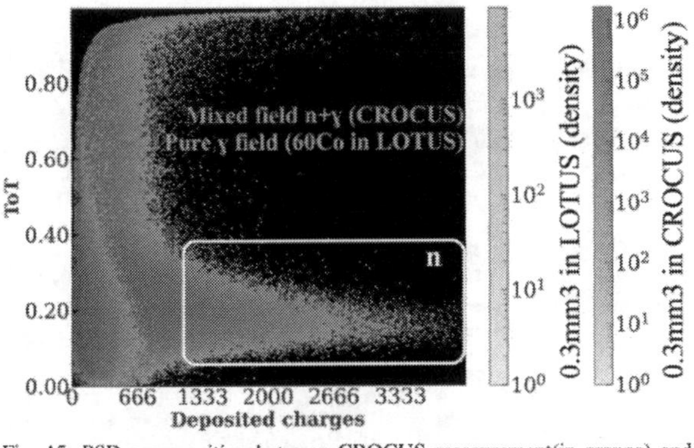

Fig. 15: PSD superposition between CROCUS measurement(in orange) and ^{60}Co (in blue) measurement with 0.3 mm^3 scintillator (bias voltages of 1.5 kV). The white frame corresponds to the neutron detection area for power monitoring.

Even if the gamma energy spectrum of Co^{60} is relatively lower than the gamma spectrum in CROCUS, we estimated that the gamma-ray source gave us sufficient complementary information. Consequently, the single cluster was attributed to neutron detection only, corresponding to the white frame in Fig.15.

Finally, we performed a comparison between the 0.3 mm^3 scintillator neutron detection (i.e., events selected from the white frame shown in the PSD plot) and the CROCUS monitoring fission chamber n°2 from 80W power to shutdown. For the ease of comparison, the two signals were normalized to the average count rate during the 80W stabilization at criticality.

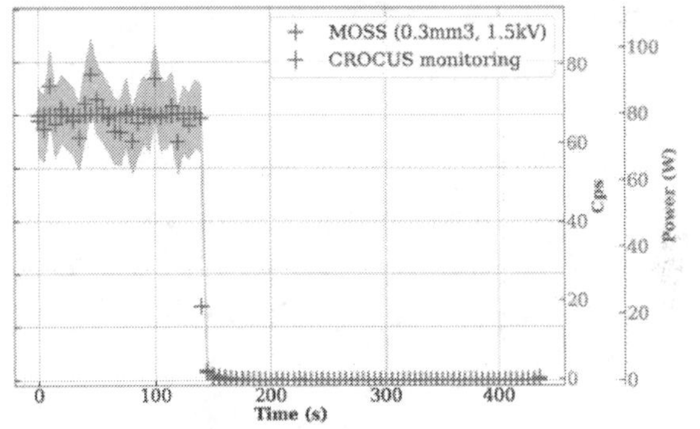

Fig. 16: Power tracking comparison between 0.3 mm^3 scintillator (1.5 kV) and CROCUS monitoring fission chamber n°2 during a shut-down from 80W to sub-critical. The dwell time of the two detectors is 5s.

According to previous research [38], [39], after shutdown, the delayed gamma-ray background in CROCUS corresponds to about 30% of the counts observed at 80W. Figure 16 displays the results from the shutdown measurement. The 0.3 mm^3 scintillator accurately follows the fission chamber profile, and does not plateau at 30% as would be expected with delayed gamma-rays. This provides further evidence that the white frame region corresponds to pure neutron counts.

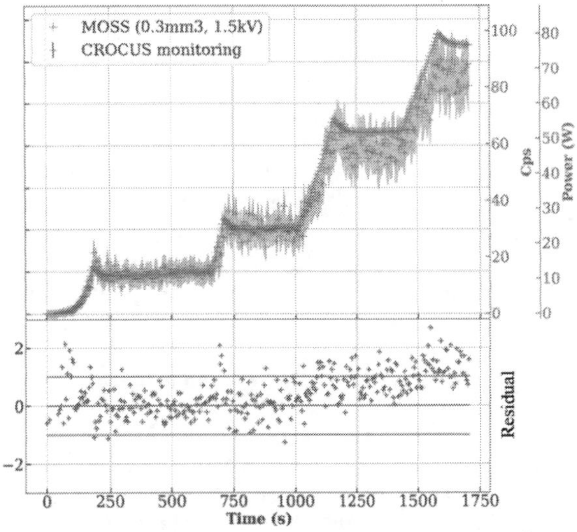

Fig. 17: Power steps comparison between the repetition of the 0.3 mm^3 fiber-coupled scintillator and CROCUS monitoring fission chamber n°2 during different power steps from 0 to 80W. Dwell time of the two detectors is 5s.

In addition to this measurement, we created a repetition of the previous detector (same volume of the scintillator) and performed a measurement of 4 different power steps in CROCUS (Fig.18).

The detector response is in agreement with the fission chamber (within 1 sigma) up to 25 W. However, above 25 W we observe a significative difference which is probably caused by the overloading of the digitizer with non neutron events (pile-ups). Future test will be carried out with an energy cut-off to prevent this overloading. Following these tests, monitoring was performed at different power levels in CROCUS to observe the detectors' behavior for a variety of scintillator sizes, and especially linearity, as shown in Figure 18).

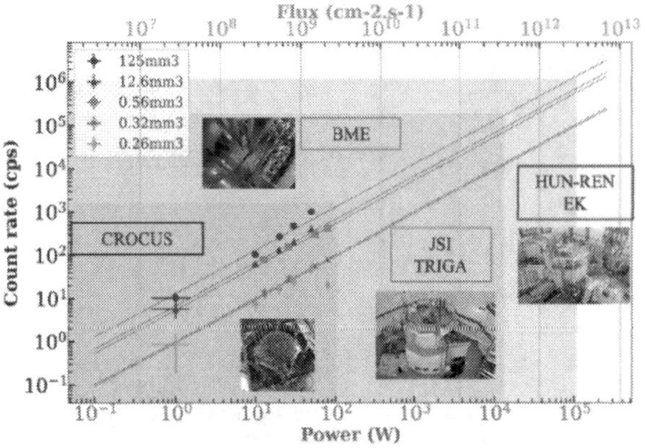

Fig. 18: Count rates of different sizes of scintillators as a function of power and flux in CROCUS South-East experimental channel. The colored regions correspond to the reactors' accessible neutron fluxes.

We tested different sizes of scintillator for the power steps experiment, each coupled to a 20m fiber. They all followed a linear trend with power/flux up to 25W. At higher power levels, the two biggest scintillators (125 mm^3 and 12.6 mm^3) have shown an increase of the count rate. It is possibly caused by the misclassifications of gamma-ray events as neutron events. We can't solve this problem for the 125 mm^3 scintillators but for the 12.6 mm^3 we can change the neutron characterization window.

The 0.32 mm^3 and 0.6 $mm^3$4 detectors were showing a very small decrease. We cannot definitively conclude the cause of this decrease. One possibility is the high data throughput in the electronics not keeping up with the signal (as we record all events, including pileups and other gamma ray events) or because the scintillation light is too intense and saturates the system, making it difficult to distinguish neutrons from pile-ups and gamma rays. All the measurements are taken into account for the linear fitting in Figure 18. Finally, a decrease in the count rate is observed for the smallest detector (0.26 mm^3). It is caused by an experimental error and is not taken into account in the fitting. The smallest scintillators appear to be well-suited for the high-power experiments in other reactors within the EVEREST project. The maximum count rate readable by the CAEN digitizer is around 10^6cps. With appropriate threshold settings we could therefore extrapolate a measurement capability up to a flux of 10^{12}-10^{13} n.cm^{-2}.s^{-1}. Naturally, this extrapolation is likely to not hold once temperatures, Cherenkov light and mechanical vibrations in the higher-power reactors become significant (some 10 kW and above), and we plan to explore the neutron detection system behavior carefully to find the true limits.

V. Conclusion

In this paper we present the development of fiber-coupled organic scintillators for fast neutron detection . Through Pulse Shape Discrimination, we established two distinct zones: one for neutron events and another for gamma-ray events. EJ-276D scintillator (125 mm^3 size) was coupled via an optical fiber to a PMT and irradiated in a mixed neutron and gamma-ray field. Despite the incurring light loss from fiber coupling, the detector still successfully performs pulse shape discrimination, and therefore neutron detection.

A final detection device was created with a 0.3 mm^3 scintillator to achieve the miniature size for high-resolution measurements in research reactors. Experiments were performed in the CROCUS zero-power reactor. By producing PSD superposition plots, we confirmed that the cluster associated with neutrons was indeed due to neutron-induced events. Power monitoring and comparison with the reactor shut-down from 80W was furthermore used to validate our hypothesis. In addition to that, we can assess that the detectors can operate up to 80W. By doing a repetition of the system with a second scintillator of the same volume, we were able to reach 80W with in agreement with the fission chamber measurement. For different volumes of scintillator we assessed the linearity of counting with increasing reactor power. By extrapolation, we provide a first hint at the feasibility of these fiber-coupled scintillators to provide data in higher fluxes reactors (10^{12}-10^{13} n.cm^{-2}.s^{-1}) performing high-spatial-resolution neutron detection.

Future work includes tests in research reactors at higher power levels, namely the JSI-TRIGA and BME reactors within EVEREST, in preparation of high resolution multiphysics experiments. We plan to also compare the performance of PMTs with Silicon Photomultipliers (SiPMs) for fiber coupling. Additionally, fiber coupling with other organic scintillator such as stilbene and Organic Glass Scintillator (OGS) developed at Michighan University are in preparation.

Author's Contribution

Alexis Dupont Bembinoff, Oskari Pakari and Vincent Lamirand conceived, planned and carried out the experiments and the analysis. Alexis Dupont Bembinoff produced the analysing scripts, the scintillators-fiber couplings and the scintillator reduction. Daniel Clément handcrafted the PMT caps and fiber holders. Oskari Pakari is the head of nuclear facilities. Vincent Lamirand is the program manager of experimental research. Thomas Ligonnet helped for the data analysis. Mathieu Hursin is responsible for the EPFL contribution in the EVEREST project. Andreas Pautz is the head of the laboratory.

Acknowledgment

The authors would like to thank the CROCUS operation team for their help and patience during the measurement campaigns. The first author expresses his gratitude to Michel Saliba for his valuable guidance and insightful explanations, which enhanced his understanding of Pulse Shape Discrimination.

This project has received funding from the Horizon-Euratom Research and Training Program 2021–2027 through the EVEREST project under grant agreement n°101163288. Views and opinions expressed in this paper reflect only the authors' view, and the Commission is not responsible for any use that may be made of the information it contains.

References

[1] M. Pecchia, H. Ferroukhi, A. Vasiliev, and P. Grimm, "Studies of intra-pin power distributions in operated bwr fuel assemblies using mcnp with a cycle check-up methodology," *Annals of Nuclear Energy*, vol. 129, pp. 67–78, 7 2019.

[2] A. S. Ványi, B. Babcsány, Z. I. Böröczki, A. Horváth, M. Hursin, M. Szieberth, and S. Czifrus, "Steady-state neutronic measurements and comprehensive numerical analysis for the bme training reactor," *Annals of Nuclear Energy*, vol. 155, 6 2021.

[3] A. S. Ványi, M. Hursin, and S. Czifrus, "Analysis of transient measurements with thermal feedback and coupled trace/parcs calculations performed on the bme training reactor," *Annals of Nuclear Energy*, vol. 194, 12 2023.

[4] V. Valero, L. Ottaviani, A. Lyoussi, V. Radulović, L. Snoj, A. Volte, M. Carette, and C. Reynard-Carette, "Thermal simulations of a new sic detector design for neutron measurements in jsi nuclear research reactor," in *Materials Science Forum*, vol. 1062 MSF, pp. 619–626, Trans Tech Publications Ltd, 2022.

[5] H. Carcreff, V. Radulović, D. Fourmentel, K. Ambrožič, C. Destouches, L. Snoj, and N. Thiollay, "Nuclear heating measurements for fusion and fission relevant materials in the jsi triga reactor," *Fusion Engineering and Design*, vol. 179, 6 2022.

[6] T. A. Cavalieri, P. de Tarso Dalledone Siqueira, J. M. B. Shorto, and H. Yoriyaz, "Thermal neutron dose measurements using tld-100 detectors in the ipen/mb-01 reactor core," *Applied Radiation and Isotopes*, vol. 208, p. 111302, 6 2024.

[7] T. Mager, J.-B. Valentin, V. Lamirand, T. Ligonnet, W. D. Shin, J.-M. Fürbringer, M. Hursin, and A. Pautz, "Design of an instrumented fuel rod for intra-pin neutron flux measurements in the crocus experimental reactor," *EPJ Web of Conferences*, vol. 288, p. 04007, 2023.

[8] T. Mager, O. Pakari, M. Hursin, T. Ligonnet, A. Pautz, and V. Lamirand, "Intra-pin reaction rate measurements in the crocus experimental reactor," *IEEE Transactions on Nuclear Science*, vol. 71, pp. 1047–1055, 5 2024.

[9] T. Mager, M. Hursin, K. Vaughn, S. Choi, B. Kochunas, T. Ligonnet, V. Lamirand, O. Pakari, C. Fiorina, and A. Pautz, "Comparison of high-fidelity transport code predictions against the nectar intra-pin reaction rate measurements," *Annals of Nuclear Energy*, vol. 219, 9 2025.

[10] C. MORI, T. OSADA, K. YANAGIDA, T. AOYAMA, A. URITANI, H. MIYAHARA, Y. YAMANE, K. KOBAYASHI, C. ICHIHARA, and S. SHIROYA, "Simple and quick measurement of neutron flux distribution by using an optical fiber with scintillator," *Journal of Nuclear Science and Technology*, vol. 31, pp. 248–249, 1994.

[11] F. Vitullo, "Miniature and minimalistic neutron detectors for online high-resolution experiments in the zero-power reactor crocus," tech. rep.

[12] F. Vitullo, V. Lamirand, K. Ambrožič, L. Braun, D. Godat, P. Frajtag, and A. Pautz, "Design of a 150-miniature detectors 3d core-mapping system for the crocus reactor," *EPJ Web of Conferences*, vol. 253, p. 04023, 2021.

[13] F. Vitullo, V. Lamirand, J. B. Mosset, P. Frajtag, O. Pakari, G. Perret, and A. Pautz, "A mm3 fiber-coupled scintillator for in-core thermal neutron detection in crocus," *IEEE Transactions on Nuclear Science*, vol. 67, pp. 625–635, 4 2020.

[14] M. Saliba, V. Lamirand, W. Monange, F. Vitullo, M. Hursin, A. Pautz, and O. Pakari, "Measurement of the prompt neutron decay constant in a zero-power reactor using a novel 3d detector system," 2025.

[15] F. Vitullo, V. Lamirand, J.-B. Mosset, P. Frajtag, O. Pakari, P. Gregory, and A. Pautz, "Developing and testing a miniature fiber-coupled scintillator for in-core neutron counting in crocus," *EPJ Web of Conferences*, vol. 225, p. 04018, 2020.

[16] C. Rochat, "Extension of miniature neutron scintillators features in crocus: novel acquisition scheme and extended flux range," jul 2021.

[17] T. Yagi, T. Misawa, C. H. Pyeon, and S. Shiroya, "A small high sensitivity neutron detector using a wavelength shifting fiber," *Applied Radiation and Isotopes*, vol. 69, pp. 176–179, 1 2011.

[18] T. Yagi, H. Unesaki, T. Misawa, C. H. Pyeon, S. Shiroya, T. Matsumoto, and H. Harano, "Development of a small scintillation detector with an optical fiber for fast neutrons," *Applied Radiation and Isotopes*, vol. 69, pp. 539–544, 2 2011.

[19] F. B. Darby, O. V. Pakari, M. Y. Hua, V. Lamirand, S. D. Clarke, A. Pautz, and S. A. Pozzi, "Neutron-gamma noise measurements in a zero-power reactor using organic scintillators," *IEEE Transactions on Nuclear Science*, vol. 71, no. 5, pp. 1033–1040, 2024.

[20] O. V. Pakari, A. Lucas, F. B. Darby, V. P. Lamirand, T. Maurer, M. G. Bisbee, L. R. Cao, A. Pautz, and S. A. Pozzi, "Gamma-ray spectroscopy in low-power nuclear research reactors," *Journal of Nuclear Engineering*, vol. 5, pp. 26–43, 1 2024.

[21] E. Technologies, "https://eljentechnology.com/products/plastic-scintillators/ej-276."

[22] K. D. Ngo, C. Cazzaniga, M. Paoletti, D. Rigamonti, M. Kastriotou, C. Frost, M. Tardocchi, J. Sykora, S. Mann, B. Lutz, and R. Nolte, "Fast neutron response characterization of an ej-276 plastic scintillator for use as a neutron monitor," *Nuclear Instruments and Methods in Physics Research Section A: Accelerators, Spectrometers, Detectors and Associated Equipment*, vol. 1051, p. 168216, 6 2023.

[23] E. V. Ryabeva, I. V. Urupa, E. E. Lupar, V. V. Kadilin, A. V. Skotnikova, Y. A. Kokorev, and R. F. Ibragimov, "Calibration of ej-276 plastic scintillator for neutron–gamma pulse shape discrimination experiments," *Nuclear Instruments and Methods in Physics Research, Section A: Accelerators, Spectrometers, Detectors and Associated Equipment*, vol. 1010, 9 2021.

[24] P. Pant, K. Banerjee, P. Roy, R. Shil, and A. K. Saha, "Characterization of ej-276d plastic scintillator and its comparison with ej-299-33a and bc-501a," *Journal of Instrumentation*, vol. 19, 10 2024.

[25] E. V. Ryabeva, D. A. Molodtsev, I. V. Urupa, R. F. Ibragimov, V. V. Gaganov, E. E. Lupar, I. S. Vershinin, and D. I. Savin, "Ej-276 based neutron spectrometer with neutron–gamma pulse shape discrimination capability," *Nuclear Instruments and Methods in Physics Research Section A: Accelerators, Spectrometers, Detectors and Associated Equipment*, vol. 1034, p. 166791, 7 2022.

[26] Mitsubishi, "https://fiberfin.com/product/ff-sh-8001-3-0/."

[27] Mitsubishi, "Sh-4001 - ff-sh-4001-data-sheet."

[28] Hamamatsu, "Photomultiplier tube assembly h3178-51."

[29] CAEN, "Dt5730 / dt5730s 8 channel 14 bit 500 ms/s digitizer," 2024.

[30] CAEN, "Compass multiparametric daq software for physics applications caen tools for discovery user manual um5960," tech. rep., 7 2024.

[31] G. F. Knoll, *Radiation Detection and Measurement, 3rd ed.* John Wiley and Sons, 3rd edition ed., 2000.

[32] E. L. Brunetto, F. Vitullo, V. Lamirand, K. Ambrožič, D. Godat, M. Buck, G. Pohlner, J. Starflinger, and A. Pautz, "High resolution measurements with miniature neutron scintillators in the sur-100 zero power reactor," *EPJ Web of Conferences*, vol. 253, p. 04029, 2021.

[33] T. Guilbaud, "Honeycomb, an effective web tool to visualize and edit lattices," No. 1 in Transactions of the American Nuclear Society; 130, p. 1052–1053, American Nuclear Society, 2024.

[34] A. Laureau, A. Gruel, V. Lamirand, T. Ligonnet, A. Sardet, A. Pautz, and al Analysis, "Analysis of the preliminary campaign for the petale program," *EPJ Nuclear Sci. Technol*, vol. 9, p. 25, 2023.

[35] A. Bozkurt, "Monte carlo calculation of proton stopping power and ranges in water for therapeutic energies," in *EPJ Web of Conferences*, vol. 154, EDP Sciences, 9 2017.

[36] Z. Francis, S. Incerti, M. Karamitros, H. N. Tran, and C. Villagrasa, "Stopping power and ranges of electrons, protons and alpha particles in liquid water using the geant4-dna package," *Nuclear Instruments and Methods in Physics Research Section B: Beam Interactions with Materials and Atoms*, vol. 269, pp. 2307–2311, 10 2011.

[37] B. W. Wiggins, A. Favalli, M. L. Iliev, K. D. Ianakiev, and M. P. Hehlen, "Computational investigation of arranged scintillating particle composites for fast neutron detection," *Nuclear Instruments and Methods in Physics Research, Section A: Accelerators, Spectrometers, Detectors and Associated Equipment*, vol. 915, pp. 17–23, 1 2019.

[38] O. Pakari, T. Mager, V. Lamirand, P. Frajtag, and A. Pautz, "Delayed gamma fraction determination in the zero power reactor crocus," *EPJ Nuclear Sciences and Technologies*, vol. 7, 2021.

[39] M. Hursin, C. Weiss, P. Frajtag, V. Lamirand, G. Perret, P. Kavrigin, A. Pautz, and E. Griesmayer, "Testing of a scvd diamond detection system in the crocus reactor," *European Physical Journal A*, vol. 54, 5 2018.

Pile-oscillation results for stainless steel with local and 3D global measurements

Thomas Ligonnet[1,*], Axel Laureau[2], Oskari Pakari[1,3], Daniel Clément[1], Alexis Dupont-Bembinoff[1], Gilles Noguère[4], Andreas Pautz[1,3], and Vincent Lamirand[1,3]
[1]EPFL, Switzerland
[2]CNRS Grenoble, France
[3]PSI, Switzerland
[4]CEA, France
(*) thomas.ligonnet@epfl.ch

Abstract*— The BLOOM pile-oscillation program on stainless steel and minor alloying elements was successfully conducted in the CROCUS reactor at EPFL between fall 2024 and spring 2025. For this purpose, a pile-oscillator and a new instrumented channel capable of measuring thermal neutrons as close as 2 mm from the sample were developed in-house, thanks to the previous development in miniature neutron scintillators. Furthermore, the SAFFRON detector array, consisting of 160 such miniature detectors, was used to measure the neutron flux across the core and within the instrumented channel. Using SAFFRON, it was verified that the chosen oscillation location was suitable and point kinetics could be applied, as the induced perturbation did not cause globalized flux redistributions. Local perturbations were measurable up to 15 cm from the sample, setting the minimum distance beyond which the global perturbation was measurable. A total of 45 pile-oscillations were performed, including measurements of 40 different samples and five repetitions of a gold sample at different points of the campaign. For each experiment, the amplitudes of the local and global perturbations and the corresponding reactivity worth were successfully measured, with relative uncertainties as low as 0.07% for the heaviest samples local perturbation, and uncertainty on reactivity worths of 0.04 pcm Future developments will include the addition of new samples and references such as the D_2O. Further developments are also planned to be able to perform pile-oscillations at higher power.

***Keywords* — Pile-Oscillation, CROCUS, Experiments, zero-power reactor, Cross-Section.**

I. Introduction

THE BLOOM experimental program consists of pile-oscillation experiments conducted in the CROCUS zero-power reactor of the LRS at EPFL. The BLOOM program is part of LRS's broader initiative, the HARVEST-X project, which aims to produce new constraints for stainless-steel nuclear data evaluations through integral experimental results [1,2]. Within this context, BLOOM was developed to reproduce and extend the critical experiments conducted as part of the CEA-EPFL PETALE program, the other part being dedicated to transmission [3,4]. Specifically, samples directly extracted from spare reflectors used in the PETALE program were oscillated during the experimental campaign.

Pile-oscillation experiments involve repeatedly inserting and extracting samples into a nuclear reactor to periodically perturb its state [5]. Although the principle is not new, pile-oscillation experiments have regained interest because they enable lower uncertainties and a higher versatility compared to traditional critical experiments [6–8]. Furthermore, the CEA MAESTRO program demonstrated the feasibility of hybrid pile-oscillation experiments, in which both localized and generalized effects of the oscillated sample are measured. The MAESTRO program also validated the equivalence between the open-loop method, which allows reactor power to drift during measurements, and the traditional closed-loop method, in which reactivity effects are actively compensated [9,10]. BLOOM is among the experimental programs inspired by these results, and its primary experimental campaign was conducted between fall 2024 and spring 2025.

This project is the latest in the series of modulation experimental programs that takes place in the CROCUS reactor. It takes full advantage of the developments and insight gained from the previous fuel rod oscillator COLIBRI and for the rotating oscillator PISTIL [2,11,12].

In the first part of this study, the adequacy of our oscillator and detector location is verified using the EPFL miniature scintillators distributed in core in the SAFFRON detector array [13,14], ensuring that oscillation does not induce global flux redistribution. Subsequently, an instrumented channel, newly developed to measure localized effects at an unprecedented proximity, is employed. Using this channel, nine miniature detectors of SAFFRON measure the local effects of the samples at a distance of 2 mm from their surfaces, thus maximizing both the amplitude and precision of the local perturbations' measurements. Finally, the data are processed using state-of-the-art methodologies, applicable to pile-oscillation and modulation experiments, including developments performed for the CORTEX program [15,16]. The experiment with the gold reference sample was repeated five times at different intervals from the beginning to the end of the program, verifying the consistency of the experimental setup and methodologies throughout the study. In total, 45 measurements were conducted on 40 distinct samples and 25 materials, providing an extensive dataset covering both local and global core perturbations, as well as reactivity worths. Results that are of interest for current and future nuclear data validations and assimilations [17,18].

II. Setup And Methodology

A. The CROCUS Reactor

The CROCUS reactor is a zero-power research reactor located in Lausanne, Switzerland. It is a light-water pool type reactor consisting of two concentric fuel zones with low enrichment. The inner zone comprises 336 fuel rods containing uranium oxide pellets enriched to 1.806 wt.%, while the outer zone is currently composed of 180 fuel rods of metallic uranium and is enriched at 0.947 wt.%. The number of rods in this outer zone can vary depending on the experimental campaign, in this instance 4 additional rods with respect to the standard configuration compensate for the reactivity worth of the SAFFRON detector array. The reactor operates at ambient pressure and a controlled temperature of 20 °C. The maximum power authorized by its operating license is 100 W, resulting in a maximum neutron flux of 2.5×10^9 $cm^{-2}.s^{-1}$ at the core center [19,20].

The primary reactivity control mechanism of CROCUS is a spillway that regulates the reactor's water level. Its height can be adjusted with a resolution of 0.1 mm, resulting in a reactivity uncertainty of 0.4 pcm. In its typical configuration, two additional boron carbide absorber rods can also be employed to control its criticality. These rods have an insertion calibration precision of 1 mm, corresponding to an average reactivity control precision of 0.2 pcm. The reactor's permanent power monitoring system consists of two boron-coated compensated ionization chambers and two fission chambers, the last two also function as the safety monitors.

In the BLOOM experimental configuration, the northern absorber rod is removed and replaced with an instrumented oscillation channel. Four Photonis CFUL-01 fission chambers are positioned at each cardinal direction and added to the permanent instrumentation to enhance neutron counting capabilities during oscillations, and provide references in comparison with previous noise and modulation campaigns [11,21].

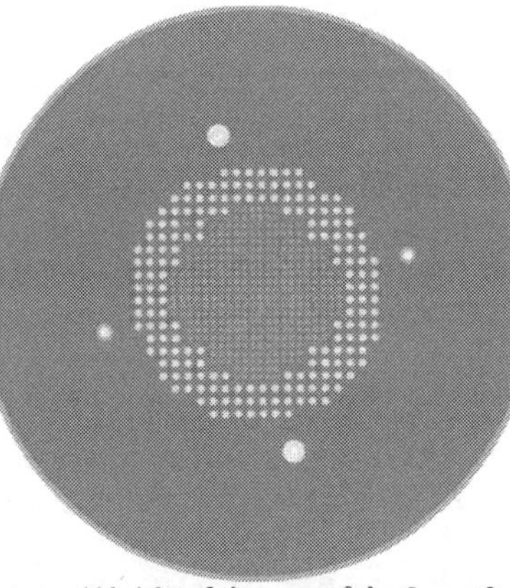

Fig. 1. Axial cut at mid-height of the core of the Serpent2 model of the CROCUS reactor. In this model the size of SAFFRON scintillators (orange squares) is magnified by a factor ten for visibility, with the exception of the ones in the instrument channel, which are visible in a separate close-up (Fig. 4).

B. The SAFFRON Detector Array

SAFFRON is an array composed of 160 miniature LiF:ZnS neutron scintillation detectors, or MiMi detectors [22]. The scintillators are coupled with optical fibers to SiPMs, which are periodically read, and each neutron detection is treated individually using a time tagging system. Of the 160 detectors, 149 are set at fixed positions within the reactor core. Most detectors are arranged in three axial layers at heights of 25, 50, and 75 cm, named L1, L2, and L3 respectively, while a few additional detectors are positioned at intermediate layers LM1 and LM2 close to control rod guide tubes. The detectors are distributed throughout the core cross-section, as shown in Fig. 1, to allow for flux map measurements. The remaining 11 fibers can be freely repositioned within the core; currently, nine of these detectors are employed in the instrumented experimental channel [13,14].

C. Instrumented Experimental Channel

The instrumented experimental channel consists of two components: the pile-oscillator and the in-air instrumented channel. A picture of the installed experimental setup is shown in Fig. 2. The pile-oscillator, referred to as POLLEN, consists of a brushless motor coupled to a wheel around which a wire is wound, allowing a displacement precision of 0.25 mm (on the left in the figure). Detailed descriptions of the motor, along with its associated software and hardware, are available in the proceedings of the previous conference edition [23]. On the right, the wire is attached to an aluminum ruler, which maintains sample alignment within the channel via the channel's cap. Oscillated samples are cylindrical, 10 mm in length and 2 to 10.3 mm in diameter. They are first inserted into an aluminum sample holder, which is then screwed onto the ruler. The sample holders are cylindrical capsules, 121 mm long with an outer diameter of 12 mm and inner dimensions of 101 mm length and a diameter ranging from 9.9 mm to 10.3 mm, ensuring proper sample centering. For samples with smaller diameters, such as gold rods, centering is achieved by replacing the holder's plugs with slightly longer plugs, partially drilled to match the dimensions of the samples. To ensure perfect alignment of the samples within the guide tube, the pulley used for the northern control rod is reused.

Fig. 2. The POLLEN pile-oscillator installed in the CROCUS zero-power reactor. On the left, motor, controller and wheel. The wire passes through the control rod pulley (on the top right) for precision guiding. The ruler connected to the wire is visible next to the double arrow that illustrates the linear displacement. The bottom of the ruler is inserted in the channel, with exiting optical fibers connected to miniature detectors set at the irradiation location.

The instrumented channel enables the measurement of local flux perturbations induced by sample insertion while maintaining its alignment during displacement. A picture of the channel instrumentation, ready for placement in the reactor core, is shown in Fig. 3.

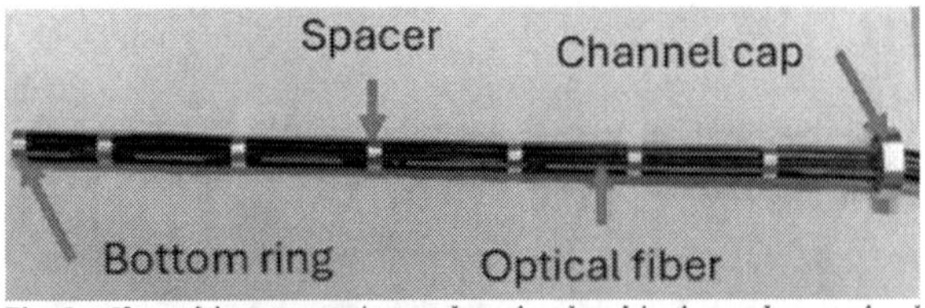

Fig. 3. Channel instrumentation ready to be placed in the north control rod guide tube. Six aluminum spacer ensure the sample alignment and straightness of the optical fibers

The instrumentation consists of an aluminum cap that fits onto the top of the control rod guide tube and a 20 mm long aluminum ring positioned at the core mid-height. The outer diameter of the ring matches the inner diameter of the guide tube, while its inner diameter is precisely manufactured to 12.2 mm to accommodate the sample holders. The ring is partially drilled with 12 holes: nine holes with a diameter of 2.2 mm accommodate miniature scintillators and their fibers, and three holes with a diameter of 2 mm serve as attachment points for three titanium structural rods. To enhance sample guidance, the 2-mm diameter holes protrude slightly toward the center of the ring, as shown in Fig. 4.

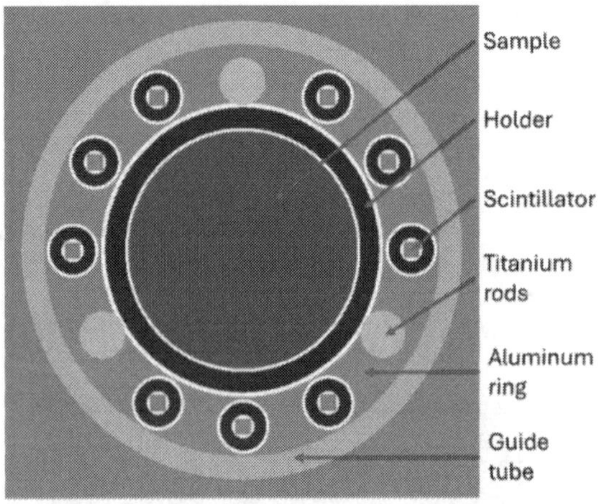

Fig. 4. Axial cross section of the instrumented channel at the reactor core mid-height, with a sample inserted. At the center is a sample encased in the sample holder. The three titanium rods, necessary for mechanical integrity, are visible in light gray. The nine orange squares are scintillators encased in their aluminum cap.

In addition to the titanium rods, the mechanical stability of the channel and the alignment of the fibers are guaranteed by six aluminum spacers, each 10 mm in length and visible in Fig. 3, into which the structural rods are embedded. The channel cap consists of two parts: a fixed component containing holes for the fibers, and a removable central plug that ensures the alignment of the sample holder and is unscrewed during sample exchange.

D. Analysis Methodology

The reactor time-dependent reactivity is calculated using inverse point kinetic, as was employed in the MAESTRO program [10,24,25]. The time-dependent reactivity $\rho_\$(t)$ is obtained using (1) where $S_d(t)$, the delayed neutron source normalized to the reactor's normalized global signal $S_G(t)$, is computed using (2).

$$\rho_\$(t) = 1 - \frac{\Lambda\, S_d(t)}{\beta\, S_G(t)} \tag{1}$$

$$\begin{cases} S_d(t) = \sum_i \lambda_i c_i(t) \\ \dfrac{dc_i(t)}{dt} = \dfrac{\beta_i}{\Lambda} S_G(t) - \lambda_i c_i(t) \end{cases} \tag{2}$$

With Λ the generation time, β the total delayed neutron fraction, c_i the i-th delayed neutron group normalized precursor concentrations, and the λ_i and β_i the respective decay constants and delayed neutron fractions.

The pseudo-square oscillation of the sample produces a corresponding pseudo-square signal in the time-dependent reactivity [9]. As the Fourier series representation of a square wave with amplitude 1 and period L is (3) [26].

$$S(x) = \frac{4}{\pi} \sum_{i=0}^{\infty} \frac{1}{2i+1} \sin\left(\frac{(2i+1)2\pi x}{L}\right) \tag{3}$$

The reactivity worth $\Delta\rho$ of the sample is measured as the full amplitude of the reactivity signal resulting from the pseudo-square oscillation. Its value is obtained by extracting the first harmonic of the time-dependent reactivity's Fourier transform $\tilde{\rho}$, as shown in (4).

$$\Delta\rho = \frac{\pi}{2} \|\tilde{\rho}(f_0)\| \tag{4}$$

Both the global and local signals, $S_G(t)$ and $S_L(t)$, are normalized by dividing each by its moving average over a window of length equal to their period T, similarly to the processing performed for during CORTEX [16]. The local perturbation $L(t)$ is then directly computed as the subtracting the global signal from the local signal [1,9,10].

$$L(t) = S_L(t) - S_G(t) \tag{5}$$

The quantities of interest – the global perturbation ΔG and the local perturbation ΔL – are defined as the amplitude of the first harmonic of $S_G(t)$ and $L(t)$ respectively. To estimate the uncertainties associated with these quantities and $\Delta\rho$, calculations are performed not on the full signal but on individual segments, each of two periods in length. The resulting set of estimates is then bootstrapped to obtain both the mean value of each quantity through the periods and the associated standard deviation [27].

Validation of the applicability of the point-kinetic assumption is performed by measuring the spectral power ratio (*SPR*) between each SAFFRON detector and one of its furthest detectors. The *SPR*, defined in (6) as the ratio between two detectors' normalized signals power, P_i and P_{ref}, is computed using the method (7) developed for the CORTEX program [15,16]. It allows comparing the relative noise distribution in a set of detectors. This method relies on cross-power spectral densities (CPSDs) with intermediate detectors j, yielding estimates $P_{i,j}$, rather than direct pairwise computations. This approach produces a set of estimates from which the variance can also be evaluated. In the absence of spatial effects, all normalized signals are expected to exhibit the same power. Consequently, the spectral power ratio (SPR) would be equal to one.

$$SPR_{i,ref} = \frac{\sqrt{P_i}}{\sqrt{P_{\text{ref}}}} \tag{6}$$

$$SPR_{i,ref} = \frac{\sqrt{P_{i,i}}}{\sqrt{P_{\text{ref,ref}}}} = \frac{1}{n}\sum_{j}^{n} \frac{\sqrt{P_{i,j}}}{\sqrt{P_{j,\text{ref}}}} \tag{7}$$

Instead of computing a simple average, weights can be applied in (7) to account difference in detector performance. In BLOOM, the chosen weighting scheme uses the total number of counts recorded by the intermediate detector j over the full experiment as its weight in the calculation of the average and standard deviation.

III. Results And Discussions

A. Point kinetic assumption verification

Verification of the absence of spatial effects in the core is carried out using signals measured during standard pile-oscillation experiments. The displacement follows a pseudo-square oscillation pattern, with a dwell time of 30 seconds at both the inserted and extracted positions, and a transition time of 2.4 seconds. In the inserted position, the center of the sample aligns with the mid-height of the core, placing it on the same horizontal plane as the SAFFRON L2 detectors and the scintillation detectors of the instrumented channel. The oscillation sequence consists of 84 periods, resulting in a total duration of slightly over an hour and a half. The reference detector for spectral power ratio calculations is the furthest mid-plane detector in SAFFRON, i.e., detector number 96 [13]. This choice is motivated by its distance from local flux perturbations caused by the sample, and by the higher count rate at the core mid-plane, which benefits from being the maximum of the cosine shaped axial flux distribution. The graph in Fig. 5 displays the results obtained with the largest gold sample used during the experimental campaign. The SPR values are plotted as a function of the distance from the perturbation and are grouped according to the respective detector plane.

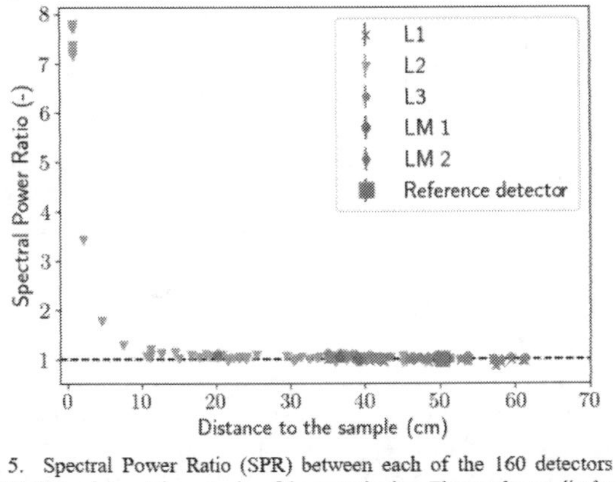

Fig. 5. Spectral Power Ratio (SPR) between each of the 160 detectors of SAFFRON and one at the opposite of the perturbation. The results are displayed as a function of the distance to the sample and grouped by the SAFFRON plane they are part of.

The results show that the local flux perturbation vanishes at a distance of 15 cm. Beyond this distance, all detectors have an SPR of 1 with respect to the reference detector, confirming that these detectors have the same modulation amplitude. It can therefore be concluded that, despite the local perturbation extending to 15 cm, the reactor continues to behave consistently accordingly to the point-kinetics model. This verification is essential for the experimental program, as variations in the spectral power ratios would indicate that the observed changes in detector count rates are partially caused by spatial redistribution of the neutron flux. In such a case, correction factors would be required to accurately retrieve the reactivity and global effect values. From these results it can be concluded that all detectors farther than 15 cm from the sample are suitable for the estimation of global perturbation.

B. Gold Sample Reactivity worth Measurements

The reactivity worth of the gold sample was measured five times between September 2024 and April 2025. The oscillation protocol was identical to that used for verifying the point-kinetics assumption: 84 pseudo-square periods, with a 30-second dwell time at each position and 2.4 seconds for each transition. Between each experiment, the sample holder was removed from the experimental channel, and the sample from the holder, while the oscillator remained untouched. Gold was chosen to perform repetitions because it is a standard in neutron activation dosimetry, and is also typically used to perform for power calibrations [20]. This makes it suitable for validating the implemented analysis methodology.

For each experiment, the time-dependent reactivity of the reactor is computed from the acquired signal using (1), the reactivity worth is then estimated using (5). The obtained values are shown in Fig. 6, alongside the reactivity worth computed through Monte Carlo simulation with the Serpent2 [28] neutron transport code and the JEFF-3.3 [29] nuclear data library.

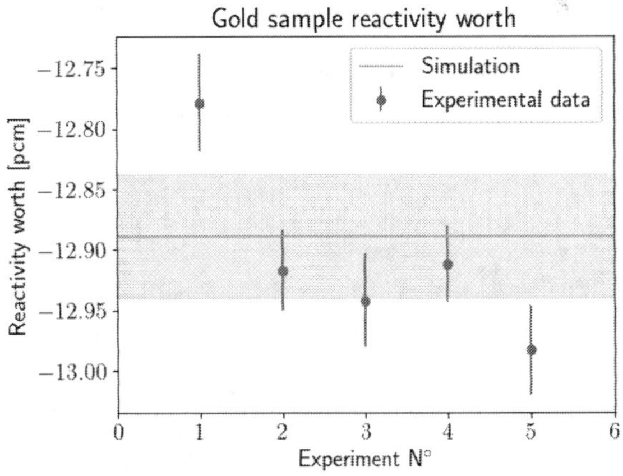

Fig. 6. Measured reactivity worth of a gold sample during five different experiments, performed between fall 2024 and spring 2025, and computed value with the Serpent2 MC code. All displayed uncertainties are 1 standard deviation.

The estimated experimental uncertainties range from 0.04 pcm to 0.05 pcm, corresponding to relative uncertainties

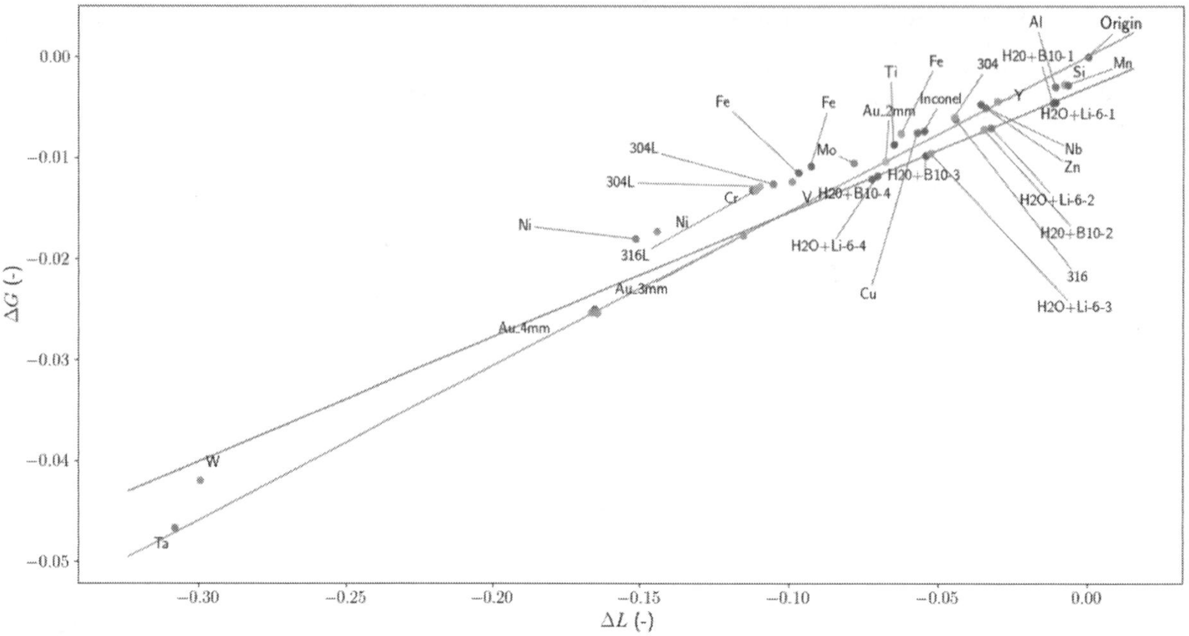

Fig. 7. Measured amplitude of the global signal ΔG vs. measured amplitude of the local perturbation ΔL for 40 experiments of the BLOOM program. Two linear regressions are also displayed: one for the measurement of gold samples with different diameters and one for the water samples with different neutron absorber concentrations.

on the order of 0.3%, as expected from the setup updates performed following the experimental design [1]. These variations are attributed to the differences in the average reactor power during the oscillations, which ranges from 800 mW to 950 mW, and to whether the power was stable or required correction for the drift in the power during the experiments due to a non-zero average of the reactivity during the oscillation. Although it is observed that the reactivity worth obtained during the measurement 1 may be an outlier, lying at more than two sigma from the experiments 3 and 5, the distance is not sufficient to rule the possibility of all of it being statistical. Moreover, the result from experiment 1 is supported by the simulations results, with which it agrees within two standard deviations. As shown in Fig. 6, all measurements are within two standard deviation agreements with the value obtained through Monte Carlo simulations, which has an uncertainty of the same order of magnitude as the experiments. This agreement support confidence in the implementation of the analysis methodology, which will later be applied to all samples in the experimental campaign.

C. *Global and local effects*

The BLOOM pile-oscillation program currently comprises a total of 45 oscillation experiments conducted on 40 different samples (25 different materials). All experiments follow the same protocol – 84 periods of pseudo-square pattern, with 30 seconds dwell times in each position and transition times of 2.4 seconds – defined to reach the targeted factor ten between measurement and nuclear data uncertainties for the iron sample [1]. The period averaged reactor power during oscillation was maintained constant at values varying between 800 mW and 950 mW.

Forty of the results are already analyzed and are shown in Fig. 7, using a layout commonly found in the literature and especially the MAESTRO program, as the amplitude of the global signal ΔG against the amplitude of the local perturbation ΔL [9]. It is observed that the measurements globally align along a single axis in the graph, showing a qualitative proportionality between the decreases in ΔG the ΔL. This result was expected, as most samples act as additional thermal neutron absorbers in the core, inducing similar behaviors in core. The orientation of the axis itself is explained by the sensitivity of the detectors, used to measure the local perturbation, to thermal neutrons. The use of fast neutron sensitive detectors would yield different results. In terms of uncertainty, the global perturbation ΔG is consistent across the experiments, with values ranging between 8×10^{-5} and 1×10^{-4}. The estimated uncertainties show no significant dependence to the mean values, resulting in relative uncertainties ranging from 0.2% to 2%, depending on the sample's reactivity worth. Similarly, the uncertainty on the local perturbation ΔL shows no dependence on the mean value either. It ranges from the 1.4×10^{-4} to 1.8×10^{-4}, yielding large range of relative uncertainties that spans from 0.07% to 2.5% depending on the sample.

While the current measurement precision is sufficient for the study of stainless-steel main elements, with especially an experimental uncertainty on iron samples ten times smaller than the nuclear data induced one, as targeted [1]. Several options are available to reduce the uncertainties on the measurements of the global effects and reactivity worth of for low-effect samples. The simplest approach would be to increase the number of periods. However, this would reduce the uncertainty by a factor two at most, since reactor operations lasting more than six hours are unlikely and the uncertainties scales with the inverse of the square root of the number of fission, and thus of the oscillation time [5]. The second option is to increase the mass of the sample to increase the amplitudes of the perturbations, as the uncertainty is observed to be independent of the amplitude. However, the limiting factor in this case is the size of the channel, which is already radially filled by most samples. Increasing the sample length is possible, but it would only affect the amplitude of the global perturbation. The last option is to increase the reactor power during pile-oscillations. Since uncertainty scales with inverse of the square root of the power, a factor 10 is theoretically achievable in the CROCUS reactor. However, the current limitation is the operational range of the detectors, which restricts the reactor power to 1 W. Operating the reactor at higher power in the future is feasible, but new instrumental developments and implementations are required.

Two linear regressions are also shown in Fig. 7. These are reproductions of two of the axes observed in the MAESTRO program, but obtained in the CROCUS reactor. The different linear trends observed in this type of graphs are hypothesized to be axes along which a macroscopic cross-section evolves. In this case it would be thermal neutron capture [9,30]. The first regression is the axis of the capture in water, and includes four ^{10}B in solutions (0.35 g/L, 0.69 g/L, 1.04 g/L, and 1.39 g/L) and four ^{6}Li solutions (0.82 g/L, 1.64 g/L, 2.46 g/L, and 3.28 g/L, each with 5% HNO_3). As shown in Fig. 8, the CROCUS measurements reproduce the expected linear trend, with a regression coefficient R^2 of 0.9993.

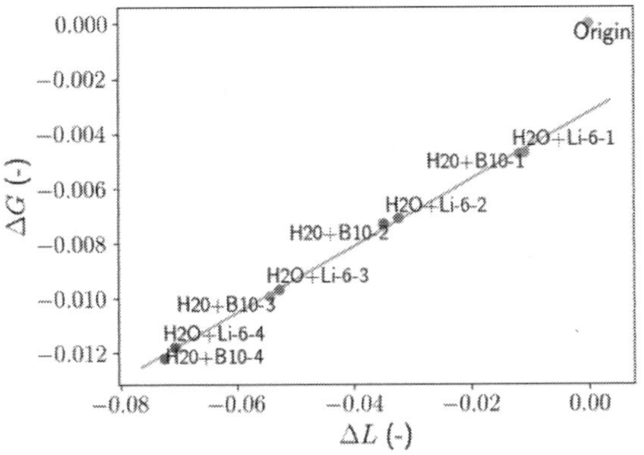

Fig. 8. Measured amplitude of the global signal ΔG vs. measured amplitude of the local perturbation ΔL for water samples of same volumes but with different absorber concentrations.

The second regression is based on the measurements using the gold samples. Three samples were measured, all of them long of 100 mm but with different radii of 4 mm, 3 mm, and 2 mm, respectively. The results are shown in Fig. 9. As with the water samples, the linearity is reproduced, with a regression coefficient R^2 of 0.9997. When extrapolated, the regression line passes near the origin of the coordinate system (i.e. unperturbed state), origin which was not part of the fitted data. In fact, it passes at a distance consistent with the previously mentioned uncertainties. This result is expected, as the linear trend in that direction corresponds to a decrease in mass up to an empty sample, i.e. sample with null radius, or, equivalently, to the total macroscopic cross-section moving toward zero. In contrast, for the water samples, only the capture cross-section changes significantly, so the regression line does not pass through the origin. These gold samples results provide additional confidence in the quality of the measurements performed during the BLOOM program.

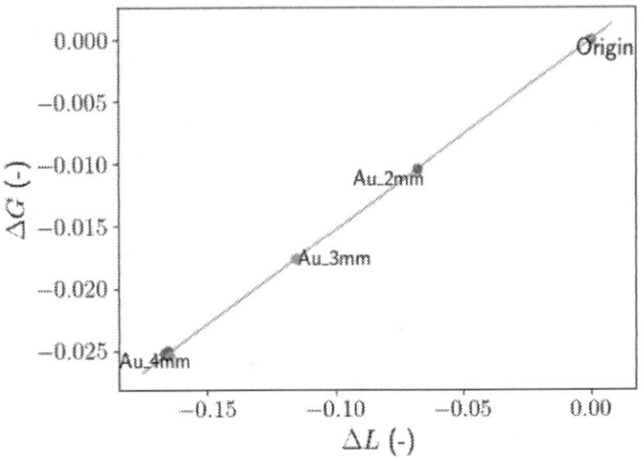

Fig. 9. Measured amplitude of the global signal ΔG vs. measured amplitude of the local perturbation ΔL for gold sample of different diameters. The origin is not included in the fitted data of the linear regression.

As mentioned previously (Sec. III.B), five repetitions were performed for the measurement of the gold sample with a diameter of 4 mm. Out of the five, only four measurements were used to perform the fit. The exclusion of the fifth point (experiment 2 in Fig. 6) is due to a difference in the sample holder used in this experiment compared to the others. In that experiment, the aluminum sample holder had an outer diameter of 12 mm, identical to the others, but an inner diameter 9.9 mm instead of the 10.3 mm used for the other samples. Thus, the holder was 0.2 mm thicker on both sides, which may affect the local perturbation ΔL measurements given the closeness of the local detectors. A close-up of these measurements, shown in Fig. 10, confirms this hypothesis. While the four measurements on the right of the graph form a consistent group, the measurement with the thicker holder (on the left) is at nine standard deviations from the average of the other four in the estimation of the local perturbation. These results demonstrate that the system is sensitive enough to detect such small differences in the measurement of the local perturbation ΔL, which, on the other hand, are not observable in the global effect ΔG.

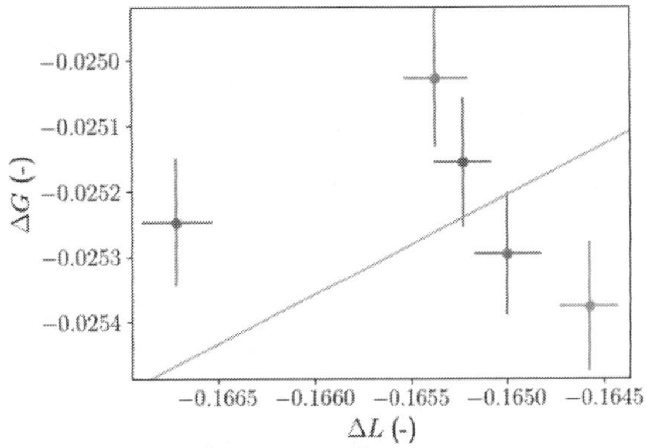

Fig. 10. Close-up on the repetitions of the measurement of the gold sample with a diameter of 4 mm. The measurement on the left side was performed with an aluminum holder thicker by 0.2 mm on both sides than the other four.

A pattern is observed in the distribution of the four measurements on the right side of in Fig. 10. All four are within two standard deviations from their average and are thus considered consistent with each other, but they also appear to be diagonally aligned. Although not yet studied, it is hypothesized that a negative correlation exists between the measurement uncertainties on the global and local effects. This may be expected, as the local perturbation over time, $L(t)$, is calculated using (5), which includes a subtraction of the global signal $S_G(t)$. These potential correlations for the gold samples, but also for all samples of the BLOOM program, will be investigated in future studies.

IV. Conclusion

The BLOOM pile-oscillation program was successfully conducted in the CROCUS reactor between fall 2024 and spring 2025. During the program, 45 oscillation experiments were carried out using 40 different samples and 25 materials, totaling over 200 hours of in-core experimentation.

As a first step, a gold cylinder with a reactivity worth 12.9 pcm, used as a reference, was oscillated and used to verify that the oscillation location was suitable for the program by checking that the flux distribution is only locally perturbed. It was observed that detectors located within 15 cm of the sample had perturbed count rates, while those farther away behaved as predicted by the point-kinetic model.

To validate the experimental and analysis methodologies, five repetitions of the gold sample oscillation experiment were performed to assess the consistency of the obtained reactivity worth. The results obtained show good agreement between themselves and with the value predicted by Monte Carlo simulations. Therefore, the implementation of the analysis methodology was validated and its robustness demonstrated.

The amplitudes of the local and global perturbations were successfully measured for each sample. The experimental uncertainties were found to show no dependence on the magnitude of the perturbations, with typically an uncertainty on the reactivity worth of 0.04 pcm. As a result, the relative uncertainty on local perturbations ranged from 2.5 % for the lightest sample to 0.07 % for the heaviest. Respectively, the uncertainty on global perturbations ranged from 2% to 0.2%. The experimental results reproduced the linear trends observed in the MAESTRO program for varying gold masses and neutron absorber concentration in water. The repetitions of the reference gold sample measurement showed that the experimental setup is sensitive enough to significantly measure the difference in local perturbation caused by a 0.2 mm variation in the thickness of the aluminum sample holder.

Next steps for the BLOOM pile-oscillation program in CROCUS include the addition of new reference samples such as D_2O. It also includes new instrumentation developments to be able to conduct new oscillations at higher power with the aim of decreasing the uncertainties on the measurements of low-reactivity samples. All results are to be included in a forthcoming NEA benchmark. Finally, new samples beyond structural materials are prospected for future pile-oscillation experiments in CROCUS.

Authors Contributions

Thomas Ligonnet designed and carried out the experiments, as well as performed the analysis and the redaction of this paper, as part of his PhD thesis.

Axel Laureau supervises Thomas Ligonnet's PhD thesis. He provided valuable feedback and help during the analysis.

Oskari Pakari is the CROCUS reactor installation chief. He contributed significantly to carrying out the experiments.

Daniel Clément participated in the design of the experimental channel and performed the mechanical production work.

Alexis Dupont-Bembinoff participated actively in the preparation of the experiments during the campaign.

Gilles Noguère provided feedback along the experimental program and enabled us to use samples from the MAESTRO program.

Andrea Pautz is Thomas Ligonnet's thesis director and the head of the LRS at EPFL.

Vincent Lamirand supervises Thomas Ligonnet's PhD thesis as co-director, and is in charge of LRS experimental research as Program manager. He is at the genesis of the experimental program and participated in the whole program.

Acknowledgment

The authors wish to thank Nicolas Weiss for the development of the pile-oscillator and Laurent Braun for his help with the electronics.

The authors also wish to thank Pierre Leconte and Benoît Geslot for their insight on the subject, and all people from CEA who took some of their time to enable us to use some of the MEASTRO program samples.

References

[1] T. Ligonnet et al., "Design of an Open-Loop Pile-Oscillation Program in the CROCUS Reactor," IEEE Transactions on Nuclear Science **71** 5, 1001 (2024); https://doi.org/10.1109/TNS.2024.3371183.

[2] V. Lamirand, "Ten springs of experiments in CROCUS," EPJ Web Conf. **288**, 04026, EDP Sciences (2023); https://doi.org/10.1051/epjconf/202328804026.

[3] V. Lamirand et al., "An Experimental Programme optimized with Uncertainty Propagation: PETALE in the CROCUS Reactor," EPJ Web Conf. **211**, O. Serot and A. Chebboubi, Eds., 03003 (2019); https://doi.org/10.1051/epjconf/201921103003.

[4] A. Laureau et al., "Uncertainty propagation for the design study of the PETALE experimental programme in the CROCUS reactor," EPJ Nuclear Sci. Technol. **6**, 9, EDP Sciences (2020); https://doi.org/10.1051/epjn/2020004.

[5] W. K. Foell, "Small-sample reactivity measurements in nuclear reactors," TID-26511, American Nuclear Society, Hinsdale, IL (1972).

[6] B. A. Baker, "Comparison of open loop and closed loop reactivity measurement techniques on the ISU-AGN-201 reactor," Ph.D., Idaho State University (2013).

[7] Y. Jiang et al., "Review of kinetic modulation experiments in low power nuclear reactors," EPJ Nuclear Sci. Technol. **6**, 55, EDP Sciences (2020); https://doi.org/10.1051/epjn/2020017.

[8] D. Bernard, "Validation of actinides nuclear cross-section using pile-oscillation experiments performed at MINERVE facility," Journal of the Korean Physical Society **59** 2, 1119, Korea, Republic of (2011).

[9] B. Geslot et al., "Innovative Hybrid Pile Oscillator Technique in the Minerve Reactor: Open Loop Versus Closed Loop," IEEE Transactions on Nuclear Science **65** 11, 2767 (2018); https://doi.org/10.1109/TNS.2018.2874690.

[10] B. Geslot et al., "A hybrid pile oscillator experiment in the Minerve reactor," Annals of Nuclear Energy **108**, 268 (2017); https://doi.org/10.1016/j.anucene.2017.04.036.

[11] V. Lamirand et al., "Analysis of the First COLIBRI Fuel Rods Oscillation Campaign in the CROCUS Reactor for the European Project CORTEX," EPJ Web Conf. **247**, 21010, EDP Sciences (2021); https://doi.org/10.1051/epjconf/202124721010.

[12] Y. Jiang et al., "PISTIL, a reactivity modulation device to probe the transfer function of the nuclear reactor CROCUS," EPJ Web Conf. **253**, A. Lyoussi et al., Eds., 04007 (2021); https://doi.org/10.1051/epjconf/202125304007.

[13] F. Vitullo, "Miniature and Minimalistic Neutron Detectors for Online High-Resolution Experiments in the Zero-Power Reactor CROCUS," PhD thesis, EPFL (2022); https://doi.org/10.5075/epfl-thesis-9492.

[14] F. Vitullo et al., "Design of a 150-miniature detectors 3D core-mapping system for the CROCUS reactor," EPJ Web Conf. **253**, 04023, EDP Sciences (2021); https://doi.org/10.1051/epjconf/202125304023.

[15] V. Lamirand et al., "CORTEX experiments – Part I: Modulation campaigns in AKR-2 & CROCUS for the validation of neutron noise codes," Annals of Nuclear Energy **211**, 110928 (2025); https://doi.org/10.1016/j.anucene.2024.110928.

[16] K. Ambrožič et al., "CORTEX experiments, Part II: Postprocessing of neutron noise time series to determine reliable mean and uncertainties," Annals of Nuclear Energy **208**, 110704 (2024); https://doi.org/10.1016/j.anucene.2024.110704.

[17] G. Noguere et al., "Interpretation of pile-oscillation measurements by the integral data assimilation technique," Nuclear Instruments and Methods in Physics Research Section A: Accelerators, Spectrometers, Detectors and Associated Equipment **629** 1, 288 (2011); https://doi.org/10.1016/j.nima.2010.11.034.

[18] A. Laureau et al., "Bayesian Monte Carlo assimilation for the PETALE experimental programme using inter-dosimeter correlation," EPJ Web Conf. **239**, 18004, EDP Sciences (2020); https://doi.org/10.1051/epjconf/202023918004.

[19] U. Kasemeyer et al., "Benchmark on Kinetic Parameters in the CROCUS Reactor," 978-92-64-99020–3, Nuclear Energy Agency of the OECD (NEA), p. 93 (2007).

[20] V. Lamirand et al., Eds., "Power calibration methodology at the CROCUS reactor," Advancements In Nuclear Instrumentation Measurement Methods And Their Applications (Animma 2019), E D P SCIENCES, Cedex A (2020); https://doi.org/10.1051/epjconf/202022504022.

[21] O. Pakari et al., "Kinetic Parameter Measurements in the CROCUS Reactor Using Current Mode Instrumentation," IEEE Transactions on Nuclear Science **65** 9, 2456 (2018); https://doi.org/10.1109/TNS.2018.2831180.

[22] F. Vitullo et al., "A mm3 Fiber-Coupled Scintillator for In-Core Thermal Neutron Detection in CROCUS," IEEE Transactions on Nuclear Science **67** 4, 625 (2020); https://doi.org/10.1109/TNS.2020.2977530.

[23] T. Ligonnet et al., "POLLEN: A Pile-Oscillator for the BLOOM Experimental Program," EPJ Web Conf. **288**, 04018, EDP Sciences (2023); https://doi.org/10.1051/epjconf/202328804018.

[24] B. Geslot, C. Jammes, and B. Gall, "Influence of the delayed neutron group parameters on reactivity estimation by rod drop analysis," Annals of Nuclear Energy **34** 8, 652 (2007); https://doi.org/10.1016/j.anucene.2007.03.005.

[25] S. TAMURA, "Signal Fluctuation and Neutron Source in Inverse Kinetics Method for Reactivity Measurement in the Sub-critical Domain," Journal of Nuclear Science and Technology **40** 3, 153, Taylor & Francis (2003); https://doi.org/10.1080/18811248.2003.9715345.

[26] E. W. Weisstein, "Fourier Series--Square Wave;" https://mathworld.wolfram.com/FourierSeriesSquareWave.html; (current as of May 2, 2025).

[27] B. Efron and R. J. Tibshirani, *An Introduction to the Bootstrap*, Chapman and Hall/CRC, New York (1994); https://doi.org/10.1201/9780429246593.

[28] J. Leppänen et al., "The Serpent Monte Carlo code: Status, development and applications in 2013," Annals of Nuclear Energy **82**, 142 (2015); https://doi.org/10.1016/j.anucene.2014.08.024.

[29] A. J. M. Plompen et al., "The joint evaluated fission and fusion nuclear data library, JEFF-3.3," Eur. Phys. J. A **56** 7, 181 (2020); https://doi.org/10.1140/epja/s10050-020-00141-9.

[30] Y. Danon, "Innovative experiments for reduction of nuclear data uncertainty," EPJ Nuclear Sci. Technol. **4**, 22, EDP Sciences (2018); https://doi.org/10.1051/epjn/2018017.

Silicon carbide neutron detectors for reactor monitoring in subcritical conditions

Vladimir Radulović[1*], Takahiro Makino[2], and Danilo Bisiach[3]
[1]Jožef Stefan Institute, Slovenia
[2]National Institutes for Quantum Science and Technology, Japan
[3]Instrumentation Technologies, d.o.o., Slovenia
(*) vladimir.radulovic@ijs.si

***Abstract*— Silicon carbide (SiC) semiconductor detectors offer key advantages for radiation detection, including a wide bandgap, high thermal conductivity and radiation hardness, making them ideal for applications in harsh environments. Engineering and optimization of dedicated neutron converter components can enable achieving high detection sensitivities, comparable to BF_3 or ^{3}He detectors of the same physical dimensions. This study explores the use of wide-area SiC Schottky Barrier Diodes (SBDs) for neutron flux measurements in a nuclear reactor in subcritical conditions. Two detectors, fabricated at the National Institutes for Quantum Science and Technology in Japan and equipped with ^{6}LiF converter layers, were tested at the Jožef Stefan Institute TRIGA reactor in Slovenia using the Libera MONACO 3 system, developed at the French Atomic and Alternative Energies Commission and manufactured by the Instrumentation Technologies company in Slovenia. Results showed excellent performance, in comparison to a Centronic FC165 fission chamber, routinely used for this purpose. SiC detectors and standalone neutron converter components also have a high educational value, allowing for visual and hands-on demonstrations of neutron detection aspects and phenomena. Such activities are implemented in the framework of the European Nuclear Experimental Educational Platform (ENEEP).**

***Keywords* —Silicon Carbide, Lithium-6, Neutron Detector, Nuclear Reactor, Criticality Measurements.**

I. Introduction

SILICON carbide (SiC) semiconductor radiation detectors are being researched in the field of radiation detection as a promising technology offering superior performance characteristics that address many limitations of traditional detector materials, thanks to inherent advantages of SiC. These are, for radiation detection in particular, its wide bandgap, high thermal conductivity and exceptional radiation hardness, which make SiC highly suitable for use in harsh environments. Several excellent review papers on the topic of radiation detection using SiC detectors have been published [1-5]. The development and deployment of SiC-based neutron detectors is opening new avenues in various applications, ranging from nuclear reactor monitoring, neutron detection in specific applications [6], as well as applications in homeland security [7].

Furthermore, SiC neutron detectors are finding applications in space exploration and high-energy physics experiments. Their ability to withstand extreme conditions such as high radiation levels and wide temperature variations makes them suitable for use in space missions, where reliable radiation detection is critical for both mission success and the protection of onboard electronics and human crew members. In high-energy physics, these detectors contribute to the study of fundamental particles and forces, providing precise measurements that are essential for advancing our understanding of the universe. As research and development continue to advance, these detectors are expected to play an increasingly vital role in ensuring safety, security, and scientific discovery.

This paper presents research into the application of SiC detectors for measurements of the neutron flux in a nuclear reactor in subcritical conditions, relevant for neutron measurements / monitoring of subcritical and zero power reactor facilities or spent fuel facilities. A first series of measurements was performed during a controlled approach to criticality to demonstrate the capability of such detectors to monitor criticality conditions. A second series of measurements was performed in which the axial neutron flux distribution was measured in steady-state subcritical conditions.

The application of SiC detectors for neutron measurements is also highly relevant in nuclear education, as was implemented at the JSI in the framework of hands-on education activities organized by the European Nuclear Experimental Educational Platform (ENEEP) [8].

II. Experimental

A. SiC Schottky barrier diodes

The detectors used in the present work were SiC-SBDs fabricated on 36 μm thick n-type 4H-SiC epitaxial layer grown on an n-type 4H-SiC substrate by Chemical Vapor Deposition (CVD) [9]. The donor concentration in the epitaxial layers were 1.7×10^{14}. Schottky contacts with 4 mm x 4 mm and and 5 mm × 5 mm were formed on the epitaxial layers using the deposition of Ni. For the ohmic contact on the backside, Ni deposition and sintering at 950 °C for 10 min in Ar were carried out. The contact quality was verified with I-V measurements. The detectors were surface-mounted onto chip carriers with wire bonded electrical contacts. The chip carriers were enclosed in 3D-printed plastic holders in order to isolate the electrical contacts under high voltage.

B. Neutron converter

A neutron converting layer based on ^{6}LiF was deposited onto the front contact of the SiC diodes. The target thickness for the ^{6}LiF deposit was 20 μm, based on our previous work [10], in order to maximize the detection efficiency for measurements in a low neutron flux corresponding to a subcritical configuration of the reactor. The converting layer was realized by dispersing 27.43 mg of ^{6}LiF powder, obtained from the Oak Ridge National Laboratory (USA) in 0.4821 g of ethanol (^{6}LiF mass fraction of approximately 5.4 %). Approximately one half of the ^{6}LiF material did not disperse and was present as precipitate. One drop of the liquid dispersion was deposited onto the diode surface, the deposit mass on the 5 mm × 5 mm diode was 14.6 mg, the deposit mass on the 4 mm × 4 mm diode was 24.8 mg. The estimated effective thickness of the deposit, assuming only half of the ^{6}LiF material dispersed, was 6 μm for the 5 mm × 5 mm diode and 17 μm for the 4 mm × 4 mm diode. Fig. 1 displays a close-up of the deposition of the converter solution onto the SiC diode surface.

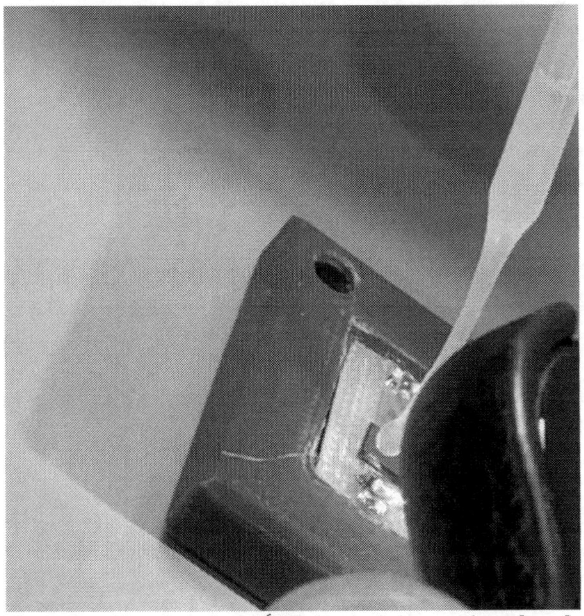

Fig. 1. Deposition of one drop of ^{6}LiF ethanol solution onto the surface of a SiC diode.

C. Measurement setup

The SiC diodes in their 3D-printed holders, equipped with the ^{6}LiF converting layers were mounted to a 3D-printed support, designed to prevent contact between the diodes and the interior surface of the irradiation channel used (and resulting damage to the diode). The detector assembly is displayed in Fig. 2. Additionally, a fission chamber (Centronic, model FC165 with a ^{235}U fissile deposit) was used in this work for comparison, also displayed in Fig. 2.

For the measurements during the approach to criticality, bottom spacers were present in both Triangular Irradiation Channels 1 and 2, designed to locate typical irradiation capsules in the core centre-line. The spacers were 238 mm in height and constructed from a top and bottom plate, shaped to fit the Triangular Irradiation Channel cross-section, joined by three aluminium tubes. A detailed description of the Triangular Irradiation Channels is given in [11]. The SiC detector assembly was lowered from the reactor platform level and rested on the spacer. For the axial profile measurements the spacer in TIC2 was removed, the SiC detector assembly was lowered to the bottom of the TIC2 and progressively raised by means of a steel cable. The assembly height was measured using a pneumatic system developed for accurate positioning of miniature fission chambers in the reactor core, used in [12] and other previous work. The system features a wheel with a rubber tread in contact with the cable, coupled to a rotary encoder. Fig. 3 displays a schematic view of the JSI TRIGA reactor core configuration and a drawing of the Triangular Irradiation Channel cross-section.

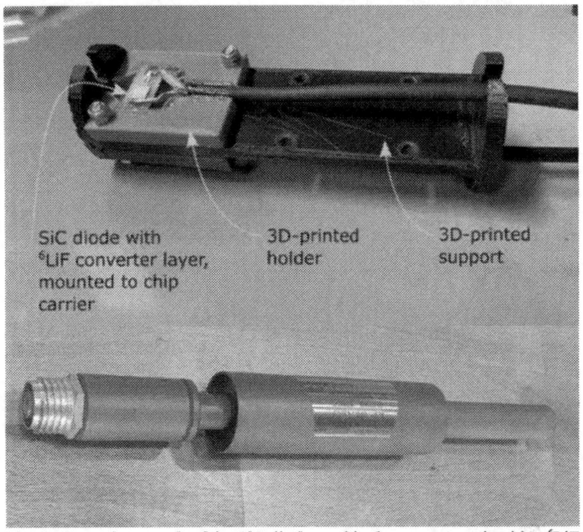

Fig. 2. Top: photograph of the SiC diode used in the present work with a ^{6}LiF converter layer, mounted to a chip carrier, enclosed in a 3D printed holder and support. Bottom: photograph of the FC165 fission chamber also used in the present work.

Pulse-height spectra resulting from the energy deposition of charge particles originating from ^{6}Li(n,t)^{4}He reactions in the coverter layer, and interacting in the active 4H-SiC material were acquired using the Libera Monaco 3 system, developed by the French Atomic and Alternative Energies Commission (CEA) [13] and industrialized by the Instrumentation Technologies company in Slovenia [14]. The system integrates high voltage, signal preamplification and processing stages, and incorporates three acquisition modes working in parallel: pulse mode, Cambpell (or variance) mode and current mode. The Monaco system was developed for acquisition of signals from fission chambers in nuclear reactors, but enables signal acquisition for different detector types as well, such as ionization chambers (current mode only) and self-powered neutron detectors (current mode only). In this work, its ability of signal acquisition from semiconductor detectors was tested with excellent results.

The JSI TRIGA reactor is equipped with four control rods, the Pulse rod (P), equipped with a pneumatic system used to eject the rod in pulse operation mode, the Safety rod (S), the

Compensating rod (C), and the regulating rod (R), equipped with an automatic system for operation in steady-state mode. The reactor core configuration including the locations of the control rods and the in-core irradiation channels is displayed in Fig. 3.

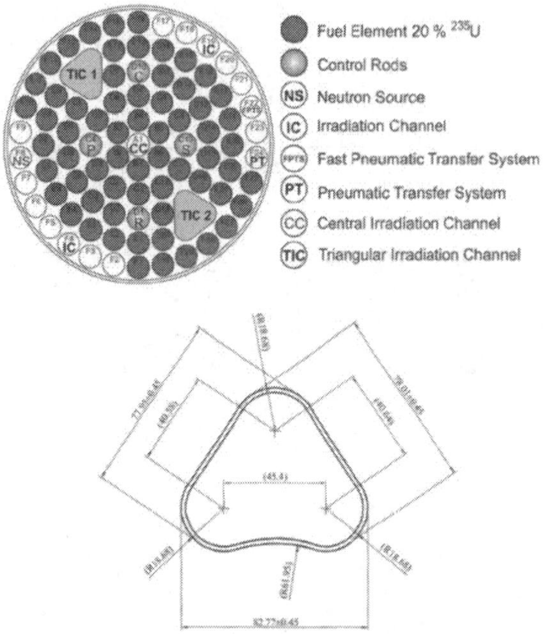

Fig. 3. Top: JSI TRIGA reactor core configuration (schematic). Bottom: cross-section of the Triangular Irradiation Channels (technical drawing).

D. *Measurements*

Measurements during the controlled approach to criticality were started with the reactor in a deeply subcritical state, with all rods inserted. Measurement steps were defined by progressive withdrawal of the control rods. Pulse height spectra from the SiC diode were recorded, displayed in Fig. 4. The spectra recorded from the SiC detectors, displayed in blue and green color in Fig. 4 appeared consistent for the measurement steps, as was the case for the spectra recorded from the FC165 fission chamber, displayed in red in Fig. 4. Detector count rates were determined by summation of the events in the shaded region of interest for the two SiC detectors as well as the FC165 fission chamber.

III. RESULTS AND DISCUSSION

Tables I and II report the experimental data obtained during the controlled approach to criticality. Table 1 reports the control rod positions and the cumulative inserted reactivity for each measurement step. Table II reports the measured count rates on the two SiC detectors as well as the FC165 fission chamber, and the computed (inverse) subcritical multiplication factors.

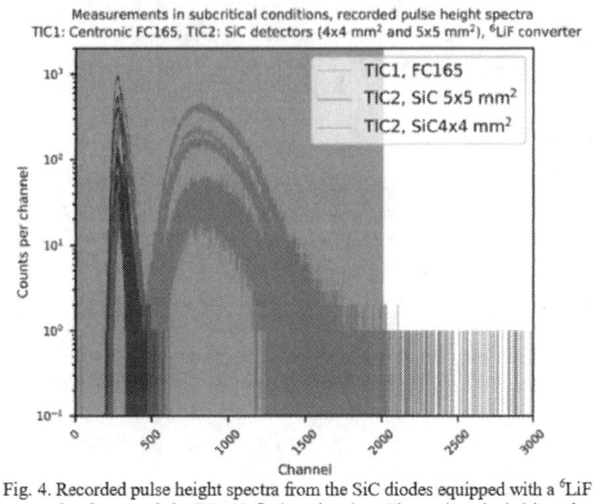

Fig. 4. Recorded pulse height spectra from the SiC diodes equipped with a ^{6}LiF converting layer and the FC165 fission chamber. The region shaded in color represents the signal considered in the present work, attributable to charged particles originating from neutron events. Spectra originating from the SiC detectors are displayed in blue and green, spectra originating from the FC165 fission chamber in red.

TABLE I
APPROACH TO CRITICALITY – CONTROL ROD POSITIONS AND CALCULATED INSERTED REACTIVITY FOR EACH MEASUREMENT STEP.

Meas. step (*i*)	T. rod position [step]	C. rod position [step]	S. rod position [step]	R. rod position [step]	Inserted reactivity [pcm]
0	901	900	900	901	0
1	451	900	900	901	1956.7
2	1	900	900	901	2962.2
3	1	900	450	901	5874.0
4	1	900	200	901	6740.3
5	1	700	200	901	7379.3
6	1	600	200	901	8007.2
7	1	500	200	901	8672.0
8	1	500	200	804	8845.5
9	1	500	200	751	9052.5
10	1	500	200	704	9275.1
11	1	500	200	675	9427.3
12	1	500	200	660	9510.1
13	1	500	200	642	9612.5
14	1	500	200	631	9676.6
15	1	500	200	620	9741.6
16	1	500	200	610	9801.3

The measurements performed give a good indication on the achievable detection sensitivity of SiC detectors. Observing the measured count rates in Table II, in particular during the last measurement steps, we see that they are of the same order of magnitude as the fission chamber count rates, despite the fact that there is a large difference in size in the active volume of the detector in favour of the fission chamber. This observation underscores the interest in researching SiC detectors for neutron applications, especially in case high detection efficiency is required.

TABLE II
APPROACH TO CRITICALITY – MEASURED COUNT RATES AND CALCULATED SUBCRITICAL MULTIPLICATION FACTORS FROM TE TWO SIC DETECTORS AND THE FC165 FISSION CHAMBER.

Meas. step (i)	SiC 5x5 mm² [cps]	SiC 5x5 mm² I_0/I_i	SiC 4x4 mm² [cps]	SiC 4x4 mm² I_0/I_i	FC [cps]	FC I_0/I_i
0	4.3	1	5.17	1	32.33	1
1	5.27	0.8165	6.56	0.7884	36.04	0.897
2	6.23	0.6914	7.59	0.6817	42.37	0.763
3	11.72	0.3672	13.59	0.3805	55.25	0.5851
4	14.47	0.2975	17.76	0.2912	65.52	0.4935
5	19.09	0.2256	23.25	0.2225	82.51	0.3919
6	25.27	0.1704	30.39	0.1702	110.17	0.2935
7	38.03	0.1132	47.18	0.1096	169.96	0.1902
8	44.65	0.0964	54.5	0.0949	188.71	0.1713
9	54.9	0.0784	65.11	0.0794	221	0.1463
10	70.12	0.0614	85.97	0.0602	269.31	0.1201
11	86.13	0.05	100.49	0.0515	322.44	0.1003
12	94.34	0.0456	111.32	0.0465	348.25	0.0928
13	114.31	0.0377	139.2	0.0371	416.88	0.0776
14	133.92	0.0321	161.85	0.032	480.22	0.0673
15	157.9	0.0273	190.29	0.0272	555.69	0.0582
16	185.75	0.0232	223.09	0.0232	654.14	0.0494

Fig. 5 displays the 1/M diagrams derived from the measurements performed with the two SiC diodes and the FC165 fission chamber. In subcritical conditions, the neutron population N in a nuclear reactor is proportional to the neutron source strength S (neutrons / s), in the present case a Ra-Be startup source, located in the core periphery, and the subcritical multiplication factor M:

$$N = SM = S\frac{1}{1-k_{eff}}, \quad (1)$$

where k_{eff} is the effective multiplication factor. If the measured signal I is proportional to the neutron population, the 1/M diagram is drawn by observing the ratio I_0/I_i, the indices 0 and i referring to the zero-th and i-th measurement step, as a function of the change made to the system, e.g. the progressive withdrawal of neutron absorbers [15]. The I_0/I_i ratios in Fig. 5 are displayed as a function of cumulative inserted reactivity, computed from the measured control rod reactivity worth tables. In theory, the ratio should depend linearly on the cumulative inserted reactivity, starting at the value 1 and approaching 0 as the reactor approaches criticality. As can be seen, the I_0/I_i diagrams from measurements using both SiC detectors are fully in line with expectations, whereas the I_0/I_i values from measurements using the fission chamber are more scattered initially, and appear to follow a linear dependence closer to criticality.

Similar measurements are performed routinely in the "Critical Experiment" exercise in the framework of education activities at the JSI TRIGA reactor. During the exercise, students / trainees observe the variation of the I_0/I_i ratio and make on-the-spot estimates on the amount of reactivity required to reach criticality. The indications obtained using the FC165 fission chamber in this work are qualitatively very similar to results from previous measurements, i.e. there are significant deviations from linearity far from criticality, and the resulting estimates on the amount of reactivity required to reach criticality are often unreliable. This can be explained by the position of the detectors used for the measurements. In principle, the closer a detector is located to the neutron source, the larger is the fraction of the signal originating from the neutron source and the lower is the sensitivity of the measurement setup to neutrons originating from the multiplication process in the fuel. Therefore, in deeply subcritical systems, more reliable indications on the criticality of a reactor are obtained by detectors located further away from the neutron source. Closer to criticality however, the neutron population increases significantly, and can become dominant over the contribution due to the neutron source. This is the reason why closer to criticality the reliability of the detector signals improves. Another important point to be considered during such measurements is the signal stabilization time. Equation (1) follows from a more general expression relating to successive neutron generations in the multiplication process:

$$N = S + k_{eff}S + k_{eff}^2 S + k_{eff}^3 S + \cdots, \quad (2)$$

which converges for $k_{eff} < 1$. Each neutron generation is separated in time by the mean generation time Λ, of the order of several 10 µs, depending on the core configuration [16]. In a deeply subcritical system the higher order terms vanish quickly, meaning a short signal stabilization time after each successive change in the system. However, close to criticality, higher-order terms do become important, and in order to obtain reliable measurements, sufficient time for signal stabilization must be ensured. Stabilization times of the order of minutes are commonly observed.

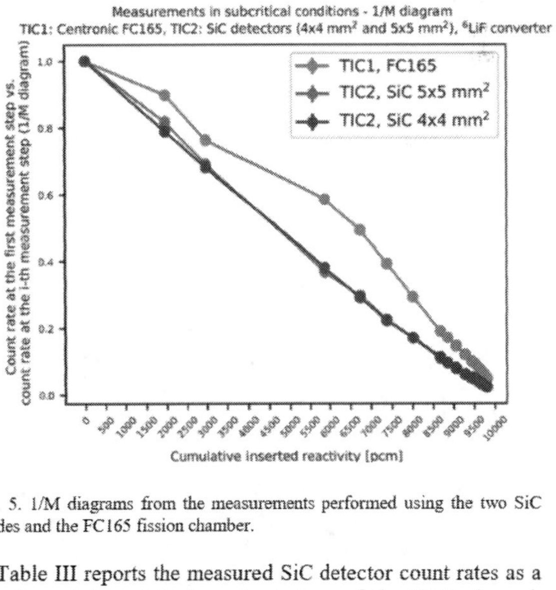

Fig. 5. 1/M diagrams from the measurements performed using the two SiC diodes and the FC165 fission chamber.

Table III reports the measured SiC detector count rates as a function of the height from the bottom of the TIC2 channel, Fig. 6 displays the measured axial distribution of the SiC detector signal in the TIC2 channel. Qualitatively, the observed distributions are consistent with previous measurements of fission rates using miniature fission chambers in collaboration with the French Atomic and Alternative Energies Commission (CEA), published as a benchmark case in the International Reactor Physics Benchmark Experiment Handbook [12]. The measured distributions display a central cosine-shaped maximum corresponding to the central fuel region, and tails corresponding to the top and bottom axial graphite reflectors.

TABLE III
MEASURED SIC DETECTOR COUNT RATES AS A FUNCTION OF HEIGHT FROM THE BOTTOM OF THE TIC2 CHANNEL.

Height above bottom in TIC2 [mm]	SiC 5×5 mm² signal [cps]	SiC 4×4 mm² signal [cps]
0	7.43	13.16
19.5	9.12	15.29
39.7	9.85	18.10
59.2	11.46	21.42
80.3	13.12	24.54
100.1	14.39	27.41
121.9	15.16	28.83
139.8	17.39	31.56
160.2	18.11	32.95
185.3	20.06	35.24
210	21.67	38.58
233.4	21.46	38.96
259.4	23.85	39.78
284.1	23.95	41.93
303.2	23.42	42.59
325.7	23.62	40.93
350.7	23.55	40.67
375.1	22.19	37.43
400.1	20.72	35.87
425.8	20.23	33.12
474.8	15.63	26.59
502.1	14.04	23.05
524.6	12.53	18.63
551.5	10.76	17.10
574.6	8.77	13.43
603.2	7.03	9.85

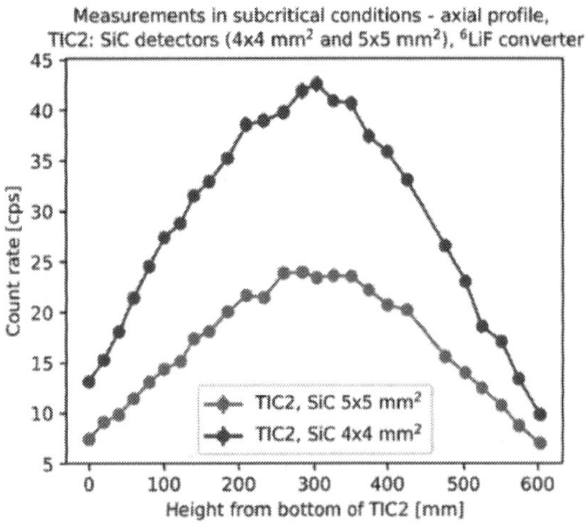

Fig. 6. Axial profiles measured using the two SiC diodes.

The measurements presented in this work clearly demonstrate the capability of SiC detectors to be used for measurements in a subcritical reactor, especially in the context of education activities, with several advantages. The detectors are less costly than typical neutron detectors, either fission chambers or gas detectors (^3He, BF_3) and do not introduce the need for handling radioactive materials as in the case of fission chambers. Additionally, they are of smaller dimensions compared to typical fission chambers or gas detectors, although miniaturized fission chambers are commonly being used. This enables achieving high spatial resolution, of the order of mm to cm, which is particularly important for neutron flux mapping experiments. Optical-fibre based neutron detectors equipped with a neutron-sensitive scintillating material, e.g. ZnS:^{6}LiF, as first developed by Mori et al. [17] and utilized e.g. in large numbers in a recent experiment at the CROCUS reactor at EPFL [18] are a comparable alternative with qualitatively similar characteristics.

For education purposes, SiC detectors equipped with standalone (removable) neutron converters introduce a series of possibilities, enabling observation and deeper understanding of a plurality of phenomena, e.g.

- measurement of I-V characteristics of SiC
- measurement of SiC detector response to α radiation, using e.g. ^{241}Am α sources
- demonstration / meaurement of the range of α radiation in air / vacuum, using a vacuum chamber and internal structures to locate the SiC detector and α source
- Energy calibration of the SiC detector setup using α sources, demonstration of α spectrometry
- Measurements of the energy(es) of secondary charged particles emitted in neutron interactions, using thin neutron converting layers containing ^{6}Li or ^{10}B isotopes
- Measurements of neutron detection efficiency, on the basis of information on the neutron flux.

IV. CONCLUSIONS

The results of this study demonstrate that silicon carbide (SiC) Schottky Barrier Diodes (SBDs) are highly effective for neutron flux monitoring in subcritical reactor conditions. Their robustness, wide bandgap, and radiation hardness enable reliable operation in environments where traditional detectors may fail or degrade over time. Measurements conducted at the Jožef Stefan Institute TRIGA reactor, using the Libera Monaco 3 data acquisition system, confirmed the capability of SiC detectors to track neutron population changes and provide spatial flux distribution data in low-power reactor scenarios. Their sensitivity and stability in subcritical conditions support their potential use in various nuclear facilities, including research reactors, zero-power systems, and spent fuel monitoring applications. These findings contribute to the growing body of evidence supporting SiC as a next-generation material for advanced radiation detectors.

In addition to their technical performance, SiC detectors offer unique benefits for nuclear education and training. The use of visible and detachable neutron converter layers allows for intuitive demonstrations of neutron interactions, particle attenuation, and (self)absorption effects, enhancing hands-on learning. Their visual accessibility and experimental flexibility make them well-suited for instructional settings, as already demonstrated through activities within the European Nuclear Experimental Educational Platform (ENEEP). This dual utility—operational and educational—positions SiC detectors as a valuable tool for both reactor monitoring and the advancement of nuclear science education.

ACKNOWLEDGMENT

The authors would like to acknowledge financial support from from the Slovenian Research and Innovation Agency - ARIS (research core funding No. P2-0073 - Reactor Physics, research project No. BI-JP/23-25-001 – Structure optimization of SiC based pixel radiation detectors for nuclear monitoring applications). The authors would like to acknowledge Ms. Lisa Box for the support rendered in the performance of the measurements at the JSI TRIGA reactor.

The detector used in this study was developed through a joint research project and was kindly provided by the Central Research Institute of Electric Power Industry (CRIEPI), Japan.

REFERENCES

[1] F. Nava, G. Bertuccio, A. Cavallini, and E. Vittone, Measurement Science and Technology 19, 102001 (2008).154
[2] N. B. Strokan, A. M. Ivanov, and A. A. Lebedev, "Sic nuclear radiation detectors," in SiC Power Materials: Devices and Applications, edited by Z. C. Feng (Springer Berlin Heidelberg, Berlin, Heidelberg, 2004) pp. 411–445.
[3] J. Coutinho, V. J. Torres, I. Capan, T. Brodar, Z. Ereš, R. Bernat, V. Radulović, K. Ambrožič, L. Snoj, Željko Pastuović, A. Sarbutt, T. Ohshima, Y. Yamazaki, and T. Makino, Nuclear Instruments and Methods in Physics Research Section A: Accelerators, Spectrometers, Detectors and Associated Equipment 986, 164793 (2021).
[4] M. De Napoli, Frontiers in Physics 10 (2022), 10.3389/f-phy.2022.898833.166
[5] F. H. Ruddy, L. Ottaviani, A. Lyoussi, C. Destouches, O. Palais, and C. Reynard-Carette, IEEE Transactions on Nuclear Science 69, 792 (2022).
[6] Y. Jiang, J. Wu, Z. Li, X. Fan, and J. Lei, Nuclear Instruments and Methods in Physics Research Section A: Accelerators, Spectrometers, Detectors and Associated Equipment 921, 14 (2019).
[7] V. Radulović, Y. Yamazaki, Željko Pastuović, A. Sarbutt, K. Ambrožič, R. Bernat, Z. Ereš, J. Coutinho, T. Ohshima, I. Capan, and L. Snoj, Nuclear Instruments and Methods in Physics Research Section A: Accelerators, Spectrometers, Detectors and Associated Equipment 972, 164122 (2020).
[8] Štefan Čerba, J. Haščík, J. Lüley, B. Vrban, V. Radulović, A. Jazbec, L. Snoj, L'ubomír Sklenka, M. Miglierini, S. Czifrus, A. Tormási, M. Cagnazzo, H. Böck, and M. Villa, Nuclear Engineering and Design 420, 112973 (2024).
[9] M. Ito, L. Storasta, and H. Tsuchida, Appl. Phys. Express., 1, (2008) 015001-1-3.
[10] R. Bernat, L. Bakrač, V. Radulović, L. Snoj, T. Makino, T. Ohshima, Pastuović, and I. Capan, Materials 14 (2021), 10.3390/ma14175105.185
[11] T. Goričanec et al., On the Optimisation of Large Sample In-Core Irradiation Channel in the JSI TRIGA Reactor, Proceedings, International Conference Nuclear Energy for New Europe (NENE), September 7-10, 2020, Portorož, Slovenia.
[12] Žiga Štancar, L. Snoj, and L. Barbot, International Handbook of Evaluated Reactor Physics Benchmark Experiments TRIGA-FUND-RESR-002, NEA/NSC/DOC(2006)1, OECD NEA (2017).
[13] L. Barbot, D. Fourmentel, G. de Izarra, C. Destouches, J.-F Villard, Y. Moline, and H. Hamrita, in ANIMMA 2019, ANIMMA 2019 – Advancements in Nuclear Instrumentation Measurement Methods and their Applications, Vol. 225, CEA (EPJ193 Web Conf, Portoroz, Slovenia, 2019) p. 03012-
[14] Barbot, Loic, De Izarra, Gregoire, Destouches, Christophe, Bisiach, Danilo, Cargnelutti, Manuel, Zorzut, Sebastjan, Škabar, Matjaž, Bardorfer, Aleš, Makovac, Aurora, Trnovec, Jure, and Paglovec, Peter, EPJ Web Conf. 288, 04019 (2023).
[15] E. Fermi, Experimental Production of a Divergent Chain Reaction, United States Atomic Energy Commission, AECD-3269, January 4, 1952.
[16] Luka Snoj, Andrej Kavčič, Gašper Žerovnik, Matjaž Ravnik, Calculation of kinetic parameters for mixed TRIGA cores with Monte Carlo, Annals of Nuclear Energy, Volume 37, Issue 2, 2010, Pages 223-229.
[17] C. Mori et al., "Simple and quick measurement of neutron flux distribution by using an optical fiber with scintillator," J. Nucl. Sci. Technol., vol. 31, no. 3, pp. 248–249, Mar. 1994.
[18] F. Vitullo et al., A mm3 Fiber-Coupled Scintillator for In-Core Thermal Neutron Detection in CROCUS, IEEE Transactions on Nuclear Science, Vol 67, No. 4, April 2020.

Modern Measurements of Fission Reactor Power Spectral Density

Gilles Caulier*, Romain Boffy, Maxime Dutertre, and Jean-Paul Goossens
CEA, DES, IRESNE, Department of Reactor Studies,
Cadarache F-13108 Saint-Paul-Lez-Durance, France
(*) gilles.caulier@cea.fr

***Abstract*—Power spectral density measurement techniques allow to derive information on nuclear reactors. In specific conditions, one can obtain reactor's kinetic parameters. At IRESNE institute, located at CEA Cadarache, the researchers of the Department of Reactor Studies have used for many years X-MODE and SPECTRON Data Acquisition systems (DAQ) to measure power spectral densities. Due to electronic components, computer, and software obsolescence, the Department started a refurbishment program. The new SCHMITT DAQ replaces X-MODE. It is based on a CAEN R7771 neutron pulse train recorder device, an industrial Linux computer, and a custom-made software. The latter, developed by the Department, is in charge of acquisition setting, data recording, post processing, and visualization. The paper describes the development of the SCHMITT DAQ. The software is written in an object-oriented approach and based on Qt framework for enhanced code portability and to allow using new standard software technologies. This new DAQ has been successfully completed and is ready to for a future qualification and usages. Current developments aim at including support for advanced nuclear data post-processing.**

***Keywords* — Neutron Noise; Reactor Kinetic; PSD**

I. INTRODUCTION

FOR safety reasons, it is important to predict correctly the kinetic behaviour of LWRs in nominal or incidental situations. When a nuclear reactor is near criticality, the delayed neutrons kinetic drive the neutron production. Integral parameters, such as the reactor reactivity, are calculated by computational tools which make use of a nuclear data library. To compare calculation results to the experimental values, it is possible to use nuclear research reactors at low power in well characterized configurations.

In the past years, the IRESNE institute, which is part of the French Alternative Energies and Atomic Energy Commission (CEA) at Cadarache, has been leading research activities focusing on producing better evaluations of delayed neutrons nuclear data, by means of summation calculations [1], microscopic measurements [2] or integral measurements in research reactors [3 – 6]. Some of these integral measurements relied on the power spectral density measurement by mean of "pile noise" techniques. The analysis of power spectral density allows to derive reactor's kinetic parameters such as the prompt decay constant, α_p, or the effective delayed neutron fraction, β_{eff}.

The two legacy DAQs often employed in pile noise experiments by IRESNE team were X-MODE [7] and SPECTRON [8]. The former was dedicated to experiments in counting mode and the latter to experiments in current mode. This paper presents the replacement of X-MODE by the new SCHMITT DAQ. The section II presents the new SCHMITT DAQ, its hardware and its software architecture, section III presents some elements of in-laboratory qualification, and section IV draws the key elements of this modernisation program as well as future developments.

II. THE NEW SCHMITT DAQ

A. Hardware architecture

The portable measurement system SCHMITT inherits its operating principles from the X-MODE system designed in the early 2000s. Due to its obsolescence, a redesign study was initiated at the end of 2022. The SCHMITT system is intended to replace and improve the X-MODE acquisition system. Its operational context involves acquiring TTL signals during reactor experiments. TTL is a logic signal in tension resulting from the discrimination of an amplified pulse coming from a detector. Figure 1 shows the X-MODE system on the left and the SCHMITT system on the right.

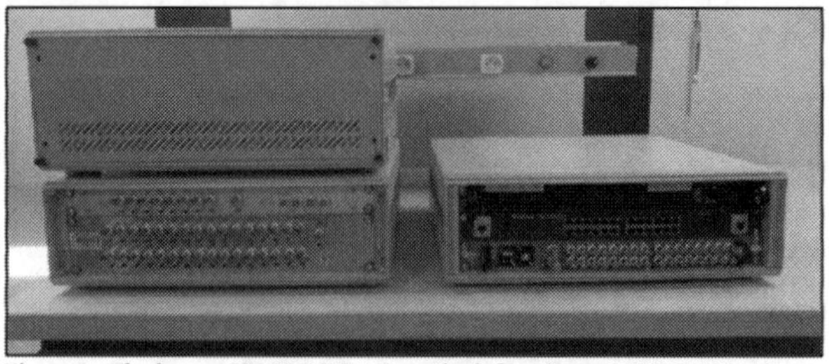

Figure 1. The legacy X-MODE DAQ on the left and new SCHMITT DAQ on the right.

Given that dead time in such experiments must remain below a certain threshold, these experiments typically involve low-power measurements or less sensitive detectors. The X-MODE system had a limitation of approximately 200 kc/s, whereas SCHMITT's limit is set at 3 Mc/s. Figure 2 illustrates the hardware architecture of SCHMITT. The complete measurement system comprises detectors inserted in the reactor, signal shaping electronics, the counting and timestamping unit CAEN model R7771 [9], and a Linux PC for acquisition control. The input signals to the R7771 module must be in a 0-5 V TTL logic form.

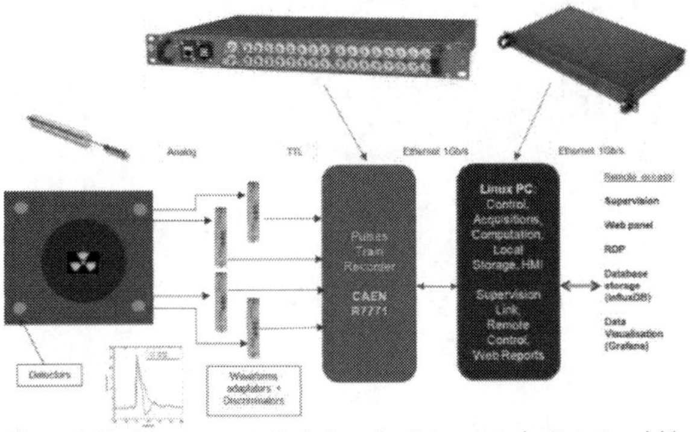

Figure 2. The Measurements Chain from the Detectors to the Data Acquisition Unit.

B. *Software architecture*

The project uses a modular architecture based on Qt [10], C++, and CMake. It includes a vast collection of unit tests and extensive data collection, with all dependencies being open-source. The application is distributed as an AppImage for Linux and comprises approximately 100,000 lines of source code, requiring about 22 person-months for its development. The Human-Machine Interface (HMI) is designed to be streamlined and intuitive with well-separated experimental functions. It uses sidebar tabs and stacked views throughout. The graphs support zooms and pan, and the interface is ready for internationalization to facilitate translations. Figure 3 shows the main view of the DAQ software.

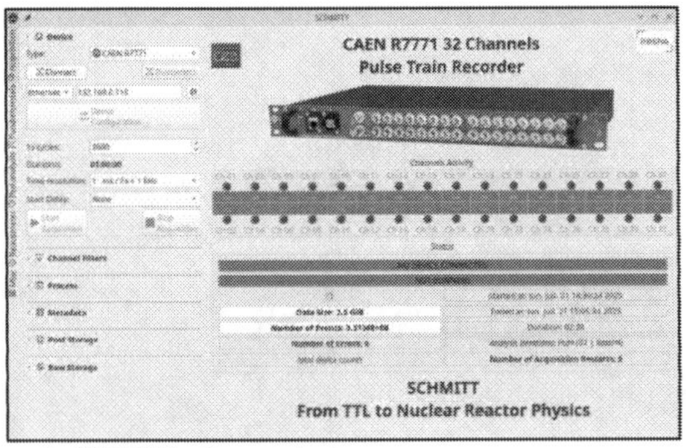

Figure 3. SCHMITT software main view.

The system offers advanced computing capabilities, including multi-core and multi-thread processing, along with a continuous workflow for data analysis and the generation of experiment graphs. It offers several modes, such as device measurements, a neutron noise simulator, a divergence simulator, and the ability to replay previous measurements. Additionally, it is capable of handling a variety of nuclear data, thus offering great flexibility in its use.

The key features related to computations includes count rate measurement, multiplicity, and the analysis of time intervals. It also presents several homemade algorithms, such as power spectral density (PSD) methods, including automatic fit. Another tool can be used to compute the inline nuclear reactivity with error bar indicators to assess measurement uncertainty.

The system supports background processing of big data as well as multi-threaded processing. It manages local files, whether they are textual or binary, and may record data in an InfluxDB database. Regarding the network, it allows remote control via VNC, web report generation, and has a supervision interface. Additionally, remote data storage is also done via InfluxDB.

Figure 4 shows the main components of the software architecture. It comprises nine primary components, each operating within separate threads to ensure efficient processing. The input data for the system can be sourced from R7771, pile noise generator, divergence simulator, or recorded files. A robust communication pipeline facilitates the data stream between the various analysis components, ensuring seamless data flow and interaction. Central to this architecture is a controller that manages the overall data workflow, orchestrating the operations and interactions between the components to maintain a coherent and efficient processing environment.

The input data can be pipelined to three computation components, namely "Fundamentals", "PSD" and "Reactimeter. The Fundamentals component comprises count rate, time-interval and multiplicity analysis. The PSD component proceeds to power spectral density analysis, a.k.a. Cohn-alpha analysis. APSD or CPSD may be chosen depending on one's needs. Finally, the Reactimeter component computes the reactivity using a linear solver applied to the logarithm of count rate.

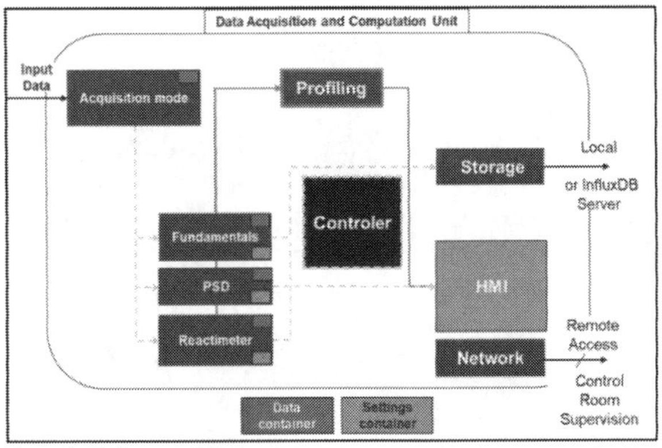

Figure 4. Software architecture

The controller is designed to manage multiple working threads, capable of handling more than 20 threads in parallel to ensure efficient processing. The pipeline facilitates data propagation in discrete blocks or packets, with a slight delay of a few seconds to manage the data flow effectively. This pipeline is built on the Qt inter-thread communication framework, utilizing the signal/slots mechanism to coordinate data transmission between threads. Additionally, the pipeline employs a Flip/Flop input buffer to further optimize data handling and processing.

Figure 5 provides a summary on the parallelized computation architecture for kinetic profiles of a reactor. It features an input buffer that receives temporal data packets for analysis. When the buffer is filled, every second, the data is sent to update the curves, initiate the power spectral density calculations, and perform archiving on the disk. The PSD buffers stack the results of the profiles, which will be reused in subsequent computation cycles. The buffers consist of two flip-flop First-in-first-out (FIFOs) storing packets protected by mutexes. The first buffer stores incoming data, while the second stores data to be processed. This kind of architecture allows to proceeds a large amount of data buffered in the memory. Note that architecture is also the same for all other computation modules.

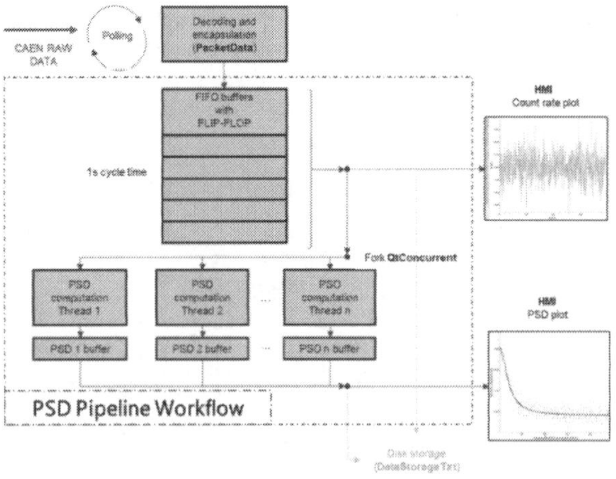

Figure 5. Details of the PSD Pipeline Workflow

III. Results

The Figure 6 shows a linearity curve obtained from a count rate measurement. A TTL pulser was used as input with a rate from 10 c/s to 3×10^6 c/s. The pulser frequency was set constant, *i.e.* without Poisson distribution. The SCHMITT was able to follow the pulser without bias on the whole frequency range. It shows comparable performances when post-treating the recorded timestamps.

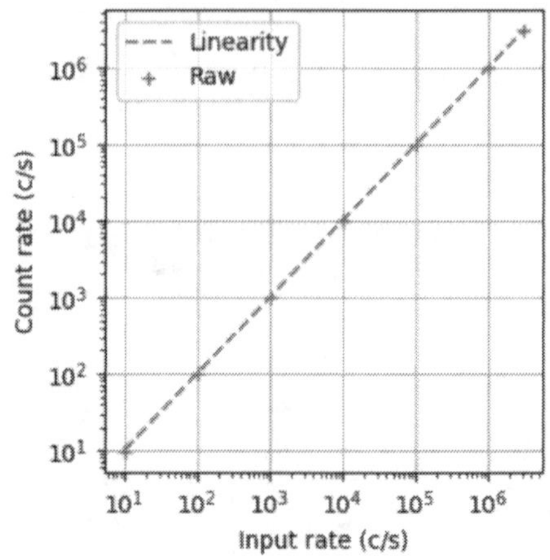

Figure 6. Count rate plot with a constant frequency input.

Figure 7 shows the PSD tab during a pile noise simulation with PSD analysis activated. The pile noise generator was set with 10 kc/s of white noise on channels 1 and 2, with an additional 1 kc/s of correlated noise to mimic the correlated signal coming from a reactor [11]. The relative amount of correlated noise was set relatively high to minimize to simulation time. The simulated prompt decay constant was set to 250 s^{-1}. The "PSD component" was set to compute APSD from channels 1 and 2, orange and green curves respectively, and the CPSD from them, purple. After each computation, the software proceeds to data fitting with the following Lorentz-like function, Y_{PSD}, from which it derives α_p

$$Y_{PSD}(f) = \frac{A}{1+\left(\frac{2\pi f}{\alpha_p}\right)^2} + C$$

The fitting procedure converges correctly and the derived α_p compares relatively well with the noise generator settings.

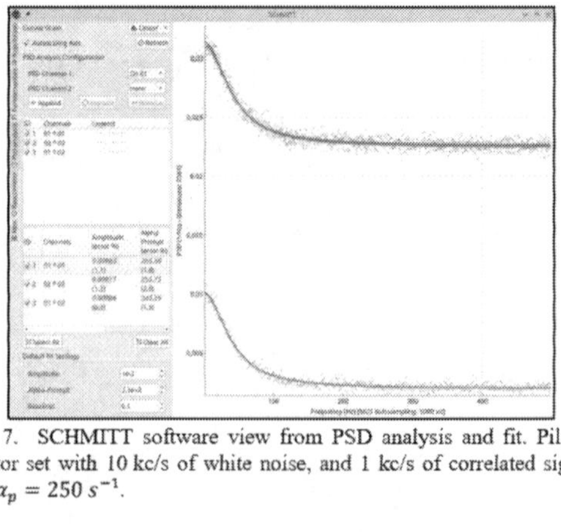

Figure 7. SCHMITT software view from PSD analysis and fit. Pile noise generator set with 10 kc/s of white noise, and 1 kc/s of correlated signal on which $\alpha_p = 250\ s^{-1}$.

Table 1 gathers computation performances with the Reactimeter and PSD components. The PSD process with two channels consists in three individual PSD computations, as in Figure 7, and the PSD process with four channels is the double. One has to recall that a data packet has to be processed every second. The DAQ is able to process the requested PSDs quickly enough to keep up with the data input rate.

TABLE I
SOME PERFORMANCES OF THE SCHMITT DAQ.

Channels	Count rate per channel	Process	Time[a]
1	10 kc/s + 10 pcm	Reactimeter	0.040 ms
2	100 kc/s	PSD[b] (F_s = 1 kHz)	2.7 ms
4	100 kc/s	PSD (F_s = 1 kHz)	3.8 ms

[a] On desktop Linux machine 12 cores, 2 threads / core, 3.8 GHz
[b] PSD = Power Spectral Density, F_s = sampling frequency

IV. CONCLUSION AND PERSPECTIVES

The SCHMITT data acquisition system is built with of-the-shelf hardware components that are a R7771 CAEN pulse train recorder device and a Linux industrial computer. The acquisition software is CEA-made and based on a modular architecture based on Qt, C++, and Cmake. In addition to the data recording, it may proceed to online analysis such as count rate, time interval distribution, multiplicity, reactimeter and power-spectral density. The SCHMITT DAQ replaces the X-MODE DAQ, developed in 2000's, with an increased maximum count rate and online PSD computation capability.

The future developments will focus on the integration of a Rossi-α algorithm in order to be able to proceed simultaneously to pile noise analysis in the time domain (Rossi-α) and in the frequency domain (Cohn-α). Finally, thanks to the modular architecture of the software it is envisaged to support other hardware such as SPECTRON which is dedicated to low noise current measurement.

ACKNOWLEDGEMENTS

This study was funded by the French research institute CEA, EDF and Framatome.

REFERENCES

[1] P. Leconte *et al.*, "New delayed neutron group constants and covariances for LWR core applications, combining summation calculations and integral experiments," *Ann Nucl Energy*, vol. 139, no. 139, p. 107250, May 2020, doi: 10.1016/j.anucene.2019.107250.

[2] B. Geslot *et al.*, "Measuring the delayed neutrons multiplicity and kinetic parameters for the thermal induced fission of ^{235}U, ^{239}Pu and ^{233}U," *EPJ Web Conf*, vol. 253, p. 01004, Nov. 2021, doi: 10.1051/epjconf/202125301004.

[3] E. Gilad *et al.*, "Experimental estimation of the delayed neutron fraction β_{eff} of the MAESTRO core in the MINERVE zero power reactor," *J Nucl Sci Technol*, vol. 52, no. 7–8, pp. 1026–1033, Aug. 2015, doi: 10.1080/00223131.2015.1038331.

[4] B. Geslot *et al.*, "A pile noise experiment in the reference core of the nuclear research reactor LR-0," *Ann Nucl Energy*, vol. 189, p. 109833, Sep. 2023, doi: 10.1016/j.anucene.2023.109833.

[5] R. Boffy, G. Truchet, B. Geslot, G. de Izarra, and A. Sardet, "Measurement of delayed neutrons in a thermal nuclear reactor by means of a long run pile noise experiment in sub-critical state," *EPJ Web Conf*, vol. 284, p. 15003, May 2023, doi: 10.1051/epjconf/202328415003.

[6] R. Boffy and et. al, "A pile noise experiment in current mode in the 37 fuel assemblies core configuration of the nuclear research reactor LR-0," *Submitted to Annals of Nuclear Energy*, 2025.

[7] B. Geslot, C. Jammes, G. Nolibe, and P. Fougeras, "Multimode Acquisition System Dedicated to Experimental Neutronic Physics," in *2005 IEEE Instrumentationand Measurement Technology Conference Proceedings*, IEEE, 2005, pp. 1465–1470. doi: 10.1109/IMTC.2005.1604394.

[8] G. de Izarra *et al.*, "SPECTRON, a neutron noise measurement system in frequency domain," *Review of Scientific Instruments*, vol. 86, no. 11, p. 115111, Nov. 2015, doi: 10.1063/1.4935250.

[9] CAEN, "R7771B Neutron Pulse Train Recording Device." Accessed: May 16, 2025. [Online]. Available: https://www.caen.it/products/r7771b/

[10] Qt Group, "Qt." Accessed: Jul. 21, 2025. [Online]. Available: https://www.qt.io/

[11] C. E. Cohn, "A Simplified Theory of Pile Noise," *Nuclear Science and Engineering*, vol. 7, no. 5, pp. 472–475, May 1960, doi: 10.13182/NSE60-4.

Interest of optimizing the response of a high-temperature ultrasonic transducer by integrating a porous aluminium backing element for the detection of structures immersed in liquid sodium

G. Chabanol[1], F. Baudry[1], E. Rosenkrantz[1], D. Laux[1], A. Paveyranne[1,2], JY. Ferrandis[1*]

[1] IES, Univ Montpellier, CNRS, Montpellier, France
[2] CEA-DES-IRESNE, Cadarache, Saint-Paul-Lez-Durance, France
(*) jean-yves.ferrandis@umontpellier.fr

Abstract—**With the emergence of major applications in harsh environments (such as the new generation of nuclear reactors and power plants, under sodium visualization reactors...), the scientific and industrial community expresses a real need for developing a range of instrumentation dedicated to monitor, control, and check resistance to ageing and damage of a large panel of structures. Ultrasonic transducers are particularly well adapted to address this issue. Up to date, there are no commercial solutions for this type of application and only a few prototypes of ultrasonic transducers exists. Thanks to the use of a specific backing material produced by uniaxial pressing of an aluminium powder containing a small amount of wax followed by sintering, we designed two ultrasonic prototypes transducers dedicated to continuous high temperature experiments. The first one is a PZ27 based 2 MHz piezo-element and the second one a Cerium-modified NBT based 10 MHz. These sensors, with a thickness of 5 mm, tested on a steel sample from room temperature to 140°C (for PZ27) and 335°C (for NBT), turned out to be entirely consistent with what one might expect for echographic time-resolved measurements. Furthermore, temperature-dependent velocity propagation relationships obtained on steel were consistent with the literature. Given their small thickness, they might be easily integrated into various experiments in harsh environments, where space is limited: measurements in sodium reactors, nuclear fuel assemblies, narrow zones in aeronautics or petrochemical industry etc.**

Keywords — **High temperature, Ultrasounds, Pulse echo, Thin transducers, Harsh environment**

I. Introduction

In order To carry out ultrasonic measurements on thin samples when it is impossible to use very high frequencies (for instance because of sample attenuation), one solution is to optimize the backing, a material added on the rear face of the piezo-element. This material has a damping effect and therefore reduces the duration of ultrasonic echoes in the studied material, thus increasing the temporal resolution of A-scans. However, it is important to avoid any round trip in the backing which could interfere with the echoes in the studied sample. The backing must therefore be highly attenuating. Finally, the backing must have a well-chosen impedance with respect to the impedance of the piezo-element. If its impedance is too low, it is useless. On the contrary, if its impedance is too high, the backing dampens the signals too much, ultimately rendering the transducer ineffective. For the classic piezo-elements used in high-temperature applications an impedance between 5 and 15 MRayl is suitable [1]. The subject is widely discussed in literature [2,3]. For example, factory-made porous steels with a thickness of 2 cm have been proposed [4]. In our previous paper [1], we proposed an experimental protocol using uniaxial pressing of aluminium powder containing a very small amount (2-4 %) of wax, to produce highly attenuating 4 mm backings with an impedance that is perfectly acceptable for the use with classical piezo-elements used in industry. Thanks to the pressing method and the addition of wax, the backing is actually lamellar and highly anisotropic, leading to the desired mechanical properties. In this new paper, we coupled such backing materials to a simple 2 MHz-PZ27 piezo-element and to a Cerium-modified NBT ($Na0.5\ Bi(4.5{-}x)Ce(x)Ti4015$) screen printed 10 MHz piezo-element settled on a 760 µm alumina substrate manufactured in our team dedicated to high temperature tests. All details concerning these Cerium-modified NBT piezo-element manufacturing can be found in [5]. Ultrasonic velocity measurements were then carried out on a 1 cm 304 steel sample up to 140°C for PZ27 sensor and up to 335°C in the NBT case in order to check the capital role of our backings, their temperature resistance and the quality of the ultrasonic signals during the temperature increase. After a detailed presentation of the piezo-elements coupling to backing, heating procedure, experimental high temperature test benches, signal acquisition and data processing, the law obtained between ultrasonic velocity in steel and temperature will be discussed regarding literature data to validate the potentialities of our transducers in hostile environments.

II. Experimental benches, data acquisition, and signal processing

To achieve the bond between the piezo-element and the backing, the characteristics of the adhesive must be precisely chosen, particularly in terms of acoustic impedance. Indeed, if the impedance of the adhesive is too different from that of the backing, the expected effect of the backing may be strongly impacted. Ideally, an adhesive material having a high attenuation and an impedance equal to those of the backing has to be used. In this case, the adhesive layer behaves like an extra-layer of backing. Obviously, the adhesive material has to be temperature-resistant and must have a thermal expansion coefficient close to that of the backing and the piezo-element. Ceramic bonding cements or epoxy like adhesives can fill this role. To guide our choice, several ceramic cements adhesives (Resbond 940, 989, 919, 907GF) and a high expansion epoxy like (Thermeez 7030) produced by Cotronics were characterized. For each adhesive, we manufactured disks with a diameter of 16 mm and a thickness of 5.5 mm. These disks were characterized with a transmission method at 5 MHz using the measurement bench described in reference [6] and validated in the framework of fine ultra-wideband characterization of poly(methyl metacrylate) (PMMA). The results in terms of acoustic impedance are given in Table 1.

TABLE I
ACOUSTICAL IMPEDANCE OF SOME RESBOND AND THERMEEZ HIGH TEMPERATURE ADHESIVES

Cotronics adhesive type	Acoustical impedance (MRayl) ± 0.5
Resbond 940	5.1
Resbond 989	9.1
Resbond 919	9
Resbond 907GF	5.9

The aluminium backing disks used in our experiments had an acoustical impedance equal to 6 MRayl. So, considering the results presented in Table 1, all the adhesives can be used because their acoustical impedance is in the same order of magnitude of backing impedance. Rigorously, the adhesives to be chosen are, in order of preference (if acoustical impedance is the major criterion): Resbond 907RF, Resbond 940, Thermeez 7030, Resbond 919 and Resbond 989. Finally, Thermeez 7030 adhesive was chosen not only for its impedance, but also for its thermal characteristics and ease of use: it can withstand temperatures up to 980°C (1800°F), and is easily applicable to small surfaces at room temperature. Furthermore, its preparation is very simple compared to the other adhesives. It only consists in mixing adhesive powder with water. At last, the attenuation of this ceramic paste has been estimated at around 40 dB/cm at 5MHz which is a perfectly acceptable value as mentioned in our previous work [1]. In practice, 100 doses of 7030 powder were mixed 20 doses of water. The obtained paste was then applied between the piezo-element and the backing disk. The assembly, firmly tightened with a 200 grams mass was heated 4 hours at 65°C.

The final thickness of these probes was estimated at 5 mm using a caliper. Before temperature measurements, the (backing+piezo-element) assembly was then coupled to the steel sample using a high-temperature (100°C-520°C) coupling paste supplied by Action-NDT. To ensure that the sample-transducer coupling remained stable throughout the test, a small device with a screw was used for the two sensor's tests and is displayed in Fig. 1 for PZ27 system. For PZ27 based sensor, temperature-dependent measurements up to 140°C we used a PID-controlled hot table. Concerning higher temperature tests for NBT based sensor, a Carbolite RWT 100 furnance (up to 1100 °C) was employed (See Fig. 2). In both cases, even if a temperature is imposed by the heating device, a thermo-couple was added in the vicinity of the ultrasonic sensor to obtain the real temperature. Furthermore, after any temperature change we followed the echoes arrival time until they reached a plateau. Indeed, as ultrasonic velocity is time dependant, if echoes do not temporally move this means that a stable temperature has been reached.

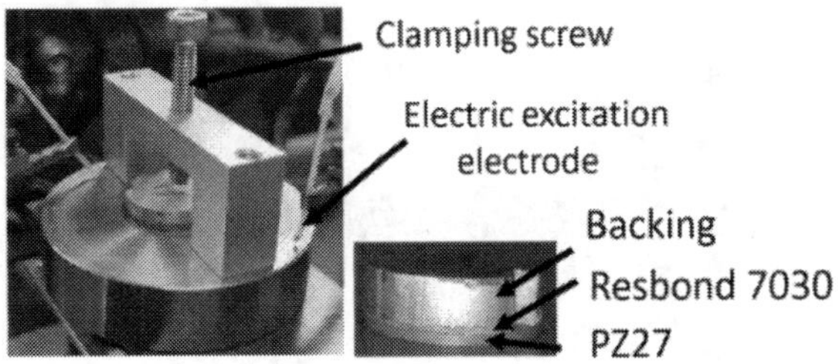

Fig. 1. PZ27 piezo-elements coupled to aluminium backing and steel sample before heating

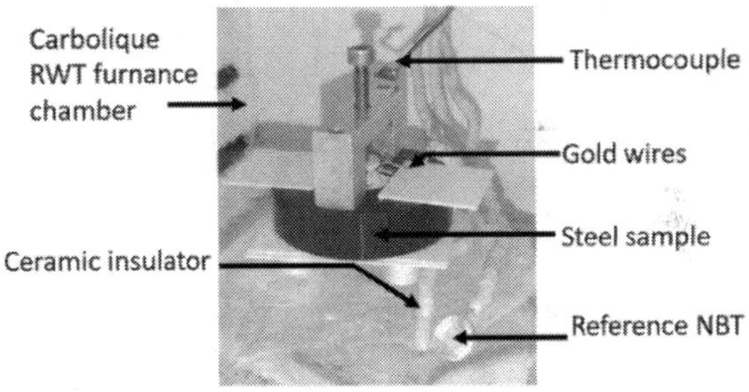

Fig. 2. NBT piezo-element coupled to aluminium backing and steel sample after heating.

III. Experimental bneches, data acquisition, and signal processing

A. Effect of backing

In order to illustrate the effect of backing on echoes duration and temporal resolution, echograms with and without backing are displayed in Fig. 3 and Fig. 4.

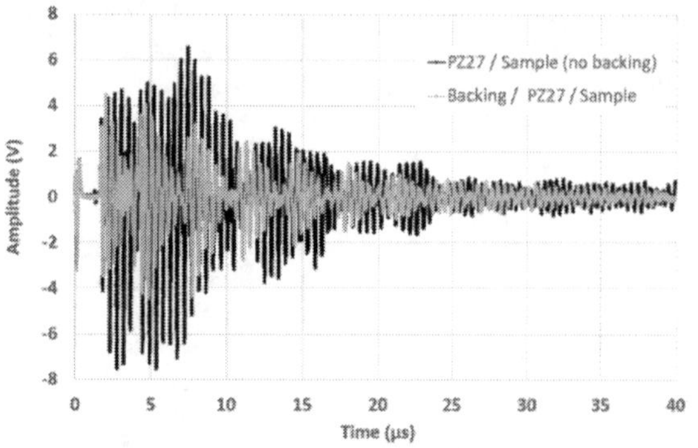

Fig. 3. Echograms for PZ27 device with and without backing at room temperature

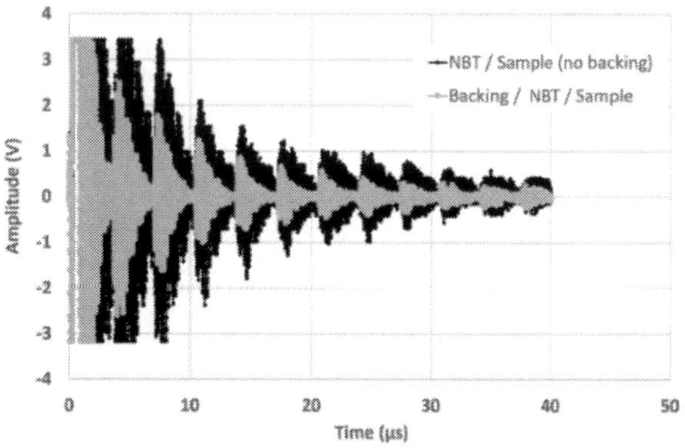

Fig. 4. Echograms for NBT device with and without backing at room temperature

As expected and clearly demonstrated in our previous work [1] with simulations, the presence of backing lead to echoes duration reduction and consequently to temporal resolution increase, overlapping between round-trips being reduced. Piezo-element coupling lead to echoes amplitude reduction which could have a deleterious effect for attenuating samples investigation. Nevertheless, the choice of an impedance around 5 MRayl limits this effect and in fine, the echograms are not impacted a lot in term of amplitude.

B. Ultrasonic velocity in steel versus temperature

Ultrasonic velocity, measured as described in [7-9] versus temperature with an accuracy of ± 30 m/s is displayed (in the range 25°C-140°C) for PZ27 device in Fig. 5 and for NBT device (in the range 30°C-330°C) in Fig. 6. Error bars are not plotted in these figures for more clarity. In both cases a linear trend was observed an led to the regressions (1) and (2). The mean behaviour law (3) can be given for ultrasonic velocity in steel sample versus temperature :

$$V_{STEEL(PZ27)} = 5913 - 0.636.T(°C) \quad (1)$$

$$V_{STEEL(NBT)} = 5926 - 0.623.T(°C) \quad (2)$$

$$V_{STEEL(MEAN)} = 5919.5 - 0.629.T(°C) \quad (3)$$

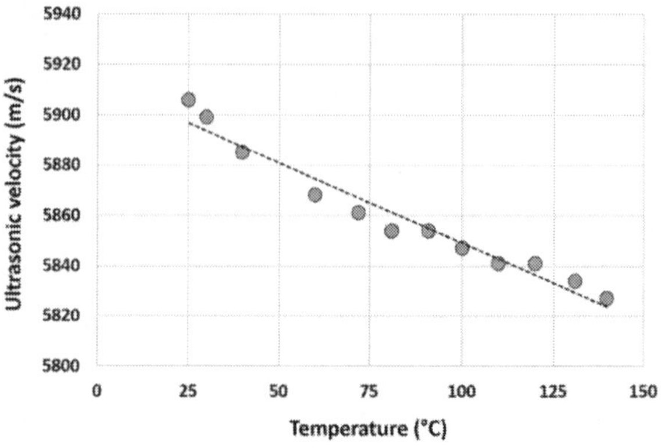

Fig. 5. Ultrasonic velocity in steel sample obtained with PZ27 device up to 140°C

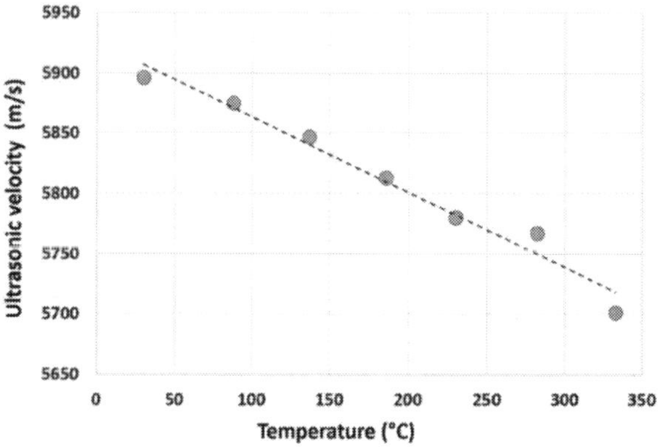

Fig. 6. Ultrasonic velocity in steel sample obtained with NBT device up to 335°C

Our results are in line with literature [10,11] in which linear trends are also reported with slopes between -0.28 and -0.68 m/s/°C. Our value of velocity at room temperature (5919 m/s) is also in line with values reported in the previous references. Regarding Curie's temperature of both PZ27 (350°C) and Cerium-modified NBT (600°C), results up to 250°C in PZ27 case and 450°C for NBT could have been expected. Indeed, it is generally admitted that piezo-elements are enough efficient up to 0.75TCurie. After experiments and return to room temperature signals have recovered the amplitudes measured before test. So, no degradation of piezo-elements occurred and no electrical connections were damaged. So, the integrity of sensors is not responsible for the temperature limitations observed. But one has to bear in mind that with temperature increase the mechanical quality factor of the piezoelement decreases and that the ultrasonic attenuation in steel increases [11, 12] leading to smaller amplitude echoes during heating.

IV. Conclusions

Results obtained in this communication clearly demonstrate the capital role of the backing material and that aluminium backings produced by uniaxial pressing of an aluminium powder containing a small amount of wax followed by sintering are perfectly adapted for high temperature time resolved measurements. Nevertheless, measurements were limited to 140°C for PZ27 device and 334°C for NBT system which is a little bit under what could have been expected regarding Curies temperature. These limitations, attributed to piezoelement quality factor and to ultrasonic attenuation in steel is perhaps also induced by the degradation of the coupling between piezo-element and the sample. Indeed, in the framework of these experiments we used a coupling paste. Even if it is theoretically given to ensure a coupling up to 520°C nothing indicates that its performance in term of coupling quality are constant during temperature elevation. It could be imagined to couple the sensor to the sample with the Thermeex 7030 adhesive but, in this case, it would not be possible to use the sensor for other samples, the sensor being definitively glued to the sample. In this communication, ultrasonic velocity measurement in steel was just used to test our system and to demonstrate the potentialities of our transducers. For practical investigation in hostile media and in particular for ultrasonic investigations under Sodium for RNR reactors study, these transducers are particularly adapted to evaluate distance between metallic components, estimate metallic components thicknesses… during maintenance operations for instance when reactor is stopped but remains hot. Indeed, their very small thickness compared to recent works using porous steel backings [13] is very promising to operate in reduced volumes. Further works will be dedicated to reach temperatures up to 500°C-600°C thanks to the use of a best coupling agent.

Acknowledgements

This work was supported by "La Region Occitanie - Pyrénées - Mediterranée" as part of a PhD grant.

References

[1] Chabanol G, Laux D, Rosenkrantz E, Ferrandis JY. Uniaxial pressing of aluminium powder to design thin lamellar high damping backing materials for high-temperature ultrasonic probes. Applied Acoustics 2024;221. https://doi.org/10.1016/j.apacoust.2024.110027

[2] Rymantas K, Vaskeliene V. High temperature ultrasonic transducers: a review. Sensors2021;21(9). https://doi.org/10.3390/s21093200.

[3] Dong Y, Yue Q, Cheng D, Zhou Z, Liang R, Dong X. Ultrasonic transducer with BiScO3-PbTiO3-based ceramics of operating temperature over 400° C. Sensors and Actuators A: Physical 2022;340. https://doi.org/10.1016/j.sna.2022.113528.

[4] Nguyen DC, Navacchia F, Tran-Huu-Hue LP, Feuillard G. Ultrasonic characterization, simulation of porous metal in the interest of high frequency applications. e-Journal of Nondestructive Testing 2022;27(12). https://dx.doi.org/10.58286/27525

[5] Baudry F, Rosenkrantz E, Combette P, Fourmentel D, Destouches C, Bonzon PV, Paumel K, Ferrandis JY. Design, manufacture and temperature characterization of screen-printed acoustic sensors for monitoring the release of fission gases in experimental fuel rods of the Jules Horowitz reactor. IEEE Transactions on Nuclear Science 2024;71(5). https://doi.org/10.1109/TNS.2024.3368618

[6] Laux D, Chabanol G, Sapey G, Ferrandis JY, Rosenkrantz E. Shear and Longitudinal attenuations and quality factors of poly(methyl metacrylate) (PMMA) from 20 kHz to 12 MHz investigation with ultrasonic spectroscopy. Ultrasonics 2023;134. https://doi.org/10.1016/j.ultras.2023.107104

[7] Rosenkrantz E, Ferrandis JY, Lévêque G, Baron D. Ultrasonic measurement of gas pressure and composition for nuclear fuel rods. Nuclear Instruments and Methods in Physics Research Section A: Accelerators, Spectrometers, Detectors and Associated Equipment 2009;609(3). https://doi.org/10.1016/j.nima.2009.02.028

[8] Sinha DN, Kaduchak G. Noninvasive determination of sound Speed in liquids. In: Levy M, Bass H, Stern R (Eds.). Published in Modern Acoustical Technics for the Measurement of Mechanical Properties. Academic Press 2001.

[9] Laux D, Gibert O, Ferrandis JY, Rosenkrantz E, Mograme MA, Prades A. In-pipe coconut water rheological characterization with ultrasonic waves. Journal of Food Engineering 2018;235. https://doi.org/10.1016/j.jfoodeng.2018.04.030

[10] Ginzel E, Ginzel R. Approximate dV/dT values for some materials. e-Journal of Nondestructive Testing 2017;22(9). https://www.ndt.net/?id=21545

[11] Tai JL, Sultan MTH, Łukaszewicz A, Shahar FS, Tarasiuk W, Napiórkowski J. Ultrasonic Velocity and Attenuation of Low-Carbon Steel at High Temperatures. Materials. 2023; 16(14). https://doi.org/10.3390/ma16145123

[12] Kube CM. Attenuation of laser generated ultrasound in steel at high temperatures;
comparison of theory and experimental measurements. Ultrasonics 2016;70. https://doi.org/10.1016/j.ultras.2016.05.009

[13] Feuillard G, Nguyen DC, Lethiecq M, Jean M, Navacchia F. Porous Metal Backing for High-Temperature Ultrasonic Transducers. Acoustics. 2024; 6(4). https://doi.org/10.3390/acoustics6040058

Gamma and neutron dose rate measurements around the KATANA water activation loop at JSI TRIGA reactor

Domen Kotnik[1, *], Julijan Peric[1, 2], Domen Govekar[1, 2], and Igor Lengar[1]
[1]Reactor Physics Department, Jožef Stefan Institute, Slovenia
[2]Faculty of Mathematics and Physics, University of Ljubljana, Slovenia
(*) domen.kotnik@ijs.si

***Abstract*— KATANA water activation facility, which serves as a well-defined high-energy gamma and neutron source, has been successfully constructed at JSI TRIGA reactor at the end of 2023. During the commissioning phase γ and neutron dose rate measurements were conducted around the KATANA to create a comprehensive dose rate map under steady-state reactor operation with activated water flowing throughout the circuit. Dose rates, quantified as dose rate equivalents of H*(10)/time, were measured using a certified neutron probe and two γ detectors. The peak γ dose rate observed was up to 5 mSv/h at the close vicinity of the main observation part of the circuit (Snail head), with neutron contributions markedly lower, by more than three orders of magnitude. Due to these elevated dose rates, the experimental zone within the concrete walls has been designated as a red zone, subject to stringent access restrictions. Outside the labyrinth, however, the dose rates remained below the limit of 10 µSv/h, indicating no need for additional shielding. Experimental data were furtherly compared with the computational results using a simplified convectional method excluding detailed fluid dynamics. C/E comparison showed that the gamma dose rates were underestimated by a factor of 2–5, while neutron dose rates near the source (Snail No. 1) were overestimated by a factor of 1.8. These findings highlight the complexity of water activation analysis at KATANA and the need for accurate modelling. Future work should incorporate advanced, state-of-the-art fluid activation codes and refined experimental methods to increase precision and support benchmark-quality experiments.**

***Keywords* —dose field measurements, N16, KATANA, water activation, Fusion, ITER.**

I. Introduction

WATER as a primary coolant will play an important role in the performance of fusion reactors, as it causes an ionising radiation field throughout the facility after its irradiation and activation and requires improved shielding for instrumentation and personnel. The activation of water consists primarily of the activation of oxygen, resulting in the generation of activated ^{16}N, ^{17}N and ^{19}O, which furtherly decay releasing high energy gammas and neutrons. The threshold energy for the primary water activation reaction, in particular, $^{16}O(n,p)^{16}N$, is about 10.5 MeV. Consequently, neutrons in fusion reactors lead to a water activity approximately five orders of magnitude higher than in fission reactors of comparable power [1].

To support ITER [2][3], the KATANA irradiation facility [4][5][6], which utilises a closed-water activation loop, was successfully licenced, built and commissioned at the end of 2023 at the JSI TRIGA research reactor in Slovenia [7]. The KATANA serves as a well-defined and stable high-energy (6 MeV - 7 MeV) gamma and ~1 MeV neutron source [8]. Such a high-energy irradiation facility will enable various experiments based on water activation, e.g. shielding experiments using ITER-relevant materials, investigation of the response of detectors to high-energy γ radiation, investigation of short-lived moving radiation sources, integral cross-section measurements, dose rates and spectrum measurements, etc. The ultimate goal of KATANA is to perform benchmark-quality experiments, e.g. experimental validation of fluid activation codes (i.e. RSTM [9] [10], FLUNED [11], GammaFlow & ActiFlow [12]), and to establish itself as a reference facility for the calibration of high-energy γ detectors [13], which will significantly support the operation of ITER, DEMO [14] and other future water-cooled fusion reactors.

To ensure safe operational conditions, it is critical that dose rates within the reactor hall remain sufficiently low, i.e. below the prescribed limit value of 10 µSv/h, during KATANA and full-power operation of the TRIGA reactor. The aim of this work is to create a comprehensive gamma and neutron dose rate map around KATANA facility under steady-state reactor operation with activated water flowing throughout the circuit by performing gamma and neutron dose rate measurements at multiple locations. In addition, an experimental data were furtherly compared with the computational results using a simplified convectional method excluding detailed fluid dynamics.

The paper is structured as follows: Section 2 presents a brief description of the KATANA irradiation facility, Section 3 presents the performed gamma and neutron dose rate measurements, and Section 4 presents the computational analysis and comparison with the experimental data.

II. KATANA WATER ACTIVATION LOOP

A. Water activation

The short-term activation of water consists primarily of the activation of oxygen isotopes by the reactions $^{16}O(n,\ p)^{16}N$, $^{17}O(n,\ p)^{17}N$ and $^{18}O(n,\ \gamma)^{19}O$ [15], with a very small contribution from the activation of dissolved gases, corrosion products and additives. Activated nitrogen and oxygen nuclides subsequently decay, releasing γ rays and neutrons with different energies. The (n,p) reactions on ^{16}O and ^{17}O are threshold reactions, while the (n,γ) reaction on ^{18}O already occurs at thermal energies. The most important water activation reactions and the corresponding decay properties are listed in Table I. The main contribution to the activity of water during operation is the activated isotope ^{16}N due to the high natural abundance of the isotope ^{16}O and high cross-section for the (n,p) reaction.

TABLE I
ACTIVATION PRODUCTS OF WATER [15].

Reaction	$t_{1/2}$ [s]	Major decay products	Threshold energy	Natural abundance
$^{16}O\ (n,p)\ ^{16}N$	7.13	γ: 6.13 MeV (67 %) γ: 7.12 MeV (5 %)	~ 10.5 MeV	99.76 %
$^{17}O\ (n,p)\ ^{17}N$	4.17	n: 0.38 MeV (35 %) γ: 0.87 MeV (3 %) n: 1.17 MeV (53 %) n: 1.70 MeV (7 %)	~ 9 MeV	0.04 %
$^{18}O\ (n,\gamma)\ ^{19}O$	26.88	γ: 0.11 MeV (3 %) γ: 0.20 MeV (96 %) γ: 1.36 MeV (50 %) γ: 1.44 MeV (3 %)	< 1 eV	0.2 %

B. KATANA irradiation facility

KATANA water circuit, schematically shown in Fig. 1, consists of a pipe loop that is partially inserted into the radial piercing port (RPP). The main components are the inner irradiation Snail, two outer observation Snails and transport pipes. All important components, sensors and instruments for controlling and operating the circuit, e.g. pump, valves, Coriolis flow metres, temperature and pressure sensors, etc., are located outside the RPP (shown in black). A three-way valve can be used to fully or partially adjust the water distribution between the primary (shown in red) and secondary loops (shown in blue), referred to as Configuration 1 and Configuration 2, respectively, further in the text. The primary loop is designed to achieve the fastest possible transport of activated water in order to achieve a high total activity of ^{16}N and ^{17}N within the outer observation part (i.e. outer Snail No. 1), resulting in a lower experimental uncertainty, e.g. for performing shielding experiments with ITER-relevant materials. The secondary loop (or so-called delay loop) extends the travel time of the activated water and favours a $^{16}N/^{19}O$ isotope ratio in favour of ^{19}O. The use of these two different circuit components, each optimised for its specific function, improves the overall performance and accuracy of the system. In addition, the loop branching also enables the investigation of more complex scenarios such as the simulation of the mixing of the activated water flow from different pipes and/or the investigation of leakage scenarios.

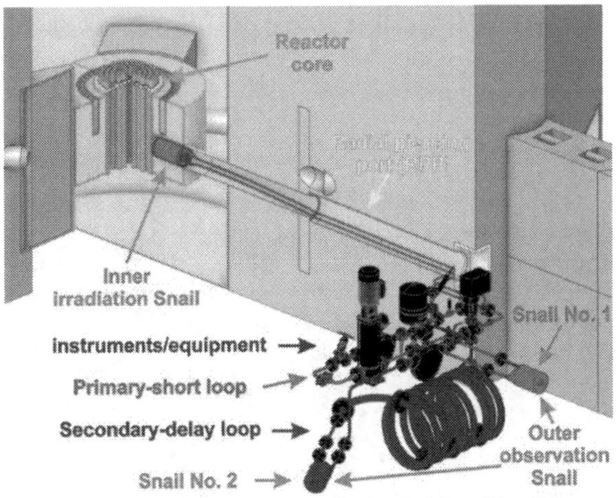

Fig. 1. Schematic of the KATANA water circuit only (adapted from [5]).

To minimise occupational radiation exposure inside the reactor hall, two special γ & neutron shielding plugs are used inside the RPP together with a 40 cm thick and 180 cm high concrete wall around the experimental area. The entire KATANA is controlled and monitored via the control panel, located outside of the labyrinth-shaped concrete wall. A fully constructed and commissioned KATANA facility is shown in Fig. 2.

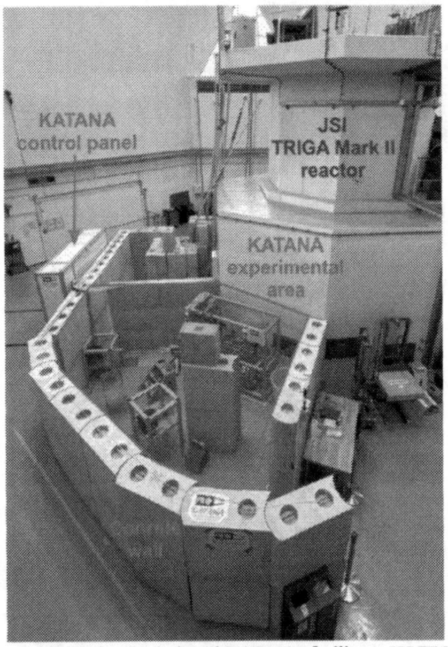

Fig. 2. Constructed and commissioned KATANA facility at JSI TRIGA research reactor (figure taken from [5]).

III. DOSE RATE MEASUREMENT

During the commissioning phase, and with a focus on safety, γ and neutron dose rate measurements were conducted around the KATANA circuit (within the experimental area enclosed by concrete walls) and across the reactor hall to create a comprehensive dose rate map under steady-state reactor operation (250 kW) with activated water flowing throughout the circuit (flow rate 0.4 L/s). Dose rates, quantified as dose rate equivalents of H*(10)/time, were measured using a certified neutron probe (Berthold LB 6411) and two γ detectors: a pressurized ionization chamber (Fluke Victoreen 451P-DE-SI-RYR) and a scintillator probe (Automess 6150AD-b/H). All three detectors, with the reported uncertainty of 25 %, are shown at the bottom left side of Fig. 3.

A total of 21 measurement positions (MPs), as shown in Fig.3, were strategically selected to characterize dose rate distributions at key locations around the KATANA facility. These included positions along the line of sight from the reactor pool plug (RPP), around the outer Snails No. 1 and 2, and at various points inside, outside, and at the entrance of the concrete shielding wall.

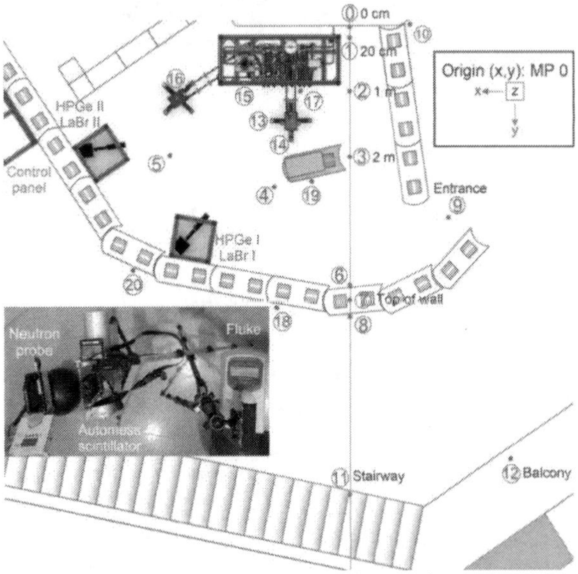

Fig. 3. 21 measurement positions for gamma and neutron dose rates around KATANA facility, covering both the shielded experimental area and selected locations in the reactor hall.

To capture peak dose rate values, most MPs were positioned at a height of 86 cm above the ground—corresponding to the vertical centreline of the RPP and the height of the primary circuit loop (Snail No. 1). Exceptions to this height included MP 17, located at the centre of the lower winding of the secondary loop inside the labyrinth wall, and several positions situated externally within the reactor hall: MP 7 (top of the concrete wall), MP 11 (stairway) and MP 12 (balcony).

Detailed spatial coordinates (x, y, z) of all MPs, with an associated positional uncertainty of ±5 cm and the coordinate origin defined at ground level beneath MP 0, are provided in Table II, along with the corresponding dose rate measurements. Due to differing detector sensitivities across radiation field intensities, gamma dose rate measurements within the concrete shielding (MPs 0–7, 13–17, and 19) were performed using a Fluke ionization chamber. Conversely, dose rates at all other reactor hall locations (MPs 8–12, 18, and 20) were measured using an Automess scintillator probe, as indicated with an asterisk (*) in the tables.

As expected, the highest gamma dose rate—reaching up to 5 mSv/h—was measured near the outer Snail No. 1. The dose rate at the inlet side (MP 13) was almost 50 % higher than at the back side (MP 14), where the activity of the water is significantly reduced due to radioactive decay along its flow path through the Snail. A comparatively elevated gamma dose rate of 0.6 mSv/h was also recorded near the pump (MP 15), which can be attributed to the substantial volume of activated water in that region (approximately 4 litres).

Despite these localized peaks, all measurement positions located outside the experimental zone (i.e., beyond the concrete shielding wall) exhibited dose rates below the regulatory threshold of 10 µSv/h. The contribution of neutrons to the overall dose rate is more than three orders of magnitude lower than that of gamma radiation and is therefore considered negligible for the purposes of radiological assessment. Nevertheless, neutrons (originating from ^{17}N) are easily measured by using time dependent (^{3}He and BF_3) detectors and can be used to characterise KATANA characteristics [8].

TABLE II
EXPERIMENTAL AND CALCULATED DATA OF GAMMA AND NEUTRON DOSE RATE FOR 21 LOCATIONS AROUND KATANA (CONFIGURATION 1, 250 KW, FLOW RATE 0.4 L/S). CALCULATIONS (WITH C/E VALUES) CONSIDERED SINGLE AND DOUBLE SOURCE LOCATION OF ACTIVATED WATER.

LOCATION				Experiment		SINGLE SOURCE (Snail no.1)				DOUBLE SOURCE (Snail no.1 + pump)			
						Calculation		C/E		Calculation		C/E	
MP	X	Y	Z	γ	neutron	γ	neutron	γ	neutron	γ	neutron	γ	neutron
[/]	[cm]	[cm]	[cm]	[μSv/h]	[μSv/h]	[μSv/h]	[μSv/h]	[/]	[/]	[μSv/h]	[μSv/h]	[/]	[/]
0	0	0	0	132	11	7.2	3.47E-02	0.05	0.003	10.6	4.31E-02	0.08	0.004
1	0	20	1	30	9	10.1	4.83E-02	0.34	0.005	13.5	5.68E-02	0.45	0.006
2	0	100	2	120	0.8	22.6	9.07E-02	0.19	0.11	26.2	1.00E-01	0.22	0.13
3	0	200	3	50	0.5	8.3	2.54E-02	0.17	0.05	9.8	2.89E-02	0.20	0.06
4	-121	246	4	122	0.4	19.9	4.95E-02	0.16	0.12	22.9	5.71E-02	0.19	0.14
5	-286	197	5	54	0.1	6.2	2.82E-02	0.1	0.28	9.6	3.71E-02	0.18	0.37
6	0	398	6	3	0.2	0.7	4.36E-03	0.22	0.02	0.8	5.34E-03	0.26	0.03
7	0	418	7	2	0	0.2	1.27E-03	0.09	N/A	0.2	1.61E-03	0.12	N/A
8	0	444	8	0.4*	0	0.1	3.18E-04	0.29	N/A	0.1	4.29E-04	0.33	N/A
9	156	293	9	1.3*	0	0.2	1.28E-03	0.16	N/A	0.3	1.63E-03	0.22	N/A
10	96	-2	10	2.8*	0	0.9	1.00E-03	0.33	N/A	1.1	1.24E-03	0.41	N/A
11	0	714	11	1.7*	0	0.1	7.30E-04	0.04	N/A	0.3	1.37E-03	0.17	N/A
12	256	661	12	1.4*	0	0.0	5.49E-04	0.02	N/A	0.0	7.20E-04	0.03	N/A
13	-107	148	13	5000	3	1388.3	3.66	0.28	1.22	1397.7	3.68	0.28	1.23
14	-94	164	14	3400	2	1231.4	3.53	0.36	1.77	1236.8	3.54	0.36	1.77
15	-167	87	15	600	1.1	26.8	9.70E-02	0.04	0.09	109.1	2.34E-01	0.18	0.21
16	-273	111	16	90	0.3	7.5	3.41E-02	0.08	0.11	14.8	5.04E-02	0.16	0.17
17	-79	100	17	320	1.1	49.0	1.59E-01	0.15	0.14	56.0	1.74E-01	0.17	0.16
18	-116	430	18	0.6*	0	0.1	4.00E-04	0.06	N/A	0.2	6.01E-04	0.37	N/A
19	-62	237	19	9	0	5.8	9.23E-03	0.65	N/A	6.3	1.05E-02	0.70	N/A
20	-344	372	20	1.7*	0	0.5	7.71E-04	0.30	N/A	0.7	1.03E-03	0.42	N/A

Remarks: a) N/A: experimental value is 0. b) C/E values between 0.2 – 5 are highlighted.
c) Relative statistical uncertainty for all MP is below 2 %. d) All dose rate detectors have stated measurement uncertainties of 25 %.
e) γ dose rate measurements noted with the (*) were performed with Automess scintillator detector (others with Fluke detector).

IV. COMPARISON OF COMPUTATIONAL AND EXPERIMENTAL DATA

A. Computational model

Experimental data described above were also compared with the computational results. To perform that, an MCNP model containing the TRIGA reactor core and the inner part of the KATANA structure, which was used for the calculation of the RR density profiles of ^{16}N, ^{17}N and ^{19}O inside the inner irradiation Snail (see article [5]), was updated to include the outer part of the KATANA facility and the reactor hall. Once again, the detailed CAD model (schematically shown in Fig. 3) was simplified in order to be converted to the MCNP file using the GEOUNED [16]program.

In the second step, the source of activated water was defined, which is evenly distributed within the outer Snail No. 1, i.e. the main source of activated water for the primary loop - Configuration 1, without taking into account contribution of activated water from the rest of the circuit. The intensity of the source was determined based on the calculated saturated activity values of ^{16}N (7.01×10^7 Bq; 77.675 %), ^{17}N (1.15×10^4 Bq; 0.017 %) and ^{19}O (9.54×10^6 Bq; 22.307 %) inside the Snail No. 1 and taking into the account the dominant released γ rays and neutrons (see Table I).

The γ and neutron dose rate values were calculated as an ambient dose equivalent H*(10) by using flux-to-dose rate conversion factors (ICRP-21 [17]and NCRP-38 [18]).

B. Computational results: single source location

γ and neutron dose rates were calculated at selected twenty-one measurement positions (MP: 0 - 20) and on a global mesh area, i.e. meshtally, around the KATANA facility, providing a global dose field distribution and insight into the important streaming paths. Results of the global γ (top) and neutron (bottom) dose field distributions are presented in Fig. 4 respectively, in a horizontal cross-section view (XY plane) at the height of the centre of the outer Snail No. 1 (Z = 86.4cm). In addition, the contour lines have been added in red for better clarity. The maximum achieved γ and neutron dose rate inside the Snail No.1 are 6.7 mSv/h and 23.8 μSv/h, respectively, which are gradually decreased for more than four orders of magnitude throughout the area. The main factor is the radial (R^2) dependence on the distance from the source, whereby the surrounding shielding concrete wall further decreases the radiation.

From the safety point of view, the dose rate outside the wall, i.e. outside the experimental area, is well below the limit value of 10 µSv/h, whereby the contribution of the neutrons to the total dose rate is negligible and more than 2 orders of magnitude smaller than that from gammas. However, some unique streaming characteristics can be observed between them, due to the nature of the interaction between gammas and neutrons with matter. Specifically, for gammas, streaming paths ("rays") with higher dose rate values are observed from the concrete shielding wall onwards. The reason for this phenomenon is the extra empty space (can be seen in Fig. 2. and Fig. 3) at the top of each shielding block that is used for handling. The effect is even more pronounced because the source (Snail No. 1), located at a height of 86.4 cm, is nearly aligned with the top of the first row of the shielding wall (90 cm). This effect was not known before the analysis. However, it was unintentionally taken into account when selecting the locations of MP 18 and MP 20, which were chosen based on the highest measured γ dose rate in the vicinity of the corresponding location, i.e. the tallies (spheres) for MP 18 and MP 20 match the beam paths in Fig. 4 on the top.

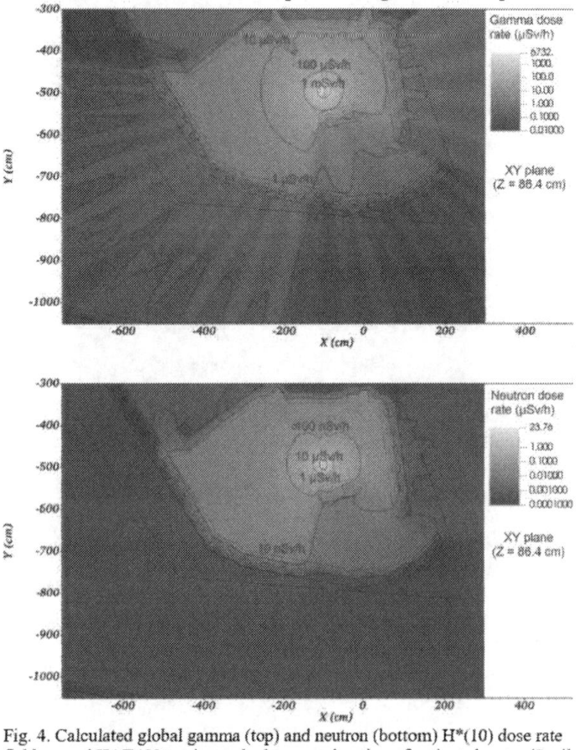

Fig. 4. Calculated global gamma (top) and neutron (bottom) H*(10) dose rate field around KATANA using a single source location of activated water (Snail No. 1) in a horizontal cross-section view (Configuration 1, flow rate 0.4 L/s, full reactor power 250 kW). Contour lines are shown in red.

Results of the calculated gamma and neutron dose rate at locations of the 21 MPs throughout the reactor hall are presented in Table II. The relative uncertainty, which only takes into account the statistical uncertainty of the Monte Carlo simulation, is well below 1% for the majority of MP and up to 2% for those outside the concrete wall. In addition, the C/E values (calculation vs experiment) are included, whereby experimental data have the stated 25 % measurement uncertainty. C/E values between 0.2 - 5 (difference by a factor of 5) are highlighted for better clarity.

The C/E values for gammas are for all MPs below 1 (underestimation), whereby eight of them show a value within a factor of 5 (highlighted). These MPs, in which the contribution from the activated water inside the outer observation Snail No. 1 is the largest, are located in the immediate vicinity of the source (MP 13 and MP 14) or are located behind the shielding wall. Results for other MPs differed even more due to the higher contribution of the activated water from the remaining part of the circuit.

The C/E results for neutrons show an even larger deviation, although C/E values for MPs outside the wall are not available due to extremely low (not measurable) dose rate values. A really good agreement and a slight overestimation (less than a factor of 1.8) are observed near the source (MP 13 and MP 14). Extremely large deviations, underestimation up to a factor of 300, are observed for MPs: [0, 1, 2, 3 and 6]. However, this deviation can easily be neglected since all these MPs are in direct line of sight to the RPP and the primary contribution to the total neutron dose rate originates from the neutrons emitted by the reactor core (which scatter along the empty instrumentation pipe of the inner part of KATANA) rather than from those produced by activated water (^{17}N).

C. Computational results: double source location

The C/E values indicate that in some areas, the contribution of the activated water from the remaining part of the circuit is not negligible and must be taken into account as well. The second most relevant component in the circuit (besides Snail No. 1) is the pump with the high water volume (estimated at 4 litres). To further improve the C/E values, an additional global dose field rate calculation was performed using a double source location of activated water, i.e. inside Snail No. 1 and the pump. The source intensity for the pump was determined on a similar basis as for the Snail, i.e. calculated saturated activity, considering a longer decay/transfer time (the pump is located further along the circuit) and a larger water volume as: 3.99×10^7 Bq (^{16}N: 64.291 %), 3.54×10^3 Bq (^{17}N: 0.008 %) and 1.05×10^7 Bq (^{19}O: 35.701 %).

The results of the global γ (top) and neutron (bottom) dose field distributions around KATANA, based on a double source location of activated water (Snail No. 1 and the pump), are presented in Fig. 5 (horizontal cross-section: XY plane at Z = 86.4 cm) and Fig. 6 (vertical cross-section: XZ plane at Y = −493 cm, corresponding to the centre of Snail No. 1), with contour lines included for improved clarity. The impact of the second source, i.e. the pump, is not negligible, as increased dose rates can be clearly observed by the outward shift of the contour lines from both sources. This effect is particularly evident in areas closer to the pump or in regions that were previously well shielded from Snail No. 1 but are now in direct line of sight from the pump. In addition, the calculated relative statistical uncertainties of the global meshtally distribution for neutrons and gammas are provided on the right side in Fig. 5 and Fig. 6. The relative statistical uncertainty is both for neutrons and gammas practically below 1 % inside the experimental area and below 5 % in the remaining reactor hall. However, the reader should be aware that these are only statistical uncertainties of the Monte Carlo simulation whereby uncertainties of the RRs, geometry, etc. are not yet included. This will be an important aspect to evaluate for the future work.

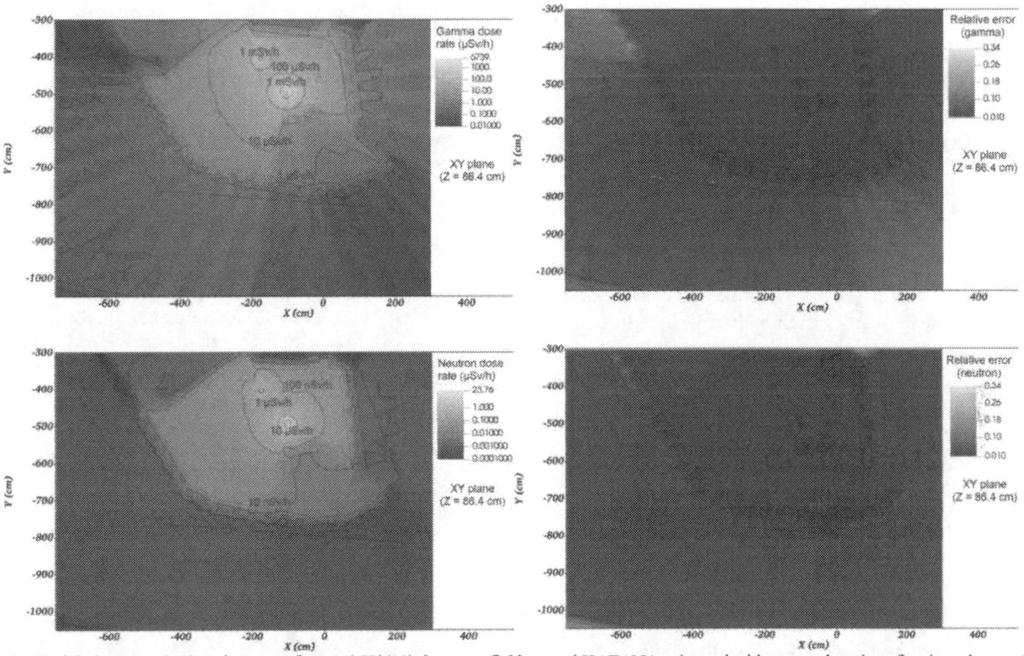

Fig. 5. Calculated global gamma (top) and neutron (bottom) H*(10) dose rate field around KATANA using a double source location of activated water Snail No. 1 + pump) in a horizontal cross-section view (Configuration 1, flow rate 0.4 L/s, full reactor power 250 kW). Contour lines are shown in red.

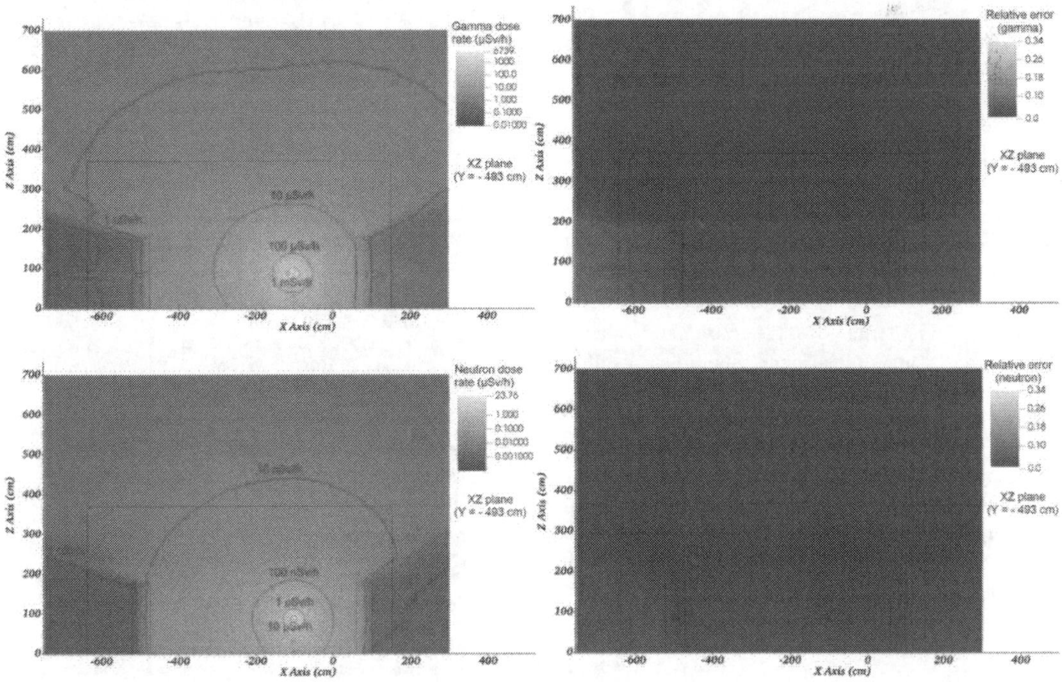

Fig. 6. Calculated global gamma (top) and neutron (bottom) H*(10) dose rate field around KATANA using a double source location of activated water Snail No. 1 + pump) in a vertical cross-section view (Configuration 1, flow rate 0.4 L/s, full reactor power 250 kW). Contour lines are shown in red.

Furthermore, results at the dedicated MPs, using both sources (Snail No. 1 and the pump), are presented in Table II as well. Improvement of C/E values is observed, especially at MPs near the pump, i.e. MP: 15, 16, 5, 18 and 20. Results indicate that considering the extra source of the activated water leads to the right path towards C/E being closer to unity (resulting in a higher number of highlighted values).

V. Conclusion

For radiological safety assessment, gamma and neutron dose rates were measured around the KATANA facility during steady-state operation of the JSI TRIGA reactor at full power. The highest gamma dose rate, reaching up to 5 mSv/h, was measured near the main observation part of the circuit, i.e. outer Snail no. 1 – short loop, with an additional local increase near the pump attributed to the higher water volume in that area. Outside the concrete shielding labyrinth, however, dose rates remained below the regulatory limit of 10 μSv/h, indicating that no further shielding is necessary. The neutron contribution to the total dose rate was significantly lower—more than three orders of magnitude lower compared to gamma contribution.

In addition, experimental results were supported with the calculations using a simplified convectional method excluding detailed fluid dynamics. A comparison between calculated and experimental (C/E) results revealed that gamma dose rates were underestimated by a factor of 2 to 5, while neutron dose rates near the primary source (Snail No. 1) were overestimated by a factor of 1.8. Including the pump as a secondary source improved agreement between simulation and measurement, emphasizing the need for accurate source modelling.

The C/E comparison of the dose rate values provided valuable insights into the water activation analysis of the KATANA facility and highlighted both the complexity and the challenges of carrying out such an absolute comparison. Further work, in particular the implementation of the use of state-of-the-art fluid activation codes (such as RSTM, FLUNED, GammaFlow & ActiFlow) to more precisely define and implement the distribution of activated water not only within the Snail No. 1 and the pump but also throughout the rest of the circuit and/or different loop configuration (Configuration 1, 2 or 3). In addition, such codes also allow for a more detailed source modelling, so that a mesh-based source distribution can be defined more correctly along the Snails (so far, the source has been distributed uniformly over the entire water volume of the Snail or the pump, thus implementing an additional simplification/uncertainty to the model).

Furthermore, additional research and the improvement of experimental results (either by using dose rate detectors with lower measurement uncertainty or by measuring the neutron and gamma flux with absolutely calibrated neutron and gamma detectors which have much lower uncertainty, i.e. HPGe) are extremely important and an essential part towards the goal of performing benchmark quality experiments at KATANA with a well-defined and low experimental and computational uncertainty.

Acknowledgment

The authors acknowledge the support of the Slovenian Ministry of Education, Science and Sport (project codes P2-0405 Fusion technologies; P2–0073 Reactor physics; PR-12326/PR-12842, Training of young researchers, NC-0022 Water activation in nuclear reactors). This work has been carried out within the framework of the EUROfusion Consortium, funded by the European Union via the Euratom Research and Training Programme (Grant Agreement No 101052200 — EUROfusion). Views and opinions expressed are, however, those of the author(s) only and do not necessarily reflect those of the European Union or the European Commission. Neither the European Union nor the European Commission can be held responsible for them.

References

[1] A. Žohar, L. Snoj, "On the dose fields due to activated cooling water in nuclear facilities," Prog. Nucl. Energy, vol. 117, 2019, Art. no. 103042. https://www.sciencedirect.com/science/article/pii/S0149197019301374
[2] N. ITER organization ITER, homepage (2024). Available: http://www.iter.org. Accessed on: 1.7.2025
[3] X. Litaudon et al., "Eurofusion contributions to iter nuclear operation," Nuclear Fusion, 2024
[4] D. Kotnik, A.K. Basavaraj, L. Snoj, I. Lengar, "Design optimization of the closed-water activation loop at the JSI irradiation facility," Fusion Eng. Des., vol. 193, 2023, Art. no. 113632. [Online]. Available: https://www.sciencedirect.com/science/article/pii/S0920379623002168
[5] Domen Kotnik, Julijan Peric, Domen Govekar, Luka Snoj, Igor Lengar, "KATANA water activation facility at JSI TRIGA, Part I: Final design and activity calculations," Nucl. Eng. Technol., 2024. [Online]. Available: https://doi.org/10.1016/j.net.2024.09.036
[6] . Peric and D. Kotnik, "KATANA: webpage," 2024. Accessed: 2025-07-22.
[7] L. Snoj, D. Kotnik et al., A Half-Century of Nuclear Research, Education and Training: Story of the JSI TRIGA Reactor, Annals of Nuclear Energy (2025). https://doi.org/10.1016/j.anucene.2024.111122
[8] J. Peric, D. Kotnik, L. Snoj, V. Radulović, Neutron emission from water activation: Experiments and modeling under fusion-relevant conditions at the KATANA facility, Fusion Engineering and Design, Volume 216, 2025, https://doi.org/10.1016/j.fusengdes.2025.115052.
[9] R. Pampin, E. Masia, F. Cau, D. Kotnik, R. Villari, A. Portone, Insights from applications of the RSTM tool for coupled CFD-activation fluid simulation, Fusion Engineering and Design, Volume 203, 2024, https://doi.org/10.1016/j.fusengdes.2024.114466.
[10] Eduardo Masiá, et. al., Application of the radio species transport model to the JSI water activation loop, Fusion Engineering and Design, Volume 219, 2025, https://doi.org/10.1016/j.fusengdes.2025.115264.
[11] M. De Pietri, J. Alguacil, E. Rodríguez and R. Juárez, Development and validation in water of FLUNED, an open-source tool for fluid activation calculations, Computer Physics Communications 291, 108807 (2023). https://doi.org/10.1016/j.cpc.2023.108807
[12] T. Berry, C. Nobs, A. Dubas, R. Worrall, T. Eade, J. Naish and L. Packer, Integration of fluid dynamics into activation calculations for fusion, Fusion Engineering and Design 173, 112894 (2021). https://doi.org/10.1016/j.fusengdes.2021.112894
[13] D. Govekar, J. Peric, D. Kotnik, V. Radulovic. Calibration of HPGe detectors and usage of prompt gamma rays to extend detector efficiency curve in above 2 MeV range.
[14] EUROfusion, "The demonstration power plant: DEMO", https://euro-fusion.org/programme/demo/, (2024), accessed: 2025-05-06.
[15] D. Brown, et.al., ENDF/B-VIII.0: The 8th Major Release of the Nuclear Reaction Data Library with CIELO-project Cross Sections, New standards and Thermal Scattering Data, Nuclear Data Sheets 148, 1 (2018), special Issue on Nuclear Reaction Data.
[16] J. García, J. P. Catalán and J. Sanz, Development of the automatic void generation module in GEOUNED conversion tool, Fusion Engineering and Design, 168, 112366 (2021).
[17] ICRP, Data for Protection Against Ionizing Radiation from External Sources: Supplement to ICRP Publication 15, ICRP Publication 21 (Pergamon Press, Oxford, 1973).
[18] H. Rossi et al., Protection against neutron radiation, NCRP-38, National Council on Radiation Protection and Measurements (1971).

Reactor test of beryllium as self-powered neutron detector emitter for fast neutron flux measurement

Elsa Dupin[1], Loïc Barbot[1,*], Stéphane Normand[2], Jean-Marc Fontbonne[3], Nicolas Thiollay[1] and Christophe Destouches[1]
[1]CEA, DES, IRESNE, DER, Cadarache, F-13108 Saint Paul-lez-Durance, France
[2]CEA DAM Île-de-France Bruyères-le-Châtel, 91297 Arpajon cedex, France
[3]Laboratoire de Physique Corpusculaire, ENSICAEN, 6 Bd Maréchal Juin, 14000 Caen, France
(*) corresponding author loic.barbot@cea.fr

***Abstract*— Neutron flux in nuclear reactors causes damages to material components. Fast neutrons, above 1 MeV, modify mechanical and structural properties of materials. Some experiments carried out in research reactors study the ageing of materials exposed to high neutron fluxes. In these experiments, the selective on-line monitoring of fast neutron flux in a water-pool type reactor environment remains a measurement challenge. The aim of this work is to experimentally validate the use self-powered neutron detector technology to monitor fast neutron flux in a water-cooled reactor.**

After a systematic review of the periodic element table, beryllium appears to be the most suitable emitter material for measurements sensitive and selective to fast neutrons. Due to its threshold cross section to neutrons and its low interaction probability with gamma rays, beryllium seems the appropriate candidate as emitter material for reactor environment use. Beryllium SPND prototypes were designed and manufactured.

In June 2024, beryllium SPND prototypes have been tested in the Slovenian TRIGA Mark II reactor (Jožef Stefan Institute - JSI). These specific reactor tests in well-characterized irradiation conditions could enforce fast neutron SPNDs as a new option for on-line monitoring of technological irradiations.

***Keywords* — Self-powered neutron detector, SPND, fast neutron, neutron flux measurement.**

I. Introduction

In material testing reactors (MTR), like future Jules Horowitz Reactor (JHR), accelerated ageing of nuclear reactor components is achieved through high neutron flux exposition as anticipation of structural changes and potential embrittlement, fast neutrons being the main cause of steel damages during mixed neutron and gamma irradiations due to atom displacements. Currently, on-line measurement of fast neutrons flux in a water pool-type reactor environment remains a challenge to monitor long-term irradiation experiments. The CEA Fast Neutron Detection System (FNDS) [1], composed of two fissions chambers coupled with an evolution code is the one of the very few existing system for on-line monitoring of fast neutron fluxes in research reactors.

Self-Powered Detectors (SPDs) have been developed since the 1960s for on-line measurement of thermal neutron flux [2-4], and recently for γ ray flux measurement [5]. This work falls in the frame of an attempt to drive the SPD technology toward a fast neutron selective self-powered neutron detector (SPND).

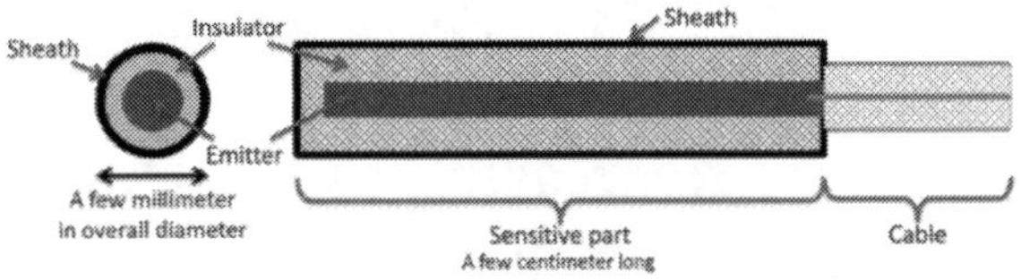

Fig. 1. Standard Self-Powered Detector scheme.

SPNDs are coaxial detectors that operate without high voltage, generally composed of three components: an emitter, an insulator and a sheath. Operating only for high fluxes, SPNDs are ideally adapted to on-line monitoring of irradiation experiments. Material choice and, in particular, emitter choice favor interactions with specific particles of interest. In our case, SPND material design must maximize fast neutron contribution.

A preliminary phase consisted of designing a detector mostly sensitive to fast neutrons in an environment where thermal neutrons and γ rays are predominant [6]. Using an in-house SPND simulation tool [7] and on purpose advanced model based on the Shockley-Ramo theorem [8], a SPND design was developed with high enough confidence meriting prototype fabrication and the set-up of a dedicated experimental campaign in the JSI TRIGA Marl II reactor.

Tests first qualified the irradiation locations with different measurements techniques such as dosimetry, standard SPD measurements and axial profiles with fission chambers and ionization chamber. It also validates the irradiation conditions (thermal neutrons, fast neutrons, and gamma rays) in support of dedicated TRIPOLI4® [9] calculations for each experimental location.

Then, the study focuses on performance analysis of the beryllium SPND prototypes in a reactor environment. The analysis yields encouraging outcomes, particularly following a comparison with SPND numerical modelling. These results permit an evaluation of beryllium benefits as an emitter.

II. Design and Fabrication of Fast Neutron Sensitive Prototypes

A. Emitter material identification

Fig. 2 presents in lethargy unit neutron and γ spectra in the Slovenian TRIGA Mark II reactor MP17 experimental location, used to design a SPND sensitive to neutrons with an energy above 1 MeV. In this location, both total neutron flux level γ-ray flux level are roughly $2\cdot10^{13}$ p·cm^{-2}·s^{-1} (particles integrated over the respective whole energy spectrum).

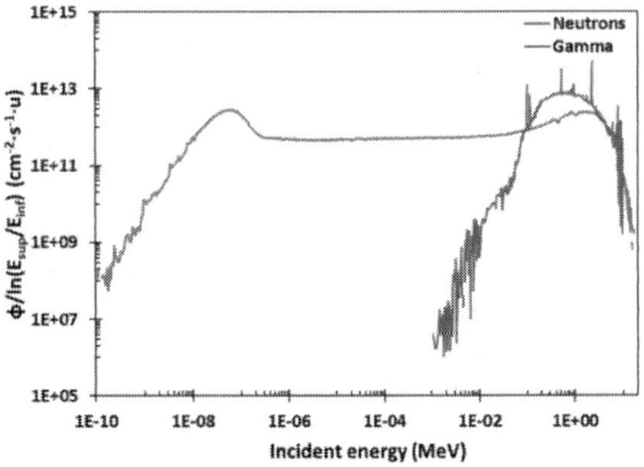

Fig. 2. Energy spectra in lethargy unit, in Slovenian TRIGA Mark II reactor MP17 position (calculated), with a neutron and γ ray total flux level of ~$2\cdot10^{13}$ cm^{2}·s^{-1} each.

Main constraint is thermal neutron presence, whose capture cross-sections are generally much larger than fast neutron ones. The (n, γ) reaction is usually the major one in a standard SPNDs. Cross-sections are generally of up to several hundred barns for neutrons with an energy below 0.625 eV (thermal neutrons), whilst only a few 0.1 barn at most for neutrons with an energy around 1 MeV (fast neutrons). γ rays from the reactor environment may also create a significant additional contribution to the SPND current.

The few main criteria for a good emitter material candidate are metallic shape (good charge conductivity) and generator of a β^- emitter with a relatively short half-life and a sufficient enough energy to leave the emitter and reach the insulator. At the end, materials and associated dimensions have to integrate these constraints while making the SPND output signal mainly representative to the fast neutron flux level.

Based on a systematic review of the periodic element table [6], the most promising material as SPND emitter is beryllium, only composed of one isotope: ^{9}Be. It is also confirmed by a literature survey [10-12], but applications were for fusion or fast reactor. Beryllium, with its (n, α) threshold interaction (667 keV) is creating ^{6}He, emitting 1.6 MeV β^- particles with 808 ms half-life [13]. No other neutron interactions generate any charge significant to the SPND signal. This is confirmed on Fig. 33 which presents the convoluted neutron reaction rate on beryllium in the TRIGA Mark II reactor, calculated with equation (1).

$$R_n(E_n) = N_{mat}\cdot\sigma(E_n).\phi(E_n) \quad (1)$$

Where $R_n(E_n)$ is the neutron reaction rate at neutron energy E_n [cm^{-3}·s^{-1}];
N_{mat} is the bulk density of studied isotope [cm^{-3}];
$\sigma(E_n)$ is the cross-section at E_n [cm^{2}];
$\phi(E_n)$ is the neutron flux at E_n [cm^{-2}·s^{-1}].

Using cross-sections from ENDF/B-VIII library [14], reaction rates are calculated for interactions that produce radioactive β isotopes with short enough half-lives. Fig. 3 confirms the absence of any neutron interaction in the thermal and epithermal energy range below 667 keV.

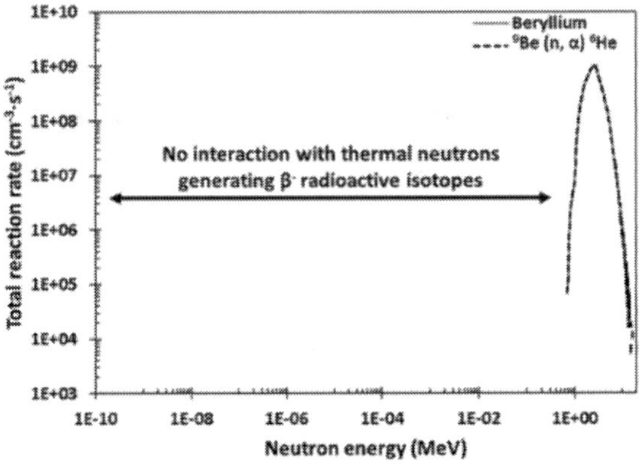

Fig. 3. Neutron reaction rate on beryllium, calculated using TRIGA Mark II neutron spectrum. No neutron interaction in the thermal and epithermal energy range.

Literature does not mention any reactor tests with a Be emitter SPND. However, Raj [11] presented and tested this emitter under a 14 MeV neutron generator. Beryllium is believed to be an interesting emitter material for water pool-type reactor application providing an optimized design.

B. SPND prototype design

To validate this innovative material emitter, the other SPND components (insulator and sheath) are chosen from materials commonly used in SPND fabrication: Inconel600® sheath and magnesia insulator for their respective low cross-sections to thermal neutrons

Starting from the CEA MATiSSe simulation tool [7, 15], an advanced model based on the Shockley-Ramo theorem [16, 17] was developed to deal with low sensitivity SPND case where multiple weak contributions are competing against each other and giving rise to a non-monotonous electric field in the insulator [8]. This was used for a parametric study of the beryllium emitter geometry to determine an optimum beryllium rod diameter.

Fig. 34 presents parametric results of the total SPND current to expect and its main contributions for the emitter diameter variation from 2 to 6.5 mm. Configuration with a ∅4 mm beryllium emitter is found to be the most relevant as thermal neutron and gamma contributions tend to cancel each other out, resulting a total current of almost 3.8 nA (20 cm long emitter), supposedly due at 93.4% to fast neutrons.

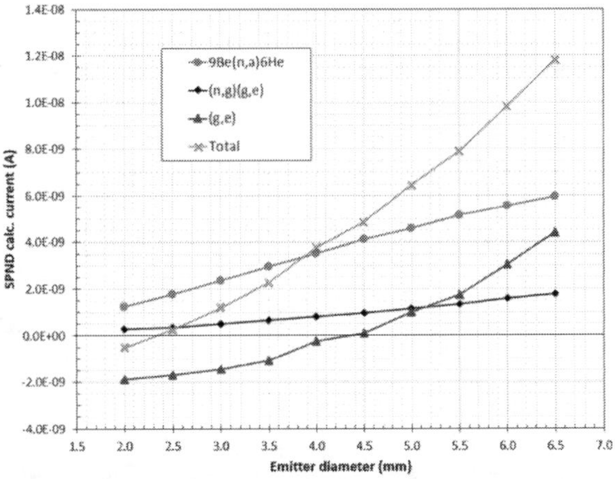

Fig. 4. Beryllium emitter SPND total current and main contributions calculated for different emitter diameter values from 2 to 6.5 mm.

In conclusion to the design phase, beryllium SPND prototypes were proposed with the global geometry presented on Fig. 35, including a ∅4 mm beryllium emitter, a 0.9 mm thick MgO insulator and a 0.2 mm thick Inconel600® sheath, making a ∅6.6 mm overall diameter detector.

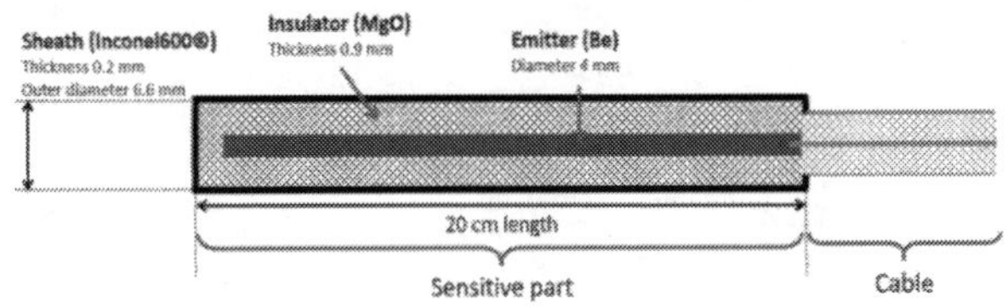

Fig. 5. Beryllium SPND design with a ∅4 mm and overall diameter of 6.6mm.

C. *Be SPND prototype fabrication*

Beryllium was identified as the best option for a fast neutron selective SPND, nevertheless it's a material with many constraints when it comes to processing. Considered as double use good with regards to nuclear non-proliferation regulations, even raw material requires export control license and end-user certificate. Being also carcinogenic material, machining and handling requires precautions and glove box use.

Five Be SPND prototypes were finally assembled by the Photonis – Exosens Group company (Brive-la-gaillarde, France) using ∅1 mm Inconel600® MI cable from the Thermocoax company (Caligny, France).
Three Be SPND prototypes were tested in the JSI TIGA Mark II reactor.

III. Reactor Tests

Be SPND prototypes were irradiated through incore reactor tests during a dedicated CEA experimental campaign, along with side neutron and gamma monitors for irradiation location thorough characterization.

A. *Experimental Location Characterization*

Three incore experimental locations (A1, C3 and F6) were picked within the TRIGA Mark II reactor core for the tests (see Fig. 6). They present identical geometry (Ø34.5mm aluminum tubes) but rather different neutron and gamma flux levels and spectra. Those were assessed using a combination of activation foils, U235 and U238 CEA miniature fission chambers, miniature ion chamber and CEA unfolding method.

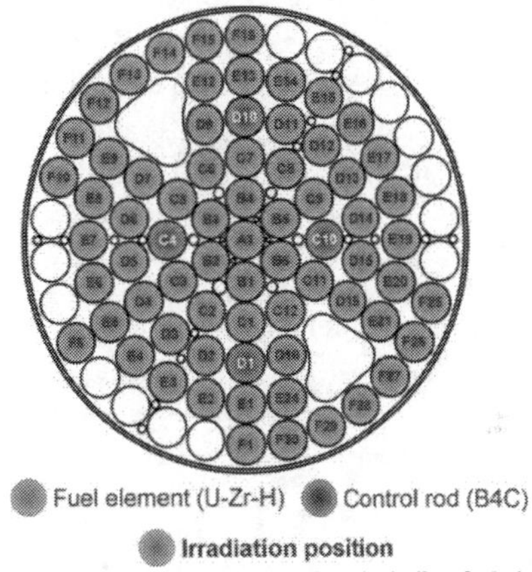

Fig. 6. Sketch of the TRIGA Mark II top view, including fuel elements (red), control rods (grey) and experimental locations (blue).

Absolute neutron (including three energy bins) and gamma flux levels along with neutron and gamma flux profiles were then achieved in the three experimental locations. All data are presented on Fig. 7 to 9 and Table I. Axial profiles ware also performed to get integrated neutron and gamma flux values to be used for Be SPND prototype analysis over their 20 cm emitter length.
Delayed gamma flux proportion to the total gamma flux was experimentally estimated using the reactor SCRAM (*Safety Control Axe Man*) operation mode.

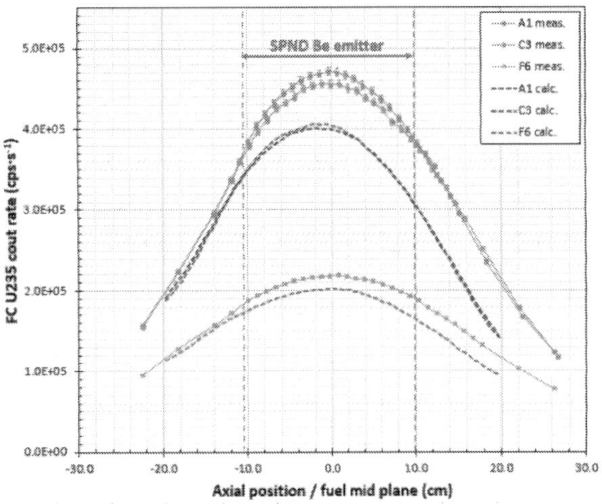

Fig. 7. U235 miniature fission chamber axial profiles in A1, C3 and F8 experimental locations (experiment to calculation comparison).

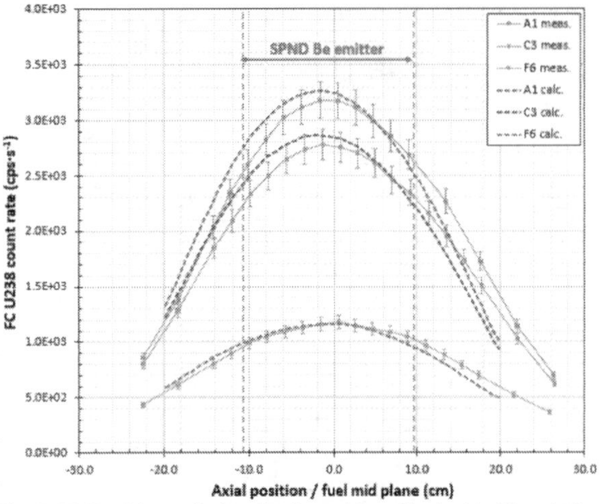

Fig. 8. U238 miniature fission chamber axial profiles in A1, C3 and F8 experimental locations (experiment to calculation comparison).

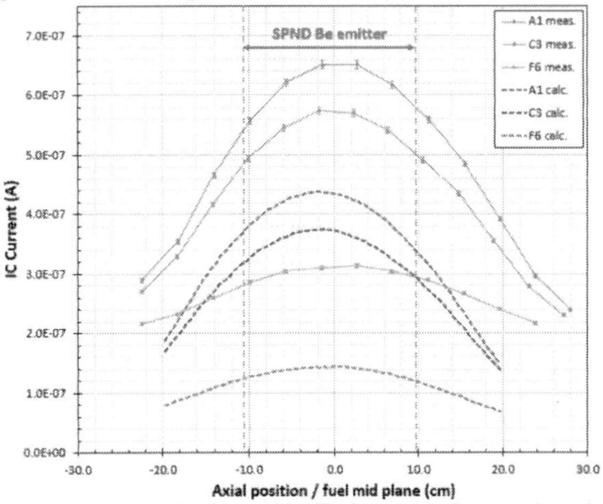

Fig. 9. Miniature ion chamber axial profiles in A1, C3 and F6 experimental locations (experiment to calculation comparison).

TABLE I. NORMALIZED NEUTRON AND GAMMA FLUXES IN THE A1, C3 AND F6 EXPERIMENTAL LOCATIONS OF THE TRIGA MARK II REACTOR

Exp. location	**A1**	**C3**	**F6**	
Power		250kW		
$\phi_{n,total\ norm.}$	$2.25\cdot10^{13}$	$1.95\cdot10^{13}$	$7.62\cdot10^{12}$	($n.cm^{-2}.s^{-1}$)
$\phi_{<0.630eV}$	37.4%	42.5%	55.0%	
$0.630eV<\phi<1MeV$	25.9%	24.7%	20.1%	
$\phi_{>1MeV}$	36.7%	32.8%	24.9%	
$\phi_{\gamma,total\ norm.}$	$4.27\cdot10^{13}$	$3.70\cdot10^{13}$	$1.40\cdot10^{13}$	($\gamma.cm^{-2}.s^{-1}$)
$\%\phi_{\gamma,delayed}$	41.9%	40.4%	41.6%	

B. Standard SPD Tests

Along with the three Be SPND prototypes, some CEA standard SPDs were also tested in the A1, C3 and F6 experimental locations to confirm the neutron and gamma data consistency using CEA models and tools for SPND simulations [6, 8]. Standard SPDs, including rhodium, cobalt and tubular bismuth (SPGD) emitters, full geometries are presented on Fig. 10.

Measured and calculated currents are compared in Table II with C/E (calculation to experiment) ratios consistent with previous results [15]

TABLE II. STANDARD SPD C/E COMPARISONS IN THE TRIGA MARK II REACTOR

SPND	**Rhodium**	**Cobalt**	**Bismuth**	
Location		A1		
$I_{measured}$	38.8±0.1	6.45±0.06	10.7±0.4	(nA)
$I_{calculated}$	35.3*	5.30*	12.6*	(nA)
C/E	0.91	0.82	1.18	

* *Statistical uncertainties < 1%*

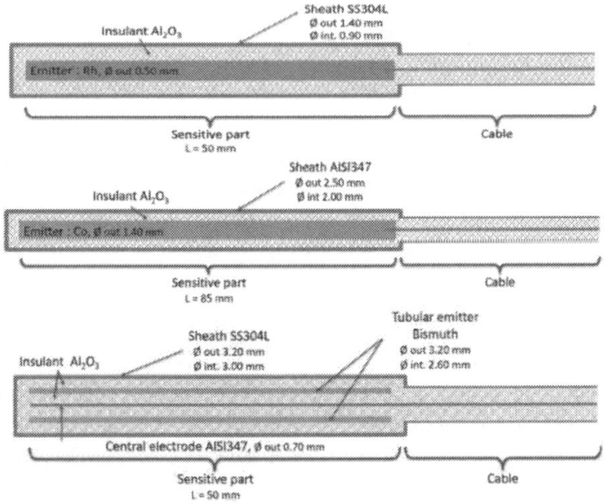

Fig. 10. Geometries of standard CEA SPNDs (rhodium, cobalt and bismuth emitters respectively)

C. Beryllium SPND Prototype Tests

The three Be SPND prototypes are each mounted in individual irradiation rigs (Fig. 11) made out of an aluminum tube. It is equipped with a large clearance to limit neutron and gamma attenuation, and some winglets at 120° for a precise and reproducible axial positioning in the center of experimental locations. The irradiation rig height is designed to lay SPND emitters at the fuel mid plane.

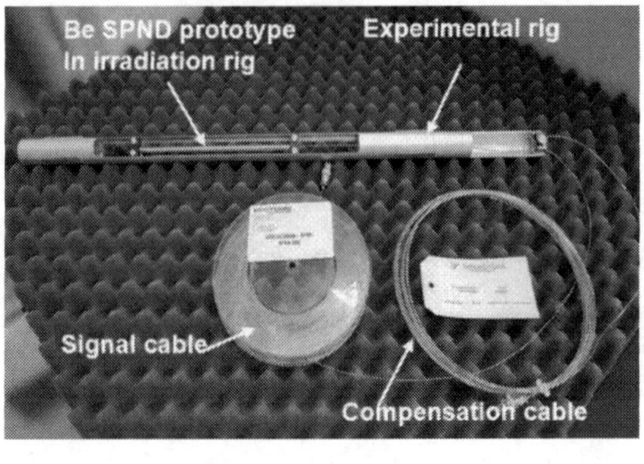

Fig. 11. Beryllium SPND irradiation rig

Fig. 11 also shows a compensation cable used along with one of the prototypes and stopping before the sensitive part to enable a posteriori cable current correction.

The three Be SNPD prototypes were irradiated and shuffled within the three irradiation locations several times to test their reproducibility as well as their sensitivities to the three different mixed fields. The TRIGA Mark II reactor was operated on purpose at different power steps to check prototype linearity and during different transient operations (ramp-up, SCRAM, cooling…) to address the prototype behavior and time response.

As example, Fig. 12 shows the three prototype currents over time during three reactor power steps (50, 150 and 250 kW).

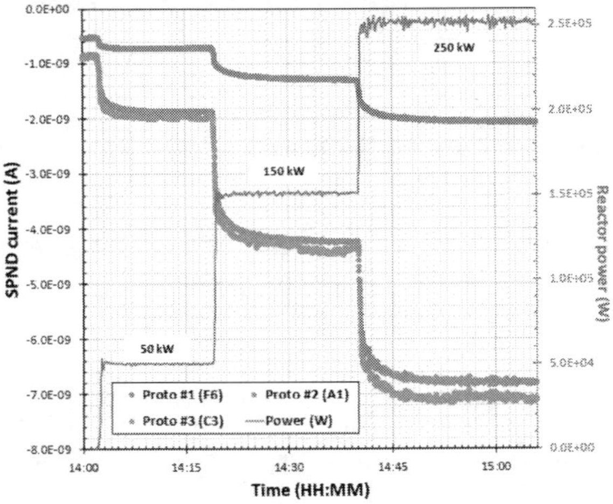

Fig. 12. Beryllium SPND prototype currents during reactor power steps.

For the two central core locations, Be SPND currents stabilized around a negative value of -7nA, to be compared to the ~4 nA expected value (see section II.B).

The rest of this document will focus on explaining what went wrong although it is still demonstrating the value of beryllium for a fast neutron sensitive SPND.

IV. Unexpected Result Analysis

A. False representativeness of sizing data

The complete design phase of the Be emitter SPND was achieved using commonly used neutron and gamma data coming from numerous past (mostly thermal neutron oriented) experiments in rather close incore irradiation locations in the TRIGA Mark II reactor, which was never an issue while designing new experiments/detectors.

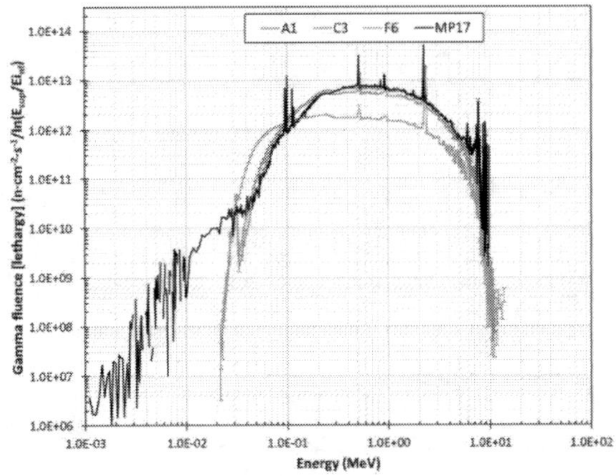

Fig. 13. Comparison of sizing gamma spectrum in MP17 to experiment gamma spectra (all calculated).

Unfortunately, attention was not paid enough to the gamma contribution and therefore to the gamma flux level and spectra definition. Up top now, gamma contribution to SPNDs was not in competition with the neutron one, it was addressed only for accuracy purposes. Here, the design phase demonstrated almost balanced fast neutrons and gamma contributions.

Fig. 13 shows differences between sizing gamma spectrum and actual gamma spectra determined during the tests, that probably explain the unexpected stronger and negative values of the prototype currents.

B. Use of transient operation data recordings

Thorough analysis of Be SPND recordings during SCRAM transients highlights a recurrent behavior of SPND currents always few seconds after reactor shut downs.

Fig. 14, as example, is showing a negative bump after the current sharp transient due to the SCRAM. Current being negative, the negative bump is due to a decreasing with time positive contribution, which is then progressively counterbalanced by a slowly appearing new contribution.

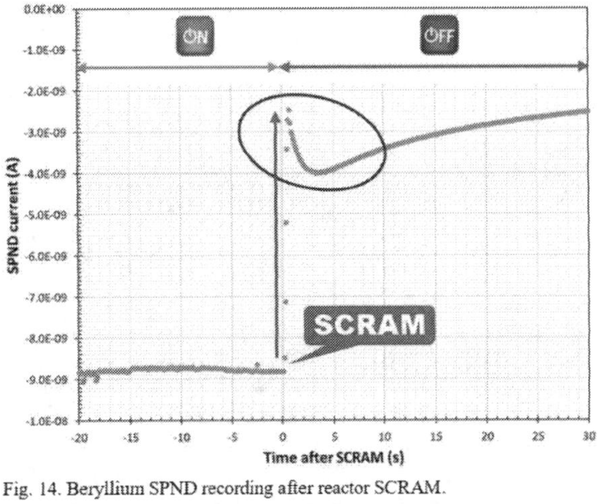

Fig. 14. Beryllium SPND recording after reactor SCRAM.

Studying the superimposition of theoretical contributions to the beryllium SPND, it was found that the observed bump could reasonably due to the ^{9}Be contribution combined with delayed gamma coming from the reactor core and activated structures, as illustrated on Fig. 15.

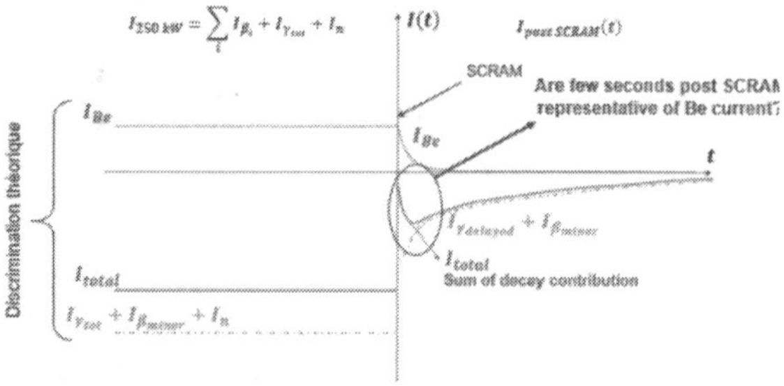

Fig. 15. Theoretical contributions to the beryllium SPND.

Using the Wigner-Way formula [18-20] which describes the decay shape of delayed gamma after a SCRAM, the commented equation on Fig. 16 is developed to model the decay of a Be SPND signal after a SCRAM. It is then possible to dig out the "measured" ^{9}Be currents representative of the fast neutron contribution. Delayed gamma contribution mentioned corresponds to the mean value of delayed gamma percentages experimentally estimated in Section III.A (41.3%).

$$\lambda_{\beta_i} = \frac{\ln(2)}{T_{1/2\,\beta_i}}$$

$$I_{post\,SCRAM}(t) = \underbrace{\sum_i^{t=0\,s} I_{\beta_i} \cdot \exp(-\lambda_{\beta_i} \cdot t)}_{\beta^{-}\ \text{contributions}} + \underbrace{I_{\gamma_{ret}} \cdot \left(t^{-0,2} - (T_{250kW} + t)^{-0,2}\right)}_{\text{Delayed } \gamma \text{ contribution}}$$

$$I_{\gamma_{tot}} = \frac{1}{41,3\%} \cdot I_{\gamma_{ret}}$$

Fig. 16. Model developed to describe the decay shape after SCRAM, applied with experimental values. In green, the Wigner-Way formula for γ decay shape [18-20]. In red, contribution currents at t = 0s (SCRAM triggered).

Measured and calculated (using CEA simulation tools) ^{9}Be absolute contributions to the SPND currents are compared for the three experimental locations, showing C/E ratio close to 1 in every location. Moreover, the ^{9}Be contribution value is also close to the 3.8 nA expected in section II.B.

TABLE III
CALCULATED TO MEASURED ^{9}BE CONTRIBUTION COMPARISONS IN DIFFERENT EXPERIMENTAL LOCATIONS. SIGNAL AND FAST NEUTRON RATIOS COMPARISONS

	A1			C3			F6		
	C	E	C/E	C	E	C/E	C	E	C/E
SPND current coming from 9Be (n, α) 6He									
I_{Be} (nA)	3.7	3.2	**1.2**	2.9	2.8	**1.05**	0.84	0.79	**1.06**
Irradiation location ratios									
$\frac{I_{Be_{loc}}}{I_{Be_{A1}}}$	1	1		0.78	0.85		0.23	0.24	
$\frac{\phi_{N_{rap_{loc}}}}{\phi_{N_{rap_{A1}}}}$			**1**			**0.78**			**0.23**

Looking also closely to the ratios between experimental locations in terms of ^{9}Be currents and fast neutron fluxes, they are very consistent among each other.

Gathering all those partial experimental results help the authors to claim that this dedicated experimental campaign validate the use of beryllium as emitter for inline fast neutron flux monitoring.

V. CONCLUSIONS

The initial objective of this work was to design a fast neutron selective SPND in a water reactor environment (high thermal neutrons and γ mixed field). Beryllium SPND prototypes were designed and manufactured including a ^{9}Be emitter, a MgO insulator and an Inconel600® sheath. Qualification tests were conducted at the JSI TRIGA Mark II reactor to evaluate beryllium material suitability.

The proposed design does not give the expected results, with prototype signal mostly sensitive to gamma flux. However, the use of beryllium as SPND emitter is experimentally validated. The signal study at the SCRAM trigger demonstrated the possibility to discriminate different contributions to the total SPND signal and to extract a beryllium contribution which follows the fast neutron flux as expected.

This work thus demonstrates the potential of beryllium as an emitter for online fast neutron measurements, although design optimization work remains to be done to remove the gamma component of the signal.

ACKNOWLEDGMENT

The PhD thesis work of Elsa Dupin was financially supported by the CEA Gen23 program, in the frame of the instrumentation project (INSNU) to develop innovative and advanced nuclear and non-nuclear instrumentation in preparation of the CEA Jules Horowitz Reactor (JHR) commissioning.

REFERENCES

[1] D. Fourmentel, J. F. Villard, C. Destouches, B. Geslot, L. Vermeeren, and M. Schyns, "In-pile qualification of the fast-neutron-detection-system," in *ANIMMA*, Belgium, 2017.

[2] J. W. Hilborn, "Self-Powered Neutron Detectors for Reactor Flux Monitoring," *Instrumentation and Measurements*, vol. 22, no. 2, pp. 69–74, Feb. 1964.

[3] M. Grin, "Collectrons, Self-Powered Neutron Flux Detectors - Part I : Theoretical considerations," Commission of the european communities, EUR4775e, 1972.

[4] O. Strindehag, "Self-Powered Neutron and Gamma Detectors for In-Core Measurements," Studsvik, Nyköping, Sweden, AE-440, Nov. 1971.

[5] H. Carcreff, L. Barbot, and L. Vermeeren, "Gamma radiation detector, and associated method for making same and uses thereof," WO 2010/103049 A1, Sep. 16, 2010

[6] E. Dupin, L. Barbot, S. Normand, JM. Fontbonne, N. Thiollay, "Design and reactor test of a selective fast neutron self-powered detector for applications in water-pool type reactor," Poster presented in 2023 8th International Conference on Advancements in Nuclear Instrumentation Measurement Methods and their Applications (ANIMMA), Lucca, June 2023.

[7] L. Barbot, V. Radulovic, V. Dewynter-Marty, F. Malouch, and F. Lopez, "Experimental validation of a Monte Carlo based toolbox for Self-Powered Neutron and Gamma Detectors simulation in the OSIRIS MTR," presented at the PHYSOR 2016 - Unifying Theory and Experiments in the 21st Century, Sun Valley, Idaho, USA, May 2016.

[8] E. Dupin, L. Barbot, S. Normand, J-M. Fontbonne, N. Thiollay, C. Destouches, "Theory and modelling for the design of a selective fast neutron self-powered detector for applications in water pool type reactors", presented at the 2023 IEEE Nuclear Science Symposium (November 4-11 2023, Vancouver, Canada)

[9] E. Brun, F. Damian, C.M. Diop, E. Dumonteil, F.X. Hugot, C. Jouanne, Y.K. Lee, F. Malvagi, A. Mazzolo, O. Petit, J.C. Trama, T. Visonneau, A. Zoia, Tripoli-4®, CEA, EDF and Areva reference Monte Carlo code, Annals of Nuclear Energy no 82, 2015, p. 151-160

[10] J. W. Jr. Upton, D. P. Brown, and W. G. Spear, "In-core, Self-powered fast neutron flux monitors," presented at the American Nuclear Society Power Division Topical Meeting, Williamsburg, Virginia, Aug. 1970, p. 13. Accessed: Mar. 08, 2022. [Online]. Available: https://www.osti.gov/servlets/purl/4059106

[11] P. Raj, "Development and Testing of Self–Powered Detectors for Nuclear Measurements in Fusion Reactors," PhD Thesis, Department of Mechanical Engineering, Karlsruhe Institute of Technology, 2019. doi: 10.5445/IR/1000096884.

[12] S. Jianxiong *et al.*, "Reactor core fast neutron flux self-powered detector," 2020CN-U3222087, Dec. 28, 2020

[13] Nuclear Energy Agency, "JANIS Web." Accessed: Dec. 17, 2021. [Online]. Available: https://www.oecd-nea.org/janisweb/

[14] D. A. Brown *et al.*, "ENDF/B-VIII.0: The 8th Major Release of the Nuclear Reaction Data Library with CIELO-project Cross Sections, New Standards and Thermal Scattering Data," *Nuclear Data Sheets*, vol. 148, pp. 1–142, Feb. 2018, doi: 10.1016/j.nds.2018.02.001.

[15] L. Barbot *et al.*, "Calculation to experiment comparison of SPND signals in various nuclear reactor environments," in *2015 4th International Conference on Advancements in Nuclear Instrumentation Measurement Methods and their Applications (ANIMMA)*, Lisbon: IEEE, Apr. 2015, pp. 1–7. doi: 10.1109/ANIMMA.2015.7465558.

[16] S. Ramo, "Currents Induced by Electron Motion," *Proc. IRE*, vol. 27, no. 9, pp. 584–585, Sep. 1939, doi: 10.1109/JRPROC.1939.228757.
[23]W. Shockley, "Currents to Conductors Induced by a Moving Point Charge," *Journal of Applied Physics*, vol. 9, no. 10, pp. 635–636, Oct. 1938, doi: 10.1063/1.1710367.

[17] W. Shockley, "Currents to Conductors Induced by a Moving Point Charge," *Journal of Applied Physics*, vol. 9, no. 10, pp. 635–636, Oct. 1938, doi: 10.1063/1.1710367.

[18] T. Van Thuong, O. L. Tashloykov, S. M. Glukhov, D. E. Shumkov, and Yu. V. Volchikhina, Experimental and theoretical justification of passive heat removal system for irradiated fuel assemblies of the nuclear research reactor in a spent fuel pool, Nuclear Engineering and Technology **55**, 2088 (2023).

[19] M. Arkani and M. Gharib, Residual heat estimation by using Cherenkov radiation in Tehran Research Reactor, Nuclear Instruments and Methods in Physics Research Section A: Accelerators, Spectrometers, Detectors and Associated Equipment **596**, 417 (2008).

[20] S. Kadalev, Assessment of residual heat generation at cyclic operation of a research reactor as input data for thermal hydraulic calculations and safety analysis, Annals of Nuclear Energy **72**, 182 (2014).

Towards the Direct Observation of Xenon-135 Poisoning in a Zero-Power Reactor via Gamma Spectroscopy

Oskari Pakari[1,2,*], Andrew Lucas[3], Flynn Darby[3], Michel Saliba[1], Vincent Lamirand[1,2], Andreas Pautz[1,2], Sara Pozzi[3]
[1]École Polytechnique Fédérale de Lausanne (EPFL), 1015 Lausanne, Switzerland
[2]Paul Scherrer Institut (PSI), 5232 Villigen, Switzerland
[3] Department of Nuclear Engineering and Radiological Sciences, University of Michigan, Ann Arbor, Michigan, 48105, USA
(*)oskari.pakari@epfl.ch

***Abstract*—Xenon-135 (Xe-135) is a high-yield fission product with high neutron-capture cross section. it is a commonly encountered reactor reactivity poison. Direct observation of Xe in reactors typically relies on off-gas measurement techniques, which can be challenging or costly to implement and often require high-flux reactors. Gamma spectroscopy of irradiated U or Pu samples provides an alternative method for obtaining fission yields and population estimates. When predicting the Xe populations of a given reactor, we rely on calculations and models to estimate the instantaneous reactivity worth and post-shutdown dynamics of Iodine-135 (I-135) and Xe-135 decay that leads to the so-called 'Xenon pit'. In this work, developing towards a direct estimate of the total Xe-135 population in an operating reactor, we measured the gamma-ray emissions of Xe-135 and I-135 emanating from the CROCUS zero-power reactor using a high-purity germanium detector in an irradiation channel. CROCUS is a uranium fueled, light water moderated, zero-power reactor operated by the Laboratory of Reactor Physics and Systems Behavior at EPFL, Switzerland. We placed an Ortec GEM-15180-P detector in the irradiation beam port of CROCUS at roughly 4.3 meters from the core center. We measured the gamma ray spectra using a CAEN DS5730 digitizer to obtain gamma-ray time stamps and pulses for the start-up, at-power operation for 1 hour at 20 W, and subsequent shutdown of the reactor for 100 hours. At power, the spectrum is dominated by the gamma ray emission from fission and neutron capture in water and aluminum. Following shutdown, the spectrum transforms, displaying a dense set of characteristic line emissions. To observe Xe-135, we used its 250 keV emission, whilst for I-135 we observed the 1260 keV emission. We binned the spectra in time for 60-minute intervals to accumulate sufficient counts under the photopeak of interest. The time dependent Xe-135 population was observed to follow the expected 'Xenon pit' shape, i.e. we were able to directly observe the Xe-135 and I-135 populations in the reactor. These preliminary data are currently only processed into count rates over time. Future work includes determining the detector efficiency in counts per emission in the fuel to finally estimate the whole-core Xe-135 and I-135 populations. Given the published IRPhE geometry of CROCUS, this dataset holds potential as a validation benchmark for reactor poison prediction models.**

***Index Terms*—Gamma spectroscopy, Nuclear Research Reactor, Xenon-135.**

I. Introduction

NEUTRON-induced fission produces prompt neutrons that, given suitable reactor geometry, can sustain a chain reaction, whose rate is controlled by inserting neutron absorbers—control rods or soluble poisons—to maintain safe, stable operation. In addition to engineered poisons, the fission product ^{135}Xe (thermal absorption cross section $\sim 2.6\times10^6$ b), also generated via decay of the fission product ^{135}I, exerts a strong, time-dependent reactivity feedback that gives rise to the so-called "xenon pit" or "iodine pit" following power changes [1], [2].

Conventional reactor neutron detectors are used to infer ^{135}Xe levels indirectly via calculations, whereas in-situ gamma-ray spectroscopy can identify the 249.79 keV line of ^{135}Xe directly. High-purity germanium (HPGe) detectors fulfill the energy resolution requirements to detect this emission, but require fast digitizers with pile-up rejection and operation in low-flux (zero-power) environments to avoid damage and saturation [3], [4]. Recent experiments using HPGe detectors in low-power reactors include measurements at LR-0 [5] and CROCUS [6]. In both, the integral gamma ray spectra of long multi-hour measurements were analyzed.

Herein, we report time-resolved HPGe measurements of ^{135}Xe and ^{135}I after shutdown of the CROCUS zero-power reactor after a 20 Wh irradiation. Using fast digitizer electronics, we achieve $\sim 2\times10^4$ cps immediately after shutdown, yielding sufficient statistics to track xenon dynamics. In Section II we detail the setup and analysis and in Section III we present the results.

II. Methods

A. Decay-Chain Kinetics

After shutdown of the reactor, the iodine ($I(t)$) and xenon ($X(t)$) populations follow these equations:

$$I(t) = I_0\, e^{-\lambda_I t}, \tag{1}$$

$$X(t) = \frac{\lambda_I I_0}{\lambda_X - \lambda_I}\left(e^{-\lambda_I t} - e^{-\lambda_X t}\right) + X_0\, e^{-\lambda_X t}, \tag{2}$$

Fig. 1. Picture of the CROCUS zero-research reactor, uranium fueled and water moderated with a maximum power of 100 W [14].

where I_0, X_0 are the populations of ^{135}I and ^{135}Xe at the moment of shutdown [7]. As ^{135}Xe is produced by ^{135}I even after shutdown, the resulting transient equilibrium leads to an increase in population at first for ca. 11 hours and then to a decay.

B. CROCUS Reactor

Measurements were carried out on the EPFL CROCUS reactor (100 W, max. thermal flux 10^9 n/cm^2/s), a light-water–moderated, zero-power research facility documented in the IRPhE handbook [8] (see Figure 1). CROCUS has been extensively studied via other gamma-ray experiments [9]–[11], yet to date no time-resolved gamma spectroscopy was attempted. The core (Ø 58 cm x 100 cm) contains 336 UO_2 rods (1.8 wt.% ^{235}U, 12.6 mm diameter, 1.8 cm pitch) and up to 180 U-metal rods (0.9 wt.% ^{235}U, 17 mm diameter, 2.9 cm pitch) in aluminum cladding. The active height is 100 cm; the core sits in a 130 cm-diam. aluminum tank (1.2 cm wall). Reactivity is adjusted via water-level control with 0.1 mm precision (0.4 pcm). The neutron flux at the detector location was estimated to be $\sim 10^4$ n/cm^2/s, which should not significantly damage the detector [12], [13].

C. Detector and Data Acquisition

We used an Ortec GEM-15180-P coaxial HPGe (57.2 mm x 50 mm) in PopTop configuration, operated at 3.6 kV with 0.235 % resolution (FWHM) at 662 keV. Signals were preamplified by a Canberra DSA-1000 and digitized by a CAEN DS5730 (DPP-PHA firmware) in list mode, with readout via USB and CoMPASS software [15], [16]. The detector was placed 4.3 m from core center, 1 m behind the beam-channel opening [17], see Figure 2. A 100 keV lower-level discriminator suppressed low-energy background, yielding $\sim 2 \times 10^4$ cps immediately after reactor shutdown.

D. Energy Calibration and Peak Fitting

Energy was calibrated assuming a linear relation,

$$E = Ax + B,$$

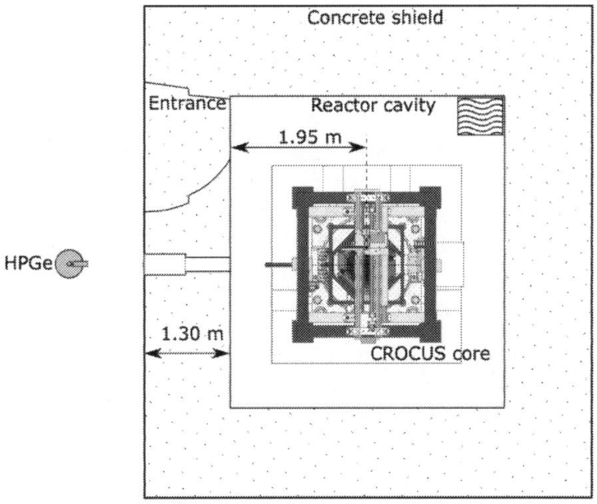

Fig. 2. Top view of the experimental setup: The HPGe was set at ca. 1 meter from the external surface of the reactor shielding, a total of 4.25 meters from the reactor core center.

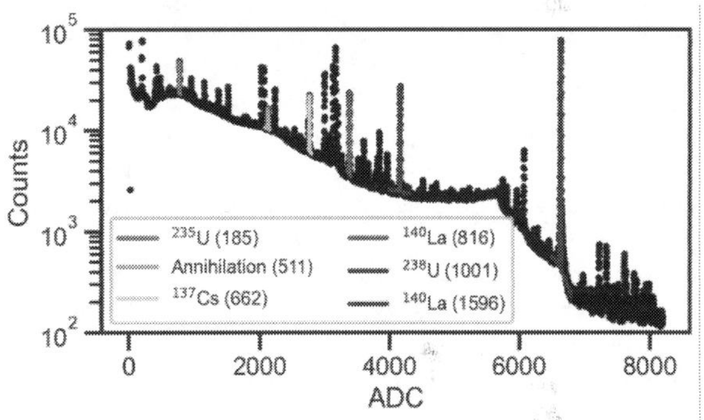

Fig. 3. Energy calibration of the gamma spectrum was achieved by using prominent emission lines identified in previous experiments [6].

where x is the ADC channel. Six photopeaks between 185 and 1600 keV (see Figure 3) were fitted with

$$G(E) = A\exp\left[-\frac{(E-\mu)^2}{2\sigma^2}\right] + aE + b$$

using trust-region nonlinear least squares. The fitted μ was then used for linear fitting across the ADC range. The calibration residuals have RMSE ¡ 0.4 keV, confirming a linear response in the examined region.

E. Spectrum treatment

Each gamma ray pulse was registered with a time stamp and a pulse integral. The histograms for time-dependent analysis were created with a 0.3 keV bin width and 1-hour wide time bins. The thus created gamma ray spectra were processed with a SNIP-type algorithm [18] to remove the background under photopeaks (see Figure 4). The peaks of interest (249.8 keV for Xe-135, 1260.5 keV for I-135) were then integrated via trapezoidal sum of the histogram in a three sigma range. For the peaks of interest herein, no neighboring peaks or overlapping peaks of concern were encountered.

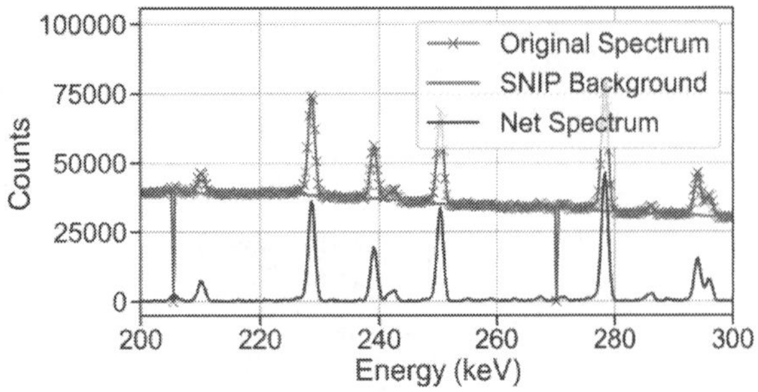

Fig. 4. Example of the gamma spectrum background removal via SNIP algorithm [18].

III. Results

In Figure 5 we display examples of the observed gamma ray spectra for different times after shutdown of the reactor. In the first hour immediately after shutdown the spectrum is dominated by delayed neutron-induced fission gamma rays and the 2.22 MeV emission from H(n,γ) and its single- and double-escape peaks. It is worth noting that 24 hours after shutdown, the count rate is higher than in the first hour. This is due to the fact that in the first 7 hours, the count rate was so high, that pile-up rejection and dead-time dominated and most pulses remained unregistered. In order to measure isotopes with short half-lives, one would need to operate at even lower reactor powers prior to shutdown, or change the shielding geometry.

Following the post-processing outlined in Section 2, we show in Figure 6 the time-dependent evolution of the Xe-135 and I-135 peaks. The expected 'xenon pit' shape, peaking at roughly 11 hours is observed, whilst ^{135}I decays exponentially with the expected half life of 6.57 h.

IV. Conclusion

In this work, we present the first analysis of time-dependent HPGe measurements of the CROCUS zero-power reactor immediately after shutdown until 90 hours thereafter. We specifically investigated the emissions of Xe-135 and I-135 at 249.8 keV and 1260.5 keV, respectively, to directly observe their populations in the reactor. Due to the high gamma-ray flux immediately after shutdown after a 1 at 20 W critical operation, a quantitative estimate of the populations was only possible after some 7 hours after shutdown. The Xe-135 count rates follow the expected transient equilibrium behavior. This observation is a first step towards the determination of the reactor-wide Xe-135 population to provide a new validation dataset for reactor poison predictors and other reactor physics codes.

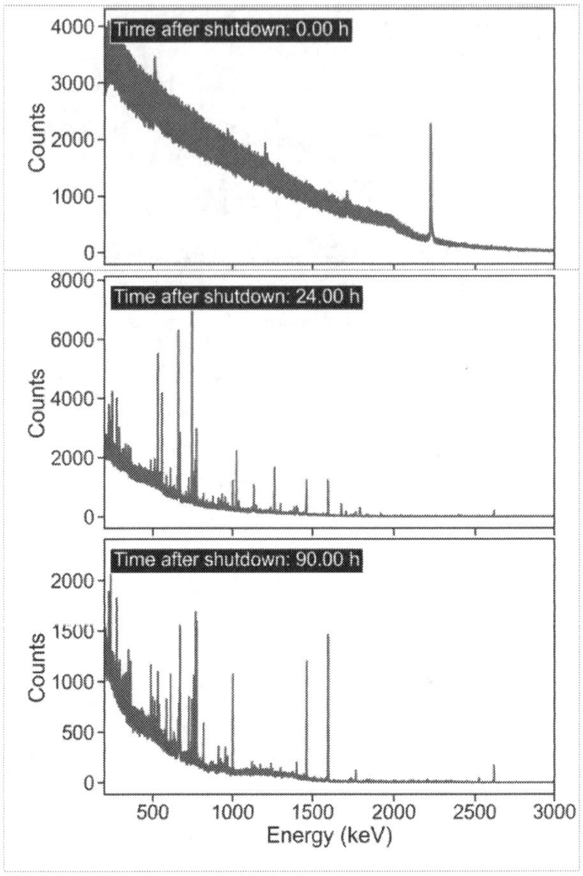

Fig. 5. Examples of measured gamma ray spectra over one hour. **Top:** Right after shutdown. **Middle:** 24 hours after shutdown. **Bottom:** 90 hours after shutdown.

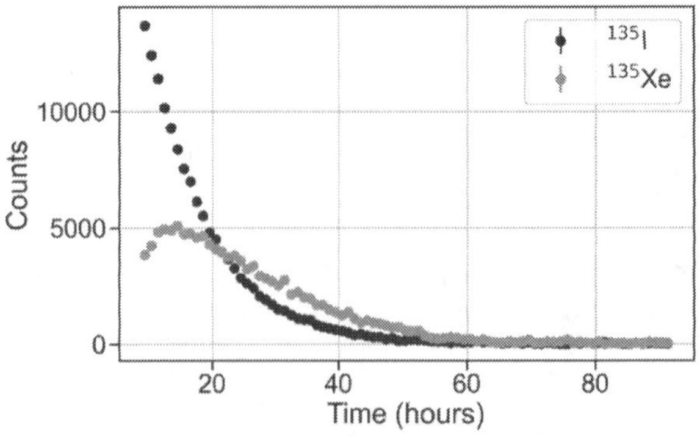

Fig. 6. First evaluation of the count rates of the 249.8 keV and the 1260.5 keV peak time evolutions.

References

[1] D. Fundamentals, "Nuclear physics and reactor theory," DOE-HDBK-1019/1-93, Tech. Rep., 1993.
[2] B. Mercier, D. Yang, Z. Zhuang, and J. Liang, "A simplified analysis of the chernobyl accident," *EPJ N - Nuclear Sciences & Technologies*, vol. 7, p. 1, 2021, publisher: EDP Sciences. [Online]. Available: https://hal.science/hal-03117177

[3] B. Singh, A. A. Rodionov, and Y. L. Khazov, "Nuclear data sheets for a = 135," *Nuclear Data Sheets*, vol. 109, no. 3, pp. 517–698, 2008. [Online]. Available: https://www.sciencedirect.com/science/article/pii/S0090375208000094
[4] S. Schmidt, T. Heftrich, J. Glorius, G. Hampel, R. Plag, R. Reifarth, Z. Slavkovská, K. Sonnabend, C. Stieghorst, N. Wiehl *et al.*, "Behavior of a trapezoid-based data acquisition system up to 100 khz and beyond," *Nuclear Instruments and Methods in Physics Research Section A: Accelerators, Spectrometers, Detectors and Associated Equipment*, vol. 768, pp. 192–196, 2014.
[5] T. Czakoj, M. Košťál, E. Novák, J. Šimon, M. Schulc, Z. Matěj, F. Mravec, F. Cvachovec, T. Urban, and E. Losa, "Measurement of gamma field inside the biological concrete shielding of vver-1000 mock-up at the lr-0 reactor," *Annals of Nuclear Energy*, vol. 211, p. 110999, 2025.
[6] O. V. Pakari, A. Lucas, F. B. Darby, V. P. Lamirand, T. Maurer, M. G. Bisbee, L. R. Cao, A. Pautz, and S. A. Pozzi, "Gamma-ray spectroscopy in low-power nuclear research reactors," *Journal of Nuclear Engineering*, vol. 5, no. 1, pp. 26–43, 2024.
[7] G. Žerovnik, B. Levpuščček, and L. Snoj, "Parametric study of the influence of 135xe build-up on required excess reactivity for load-following operations," *Nuclear Engineering and Design*, vol. 424, p. 113297, 2024.
[8] J. Paratte, R. Früh, U. Kasemeyer, M. Kalugin, W. Timm, and R. Chawla, "A benchmark on the calculation of kinetic parameters based on reactivity effect experiments in the CROCUS reactor," *Annals of Nuclear Energy*, vol. 33, no. 8, pp. 739–748, 2006. [Online]. Available: https://www.sciencedirect.com/science/article/pii/S0306454905002112
[9] O. Pakari, "Experimental and numerical study of stochastic branching noise in nuclear reactors," *PhD thesis, EPFL*, 2020.
[10] O. Pakari, T. Mager, P. Frajtag, A. Pautz, and V. Lamirand, "Gamma noise to non-invasively monitor nuclear research reactors," *Scientific Reports*, vol. 14, no. 1, p. 8409, 2024.
[11] O. Pakari, V. Lamirand, T. Mager, A. Laureau, P. Frajtag, and A. Pautz, "First in-core gamma spectroscopy experiments in a zero power reactor," in *EPJ Web of Conferences*, vol. 253. EDP Sciences, 2021, p. 04022.
[12] R. G. Gordon and O. Hemingway John, "Practical gamma-ray spectrometry," *Sussex, ENGLAND: Wiley*, 2008.
[13] V. Borrel, B. Kandel, F. Albernhe, P. Frabel, B. Cordier, G. Tauzin, S. Crespin, R. Coszach, J. Denis, and P. Leleux, "Fast neutron-induced damage in integral n-type hpge detectors," *Nuclear Instruments and Methods in Physics Research Section A: Accelerators, Spectrometers, Detectors and Associated Equipment*, vol. 430, no. 2-3, pp. 348–362, 1999.
[14] J. Paratte, R. Früh, U. Kasemeyer, M. Kalugin, W. Timm, and R. Chawla, "A benchmark on the calculation of kinetic parameters based on reactivity effect experiments in the CROCUS reactor," *Annals of Nuclear Energy*, vol. 33, no. 8, pp. 739–748, may 2006.
[15] J. Keightley, C. Bobin, J. Bouchard, M. Capogni, S. Loreti, and M. Roteta, "Recent advances in digital coincidence counting for radionuclide metrology," in *2013 3rd International Conference on Advancements in Nuclear Instrumentation, Measurement Methods and their Applications (ANIMMA)*. IEEE, 2013, pp. 1–7.
[16] O. V. Pakari, A. Lucas, F. B. Darby, V. P. Lamirand, T. Maurer, M. G. Bisbee, L. R. Cao, A. Pautz, and S. A. Pozzi, "Gamma-ray spectroscopy in low-power nuclear research reactors," *Journal of Nuclear Engineering*, vol. 5, no. 1, pp. 26–43, 2024.
[17] M. Saliba, E. Lehmann, P. Trtik, V. Lamirant, A. Pautz, and O. Pakari, "Computations for a new neutron imaging station at the CROCUS zero power reactor," 2025, in Proceedings of ANIMMA 2025.
[18] M. Caccia, A. Ebolese, M. Maspero, R. Santoro, M. Locatelli, M. Pieracci, and C. Tintori, "Background removal procedure based on the snip algorithm for γ–ray spectroscopy with the caen educational kit," *Educational Note*, vol. 2, p. ED3163, 2013.

Predicting Neutron Noise Detector Responses for Zero-Power Molten Salt Reactor Experiments

Oskari Pakari[1,2,*], Julia Świątkowska[1], Michel Saliba[1]
[1]École Polytechnique Fédérale de Lausanne (EPFL), 1015 Lausanne, Switzerland
[2]Paul Scherrer Institut (PSI), 5232 Villigen, Switzerland
(*) oskari.pakari@epfl.ch

***Abstract*—We use Monte Carlo simulations to predict the responses of neutron noise experiments in zero-power Molten Salt Reactors (MSRs). As for conventional reactors, the initial testing of a newly constructed reactor often comprises zero-power tests and criticality experiments. In zero-power environments, neutron noise experiments can be part of the experimental portfolio to provide experimental data on integral kinetic parameters such as the prompt neutron decay constant or the effective delayed neutron fraction. Neutron noise experiments exploit the statistical correlations in neutron detector signals via noise analysis methods, such as the Rossi-α or PSD method. The results are then used for code validation, nuclear data feedback or safety monitoring purposes. For MSRs, the flowing fuel and the therefore flowing precursors leads to a unique dynamic behavior that could be experimentally studied with neutron noise techniques. For example, β_{eff} depends on the fluid velocity: For higher velocities the precursors are swept out of the core and therefore might not contribute to the reactivity, and in many MSR designs β_{eff} therefore goes down with higher flow rate. In this work, we present simulations using the MSRE model in Serpent 2 to predict neutron noise curves for different flow rates. Our approach is new in its use of explicit Monte Carlo modeling for noise experiments in MSRs, with a simplified treatment of the precursor flow. Our results indicate that noise experiments in zero-power MSRs are feasible and that noise experiments could be used to measure recirculation time and kinetic parameters.**

***Keywords* — Neutron Noise, Monte Carlo Neutron Transport, Prompt Decay Constant.**

I. Introduction

ACCURATE understanding of reactor kinetics is essential for ensuring nuclear reactor safety and efficiency. Two important parameters in transient analysis are the prompt neutron decay constant α and the effective delayed neutron fraction β_{eff}, which are part of the so-called kinetic parameters of a reactor. The kinetic parameters are coefficients of the differential equations that describe the time-dependent neutron population, often used in the point kinetics approximation [1]. Experimental determination of α and β_{eff} is often necessary when nuclear data uncertainties of system materials are large, and the values then serve as an integral reference quantity for validating reactor simulation tools [2], [3].

Neutron noise techniques to determine kinetic parameters in fissile systems have been studied since the early days of nuclear engineering. The foundational work by Feynman [4] introduced methods for determining integral reactor parameters from detector signal fluctuations in steady-state fissile assemblies. More recently, neutron noise experiments have been conducted at various facilities, including MINERVE [5] [6], VENUS at SCKCEN [7], IPEN in Brazil [8], and critical assemblies at Los Alamos National Laboratory's NCERC facility [9].

Molten Salt Reactors (MSR), which are advanced Generation IV reactor designs, some of which are in advanced licensing stages, have flowing fuel. This introduces unique dynamic behavior, notably due to the phenomenon of delayed neutron precursors flowing out of the active fissioning volume. The precursors can decay in the surrounding piping, or even re-enter and decay in the reactor. Overall, this leads to a drop in the effective delayed neutron fraction β_{eff} with higher flow rate [10] [11].

To find kinetic parameters via simulation, deterministic methods have traditionally been employed to estimate the prompt neutron decay constant (α) [12] [13]. Monte Carlo codes are also frequently used for this purpose, such as the Iterated Fission Probability (IFP) method [14]. Alternative approaches involve neutron-noise-based simulations that track individual fission events to estimate α [15]. Such analyses typically utilize MCNPX-PoliMi [16] or MCNP6.3 with PTRAC capabilities [17].

Herein, we present an investigation into the feasibility of simulating of noise experiments in MSR geometries using Monte Carlo codes. Most Monte Carlo codes do not support time dependent geometries or require coupling to fluid solvers to account for the flowing fuel. To simplify, we propose to assume a uniform velocity field and to 'transport' precursors inside the simulation to approximate the flow. The thus recorded time stamps from an external source calculation are processed with common noise analysis tools, as outlined in the Methods in Section II. The results are presented in Section III.

II. Methods

A. Serpent 2

We use the Monte Carlo neutron transport code Serpent 2 [18], widely used in reactor physics analyses and radiation shielding calculations as well as reactor kinetic parameters via IFP [19].

B. Event Tracking in Monte Carlo

The event-tracking capability in Serpent 2, which internally records only incident neutrons causing fission events, was custom-developed and is not available in the official Serpent release [20]. Although still undergoing internal testing, this feature has already been validated with experimental data from the CROCUS reactor [15].

C. Model I: The CROCUS zero-power reactor

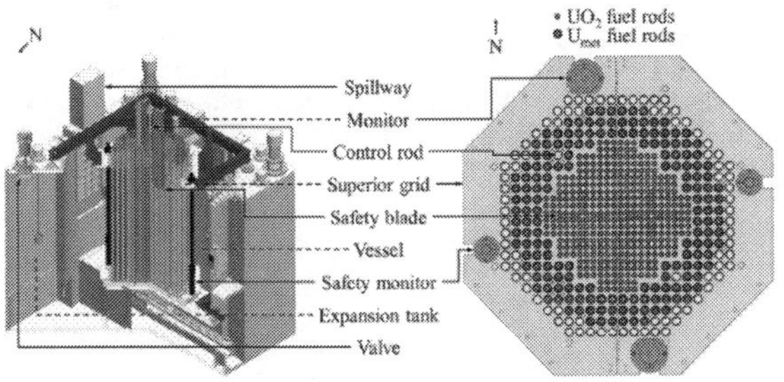

Fig. 1. Isometric view of the CROCUS reactor internals and top fuel grid.

The EPFL CROCUS reactor is a zero-power research reactor [21], functioning as a critical assembly fueled by uranium and moderated with light water (see Figure 1). Primarily dedicated to educational and research applications, CROCUS has a maximum operational power of 100W. The reactor geometry is well-documented and included in the Nuclear Energy Agency's International Handbook of Evaluated Reactor Physics Benchmark Experiments [22] and has well-characterized kinetic parameters (both in experiment [23], [24], [25] and from simulation [20]) and is thus used here as a reference model.

D. Model II: Molten Salt Reactor Experiment (MSRE)

The Molten Salt Reactor Experiment (MSRE) was an experimental reactor built and operated at Oak Ridge National Laboratory from 1965 to 1969. It utilized a liquid fluoride salt fuel, composed primarily of uranium fluoride dissolved in a mixture of lithium, beryllium, and zirconium fluorides, with graphite serving as a moderator. Operating at approximately 8 MW thermal power, the MSRE demonstrated key features of molten salt reactor technology, such as online fuel processing, continuous circulation of liquid fuel, and inherent safety characteristics. The reactor's geometry, materials, and operational data are thoroughly documented and widely employed as a benchmark for reactor physics modeling and validation studies. Here we use the available Serpent 2 MSRE model kindly provided by UCB, see also Figure 2 [26].

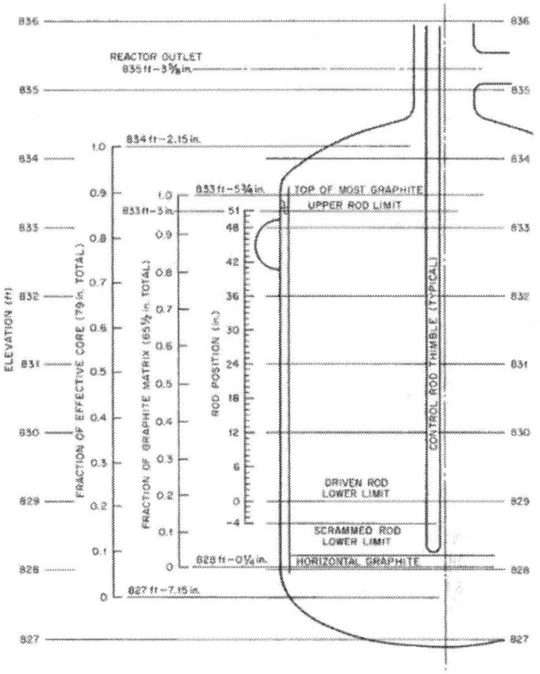

Fig. 2. Side view of the MSRE geomtery.

E. Simulation Workflow

For both models, simulations were performed following this procedure:

- Two-step simulation procedure:
    - Perform k-static calculation to determine k_{eff} and spatial fission source distribution.
    - Initialize external source simulation with spatially converged source and record individual fission events.
- Recorded time stamps analyzed via Rossi-alpha method or Power spectral density (PSD) method. Details on how the theory behind the methods can be found e.g. in [15]. α is extracted via curve fitting. β_{eff} was only determined from the PSD method, using

$$\beta_{eff} = \rho + \sqrt{\frac{2D_v}{F\,A}}$$

  F was estimated from the synthetic time series of fission time stamps, ρ was the reactivity compared to k_{eff}=1 from the simulation output file, the Diven factor D_v was assumed to be ~0.806 [27], and A the amplitude term of the PSD.
- Serpent 2 also gives for each simulation the IFP method results for comparison.
- To approximate the flowing conditions, we assume a uniform velocity field and a total loop length of ca. 6 m. Using the MSRE experimental report [28], we find a velocity of 0.22 m/s for the zero-power criticality test. Each time a delayed precursor is sampled, the newly sampled neutron is also displaced in z by the distance it could travel

at 0.22 m/s during its decay time. If the new location of emission is outside the reactor (i.e. in the pipe), the neutron is killed. This check is performed as total distance traveled modulus the total circuit length, and the remainder needs to be smaller than the reactors active height.

III. Results

In Figure 3 we show as reference case the critical CROCUS geometry Rossi-α curve extracted from some 10000 time stamps of fissions. The fitted value for α falls into the expected range and serves as a first sanity check that the method functions as intended.

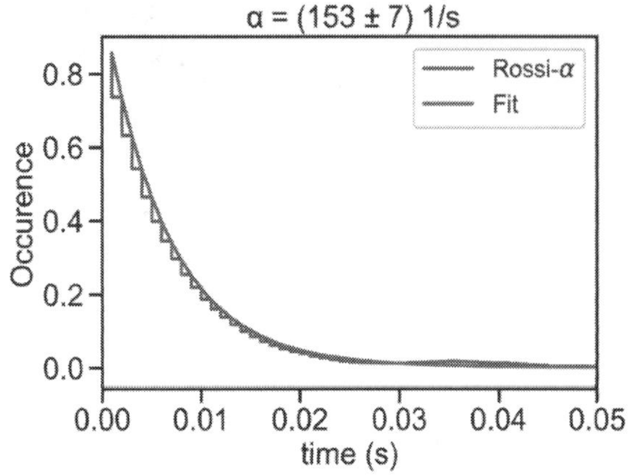

Fig. 3. Rossi-α curve from simulated fission time stamps in the CROCUS geometry. The fitted α is reported above, which is consistent with previous simulations and experiments.

In Figure 4 we display the PSD curve for the same geometry. To achieve it, the time stamps were grouped by history number and then each history distributed in time via a Poissonian sampler in python. This synthetic time series was then processed via scipy signal welch into a PSD. The time stamps were distributed in time up to 1000s, the segment length was 2^{11} and the bin width was set such that the sampling rate was 1 kHz. The PSD curve also follows the expected behavior and yields a similar α compared to the Rossi-α.

Next we display the results for the MSRE model. We compare the simulated PSD to available experimental data [28] in Figure 5 for the static MSRE. The simulated PSD suffers from poor statistics, but the overall shapes (and thus α values) agree to a satisfactory degree for a first attempt.

In Figure 6 we display the Rossi-α curves for the MSRE geometry with 'flowing' precursors as outlined in Section II. As the precursors re-enter the reactor at roughly predictable times, around the average recirculation time, we expect more correlations around that time. Indeed, we observe distinct peaks in the Rossi-α curves at the recirculation time (at its harmonics). This effect is similar to peaks in cross-correlation curves that observe a time lag [29]. This means that noise

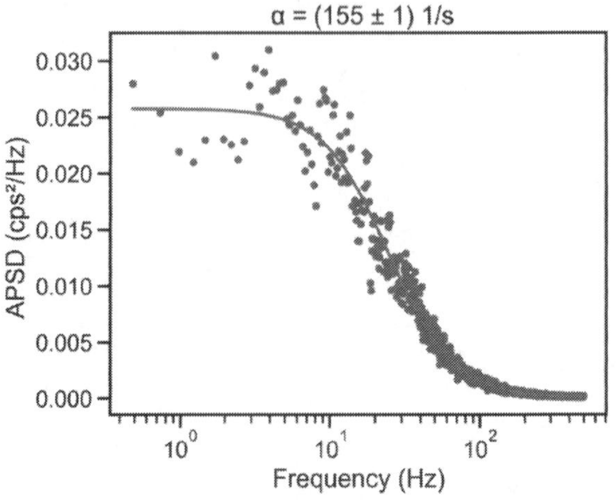

Fig. 4. Power spectral density curve from simulated fission time stamps in the CROCUS geometry. In order to create a signal in time, the recorded time stamps were distributed in time via a Poissonian sampler per history, i.e. a given time shift was applied to all fissions belonging to the same history.

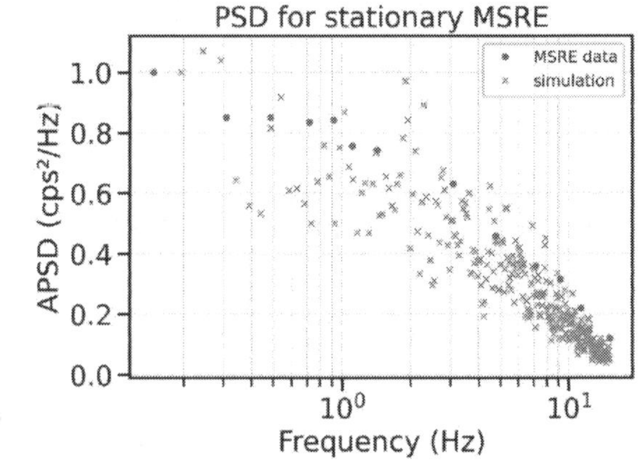

Fig. 5. Simulated power spectral density for the stationary MSRE geometry vs. experimental data *[28]*.

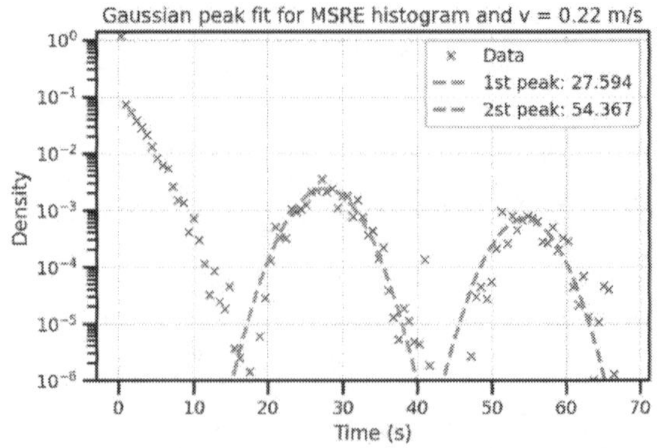

Fig. 6. Rossi-α curve for the MSRE geometry with shifted precursor emission locations using a uniform velocity field of 0.22 m/s. In these conditions, the recirculation time is 27 s. At the recirculation time (and its multiple) peaks appear in the autocorrelation. The mean corresponds to the recirculation time, whilst the width should roughly reflect the reactor's active volume height.

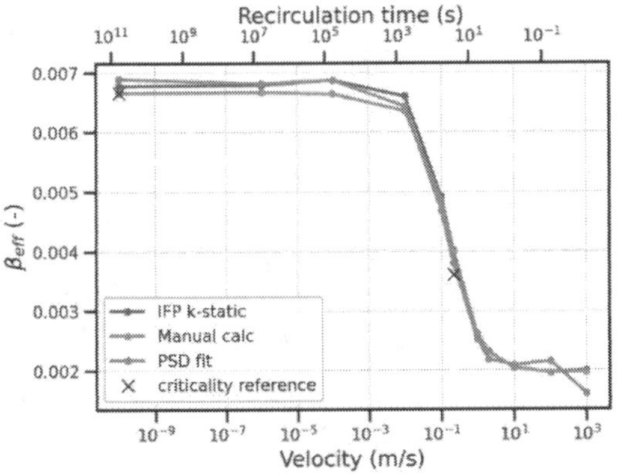

Fig. 7. Extracted β_{eff} values for different methods and different uniform flow velocities. We compare IFP vs. manual calculation (counting delayed neutrons inside the code) vs. extracting the value from the time stamp PSD. We also show the experimental values reported for the MSRE (x). All methods capture the essential feature: βeff drops with higher flow rate.

experiments in zero-power MSRs are likely to be useful to determine or verify the recirculation time.

Finally, in Figure 7 we present a comparison of the extracted β_{eff} values for the different methods: IFP, manual delayed neutron counting inside the code, and the fitting of the noise PSD. We also included experimental data points [28]. The expected drop in β_{eff} is observed by all methods, and we confirm the possibility to simulate the effect with Monte Carlo calculations and our simplified transport of precursors.

IV. Conclusions

In this work, we present the first tests of applying neutron noise analysis to a MSR Monte Carlo model to observe the unique dynamic behavior of recirculating precursors. The MSRE experiments and other work predicted a drop in β_{eff}, which we were able to reproduce from PSD fitting and IFP. The Rossi-α analysis showed peaks in the correlation histogram at the recirculation time, indicating that neutron noise could be used to verify flow velocity in MSRs. These peaks were also predicted by other simulations [10], [30], but were not observed in the MSRE [28]. Future zero-power MSRs could therefore benefit from neutron noise experiments to determine their kinetic parameters and verify if the recirculation time peak is visible or not.

References

[1] M. M. R. Williams, Random processes in nuclear reactors, Elsevier, 1974.

[2] R. D. Mosteller, "Validation suite for mcnp," 2002.

[3] B. C. Kiedrowski, F. B. Brown, N. A. Gibson, A. S. Bennett and M. A. Gonzales, "MCNP 6 Shielding Validation Suite: Past, Present, and Future," *Transactions of the American Nuclear Society,* vol. 105, p. 559–561, 2011.

[4] R. P. Feynman, F. D. Hoffmann and R. Serber, "Dispersion of the neutron emission in U-235 fission," *Journal of Nuclear Energy (1954),* vol. 3, p. 64–IN10, August 1956.

[5] G. Perret, B. Geslot, A. Gruel, P. Blaise, J. Di-Salvo, G. De Izarra, C. Jammes, M. Hursin and A. Pautz, "Kinetic parameter measurements in the MINERVE reactor," *IEEE Transactions on Nuclear Science,* vol. 64, p. 724–734, 2016.

[6] B. Geslot, A. Gruel, A. Pepino, J. Di Salvo, G. de Izarra, C. Jammes, C. Destouches and P. Blaise, "Pile noise experiment in MINERVE reactor to estimate kinetic parameters using various data processing methods," in *2015 4th International Conference on Advancements in Nuclear Instrumentation Measurement Methods and their Applications (ANIMMA)*, 2015.

[7] X. Doligez, A. Billebaud, S. Chabod, T. Chevret, D. Fourmentel, A. Krasa, A. Kochetkov, F. R. Lecolley, J. L. Lecouey, G. Lehaut and others, "Effective delayed neutron fraction measurement in the critical VENUS-F reactor using noise techniques," in *2015 4th International Conference on Advancements in Nuclear Instrumentation Measurement Methods and their Applications (ANIMMA)*, 2015.

[8] A. dos Santos and R. Diniz, "A Noise Analysis Approach for Measuring Effective Delayed Neutron Parameters in the IPEN/MB-01 Reactor," in *AIP Conference Proceedings*, 2005.

[9] J. Goda, C. Bravo, T. Cutler, T. Grove, D. Hayes, J. Hutchinson, G. McKenzie, A. McSpaden, W. Myers, R. Sanchez and others, "A new era of nuclear criticality experiments: The first 10 Years of Godiva IV operations at NCERC," *Nuclear Science and Engineering,* vol. 195, p. S55–S79, 2021.

[10] L. Bureš, "Comparison of Analytical Models for Frequency Response and Linear Stability of Molten Salt Reactors," *Nuclear Science and Engineering,* vol. 0, p. 1–24, 2025.

[11] M. Aufiero, M. Brovchenko, A. Cammi, I. Clifford, O. Geoffroy, D. Heuer, A. Laureau, M. Losa, L. Luzzi, E. Merle-Lucotte and others, "Calculating the effective delayed neutron fraction in the Molten Salt Fast Reactor: Analytical, deterministic and Monte Carlo approaches," *Annals of Nuclear Energy,* vol. 65, p. 78–90, 2014.

[12] J. M. Hykes and R. M. Ferrer, "Validation of CASMO5 neutron kinetics parameters," *Annals of Nuclear Energy,* vol. 136, p. 107015, 2020.

[13] M. I. Radaideh, W. A. Wieselquist and T. Kozlowski, "A new framework for sampling-based uncertainty quantification of the six-group reactor kinetic parameters," *Annals of Nuclear Energy,* vol. 127, p. 1–11, 2019.

[14] N. Terranova and A. Zoia, "Generalized Iterated Fission Probability for Monte Carlo eigenvalue calculations," *Annals of Nuclear Energy,* vol. 108, p. 57–66, October 2017.

[15] O. V. PAKARI, "Experimental and numerical study of stochasticbranching noise in nuclear reactors", PhD thesis, 2020.

[16] S. A. Pozzi, E. Padovani and M. Marseguerra, "MCNP-PoliMi: a Monte-Carlo code for correlation measurements," *Nuclear Instruments and Methods in Physics Research Section A: Accelerators, Spectrometers, Detectors and Associated Equipment,* vol. 513, p. 550–558, 2003.

[17] J. A. Kulesza, T. R. Adams, J. C. Armstrong, S. R. Bolding, F. B. Brown, J. S. Bull, T. P. Burke, A. R. Clark, R. A. A. Forster III, J. F. Giron and others, "MCNP® code version 6.3.0 theory & user manual," Los Alamos, New Mexico, 2022.

[18] J. Leppänen, M. Pusa, T. Viitanen, V. Valtavirta and T. Kaltiaisenaho, "The Serpent Monte Carlo code: Status, development and applications in 2013," *Annals of Nuclear Energy,* vol. 82, p. 142–150, August 2015.

[19] J. Leppänen, M. Aufiero, E. Fridman, R. Rachamin and S. van der Marck, "Calculation of effective point kinetics parameters in the Serpent 2 Monte Carlo code," *Annals of Nuclear Energy,* vol. 65, p. 272–279, March 2014.

[20] M. Saliba, V. Lamirand, W. Monange, F. Vitullo, M. Hursin, A. Pautz and O. Pakari, "Measurement of the Prompt Neutron Decay Constant in a Zero-Power Reactor Using a Novel 3D Detector System," *Nuclear Science and Engineering,* p. 1–9, March 2025.

[21] J. M. Paratte, R. Früh, U. Kasemeyer, M. A. Kalugin, W. Timm and R. Chawla, "A benchmark on the calculation of kinetic parameters based on reactivity effect experiments in the CROCUS reactor," *Annals of Nuclear Energy,* vol. 33, p. 739–748, May 2006.

[22] J. D. Bess, T. Ivanova, I. Hill and L. Scott, "The 2019 Edition of the IRPhEP Handbook," *Trans. Am. Nucl. Soc.,* vol. 121, p. 1565–1568, November 2019.

[23] O. Pakari, T. Mager, P. Frajtag, A. Pautz and V. Lamirand, "Gamma noise to non-invasively monitor nuclear research reactors," *Scientific Reports,* vol. 14, p. 8409, 2024.

[24] O. Pakari, V. Lamirand, G. Perret, P. Frajtag and A. Pautz, "Kinetic Parameter Measurements in the CROCUS Reactor Using Current Mode Instrumentation," *IEEE Transactions on Nuclear Science,* vol. 65, pp. 2456-2460, 2018.

[25] F. B. Darby, O. V. Pakari, M. Y. Hua, V. Lamirand, S. D. Clarke, A. Pautz and S. A. Pozzi, "Neutron-gamma noise measurements in a zero-power reactor using organic scintillators," *IEEE Transactions on Nuclear Science,* vol. 71, p. 1033–1040, 2023.

[26] D. Shen, G. Ilas, J. J. Powers and M. Fratoni, "Reactor physics benchmark of the first criticality in the molten salt reactor experiment," *Nuclear Science and Engineering,* vol. 195, p. 825–837, 2021.

[27] O. Pakari, V. Lamirand, G. Perret, L. Braun, P. Frajtag and A. Pautz, "Current mode neutron noise measurements in the zero power reactor CROCUS," in *EPJ Web of Conferences*, 2018.

[28] R. C. Steffy Jr, "EXPERIMENTAL DYNAMIC ANALYSIS OF THE MSRE WITH sup 233 U FUEL.," 1970.

[29] C. M. Kostelac, F. B. Darby, V. Lamirand, S. A. Pozzi, M. Saliba and O. V. Pakari, "Synchronization of Neutron and Gamma Detector Signals for Noise Studies," *submitted to: Nuclear Instruments and Methods in Physics Research Section A: Accelerators, Spectrometers, Detectors and Associated Equipment,* 2025.

[30] I. Pázsit and A. Jonsson, "Reactor kinetics, dynamic response, and neutron noise in molten salt reactors," *Nuclear Science and Engineering,* vol. 167, p. 61–76, 2011.

Simulations of a new neutron imaging station at the CROCUS zero power reactor

Michel Saliba[1,*], Eberhard Lehmann[2], Pavel Trtik[2], Vincent Lamirand[1], Andreas Pautz[1,3], and Oskari Pakari[1,3]
[1]EPFL, Laboratory of Reactor Physics and System Behavior, 1015 Lausanne, Switzerland
[2]PSI Center for Neutron and Muon Sciences, 5232 Villigen, Switzerland
[3]PSI Center for Nuclear Engineering and Sciences, 5232 Villigen, Switzerland
(*) michel.saliba@epfl.ch

Abstract — **Neutron imaging (NI) complements X-ray radiography by exploiting the strong interaction of neutrons with several light elements and a high penetration for many heavy elements. Hitherto no dedicated NI beamline currently exists at the zero-power CROCUS reactor. Establishing such a station would create a low-cost platform for student training, materials R&D, and proof-of-principle studies that feed larger facilities. Using the Serpent 2 Monte Carlo neutron transport code, we explored how to modify an existing zero power research nuclear reactor into a useful neutron source for NI. Three measures proved decisive: (i) inserting a 70 mm-diameter air (or vacuum) tube through the 300 mm water reflector, boosting channel flux by approximately factor of fifty (ii) retaining a thin water layer between fuel and tube to convert fast neutrons to thermal ones without excessive attenuation; and (iii) employing low-scatter aluminum windows and vacuum flight tubes to preserve beam intensity over source-to-detector ~5m distance. A resolution-versus-flux map identified an optimal layout that balances image sharpness and count rate for the detector planned for first radiographs in 2025. The outcomes of the simulations show that a reflector modification can deliver performance comparable to demonstrator stations at VR-1 in Czechia and AKR-2 in Germany, and they underline CROCUS's potential as a national testbed for neutron radiography, and potentially even low-resolution computed tomography.**

Keywords —**Neutron imaging, Zero-power reactor, Monte Carlo neutron transport.**

I. Introduction

Research reactors, particularly zero-power facilities, play an important role in nuclear-engineering education [1], code validation [2], and small-scale R&D [3], [4], [5], [6]. CROCUS, a 100 W water-moderated UO_2-$U_{metallic}$ research reactor at EPFL, has served these functions since the early 1980s [7]. To date it has been used mainly for pile oscillation measurements [8], benchmark experiments [7], detector testing [9], and hands-on training of hundreds of students and young professionals. Maintaining such reactors is increasingly challenging as budgets tighten and ageing facilities face shutdown; expanding their experimental portfolio is therefore essential for long-term sustainability [10]. Neutron imaging (NI) [11], a non-destructive radiographic technique sensitive to several light elements such as hydrogen, lithium or boron offers an attractive, modern use case. Upon this call, few facilities in Europe have started designing and running NI experiments such as the VR-1 zero power reactor in Czechia [12] and AKR-2 zero power reactor in Germany [13], [14]. Installing an NI beamline at CROCUS would (i) broaden the reactor's academic impact by supporting experiments over a broad palette of scientific domains, such as materials science [15], cultural-heritage studies [16], and advanced fuel research [17]; (ii) provide a unique Swiss platform for NI at reactor energies, complementing the high-flux spallation source SINQ [18]; and (iii) supply compelling justification for continued operation and upgrades. This paper presents Monte Carlo simulations to assess and optimize a practical NI station for the CROCUS reactor, demonstrating that a reflector modification can provide a useful thermal neutron flux and spatial resolution without compromising reactor safety or usability. The design goals of the NI setup are to maximize thermal neutron flux while minimizing fast neutrons and gamma radiation, achieve a field of view (FOV) of approximately 15 × 15 cm², maintain a reasonable collimation ratio (L/D) suitable for industrial applications, incorporate a variable aperture for flexibility, and ensure the overall system is not complicated for the installation and user-friendly for operation. The foreseen detector system is based on a scintillator coupled with a CCD (Charge-Coupled device) camera [19]. The structure of this paper is as follows: Section II introduces the CROCUS zero-power research reactor and the Serpent 2 Monte Carlo neutron transport code used in this study. Section III outlines the steps taken to enable neutron imaging at CROCUS, including proposed design modifications, simulation results, and the final layout. Section IV summarizes the findings and suggests directions for future development.

II. Tools and Material

In this section, we present the characteristics of the zero-power reactor CROCUS and the Monte Carlo neutron transport code Serpent 2 used to conduct the optimization in this study.

A. The CROCUS zero power reactor

The CROCUS reactor at EPFL is a zero-power research facility designed for educational and experimental use [2], [7], [8]. Operating at a maximum thermal power of 100 W,

CROCUS functions as a critical assembly using uranium fuel and light water as a moderator. The reactor features a concentric dual-zone core configuration, with uranium oxide fuel in the inner zone and metallic uranium fuel in the outer zone, as shown in Fig 1. CROCUS is instrumented with four main monitoring detectors: two compensated ionization chambers placed to the north and south of the core for general monitoring, and two east-west fission chambers calibrated for power measurements [20]. Additionally, the reactor supports various neutron detection systems, including ^{3}He proportional counters and scintillation detectors [9]. The reactor's diameter is 60 cm and has a height of one meter.

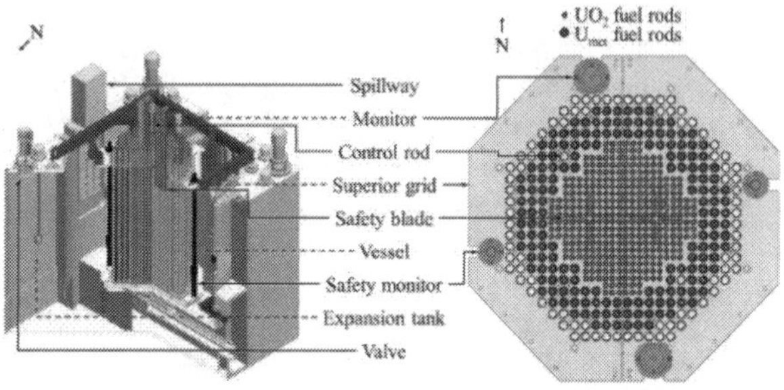

Fig 1- (left) Isometric view of the CROCUS reactor vessel and its internal components.(right) Top grid of the reactor shown in detail.

The reactor is enclosed in a 1.3 m thick concrete shielding structure, as shown in Fig 2 ,that serves as radiological protection during operation. However, a horizontal channel has been built into the shielding at the core's mid-height, specifically for experimental purposes. This channel enables placing the neutron imaging detector outside the concrete shielding, taking advantage of the increased distance for better image formation. All subsequent calculations are performed along the axis of this experimental channel.

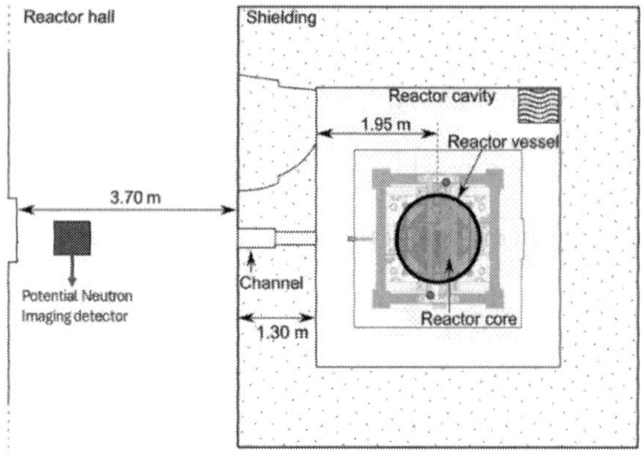

Fig 2- Top view of the reactor cavity and shielding in the reactor hall; proposed detector location shown.

B. Serpent 2 Monte Carlo neutron transport code

The Serpent 2 code [21], a Monte Carlo neutron transport code widely used in reactor physics and radiation shielding studies, was used in this work. In addition to its primary use for generating group cross-section libraries for diffusion codes, Serpent 2 can also predict kinetic parameters [22], includes variance reduction techniques [23], and has been validated for shielding applications [24]. These features make it a suitable candidate for modeling the CROCUS zero power reactor and performing calculations relevant for the implementation of neutron imaging within the CROCUS facility.

III. Enabling Neutron Imaging in CROCUS

In this section, we present the proposed modification to be implemented in the CROCUS reactor to achieve the best possible neutron imaging within the facility's limitations.

A. Channel within reflector

The reflector in the CROCUS reactor consists of approximately 300 mm of water, which significantly attenuates both thermal and fast neutrons. To increase the fraction of thermal neutrons passing through the experimental channel, an air-filled channel with an inner diameter of 70 mm was inserted into the reflector, as shown in Fig 3.

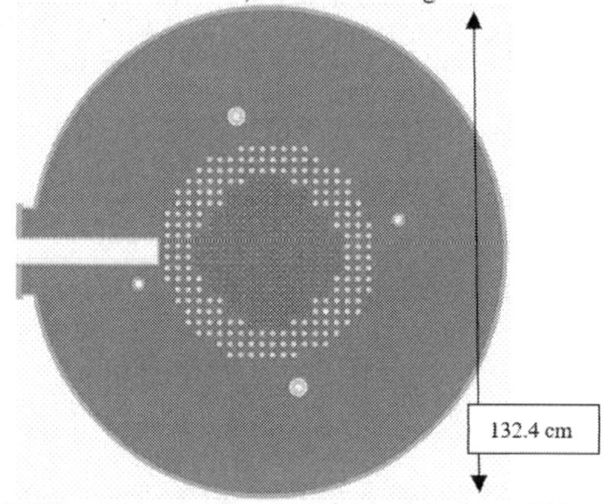

Fig 3 - CROCUS model, horizontal cut at mid-height, in the Serpent 2 Monte Carlo code, featuring an air channel inserted into the reflector.

Simulations were performed with several tallies placed at different distances from the core center, all located at the mid-height of the reactor and aligned with the air channel in the reflector. Notably, this air channel is concentric with the experimental channel in the concrete shielding. A second simulation was conducted using the same tally positions but without the air channel. The results, shown in Fig 4, indicate a 50-fold increase in neutron flux with the air channel in place, highlighting the importance of this modification for maximizing the flux outside the reactor vessel for neutron imaging applications.

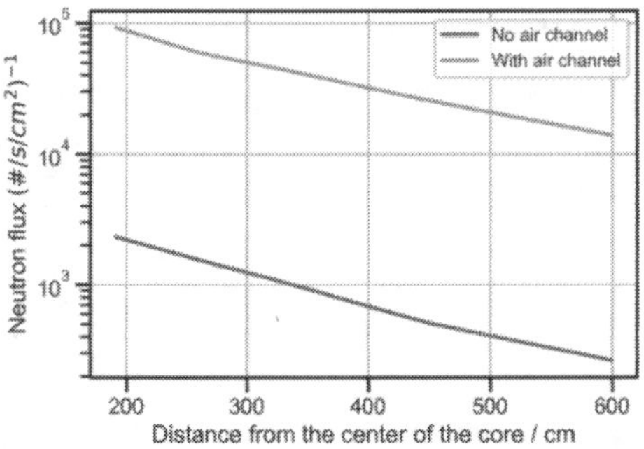

Fig 4 - Neutron flux outside the reactor as a function of distance from the core center. Green line represents results with the air channel inside the reflector; red line corresponds to the results without the air channel.

B. *Material used inside the channel inside the reflector*

To enhance the neutron flux exiting the core, specifically the thermal neutron component, we investigated replacing the air in the channel with polyethylene or heavy water. The goal was to assess whether these materials could improve thermalization of fast neutrons and thereby increase the thermal neutrons reaching the outside.

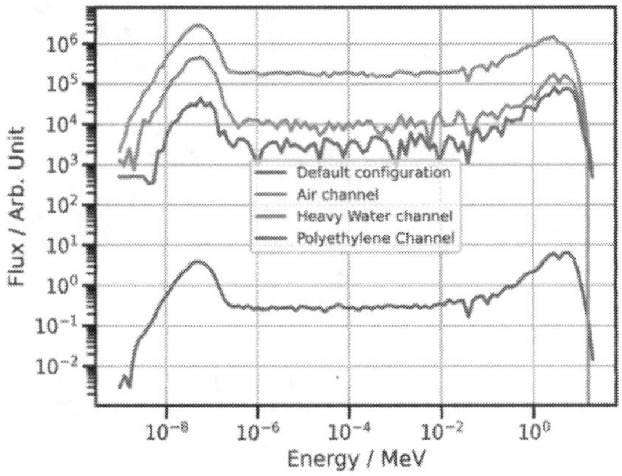

Fig 5- Neutron spectra measured outside the reflector channel for different channel materials. The default configuration refers to the case with no channel present in the reflector.

Several simulations were conducted using different materials within the reflector channel. A tally positioned 10 cm outside the channel was used to evaluate the resulting neutron spectrum for each case. As shown in Fig 5, air yields the highest thermal neutron flux. While heavy water improves thermalization of fast neutrons, it also scatters already thermalized neutrons, reducing overall thermal flux. The latter case is similar for polyethylene. Given the limited flux available in a zero-power reactor, air is the recommended material for the channel. Although a vacuum could further reduce scattering, it is not considered due to mechanical complexity.

C. *Gap between the fuel and the channel*

Reactor cores with reflectors typically exhibit a peak in thermal neutron flux at the core–reflector interface [25]. By adjusting the gap between the air channel and the core, it may be possible to exploit this effect. As shown in Fig 6, the gap was modeled in Serpent 2 (S2), and several cases were simulated, starting with the channel in direct contact with the fuel and progressively increasing the distance. A tally was placed outside the channel to measure total, thermal, and fast neutron fluxes. Additionally, a tally, 10 cm outside the reactor in front of the channel, was included to monitor the reaction rate of ^{6}Li, as the planned neutron imaging detector will use a ^{6}LiF:ZnS based scintillator. As illustrated in Fig 7, the thermal neutron flux remains relatively constant within the first 20 mm of gap distance before starting to decline. In contrast, the fast neutron flux decreases linearly as the gap increases. As the total flux is the sum of both components, it reflects a combination of these trends. The key performance metric is the ^{6}Li reaction rate, which, as shown in Fig 8, remains stable within the first 20 mm, mirroring the thermal flux behavior. The observed variation falls within the statistical uncertainty of the Monte Carlo simulations. Considering mechanical constraints, a gap of 10 mm between the fuel and the channel is recommended as it preserves thermal flux while ensuring structural feasibility.

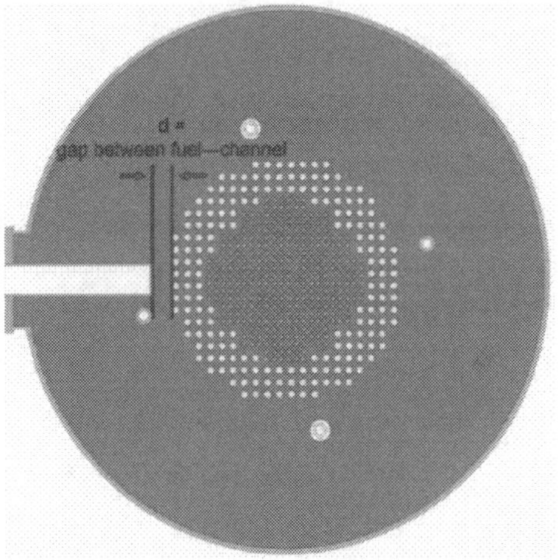

Fig 6 - Model of the CROCUS reactor in Serpent 2, showing a horizontal cross section at core mid-height and highlighting the gap between the fuel region and the channel.

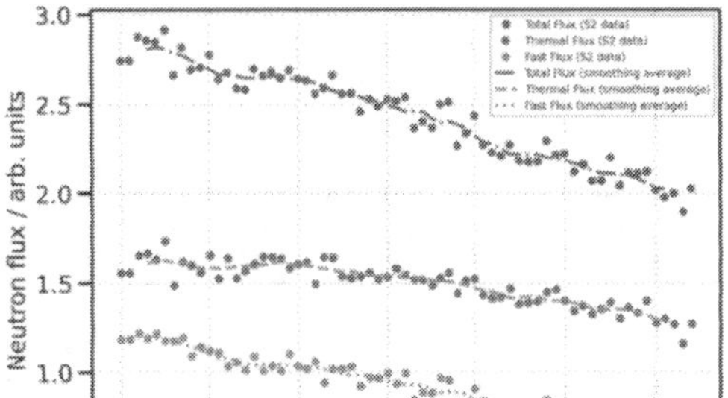

Fig 7 Total, thermal, and fast neutron flux as a function of the gap distance between the reactor channel and the fuel. The flux tally was positioned outside the vessel, directly in front of the channel.

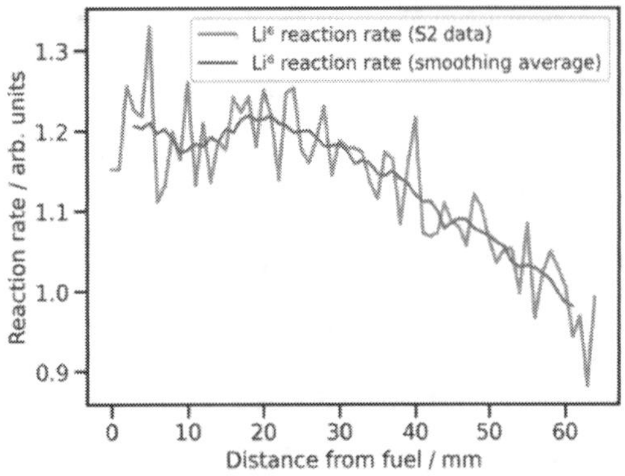

Fig 8 - Lithium-6 capture reaction rate in function of the gap distance. Tally was also placed outside the vessel and in front of the channel.

D. Simulation of flight tubes

In this section, we investigate the effect of air on beam intensity, focusing on its impact on thermal and fast neutron flux. To simulate this, a 6 m-long, 10 cm-diameter cylinder—matching the dimensions of the experimental setup—was modeled. The source energy spectrum was taken from the detailed CROCUS simulation at the exit of the reflector channel and placed at one end of the cylinder, with a neutron tally at the opposite end. The cylinder was filled with air containing water vapor at varying relative humidity levels. The uncertainties shown on the graphs are directly extracted from the tally information given by the Monte Carlo code. The first test examined how relative humidity affects neutron flux. As shown in Fig 9, no significant change was observed in total, thermal, or fast neutron flux.

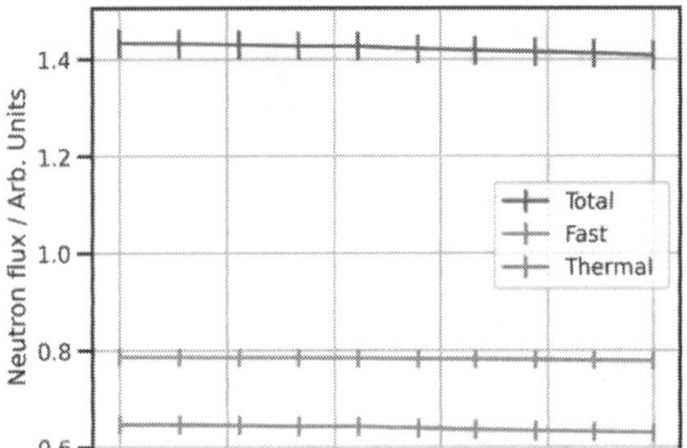

Fig 9 - Total, fast and thermal flux in function of the relative humidity of air.

Next, we investigated the effect of air pressure on neutron flux. Reducing air pressure lowers air density, which in turn decreases the total macroscopic cross section. As shown in Fig 10, a noticeable reduction in neutron flux occurs when the air pressure exceeds 100 mbar.

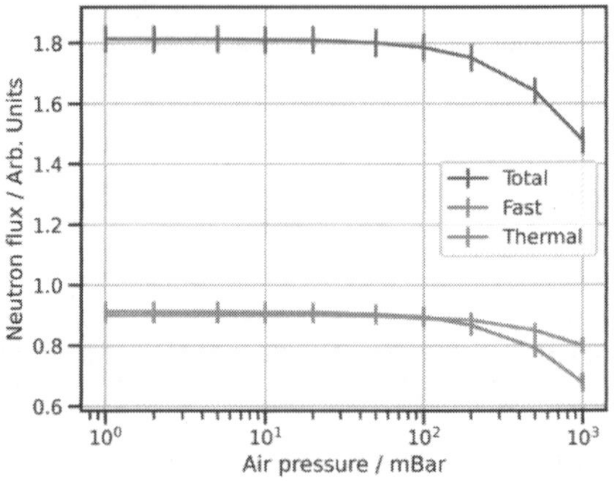

Fig 10 - Total, fast and thermal flux in function of the air pressure.

Additionally, as shown in Fig 11, hydrogen contributes only a small fraction (10–16%) to the total macroscopic cross section. This explains why changes in humidity had little impact on neutron flux compared to variations in air pressure.

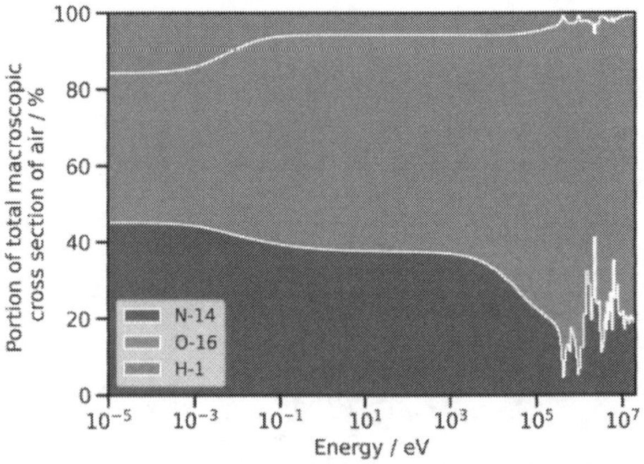

Fig 11 - Relative contribution of each nuclide in air to the total macroscopic cross section for a relative humidity of 78%.

For a zero-power reactor like CROCUS, a flux reduction of up to 20% is observed at ambient pressure compared to 50 mbar, as shown in Fig 10 . To mitigate this loss, aluminum flight tubes can be installed along the beam path, i.e., a tube from the exit of the reactor's vessel until the camera's field of view, allowing air to be passively or actively evacuated to maintain a pressure below 100 mbar.

E. Design proposal – Final layout proposal

After exploring various design optimizations to enable successful neutron imaging at the CROCUS zero-power research reactor, we propose a consolidated experimental layout. Table I summarizes the key modifications and design choices discussed in the previous sections. A schematic overview, see Fig 12, of the proposed setup is presented afterward to illustrate how the facility would be conFigd for neutron imaging experiments.

TABLE I
DETAILS OF THE FINAL DESIGN PROPOSAL

Object	Modification
Air Channel in the reactor's reflector	70 cm diameter, 41 cm length, till it exits the reactor's vessel. Air would maximize the absolute thermal flux
Gap between fuel and air channel	1 cm maximizes the ^{6}Li absorption reaction rate
Flight tubes	Insertion of flight tubes would increase the flux by 20%

The length of the flight tube depends on the required L/D ratio [14], which determines the geometric resolution of the radiographic image, and the desired neutron flux needed to achieve a reasonable irradiation time. As illustrated in Fig 13, neutron flux and L/D ratio are inversely related.

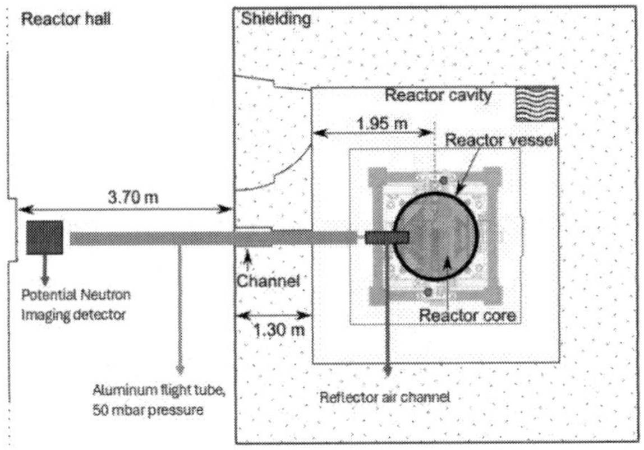

Fig 12 - Schematic showing the final design proposed to perform the first neutron imaging experiment at the CROCUS zero power reactor.

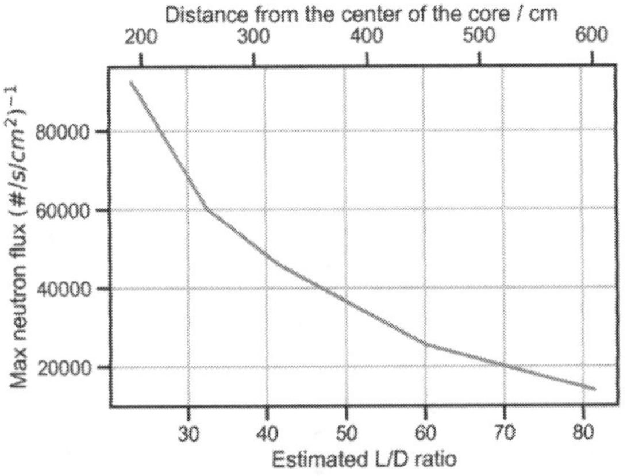

Fig 13 - Calculated neutron flux as a function of the estimated L/D ratio for the neutron imaging setup. The corresponding distance to the center of the CROCUS reactor core is also shown on the top axis. Simulations were performed in Serpent 2 assuming a maximum reactor power of 100 W. The imaging setup can be placed as close as 330 cm from the core center. The L/D ratio was estimated based on the diameter of the inner channel within the collimator at various possible setup positions.

The L/D ratio is estimated based on the diameter of the inner channel in the reflector and the distance L from the center of the core. To maximize neutron flux from the core, the initial experiments will use the full channel diameter without additional apertures. The L/D ratio can be adjusted by changing the flight path length, i.e., the distance from the core to the imaging setup. The setup can be placed as close as 330 cm from the core, but for the first experiments, the detector will be positioned at the maximum distance of 600 cm. At this location, the estimated neutron flux from Monte Carlo simulations is approximately 13,000 #/s/cm^2, corresponding to an L/D ratio of about 80.

IV. CONCLUSIONS

This paper presents Monte Carlo simulations aimed at assessing the feasibility of establishment of a neutron imaging beamline at the CROCUS zero power research reactor at EPFL, Switzerland. The proposed reflector modification, involving the insertion of a 70 mm-diameter air-filled channel through the reactor's water reflector, which enhances thermal neutron flux at the imaging location by approximately factor of fifty. Air was identified as the optimal material for the channel, outperforming alternatives such as polyethylene and heavy water due to its minimal neutron scattering and absorption. A gap of 1 cm between the channel and the reactor fuel is recommended, balancing mechanical feasibility and maintaining high thermal neutron flux.

Further flux improvement can be achieved by incorporating vacuum flight tubes at reduced pressure (~50 mbar), which decreases neutron scattering and boosts thermal neutron flux by an additional 20%. A practical NI design proposal is provided, incorporating these optimized features to ensure maximum neutron flux and appropriate image resolution for industrial and research applications.

The findings underline the feasibility of implementing a dedicated neutron imaging station at CROCUS, promising a valuable national platform for training, materials research, and neutron imaging demonstrations. Future work will include the experimental validation of these simulations.

REFERENCES AND FOOTNOTES

A. References

[1] R. Chawla, G. Girardin, and O. P. Joneja, "Reactor Experiments Course of the Swiss Nuclear Engineering Master," presented at the 18th International Conference on Nuclear Engineering, American Society of Mechanical Engineers Digital Collection, Apr. 2011, pp. 323–330. doi: 10.1115/ICONE18-29183.

[2] T. Mager, O. Pakari, M. Hursin, T. Ligonnet, A. Pautz, and V. Lamirand, "Intra-Pin Reaction Rate Measurements in the CROCUS Experimental Reactor," *IEEE Transactions on Nuclear Science*, vol. 71, no. 5, pp. 1047–1055, May 2024, doi: 10.1109/TNS.2024.3370371.

[3] O. Pakari, T. Mager, P. Frajtag, A. Pautz, and V. Lamirand, "Gamma noise to non-invasively monitor nuclear

research reactors," *Sci Rep*, vol. 14, no. 1, p. 8409, Apr. 2024, doi: 10.1038/s41598-024-59127-y.

[4] V. Lamirand *et al.*, "CORTEX experiments – Part I: Modulation campaigns in AKR-2 & CROCUS for the validation of neutron noise codes," *Annals of Nuclear Energy*, vol. 211, p. 110928, Feb. 2025, doi: 10.1016/j.anucene.2024.110928.

[5] V. Lamirand, "Ten springs of experiments in CROCUS," *EPJ Web Conf.*, vol. 288, p. 04026, 2023, doi: 10.1051/epjconf/202328804026.

[6] M. Saliba *et al.*, "Measurement of the Prompt Neutron Decay Constant in a Zero-Power Reactor Using a Novel 3D Detector System," *Nuclear Science and Engineering*, vol. 0, no. 0, pp. 1–9, doi: 10.1080/00295639.2025.2456411.

[7] J. M. Paratte, R. Früh, U. Kasemeyer, M. A. Kalugin, W. Timm, and R. Chawla, "A benchmark on the calculation of kinetic parameters based on reactivity effect experiments in the CROCUS reactor," *Annals of Nuclear Energy*, vol. 33, no. 8, pp. 739–748, May 2006, doi: 10.1016/j.anucene.2005.09.012.

[8] T. Ligonnet, A. Laureau, P. Simbsler, A. Pautz, and V. Lamirand, "Design of an Open-Loop Pile-Oscillation Program in the CROCUS Reactor," *IEEE Transactions on Nuclear Science*, vol. 71, no. 5, pp. 1001–1009, May 2024, doi: 10.1109/TNS.2024.3371183.

[9] F. Vitullo *et al.*, "A mm 3 Fiber-Coupled Scintillator for In-Core Thermal Neutron Detection in CROCUS," *IEEE Transactions on Nuclear Science*, vol. PP, pp. 1–1, Mar. 2020, doi: 10.1109/TNS.2020.2977530.

[10] "Research Reactors." Accessed: Jul. 10, 2025. [Online]. Available: https://www.iaea.org/bulletin/64-4

[11] M. Strobl and E. Lehmann, *Neutron Imaging: From applied materials science to industry*. IOP Publishing, 2024. Accessed: Jul. 10, 2025. [Online]. Available: https://iopscience.iop.org/book/edit/978-0-7503-3495-2

[12] J. Matouskova, B. Schillinger, and L. Sklenka, "New Neutron Imaging Facility NIFFLER at Very Low Power Reactor VR-1," *Journal of Imaging*, vol. 9, no. 1, Art. no. 1, Jan. 2023, doi: 10.3390/jimaging9010015.

[13] C. Lange and N. Bernt, "Neutron imaging at the low flux training and research reactor AKR-2," *Nuclear Instruments and Methods in Physics Research Section A: Accelerators, Spectrometers, Detectors and Associated Equipment*, vol. 941, p. 162292, Oct. 2019, doi: 10.1016/j.nima.2019.06.033.

[14] R. Hübscher, N. Bernt, and C. Lange, "L/D study via low-flux neutron imaging with TRAPY," *Nuclear Instruments and Methods in Physics Research Section A: Accelerators, Spectrometers, Detectors and Associated Equipment*, vol. 1040, p. 167294, Oct. 2022, doi: 10.1016/j.nima.2022.167294.

[15] J. Torres *et al.*, "A compact furnace to support in situ neutron imaging of hydrogen dynamics in yttrium hydride moderators," *MRS Advances*, vol. 9, no. 16, pp. 1227–1233, Nov. 2024, doi: 10.1557/s43580-024-00932-y.

[16] E. H. Lehmann, "Using neutron imaging data for deeper understanding of cultural heritage objects experiences from 15+ years of collaborations," *Journal of Archaeological Science: Reports*, vol. 19, pp. 397–404, Jun. 2018, doi: 10.1016/j.jasrep.2018.02.046.

[17] P. Trtik, R. Zubler, W. Gong, R. Grabherr, J. Bertsch, and L. I. Duarte, "Sample container for high-resolution neutron imaging of spent nuclear fuel cladding sections," *Review of Scientific Instruments*, vol. 91, no. 5, p. 056103, May 2020, doi: 10.1063/1.5143226.

[18] W. E. Fischer, "SINQ—the spallation neutron source, a new research facility at PSI," *AIP Conference Proceedings*, vol. 392, no. 1, pp. 1119–1122, Feb. 1997, doi: 10.1063/1.52691.

[19] E. Lehmann, P. Trtik, and D. Ridikas, "Status and Perspectives of Neutron Imaging Facilities," *Physics Procedia*, vol. 88, pp. 140–147, Jan. 2017, doi: 10.1016/j.phpro.2017.06.019.

[20] V. Lamirand, A. Laureau, O. Pakari, P. Frajtag, and A. Pautz, "Power calibration methodology at the CROCUS reactor," *EPJ Web Conf.*, vol. 225, p. 04022, 2020, doi: 10.1051/epjconf/202022504022.

[21] J. Leppänen, V. Valtavirta, A. Rintala, and R. Tuominen, "Status of Serpent Monte Carlo code in 2024," *EPJ Nuclear Sci. Technol.*, vol. 11, p. 3, 2025, doi: 10.1051/epjn/2024031.

[22] J. Leppänen, M. Aufiero, E. Fridman, R. Rachamin, and S. van der Marck, "Calculation of effective point kinetics parameters in the Serpent 2 Monte Carlo code," *Annals of Nuclear Energy*, vol. 65, pp. 272–279, Mar. 2014, doi: 10.1016/j.anucene.2013.10.032.

[23] J. Leppänen, "Response Matrix Method–Based Importance Solver and Variance Reduction Scheme in the Serpent 2 Monte Carlo Code," *Nuclear Technology*, vol. 205, no. 11, pp. 1416–1432, Nov. 2019, doi: 10.1080/00295450.2019.1603710.

[24] S. Häkkinen, "serpent 2 Validation for Radiation Shielding Applications," *Journal of Nuclear Engineering and Radiation Science*, vol. 8, no. 042001, Mar. 2022, doi: 10.1115/1.4051614.

[25] J. R. Lamarsh and E. Melkonian, "Introduction to Nuclear Engineering," *Physics Today*, vol. 30, no. 6, p. 53, Jun. 1977, doi: 10.1063/1.3037597.

The basis for a feasibility study for a high-power cyclotron to be used as a multi-purpose irradiation facility

G. Ragonesi[1*], M. Croia[1], L. Falconi[1], N. Burgio[1], A. Santagata[1], A. Pietropaolo[2], M. De Santis[3].
[1]ENEA Casaccia Research Center NUC-IRAD-RNR, Rome Italy
[2]ENEA Frascati Research Center, NUC-FUSEN-TEN, Rome Italy
[3]University of Campania "Luigi Vanvitelli", Caserta Italy
(*) giuseppe.ragonesi@enea.it

Abstract—**The primary objective of this work is to establish the groundwork for the development of a new experimental facility integrating a particle accelerator to be installed in ENEA Casaccia research centre. More specifically, this study has taken into consideration a high beam repetition frequency and a high beam integrated current accelerator.**

Preliminary studies have shown that the finest choice for an accelerator with high beam current and simplicity of installation and management is a proton cyclotron capable of reaching, for our research scopes, a final beam energy of at least 70 MeV.

To increase the operating flexibility, the accelerator shall supply at least two beamlines and should be capable of accelerating different kinds of ions, in addition to protons. The first beamline to generate neutrons by bombarding a metallic target with protons for neutron applications. The second one directly bombards various kinds of targets with protons or alpha particles, covering a wide range in beam energy and currents.

The facility would be of interest to aerospace, nuclear medicine, materials science, radiobiology, medical applications, detector calibration, and fusion physics.

Keywords — **Particle accelerator, Monte Carlo simulation for particle transport in matter, FLUKA, cyclotrons, radioisotope production with cyclotrons, accelerator-driven neutron activator.**

I. INTRODUCTION

INDUSTRIAL cyclotrons, which accelerate protons, have demonstrated their robustness and technological maturity in producing a wide range of radioisotopes for nuclear medicine. These radioisotopes, once bonded with suitable biomolecules, can be used as diagnostic and therapeutic agents against cancer and cardiocirculatory diseases [1]. On the radiopharmaceutical market, cyclotrons of various sizes are available, ranging from 9 to 70 MeV in energy and with currents of up to 750-800 μA (offers are also available for currents of up to 1500 μA) [2]. There are also cyclotrons designed for hadrontherapy, whose energy can reach 230–250 MeV; however, for their mission, they have a current of the order of nA [3]. Our goal is to produce a feasibility study for the creation of a multi-purpose, cyclotron-based irradiation facility [4-6] for various activities related to material activation and radiation damage, including aerospace, fission and fusion technologies, molecular and crystal structures, quality test design for industry, radiobiology, and pharmaceutical applications. Such a facility should be classified as a regional testing facility in which new ideas, proof of concept and training will be carried out at a relatively low cost, where they are posed on trial and, once passed a robust experimental campaign, transferred to the European Central facilities to be further investigated and exploited. The feasibility study aims to identify, via FLUKA [7], PHITS [8] and beam dynamics simulation codes such as GPT [9], and from interaction with stakeholders, the main interesting parameters for such a multi-purpose facility, defining the kind of particle to be accelerated starting from protons, deuterons and alpha, the respective proper range of currents, the targets' material necessary to obtain further particles as secondary emission field (neutrons, photons), and the main issues related to radioprotection and post-irradiation handling. Based on the outcomes of the present work, an interlocution with the cyclotron manufacturer will be initiated to identify (or design) the cyclotron that fits our purposes. Additionally, a discussion with the potential stakeholders will also be initiated.

II. CYCLOTRON TYPE SELECTION

WE are currently evaluating the type of particles to be accelerated. As already stated in the introduction, the cyclotrons having the maximum beam energy of 70 MeV were supplied by high maximum current that allows a very flexible range of activities. Consequently, we focused on two types of machines whose beam characteristics are reported in Table I.

TABLE I
COMPARISON OF THE MAIN PROPERTIES OF THE TWO TYPES OF CYCLOTRONS

Accelerated Particles	Machine 1	Machine 2 [a]
Protons	Beam Energy 30÷70 MeV Average current: 750 μA	Beam Energy 30÷70 MeV Average current: 750 μA
H^- ions	Beam Energy 30÷70 MeV Average current: 750 μA	Beam Energy 30÷70 MeV Average current: 750 μA
Deuterons		Beam Energy 35 MeV Average current: 50 μA
Alpha		Beam Energy 70 MeV Average current: 50 μAe

The first is a cyclotron capable of accelerating protons and H^- ions within an energy range of 30 to 70 MeV and an average current of 750 μA. Additionally, the second machine would also enable the acceleration of other particle and ion species, including deuterons up to a final energy of 35 MeV with an average current of 50 μA, and alpha particles up to 70 MeV with an average current of 50 μA. As reported in Table I, both machines span a proton energy range of 30–70 MeV, with the same maximum average current of 750 μA. Machine 2 can also accelerate deuterons up to 35 MeV and alpha particles up to 70 MeV with a maximum averaged current of 50 μA. The final selection of the most suitable accelerator for ENEA interests, among the two reported in Table I, will be the object of future work. However, preliminary computational work was conducted using the Monte Carlo FLUKA code to investigate the neutron generation by proton beams, which is reported in the following paragraph. Whereas, the study of applications that use the direct interaction of beam particles with targets to produce isotopes, test materials, or detectors will be deferred to a future phase.

III. FLUKA Simulations of Beam–Target Interactions for Neutron Generation

A first preliminary FLUKA simulation set was executed using cylindrical targets of Beryllium, Tungsten, and Rhenium with a proton beam as source, focusing on high-energy neutron production via the (p,n) reaction [3, 5]. The geometrical layout is straightforward:

1. We selected the same radius of 2.5 cm for all the targets. To ensure that the Bragg Peak of each material is included in the target body, we changed the cylinder's thickness using 10 cm for Beryllium and 0.96 cm for Tungsten and Rhenium;
2. A transverse Gaussian shape ($\sigma_{r,fwhm}$ = 1.5 cm), perfectly collimated proton beam impinges on one of the circular surfaces of the target and the transport and interaction of the proton within the target were simulated. The neutron-to-proton ratio per source, the energy deposited in the target volume, and its spatial distribution were estimated;
3. The target region is surrounded by a spherical vacuum region (radius = 10000 cm), and the escaping neutron flux is sampled on its spherical boundary;
4. All simulations have been carried out at the beam energies of 70, 120, and 200 MeV, and the results have been normalized at the maximum available proton current of each energy, which are respectively 750, 20, and 0.8 μA.

Fig. 1 compares the resulting spatial distributions of the deposited energy (GeV/cm³) for Beryllium, Tungsten, and Rhenium cylindrical targets at a proton energy of 70 MeV. It is apparent that, according to their bulk densities and atomic numbers, Beryllium has a more extended spatial distribution of the deposited energy density than Rhenium and Tungsten.

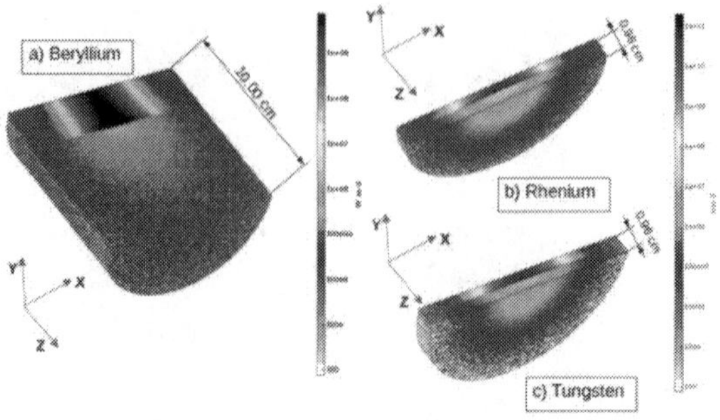

Fig. 1. Estimated distribution of the power density [W/m³] in a spatial mesh for Beryllium (a), Rhenium (b), and Tungsten (c) targets bombarded with a 70 MeV beam and a transverse Gaussian shape ($\sigma_{r,fwhm}$ = 1.5 cm) proton beam during FLUKA simulations.

To better highlight the Bragg Peaks position in each target, we integrated the energy deposited by the proton beam at different depths around the axis of the cylindrical target, varying the depth at various points on the given spatial mesh. We reported the power densities versus depth along the target axis in Fig. 2. As expected, Beryllium, Rhenium, and Tungsten show different Bragg locations. Rhenium (Z=75, bulk density 21.02 g/cm³) and Tungsten (Z=74, bulk density=19.3 g/cm³) have very close Bragg peaks (0.44 and 0.43 cm respectively) compared with Beryllium (Z=4, bulk density =1.85 g/cm³), which is located at 2.67 cm from the beam impinging surface.

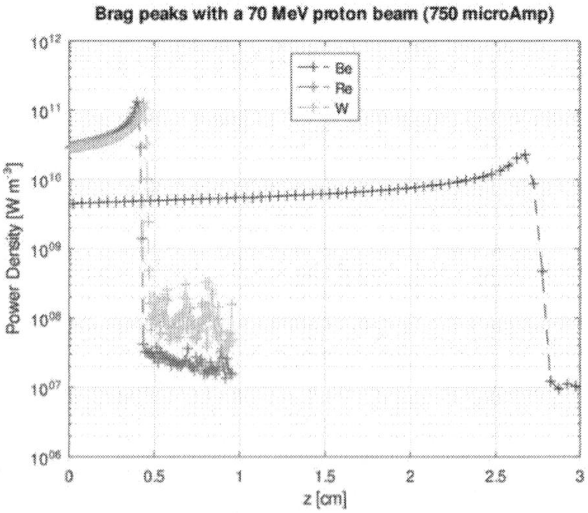

Fig. 2. Power density vs depth of interactions of proton beams with an energy of 70 MeV in Beryllium (Be-black curve), Rhenium (Re-red curve) and Tungsten (W-Green curve) estimated during the FLUKA simulations. The curves are obtained by integrating the deposited energy contribution of each mesh voxel along the x and y axes (see Fig. 1) of the FLUKA reference system and reported along the z axis (coincident with the beam propagation direction).

Fig. 3 presents the comparisons of the energy differential distributions of the escaping neutron flux intensities (neutron spectra) collected at the boundary of the void spherical region containing the target and constituting the boundary of the simulation layout (see point 3 of the layout description). The neutron spectrum generated by the Beryllium target has a significant fraction of its intensity below 100 keV, due to the slowing down through elastic scattering and (n,2n) reaction with the Beryllium light nuclei. At energies above 100 keV, the spectrum shape exhibits a shoulder compatible with evaporative neutron emission in the 0.7-1 MeV region. Conversely, Rhenium and Tungsten, being heavier nuclei, show a less thermalized spectrum with very similar evaporative peaks. Noteworthy is that, above 5-6 MeV up to 65 MeV, the high-energy neutron emission is higher for Beryllium than for the other two targets.

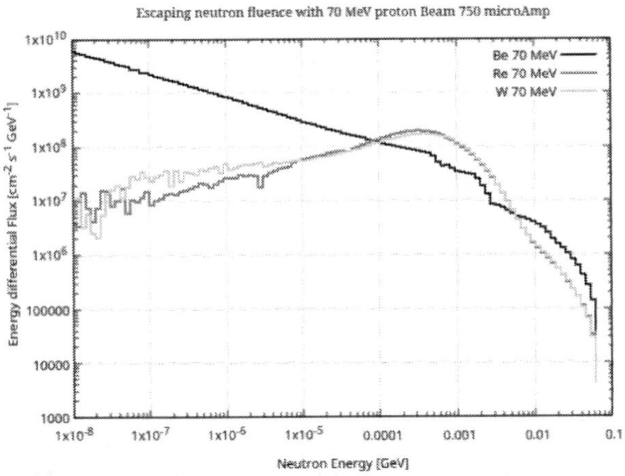

Fig. 3. FLUKA outcomes for the energy distribution of the neutron flux intensities generated on the spherical surface (see point 3 of the layout) for Beryllium (Be-black curve), Rhenium (Re-red curve), and Tungsten (W-green curve) using a 70 MeV proton beam.

Fig. 4 compares the estimates of the neutron-to-proton ratio (n/p) for the three considered targets when bombarded at the three selected beam energies.

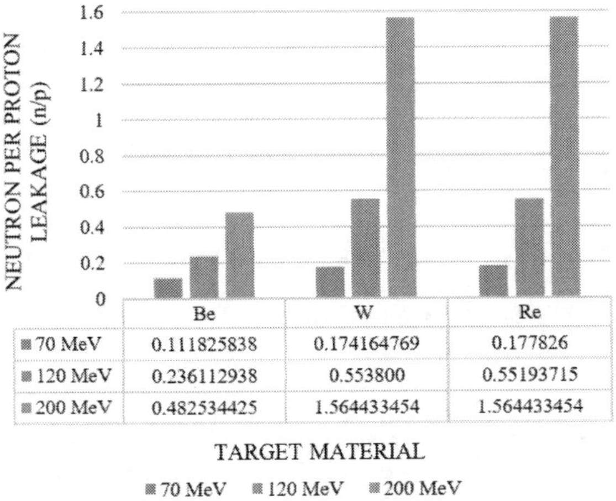

	Be	W	Re
70 MeV	0.111825838	0.174164769	0.177826
120 MeV	0.236112938	0.553800	0.55193715
200 MeV	0.482534425	1.564433454	1.564433454

Fig. 4. FLUKA estimates for the neutron per proton ratio of Beryllium, Tungsten, and Rhenium metallic targets at the three considered beam energies (70, 120, 200 MeV).

The (n/p) is a parameter used to quantify and compare the intrinsic capability of a target to generate neutrons under proton bombardment. As reported in Fig. 4, higher beam energy increases the n/p, and the Tungsten and Rhenium targets, due to their higher Z and bulk densities, prevail over Beryllium in neutron production.

However, as evidenced in the introduction to this work, the available cyclotron designs show that the maximum proton beam current decreases as the proton energy increases. Consequently, as shown in Fig. 5, the neutron flux reaches its maximum at 70MeV for all examined targets, considering the highest beam current available for each energy.

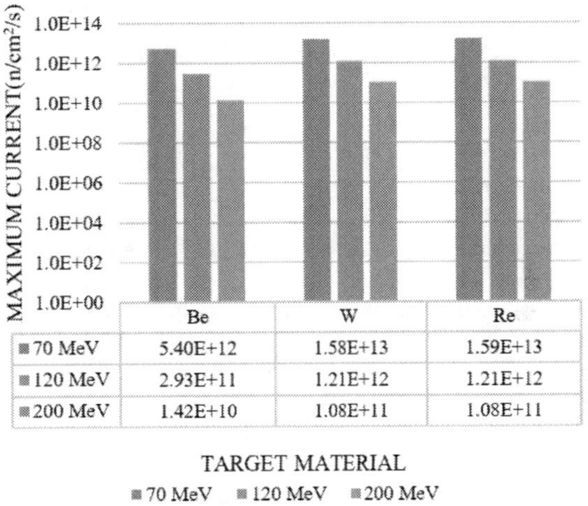

	Be	W	Re
70 MeV	5.40E+12	1.58E+13	1.59E+13
120 MeV	2.93E+11	1.21E+12	1.21E+12
200 MeV	1.42E+10	1.08E+11	1.08E+11

Fig. 5. Comparison of the FLUKA estimates for the neutron flux normalised at the maximum available beam current at 70, 120, 200 MeV of proton beam energy.

Another critical parameter is the power densities developed during the bombardment. Fig. 6 compares the power density of Beryllium, Tungsten, and Rhenium targets, showing that the deposited power densities of Tungsten and Rhenium are a factor of four to an order of magnitude higher than those of Beryllium, which exhibits lower power density at all beam energies.

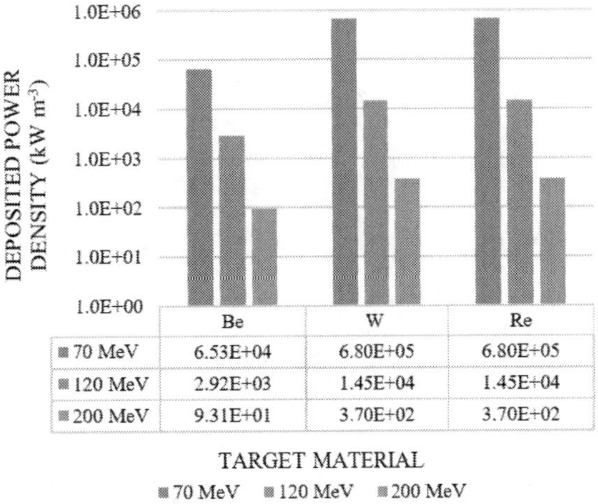

	Be	W	Re
70 MeV	6.53E+04	6.80E+05	6.80E+05
120 MeV	2.92E+03	1.45E+04	1.45E+04
200 MeV	9.31E+01	3.70E+02	3.70E+02

Fig. 6. Comparison of the power densities (kW/m³) estimated in all the considered targets at the three beam energies and normalised at respective maximum currents.

Table II summarises further FLUKA simulations that evaluated the target activation contact doses induced during a reference bombardment period.

TABLE II
REFERENCE BOMBARDMENT PERIOD

Period	Beam energy 70 MeV	Beam energy 120 MeV	Beam energy 200 MeV
#1	6 h bombardment @ 750 μA	6 h bombardment @ 20 μA	6 h bombardment @ 0.8 μA
#2	18 h cooling	18 h cooling	18 h cooling
#3	6 h bombardment @ 750 μA	6 h bombardment @ 20 μA	6 h bombardment @ 0.8 μA
#4		1 month of cooling	

As reported in Table II, the targets were bombarded for 6 hours, then left to cool for 18 hours, and again bombarded for another 6 hours. Then, the decay of the radiative source associated with the target activation is simulated for 1 month, during which the contact dose at the End Of Bombardment (EOB) and at 1 month are estimated.

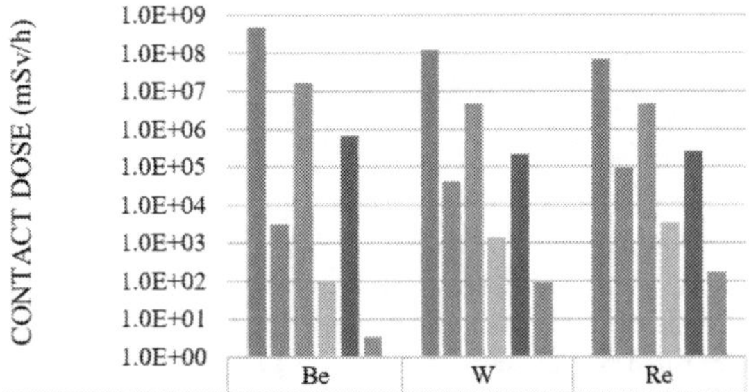

	Be	W	Re
70 MeV EOB	4.56E+08	1.22E+08	6.76E+07
70MeV 1Month	3.13E+03	4.18E+04	9.33E+04
120 MeV EOB	1.65E+07	4.70E+06	4.62E+06
120 MeV 1Month	9.65E+01	1.38E+03	3.34E+03
200 MeV EOB	6.80E+05	2.16E+05	2.63E+05
200 MeV 1 Month	3.38E+00	9.24E+01	1.70E+02

TARGET MATERIAL

70 MeV EOB | 70MeV 1Month | 120 MeV EOB | 120 MeV 1Month | 200 MeV EOB | 200 MeV 1 Month

Fig. 7. Targets estimated contact doses at EOB and after 1 month from EOB.

Fig. 7 compares the estimated doses at EOB and after one month of cooling: the Beryllium target shows a higher contact dose at EOB but it reaches the lowest value after one month of cooling. Despite the Beryllium contact dose at 1 month remains higher in absolute terms, it can be more easily managed with an appropriate cask and cooling period than the ones developed in the Tungsten and Rhenium target. Moreover, the Beryllium shows fewer mechanical and cooling problems than the other two materials.

IV. CONCLUSIONS

THIS work represents the first approach aimed at considering the possibility of installing a new irradiation facility at the ENEA Casaccia Research Centre to expand and enrich the agency's portfolio of services.

The study revealed that increasing the final beam energy leads to a significant decrease in the extracted beam current.

In particular, by examining the results obtained from simulations carried out to evaluate the available options for the target required for the (p,n) reaction, we found that the configuration that allows for the highest overall neutron flux is 70 MeV with a beam current of 750 μA, while the material presenting the fewest complications related to energy deposition and contact dose is Beryllium.

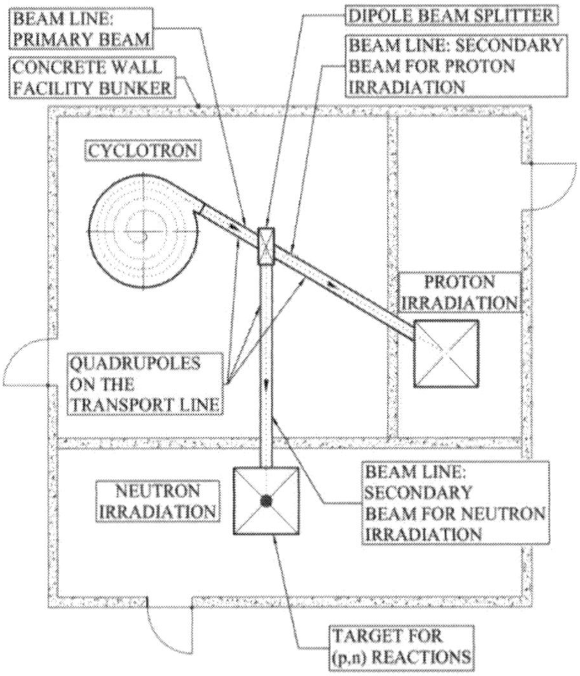

Fig. 8. A tentative layout for the cyclotron facilities with two beam lines, one dedicated to the neutron generation and the second for irradiation with protons and other ions

Based on similar facilities and data available in the literature, we developed a preliminary layout that envisions the installation of a cyclotron with an extraction point splitted in two beam lines by a dipole.

As shown in Fig. 8, the bunker contains the cyclotron room and two irradiation rooms, each served by an individual beamline. These beamlines depart from a dipole beam splitter, which supplies the two beamlines. The first one is for proton (and other ions) irradiation of targets for radioisotope production, detector calibration, and radiobiology experiments. The second one is dedicated to generating neutrons for irradiation experiments using this type of particle.

Future work will focus on refining specific simulation models, particularly understanding the target geometries and constituent materials related to beam dynamics. In parallel, a series of simulations will begin, focusing on the evaluation of radioisotopes producible through the facility and the particles intended for use. This activity will facilitate engagement with potential stakeholders and enable the completion of necessary assessments for selecting the appropriate cyclotron.

REFERENCES

[1] Y. Wang et al., "Production Review of Accelerator-Based Medical Isotopes", Molecules 2022, 27, 5294. https://doi.org/10.3390/molecules27165294;
[2] W. Kleeven, et al. "Recent development and progress of IBA cyclotrons." Nuclear Instruments and Methods in Physics Research Section B: Beam Interactions with Materials and Atoms 269.24 (2011): 2857-2862;
[3] A. Geisler et al. "Commissioning of the ACCEL 250 MeV Proton Cyclotron" in Proc. of 18th Int. Conf. on Cyclotrons and their Applications, Giardini Naxos, Italy. 2007. p. 9-14.
[4] N. Burgio et al., "Experimental Results of an Accelerator-Driven Neutron Activator for Medical Radioisotope Production" in IEEE Transactions on Nuclear Science, vol. 58, no. 2, pp. 445-450, April 2011, doi: 10.1109/TNS.2011.2112377;
[5] N. Gibson et al., "Radiolabelling of engineered nanoparticles for in vitro and in vivo tracing applications using cyclotron accelerators", Arch Toxicol 85, 751–773 (2011). https://doi.org/10.1007/s00204-011-0701-6;
[6] L. Maciocco, S. Avila, S. Buono, N. Burgio, F. Haddad and K. Abbas, "285 Experimental Qualification of the Theranean Accelerator-driven Neutron Activator for the Production of Activated Nanoparticles for Cancer Treatment", Radiotherapy and Oncology, Volume 102, Supplement 1, 2012, Page S150, ISSN 0167-8140, https://doi.org/10.1016/S0167-8140(12)70248-7;
[7] G. Hugo et al, "Latest FLUKA developments", EPJ Nuclear Sciences & Technologies **10** (2024).
[8] T. Sato et al, "Features of Particle and Heavy Ion Transport Code System PHITS Version 3.02", J. Nucl. Sci. Technol. 55, 684-690 (2018)
[9] M.J. de Loos, S.B. van der Geer, "General particle tracer: a new 3D code for accelerator and beam line design". https://accelconf.web.cern.ch/e96/PAPERS/THPG/THP001G.PDF

Design of a Liquid Neutron Filter Device at the JSI TRIGA Reactor for Fission Rate Measurements under Simulated High-Temperature Conditions

Blaž Levpuščček[1,2,*], Vladimir Radulović[1,2], Andrej Trkov[1,2],
Gilles Noguère[3], Olivier Serot[3], Christophe Destouches[3], Kieran Kessaci[3]
[1] "Jožef Stefan" Institute, Slovenia
[2] University of Ljubljana, Faculty of Mathematics and Physics, Slovenia
[3] CEA, DES, IRESNE, DER, France
(*) blaz.levpuscek@ijs.si

***Abstract*—A novel experimental setup has been developed for improving nuclear data through fission reactor rate measurements using a borated liquid filter. The experimental device involves simulating a temperature shift of the thermal peak in the neutron flux spectrum by varying the boric acid concentration in a specially designed container filled with borated water. Monte Carlo neutron transport simulations were conducted to optimize the filter geometry, evaluate fission chamber response, and determine the most suitable beam port location for implementation at the Jožef Stefan Institute TRIGA Mark II reactor. Among the options, the thermal column beam port was selected for its favorable neutron spectrum and reduced radiation streaming, despite the radial port offering more uniform flux conditions. The device enables measurements using dual fission chamber setup for comparative measurements under similar irradiation conditions, with an additional channel for dosimetry. Results show that the thermal spectrum can be shifted to correspond to an equivalent temperature from 30°C to 100°C. Boron concentration will be precisely controlled via a combination of volumetric analysis and neutron absorption measurements using an external neutron source. Construction of the experimental system is underway, with the first measurements expected in early 2026.**

***Keywords*— boric acid, boron concentration measurement, liquid neutron filter, neutron spectrum shaping, temperature feedback effects, thermal neutron peak shift**

I. Introduction

Accurate nuclear cross-section data is essential for advancing reactor simulations, improving fuel cycle performance, and ensuring nuclear safety. However, current nuclear data libraries such as ENDF/B [1] and JEFF [2] exhibit inconsistencies, particularly in actinide evaluations, which can lead to compensating errors in reactivity predictions [3], [4]. A major source of uncertainty lies in the modeling of temperature-dependent effects on fission cross sections, especially for heavy nuclides that significantly contribute to temperature feedback effects in nuclear reactors.

The temperature reactivity coefficient, which describes the sensitivity of the reactor to temperature change, is strongly influenced by the spectral shift of the thermal neutrons and the thermal cross-section shape of fissile actinides. Recent experiments—such as those performed in the MISTRAL campaign at the EOLE zero-power reactor—have shown notable discrepancies between calculated and experimental temperature reactivity coefficients, reaching up to 1.0 pcm/°C in MOX fuel [5]. These deviations have highlighten the need for experimental validation tools that can simulate temperature-dependent spectral effects with high resolution.

This work proposes a novel approach to address this need through the design of a borated liquid neutron filter. By varying the concentration of boric acid in water, the device can induce controlled shifts in the thermal neutron spectrum, effectively simulating the effect of temperature changes without altering the physical temperature of the system. The spectral shift is driven by the 1/v shape of the absorption cross-section of ^{10}B, which removes slower neutrons and results in a shift of thermal peak of neutron flux spectrum to higher effective temperatures.

The paper presents the conceptual design of the liquid filter system, optimization of its geometry and location using MCNP calculations, and a proposed method for precise boron concentration monitoring using an external AmBe source. This work lays the foundation for experimental implementation, with construction and commissioning planned in the near future.

Section 2 introduces the conceptual design of the experimental device and outlines the computational model used. Section 3 presents the results of the MCNP calculations, including optimization of the filter geometry and selection of the most suitable beam port for the experimental setup. Section 4 discusses possible methods for boric acid concentration measurement and identifies the configuration offering the highest sensitivity. The experimental implementation is described in Section 5, followed by conclusions and future work in Section 6.

II. Liquid Filter Design Description

A. Concept Description

The liquid neutron filter is designed to continuously and precisely shift the thermal peak of the neutron flux spectrum at the measurement location, while keeping the rest of the experimental setup unchanged. The shift can be performed while the reactor operates at steady power, ensuring that the position of the device and detectors remains fixed, providing consistent and comparable measurements across different spectral conditions. The device is intended for deployment in a well-thermalized beam port of the TRIGA Mark II reactor at the Jožef Stefan Institute (JSI).

A schematic representation of the liquid neutron filter is shown in Fig. 1. The core component of the device is a container filled with boric acid solution, which serves as a neutron filter. The fundamental principle behind its operation relies on the 1/v neutron absorption cross-section behavior of ^{10}B. Since it has a significantly higher absorption probability for low-energy neutrons, increasing the concentration of boron in the solution disproportionately removes more lower-energy neutrons, effectively shifting the neutron spectrum toward higher energies. This mechanism allows for precise and controllable spectral modifications without physically altering reactor or experimental conditions.

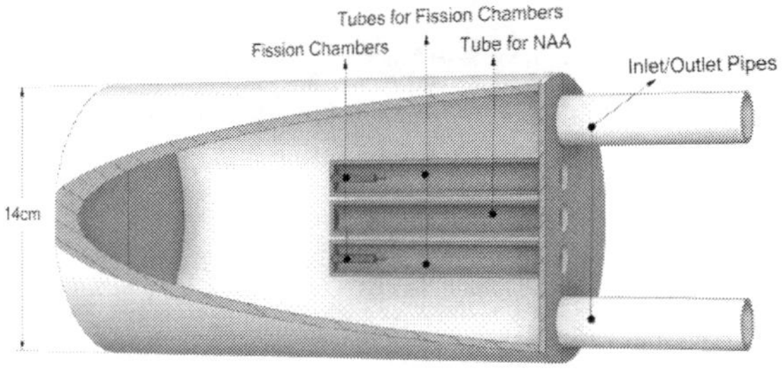

Fig. 1. Cross section view of the liquid filter device and its main internal components.

The primary function of the liquid neutron filter is to enable relative fission rate measurements using fission chambers. The device is equipped with three equivalent access tubes (marked in blue and green in Fig. 1), providing direct access to the neutron field inside the liquid filter. The two outer tubes (blue in Fig. 1) are designated for housing fission chambers. One chamber contains a reference fissile deposit, ^{235}U, which serves as a standard for normalizing fission rate measurements. The second chamber is used for measuring actinides of interest (e.g., ^{233}U, ^{238}U, ^{237}Np, ^{239}Pu, ^{241}Pu, ^{241}Am). By monitoring relative fission rates between these two chambers at different boron concentrations, the variation in fission cross-sections as a function of the thermal neutron peak temperature can be determined.

In addition to fission chamber measurements, the liquid neutron filter includes a dedicated tube for dosimetry measurements. This feature allows the irradiation of samples within the neutron filter, enabling direct characterization of neutron flux properties and verification of spectral modifications induced by varying boron concentrations. The integration of dosimetry strengthens the experimental methodology by providing an independent validation of neutron spectrum shifts, complementing the fission chamber measurements and enhancing overall measurement reliability.

B. Computational Model Description

The design optimization calculations for the liquid neutron filter were performed using the Monte Carlo N-Particle (MCNP) code [6], using the ENDF/B-VIII.0 nuclear data library [1]. The primary focus of the optimization was on neutron transport behavior, while also considering experimental feasibility and ease of implementation.

Since boric acid solubility is highly temperature-dependent [7], a constant water temperature of 20°C was assumed throughout the system. At this temperature, the maximum achievable concentration of boric acid was set to w(H_3BO_3) = 5 %.

To optimize the device geometry, initial calculations were performed using a simplified neutron source model. In this approach, only the device and the neutron source were modeled (without reactor geometry). The energy spectrum of the neutron source was extracted from the center of the tangential beam port at the JSI TRIGA. This simplification significantly reduced computational time, allowing for an efficient exploration of optimal filter dimensions and configurations.

Once the optimal filter geometry was determined, the full MCNP model of the TRIGA reactor was used to assess the device's performance in realistic conditions. The liquid neutron filter was inserted into all currently available beam ports at the JSI TRIGA reactor - the tangential, radial and thermal column beam port. These locations are highlighted in Fig. 2. The filter was inserted in different orientations with the goal to achieve the highest similarity of the neutron flux inside of the fission chambers.

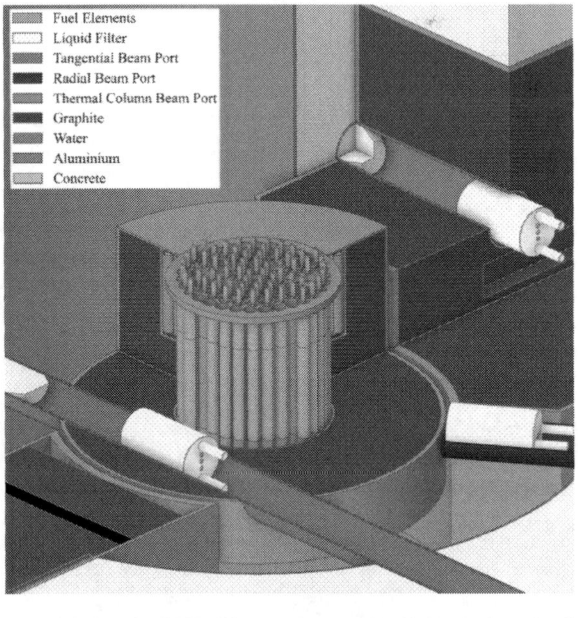

Fig. 2. Model of a JSI TRIGA research reactor. Liquid filter is shown in all different possible beam ports.

Due to the significant neutron moderation and attenuation caused by the water and graphite shielding in the radial and thermal column beam ports, a variance reduction technique was necessary to improve simulation efficiency. For this purpose, the ADVANTG code [8] was employed to generate weight windows and source biasing to improve the statistics of the neutron transport calculations.

C. Fission Chambers

The experimental device is designed to be able to facilitate multiple different fission chambers inside of its measurement tubes. The tubes are designed to be able to facilitate the widest fission chamber available. To achieve precise fitting of the smaller fission chambers the aluminum spacers will be used to guarantee that the fission chambers are placed at the exactly the same location each time they are inserted. Table I contains fission chambers and their fissile nuclides that were used in the calculations.

TABLE I
FISSION CHAMBERS USED IN THE CALCULATIONS.

Fission Chamber Type	Fissile Nuclide	Diameter [mm]
CF8Rgr	^{238}U	8
CF8Rgr	^{235}U	8
CF8Rgr	^{237}Np	8
CF4	^{235}U	4
CF4	^{241}Am	4
CF4	^{238}U	4
CFP12	^{233}U	12
CFP12	^{235}U	12
CFP12	^{239}Pu	12

III. LIQUID FILTER DESIGN AND OPTIMAL POSITION RESULTS

A. Optimal Geometry

A variety of different designs were tested to determine which configuration provides the best performance in terms of reaction rate ratios and neutron flux symmetry at the fission chamber locations. The tested layouts of the measurement tubes inside the liquid neutron filter are shown in Fig. 3.

Each design offers a distinct approach. The opposite-tube configuration provides the highest axial symmetry, ensuring similar neutron flux spectrum at both fission chamber locations. The slightly shifted tube configuration allows for measurements to be performed at two closely spaced positions. The three-tube configuration, while sacrificing proximity between fission chambers, introduces the advantage of an additional measurement tube. Furthermore, this design simplifies the experimental implementation, as it eliminates the need to route fission chamber wiring around the entire liquid filter, which is a significant benefit given the limited space within the reactor beam ports. From a separate calculation it was concluded, that the device should be as wide as possible to maximize the sensitivity to the boron concentration changes.

The reaction rate ratios between different actinides of interest and ^{235}U as a function of boric acid concentration are shown in Fig. 4. The ratio $\frac{^{235}UFC_1}{^{235}UFC_2}$ serves as an indicator of neutron flux symmetry between the two fission chamber positions. Ideally, this ratio should be equal to unity for all concentrations of dissolved boric acid. The results show that

Fig. 3. Cross section view of the liquid filter with different tested inner tubes configurations.

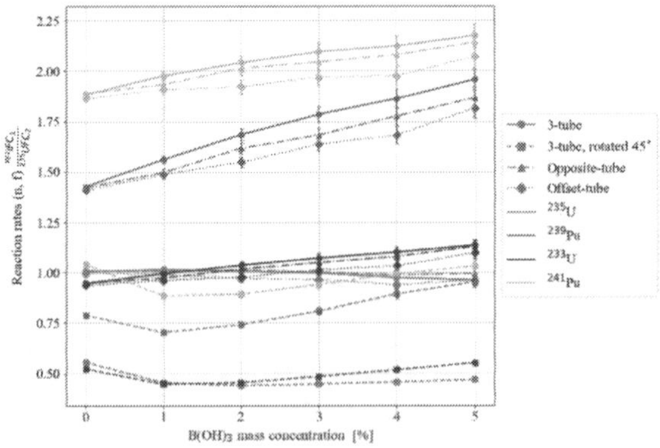

Fig. 4. Reaction rate ratios between fission chamber FC_1 with fissile nuclide of interest ^{xyz}i and a reference fission chamber FC_2 with ^{235}U.

the opposite-tube configuration achieves the closest agreement with unity, indicating the most similar neutron flux spectrum at both fission chambers. However, deviations in the other designs are not significant, as they remain within standard deviation.

Among the tested configurations, the three-tube design yields the highest reaction rate ratios. However, this design has a notable drawback: the increased distance of the fission chambers from the central axis of the liquid filter makes the measurements more susceptible to uncertainties arising from the unsuspected rotation of experimental device inside a neutron field. If the device experiences any unintended rotational shift, the reaction rate ratios can change significantly. This effect is demonstrated in Fig. 4, where a 45° rotation of the three-tube design results in a noticeable shift in the reaction rate ratios, emphasizing the importance of precise device positioning inside of the beam ports.

B. Optimal Beam Port

Figure 5 presents the thermal neutron flux (0–5 eV) within the liquid filter at three possible beam port locations. Clear differences are observed among the ports in terms of neutron spectrum, intensity, and spatial distribution. Notably, the tangential beam port exhibits a pronounced spatial tilt in the neutron flux at the fission chamber locations. This asymmetry arises from the beam port's position slightly below the reactor core's centerline. In contrast, the radial beam port shows a nearly symmetrical flux profile at the fission chamber locations, as both detectors are situated at the same vertical position. This uniformity is advantageous for calibration and performance stability.

Importantly, the steep neutron flux gradient within the liquid filter in the radial beam port enables precise tuning of the fission chamber response during the device design phase. However, the radial beam port poses significant experimental challenges due to its direct alignment with the reactor core. This results in substantial streaming of radiation during reactor operation, necessitating extensive radiation shielding, which may be cost-prohibitive.

The thermal column beam port, by comparison, offers a more favorable experimental environment. Its tangential orientation and distance from the reactor core significantly reduce direct radiation streaming. Although it is positioned slightly above the core centerline—introducing a mild flux asymmetry—this effect is minimal and can be further mitigated by rotating the experimental device by a few degrees. Given these advantages and acceptable neutron flux characteristics, the thermal column beam port was selected as the primary location for the experimental setup. However, due to the standardized port diameters, the device design remains compatible with installation in the radial beam port if adequate shielding measures can be implemented.

Figure 6 shows the change in the neutron flux spectrum per unit of lethargy in the thermal column beam port. As the boric acid concentration increases, the thermal peak diminishes in magnitude and shifts toward higher energies.

To quantify the spectral shift in terms of equivalent temperature, Fig. 7 illustrates the energy (and corresponding temperature) of the thermal peak as a function of boric acid concentration. A linear fit is applied to the data, accounting for uncertainties from the MCNP simulation. The slope of the fitted line is 1.21 ± 0.03 meV per weight percent of boric acid. The fit matches the trend well at lower concentrations, although uncertainties increase at higher concentrations due to the stochastic nature of Monte Carlo calculations. Simulated equivalent neutron temperatures range from approximately 30°C to 100°C, covering the desired temperature range.

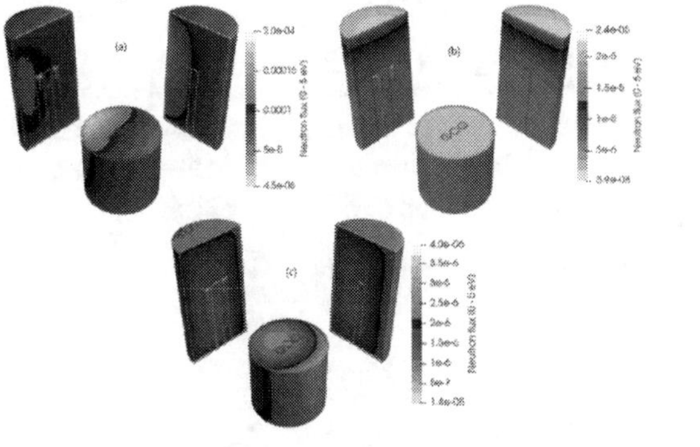

Fig. 5. Heat maps of the thermal neutron flux (0–5 eV) inside the liquid filter at $w(B(OH)_3) = 0\%$ for different beam port positions: (a) tangential, (b) radial, and (c) thermal column.

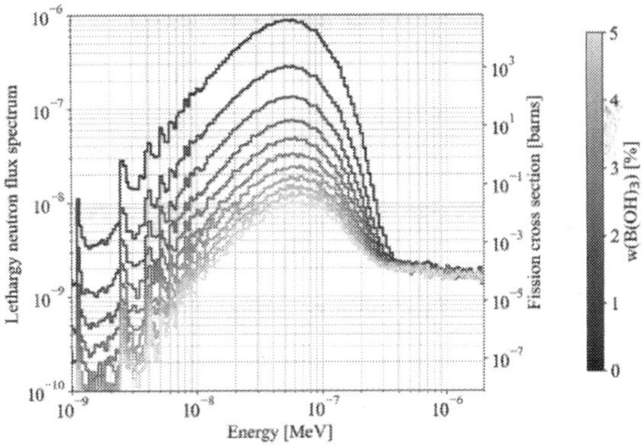

Fig. 6. Thermal peak of the neutron flux spectrum per unit lethargy in fission chambers located in the thermal column beam port for varying boric acid concentrations.

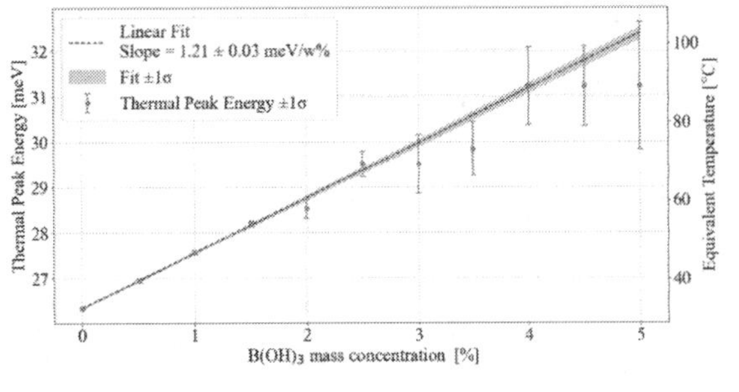

Fig. 7. Energy and equivalent temperature of the thermal neutron flux peak as a function of boric acid concentration in the thermal column beam port.

Finally, Fig.8 shows the ratio of reaction rates between two fission chambers—^{235}U one serving as a reference and the other containing nuclides of interest. Upper figure (a) displays the ratios for various nuclides on a logarithmic scale, while figure (b) provides a zoomed-in view of selected nuclides on a linear scale. Under ideal conditions, the ratio $\frac{^{235}U\ FC_1}{^{235}U\ FC_2}$ should be unity. Deviations from this ideal behavior primarily result from statistical uncertainties in the Monte Carlo calculations at higher boron concentrations.

The results highlight the system's sensitivity to changes in fission cross-section as a function of neutron spectrum modification. Notably strong responses are observed for ^{241}Am, ^{237}Np, and ^{238}U. While the sensitivity to plutonium isotopes (^{239}Pu and ^{241}Pu) is somewhat less pronounced, the change in reaction rate ratio still reaches approximately 20 % between boron-free and saturated conditions, offering a reliable mean to quantify the differences in the fission cross-section.

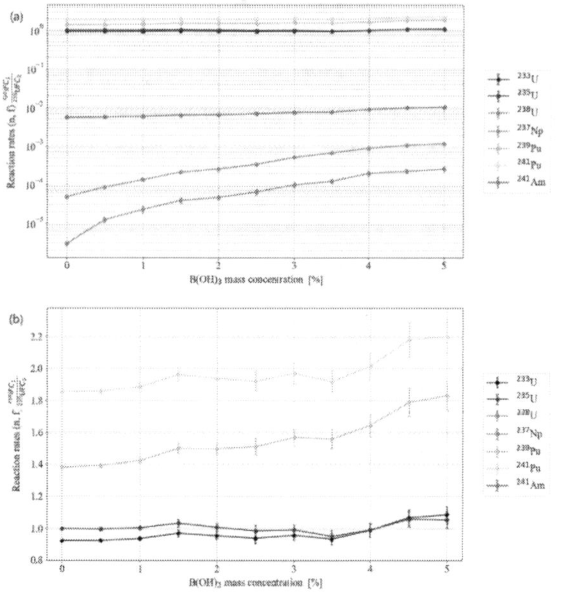

Fig. 8. Ratio of reaction rates in ^{235}U fission chambers for different boric acid concentrations in the thermal column beam port. (a) Logarithmic scale showing all nuclides of interest. (b) Linear zoom highlighting selected isotopes.

IV. Boron Concentration Measurement

A. Measuring principle

A critical requirement for ensuring the reliability of the proposed experimental setup is the precise determination of boron concentration within the liquid filter. Currently, two methods are proposed for quantifying this concentration.

The first is a straightforward volumetric approach, in which the boron concentration, C_B, is calculated from the known volume of water, V_w, in the system and the mass of added boric acid, $m_{\mathrm{B(OH)_3}}$, using the expression:

$$C_B = \frac{M_B}{M_{\mathrm{B(OH)_3}}} \cdot \frac{m_{\mathrm{B(OH)_3}}}{V_w}, \qquad (1)$$

where M_B and $M_{\mathrm{B(OH)_3}}$ denote the molar masses of boron and boric acid, respectively.

The second method utilizes an external neutron source in combination with a neutron detector. The principle of this technique is illustrated in Fig. 9. Neutrons emitted from the source are moderated and subsequently absorbed in the pipe system containing water and dissolved boric acid. Neutrons that are not captured via the (n, α) reaction on ^{10}B are detected by the neutron detector. As the boron concentration increases, a greater proportion of neutrons are absorbed, leading to a decrease in the detector count rate. This inverse relationship between boron concentration and neutron count rate is described by a calibration function that correlates the measured count rate with the corresponding boron concentration.

Similar measurement techniques have been employed and extensively studied in pressurized water reactors, where continuous monitoring of boron levels is crucial for reactor control and safety [9].

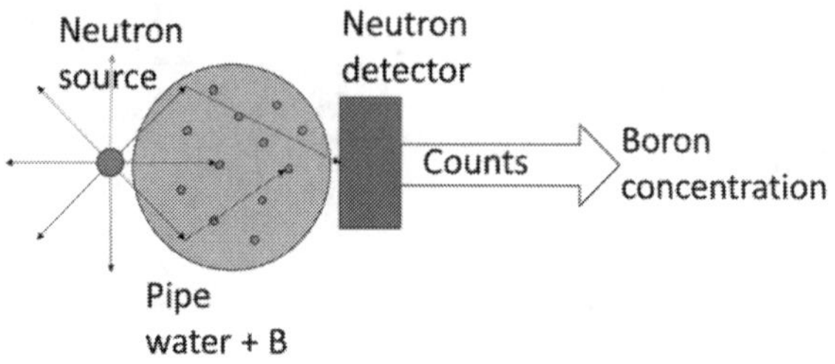

Fig. 9. Schematic of boron concentration measurement using an external neutron source.

B. Model description

The simulations were carried out using the Monte Carlo neutron transport code MCNP [6], in combination with the ENDF/B-VIII.0 nuclear data library [1]. A segment of the external measurement system was modeled, incorporating the AmBe neutron source, a ^{3}He neutron detector, and a pipe section. The diameter of the pipe section was varied in order to identify the optimal configuration for boron detection sensitivity.

The primary objective of these calculations was to investigate how variations in pipe diameter influence the neutron count rate detected by the ^{3}He detector at different boron concentrations. The geometry and components included in the simulation model are depicted in Fig. 10.

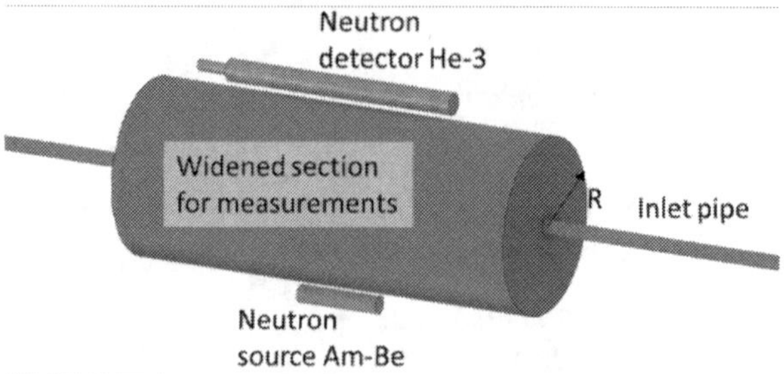

Fig. 10. Simulation model used in MCNP to evaluate the influence of the radius R of the widened pipe section on the neutron count rate.

C. Results

The primary outcomes of the MCNP calculations are presented in Fig. 11. As anticipated, increasing the boron

https://doi.org/10.1051/epjconf/202533804014

concentration within the pipe results in a reduction of the neutron signal detected by the ^{3}He detector. This decrease becomes more pronounced with larger pipe diameters. The observed trend arises from a combination of effects, including changes in geometry as well as variations in neutron interaction rates—specifically moderation in water and absorption by ^{10}B.

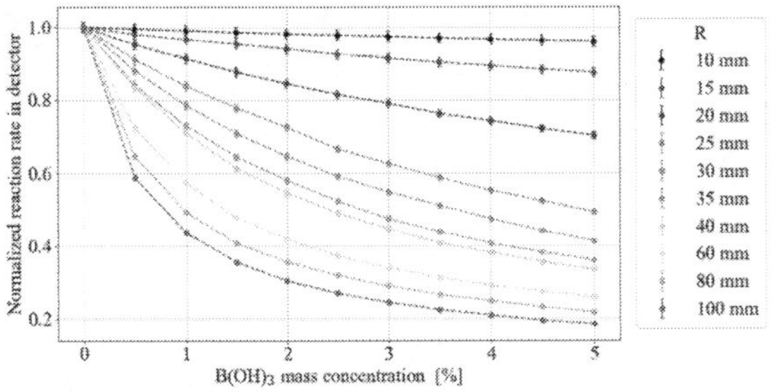

Fig. 11. Normalized reaction rates in the ^{3}He detector at varying boric acid concentrations for different pipe radii R.

As the pipe radius increases, the volume of moderating and absorbing medium between the neutron source and detector also increases. This leads to two dominant effects: (1) enhanced neutron moderation by water, and (2) increased neutron absorption by dissolved ^{10}B. The AmBe neutron source emits primarily fast neutrons, with energies around 3 MeV. As these fast neutrons pass through a greater volume of water, a larger fraction are thermalized. This thermalization has a dual consequence: more thermal neutrons are captured by ^{10}B nuclei (via the (n, α) reaction), while the fraction of neutrons that reach the ^{3}He detector and are detected also increases, due to its sensitivity to thermal neutrons.

Figure 12 illustrates the sensitivity of the detector response to changes in boron concentration by plotting the derivative of the normalized reaction rate with respect to concentration. It can be seen that a larger pipe radius enhances sensitivity at lower boron concentrations, while reducing sensitivity at higher concentrations. Conversely, a narrower pipe exhibits lower sensitivity in the low-concentration regime, but maintains better response at higher concentrations. This trade-off is a key consideration in optimizing system geometry for the desired measurement range.

In addition to neutron moderation and absorption, geometrical effects play a significant role in determining the detector signal. In the modeled configuration, the neutron source and the detector are placed on opposite sides of the pipe. As the pipe diameter increases, the distance between the source and detector also increases. This leads to a geometric reduction in the number of neutrons reaching the detector due to a smaller solid angle subtended by the detector, as well as increased scattering in the water-filled pipe.

These effects are quantified in Fig. 13, which shows the change in detector response as a function of pipe radius under three different conditions. The "void" case (green points), in which the pipe contains no material, isolates the pure geometrical contribution to signal reduction. Because there is no moderation in this case, the reaction rate is significantly lower and is normalized independently. The red and blue points correspond to cases with boric acid concentrations of 0% and 5%, respectively, and demonstrate the combined influence of geometry, scattering, and absorption. The suppression of the detector signal with increasing pipe radius in these cases is more severe, highlighting the compounding impact of neutron scattering and absorption processes in addition to geometric attenuation.

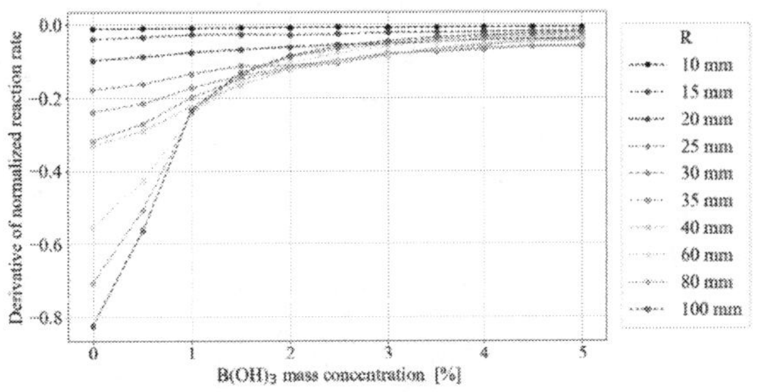

Fig. 12. Sensitivity of the ^{3}He detector to boron concentration, expressed as the derivative of the normalized reaction rate, at varying pipe radii R.

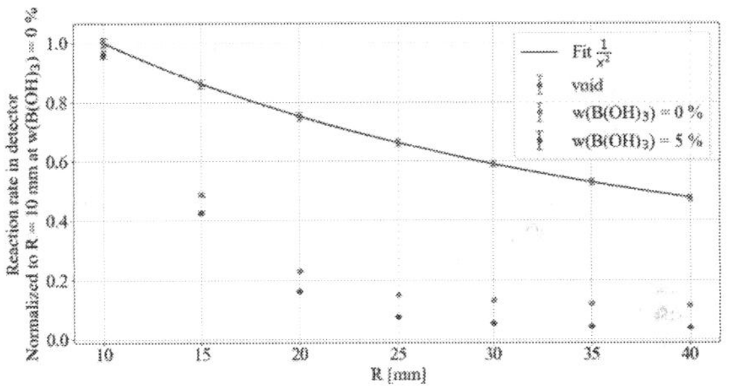

Fig. 13. Reaction rate in the ^{3}He detector as a function of pipe radius for three cases: void (green), boric acid concentration $w(\mathrm{B(OH)_3}) = 0\%$ (red), and $w(\mathrm{B(OH)_3}) = 5\%$ (blue). The void case is normalized separately and illustrates the geometric effect in the absence of material.

V. Experimental setup

Based on the results of the simulation study, the thermal column beam port was selected as the optimal experimental location for the liquid filter system. The planned external setup is depicted in Fig. 14. The device is positioned within the thermal column beam port, with inlet and outlet pipes connected to an external loop for boron concentration regulation.

Due to the absence of a standard radiation plug—removed to accommodate the piping—additional shielding blocks are placed at the beam port entrance to ensure radiological safety. The external section of the system includes a water tank where boric acid will be added, a circulation pump to drive the solution through the filter, and a boron concentration measurement system based on an external neutron source and detector configuration.

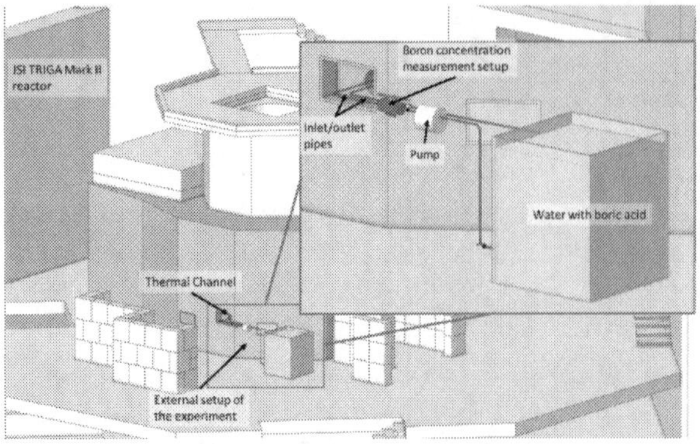

Fig. 14. External experimental setup located in the TRIGA reactor hall at JSI.

VI. Conclusion

The present study marks a significant step toward the development of a novel experimental setup aimed at improving nuclear data through fission reactor rate measurements. The novelty of the device lies in its ability to simulate temperature-induced shifts in the thermal neutron spectrum by varying the concentration of boric acid within a so-called liquid filter. The study focused on optimizing the design of this borated liquid filter, with careful attention to both measurement quality and experimental feasibility.

The finalized design enables comparative measurements between two fission chambers under identical irradiation conditions and incorporates an additional measurement tube for dosimetry measurements. Furthermore, a detailed assessment of beam port configurations led to the selection of the thermal column beam port as the optimal experimental location. Although the radial beam port provides the most uniform neutron flux distribution across both fission chambers, its unfavorable radiation streaming conditions during reactor operation present substantial challenges. Consequently, the thermal column beam port was selected as the primary experimental site due to its more manageable radiological conditions and adequate spectral characteristics.

The device is capable of simulating a thermal neutron spectrum shift equivalent to a temperature variation from 30°C to 100°C. Achieving precise control over this spectral shift requires accurate determination of boron concentration, which will be ensured through a combination of a volumetric method and neutron absorption measurements using an external AmBe source.

To optimize sensitivity across different boric acid concentration ranges, pipe diameter must be tailored accordingly. Wider pipes ($2R$ = 20 cm) are more suitable for low boron concentrations, while narrower pipes ($2R$ = 5 cm) provide better sensitivity at higher concentrations.

The next phase of the project will focus on the construction and commissioning of the experimental device, with initial measurements anticipated in early 2026. Additional tasks will include detailed radiation field simulations during reactor operation and possible activation characterization of the fission chambers, to ensure safe experimental conditions.

Acknowledgment

The authors acknowledge the financial support from the Slovenian Research Agency (research core funding No. P2-0073 (B), research project No. NC-24003).

References

[1] D. A. Brown, M. B. Chadwick, R. Capote, A. C. Kahler, A. Trkov, M. W. Herman, *et al.*, "ENDF/B-VIII.0: The 8th major release of the nuclear reaction data library with CIELO-project cross sections, new standards and thermal scattering data," *Nucl. Data Sheets*, vol. 148, no. 1, pp 1-142, Feb. 2018, 10.1016/j.nds.2018.02.001.

[2] A.J.M. Plompen, O. Cabellos, C. De Saint Jean, M. Fleming, A. Algora, M. Angelone, *et al.*, "The joint evaluated fission and fusion nuclear data library, JEFF-3.3," *Eur. Phys. J.*, vol. 56, no. 181, Jul. 2020, 10.1140/epja/s10050-020-00141-9.

[3] D. Neudecker, J. Alwin, A. R. Clark, T. Cutler, N. Gibson, M. J. Grosskopf, *et al.*, "Uncovering Where Compensating Errors Could Hide in ENDF/B-VIII.0," *EPJ Web of Conf.*, vol. 284, May. 2023, 10.1051/epjconf/202328416003.

[4] E. Bauge, G. Bélier, J. Cartier, *et al.*, "Coherent investigation of nuclear data at CEA DAM: Theoretical models, experiments and evaluated data," *Eur. Phys. J. A*, vol. 48, no. 113, Aug. 2012, 10.1140/epja/i2012-12113-7.

[5] J. P. Scotta, G. Noguere, D. Bernard, J. I. M. Damian and A. Santamarina, "Impact of the thermal scattering law of H in H2O on the isothermal temperature reactivity coefficients for UOX and MOX fuel lattices in cold operating conditions," *EPJ Nuclear Sci. Technol.*, vol. 2, no. 28, Jun. 2016, 10.1051/epjn/2016020.

[6] M. E. Rising, J. C. Armstrong, S. R. Bolding, F. B. Brown, J. S. Bull, T. P. Burke, *et al.*, "MCNP® Code Version 6.3.0 Release Notes," LANL, Los Alamos, NM, USA, Rep. LA-UR-22-33103, 2023.

[7] K. Crapse and E. Kyser, 'Literature review of boric acid solubility data," SRS, Aiken, SC, USA, Rep. SRNL-STI-2011-00578 , 2011.

[8] S. W. Mosher, S. R. Johnson, A. M. Bevill, A. M. Ibrahim, C. R. Daily and T. M. Evans, "ADVANTG an automated variance reduction parameter generator, Rev. 1," ORNL, Oak Ridge, TN, USA, Rep. ORNL/TM-2013/416, 2015.

[9] N. Tabti, A. Sari and D. Tromson, "Towards the enhancement of nuclear reactor safety: physical phenomena involved in the measurement accuracy of boron concentration online monitoring systems" *IEEE Transactions on Nuclear Science*, vol. 71, no. 5, Dec. 2024, 10.1109/TNS.2024.3351481.

https://doi.org/10.1051/epjconf/202533804015

Investigation of Noise Detection Systems in the CROCUS Reactor from SANDcaStLE

F.B. Darby[1,2,*], C. Kostelac[2], Michel Saliba[3], V. Lamirand[3,4], S.D. Clarke[1], A. Pautz[3,4], S.A. Pozzi[1], and O.V. Pakari[1,3,4]

[1] Department of Nuclear Engineering and Radiological Sciences, University of Michigan, Ann Arbor, 48109 MI, USA
[2] Los Alamos National Laboratory, Los Alamos, 87545 NM, USA
[3] Laboratory of Reactor Physics and Systems Behaviour, Ecole Polytechnique Fédérale de Lausanne, 1015 Lausanne, Switzerland
[4] Nuclear Energy and Safety Division, Paul Scherrer Institut, 5232 Villigen PSI, Switzerland
[*] fdarby@umich.edu

***Abstract*—The need for accurate, non-invasive methods to monitor reactor kinetics motivated SANDcaStLE (Simultaneous Analysis of Noise Detectors and Spectroscopy of a Low-power Experiment) at the CROCUS zero-power reactor. This study compares various neutron and gamma-ray detection systems, including fission chambers, a novel 3D neutron detection array called SAFFRON, and inorganic scintillators, to estimate the prompt neutron decay constant (α) from the fission "noise" at 8.92 mW critical. Our results show that gamma-ray measurements are highly effective, achieving a relative uncertainty of less than 15% in just one minute, an improvement over the three minutes required by ex-core fission chambers. Additionally, the SAFFRON array could detect fission noise despite its low total efficiency. The SAFFRON array was distributed through the radius and height of the core, suggesting the importance of detector spatial distribution in noise analysis. The fission chamber and inorganic scintillator systems agreed with the Serpent 2 IFP simulation α within 10 minutes of measurement time.**

***Keywords*—prompt neutron decay constant, fission noise, neutron noise, gamma noise, fission chambers, inorganic scintillators, SAFFRON**

I. Introduction

In nuclear reactors, "noise" refers to the small, random fluctuations in the neutron and gamma-ray populations that arise from fission chains. This phenomenon is similar to how a whale's calls, or "vocalizations," exist within the broader, often noisy, soundscape of the ocean. Just as bioacousticians study whale calls to understand their behavior, reactor physicists can analyze reactor noise to non-invasively infer the steady-state kinetic parameters, such as the prompt neutron decay constant (α), that define the temporal fluctuations arising from steady-state fission chains. These parameters are essential for accurately modeling reactor time-dependent behavior, validating computational methods, and ensuring safe and reliable operations [1]–[3].

While neutron noise measurements are traditionally used, gamma-ray noise measurements have not been studied as extensively, and may provide the opportunity for improved signal statistics [4]. On average, more gamma rays are emitted during fission than neutrons (7-10 gamma rays versus 2-3 neutrons for ^{235}U per fission). However, the theory for interpreting gamma-ray noise is not as well-developed as it is for neutron noise, necessitating investigation. Current theory allows for estimation of the prompt neutron decay constant and reactivity with gamma-ray noise [4]–[8], but estimation of the delayed neutron fraction and mean generation time has not been addressed. In this work, we compare various detector types, such as neutron-sensitive fission chambers and gamma-ray-sensitive scintillators, to cross-validate our results and enhance our understanding of both types of radiation correlations arising from fission chains.

SANDcaStLE (**S**imultaneous **A**nalysis of **N**oise **D**etectors and **S**pectroscopy of a **L**ow-power **E**xperiment) was designed to compare multiple detection systems simultaneously [9]. The spectroscopy work from this experiment is addressed in [10]. This work analyzes the α estimates from three systems, including a new three-dimensional (3D) neutron detection system called SAFFRON (an acronym for Scintillator Array with optical Fibers For high-Resolution Online Neutron counting), gamma-ray sensitive scintillators, and fission chambers [4], [5], [11]–[13].

II. Methods

A. CROCUS Zero-power Reactor

The CROCUS zero-power research reactor, operated by the Laboratory for Reactor Physics and Systems Behaviour (LRS) at the Swiss Federal Institute of Technology Lausanne (EPFL), is a two-region, water-moderated uranium core with a maximum operating power of 100 W. The cylindrical CROCUS core, shown in Figure 1, is approximately 58 cm in diameter and 100 cm in height, surrounded by a water-tight aluminum vessel, and all housed within a 1.3 m thick concrete shield. The central zone contains 336 uranium dioxide (UO_2) fuel rods, enriched to 1.806 wt% and arranged in a square lattice. The peripheral zone has up to 176 thicker uranium metal (U_{met}) fuel rods, enriched to 0.947 wt%, also in a square lattice with a wider pitch. Core criticality is achieved by precisely raising the water level from below, with a control accuracy of 0.1 mm. During operation, the water level was nominally held at 960 mm out of 1 m of the water fill height, and the detector measurements occurred during a two-hour, 8.9 $\pm$ 0.6 mW [14] steady-state critical operation.

B. Detection Systems

SANDcaStLE employed several detection systems to gather data on neutron and gamma radiation arising from the fission

Fig. 1. The CROCUS reactor core.

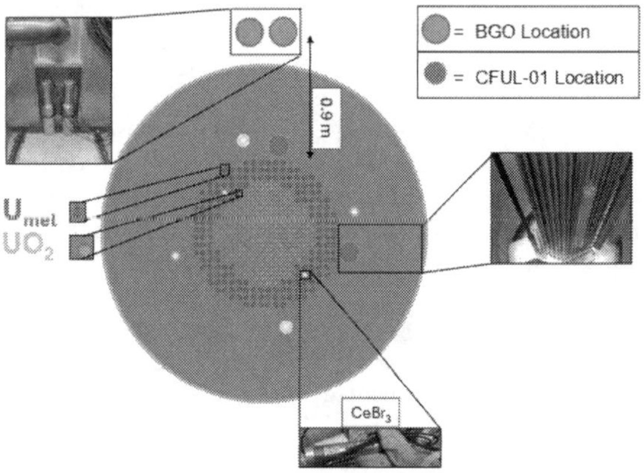

Fig. 2. The axial placement of two detection systems of 1) gamma-ray detectors, including two bismuth germanate (BGO) detectors 0.9 m north of the north edge of the CROCUS core and one $CeBr_3$ detector in the southeast control rod position, and 2) two CFUL-01 fission chambers in contact with the north and east edges of the core.

noise. These systems were positioned in, on, and outside the CROCUS reactor. In Figure 2, four CFUL-01 (^{235}U) fission chambers [15] were placed around the core perimeter to detect ex-core neutrons, two large (12.7 cm diameter by 25.0 cm length) bismuth germanate (BGO) inorganic scintillators were placed outside the north edge of the vessel to detect ex-vessel gamma rays, and one $CeBr_3$ inorganic scintillator (1.3 cm diameter by 5.1 cm length) was placed in the southeast control rod position to detect gamma rays in the core reflector. Figure 3 shows the SAFFRON 3D detector system [11]–[13], [16]–[18], consisting of 160 ZnS scintillators doped with ^{6}Li, each with a sensitive area of 1 mm^2 and a thickness of 0.2 mm. The scintillators are arranged in three main axial planes (15, 50, and 85 cm) and are coupled to optical fibers for readout outside the reactor shielding. SAFFRON is proven to be neutron noise sensitive despite the total efficiency being below the expected empirical limit observed for fission chambers ($\epsilon = 4 \times 10^{-5}$) [11], [19].

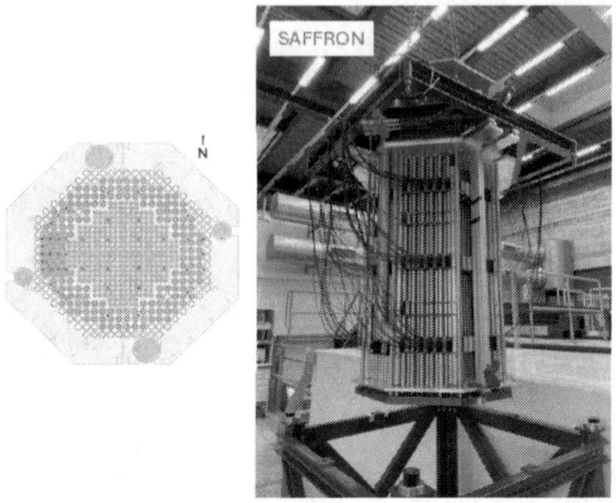

Fig. 3. The arrangement of the SAFFRON array. The SAFFRON array consists of 160 sheets of ZnS scintillator doped with ^{6}LiF, with a sensitive area of 1 mm^2 and a thickness of 0.2 mm, all coupled to optical fibers. The measurement system is distributed in two axial rings of the core and one ring in the reflector. The axial positions are distributed on three primary levels about the height of the core [11], [12].

C. Noise Analysis

The auto power spectral density (APSD) method was used to analyze the collected signals. This technique, also known as the Cohn-α method [20], [21], relies on analyzing the frequency content of the detector signals. The power spectral density of a signal is estimated by computing the periodogram, which is the squared magnitude of its Fourier transform. APSD is calculated by summing the relevant detector signals in counts or current per second and applying the Welch method [6], [22] with a sampling frequency of 2 kHz, a window length of 2^{11} samples, and 50% overlapping Hann window filter. The corresponding analytical expression is derived from the point-kinetics model [1] and results in a Lorentzian curve with a cutoff frequency that is directly proportional to the prompt neutron decay constant:

$$\begin{aligned} G_{ii}(\omega) &= \int_{-\infty}^{\infty} dt \cdot e^{-i\omega t} P_{ii}(\tau) \\ &= \epsilon_i F_0 + \frac{\epsilon_i^2 F_0 D_\nu}{(\beta_{\text{eff}} - \rho)^2} \cdot \frac{1}{1 + \omega^2/\alpha^2}, \end{aligned} \tag{1}$$

where $G_{ii}(\omega)$ is the APSD distribution, $P_{ii}(\tau)$ is the autocorrelation distribution, ϵ_i is the detection system total efficiency, F_0 is the total fission rate of the measured reactor, D_ν is the "Diven" factor for neutron multiplicity, β_{eff} is the effective delayed neutron fraction, ρ is the reactivity of the system ($\rho = 0$ at critical), and α is the prompt neutron decay constant. The equation is derived for neutron signal, but the cutoff frequency is shown to be consistent in gamma-ray measurement analysis [4]–[6]. We fit the APSD distribution with:

$$G_{ii}(f) = A + B/(1 + f^2/f_c) \tag{2}$$

where the prompt neutron decay constant is related to the break frequency (f_c) of the fitted Lorentzian curve by $\alpha = 2\pi f_c$.

III. Results

Figure 4 shows the APSD for the fission chambers transformed from the sum of the current-mode signals, with noticeable contamination peaks at multiples of 50 Hz due to AC leakage, which are ignored in the fitting process. The BGO and $CeBr_3$ APSD was computed from the sum of the count histograms from the time-tagged detection events and resulted in a smooth Lorentzian distribution. The APSD fit yielded α estimates consistent between the fission chamber and gamma-ray detection systems, and aligned with the Serpent 2 code predictions using the ENDF/B-8.0 library [5]. Both systems converged to about 152 s^{-1}, as shown in Table I, and agree with the Serpent 2 prediction within 10 minutes of measurement time. The APSD fit for the SAFFRON array showed a clear Lorentzian shape; however, the fit did not converge to the expected simulated value, with the mean measurement value being just above one standard deviation higher than the mean simulated value.

The signal-to-noise ratio shown in Table I is estimated as the A/B ratio of the Equation 2 fit for each system. Notably, the gamma-ray detection system has the highest signal-to-noise ratio, three times larger than the fission chambers and 300 times larger than SAFFRON. This is consistent with the precision of each system, where a higher signal-to-noise ratio corresponds with a more precise α estimate.

In Figure 5 and Table II, the gamma-ray detector system achieved a three-standard-deviation relative uncertainty converged below 15% in just 1 minute of measurement time, while the fission chambers achieved this in 3 minutes, and SAFFRON achieved this in 107 minutes. As shown in Figure 6, the SAFFRON array did not converge to the simulated α after 120 minutes of measurement.

IV. Conclusion

SANDcaStLE successfully demonstrated the advantages of using gamma-ray measurements for determining the prompt neutron decay constant (α) in the CROCUS reactor. The APSD of the gamma-ray detector system converged to a value with 15% relative uncertainty in just one minute, which was three times faster than the neutron-sensitive fission chambers. Furthermore, the study compared these systems to the SAFFRON array. Despite having an efficiency below the expected empirical limit for noise measurements, the unique 3D distributed geometry of SAFFRON allowed for a reasonable estimation of the prompt decay constant. These findings provide initial noise analysis results from SANDcaStLE. Future work will include estimating β_{eff} from the SAFFRON system and conducting comparative simulations of the gamma-ray signal to understand the gamma-ray amplitude term.

Acknowledgment

This work was supported in part by the Consortium for Monitoring, Technology, and Verification under the Department of Energy National Nuclear Security Administration award number DE-NA0003920.

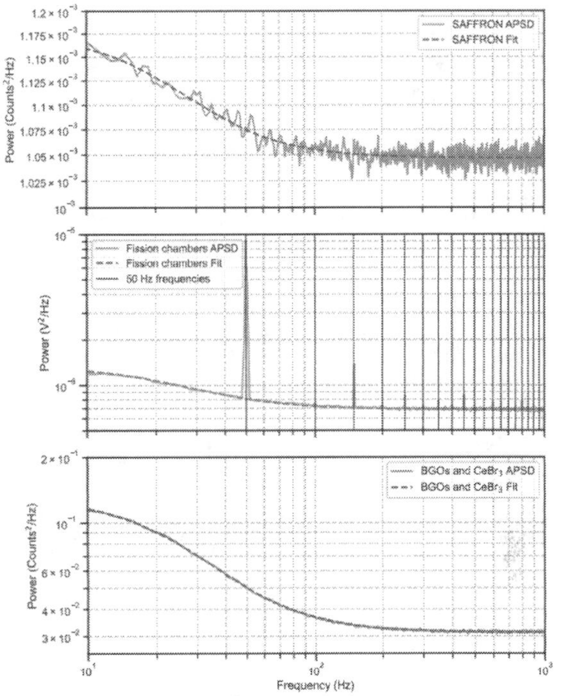

Fig. 4. The APSD of the SAFFRON (top), fission chamber (middle), and gamma-ray (bottom) detection systems after 120 minutes of measurement and corresponding Equation 2 fits for each APSD. The gamma-ray APSD was derived from the summed signal of the two BGO and one $CeBr_3$ detectors.

References

[1] J. Lewins, *Nuclear Reactor Kinetics and Control*, en. Amsterdam, Netherlands: Elsevier, Oct. 2013.

[2] M. Johnson, S. Lucas, and P. Tsvetkov, "Modeling of reactor kinetics and dynamics," Tech. Rep., Sep. 2010.

[3] F. W. Shabangu, D. Yemane, G. Best, and B. J. Estabrook, "Acoustic detectability of whales amidst underwater noise off the west coast of south africa," en, *Mar. Pollut. Bull.*, vol. 184, no. 114122, p. 114 122, Nov. 2022.

[4] O. Pakari, T. Mager, P. Frajtag, A. Pautz, and V. Lamirand, "Gamma noise to non-invasively monitor nuclear research reactors," en, *Sci. Rep.*, vol. 14, no. 1, p. 8409, Apr. 2024.

[5] O. V. Pakari, V. Lamirand, T. Mager, P. Frajtag, and A. Pautz, "High accuracy measurement of the prompt neutron decay constant in CROCUS using gamma noise and bootstrapped uncertainties," en, *Ann. Nucl. Energy*, vol. 170, p. 109 011, Jun. 2022.

[6] F. B. Darby, S. J. Tyler, O. V. Pakari, *et al.*, "Welch method and bootstrapping applied to subcritical gamma noise," *Nuclear Science and Engineering*, vol. 0, no. 0, pp. 1–8,

[7] F. B. Darby, O. V. Pakari, M. Y. Hua, *et al.*, "Neutron-gamma noise measurements in a zero-power reactor

TABLE I
α ESTIMATES AT 120-MINUTE MEASUREMENT TIME.

Detector System	α Value (s^{-1})	3σ Uncertainty (s^{-1})	Relative Uncertainty (%)	Signal-to-noise (A/B) Ratio
SAFFRON	167.0	23.0	13.8	1×10^{-1}
Fission chambers	151.7	3.5	2.3	1
BGOs and $CeBr_3$	152.6	1.1	0.7	3

TABLE II
α ESTIMATES WHEN THE THREE STANDARD DEVIATION UNCERTAINTY CONVERGED AT OR BELOW 15% RELATIVE UNCERTAINTY.

Detector System	Time to Achieve < 15% Uncertainty (min)	α Value (s^{-1})	3σ Uncertainty (s^{-1})	Relative Uncertainty (%)
SAFFRON	107	163	24	15
Fission chambers	4	155	14	9
BGOs and $CeBr_3$	1	158	12	7

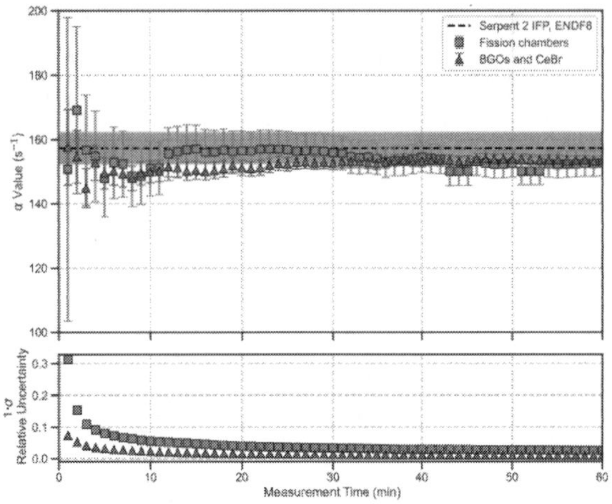

Fig. 5. The cumulative α estimates from APSD of the gamma detector system and fission chamber systems for the first 60 minutes. Three-standard-deviation uncertainty derived from a least-squares fit with Equation 2 is included for each estimate to emphasize the fast convergence of the gamma-ray system.

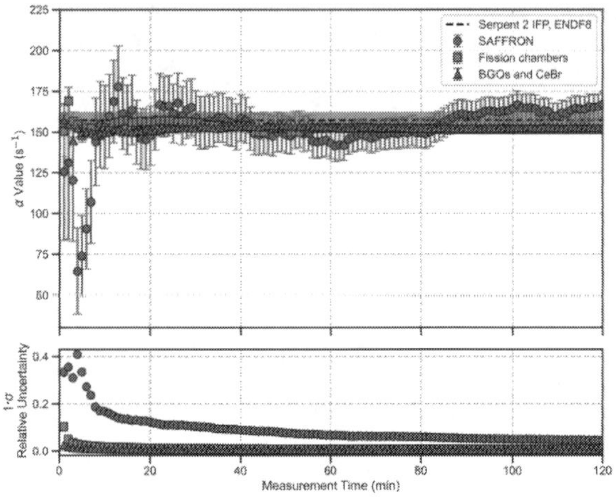

Fig. 6. The cumulative α estimates from APSD of each system as a function of measurement time. One-standard-deviation uncertainty derived from a least-squares fit with Equation 2 is included for each estimate.

using organic scintillators," *IEEE Trans. Nucl. Sci.*, vol. 71, no. 5, pp. 1033–1040, May 2024.

[8] O. V. Pakari, "Experimental and numerical study of stochastic branching noise in nuclear reactors," en, Ph.D. dissertation, EPFL, Lausanne, 2020.

[9] F. Darby, A. Lucas, and O. Pakari, "Simultaneous analysis of noise detectors and spectroscopy of a low-power experiment (SANDcaStLE)," Jun. 2025.

[10] O. Pakari, A. Lucas, F. Darby, *et al.*, "Towards the direct observation of xenon-135 poisoning in a zero-power reactor via gamma spectroscopy," in *Proceedings of ANIMMA*, 2025.

[11] M. Saliba, V. Lamirand, W. Monange, *et al.*, "Measurement of the prompt neutron decay constant in a zero-power reactor using a novel 3D detector system," en, *Nucl. Sci. Eng.*, pp. 1–9, Mar. 2025.

[12] F. Vitullo, V. Lamirand, D. Godat, and A. Pautz, "Simultaneous inter-calibration of 160 MiMi neutron detectors," *EPJ Web Conf.*, vol. 288, p. 04 023, 2023.

[13] F. Vitullo, "Miniature and minimalistic neutron detectors for online high-resolution experiments in the zero-power reactor CROCUS," en, Ph.D. dissertation, EPFL, Lausanne, 2022.

[14] V. Lamirand, A. Laureau, O. Pakari, P. Frajtag, and A. Pautz, "Power calibration methodology at the CROCUS reactor," *EPJ Web Conf.*, vol. 225, p. 04 022, 2020.

[15] O. Pakari, V. Lamirand, G. Perret, P. Frajtag, and A. Pautz, "Kinetic parameter measurements in the CROCUS reactor using current mode instrumentation," *IEEE Trans. Nucl. Sci.*, vol. 65, no. 9, pp. 2456–2460, Sep. 2018.

[16] F. Vitullo, V. Lamirand, J.-B. Mosset, *et al.*, "Developing and testing a miniature fiber-coupled scintillator for in-core neutron counting in CROCUS," *EPJ Web Conf.*, vol. 225, p. 04 018, 2020.

[17] F. Vitullo, V. Lamirand, J.-B. Mosset, *et al.*, "A mm^3 fiber-coupled scintillator for in-core thermal neutron detection in CROCUS," *IEEE Trans. Nucl. Sci.*, vol. 67, no. 4, pp. 625–635, Apr. 2020.

[18] F. Vitullo, V. Lamirand, K. Ambrožič, *et al.*, "Design of a 150-miniature detectors 3D core-mapping system for the CROCUS reactor," *EPJ Web Conf.*, vol. 253, p. 04 023, 2021.
[19] O. Pakari, V. Lamirand, G. Perret, *et al.*, "Investigation of spatial effects on neutron noise measurements in the zero power reactor CROCUS," *PHYSOR 2018: Reactor physics paving the way towards more efficient systems*, 2018.
[20] C. E. Cohn, "A simplified theory of pile noise," en, *Nucl. Sci. Eng.*, vol. 7, no. 5, pp. 472–475, May 1960.
[21] C. E. Cohn, "Determination of reactor kinetic parameters by pile noise analysis," en, *Nucl. Sci. Eng.*, vol. 5, no. 5, pp. 331–335, May 1959.
[22] P. Welch, "The use of fast fourier transform for the estimation of power spectra: A method based on time averaging over short, modified periodograms," *IEEE Transactions on audio and electroacoustics*, vol. 15, no. 2, pp. 70–73, 1967.

Multi-Particle Irradiation of CubeSat Components using the ENEA Distributed Facility: Experimental Setup and Radiation Response Analysis

Michele Croia[1,*], Nunzio Burgio[1], Alfonso Santagata[1], Alessandro Ampollini[2], Maria Denise Astorino[2], Giulia Bazzano[2], Barbara Bianchi[1], Matteo Cesaroni[1], Andrea Colangeli[2], Luca Falconi[1], Nicola Fonnesu[2], Stefano Loreti[2], Paolo Nenzi[2], G. Pagano[2], Antonino Pietropaolo[2], Pierpaolo Ricci[1], Giuseppe Ragonesi[1], Concetta Ronsivalle[2], Luigi Scaramuzzo[1], Emiliano Trinca[2], Androniki Vezyri[3]
[1] ENEA Casaccia Research Center, Rome (Italy)
[2] ENEA Frascati Research Center, Frascati (Italy)
[3]University of Ioannina, Ioannina (Greece)
(*) michele.croia@enea.it

Abstract— **This work presents the first experimental outcomes of a test campaign on CubeSat components, conducted at the ENEA distributed irradiation facility. The facility integrates several particle sources, including the TAPIRO fast nuclear reactor (fast neutrons), the FNG facility (14 MeV and 2.5 MeV neutrons), the REX linac (5 MeV electrons), and the TOP-IMPLART linac (protons up to 71 MeV), enabling the possibility to perform multi-particle irradiations. A dummy 1-unit CubeSat equipped with a RADFET dosimeter, like the one that flew on the ABCS mission, was exposed to various radiation fields. The primary aim of the irradiation was to calibrate the simulation code, thereby reducing discrepancies between the simulated doses and those measured during the mission. During the irradiation campaign, an intensive optimisation of the satellite model using FLUKA was performed, resulting in a reduction of the discrepancies between the satellite mission data and simulated doses, from 27% to the current value of 12%. This relevant update shifts the primary focus of investigation from discrepancy reduction to response analysis of silicon-based dosimeters. The distributed facility, allowing a certain degree of decoupling of the irradiating particles, may help to clarify the contribution of each particle generated in the shower and delivered to the detector. In this paper, we present the irradiation setup and the initial irradiation results. These findings enable a more in-depth analysis of the response of silicon-based dosimeters and the enhanced calibration of predictive simulation models. Future steps include proton irradiation at TOP-IMPLART to complete the test matrix.**

Keywords —**Nuclear research reactor, Particle accelerator, Neutron generator, Irradiation Facility, Aerospace, Dosimeter, Science material, Monte Carlo simulations.**

I. Introduction

The ENEA distributed irradiation facility offers a versatile environment for irradiating samples and conducting experiments with various particles, including neutrons, protons, and electrons. This facility comprises several sub-facilities: the RSV TAPIRO (Reattore Sorgente Veloce - Taratura Pila Rapida a Potenza 0 - Fast Pile Calibration at Zero Power) fast nuclear reactor as a fission neutron source, the TOP-IMPLART (Terapia Oncologica con Protoni–Intensity Modulated Proton Linear Accelerator for RadioTherapy) linear accelerator capable of producing proton beams up to 71 MeV, the FNG (Frascati Neutron Generator) facility, which generates neutrons of 2.5 MeV and of 14 MeV, and the REX ((Removable Electron to X-ray source) linear accelerator for electron beams up to 5 MeV. Together, these facilities support a broad range of irradiation experiments and are particularly valuable for studies that require exposure to multiple particle types on the same batch of samples. In a previous work [1], we evaluated the energetic spectra at 6000 km altitude from the Earth's surface using the SPENVIS code, obtaining such quantities for trapped particles (TP) and Galactic Cosmic Rays (GCR). The primary purpose was to compare the mean energy of TP and GCR with that of our facilities. We found that our distributed facility can represent the trapped electron (TP-e) and proton (TP-H) populations, which are the dominant radiation terms, up to the inner Van Allen Belt. In another work [2], taking as reference the irradiation condition of the ABCS mission and with the aid of some FLUKA simulations, we presented the design of a set of experiments to be carried out on our distributed facility with the primary purpose of “check and calibrate” the FLUKA simulations to reduce the 27% discrepancies between the predicted and measured dose rate during ABCS mission [3]. However, during the realisation of the experimental campaign, a partial modification of the ABCS FLUKA model and a punctual re-evaluation of the orbital radiative sources at the mission epoch led to a substantial reduction in the discrepancy to 12% [4]. This relevant update shifts the primary focus of investigation from discrepancy reduction to response analysis of silicon-based dosimeters.

The current work deals with the presentation and discussion of the first results of such irradiation experiments carried out with electrons (REX), neutrons (TAPIRO and FNG), using as a target a dummy 1 unit CubeSat mounting a RADFET chip like the one used in the ABCS mission to compare the dose measured under irradiation in the distributed facility. We are currently working on chip irradiation at the TOP-IMPLART proton accelerator. From such experimental outcomes, supported by further simulations in which more refined modelling and numerical techniques will be implemented, we will be confident in enhancing our design tools to support the radiometric characterisation of components and systems used, at least in near-Earth spatial missions.

II. THE DISTRIBUTED IRRADIATION FACILITY

The Distributed Irradiation Facility, operated by ENEA, comprises four distinct systems: the RSV TAPIRO nuclear research reactor, the FNG neutron generator, the TOP-IMPLART linear proton accelerator, and the REX linear electron accelerator. We report the sub-facilities description following the one made in [2].

A. The RSV TAPIRO fast nuclear research reactor

The RSV TAPIRO is a neutron source based on the fission process of a highly enriched uranium-molybdenum alloy. The neutron energy spectrum and flux intensity vary depending on the irradiation location within the reactor, allowing the facility to meet a range of experimental requirements. Near the reactor core, a spectrum similar to that of fission neutrons can be achieved, with a flux intensity of 4×10^{12} cm^{-2} s^{-1} at the maximum operating power of 5 kW. Other irradiation positions offer access to lower-energy neutron spectra, including epithermal and thermal neutrons, particularly within the Thermal Column Cavity (TCC). Further technical details on the reactor and its applications can be found in [5]. A schematic of RSV TAPIRO with the labelled experimental channels is presented in Fig. 1.

Fig. 1. Schematic 3-D layout of the RSV TAPIRO nuclear research reactor and its experimental channels.

B. The Frascati Neutron Generator – FNG

The Frascati Neutron Generator (FNG) is a compact neutron source based on accelerator technology, currently in operation at the ENEA Frascati Research Centre [6-8]. It utilises both Deuteron-Deuteron (D–D) and Deuteron-Tritium (D–T) fusion reactions to generate nearly monochromatic neutron beams with energies of approximately 2.5 MeV and 14 MeV, respectively. The neutron emission rates reach up to 10^9 s^{-1} for D–D reactions and 10^{11} s^{-1} for D–T reactions. A three-dimensional schematic overview of the FNG system, highlighting the main components, is illustrated in Fig. 2.

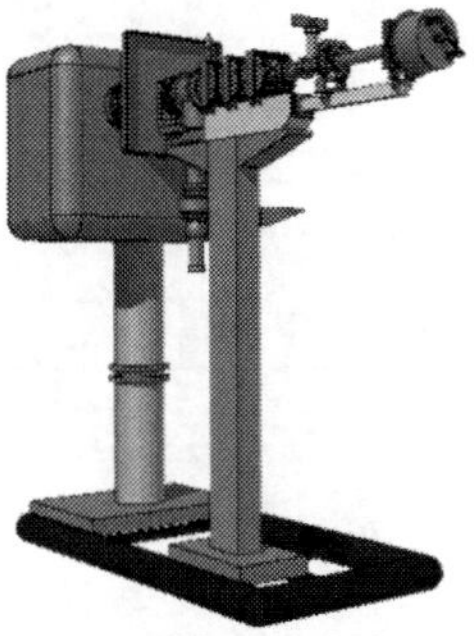

Fig. 2. Schematic 3-D layout of the FNG facility.

In D-T mode, the 14 MeV neutron calibration relies on the use of a ^{93}Nb activation foil. The reaction used is ^{93}Nb(n,2n)^{92m}Nb. A gamma line characterises the activation spectrum of ^{92m}Nb at 934 keV with a relative intensity of 100%. This is then measured using an absolutely energy-calibrated HPGe. The D-D neutron emission is measured at ±3.5% uncertainty. In D-T mode, the 14 MeV neutron emission rate is also determined using the foil activation technique, where a ^{115}In foil is irradiated and the gamma activation following the ^{115}In (n,n') ^{115m}In reaction channel is measured using the same HpGe. The uncertainty associated with D-D neutron emission is measured at ±7%.

C. The TOP-IMPLART proton linac

The TOP-IMPLART particle accelerator [9] is a technology demonstrator of a pulsed, fully linear machine for proton therapy under development at ENEA Frascati Research Centre in collaboration with the Italian Institute of Health (ISS) and the Oncological Hospital Regina Elena-IFO. It is a unique and compact accelerator, utilising S-band accelerating structures to achieve energies as low as 7 MeV, and one of the few proton sources available in Italy above 60 MeV. It comprises a commercial injector, by ACCSYS-Hitachi, and eight side-coupled DTL structures, patented by ENEA, which accelerate the proton beam to a maximum energy of 71 MeV and maximum intensities of 4.5×10^8 protons per pulse. A set of two scanning magnets allows irradiation of an area up to 10 cm × 10 cm. Fig. 3 shows a schematic layout of the accelerating line, from the injector up to reach the final energy of 71 MeV.

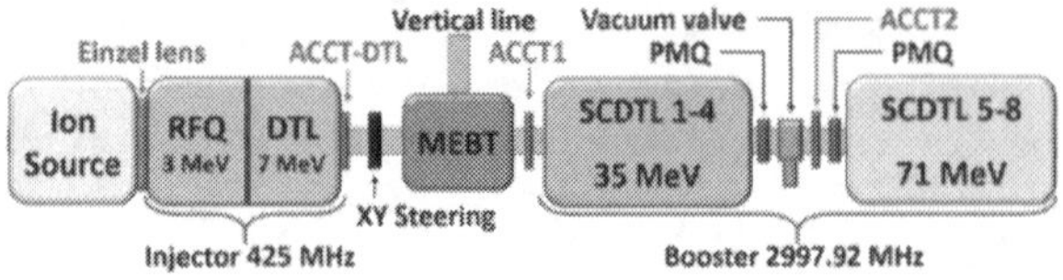

Fig. 3. Layout of the TOP-IMPLART accelerating line. The first extraction point is in the vertical line (7 MeV). At the end of the entire line (71 MeV) are the two scanning magnets.

Table I summarises the beam characteristics measured at the linac exit, just upstream of the two scanning magnets.

TABLE I
TOP-IMPLART BEAM PARAMETERS AT TARGET POSITION

Quantity	Values
Pulse length	2.4 μs
Pulse repetition frequency	25 Hz
Protons per pulse (max)	4.5×10^8
Pristine energy on target	61 – 70 MeV
Pencil beam spot size (FWHM)	25 mm
Maximum proton flux on a 10 cm × 10 cm area	1.1×10^8 $cm^{-2}s^{-1}$

Beam parameters at the end of the beam line.

D. *The REX electron linac*

The REX (see Fig. 4) irradiation facility [10], operating at the ENEA Frascati Research Centre, is based on a 5 MeV S-band on-axis coupled electron linear accelerator. It can provide two types of radiation: electrons and photons. It is equipped with a removable conversion unit for secondary X-Ray emission, generated by bremsstrahlung from the 5 MeV electron beam. The beam is produced in short pulses (~3 µs) with a maximum repetition rate of 20 Hz. Irradiation occurs in air within a 40 x 40 x 80 cm^3 lead-shielded chamber.

The beam size, flux and homogeneity vary within the irradiation chamber. In electron mode, the typical spot size is between 2 and 4 cm in diameter, with ±10 % homogeneity, and an electron flux greater than 10^{11} electrons cm^{-2} sec^{-1}. Electron energy can vary in the range of 5 to 3.5 MeV; a lower flux can be obtained by reducing the pulse repetition frequency down to 1 Hz.

The electron beam intensity is monitored online using an AC current transformer, which is calibrated against a Faraday collector. Dose mapping inside the irradiation chamber is performed with radiochromic films and alanine pellets.

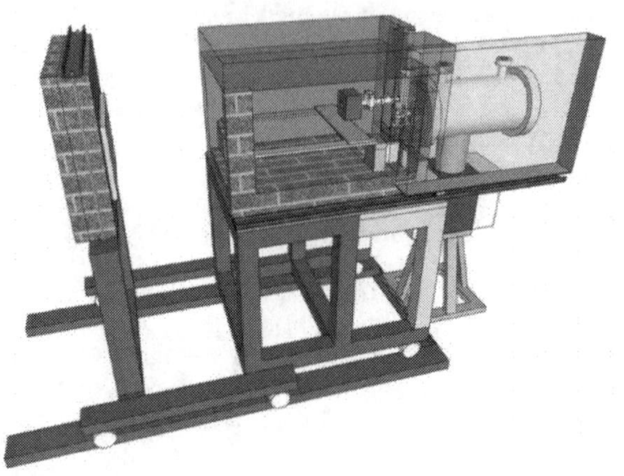

Fig. 4. Schematic 3-D layout of the REX accelerator and the irradiation chamber.

III. IRRADIATION AIMS AND STARTING EXPERIMENTAL SETUP

We had a strong interest in decoupling the response of silicon-based dosimeters under the effect of a mixed radiation field, as the one in the Van Allen Belt.

The distributed facility, which allows for a certain degree of representativeness of such radiative sources, may help to clarify the contribution of each primary and secondary particle generated in the shower through the satellite structure and delivered to the detector.

The first action was the realisation of four CubeSat units, one for each facility, with 2 mm-thick aluminium walls, to achieve a certain degree of particle shower representativeness similar to that encountered by the ABCS payload [2].

Each satellite replica hosts the same RADFET model used during the mission. The VARADIS VT02 has been installed on a removable Printed Circuit Board (PCB). Each VT02 contains two RADFETs; therefore, in each irradiation, we had two different measurements available. After each irradiation step, the VT02 was removed from the PCB for dose measurements. Fig. 5 shows the VT02 installed on its PCB before the irradiation. Each CubeSat is equipped with a small hole to implement a future remote measurement system to be used online during irradiations, thereby avoiding the need to switch off facilities and radiation protection problems due to the activation of the payload and RADFETs. In all irradiations, we grounded the VT02 through the side hole in the CubeSat. All operations and VT02 installations have been carried out in accordance with Varadis recommendations [11].

 https://doi.org/10.1051/epjconf/202533804016

Fig. 5. The removable PCB, inserted into the slot within the CubeSat, with the VT02 installed. Visible on the right is the cable that grounds the VT02.

To evaluate the acquired doses during device irradiation, we measured the RADFET voltage, as described in [11, 12]. This assessment relies on the established calibration relationship between the acquired dose and the RADFET degradation:

$$Dose[rad(Si)] = \left(\frac{\Delta V_{th}}{a}\right)^{\frac{1}{b}}, \qquad (1)$$

According to the Varadis recommendations:

- For the Voltage measurements at room temperature, the RADFETs were powered with a current of 10 μA to avoid temperature dependence.
- Varadis supply the a and b coefficients and were chosen adequately for the dose range.

IV. RESULTS AND DISCUSSION

According to FLUKA simulation outcomes [2], we programmed the exposure time of the CubeSats to the different facility radiation fields offered by the distributed facility. However, the first irradiation time steps have been carried out for a very short time to verify the response.

A. Irradiation @ RSV TAPIRO

Due to the dimensions of the Satellite, we decided to carry out the irradiations in the empty TTC (see Fig. 1 and [5]), positioning the satellite in the part of the TCC closest to the reactor core, directly in contact with the external reflector. Fig. 6 shows the CubeSat inside the TCC before the irradiation.

Fig. 6. TAPIRO Thermal Column, inside which is visible the CubeSat that we positioned, next to the external reflector, during the irradiation.

This irradiation layout enabled us to utilise the maximum neutron flux within the thermal column, and in the absence of graphite, to employ a hard neutron spectrum. During the irradiation, the reactor was maintained critical at a power of 50 W. To measure the progressive delivered dose, we shut down the reactor at each programmed time step and, after one night of cooling, removed the PCB slot from the TCC to measure the voltage variation of each RADFET in the irradiated VT02. After the measurements, the VT02 and the PCB were inserted into the CubeSat interior in the TCC. Once the TCC door has been closed, the TAPIRO ramp-up again to criticality, reaching 50 W of power. Due to the low final power, the dose acquired during the rise is negligible compared to the steady state period of irradiation.

Using (1) and the a and b coefficients provided by Varadis for that specific dose range, we calculated the acquired dose values. Fig. 7 shows the results of the irradiations.

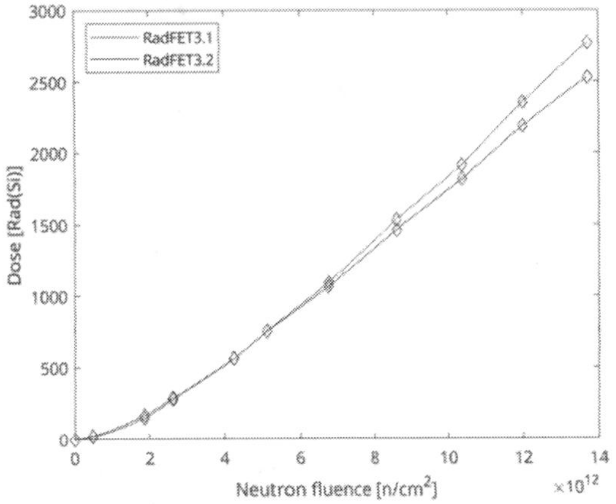

Fig. 7. Doses vs neutron fluence trends during the VT02 irradiation campaign @ TAPIRO. Since the VARADIS VT02 is equipped with two RADFETs, there are two sets of measurements.

B. Irradiation @ REX

During irradiation with electron beams, we positioned the satellite face at approximately 40 cm from the linac exit flange; in this way, the RADFET is located 45 cm from the linac exit. In the selected accelerator configuration, the beam peak energy was 4.5 MeV while the beam average energy was 3.8 MeV. Considering the measured beam current in the pulse of 92 mA, and the pulse repetition frequency of 1 Hz, the delivered charge is Q = 92 × t_{irr}. Fig. 8 shows the experimental layout.

Fig. 8. Experimental layout @ REX: on the left the CubeSat, on the right the end of the linac beamline.

To have more experimental points, we split the measurement into different sub-irradiations, switching off the linac each time. In Fig. 9, the experimental results, measured using the same procedure as for the TAPIRO irradiation, are presented.

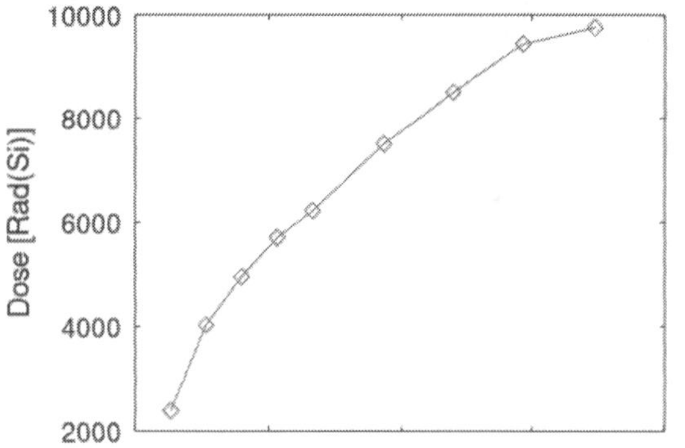

Fig. 9 Dose vs electron charge during the campaign @ REX. Due to the early failure of one of the RADFETs, only the results for the working one have been reported.

Unfortunately, after the first irradiation, one of the two RADFETs presented a fault. We continued the irradiations with only one working device; here we present the results for this single RADFET.

Due to the high beam currents, the delivered doses were extremely high in a very short time. Compared with the RADFET dams with neutrons (see Figs. 7 and 9), the trend of the curve in the plot shows a rise toward saturation. In addition to future considerations, we note that such an electron current availability will help test the saturation limit of the RADFETs that are critical for long-term missions.

C. Irradiation results @ FNG

During irradiations at the FNG facility, to cover a relatively different range of neutron energies compared to TAPIRO, we selected the D-T fusion reactions to deliver neutrons to the CubeSat with an energy of 14 MeV.

In Fig. 10, the measured voltage response after the irradiation.

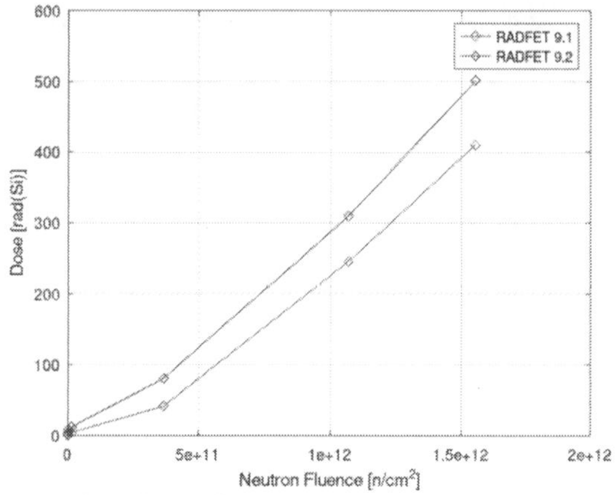

Fig. 10. Experimental results @ FNG.

Fig. 10 reports all the doses obtained in both RADFETs at each time step. The first four irradiation steps are very short and are carried out at low fluences, appearing in the plot very close to zero dose. They should be considered as a set of trial irradiations, which were executed at a distance of 5 cm from the neutron source, resulting in negligible doses. Starting from the fifth measurement, we shifted the CubeSat to a lower distance of 1.5 cm from the generator. Additionally, we increased the irradiation time per step, resulting in a clear increase in the delivered dose, which reached approximately 500 Rad (Si) in 6 hours.

V. CONCLUSIONS AND FUTURE WORK

The neutron irradiations show that the spectral difference between FNG and TAPIRO leads to different quantitative dose responses from the VT02 device and deserve an accurate comparison with simulation to achieve goals in terms of designing more accurate measurements to be carried out on our facility and in terms of design tools for new aerospace missions. The REX capability to execute irradiation with very high electron fluxes, allows the exploration of the lower and higher limits for dosimeters to be used in long-term space missions. The forthcoming proton irradiation campaign will complete the test matrix, yielding a comprehensive dataset to disentangle and attribute in-orbit damage to specific particle types.

REFERENCES

[1] N. Burgio, et al. "Modelling the interaction of the Astro Bio Cube Sat with the Van Allen's Belt radiative field using Monte Carlo transport codes." *Radiation Detection Technology and Methods* 6.2 (2022): 262-279.
[2] Croia, Michele, et al. "Exploring the ENEA Casaccia and ENEA Frascati irradiation capabilities: Status and perspectives." EPJ Web of Conferences. Vol. 288. EDP Sciences, 2023
[3] N. Burgio, "Radiometric impact assessments and shielding of CubeSat class satellites in interaction with orbital radiation fields and representativity of the calibration procedure and radiation damage tests in the ground facility." PhD thesis, DIAEE, Sapienza University of Rome, Rome, Italy, 2023.
[4] N. Burgio et al. "Comparison of the measured and simulated dose responses within the AstroBio CubeSat payload during its mission in the Van Allen Belt", Radiation Detection Technology and Methods, https://doi.org/10.1007/s41605-024-00508-2.
[5] N. Burgio, L. Cretara, M. Frullini, A. Gandini, V. Peluso, A. Santagata, Nucl. Eng. Des. (2014). https://doi.org/10.1016/j.nucengdes.2014.03.040
[6] M. Martone, *et.al.* J. Nucl. Mat. **212-215**, 1661 (1994);
[7] M.Angelone et al., Rev. Sci. Instr. **67**, 2189 (1996);
[8] A. Pietropaolo et al. "The Frascati Neutron Generator: A multipurpose facility for physics and engineering", IOP Conf. Series: Journal of Physics: Conf. Series 1021 (2018) 012004, doi:10.1088/1742-6596/1021/1/012004.
[9] V. Surrenti et al. Proc. 14th International Particle Accelerator Conference, 1516-1520, TUPM092 (2023)
[10] G.Bazzano et al., Proc. 14th International Particle Accelerator Conference, 5161-5164, THPM127 (2023)
[11] https://www.varadis.com/products/
[12] A. Holmes-Siedle, and L. Adams. "RADFET: A review of the use of metal-oxide-silicon devices as integrating dosimeters." International Journal of Radiation Applications and Instrumentation. Part C. Radiation Physics and Chemistry 28.2 (1986): 235-244.

https://doi.org/10.1051/epjconf/202533804017

A step-by-step approach to validate a new thin-film deposited heating element integrated into a calorimeter from laboratory to irradiation conditions in the JSI TRIGA reactor

Adrien Volte[1*], Vladimir Radulović[2], Florian Baudry[3], Alain Giani[3], Jérémy Rebaud[1], Michel Carette[1], Abdallah Lyoussi[4], Luca Snoj[2], Jean-Yves Ferrandis[3] and Christelle Reynard-Carette[1]

[1]Aix Marseille Univ, Université de Toulon, CNRS, IM2NP, Marseille, France
[2]Reactor Physics Division, Jožef Stefan Institute, Ljubljana, Slovenia
[3]IES, University of Montpellier, CNRS, Montpellier, France
[4] CEA/DES/IRESNE/DER, Section of Experimental Physics, Safety Tests and Instrumentation, Cadarache, F-13108, Saint Paul-lez-Durance, France

(*) corresponding author: adrien.volte@univ-amu.fr

Abstract—**This paper presents a step-by-step approach, from laboratory to irradiation conditions, to validate a new thin-film deposited heating element integrated into a single-cell calorimeter. First, the specifications of the new multi-thin-film heating element are given, and the main characteristics and performance, obtained thanks to a comprehensive characterization protocol, are detailed. Then, the sensor, which corresponds to a CALORRE-type single-cell calorimeter is presented, and the integration of this new multi-thin-film heating element into the calorimeter is described by focusing on the specific optimization realized. Finally, the irradiation campaign realized in the Jožef Stefan Institute TRIGA Mark II reactor in Slovenia is detailed. The assembly and the irradiation channel are presented. The calorimeter was located in the Triangular Irradiation Channel #1 with natural convection (dry air channel and ambient temperature), and the axial position corresponded to the median plane of the fuel elements. The new multi-thin-film heating element was qualified up to 3.2 × 10^{17} n.cm^{-2} and the evolution of its resistance was compared before and after 3 days of irradiation. Absolute temperatures at specific points were measured as a function of the reactor power from 12.5 kW to 250 kW.**

Keywords — **Calorimetry, Sensor, Calibration, Irradiation**

I. Introduction

THE intense mixed neutron and photon fluxes existing in nuclear environments, such as research reactors, lead to energy deposition within materials. This energy deposition is called absorbed dose rate or nuclear heating rate in the case of intense value. This latter quantity is a key parameter for the design and interpretation of the experiments, conducted using major facilities such as Material Testing Reactor (MTR), for safety and security reactor operations, optimization of materials, fuels and radioactive wastes as well as testing new concepts of reactor. Due to the aging of the major research facilities, particularly the MTRs, the Jules Horowitz Reactor is currently under construction at the CEA Cadarache center in the south of France. It will produce extreme and harsh conditions: high fast neutron flux of 5.5x10^{14} n.cm^{-2}.s^{-1} (from 1 MeV) leading to a high accelerated aging (up to 16 dpa/year) and a high absorbed dose rate (up to 20 W.g^{-1} in aluminum) [1, 2]. These unequalled conditions in Europe, launched new collaborative research topics in 2009 between Aix-Marseille University and the CEA thanks to the joint laboratory LIMMEX - Laboratory for Instrumentation and Measurement in EXtreme Environments) and its IN-CORE program (Instrumentation for Nuclear Radiations and Calorimetry online in REactor).

One of the objectives is to develop and design innovative sensors and measurement methods dedicated to the on-line quantification of the nuclear heating rate using non-adiabatic calorimeters. The associated challenge is to increase the measurement range (from very low to high nuclear heating rate values), while working on the miniaturization of the calorimeters, the optimization of their metrological performances (sensitivity, linearity, response time, etc.), the development of new calibration and measurement methods and the measurement of the associated thermal properties as a function of the temperature [3-8]. Consequently, a new research program within the framework of the LIMMEX laboratory, called MICRO-CALOR, began in 2020 to design and study miniaturized single-cell calorimeter integrating new thin-film heating element (sensor patented by AMU and the CEA), under laboratory and real conditions [9, 10].

This paper will be dedicated to a step-by-step approach to study a new multi-thin-film deposited heating element integrated inside a specific CALORRE single-cell calorimeter. This CALORRE single-cell calorimeter was already tested and validated during a previous irradiation in the Triangular Irradiation Channel #1 (TIC) inside the JSI TRIGA Mark II reactor) but by using a classical heating element. This approach is realized from the design of this multi-thin-film heating element to its characterization under laboratory conditions and irradiation conditions in the dry TIC.

The first part will focus on the new multi-thin-film heating element. This part will describe its design and fabrication achieved using screen-printing technique. Then, its chemical, dimensional, electrical, and thermal characterizations out-of-calorimeter under non-nuclear conditions will be given. The second part of the paper will briefly present the CALORRE single-cell calorimeter. A same configuration was irradiated previously (in 2023) but it was adapted and optimized to integrate the new multi-thin-film heating element discussed in the first part. The last part will present the irradiation inside the dry Triangular Irradiation Channel (TIC) of the JSI TRIGA reactor at the median plane of the fuel. Firstly, the JSI TRIGA reactor core conditions and the TIC channel, as well as the nuclear-hardened CALORRE assembly integrating the new multi-thin-film heating element, will be slightly detailed. Secondly, the evolution of the new multi-thin-film heating element resistance before and after irradiation will be provided to confirm the stability and robustness of this heating element. Finally, the absolute temperatures measured at specific points and temperature difference obtained in steady state as a function of the reactor power will be shown and discussed.

II. CHARACTERIZATION OF A NEW MULTI-THIN-FILM DEPOSITED HEATING ELEMENT OUT-OF-CALORIMETER

A. Design and fabrication using the screen-printing technique

The multi-thin-film heating element corresponds to a new multilayer heating element realized thanks to the screen-printing technique (at the IES - Institute of Electronic and System of Montpellier). It is deposited on a first sample of alumina (18 mm in height, 10 mm of width and 1 mm of thickness) and is composed of 4 contact pads, a resistive layer with a specific pattern and a protective layer which are successively deposited on the alumina sample. The deposition of the layers required specific protocols, masks and thermal treatments not detailed in this paper. Then, 4 electrical wires are welded on contact pads and a second sample made of Alumina (15 mm in height, 10 mm of width and 1 mm of thickness) integrating the T_{hot} thermocouple is used to obtain a "core" (cf. Figure 1).

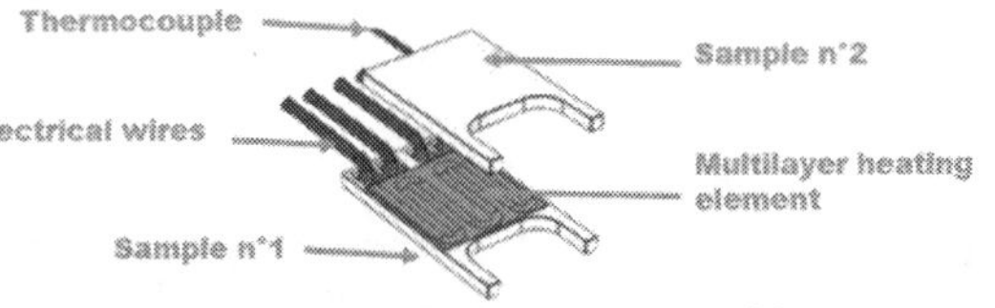

Fig. 1. Schematic of the multi-thin-film heating element and the core prototype

B. Characteristics and performance

Once the prototype cores containing the new multi-thin-film heating elements were produced, the new multi-thin-film heating elements were characterized under laboratory conditions by using several prototypes corresponding to different steps of the fabrication.

First, the surface structure and chemical compositions of the deposited thin-film were analyzed by means of Scanning Electron Microscopy (SEM) and Energy Dispersive X-ray Spectroscopy (EDX). The main result is that no organic material was identified. The composition of the protective layer is mostly composed of Bismuth and Oxygen, the resistive layer of Palladium, Silver and Oxygen and the contacts pads of Silver. Then, the thicknesses of the contact pads, the resistive layer and the protective layer were determined thanks to profilometric measurements. The thicknesses are reproducible: 11.15 µm +/- 1.8 µm for the contact pads, 7.0 µm +/- 1.6 µm for the resistive layer and 7.8 µm +/- 1.4 µm for the protective layer. The linear resistance was quantified using a 4-point station while moving along the resistive track pattern. The linear resistance varies from 0.186 $\Omega.mm^{-1}$ to 0.207 $\Omega.mm^{-1}$ showing that the thin-film deposits are homogenous. The resistance as a function of temperature was studied using a climatic chamber from 23 °C to 73 °C. The Temperature Coefficient Resistance obtained is low and varies from 69.07 $ppm.°C^{-1}$ to 68.49 $ppm.°C^{-1}$. Finally, an electrical current, up to 0.262 A, was injected inside the multi-thin-film heating element, corresponding to a current density of 46.6 $A.mm^{-2}$ during 12 tests (16 steps of current increments with a duration of 30 minutes each). This ageing process has stabilized the resistance response.

This characterization allows the complete validation of the new multi-thin-film heating elements: electrical performances, design, geometry, robustness, durability, future compatibility with nuclear environment.

III. CALORRE SINGLE-CELL CALORIMETER

The CALORRE single-cell calorimeter is made of Aluminum 6061 and is composed of four parts. The first part is an essential part: the sample, in which the nuclear heating rate is measured. The cylindrical sample, manufactured in a single block with the head of the calorimetric cell, has an external diameter of 16 mm and a height of 46.2 mm (cf. Figure 2). It has a parallelepiped hole to host the core.

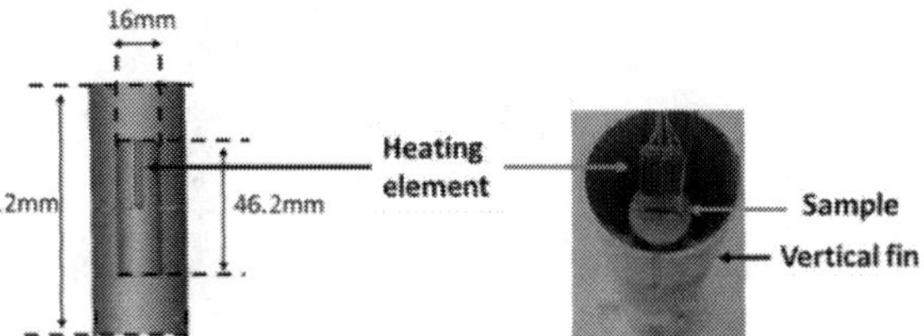

Fig. 2. Schematic (on the left-hand section) and picture (on the right-hand section) of the CALORRE-type single cell calorimeter.

The second part is the cell structure including the horizontal and vertical fins. The sample is surrounded by the horizontal fin composed of 4 horizontal metal sectors with internal and external diameters of 16 mm and 32 mm, respectively, and a thickness of 1 mm. These sectors are positioned in the median plane of the sample. This horizontal fin is surrounded by a cylindrical annular vertical fin with a thickness of 1 mm. This vertical fin also serves as an envelope similar to that of the CALORRE MIT configuration to suppress thermal contact resistances [11]. The third part corresponds to temperature measurement by means of K-type nuclear-hardened thermocouple. The hot temperature T_{hot}, is placed within the core containing the new multi-thin-film heating element (as presented in the previous section) and is inserted 5 mm into the sample. The second thermocouple, T_{cold}, is positioned on the vertical fin at the same axial height as that of the T_{hot} thermocouple. The fourth part corresponds to a 4-wire connected multi-thin-film heating element integrated in a core which is put at the center of sample (cf. Figure 2).

IV. Irradiation inside the JSI TRIGA Reactor

A. Irradiation channel and calorimeter assembly

The JSI TRIGA Mark II reactor is a conventional 250 kW TRIGA research reactor that is used for several applications, from the Neutron Activation Analysis, the Education and Training to research activities [12, 13]. As mentioned, the Triangular Irradiation Channel (TIC #1) was chosen to qualify the new multi-thin-film heating element (cf. Figure 3) as a previous irradiation was realized in this channel with a similar CALORRE single-cell calorimeter integrating a classical heating element.

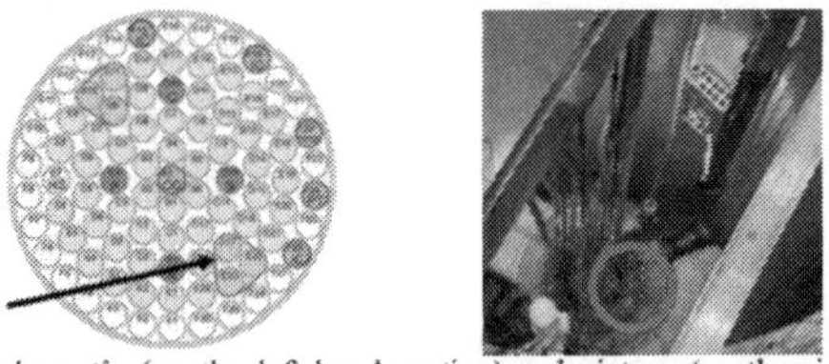

Fig. 3. Schematic (on the left-hand section) and picture (on the right-hand section) of the JSI core and location of the TIC #1

The TIC #1 has approximately an internal diameter of 5.3 cm. It is a dry air channel with an ambient temperature and natural convection. This shape and these geometric dimensions were considered when designing the calorimeter assembly.

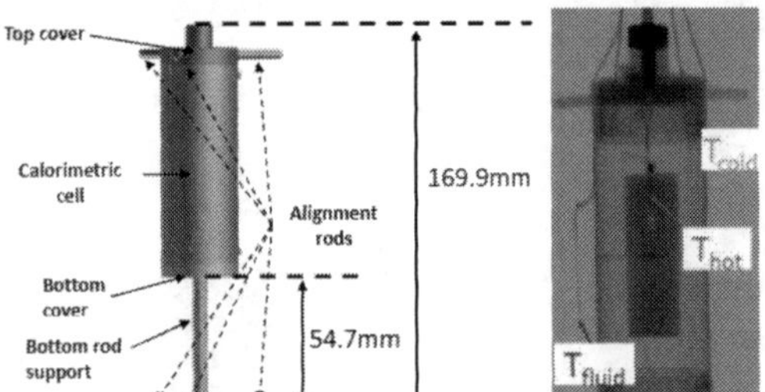

Fig. 4. Schematic of the calorimeter assembly (on the left-hand section) and microtomography picture (on the right-hand section)

The first component of the calorimeter assembly is also made of Aluminum 6061 and is composed of two covers (the bottom and top covers) positioned on either side of the calorimetric cell. There are 3 alignment rods on each cover that set the calorimeter at the center of the TIC. The bottom cover serves to position the calorimeter at the desired location from the bottom of the channel thanks to a rod of 55.7 mm in height (including 1 mm inside the support structure) and 6 mm in diameter. Another K-type thermocouple was added to measure the fluid temperature (cf. Figure 4).

B. Qualification of the new multi-thin-film heating element

The new thin-film heating element was irradiated during 20 hours and for different reactor powers (from 12.5 kWth to 250 kWth). In order to validate the use of the new multi-thin-film heating element, the change in resistance as a function of the electrical power injected during the calorimeter calibration before and after irradiation was studied with the reactor shut down. As shown in Figure 5, the maximum difference of resistance between pre-irradiation and post-irradiations is very low: 8 mΩ. This negligible difference demonstrates the robustness of the new multi-thin-film heating element up to a total fluence of 3.2 10^{17} n.cm^{-2}.

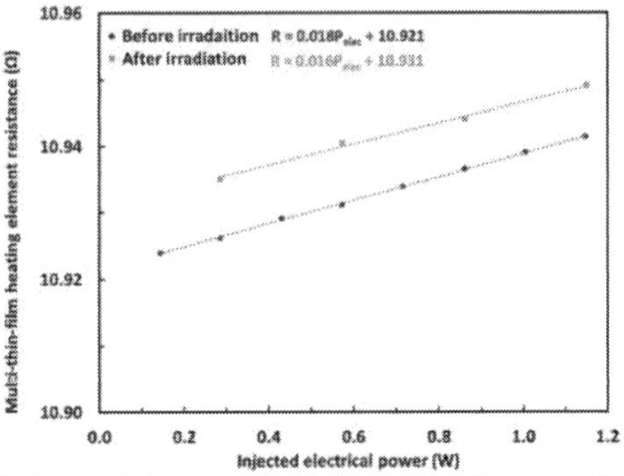

Fig. 5. Resistance of the new multi-thin-film heating element as a function of the injected electrical power obtained inside the TIC #1 before and after 20 hours of irradiation

C. Temperature measurements

During the first day of the irradiation campaign, 5 reactor powers were imposed: 50, 100, 150, 200 and 250 kW. The temporal response of the 3 temperatures: T_{hot}, T_{cold} and T_{fluid} is shown in Figure 6. The response time obtained depends on the reactor power. It varies from 40 minutes at 50 kW to 60 minutes at 250 kW. The maximum temperatures reached at 250 kW are 101.6 °C, 93.8 °C, and 54.0 °C for T_{hot}, T_{cold} and T_{fluid} respectively.

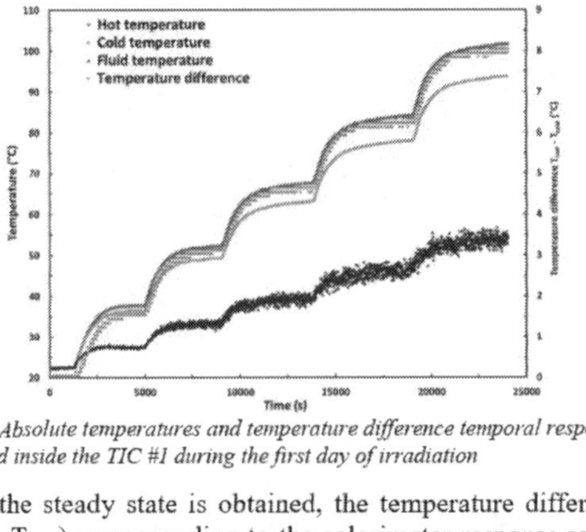

Fig. 6. *Absolute temperatures and temperature difference temporal response obtained inside the TIC #1 during the first day of irradiation*

Once the steady state is obtained, the temperature difference ($T_{hot} - T_{cold}$) corresponding to the calorimeter response can be calculated by doing an average for each temperature. Figure 7 gives the curve of the calorimeter response as a function of the reactor power. The behaviour of the calorimeter was not modified by the integration of the new multi-thin-film heating element.

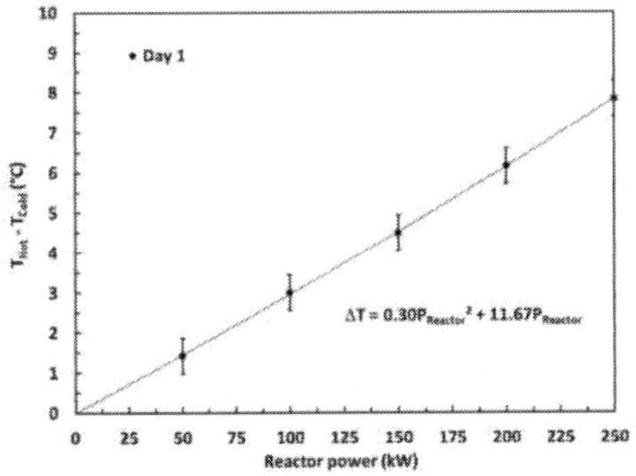

Fig. 7. *Steady calorimeter response as a function of the reactor power obtained inside the TIC #1 during the first day of irradiation*

V. Conclusions

A new multi-thin-film heating element was validated from laboratory to irradiation conditions. First, the design criteria were identified, then prototypes were fabricated by multilayers using screen-printing and finally were characterized under laboratory conditions. The obtained key parameters in terms of chemical composition, dimension, thermal, and electrical properties and performances allowed the validation of this new type of multi-thin-film heating element in term of design, fabrication and resistance value. After the complete characterization under laboratory conditions, this new multi-thin-film heating element was successfully integrated inside a CALORRE single-cell calorimeter made of Aluminum 6061 (same configuration that was previously irradiated in 2023) and qualified for the first time during 3 days of irradiation inside the TIC #1 of the JSI TRIGA reactor. The multi-thin-film heating element resistance was studied before and after irradiation and its robustness was verified after a fluence of $\Phi = 3.2\times 10^{17}$ n·cm^{-2}. The irradiation induces a non-significant drift of the multi-thin-film heating element resistance value. At the same time, the CALORRE calorimeter integrating for the first time such a multi-thin- heating element, was studied. The calorimeter response given by the temperature measurements at specific points under steady state for several reactor powers confirmed a good behavior of the sensor including the new multi-thin-film heating element.

Future work will be related to the determination of the nuclear heating rate by analyzing the data obtained during the irradiation campaign and 3-D numerical thermal and electrical simulations considering the new multi-thin-film heating element under laboratory and irradiation conditions.

References

[1] A. Lyoussi, D. Fourmentel, J-F. Villard, J-Y. Malo, P.Guimbal, H. Carcreff, C. Gonnier, G. Bignan, J-P. Chauvin, C. Reynard-Carette, J. Brun, O. Merroun, M. Carette, M. Muraglia, A. Janulyte, Y. Zerega and J. Andre, "Advanced methodology and instrumentation for accurate on line measurements of neutron, photon and nuclear heating parameters inside Jules Horowitz MTR Reactor", Proceedings of Conferences RRFM and IGORR 2012, pp. 21-25, Prague, Czech Republic, 18-22 March 2012.

[2] G. Bignan, X. Bravo and P.M. Lemoine, "The Jules Horowitz Reactor: A new high Performances European MTR (Material Testing Reactor) with modern experimental capacities: Toward an International Centre of Excellence", Proceedings of Conferences RRFM and IGORR 2012, pp. 21-25, Prague, Czech Republic, 18-22 March 2012.

[3] C. Reynard-Carette, G. Kohse, J. Brun, M. Carette, A. Volte, A. Lyoussi, "Review of nuclear heating measurement by calorimetry in France and USA", Proc. ANIMMA 2017, Jun. 2017, EPJ Web of Conferences 170, 04019 (2018).

[4] J. Brun, C. Reynard-Carette, A. Lyoussi, C. De Vita, M. Carette, M. Muraglia, D. Fourmentel, P. Guimbal, J.-F. Villard, "Comparison of the thermal calibration response of two calorimetric cells dedicated to nuclear heating measurements", IEEE Trans. Nucl. Sci., vol. 61, no. 4, pp. 2254-2261, Aug. 2014.

[5] M. Carette, A. Lyoussi, J. Brun, C. Reynard-Carette, J.F. Villard, « Eprouvette pour mesure d'échauffements nucléaire dans un réacteur nucléaire, et cellule calorimétrique comprenant au moins une telle éprouvette », FR1553136A, 2015-04-10.

[6] A.Volte, C. Reynard-Carette, A. Lyoussi, J. Brun, M. Carette, "Study of the Response of a New Compact Calorimetric Cell for Nuclear Heating Rate Measurements", IEEE Trans. Nucl. Sci., vol. 65, no. 9, pp. 2461-2470, Sept. 2018.

[7] A.Volte, J. Brun, A. Lyoussi, M. Carette, C. Reynard-Carette, "Qualification of a New Differential Calorimeter Configuration Dedicated to Nuclear Heating Rates up to 20 W.g-1", IEEE Trans. Nucl. Sci., vol. 67, no. 11, pp. 2405-2414, Nov. 2020.

[8] A.Volte, M. Carette, A. Lyoussi, G. Kohse, J. Rebaud, V. Valero, C. Reynard-Carette, "Review of CALORRE Calorimeter Characterizations Under Laboratory and Irradiation Conditions", IEEE Trans. Nucl. Sci., vol. 69, no. 4, pp. 840-848, April 2022, doi: 10.1109/TNS.2022.3150148.

[9] C. Reynard-Carette, A. Volte, A. Lyoussi, M. Carette, « Noyau de mesure pour la mesure d'échauffement nucléaire en réacteur nucléaire et capteur calorimétrique intégrant un tel noyau de mesure », Patent 21212899.5, 2021-12-07.

[10] J. Rebaud, A. Volte, M. Carette, A. Lyoussi, C. Reynard-Carette, "Experimental and numerical results for a new single-cell calorimeter dedicated to nuclear heating rate measurement", ANIMMA 2023, 12-16 June 2023, Lucca, Italy, EPJ Web of Con-ferences 288, 04008 (2023), https://doi.org/10.1051/epjconf/202328804008.

[11] A. Volte, S. Hauptman, D. Carpenter, M. Carette, A. Lyoussi, G. Kohse, C. Reynard-Carette, "Numerical and Experimental Characterization of a New Reduced-Height CALORRE Differential Calorimeter for CALOR-I Irradiation in MITR", in IEEE Trans-actions on Nuclear Science, vol. 71, no. 5, pp. 1134-1143, May 2024, doi: 10.1109/TNS.2024.3388518.

[12] L. Snoj, G. Zerovnik, A. Trkov, Computational analysis of irradiation facilities at the JSI TRIGA reactor, Appl. Radiat. Isot. 70, 3 (2012).

[13] H. Carcreff, V. Radulovi´c, D. Fourmentel, K. Ambrozic, C. Destouches, L. Snoj, N. Thiollay, Nuclear heating measurements for fusion and fission relevant materials in the JSI TRIGA reactor, Fusion Eng. Des. 179, 113136 (2022).

Qualification of miniature fission chambers at high temperature in OSU Research Reactor

Grégoire de Izarra[1,*], Loïc Barbot[1], and Kevin Tsai[2]
[1]Instrumentation Sensors and Dosimetry Laboratory, CEA Cadarache, St-Paul-lez-Durance, France
[2]Idaho National Laboratory, Idaho Falls, ID, USA
(*) gregoire.deizarra@cea.fr

Abstract—**With the renewed interest in Sodium Fast Reactors and Advanced Small modular reactors such as Molted Salt ones, there is a need for high temperature (600 °C) neutron instrumentation. In this context, two types of miniatures fission chambers were identified for testing, a CFUE43 dedicated to high temperature conditions and a CF3, the reference fission chamber for in core measurement at CEA. An irradiation was performed at the Ohio State University Research Reactor, in a high temperature irradiation rig that might reach temperatures up to 1200 °C. The rig was installed against the reactor casing, allowing the experimental load to see a maximum neutron flux of about 10^{12} n/cm²/s at full power.**

During three days, the fission chambers were irradiated at different temperature steps, from 20 °C up to 820 °C. Electronic pulse shape, leakage current and counting mode were monitored in parallel thanks to the libera MONACO 3 measurement system. CFUE 43 cannot be operated in counting or Campbell mode because of the low electromagnetic immunity of its mineral cable. CF3 fission chamber can be operated in counting mode above 600 °C even if its leakage current is high, of the order of 10 µA. Failure of CF3 with temperature was monitored. It can be explained with simple models.

Keywords —**Enter key words or phrases.**

I. INTRODUCTION

Between the 1970s and the late 1980s, the CEA developed several fission chambers operating at high temperature to instrument French sodium-cooled fast reactors (SFR) such as Phénix and SuperPhénix. These developments, carried out in collaboration with the Photonis company (now Exosens), led to the marketing of ex-core detectors, the CFUC, as well as miniature detectors such as the CFUE. These developments were put on hold in the late 90s before being restarted in the 2010 decade through the ASTRID project [1]. At the end of the ASTRID project, it was decided to pursue the work on high temperature fission chambers to maintain the acquired knowledge and for use in specific research programs such as the SAIGA experiment [2] or future advanced reactors.

As part of a collaboration between CEA and INL through the US Department Of Energy, it was possible to test neutron detectors under irradiation at high temperature in the Ohio State University Research Reactor (OSURR).

It was decided to focus the tests on two type of miniature fission chambers: a modified CFUE43 from Exosens designed for sodium fast reactor conditions, and a CF3, which is the reference detector for in-core neutron flux measurement at CEA. The irradiation campaign was tailored first to assess the performance of these sensors at a temperature of 600 °C and then to monitor their failure and their maximum operating temperature.

In this paper, we describe the irradiation campaign. After a brief description of the detectors and the associated electronic measurement system, the main characteristics of the OSU reactor and high temperature irradiation rig are given. A presentation of the main experimental results and their interpretation is given. At last, several improvements to these fission chambers for high temperature application are envisaged.

II. MATERIAL AND METHODS

A. Fission chambers

Two fission chamber models were tested: one CF3 and two CFUE43. The CF3 is a wide range fission chamber designed and built by the CEA for irradiation in PWR in-core conditions. The CFUE43 were purchased from Exosens. This version of CFUE operate specifically in current mode, which is not in the first place attractive for our application because it is not possible to distinguish the neutron contribution from the gamma-induced and the leakage current. We therefore chose to adapt the filling gas pressure in order to maximize the amplitude of the pulses and to exploit CFUE fluctuation and pulse modes. The characteristics of these chambers are summarised in **Erreur ! Source du renvoi introuvable.**.

TABLE I
CHARACTERISTICS OF THE TESTED FISSION CHAMBERS

	CFUE43 N°200 & 201	CF3 N°2349
Fissile deposit	170 µg of ^{235}U	1 µg of ^{235}U
Sensitive length	15 mm	10 mm
Diameter	7 mm	3 mm
Inter-electrodes gap	0.3 mm	0.25 mm
Filling gas	Ar	Ar+4%N_2
Cable	15m, Thx 1 I I 30 Mg	15m, Thx 1 C CaC 22Si
Measurement mode	Counting, Campbell	Counting

The CFUE43 detector is suitable for high-temperature operation: the mineral insulation used in the cable is MgO powder, while the filling gas is pure argon. Argon/nitrogen mixture used in CF3 increase electron drift speed but it is known to change under high-temperature irradiation (nitriding of detector structures).

B. Measurement system

A Libera MONACO 3 system was used during the whole irradiation to polarize the detectors, record signals and process them in real time [3]. This four channels instrument can operate simultaneously in counting, Campbell and current mode at a maximum sampling rate of 1 kHz. This is valuable for the qualification of fission chambers at high temperature: the evolution of leakage current with temperature is monitored while the counting and Campbell mode give the fission rate.
Automatic discrimination and saturation curve generation are available to check the good condition of the detector at the start of the experiment. At last, an oscilloscope mode is also available to monitor electronic pulses generated by the fission chambers. However, due to the 20 MHz bandwidth of the preamplifier, it is not possible to capture the exact duration and shape of the pulses.

C. Reactor and irradiation rig

The OSURR reactor is a pool-type research reactor built in 1960. Originally fuelled with highly enriched fuel plates, the reactor was converted to low-enriched fuel in the 1980s, then its maximum licensed power was increased from 10 kW up to 500 kW in 1992. The depth of the pool is 6 m (20 feet), while the altitude of the core centre is 1.2 m. The typical height available for sample irradiation is around 60 cm (2 feet).

Fig. 1. Picture of the OSU research reactor, with a 23.7 cm diameter dry channel installed against the core casing.

The reactor has removable ex-core dry channels, 16.5 and 23.7 cm in diameter, which are positioned against the core housing to irradiate large elements **Erreur ! Source du renvoi introuvable.**. At the tubes' location, the neutron flux is of the order of 10^{12} n.cm^{-2}.s^{-1}. Furnaces and cryostats have been developed at the NRL (Nuclear Reactor Laboratory - OSU) to equip those ex-core tubes and allow to perform irradiations at temperatures between 4 K and 1500 K [3].
To carry out irradiations at high temperature, the OSURR facility provided us a furnace composed of a cylinder of insulating material buster M35 and a silicon carbide tubular heating element which is positioned at the bottom of the 23.7 cm diameter aluminium dry tube.

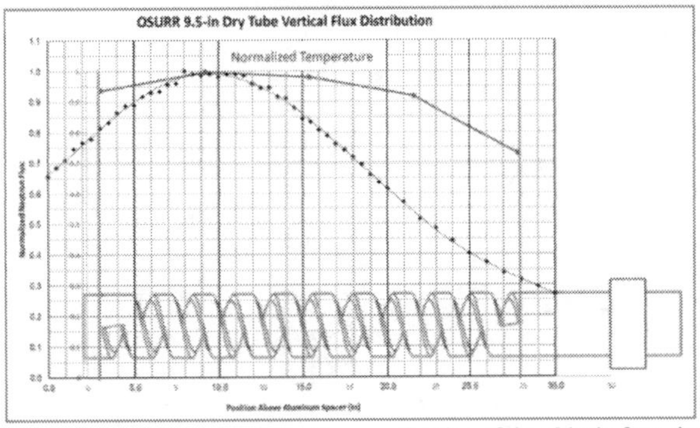

Fig. 2. Temperature and neutron flux as a function of the altitude from the bottom of the SiC Heating element. The twisted SiC element is depicted in the picture for more clarity.

The fission chambers were first encapsulated in closed end steel tubes sealed to prevent leakage in case of failure. They were then inserted into a silica tube with an internal diameter of 50 mm. Fission chambers altitudes were set to have the centre of the sensitive length in the maximum neutron flux (Fig. 2). In order to monitor the temperature at the level of the detectors, three type K thermocouples were positioned in the silica tube at altitudes of 3, 6 and 9 inches from the bottom of the heating elements.

D. Irradiation schedule

The experiment, which took place over three days, allowed the detectors to be irradiated at different temperatures and power levels.
Each day started with the detector at room temperature and the reactor power at 100 W to check the proper operation of the detectors with saturation, discrimination curve and pulse shape analysis. Then measurements were recorded at reactor power of 0.1, 1, 10, 100 and 200 kW to verify the linearity of the measurement modes. Once the nominal operation of the detectors was proven, the furnace was heated up to the target temperature while keeping the reactor power at 1 kW. Few temperatures target were planned for this experimental campaign: 350 °C, 600 °C, 650 °C and 750 °C. After the temperature was stabilized, few reactor power steps were recorded to monitor the measurement modes linearity. Leakage current and mean pulse shape were also recorded.
A part of the third irradiation day was dedicated to the study of the maximum operating temperature and then the failure of the fission chambers. To do so, the reactor power was kept constant at 1 kW and the furnace temperature was raised slowly from 750 °C up to the failure point.

III. Experimental results

A. Results at room temperature

Before starting the reactor, raw signals were recorded with the FC polarized to evaluate the noise level. As shown in figure 3, the noise level in the CF3 is acceptable, whereas the

CFUE43 pic to pic noise level is larger than the expected pulses amplitude. This might be due to the mineral cable used in the CFUE43 which has a high transfer impedance. As a result, counting and Campbell mode are not usable with CFUE43. Only the current mode is immune to high frequency noise.

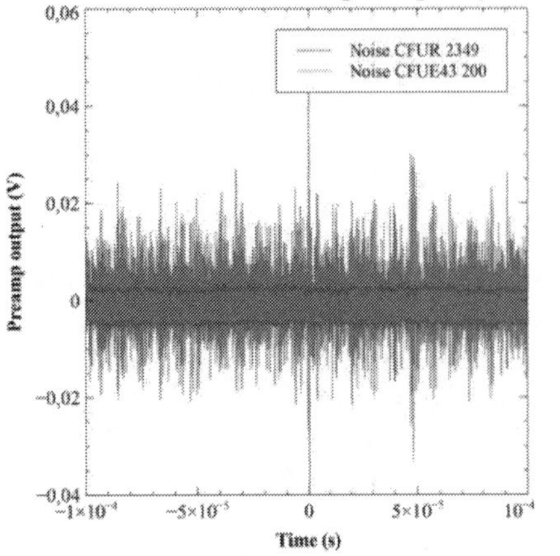

Fig. 3. Noise level recorded with the Libera MONACO 3 system on the CF3 N°2349 and on the CFUE43 N°200. The reactor was shutted down for this measurement.

The reactor was started and stabilized at 100 W to record the CF3 discrimination curve and mean pulse shape (Fig.4). At 1 kW, we noticed CFUE43 current mode seems usable.

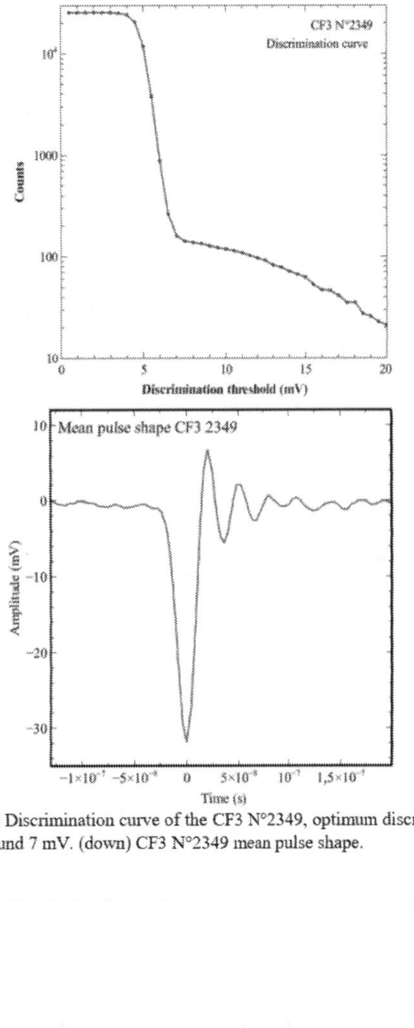

Fig. 4. (up) Discrimination curve of the CF3 N°2349, optimum discrimination level is around 7 mV. (down) CF3 N°2349 mean pulse shape.

Few reactor power steps were performed to monitor the linearity of the fission chambers. The count rate against reactor power recorded with the CF3 is depicted in figure 5.

The maximum count rate of the CF3, 2.78 10^5 cps, is too low to take advantage of the Campbell mode.

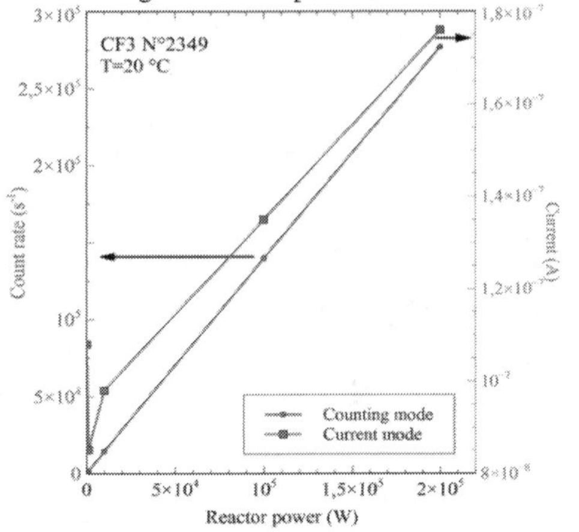

Fig. 5. CF3 N°2349 count rate versus reactor power at room temperature. The mean current measurement is also depicted for comparison with leakage current.

B. *High temperature results*

Because of the SiO_2 mineral cable, the CF3 leakage current is rather high above 600 °C, of the order of 10 μA. Nevertheless, the counting mode is still usable (Fig 6), even if the maximum count rate decreases slightly with the temperature. Between room temperature and 600 °C, the count rate decreased by 4 %.

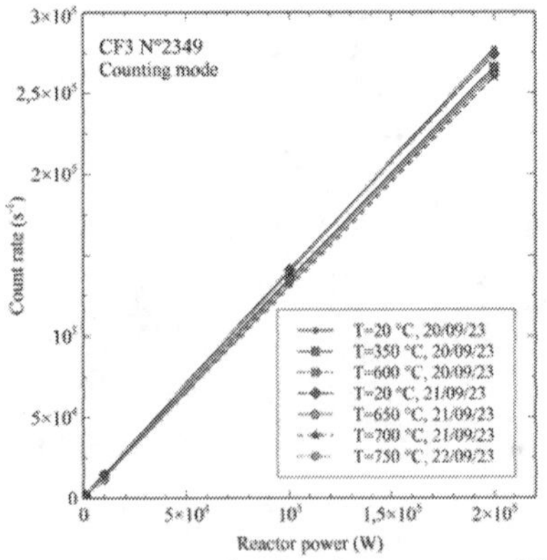

Fig. 6. CF3 N°2349 count rate as a function of reactor power at different furnace temperature. The count rate decreases slightly with the temperature.

During the high temperature power steps, the pulses generated by the CF3 were monitored using the built-in oscilloscope feature of the libera MONACO 3 system. Only one pulse per second is recorded which is not sufficient to extract precise quantitative results from the data. However, it was possible to analyse the FWHM and the amplitude of the recorded pulse to get tendencies. The pulse shape analysis is available in figure 7. Between room temperature and 650 °C, there is no change of the pulse shape. At 750 °C, the pulses start to widen drastically, indicating a change of filling gas transport properties. For the last part of the irradiation, we decided to monitor the failure of the CF3, the temperature was raised slowly while keeping a constant reactor power of 1 kW. At 820 °C, a drastic decrease of pulse duration was noticed along with really narrow pulses. By decreasing the high voltage, the narrow pulses disappeared. To our opinion, the decrease of the pulse duration is due to the failure of the gas thigh feedthrough: A part of the filling gas moved in the mineral cable. This leaded to an increase of reduced electric field and thus to an increase of the electron drift velocity. The really narrow pulses might be due to partial discharge, a well-known phenomenon that appears in fission chambers at high temperature.

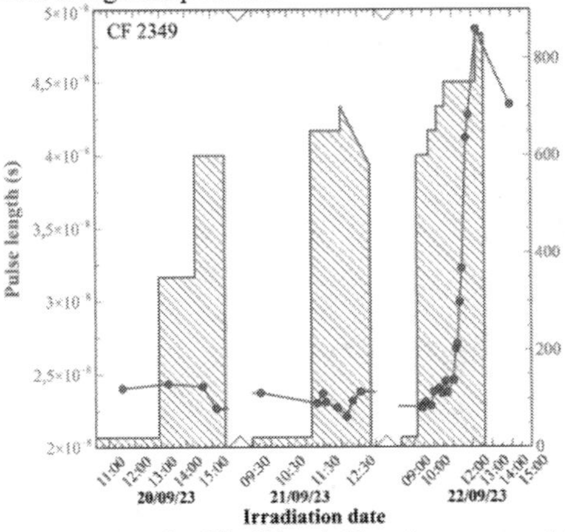

Fig. 7. CF3 N° 2349 pulse full width at half maximum measured during the irradiation campaign.

IV. Data analysis

It was shown that performances at high temperature of CF3 fission chambers are slightly different than the one at room temperature. To explain the decrease of count rate and the change of pulse shape with temperature, we developed simple models. The count rate change with temperature might be due to a change of neutron temperature that impact neutron reaction rate. To prove this hypothesis, a furnace model was built with PHITS and JENDL5 [5] library. Using a room temperature thermal neutron source, cross sections at 300 K, 900 K and thermal scattering law for SiC and SiO_2, we were able to compute the neutron spectrum and the fission rate at the centre of the furnace (Fig. 8). With this simple model, a decrease of neutron reaction rate of 4% was computed between 300 K and 900 K, which is in agreement with the experiment.

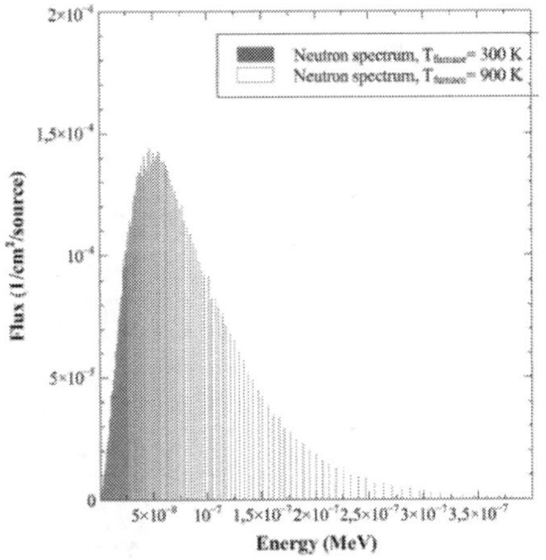

Fig. 8. Computed neutron spectrum inside the furnace, when the irradiation rig is at 300 K (red) and 900 K (blue).

The change of CF3 pulse shape at high temperature might be due to a change in the filling gas composition. Gas mixtures with nitrogen are not used at high temperature since it disappear under the combined action of temperature and radiation in nitriding reaction of the fission chamber's structures. To prove the removal of nitrogen from the filling gases, we computed the electron drift velocity in various argon/nitrogen mixtures [6] (Fig.9). Using the full width at half maximum, and assuming the electronic transfer function has no impact on FWHM differences, the longest pulses recorded at 750 °C correspond to an electron drift velocity of about (10 ± 3) 10^3 m/s. This corresponds to argon with a molar content of nitrogen between 0 and 0.5 % of nitrogen.

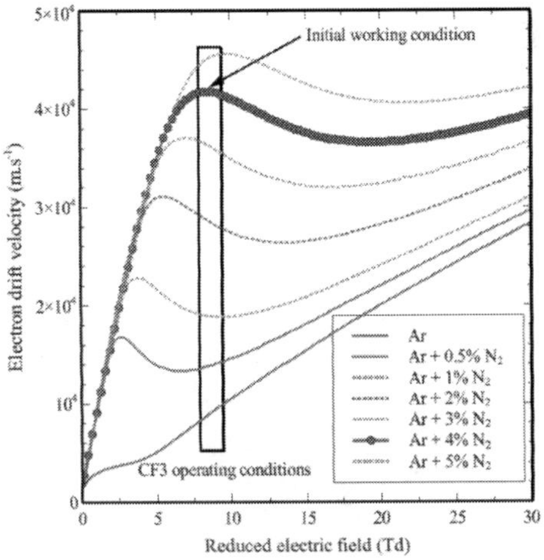

Fig. 9. Electron drift velocity in various argon/nitrogen mixture.

V. Conclusions

During the irradiation, it was not possible to test CFUE43 fission chamber in pulse and Campbell mode. This is probably due to the high transfer impedance of the 3 mm mineral cable.

https://doi.org/10.1051/epjconf/202533804018

The mineral cable should be upgraded to a high immunity one with an outer copper layer to improve the electromagnetic noise immunity of the detector. Even if the CF3 detector was not designed for high temperature operation, the experiment proved that pulse mode worked properly up to 750 °C. Above 750 °C temperature, the combined effect of radiation and temperature remove the nitrogen from the filling gas through nitriding reaction with the metallic structure of the detector. The Campbell mode was not tested because of the low fissile mass and the filling gas composition. By increasing the fissile mass deposit and filling the detector with pure argon, it should be possible to test the Campbell mode in the OSU research reactor.

REFERENCES

[1] C. Jammes, P. Filliatre, B. Geslot, T. Domenech, S. Normand, "Assessment of the High TemperatureFission Chamber Technology for the French Fast Reactor Program" IEEE TNS, vol 59, no 4, pp 1351–1359:2012.

[2] F. Payot, R. Clavier, F. Charollais, D. Cambet, E. Martin-Lopez, L. Trotignon, E. Dufour, F. Serre, E. Batyrbekov, A. Vurim, A. Pakhnits, V. Baklanov, V. Vityuk, "The SAIGA in-pile experimental programm to qualify the SIMMER calculation tool is SFR Severe Accident Conditions", in: Proceedings of the FR22 conference, 2022, Vienna.

[3] L. Barbot, G. de Izarra, C. Destouches, D. Bisiach, M. Cargnelutti, S. Zorzut, M. Skabar, A. Bardorfer, A. Makovac, J. Trnovec, P. Paglovec, "Qualification of the industrialized libera MONACO 3 digital acquisition system dedicated to fission/ion chamber measurements in research reactors" in Procedings of the ANIMMA conference, 2023, Lucas

[4] A. Kauffman, K. Herminghuysen, M. Van Zile, S. White, J. Hatch, A. Maier, L. R. Cao, "Review of research and capabilities of 500 kw research reactor at the ohio state university", ANE 206: 2024

[5] O. Iwamoto, N. Iwamoto, S. Kunieda, F. Minato, S. Nakayama, Y. Abe, K. Tsubakihara, S. Okumura, C. Ishizuka, T. Yoshida, S. Chiba, N. Otuka, J. Sublet, H. Iwamoto, *et al*, "Japanese evaluated nuclear data library version 5: Jendl-5", JNST vol 60 no 1, pp 1-60, 2023.

[6] G. J. M. Hagelaar, L. C. Pitchford, "Solving the Boltzmann equation to obtain electron transport coefficients and rate coefficients for fluid models", Plasma Sources Sci. Technol. Vol 14, pp 722–733:2005.

Development of a wireless transmitter for in-core monitoring of reactor irradiation experiments

A.V. Fedorov*, T.J. Schröder, A.S. Booij, D.N. Verbruggen, P.R. Hania, E. de Visser – Týnová
NRG PALLAS, The Netherlands
(*) fedorov@nrg.eu

***Abstract*— A wireless in-core radio transmitter which is currently under development at NRG-PALLAS will be able to amplify a signal measured inside a reactor core and transmit the signal to a receiver located outside the core without using instrumentation cables. The transmitter uses vacuum tubes as active components as these are judged – unlike transistors based on semi-conductors – to be able to withstand in-core radiation field. The transmitter is self- powered by means of a thermo-electric generator (TEG) utilizing nuclear heating present in the core.**

A selection of electronic components envisaged to be used in the transmitter was irradiated in steps in a gamma field of 20 kGy/h reaching an accumulated irradiation dose of 720 kGy. The irradiation took place in the spent fuel pool of the High Flux Reactor (HFR). Relevant electronic properties of the components were measured before and after every irradiation step. The tested components included three types of miniature low anode voltage vacuum tubes, a variety of capacitors and resistors of different nominals, structures and compositions, and a bismuth telluride TEG unit.

In a next step electronic characteristics of vacuum tubes and capacitors were measured on-line during irradiation. These type of experiments were focused on the performance characteristics of the components in strong gamma field.

In all irradiation experiments the pre-selected electronic components behaved reasonably well and did not show any significant degradation of properties.

Details on the irradiation conditions and results of the measurements are presented.

***Keywords* — wireless in-core radio transmitter, self- powered, radiation resistant, vacuum tubes**

I. Introduction

Continued development of new experimental methods and measuring techniques is crucial for successful conduction of irradiations performed at the High Flux Reactor (HFR) and in perspective in the future Pallas reactor which is currently under construction. In this respect self-powered wireless sensors that can precisely monitor process variables inside an operating nuclear reactor core, while withstanding its harsh radiation environment, are of great potential benefit. This technology has the potential to significantly reduce amount of cables and number of penetrations in the reactor vessel, provide increased flexibility for installing various sensors, in particular in locations which are difficult to access through conventional methods. Wireless sensors are applicable for Material Research Reactors, Nuclear Power Plants and other nuclear facilities including fuel fabrication facilities and spent fuel storage installations [1][2][3][4].

In support of future irradiation experiments, NRG-PALLAS has started the RADIO (iRrADIatIon technOlogy) project aiming to evaluate technical options of wireless transmission out of the core of a nuclear reactor during operation. The transmitter will be able to amplify a signal measured inside a reactor core and transmit the signal to a receiver located outside the core without using instrumentation cables. The transmitter uses vacuum tubes as active components as these are judged to be better suited to withstand in-core radiation field than transistors based on semi-conductors. The transmitter is self- powered by means of a thermo-electric generator (TEG) utilizing nuclear heating present in the core.

A similar approach was employed in the joint effort of Westinghouse and the Penn State University aiming to develop a self-powered wireless real-time reactor power sensor to measure core power peaking factor. The sensor uses a vacuum micro-electronic transmitter as active component which is powered by rhodium self-powered neutron sensor. The transmitter was tested in the Penn State TRIGA reactor reaching the neutron dose 8.97×10^{17} n/cm^2 and gamma dose 725 MRad (7.25MGy) [6][7].

The development of the self-powered wireless transmitter at NRG-PALLAS faced a number of challenges from which the main are related to the following aspects:

- **Radiation resistance of the electronic components**: wireless transmitter operated in radiation fields have to fulfil requirements on radiation resistance. In many cases, the requirements on radiation resistance of the used components raises the need to develop customized solutions. The results of the radiation tests performed within the project are presented in Section III.
- **Low frequency data transmission:** It was realized that to bridge the reactor vessel containment, application of very low frequencies (VLF; <30 kHz) will be required. The VLF range is technical challenging for several reasons, including low antenna sensitivity, large interferences in industrial

environments, and spatial limitations that hampers the implementation of efficient antennas [14]. Because the VLF range is usually avoided for communication purposes and only very few commercial solutions are available, often using large amounts of energy, the deployment of VLF technology in a reactor environment requires customized technical solutions for the electronic hardware, and careful evaluation and testing of feasible transmission frequencies. Application of low frequencies solutions through solid, electrically highly conductive media necessitates the use of magnetic fields which have a complex propagation behavior under near field and extended near field conditions. For the successful implementation of a low frequency transmitter located behind a significantly shielding barrier, and with a constrained supply of power, a good quantitative understanding of field generation and propagation through electrically conductive media is important. Results of the studies performed on magnetic field propagation in water and metals are presented in Section IV.

• **Technology development and integration:** an important aspect and key driver for the approach followed in this project is that none of the necessary components of the transmission chain is covered by the technological mainstream and the vast variety of technical solutions offered on the market. The development of an autonomous operating wireless sensor unit for the intended application is a complex and unique task, requiring multidisciplinary knowledge and experience in working fields that goes beyond material science and irradiation experiments: besides a good understanding of analog technology and application electronics in general, knowledge on communication technology and signal processing, and the physics of magnetic field propagation is necessary. Section V is focused on understanding the requirements of wireless in-core monitoring devices and translate this into creative and innovative solutions that allow to monitor under extreme harsh environmental conditions.

II. Experimental

A. Gamma Irradiation Facility

Radiation resistance of the electronic components were tested in the gamma field in the HFR spent fuel pool. A dedicated Gamma Irradiation Facility (GIF), shown in Fig. 1 and Fig. 2, was developed that allows to perform in-situ measurements of electronic components during the irradiations.

The irradiation experiments in the GIF facility were performed in two steps: first an initial screening of the radiation resistance of selected components was performed. For the components that passed, in a second step the behaviour under irradiation was studied in more detail. Together with these tests, a workflow and the necessary tooling was implemented and built, allowing to irradiate and test components efficiently under well-defined conditions.

Three irradiations were carried out. The irradiation parameters are shown in Table I. Irradiation A was performed in multiple steps (A1, A2 and A3). The irradiation B was performed in one step. For both irradiations the electronic characteristics of the irradiated components were measured before and after each irradiation step. These irradiations were focused on permanent irradiation damage which increases with the dose and on its effect on electronic characteristics of the components under study. The maximum irradiation gamma dose achieved in these tests was 725 kGy.

Differently, irradiation test C was focused on measurements of characteristics of the electronic components during irradiation. The gamma dose rate in the irradiation C was 17.3 kGy/h, which corresponds to 4.8×10^{-3} W/g. A typical gamma heating in HFR is about 5 W/g, thus irradiations in the GIF facility can only discard electronic components with poor performance under irradiation conditions, but will not guarantee their operation in an in-core position.

Fig. 1 *Gamma irradiation facility with instrumentation cable in the HFR spent fuel pool.*

Fig. 2 *Gamma irradiation facility in the irradiation position surrounded by spent fuel elements.*

TABLE I
PARAMETERS OF THE IRRADIATION EXPERIMENTS

No	Instrumented	Dose rate [kGy/h]	Accumulated dose [kGy]
A1	No	15.9	30
A2	No	15.9	90
A3	No	15.9	690
B	No	21.0	725
C	Yes	17.3	26

The electronic components under study included vacuum tubes, capacitors and resistors of different nominal values and different material composition and build, antenna wire with enamel insulation and a commercial thermo-electric generator. Electronic components are usually classified as passive or active components. Active components in this project are the vacuum tubes, while passive components include capacitors and resistors. The vacuum tubes chosen for the program were the miniature pentode tubes 1Zh37B and 1P24B (in two different making), with rod structure [8]. The rod structure of these tubes in contrast to commonly used grids offers extra robustness against vibrations. Both tubes run at low anode voltage, ~10 V, which is a key feature enabling to power the transmitter with a low voltage TEG.

In the pre-selection of passive components most attention was focused on the materials in use and their radiation resistance. Materials as polymer film, paper and liquid electrolytes were directly excluded. Capacitors pre-selected for the irradiation tests contained dielectric materials such as glass, ceramic, mica, and oxide layers (As exception, one capacitor with polystyrene film as dielectric layer was present in the program). The main characteristics measured in the tests are the capacitance and leakage current. The nominal capacitance varied from 10 nF to 10 μF. The capacitance was determined over a larger bandwidth to avoid problems with interferences on the long signal wiring used.

A variety of resistors of different makings with resistance values between 47 Ohm and 10 MOhm was preselected for tests. Both 'traditional' thru-hole-types with leads on both ends, and smaller 'Surface Mounting Devices' (SMD) without leads were included. Variety of tested resistive element included carbon, metal film and wire.

All electronic components were soldered on a Printed Circuit Board (PCB) as shown in Fig. 5. For the measurements, both standard apparatus and in-house made dedicated electronics were used.

The irradiation tests have also included a commercial 40 × 40 mm bismuth telluride thermo-electric generator (TEG). The measured thermo-electric properties included thermo-electric force and internal resistance.

B. Measurements of electro-magnetic field attenuation

To characterize frequency- and distance-dependent attenuation of magnetic field in water, a large bandwidth transmitter operating between 200 Hz and 9 kHz was developed. The transmitter is powered by low-voltage batteries, and provides an output voltage > 4V to the low ohm, inductive load of the transmitter antenna. Because these tests do not involve radiation conditions the test-transmitter uses conventional semiconductor based electronics. The test-transmitter uses a D-class amplifier and a 40 cm diameter magnetic loop antenna. The receiving antenna is a 2 meter diameter loop antenna with a pre-amplifier and a data logger.

The underwater test was carried out in a test basin at NRG-PALLAS. The basin is 6 x 6 m wide and has a water depth of 4 m. Fig. 3 gives an overview of the experimental set-up with the test transmitter located in the middle of the basin and the receiver antenna floating in a coaxial position on the water, and the receiver electronics located on a desk beside the basin. During the measurements, the test transmitter was lowered in intervals of 50 cm, from 50 cm below the water line down to 4 m on the basin's floor. At each depth, signals from the transmitter between 200 Hz and 9 kHz were recorded.

Fig. 3. General set-up for underwater testing. The transmitter antenna is located close to the basin floor, and the receiver antenna is floating on top.

To study magnetic field propagation through a highly conductive slab of aluminium with a relevant thickness a desktop experiment was performed (see Fig. 4). In this experiment, the frequency-dependent signal transmission between two coaxial coils of 40 mm diameter was measured, separated by an aluminium plate of 200 × 400 × 4 mm.

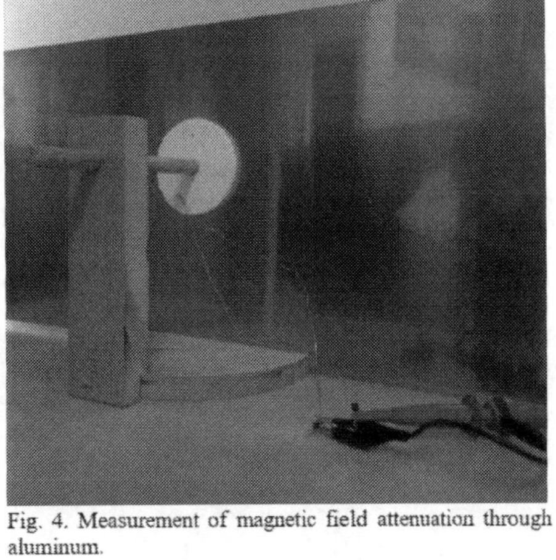

Fig. 4. Measurement of magnetic field attenuation through a 4 mm slap of aluminum.

Furthermore, in preparation of future transmitter experiments, vertical and radial magnetic background noise and interferences in the HFR has been measured and characterized on several locations foreseen for the receiving antenna.

III. RESULTS OF IRRADIATION TESTS

Material qualification experiments presented in this Section are focused on selection of suitable active and passive electronic components, and determination of the radiation resistance and performance of these components in a strong gamma field.

A. *Vacuum tubes*

PCB adapters with electronic components under study before and after irradiation are shown in Fig. 5. The borosilicate glass of the vacuum tubes have shown a red-brownish coloring, already clearly visible after the first 30 kGy of irradiation. This effect of radiation on glass is known in literature (e.g. [9][10]), but no impairment of the glass bulb's function was found in the irradiations.

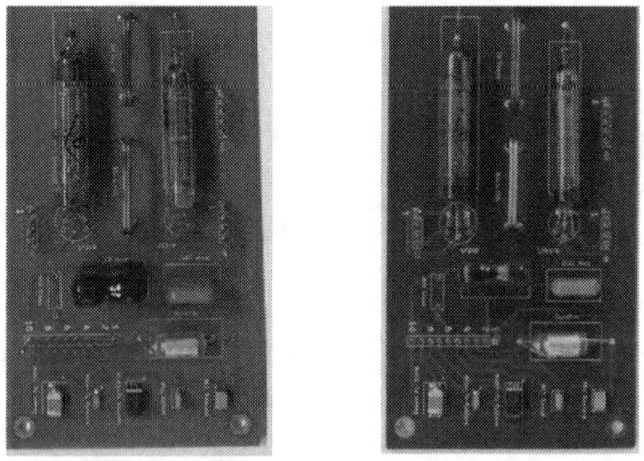

Fig. 5. PCB adapters with the vacuum tubes and capacitors under testing before irradiation (left) and after irradiation, 690 kGy (right).

The vacuum tubes were characterized by measuring the anode current as a function of grid voltage, at constant anode-cathode voltage (transfer function) [11][12]. Fig. 6 shows the transfer function measured for the vacuum tube 1P24B during and after irradiation with 17.3 kGy/h at 9 V anode voltage. The vacuum tubes showed only small deviations during and after irradiation, compared to production tolerances.

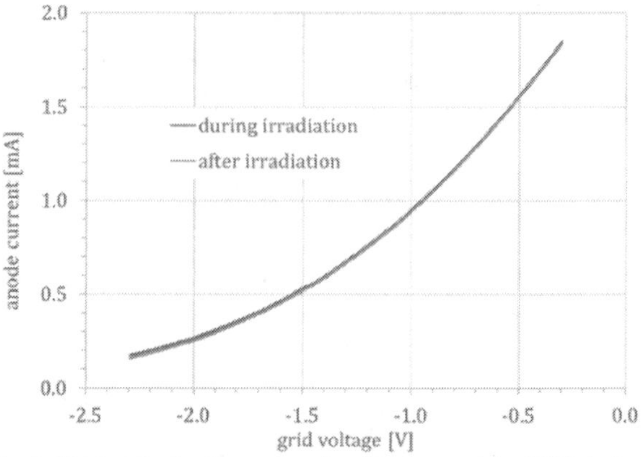

Fig. 6. The transfer function measured on the vacuum tube 1P24b during irradiation with a dose rate of 17.3 kGy/hr and after irradiation with about 26 kGy for an anode voltages of 9 V.

Two other important characteristics measured in the tests are filament current and saturated anodic current. No significant changes of the filament currents were found after irradiation at 725 kGy. The anodic current, however, was significantly decreased, by 21%, in one of the vacuum tubes (1P24B), pointing to chemical alterations of the filament's surfaces that decreases the electron emission efficiency. More detailed testing showed that this decrease is partially reversible by heating, raising the question if this decrease would also occur if the tube is operated during irradiation. The irradiation C with on-line measurements has strengthened this assumption: here, no decrease of the anodic current was found.

B. *Capacitors*

For a selected set of capacitors, the capacitance and leakage currents were measured. The capacitance was determined over a larger bandwidth.

Table II summarizes the measured capacity before and after 30, 90 and 690 kGy of irradiation. No significant change of the capacitance was found for two mica capacitors and one of the MLC (Multilayer Ceramic Capacitor) capacitor (MLC-1). The capacity of the polystyrene capacitor increased slightly (about 1%) after 690 kGy of irradiation. The MLC-2 showed a decrease of about 5% of its capacity. For the Ta-type capacitor a significant increase was found for all three tested versions already after 30 kGy. After 690 kGy, the MnO_2-Ta-2 showed an increase of more than a factor of 5. The capacitance of the NbO-Ta capacitor increased by more than a factor of two. Despite the strong capacitance increase of some capacitor types, none of the capacitors showed a voltage breakthrough.

The objective of this irradiation tests was not to get understanding of the mechanisms responsible for degradation of the properties but rather to discard electronic components which behaved poorly in the tests from the further investigation.

TABLE II
MEASURED CAPACITY BEFORE AND AFTER 30, 90 AND 690 KGY OF IRRADIATION

Type	Reference	30 kGy	90 kGy	690 kGy
Mica-1	9.89 nF	9.87 nF	9.89 nF	9.90 nF
Mica-2	9.95 nF	9.94 nF	9.96 nF	9.97 nF
Polystyrene	10.1 nF	10.1 nF	10.1 nF	10.2 nF
MLC-1	99.6 nF	99.6 nF	99.6 nF	99.6 nF
MLC-2	91.5 nF	88.6 nF	87.4 nF	86.6 nF
MnO_2-Ta-1	10.7 µF	13.6 µF	23.1 µF	53.4 µF
MnO_2-Ta-2	10.2 µF	10.9 µF	11.9 µF	16.2 µF
NbO-Ta	11.0 µF	12.2 µF	15.2 µF	24.4 µF

The capacitors which showed good performance in these irradiation tests, mica, polystyrene and MLC, were tested further in the irradiation C with on-line measurements. A capacitance measurement in broad frequency range performed on the MLC capacitor is shown in Fig. 7 and shows a good agreement between the measurement during irradiation and two reference measurements, before the irradiation and after. Similar results were obtained for mica and polystyrene.

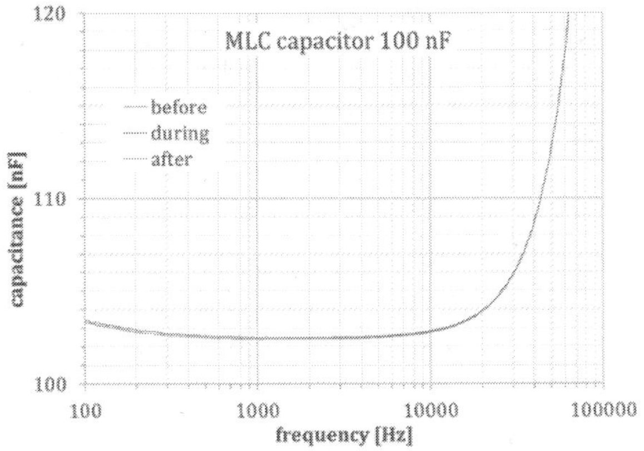

Fig. 7. Measured capacitance of the MLC capacitor measured before, during, and after irradiation.

The leakage currents measured on the tested capacitors during irradiation were ≤4 nA/V, which is sufficiently low for practical application, e.g. in an RCL-tank or as bias-blocking device. Unfortunately, but not unexpected, the larger capacitors with capacitances in the µF-range failed the irradiation tests and an alternative solution has to be found.

C. Resistors

The selected resistor types showed no effects during the irradiation with 690 kGy in the GIF facility, allowing to select from a wide variety of resistor types and designs in the next steps. The results of the measurements are shown in Table III.

TABLE III
MEASURED RESISTANCE BEFORE AND AFTER IRRADIATION

Type	Leads	Reference	690 kGy
Metal film-1	through hole	10 MΩ	9.99 MΩ
Metal film-2	through hole	47.2 Ω	47.3 Ω
Metal oxide	through hole	10.04 MΩ	10.03 MΩ
Thick film	SMD	10.03 MΩ	10.03 MΩ
Thin film	SMD	47.4 Ω	47.4 Ω
Carbon film	through hole	10.13 MΩ	10.14 MΩ
Thin film	SMD	10.01 MΩ	10.01 MΩ
Thick film	SMD	47.2 Ω	47.3 Ω
Thin film	SMD	47.4 Ω	47.4 Ω

D. TEG

After the irradiation the ceramic surface of the TEG showed a slight yellowish-grey colouring. The measured thermoelectric force after irradiation was determined as 87.9 mV/°C without load, this is within 4% of the value given in the datasheet (91.2 mV/°C for a ΔT of 170°C). 53.4 mV/°C was measured with a 10 Ohm load, resulting in an estimated internal resistance under load conditions of about 6.5 Ohm, which is lower than the 9.5 Ohm ±15% given in the datasheet, but consistently within 4% of the thermal force given in the datasheet, considering the estimated internal resistance correct. Thus, it can be concluded that the TEG's thermal force after the irradiation is well within the value provided in the datasheet, considering that thermal coupling of the device can introduce experimental errors of 10%. Finally, the thermo-electric performance of the TEG was within specifications after irradiation, pointing to limited or no permanent damage.

The PCB showed no impairment of its basic function and is found to be suitable for the applied radiation conditions. The observed colouring of the solder mask does not impair the performance of the PCB. For future irradiation experiments with a much higher irradiation dose, however, it could be considered to manufacture a PCB without the solder mask.

IV. PROPAGATION OF ELECTROMAGNETIC FIELD IN WATER AND METALS

Signal from an in-core radio transmitter has to propagate through the wall of the reactor core and several meters of water of the reactor pool. In case of HFR that would be 5 cm of aluminum (AA5451) and 8 meter of fresh water. Therefore, propagation of radio waves in low frequency range through relevant media (water, air, aluminum, Zircaloy, stainless steel) was essential part of the investigation.

Propagation of electro-magnetic field in water has been studied in the test basin at NRG-PALLAS site as described in Section II.B. Fig. 8 summarizes the outcome of the measurements as a function of the transmission distance. Within the investigated frequency range the observed attenuation of the signal strength is essentially due to the geometry factor, and no significant signal attenuation in ultrapure water of the basin was observed.

 https://doi.org/10.1051/epjconf/202533804019

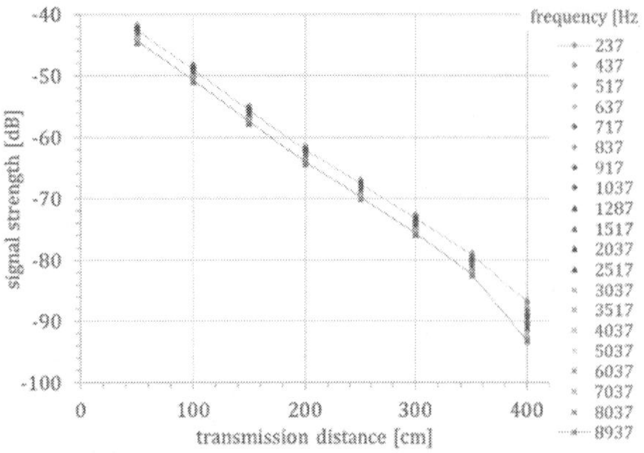

Fig. 8. Signal strength as a function of the transmission distance for frequencies between 200 Hz and 9 kHz

Propagation of electro-magnetic field in aluminum has been studied using the setup described in Section II.B. Attenuation as a function of the frequency is shown in Fig. 9.

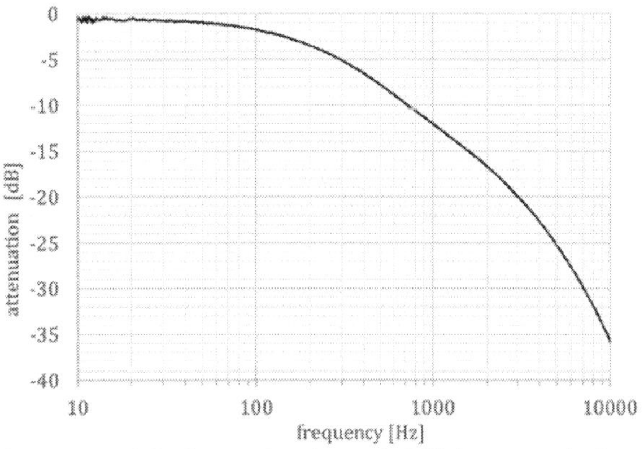

Fig. 9. Measured signal attenuation of magnetic fields by a 4 mm aluminum slab.

The propagation behaviour's dependency on the slab's thickness (t) is roughly defined by the product of $t{\cdot}\sqrt{\omega}$, allowing to scale the thickness of the slab by invers scaling of the frequency (ω): upscaling of the above measurement by 12:1 would then represent a slab thickness comparable to the 5 cm of the HFR reactor wall [13][14].

This approach allows to make first principal analyses of the signal attenuation to be expected in the HFR, or other reactor designs as the future Pallas reactor. Fig. 10 shows the frequency-dependent attenuation of magnetic fields through 5 cm of aluminium as present in the HFR, and through 1 - 5 cm of SS 301 or Zircaloy-4. The results clearly show that reactor vessels made from less conductive metals than aluminium allow transmitting at technically more suitable frequencies, with less than 40 dB attenuation at 1 kHz.

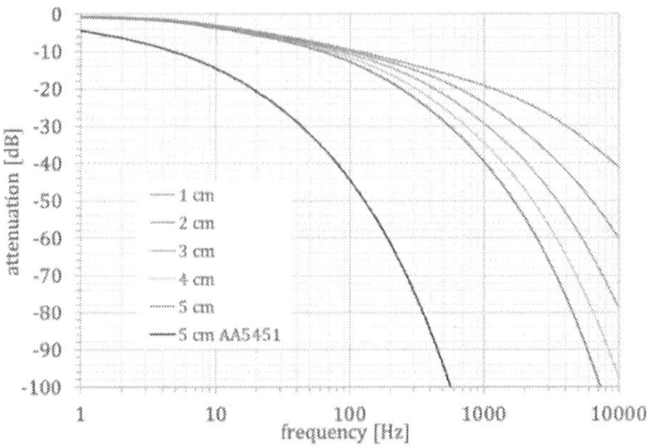

Fig. 10. Calculated signal strength for 1 to 5 cm of SS 301 or Zircaloy and 5 cm of AA5451 for the frequency range 1 Hz to 10 kHz.

V. Wireless sensor unit

A wireless monitoring solution consists of several parts. In the first step of the transmission chain, data from a sensor is coded by modulating its information on a fixed carrier frequency. This signal is usually a voltage signal that needs to be amplified to provide sufficient current to the transmitting antenna. The amplified signal is then injected into the transmitter antenna, located at wall of the reactor vessel. The magnetic field build up by the transmitter antenna propagates through the reactor wall and induces a voltage in the receiver antenna. The voltage is then amplified and filtered, and the sensor data is extracted from the resulting signal thereafter.

The most critical element of the above set-up is the in-core transmitter, as the receiver is located outside the reactor basin and its technology can rely on a wide variety of available hardware and software solutions. The in-core transmitter can be envisioned as an autonomous operating Wireless Sensor Unit (WSU), with the core components depicted in Fig. 11.

The essential part of the wireless sensor unit is the oscillator providing the carrier frequency. In 2020 a proof-of-principle lab experiment was realised in which an oscillator built on a vacuum tube 1Zh37B was powered by a bismuth telluride commercial TEG. The TEG used the heat produced by a halogen bulb.

Later a prototype of the VLF (8.68 kHz) transmitter was developed shown on Fig. 12. This transmitter operated on the 1P24B vacuum tube and used an 55 mm diameter antenna coil. The transmitter is enclosed in a watertight containment. This prototype was successfully used for the underwater tests along with the large bandwidth transmitter described above.

Several radiation resistant sensors for in-core monitoring have been considered for the WSU. A variety of sensor is discussed in e.g. [2]. The sensor of interest for the current stage of the RADIO project is the thermocouple. This sensor type generates a temperature-dependent voltage. A standard K-type, nickel-alloy based chromel-alumel thermocouple is a passive sensor producing an open load voltage of about 40 μV/°K. This voltage needs to be amplified by an instrumentation amplifier with a high input impedance, to provide sufficient output voltage to modulate the oscillator, either by FM or AM methods. At later stage a broad range of sensors will be considered in the project. The list will include Self-Powered

 https://doi.org/10.1051/epjconf/202533804019

Neutron Detectors (SPND), Linear Variable Differential Transformer (LVDT), pressure transducers, and gamma thermometers.

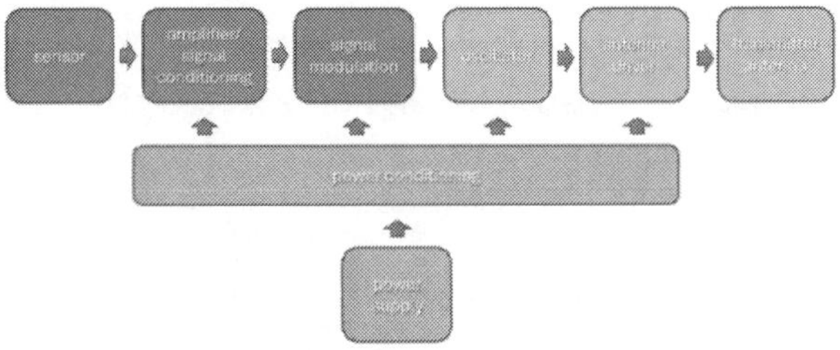

Fig. 11. Principal components of a wireless sensor unit (WSU)

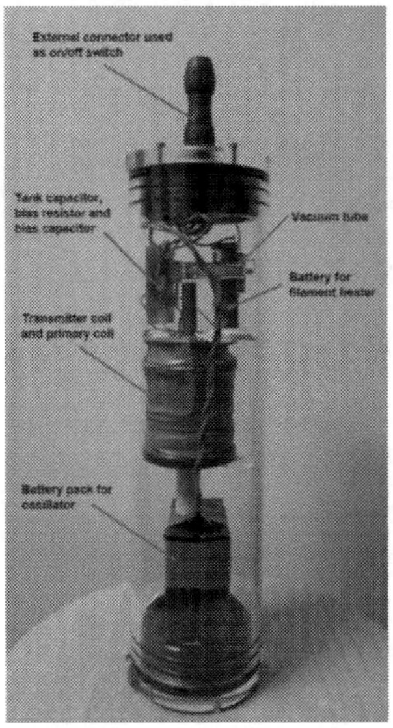

Fig. 12. Prototype low frequency oscillator design in a waterproof enclosure.

Another important element of the WSU is a TEG which provides electric power to the transmitter. The thermo-electric effect is widely applied in industry and consumer applications, with a large range of parts available on the market, either as TEG, or a Peltier cooling element. These commercial units can provide sufficient power for the application at hand, however, available heat, even if present in large quantities, can only be used if a sufficient large temperature difference can be maintained over both sides of the TEG, which can be limiting in small confined spaces [15]. The commercially available parts are usually based on bismuth telluride, as it provides one of the highest thermo-electric forces in the temperature range of interest (~ 280 μV/°C). These parts come in standard size, e.g. 30×30 or 40×40 mm, with typically between 100 and 250 thermocouples and source resistances between 2 and 20 Ohm. TEGs with other semiconducting materials are available as well, e.g., half-Heusler materials, lead telluride (PbTe) and silicon-germanium (SiGe) , but are not considered to have more favourable thermo-electric properties than the widely available bismuth telluride based TEGs [15]. Despite good performance of a commercial bismuth telluride in the gamma field of the GIF facility it is still uncertain if semiconducting material would work under strong radiation fields as present in the HFR core. Alternatively, all-metal TEGs are considered. Nickel-alloy based thermocouples as chromel-alumel or chromel-constatan are a radiation resistant and widely used in nuclear instrumentation, but have smaller thermo-electric forces: about 40 and 60 μV/°C, respectively.

To produce the same output voltage as a bismuth-telluride based TEG, four to seven times more thermocouples are necessary. Alumel, chromel and constantan have a slightly higher thermal conductivity (about 30, 19, and 20 W/m·K, respectively), compared to bismuth telluride, and an about one order of magnitude lower electrical resistivity (about 0.3, 0.7 and 0.5 μOhm·m for alumel, chromel, and constantan, compared to roughly 6 to 9 μOhm·m for bismuth telluride). The latter is relevant, because integration of a larger number of thermocouples on the same surface area will relevantly increase the source resistance of the TEG. However, since for obvious reasons thermo-couple-based TEGs are not available at the market, such a device must be designed and probably even built in-house.

Power conditioning is achieved by careful load matching of TEG and the oscillator. When combined with an instrumentation amplifier, additional power conditioning might be necessary.

VI. Conclusions

A wireless in-core radio transmitter is currently under development at NRG-PALLAS. The transmitter will use vacuum tubes as active components and will be powered by means of a thermo-electric generator utilizing nuclear heating present in the core. The concept has been demonstrated in the proof-of-principle laboratory experiment.

In perspective of preparation of the future in-core irradiation test at HFR various components of the radio transmitter have been tested in the gamma irradiation facility in the spent fuel pool. Several irradiation tests have been carried from which one with on-line measurements.

The following conclusion on the suitability of the tested components and materials under gamma irradiation are made:

- The tested vacuum tubes showed good results during and after the irradiation and can be used in the future in-core irradiation, disregards the implementation version. It is unclear whether the reddish-brown colouring of the glass goes along with an embrittlement and may result in a potential failure of the tube's glass enclosure during harsher environmental conditions (heat, pressure, vibrations, neutron irradiation).
- Mica capacitors: both tested capacitors showed no change of the capacity after 690 kGy of irradiation, and no visual alterations were found.
- The polystyrene capacitor showed only a minor increase of the capacity after 690 kGy of irradiation.
- No changes were found at least for one type of a MLC capacitor.
- The three tested Ta-types are found unsuitable for the operation in a radiation field.
- No alterations were found for the tested wires and coils, showing that PU insulation and aliphatic polyamide bond

coatings are suitable isolation and bonding materials for the applied radiation conditions.

- It can be concluded that all tested resistors did not show any measurable alternation in the measured resistance.
- The Bismuth/Telluride TEG demonstrated perfect radiation resistance in contrast to the expectations. Both measured parameters, the Seebeck coefficient and internal resistance, did not show any degradation after the irradiation.

It was demonstrated that the highly electrically conductive wall of a reactor vessel presents a relevant barrier that determine the frequency range to be used and that can only be bridged by wireless technology at technical unfavourable very low frequencies (VLF). The results so far allow making first principal analyses of the signal attenuation to be expected in the HFR, or other reactor designs as the future Pallas reactor.

The results clearly show that reactor vessels made from less conductive metals than aluminium allow transmitting at technically more suitable frequencies, with less than 40 dB attenuation at 1 kHz.

It has been shown that in ultrapure water of the HFR basin transmission of magnetic waves proceeds without relevant attenuation.

Analyses of different components of the in-core Wireless Sensor Unit is presented. The current status of the different components of the WSU is summarized as follows:

- **Sensor:** As primary sensor for the current project, a thermocouple is considered, as routinely used in-core at the HFR and other reactors.
- **Signal modulation:** For signal modulation both AM and FM are been considered. No final selection of the method has been made.
- **Power supply:** Currently a commercial bismuth-telluride based TEG is selected as power source for the WSU. A TEG-powered oscillator built on vacuum tubes was successfully tested. Failure or insufficient performance would eventually leading to the need to design an all-metal TEG.
- **Transmitter antenna:** Several antenna designs have been tested. The designs include a proper matching to the oscillator, and considers the limited capacitance of available radiation resistant capacitors, and the limited space available in the GIF and HFR in-core positions.
- **Antenna driver:** The currently used oscillator design provides sufficient output current to directly drive the transmitter antenna.

Acknowledgement

This research programme has been supported by the Dutch ministry of economic affairs and climate.

References

[1] V. Agarwal,, J.A. Smith, K.A. Manjunatha, "Assessing of Wireless Communication Technologies", Presentation on the Advanced Sensors and Instrumentation Annual Webinar, 29.10.2020 - 5.11.2020.

[2] D. Xiang, "A Self-Powered, Wireless Sensor System for Remote and Long-Term Monitoring of Internal Conditions of Spent Nuclear Fuel Dry-Storage Casks", Presentation on the Advanced Sensors and Instrumentation Annual Webinar, 29.10.2020 - 5.11.2020

[3] L. Zuo, "Self-Powered Wireless Through-wall Data Communication for Nuclear Environments", Presentation on the Advanced Sensors and Instrumentation Annual Webinar, 29.10.2020 - 5.11.2020.

[4] T.J. Schröder, E. Rosca-Bocancea, "Current state of the art of wireless data transmission systems for repository monitoring", The 2nd international conference on Monitoring in Geological Disposal of Radioactive Waste, 9-11 April 2019, Paris, France.

[5] T. J. Schröder, E. Rosca-Bocancea, "Wireless Data Transmission Demonstrator: from the HADES to the surface", MoDeRn Deliverable D.3.4.2, 2013.

[6] J. Turso, J. V. Carvajal, S.C. Stafford, M.D. Heibel, P.M. Sirianni, M.M. Heagy, G. Corak, R.W. Flammang, N.G. Arlia, K. Unlu, 2020, "Toward the implementation of self-powered, wireless, real-time reactor power sensing", Annals of Nuclear Energy 139, 2020.

[7] J. Carvajal, "Wireless Reactor Power Distribution Measurement System Utilizing an In-Core Radiation and Temperature Tolerant Wireless Transmitter and a Gamma-Harvesting Power Supply", DoE Advanced Sensors and Instrumentation Annual Webinar, 2020.

[8] B. Sukhanov, A. Kireev, "Rod Tubes, Principle and design", Translation of an original Russian article published in July 1960 in the Russian Radio Magazine "Radio" by Dmitri Faguet; English proof-reading Joe Sousa, October 2011.

[9] C.C. Phifer, "Radiation and Hydrogen in Glass", Sandia report SAND90 - 0950 - UC-404, 1990.

[10] D. Ehrt, P. Ebeling, "Radiation defects in borosilicate glasses", Glass Technol., 44 (2), p.46–49, 2003.

[11] H.J. Reich, "Theory and Application of Eletron Tubes", 2nd ed., McGraw-Hill, 1944.

[12] J. Deketh, "Fundamentals of Radio-Valve Technique". N.V. Philips' Gloeilampenfabrieken, 1949.

[13] Dood, C.V., W.E. Deeds, 1968, Analytical Solutions to Eddy-Current Probe-Coil Problems, Journal of Applied Physics, 39(6), p.2829-2838.

[14] T.J. Schröder, "Low frequency data transmission by magnetic induction", NRG note 24831/21.229319, 15 December 2021.

[15] T.J. Schröder, E. Rosca-Bocancea, J. Hart, "Thermal energy harvesting from High-Level Waste", The 2nd international conference on Monitoring in Geological Disposal of Radioactive Waste, 9-11 April 2019, Paris, France.

Evaluation of 4H-SiC sensor performances for thermal neutron flux measurements

Olivier Llido[1,*], Christophe Destouches[1], Abdallah Lyoussi[1], Laurent Ottaviani[2], Christelle Reynard-Carette[2], Heiko Jasper[3], Patrick Weidenauer[3]
[1] CEA, DES, IRESNE, DER, Cadarache F-13108, Saint-Paul-Lez-Durance, 13108, France
[2] Aix Marseille Univ, Université de Toulon, CNRS, IM2NP, Marseille, FRANCE
[3] Framatome GmbH Erlangen/Karlstein Germany
(*) olivier.llido@cea.fr

***Abstract*— Silicon Carbide (SiC) based neutron sensors present very interesting properties for neutron flux measurements: selectivity between neutron and gamma radiations, response linearity, radiation and temperature hardness, ... In 2021, CEA, AMU (Aix-Marseille University), within the framework of their joint laboratory LIMMEX, and FRAMATOME started a study to explore 4H-SiC diodes as possible thermal neutron detectors by using ^{10}B converter. Following a first exploratory measurement campaign of boron implanted 4H-SiC p-n diode sensors performed in 2021 in the Slovenian TRIGA reactor facility (Jožef Stefan Institute), new diodes have been manufactured with the aim of improving their thermal neutron sensitivity: larger surface (about 1 cm²) and a 1 µm thick layer of B_4C coating.**

These diodes have been irradiated in the same reactor in March 2023. A degradation of diode responses was observed as neutron fluence becomes significant. Investigations of the possible phenomena at the origin of the observed degradation of detector performances raise suspicions of defects induced linked to $^{10}B(n, \alpha)^7Li$ reaction product impacts which deposit their whole energy in the diode depleted area. A literature review confirmed this hypothesis. A model, based exclusively on carbon vacancies concentration, was proposed to simulate degradation of detector performances. It gives a good estimate of the typical fluence of reaction products that leads to a loss of counting rate (about 10^9 particles.cm^{-2}).

***Keywords* — Diode, p-n junction, thermal neutrons, ion implantation, boron carbide B_4C, silicon carbide 4H-SiC, carbon vacancy**

I. Introduction

This document is the summary of a several-year project involving the development and characterization of thermal neutron flux detectors based on 4H-SiC semiconductor. This work was carried out in partnership with FRAMATOME. Detectors were designed, developed and tested both by CEA and IM2NP of Aix-Marseille University, within the framework of their joint laboratory LIMMEX.

The main objectives were to design a compact detector as sensitive as possible and evaluate its performances and limits especially in term of neutron flux and fluence.

Silicon carbide has some very interesting properties for radiation detection:

- It is a wide bandgap semiconductor, which means it can be used as a detector with low thermal noise and can be operated at high temperatures (up to 700 °C).
- It has high breakdown voltage, high saturated drift velocity.
- It is known to be more resistant to radiations than other semiconductors.

It is inherently sensitive to fast neutrons, thanks to scattering and threshold reactions on carbon and silicon atoms. The main reactions are listed in the table I below.

TABLE I
MAIN FAST NEUTRON REACTIONS IN SIC

Reaction	Threshold energy (Mev)
Elastic scattering $^{12}C(n,n')^{12}C$	0
Elastic scattering $^{28}Si(n,n')^{28}Si$	0
Inelastic scattering $^{12}C(n,n')^{12}C^*$	4.8
Inelastic scattering $^{28}Si(n,n')^{28}Si^*$	1.8
$^{12}C(n,\alpha)^9Be$	6.2
$^{28}Si(n,\alpha)^{25}Mg$	2.7
$^{12}C(n,p)^{12}B$	13.6
$^{28}Si(n,p)^{28}Al$	4.0

To be sensitive to thermal neutrons, it is necessary to use a converter material to generate secondary charged particles. ^6Li and ^{10}B are commonly used to this purpose. Their reactions with thermal neutrons (25 meV) are described by the following equations:

$$\underset{(941\,b)}{^6Li + n_{th}} \rightarrow \underset{2.73\text{ MeV}}{^3H} + \underset{2.04\text{ MeV}}{^4He} \quad (1)$$

$$\underset{(3838\,b)}{^{10}B + n_{th}} \rightarrow \begin{cases} \underset{0.84\text{ MeV}}{^7Li} + \underset{1.47\text{ MeV}}{^4He} + \underset{0.477\text{ MeV}}{\gamma} & (93.7\,\%) \\ \underset{1.01\text{ MeV}}{^7Li} + \underset{1.78\text{ MeV}}{^4He} & (6.3\,\%) \end{cases} \quad (2)$$

One of the objectives of this study was to achieve the highest possible sensitivity, which is why a ^{10}B converter was chosen because its capture cross-section (3838 b) is about 4 times greater than that of the ^6Li (941 b).

II. Detector designs and manufacturing

Detectors were designed based on a silicon carbide p-n junction with a n-doped layer of 20 µm thick (see Fig. 1.) A depleted zone appears at the interface between the p-doped layer and the n-doped layer. When the applied voltage is sufficiently high (> 70 V), the n-doped layer is fully depleted and forms the active volume of the detector. Two designs were developed: one with an ionic implantation of ^{10}B and the other with a layer of boron carbide B_4C.

A. Ion-implanted diodes

Previous studies [1][2] carried out jointly by CEA and Aix-Marseille University in the framework of the I_SMART project, lead them to propose, develop, patent and qualify a 4H-SiC detector with implanted ^{10}B as converter. The good behavior of this kind of detector under thermal neutron flux has been experimentally demonstrated up to a neutron fluence of 2×10^{13} $n\cdot cm^{-2}$[3].

Based on this knowledge a first design of detectors was developed in 2021. Fig. 1. (a) shows that a 1 µm thick layer of aluminum was added to the top of the diode. ^{10}B ions were implanted in this layer with an areal density of 5×10^{15} $atoms\cdot cm^{-2}$. Two sizes of diodes were made: 2 mm × 3 mm and 4 mm × 5 mm. The maximal achieved sensitivity was 2.6×10^{-6} $count\cdot s^{-1}/n\cdot cm^{-2}\cdot s^{-1}$.

B. B_4C layer diode

In 2023, a second detector design was developed and tested. The purpose was to increase its sensitivity. Fig. 1. (b) shows that the ion-implanted layer was replaced by a 1 µm-thick layer of B_4C with a natural ^{10}B isotopic concentration (~20 %). The size of the diode has been increased to 9 mm ×11 mm. This converter layer offers a much higher areal density of ^{10}B (2.2×10^{18} $atoms\cdot cm^{-2}$). So, the overall sensitivity has been increased by a factor of about 1960.

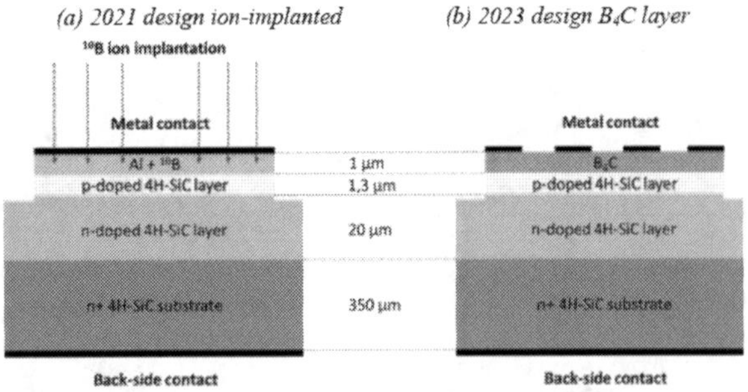

Fig. 1. Designs of detectors. Both are based on a silicon carbide p-n, junction. (a) Ion-implanted design made in 2021 and (b) boron carbide layer design made in 2023.

TABLE II
CHARACTERISTICS OF DETECTORS

	2021 design ion-implanted	2023 design B_4C layer
^{10}B areal density [$atoms\cdot cm^{-2}$]	5×10^{15}	2.2×10^{18}
Size	2 mm ×3 mm 4 mm × 5 mm	9 mm × 11 mm
Sensitivity [$count\cdot s^{-1}/n\cdot cm^{-2}\cdot s^{-1}$]	1.3×10^{-6} 2.6×10^{-6}	5.1×10^{-3}

III. Experimental facilities

Both detectors have been irradiated in the TRIGA reactor of the Joseph Stephan Institute in Slovenia, but not in the same location. In 2021 the ion-implanted diodes were placed in the Tangential Channel and in 2023 the diode with the B_4C converter layer was placed in the Thermal Column (see Fig. 2.). We made this choice for two reasons. Firstly, as the B_4C converter layer diode is much more sensitive than the ion-implanted one, it can operate at a lower neutron flux. Secondly, in the thermal column, the neutron spectrum is more thermalized. This is more suitable for our study, because fast neutrons are more likely to induce damage to the diode (see characteristics of irradiation positions in Table III).

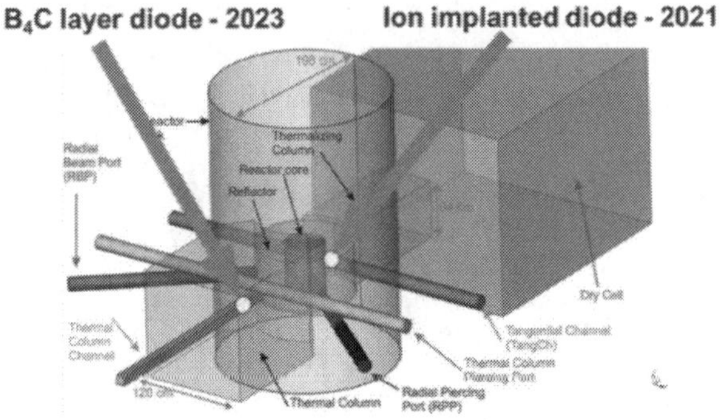

Fig. 2. Irradiation position of detectors in the TRIGA reactor of JSI.

TABLE III
CHARACTERISTICS OF IRRADIATION POSITIONS

	2021 design ion-implanted	2023 design B_4C layer
Irradiation location	Tangential channel	Thermal column channel
Conversion factor [$n_{th}\cdot cm^{-2}\cdot s^{-1}$/Wreactor]	6.7×10^{6}	1.28×10^{5}
Proportion of thermal neutron in the spectrum	67 %	98 %

To perform data acquisition during irradiation campaigns, we used the same measurement chain (see Fig. 3.), with the following elements:

- A preamplifier Cividec C2 HV (bandwidth 1 MHz – 2 GHz) power in 12 V.
- A high voltage power supply DT5533E from CAEN for detectors polarization.
- An Agilent MSO9104A oscilloscope capable of record pulse train with 512 MS.
- From detector to current amplifier: a BNC cable R285427, 10 m, 50 Ω, 1 nF.
- From current amplifier to the oscilloscope: a BNC cable R285427000, 2m, 50 Ω, 200 pF.

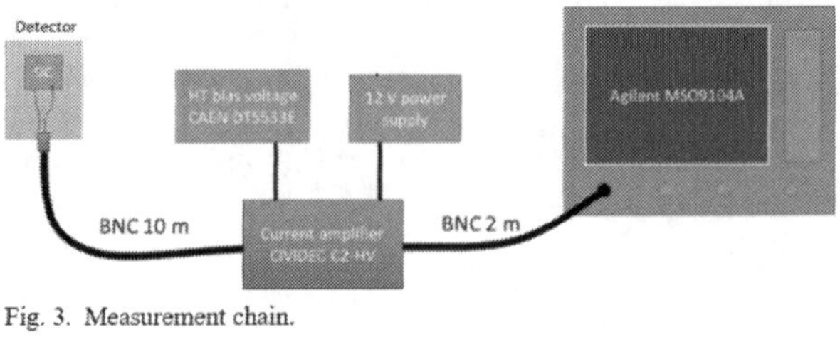

Fig. 3. Measurement chain.

IV. EXPERIMENTAL OBSERVATIONS

A. Raw signals observations

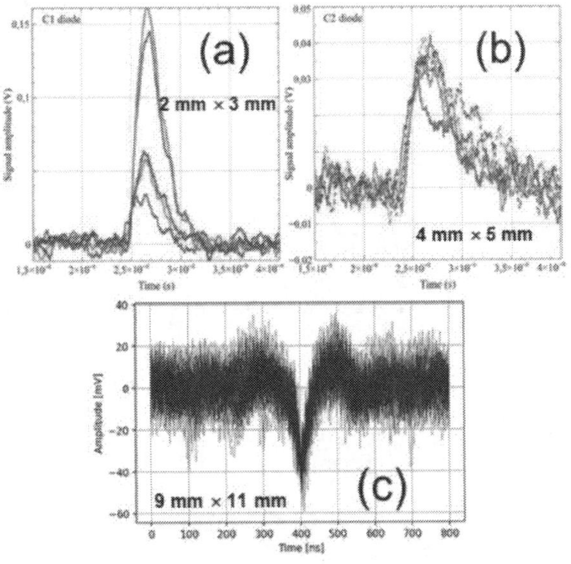

Fig. 4. Raw signals observed with the 3 diodes. (a) ion-implanted diode of 2 mm × 3 mm, (b) ion-implanted diode of 4 mm × 5 mm, (c) B_4C layer diode of 9 mm × 11 mm.

Fig. 4 shows examples of raw signals recorded in 2021 and 2023 with ion-implanted diodes and B_4C layer diode respectively. Pulses from ion-implanted diodes are sharp and well defined. In the contrary, raw signals from the B_4C layer diode are very noisy, and it was very difficult to get a signal out of the noise. Two reasons should be considered to explain this deterioration:

- First, the size of diode. The self-capacitance of a diode can be approximated according to (3) where C is the diode capacitance [F], ε is the dielectric constant of the material [$F\cdot m^{-1}$], S is the diode area [m^2] and e is the depletion layer thickness [m].

$$C = \frac{\varepsilon \times S}{e} \quad (3)$$

  The self-capacitance increases with the surface area of the diode. It acts as a low-pass filter and spreads pulses. It can be seen in Fig. 4 that the larger the diode, the smaller and wider the pulse. In Fig. 4 (a) the pulse duration is about 5 ns, in Fig. 4 (b) 15 ns and in Fig. 4 (c) 100 ns. So, increasing the size of the diode reduces the signal amplitudes.
- Secondly, boron carbide B_4C is also a semiconductor. A diode structure may appear at the contact between boron carbide and silicon carbide. This structure is likely to generate a high leakage current and therefore increase noise amplitude. This aspect could be improved with a new diode design, paying particular attention to preventing leakage currents, but this has not been studied yet.

These two points leads to a reduced signal to noise ratio.

B. Detector responses to a gradual increase in reactor power

In order to study response of detectors to reactor power, several measurements were carried out by increasing the power in stepwise increments. Each step lasted about 5 to 10 min. An increase in the counting rate with reactor power is expected, hopefully with a linear relationship. Fig. 5. (a) shows the result obtains with the two ion-implanted diodes. A very good linearity of the count rate with reactor power is observed over 4 decades. The slope of the straight lines is used to determine the sensitivity of detectors: 1.3×10^{-6} count$\cdot s^{-1}/n\cdot cm^{-2}\cdot s^{-1}$ and 2.6×10^{-6} count$\cdot s^{-1}/n\cdot cm^{-2}\cdot s^{-1}$. The behavior of the B_4C layer diode observed in Fig. 5 (b) is surprisingly very different. There is a fast collapse of count rate as the reactor power increases. In addition, we do not observe any recovery of the count rate when reactor power decreases.

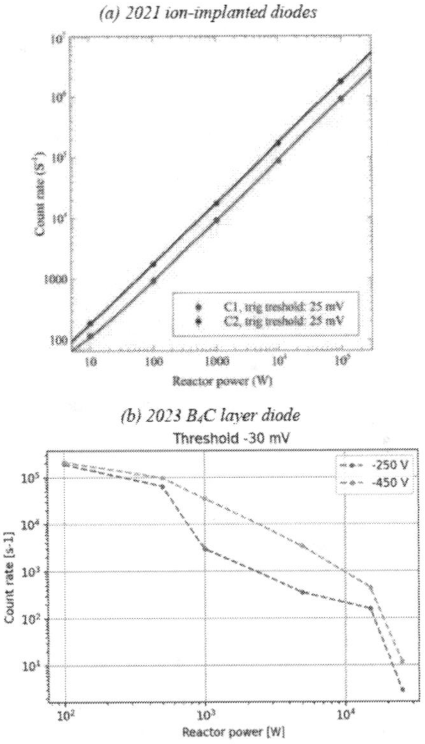

Fig. 5. Evolution of count rate with reactor power. (a) Response of ion-implanted diode, (b) response of B_4C layer diode.

C. Detector responses at a steady state power of the reactor

Fig. 6 (b) shows an exponential collapse of the count rate of the B_4C layer diode when the reactor power is at a steady state of 15 kW. The decay time constant is about 105 s. For ion-implanted diodes, in 2021 an irradiation of one hour at 250 kW was conducted. Fig. 6 (a) shows that an exponential collapse in the count rate also occurs, but at a much slower rate. Here, the decay time constant is about 15.6 min. It may explain why no degradation is visible in Fig. 5. (a), because the decay time constant is much longer than the time spent at each power step. Furthermore, a correlation between neutron flux and decay time can be expected: the higher the neutron flux, the shorter the decay time.

(a) 2021 ion-implanted diodes

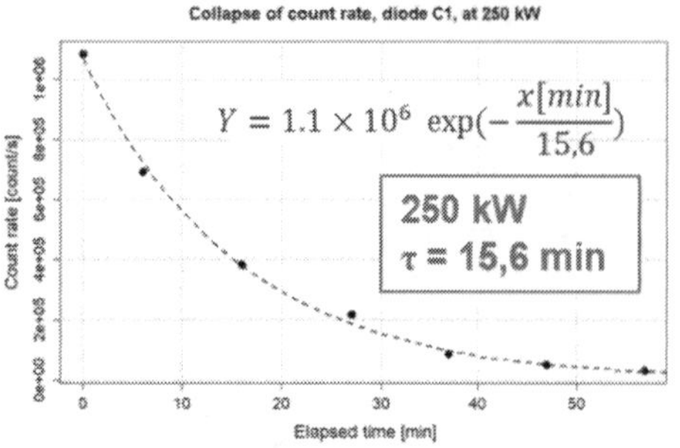

(b) 2023 B_4C layer diode

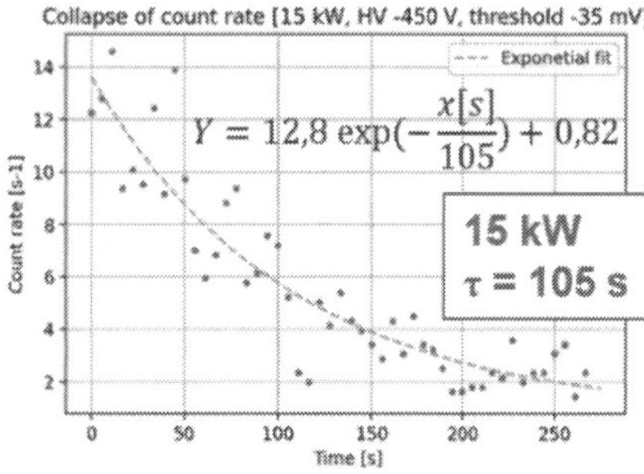

Fig. 6. Collapse of count rate at a steady state power of the reactor. (a) Response of ion-implanted diode, (b) response of B_4C layer diode.

In both cases, an exponential collapse in count rate is observed as the neutron flux becomes significant. This behavior suggests that a degradation of the diode occurs under irradiation. It is widely reported in literature that silicon carbide diodes are very resistant to high neutron fluence, up to 10^{15} (1 MeV) $n \cdot cm^{-2}$ [4]. Such a fluence level is achieved within an hour at 250 kW for the ion implanted diode, but for the B_4C layer diode at 15 kW, neutron fluence is still far below this limit. Moreover, we observe a much faster collapse for the B_4C layer diode with a lower neutron flux than for the ion-implanted diode. So, the neutron flux alone cannot explain the degradation we observed.

The fact that degradation rates are so different suggests that it might be a link with the concentration of ^{10}B (which has been significantly increased). Here we formulate the hypothesis that **the degradation is not due to neutron fluence but rather to the fluence of (n, α) reaction products (i.e. secondary charged particles) which stop in the detector active volume.**

Using equations (4-6) and the corresponding useful values listed in Table IV, we calculated for both diode the reaction products fluence corresponding to the observed decay time constant. Results are very close: 2×10^9 particle$\cdot cm^{-2}$ for the ion-implanted diode and 1×10^9 particle$\cdot cm^{-2}$ for the B_4C layer diode. This is an interesting clue that supports our hypothesis.

$$\begin{matrix}\text{Reaction products} \\ \text{fluence}\end{matrix} = \begin{matrix}\text{decay} \\ \text{time}\end{matrix} \times \begin{matrix}\text{reaction} \\ \text{rate}\end{matrix} \quad (4)$$

$$\begin{matrix}\text{Reaction products} \\ \text{fluence}\end{matrix} = \begin{matrix}\text{decay} \\ \text{time}\end{matrix} \times \begin{matrix}\text{neutron} \\ \text{flux}\end{matrix} \times \begin{matrix}\text{detector} \\ \text{sentivity}\end{matrix} \quad (5)$$

$$\begin{matrix}\text{Reaction products} \\ \text{fluence}\end{matrix} = \begin{matrix}\text{decay} \\ \text{time}\end{matrix} \times \begin{matrix}\text{reactor} \\ \text{power}\end{matrix} \times \begin{matrix}\text{conversion} \\ \text{factor}\end{matrix} \times \begin{matrix}\text{detector} \\ \text{sentivity}\end{matrix} \quad (6)$$

TABLE IV
USEFUL VALUES TO CALCULATE THE REACTION PRODUCTS FLUENCE CORRESPONDING TO THE OBSERVED COUNT RATE DECAY TIME CONSTANT

	2021 ion-implanted diode	2023 B_4C layer diode
Reactor power	250 kW	15 kW
Conversion factor [$n_{th} \cdot cm^{-2} \cdot s^{-1}$/Wreactor]	6.7×10^6	1.28×10^5
Thermal neutron flux [$n_{th} \cdot cm^{-2} \cdot s^{-1}$]	1.7×10^{12}	1.9×10^9
Detector sensitivity [$count \cdot s^{-1}/n \cdot cm^{-2} \cdot s^{-1}$]	1.3×10^{-6}	5.1×10^{-3}
Decay time constant	15.6 min	105 s
Reaction products fluence related to the decay time [$particle \cdot cm^{-2}$]	**2×10^9**	**1×10^9**

V. RADIATION DAMAGE IN SILICON CARBIDE

In order to better understand the degradation that we observed in the detectors, here are some results from literature reviews.

A. Effects of irradiation in silicon carbide [4][5]

Defects in silicon carbide have been thoroughly studied for many years by various teams around the world. Some focus on defects created during the diode manufacturing process (as-grown defects). Others are more specifically interested in defects generated during the use of these components, particularly under irradiation.

Defects generated under irradiation are added to as-grown defects. Energetic particles such as electrons, protons, alphas, neutrons, γ-ray photons and heavy ions can cause two types of radiation damages in detector materials:

- surface damages due to ionizing radiation loss (IEL);
- bulk damages due to non-ionizing energy loss (NIEL).

Surface damage are due to accumulation of charge in the surface passivation layer. They affect the breakdown behavior and the inter-electrode capacitance, resulting in higher electric noise. Bulk damages can be created by the collision of energetic particles with atoms in the crystal lattice. When a lattice atom is pushed away from its equilibrium position, vacancies and interstitial atoms can form. These are the so-called primary radiation defects. The majority of these primary defects recombine, whereas the others form deep levels in the energy gap. They may stay isolated as point defects: vacancies, interstices, anti-sites, impurity substitutions (see Fig. 7). Some of them may interact with each other (Frenkel pair), or with impurity atoms (vacancy-impurity complex) and form clusters. These mechanisms mainly occur during the annealing phase, resulting in secondary radiation defects. Bulk damages refer to both primary and secondary defects of this nature.

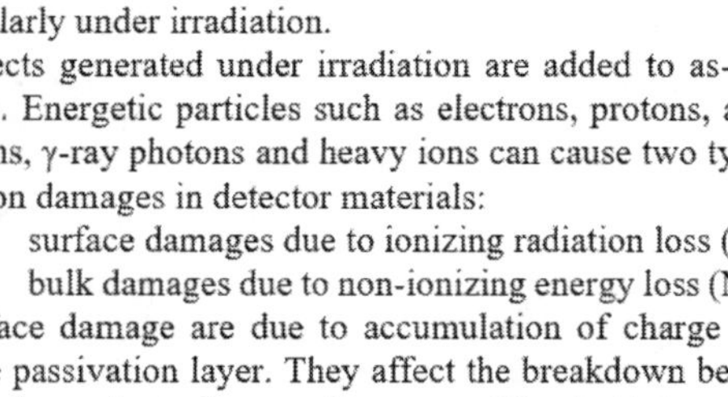

Fig. 7. Schematic diagram of the different types of point defect in a crystal lattice. Reproduced from [6].

Bulk damages, whether in the form of point defects or clusters, are responsible for the following:

- A decrease in signal amplitude due to charge carrier trapping.
- A rise in ohmic resistance due to a decrease in free carrier concentration.
- A shortening of carrier lifetime and diffusion length due to an increase in the concentration of recombination centers.
- An increase in the leakage current (and so noise background) due to the formation of charge-donor states.
- A change of effective doping concentration.

All these effects contribute to decrease the charge collection efficiency, to increase the noise and thus to lower the signal/noise ratio and consequently to deteriorate the detector performances.

B. Defects characterization

Various techniques are available to characterize defects in SiC. Among them, DLTS (Deep-Level Transient Spectroscopy) is one of the most commonly used. Here we draw particularly on the results presented in [7] and obtained using this characterization technique.

Fig. 8 (a) shows that in as-grown SiC, prior any irradiation, two deep levels appear to be dominant in DLTS measurements: $Z_{1/2}$ (EC - 0.64 eV) and $EH_{6/7}$ (EC - 1.84 eV). The $Z_{1/2}$ defect is the main contributor to the reduction in charge carrier lifetime. It is sometime referred as the "lifetime killer". Its capture cross section is quite large: 10^9-10^{10} barn [7][8]. It is the most important deep level in SiC. It has been studied for a long time, but it was only in 2012 that it (as the $EH_{6/7}$ defect) became clearly linked to carbon vacancy [9]. The density of deep levels in lightly doped as-grown SiC epitaxial layers is typically 10^{12} - 10^{13} cm^{-3} [10] although it appears that, under certain conditions, it can be reduced to 3×10^{10} cm^{-3} [11]. Lastly, it is extremely stable in temperature (as is the $EH_{6/7}$ defect), since temperatures between 1,400 °C and 1,700 °C are required to start an annealing process [4][11].

Under irradiation (Fig. 8 (b)), the concentration of defects observed in as-grown SiC ($Z_{1/2}$ and $EH_{6/7}$) increases. Beside these defects, new ones are appearing. The three most important are labeled ET1 (EC - 0.35 eV), ET2 (EC - 0.65 eV) and EH3 (EC - 1.06 eV) [7]. Defects concentration increases with radiation fluence. The type of generated defects doesn't seem to depend on the nature of the radiation (electrons, neutrons, ions ...), however there may be a difference in the introduction rate of each defect depending on the type of radiation and particle energy.

Fig. 8 (c) shows that apart from $Z_{1/2}$ and $EH_{6/7}$, defects introduced by irradiation appear to be annealable at rather reasonable temperatures (up to 425 °C). However, various recombination mechanisms seem to lead to an increase in concentration of $Z_{1/2}$ and $EH_{6/7}$ defects [7][12]. After an annealing, almost only $Z_{1/2}$ and $EH_{6/7}$ defects are found in SiC, in concentrations higher than those of as-grown material. This is consistent with observations reported in the literature [13] that is to say that after an annealing phase, SiC-based components do not fully recover their pre-irradiation electrical properties.

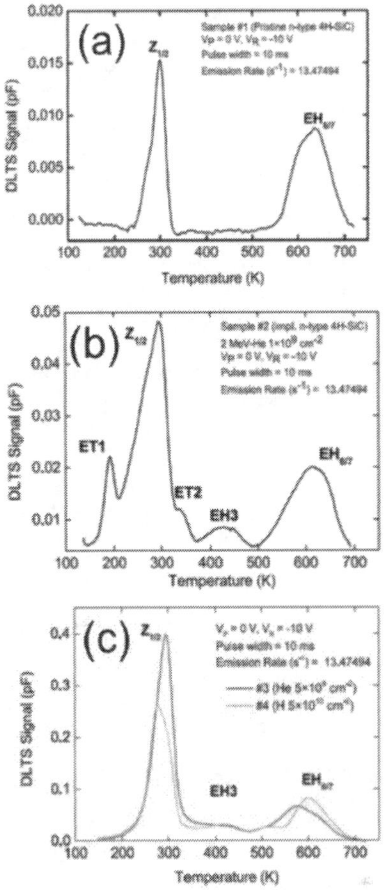

Fig. 8. DLTS spectra reproduced from [7]. (a) Spectrum of a as grown SiC sample. (b) Spectrum of SiC sample after irradiation with alpha particles of 2 MeV at a fluence of 1×10^9 cm^{-2}. (c) Spectrum of a SiC sample after irradiation and annealing up to 700 K.

VI. Simulations

A. Numerical model

Carbon vacancies appears to be the primary defect that we have to deal with. To better understand the generation of carbon vacancies in the detector and their effect on its operation we performed a simulation with the SRIM and TRIM software [14]. They are reference tools for simulating the interaction of ions with matter. More specifically, they can be used to calculate the deposited energy by ionization and collision, and to estimate the defects generated in the material. We have modeled a detector with a 1 µm thick layer of B_4C next to a 20 µm thick layer of SiC. We performed a simulation of 10,000 ^{10}B (n, α) reactions uniformly distributed in the B_4C layer and over an area of 10 µm × 10 µm. ^{7}Li and ^{4}He particles are isotropically emitted and at 180 ° to each other. To study defect generation, the "full damage cascades" model was used. For the data processing, in order to avoid edge effects and obtain a semi-infinite volume, we assumed that any particle leaving the volume via one of the side walls would immediately re-enter via the opposite wall, maintaining the same direction and energy.

B. Distribution of carbon vacancies and charge carriers

A first result is that ranges of reaction products are shorter than the thickness of the depletion zone (20 μm):

- 1.6 to 1.8 μm for ^{7}Li;
- 3.4 to 4.2 μm for ^{4}He.

The consequence is that these particles stop in the active volume of the detector and induce damages in it.

Fig. 9 shows overall distributions of carbon vacancies and charge carrier creation in the detector. In silicon carbide, carbon vacancies are created in a decreasing continuum. Charge carrier production distribution in silicon carbide follows the same trend as the one for carbon vacancies. It is maximal where vacancy concentration is maximal too. So, it doesn't appear to be possible to optimize the signal-to-damage ratio by adjusting the detector active area thickness. This is a harmful configuration because charge carriers are created in a heavily damaged area. So, there is a high probability that they recombine or be trapped before drifting and inducing a signal. Moreover, a large part of the peak charge carrier production takes place in the converter layer, outside the active volume of the detector, and so do not contribute to signal formation.

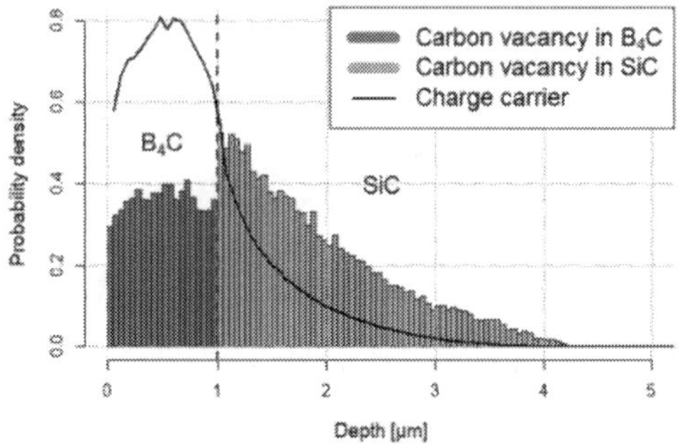

Fig. 9. Distribution of vacancies and charge carriers in the detector.

C. Density of carbon vacancies

Thanks to simulation, we have access to every carbon vacancy position. Most of them are created near the ion stopping point. They are concentrated in a small volume and so crystal properties are locally modified. We want to estimate the average local concentration of carbon vacancies at a stopping point. We assume that they are Gaussian distributed around this point. For each particle, to estimate the impacted volume, we look for the rotational ellipsoid that encompasses 68 % of these vacancies (see Fig. 10). The 68 % value corresponds to a dispersion of plus or minus a standard deviation generalized in 3 dimensions.

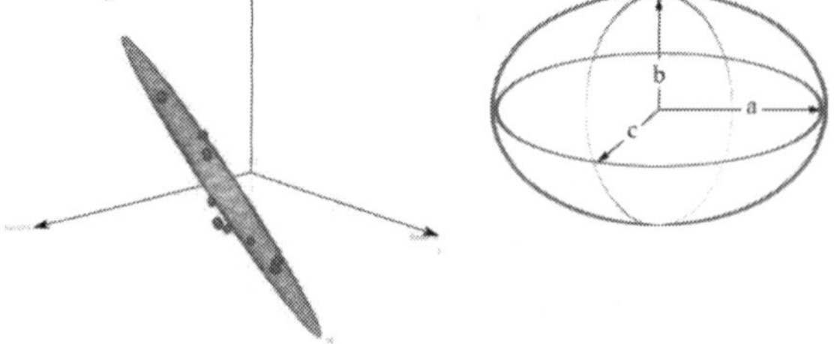

Fig. 10. Example of an ellipsoid encompassing carbon vacancies.

The ellipsoid orientation corresponds to the eigenvectors of the vacancies cloud of points. Dimensions along each axis are related to eigenvalues by the relation:

$$a = \sqrt{3.5 \times \lambda_a} \tag{7}$$

Where a is the dimension along an axis and λ_a the eigenvalue associated with this axis. The 3.5 factor corresponds to the value obtained with the distribution function in dimension 3 to reach the 68 % target. Ellipsoid volume is calculated using the relation:

$$V = \frac{4}{3} \times \pi \times a \times b \times c \tag{8}$$

As we know how many vacancies a particle generates, we can calculate the local concentration of carbon vacancies for each particle. Fig. 11 shows the distribution of so evaluated carbon vacancy concentration. In the following, we will consider the median value of the concentration: 1.7×10^{16} vacancy·cm^{-3}.

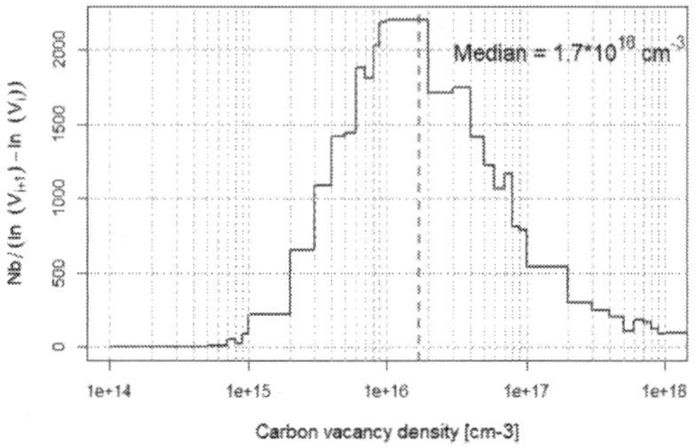

Fig. 11. Distribution of carbon vacancy densities at stopping point.

Fig. 12 shows influence of carbon vacancy concentration on charge carrier lifetime. At low carbon vacancy density, the charges carrier lifetime is almost a constant. We can estimate that, for a 20 μm thick epitaxial layer diode, at low carbon vacancy density (prior irradiation), the order of magnitude of charge carrier lifetime is about 1 μs. When the carbon vacancy density is relatively high (higher than about 10^{13} vacancy·cm^{-3}), the inverse of the carrier lifetime is proportional to the carbon vacancy density [11]. So, for a concentration of 1.7×10^{16} vacancy·cm^{-3}, charge carrier lifetime is locally reduced by a factor 1.7×10^{3}. This is a big difference, so it is therefore sensible to assume that a particle stopping point is a dead point in the detector.

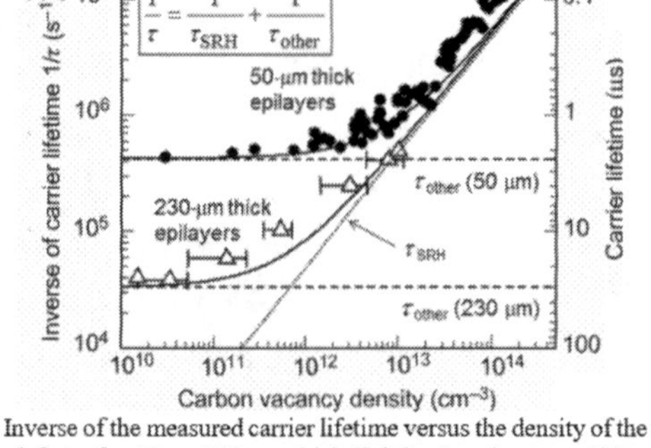

Fig. 12. Inverse of the measured carrier lifetime versus the density of the carbon vacancy defects for 50 or 230 μm thick lightly doped n-type SiC epitaxial layers. Reproduced from [11].

D. *Count rate evolution with reaction product fluence*

If a particle stopping point becomes a dead point in the detector, the count rate should be proportional to the non-impacted surface area. If we consider that charge carriers drift rectilinearly towards detector faces, the surface area impacted by an ion implantation is the projection onto the upper face of the detector of the encompassing volume of carbon vacancy we previously determined (see Fig. 13). Fig. 14 shows the distribution of these surface areas. In the following, we will consider the median value: 0.026 µm^2. It represents the typical detector surface that become inoperative when a particle passes through it.

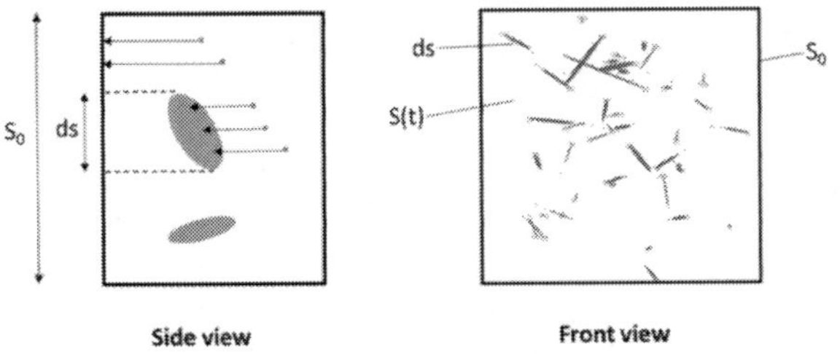

Fig. 13. Illustration of impact point projection on detector surface.

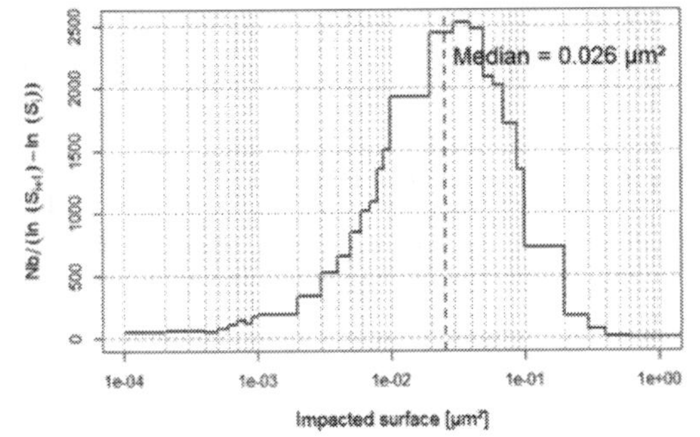

Fig. 14. Distribution of the impacted surface by an ion implantation.

The intact surface area evolves as follows (see Fig. 14 for references):

$$dS(t) = -ds \times \tau \times dt \qquad (9)$$

Where:

- *S(t)* is the intact surface area of detector at time *t*.
- *ds* the small surface element impacted by the passage of a particle, here 0.026 µm^2.
- τ the reaction rate in the detector (i.e. the number of reaction products that reach the detector active zone per time unit).

There is the relation with the reaction product fluence rate:

$$\tau = fluence\ rate\ (t) \times S(t) = \frac{d\ fluence(t)}{dt} \times S(t) \qquad (10)$$

So, it can be written:

$$dS(t) = -ds \times S(t) \times dfluence(t) \qquad (11)$$

$$\frac{dS(t)}{S(t)} = -ds \times dfluence(t) \qquad (12)$$

By integration with have the following relation, where C is a constant:

$$ln\ (S(t)) = -ds \times fluence(t) + C \qquad (13)$$

$$S(t) = exp(-ds \times fluence(t) + C)$$
$$S(t) = e^{C} \times e^{-ds \times fluence(t)} \qquad (14)$$

At the beginning ($t = 0$) we have *fluence(0)* = 0 cm^{-2} and S(0) = S_0, therefore $e^C = S_0$.
Finally, with have the relation of the intact surface area with the reaction product fluence, where S_0 is the detector initial surface:

$$S(t) = S_0\ e^{-ds \times fluence(t)} \qquad (15)$$

Here we have the appearance of a typical fluence of collapse $\frac{1}{ds}$. This is a decay constant. In our case its value is 3.8×10^9 particles·cm^{-2}. It can be noticed that its order of magnitude is in good agreement with what we had previously estimated from experimental data in section IV.C: 2×10^9 particle·cm^{-2} for the ion-implanted diode and 1×10^9 particle·cm^{-2} for the B_4C layer diode. This is a very interesting result for a fairly simple model.

This model, based exclusively on carbon vacancies, gives a good estimate of the typical fluence of reaction products that leads to a degradation of detector performance. It can also be noted that it slightly overestimates the typical fluence value, leading to a slower count rate degradation than that observed in Fig. 6. This is consistent with the fact that we didn't consider effects of other defects generated under irradiation. We also didn't take into account the possibility that when two impact points are close to each other, their encompassing ellipses may overlap, resulting in a larger impacted area.

Decrease in intact detector surface area results in a loss of count rate and a loss of signal amplitude. As the signal-to-noise ratio decreases, it becomes increasingly difficult to correct the response function to maintain a linearity between neutron flux and count rate. When the typical fluence value is reached, only 37 % of detector surface remains intact. From our point of view, it doesn't seem reasonable to exceed this typical fluence of reaction products.

VII. Conclusions

For this studies, 2 types thermal neutron flux detectors have been designed and tested. There are both based on 4H-SiC p-n junction and use ^{10}B as converter. In the first design ^{10}B is ion-implanted while the second use a B_4C layer as converter. A wide range of sensitivity can be achieved by using either one or the other.

Tests performed under neutron flux show that the signal to noise ratio of the B_4C layer diode is very poor. Improvements can be made to its design and manufacturing process. An excellent linearity of count rate with neutron flux was observed for the ion implanted diode but it is collapsing for the B_4C converter diode. For both diodes, we observed a collapse in the count rate at a steady state of reactor power. We attribute this collapse to the fluence of the reaction products, which stops in the active volume of the detector. The reaction product fluence is in the same order of magnitude for both detectors:

1×10^9 particle·cm^{-2}.

We proposed a model, based exclusively on carbon vacancies concentration, to simulate degradation of detector performances. It gives the same order of magnitude of the typical fluence of reaction products that leads to a loss of counting rate (10^9 particles.cm^{-2}).

From our perspective 4H-SiC diode with ^{10}B converter might be suitable for low neutron flux measurement, but we do not recommend using them above a fluence of reaction products of 1×10^9 particle·cm^{-2}.

References

[1] F. Issa, *et al.* "4H-SiC neutron sensors based on ion implanted 10B neutron converter layer," 2015 4th International Conference on Advancements in Nuclear Instrumentation Measurement Methods and their Applications (ANIMMA), Lisbon, Portugal, 2015, pp. 1-5, doi: https://doi.org/10.1109/ANIMMA.2015.7465544

[2] F. H. Ruddy, L. Ottaviani, A. Lyoussi, C. Destouches, O. Palais and C. Reynard-Carette, "Silicon Carbide Neutron Detectors for Harsh Nuclear Environments: A Review of the State of the Art," in IEEE Transactions on Nuclear Science, vol. 69, no. 4, pp. 792-803, April 2022, doi: https://doi.org/10.1109/TNS.2022.3144125

[3] F. Issa, *et al.* "Radiation Silicon Carbide Detectors Based on Ion Implantation of Boron," in IEEE Transactions on Nuclear Science, vol. 61, no. 4, pp. 2105-2111, Aug. 2014, doi: https://doi.org/10.1109/TNS.2014.2320943

[4] F Nava, *et al.* 2008, "Silicon carbide and its use as a radiation detector material", Meas. Sci. Technol. 19 102001, doi: 10.1088/0957-0233/19/10/102001

[5] De Napoli M (2022), "SiC detectors: A review on the use of silicon carbide as radiation detection material". Front. Phys. 10:898833. doi: 10.3389/fphy.2022.898833

[6] Q. Potiron, "Advanced solid state neutron detectors for fast neutron measurement and monitoring in fusion and fission environments", Ph.D. dissertation, Aix Marseille Univ, Marseille, France, 2024. p. 57.

[7] Pastuović Ž, Siegele R, Capan I, Brodar T, Sato SI, Ohshima T. "Deep level defects in 4H-SiC introduced by ion implantation: the role of single ion regime". J Phys Condens Matter. 2017 Nov 29;29(47):475701. doi: 10.1088/1361-648X/aa908c. PMID: 28972198.

[8] Brodar, T, *et al.* "Depth Profile Analysis of Deep Level Defects in 4H-SiC Introduced by Radiation". Crystals 2020, 10, 845. https://doi.org/10.3390/cryst10090845

[9] N.T. Son, X.T. Trinh, L.S. Løvlie, B.G. Svensson, K. Kawahara, J. Suda, T.Kimoto, T. Umeda, J. Isoya, T. Makino, T. Ohshima, E. Janzén, "Negative-U system of carbon vacancy in 4H-SiC", Phys. Rev. Lett. 109 (18) (2012) 187603, http://dx.doi.org/10.1103/physrevlett.109.187603.

[10] José Coutinho, *et al*, "Silicon carbide diodes for neutron detection", Nuclear Instruments and Methods in Physics Research Section A: Accelerators, Spectrometers, Detectors and Associated Equipment, Volume 986, 2021, 164793, ISSN 0168-9002, https://doi.org/10.1016/j.nima.2020.164793

[11] Tsunenobu Kimoto,2 - SiC material properties, Editor(s): B. Jayant Baliga, Woodhead Publishing, Wide Bandgap Semiconductor Power Devices, 2019, Pages 21-42, ISBN 9780081023068, https://doi.org/10.1016/B978-0-08-102306-8.00002-2.

[12] Hazdra, P. and Vobecký, J. (2019), "Radiation Defects Created in n-Type 4H-SiC by Electron Irradiation in the Energy Range of 1–10 MeV". Phys. Status Solidi A, 216: 1900312. https://doi.org/10.1002/pssa.201900312

[13] Muhammad Usman, *et al*, 2010 Phys. Scr. 2010 014012 "Effect of 3.0MeV helium implantation on electrical characteristics of 4H-SiC BJTs" doi: 10.1088/0031-8949/2010/T141/014012

[14] Ziegler, J.F.; Ziegler, M.D.; Biersack, J.P. "SRIM—The stopping and range of ions in matter (2010)". Nucl. Instrum.Methods Phys. Res. Sect. B Beam Interact. Mater. Atoms. 2010, 268, 1818–1823, https://doi.org/10.1016/j.nimb.2010.02.091

EU OFFERR Project. IRRADCOEFH Gamma Tolerance Test of Thermal CoefH innovative Sensors

Olivier Braillard[1,2 (*)], Stephane Gaillot[2 (*)]
Martin Mareček[3], Michal Cihlář[3], Lukas Prochazka[3], Jan Prehradny[3], Marek Miklos[3]

(1) CEA, DES, IRESNE, Nuclear Technology Department, Cadarache. 13108 Saint Paul lez Durance. France
(2) Improve-Heat company, Aix en Provence, France
(3) Centrum VÝZKUMU ŘEŽ S.R.O., Hlavní 130, Řež, 250 68 Husinec, Czech Republic

(1,2*) corresponding.authors : stephane.gaillot@cea.fr , olivier.braillard@improveheat.com
(3) other authors: Martin Mareček@cvrz.cz, Michal Cihlář@cvrez.cz, Lukas Prochazka@cvrez.cz, jan.prehradny@cvrez.cz, Marek.Miklos@cvrez.cz

Abstract — **In the context of experiments aimed at characterizing the thermal exchanges between fluids, the determination of the h thermal exchange coefficient at the wall is requested.**
In this objective, the innovative COEFH thermal sensor has been developed and optimized to precisely measure this coefficient. The robustness of its measurement is obtained by performing various tests of the COEFH sensor's resistance to severe environmental conditions (pressure, temperature, irradiation...). In this case, the tests presented in this paper aim to check the gamma dose tolerance resistance of the sensor (at room temperature, without pressure).
The framework of this action is the European OFFERR program underway over the period [2022-2026] and which aims to carry out R&D actions in the facilities identified in the OFFERR network (https://snetp.eu/offerr/). The present action is called IRRADCOEFH.
This action was carried out in 2024 April-May through a collaboration between the CEA-DES-IRESNE (France) and the CVR located in the Czech Republic.
Tests performed with CVR's gamma irradiator (*) allowed the electrical characteristics of the COEFH sensor to be continuously tested up to a gamma dose of 50 kGy. These tests were carried out over a period of 3.5 days. The first results obtained show a good behavior of the sensor in the face of gamma irradiation without any observed degradation of its characteristics.
These tests made it possible to validate the use of the sensor under gamma radiative environment.
After a general description of the OFFERR project, the paper describes the irradiation tests performed with the COEFH sensor at the CVR and provides the first results obtained.

Note: *the presented results were obtained using the CICRR infrastructure, which is financially supported by the Ministry of Education, Youth and Sports - project LM2023041.*

Keywords — **Innovative Sensor. Thermal CoefH sensor, Gamma dose tolerance test, CEA, CVR, EU OFFERR Project**

I. INTRODUCTION

As part of the European OFFERR Project for the period [2022-2026] aimed at carrying out collaborative R&D projects in support of the nuclear industry [R1], [R2] the action called IRRADCOEFH [R3] presented in this paper aims to verify the resistance under gamma irradiation of an innovative sensor for measuring heat exchange coefficient at the wall. The results obtained [R4],[R5] make it possible to specify the field of use of the sensor in a gamma radiative environment. This project was carried out between CEA-DES-IRESNE (France) and CVR (Czech Republic) in the spring of 2024.

II. SENSORS DESCRIPTION

The COEFH sensor allows the characterization of thermal coefficients, using miniaturized thermocouples or resistive tracks implanted on a ceramic substrate [R4].
Different models used are described hereafter:

STANDARD COEFH

COEFH sensors are specially designed to measure experimentally all the thermal parameters at the interface of a convective heat transfer; wall temperatures (three measurements), fluid temperature, wall temperature (at the interface), heat flux density, and the exchange coefficient (h factor). This last parameter is original. Few sensor manufacturers propose this experimental measurement.
This feature has been patented since 2008.

THIN COEFH

The new sensor uses the technology of thin layers in ceramic (Al2O3) cofired at T=1600°C with platinum micro-tracks inserted. Its manufacturing process, like the components of the nanotechnology, is carried out in the form of a wafer, thus accessing to a series production. Another particularity, the sensor is fixed by glue to the sensible surface, no more machining of the wall, thus the installation on a thin wall is possible as long as its thickness is compatible with that of the sensor. This is precisely the reason for the drastic reduction pathway in its thickness from 6 mm to 0.6 mm for the new generation.

Fig.1 view of the standard COEFH sensor

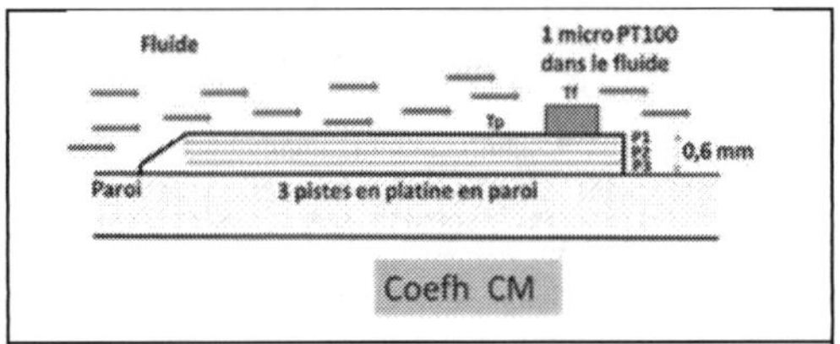

Fig.2 new generation of Coef h CM with 3 platinum tracks (P1, P2, and P3) inside the body sensor and Tf (PT100 or PT1000) outside the body sensor

In the IRRADCOEFH action, two versions of the sensor will be tested: COEFH, the standard model and the THINCOEFH which stands for an optimized sensor design based on ceramic substrate.

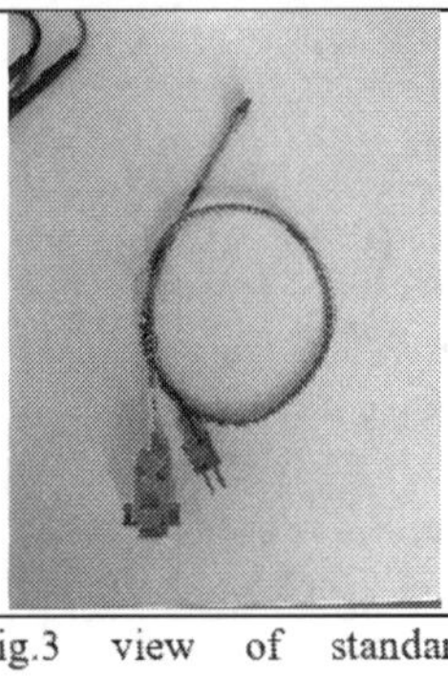

Fig.3 view of standard COEFH Sensor

Fig.4 view of optimised thin COEFH sensor

III. Gamma tolerance test objectives

The objective of these tests is to check the sensors with a gamma dose of 10 kGy (at minimum).
These tests will be performed at room temperature.
The total duration of the irradiating experiment will be about one week. The control of the sensor will be given by verifying electrical resistivity's measurements during the irradiation phases. Nevertheless, if the sensors are not off after this first phase, higher gamma tolerance tests will be achieved.
In a first phase, CVR will prepare the irradiation campaign by performing dosimetry measurements in the test chamber using alanine detectors.

IV. CVR irradiation facility overview

The irradiation has been performed in the Gamma Irradiation Facility. The Gamma Irradiation Facility uses 60Cobalt with activity of about 67 TBq as a source of ionizing radiation. The source is enclosed in a stainless-steel tube with an erecting mechanism. The stainless-steel tube with source is retracted during the manipulations. For irradiation, the source is erected to the irradiation chamber. The irradiated samples are placed on a moveable irradiation table. The moveable irradiation table can slide out of the irradiation chamber for easier manipulation with irradiated samples. The movable irradiation table can be equipped with additional apparatus, e.g., supplementary shielding, high vacuum chamber, inert atmosphere (Ar, N_2, …) chamber, cooling or heating system (in range between T=196 °C and T=400 °C), or various online monitoring systems.

Fig.5 view of CVR irradiation chamber

Fig.6 view of Irradiation Chamber (internal part)

V. SENSORS IRRADIATION TEST DESCRIPTION

COEFH Sensors were located on the moveable irradiation table as shown in fig.7 & 8.

Fig.7 location of sensors on the the moveable irradiation table

Fig.8 overview of the moveable irradiation table with sensors in place

The sensors were connected throughout the irradiation chamber's shielding to the measuring station. During the instalment, it was found out that some channels give incorrect results for the measurement of resistivity.

Extension cables for both, online and before and after the irradiation measurements were used. The cables were connecting the measuring device and the Sensor 1 bus. The values of the resistance of the extension cables are shown in Table 1. The measured values include both directions, to and from the Sensor 1 bus. Note that the values for P2 Transparent and P3 Light purple are lower. That is caused by the use of power cables with bigger diameter.

EXTENSION CABLES RESISTANCE				
Color	Tf(ohm)	P1(ohm)	P2(ohm)	P3(ohm)
Brown	1,25			
Black	1,15			
Grey		1,12		
White		1,16		
Red			1,16	
Transparent			0,73	
Blue				1,17
Light Purple				0,7

Table 1: Values of resistance of the extension cables connecting the measuring device and the Sensor 1 bus.

VI. DOSIMETRY

Before the experiment an offline dosimetry using alanine was performed. Alanine dosimetry is a common industrial process for use in gamma, electron beam, or X-ray applications that require a wide dose range, e.g., medical device sterilization or food irradiation. Alanine pellets were placed in the same place as the sensors in the following experiment. Alanine pellets were irradiated and sent for gamma dose measurements.

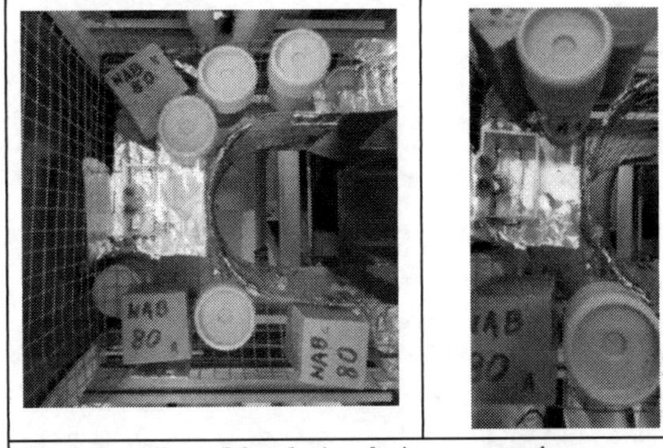

Fig.9: Location of the alanine dosimetry samples.

The alanine dose measurement was performed by ÚJV Řež's Testing Laboratory of the Radiation Chemistry and Environmental Qualification Department. The dose rate at the place of alanine pellets, i.e., at the place of sensors in the following experiment, was determined to be 0,605 kGy/hour with k=2 (95 % probability) uncertainty of 0,019 kGy.

Alanine standards used were produced by Far West Technology. The Electron Paramagnetic Resonance spectrometer was e-scan Bruker Biospin, type ES 200A, serial No SC0260

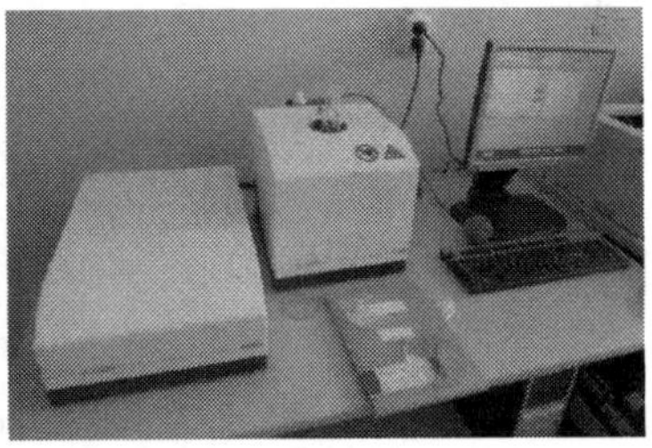

Fig.10 : EPR spectrometer e-scan Bruker Biospin, type ES 200A used for the alanine dosimetry

VII. IRRADIATION PROCESS

During the irradiation, the sensors were located in a distance of about 15 cm from the source.
The dose-rate at the place was determined to be 0,605 kGy/hour measured using alanine dosimetry (Fig.9) as mentioned above. Therefore, in order to gain a dose of 50 kGy, the irradiation was set to 82 hours and 38 minutes (Three days, 10 hours, and 38 minutes).
The irradiation started on 8th April at 20:50 PM and ended on 12th April at 7:40 AM.
There were no breaks in the irradiation – the irradiation was carried out in one continuous session.
The absorbed dose was approximately 50,1 kGy with k=2 (95 % probability) uncertainty of about 1,6 kGy.

VIII. ON-LINE MEASUREMENTS

The values of resistance for temperature sensor, Sensor 1, and Sensor 2 were measured online during the irradiation phase. As the measurement was carried out online the channels and pins could not be measured individually and independently. Therefore, due to the problems encountered and described in the previous chapters, the measured resistances were :
Tf Brown, Tf Black, S1 P1 White, S1 P2 Red, S1 P2 Transparent, S1 P3 Blue, S2 P1, and S2 P3.
The measured values are shown in fig.11 & 12
All the data are enclosed in a specific file.
For the online measurement, COMET MS6D data logger was used.

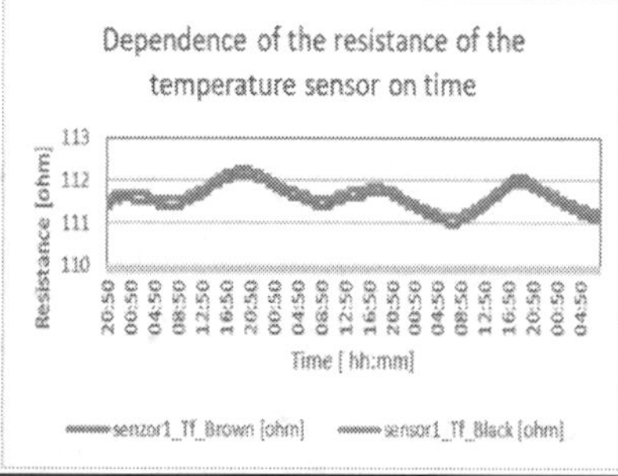

Fig.11 a graph of the dependence of the resistance of the temperature sensor on time.

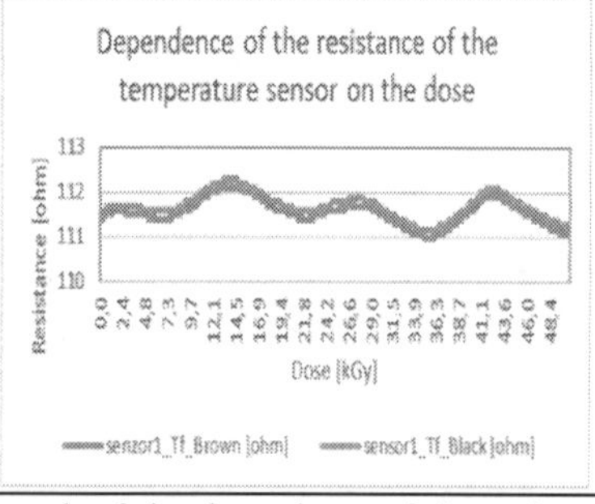

Fig.12 a graph of the dependence of the resistance of the temperature sensor on the absorbed dose

In these graphs, it is easy to see, that the strongest influence is caused by the day-night cycle and the corresponding slight changes in the temperature.

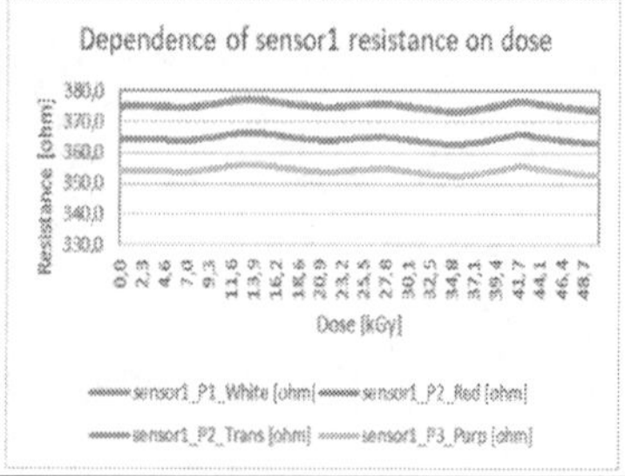

Fig.13 a graph of the dependence of the resistance of the sensor1 on the absorbed dose

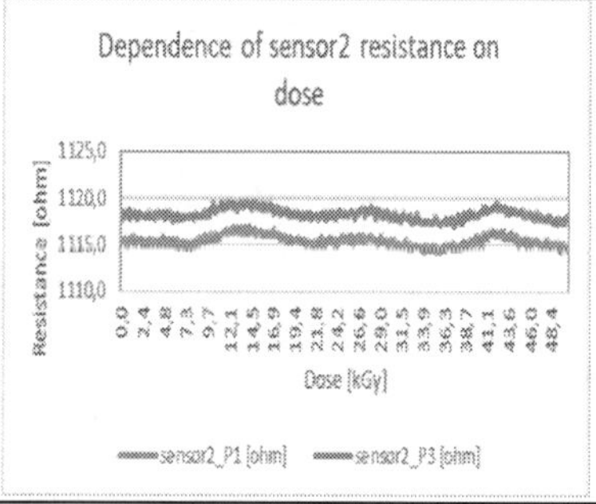

Fig.14 a graph of the dependence of the resistance of the sensor2 on the absorbed dose

IX. ON-LINE MEASUREMENTS

The values of resistance for individual channels of Sensor 1 were measured before and after the irradiation. Each channel and pin were measured individually and independently. For the before and after the irradiation measurements the Keithley 2001 Series: 7.5 Digit Multimeter with Scanning was used.

RESISTANCES BEFORE IRRADIATION				
Color	Tf (ohm)	P1 (ohm)	P2 (ohm)	P3(ohm)
Brown	114,6			
Black	114,9			
Grey		370,6		
White		370,4		
Red			381,9	
Transparent			381,5	
Blue				360,0
Light Purple				360,0

Table 2: values of resistance measured individually and independently for each channel and pin of Sensor 1 before the irradiation.

RESISTANCES AFTER IRRADIATION				
Color	Tf (ohm)	P1 (ohm)	P2 (ohm)	P3(ohm)
Brown	114,05			
Black	114,09			
Grey		368,76		
White		368,67		
Red			380,06	
Transparent			379,81	
Blue				358,72
Light Purple				358,28

Table 3: values of resistance measured individually and independently for each channel and pin of Sensor 1 After the irradiation.

X. CONCLUSION

The Sensor 1 and Sensor 2 were irradiated in the CVR Gamma Irradiation Facility using 67 TBq 60Co source. The sensors were in a distance of about 15 cm from the source with the dose-rate at the place of 0,605 kGy/hour. The irradiation started on 8th April at 20:50 PM and ended on 12th April at 7:40 AM. The absorbed dose was approximately 50,1 kGy with k=2 (95 % probability) uncertainty of about 1,6 kGy.
As it can be seen from both, the online measured data and before and after the irradiation data, it seems that the resistance of the sensors was not influenced by the irradiation. The variation in the measured resistance was probably caused by the day-night cycle and the corresponding changes in the temperature.

XI. PERSPECTIVES

At the end of these tests, the standard COEFH and the optimized thin COEFH sensors gave interesting results on the ability of the sensors to operate under a gamma radiation environment. In addition, the resistances of the platinum wires of the Thin COEFH sensor and those of the thermocouples of the standard sensor were not affected by irradiation. Note that neutrons tolerance tests have not been checked. Nevertheless, these first results open new perspectives on use of the COEFH sensors in Gamma Radiative environments.

XII. REFERENCES

[R1] EU OFFERR Project. https://snetp.eu/offerr/

[R2] EU OFFERR Brochure. https://snetp.eu/offerr-brochure/

[R3] EU OFFERR Project. IRRADCOEFH proposal. Fast Track. 23 Sept Call

[R4] EU OFFER Project. IRRADCOEFH. CEA Report. Gamma Tolerance Tests on Innovative Sensor (CoefH). O.Braillard & Al.

[R5] EU OFFER Project. IRRADCOEFH. CVR Report. Gamma Tolerance Tests on Innovative Sensor (CoefH). Report n° 04-2024. Martin Mareček, Michal Cihlář

Photon flux characterization of a new electron LINAC in the CINPHONIE irradiation facility

Benoit Geslot*, Nicolas Estre, Emmanuel Payan, Daniel Eck, Frédéric Moutet, Alix Sardet, Malo Lebreton, Cédric Carasco, Bertrand Perot
CEA, DES, IRESNE, DTN, SMTA, Nuclear Measurement Laboratory, Cadarache, F-13108 St Paul Lez Durance, France
(*) corresponding author: benoit.geslot@cea.fr

***Abstract*— Electron linear accelerators (LINACs) are versatile and powerful X-ray sources, that can be used in medical radiotherapy as well as in various industrial applications including non-destructive testing, imaging and security inspection. LINACs accelerate electrons by passing them through a series of oscillating electric fields within a vacuum tube. These high-energy electrons are then directed towards a metallic target, producing X-rays (bremsstrahlung radiation) when they decelerate upon impact. In the field of non-destructive radioactive waste characterization, high-energy photon imaging (radiography, tomography) is used on large cemented radiological waste containers, with a volume of the order of 1 m^3, to check their integrity and assess their content [1][2]. However, for such packages, passive gamma-ray spectroscopy, passive neutron counting and even active neutron interrogation fail in measuring nuclear materials, like plutonium and uranium. Therefore, high-energy photon interrogation techniques is under study to detect and quantify nuclear materials through the detection of induced-photofission particles. For the past years, CEA has been developing high-energy imaging [3] and photon interrogation techniques in CINPHONIE irradiation bunker (CHICADE facility, CEA IRESNE, Cadarache, France). CINPHONIE was recently upgraded with a K15 Varex accelerator that can reach a maximum dose rate of 130 Gy/min at 1 m from the X-ray target [4]. For advanced techniques (high-energy photon and photoneutron activations, photofission, bi-energy imaging), it is paramount to simulate precisely the irradiation field. For that purpose, a numerical model of the LINAC internals was built (with MCNP 6.3). It aims at simulating photon and neutron fields in view to calculate dose rates and reaction rates in irradiation samples, waste packages, but also in the whole casemate. A thorough characterization campaign was carried out to validate and calibrate this MCNP model against various experiments, including dose rate measurements in a water tank and delayed gamma-ray spectroscopy of thin metal foils activated in the X-ray field. These experimental results were used to fine-tune the electron source energy distribution as well as to estimate the average beam current. Its high-energy part is indeed particularly crucial for photofission and bi-energy studies.**

Keywords **—Photo-activation, Active photon interrogation, LINAC, Bremsstrahlung, CINPHONIE**

I. Introduction

The Nuclear Measurement Laboratory (CEA Cadarache) has long been developing non-destructive measurements in the context of inspection and characterization of nuclear waste packages [1], such as passive gamma spectrometry (global or segmented), passive neutron counting, high energy imaging (radiography, tomography), active neutron interrogation (prompt or delayed neutron measurement) or active photon interrogation [5].

The CINPHONIE irradiation bunker, located in the CHICADE facility (Cadarache, France) is dedicated to inspecting large and dense objects, such as 870 L cemented waste drums. Coupled to a motorized 3-axis bench, objects of dimensions up to 1.6 m in diameter and 5000 kg in mass can be handled. CINPHONIE was recently equipped with a new high energy X-ray source: a Varex K15 LINAC [2]. The LINAC was installed and commissioned in June 2024. The stability of the dose rate, as well as the reproducibility of irradiation conditions have been greatly improved compared to previous setups with a legacy SATURNE LINAC [4].

The K15 can be operated in two irradiation modes. In the so-called "high energy mode", the maximum dose rate reaches 130 Gy/min (at 1 m from the Bremsstrahlung target). In this mode, the photons have sufficient energy to induce activation by (γ,xn) reactions on the surrounding materials. In the "low dose rate" mode, the beam energy is lowered (around 9 MeV) so as to drastically reduce photo-activation. The maximum dose rate in this mode is 23 Gy/min @1 m.

To correctly predict dose rates (gamma and neutron), material activation and reaction rates of interest, such as photo-fission in nuclear material, Monte Carlo simulations with MCNP 6.3 are extensively used. A realistic model based on the geometry and materials of the LINAC interns (Bremsstrahlung target, collimator and shielding) was built and used to simulate the X-ray field (spectrum and geometry). This article is focused on the experiments that were carried out to validate and calibrate the MCNP model.

One of the main parameters of the simulations is the source definition and more specifically the average electrons' energy. Varying the electron energy or, to a minor extent, the distribution of energy can drastically change the results. In order to compare simulation results to experiments, the beam current needs also to be estimated.

 https://doi.org/10.1051/epjconf/202533804022

II. Description of the experiments

A. Experimental setup

Calibration tests were carried out with the objective of estimating the electron average beam's energy and current. First, the total dose rate deposited in a water tank was measured using an ionizing chamber (PTW equipped an electrometer). By fitting the attenuation curves along the beam's axis, it was possible to estimate the linear attenuation coefficient and to compare it to simulation results at various beam's energies. This was done in both operating modes.

Complementarily, photo-activation experiments were carried out with the objective to obtain more information on the high energy part of the X ray field. Photo-activation reactions like (γ,xn) are only possible when the photon's energy is above the minimum binding energy of nucleons (typically a few MeV). When reaching this threshold, the nucleus enters the Giant Dipole Resonance (GDR), which may result in the emission of neutrons and gamma rays.

By irradiating metal foils of gold, nickel and depleted uranium, short-lived radionuclides were produced by (γ,n) and (γ,f) reactions and then measured on a gamma spectrometry bench equipped with an hyper pure germanium detector and a Canberra acquisition system with Genie-2000 software. Interestingly, the four reactions of interest have different energy thresholds (see table I). This explains why their sensitivity to the beam's energy vary greatly.

Table I
Reactions of interest for photo-activation experiments, energies and intensities of gamma rays (source: LNHB)

Reaction	Threshold (MeV)	Radionuclide	Half-life	Energy (keV)	Intensity (%)
^{197}Au(g,n)	8.5	^{196}Au	6.18 d	333	22.9
				355	87
^{58}Ni(g,n)	12	^{57}Ni	1.5 d	127	16
				1377	81.2
^{238}U(g,n)	6.5	^{237}U	6.75 d	208	21.3
^{238}U(g,F)	5	^{135}I	6.57 h	1131	22.6
				1260	28.7
^{238}U(g,F)	5	^{92}Sr	2.65 h	1384	90

The metal foils were mounted on a 3D-printed plastic holder, the center of which was aligned with respect to the beam's axis (X axis). The measurement's reference point was located at 104.4 cm from the X-ray target (referred to as "position 1"). It corresponds to the aluminum door mounted on the K15 collimation system. This door was customized with a central housing to accommodate the plastic holder. Several irradiations were performed in position 1 to check the reproducibility and consistency in the results.

A second configuration located further away from the target (X = 242 cm, referred to as "position 2") was explored with the objective of mapping the gold and nickel reaction rates on a square grid with a pitch of 3 cm.

Prior to the irradiations, the metal foils were weighted to account for small variations in their mass. Then they were put in small 3D-printed "pill box" loaded with 3 nickel disks (each 1 mm in thickness, ~1.55 g) on top of which a thin gold foils was taped (0.1 mm in thickness, ~95 mg).

After irradiation, each set of disks was measured altogether, putting the gold foils on the bottom side, the closest to the HPGe diode.

The metallic uranium 238 disk (0.1 mm in thickness, ~100 mg) was irradiated and measured separately, in a vinyl bag for radioprotection. The bag was taped on the plastic holder, so as to center the uranium disk with respect to the X axis.

In both setups, the metal foils were arranged symmetrically with respect to the beam axis (X axis), either diagonally or vertically (or in both directions in the case of configuration #6, see table II). Symmetrical positions were chosen to check the alignment of the beam with the X axis.

Table II
List of irradiation configurations

	Duration	Position	Number of disks (gold, nickel, U)	Setup	Average dose rate (Gy/min) *
#1	1 h	1	1 – 1 – 0	Central	127.9
#2	1 h	1	3 – 3 – 0	Diagonal	126.1
#3	1 h	1	3 – 3 – 0	Diagonal	129.3
#4	1 h	1	0 – 0 – 1	Central	131.5
#5	1 h	1	3 – 3 – 0	Vertical	131.0
#6	2 h	2	10 – 10 – 0	Diagonal & Vertical	132.4

* The dose rate and dose were recorded by the LINAC's monitoring system.

B. Monte Carlo simulations

The MCNP 6.3 particle transport code, associated with the IAEA PD-2019 photo-nuclear cross sections library, was used to calculate photon spectra (illustration on Fig. 1, left) and photo-activation reaction rates inside the metal disks (using F4 tallies with flux modifiers FM card).

After analyzing configurations #2 and #3, a small shift of the position of the maximum flux was discovered. To account for this bias in the simulation, the beam axis was shifted by +5 mm vertically at X = 104 cm and by +12 mm at X = 242 cm (illustration on Fig. 1, right).

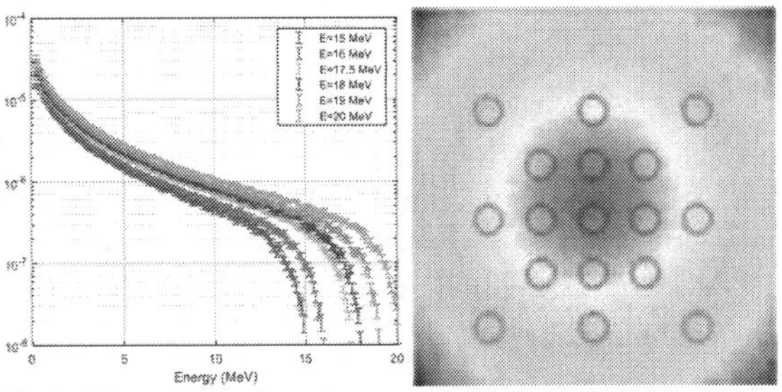

Fig. 1: *Photon spectra for various energy of the beam (left). Flux mesh in the Y-Z plane for configuration #6 (right): the black circles represent the activation foils containers.*

III. Results and discussion

A. Total dose rate in water

The photon dose rate deposited in water is sensitive to the whole photon spectrum, which is mainly composed of photons below 5 MeV. MCNP F6 tallies were used to calculate the total dose rate in the water tank. The volume of cells was chosen equal to the sensitive volume of the ionization chamber.

By varying the beam's energy, several attenuation curves were produced. They are displayed on Fig. 2, along with the measurement data points (square markers). All curves show the same behavior: a short increase followed by a typical exponential decay. By dividing measurements dose rates with the beam current (108 µA, estimated based on information issued by the monitoring system), it was possible to directly compare simulations and experiments. It appears that they fall in between simulations obtained with 17.5 MeV and 18 MeV for the beam's energy.

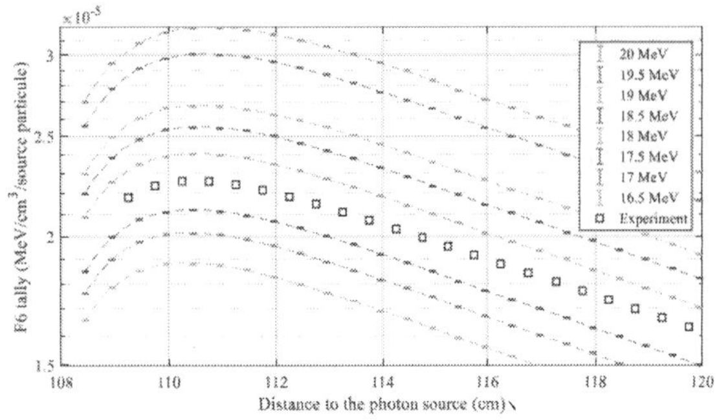

Fig. 2. Dose rate vs. distance in water calculated with MCNP (F6 tally) and measurement (scaling factor with a beam current of 108 µA)

B. Photo-activation results

After the irradiation, the metal foils were placed on the gamma spectrometry bench for counting. Then, the analysis process was as follows: counting spectra were processed to obtain integral counts in the photopeaks of interest (see table I). Knowing the detector's efficiency ε (based on an MCNP 6.3 model), the decay constant λ of the radionuclide, the intensity of gamma rays I_γ and the mass of the metal disk M, the measurement C could be converted into an activity per mass unit A (in Bq/g), as expressed in equation (1).

$$A = \frac{\lambda}{M}\frac{1}{\varepsilon}\frac{C}{I_\gamma\cdot\left(1-e^{-\lambda T_i}\right)\cdot e^{-\lambda T_c}\cdot\left(1-e^{-\lambda T_m}\right)}, \quad (1)$$

In the previous equation, T_i, T_c and T_m are the durations of the irradiation, cooling down and measurement phase. By dividing A by electronic charge q_e and by the reaction rate simulated with MCNP, an estimation of the average beam's current I_X can be obtained. In equation (2), the MNCP tally for reaction X is referred to as RR_X and expressed in s^{-1}/g/source particle.

$$I_X(E) = \frac{1}{q_e}\frac{A}{RR_X(E)} \quad (2)$$

RR_X strongly depends on the beam's energy E. By varying E from 16 MeV to 20 MeV, a list of beam's currents was obtained. All curves should cross at the LINAC operating point that best reflects the measurements. Results displayed on Fig. 3 were produced based on the measurements obtained in configurations #1 and #4. As expected, the curve associated with the ^{58}Ni(γ,n) reaction is the most sensitive to the beam's energy, since the threshold of the reaction is the highest (12 MeV). On the contrary, the reaction ^{238}U(γ,n) gives the flattest curve.

The four reactions give consistent estimations of the average for a beam's energy sitting between 18 MeV and 18.2 MeV. This is in good agreement with the estimation obtained from the dose rate measurements.

Uncertainties associated with the gamma spectrometry measurements are so low (below 1 %) that they cannot be seen on the graph, but there are other sources, as listed below:

- LINAC's reproducibility (around 2 %);
- Efficiency of the detector (typically around 5 %);
- Geometry errors associated to the irradiation setup (position of the metal foils and distance to the X-ray source);
- Geometry errors during gamma spectrometry measurements;
- Photo-nuclear cross sections (around 10 % to 15 %).

The uncertainty breakthrough is clearly dominated by the photo-nuclear cross sections. Under the assumption that the four cross sections are independent from each-others, it is hopefully possible to benefit from the average of the four estimations.

A best estimate of the beam's current was obtained by averaging of the 8 data points lying between 18 MeV and 18.2 MeV. The associated uncertainty was obtained based on the data point experimental standard deviation (square box in Fig. 3). The beam current was estimated at (94 ± 5) µA.

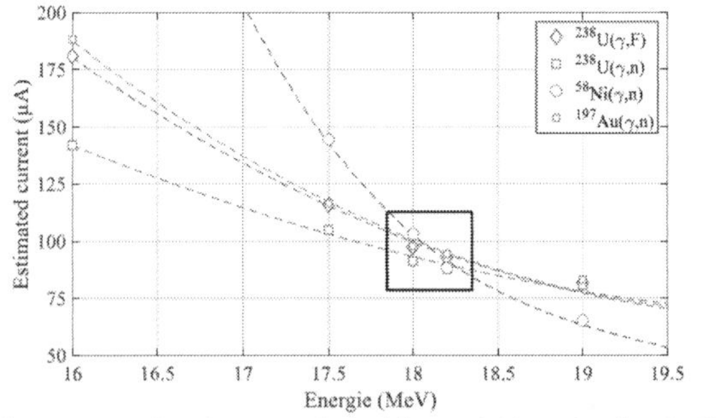

Fig. 3. Average beam's current obtained from activities and analyzed with various beam's energy simulations. Polynomial fit curves are illustrative.

IV. CONCLUSIONS AND OUTLOOKS

A new X-ray source was installed in the CINPHONIE irradiation casemate and commissioned in 2024. Improved irradiation conditions, in terms of dose rate and reproducibility, are expected from this upgrade.

An MCNP 6.3 model was built and validated against experiments. Dose rate attenuation in water and photo-activation of metal foils were carried out. This article focused on the photoactivation results. Thanks to four reactions of interest, it was possible to estimate two key parameters: the electron beam's average energy and the beam's current. The uncertainty breakdown is dominated by the uncertainties on photo-nuclear cross sections. The beam's energy was estimated at (18.1 ± 0.1) MeV. The best estimate of the beam's current was (94 ± 5) µA.

In future works, a methodology based on Bayesian inference will be used to extract more information from the photo-activation measurements. Indeed, two sets of experimental data were produced (for the irradiation setups) and they should be analyzed altogether within an enlarged mathematical model. This model will allow estimating extra parameters, such as the distance to the X-ray source and the position of the maximum of the X-ray field. Thanks to the Bayesian framework, uncertainties, especially those associated to the photo-activation cross sections, will be natively included into the estimation process.

References

[1] B. Pérot, *et al.*, "The characterization of radioactive waste: a critical review of techniques implemented or under development at CEA, France," EPJ Nuclear Sciences & Technologies, 4, 3 (2018). doi.org/10.1051/epjn/2017033

[2] N. Estre, *et al.*, "Performances of 9 to 15 MV Computed Tomography of large objects in industrial or nuclear field," 10th International Symposium on Digital Industrial Radiography and Computed Tomography (DIR 2025), Paris, 1-3 July 2025

[3] M. Kistler, *et al.*, "Simulated performances of very high energy detectors for non-destructive computed tomography characterization of large objects," IEEE Transactions on Nuclear Science, 65(9):2527–2532, 2018.

[4] N. Estre, *et al.*, "High-Energy X-Ray Imaging Applied to Nondestructive Characterization of Large Nuclear Waste Drums," IEEE Transactions on Nuclear Science, 62(6), 3104–3109, (2015). doi.org/10.1109/TNS.2015.2498190

[5] M. Delarue, *et al.*, "New measurements of cumulative photofission yields of ^{239}Pu, ^{235}U and ^{238}U with a 17.5 MeV Bremsstrahlung photon beam and progress toward actinide differentiation," Nuclear Instruments and Methods in Physics Research Section A: Accelerators, Spectrometers, Detectors and Associated Equipment, 1040, 167259, (2022). doi.org/10.1016/J.NIMA.2022.167259

Report on TESCA irradiation of optical sensors and tests of glasses

Guy Cheymol[1*], Marion Agoyan[2], Ayoub Ladaci[1], Florence Martin[3] and Christophe Destouches[3]
[1] Université Paris-Saclay, CEA, Service de Recherche en Matériaux et procédés Avancés, 91191, Gif-sur-Yvette, France
[2] Univ. Lyon, UJM, CNRS, IOGS, Laboratoire Hubert Curien, UMR 5516, 18 rue Prof. B. Lauras, 42000 Saint-Etienne, France
[3] CEA, IRESNE, Cadarache, DER/SPESI, F-13108 Saint-Paul-lez-Durance, France
(*) guy.cheymol@cea.fr

***Abstract*—The TESCA irradiation campaign, conducted from November 16 to December 17, 2023 (one reactor operating cycle) in the SCK•CEN BR2 research reactor, enabled testing of optical sensors at different development stages, under intense radiation: fast neutron fluence (E > 1 MeV) of 1 to 2 x 10^{19} neutrons/cm^2 and gamma dose of about 5 GGy. The temperature ranged between 100°C and 500°C. TESCA irradiation enabled on-line testing of Fabry-Perot extensometers, and miniature, non-contact, high-temperature optical pyrometry sensors. We also report an online measurement of radiation-induced attenuation on a sapphire sample. Because all sensors and measurements are temperature-sensitive, it was crucial to either maintain stable temperatures or accurately monitor them to assess the impact of radiation on sensor response stability at constant load or on glass parameters. Brief temperature jumps were also applied to test sensors accuracy. An overview of results from monitoring the various sensors and experiments throughout the irradiation cycle are provided. A well thermally stabilized Fabry–Perot extensometer maintained high stability, with a drift below 1 µm under constant load. It also ensured accurate thermal expansion measurements during temperature jumps. The pyrometric sensors operated effectively under intense irradiation at steady temperature: the collected blackbody emission spectrum was preserved, as expected with constant temperature. The unusable spectral range around the OH peak will be extended to [1250–1430 nm]. In contrast, the RIA setup showed susceptibility to temperature variations and mechanical vibrations, indicating the need for improved mechanical robustness.**

***Keywords* —Fabry-Perot extensometer, Glasses under irradiation, Irradiation of optical sensors, Optical pyrometry sensors, Optical sensors in Material testing reactor.**

I. Introduction

NUCLEAR energy remains a key solution to meet future energy demands while limiting carbon dioxide emissions and other atmospheric pollutants. In this context, research and development efforts increasingly rely on highly instrumented experiments conducted in Material Testing Reactors (MTRs). These experiments are crucial for extending the lifetime of nuclear reactors, optimizing fuel management, and enhancing the safety of both current and next-generation nuclear power plants.

Optical Fiber Sensors (OFS) offer several advantages for instrumentation within MTR irradiation devices. Their ability to enable remote measurements, multiplex multiple sensing points, and resist electromagnetic interference makes them particularly attractive. Moreover, their compact size is ideal for the constrained spaces typical of irradiation setups, minimizing perturbation of the thermal and radiation fields near the tested material samples. OFS can be used to monitor a wide range of physical parameters through various optical principles, even in harsh environments [1].

However, in order to qualify these sensors for nuclear applications, it is essential to assess their performance under irradiation and realistic operating conditions.

This article presents some results from the TESCA irradiation campaign, which enabled the testing of optical sensors—at different stages of development—under intense radiation and elevated temperatures (100 to 500 °C). In addition, several optical glasses intended for integration into future confocal chromatic sensors were evaluated. Both online Radiation-Induced Attenuation (RIA) measurements and optical path variation measurements were conducted, along with post-irradiation assessments. In this paper, we present an example of online RIA measurement results obtained for a sapphire sample.

The following sections describe the experimental setup, test conditions, materials and sensors used, followed by a presentation and discussion of key results. The article concludes with a summary of the findings and perspectives.

II. Description of Experiment

A. *Irradiation and Test Device*

The irradiation was conducted in the BR2 reactor at the SCK•CEN site in Mol, Belgium. The experiments were placed in the reflector region, inside the Liberty test device, which consists of five elongated "needles" (Figure 1). Each needle has an outer diameter of 18 mm and a length slightly under 6 meters, with its lower section located in the reactor core. The lower part of each needle, aligned with the reactor's midplane, contains a heating core that enables temperature regulation over a vertical span of approximately 40 cm. This heating core limits the experimental volume to a cylindrical region with a diameter of 9 mm and a length of 40 cm. From each needle emerge a metallic tube containing the fibers that carry the probe beam and the optical signal, as well as the electrical wires of the heating core and five thermocouple (TC) wires. The experimental zone within each needle is maintained in a helium atmosphere.

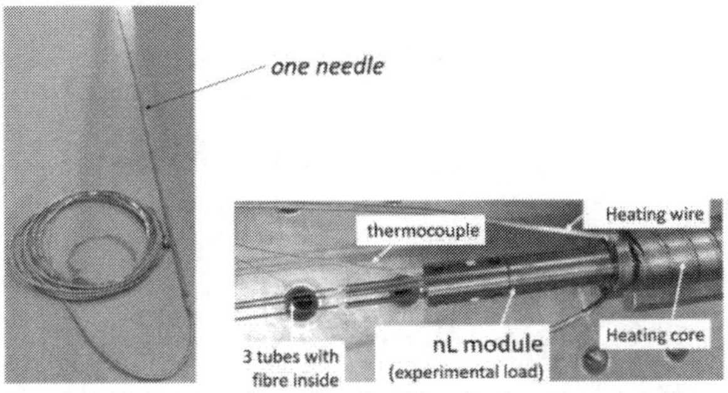

Figure 1: Left: a needle and a flexible tube housing relay fibers, thermocouples and heating wires; right: picture taken during insertion of a part of the experimental load in the heating core.

The irradiation period lasted 30 days. The fast neutron flux (E > 1 MeV) ranged from 5×10^{12} to 1×10^{13} n/cm²/s, depending on the specific needle and vertical positioning, resulting in a total fluence between 1×10^{19} and 2×10^{19} n/cm². The associated gamma heating was approximately 2 kGy/s, leading to a total dose of 5 GGy.

B. *Sensors and Tests Conducted*

We tested hardened Fabry-Perot (FP) extensometers (Figure 2, left), designed to operate at high temperatures (up to 400 °C), in MTRs. Previous work [2] described the sensor and its performance at high temperature without irradiation. Here the objective is to test the sensor response under high neutron and gamma irradiation, at high temperature, at constant stress and also to test the sensor response when change in stress is produced by deliberate temperature shifts generating thermal expansion of the support.

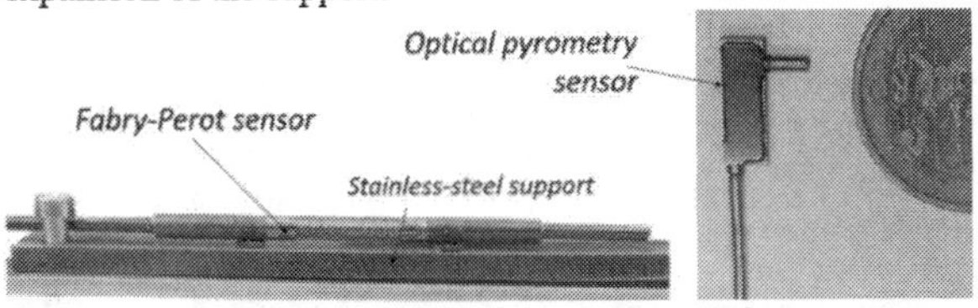

Figure 2: Left: FP sensor welded onto stainless steel support (gauge length: 15 mm); right: pyrometry sensor designed to collect IR emission at 90° to the fiber direction.

Miniature, non-contact, high-temperature optical pyrometers (Figure 2, right), developed for Loss Of Coolant Accident (LOCA) simulation experiments [3,4], were also tested. Three such sensors faced highly oxidized surfaces held at controlled temperature during most of the irradiation cycle, with some intentional temperature jumps introduced. In multispectral pyrometry, once calibration has been completed, it is important to minimize differential spectral variations in the collected signal, other than those caused by changes in the surface temperature, because they can be interpreted as temperature variations. The main objective here is to study the differential spectral variations of the received optical signal at constant temperature that may result from irradiation, particularly RIA in the optical fiber.

Additionally, tests were conducted as part of the development of a confocal chromatic displacement sensor, intended for in-core measurements such as fuel rod swelling [5]. To evaluate RIA in candidate optical glasses for this sensor, an assembly with seven miniature optical lines was constructed (Figure 3, top). Each line consisted of a radiation-hardened fiber whose end face was positioned at the center of curvature of a polished stainless steel concave mirror, with a small glass sample (1.4 × 1.4 × 3 mm³) placed in between. RIA was determined by the decrease in optical signal during irradiation, after subtracting attenuation in the fiber and losses from mirror reflection. These baseline losses were measured on a reference line without any glass sample.

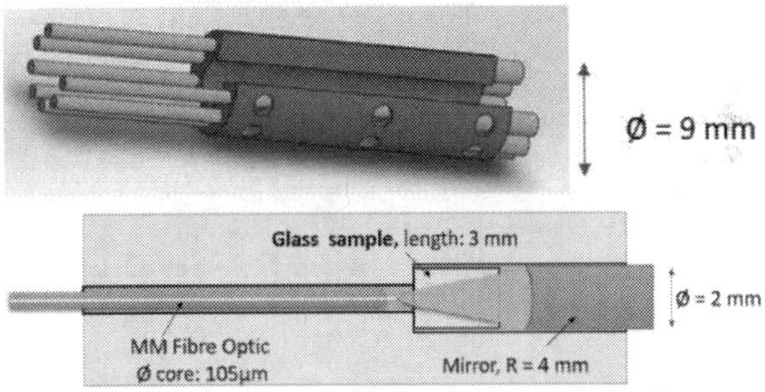

Figure 3: Top: assembly with 7 independent lines; bottom: diagram of a single line of RIA glass sample measurement.

C. *Temperature*

Since all tested sensors exhibit temperature sensitivity, maintaining thermal stability was essential for evaluating radiation effects on sensor response. Given the variability of nuclear heating, supplementary electrical power was required to regulate the temperature. Heating wires embedded in the core were connected to the BR2 facility's temperature control system. Five thermocouples were installed per needle, with one selected as the reference for active regulation.

Figure 4 illustrates the temperature evolution in one needle: needle A dedicated to Fabry-Perot sensors. The control thermocouple (TC1) successfully maintained the setpoint temperature, demonstrating the effectiveness of the regulation system. However, other thermocouples located further from the control point showed significant deviations and fluctuations during the irradiation cycle, highlighting temperature inhomogeneities.

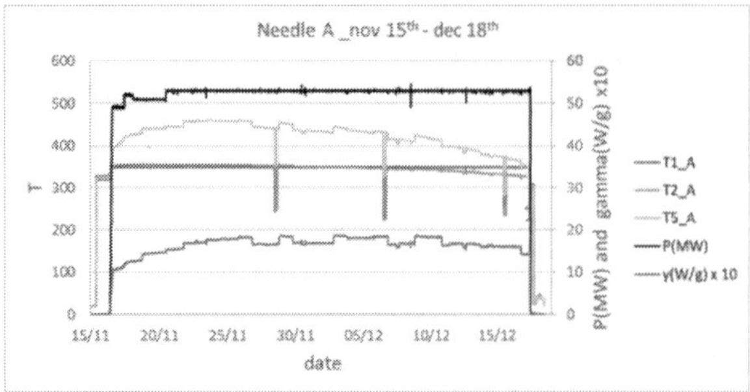

Figure 4: Temperatures at various points, gamma heating and reactor power on needle A dedicated to Fabry-Perot sensors. The heights of the TCs relative to the core median plane are respectively: +150 mm (TC1), +80 mm (TC2) and -130 mm (TC5).

III. Results and Analysis

A. Fabry-Perot Extensometer

Figure 5 presents the evolution of temperature and optical cavity length (Lc) of Fabry-Perot sensor no. 1 during the irradiation cycle. At constant temperature, variations in Lc attributable to the experiment—particularly to irradiation—remain below 1 µm. The associated fast neutron fluence (>1 MeV) reaches approximately 1.5 × 10^{19} n/cm^2 at the end of the irradiation cycle.

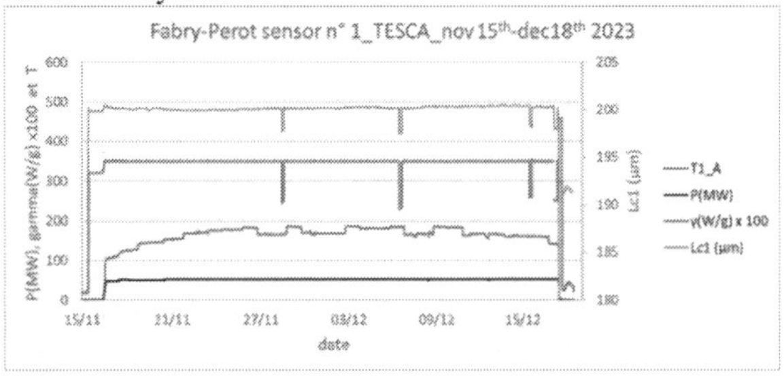

Figure 5: Optical cavity length (Lc) of Fabry-Perot sensor no. 1, temperature, gamma heating, and reactor power during the irradiation cycle.

The temperature sensitivity of the optical cavity length is due to the thermal expansion of both the sensor and the stainless-steel support [2]. After accounting for the sensor's own expansion, the device effectively measures the expansion of the support structure. It is therefore a validation of the capacity to measure small extensions that we aim for by making 3 temperature jumps of approximately -100 °C during irradiation. The measured cavity length changes are consistent with those recorded during post-manufacturing tests (+2.6 µm/100 °C), confirming the preservation of measurement accuracy.

Additionally, the signal level showed strong radiation resistance, with attenuation remaining below 3 dB throughout the cycle. The sensor was built with a single-mode (SM) radiation-hardened fiber made from a preform provided by Exail (formerly iXblue).

B. Pyrometric Sensor.

Figure 6 displays normalized spectra collected by a pyrometric sensor at several time points when the corresponding thermocouple indicated temperatures close to 500 °C (±2 °C). Spectral normalization facilitates comparison of the signal shape, in line with multispectral pyrometry principles. A variable peak at 1280 nm—attributable to oxygen fluorescence [6]—was observed. To reduce its impact on temperature measurements, the spectral exclusion window (1300–1430 nm, corresponding to the OH peak) [3,4] should be extended to include the 1280 nm region. Outside this range, differential attenuation appears negligible. The spectrum of the collected blackbody emission is preserved, as expected since the temperature is constant. This means that the measurement is not affected by the irradiation environment encountered in this experiment.

However, due to issues during assembly, the initial black-body-based calibration—which provided a temperature reference for the spectral data—was lost. Therefore, we are unable to associate specific temperatures with the recorded spectra, which complicates the precise and quantified assessment of the irradiation-induced error.

The same is true for temperature jumps of about 30°C; but we still observed variations in intensity, at a given wavelength, in agreement to within a degree of the evolution expected according to Planck's laws.

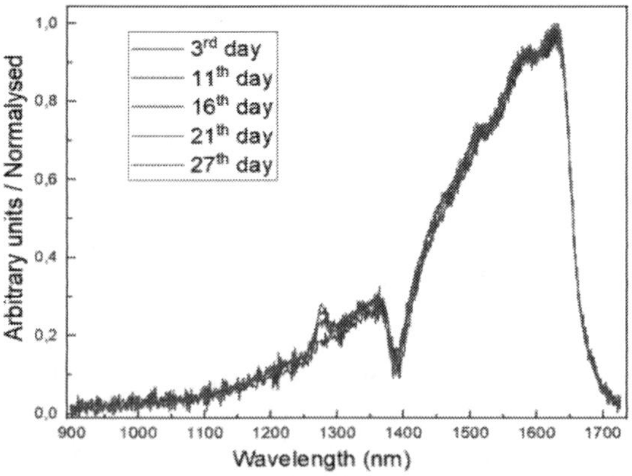

Figure 6: Normalized spectra from the pyrometric sensor at different time points. Fast neutron fluence ~10^{19} n/cm^2 for the last timepoint.

C. RIA of glass samples

Two main challenges were encountered: loss of measurement channels due to fiber breakage along the transport lines, and unexpected signal fluctuations caused by slight sample displacements within their housings. These displacements likely come from mechanical vibrations and unintended temperature changes. A slight defocusing considered to counter this effect was difficult to implement. This reduces the confidence in the signal intensity and, by extension, the derived Radiation-Induced Attenuation (RIA) values. Nevertheless, Figure 7 presents RIA data for a double-pass sapphire crystal (6 mm total length), obtained by subtracting the reference signal measured on a channel without a sample.

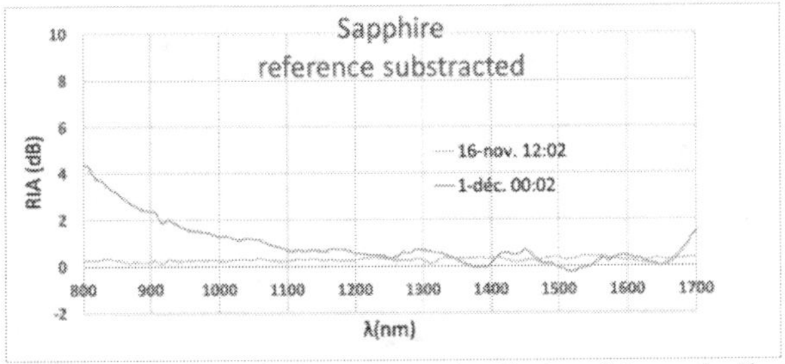

Figure 7: RIA(λ) at the beginning and midpoint of the cycle (fast neutron fluence ~0.8 10^{19} n/cm²) for the sapphire sample alone (double-pass, 6 mm length), derived by subtracting the RIA of the reference channel. Temperature during the period: between 110 and 270°C (average: ~ 150°C).

Although the absolute values remain uncertain, the spectral profile can be considered reliable, as it indicates stronger attenuation in the lower part of the spectrum—consistent with post-irradiation measurements performed later (to be presented in a forthcoming article). Similar increases in RIA at wavelengths below 1 µm have been reported in [7] for sapphire optical fibers irradiated at comparable temperatures but under lower fluences. Reference [8] also presents post-irradiation, thermally-annealed absorption data for sapphire samples, again obtained at lower neutron fluences. These differences in irradiation conditions complicate any direct comparison of RIA levels.

IV. Conclusions

We have presented the on-line testing of Fabry-Perot extensometers, miniature non-contact high-temperature optical pyrometric sensors, and radiation-induced transmission changes in optical glass samples during the TESCA irradiation experiment in the BR2 reactor at SCK•CEN. The experimental setup and test conditions were described. Among the various parameters, temperature control is often particularly critical for evaluating radiation effects on sensor responses. Achieving both homogeneous and stable thermal conditions in such environments remains a significant challenge.

Under fast neutron fluences (E > 1 MeV) of 1–2 × 10^{19} n/cm² and total gamma doses of approximately 5 GGy, the RIA observed in radiation-hardened fibers - manufactured by Exail (formerly iXblue) - was sufficiently low to ensure satisfactory sensor performance. Specifically, an RIA of ~3 dB is well within acceptable limits for interferometric measurements using the Fabry-Perot sensor. Furthermore, the low differential RIA observed in 400 µm core multimode fibers at temperatures around 500 °C had minimal impact on multispectral pyrometry. The effective double-pass fiber length exposed to ~70% of the maximum flux is estimated at about 50 cm.

The Fabry-Perot extensometer demonstrated excellent stability under high radiation, showing a drift of less than 1 µm under constant load and providing reliable measurements of thermal expansion. Pyrometric measurements also remained functional under intense irradiation, with an extension of the unusable spectral range to [1250–1430 nm].

However, the RIA measurement setup exhibited sensitivity to temperature fluctuations and mechanical vibrations, indicating that its mechanical robustness should be improved.

Finally, we performed on-line optical path measurements on glass samples, the results of which will be detailed in a forthcoming publication. These on-line observations will be complemented by post-irradiation characterizations of additional glass samples irradiated in dedicated containers and retrieved approximately one year after the TESCA experiment.

The Fabry–Perot extensometer and the pyrometric sensor are suitable for use in MTR irradiation devices, given their demonstrated resistance to radiation under the tested conditions.

Acknowledgment

This work was supported by the French Alternative Energies and Atomic Energy Commission (CEA), Division of Nuclear Energy. The authors thank the staff of SCK.CEN BR2 and in particular Andrei Goussarov for their technical assistance in the preparation and conduct of the irradiation test.

References

[1] Y. Deng and J. Jiang, "Optical Fiber Sensors in Extreme Temperature and Radiation Environments: A Review," in IEEE Sensors Journal, vol. 22, no. 14, pp. 13811-13834, 15 July15, 2022, doi: 10.1109/JSEN.2022.3181949.

[2] G. Cheymol, *et al.*, "High-Temperature Measurements With a Fabry–Perot Extensometer," in IEEE Transactions on Nuclear Science, vol. 67, no. 4, pp. 552-558, April 2020, doi: 10.1109/TNS.2020.2975707.

[3] B. Bouvry *et al.*, "Multispectral pyrometry for surface temperature measurement of oxidized Zircaloy claddings," Infrared Physics and Technology, vol. 83, pp. 78-87, 2017. https://doi.org/10.1016/j.infrared.2017.04.013

[4] A. Ladaci, *et al.*, "Measurement of reactor core temperature using multispectral infrared pyrometry in accidental conditions." EPJ Web of Conferences. Vol. 225. 2020. ANIMMA 2019. https://doi.org/10.1051/epjconf/202022508005.

[5] M. Agoyan, *et al.*, "Toward confocal chromatic sensing in nuclear reactors: In situ optical refractive index measurements of bulk glass." IEEE Transactions on Nuclear Science 69.4 (2022): 722-730.

[6] B Brichard *et al.*, « Fibre-optic gamma-flux monitoring in a fission reactor by means of Cerenkov radiation", 2007 Meas. Sci. Technol. 18 3257 DOI 10.1088/0957-0233/18/10/S32.

[7] C.M. Petrie *et al.*, "In situ reactor radiation-induced attenuation in sapphire optical fibers heated up to 1000 °C", Nuclear Instruments and Methods in Physics Research Section B: Beam Interactions with Materials and Atoms, Volume 342, 2015, Pages 91-97, ISSN 0168-583X, https://doi.org/10.1016/j.nimb.2014.09.014.

[8] B.W. Morgan *et al.*, "Optical Absorption of Fused Silica and Sapphire Exposed to Neutron and Gamma Radiation with Simultaneous Thermal Annealing", Journal of Nuclear Materials, Volume 570, November 2022, 153945. https://doi.org/10.1016/j.jnucmat.2022.153945.

Characterization of low output portable neutron generators operated at FNSPE CTU in Prague

Tomas Bily[1,*], Ondrej Huml[1], Ondrej Novak[1], Pavel Suk[1], Filip Fejt[1], Jan Frybort[1], Zhao-Ming Pan[1] and Jan Rataj[1]
[1]Czech Technical University in Prague, Faculty of Nuclear Sciences and Physical Engineering, Czech Republic
(*) tomas.bily@fjfi.cvut.cz

***Abstract*—Low output portable DD and DT neutron generators are used at Faculty of Nuclear Sciences and Physical Engineering of the Czech Technical University in Prague for variety of research and educational applications. The paper summarizes some acquired experience related to characterization of these devices. Specifically, it includes determination of the total neutron generator emission, dependence of neutron emission on operational parameters, and characterization of pulsed mode operation. Whereas for ca. 1E8 n/s DT neutron generator, the total neutron generator emission has been estimated via foil activation technique, for 7E6 DD neutron generator the manganese bath technique has been applied. Further, the pulsed mode operation has been analysed over wide range of frequencies utilizing in-house developed set-up based on Red Pitaya Stem-lab. The verification of the set-up for the purpose is provided as well as the discussion of using epithermal (cadmium covered thermal neutron gas-filled detector) and fast neutron detector (of diamond type) for the purpose. Finally, the discussion is provided on measurement of neutron output dependence on operational parameters, as well as on self-activation of the devices and characteristic X-ray produced during their operation.**

***Keywords* —portable neutron generators, manganese bath technique, pulsed neutron source characterisation.**

I. Introduction

Portable electronic neutron generators have found many applications in training, research and industry [1]. At CTU in Prague, they are used for studies related to nuclear material characterisation techniques [2], to drive a subcritical reactor [4], and study its reactivity [4], or its kinetic parameters [5], or as neutron irradiator for training purposes [6] and for training purposes.

Optimizing their use for applications is connected with understanding of their characteristics which depends on device design. As they are electronic devices it is worth to check their parameters by measurements. The parameters to be verified typically include the total emission rate (and eventually its space profile, or flux at given irradiation position), dependence of emission rate on the set-up of high voltage and beam current, characterisation of pulse mode operation, i.e. of time profile of neutron emission during neutron pulses and stability of pulsing frequency, and as well the byproduct radiation caused, i.e. x-rays and self-activation.

II. Materials and Methods

A. Portable neutron generators at CTU in Prague

At CTU in Prague, two portable neutron generators from Thermofisher Scientific are used: the DD neutron generator model P-385 [7] and DT neutron generator of model MP320 [8]. Both devices are rugged, portable electronic devices designed for laboratory or field applications.

The systems are comprised of two main components: the accelerator assembly and the electronics enclosure. These components are connected to each other via a set of cables that are used to send control and monitoring signals. The accelerator assembly and the electronics enclosure are usually mounted separately; however, they may be physically connected using a mounting plate or other hardware, if desired. A warning lamp and HVPS Disable box must also be connected to the system for the system to operate. The High Voltage Power Supply (HVPS) used to accelerate ions is integrated into the accelerator assembly. The system can be completely controlled using a standard RS 232 serial communications interface with a simple terminal program or with the supplied LabWindows Graphical User Interface (GUI) program. Technical parameters of both devices are summarized in table I.

TABLE I
BASIC TECHNICAL DATA OF USED NEUTRON GENERATORS [7, 8]

Parameter	DD model P385	DT model MP320
Neutron yield (n/s)	7E6	1E8
Neutron energy (MeV)	2.4	14.1
Max. accelerator voltage (kV)	130	90
Max. beam current (μA)	70	70
Frequency range	250 Hz to 20 kHz, continuous	250 Hz to 20 kHz, continuous
Duty Cycle	5% to 100%, 5 μs pulse width minimum	5% to 100%, 5 μs pulse width minimum
Tube length/diameter (mm)	700 /101.6	558.8/120.6

B. Manganese bath and foil activation techniques

For very low output portable neutron generator, it seems reasonable to select different approaches for DD and DT tubes to determine the total neutron output. For DT generators the combination of higher neutron production and 14 MeV neutrons enabling use foil activation technique utilizing properly selected threshold reactions. Contrary, 2.45 MeV neutrons from DD generators has not high enough energy to utilize threshold reactions (115In(n,n')115m might be one of very few candidates) and together with significantly lower neutron output it would be doubtful to get reasonable precision even with thick activation foils. Thus, two experimental methods were used to measure the intensity of neutron production by the neutron generator: foil activation method for DT neutron generators and manganese bath method for DD neutron generator. Both methods are based on the similar principle, with the help of a suitable gamma spectrometric system, the activity, respectively neutron reaction rate of the material irradiated by the investigated neutron source is determined. Contrary to utilizing threshold reactions, the manganese bath technique is based on moderation and absorption of vast majority of neutrons emitted from the source and measurement of volumetric activity of the bath. For both methods the irradiated materials were analysed in the NAA laboratory equipped with the semiconductor HPGe (high purity germanium) detector utilised for gamma spectrometric measurements. It is Canberra-Packard HPGe detector (model GC2518) with relative detection efficiency of 25 % and energy resolution of 1.8 keV. Analysis of gamma spectra was performed using Genie 2000 Spectroscopy Software. The matters of absolute detection efficiency ε_γ calibration for relevant source-to-detector distances were performed earlier [9, 10], for thick foils the Labsocs SW was use for the purpose [11]. The reaction rate (RR) which equals to saturated activity can be obtained using the equation ([7]):

$$RR = \frac{S_\gamma \lambda^{t_{real}/t_{live}}}{N_0 \varepsilon_\gamma I_\gamma (1-e^{-\lambda t_{irr}}) e^{-\lambda t_{cool}} (1-e^{-\lambda t_{irr}})}, \quad (1)$$

where S_γ is the area under the full energy peak (FEP) provided by gamma spectrometric system, λ the decay constant of radionuclide, ε_γ the absolute detection efficiency for the FEP, I_γ the intensity per decay, t_{irr} and t_{cool} the irradiation and cooling time, t_{real} and t_{live} the real and live time of the spectrometric measurement and both times are also provided by gamma spectrometric system.

In foil activation technique, the neutron generator emission rate was determined through activation foil saturated activity and comparison with expected activity at given position as calculated by the MCNP6 code [12] and taking into account the spatial distribution of neutron emission based on [13]. Three foil materials were used Al, Nb and Cu. The foils were put to the distance of 7.5 cm from outer tube of neutron generator at 90° from target plane. During measurement, foils were put directly on HPGe surface. Tab. II shows the reaction channels used as well as the products half-lives and energies and intensities of measured gamma lines. Fig. 1 shows the cross sections for these reactions.

Obviously, the most desirable would be reactions with threshold rather closely below 14 MeV to avoid contributions from scattered neutrons, with high cross section and reasonable half-live to get enough counts at the detector, and flat profile of cross-section around 14 MeV. Many candidates have been tested for the purpose, see e.g. [14]. In this work we used Al-27 having flat cross section around 14 MeV and short half-live, copper foil which is considered to be kind of standard for the purpose [15, 16, 17], and Nb-93 with high threshold. Foils had diameter of 25 mm, and thickness of 1 mm. The aluminium foil had weight of 1.38 g and purity of 99.999 %, copper foil weighted 4.4 g with purity of 99.9%, and niobium foil weighted 4.21 g and had purity of 99.9%.

TABLE II
PARAMETERS OF USED REACTIONS

Reactions	Product half-live (s)	E_γ (keV)	Intensity %
Al27(n,α)Na24	53989	843.7	71.8
Al27(n,p)Mg27	567	1368	100
Nb93(n,2n)Nb, m92	10.15 days	934.4	99.07
Cu65(n,2n)Cu64	45724	511(annih.)	35.2

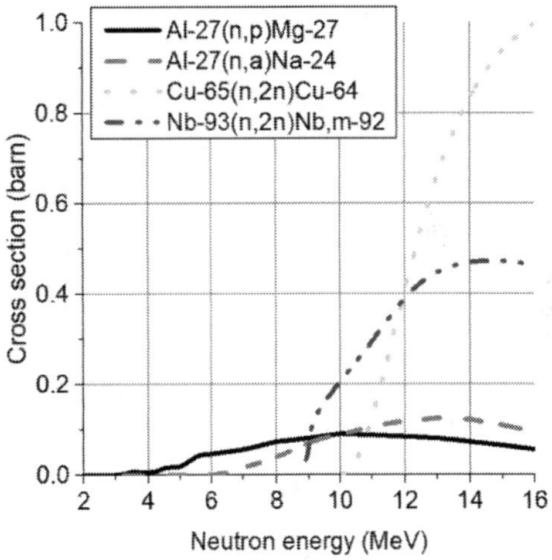

Fig. 1. Reaction cross sections used for DT generator neutron emission rate determination

For the estimation of neutron emission rate we used the determination of saturated activity of the foils based on measurement, calculation of the foil saturated activity per one source neutron in MCNP with the assumption of isotropic emission and neglecting generator internals and surrounding material, and the difference in neutron flux at 90° between models based on isotropic neutron emission and that taking into account generator internals as published in [13] (the difference for the given position reaches ca 20 %).

The manganese bath technique is a proven technique used usually for calibrating radionuclide neutron sources, e.g. [18, 19, 20], although to the knowledge of the authors its use for determination of neutron generator emission rate has not been published. It is based on the determination of the activity of Mn-56 produced by the irradiation of manganese sulphate with neutrons emitted from calibrated source. The resulting emission rate of the source is determined from the manganese activity and the efficiency of the manganese bath. The manganese bath utilized for the experiment consists of a spherical plexiglass tank filled with an aqueous solution of $MnSO_4$. The aqueous solution of $MnSO_4$ has the density 1.21 g/cm^3 and the concentration of manganese sulphate 0.215 g per gram of aqueous solution. The total volume of the bath is 163.3 litres. The bath (see Fig. 2) was designed as a simple device; It consists of two plexiglass hemispheres, which are water-tightly connected by screws and silicone glue. The sources are manually introduced into the centre of the bath using a dry plexiglass channels. The channels are inserted through the neck at the top of the bath. The channels are optional according to the neutron source size, so that it is placed in the centre of the bath. The bath can be moved using the transport trolley on which is mounted. Mixing of the activated solution is performed by bubbling compressed air through the tank. The dump valve located at the bottom part of the tank serves for sampling of the activated solution. Activated solution is taken into the Marinelli bakers.

The neutron generator tube was inserted into the bath using a special dry channel, so the position of the generator target corresponds to the centre of the manganese bath. The generator produced neutrons inside the bath for more than 21.5 hours. Substantial number of the source neutrons were captured by manganese nuclei forming Mn-56, which β decays to Fe-56 with a half-life of 2.579 hours:

$$Mn55 + n \rightarrow 56Mn \rightarrow 56Fe + \gamma + \beta - , \qquad (2)$$

The energy of gamma rays emitted during the decay process of Mn-56 ranges from 846.77 keV to 3369.60 keV. The activated solution of $MnSO_4$ was taken into two Marinelli bakers after the source switching off.

The activity of the samples was determined based on the measurement by HPGe gamma spectrometric system using the equation 1. The analysis of the irradiated samples was done for three energies of gamma radiation with the highest intensity I_γ these are 846.771 keV (I = 98.9 %), 1810.772 keV (I = 27.2 %) and 2113.123 keV (I = 14.3 %). The resulting activity A_{bath} was determined as a weighted average of the activities corresponding to the observed gamma-ray energies:

$$A_{bath} = \frac{\sum_{i=1}^{n} \frac{A_i}{s_i^2}}{\sum_{i=1}^{n} \frac{1}{s_i^2}}, \qquad (3)$$

where A_i is the activity determined for the gamma-ray energy E_γ^i and s_i is corresponding standard deviation. The neutron emission rate B (s^{-1}) was obtained from the activity A_{bath} and bath efficiency ε_{bath} using the equation:

$$B = \frac{A_{bath}}{\varepsilon_{bath}}, \qquad (4)$$

ε_{bath} reflects neutron losses by absorption (except for manganese) and by leakage from the bath. It can be described by equation (ref):

$$\varepsilon_{bath} = f\,(1 - O - S - L)\,, \qquad (5)$$

The coefficient f is the fraction of thermal neutrons captured by manganese as opposed to those captured by other nuclei. The coefficient O represents the fraction of neutrons lost by (n,α) reactions in oxygen, and (n, α) and (n,p) reactions in sulphur. The coefficient S represents the fraction of source neutrons absorbing by the construction materials surrounding the source (stainless steel capsule, dry channel). The coefficient L describes the fraction of source neutrons escaping the bath. The efficiency of the bath was determined based on the simulation using MCNP code.

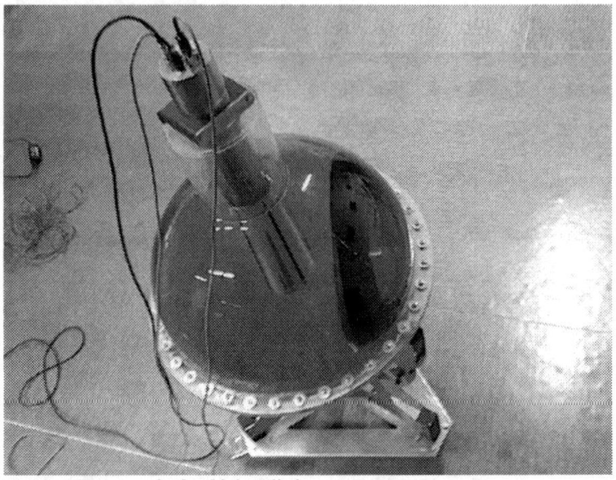

Fig. 2. Manganese bath with installed neutron generator

C. Measurement of dependence on operational parameters

Measurement have been performed for DD neutron generator with fast neutron detector (ZnS(Ag) encapsulated in resin, i.e. using proton recoil reaction) and DA310 analyser. The detector was operated in pulsed mode and have been located 10 cm from the source at 90° from deuteron beam at target plane. Dependence on beam current and high voltage has been measured in the range from 20 μA to 70 μA, and from 60 kV to 130 kV respectively. The time of measurement has been adjusted to reach relative statistical uncertainty of all values at 1%.

D. Characterisation of pulse mode operation

For characterisation of neutron generators operation in pulsed mode Red Pitaya – STEMlab device with in-house SW that evaluates TTL signal from NG ion source is used, either together with output of neutron detector, or standalone to characterize stability of pulse frequency. Diamond detectors have been used for characterisation of whole pulsing range of DT neutron generator [21]. Alternatively, low frequency range (250 Hz and lower) can be also characterized by cadmium covered thermal neutron detector, typically gas-filled one [22].

E. Self-activation, x-ray production

Self-activation and x-ray production of neutron generators have been evaluated using portable HPGe detector. Dose rates after operation have been determined using Rados RDS-120 radiation survey meter.

III. RESULTS

A. *Foil activation technique to determine the emission rate*

The measurement was performed at maximal allowable parameters of the tube, i.e. 90 kV and 70 μA. Parameters and results of the foil activation measurement are summarized in Table III.

TABLE III
PARAMETRES AND RESULT OF THE FOIL ACTIVATION MEASUREMENT (UNCERTAINTIES IN BRACKETS)

	Al27 (n,α)	Al27 (n,p)	Al27 (n,α)	Al27 (n,p)	Nb93 (n,2n)	Cu65 (n,2n)
Activation time (s)	3600	3600	3600	3600	3600	3600
Cooling time (s)	300	300	300	300	300	7588
Measurement time (s)	4417	4417	75610	75610	487.8k	176.7k
Net peak area (counts)	2295 (50)	4824 (70)	24942 (160)	4882 (100)	63275 (252)	4882 (100)
Saturated activity (Bq)	462 (21)	306 (16)	445 (18)	308 (17)	1161 (51)	308 (17)
Estimated generator emission rate (n/s)	1.62E8	1.65E8	1.56E8	1.66E8	1.42E8	1.64E8

Regarding uncertainties, the uncertainty of saturated activity determination reaches 5 % (mostly caused by uncertainty in efficiency calibration), calculation uncertainty is about 1 %, uncertainty in determination ratio between neutron emission from isotropic model and model with neutron generator internals taken from [13] was estimated to 10 %. Another 10 % uncertainty is estimated to be connected with foil positioning. Thus, the total uncertainty of neutron generator emissions was estimated to be 15 % at 1 σ. However, it should be noted that the estimation does not include a possible systematic error caused scattering on surrounding materials which was not evaluated (the nearest shielding wall was ca 30 cm from the foil). The value corresponds with the results presented in [13] where for the same type of generator the neutron emission was determined for 90 kV and 60 μA to be ca. 1.56E8 n/s, which assuming linear dependence on beam current would give some 1.82E8 n/s at the parameters used by us (90 kV, 70 μA). Obviously, the two results cannot be easily compared due to lack of data on previous operation, tritium decrease due to radioactive decay, or production uncertainty in the targets. The producer certificate gives emission rate of 9.02E7 n/s for settings of 80 kV and 60 μA which should correspond to ca. 1.49E8 n/s at 90 kV and 70 μA. Finally, the measurement has been performed at the 5 year old tube, so that tritium decay should be considered. Authors do not have study related to their tube, however based on the study [23] performed at different model of DT generator from the same producer, the neutron yield can be considered stable for the first ca 10 years and in the first 4.5 years it can even slightly rise. Taking this into account together with uncertainty of producer check-out record of 25 % and our uncertainty, the measured values correspond with the expectations.

B. *Manganese bath technique to determine the emission rate*

The results of the experiment are summarized in Table IV. The table contains the determined activity of the manganese bath and the corresponding emission rate, including a comparison with the reference value provided by manufacturer. These values were determined for both collected samples of activated manganese sulphate solution. The resulting values, determined by a weighted average, are given in the last row of the Table IV.

TABLE IV
RESULTS OF THE MEASUREMENT BY MANGANESE BATH METHOD

Sample	A_{Bath} (Bq)	s (%)	B (s-1)	s (%)	1 – C/E (%)
1	1.92E+06	0.44	7.40E+06	0.44	-5.73
2	1.86E+06	0.44	7.20E+06	0.44	-2.79
Average	1.89E+06	0.31	7.30E+06	0.31	-4.23

C. *Measurement of dependence on operational parameters of DD generator*

The results are shown in the fig. 3 and 4 and in the normalized form in the fig. 5 and 6. Values were normalized to the minima of each row. Obviously, the rise of neutron emission to 90° with high voltage is slightly faster for higher values of beam current. The rise with beam current is slightly higher for lower values of high voltage. The dependences were fitted with an appropriate function: linear for dependence on beam current and power for the case of high voltages. The obtained fit parameters are summarized in tables V and VI.

TABLE V
FIT PARAMETRES OF NEUTRON EMISSION TO 90° FOR DEPENDANCE ON HIGH VOLTAGE ($C * HV^D$)

Beam current (μA)	c (kV^{-1})	d (-)
20	4.90E-05	2.72
25	4.91E-05	2.78
30	4.86E-05	2.83
35	4.79E-05	2.87
40	4.52E-05	2.91
45	4.03E-05	2.96
50	3.99E-05	2.99
55	4.14E-05	3
60	4.57E-05	3
65	4.22E-05	3.03
70	3.92E-05	3.06

TABLE VI
FIT PARAMETRES OF NEUTRON EMISSION TO 90° FOR DEPENDANCE ON BEAM CURRENT (A * BC + B)

High voltage (kV)	a (μA^{-1})	b (-)
60	0,14	0,49
70	0,25	0,39
80	0,4	0,24
90	0,58	-0,51
100	0,82	-2,01
110	1,05	-2,57
120	1,34	-4,51
130	1,71	-8,05

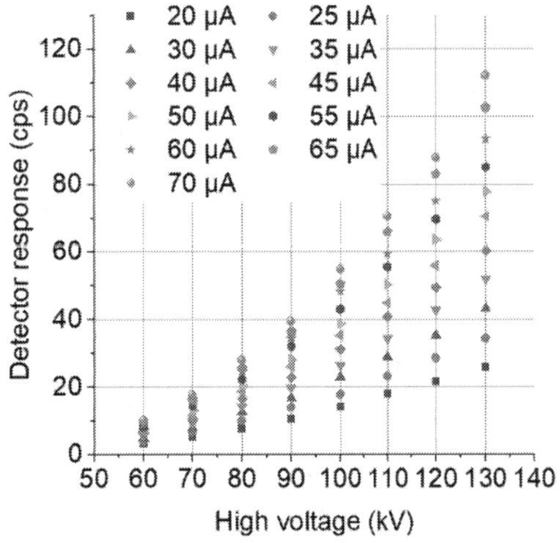

Fig. 3. Dependence of neutron output at 90°on high voltage

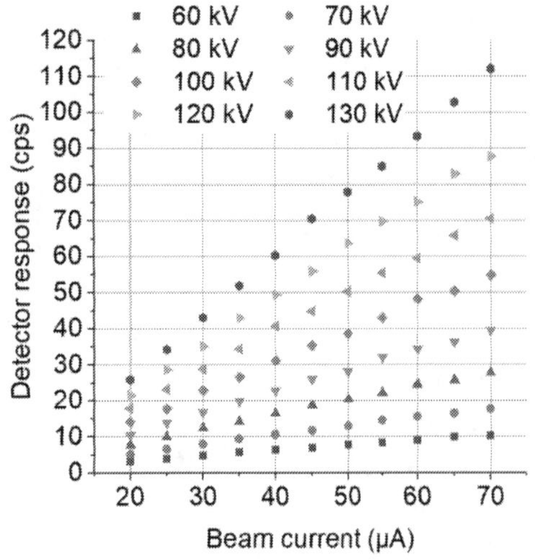

Fig. 4. Dependence of neutron output at 90°on beam current

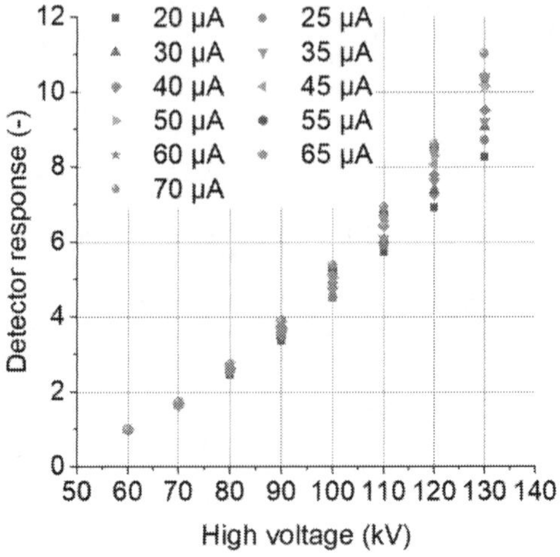

Fig. 5. Normalized dependence of neutron output at 90° on high voltage

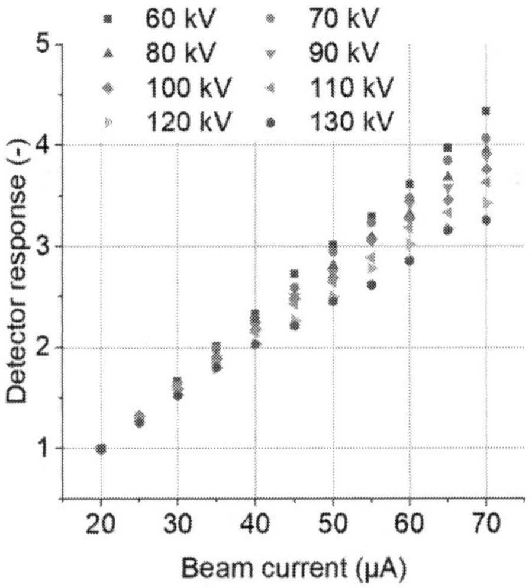

Fig. 6. Normalized dependence of neutron output at 90° on beam current

D. Characterisation of pulse mode operation

Typical result for neutron emission during pulsed mode operation for several frequencies of DT MP320 neutron generator measured via diamond detector for set of frequencies and fixed duty cycle, HV and BC settings is shown in the fig. 7. Red curve indicates idealized rectangular pulse for set pulse parameters; the black is measured one.

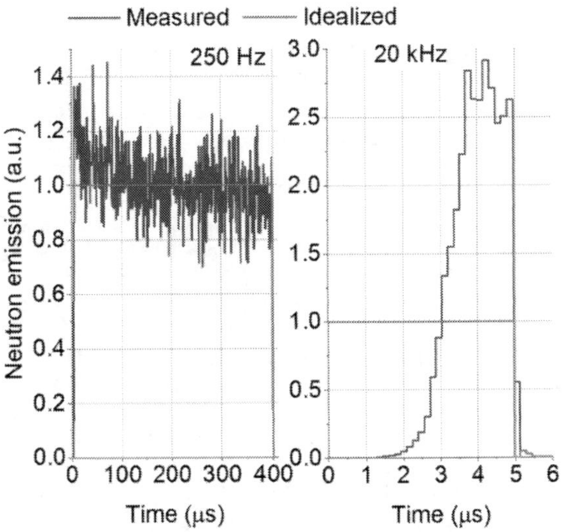

Fig. 7. Typical neutron pulse emission for D-T neutron generator at nominal HV and BC and frequency of 250 Hz and 20 kHz. Adapted from [21].

Typical results of frequency stability measurement are given in fig. 8 and 9 for 100 Hz and 20 kHz respectively. E.g. for 100 Hz operation in 99.62 % of cases the next TTL signal came after 10 000 µs ± 1 µs, however in few cases significant deteriorations has been observed. For 20 kHz operation in 99.39 % cases the next TTL signal came after 50 µs ± 1 µs, however in few cases significant deteriorations has been observed.

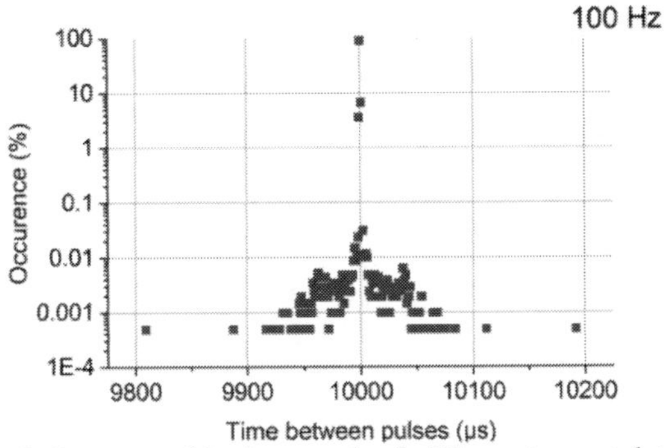

Fig. 8. Occurrences of time between pulses for DT generator operated at 100 Hz and DF 10 %

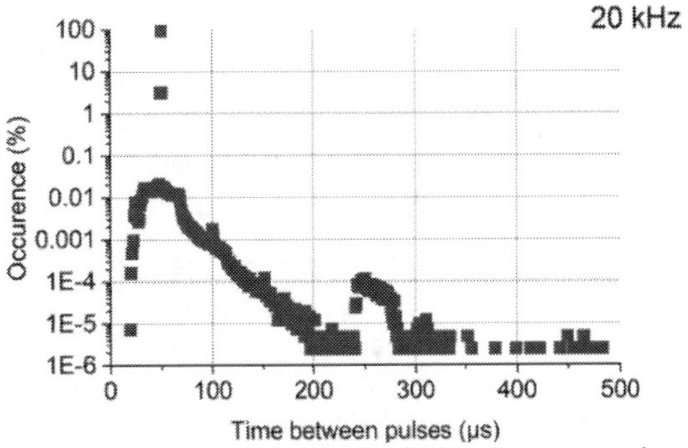

Fig. 9. Occurrences of time between pulses for DT generator operated at 20 kHz and DF 10 %

E. Self-activation, x-ray production

Whereas for ca. 7E6 DD neutron generator the self-activation was found negligible, for ca. 1E8 DT generator weak activation is measurable. E.g. after 1-hour long irradiation, the dose rate at 30 cm from the tube reaches ca. 5 µSv/h at 5 minutes after stop, and ca. 1.5 µSv/h at 30 minutes after stop. Visible gamma lines originates mainly from aluminium and iron activation corresponding to Al-27(n,g), Al-27(n,p) , Al-27(n,α), and Fe-56(n,p) reactions. Although the capture reactions may significantly depend on neutron generator environment, the dose rate results correspond to information provided by producer [24].

During neutron generator operation X-rays of about 80 keV are produced which may need to be handled for some measurements and also it need to be taken into account in radioprotection considerations. However, it can be easily shielded by few mm of lead (see Fig. 10). The results were consistent with findings of [25].

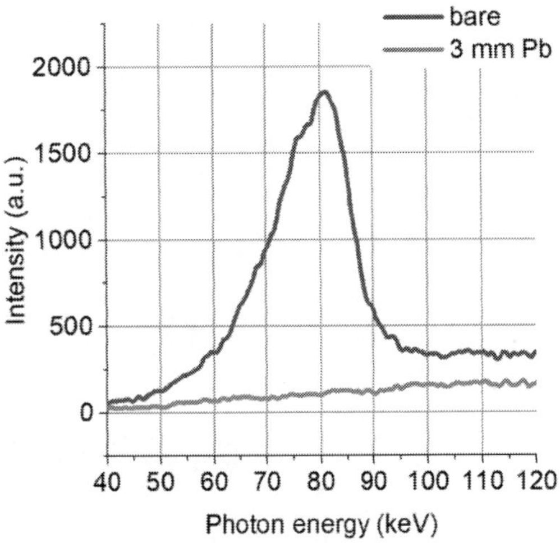

Fig. 10. X-ray peak for bare MP320 DT neutron generator operated at 20 µA and 80 kV and its suppression by 3 mm of lead.

IV. Conclusions

The detailed understanding of neutron generator behaviour under specific operating conditions enhances their applicability both for training and research purposes and makes results of any application more reliable. The paper summarizes techniques used for characterisation of DD and DT portable neutron generators used at CTU in Prague. To measure neutron generator emission rate, foil-based technique and manganese bath technique were used. Foil-based technique is well established for DT neutron generators; however, its application requires knowledge on spatial distribution of outgoing neutrons. Contrary, the paper shows good applicability of using the manganese bath technique for neutron emission rate determination for low output DD generators. In fact, the applicability of manganese bath technique is not limited to generators of DD type, but the limitation lies rather in the tube geometry (e.g. due to geometry of cable inputs it cannot be used for MP 320 type generator). Further, paper shows typical results of dependence of neutron output on operational parameters (high voltage and beam current set-up). Next, it shows the approach to characterize pulsed mode operation, i.e. the time profile of neutron emission during pulsed mode operation and stability of pulsing frequency. Finally, self-activation and x-ray production was measured and found consistent with producer-provided data.

Acknowledgment

The work was supported by the project Large Research Infrastructures of the Ministry of Education, Youth and Sports of the Czech Republic LM2023073.

References

[1] IAEA "Neutron generators for analytical purposes." IAEA radiation technology reports series, ISSN (2012): 2225-8833
[2] Bílý, T.; Huml, O.; Belko, B. "Delayed neutron counting of uranium pins using moderated neutrons from portable DD neutron generator in high frequency mode", Applied Radiation and Isotopes. 2023, 191 ISSN 0969-8043.
[3] J. Rataj, F. Fejt, J. Frybort,et al., "The project of vr-2 subcritical assembly," NUCLEAR ENGINEERING AND DESIGN, vol. 386, JAN 2022.
[4] Herčík, M.; Rataj, J.; Huml, O. "Fejt, F. Reactivity measurement using pulsed neutron source methods on VR-1 and VR-2 reactors and their verification by SERPENT code", Annals of Nuclear Energy. 2025, 219 1-16. ISSN 1873-2100.
[5] T. Bily, J. Rataj, O. Huml, et al, "Effect of kinetics parameters on transients calculations in external source driven subcritical vr-1 reactor," Annals of Nuclear Energy, vol. 123, pp. 97 - 109, 2019.
[6] J. Rataj, P. Suk, T. Bily, et al., "Characterisation of neutron field in the polyethylene neutron irradiator," APPLIED RADIATION AND ISOTOPES, vol. 168, FEB 2021.
[7] Thermo Fisher Scientific Inc., „MP320 Neutron Generator Operation Manual", Manual P/N 120004, 2010
[8] Thermo Fisher Scientific Inc., "P 385 Neutron Generator Operation Manual", Tech. Rep. Manual P/N 120006-A 062907, 2010.
[9] M. Stefanik, L. Sklenka, O. Huml, et al., "Activation analysis of tibetan coins and thermal neutron flux measurement at the vr-1 training reactor," Radiation Physics and Chemistry, vol. 155, pp. 304 - 309, 2019. IRRMA-10.
[10] J. Rataj, T. Bily, O. Huml, et al., „Verification of the mcnp model of manganese bath," in 2020 21st International Scientific Conference on Electric Power Engineering (EPE), pp. 1-5, 2020.
[11] F. Bronson "Validation of the accuracy of the LabSOCS software for mathematical efficiency calibration of Ge detectors for typical laboratory samples." Journal of Radioanalytical and Nuclear Chemistry 255.1 (2003): 137-141.
[12] J. A. Kulesza, et al. "MCNP® code version 6.3. 0 theory & user manual". No. LA-UR-22-30006. Los Alamos National Laboratory (LANL), Los Alamos, NM (United States), 2022.
[13] R. Remetti, L. Lepore, N.Cherubini, "Development and experimental validation of a monte carlo modeling of the neutron emission from a dt generator." Nuclear Instruments and Methods in Physics Research Section A: Accelerators, Spectrometers, Detectors and Associated Equipment 842 (2017): 7-13.
[14] D. Lee, et al., "Determination of the Deuterium-Tritium (DT) Generator Neutron Flux using Multi-foil Neutron Activation Analysis Method". No. INL/EXT-19-53812-Rev000. Idaho National Lab.(INL), Idaho Falls, ID (United States), 2019
[15] R. L. Heath, "The texas convention on the measurement of 14-MeV neutron fluxes from accelerator," in Proc. Int. Conf. Modern Trends Activation Anal., College Station, TX, USA, Apr. 1965
[16] International Union of Pure and Applied Chemistry, "Recommended procedure for the measurement of 14-MeV neutron fluxes from accelerators for activation analysis," Pure Appl. Chem., vol. 49, pp. 399–401, 1977.
[17] R. Uhlář, et al. "Modified Texas convention method for fast neutron flux measurements." In IEEE Transactions on Nuclear Science 67.1 (2019): 382-388.
[18] N. J. Roberts, N. N. Moiseev, M. Kralik. "Radionuclide neutron source characterization techniques." Metrologia 48.6 (2011): S239.
[19] N. J. Roberts, et al. "International key comparison of measurements of neutron source emission rate (1999–2005): CCRI (III)-K9. AmBe." In Metrologia 48.S 251 (2011): 06018.
[20] J. Lüley, et al. "On uncertainties in the manganese sulphate bath techniques." In Radiation Protection Dosimetry 198.9-11 (2022): 634-640.
[21] T. Bily, O. Novak, O. Huml, Pan, Z.-M., "Characterisation of pulse-mode operation of D-T portable neutron generator by diamond detector", in Fusion Engineering and Design. 2024, 208.
[22] T. Bily, O. Huml. "Characterization of neutron emission during pulse mode of low output electronic neutron generator," in EPJ Web of Conferences. Vol. 253. EDP Sciences, 2021.
[23] J. Jeon, R. J. Goldston, E. P. Gilson, Neutron Yield of Thermo Scientific P385 DT Neutron Generator vs. Current and Voltage. IEEE Transactions on Nuclear Science, 2025.
[24] Thermofisher Scientific, "Shielding considerations as applied to fast neutron generator applications", Technical Bulletin 104, 2020
[25] C. J. Wharton, et al. "X-ray measurements of a thermo scientific P385 DD neutron generator." in AIP Conference Proceedings. Vol. 1336. No. 1. American Institute of Physics, 2011.

Oscillation experiment design at the VENUS-F fast reactor

Federico Di Croce[1,2,*], Federico Grimaldi[1,2], Antonin Krása[1], Pierre-Etienne Labeau[2], Guido Vittiglio[1], and Jan Wagemans[1]
[1]SCK CEN Belgian Nuclear Research Centre, Boeretang 200, Mol, 2400, Antwerp, Belgium
[2] Université libre de Bruxelles, Avenue Franklin D. Roosevelt 50, Brussels, 1050, Belgium
(*) federico.di.croce@sckcen.be

***Abstract*— Improvement of nuclear data for fast and epithermal neutron interactions is crucial to the deployment of advanced nuclear reactor technologies. Integral experiments in VENUS-F, the zero-power fast spectrum reactor operated at SCK CEN, can play a major role in addressing this challenge. With this perspective, sample reactivity worth experiments applying the oscillation technique are being implemented in VENUS-F. In this work, we present the design of such experiments and the developed processing tools. This work precedes the analysis of actual oscillation experiments of a test B_4C sample. Firstly, Monte Carlo simulations are performed to calculate the reactivity worth of the sample and its local perturbation effect. Then, synthetic experimental output signals are produced and processed to showcase the analysis procedure. The oscillation parameters are varied to assess their impact on the measurement accuracy and identify the optimal experimental configuration. Moreover, the developed processing routines are validated for the synthetic cases and compared. This work outlines the experiment design process, it demonstrates the reliability of the developed routines, and underlines their limitation under certain oscillation conditions.**

***Keywords* — VENUS-F, sample reactivity worth, oscillation experiments, Fast reactor, Inverse point kinetics**

I. INTRODUCTION

INTEGRAL experiments in zero-power reactors are used to support the design of advanced reactor technologies [1]. Work still needs to be done to improve the nuclear data of key materials in the fast and epithermal energy range, which are the largest uncertainty sources in advanced reactors neutronic simulations [2]. To address this challenge, integral experiments are envisaged at VENUS-F [3], the zero-power fast spectrum research reactor operated at SCK CEN. They include sample reactivity worth experiments, which allow to validate nuclear data by measuring the reactor reactivity response to the insertion of a sample. The measured response is highly sensitive to the sample properties and its agreement with Monte Carlo calculations directly reflects the quality of the sample nuclear data.

Among the available experimental techniques, pile-oscillation is a very powerful method to reduce the experimental uncertainty when dealing with small samples [4]. This technique consists of the repeated insertion and extraction of a sample in the reactor generating a periodic perturbation.

This work describes the possible settings for oscillation experiments at VENUS-F and the developed processing routines. The oscillation experiment design of a reference test B_4C sample is considered. First, the perturbation induced by the sample is modeled. This is used to produce synthetic experimental data mimicking the experiment outputs. Trapezoidal reactivity oscillations are assumed and the wave parameters are varied to investigate their impact on the experimental results. To do so, the synthetic data are processed with the implemented processing routines. The ultimate goal of this work is to describe the pipeline developed for oscillation experiments design and analysis, highlighting how data processing and oscillation settings influence the precision and accuracy of the measured sample reactivity worth.

II. VENUS-F SET-UP FOR OSCILLATION

VENUS-F is the zero-power fast spectrum research reactor operated at SCK CEN since 2011. It operates at ambient temperature and it has a maximum nominal power of 500 W. The core is a 12x12 square lattice, featuring fuel assemblies as well as reflector and absorber assemblies. The fuel is metallic uranium (enrichment: 30% w.t.) and there are plates and rodlets of lead and bismuth as coolant simulator. The reactor has an active core height of 60 cm and it features lead reflectors below, above and around the active core. The extensive description of the current critical core configuration is found in [5].

Oscillation experiments are performed in this configuration. The sample is oscillated in one of the experimental assemblies in the core center from the core midplane to the reactor top. The chosen sample is a stack of B_4C pellets of well-known composition and largely available at SCK CEN. The net negative reactivity effect is largely determined by the ^{10}B (n,α) reactions, which has an accurately known cross section.

The measurement system consists of ^{235}U fission chambers placed in the core periphery to monitor the neutron flux amplitude evolution during the experiment. They provide count rate ranging from 10000 to 1000000 c/s count rate depending their mass and the neutron flux spectrum at the detector location.

The test experiments employ one of the available control rod motors to move the sample from the top (OUT) to the midplane

(IN) of the reactor. It allows for trapezoidal movements with variable stable time in IN/OUT position (t_{dwell}). It has a maximum speed of 3 cm/s, limiting the insertion/extraction velocity (t_{trans}) to 27 s. The motor constraints are used to define feasible reactivity shapes for the synthetic signals.

III. COMPUTATIONAL METHODS

This section introduces the computational methods for reactor response calculation with Monte Carlo and for oscillation experiment simulation with point kinetics.

A. Monte Carlo simulations

The VENUS-F configuration in which oscillation experiments are performed is modeled in the Serpent2 Monte Carlo code [6]. Model verification is presented in [5].

The difference of two static simulations with the sample IN (2) and OUT (1) provides the perturbation of the neutron flux due to the sample and the eigenvalue ($1/k_{\text{eff}}$) difference of the two states is used to estimate the sample reactivity worth $\Delta\rho$ as

$$\Delta\rho = \frac{k_{\text{eff,2}} - k_{\text{eff,1}}}{k_{\text{eff,2}}\, k_{\text{eff,1}}}. \quad (1)$$

Despite being accurate, the difference of close eigenvalues is slow to converge when precision higher than 1 pcm is desired. Perturbative techniques [7] offer a valid alternative, determining reactivity variation as function of the perturbation in the transport equation operators. The exact perturbation theory formalism yields

$$\Delta\rho = \frac{\left\langle\varphi_1^+\middle|\delta\hat{H}\varphi_2\right\rangle_V - \left\langle\varphi_1^+\middle|\frac{1}{k_{\text{eff,2}}}\delta\hat{F}\,\varphi_2\right\rangle_V}{\left\langle\varphi_1^+\middle|\hat{F}_1\,\varphi_2\right\rangle_V}, \quad (2)$$

where φ_1^+ is the adjoint flux of the unperturbed system, φ_2 is the forward flux of the perturbed system, and the brackets denote integration over the whole phase space. $\delta\hat{H}$ and $\delta\hat{F}$ are the perturbation of the transport ($\hat{H}$) and fission ($\hat{F}$) operator due to the sample. The operators differ from zero only in the sample volume, which makes it the numerator integration boundary. The denominator is a normalization on the importance of the neutron source. The first order approximation of this equation is widely implemented in Monte Carlo codes for perturbative and sensitivity studies [8], [9], [10] while exact formulation is only available in a custom version of TRIPOLI4 [11]. For experiment design, we rely on the 1st order approximation implemented in the Serpent2 XGPT routine [12], that consists of a rapid method to obtain more precise estimations than with the eigenvalue difference method. The XGPT approximation holds when dealing with small samples in fast systems, where the flux perturbation in the sample volume is smaller than in thermal ones [13]. The gain in precision with respect to the eigenvalue difference method highlights the advantage of this approach.

B. Point kinetics modeling

The neutron flux amplitude variation due to any reactivity insertion can be modeled with the point kinetics equations [14]. Both numerical solution of such equations and the transfer function formalism are exploited.

The numerical solution of the coupled system of ordinary differential equations (ODEs) involves time discretization. The knowledge of the reactivity evolution and the effective delayed neutron parameters is required. The prompt and delayed neutrons acting on different time scales make the solution matrix stiff and the solution numerically unstable, unless very fine time discretization is taken. The system of equations is solved using a fully implicit Radau method of fifth order [15], implemented in *SciPy* [16].

The point kinetics equations with oscillating reactivity insertions can also be solved in the frequency domain by assuming linear perturbations of the neutron flux amplitude, the delayed neutron precursor concentration, and the reactivity. A full derivation of this approach is provided in [17]. The final expression obtained is referred to as the zero-power transfer function (ZPTF) of the reactor

$$\frac{\delta\hat{N}(j\omega)}{\delta\hat{\rho}(j\omega)} = \hat{G}(j\omega) = \frac{N_0}{\Lambda j\omega + \beta - \sum_g^G \frac{\lambda_g \beta_g}{j\omega + \lambda_g}}, \quad (3)$$

which defines the relationship between the perturbation of the neutron flux amplitude $\delta\hat{N}(j\omega)$ and the input reactivity $\delta\hat{\rho}(j\omega)$ in the complex frequency domain ($j\omega$). $\hat{G}(j\omega)$ is the zero power transfer function, function of the effective delayed neutron parameters, the delayed neutron fraction β_g and the associated decay constant λ_g for the precursors group g in the G-group approximation, the effective prompt neutron generation time Λ and the initial neutron flux N_0.

As compared to the ODE solution, which is only limited by point kinetics validity, the ZPTF formalism holds in the range of small perturbations, worsening in case of reactivity much larger than 1 pcm as will be shown in Section VI.A.

IV. EXPERIMENT PROCESSING METHODS

In addition to methods for simulating experiment outputs, a procedure for raw experiment data processing is developed. The experiment output consists of the detector count rate evolution. Assuming the detector only sees the global effect of the reactivity insertion, its response is proportional to the neutron flux amplitude, from which we reconstruct the system reactivity. The net sample reactivity worth is then the difference between the reactivity when the sample is in position IN and OUT.

The data processing can be performed in the time domain, by implementation of inverse point kinetics or in the frequency domain making use of the ZPTF formalism.

A. Inverse point kinetics (IPK)

The inverse point kinetics [18] expresses the reactivity $\rho(t)$ as function of N(t) and its derivative is

$$\rho(t) = \frac{\Lambda}{N(t)}\frac{dN(t)}{dt} - \frac{1}{N(t)}\sum_g^G \beta_g \left(N_0 e^{-\lambda_g t} + \lambda_g \int_0^t N(t') e^{\lambda_g (t'-t)} dt\right) + \beta. \quad (4)$$

The integral in the equation is commonly referred to as the delayed neutron source [19]. The discretized version of the equation is implemented to get the reactivity with the same time binning in which the detector count rate, proportional to the neutron flux amplitude N(t), is measured. The derivative is approximated using the backward differentiation scheme. Furthermore, we recall that

$$\frac{1}{N(t)}\frac{dN(t)}{dt} = \frac{d\left(\log(N(t))\right)}{dt},$$

which allows to rewrite the first term of Eq. 4 as

$$D_j = \frac{\log\left(\frac{N_j}{N_{j-1}}\right)}{t_j - t_{j-1}}. \quad (5)$$

The logarithmic expression of the derivative is preferred since it features potentially better numerical stability than the standard differentiation scheme

$$\frac{1}{N_j}\frac{N_j - N_{j-1}}{t_j - t_{j-1}},$$

despite having the same order of accuracy.

Numerical integration could be performed for the third term of the expression but the standard trapezoidal rule does not give accurate results because of the exponential term. Although more advanced integration methods can be applied to reduce the integration error, a more practical solution was applied following [20], calculating the integral analytically by prescribing the N(t) evolution within a single integration step. Exponential evolution is assumed. It can be expressed as

$$\tilde{N}(t) = N_j e^{D_j(t-t_j)} \text{ with } t_j < t < t_{j+1}, \quad (6)$$

which leads to a discretized form of the integral for time steps $t_1, t_2, \ldots t_k$ as:

$$I_1 = \int_0^{t_1} N(t')e^{\lambda_g(t'-t_1)}dt = N_0\int_0^{t_1} e^{D_1 t'}e^{\lambda_g(t'-t_1)}dt = \frac{N_1 - N_0 e^{-\lambda_g t_1}}{D_2 + \lambda_g}$$

$$I_2 = \int_0^{t_1} N(t')e^{\lambda_g(t'-t_2)}dt + \int_{t_1}^{t_2} N(t')e^{\lambda_g(t'-t_2)}dt = I_1\, e^{-\lambda_g(t_2-t_1)} + \frac{N_2 - N_1 e^{-\lambda_g(t_2-t_1)}}{D_2 + \lambda_g}$$

$$I_k = I_{k-1}\, e^{-\lambda_g(t_k-t_{k-1})} + \frac{N_k - N_{k-1}e^{-\lambda_g(t_k-t_{k-1})}}{D_k + \lambda_g}. \quad (7)$$

Expression (7) is convenient since it eliminates the dependence on the full preceding history, replacing it with the integral computed in the previous time step. The final form of the discrete inverse point kinetics (4) using the terms derived in (5), (7) is:

$$\rho_j = \beta + \Lambda D_j + \frac{1}{n_j}\sum_g^G \beta_g\left(N_0 e^{-\lambda_g t_j} + \lambda_g I_j\right). \quad (8)$$

The formula is applied to retrieve the reactivity evolution from measured or simulated data. The VENUS-F effective delayed neutron parameters calculated through adjoint weighting in Serpent2 [21] are used.

With this formula, the raw oscillation experiment output signal, consisting of the detector count rate evolution array, is processed into the reactivity evolution array, with the same time discretization. The sample reactivity worth is extracted from the full reactivity array by identifying the time intervals when the sample is in position OUT (ρ_{OUT}) and IN (ρ_{IN}). For each oscillation cycle *i*, estimations of the distribution of $\rho_{OUT,i}$ and $\rho_{IN,i}$ are obtained. The in-cycle reactivities IN and OUT are characterized as the mean of the distribution and the associated standard error. Then, the in-cycle reactivity difference $\Delta\rho_i$ and its variance $u(\Delta\rho_i)^2$ is calculated, assuming no correlation. The averaged reactivity worth between the different cycles is computed as the weighted average of the in-cycle quantity:

$$\Delta\rho = \frac{\sum_i \frac{\Delta\rho_i}{u(\Delta\rho_i)^2}}{\sum_i \frac{1}{u(\Delta\rho_i)^2}}, \qquad u(\Delta\rho) = \frac{1}{\sqrt{\sum_i \frac{1}{u(\Delta\rho_i)^2}}}. \quad (9)$$

The compatibility of the data is tested with a χ^2 test as suggested in [22]. If the spread of $\Delta\rho_i$ is not explained by the uncertainty (reduced $\chi^2 > 1$), the propagated uncertainty is increased by a factor

$$C = \sqrt{\frac{1}{M-1}\sum_i \frac{(\Delta\rho_i - \Delta\rho)^2}{u(\Delta\rho_i)^2}}, \quad (10)$$

accounting for the spread of the individual $\Delta\rho_i$, where M is the number of oscillation cycles. This approach ensures that the final uncertainty is consistent with the individual uncertainty of each $\Delta\rho_i$, corrected in case of larger spread of the mean values. This correction is then applied when the in-cycle uncertainty is underestimated, or when there is a drift in the reactivity between cycles. In the former case, the method is valid and correctly adjusts the final uncertainty. In the latter case, instead, it averages data that should be disregarded, leading to a misleading result. On the other hand, a reduced χ^2 value significantly lower than one indicates that the in-cycle uncertainty is overestimated, resulting in an overestimation of the final uncertainty. For this reason, the standard error of the mean of $\Delta\rho_i$ is now considered as a potentially more appropriate estimate of the final uncertainty, as better representative of the repeatability of the measurements. Moreover, applying a linear fit to the $\Delta\rho_i$ values to detect significant drifts over cycles could be useful to identify and exclude inconsistent data.

B. *Fourier Transform (FT)*

In Eq. (3), the ZPTF expresses the functional relationship between the input reactivity and the output neutron flux amplitude (or detector count rate) in the frequency domain. Hence, it can be used to reconstruct the reactivity from the transformed output signal $\hat{N}(j\omega)$. The reactivity $\hat{\rho}(j\omega)$ is expressed as the ratio of $\hat{N}(j\omega)$ and the ZPTF. This approach is valuable since prior knowledge of the relevant frequencies allows to get rid of the noise in the reconstructed reactivity. Oscillation experiments analysis usually focuses on the first harmonic [23], identified by its dominant amplitude and its frequency being the inverse of the oscillation period.

The first step of the procedure prior to transforming the measured signal N(t) is the drift correction. A reactor power drift always appears when oscillating in open loop, even for perfectly symmetric reactivity oscillations between $-\frac{\Delta\rho}{2}$ and $+\frac{\Delta\rho}{2}$; the greater the asymmetry, the more pronounced the drift becomes. This is not a principal issue, but it is recommended to correct for such drift because it introduces additional spurious components in the transformed signal. An empirical method that divides the signal by the fitting second order polynomial is used, as proposed in [23]. Once the signal is normalized, its Fourier Transform $\hat{N}(j\omega) = FFT[N_t]$ is computed using *SciPy* implemented Fast Fourier Transform algorithm (scipy.fft) and the corresponding reactivity frequency spectrum $\hat{\rho}(j\omega)$ is retrieved by dividing it by the ZPTF derived in Eq. (3).

The Inverse Fourier Transform implemented in SciPy (scipy.ifft) can then be used to reconstruct the reactivity time array, equivalent to what was obtained in Eq. (5), or directly evaluate the reactivity amplitude ($\Delta\rho$) from applying suitable correction coefficients to the $\hat{\rho}(j\omega)$ relevant harmonics.

Every Fourier coefficient c_n of the transformed signal is a complex number defined by the integral

$$c_n = \frac{1}{T}\int_0^T f(t)e^{j2\pi\frac{n}{T}t}\,dt\,, \tag{11}$$

where *f(t)* is the original signal, *T* is its period and *n* is the rank of the coefficient. Therefore, given a fixed reactivity shape $a(t)$ of unknown amplitude $\Delta\rho$,

$$\rho(t) = \Delta\rho\; a(t),$$

one can solve the integral (11) and get a relationship between the amplitude and each of the Fourier coefficients.

One can then write an explicit relationship of $\Delta\rho$ as

$$\Delta\rho = \frac{c_n}{\frac{1}{T}\int_0^T a(t)e^{j2\pi\frac{n}{T}t}\,dt}. \tag{12}$$

The analysis is performed on the fundamental harmonic consistently with the approach used in [23], since it contains most of the signal's power.

The main limitation of this method is that it relies on prior knowledge of the reactivity shape, which is not available in the transient from position IN to OUT. If sinusoidal oscillations are performed, the reactivity shape will differ from a sinusoid because of the non-linear relationship between the sample position in the core and its reactivity effect in that position. This is due to the axial change of the sample importance to k_{eff} and of the neutron flux.

The only case of square oscillations with negligible transition time guarantees that the reactivity shape is the same as the sample position. In practical cases, the transition time should be reduced to limit the signal distortion. The proportionality coefficient $\frac{c_1}{\Delta\rho}$ has been calculated for the fundamental component of both square and trapezoidal waves and presented in Table I.

TABLE I
PROPORTIONALITY COEFFICIENT $\frac{c_1}{\Delta\rho}$ OF THE FUNDAMENTAL HARMONIC FOR DIFFERENT OSCILLATION WAVES

Oscillation wave	$\frac{c_1}{\Delta\rho}$ [-]
Square	$2/\pi$
Trapezoidal (t_{dwell}, t_{trans})[a]	$\frac{2}{\pi^2}\frac{2(t_{\text{trans}}+t_{\text{dwell}})}{t_{\text{trans}}}\sin\left(\pi\frac{t_{\text{trans}}}{2(t_{\text{trans}}+t_{\text{dwell}})}\right)$

[a] Trapezoidal wave parameters; t_{dwell}= dwell time [s], t_{trans}= transition time [s].

The Fourier Transform based analysis (FT) has implementation advantages since minor manipulation of the signal is required, while IPK requires signal cutting and averaging. On the other hand, the FT method requires knowledge of the reactivity shape and does not provide direct estimation of the measurement uncertainty as for IPK. Sampling of the input data can be performed to attribute an uncertainty estimation to the obtained reactivity worth. A more practical approach used in actual measurements consists in averaging among different detectors, or taking their cross-correlation [24].

V. SYSTEM PERTURBATION MODELING

The sample is simulated as a B_4C cylinder without holder and connection rod. Position IN is at the VENUS-F active core midplane in one of the central fuel assemblies. More detailed information of the core configuration is found in [5].

A. Local parameters perturbation

Simulation difference of the system with/without the sample provides the relevant quantities of interest such as the neutron flux perturbation in the sample volume and close to it. These quantities are specific for the sample in use and this study must be repeated for each sample. The neutron flux perturbation in the sample volume mainly relates to the applicability of the 1st order perturbation theory for the reactivity estimation while the perturbation of thermal (^{235}U fission rate) and fast (^{238}U fission rate) neutron flux in the proximity of the sample defines the locality of the perturbation and the possibility to measure any local signal changes. The neutron flux perturbation caused by the sample in the sample volume is reported as in [13], weighting the neutron flux on the sample reactions cross-sections.

TABLE II
FLUX PERTURBATION IN B_4C SAMPLE CALCULATED WITH SERPENT XGPT

Reaction	Flux Perturbation [%]
^{10}B (n,α)	-0.42% ± 0.15%
^{10}B (n,el)	-1.10% ± 0.11%
^{nat}C (n,el)	-0.79% ± 0.11%
^{11}B (n,el)	-0.83% ± 0.11%

Table II shows the flux perturbation for the relevant reactions. The perturbation is negative, limited to 1% for every reaction meaning the perturbation is small enough to allow using the 1st order perturbation theory with negligible bias. Despite being the most important in terms of reactivity effect, the (n,α) reaction in ^{10}B causes lower flux perturbation than the scattering reactions. This underlines the potential significance of scattering effects at local level and the importance of simulating the response of local detectors when samples with multiple effects are oscillated.

The local effect of the B_4C sample is investigated placing a virtual detector wrapped around the sample to magnify its response change. With this configuration a negligible perturbation of the fission rate of ^{235}U and a reduction of the fission rate of ^{238}U of around 1% are observed introducing the sample. This effect is due to the removal of neutrons at energies above the ^{238}U fission threshold, by both scattering and absorption in the sample. Moving further from the sample, outside the experimental channel, no significant change in ^{235}U, ^{238}U fission rates is detectable. The analysis of the sample's local effect provides valuable insights, even though these observations are sample-specific and not always applicable. Firstly, the local effect is limited to regions very close to the

https://doi.org/10.1051/epjconf/202533804025

oscillation location and this gives flexibility in positioning detectors for reactivity measurement. Secondly, despite its small magnitude, local effects could still be measured in purely fast spectrum. Using detectors with different response functions can be especially interesting to catch perturbations of the neutron flux in specific energy regions, relatable to different reactions in the sample.

B. Perturbation of reactivity

B_4C sample reactivity effect is calculated both by eigenvalue difference method and Serpent2 XGPT. The results agree with one another and are shown in Table III.

TABLE III
B_4C SAMPLE REACTIVITY WORTH CALCULATED WITH SERPENT2

Method	$\Delta\rho$ [pcm]
eigenvalue difference	-6.0 ± 1.2
XGPT	-6.8 ± 0.3

The reactivity effect is of the order of 7 pcm, principally due to $^{10}B(n,\alpha)$; only Monte Carlo uncertainties are reported. The gain in precision of XPGT compared to eigenvalue difference is observed. Therefore, the XGPT sample reactivity worth estimation is fed ODE solver to produce synthetic data.

VI. SYNTHETIC DATA ANALYSIS

A. Production of Synthetic data

The synthetic data are produced to mimic oscillation conditions achievable with the current VENUS-F oscillator setup. A trapezoidal reactivity wave is assumed, proportional to the sample position. The actual sample reactivity profile from core midplane (IN) to OUT is not considered at this stage.

The produced wave has the following parameters:

- Dwell time: 30 s;
- Transition time: 27 s, corresponding to the time to completely withdraw the sample from active core midplane at the maximum speed of the current motor;
- Starting position: intermediate for symmetric oscillations between $-\frac{\Delta\rho}{2}$ and $+\frac{\Delta\rho}{2}$;
- Sampling frequency: 10 Hz is the maximum acquisition frequency of the current DAQ. The time bin is therefore Δt = 0.1 s;
- Wave amplitude: 6.8 pcm from sample reactivity worth XGPT calculations;
- Number of cycles: 30 for about one hour experiment.

The initial detector count rate is set to 1×10^5 c/s, which is identified in real experiments as an optimal tradeoff between measurement precision and detectors dead time [25].

The signals produced by both the ODE and ZPTF method are shown in Figure 1. The ODE-generated signal exhibits a positive drift in the neutron flux (or count rates), consistent with the behavior observed in [26], although that study focused on reactivity oscillations of higher magnitude. The positive drift is not captured by the ZPTF solution, which consider small perturbations of $\rho(t)$ and $N(t)$ around an equilibrium point [17]. This is shown in the third subplot of Figure 1, where the

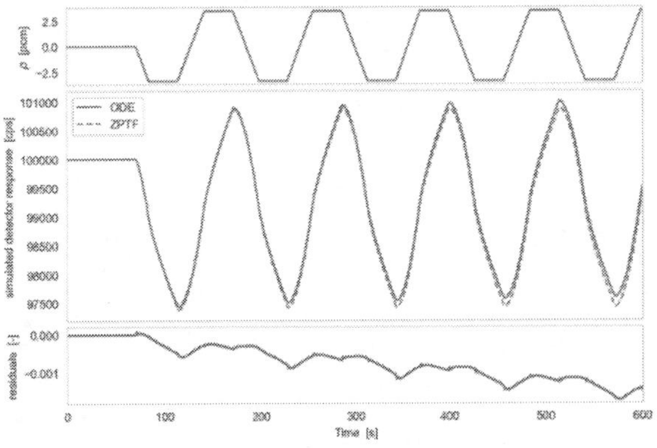

Fig. 1. Synthetic data from trapezoidal reactivity insertion. Top: assumed reactivity shape. Center: simulated response by ODE and ZPTF. Bottom: residuals as relative difference of ZPTF with respect to ODE.

residuals increase linearly over time. The ODE-generated data are used as synthetic signal for the following analysis.

The detectors output in real oscillation experiments is rather different from the simulated one due to noise. This is the combination the counting statistic (Poisson noise) and the intrinsic neutron noise, limiting the reactivity measurement accuracy [4],[27]. Additional electronic noise is introduced when detector signals are acquired in current mode. In the synthetic signals produced, only the Poisson component of the noise is applied. With this strategy, diverse synthetic cases are produced to study the sensitivity of the measured reactivity to any experimental setup parameter and to compare the processing tools available.

B. Processing of the synthetic signal

Inverse point kinetics is applied to the synthetic signal of Figure 1 to evaluate the reactivity distribution within each oscillation cycle. The reactivity distributions from every oscillation cycle are averaged to enhance the precision of the reactivity evaluation, following the procedure described in Section IV.A. Fourier Transform analysis is performed on the same signal. The sample reactivity worth is reconstructed from the magnitude of the fundamental harmonic following the methodology described in Section IV.B.

Inverse point kinetics reactivity reconstruction and cycle-averaging leads to a reactivity amplitude $\Delta\rho = -6.805 \pm 0.043$ pcm. The reactivity amplitude obtained from the first harmonic of the reconstructed reactivity spectrum is consistent (-6.797 pcm).

C. Sensitivity to the oscillation parameters

The impact of perturbing any of the oscillation parameters previously defined and their impact on the final experimental result is analyzed in this section. Varying dwell and transition times, while keeping the number of cycles constant also impacts the overall reactivity worth evaluation and its uncertainty.

Figure 2 shows the mean and uncertainty as function of the dwell time, in the top and bottom plot respectively. The marker styles indicate different transition times while the colors distinguish the processing methods. The best estimate results converge to the expected reactivity of -6.8 pcm. For each dwell time, they spread randomly around the expected value, with no visible trend due to the transition time and the processing

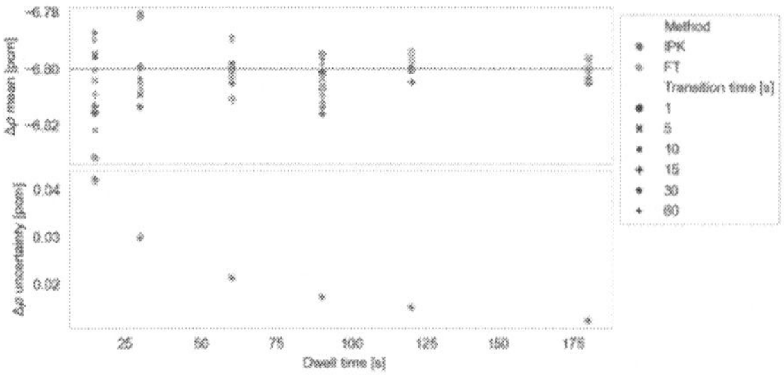

Figure 2: Sensitivity of Δρ best estimate and uncertainty to dwell and transition time variation.

method. The spread of the points reduces increasing the dwell time. This trend reflects in the uncertainty, which smoothly decreases for increasing dwell times: the accuracy of the evaluated reactivity within each cycle improves as more points are averaged in longer cycle times. The uncertainty reduction is proportional to the dwell time (t_{dwell}) as $1/\sqrt{t_{\mathrm{dwell}}}$. On the other hand, the results are rather unsensitive to the transition time: this part of the transient is neglected in the IPK analysis and it is not relevant in the first harmonic analysis, as long the correct correction for the actual reactivity shape is computed (see Table I).

Figure 3 shows the impact of the ratio dwell time / number of cycle to the measurement accuracy. Several synthetic experiments are produced varying dwell time and number of cycles together, while keeping the transition time equal to 1 s since we previously observed negligible impact. The resulting $\Delta\rho$ (by IPK and FT) is plotted as function of the effective experiment time $t_{\mathrm{eff}} = 2\,M\,t_{\mathrm{dwell}}$, where M is the number of cycles and t_{dwell} is the dwell time. t_{eff} is proportional to the number of total observations collected.

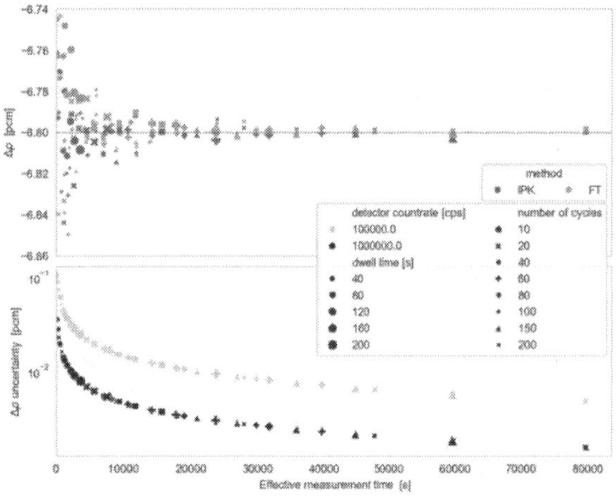

Figure 3: Δρ variation as function of the effective measurement time, varying t_{dwell} and the number of cycles. Uncertainty comparison between detectors with different count rates.

The best estimate results (Figure 3 top) are shown for a dummy detector with an initial count rate of 1×10^5 c/s. The uncertainty (Figure 3 bottom) is presented for both a detector with initial count rate of 1×10^5 c/s and another detector with one order of magnitude larger count rate. The uncertainty is only quantified for the IPK method as explained in Section IV.A.

Increasing t_{eff}, the evaluated reactivity worth smoothly converges to the expected value of -6.8 pcm and the two processing methods perform similarly. The visible discrepancy for $t_{\mathrm{eff}} < 1 \times 10^5$ s is covered by the uncertainty. The different combination of dwell times and number of cycles does not impact the estimated reactivity and its uncertainty: any combination leading to the same t_{eff} results in the same uncertainty on Δρ, see bottom of Figure 3. Moreover, the uncertainty reduces smoothly as $1/\sqrt{t_{\mathrm{eff}}}$. This directly relates to how it is estimated. Assuming the same number of IN/OUT observations in cycle i the uncertainty in that cycle is:

$$u(\Delta\rho_i) = \frac{1}{\sqrt{\frac{t_{\mathrm{dwell}}}{\Delta t}}}\sqrt{u^2(\rho_{OUT,i}) + u^2(\rho_{IN,i})}, \qquad (13)$$

where $t_{\mathrm{dwell}}/\Delta t$ is the number of observations. If this uncertainty stays the same from cycle to cycle – which is reasonable if the drift in N(t) is negligible – the uncertainty propagated to the experiment is

$$u(\Delta\rho) = \frac{C}{\sqrt{M}}u(\Delta\rho_i) = \frac{C}{\sqrt{M\frac{t_{\mathrm{dwell}}}{\Delta t}}}\sqrt{u^2(\rho_{OUT,i}) + u^2(\rho_{IN,i})} = \frac{C}{\sqrt{\frac{t_{eff}}{2\,\Delta t}}}\sqrt{u^2(\rho_{OUT,i}) + u^2(\rho_{IN,i})}, \qquad (14)$$

where C is the correction factor in (10), applied in case the reduced χ^2 is larger than one. Thus, the Δρ uncertainty consists of two components: one related to the measurement time, and the other to the reactivity fluctuations themselves. The latter, in turn, arises from both the Poisson noise in the detector and the inherent neutron noise. In this synthetic case, the neutron noise component was not introduced in the signal and therefore the uncertainty is underestimated compared to a real experiment. In real cases, the Poisson noise can be reduced by increasing the detector counting statistics, and the neutron noise by increasing the reactor power [27].Therefore, the increase of power in oscillation experiment is beneficial to minimize both uncertainty sources. The principal limitation to power increase is posed by the dead time of the detectors. To overcome this, measurements in current mode could be employed.

The last feature we want to investigate is the impact of performing non-symmetric oscillations. Open loop oscillation experiments [28], [29] are usually symmetric between subcriticality and supercriticality, targeting to oscillate between $-\frac{\Delta\rho}{2}$ and $+\frac{\Delta\rho}{2}$ and correcting for the power drift. The impact of oscillations starting from different position is studied through dedicated synthetic signals. The oscillation setup of the reference case is considered, and the results are summarized in Table IV.

TABLE IV
SENSITIVITY OF THE $\Delta\rho$ [PCM] TO OSCILLATION STARTING POSITION DEPENDING ON THE PROCESSING ROUTINE

Oscillation Range	IPK	FT
CRIT-SUBCRIT (0 ; $-\Delta\rho$)	-6.77 ± 0.05	-6.22

$(-\frac{2\Delta\rho}{3};\frac{\Delta\rho}{3})$	-6.81 ± 0.04	-6.74
$(-\frac{\Delta\rho}{2};\frac{\Delta\rho}{2})$	-6.81 ± 0.03	-6.80
$(-\frac{\Delta\rho}{3};\frac{2\Delta\rho}{3})$	-6.79 ± 0.03	-6.86
CRIT-SUPCRIT $(0;\Delta\rho)$	-6.78 ± 0.02	-5.99
$(\frac{2\Delta\rho}{3};-\frac{\Delta\rho}{3})$	-6.80 ± 0.03	-6.81
$(\frac{\Delta\rho}{2};-\frac{\Delta\rho}{2})$	-6.84 ± 0.03	-6.81
$(\frac{\Delta\rho}{3};-\frac{2\Delta\rho}{3})$	-6.80 ± 0.04	-6.79

The results with IPK are always consistent with each other and with the expected reactivity. The uncertainty change is driven by the drift: if N(t) drifts positively the Poisson noise reduces while if it drifts negatively the Poisson noise increases and so does the uncertainty, the extreme cases are CRIT-SUPCRIT and CRIT-SUBCRIT, respectively. On the contrary, FT results dramatically worsen in the extreme cases. The fundamental assumption in the derivation of the ZPTF, already discussed in Section VI.A, is violated when oscillations from critical to supercritical or critical to subcritical are performed. In these cases, the Fourier Transform based processing gives poor results and its applicability is limited.

VII. Conclusions

This study presents the workflow to design oscillation experiments at VENUS-F. It precedes the analysis of the test oscillation experiments of the B_4C sample in the VENUS-F core center.

Firstly, the sample reactivity worth and neutron flux perturbation are calculated with the Serpent Monte Carlo code. The local perturbation caused by the sample is limited and we observed it is mostly due to scattering, as opposed to the sample reactivity effect which is driven by (n,α) reaction in ^{10}B. The change in the virtual detectors fission rate is strongly dependent on the deposit, which highlights the importance of selecting detectors of different energy sensitivities as local detectors. Moreover, the local response is strongly sample-dependent and a dedicated analysis must be carried out for each sample.

The calculated B_4C sample reactivity worth was fed to point kinetics solver to produce synthetic data. They were used to showcase and verify the developed processing pipelines and study the sensitivity of its output to the oscillation parameters. Inverse point kinetics and Fourier Transform processing routines gave results consistent with one another and with the calculated $\Delta\rho$. Only in the case of asymmetric reactivity oscillation, Fourier Transform processing failed to predict the expected $\Delta\rho$. This observation relaxes the constraint of having perfectly symmetric oscillations during the experiments, which is challenging to achieve. The transition time has no effect on the $\Delta\rho$ as well as the ratio between dwell time and number of cycles. The best estimate and uncertainty convergence is only driven by the effective measurement time. Moreover, the uncertainty is proportional to the reactivity fluctuation within the cycle that is a combination of Poisson and neutron noise. Therefore, setup optimization should aim to minimize reactivity fluctuations by operating at the highest possible power while carefully managing the impact of detector dead time and temperature effect, which could otherwise degrade the measurement quality. A viable option would be to perform measurement in current mode, where the detector dead time is not a problem but electronic noise plays a larger role.

This work is the first phase of the effort to conduct oscillation experiments at VENUS-F. The next phase will focus on the analysis of the first test experiments, where the proposed design choices were applied on a real case. Future plans include experiments using new equipment – such as a faster motor to enable quasi-square-wave oscillations and additional detectors to acquire local signals. A full scale experimental campaign will then be carried out oscillating fissile materials, such as PuO_2.

References

[1] G. Palmiotti, P. Blaise, and F. Mellier, "Massimo Salvatores: integral experiments and their use for the validation of nuclear data and the neutronic design of advanced nuclear systems," *EPJ Nuclear Sciences & Technologies*, vol. 7, p. 11, Jun. 2021, doi: 10.1051/epjn/2021007.

[2] S. Panizo *et al.*, "Sensitivity and uncertainty analyses for advanced nuclear systems (ALFRED, ASTRID, ESFR and MYRRHA)," *Progress in Nuclear Energy*, vol. 172, p. 105207, Jul. 2024, doi: 10.1016/J.PNUCENE.2024.105207.

[3] A. Kochetkov *et al.*, "Integral experiments in the VENUS-F reactor," *Ann Nucl Energy*, vol. 157, p. 108230, Jul. 2021, doi: 10.1016/J.ANUCENE.2021.108230.

[4] W. K. Foell, *Small-sample reactivity measurements in nuclear reactors*. 1972. Accessed: Feb. 27, 2024. [Online]. Available: https://www.osti.gov/biblio/4351459

[5] F. Grimaldi *et al.*, "The CoRREx neutron spectrum filtering campaign at VENUS-F for calculation-to-experiment discrepancy interpretation," *Ann Nucl Energy*, vol. 219, p. 111425, Sep. 2025, doi: 10.1016/J.ANUCENE.2025.111425.

[6] J. Leppänen, V. Valtavirta, A. Rintala, and R. Tuominen, "Status of Serpent Monte Carlo code in 2024," *EPJ Nuclear Sciences & Technologies*, vol. 11, p. 3, Jan. 2025, doi: 10.1051/epjn/2024031.

[7] Bell G. and Glasstone S., *Nuclear reactor theory*. Van Nostrand Reinhold Company, 1970.

[8] B. Liu, X. Lv, W. Zhao, K. Wang, J. Tu, and X. Ouyang, "The comparison of MCNP perturbation technique with MCNP difference method in critical calculation," *Nuclear Engineering and Design*, vol. 240, no. 8, pp. 2005–2010, Aug. 2010, doi: 10.1016/j.nucengdes.2010.04.022.

[9] M. Aufiero *et al.*, "A collision history-based approach to sensitivity/perturbation calculations in the continuous energy Monte Carlo code SERPENT," *Ann Nucl Energy*,

vol. 85, pp. 245–258, Nov. 2015, doi: 10.1016/J.ANUCENE.2015.05.008.

[10] X. Peng, J. Liang, A. Alhajri, B. Forget, and K. Smith, "Development of continuous-energy sensitivity analysis capability in OpenMC," *Ann Nucl Energy*, vol. 110, pp. 362–383, Dec. 2017, doi: 10.1016/J.ANUCENE.2017.06.061.

[11] G. Truchet and P. Leconte, "Small sample reactivity worths calculation exact perturbation theory and Monte Carlo transport," in *M&C019*, 2019. [Online]. Available: https://hal.science/hal-02411094

[12] M. Aufiero, M. Martin, and M. Fratoni, "XGPT: Extending Monte Carlo Generalized Perturbation Theory capabilities to continuous-energy sensitivity functions," *Ann Nucl Energy*, vol. 96, pp. 295–306, Oct. 2016, doi: 10.1016/j.anucene.2016.06.012.

[13] F. Di Croce, P. Leconte, F. Grimaldi, A. Krása, P.-E. Labeau, and J. Wagemans, "Small Sample Reactivity Worth Calculation: Comparison of Serpent2 and TRIPOLI4 Perturbation Techniques," in *International Conference on Mathematics and Computational Methods Applied to Nuclear Science and Engineering (M&C 2025)*, Illinois: American Nuclear Society, 2025, pp. 1156–1165. doi: 10.13182/xyz-47168.

[14] G. R. Keepin, *Physics of Nuclear Kinetics*, 1st ed. Addison-Wesley, 1965.

[15] E. Hairer and G. Wanner, *Solving Ordinary Differential Equations II*, vol. 14. Berlin, Heidelberg: Springer Berlin Heidelberg, 1991. doi: 10.1007/978-3-662-09947-6.

[16] P. Virtanen *et al.*, "SciPy 1.0: Fundamental Algorithms for Scientific Computing in Python," *Nat Methods*, vol. 17, pp. 261–272, 2020, doi: 10.1038/s41592-019-0686-2.

[17] Hetrick David L, *Dynamics of nuclear reactors* . Chicago: University of Chicago press, 1971.

[18] V. M. Monteiro, Á. Calebe Ribeiro Lima, A. S. Martinez, and A. C. Gonçalves, "A differential equation for inverse point kinetics," *Nuclear Engineering and Design*, vol. 421, p. 113094, May 2024, doi: 10.1016/j.nucengdes.2024.113094.

[19] B. Geslot, C. Jammes, and B. Gall, "Influence of the delayed neutron group parameters on reactivity estimation by rod drop analysis," *Ann Nucl Energy*, vol. 34, no. 8, pp. 652–660, Aug. 2007, doi: 10.1016/j.anucene.2007.03.005.

[20] A. dos Santos, "A reactivity meter with uncertainties," *Ann Nucl Energy*, vol. 202, p. 110463, Jul. 2024, doi: 10.1016/J.ANUCENE.2024.110463.

[21] J. Leppänen, M. Aufiero, E. Fridman, R. Rachamin, and S. Van Der Marck, "Calculation of effective point kinetics parameters in the Serpent 2 Monte Carlo code," *Ann Nucl Energy*, vol. 65, pp. 272–279, Mar. 2014, doi: 10.1016/J.ANUCENE.2013.10.032.

[22] M. Tanabashi *et al.*, "Review of Particle Physics," *Physical Review D*, vol. 98, no. 3, p. 030001, Aug. 2018, doi: 10.1103/PhysRevD.98.030001.

[23] B. Geslot, A. Gruel, P. Walczak, P. Leconte, and P. Blaise, "A hybrid pile oscillator experiment in the Minerve reactor," *Ann Nucl Energy*, vol. 108, pp. 268–276, Oct. 2017, doi: 10.1016/j.anucene.2017.04.036.

[24] T. Ligonnet, A. Laureau, P. Simsbler, A. Pautz, and V. Lamirand, "Design of an Open-Loop Pile-Oscillation Program in the CROCUS Reactor," *IEEE Trans Nucl Sci*, pp. 1–1, 2024, doi: 10.1109/TNS.2024.3371183.

[25] International Commission on Radiation Units and Measurements, "Particle Counting in Radioactivity Measurements," Bethesda, MD, 1994. [Online]. Available: https://www.icru.org/report/particle-counting-in-radioactivity-measurements-report-52/

[26] B. D. Ganapol, "A highly accurate algorithm for the solution of the point kinetics equations," *Ann Nucl Energy*, vol. 62, pp. 564–571, Dec. 2013, doi: 10.1016/j.anucene.2012.06.007.

[27] C. E. Cohn, "A Simplified Theory of Pile Noise," *Nuclear Science and Engineering*, vol. 7, no. 5, pp. 472–475, May 1960, doi: 10.13182/NSE60-4.

[28] B. Geslot, A. Gruel, S. Bréaud, P. Leconte, and P. Blaise, "Innovative hybrid pile oscillator technique in the Minerve reactor: open loop vs. closed loop," *EPJ Web Conf*, vol. 170, p. 04009, Jan. 2018, doi: 10.1051/epjconf/201817004009.

[29] G. R. Imel, B. Baker, T. Riley, A. Langbehn, H. Aryal, and M. L. Benzerga, "Study of the open loop and closed loop oscillator techniques," in *ANIMMA2015*, IEEE, Apr. 2015, pp. 1–8. doi: 10.1109/ANIMMA.2015.7465538.

Irradiation Rig Design and Preliminary Dosimetry Results for CORANI Phase 1 at HANARO Reactor

Junesic Park[1,*], Seong Woo Yang[1], Sung Jae Park[1], Byung Gun Park[1], Jong Woo Kim[1], Kishore B. Dasari[2], Christophe Destouches[3], Loïc Barbot[3], Christophe Domergue[3], and Hervé Philibert[3]
[1]Korea Atomic Energy Research Institute (KAERI), Korea
[2] Gyeongsang National University (GNU), Korea
[3] CEA, DES, IRESNE, DER, Cadarache, F-13108 Saint Paul-lez-Durance, France
(*) jp@kaeri.re.kr

Abstract— **The CORANI (KAERI/CEA COllaboration for Research reactor Application of Neutron dosimetry and Instrumentations) project is a collaborative effort between KAERI and CEA aimed at validating advanced neutron instrumentation and dosimetry techniques at the High-Flux Advanced Neutron Application Reactor (HANARO). The Phase 1 campaign focuses on neutron field characterization using self-powered neutron detectors and fission chambers, complemented by activation-foil dosimetry. This paper presents the feasibility study results for CORANI Phase 1 to confirm the reliability of the adopted methodologies and provide a solid basis for experimental conditions and irradiation rigs. The preliminary results demonstrate well agreement between the two different measurement systems and confirm the operational parameters for the forthcoming Phase 1 campaign scheduled for the fourth quarter of 2025.**

Keywords — **HANARO, JHR, CORANI project, neutron dosimetry, SPND, fission chamber, irradiation rig**

I. Introduction

Korea Atomic Energy Research Institute (KAERI) and the French Atomic Energy Commission (CEA) are jointly undertaking the CORANI (KAERI/CEA COllaboration for Research reactor Application of Neutron dosimetry and Instrumentations) project, a initiative aimed at advancing nuclear instrumentation technologies for application in research reactors. The primary objective is to test and validate innovative neutron and gamma measurement systems—intended for application in the Jules Horowitz Reactor (JHR)—under realistic irradiation conditions at the High-Flux Advanced Neutron Application Reactor (HANARO) in Korea.

This collaboration arises from the increasing demand for advanced fuel testing, particularly for Small Modular Reactor (SMR) fuels and Accident Tolerant Fuels (ATF), where precise, real-time monitoring of parameters such as fuel centerline temperature, internal pressure, deformation, neutron flux, and fission product activity is essential during irradiation experiments. The project is divided into two campaign phases. Phase 1 focuses on characterizing the neutron field in specific irradiation positions of HANARO through the deployment of commercial self-powered neutron detectors (SPNDs) and CEA miniature fission chambers (FCs), complemented by activation foil dosimetry. Phase 2, to follow, will extend the work to gamma heating measurements, temperature monitoring, and advanced dosimetry techniques.

The Phase 1 campaign is scheduled to be conducted in the fourth quarter of 2025. This feasibility study describes the procedures and results of the preliminary experiments conducted for the CORANI project. These results lead to the overall Phase 1 experimental plan, including the design of the irradiation rig.

II. Preliminary Experiment

A. HANARO Overview and Irradiation Hole Selection

HANARO is a 30 MW_{th} open-pool type research reactor at the Korea Atomic Energy Research Institute that uses heavy water as a reflector. The fuel region lies on the reactor's radial centerline, at the center of the reflector tank. The active fuel length is 70 cm (z = −35 to +35 cm). It provides seven core-region vertical irradiation holes and 17 reflector-region vertical irradiation holes ("IP" holes). The IP holes allow on-power insertion and removal of irradiation rigs, making them appropriate for our approach of conducting activation-foil analysis in advance of or in parallel with the online SPND/FC measurements to strengthen the sensor validation.

The IP holes have a 6-cm diameter, pass vertically through the reflector tank, and operate without forced convection (natural convection only). They are distributed at radial distances of 48–81 cm from the reactor centerline. Evaluations with the HANARO MCNP model indicate maximum total neutron flux ranging from 2.4×10^{13} to 2.0×10^{14} $n \cdot cm^{-2} \cdot s^{-1}$, with a ratio of thermal (Φ_{th}) to fast (Φ_f) in the range of about 22 to 3000. To verify the radiation tolerance of the sensors, we selected IP5 as the irradiation site among the IP holes because it offers relatively high thermal neutron flux and is readily accessible during power operation. Fig. 1 shows a cross-sectional view of the HANARO core.

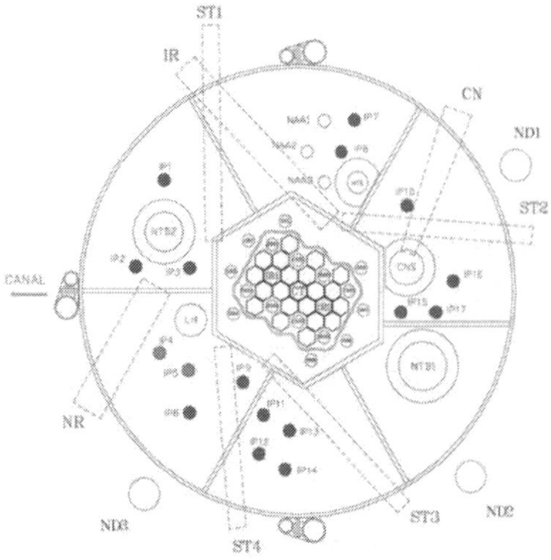

Fig. 1. A cross-sectional view of the HANARO core. Gray shading indicates the IP holes; red shading highlights IP4 and IP5 relevant to this study.

B. Preliminary Experiment: Methods and Results

As a preliminary experiment, we evaluated the neutron flux at the IP4 irradiation hole using various activation foils. The activation foils used are listed in Table I. The foils were fabricated by CEA-LDCI Lab (Dosimetry, Sensor and Instrumentation Lab) at the MADERE facility [1,2]. The foils were irradiated at IP4, retrieved and measured by gamma-ray spectrometry. Since IP5 was unavailable at the time of the preliminary experiment, the irradiation was performed at IP4, a nearby position with similar specifications. Prior to irradiation, foils were first loaded into a primary aluminum container (28-mm inner diameter, 80-mm height, 1.5-mm thickness). Each primary container was then placed inside a secondary container (34-mm inner diameter, 100-mm height, 2-mm thickness), which was helium-sealed. Four sealed containers were prepared, each containing three to four activation foils of different types. During irradiation, the four containers labeled as Lv1, Lv2, Lv3, and Lv4 were stacked from bottom to top in the IP carrier. The axial positions of the activation foils corresponded to $z = -27.3, -11.5, +4.2$, and $+19.95$ cm relative to the fuel midplane ($z = 0$ cm). Equivalently, measured from the bottom of the 70 cm active fuel length, the normalized positions were 0.110, 0.336, 0.560, and 0.785. The reactor operated at 13 MW_{th} for 6 hours, with the control-rod position at approximately 345 mm. After irradiation, the dosimetry containers were cooled for 26 days, transferred to the hot-cell area, disassembled, and the foils were recovered for measurement.

Measurements were performed using CEA's Mobile X-γ spectrometry system [1] and the gamma-ray spectroscopy in the HANARO Pneumatic Transfer System (PTS) facility [3]; during source counting, a low-Z spacer was used to maintain a fixed source-to-detector distance from the HPGe endcap. Energy and efficiency calibrations employed a multi-nuclide standard source and ^{152}Eu. Data analysis followed each institute's standard nuclear-data practice: KAERI used NuDat 3.0 [4], and CEA primarily used Nucléide–LARA [5]. For ^{119m}Sn, both institutes—by exception—used NuDat 3.0.

TABLE I
SPECIFICATIONS OF THE ACTIVATION FOILS FOR PRELIMINARY EXPERIMENT

Material	Diameter (mm)	Thickness (mm)
Al–Co	4	0.1
Al–Sc	4	0.1
Fe	3	0.5
Ni	4	0.1
Ag	4	0.1
^{119}Sn	7	0.1

In the Al–Co alloy, the Co content is 1 wt%, and in the Al–Sc alloy, the Sc content is 2 wt%.

Table II shows the results of the preliminary experiment. The selected experimental conditions provided activity levels sufficient for reliable quantification with both measurement systems of KAERI and CEA. Overall, the activity values evaluated by both systems were within or near the combined uncertainty; residual discrepancies in ^{60}Co, ^{46}Sc, and ^{110m}Ag are consistent with differences in coincidence-summing corrections and counting geometry, whereas the large deviation observed for ^{119m}Sn is likely to result from low-energy self-absorption and sensitivity to efficiency calibration.

The Al–Co results indicate that the neutron flux profile at IP4 shows an axial trend, peaking near Lv2 and Lv3 and falling toward the ends. These data support the use of foils for Phase 1 sensor-validation calibrations.

III. EXPERIMENTAL DESIGN OF PHASE 1 CAMPAIGN

In the CORANI Phase 1 campaign, two irradiation experiments are conducted: a "short" irradiation, in which various activation foils are irradiated for a short time to measure the neutron flux in the HANARO irradiation hole IP5, and a "long" irradiation, in which multiple real-time neutron sensors are used to measure the flux signal evolution inside the hole over (at least) a full HANARO operating cycle.

In the short-irradiation experiment, activation foils of Al–Co, Ag, Fe, Ni, and Sn—used either in the bare condition or with a B_4C filter—are employed to measure the neutron flux at the center of the irradiation hole at multiple axial positions. Based on the preliminary-experiment results, the activation-foil dimensions were specified, and their detailed specifications are summarized in Table III.

In the long-irradiation experiment, three SPNDs and six FCs (fabricated at the CEA CHICADE FC workshop [6]) measure the neutron flux at the center of the irradiation hole at multiple axial positions (covering different flux levels). After the reactor operation cycle has started, the short-irradiation experiment is performed once the neutron flux has stabilized (corresponding to stable control-rod positions, typically after 1 day), and the irradiation is completed within a few hours. Subsequently, the long-irradiation experiment begins immediately afterward and continues monitoring the flux signals until the end of the reactor cycle.

TABLE II
PRELIMINARY EXPERIMENTAL RESULTS

Dosimeter and position	Radio-nuclide	Energy (keV)	Mass (mg)	KAERI		CEA		Max acceptable deviation (%)
				Specific Activity (Bq/mg)	Uncertainty (%)	Specific Activity (Bq/mg)	Uncertainty (%)	
Al-Co (Lv1)	^{60}Co	1173.23 1332.49	3.444	3.424E+03	2.324	3.567E+03	2.9	3.72
Al-Co (Lv2)			3.424	4.109E+03	2.312	4.285E+03	2.9	3.71
Al-Co (Lv3)			3.569	3.966E+03	2.320	4.096E+03	2.9	3.71
Al-Co (Lv4)			3.418	3.166E+03	2.340	3.263E+03	2.9	3.73
Al-Sc (Lv3)	^{46}Sc	889.27 1120.54	3.514	1.712E+05	2.336	1.793E+05	2.1	3.14
Fe (Lv2)	^{54}Mn	834.85	28.247	*	*	9.551E-01	6.6	*
	^{59}Fe	1099.24 1291.59	28.247	1.906E+03	2.452	1.923E+03	2.5	3.50
Ni (Lv1)	^{58}Co	810.76	11.842	5.193E+01	2.953	4.985E+01	3.0	4.21
Ni (Lv2)			10.671	6.769E+01	2.910	6.548E+01	2.8	4.04
Ni (Lv4)			11.41	4.344E+01	3.002	4.251E+01	2.9	4.17
Ag (Lv1)	^{110m}Ag	657.76 763.94	14.094	7.323E+04	3.008	7.586E+04	2.2	3.73
Ag (Lv2)			14.004	8.759E+04	3.008	9.109E+04	2.3	3.79
Ag (Lv3)			14.108	6.643E+04	3.022	6.959E+04	2.3	3.80
^{119}Sn (Lv3)	^{119m}Sn	23.875	31.198	3.770E+02	6.220	2.332E+01	11.1	12.72
		25.044 ($K_{\alpha 2}$) 25.271 ($K_{\alpha 1}$)	31.198	*	*	2.641E+01	10.2	*

Uncertainties are relative expanded (k = 2, ≈95%). Entries marked '*' reflect results for which reliable reporting thresholds were not met in this summary and are therefore excluded. Max acceptable deviation is computed as the root-sum-square of the reported expanded uncertainties (k = 2) from both facilities.

TABLE III
SPECIFICATIONS OF THE ACTIVATION FOILS FOR SHORT-IRRADIATION EXPERIMENT OF CORANI PHASE 1

Material	Density (g/cm^3)	Diameter (mm)	Thickness (mm)	Mass (mg)	Filter
Al-Co	2.70	7	0.1	10.39	Bare
Ag	10.5	3	0.1	7.42	
Al-Sc	2.70	3	0.1	1.91	
Fe	7.87	7	0.1	30.30	
^{117}Sn	7.31	3	0.1	5.17	B_4C
^{119}Sn	7.31	3	0.1	5.17	
Ni	8.90	3	1	62.92	
Fe	7.87	3	1	55.66	
Al-Co	2.70	3	1	19.08	

In the Al-Co alloy, the Co content is 1 wt%, and in the Al-Sc alloy, the Sc content is 2 wt%. The thickness of the B_4C filter is 1.5 mm.

For Phase 1, two specialized irradiation rigs were designed and manufactured: a short-irradiation rig and a long-irradiation rig. At the HANARO IP irradiation holes, insertion and withdrawal of irradiation specimens are generally performed using an IP carrier—a basket-type carrier approximately 1 m long with a 55-mm inner diameter. Each experiment-specific rig is loaded into this carrier for irradiation and retrieval. Fig. 2 shows the configuration of the short- and long-irradiation rigs.

The short-irradiation rig was developed for neutron dosimetry using activation foils. It provides nine predefined axial installation positions along the centerline; considering the container length, up to five containers can be installed per irradiation. Each sealed container accommodates activation foils together with a B_4C filter, is helium-filled to protect the contents chemically, and is hermetically sealed. Each sealed container has an 8.35-mm inner diameter and a height of 30 mm, providing sufficient space to accommodate multiple foil types. When multiple foil types are housed in a single container, disk-shaped aluminum spacers are inserted to prevent cross-contamination via inter-foil diffusion.

After irradiation, the IP carrier is withdrawn to the service pool, the rig is removed from the carrier, and all subsequent handling (e.g., extracting dosimetry containers) is performed underwater. Because operators access the rig and containers from above the water surface using long tools, the assemblies include top-side tool interfaces for positive engagement. Conducting these tasks underwater mitigates radiation exposure and prevents inadvertent reactor protection system trips during power operation. For hot-cell transfer, containers are moved underwater into a shielded transfer cask, which is then lifted out for transport.

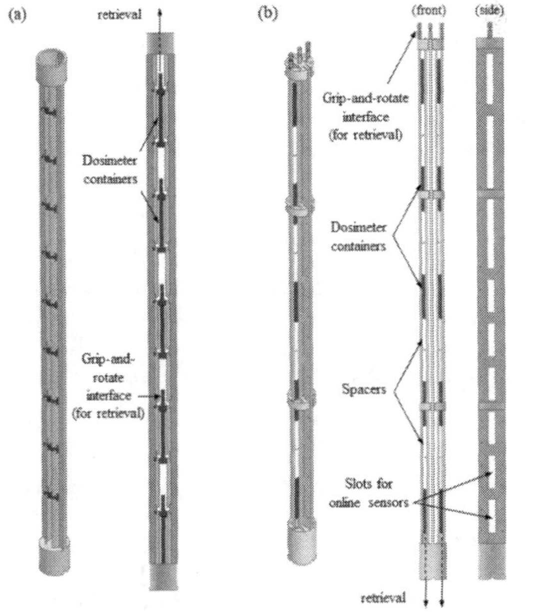

Fig. 2. Configuration of the short- (a) and long-irradiation rigs (b), designed for the CORANI Phase 1 campaign.

The long-irradiation rig mounts six FCs (enriched ^{235}U and ^{232}Th) and three SPNDs (Rh, Co, and Ag emitter type) to enable simultaneous online neutron-flux monitoring and offline dosimetry. For refined signal analysis, neutron-dosimetry containers are positioned axially on both sides of the sensor stack. The materials used for neutron dosimetry are Al–Co, Fe, Ni, and Nb; 3–4 foils are inserted per container either with a B_4C filter or bare. The foils will be arranged to measure thermal- and fast-neutron gradients and the axial profile. Following irradiation, a dedicated non-disassembly retrieval mechanism allows removal of only the dosimetry containers in the service-pool environment: a shielded transfer cask is positioned beneath the rig, and rotating a key rod at the top of the rig releases the containers through a bottom discharge port directly into the cask, leaving all installed sensors undisturbed.

IV. Conclusions

KAERI and CEA have jointly planned and are conducting the CORANI collaborative research project, with the first campaign scheduled to commence in the fourth quarter of 2025. Through the preliminary experiment, we established the detailed experimental plan for the Phase 1 campaign and designed a dedicated irradiation rig to implement it. The preliminary work also provided a valuable opportunity to compare the neutron activation analysis systems of the two institutes and the results showed good agreement between them. Consequently, this study consolidates the methodology and parameters to be used in the CORANI project.

Acknowledgment

This work was supported by the Korea government (MSIT) (2710007351). On CEA side, it was financially supported by the CEA Gen23 program, in the frame of the instrumentation project (INSNU) to develop innovative and advanced nuclear and non-nuclear instrumentation in preparation of the CEA Jules Horowitz Reactor (JHR) commissioning.

References

[1] Nicolas Thiollay, Christophe Destouches, Christophe Domergue, and David Tisseur, "The MADERE Platform: Update of the γ and X-Ray Activity Measurements Techniques," in *Proc. RPSD 2024,* Orlando, FL, USA, 2024, pp. 89–93.

[2] D. Tisseur, C. Domergue, C. Destouches, H. Philibert, A. Pepino, J. Bonora, et al., "Recent improvements in X-ray spectrometry measurements performed at the MADERE facility dedicated to reactor dosimetry requirements," EPJ Web Conf., vol. 308, Art. no. 01007, 2024.

[3] Y.-S. Chung, S.-H. Kim, J.-H. Moon, S.-Y. Baek, Y.-J. Kim, H.-R. Kim, I.-J. Park, and K.-S. Min, "New pneumatic transfer systems for neutron activation analysis at the HANARO research reactor," J. Radioanal. Nucl. Chem., vol. 278, no. 3, pp. 707–712, 2008.

[4] National Nuclear Data Center (NNDC), "NuDat 3.0: Nuclear Structure and Decay Data," Brookhaven National Laboratory, Upton, NY. [Online]. Available: http://www.nndc.bnl.gov/nudat3. Accessed on: August 19, 2025.

[5] Laboratoire National Henri Becquerel (LNHB), "Nucléide–LARA: Library for gamma and alpha emissions," CEA/LNE–CNAM. [Online]. Available: http://www.lnhb.fr/home/nuclear-data/module-lara. Accessed on: August 19, 2025.

[6] B. Geslot, F. Berhouet, L. Oriol, S. Breaud, C. Jammes, P. Filliatre, and J-F. Villard, "Development and manufacturing of special fission chambers for in-core measurement requirements in nuclear reactor," in Proc. ANIMMA 2009, Marseille, France, 2009, pp. 1–4, 10.1109/ANIMMA.2009.5503816.

Phenomenological Analysis on 1,000°C Material Irradiation Testing in HANARO using the Instrumented-Capsule(15M-03K)

Jong Woo Kim[1,*], Junesic Park[1], Ki Nam Choo[1], and Seong Woo Yang[1]
[1]Korea Atomic Energy Research Institute, Republic of Korea
(*) jw@kaeri.re.kr

***Abstract*—With the increasing demand for materials development for next-generation nuclear systems such as Molten Salt Reactors, Sodium-cooled Fast Reactors, and Very High-Temperature Reactors, high-temperature irradiation testing has become essential. This study aimed to establish irradiation technology capable of operating up to 1,000°C in the HANARO reactor. For this purpose, an instrumented capsule (15M-03K) with a double-layered thermal medium was designed, fabricated, and tested. Irradiation experiments were conducted in multiple reactor cycles between 2018 and 2024. The capsule achieved a maximum specimen temperature of 1,000°C, while stable operation was demonstrated at 800–900°C. A transient temperature drop observed at 1,000°C was promptly mitigated by helium injection, ensuring the safety of both the capsule and the reactor. Neutronic evaluations using MCNP confirmed the dominance of ~1 MeV neutrons in the specimens, with flux variations linked to control rod positions, and the accumulated fast neutron fluence was quantified for all cycles. These results demonstrate the feasibility of controlled high-temperature irradiation testing in HANARO and provide key data for material performance evaluation under advanced reactor conditions. The developed technology will contribute to future irradiation programs requiring longer durations, higher stability, and extended temperature ranges.**

***Keywords*—HANARO, high-temperature irradiation, MCNP, instrumented capsule**

I. Introduction

The HANARO, a multi-purpose research reactor, has been conducting various irradiation tests to meet the demands from academia, industry, and research institutions for nuclear materials and fuel testing.

Recently, with the advent of next-generation nuclear systems such as the Sodium-cooled Fast Reactor (SFR) and the Very High-Temperature Reactor (VHTR), designed to operate at high-temperatures, the demand for irradiation testing of structural and fuel materials has increased significantly. Therefore, it is necessary to develop technologies and devices suitable for neutron irradiation tests at high temperatures [1].

In this regard, KAERI has been conducting research on high-temperature irradiation test technology for several years [2, 3, 4, 5, 6, 7, 8, 9, 10].

This research aimed to develop a method capable of conducting irradiation tests up to 1,000°C. Accordingly, an instrumented-capsule was fabricated and irradiation tests were performed to achieve 1,000°C and verify high-temperature stability in the HANARO.

II. The Instrumented Capsule (15M-03K)

For high-temperature irradiation testing, the instrumented-capsule (15M-03K) was designed and fabricated based on neutronic and thermal analysis.

First, neutron transport calculations were performed to estimate the neutron flux and gamma heating using the MCNP6.2 code, a Monte Carlo-based simulation tool [11]. This calculation assumed an equilibrium core configuration of HANARO operating at 30MW.

Then, temperature distribution was calculated using the one-dimensional GENGTC code [12]. The material temperature in each node was calculated as follows (1),

$$T_i = T_o + \frac{q''}{4k}(R_o^2 - R_i^2) + \frac{Q}{2\pi k}\ln\left(\frac{R_o}{R_i}\right) - \frac{q'''}{2k}R_i^2\ln\left(\frac{R_o}{R_i}\right) \qquad (1)$$

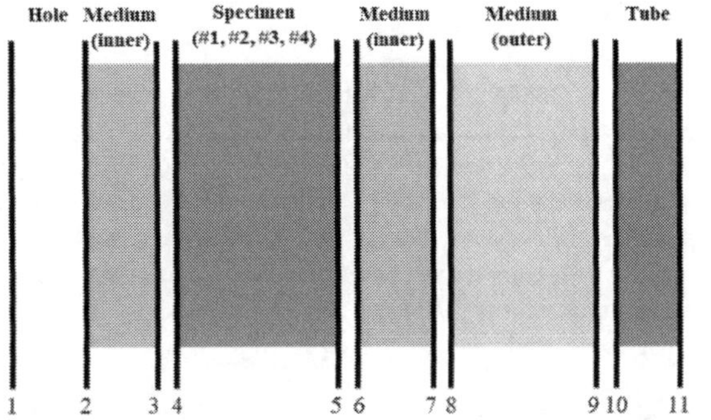

Fig. 1. One-dimensional material arrangement from the center of the capsule with corresponding numbers

The capsule features a double-layered thermal medium (Ti/Fe). Four specimens were loaded only in segment 3, while the other segments used a single-layered thermal medium (Al) without specimens.

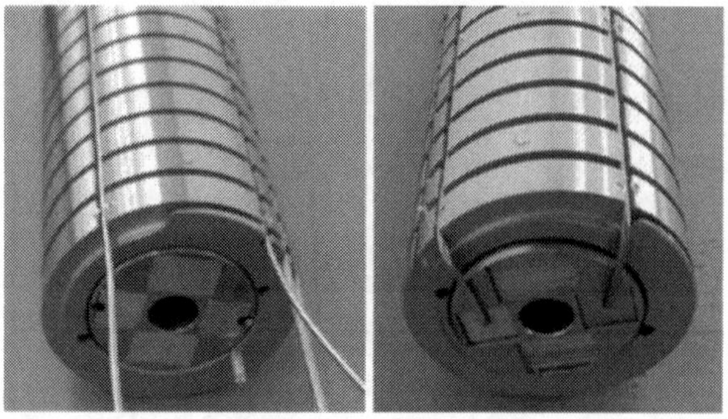

Fig. 2. Double layered thermal medium of the capsule (15M-03K)

III. Neutron Irradiation Test

The capsule underwent a series of out-of-pile tests, including loading/unloading, hydraulic, and integrity of instruments in high-temperature. And then, the capsule is loaded into the CT hole located at the central region of the core in the HANARO.

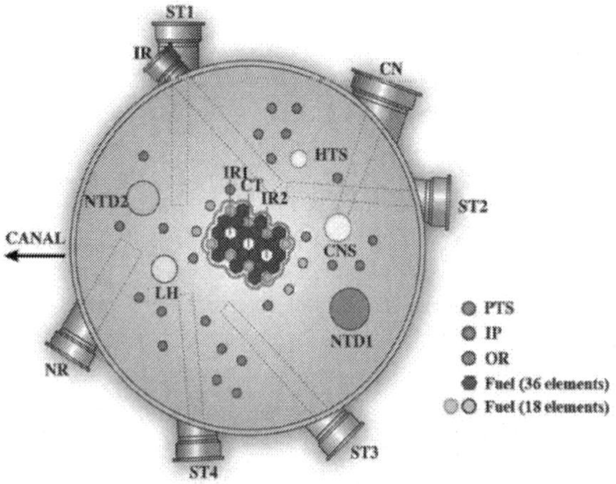

Fig. 3. Schematic diagram of HANARO core and irradiation hole

The capsule was connected to the test control system installed in HANARO. Through this system, the heater power and helium pressure inside the capsule could be precisely adjusted in real time to achieve and maintain the target temperature. In addition, the system provided a comprehensive monitoring interface, displaying key parameters such as temperature, heater power, helium pressure, reactor power, and control rod height in real time.

The first irradiation test was conducted in December 2018, during the 99th cycle of operation at 30MW, where the maximum temperature recorded was 791°C. Although an unexpected reactor shutdown led to the early termination of the test, the feasibility of high-temperature irradiation testing was confirmed. After the preliminary test in 2018, the high-temperature irradiation test was resumed in 2024 with the following objectives.

A. 1st Irradiation Test in 2024

The irradiation test was performed during HANARO 108-1 cycle of operation at 27MW. The objective of test was to reach the temperature up to 1,000°C.

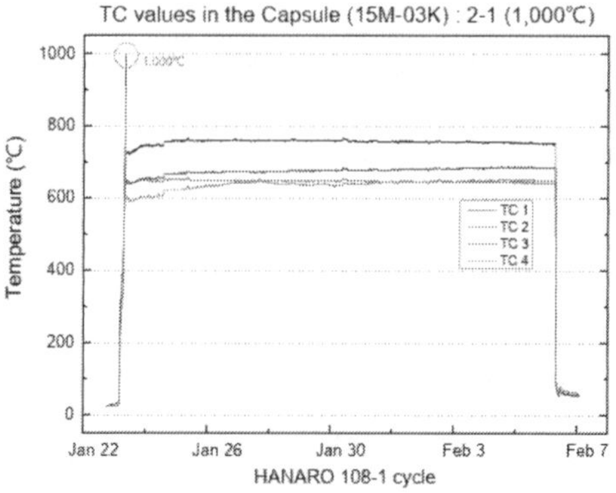

Fig. 4. Temperature variations inside of the capsule (15M-03K) during HANARO 108-1 cycle

B. 2nd Irradiation Test in 2024

The irradiation tests were performed during HANARO 108-3 cycle (27MW), 109-1 cycle (27MW), and 109-2 cycle (25MW), respectively. The objective of test was to maintain stable temperatures of 800°C, 850°C, and 900°C during successive cycles.

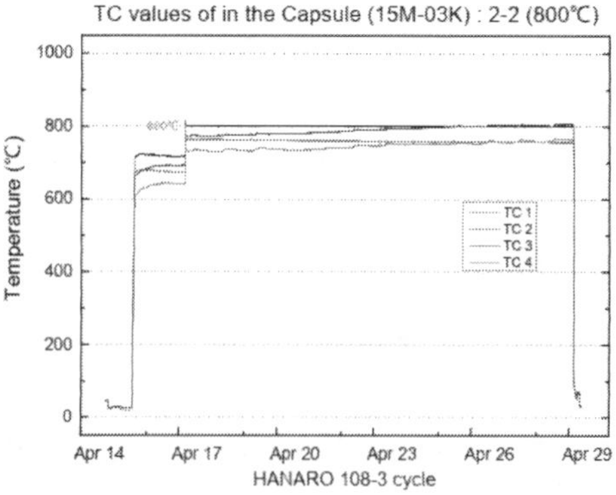

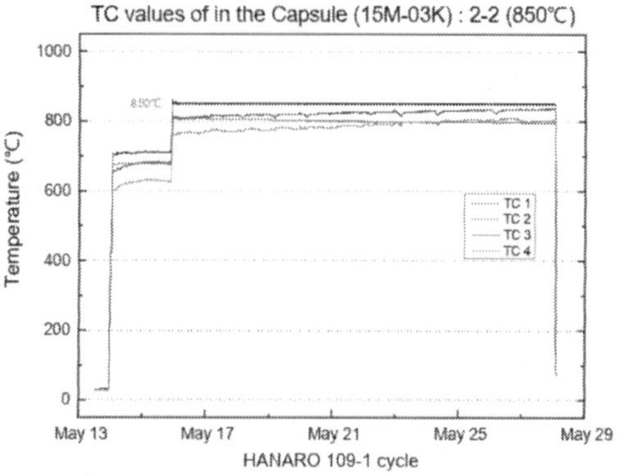

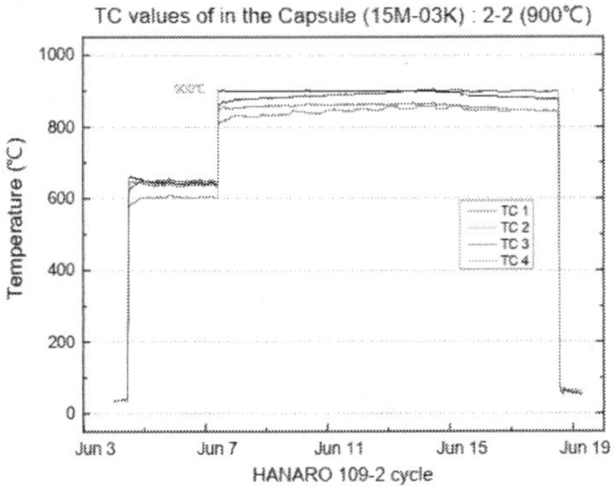

Fig. 5. Temperature variations inside of the capsule (15M-03K) during HANARO 108-3, 109-1, 109-2 cycle of operation.

IV. Phenomenological Analysis

During the 1st irradiation test conducted in 2024, the temperature unexpectedly dropped after initially reaching 1,000℃. Since the exact cause of the temperature drop was not clearly identified, helium gas was immediately injected into the capsule. This action reduced the capsule temperature to the range of 600–700℃, thereby ensuring the safety of both the capsule and the reactor. This behavior is attributed to potential reactions between specimen impurities and the inner thermal medium. That is, the temperature of the inner thermal medium (Ti) is limited to below 942℃ due to eutectic reaction [13].

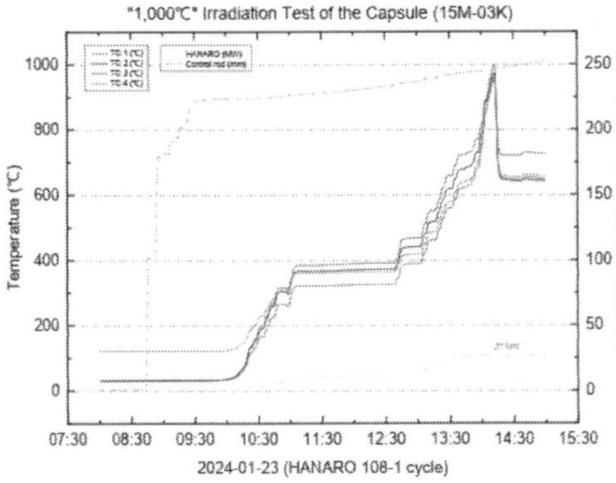

Fig. 6. 1,000℃ Irradiation Test of the capsule (15M-03K)

In the 2nd irradiation test performed in 2024, it was confirmed that the capsule integrity was not affected by the anomalous behavior observed near 1,000℃. Furthermore, the stability of high-temperature irradiation testing was successfully demonstrated at temperature below 942℃, validating the reliability of the system under sustained high temperature conditions.

The neutron behavior and irradiation exposure during reactor operation were evaluated through neutron transport calculations using the MCNP code. The capsule and specimen geometry were modeled in MCNP, as illustrated in Fig. 7.

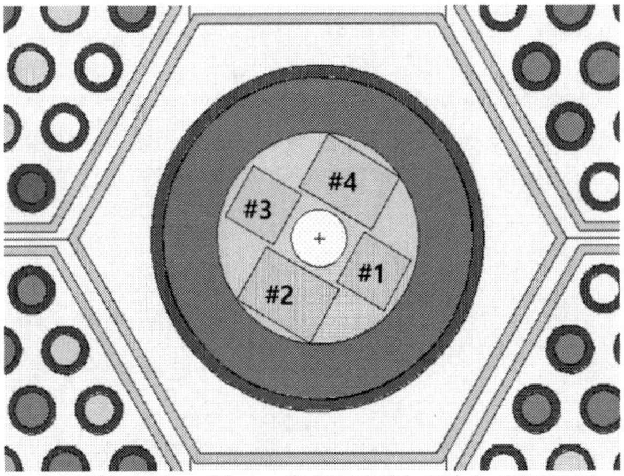

Fig. 7. MCNP modeling of the capsule (15M-03K) in CT hole of HANARO

During reactor operation, the control rod positions vary with time. By comparing the neutron flux at the specimen locations within the capsule, noticeable changes were observed in accordance with the control rod movement, as shown in Fig. 8.

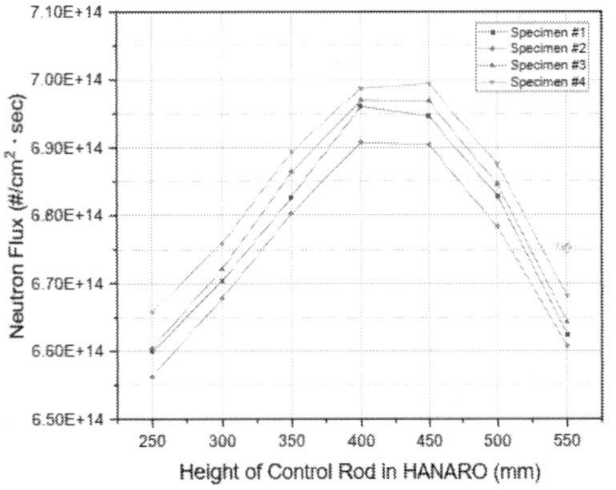

Fig. 8. Neutron flux variation for specimens as a function of control rod

Due to the structural characteristics of the reactor core, slight differences in neutron flux were observed among the specimens. The neutron flux reached its maximum when the control rod position was in the range of 400–450mm.

In contrast, the neutron energy spectrum remained relatively constant throughout the reactor operation cycle, with neutrons around 1MeV being dominant in the specimens, as shown in Fig. 9.

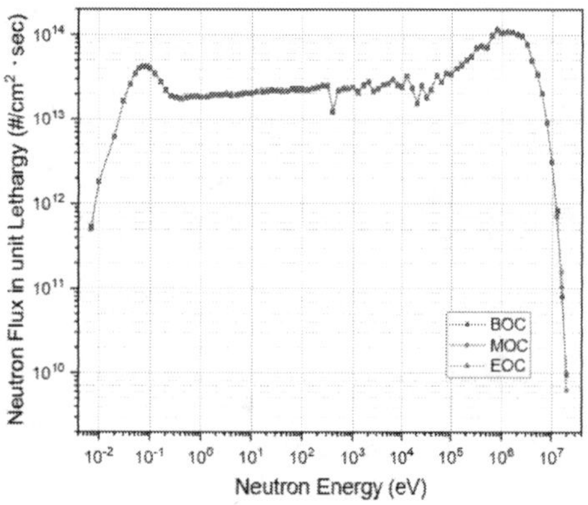

Fig. 9. Neutron energy spectrum throughout the reactor operation cycle

The fast neutron fluence on the specimens was evaluated for each HANARO operation cycle, and the results are summarized in Table I.

TABLE I
FAST NEUTRON FLUENCE ON SPECIMENS OF THE CAPSULE (15M-03K)

HANARO Operation	#1	#2	#3	#4
108-1	1.26E+20	1.28E+20	1.28E+20	1.27E+20
108-2	6.21E+17	6.35E+17	6.32E+17	6.31E+17
108-3	1.25E+20	1.28E+20	1.27E+20	1.26E+20
109-1	1.25E+20	1.25E+20	1.24E+20	1.25E+20
109-2	1.10E+20	1.11E+20	1.10E+20	1.11E+20
Total	4.87E+20	4.92E+20	4.90E+20	4.90E+20

V. CONCLUSIONS

In this study, the instrumented capsule (15M-03K) was successfully designed and fabricated for high-temperature irradiation testing in HANARO. Both neutronic and thermal analyses were performed to support the capsule design, and the out-of-pile verifications confirmed its operability under high-temperature conditions.

Through a series of irradiation tests, the capsule achieved a maximum specimen temperature of 1,000°C, demonstrating the feasibility of controlled high-temperature irradiation in HANARO. Although an unexpected temperature drop was observed after reaching 1,000°C in the first 2024 test, the safety of the capsule and the reactor was ensured through prompt helium injection. Subsequent tests confirmed capsule integrity and validated stable irradiation performance at temperatures below 942°C.

The neutronic evaluation using MCNP showed that neutron flux varied with control rod positions, while the energy spectrum remained relatively constant with ~1MeV neutrons dominating. The accumulated fast neutron fluence on the specimens was quantified for each cycle, providing essential data for high-temperature material behavior studies.

These results verify the effectiveness of the developed instrumentation technology for high-temperature irradiation testing. Future work will pursue enhanced stability, longer irradiation durations, and higher operational temperatures to further support advanced reactor materials development.

ACKNOWLEDGMENT

This work was supported by the Korean government (Ministry of Science and ICT) (2710007351).

REFERENCES

[1] J. Y. Park, "Possibility of Fusion Applications of Current Materials R&D for GEN-IV Nuclear Energy System in KAERI", Korea-Japan Blanket Workshop, 2009.
[2] M. S. Cho et. al., "Development of a Capsule with Double Layered Thermal Media", Korea Nuclear Society Autumn Meeting, 2010.
[3] M. S. Cho et. al., "Development of High Temperature Irradiation Technologies", Korea Nuclear Society Spring Meeting, 2011.
[4] Gillemot, Ferenc, et al., "Structural Material Investigation in the High Temperature Irradiation facility of the Budapest Research Reactor", Research Reactor Application for Materials under High Neutron Fluence. Proceedings of an IAEA Technical Meeting (TM-34779). No. IAEA-TECDOC-1659, 2011.
[5] M. S. Cho et. al., "Design, Fabrication and Test Report of the Material Capsule(08M-10K) with Double Thermal Media for High-temperature Irradiation", KAERI/TR-3890/2009, Korea Atomic Energy Research Institut, Jan. 2010.
[6] M. S. Cho et. al., "Design, Fabrication and Test Report on Instrumented Capsule (09M-09K) with Ti Thermal Media for High Temperature Irradiation Test", KAERI/TR-4109/2010, Korea Atomic Energy Research Institute, Aug. 2010.
[7] M. S. Cho et. al., "Design, Fabrication and Test Report on the Instrumented Capsule (11M-22K) with Ti/Graphite Thermal Media for High Temperature Irradiation Test", KAERI/TR-4879/2013, Korea Atomic Energy Research Institute, Jan. 2013.
[8] M. S. Cho et. al., "Design, Fabrication and Test Report on Double Layer Structured Instrumented Capsule (13M-01K) for High Temperature Irradiation Test", KAERI/TR-6836/2017, Korea Atomic Energy Research Institute, Jul. 2017.
[9] M. S. Cho et. al., "Development of Capsule Technology for High Temperature Irradiation – Double Layer Structured Instrumented Capsule (13M-01K/15M-03K) Design, Fabrication and Test Report", KAERI/TR-8596/2021, Korea Atomic Energy Research Institute, Apr. 2021.
[10] J. W. Kim et. al., "Report on the Second Test of an Instrumented Capsule(15M-03K) with Double Thermal Medium for High-Temperature Irradiation Test", KAERI/TR-10682/2024, Korea Atomic Energy Research Institute, Dec. 2024.
[11] D. B. Pelowitz., MCNP6TM USER'S MANUAL Version 1.0, LA-CA-13-00634, Los Alamos National Laboratory, 2013
[12] H. Someya et. al., "GENGTC-JB: A Computer Program to calculate temperature distribution for cylindrical geometry capsule", JAERI-M 87-148, Japan Atomic Energy Research Institute, 1987.
[13] Choo. K. N. Choo, Irradiation Test of High Temperature Irradiation Capsule(15M-03K) up to 1000°C in HANARO, Transactions of the Korean Nuclear Society Spring Meeting, Jeju, Korea, 9-10 May 2024.

The NEREA Python Package for Integral Experiment Analysis: a Fission Chamber Calibration Case Study

Federico Grimaldi[1,2,*], Federico Di Croce[1,2], Antonin Krása[1], Jan Wagemans[1], and Guido Vittiglio[1]
[1]SCK CEN Belgian nuclear research centre, Boeretang 200, Mol, 2400, Antwerp, Belgium
[2]Université libre de Bruxelles, Avenue Franklin D. Roosevelt 50, Bruxelles, 1050, Belgium
(*) federico.grimaldi@sckcen.be

Abstract—**NEREA is a Python package designed for integral experiment processing in nuclear reactors. In this work we present the main features of NEREA to date, including data processing for pulse height spectrum analysis, fission chamber calibration, fission rate traverse measurement processing, control rod calibration and reactivity worth calculation. NEREA offers high control of the processing flow yet keeping it smooth and user-friendly thanks to is object-oriented conception.**

We showcase the capabilities of NEREA presenting the calibration of a large set of fission chambers at the BR1 MARK-III converter. NEREA allows for analysis of the impact of processing choices and for considerations over several measurements. Strengths and weaknesses of the implemented calibration methodology are also discussed, highlighting room for future work on this. This is particularly true when the measured pulse height spectrum is characterized by low peak-to-valley ratio, making the choice of a proper discrimination level challenging.

Keywords —**Fission Chambers, MARK-III converter, NEREA Python API, Integral Experiments.**

I. Introduction

VENUS-F is a fast spectrum zero power research reactor operated at the Belgian Nuclear Research Centre SCK CEN. Integral experiments have been performed there in support of nuclear data improvement [1] [2] [3] [4] and code development [5]. Research on fast heavy-metal-cooled reactors has recently renewed the interest in VENUS-F critical configurations, with new experimental campaigns conducted in the OFFERR European Commission framework. Moreover, since 2021, students have been more and more involved in the research activities at VENUS-F with Ph.D., Master and Bachelor theses.

With the new activities and the data produced, a renewal of the data processing codes was performed. This work presents the NEREA Python package developed for this purpose, with its assumptions and implementation choices. As an application case study, this paper also presents several results of a fission chamber calibration campaign carried out in January and February 2025 in the MARK-III facility of the BR1 reactor, operated at SCK CEN [6].

II. Structure of NEREA

NEREA is an object-oriented Python package available publicly on GitHub [7] which structure is schematically reported in Fig. 1. On the experiment side, a few data classes store the acquired raw data and allow for their preprocessing. Such classes can then be passed to *Experimental* classes, where the actual processing is performed.

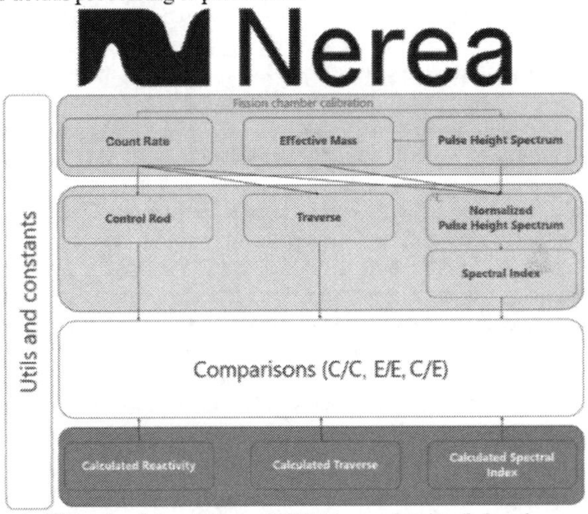

Fig. 1: Schematic of the structure of NEREA: raw-data-interfacing classes (pink), Experimental classes (light blue) and Calculated classes (dark blue).

On the calculation side, NEREA provides the user with *Calculated* classes to then enable for calculation-to-experiment comparisons. NEREA wraps *serpenTools* [8] to interface with the Serpent Monte Carlo code [9]. *Comparison* classes can couple *Experimental* and *Calculated* instances and compare them. Through the whole data handling performed by NEREA (Neutron Energy-integrated Reactor Experiment Analysis), best estimate and uncertainty values are reported one by another in *pandas.DataFrame* [10] instances.

A. Raw data interfacing classes

Each raw data interfacing class comes with one or more

classmethods that allow its creation from common raw data file formats. Attributes identifying the data source (e.g., processed detector, its deposit, …) are implemented in NEREA and their consistency is checked through the data processing.

1) The CountRate class

nerea.CountRate stores time-dependent reaction rate data. *nerea.CountRate* handles time through the *datetime* module of Python.

Methods for timeseries integration and averaging are implemented and used in the processing of spectral index measurements (see section B.2). NEREA also implements methods to find the asymptotic exponential counts in a timeseries, to fit them with an exponential and finally to get the asymptotic period measured by the processed detector.

The *nerea.CountRates* class can import multiple time-dependent data from a single output file and select that with the most counts.

2) The PulseHeightSpectrum class

nerea.PulseHeightSpectrum instances can be instantiated reading from GENIE 2000 "TKA" files. Any *nerea.PulseHeigthSpectrum* instance retains information on the counted pulses in each voltage channel and implements methods to smooth and re-bin the acquired curve.

Currently, NEREA enables processing of the measured pulse height spectra with the R-channel effective mass methodology presented in [11] [12] [13] [14]. The R-channel methodology bases the identification of the discrimination level for pulse height integration on a detector calibration measurement. Such a calibration shall be performed in a similar neutron flux spectrum as that of its application. The application and calibration spectrum are re-binned to align at the R-channel (i. e., the channel with half-maximum counts right to the maximum itself). The discrimination level for application pulse height spectrum integration is then chosen as a fraction of the R-channel (see section B1)). Several discrimination levels are considered and the pulse height spectrum integral is calculated from each of them. If the same discrimination approach is used for both calibration and measurement acquisitions, they can be normalized to one another as described in section B.1).

Integration from custom discrimination levels and normalization by fission chamber measured mass (rather than by its effective mass coming from the detector calibration) can also be performed with NEREA. Fig. 2 shows the output of the *plot* method of this class.

The *integrate* method of the *nerea.PulseHeightSpectrum* class integrates the spectrum data from a set of discrimination levels passed by the user. This is then used in the *nerea.SpectralIndex* class for appropriate reaction rate mass normalization.

Integration can be performed taking the raw (default) or smoothed data. In such case, data renormalization can be performed within NEREA, since some smoothing methods may not be conservative of the pulse height spectrum integral.

The *calibrate* method implements the R-channel methodology to compute the fission chamber effective mass processing a calibration pulse height spectrum.

3) The EffectiveMass class

Spectral indices measured with fission chambers are processed by NEREA following the effective mass approach reported in [11] [12] [13] [14]. This consists in associating an equivalent

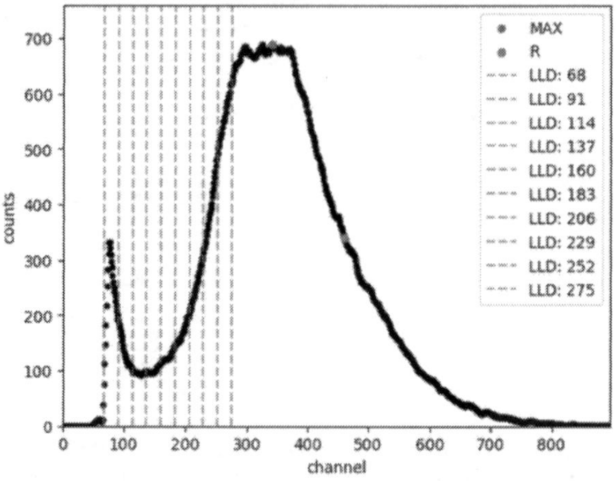

Fig. 2: Pulse height spectrum of fission chamber 19 after 10-channel moving average smoothing. In green the maximum of the spectrum found by NEREA and in red the corresponding R. Dashed lines indicate the discrimination levels (equi-spaced from 0.15 to 0.6R)

fission chamber deposit mass to several discrimination levels from which the measured pulse height spectrum is integrated. More considerations on this approach are reported in section IV.

Fission chamber mass data (either from deposit mass measurement as in [15] [16] [17] or from calibration as in section 3) are stored in the *nerea.EffectiveMass* class. The effective mass is processed and treated by NEREA assuming it has units of mass [μg].

B. Experimental classes for experiment processing

1) The NormalizedPulseHeightSpectrum class

The *nerea.NormalizedPulseHeightSpectrum* takes a *nerea.PulseHeightSpectrum* instance and normalizes it to a *nerea.EffectiveMass* and to a *nerea.CountRate* instance. The normalization is performed in steps:

1. Normalization per unit effective mass accounting for the discrimination levels used during calibration data processing;
2. To exclude the measurement noise, a plateau of the pulse height spectrum normalized per unit effective mass as a function of discrimination level is found (1% default tolerance, user customizable);
3. The plateau value is then normalized per unit live measurement time and per average count rate in the power monitor data stored in a *nerea.CountRate* instance.

The processed data can be plotted via the plot method (see Fig. 3). This displays information on the effective mass (top left), on the power monitor count rate (top right), on the processed pulse height spectrum (bottom left) and on the calculated reaction rate per unit mass and time (bottom right).

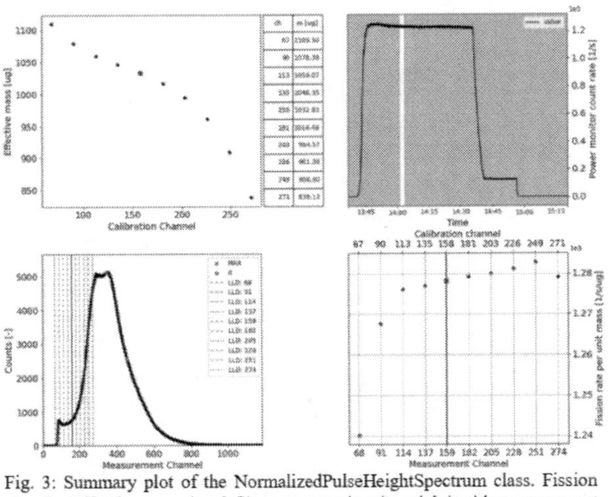

Fig. 3: Summary plot of the NormalizedPulseHeightSpectrum class. Fission chamber effective mass (top left), power monitor (top right) with measurement window in white, pulse height spectrum (bottom left) and fission rate per unit mass (bottom right). The chosen discrimination level is highlighted in blue.

2) The SpectralIndex class

A spectral index is defined as the ratio of two reaction rates. In a neutron flux φ, considering fission cross sections σ_f for ^{235}U and ^{238}U, for example, the $^{238}U/^{235}U$ spectral index would be defined as

$$SI\left(\frac{^{238}U}{^{235}U}\right) = \frac{\int d\tau\ \varphi\ \sigma_f^{238}}{\int d\tau\ \varphi\ \sigma_f^{235}},$$

where τ indicates all the dimensions of the phase space considered for integration in the fission chamber deposit. The *nerea.SpectralIndex* instances are initiated with two *nerea.NormalizedPulseHeightSpectrum*: one for the spectral index numerator and one for its denominator. A correction is subtracted from the ratio to account for impurities in the detector deposits as in [13]. Different processing choices (e.g., in terms of pulse height spectrum smoothing and of considered discrimination levels) can be passed to the numerator and denominator separately.

3) The Traverse class

NEREA is suited for processing of reactor experiments beyond spectral index measurements. The *nerea.Traverse* class takes a list of measured *nerea.CountRate*, averages their count rate over the measurement real time and normalizes them to any user-specified element in the list, corresponding to a traverse point.

4) The ControlRodCalibration class

The *nerea.ControlRodCalibration* class is initiated with a critical height and with a *nerea.CountRate* instance where calibration data are stored. Various variants of control rod calibration with positive period method (integral and differential, with and without compensation [18]) are implemented in NEREA.

From the fitted asymptotic period and the In-Hour equation, the *nerea.ControlRodCalibration* allows for calculation of the reactivity associated with each step of the control rod movement. The integral and differential control rod calibration curves are then produced and fitted. Using such fitted curves, the reactivity worth associated with any control rod movement can be computed with the *get_reactivity_worth* method, passing the initial and final control rod position

C. Calculated and Comparison classes

NEREA interfaces with calculation data with *Calculated* classes. Respectively, the *nerea.CalculatedSpectralIndex* and *nerea.CalculatedTraverse* handle calculated values for the spectral indices and traverse and can be initiated with *classmethods* wrapping *serpentTools*.

NEREA also allows for comparison of calculated (C) to experimental (E), experimental to experimental and calculated to calculated data. This is done via *Comparison* classes, which implement calculation of C/E, E/E and C/C figures of merit. The *nerea.FrameCompare* class enables comparison of any single row *pandas.DataFrame* instances with one another, given that they come with *value* and *uncertainty* columns, which is the minimal format of any output of NEREA.

D. Considerations on the uncertainty propagation in NEREA

NEREA assumes Poisson-distributed random variables for any of the count-related quantities processed in the data-interfacing classes. The uncertainties are then propagated with the first-order momentum propagation equation, assuming no correlations. To allow for a more complete uncertainty analysis, NEREA also returns the portion of output variance associated to any input quantity involved in the processing. Such a variance portion is returned in absolute terms.

For more sophisticated analyses (e.g., due to experiment specificity or processing further than the implemented comparisons), one may require NEREA to output more than the default. This is done with the *long_output* argument of any *processing* method. With this, NEREA outputs also the best estimate of each processing input and the associated variance. One can then propagate uncertainties beyond NEREA's current implementation and explicitly account for correlations.

E. NEREA use example

```
import nerea
import pandas as pd

composition1 = pd.DataFrame({"value": 100, "uncertainty": 0}, index=["U235"])
pm_cal1 = nerea.CountRate.from_ascii("PM1.txt", detector=1, deposit_id="U235")
em1 = nerea.PulseHeightSpectrum.from_formatted_TKA("cal1.TKA").calibrate(
                              nerea.constants.KNBS, composition1,
                              monitor=pm_cal1, one_group_xs=nerea.constants.XS_FAST)
composition2 = pd.DataFrame({"value": 100, "uncertainty": 0}, index=["U238"])
pm_cal2 = nerea.CountRate.from_ascii("PM2.txt", detector=2, deposit_id="U235")
em2 = nerea.PulseHeightSpectrum.from_formatted_TKA("cal2.TKA).calibrate(
                              nerea.constants.KNBS, composition2,
                              monitor=pm_cal2, one_group_xs=nerea.constants.XS_FAST)
phs1 = nerea.PulseHeightSpectrum.from_formatted_TKA("meas1.TKA")
phs2 = nerea.PulseHeightSpectrum.from_formatted_TKA("meas2.TKA")
pm1 = nerea.CountRate.from_ascii("M1.txt", detector=1, deposit_id="U235")
pm2 = nerea.CountRate.from_ascii("M2.txt", detector=1, deposit_id="U235")
si = nerea.SpectralIndex(nerea.NormalizedPulseHeightSpectrum(phs1, em1, pm1),
                         nerea.NormalizedPulseHeightSpectrum(phs2, em2, pm2))
c = nerea.CalculatedSpectralIndex.from_detectors('model_det0.m', ['d1', 'd2'])
## C/E - 1 [%] can be computed returning pandas.DataFrame
## with `value` and `uncertainty` columns
ce = nerea.CoverE(c, si).process(minus_one_percent=True)
```

Code 1: NEREA code to process spectral index C/E.

Code *1* reports a code snippet that can be used to process spectral index C/E results with NEREA. Fission chamber composition aside, most of it consists of data import (pulse height spectra and monitor fission chamber count rates). Formatted file import does not differ substantially from the import functions described in the previous section: the only difference is that the object metadata are read from the file name directly in this case rather than being passed by the user.

All the processing is handled in the *calibrate* method used to get the fission chamber effective mass, and in the *process* method, that calculates the C/E in the last row. Any processing option (e.g., tolerances, fitting and smoothing parameters, …) can be passed as a keyword argument to these functions. Iterating scripts analogous to that presented in Code *1* have been used to process the data and produce the results presented in the next section stopping at the effective mass calculation.

III. Fission Chamber Calibration At BR1 MARK-III

In January and February 2025, several fission chambers of different designs were calibrated in the MARK-III converter at BR1, where a ^{235}U prompt fission is achieved [6]. The calibration was performed following the R-channel methodology (see section II.A.2). Table I reports a list of the calibrated fission chambers with their design and the associated effective mass at 0.4 R.

TABLE I
EFFECTIVE MASS OF THE CALIBRATED FISSION CHAMBERS (PROCESSED AT 0.4 R). * INDICATES PULSE HEIGHT SPECTRA THAT CAN HARDLY BE USED FOR SPECTRAL INDEX MEASUREMENT (SEE APPENDIX).

ID	Deposit	Effective Mass [μg] (0.4 R)	Design
1	^{232}Th	886*	4 mm diameter by CEA
2	^{234}U	250	
3	^{235}U	87	
4	^{236}U	466	
5	^{238}U	500	
6	^{238}U	430	
7	$^{nat.}$U	418	
8	$^{nat.}$U	544	
9	^{237}Np	102	
10	^{239}Pu	22	
11	^{240}Pu	261	
12	^{240}Pu	75	
13	^{242}Pu	191	
14	^{241}Am	125	
15	^{232}Th	2058	8 mm diameter by CEA
16	^{238}U	2626	
17	^{237}Np	364*	
18	^{237}Np	321*	
19	^{235}U	1012	CFUF by PHOTONIS
20	^{239}Pu	20	Built-in extension

Each fission chamber was connected to an extension measured to align the fission chamber deposit midplane with the center of the MARK-III converter. The electronic chains of the measurement consisted of a high voltage power supply (+ 310 V), a charge-sensitive pre-amplifier (CANBERRA 2006), a spectroscopic amplifier (CANBERRA AFT 2025) and a multiport module. The data were acquired with the GENIE 2000 spectroscopy software. The calibration coefficient to correlate the power monitor counts with the absolute neutron fluence in the experiment site was measured through Ni foil irradiation. The Ni foil activation was measured with a HPGe detector with an uncertainty of about 2%. As already reported in [3] [13] this is a major contributor to the effective mass uncertainty. Calibrating fission chamber in a single campaign (i.e., with a single calibration factor) makes the processed spectral index independent of the calibration factor itself, as it appears at the spectral index numerator and denominator.

A. A chronological overview

Fission chambers calibration data of campaigns at BR1 MARK-III since 2010. These campaigns featured calibration of some of the fission chambers reported in this study, as reported in Table II. No impurity correction was found in the calibration files earlier than 2025: the uncorrected effective masses are then reported for those.

As it is visible in Table II, the measured effective mass changes with time. This can be attributed to several causes, among which, fluctuations in the calibration factor, slight changes in the fission chamber design characteristics (e.g., changes in filling gas pressure), changes in the used electronics and differences in the processing. A similar time-wise comparison is also reported in [11]. Fission chamber calibration also calibrates for the data processing choices, to some extent. This means that the observed differences in effective mass could reflect similarly on the measured count rate resulting in limited changes to the final spectral index value. The recommendation we give is to reperform detector calibration before each experimental campaign to maximize consistency.

TABLE II
EFFECTIVE MASS [MG] MEASURED OVER THE YEARS AT 0.4 R WITH CALIBRATIONS AT BR1 MARK-III. *: PULSE HEIGHT SPECTRA HARDLY USED FOR SPECTRAL INDEX MEASUREMENT (SEE APPENDIX); **: REPORTED AT 0.5 R BECAUSE OF HIGH NOISE.

	Year					
	2010	2013	2016	2020	2022	2025
ID	Effective mass [μg]					
1						886*
2	206				214	250
3		84	86			87
4			442		449	466
5		488	488			500
6			419			430
7			391			418
8		528	538			544
9		101				102
10			21			22
11	232**					238**
12		72				75
13	170					191
14						125
15						2058
16						2626
17						364*
18		100*				321*
19			1217			1012
20			21			20

B. *Impact of processing choices*

From now on, we consider fission chamber 19 as a reference case study, which pulse height spectrum is shown in Fig. 2. The default implementation of NEREA features moving average (MA) smoothing over 10 channels. Fig. 4 shows a comparison with another suitable smoothing option: exponential weighted mean (EWM) [19] (span=10). Visually, the resulting spectra differ very little. The pulse height spectrum can also be fitted to find its maximum and R channel. NEREA implements the possibility to fit the spectrum from any channel before the maximum to the first zero on its right. The pulse height spectrum fitted with a polynomial of order 5 is shown in Fig. 4 (FIT). Before channel 270, the raw data are concatenated to the fitted ones.

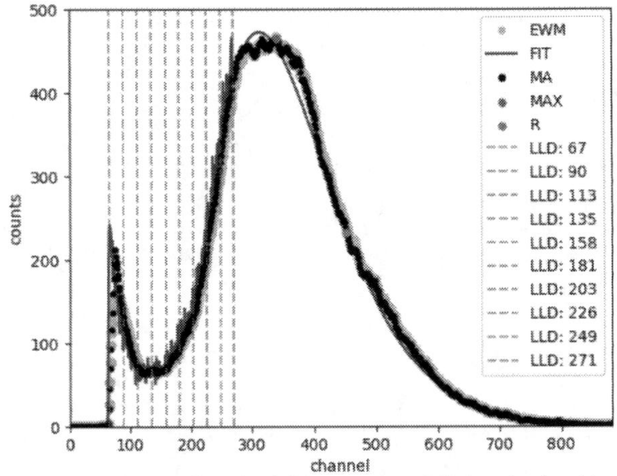

Fig. 4: Comparison of the pulse height spectrum of fission chamber 19 smoothed with moving average (black), exponential weighted mean (orange) and the fitted one (blue).

TABLE III
EFFECTIVE MASS OF FISSION CHAMBER 19. PULSE HEIGHT SPECTRUM PROCESSED WITH DIFFERENT SMOOTHING OPTIONS. RESULTS REPORTED AT 0.4 AND 0.6 R.

Smoothing method	Window, Span or Order	Effective Mass [μg] (0.4 R)	Effective Mass [μg] (0.6 R)
None	n.a.	1017.7	854.2
Moving	5	1013.5	825.4
Average	10	1012.5	821.3
	-5 to +5	1014.6	829.4
	15	1011.6	821.3
Exponential	5	1015.4	833.1
weighted	10	1012.5	821.3
mean	15	1011.6	814.0
Mean		1013.5	827.5
Absolute standard deviation		2.3	12.2

From now on, all the pulse height spectrum integrations are performed finding the R-channel of the smoothed spectrum and integrating the raw data from the so-found discrimination levels. The effective mass emerges to be quite insensitive to the smoothing options (see Table III). A larger standard deviation is found for effective mass values at 0.6 R then at 0.4 R. The effective mass obtained fitting the pulse height spectrum is rather sensitive to the fit choice. Even if the fitted integral counts in the spectrum differ by about 0.1 % from the raw ones, the discrimination levels considered to determine the effective mass easily fall in the region of transition between raw data and fitted data (see Fig. 4). The effective mass is then quite sensitive to the choice of the channel where the fit starts. Low polynomial orders or low channels to start the fit from can distort the spectrum significantly. For this reason, the results of the pulse height spectrum fit are not reported in Table III.

Discrimination level 0.4 R (channel 181) is in the pulse height spectrum valley between noise and fission signal. Changing the discrimination level by one channel there impacts the spectrum integral less than around 0.6 R (channel 271), which is much closer to the pulse heigh spectrum maximum. The larger sensitivity of the integrated PHS to the discrimination level at 0.6 R than at 0.4 R explains the larger standard deviation reported in Table III.

These results depend on the measured pulse height spectrum (i.e., on the measured fission chamber). In general, the more distinctly the fission signal stands out from the measurement noise (i.e., the higher the peak-to-valley ratio) and the broader the valley region (relative to the overall spectrum), the less impact the smoothing method will have on the results.

C. *Comparison of successive acquisitions*

Fig. 5 shows the pulse height spectra acquired in three successive acquisitions with fission chamber 19. Nor the reactor nor the electronics were shut down between the acquisitions and no change in the fission chamber position or acquisition setting happened.

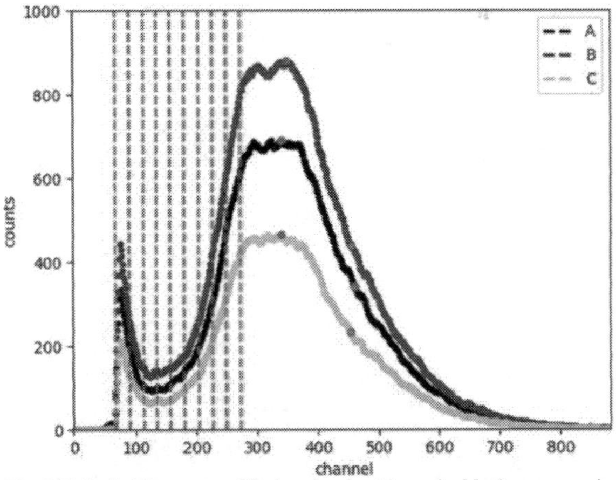

Fig. 5: Pulse height spectra of fission chamber 19 acquired in three successive acquisitions (A, B, C) after smoothing.

Table IV reports the effective mass measured in each of the three acquisitions for each of 10 discrimination levels computed as a fraction of the R channel (note that this requires rounding by NEREA). The standard deviation of the measured effective masses (A, B and C) ranges from about 0.5%, at 0.4 R, to 1% at 0.6 R. The spread of the measurements is not fully explained by the uncertainty propagated from the fission chamber counting uncertainty (0.2% - 0.3%) and the power monitor counting uncertainty (< 0.05% - the difference in power monitor count rate between the three acquisition goes up to 0.2%).

TABLE IV
EFFECTIVE MASS [MG] OF FISSION CHAMBER 19 AFTER ACQUISITIONS A, B AND C FOR EACH OF 10 DISCRIMINATION LEVELS FRACTIONAL OF THE R-CHANNEL.

R		A		B		C
	ch	Effective Mass [μg]	ch	Effective Mass [μg]	ch	Effective Mass [μg]
0.15	68	1095.6	68	1104.8	67	1107.8
0.20	91	1061.6	90	1069.7	90	1073.5
0.25	114	1045.9	113	1051.9	113	1056.3
0.30	137	1033.1	136	1037.9	135	1043.7
0.35	160	1019.4	158	1023.6	158	1030.1
0.40	183	1001.7	181	1006.2	181	1012.5
0.45	206	977.7	204	982.0	203	989.4
0.50	229	940.8	227	944.0	226	952.9
0.55	252	882.8	249	889.6	249	896.1
0.60	275	804.9	272	811.9	271	821.3

Table V shows how the effective mass results of acquisitions A, B and C compare. Sensitivity to the pulse height spectrum processing choice is here observed. This is particularly relevant fitting the spectrum as the number of channels fitted before the maximum and the polynomial order (first and second value in parenthesis in Table V) give rather oscillating results. On the other hand, integrating from the same channel discriminator (independently of R), the spread of the measured effective masses reduces, as shown in Table V.

TABLE V
DIFFERENCE IN EFFECTIVE MASS INTEGRATING FROM R-CHANNEL-RELATED AND ABSOLUTE DISCRIMINATION LEVELS.

Spectrum processing and discriminator criterion		$\frac{\text{Effective Mass B}}{\text{Effective Mass A}} - 1$	$\frac{\text{Effective Mass C}}{\text{Effective Mass A}} - 1$
0.6 R	MA	0.87 %	2.04 %
	EWM	1.36 %	2.04 %
	FIT (100, 7)	1.34 %	2.51 %
	FIT (150, 7)	- 0.07 %	2.05 %
	FIT (150, 9)	0.38 %	1.60 %
Ch. 275	Raw data	- 0.53 %	0.19 %

While one could argue that the observed effective mass standard deviation between acquisitions A, B and C is of a similar order of magnitude as the counting uncertainties until channel 0.4 R, after that this is not the case.

We attribute the difference in measured effective mass between acquisitions A, B and C to the R channel calibration methodology. As it was already observed, the sensitivity of the pulse height spectrum integral to the associated discrimination level (i.e., to the integration lower bound) increases moving towards the pulse height spectrum maximum. When the calibration methodology results in different discriminators (see Table V), the measured effective mass is affected in a way that is not reflected by any of the considered uncertainties.

IV. DISCUSSION ON THE R-CHANNEL CALIBRATION METHODOLOGY

Adoption of the R-channel to compare pulse height spectra comes with many advantages and some shortcomings. Fundamentally, the R-channel methodology consists of a pulse height spectrum re-binning (and consequent reshaping). When both the calibration and the measurement spectrum are re-binned to the same fractions of R, one can look at them in the same domain and compare when the pulse height spectrum integrals are proportional with a constant factor. This has mostly two advantages:

- the processing methodology is inherently rather independent of the acquisition chain amplification gain, as the R re-binning corrects for it;
- when the pulse height spectrum valley is well identified, the pulse height spectrum integral ratio is rather stable in that region.

These two advantages should not be underestimated: the experimental procedures and practices are simplified by the R-channel methodology, as there is no need to calibrate each fission chamber with each amplifier – pre-amplifier couple. Moreover, one could argue that a fission chamber with poor peak-to-valley ratio is not well suited for spectral index measurements and should rather be used for bare count rate comparisons as in fission traverse measurements.

On the other hand, it was observed that the R-channel methodology suffers from uncertainties associated with the determination of the R channel as this does not only require good statistics in the pulse heigh spectrum integral, but also in the individual channels of the maximum and of R. This can be quite hard to achieve with low-counting fission chambers and gives a visible effect also on the effective mass measured with fission chamber 19 (about 10^5 counts after 0.6 R).

While alternatives to the R-channel calibration methodology are being considered, the recommendation we give is to perform multiple calibration and measurement acquisitions, possibly varying the electronics. Then the spread of the measured effective masses could be taken as an agnostic estimate of the calibration uncertainty, accounting not only for counting statistics uncertainty, but also for the inherent uncertainties associated with the calibration methodology.

V. CONCLUSIONS

The NEREA Python package for integral reactor experiments is presented in this paper. The living version of the code is available on GitHub [7] and the code is available on PyPi. NEREA provides a user with classes to process fission chamber acquisitions for spectral index, fission rate traverse and control rod calibration measurements.

To show the versatility of NEREA and the degree of processing control it allows, the results of a fission chamber calibration campaign performed at BR1 are presented. The results highlight

a benefit in fission chamber recalibration before each experimental campaign. The little-to-no impact of the pulse height spectrum smoothing choices implemented in NEREA on the fission chamber effective mass results was shown for a fission chamber with good signal-to-noise discrimination.
An analysis of the R-channel calibration methodology implemented in NEREA is also presented. This shows that, besides its advantages, such a methodology comes with uncertainties that are not fully reflected by the observed counting statistics. While alternative options are under investigation, it is recommended to perform multiple independent calibration measurements and derive a calibration uncertainty after statistical considerations on the repeated measurement set.

ACKNOWLEDGMENT

This work was supported financially by Association Vinçotte Nuclear (AVN).

REFERENCES

[1] A. Krása et al, "Comparative study on neutron data in integral experiments of MYRRHA mockup critical cores in the VENUS-F reactor", *EPJ Web of Conferences*, 2017.

[2] A. Kochetkov, et al., "Integral experiments in the VENUS-F reactor", *Annals of Nuclear Energy*, 2021.

[3] F. Grimaldi et al., "The CoRREx neutron spectrum filtering campaign at VENUS-F for calculation-to-experiment discrepancy interpretation", *Annals of Nuclear Energy*, 2025.

[4] A. Kochetkov et al., "The Lead-Based VENUS-F Facility: Status of the FREYA Project", *EPJ Web of Conferences*, 2016.

[5] E. Fridman et al., "Modeling of FREYA fast critical experiments with the Serpent Monte Carlo code", *Annals of Nuclear Energy*, 2017.

[6] J. Wagemans et al., "The 235U Prompt Fission Neutron Spectrum in the BR1 Reactor at SCK•CEN", EPJ Web of Conferences, 2016.

[7] F. Grimaldi, "NEREA GitHub", 2025. [Online]. Available: https://github.com/GrimFe/NEREA. [Accessed 2025].

[8] A. Johnson et al., "serpentTools: A Python Package for Expediting Analysis with Serpent", *Nuclear Science and Engineering*, 2019.

[9] J. Leppänen et al., "The Serpent Monte Carlo code: Status, development and applications in 2013", *Annals of Nuclear Energy*, 2015.

[10] W. Mckinney, "Data Structures for Statistical Computing in Python", Proceedings of the 9th Python in Science Conference, 2010.

[11] N. Blanc De Lanaute, "Développement et optimisation de méthodes de mesures neutroniques par chambre à fission auprès de réacteurs expérimentaux. Maîtrise , traitement et réduction des incertitudes", PhD thesis, 2012.

[12] B. Geslot et al., "Method to Calibrate Fission Chambers in Campbelling Mode", *International Conference on Advancements in Nuclear Instrumentation, Measurement Methods and their Applications (ANIMMA)*, 2011.

[13] V. Lamirand et al., "Miniature fission chambers calibration in pulse mode: interlaboratory comparison at the SCK•CEN BR1 and CEA CALIBAN reactors" *International Conference on Advancements in Nuclear Instrumentation, Measurement Methods and their Applications (ANIMMA)*, 2013.

[14] A. Kochetkov et al., "Calibration of Large Photonis Fission Chambers in Standard Neutron Fields of the BR1 Reactor", IEEE TNS, 2022.

[15] V. Doolin and V. Mozhaev, "Method of fission chamber efficiency determination using uncorrelated neutron background measurement", Nuclear Instruments and Methods, 1972.

[16] J. A. Grundl et al., "Measurement of absolute fission rates," *Nuclear Technology*, 1975.

[17] M. Obu, "Preparation and Characteristics of Fission Chambers", 1981.

[18] T. Misawa et al., Nuclear Reactor Physics Experiments, Kyoto University Press, 2010.

[19] pandas, "pandas.Series.ewm," [Online]. Available: https://pandas.pydata.org/docs/reference/api/pandas.Series.ewm.html. [Accessed 2025].

[20] J.-P. Hudelot, " Developpement amelioration et calibration des mesures de taux de reaction neutroniques: elaboration d'une base de techniques standards", PhD thesis, 1998.

APPENDIX

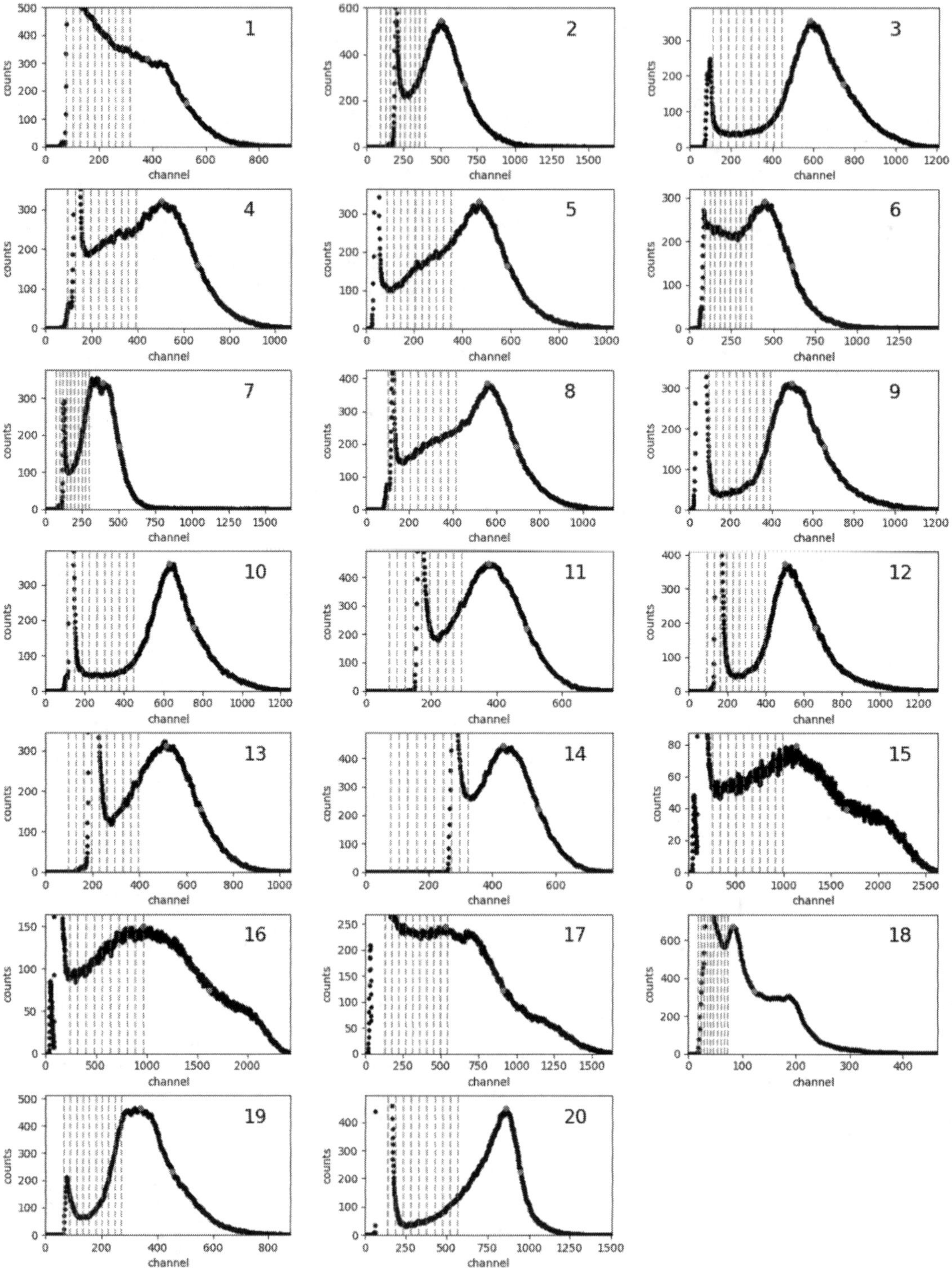

Fig. 6: Measured calibration pulse height spectra. The maximum (green) and R (red) points are highlighted. Red dashed lines indicate 10 discrimination levels as a fraction of the R-channel.

Subcritical Neutron and Gamma Noise Measurements at the Seven Percent Critical Experiment (7uPCX)

Cole Kostelac[1,2*], George McKenzie[1], Robert Weldon[1], Nicholas Whitman[1], Jesson Hutchinson[1], Ayodeji Alajo[2]
[1]Los Alamos National Laboratory, United States
[2]Missouri University of Science and Technology, United States
(*) kostelac@lanl.gov

***Abstract*— As part of a collaborative international effort organized by Lawrence Livermore National Laboratory (LLNL), with key participants from L'Institut de radioprotection et de sûreté nucléaire (IRSN), Los Alamos National Laboratory (LANL), and Sandia National Laboratories (SNL), a series of high-multiplication subcritical neutron and gamma noise measurements were planned and executed. The primary aim of this research was to advance detector technology, assess the validity of gamma noise for subcriticality measurements, and nuclear criticality safety, focusing on collecting list-mode or time-series data from various reactor configurations with multiplication values ranging from 20 to 310. This comprehensive dataset enabled a detailed comparative analysis of multiple detector systems and the results of both neutron and gamma noise measurements. In this work we focus on experimentally comparing the results from neutron and gamma noise measurements. We note good agreement between estimations of the prompt neutron decay constant and demonstrate the effects of changing reactor geometry on the efficiency of the differing methods.**

***Keywords* — Criticality, Neutron Noise, Gamma Noise, Detectors, Cross Power Spectral Density**

I. Introduction

The behavior of nuclear reactors and other systems hosting fission chain reactions is a topic of central importance in nuclear engineering, particularly to reactor physicists and criticality safety analysts. A key parameter in describing such systems is the effective neutron multiplication factor, k_{eff}, defined as the ratio of the number of neutrons in one generation to that in the previous one [1].

Neutrons in these systems are produced through two mechanisms: prompt neutrons, which are emitted immediately during the fission process, and delayed neutrons, which are released from the decay of specific fission products known as delayed neutron precursors. The presence of delayed neutrons is critical to the controllability of nuclear systems. When k_{eff} is equal to unity, the reactor is said to be delayed critical, meaning the chain reaction is sustained primarily by delayed neutrons. Systems with a k_{eff} greater than one is supercritical, and their neutron populations grow exponentially; the degree of supercriticality in such cases can often be inferred using inverse kinetics techniques. Conversely, when k_{eff} is less than one, the system is subcritical, and the neutron population declines over time but can achieve a steady state neutron population by use of a neutron source such as PuBe or ^{252}Cf.

Accurately determining the degree of subcriticality can be accomplished with noise methods. In subcritical systems, fission chains are inherently stochastic, driven by the probabilistic nature of neutron interactions and nuclear data. This stochastic behavior gives rise to measurable fluctuations in neutron population, which can be analyzed using techniques collectively referred to as reactor noise analysis. These methods offer a non-invasive approach to inferring important system parameters, including subcriticality, and are the focus of this work.

These methods are well understood for analyzing the fluctuations in the neutron population but can also be performed by observing similar fluctuations in the prompt gamma population [2][3][4]. Like neutrons, these prompt gamma emissions are born directly from the fission process. Since they propagate at the speed of light, these gamma emissions essentially carry a timestamp of the fission event with them when detected by a detector unlike neutrons which often scatter and must be thermalized to be detected by thermal neutron detectors such as ^{3}He proportional counters or BF_3 detectors. Additionally, gamma detectors are typically less invasive to the system being measured, the aforementioned neutron detectors detect said neutrons by a capture process. This removes a neutron from the system, thus perturbing the chain reaction. Gamma detectors used for noise analysis typically have very low neutron interaction cross sections and are less perturbing to the system they are measuring.

Due to these possible advantages of gamma noise, we compare results of noise measurements of thermal neutrons, fast neutrons, and gammas at various levels of subcriticality.

II. Theory

Of principle interest in this study is the kinetic parameter known as the prompt neutron decay constant, α.

$$\alpha = \frac{\rho-\beta_{eff}}{\Lambda}, \quad (1)$$

where ρ is the reactivity of the system, β_{eff} is the effective delayed neutron fraction and Λ is the prompt neutron generation time. Α serves as an experimentally measurable quantity directly related to the subcriticality of a multiplying system. Various so-called noise methods have been developed that use the fluctuating population of neutron and gammas measured by a detector to the α of a system. In this work, we make use of the cross power spectral density (CPSD) technique to do so. The CPSD between two detector signals measures their correlation as a function of frequency. For reactor noise, this CPSD takes the form of a Lorentzian distribution whose break frequency is proportional to α [5][6].

$$||\mathrm{CPSD}|| = \frac{A}{1+\left(\omega/\omega_b\right)^2} \qquad (2)$$

$$A = \frac{2D}{F(\beta_{eff}-\rho)} \qquad (3)$$

$$\alpha = 2\pi\omega_b \qquad (4)$$

Where D is the Diven factor, and F is the integral fission rate of the multiplying system. Although this relationship was derived for neutron noise, we will also apply it to the prompt gamma fluctuations as well.

III. EQUIPMENT

The Seven Percent Critical Experiment (7uPCX) is a critical assembly consisting of a light water moderated and reflected array of UO_2 fuel rods enriched to 6.90% ^{235}U located at Sandia National Laboratory. A high-quality benchmark of the system is included in the International Criticality Safety Benchmark Evaluation Project (ICSBEP) Handbook as LEU-COMP-THERM-078 [7]. Its highly reconfigurable fuel region allowed subcritical measurements driven by a ^{252}Cf source ranging from a multiplication of 20 to 310.

Figure 1: 7uPCX Experiment. The four MC-15 detectors are seen surrounding the water tank and EJ-309 scintillators are seen in detector drywells with orange signal cable.

These configurations were diagnosed by four ^3He MC-15 detectors placed outside the water tank of the assembly and four 2.54 cm x 2.54 cm EJ-309 liquid scintillators placed in detector drywells seen in Fig. 1. Thermal neutrons were measured by the ex-core MC-15 detectors and fast neutrons and gammas were measured by the in-core EJ-309 using pulse shape discrimination.

The MC-15 is a multiplicity detector designed by LANL consisting of 15 ^3He proportional counters embedded in HDPE. Detection events are time tagged by on board acquisition hardware with a time resolution of 100 ns [8]. The list mode data from all four MC-15s (a total of 60 ^3He detectors) were recorded for each measurement.

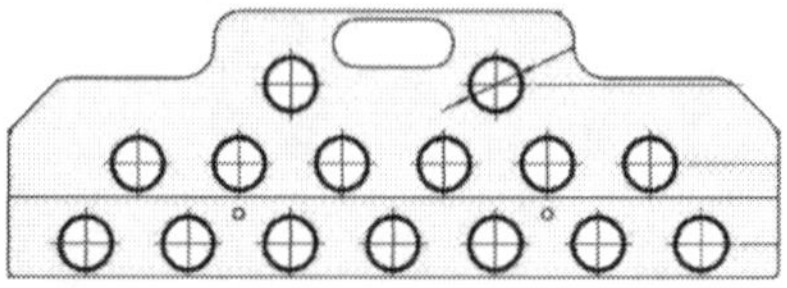

Figure 2: MC-15 detector plan view.

The EJ-309 detectors used a CAEN R1470ETD power supply module to power the detectors, and a CAEN DT5730 (500 MHz, 14 bit) digitizer was used to record a short (Q_s) and long (Q_L) integrated charge for each recorded pulse [9]. Since fast neutron and gamma induced pulses produce slightly different pulse shapes (Fig. 2), one can identify these particles by plotting Q_L against the ratio Q_S/Q_L as seen in Fig. 3.

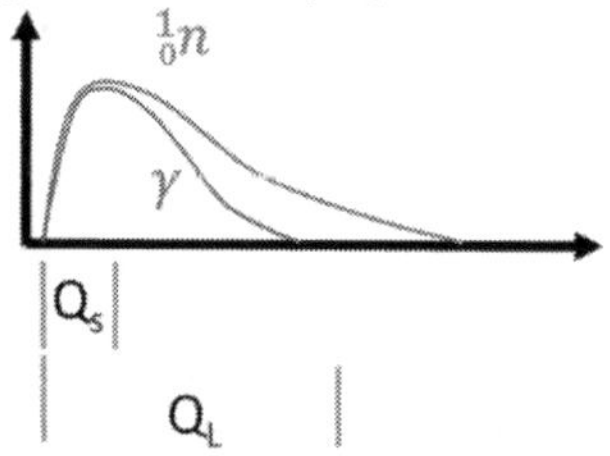

Figure 3: Notional neutron and gamma induced pulses from EJ-309 scintillators.

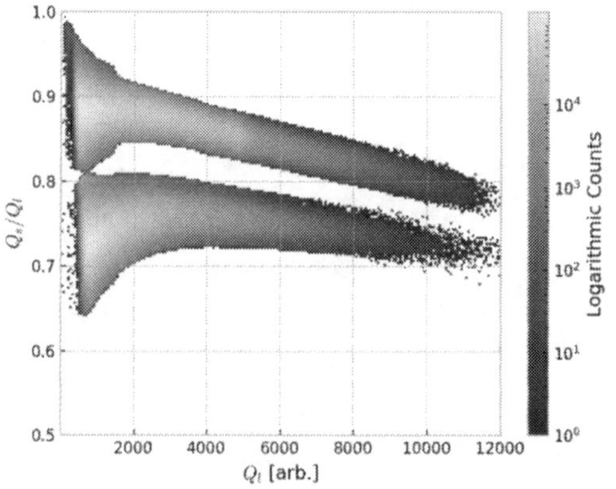

Figure 4: Pulse shape discrimination plot from EJ-309 scintillators for the 1048 rod 7uPCX configuration.

IV. METHODS

Fluctuations in the thermal and fast neutron and gamma populations were analyzed in the frequency domain using the aforementioned CPSD method to infer the prompt neutron decay constant. The CPSD between each detector type provides six unique estimations of the prompt neutron decay constant. Notably, the MC-15 and EJ-309 detector signals were acquired by different digitizers. Therefore, a scheme to synchronize the signals was developed. This entailed computing the cross correlation of the signals in order to estimate the time lag between the beginning on their acquisition. Upon the estimation of this time lag, the signals were synchronized and the CPSD was computed to estimate α.

To estimate α for each configuration first, detected pulses from each detector were binned into counts per one millisecond. The CPSD of each pair of detector signals was estimated using a Python implementation of the Welch method [10][11].

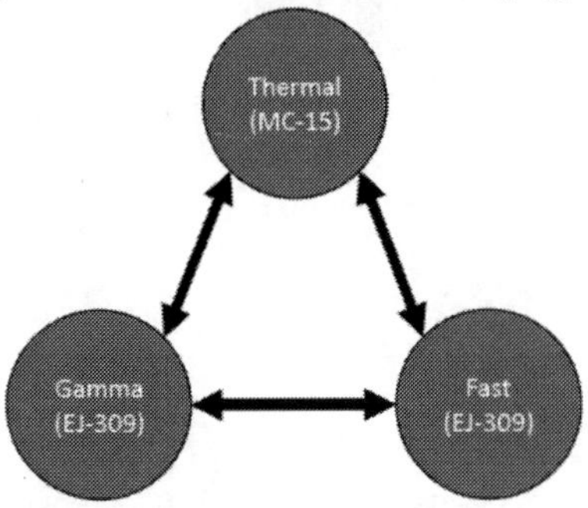

Figure 5: Combinations of three particle signals to estimate α.

V. RESULTS

Estimations of α were made for five subcritical configurations of the assembly. Notably, the CPSD of the fast neutron signal from the EJ-309 detectors did not yield a measurable α, however the CPSD between the fast and gamma signal and fast and thermal signals did.

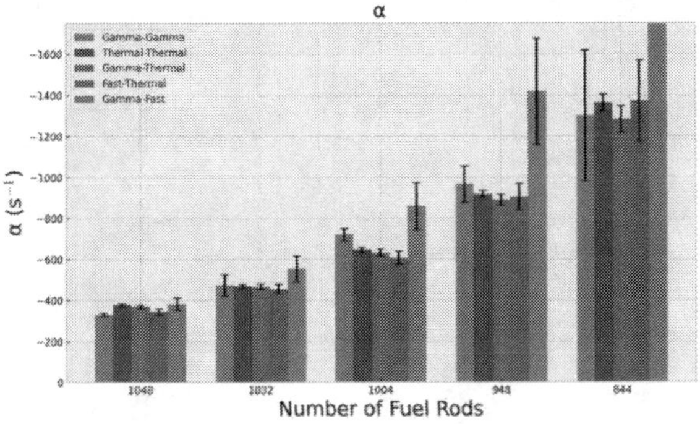

Figure 6: The prompt neutron decay constant of various subcritical configurations as a function of number of fuel rods.

The parameter $A\alpha^2$ is proportional to the inverse of the integral fission rate, F (Eq. 5). This value is plotted against the number of fuel rods in each measured configuration. As expected, this value increases as the multiplication decreases as shown in Fig. 7.

$$A\alpha^2 = \frac{2D}{F\Lambda^2} \tag{5}$$

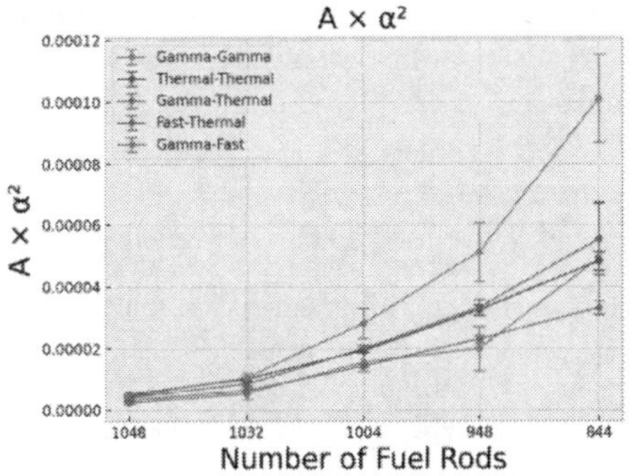

Figure 7: The $A\alpha^2$ parameter shows an increase as neutron multiplication decreases, indicating a lower integral fission rate in the system.

Multiplying the count rate of the detectors by this $A\alpha^2$ give a parameter that is proportional to the detector efficiency, ε (Eq. 6). As seen in Fig. 8, estimation using signals from the MC-15 detectors lose efficiency as fuel rods are removed. This is due to the fact that as fuel rods are removed, the distance between the MC-15 detectors and the fuel is increased, and neutrons must travel through more water to reach the detector. The effect of both lead to a decrease in the probability a neutron from the core will reach the detector, thus decreasing the detector's efficiency. This effect is not noticeable with signals using only the in-core EJ-309 detectors measuring fast neutrons and gammas.

$$A\alpha^2 c = \frac{2D\epsilon}{\Lambda^2} \tag{6}$$

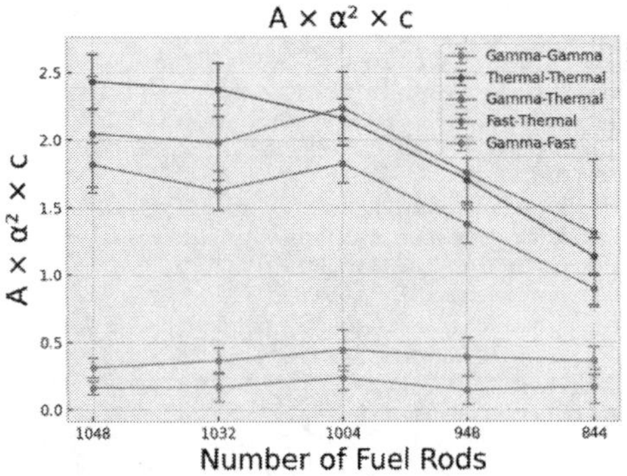

Figure 8: The effects of efficiency loss are seen in the CPSDs using the thermal neutron signal from the ex-core MC-15 detectors.

VI. CONCLUSIONS

From this work we note the good agreement between neutron and gamma noise in estimating the prompt neutron decay constant. We also demonstrate that the usage of gamma noise can be advantageous in certain scenarios where there are changes in fuel and moderator geometries that would otherwise have effects on neutron detection efficiency. Further studies should include an explicit derivation of the CPSD between thermal and fast neutrons and prompt gammas. This would allow for the estimation of β_{eff} and the integral fission rate using gamma noise.

ACKNOWLEDGMENT

Thank you to Sandia National Laboratories (SNL) for providing staffing support and consulting for these measurements: Beth Hanson, Jason Soares, Patrick Ward, Gary Harms, and David Ames. Los Alamos National Laboratory is operated by Triad National Security, LLC, for the National Nuclear Security Administration of the US Department of Energy under Contract No. 89233218CNA000001.

REFERENCES

[1] G. R. Keepin, Physics of Nuclear Kinetics. Addison-Wesley Publishing Company, 1965. [Online]. Available: https://books.google.com/books?id=mNVEAAAAIAAJ

[2] J. Orndoff, "Prompt Neutron Periods of Metal Critical Assemblies," Nuclear Science and Engineering, vol. 4, pp. 450–460, 1958.

[3] W. K. Lehto and J. M. Carpenter, "Reactor Noise Measurements Using Gamma Rays," Nuclear Science and Engineering, vol. 32, no. 1, pp. 142–146, 1968, doi: 10.13182/NSE68-A18838.

[4] O. Pakari, T. Mager, P. Frajtag, A. Pautz, and V. Lamirand, "Gamma noise to non-invasively monitor nuclear research reactors," Scientific Reports, vol. 14, no. 1, 2024, doi: 10.1038/s41598-024-59127-y.

[5] C. E. Cohn, "Determination of Reactor Kinetic Parameters by Pile Noise Analysis," Nuclear Science and Engineering, vol. 5, no. 5, pp. 331–335, 1959, doi: 10.13182/NSE59-A25605.

[6] E. Gilad, B. Geslot, P. Blaise, and C. Dubi, "Sensitivity of Power Spectral Density Techniques to Numerical Parameters in Analyzing Neutron Noise Experiments," Progress in Nuclear Energy, vol. 101, pp. 288–298, 2017.

[7] G. Harms, "Water-Moderated Square-Pitched U(6.90)O2 Fuel Rod Lattices With 0.52 Water-To-Fuel Volume Ratio (0.855 cm Pitch)," in International Handbook of Evaluated Criticality Safety Benchmark Experiments, OECD Nuclear Energy Agency, 2016, ICSBEP No. LEU-COMP-THERM-07.

[8] Multiplicity Counter-15 (MC-15) User Manual, Los Alamos National Laboratory (LANL), LA-UR--18-29563, 2018.

[9] R. A. Weldon, T. E. Culter, J. D. Hutchinson, G. E. McKenzie IV, L. Misurek, and E. B. Sorensen, Rossi Alpha Measurements – Rapid Organic (n, γ) Discrimination Detector (RAM-RODD) System Capabilities, Los Alamos National Laboratory (LANL), LA-UR-20-28902, 2020.

[10] Python Software Foundation, Python 3.8: Python Language Reference, Python Software Foundation, 2019. [Online]. Available: https://docs.python.org/3.8/

[11] P. Virtanen et al., "SciPy 1.0: Fundamental Algorithms for Scientific Computing in Python," Nature Methods, vol. 17, no. 3, pp. 261–272, 2020, doi: 10.1038/s41592-019-0686-2.

Development of the pile oscillation method for the determination of integral cross section data at the AKR-2 training and research reactor

Alexander Knospe[1,*], Juan José Gómez-Rodriguez[1], Marco Viebach[1], Carsten Lange[1] and Antonio Hurtado[1]
[1]Dresden University of Technology, Germany
(*)alexander.knospe@tu-dresden.de

***Abstract*— The development of Generation-IV reactors requires validated low uncertainty nuclear data for novel materials. To address this, the pile oscillation method for measuring integral absorption cross sections has been developed at the AKR-2 research reactor at Dresden University of Technology. The experimental setup consists of a linear motor-driven oscillator, four He-3 proportional counters, and a custom FPGA-based data acquisition system that records individual neutron events. Signals are transformed to the frequency domain, and cross-power spectral density analysis is used to determine the sample's reactivity worth while minimizing biases. The method was validated against Monte Carlo simulations using Serpent 2.2.1. Measurements on foils of various natural materials, including gold, indium, iridium, copper, hafnium, zirconium, and carbon, were performed. Excellent agreement was found for strong absorbers like indium and iridium, with calculated-to-experimental ratios deviating by less than 6%, accurately reproducing self-shielding effects. As expected, larger deviations were observed for weakly absorbing materials such as copper and carbon, confirming the method's performance characteristics. These results demonstrate that the developed experimental design, data acquisition, and analysis techniques provide a validated and reliable tool for obtaining integral cross-section data, supporting the validation of nuclear data libraries and the safety assessment of advanced reactor designs.**

***Keywords* —Pile oscillator, reactor noise, nuclear data, AKR-2.**

I. INTRODUCTION

THE advancement of Generation-IV nuclear reactor concepts, such as the chloride based Molten Salt Fast Reactor (MSFR), necessitates the development and qualification of novel materials for fuels, coolants, and structural components. The operational safety and efficiency of these advanced designs depend critically on the accurate knowledge of nuclear data for these materials, particularly because their neutron energy spectra differ significantly from those of the conventional Light Water Reactor (LWR) fleet. To address this critical need for high-fidelity experimental data, the NAUTILUS project [1] was established at the Dresden University of Technology. One of the project's primary objectives is to develop an experimental platform at the AKR-2 research reactor for the measurement of nuclear data and the validation of nuclear simulation codes. This contribution presents the development, implementation, and validation of a key experimental technique within this project: the pile oscillation method for determining integral absorption cross sections.

The method, albeit already used in the 1940s - 1960s (cf. for example [2]–[4]), has gained more attention in recent years due to advancements in signal processing, electronics, and detector developments, [5]–[7]. The method is suitable for the measurement of compound materials and by tuning the frequency and measurement time it is possible to measure very small reactivity perturbations and thus allow for small sample sizes.

This contribution details the first comprehensive validation of the pile oscillation technique for cross section measurements at the AKR-2 facility, establishing a robust methodology for future experiments. To this end, thin stacks of solid foils were oscillated between the core center and the outside of the core in a way that the neutron flux at the location of the sample can be assumed to be constant despite the relatively small core of the AKR-2 reactor and the perturbation can be assumed to be located at a point in space.

II. THEORETICAL FRAMEWORK

The dynamic response of the AKR-2 reactor to small reactivity perturbations inside the reactor is well-described by the point kinetics equations. For a small reactivity perturbation in frequency space, $\delta\rho(\omega)$, at an angular frequency ω around a critical state (where the steady-state power is P_0), the fractional change in reactor power, $\delta P(\omega)$, is related to the reactivity insertion via the zero-power reactor transfer function, $H(i\omega)$,[8]:

$$\frac{\delta P(\omega)}{P_0} = H(i\omega) \cdot \delta\rho(\omega), \tag{1}$$

$$H(i\omega) = \frac{1}{i\omega\left(\Lambda + \sum_k \frac{\beta_k}{i\omega + \lambda_k}\right)}, \tag{2}$$

with β_k being the effective delayed neutron fractions, Λ being the neutron generation time and λ_k the decay constant of the precursor group k. The reactivity worth of a small, oscillating sample is given by first-order perturbation theory, which relates the change in reactivity to the change in the neutron absorption $\delta A(\omega)$ and production operators $\delta F(\omega)$, weighted by the adjoint flux, Φ^*. For a sample where absorption is the dominant effect, the reactivity perturbation, $\delta\rho$, is proportional to the adjoint-weighted absorption reaction rate $\langle\Phi^*, \delta A(\omega)\Phi\rangle$, where $\langle\Phi^*, \delta A(\omega)\Phi\rangle$ represents the inner product of the adjoint flux, the change in the absorption operator, and the forward flux.

By performing measurements on a sample of material M1 and second sample, M2, under identical oscillation conditions, the reactor transfer function and other constants cancel out. This allows the ratio of their reactivity worths to be determined from the ratio of their measured power oscillations:

$$\frac{\delta P(\omega)_{\mathrm{M1}}}{\delta P(\omega)_{\mathrm{M2}}} \approx \frac{\delta\rho_{\mathrm{M1}}(\omega)}{\delta\rho_{\mathrm{M2}}(\omega)} = \frac{\langle\Phi^*, \delta A_{\mathrm{M1}}(\omega)\Phi\rangle}{\langle\Phi^*, \delta A_{\mathrm{M2}}(\omega)\Phi\rangle}. \quad (3)$$

Since the samples are oscillated between the same locations (the reactor center and the outside position), the adjoint flux weighting term, Φ^*, is approximately the same for both measurements and can be cancelled. This simplifies the relationship to the ratio of the absorption reaction rates, $R_{\mathrm{M}i} = \int \delta\Sigma_{\mathrm{abs},i}\Phi\, \mathrm{d}E\mathrm{d}V$:

$$\frac{\delta P(\omega)_{\mathrm{M1}}}{\delta P(\omega)_{\mathrm{M2}}} \approx \frac{\delta\rho_{\mathrm{M1}}(\omega)}{\delta\rho_{\mathrm{M2}}(\omega)} \approx \frac{R_{\mathrm{M1}}}{R_{\mathrm{M2}}}. \quad (4)$$

This relationship forms the basis of the pile oscillator technique, enabling the determination of cross section ratios between different samples investigated in this contribution. The right-hand side of (4) will be measured in the experiment while the left-hand side will be determined with the neutron transport Monte Carlo code Serpent 2.2.1, [9].

III. EXPERIMENTAL METHODS AND SIMULATION SETUP

A. The AKR-2 Research Reactor

The experiments were conducted at the Ausbildungs-kernreaktor-2 (AKR-2), a thermal zero-power training and research reactor located at the Dresden University of Technology, Germany, operated by the Chair of Hydrogen and Nuclear Energy Technology. The reactor operates at a nominal thermal power of 2 W and features a core of a diameter of 25 cm and a height of 27 cm composed of UO_2 fuel with a mean enrichment of 19.62 %, homogeneously mixed with a polyethylene moderator. For this work, the pile oscillation experiment was installed in the central horizontal channel, designated 1-2, of the reactor core. This experimental channel passes directly through the core allowing samples to oscillate radially in the reactor. The core is surrounded by a graphite reflector followed by a biological shield made from paraffin which is surrounded by heavy concrete for radiation shielding.

B. Material samples

For the experiments presented here, material samples of solid materials were laser cut from foils of different thicknesses of the natural elemental materials indium (In), iridium (Ir), gold (Au), hafnium (Hf), copper (Cu), zirconium (Zr), and carbon (C). Multiple circular pieces of the foils were stacked to form cylinders of different heights. All foils have a diameter between 12 mm and 14 mm (see Table I). The selection of materials was composed in such a way that there was a mix of stronger absorbing and weaker absorbing materials with respect to the AKR-2 spectrum.

TABLE I
SAMPLE GEOMETRIES AND INVESTIGATED MATERIALS

Material	Thickness in µm	Diameter in mm
Copper	500	14
Gold	30	14
Graphite	500	14
Hafnium	125	14
Indium	70 - 630	12
Iridium	50 - 400	13
Zirconium	500	14

C. Pile Oscillation Apparatus

The pile oscillator device, which was used before for reactor noise experiments at AKR-2 [10], [11], utilizes a high-precision linear motor axis, to which an aluminum guide rod is attached. Material samples are placed into low-mass aluminum containers, which are then inserted into the guide rod. Distance pieces ensure accurate and repeatable axial positioning of the sample container within the rod. The motion profile can be adapted by the user. For the measurements presented here, a trapezoidal motion profile was implemented oscillating the sample between the reactor core's center and a position 20 cm outside the core, which is a profile optimized for measuring reactivity changes dominated by neutron absorption. The driving time of the linear motor axis in between these termination points is approximately 0.7 s. A second empty container is placed at 20 cm from the sample container which minimizes the contribution of the containers themselves if the radial neutron flux is assumed to be symmetrical around the center. The instantaneous position of the guide rod is monitored online by a pull wire displacement sensor to verify the motion profile.

D. Neutron Detection and Data Acquisition

The neutron flux is monitored by four He-3 proportional counter detectors positioned inside the two tangential channels of the reactor at approximately 1 m from the core center in the graphite reflector (see Fig *1*). Two detector types of different lengths are used where each detector casing is equipped with a pair of ½ in diameter counter tubes of an active length of 50 mm or 100 mm, resp. The detectors feature an built-in preamplifier, shaper discriminator combination. They were custom built for AKR-2 by GBS Elektronik GmbH to ensure low dead time at sufficiently high count rate.

The detector signals are processed by a custom data acquisition (DAQ) system based on a ZYNQ™ Field-Programmable Gate Array (FPGA). This advanced DAQ system registers individual neutron detection events with timestamps synchronized over all channels, which permits user-defined dwell times and detailed statistical analysis during post-processing.

To measure the movement of the linear axis a pull wire displacement sensor signal is recorded separately on a BeagleBone® black minicomputer where the measurement is synchronized with a dedicated trigger signal between the DAQ and the BeagleBone® minicomputer.

E. Data Analysis Technique

The data analysis workflow begins with adjusting the dwell times to the registered motion profile and binning the data in post processing. This ensures that upon transforming the time-series data from the neutron detectors into the frequency domain, using spectral density estimation, the sampling frequency is optimized for the profile. The data is then drift-corrected by a cubic fit and normalized to the fitted power. To minimize biases from detector-specific noise and efficiency fluctuations, the cross-power spectral density (CPSD) is calculated between signals from all detector pairs. The amplitude of the first harmonic of the oscillation frequency is then extracted by fitting the respective peaks in the spectra to determine the sample's power oscillation amplitude, $\delta P(\omega)$. All detector combinations are used to determine the relative worths of the samples similar to the method developed in [10], [12]. The uncertainties of the experimental values are estimated by the standard deviation of the different detector combinations in the CPSD (for more detail see [10], [12]).

F. Monte Carlo Simulation for Validation

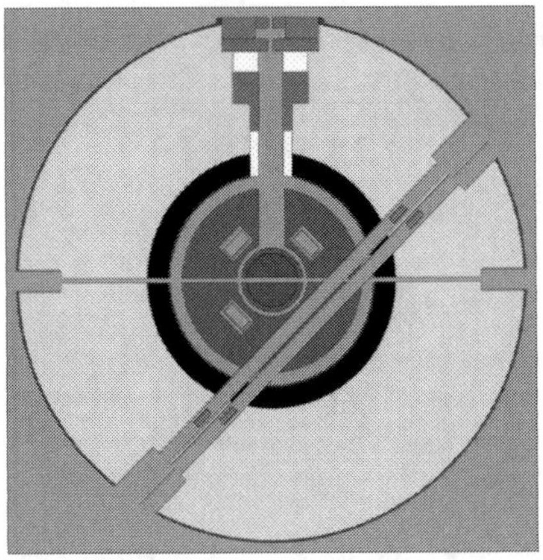

Fig 1. Horizontal cut through the core center in the used Serpent 2.2.1 model of AKR-2. Detector locations shown in red and the tangential channels are depicted while the location of the tangential channels is at ±8 cm height from the core center.

To validate the experimental method, a high-fidelity computational model of the AKR-2 reactor was used, utilizing the Monte Carlo transport code Serpent 2.2.1 with the ENDF-B/VIII.0 nuclear data library [13]. The simulations were performed using a two-step approach: first, the detailed neutron energy spectra along the experimental channel were calculated; second, these spectra were used to determine the reaction rates and resulting reactivity worths for the oscillated samples accounting for the neutron attenuation inside the sample. This approach, while an approximation, is valid under the assumption that the adjoint flux is unperturbed during the oscillation and that the trapezoidal motion profile can be assumed to be close to a rectangular shape.

IV. RESULTS AND DISCUSSION

The experimental results are presented from two distinct measurement campaigns. The first campaign focused on validating the method and investigating self-shielding effects in strong absorbers. The second campaign aimed to compare the reactivity worths of various materials with widely differing absorption cross sections. To ensure independence from the absolute values of the reactor kinetic parameters (β_k and Λ), which are sensitive to deviations in e.g. nuclear data libraries of the Monte Carlo models, the results for both campaigns are presented as reactivity ratios, and thus reaction rate ratios (see (4)). For the first campaign, results are normalized to the measurement of a single foil of the respective material. For the second campaign, ratios are calculated between the different measured materials. For all reported quantities, both experimental and simulation results were determined and subsequently compared by calculating their C/E ratios.

A. Validation with Strong Absorbers

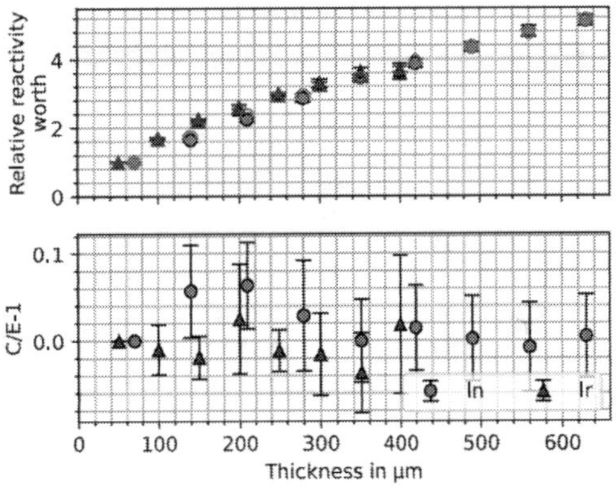

Fig. 2. Relative reactivity worth of indium and iridium w.r.t. one foil (top) and computation (C) over experiment (E) – 1 (bottom).

The method's performance was first evaluated using strongly absorbing materials, namely 1 to 9 foils of natural indium (In) with a thickness of 70 μm and 1 to 8 foils of iridium (Ir) with a thickness of 50 μm. For this campaign, a holding time of 10 s at each end position of the oscillation was sufficient to achieve good statistics due to the large reactivity signal of these materials. The experimental results, depicted in Fig. 2, show good agreement with the values predicted by Serpent simulations, with the calculated-to-experimental (C/E) ratios deviating by less than 6%, with slight deviations outside of estimated uncertainties for indium at thicknesses 140 μm and 210 μm. Furthermore, the effect of self-shielding was investigated by incrementally stacking foils of both materials. The measured relative reactivity worth demonstrates a clear non-linear dependence on sample thickness, a trend that is accurately reproduced by the Monte Carlo simulations. This confirms the experimental setup's ability to correctly capture complex self-shielding phenomena.

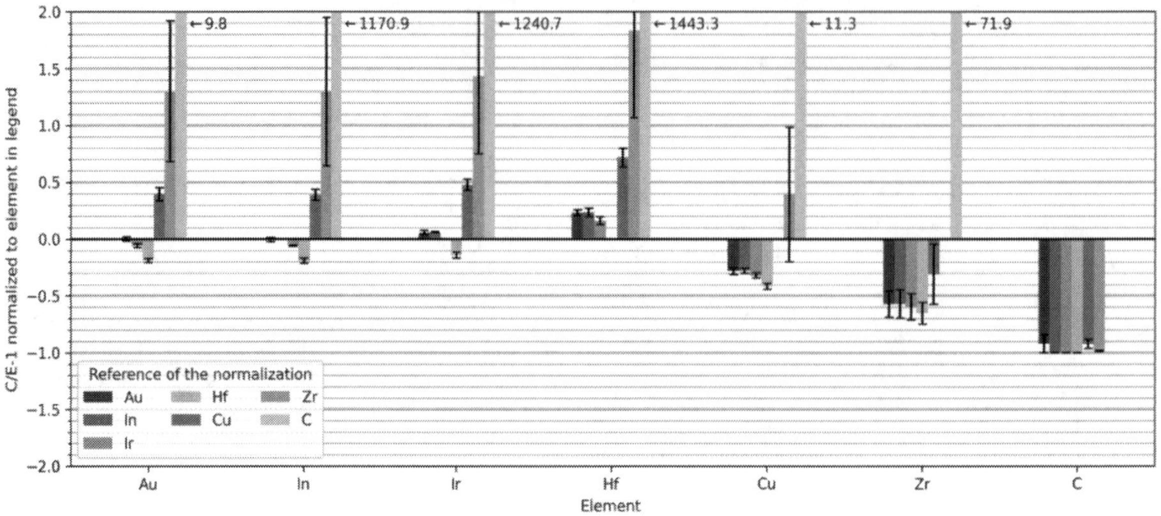

Fig. 3. Relative deviations of computational (C) to experimental (E) reaction rate ratios (R_{M1}/R_{M2}) between two materials M1 and M2. The x-axis labels correspond to the numerator (R_{M1}) while colors correspond to the denominator (R_{M2}) of the respective ratios.

B. *Measurements of Various Materials*

Following the initial validation, the technique was applied to a broader range of single-element foils, including gold (Au), copper (Cu), zirconium (Zr), hafnium (Hf) and carbon (C), in addition to indium (In) and iridium (Ir), where a holding time of 100 s was used. The results, see Fig. *3*, provide a comprehensive comparison between the measured integral absorption effects and the values predicted by simulation across materials with widely varying absorption cross sections. Results for the primarily absorbing materials (In, Ir, Au) show good agreement between the simulation and experiment with deviations below 6 %. For hafnium, there are systematic deviations of around 20 %, most likely from higher-than-expected zirconium content in the samples.

C. *Performance with Weak Absorbers*

For weakly absorbing materials such as copper, zirconium and carbon, larger deviations between experiment and simulation were observed (cf. Fig. *3*), with C/E-1 values w.r.t. the higher absorbtion materials (Au, In, Ir) reaching approximately 30% for copper, 60 % for zirconium and higher for carbon (near 100 %). This outcome is consistent with the inherent limitations of the reactivity-based pile oscillation method, where the signal-to-noise ratio diminishes for samples whose absorption cross section is very small, making their reactivity effect difficult to distinguish from statistical fluctuations in the reactor's neutron population. The very small absorbtion reaction rate of the carbon samples then leads to very high deviations for reaction rate ratios with carbon in the denominator (cf. Fig. *3*). Longer measurement times or larger sample sizes would be needed for all these materials with low absorbtion cross sections for this motion profile.

V. CONCLUSIONS AND OUTLOOK

This work has successfully demonstrated the development, implementation, and validation of a pile oscillation experiment at the AKR-2 training and research reactor. The experimental apparatus, coupled with a sophisticated FPGA-based data acquisition system and a robust CPSD analysis method, has proven to be a reliable tool for measuring integral absorption cross sections. The validation against high-fidelity Serpent simulations shows excellent agreement, particularly for strong absorbers, confirming the accuracy of both the experimental technique and the computational model of the AKR-2. This provides the Chair of Hydrogen and Nuclear Energy Technology at the Dresden University of Technology with a valuable and validated capability for obtaining integral data essential for the validation of nuclear data libraries and the safety assessment of future reactor designs.

To expand the capabilities of this facility, the experiment is currently being enhanced to enable the measurement of epithermal and thermal-spectrum-averaged cross sections. This involves the design and construction of a new pile oscillator device that incorporates a combination of cadmium shielding, to filter out thermal neutrons, and a polyethylene moderator, to locally tailor the neutron spectrum and compensate for the negative reactivity introduced by the cadmium. This next phase will allow for more detailed investigations into the energy-dependent behavior of neutron cross sections.

ACKNOWLEDGMENT

This work is part of the research project NAUTILUS funded by the German Federal Ministry of Research, Technology, and Space under the reference number 02NUK079.

REFERENCES

[1] M. Viebach, A. Knospe, C. Lange, B. Merk, and A. Hurtado, "NAUTILUS: A Project for the Development of Experimental Methods for Investigating Innovative Approaches to Nuclear Waste Management and to Nuclear Safety," in *PHYSOR 2024*, 2024, pp. 316–325.

[2] J. M. Harrer, R. E. Boyar, and D. Krucoff, *Measurement of CP-2 Reactor Transfer Function*, vol. 4373. Argonne National Laboratory, 1952.

[3] P. S. Christensen, "A description of the Pile Oscillators at DR1," 1967.

[4] H. Pomerance, "Thermal Neutron Capture Cross Sections," *Phys. Rev.*, vol. 83, no. 3, pp. 641–645, 1951.

[5] P. Leconte, P. Blaise, B. Geslot, and A. Gruel, "1959-2017: Highlights of sixty years of pile-oscillation experiments in the minerve facility for nuclear data improvement," *Int. Conf. Phys. React. PHYSOR 2018*, vol. Part F1683, pp. 1690–1702, 2018.

[6] P. Leconte *et al.*, "MAESTRO: An ambitious experimental programme for the improvement of nuclear data of structural, detection, moderating and absorbing materials - First results for ^{nat}V, ^{55}Mn, ^{59}Co and ^{103}Rh," in *2013 3rd International Conference on Advancements in Nuclear Instrumentation, Measurement Methods and their Applications (ANIMMA)*, 2013, pp. 1–9.

[7] T. Ligonnet, N. Weiss, V. Lamirand, A. Laureau, and A. Pautz, "POLLEN: A Pile-Oscillator for the BLOOM Experimental Program," *EPJ Web Conf.*, vol. 288, p. 04018, Nov. 2023.

[8] D. L. Hetrick, "DYNAMICS OF NUCLEAR REACTORS.," 1971, pp.66-77.

[9] J. Leppänen, M. Pusa, T. Viitanen, V. Valtavirta, and T. Kaltiaisenaho, "The Serpent Monte Carlo code: Status, development and applications in 2013," *Ann. Nucl. Energy*, vol. 82, pp. 142–150, 2015.

[10] V. Lamirand *et al.*, "CORTEX experiments – Part I: Modulation campaigns in AKR-2 & CROCUS for the validation of neutron noise codes," *Ann. Nucl. Energy*, vol. 211, Feb. 2025.

[11] A. Knospe, S. Hübner, K. Ambrožič, M. Viebach, C. Lange, and V. Lamirand, "CORTEX experiments, Part III: Experimental determination of the zero power transfer function of AKR-2 with reliable uncertainties," *Ann. Nucl. Energy*, vol. 209, p. 110686, Dec. 2024.

[12] K. Ambrožič *et al.*, "CORTEX experiments, Part II: Postprocessing of neutron noise time series to determine reliable mean and uncertainties," *Ann. Nucl. Energy*, vol. 208, p. 110704, Dec. 2024.

[13] D. A. Brown *et al.*, "ENDF/B-VIII.0: The 8th Major Release of the Nuclear Reaction Data Library with CIELO-project Cross Sections, New Standards and Thermal Scattering Data," *Nucl. Data Sheets*, vol. 148, pp. 1–142, 2018.

Application of Unfolding Methods on Thermal Neutron Data Measured with a Single-Chopper Time-of-Flight Setup

Vincent Melzer[1,*], Alexander Knospe[1], Louis Stein[1,‡], Marco Viebach[1], Carsten Lange[1], and Antonio Hurtado[1]
[1]Chair of Hydrogen and Nuclear Energy, Dresden University of Technology, Germany
(*) vincent.melzer@tu-dresden.de

***Abstract*—As part of the NAUTILUS research project, the thermal neutron spectrum of the Training Reactor AKR-2 of the Dresden University of Technology is being determined by means of the time-of-flight technique. However, due to the AKR-2 being a zero-power reactor and a facility with low neutron flux, to increase overall counting statistics only a single-disk chopper instead of multiple disk choppers in synchronicity is used in the time-of-flight setup. As no appropriate energy selection is possible with a single-disk chopper, a method was developed for the assessment of real flight times by unfolding measured flight-time data with a flight-time response matrix. This publication presents the current time-of-flight setup at the AKR-2 and the developed deconvolution method. Experimental and simulated flight-time data with the software McStas are shown and compared to theoretical approximations. Results of the deconvolution approach are discussed.**

***Keywords* —neutron spectrum, thermal neutrons, time-of-flight method, deconvolution, He-3 counters, McStas**

I. Introduction

THERMAL neutron time-of-flight (ToF) measurements are a fundamental technique for characterizing neutron energy spectra. Because of this, they are used in the NAUTILUS project [1] for determining the thermal neutron spectrum of the Training Reactor AKR-2 of the Dresden University of Technology in Germany.

For the selection of individual thermal neutron energies in ToF experiments, disk choppers can be utilized. The latter are disk-like devices that are placed in the neutron beam. As they are constructed of neutron-absorbing materials like cadmium or gadolinium and with slits for neutron passing, their rotation can be exploited to partition the neutron beam into individual bursts.

For the determination of neutron spectra, the synchronization of multiple disk choppers can be used to filter specific neutron energies. While this approach is effective, it becomes impractical at neutron sources with low source strength. This is because the combination of multiple choppers significantly reduces the neutron flux, resulting in long measurement times, particularly when scanning over a range of discrete energies.

This limitation can be overcome by means of ToF experiments with a single-disk chopper setup. Maintainable measurement times are achieved as the transmitted flux is maximized. However, as Niederhaus describes [2], the finite opening time of the chopper, together with neutron transmission through the closed chopper and geometric imperfections, inevitably cause the *measured flight times* to deviate from the *real flight times*. This ultimately results in a compromised energy resolution and thus the loss of a true energy selection.

To enable an assessment of the real flight times from measured flight-time data of a single-disk chopper ToF setup, a method was developed at the AKR-2 in which the distribution of measured flight times is deconvoluted by means of a flight-time response matrix to yield the distribution of real flight times. In this approach, the flight-time response matrix describes the measurement probabilities of the measured flight times for the given real flight times. It can be determined by knowledge of the chopper opening function and the microscopic cross sections of the materials in the flight path.

This publication presents the developed deconvolution method and the underlying single-disk chopper ToF setup at the AKR-2. Experimental results of an investigation of the chopper opening function and how they were used to determine a flight-time response matrix are shown. A comparison between measured flight times, McStas simulation data and theoretical approximations is performed. Finally, results of the application of the developed method and the yielded distributions of real flight times are discussed.

II. Theory

A schematic for a single-disk chopper ToF setup is depicted in Fig. 1. It is assumed that the measured flight times for a corresponding real flight time are mainly influenced by the following aspects:

1. Transmission through media

The signal strength in the stop detector is influenced by the transmission of the neutrons through the media in the flight path. This can be described by Beer-Lambert law:

‡ L. Stein, now at Helmholtz-Zentrum Dresden-Rossendorf, 01328 Dresden, Germany

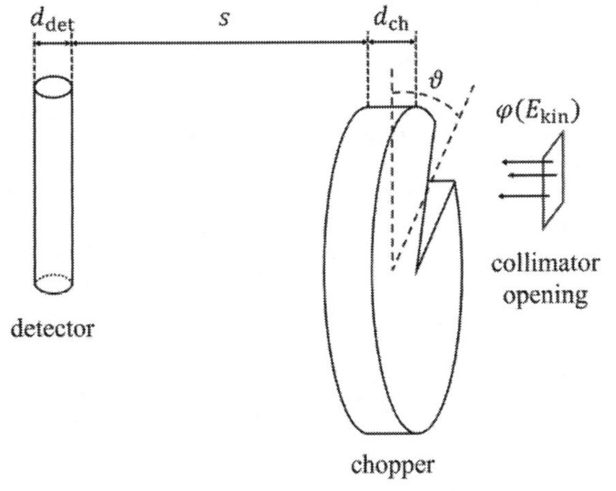

Fig. 1. Schematic of a single-disk chopper ToF setup

$$\begin{aligned}\varphi(E_{\text{kin}}, d_{\text{M}}) &= \varphi_0(E_{\text{kin}}) \cdot T_{\text{M}}(E_{\text{kin}}, d_{\text{M}}) \\ &= \varphi_0(E_{\text{kin}}) \cdot e^{-\Sigma_{\text{M}}(E_{\text{kin}})d_{\text{M}}} \\ &= \varphi_0(E_{\text{kin}}) \cdot e^{-\sigma_{\text{M}}(E_{\text{kin}})n_{\text{M}}d_{\text{M}}}\end{aligned}$$

Here, $\varphi(E_{\text{kin}}, d_{\text{M}})$ is the neutron flux of neutrons with the kinetic energy E_{kin} after transmission through a medium with thickness d_{M}. The initial neutron flux is denoted with $\varphi_0(E_{\text{kin}})$. The transmission through the medium $T_{\text{M}}(E_{\text{kin}}, d_{\text{M}})$ is calculated from the macroscopic cross section $\Sigma_{\text{M}}(E_{\text{kin}})$, the microscopic cross section $\sigma_{\text{M}}(E_{\text{kin}})$ and the particle density n_{M}.

The kinetic energy E_{kin} for thermal neutrons can be expressed in terms of their velocity v:

$$E_{\text{kin}} = \frac{1}{2} m_n v^2$$

where m_n is the neutron mass. The velocity v can be further expressed in terms of the flight time t_{path} if the length of the flight path l_{path} is known:

$$v = \frac{l_{\text{path}}}{t_{\text{path}}}$$

Therefore, Beer-Lambert law can be expressed as such:

$$\begin{aligned}\varphi(t_{\text{path}}) &= \varphi\left(E_{\text{kin}}\left(v(t_{\text{path}})\right)\right) \\ &= \varphi_0\left(E_{\text{kin}}\left(v(t_{\text{path}})\right)\right) e^{-\Sigma_{\text{M}}(E_{\text{kin}}(v(t_{\text{path}})))d_{\text{M}}}\end{aligned}$$

2. Chopper opening function

When the chopper opening moves in front of the collimator, the intensity of the signal changes. This is represented by the chopper opening function $G(\vartheta)$:

$$G(\vartheta) = \frac{A_{\text{chopper}}(\vartheta) \cap A_{\text{collimator}}}{A_{\text{collimator}}}$$

Here, $A_{\text{chopper}}(\vartheta)$ is the area of the chopper opening at the rotation angle ϑ. The area of the collimator opening is denoted by $A_{\text{collimator}}$.

If the start angle ϑ_0 corresponding to a time t_0 and the rotation frequency of the chopper f are known, the chopper opening function can be calculated for any time t:

$$G(t, t_0) = G\left(\vartheta(t, t_0)\right) \text{ with } \vartheta(t, t_0) = \frac{ft}{360°} + \vartheta_0(t_0) = \frac{f(t + t_0)}{360°}$$

Under consideration of these aspects and setting an initial angle $\vartheta_0 = \vartheta(t_0)$, the flight-time response function $R(t_{\text{real}}, t_{\text{meas}})$ for the real times t_{real} and measured times t_{meas} can be constructed as

$$\begin{aligned}&R(t_{\text{real}}, t_{\text{meas}}) \\ &= G(t_{\text{real}}, t_0 - \Delta t) \cdot e^{-\Sigma_{\text{air}}(d_{\text{ch}}+s)}\left(1 - e^{-\Sigma_{\text{det}}d_{\text{det}}}\right) \\ &+ \left(1 - G(t_{\text{real}}, t_0 - \Delta t)\right) \cdot e^{-\Sigma_{\text{ch}}d_{\text{ch}}} e^{-\Sigma_{\text{air}}s}\left(1 - e^{-\Sigma_{\text{det}}d_{\text{det}}}\right)\end{aligned}$$

with $\Delta t = t_{\text{meas}} - t_{\text{real}}$ and $t_{\text{meas}} = t_{\text{stop}} - t_{\text{start}}$ using t_{stop} and t_{start} as the recorded start and stop time. The indices of the macroscopic cross section $\Sigma = \Sigma(t_{\text{real}})$ and thickness d refer to the components of the flight path being the chopper ("ch"), the traversed air ("air") and the detector ("det").

As described in the equation, the flight-time response function is constructed from two contributions: neutrons that enter the flight path through the open chopper window, and neutrons that pass through the closed window. Because different media are present along the individual flight paths, the signal strengths in the detector differ according to the respective macroscopic cross sections. The time offset in the chopper opening function accounts for the difference between the measured flight time – derived from the recorded start and stop time – and the real flight time.

Finally, the distribution of measured flight times $S(t_{\text{meas}})$ can be expressed as a convolution of the flight-time response function $R(t_{\text{real}}, t_{\text{meas}})$ with the differential distribution of real flight times $\mathrm{d}N(t_{\text{real}})/\mathrm{d}t_{\text{real}}$:

$$S(t_{\text{meas}}) = \int_0^{\frac{1}{f}} R(t_{\text{real}}, t_{\text{meas}}) \frac{\mathrm{d}N(t_{\text{real}})}{dt_{\text{real}}} \,\mathrm{d}t_{\text{real}}$$

In the case of a discrete time binning, this expression can be restated in a tensor formulation:

$$\vec{S}(t_{\text{meas}}) = \hat{R}(t_{\text{real}}, t_{\text{meas}})\vec{N}(t_{\text{real}})$$

Here, $\vec{S}(t_{\text{meas}})$ and $\vec{N}(t_{\text{real}})$ denote the corresponding vectors of the distributions of measured and real flight times. $\hat{R}(t_{\text{real}}, t_{\text{meas}})$ is the matrix representation for the flight-time response function. It is further referred to as the *flight-time response matrix*.

It follows that the determination of the distribution of real flight times can be achieved by a deconvolution approach. Deconvolution methods are, for example, the Richardson-Lucy deconvolution [3] [4], Wiener deconvolution [5] and Tikhonov regularization [6].

III. Materials

The single-disk chopper ToF setup at the AKR-2 is depicted in Fig. 2. The neutron beam originating from the radial channel (opening #7) of the AKR-2 is focused and aligned by two collimators. It is then partitioned into individual neutron bursts by a single-disk chopper. The neutron bursts pass a boron-aluminum aperture before being registered by a ^{3}He detector in a shielding construction.

The total flight distance of the setup depends on its alignment and is approximately 3.2 m. The start and stop signals for the determination of the flight times are set by the opening and closing instances of the chopper, and the detection events inside the ^{3}He counter. They are generated in the data acquisition system from the signals sent by their individual components.

The individual components of the setup are further presented in the following sections.

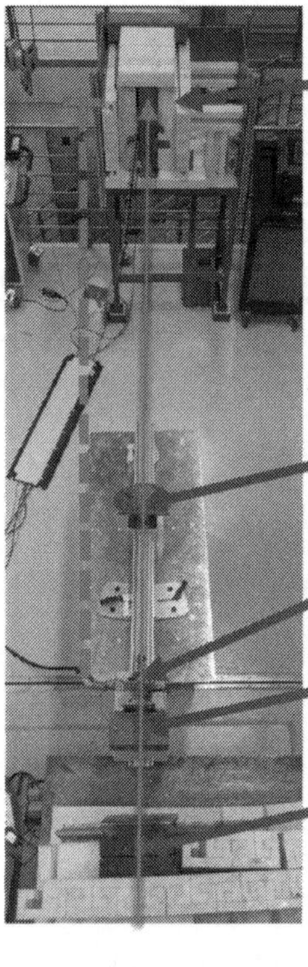

Fig. 2. Single-disk chopper ToF setup at the AKR-2

A. *Training Reactor AKR-2*

The Training Reactor AKR-2 (German: **A**usbildungs**k**ern**r**eaktor) is described in Lange and Bernt [7]. It is situated at the Chair of Hydrogen and Nuclear Energy at the Dresden University of Technology, Germany. Since its commissioning on March 22nd, 2005, it has been used for education, research, and public relations. It is a homogeneous, thermal, zero-power reactor (nominal thermal power: 2 W).

The reactor core is cylindrical, measuring 250 mm in diameter and 275 mm in critical height. It consists of two vertically separable halves; each composed of layered fuel plates made from a combination of uranium dioxide (19.8 % enrichment in uranium-235) and polyethylene. Criticality is achieved by vertically aligning the lower half with the upper half of the core. Surrounding the core is a reflector made from high-purity graphite, encased by paraffine and baryte concrete as radiation shielding. Reactor control is maintained by three cadmium control and safety rods. The neutron flux level is monitored by several fission chambers and a gamma-compensated ionization chamber. Five channels, marked with openings numbered 1 through 7, can be used at the AKR-2 for the conduct of experiments. A cross-sectional diagram of the reactor is shown in Fig. 3.

B. *Disk chopper*

The disk chopper used in the ToF setup is depicted in Fig. 4. It consists of a cadmium layer of 2.4 mm thickness positioned between two aluminum disks with thicknesses 2 mm and 3.9 mm. The chopper furthermore has a total diameter of 17 cm and features three slits with an opening angle of 6°. By means of a PID controller, the rotation frequency can be adjusted.

For ToF measurements, the opening and closing instances of the chopper are registered by a laser sensor whereas the first are used as start signals.

C. *^{3}He Detector*

The ^{3}He detector used in the setup was produced by Berthold. It has a diameter of 1.27 cm and a height of 30 cm. Its gas pressure is specified as 4 atm and its bias voltage of +1200 V is

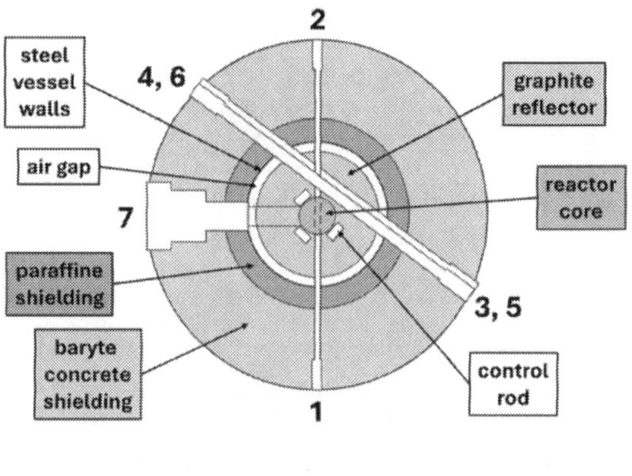

Fig. 3. Cross-sectional diagram of the AKR-2

Fig. 4. Photo of the disk chopper

applied by a "556" module by ORTEC. Gamma discrimination in the detector is achieved by a pulse-height threshold.

For ToF measurements, the pulses of detection events in the ^{3}He counter are used as a stop signal.

D. Data acquisition system

The data acquisition system of the ToF setup consists of a field-programmable gate array board ("Zedboard" by Avnet), with BNC connectors interfaced to the relevant FPGA pins via a PMOD adapter. The start and stop signals of the chopper and ^{3}He counter are separately processed on the FPGA for the generation of timestamps.

IV. Measurements

For the application of the unfolding approach to measurement data of the single-disk chopper ToF setup at the AKR-2, the flight-time response matrix had to be determined. Therefore, to investigate the chopper opening function a group of three measurements M1.1 to M1.3 with slow chopper rotation frequencies was performed.

To apply different unfolding algorithms to flight-time data, an additional long-term measurement M2 with a chopper frequency of 50 Hz was performed.

Details of the experiments conducted are listed in the following table:

Table I
Details of the Performed Measurements

Index	Flight distance	Chopper frequency	Measurement time
M1.1	3.297 m	15 Hz	2 h 42 min
M1.2	3.297 m	10 Hz	3 h 5 min
M1.3	3.297 m	1.6 Hz	1 h 14 min
M2	3.200 m	50 Hz	12 h 46 min

V. Simulations

For measurement M2, Monte Carlo particle transport simulations with McStas [8] were used to validate the distribution of measured flight times. The latter is a Monte Carlo ray-trace simulation package for neutrons in which experiments can be recreated by predefined instrument components in the software. The experimental setup of measurement M2 was remodeled (see Fig. 5). With this model, a Monte Carlo simulation was performed with $N_{\text{rays}} = 10^{17}$ neutron rays.

VI. Analysis

A. Investigation of the chopper opening function

The measured flight times of measurements M1.1 to M1.3 were sorted into a histogram and approximated by triangular functions (see Fig. 6).

Due to the slow rotation speed of the chopper, the distributions of measured flight times can be interpreted as the angle-dependent signal intensity distribution for the chopper. Therefore, the chopper opening function was further assumed to be of triangular shape. As the opening of the collimator in front of the chopper was constructed congruent to the chopper opening, the maximum value of the chopper opening function was fixed to 1. The triangle width was defined based on the chopper opening time, representing the time required for a complete open–close transition:

$$t_{\text{open}} = 2 \cdot \frac{6°}{360°} \cdot \frac{1}{f}$$

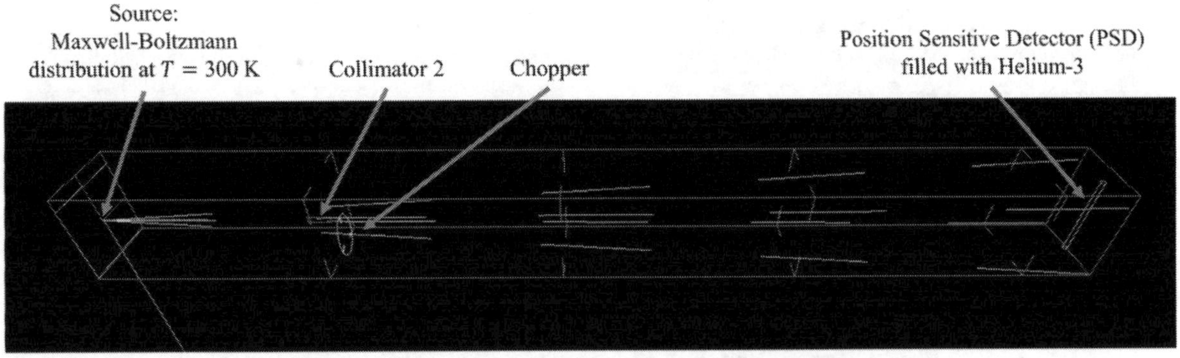

Fig. 5. McStas model of the single-disk chopper ToF setup for measurement M2

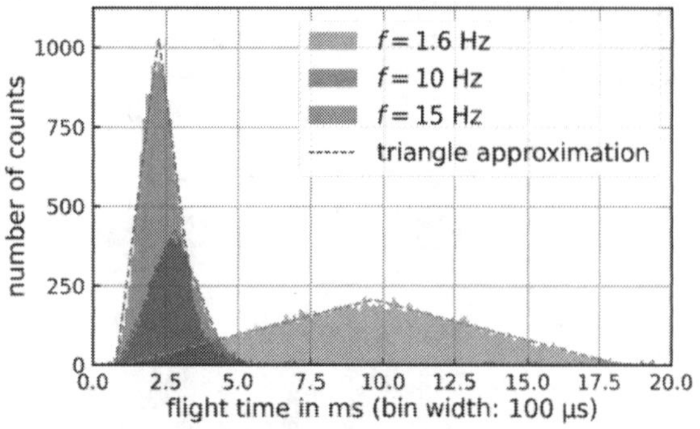

Fig. 6. Distributions of measured flight times for measurements M1.1 to M1.3

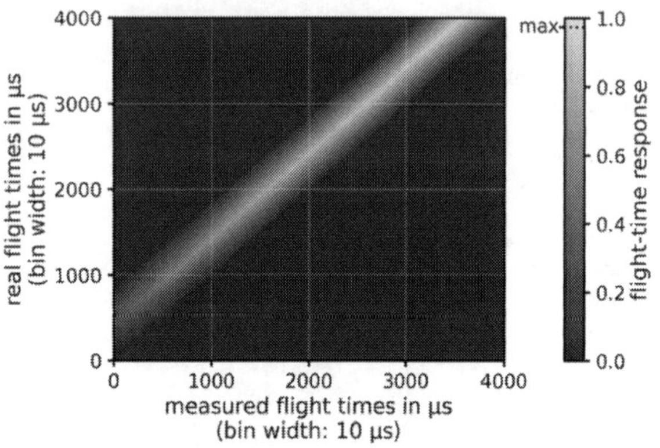

Fig. 7. Determined flight-time response matrix for the evaluation of measurement M2

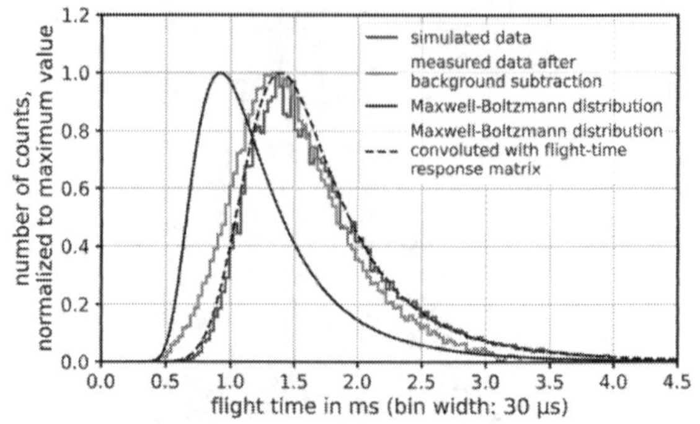

Fig. 8. Compared flight-time distributions for measurement M2 (measured data, McStas simulation data, theoretical approximation)

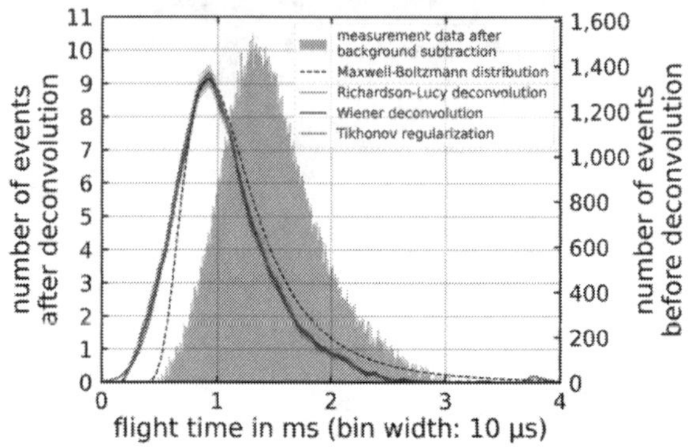

Fig. 9. Results of the application of different unfolding algorithms on the distribution of measured flight times of measurement M2

B. Flight-time response matrix

With the determined chopper opening function, the flight-time response matrix was determined for the evaluation of measurement M2. The individual entries of the matrix were approximated according to

$$R_{i,j} \approx G\left(t_{\mathrm{real},i}, t_0 - \Delta t_{i,j}\right) \cdot \left(1 - e^{-\Sigma_{\mathrm{det}}(t_{\mathrm{real},i}) d_{\mathrm{det}}}\right)$$

with $\Delta t_{i,j} = t_{\mathrm{meas},j} - t_{\mathrm{real},i}$. The determined flight-time response matrix is depicted in Fig. 7.

C. Comparison of the distribution of measured flight times with McStas simulation data and theoretical approximations

The measured flight times of measurement M2 were sorted in a histogram and a constant background was subtracted. The McStas simulation data was sorted in a histogram as well. A Maxwell-Boltzmann distribution was calculated for the temperature $T =$ 295 K. Then, it was folded with the determined flight-time response matrix for the evaluation of measurement M2. Finally, all distributions were normalized for the comparison of their individual shapes. The results are depicted in Fig. 8.

D. Application of unfolding algorithms to the distribution of measured flight times

The measured flight times of measurement M2 were sorted into a histogram. Then, a constant background subtraction was performed. Three unfolding algorithms were applied by means of the calculated flight-time response matrix for the determination of the distribution of real flight times. The applied unfolding parameters were determined by a separate parameter study:

1. Richardson-Lucy deconvolution (GRAVEL [9] [10]): Tolerance parameter: $tol_{\mathrm{Richardson}} = 2 \cdot 10^{-4}$
2. Wiener deconvolution: Multiplicative noise parameter $\alpha_{\mathrm{Wien}} = 2 \cdot 10^{-2}$
3. Tikhonov regularization: Additive noise parameter $\alpha_{\mathrm{Tikhonov}} = 20$

The measurement uncertainties of the unfolded distribution were accessed by the bootstrapping method [11]. Therefore, flight times were sampled $N_{\mathrm{sample}} = 4000$ times from the background-subtracted distribution and sorted into histograms. Then, all were deconvoluted by means of the unfolding algorithms. The statistical uncertainties of the deconvoluted distributions were calculated.

As a theoretical reference, the Maxwell-Boltzmann distribution for the temperature T = 295 K was calculated and normalized to the maximum of the unfolded distributions. The results are depicted in Fig. 9.

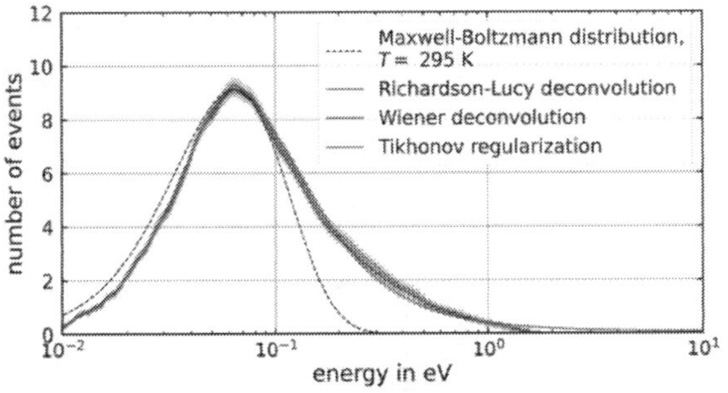

Fig. 10. Results of application of different unfolding algorithms on the distribution of measured flight times of measurement M2, after conversion to kinetic energies (nonlinear binning)

Additionally, the corresponding energy distribution was determined by a direct conversion of the flight times to their kinetic energies:

$$E_{\text{kin}} = \frac{1}{2} m_{\text{n}} \left(\frac{s}{t_{\text{real}}}\right)^2 \Leftrightarrow t(E_{\text{kin}}) = \pm s \sqrt{\frac{m_{\text{n}}}{2E_{\text{kin}}}}$$

Here, E_{kin} is the kinetic energy of the neutron, m_n is the neutron mass, t_{real} is the real flight time and s is the flight path. The results are depicted in Fig. 10. According to the conversion, the energy binning is nonlinear.

VII. Discussion

A. Chopper opening function

As depicted in Fig. 6, the approximation of the chopper opening function by a triangular approach is a suitable first approach. However, the quantitative comparison with other approximating functions has to be performed in the future.

B. Flight-time response matrix

In the determined flight-time response matrix (see Fig. 7), it is reflected that the maximum flight-time response increases with higher real flight times. This is due to the neutron cross section of ^{3}He increasing with lower kinetic energies. The overall shape of the flight-time response for different real times reflects the chopper opening function.

In the calculation of the flight-time response matrix, the contribution of the closed chopper state and of the air in the flight path was neglected.

C. Comparison of the measured, the simulated, and theoretically approximated flight-time distributions

The shapes of the theoretically approximated and simulated flight-time distributions (see Fig. 8) are in good agreement with each other. This underlines that the measured flight times can be represented by a convolution of the distribution of real flight times with a flight-time response matrix.

However, for the shape of the distribution of measured flight times a deviation can be observed. There seems to be a lack of slower thermal neutrons and an excess of faster thermal neutrons in comparison with the other distributions. It is assumed that this is for three reasons:

1. The neutron spectrum of the AKR-2 is not fully thermalized.
2. Slower thermal neutrons are more likely to be absorbed by the air in the flight path before they reach the detector.
3. The transition of the thermal neutron spectrum into the epithermal neutron spectrum is not represented in the theoretical approximation and the McStas simulation, as both are based exclusively on a Maxwell–Boltzmann distribution.

D. Results of the application of unfolding algorithms

In Fig. 9 and Fig. 10 the distributions of real flight times for the different applied unfolding algorithms agree with each other. The determined uncertainties are relatively small. As the deconvolution results deviate from a theoretical Maxwell-Boltzmann distribution, it is assumed that these incongruities stem from the mentioned reasons.

VIII. Summary and Outlook

Using a single instead of multiple disk choppers in ToF experiments offers the advantage of significantly reducing the overall measurement time required to achieve low statistical uncertainties. However, due to the finite opening time of the chopper, neutrons passing through the closed chopper window, and geometric imperfections, the measured flight times deviate from the real flight times, and a true energy selection is not possible.

This publication demonstrated that, by means of a deconvolution approach, the distribution of real flight times can still be obtained from the measured flight-time distribution acquired with a single-disk chopper ToF setup. First, the chopper opening function was investigated to determine a flight-time response matrix, which was then used to unfold the distribution of measured flight-times using different deconvolution algorithms. The resulting unfolded distributions showed good mutual agreement and small uncertainties.

Nevertheless, deviations between the distributions of measured flight times and both theoretical approximations and McStas simulations were observed. These differences are likely caused by the thermal neutron spectrum deviating from an ideal Maxwell–Boltzmann distribution.

One aspect was not discussed in the scope of this publication: Currently, this method is further validated at AKR-2 by means of single-disk chopper ToF measurements combined with transmission and Bragg scattering experiments. The results – for example, determined cross sections in the thermal region – agree well with literature data. Therefore, the authors are confident in the reliability of the presented method.

In the future, it is planned to determine the thermal spectral neutron flux from distributions of real flight times determined by the presented technique. With that, the spectrum of the AKR-2 will be experimentally characterized in the thermal region.

Acknowledgment

This work is part of the research project NAUTILUS funded by the German Federal Ministry of Research, Technology, and Space (BMFTR) under the reference number 02NUK079. The authors also want to thank the technicians Erik Findeisen and

Tobias Langnickel, the research associate Sascha Weichel as well as the student Paula Röver.

REFERENCES

[1] M. Viebach, A. Knospe, C. Lange, B. Merk, and A. Hurtado, "NAUTILUS: A Project for the Development of Experimental Methods for Investigating Innovative Approaches to Nuclear Waste Management and to Nuclear Safety" in Proc. PHYSOR 2024, San Francisco, CA, US, 2024, pp. 316-325.

[2] J. H. J. Niederhaus, "A single-disk-chopper time-of flight spectrometer for thermal neutron beams," M.S. thesis, Dept. Nuc. Eng., Penn State, State College, PA, US, 2003, pp. 24-31.

[3] W. H. Richardson, "Bayesian-Based Iterative Method of Image Restoration," J. Opt. Soc. Am., vol. 62, no. 1, pp. 55-59, Jan. 1972, 10.1364/JOSA.62.000055.

[4] L. B. Lucy, "An iterative technique for the rectification of observed distributions," Astron. J., vol. 79, p. 745, June 1974, 10.1086/111605.

[5] N. Wiener, *Extrapolation, interpolation, and smoothing of stationary time series*, 1st M.I.T. Press Paperback Edition. Cambridge, MA, US: M.I.T. Press, 1964.

[6] A. Tikhonov, V. Glasko, "Use of the regularization method in non-linear problems", U.S.S.R. Comput. Math. Math. Phys., vol. 5, no. 3, pp. 93-107, Feb. 1965, 10.1016/0041-5553(65)90150-3.

[7] C. Lange and N. Bernt, "The Education and Research Reactor AKR-2 and its Experimental Programme for Education," in Proc. PHYSOR, Cancún, MX, 2018, pp. 2705-2720.

[8] P. Willendrup, E. Farhi, E. Knudsen, U. Filges, K. Lefmann, and J. Stein. "User and Programmers Guide to the Neutron Ray- Tracing Package McStas, version 3.5," Technical University of Denmark, 2800 Kongens Lyngby, Denmark, August 2024. [Online] Available: https://www.mcstas.org/documentation/manual/mcstas-3.5.32-manual.pdf. Accessed on: June 6, 2025.

[9] M. Matzke, "Unfolding of Pulse Height Spectra, The HEPRO Program System", PTB, Braunschweig, Germany, Rep. PTB-N-19, 1994.

[10] T. D. Doležal. "Unfolding the AmBe Neutron Spectra using GRAVEL and MLEM", Air Force Institute of Technology, Department of Engineering Physics, OH, US. Rep. NENG612, 2024. [Online] Available: https://github.com/tylerdolezal/Neutron-Unfolding/blob/main/report_and_presentation/NENG612_Final_Report.pdf. Accessed on: June 6, 2025.

[11] B. Efron, "Bootstrap methods: Another look at the jackknife," Ann. Statist., vol. 7, no. 1, pp. 1-26, Jan. 1979, 10.1214/aos/1176344552.

Nuclear heating rate measurement in the MIT research reactor using a new compact CALORRE differential calorimeter

C. Reynard-Carette [1*], A. Volte[1], G. Kohse[2], D. Carpenter[2], Y. Ostrovsky[2], M. Ames[2], S. Hauptman[2], A. Lyoussi[3], M. Carette[1]
[1]Aix Marseille Univ, Université de Toulon, CNRS, IM2NP, Marseille, France
[2]Massachusetts Institute of Technology, Nuclear Reactor Laboratory, Cambridge, Massachusetts, USA
[3]CEA/DES/IRESNE/DER, Section of Experimental Physics, Safety Tests and Instrumentation, Cadarache, F-13108, Saint Paul-Lez-Durance, France

(*) corresponding author: christelle.carette@univ-amu.fr

Abstract—**This paper presents experimental results obtained thanks to a new compact prototype of CALORRE calorimeter during an irradiation campaign inside the MIT Research Reactor (MITR) in October 2024 within the framework of the CALOR-I research program led by Aix-Marseille University and involving the Nuclear Reactor Laboratory of the MIT and the CEA. The nuclear heating rate was measured for the first time in the MITR. An axial mapping of the nuclear heating rate was realized in 15 axial positions of the in-core water-loop of the MITR (B3 position) by moving the CALORRE differential calorimeter while keeping the reactor in operation (at a reactor power equal to 5 MW_{th}) for a stainless steel (ASI 316L) sample. Moreover, the nuclear heating rate was quantified versus the reactor power at a fixed axial position with and without the displacement of the compact CALORRE calorimeter. The sensor reliability and robustness were studied thanks to calibrations conducted before, during and after the irradiation campaign inside the in-core water-loop with the reactor shut down.**

Keywords — **Calorimetry, Sensor, Irradiation, Mapping, Nuclear heating rate**

I. Introduction

To design and propose innovative nuclear instrumentation, Aix-Marseille University and the CEA as part of the LIMMEX joint laboratory (Laboratory for Instrumentation and Measurement in EXtreme Environments) created in 2010 are conducting several research programs addressing key values for research reactors and tokamaks, such as fast and thermal neutron fluxes, prompt and delayed photon fluxes and intense absorbed dose rates (also called nuclear heating rate). A comprehensive approach was developed particularly regarding heat-flow calorimetric sensors by coupling experimental work with 1-D calculations and 3-D simulations (heat transfer and ray-matter interactions) from non-nuclear laboratory conditions to irradiation campaigns in nuclear environments [1-11]. Two types of calorimeter integrating at least one heating element were patented by AMU and the CEA: one differential calorimeter in 2015 (called CALORRE) [4] and one single-cell calorimeter in 2020 (called Mono-CALO) [7]. After successful tests of a first prototype of the CALORRE differential calorimeter in the MARIA reactor in November 2015 [9], different challenges associated to this sensor type were and are targeted such as: the extension of the measurement range from very low values for TRIGA reactors such as the JSI TRIGA reactor in Slovenia to high values for MTRs such as the JHR reactor under construction in France, the sensor miniaturization to design a micro-sensor in the future, and the characterization of different materials for fission and/or fusion purposes.

In this context, a research program, called CALOR-I (compact-CALORimeter Irradiations inside the MIT Research Reactor), funded by Aix-Marseille University foundation (A*Midex) and involving the Nuclear Reactor Laboratory of the MIT and the CEA, began in 2020 with four main objectives: (1) the design of a new CALORRE prototype, (2) the study of this prototype from laboratory to irradiation conditions, (3) the axial mapping of an in-core channel of the MIT Research Reactor (MITR) in terms of nuclear heating rate (never measured before), and (4) the comparison of measurement methods and sensors. By taking the feedback from the MARIA campaign into account and carrying out thermal and neutron modeling, a new CALORRE calorimeter prototype of reduced size and similar height to that of the usual single-cell calorimeters was defined and manufactured [9-11].

This paper deals with the experimental study realized on this reduced-size CALORRE prototype inside the in-core water-loop of the MITR in October 2024.

The first part of the paper presents the design of this CALORRE calorimeter with its main features and the calorimeter vehicle placed into the water-loop autoclave and its components. The second part focuses on results obtained during the MITR irradiation campaign. A brief description of the experimental operating protocols is given for the CALORRE calorimeter calibrations and the nuclear heating rate measurements. The axial profile of the nuclear heating rate in the B3 channel is shown for a reactor power of 5 MW_{th}. The measured nuclear heating rate at a specific position as a function of the reactor power is given for two operating protocols. Finally, the reliability and the robustness of the calorimeter is studied by comparing the calibration curves obtained before, during and after the irradiation campaign with the reactor shut down.

II. Calorimeter and Experimental Set-up

A. CALORRE calorimeter

The new CALORRE prototype corresponds to the most compact CALORRE differential calorimeter built and irradiated to date. This prototype is made of stainless steel (ASI 316L) with a total height equal to 88.5 mm from its nose to its watertight passage and an external diameter of 18 mm (cf. Figure 1). It is composed of two superimposed calorimetric cells with a half-horizontal fin configuration (8 metal sectors and 8 empty sectors each with an equal angle of 22.5°).

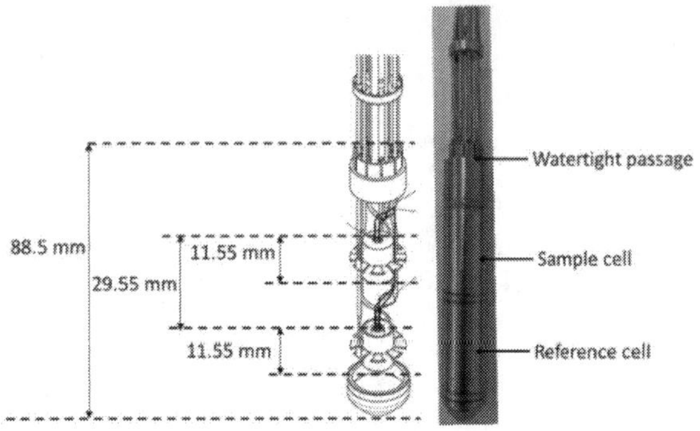

Fig. 1. Schematic diagram and picture of the CALORRE calorimeter

The upper calorimetric cell contains the sample to be characterized. The sample and head structure were manufactured into a single stainless-steel block to avoid additional thermal contact resistance. The lower calorimetric cell, which contains no sample, acts as a reference cell to remove the energy deposition in areas of the sample cell not corresponding to the sample. Each cell includes two K-type thermocouples and a heating element (4-wire made of Nickel-Chromium). Each calorimetric cell has a height equal to 11.55 mm with an external vertical fin of 31.55 mm in height and 18 mm in external diameter serving as sensor envelope by welding. The watertight assembly of the calorimeter has an inter-cell spacing of 18 mm and an inter-sample area spacing of 29.55 mm. The calorimeter is filled with N_2.

B. Experimental set-up

The experimental set-up is composed of several components (cf. Figure 2):

- the CALORRE calorimeter and a gamma thermometer
- a K-type thermocouple to measure the temperature in the water-loop close to the CALORRE calorimeter
- mechanical elements to center the calorimeter vehicle in the channel, displace it, impose a fluid flow temperature in the loop, to watertight (centering fixture, sensor tube, water inlet and outlet pipes, autoclave flange with dynamic seal for sensor tube and its displacement)
- a displacement system to change the axial position of the calorimeter
- a programmable power supply for the calibration of each CALORRE calorimetric cell
- a data acquisition system

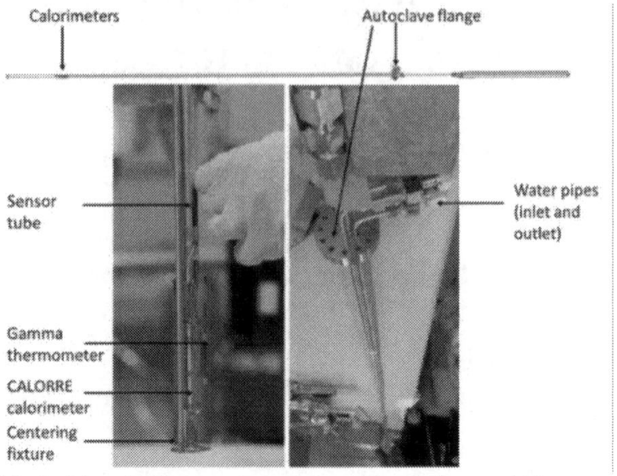

Fig. 2. Schematic diagram and two pictures of the calorimeter vehicle

III. Experimental Results

A. Axial mapping of the nuclear heating rate

The axial profile of the nuclear heating rate was obtained by moving the CALORRE calorimeter with the reactor operating at a power of 5 MW_{th} and a water-loop temperature of approximately 120 °C.

For each axial position z_i, the measurement was realized in two steps. In step a), the calorimeter is placed at the axial position z_i and the steady temperatures of the two cells are measured. In step b), the calorimeter is moved 29.55 mm upward to place the reference cell at the same axial position z_i as that of the sample cell in step a), and the steady temperatures of the two cells are measured. This protocol was repeated to determine the axial profile in 15 positions of the sample between -109.38 mm and +304.32 mm (cf. Figure 3).

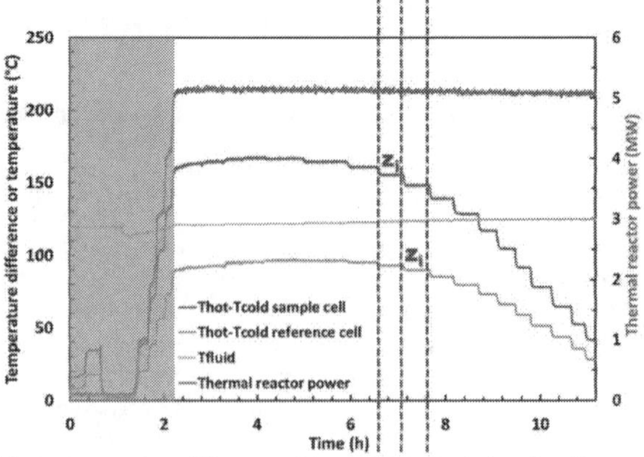

Fig. 3. Temperature differences of the CALORRE calorimetric cells, water-loop temperature and reactor power versus time for various axial positions in B3 channel of the MITR

The nuclear heating rate is calculated for each axial position z_i thanks to the two differences of steady temperatures obtained for the sample cell at the step a) and for the reference cell at step b) respectively, and by considering the calibration curves (cf. section III.C).

The axial profile is given in Figure 4 (in order to take the reactor power variation into account during the experiments, the nuclear heating rate was set to a reactor power of 5 MW_{th} whatever the axial position). The maximum value of the nuclear heating rate in stainless steel is equal to 2.04 $W.g^{-1}$ at -79.83 mm.

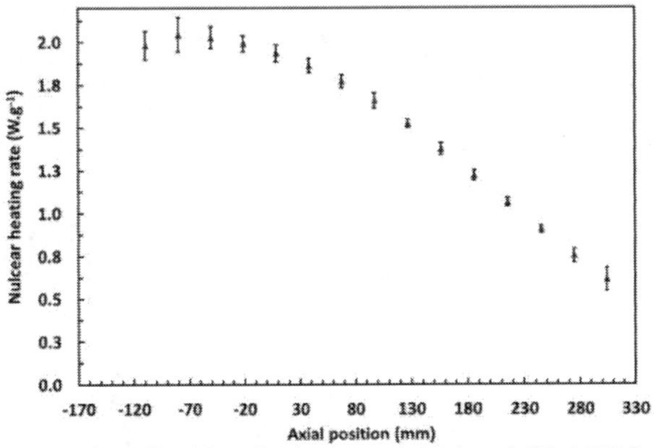

Fig. 4. Axial profile of the nuclear heating rate in B3 channel of the MITR for a power of 5 MW_{th}

B. Nuclear heating rate versus the reactor power

The nuclear heating rate measurement versus the reactor power was quantified over several days of the irradiation campaign for a specific position of the sample (-50.28 mm). The results are given in Figure 5 for two operating protocols (with or without CALORRE displacement).

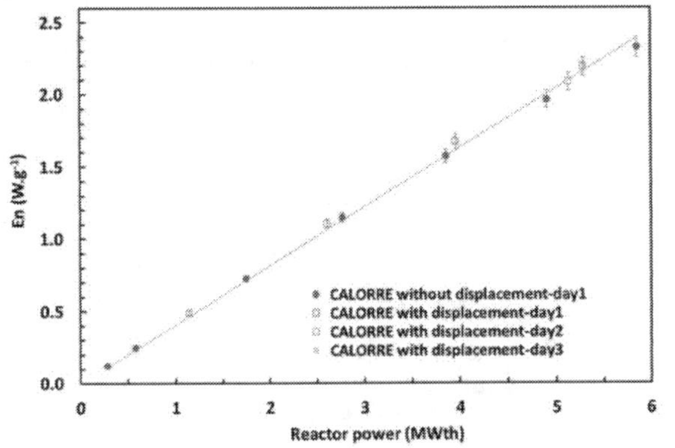

Fig. 5. Nuclear heating rate versus the reactor power at -50.28 mm (sample median plane) with or without moving the calorimeter

The empty symbols correspond to results obtained with the same operating protocol described in the previous section. The nuclear heating rate is determined by moving the calorimeter to consider the temperature differences of the two cells at the same axial position.

The solid symbols are obtained without moving the calorimeter and thus by using the temperature differences of the two superimposed cells not located at the same axial position. That is possible to apply this method without displacement because the discrepancy of the nuclear heating rate between these two protocols at -50.28 mm is very low (0.2 %) due to the reduced size of the calorimeter and the low flux gradient in this area (a study with and without moving CALORRE calorimeter was realized during the axial mapping, these results are not presented in this paper).

The nuclear heating rate was measured versus the reactor power from 284 kW_{th} to 5.9 MW_{th} and it is between 0.13 $W.g^{-1}$ and 2.32 $W.g^{-1}$. A good reproducibility was observed.

C. Reliability of the calorimeter

Another objective of the irradiation campaign was to verify the reliability and robustness of the CALORRE calorimeter by carrying out calibrations of the CALORRE calorimetric cells while the reactor was shut down at different times during the irradiation campaign. They were performed in the night before, after and during the irradiation campaign with reproducibility tests.

In that case, the calorimeter was placed in a fixed position in the upper part of the channel to avoid residual power after the reactor shut down: bottom of the CALORRE calorimeter at +355.4 mm. To calibrate each calorimetric cell, the nuclear heating rate is simulated by generating heat by means of Joule effect thanks to the heating element integrated into each cell. Increments of electrical current were injected successively into each heating element until a stable thermal state was reached for each increment (cf. Figure 6). Different water-loop temperatures were tested. The water-loop temperature for the first calibration realized in the reactor before the start of the irradiation campaign was set to 75 °C but as it was not possible to reach this value with the reactor in operation at 5 MW_{th}, new higher water-loop temperature values were used for the next calibrations: 115 °C and 120 °C.

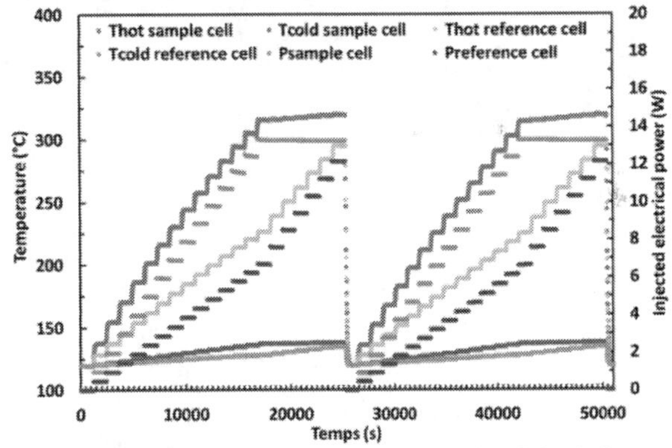

Fig. 6. Temporal response curves in temperature and injected electrical power for two successive calibrations of the two CALORRE calorimetric cells after the irradiation campaign with reactor shut down

Figure 6 shows an example of temporal curves in temperatures and injected electrical power for two calibrations realized successively (reproducibility tests) after the irradiation campaign with reactor shut down, for an imposed water-loop temperature equal to 120 °C and an electrical power range up to 13.3 W.

Figure 7 presents the results obtained for the different experimental conditions and reproducibility tests. The calibration curves obtained for the two calorimetric cells with an imposed water-loop temperature close to that reached when the reactor was in operation, have very low drift demonstrating the reliability and robustness of the CALORRE calorimeter.

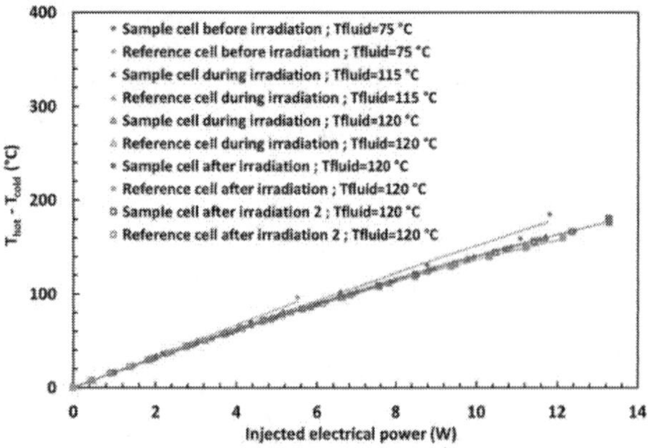

Fig. 7. Calibration curves of the two CALORRE calorimetric cells at different times during the irradiation campaign with reactor shut down

IV. Conclusions

The most compact prototype of the CALORRE differential calorimeter to date, with a height of 88.5 mm, was studied thanks to an irradiation campaign in the in-core water-loop in the B3 position of the MITR.

The CALORRE sensor allowed the quantification of the nuclear heating rate for the first time in the MITR. The measurements were realized for a sample made of stainless steel (ASI 316L) for different experimental conditions and operating protocols. A maximum nuclear heating rate of 2.04 $W.g^{-1}$ were obtained at 5 MW_{th}.

An axial profile of the nuclear heating rate was determined in 15 positions of the sample between -109.38 mm and +304.32 mm by moving the calorimeter with an increment of 29.55 mm while keeping the reactor in operation at 5 MW_{th} and thus by considering the temperature differences of the two superimposed calorimetric cells for the same axial position.

The influence of the size of this compact CALORRE calorimeter on the nuclear heating rate quantification was estimated by using the responses of the two calorimetric cells without calorimeter displacement. A low influence was observed in the zone from -109 mm to +38.37 mm.

The nuclear heating rate was measured at -50.28 mm versus the reactor power: between 0.13 $W.g^{-1}$ and 2.32 $W.g^{-1}$ from 284 kW_{th} to 5.9 MW_{th} respectively.

The reliability and the robustness of the calorimeter was verified thanks to different calibrations before, during and after the irradiation campaign with the reactor shut down.

Other experiments during this successful irradiation campaign in the MITR were conducted (and not presented in this paper) with the CALORRE calorimeter (axial profile with a higher space-resolution in a specific area, another measurement method by using one heater, reproducibility, residual power measurement...) and with the gamma thermometer. The corresponding data are currently being processed. After analyzing the experimental data, 3-D simulations will be performed taking the entire system into account in order to make future comparisons.

V. Acknowledgements

The CALOR-I project leading to this publication has received funding from the Excellence Initiative of Aix-Marseille University - A*Midex, a French "Investissements d'Avenir" programme", AMX-19-IET-013.

References

[1] A. Lyoussi, D. Fourmentel, J-F. Villard, J-Y. Malo, P.Guimbal, H. Carcreff, C. Gonnier, G. Bignan, J-P. Chauvin, C. Reynard-Carette, J. Brun, O. Merroun, M. Carette, M. Muraglia, A. Janulyte, Y. Zerega and J. Andre, "Advanced methodology and instrumentation for accurate on line measurements of neutron, photon and nuclear heating parameters inside Jules Horowitz MTR Reactor", Proceedings of Conferences RRFM and IGORR 2012, pp. 21-25, Prague, Czech Republic, 18-22 March 2012.

[2] C. Reynard-Carette, G. Kohse, J. Brun, M. Carette, A. Volte, A. Lyoussi, "Review of nuclear heating measurement by calorimetry in France and USA", Proc. ANIMMA 2017, Jun. 2017, EPJ Web of Conferences 170, 04019 (2018).

[3] J. Brun, C. Reynard-Carette, A. Lyoussi, C. De Vita, M. Carette, M. Muraglia, D. Fourmentel, P. Guimbal, J.-F. Villard, "Comparison of the thermal calibration response of two calorimetric cells dedicated to nuclear heating measurements", IEEE Trans. Nucl. Sci., vol. 61, no. 4, pp. 2254-2261, Aug. 2014.

[4] M. Carette, A. Lyoussi, J. Brun, C. Reynard-Carette, J.F. Villard, « Eprouvette pour mesure d'échauffements nucléaire dans un réacteur nucléaire, et cellule calorimétrique comprenant au moins une telle éprouvette », FR1553136A, 2015-04-10.

[5] A.Volte, C. Reynard-Carette, A. Lyoussi, J. Brun, M. Carette, "Study of the Response of a New Compact Calorimetric Cell for Nuclear Heating Rate Measurements", IEEE Trans. Nucl. Sci., vol. 65, no. 9, pp. 2461-2470, Sept. 2018.

[6] A.Volte, J. Brun, A. Lyoussi, M. Carette, C. Reynard-Carette, "Qualification of a New Differential Calorimeter Configuration Dedicated to Nuclear Heating Rates up to 20 W.g-1", IEEE Trans. Nucl. Sci., vol. 67, no. 11, pp. 2405-2414, Nov. 2020.

[7] C. Reynard-Carette, A. Volte, A. Lyoussi, M. Carette, « Noyau de mesure pour la mesure d'échauffement nucléaire en réacteur nucléaire et capteur calorimétrique intégrant un tel noyau de mesure », Patent 21212899.5, 2021-12-07.

[8] J. Rebaud, A. Volte, M. Carette, A. Lyoussi, C. Reynard-Carette, "Experimental and numerical results for a new single-cell calorimeter dedicated to nuclear heating rate measurement", ANIMMA 2023, 12-16 June 2023, Lucca, Italy, EPJ Web of Conferences 288, 04008 (2023), https://doi.org/10.1051/epjconf/202328804008.

[9] A. Volte, M. Carette, A. Lyoussi, G. Kohse, J. Rebaud, V. Valero, C. Reynard-Carette, "Review of CALORRE Calorimeter Characterizations Under Laboratory and Irradiation Conditions", IEEE Trans. Nucl. Sci., vol. 69, no. 4, pp. 840-848, April 2022, doi: 10.1109/TNS.2022.3150148.

[10] S. Hauptman, A. Volte, G. Kohse, M. Carette, A. Lyoussi, C. Reynard-Carette, "Flux and Nuclear Heating Rate Calculations for CALORRE Irradiation in MITR", ANIMMA 2023, 12-16 June 2023, Lucca, Italy,, EPJ Web of Conferences 288, 04010 (2023), https://doi.org/10.1051/epjconf/202328804010.

[11] A. Volte, S. Hauptman, D. Carpenter, M. Carette, A. Lyoussi, G. Kohse, C. Reynard-Carette, "Numerical and Experimental Characterization of a New Reduced-Height CALORRE Differential Calorimeter for CALOR-I Irradiation in MITR", in IEEE Trans-actions on Nuclear Science, vol. 71, no. 5, pp. 1134-1143, May 2024, doi: 10.1109/TNS.2024.3388518.

EPJ Web of Conferences **338**, 04033 (2025) https://doi.org/10.1051/epjconf/202533804033
ANIMMA 2025

Results of South-East Flux Trap Dosimetry Measurements for the Advanced Test Reactor Critical Facility in support of Advanced Sensors and Instrumentation Development.

M. Reichenberger[1,*], K. McCary[1], T. Holschuh[1], K. Tsai[1], T. Quist[1], and D. Ottaway[1]
[1]Idaho National Laboratory, USA
(*) Michael.Reichenberger@INL.gov

***Abstract*— Reactor dosimetry measurements are commonly used to validate simulation and modeling in nuclear reactor experiments. Numerous standard dosimeter materials exist which are commonly utilized for their sensitivity to different energy ranges of neutrons. At the Advanced Test Reactor, cobalt alloy and pure nickel wires are installed every cycle to monitor thermal- and fast-neutron fluence rates. However, there is growing interest in exploring less commonly used materials which are either more sensitive to different parts of the neutron energy spectrum, or which can incorporate multiple activation paths in a single material. Epithermal and fast-neutron energies beyond the typical 1-MeV threshold are of particular interest. Two Advanced Test Reactor Critical Facility Flux Runs took place during 2024; each flux run included supplemental dosimetry packages in the South-East Flux Trap. The focus of the dosimetry package for flux run 23-4 was to test two novel dosimetry methods that can provide simultaneous thermal and threshold (fast) sensitivity in a single dosimeter wire. A selection of 3% gold-in copper alloyed wires were available that provided sensitivity to fast and thermal neutrons through 5 different reactions. Likewise, iron offers multiple interaction pathways with sensitivity to both thermal and fast neutrons. The main question to be answered by these irradiations was if sufficient radioactivation would take place in the Advanced Test Reactor Critical Facility South-East Flux Trap during a nominal 20-minute irradiation at typical power levels (near 600 W_{th}) to allow the observation of the threshold reactions with smaller activation cross-sections than the thermal reactions, without being saturated by interfering interactions and Compton continuum during the High-Purity Germanium measurements. The results from comparing the measurement results to anticipated activity levels provide confidence in our ability to activate both traditional and novel dosimetry materials in Advanced Test Reactor Critical Facility, however not all the measured values matched with the predicted activities. This leaves further room for investigation both on the experimental and computational approaches for future irradiation experiments.**

***Keywords* —ATR-C, Reactor Dosimetry, ASI.**

I. Introduction

Reactor dosimetry measurements are commonly used to validate simulation and modeling in nuclear reactor tests. Numerous standard dosimeter materials exist which are commonly utilized with sensitivities to different energy ranges of neutrons [1]. At the Advanced Test Reactor (ATR), cobalt alloy and pure nickel wires are installed every cycle to monitor thermal- and fast-neutron fluence rates [2]. However, there is growing interest in exploring less commonly used materials which are either more sensitive to different parts of the neutron energy spectrum, or which can incorporate multiple activation paths in a single material, shown in Fig. 1. Epithermal and fast-neutron energies beyond the typical 1-MeV threshold are of particular interest.

The classical dosimeter materials shown in Fig. 1 do not provide coverage over the neutron energy range from 1 eV – 100 keV. For light-water reactors, the thermal neutron flux is of the greatest interest, however, continued interest in fast-spectrum nuclear reactor development has expanded the need for dosimetry methods in the > thermal energy regions. To this end, an irradiation campaign has been executed to demonstrate capabilities to conduct conventional dosimetry measurements and to begin investigation of novel reactions for epithermal- and fast-neutron measurements.

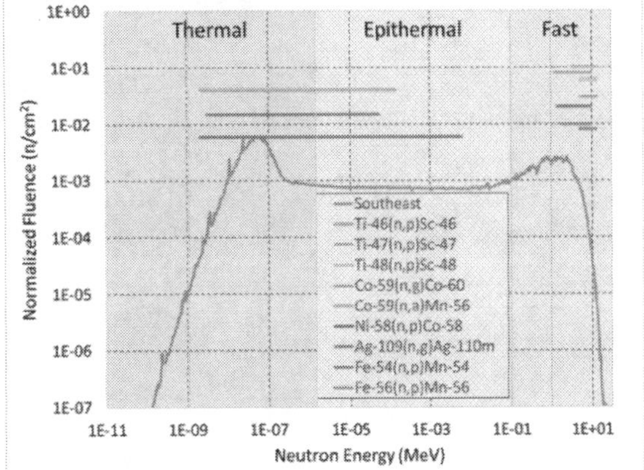

Figure 1. Distribution of traditional dosimeter interaction sensitivities utilized in flux run 24-1 relative to the normalized SEFT neutron energy profile in ATR-C.

II. METHOD

Two Advanced Test Reactor Critical Facility (ATR-C) flux runs took place during FY 2024, described in Table I. ATR-C serves as a prototypical irradiation environment to complement ATR with a similar flux distribution, but operated a much lower power (typically <1 kW_{th}). Each flux run included supplemental dosimetry packages in the South-East Flux Trap (SEFT) filler depicted in Fig. 2. For each of flux runs, 23-4 and 24-1, duplicate dosimetry packages were installed into the SE-2 and SE-4 positions. The loading for each flux run was different, as described in App. A. The dosimeter wires in all cases were encapsulated within a 0.075-in. outer diameter by 0.050-in. inner diameter pure aluminum tube, within a standard BR holder (DWG-035561-2), held in position by crimping onto the end of a 0.040-in diameter spacer wire.

The focus of the dosimetry package for flux run 23-4 was to test two novel dosimetry methods that can provide simultaneous thermal and threshold (fast) sensitivity in a single dosimeter wire, shown in Fig 3. A selection of 3% Au in Cu alloyed wires were available at the ATR Radiation Measurements Laboratory (RML). These wires provide sensitivity to fast and thermal neutrons through 5 reactions. Likewise, Fe offers multiple interaction pathways with sensitivity to both thermal and fast neutrons.

The main question to be answered by these irradiations was if sufficient radioactivation would take place in the ATR-C SEFT during a nominal 20-minute irradiation at typical power levels (near 600 W_{th}) to observe the threshold reactions that have smaller activation cross-sections than the thermal reactions without being saturated by interfering interactions and Compton continuum during the High-Purity Germanium (HPGe) measurements. The reported total uncertainty includes components from standard measurement, mass (0.00009 g), systemic (1%), and random (1%), combined by quadrature.

The focus of flux run 24-1 was related to the expanded testing of numerous threshold reactions in the SEFT position. Fe wires were irradiated in both flux runs, but the additional dosimetry in flux run 24-1 was comprised entirely of conventional materials that have been well documented by institutionally accepted standard test methods [3]. Predicted activities for all reactions were determined by numerical integration of the a-priori neutron energy spectrum with cross-section data from IRDFF-II [4] using the assumed values in Table II. A total flux of 1 x 10^9 $n \cdot cm^{2} \cdot s^{-1}$ was assumed based on scaling from 250 MW_{th} flux estimates in the ATR SEFT [5].

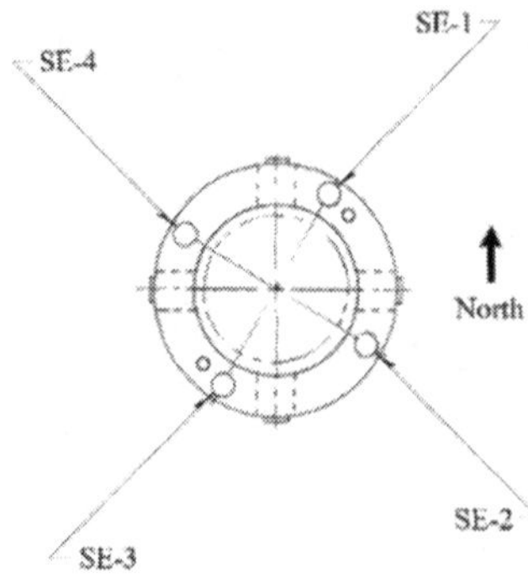

Figure 3. ATR-C SEFT Filler

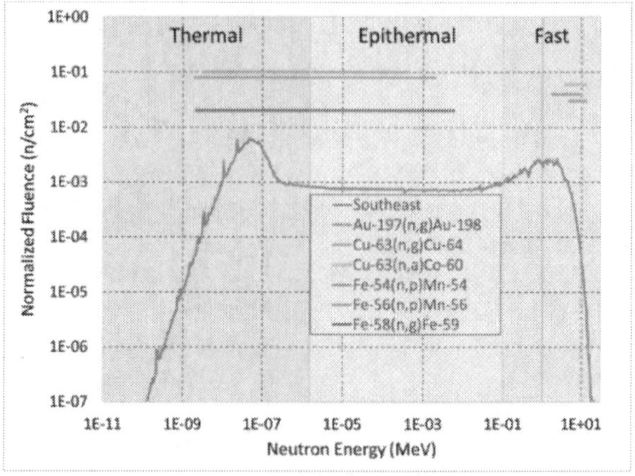

Figure 2. Distribution of traditional dosimeter interaction sensitivities utilized in flux run 23-4 relative to the normalized SEFT neutron energy profile in ATR-C.

TABLE I
ATR-C FLUX RUN SCRAM DATES/TIMES.

Flux run	SCRAM Date	SCRAM Time
23-4	05/30/2024	09:52:56
24-1	07/09/2024	09:20:28

TABLE II
ASSUMED VALUES USED IN CALCULATION OF PREDICTED ACTIVITIES.

Parameter	Value
Neutron Energy Profile	ATRC Southeast Flux Trap
Neutron Flux (n/cm2/sec)	1.00E+09
Irradiation Time (sec)	1200
Decay Time (sec)	3600

III. RESULTS AND CONCLUSIONS

THE MEASUREMENT RESULTS FROM DOSIMETER WIRES ARE SUMMARIZED IN

Table III Although they do provide confidence in our ability to activate both traditional and novel dosimetry materials in ATR-C, the measured values do not match with the predicted activities, shown in the computational to experimental (C/E) values. This leaves further room for investigation both on the experimental and computational approaches. First, the assumed a-priori neutron energy profile was a scaled version of the generic spectrum in the same position in ATR. It is well-known that differences exist between ATR and ATR-C which could impact the neutron energy spectrum, and it is recommended that a dedicated model be used in future experiments. Furthermore, several of the dosimeters required multiple measurements to capture activity from reaction products of various half-lives. This experience has already improved measurement techniques by illustrating the importance of measurement sequence, indicated in the reduced uncertainty in the iron wire activities measured in both flux runs 23-4 and 24-1. As indicated in Table 3, our predictions indicated that the iron activation between flux run 23-4 and 24-1 should be the same. These results did not match as well as expected. The measurement uncertainty and minimum detectable activity for activation in the 23-4 run was much higher. This was observed during the measurement process and improved for the 24-1 run. Further, the specific location (see Appendix A and Appendix B) was slightly different. At the core centerline region, this small difference shouldn't matter much but could contribute to the difference. The availability of numerous HPGe detectors was critical in our ability to accurately measure many of these dosimeters. It was only possible to conduct the long measurement of several of the low-activity reaction products by measuring the samples for several days. If only one or two detectors were used for these measurements the activity of other important reaction products would have decayed while competing measurements were taking place. For future irradiation experiments, careful consideration of the counting plan should be made to ensure that sufficient resources (both time and hardware) are available to measure the reactions of interest. The project planning team will have to make decisions regarding the amount and selection of dosimetry materials to balance the ability to measure the samples. This may require multiple irradiations for future experiments, expansion of counting capabilities, or reduced dosimetry. Further improvement could be made, however, by using detectors with higher counting efficiency and by performing multiple measurements of each dosimeter on numerous detectors to determine an average and standard deviation for sample activity. The method employed herein relied on only a single measurement of activity in many cases which lends the results to inheriting any biases or systemic issues that could exist in the detectors or methods. These issues could be associated, for instance, with the calibration of the detectors or placement of the sample by the measurement technicians. These problems can only be identified and resolved if sufficient time and resources are available to measure each wire multiple times.

The RML performed quality assurance and quality control checks applicable to the gamma-ray measurement for the detector system utilized for these measurements. The RML gamma-ray counting systems were "in control" for all relevant parameters.

TABLE III
PREDICTED AND MEASURED SPECIFIC ACTIVITIES FOR DOSIMETRY INTERACTIONS IN ATR-C FLUX RUNS 23-4 AND 24-1.

Element	Interaction	Predicted Activity (uCi/g)	SE-2 Measured Activity (uCi/g)	Total Uncert. (%) 2σ	C/E SE-2	SE-4 Measured Activity (uCi/g)	Total Uncert. (%) 2σ	C/E SE-4	SE-2/SE-4
Ti	^{46}Ti (n,p) ^{46}Sc [6]	1.74E-05	4.46E-05	17.86%	0.39	2.49E-04	9.51%	0.070	0.18
Ti	^{47}Ti (n,p) ^{47}Sc [6]	6.29E-04	1.34E-03	3.65%	0.47	1.65E-02	3.47%	0.038	0.08
Ti	^{48}Ti (n,p) ^{48}Sc [6]	1.74E-04	2.88E-04	7.26%	0.60	2.08E-03	4.26%	0.083	0.14
Co	^{59}Co (n,γ) ^{60}Co [7]	6.59E-02	3.97E-02	3.06%	1.66	5.39E-02	3.06%	1.22	0.74
Co	^{59}Co (n,α) ^{56}Mn [7]	7.78E-04	6.01E-01	3.17%	0.00129	8.08E-01	3.20%	0.00096	0.74
Ni	^{58}Ni (n,p) ^{58}Co [8]	6.53E-04	1.86E-03	9.37%	0.35	1.17E-03	16.36%	0.56	1.59
Ag	^{109}Ag (n, γ) ^{110m}Ag [7]	2.97E-02	1.07E-02	6.57%	2.77	1.16E-02	6.26%	2.55	0.92
Fe	^{54}Fe (n,p) ^{54}Mn [9]	1.97E-05	3.29E-05	18.73%	0.60	4.16E-05	17.00%	0.47	0.79
Fe	^{56}Fe (n,p) ^{56}Mn [9]	4.95E-03	2.29E-02	4.88%	0.22	1.84E-02	4.00%	0.27	1.24
Fe	^{58}Fe (n, γ) ^{59}Fe [9]	2.73E-04	3.25E-04	9.19%	0.84	2.59E-04	23.26%	1.05	1.25
Fe	^{54}Fe (n,p) ^{54}Mn [9]	1.97E-05	-	-	-	5.87E-05	48.03%	0.34	-
Fe	^{56}Fe (n,p) ^{56}Mn [9]	4.95E-03	-	-	-	1.93E-02	25.65%	0.26	-
Fe	^{58}Fe (n, γ) ^{59}Fe [9]	2.73E-04	3.13E-04	44.19%	0.87	3.89E-04	70.19%	0.70	0.80
Cu	^{63}Cu (n, γ) ^{64}Cu [10]	1.37E+01	1.49E+01	8.81%	0.92	1.66E+01	9.33%	0.82	0.90
Cu	^{63}Cu (n, α) ^{60}Co [10]	2.29E-07	-	-	-	3.11E-04	29.06%	0.00073	-
Au	^{197}Au (n, γ) ^{198}Au [1]	7.09E+01	6.76E+01	6.91%	1.05	8.00E+01	7.11%	0.89	0.84

APPENDIX A

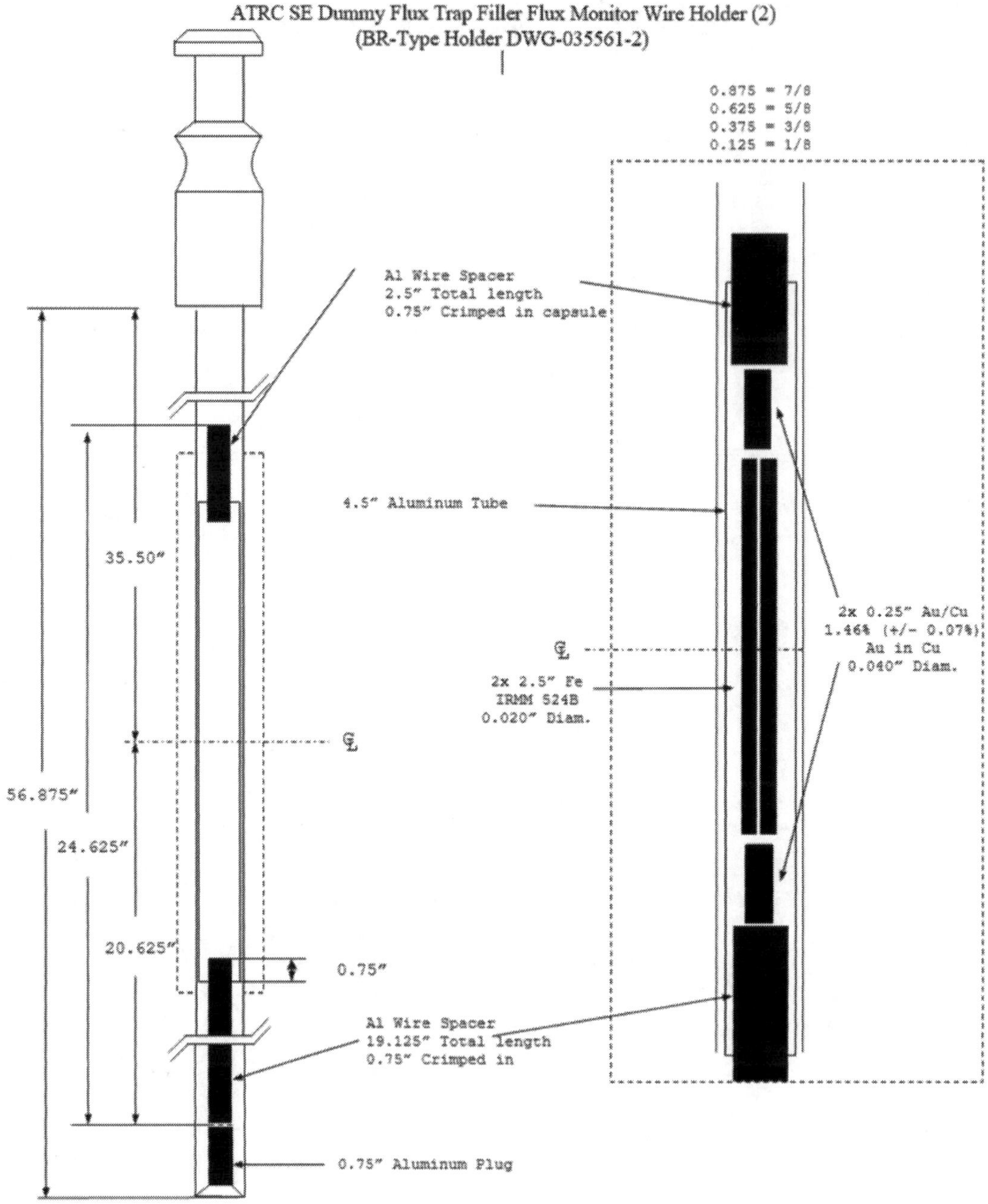
ATRC SE Dummy Flux Trap Filler Flux Monitor Wire Holder (2)
(BR-Type Holder DWG-035561-2)
0.875 = 7/8
0.625 = 5/8
0.375 = 3/8
0.125 = 1/8
Al Wire Spacer
2.5″ Total length
0.75″ Crimped in capsule
4.5″ Aluminum Tube
35.50″
2x 0.25″ Au/Cu
1.46% (+/- 0.07%)
Au in Cu
0.040″ Diam.
2x 2.5″ Fe
IRMM 524B
0.020″ Diam.
56.875″
24.625″
20.625″
0.75″
Al Wire Spacer
19.125″ Total length
0.75″ Crimped in
0.75″ Aluminum Plug

https://doi.org/10.1051/epjconf/202533804033

APPENDIX B

ATRC SE Dummy Flux Trap Filler Flux Monitor Wire Holder (2)
(BR-Type Holder DWG-035561-2)

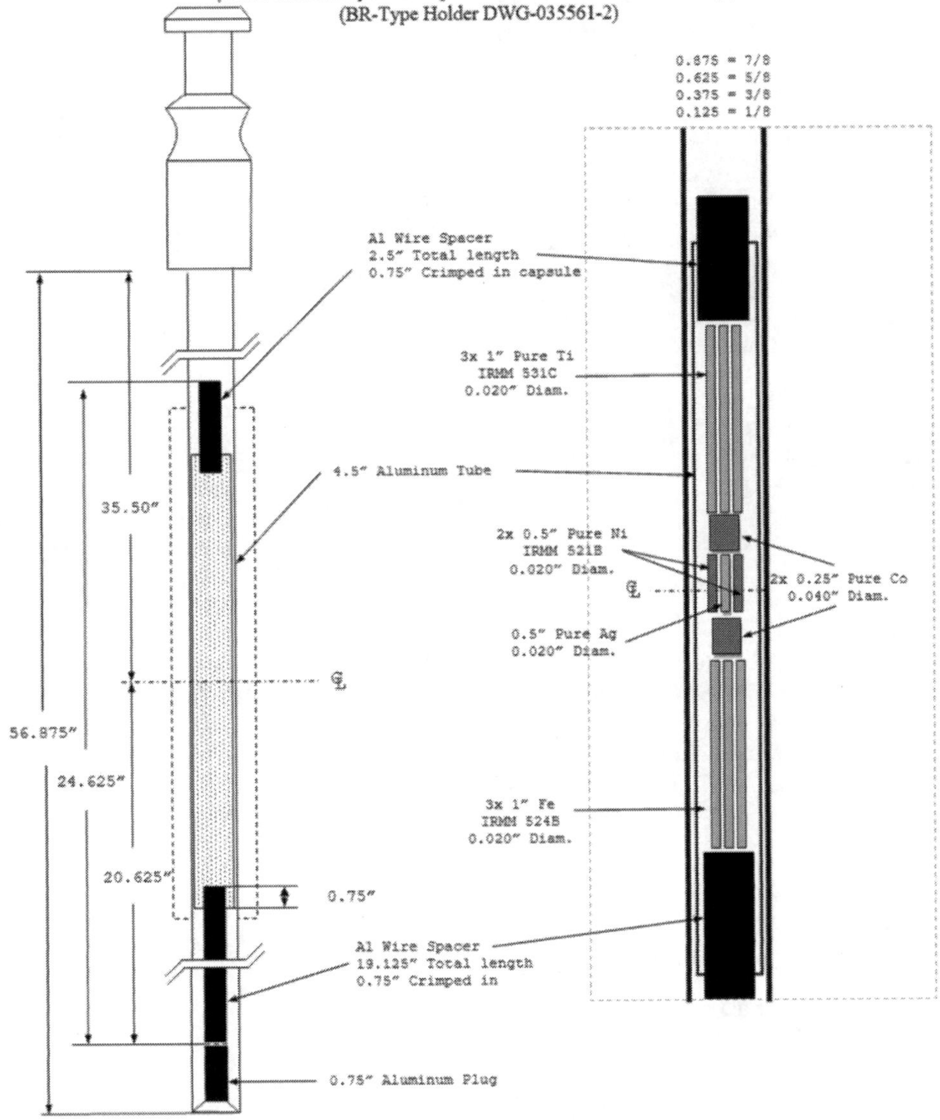

ACKNOWLEDGMENT

This work, presented at the Advancements in Nuclear Instrumentation Measurement Methods and their Applications (ANIMMA) 2025, was supported by the United States Department of Energy Contract Number: DE-AC07-05ID14517. This information was prepared as an account of work sponsored by an agency of the U.S. Government. Neither the U.S. Government nor any agency thereof, nor any of their employees, makes any warranty, expressed or implied, or assumes any legal liability or responsibility for the accuracy, completeness, or usefulness, of any information, apparatus, product, or process disclosed, or represents that its use would not infringe privately owned rights. References herein to any specific commercial product, process, or service by trade name, trademark, manufacturer, or otherwise, does not necessarily constitute or imply its endorsement, recommendation, or favoring by the U.S. Government or any agency thereof. The views and opinions of authors expressed herein do not necessarily state or reflect those of the U.S. Government or any agency thereof.

REFERENCES

[1] ASTM International Standard E262, "Standard Method for Determining Thermal Neutron Reaction and Fluence Rates by Radioactivation Techniques," 2024.

[2] C. Jones and M. Reichenberger, "Measured Thermal and Fast Neutron Fluence Rates ATR Cycles 171B MP-2," Idaho National Laboratory, 2024.

[3] ASTM International, "Standard Practice for Determining Neutron Fluence, Fluence Rate, and Spectra by Radioactivation Techniques," 2021.

[4] A. Trkov, P. Griffin, S. Simakov, L. Greenwood, K. Zolotarev, R. Capote, D. Aldama, V. Chechev, C. Destouches, A. Kahler, C. Konno, M. Kostal, M. Majerle, E. Malambu, M. Ohta, V. Pronyaev, V. Radulovic, S. Sato and M. Schulc, "IRDFF-II: A New Neutron Metrology Library," *Special issue of Nuclear Data Sheets,* vol. 163, pp. 1-108, 2019.

[5] Idaho National Laborator, "Advanced Test Reactor Users Guide," Idaho National Laboratory, Idhao Falls, ID, 2023.

[6] ASTM International Standard E526, "Standard Test Method for Measuring Fast-Neutron Reaction Rates By Radioactivation of Titanium," 2022.

[7] ASTM Internatinoal Standard E481, "Standard Test Method for Measuring Neutron Fluence Rates by Radioactivation of Cobalt and Silver," 2016. [Online]. Available: https://www.astm.org/e0481-16.html.

[8] ASTM International Standard E264, "Standard Test Method for Measuring Fast-Neutron Reaction Rates by Radioactivation of Nickel," 2019. [Online]. Available: https://www.astm.org/e0264-19.html.

[9] ASTM International Standard E263, "Standard Test Method for Measuring Fast-Neutron Reaction Rates by Radioactivation of Iron," 2018.

[10] ASTM International Standard E523, "Standard Test Method for Measuring Fast-Neutron Reaction Rates by Radioactivation of Copper," 2021.

Measurement of the Thermal Neutron Fluence Rate in the Instrument Thimble 11 Position in the Advanced Test Reactor Critical

Dani G. Ottaway[1,*], Kelly M. McCary[1], and Michael A. Reichenberger[1]
[1]Radiation Measurement Laboratory, Idaho National Laboratory, USA
(*) dani.ottaway@inl.gov

***Abstract*— The Advanced Test Reactor Critical (ATRC) is a full-scale replica of the Advanced Test Reactor (ATR) core, utilized primarily for evaluating experiment impacts on power distribution. This study focuses on neutron fluence measurements within Instrument Thimble 11 (IT-11) by means of radiometric activation of passive dosimetry packages. These packages contained indium and gold thin foils as well as gold, cobalt, and nickel wire utilized during the 23-3 and 23-4 ATRC irradiations within IT-11 to evaluate the neutron field present within the test position. Measurement variances were observed between the various dosimetry methods, likely due to self-shielding effects in dosimetry foils and potential epithermal neutron contributions. Future work includes refining self-shielding corrections, exploring neutron field directionality, and developing epithermal corrections to resolve discrepancies. The findings aim to establish a more accessible and affordable test platform for nuclear instrumentation development and testing at the ATR facility.**

***Keywords* — Advanced Test Reactor Critical, ATRC, Neutron Dosimetry, Instrument Thimble 11, IT-11, Thermal Neutron Column**

I. Introduction

THE Advanced Test Reactor (ATR) is a highly versatile, pressurized, light water cooled, beryllium moderated test reactor with a nominal power capacity of 250 Megawatts (thermal). Its unique core design utilizes a serpentine clover leaf of 40 fuel elements to create 9 different flux traps. The Advanced Test Reactor Critical (ATRC) is a full-scale, 1:1, replica of the Advanced Test Reactor core but in a pool in lieu of a pressure vessel and usually operates at a power level near 600 Watts [1]. ATRC is typically used to evaluate experiment impacts on power distribution. These measurements of core power distribution in ATRC are utilized to indicate how an experiment will perform during the full irradiation in ATR and verify experiment reactivity models. The cores were designed with space to accommodate up to 12, dry, in-pile tubes located outside the main reactor tank that were to be used as instrument thimbles. Measurements of the neutron fluence rate in Instrument Thimble 11 (IT-11) in the ATRC will be used to assess the test position for nuclear instrumentation development, research, and testing. Dosimetry packages were placed in IT-11 during the 23-3 and 23-4 ATRC irradiations containing several experiments. These experiments also established the processes needed to execute a wider variety of experiments within the instrument thimbles, as well as aided in refining the material selections for further dosimetry packages.

II. Background

The ATRC core was built with 8 dry air tubes for nuclear instrumentation. Most of the tubes are filled with instrumentation to support reactor operations but Instrument Thimbles 6 (IT-6) and 11 (IT-11) are left empty and unutilized during typical irradiations. The neutron fluence within these instrument thimbles is neither well characterized nor typically included in reactor models because they do not impact core reactivity.

Most active dosimetry instrumentation and electronics cannot survive for a prolonged operation in-core to facilitate the measurement of neutron levels because the environment inside the cores of nuclear reactors is hostile during irradiations. In-core passive dosimetry is one method to measure the neutron environment within the core during the irradiation through material radioactivation methods [2,3,4,5]. Passive dosimetry is inserted prior to irradiation and left in while the core is critical. It is then disassembled after the irradiation is completed and the material activation is measured. From this measured activity, information about the neutron fluence can be determined.

Indium can be used to measure thermal neutron fluence [6]. Its thermal neutron absorption cross section of 159 barns makes it an excellent absorber through the $^{115In}(n,\gamma)^{116}In$ reaction. ^{116}In has a 14.1 second half-life and cannot typically be measured directly, but in 79% of reactions, it produces ^{116m}In as a daughter product. 116mIn has a 54.29 minute half-life and can facilitate measurements if samples can be recovered and measured quickly [3]. However, indium has some limitations. It exhibits a high series of resonance peaks in its neutron absorption cross section, necessitating adjustments during measurements. Additionally, calculations require multiple corrections to account for factors such as the natural abundance of 115In and gamma emission probability, complicating post-measurement analysis. The short half-life of ^{116m}In further complicates analysis at secondary locations, as several half-lives may elapse before initial measurements can be performed [2,3].

Gold is also commonly utilized for thermal neutron dosimetry [6]. Being monoisotopic and having a single gamma-ray emission, gold minimizes the potential for errors compared to other dosimetry materials such as indium. The ^{197}Au(n,γ)^{198}Au reaction has an absorption cross section of approximately 98.65 barns, making gold an excellent thermal neutron absorber. However, gold has a significant resonance peak in the epithermal range, necessitating careful consideration when unfolding and quantifying the neutron spectrum [2,3]. ^{198}Au, with a half-life of 2.695 days, reaches saturation quickly and maintains sufficient activity for off-site measurements. When utilizing geometries like thin foils or wires, they can be reused for further irradiation once their activity has completely decayed. Due to gold's large cross section, it is crucial to account for self-shielding issues during calculations which are more prevalent in dosimetry foils than they are in wires [2,3].

Cobalt is a preferred thermal neutron dosimeter for many activation experiments. However, it may encounter issues with isotopic burnup if total fluence exceeds 10^{20} n·cm-2 [2,3,5]. With a half-life of 1925.3 days, cobalt can be measured long after irradiation, while still yielding quality results. Its thermal neutron absorption cross section of 37.2 barns makes it an excellent absorber through the ^{59}Co(n,γ)^{60}Co reaction. Nonetheless, cobalt has a notable drawback: it also has a large resonance peak in the epithermal range, which can affect activation measurements. Therefore, additional care or alternative measurement techniques are necessary to ensure accurate measurements [2,3,5].

Finally, nickel is a widely used fast neutron dosimeter for assessing neutron levels above 1 MeV through the ^{58}Ni(n,p)^{58}Co reaction. With a half-life of 70.85 days, ^{58}Co remains detectable after extended irradiation periods. However, in the presence of high thermal neutron fluence, there are concerns about burnup due to concurrent interactions with ^{58m}Co and ^{58}Co [4]. Additionally, most nickel contains impurities of ^{59}Co, which can lead to the production of ^{60}Co and potentially contaminate measurement results if measured alongside other ^{60}Co dosimeters [4].

III. Experiment Methods

A series of High Purity Germanium (HPGe) detectors were utilized to perform gamma spectroscopy measurements of the activated dosimeter materials. These detectors are calibrated and maintained in accordance with industry standards [7]. All dosimetry was measured in a round-robin style to ensure potential bias of any individual detector was removed.

A. ATRC 23-3

For the 23-3 ATRC irradiation a series of initial measurements were obtained utilizing gold and indium dosimetry foils. These foils (circular, roughly 0.188 in. diameter by 0.004 in. thick) were cleaned, weighed three times, placed in individual, 1.5 ml sample vials and labelled as shown in Fig. 1.

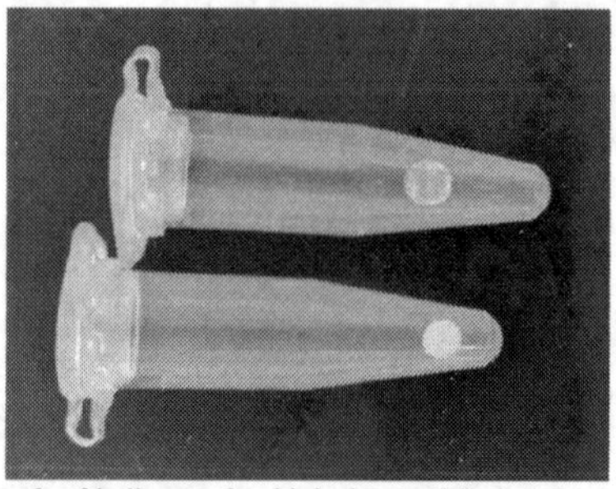

Fig. 1. Photograph of indium and gold dosimeter foils in 1.5 ml sample vials after 23-3 ATRC irradiation [8].

The sample vials were then secured to a rope at pre-determined locations with a weight at the end and lowered into IT-11 during the irradiation as illustrated in Fig. 2. Test positions spanned from the bottom of the instrument thimble, 15 inches below the fueled region of the core, to just below the reactor pool water level. After 10 minutes at the stable reactor power of 533.82 W, the dosimetry was removed and taken to the Radiation Measurement Laboratory (RML) for gamma spectroscopy measurement. At RML, the various vials were measured on several HPGe detectors, and the specific activity was determined for each dosimeter [8].

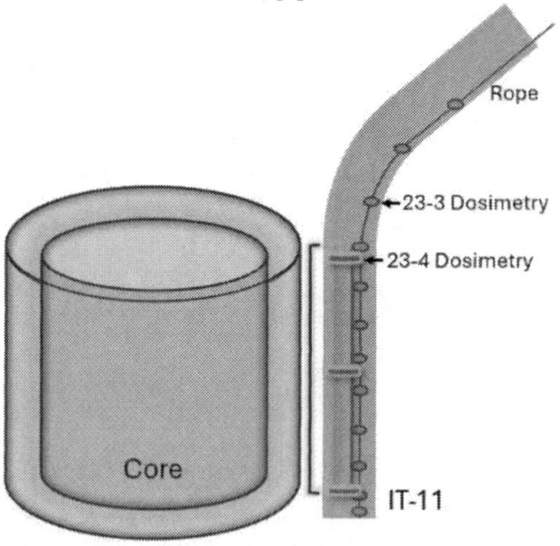

Fig. 2. Illustration of installed dosimetry packages for the 23-3 and 23-4 ATRC irradiations.

B. ATRC 23-4

Prior to the 23-4 ATRC irradiation a dosimetry package was assembled to further explore the thermal neutron fluence rate in IT-11 within the elevation of the fueled region of the core. This dosimetry package consisted of gold and indium dosimetry foils (described previously) at several elevations as well as two modified ATR B position dosimetry flux monitor wire holders containing 55" of pure gold, cobalt and nickel wires. The flux monitor wire holders were aligned by 3D printed PLA alignment discs at the top, bottom and middle of the holders. The full assembly prior to irradiation is shown in Fig. 3. An attempt was made to place the same dosimeter foils in locations equivalent to and comparable to their irradiation location during the 23-3 ATRC irradiation [8], as indicated in Fig. 2, as sufficient time passed between the irradiations.

Fig. 3. Full dosimetry package for 23-4 ATRC irradiation.

After a 20-minute irradiation at the stable reactor power of 571.92 W, the full dosimetry package was disassembled at RML and foils measured before the indium dosimetry experienced a full half-life. Dosimetry wires were removed from the modified flux monitor wire holders and cut into 1-inch segments, weighed, and placed in labelled sample vials. To prevent any accidental misplacement or confusion of dosimetry, only one wire segment was cut, moved and weighed at a time until the corresponding vial for that sample was closed and contents verified. Specific activity was determined for the dosimetry wires and foils and compiled into a data set.

IV. Results

The data from the 23-3 and 23-4 ATRC irradiations were utilized to calculate neutron fluence rates utilizing industry standard practices [2,3,4,5]. No quantifiable activity was detected on the nickel wire segments, therefore no fast neutron fluence could be determined. After all the appropriate corrections were made, the thermal neutron fluence rates were plotted to provide a profile of the fluence rate within IT-11 relative to the core elevation. The cobalt and gold wires cut into nominal 1-inch segments provide a detailed characterization of the fluence rate profile of the core as seen in Fig. 4.

V. Conclusions and Future Work

The information obtained from utilizing the various methods of determining neutron fluence have resulted in more variance than was anticipated. Thermal neutron fluence rates were between 1.0×10^{6} n·cm^{-2}·s^{-1} and 1.2×10^{7} n·cm^{-2}·s^{-1} throughout the active fueled region of the core. Thermal neutron fluence rate calculations based on dosimetry foil measurements are consistently reporting higher than the wire measurements that were performed. Repeated measurements with the same dosimeter foils in approximately the same locations are yielding different results and indium and gold foils do not agree. The cobalt and gold wires yield more similar results. This suggests that the difference may be attributed to self-shielding within the foils that is less significant in the wires. Given foils have considerably more mass, and subsequently atoms, the likelihood of a neutron interaction is heterogeneous throughout the foil as atoms on the surface interact with neutrons, react and shield nuclei deeper in the material from interactions. Wires have a far more homogenous and consistent interaction probability throughout the whole material. This has greater impact on the gold measurements as the wire has a diameter of 0.004 in. compared to the 0.118 in. diameter of the foils. Additional measurements and simulations will be performed to determine

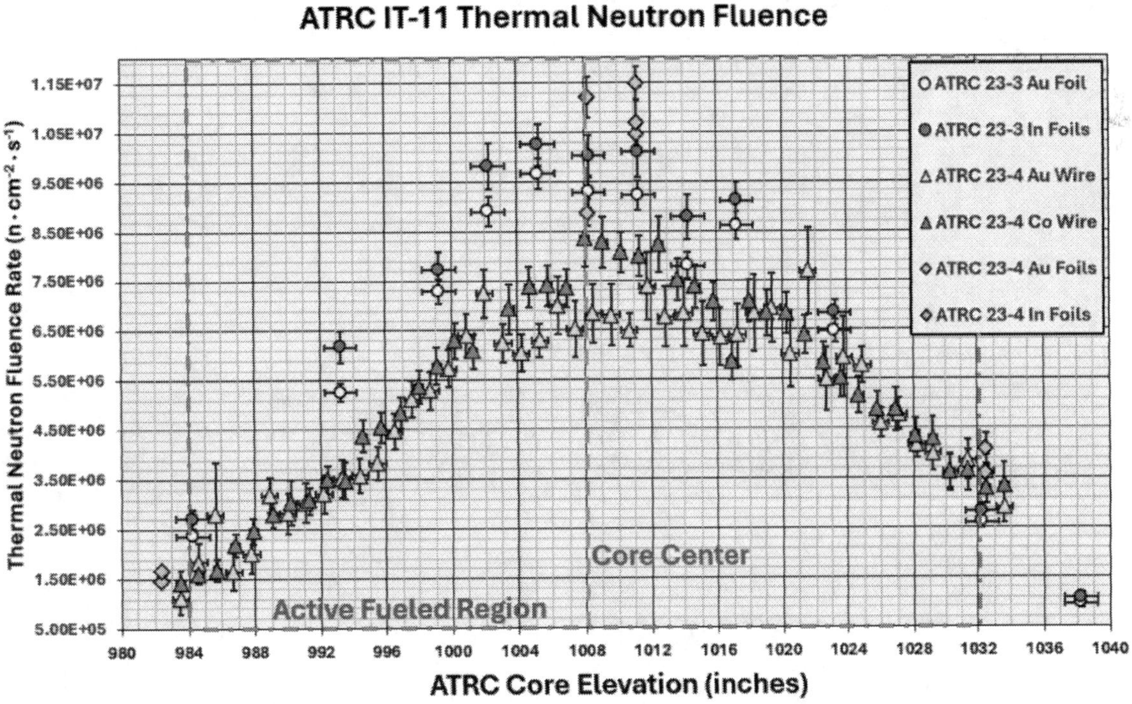

Fig. 4. Calculated thermal neutron fluence rates from the various dosimeters used in the 23-3 and 23-4 ATRC irradiations.

adequate self-shielding factors for these dosimetry foils. Questions have also arisen regarding the impact of directionality of the thermal neutron field within IT-11 and if angle of incidence has an impact on material activation and self-shielding corrections for the foils. Simulations of various neutron fields, beams and dosimeter orientation will be performed to explore this possibility.

Another possible factor is the activation contribution due to epithermal neutron activation. While it was assumed that the neutron field would be purely thermalized by the reactor tank, components and water prior to reaching IT-11, this may not be strictly true. As illustrated in figure 5, gold, indium and cobalt all have very large resonances in epithermal neutron absorption cross section.

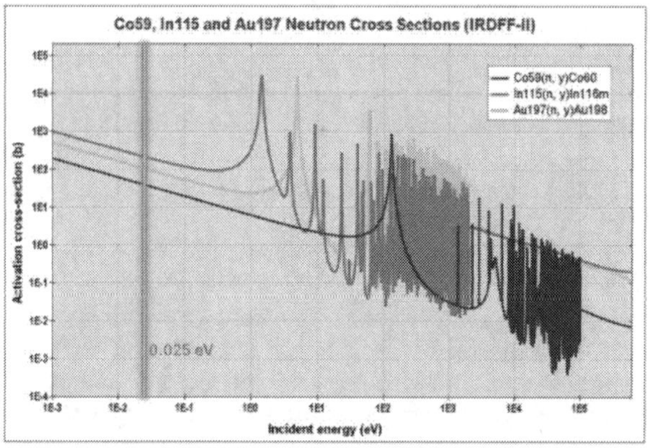

Fig. 5. Neutron absorption cross sections for cobalt, indium and gold dosimeters to illustrate epithermal neutron absorption probabilities [9,10].

Even if an epithermal field was orders of magnitude lower than the thermal neutron field, this assumption may cause inaccuracies. Testing is being developed to utilize cadmium covered dosimeters to determine an effective cadmium ratio for the dosimeter materials. By comparing the results of these measurements, we hope to develop an epithermal correction to see if it helps to resolve the differences between the various radioactivation methods.

The fast neutron field was expected to be negligible in IT-11. The absence of measurable activity in the nickel wires lends further credence to this assumption. If an opportunity is presented for prolonged irradiation in ATRC, an attempt will be made to irradiate additional fast neutron dosimetry within IT-11. Ideally, if there is a small field present, extending the irradiation time, and increasing the overall fluence, that field can activate enough material to be quantified.

References

[1] J. CAMPBELL, "ATR User Guide," Idaho National Laboratory, Idaho Falls, Idaho (2021).
[2] ASTM International Standard E261, "Standard Practice for Determining Neutron Fluence, Fluence Rate, and Spectra by Radioactivation Techniques," 2021.
[3] ASTM International Standard E262, "Standard Test Method for Determining Thermal Neutron Reaction Rates and Thermal Neutron Fluence Rates by Radioactivation Techniques," 2017.
[4] ASTM International Standard E264, "Standard Test Method for Measuring Fast-Neutron Reaction Rates by Radioactivation of Nickel," 2019.
[5] ASTM Standard E481, "Standard Test Method for Measuring Neutron Fluence Rates by Radioactivation of Cobalt and Silver," 2016.
[6] Hogg, C H. "THERMAL NEUTRON FLUX MEASUREMENTS AT THE MTR-ETR SITE.", Oct. 1960.
[7] ASTM International Standard E181, "Standard Guide for Detector calibration and Analysis of Radionuclides in radiation Metrology for Reactor Dosimetry," ASTM (2023).
[8] Ottaway, D G, et al. "Qualifying the Instrument Thimble 11 Test Position in the Advanced Test Reactor Critical," Nov. 2024.
[9] M. B. E. D. N. Soppera, " JANIS 4: An Improved Version of the NEA Java-based Nuclear Data Information System," Nuclear Data Sheets, vol. 120, pp. 294-296, 2014.
[10] A. Trkov, P. Griffin, S. Simakov, L. Greenwood, K. Zolotarev, R. Capote, D. Aldama, V. Chechev, C. Destouches, A. Kahler, C. Konno, M. Kostal, M. Majerle, E. Malambu, M. Ohta, V. Pronyaev, V. Radulovic, S. Sato and M. Schulc, "IRDFF-II: A New Neutron Metrology Library," Special issue of Nuclear Data Sheets, vol. 163, pp. 1-108, 2019.

Validation of working of acoustic sensor for fission gas release characterization devoted to Jules Horowitz reactors in harsh condition up to 350°C and 120 bar

F. Baudry[1], G. Galeron[2], E. Rosenkrantz[1], P. Combette[1], PV. Bonzom[2] C. Destouches[2], F. Martin[2], D. You[3], R. Abadie[3], JY. Ferrandis[1*]

[1] IES, Univ. Montpellier, CNRS, Montpellier, France
[2] CEA-DES-IRESNE, Cadarache, Saint-Paul-Lez-Durance, France
[3] CEA-DES-ISAS, Université Paris-Saclay, Gif-sur-Yvette, France
(*) jean-yves.ferrandis@umontpellier.fr

***Abstract*— For several years, the IES laboratory has been collaborating with the CEA to develop acoustic instrumentation for nuclear applications. As part of this collaboration, the IES acoustic team is designing a miniaturized sensor capable of measuring gas composition in situ within a fuel rod. Initial evidence of this approach dates back to 2010 with the REMORA 3 experiment, conducted in the OSIRIS reactor at CEA Saclay, which enabled acoustic measurements of fission gas release at 150 °C. The new sensors, however, are designed to operate up to 300 °C, thanks to a modified Bismuth Titanate (NBT) piezoelectric material, integrated by screen printing on an alumina substrate.**

This paper focuses on an experiment conducted in January 2024 in a high-pressure vessel at CEA Saclay, where the sensors were tested at 350 °C and 120 bar in various gas mixtures. Impedance and acoustic signal measurements were performed to evaluate sensor ageing and performance. After 79 hours, slight degradation was observed, but the sensor continued to emit acoustic signals and demonstrated its potential for reliable in situ gas composition monitoring. These results are a key step toward validating the device before irradiation tests planned for 2025.

***Keywords* — Fission gas release, High temperature, Nuclear Instrumentation, Ultrasonic sensor**

I. Introduction

THE first experimental evidence of an acoustic measurement used to determine gas composition dates back to 2010 with the REMORA 3 experiment, which measured the release of fission gases from fissile fuel. This experiment was carried out in the OSIRIS reactor located at CEA Saclay in France. The setup allowed the devices to be tested under irradiation at 150 °C [1], [2], corresponding to the maximum operating Curie temperature permitted by PZT elements.

However, the new sensors presented in this paper are designed to operate up to 300,ÄØ¬∞C. This improved performance is made possible through the use of a modified Bismuth Titanate piezoelectric material (NBT). The material is integrated using a screen-printing process on an alumina substrate [1], [3]. The fabrication process and initial characterizations were previously presented at ANIMMA 2021 [4] and ANIMMA 2023 [5], [6].

The research presented here is a direct continuation of the work introduced at the 2023 ANIMMA conference, which detailed the design of an experimental gas composition sensor for measurements in an experimental fuel rod of the Jules Horowitz Reactor (RJH).

These devices are developed by the Electronic and System Institute (IES) in Montpellier, France, specifically by the acoustics team, which also develops acoustic technologies for NDT applications in harsh environments. This sensor is intended to monitor fission gas release in experimental fuel rods. Its operational requirements are as follows:

The sensor must remain functional at temperatures of at least 350 °C

It must perform measurements at pressures of at least 60 bar

The sensor operates based on an acoustic time-of-flight measurement. An analytical model is then used to estimate the gas composition as a function of temperature, pressure, and acoustic velocity.

The manufacturing process and operating principles of the sensor were described in detail in [4]. At IES, we tested the sensor's behavior under varying pressures at room temperature [6], and these tests validated the concept of a gas composition sensor.

This article presents experiments conducted at the DRMP laboratory at the CEA Saclay site, where the sensor is tested in an autoclave at temperatures up to 350 ° C and pressures up to 125 bar.

In the first part, we will describe the sensor preparation process, followed by the experimental setup and test protocol. Finally, the third section will present the experimental measurements and results.

II. Presentation of the sensor

The sensor tested in this study was manufactured and assembled at IES, as shown in Fig. 1, and is composed of two main components. The first is a piezoelectric, or active, element that emits the acoustic wave. As shown in Fig. 1, each sensor consists of two piezoelectric elements positioned face-to-face. The second component is a cylindrical ceramic cavity. All components are assembled using alumina cement, an adhesive capable of withstanding thermal stresses up to 1000 °C in theory. The piezoelectric ceramic used is based on bismuth titanium niobate (NBT), a material extensively studied in [7]. Its Curie temperature, depending on the specific application, is approximately 640 °C.

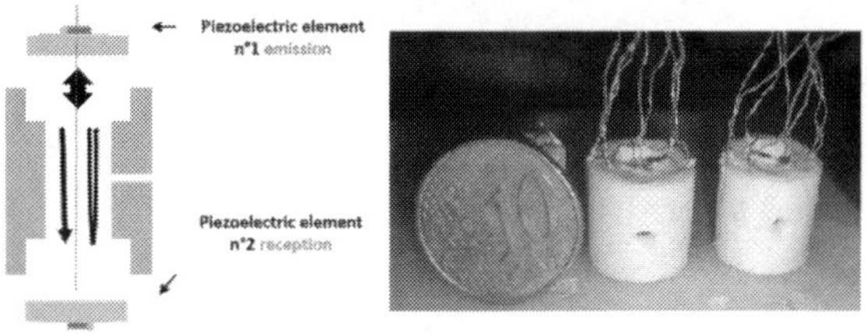

Fig 1. Transmission sensors used for the experiments

The sensor has four connection channels, two for each acoustic element. However, due to technical constraints of the experiment, the sensor had to be connected using only three wires, meaning that both piezoelectric elements share a common ground. Fig. 2 shows the sensor prepared for testing, with wire extensions. These extensions allowed for improved connectivity between the sealed electrical contacts and the sensor.

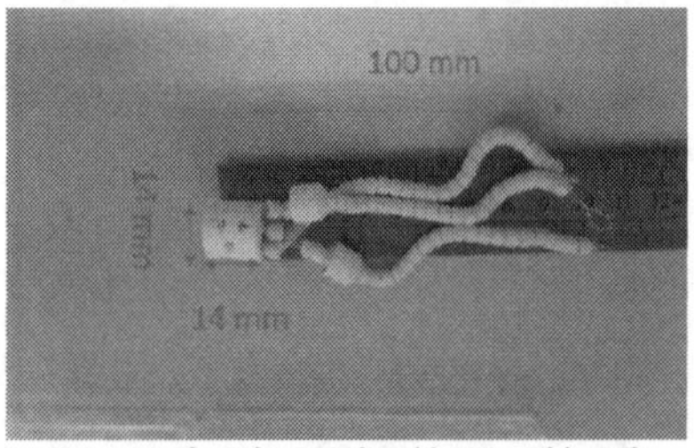

Fig 2. Gas sensor ready to be tested and integrated into the autoclave

Now we're going to go into detail about the experiment and the experimental protocol.

III. Experimental set-up and measurements

The aim of the experiments was to characterize the sensor under thermal, pressure, and gas mixture conditions similar to those found in an experimental fuel rod. These conditions can be reproduced in an autoclave.

A. Material and Methods

The autoclave at DRMP is a device capable of containing a gas mixture at temperatures up to 350 °C and pressures up to 120 bar. A schematic view of the autoclave is shown in Fig. 3. It consists of two sections: a cold section and a hot section. The cold section houses the pressure gauges, gas mixture cylinders, and safety devices. It is connected to the hot section via a pipe, and it is not possible to isolate the two sections from each other. The hot section is equipped with a temperature-controlled heating collar, which raises the chamber to the desired temperature. A type K thermocouple placed inside the chamber is used to measure the temperature.

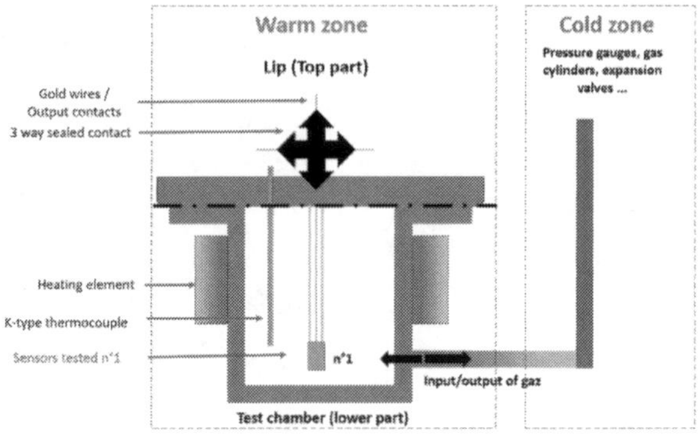

Fig 3. Schematic cut view of the autoclave

The sealed contacts allowed us to connect our instrumentation and monitor the acoustic sensors. The sensor is integrated into the autoclave in three stages, as shown in Fig. 4. It is connected to the sealed contacts via pure gold wires with a diameter of 300 µm. Additionally, the sensor is suspended to avoid any mechanical coupling with the internal wall of the chamber.

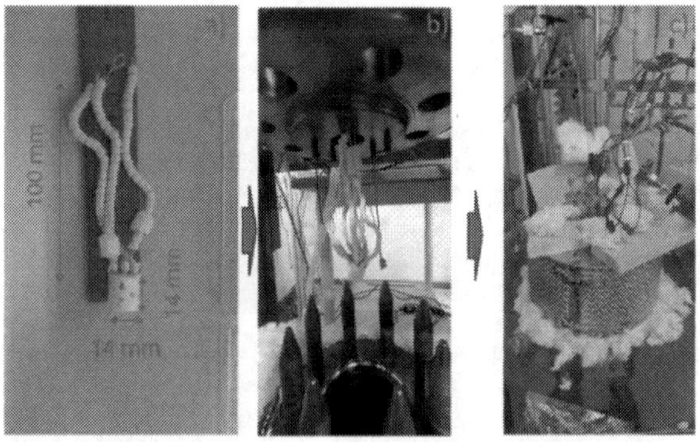

Fig 4. (a) Presentation of the sensor, (b) Sensor before closing the autoclave, (c) Autoclave closed and insulated ready to operate with the instrumentation connected

Once the autoclave is closed and sealed, a number of preparatory steps are required to ensure it is leak-tight. The autoclave then undergoes a helium purging process, which consists of increasing the pressure, then reducing it, and repeating this cycle three times. This process results in a helium concentration of over 99.9% within the autoclave.

For this experiment, two different acquisition systems were used. The first is dedicated to acquiring acoustic data, generally in the form of echograms, as shown in Fig. 5a. From these echograms, the acoustic velocity in the gas can be estimated, which in turn allows for an estimation of the gas composition.

This acquisition chain consists of a Tektronix TDS 3032 oscilloscope and an Olympus Panametrics 5800 acoustic signal conditioner, with data acquisition performed on a laptop computer running dedicated software. The second acquisition system uses a Keysight E4990A impedance analyzer, which enables monitoring of impedance changes over time and as a function of temperature (Fig. 5b).

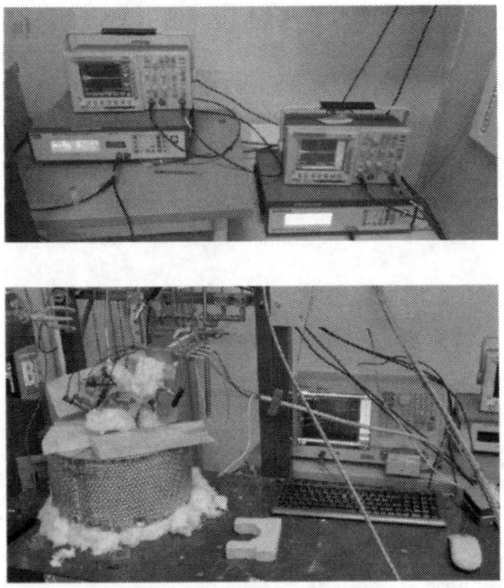

Fig 5 : (a) Photograph of the acoustic test benches used to measure velocity, (b) Impedance analyser Keysight E4990A used to measure real and imaginary impedance

B. *Experimental environment*

The experiment was conducted over a period of four days, during which various gas mixture compositions were evaluated. The experiments presented in this article were performed at distinct temperature levels: 30, 100, 250, and 350°C. These conditions are further illustrated by the impedance measurements displayed in Fig. 6. The gas utilized was sourced from a Messer Group cylinder, consisting of 23.6% xenon, 2.04% krypton, and the balance helium. Hereafter, this will be referred to as the He-Xe-Kr mixture.

Subsequently, gas composition measurements will be conducted at 250°C and 350°C under isothermal conditions.

IV. Results and discussion

A. *Impedance vs. Time*

To study the sensor's evolution, its impedance was measured in situ throughout the experiment. Measurements were conducted at various temperatures, ranging from 30°C to 350°C. The sensor consists of two opposing piezoelectric elements, designated as number 1 and number 2 in Fig. 1. Their temperature-dependent impedances are presented in Fig. 6. The chosen frequency window enables the observation of the first two impedance peaks (approximately 5 MHz and 9 MHz). These peaks are investigated due to the enhanced efficiency of the [1; 10] MHz bandwidth for acoustic emission in gases.

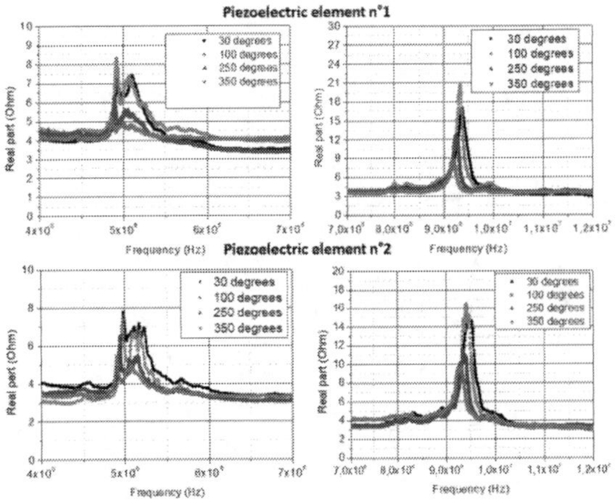

Fig 6. Measurement of real parts of electrical impedance as a function of temperature between [4; 7] MHz and [7; 12] MHz of piezoelectric elements number 1 and number 2

The measurements indicate that the piezoelectric elements responses change with temperature. This observed decrease is linked to a reduction in the coupling coefficient parameter kt, but it does not indicate a degradation of the piezoelectric elements. This phenomenon has been consistently observed in previous laboratory experiments, and post-test analysis confirmed that the acoustic elements suffered no degradation [6].

B. *Measurement at 250 degrees in He-Xe-Kr mixture*

In this experiment, the autoclave was held at a constant temperature of 250°C and a pressure of 124 bar. Initially, an He-Xe-Kr mixture, comprising 10% xenon and 0.9% krypton, was injected into the chamber. Subsequently, helium was introduced into the autoclave over an 11-hour period to progressively decrease the xenon concentration. To ensure isobaric conditions were maintained, a deliberate leak was created to offset the helium injection. Fig. 7 presents the velocity measurements recorded throughout the experiment, demonstrating the gradual convergence of the echoes over time.

Furthermore, at the experiment's onset, only two echoes are visible. However, after 10 hours, a multitude of echoes appear with increased amplitudes. Experimentally, we found that decreasing the xenon concentration reduces attenuation. Fig. 8 displays the velocity measurements extracted from the echograms between 7 PM and 6 AM. After 10 hours, the velocity measured in the autoclave approaches the speed of sound of helium at 124 bar and 250°C, which is 1382 m/s (data from the NIST institute).

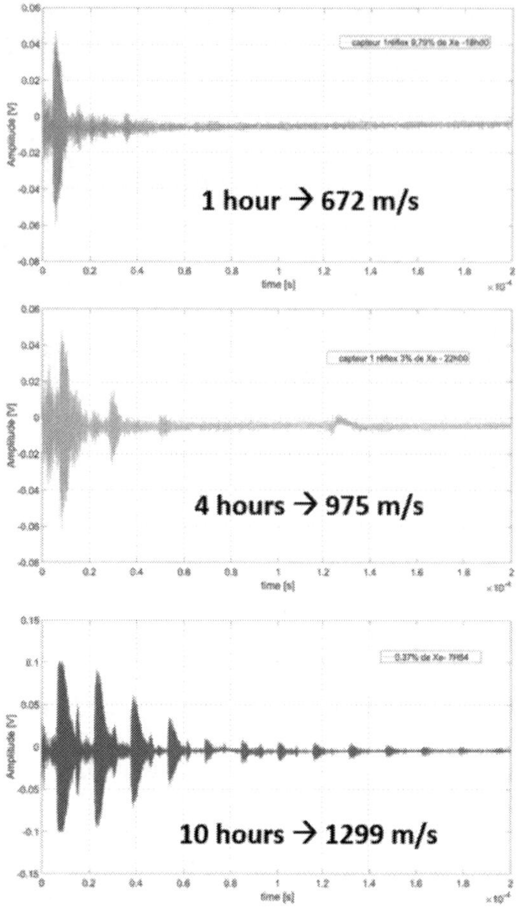

Fig 7. Three echogram measurements taken at 1 hour, 4 hours and 10 hours respectively

As mentioned previously, by measuring acoustic velocity, temperature, and pressure, we can estimate the xenon composition. This estimation has been detailed in several articles [1, 2] and is based on the following equation:

$$C = \sqrt{\frac{\gamma R T}{x M_{Xe} + (1-x) M_{He}} (1 + \beta_{mix} P + \gamma_{mix} P^2 + \cdots)} \quad (1)$$

with C the acoustic velocity, T the temperature and P the pressure, M_{Xe} and M_{He} the molar masses of the different gases, R the perfect gas constant, γ which in our case is equal to 5/3 and β_{mix} and γ_{mix}, the virial coefficients corresponding to the gas. From these parameters, it is then possible to trace the x parameter, which is the mole fraction of xenon. These parameters can then be used to determine the x parameter, which is the mole fraction of xenon. Using the measurements from Fig. 8 and Equation (1), an estimate of the xenon composition is provided in Fig. 9.

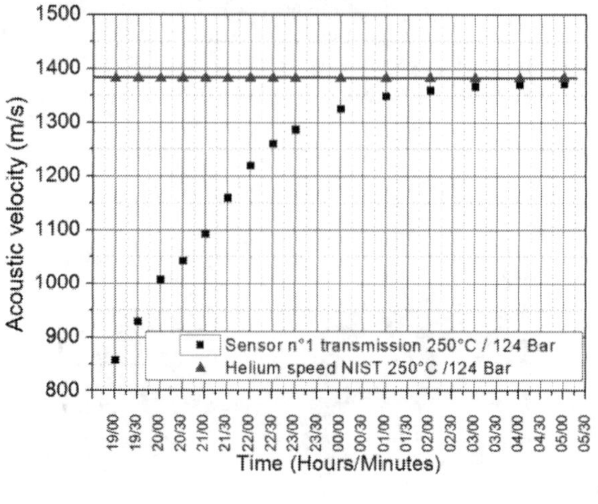

Fig 8.: Acoustic speed measurements between 7 PM and 5:30 AM, i.e. for 10 hours and 30 minutes

Uncertainties in xenon composition are directly linked to uncertainties in temperature measurements. This demonstrates the sensor's ability to track variations in gas composition at 250°C. However, these measurements are estimations, and no mass spectrometry was performed.

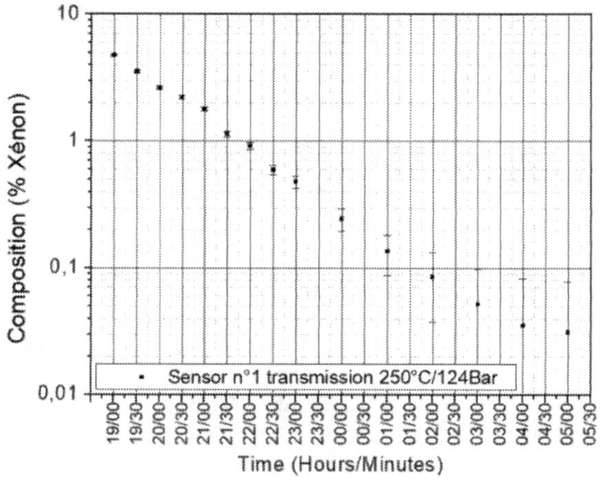

Fig 9. Composition measurements between 7 PM and 5:30 AM, i.e. for 10 hours and 30 minutes

C. Measurement at 350°C in pure He

The autoclave is now at a temperature of 350°C, containing a helium mixture of over 99%. The advantage of this mixture is its low attenuation, which allows us to test the minimum operational pressure. Fig. 10 shows three echograms for three different pressure values: 119 bar, 90 bar, and 60 bar.

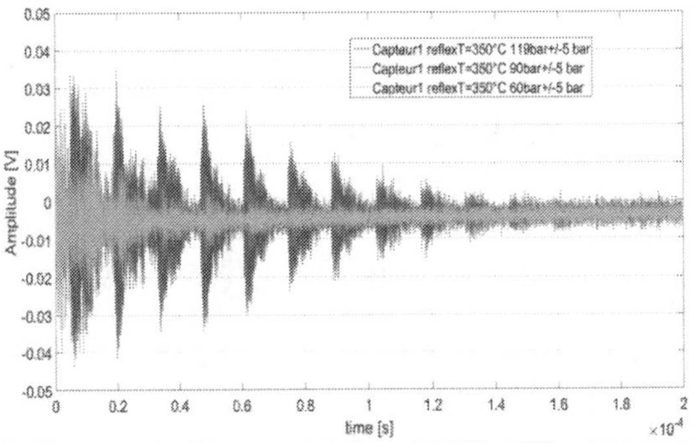

Fig 10. Measurements of the sensor in helium at 350°C at different pressures

We found that the sensor was operational at 350°C and successfully performed an acoustic measurement. This demonstrates that both requirements have been met: first, operating at 350°C, and second, being able to measure at a pressure of at least 60 bar.

V. Conclusions

The gas composition sensor developed by IES in partnership with CEA IRESNE successfully performed an initial measurement of changes in gas composition. The temperature and pressure conditions, 250°C and 124 bar respectively, were representative of experimental conditions in Material Testing Reactor experiments. We studied concentrations ranging from 10% xenon to less than 1%, and all sensor components withstood the thermal stresses. The temperature was then increased to 350°C, and we tested a mixture with a low helium content. Encouragingly, many echoes are visible at pressures between 120 and 60 bar.

Post-test analysis confirmed that the acoustic elements suffered no degradation. The sensors remained operational after a total duration of 34 hours at 250°C and 10 hours at 350°C. The radiation reliability of the final sensor remains to be tested.

These encouraging results suggest the sensor's potential application as a fission gas release characterization device within the test loops of the forthcoming RJH reactor, particularly the Madison loop.

Acknowledgment

This work was financed by a partnership between IES and the CEA IRESNE institute and more specifically the LISM laboratory and the INSNU project. The authors would like to thank the CTM (Montpellier Micro and nanoelectronics Technology Centre). Finally, we would like to warmly thank Vincent Lechien (technician at Montpellier University) for his support on screen printing methods.

References

[1] D. Fourmentel, Jf. Villard, Jy. Ferrandis, F. Augereau, E. Rosenkrantz, et M. Dierckx, « Acoustic sensor for in-pile fuel rod fission gas release measurement », in *2009 1st International Conference on Advancements in Nuclear Instrumentation, Measurement Methods and their Applications*, Marseille, France: IEEE, juin 2009, p. 1-5. DOI : 10.1109/ANIMMA.2009.5503679.

[2] E. Rosenkrantz, J. Y. Ferrandis, F. Augereau, T. Lambert, D. Fourmentel, et X. Tiratay, « An Innovative Acoustic Sensor for In-Pile Fission Gas Composition Measurements », IEEE Trans. Nucl. Sci., vol. 60, no 2, p. 1346-1353, avr. 2013, DOI : 10.1109/TNS.2013.2252624.

[3] O. Gatsa, P. Combette, D. Fourmentel, C. Destouches, et J. Y. Ferrandis, « High temperature ultrasonic sensor for fission gas characterization in MTR harsh environment », EPJ Web of Conferences vol. 170, 04008, 2018.

[4] F. Baudry, E. Rosenkrantz, P. Combette, D. Fourmentel, C. Destouches, et J. Y. Ferrandis, « Design of an acoustic sensor for fission gas release characterization devoted to JHR environment measurements », EPJ Web Conf., vol. 253, p. 04028, 2021, DOI : 10.1051/epjconf/202125304028

[5] F. Baudry et al., « Characterization and performance of an acoustic sensor for fission gas release characterization devoted to JHR environment measurements », EPJ Web Conf., vol. 288, p. 05001, 2023, DOI : 10.1051/epjconf/202328805001.

[6] F. Baudry *et al.*, « Design, Manufacture and Temperature Characterization of Screen-Printed Acoustic Sensors for Monitoring the Release of Fission Gases in Experimental Fuel Rods of the Jules Horowitz Reactor », *IEEE Trans. Nucl. Sci.*, vol. 71, n° 5, p. 1100-1108, mai 2024, DOI : 10.1109/TNS.2024.3368618.

[7] Y. Y. Feng et al., « The High-Performance of Cerium Doped Sodium Bismuth Titanate Aurivillirs Phase Material », Adv. Mater. Res., vol. 284-286, p. 1389-1392, juill. 2011, DOI : 10.4028/www.scientific.net/AMR.284-286.1389.

Non-destructive measurement of fuel rods inner pressure and composition: numerical consideration of different spring dimensions.

D. Ayache[1], S. Gillet[2], L. Ramiandrisoa[3], C. Reece[2], JY. Ferrandis[1,*]
[1] IES, University of Montpellier, CNRS, Montpellier (France)
[2] DCN/PEL/DER, EDF, Saint-Denis (France)
[3] R&D/PRISME, EDF, Chatou (France)
(*) jean-yves.ferrandis@umontpellier.fr

***Abstract*—The technique presented in this paper enables non-destructive measurement of both the internal pressure and gas composition within fuel rods used in pressurised water reactors. The sensor consists of a piezoelectric element shaped as a tile, which can be applied directly on the fuel rod cladding in the plenum region, where fission gases accumulate. Acoustic waves generated by the sensor propagate inside the cladding and the reflections occurring within the gas are detected. The time-of-flight between measured echoes is used to retrieve the composition of the internal gas mixture. Pressure is related to the amplitude of the acoustic signal and is determined through a calibration process that links theoretical amplitude to experimental measurements. The developed theoretical model significantly reduces the calibration time by taking into account the losses and signal attenuation caused by the presence of a spring located within the measurement zone. This paper presents the construction of a theoretical model that considers the effects of the pressure and gas composition, as well as the shadowing and diffraction phenomena induced by the spring, all of which influence the amplitude of the measured acoustic signal.**

***Keywords* — Acoustic, Non-Destructive Testing, Fuel rods, Pressurized Water Reactors.**

I. Introduction

This work presents the development of an innovative, non-destructive acoustic technique for in situ measurement of internal pressure and gas composition within the upper plenum of standard pressurised-water reactor (PWR) fuel rods. The proposed technique is designed to be cost-effective, straightforward to implement and suitable for deployment directly at nuclear power plant (NPP) sites.

Internal rod pressure is one of many critical parameters of engineering and safety criteria, such as cladding corrosion, against which the acceptable burn-up limits for nuclear fuel are defined. Accurate pressure monitoring is essential to prevent localised swelling or mechanical degradation of the cladding. Additionally, internal pressure serves as a key indicator of the overall behaviour of the fuel rod during operation, transport and long-term storage and can offer insights into gas desorption dynamics.

A strong correlation exists between internal pressure and gas composition. The latter reflects the release of noble fission gases—primarily helium (He), krypton (Kr) and xenon (Xe)—into the rod's free volume [1], [2]. Acoustic techniques enable a direct inventory of these released gases, thus providing valuable information about the fuel performance and fission gas release behaviour.

Such data are also essential for code validation and multi-physics simulation tools used in nuclear engineering [3]. However, current methodologies in France require the extraction of fuel rods followed by their transport to specialised hot laboratories using dedicated casks. These operations are expensive, time-consuming and involve radiation risks. Once in the laboratory, the rods are pierced and cannot be used anymore, resulting in the generation of nuclear waste.

The development of a non-destructive, on-site measurement, would prevent the need for complex logistics and both operational costs and nuclear waste production. Furthermore, it would enable larger measurement campaigns across fuel assemblies, thus supporting the creation of statistically robust datasets and the development of probabilistic models. Finally, this technique could provide an alternative approach to the detection of leaking fuel rods, which is currently performed using ultrasonic inspection methods across the French nuclear fleet.

EDF, French energy supplier, and IES (University of Montpellier, CNRS) have collaborated for more than 20 years to develop a methodology for on-site measurements. The first generation of the developed system was successfully used on irradiated fuel rods in the LECA-STAR facility at the CEA CADARACHE Centre [4]. The latest improvements of the system have demonstrated the feasibility of the measurement for pressure below 40 bar in pure helium (100% He), covering a wider range of the fuel cycle.

The challenge of the measurement lies in the different parameters, which can evolve during the fuel cycle and need to be considered. The spring used to mechanically maintain the fuel pellets inside the rod constitutes one of the most important parameters to consider as it can significantly affect the pressure measurement. There are two different approaches to assess the effect of the spring. The first is to include it directly in the calibration process that links the theoretical model to the experiments. However, this requires having specific calibration for each existing spring configuration. Given the variation of spring geometry between manufacturers and fuel assemblies, this would make the calibration process complex and time-consuming. The second approach is to describe the effect of the spring using explicit analytical equations within the theoretical model. This allows the calibration process to be decoupled from the different spring dimensions that can be encountered. The consideration, by the numerical model, of the different types of springs leading to the validation of the system in laboratory conditions will be presented in this paper.

II. Presentation of the system

The next section will provide a brief description of the acoustic sensor and the principle of the measurement. More detailed information can be found in [5-7].

A. Acoustic sensor design

The acoustic transducer has a tile shape designed as concentric to the fuel rod. Fig. 1 shows a representation of the device. The sensor is the complete system composed of different layers including a backing, a piezoelectric element and a coupling material with the rod cladding.

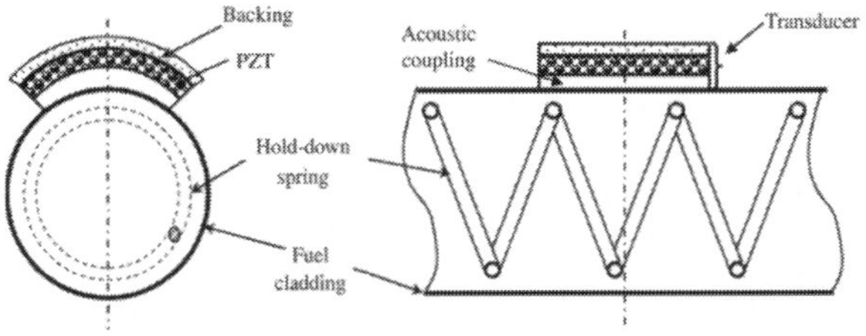

Fig. 1. Schematic representation of the acoustic device. Fuel rod inner diameter D = 8.3 mm and thickness e = 0.57 mm [5].

The design and the optimisation of the acoustic device layers are defined through the stacking model presented in [8] which enables the control of the thickness of the different layers, the electromechanical properties of the piezoelectric element and the acoustic velocity, the density and the attenuation of the layers.

The sensor can be operated in two frequency ranges: frequency 1 (f_1) around 3–4 MHz and frequency 2 (f_2) around 7–8 MHz, each representing different gas concentration ranges. For gas mixtures containing more than 5% Xe/He (i.e., 5% Xe and 95% He), the measurement is performed at f_1. For mixtures containing less than 5% Xe/He, it is more efficient to work at f_2. Indeed, since the speed of sound is higher in He, it is challenging to identify the first echoes in the measured signal at f_1, as they tend to be lost in the background noise.

Several mounting configurations can be used depending on the application (Fig. 2):

- A "clamp", used in testing and qualification phase in laboratory
- A "weight", specifically designed for hot lab applications. It consists of two metallic masses tied to the sensor. This auto positioning system ensures good coupling between the sensor and the rod.
- A "saber", with a miniaturized version of the sensor to perform, that can be mounted on a motorized arm to perform measurements on peripheral but also fuel rods positioned inside the assembly.

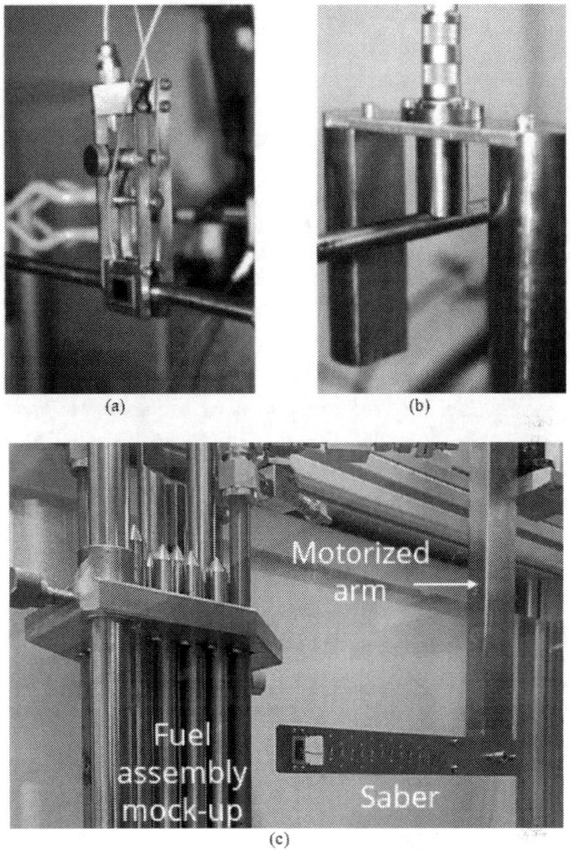

Fig. 2. Different types of mounting configurations for the sensor (a) "clamp", (b) "weight" and (c) "saber".

B. Principle of the measurement

Acoustic waves, generated by the piezoelectric transducer, propagate through the rod cladding and inside the gaseous mixture. Therefore, a modification in the frequency spectrum of the vibration of the gas can be observed. The acoustic impedance of a gas in a rigid plane cavity is described as a periodic function. The resonances of the gas are separated with a frequency Δf described in (1):

$$\Delta f = \frac{c_{gas}}{2D} \tag{1}$$

where c_{gas} is the sound velocity in $m.s^{-1}$ and D is the inner diameter of the rod in m.

Then, the sound velocity can be computed from (1) to deduce the molar mass M of the gas mixture (2):

$$M = \frac{\gamma RT}{c_{gas}^2} = xM_{Xe} + (1-x)M_{He} \tag{2}$$

where x is the molar fraction of xenon in a gas mixture composed of helium and xenon, γ is the adiabatic index equal to $\frac{5}{3}$ for a monoatomic gas and T is the temperature in K.

In our application, the composition of the gaseous mixture involves a small amount of krypton. In order to retrieve its composition, the molar mass is described as in (3):

$$M = xM_{Xe} + \frac{x}{\frac{x}{X_{Kr}}} M_{Kr} + \left(1 - x - \frac{x}{\frac{x}{X_{Kr}}}\right) M_{He} \qquad (3)$$

In (3), the ratio of the concentration of xenon to the concentration of krypton is known and assumed to be equal to 17 or 10 for UO_2 or MOX fuel. This assumption is true after the first fuel cycle and results from measurements carried out by the CEA on several fuel rods and is confirmed by previous studies [9].

The gas mixture composed of He-Xe-Kr is not a perfect gas, which implies that the sound velocity increases with the pressure. Therefore, as the pressure range during the tests varies from 0 to 80 bar, some corrections must be applied to the sound velocity as described in (4):

$$c_{gas}^2 = \frac{5RT}{3M}\left(1 + \left(2b - \frac{a}{RT^{\frac{3}{2}}}\right)\frac{P}{RM}\right) \qquad (4)$$

where P is the pressure and the adjustable quantities a and b are obtained from [10].

In Fig. 3, an echogram, i.e., the reflection of the acoustic wave inside the medium, can be observed.

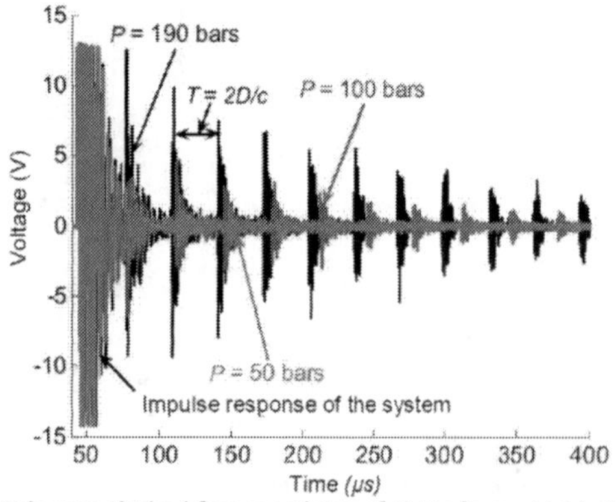

Fig. 3. Echogram obtained for a gas mixture of 40 % of argon (Ar) in He and for pressure equal to 50 bar, 100 bar and 190 bar [11].

The echograms presented in Fig. 3 characterize a gas mixture of 40 % Ar/He. The period between two echoes, usually called the time-of-flight (ToF), is used to calculate the sound velocity and retrieve the gas composition as previously explained. Moreover, it can be observed that the amplitude of the different echoes increases with the pressure, which makes it possible to determine the pressure through a calibration process.

The pressure is not the only parameter affecting the amplitude of the measured signal. In the next section, the effect of the gas composition and the presence of the spring in the plenum part on the measured amplitude will be described.

III. Parameters affecting the signal amplitude

The theoretical amplitude Amp_{th} of the signal is defined as in (5):

$$Amp_{th} = P \times \sqrt{M} \times Absorption^{je} \times Losses^{je} \qquad (5)$$

where P is the pressure in bar, M is the molar mass of the gas in g.mol^{-1}, $Absorption$ represents the losses induced by the gas as an absorption medium and $Losses$ defines the losses induced by the presence of a spring. The losses in the system depend on the echo order je.

A. Effect of the pressure

The effect of the pressure on the signal amplitude can be observed in Fig. 4. The echograms are obtained in pure helium for 3 different pressures: 80 bar, 40 bar and 30 bar. In Fig. 4 the envelope of each echo is obtained after signal processing including double FFTs, signal filtering at f_1 or f_2 and correlation with theoretical echoes implemented to increase the noise rejection. In this case, the measurement is performed at $f_2 = 7.95$ MHz. The experimental setup is detailed in part IV.

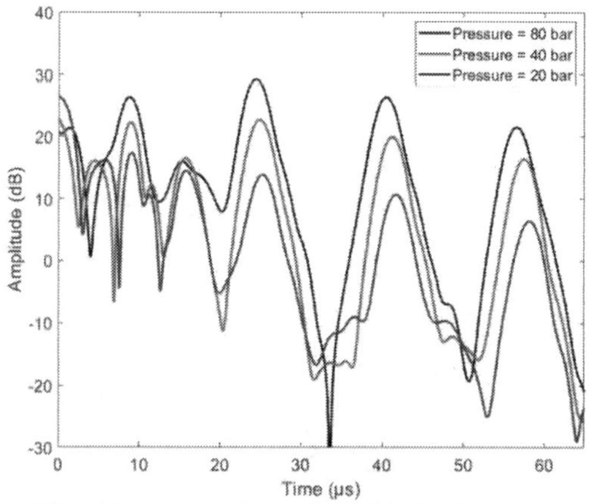

Fig. 4. Effect of the pressure on the amplitude of the signal. Echograms obtained for 100% He at different pressures: 80 bar (black curve), 40 bar (red curve) and 20 bar (blue curve). Measurements performed at ambient room temperature (20.5 °C) and the sensor operated at $f_2 = 7.95$ MHz.

The sound velocity is higher in pure He which complicates the separation of the first echo from the noise. For the second echo of the signal observed at 25 μs, the amplitude of the signal decreases from about 30 dB at 80 bar to 13 dB at 20 bar. This shows that for a stable composition, the amplitude of the measured signal decreases when the pressure decreases.

B. Effect of the gas composition

Fig. 5. shows the effect of the gas composition and the absorption induced by the gas. The echograms are obtained at 70 bar in two different gas mixtures: 20% Xe/He and 10% Xe/He. In this case the sensor is operated at $f_1 = 3.68$ MHz.
The gas composition has an effect on the period and the amplitude of the signal that varies even for a stable pressure.

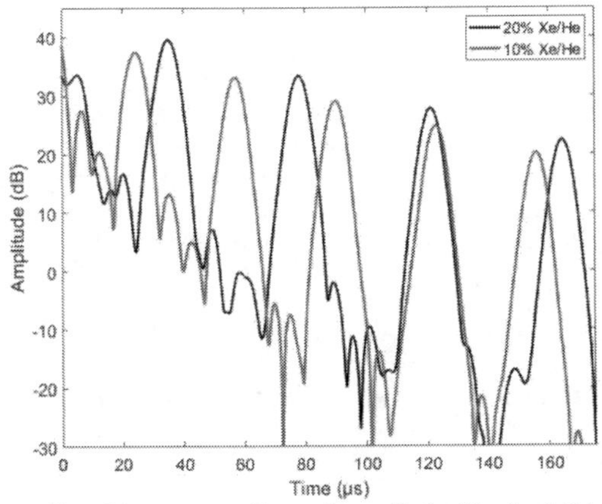

Fig. 5. Effect of the gas composition on the amplitude of the signal. Echograms obtained at 70 bar for different gas mixtures: 20% Xe/He (black curve), 10% Xe/He (red curve). Measurements performed at ambient room temperature (20 °C) and the sensor operated at f_1 = 3.68 MHz.

C. *Effect of the spring*

The echograms are obtained in pure He at 70 bar with f_2 = 7.95 MHz. In Fig. 6. three different configurations are presented: without a spring, with a spring type A and a spring type B. The pitch g of the spring is about 7.8 mm and 4.5 mm and the wire diameter d is equal to 1.2 mm and 1.5 mm, respectively, for spring A and spring B.

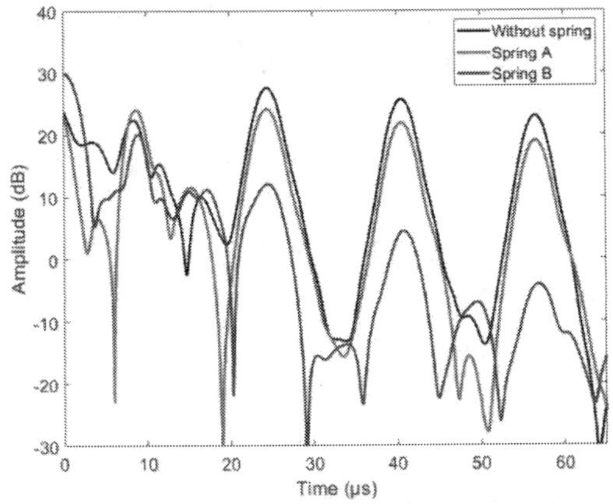

Fig. 6. Effect of the presence of a spring in the measurement zone on the amplitude of the signal. Echograms obtained at 70 bar in pure He in three configurations: without spring (black curve), with spring A (pitch g = 7.8 mm, wire diameter d = 1.2 mm) (red curve) and spring B (pitch g = 4.5 mm, wire diameter d = 1.5 mm) (blue curve). Measurements performed at ambient room temperature (20.5 °C) and the sensor operated at f_2 = 7.95 MHz.

In Fig. 6. it can be observed that for the same experimental condition of pressure and composition, the amplitude of the signal decreases with the presence of the spring. Moreover, spring A is less compressed than spring B. The difference observed between the measurements performed on two different types of springs shows that the effect of the springs on the signal amplitude depends on their geometrical parameters.

If the experimental amplitude Amp_{exp} is defined as in (6) :

$$Amp_{exp} = S \times Amp_{th} \quad (6)$$

where S is a constant, then the losses, defined in (5), can either be a constant for each configuration of spring or defined by explicit equations taking into account the geometry of each spring. In the latter case, S would be a constant exclusively related to the type of sensor employed for the measurement.

1) Shadow effect

Part of the acoustic waves that are transmitted through the rod cladding will not propagate inside the gas due to the presence of the spring. In this case, the spring acts as a screen, blocking the propagation of the emitted signal and resulting in a reduction of the intensity of the waves. This phenomenon does not depend on the order of the echo je, and can be described as function of the geometrical parameter of the spring:

$$Shadow = 1 - \frac{2d}{g} \quad (7)$$

2) Diffraction effect

Diffraction occurs when the acoustic waves propagating inside the rod hit the surface of the spring. This phenomenon is function of the wavelength λ, thus the frequency, and the geometrical parameters of the spring:

$$Diff = \frac{1}{1 + \left(\frac{\lambda}{cste \times \left(\frac{g}{2} - d\right)}\right)^3 + \left(\frac{\lambda}{cste \times l}\right)^3} \quad (8)$$

where $cste$ is a constant equal to 190 obtained through empirical work and l is the width of the acoustic device approximately 10 mm.

D. *Calibration and measurement processes*

By injecting (7) and (8) in (6), the Amp_{exp} is defined as:

$$Amp_{exp} = S \times P \times \sqrt{M} \times Shadow \times (Absorption \times Diff)^{je} \quad (9)$$

During the calibration process, the pressure P, the gas composition, related to M and the spring dimension is known. Thus, the model can be used to calculate the theoretical amplitude of the signal. It is then possible to determine the constant S, by adjusting experimental and theoretical values for a few known values of pressure and composition.

During the measurements, S is known and related to the sensor properties. By inserting the spring dimensions in the theoretical model, the losses can be calculated and the pressure and gas composition retrieved.

IV. Experimental measurement

The experimental setup is composed of a SEPEMA generator developed by SONAXIS company. This equipment is used as a pulse generator and a signal receptor and is capable of

performing data acquisition and processing. A single pulse of few nanoseconds targeting f_1 frequency range is sent to the sensor. The complete echogram is collected and signal processing is performed at f_1 and f_2.
A homemade gas mixing system using calibrated gas bottles is used to fill the rod-mock available in our laboratory. A manometer is used to control and adjust the pressure inside the rod. The sensor is clamped over a rod with water as a coupling medium. The rod mock-up is long enough to contain two different types of spring: spring A, previously described, and spring C (pitch $g = 5.9$ mm and wire diameter $d = 1.5$ mm).

During the calibration process, measurements are performed at two different compositions and, in each case, for two different pressures. Therefore, the first measurements are conducted at 20% Xe/He for 70 bar and 50 bar. The second part of the process is conducted in pure helium for 70 bar and 50 bar. For each configuration of composition/pressure, the measurement is performed 3 times, where each represents the mean of 10 measurements performed at 10 different sensor positions. The calibration process is conducted on the zone containing the spring C, and the sensitivity S of the sensor is retrieved.

Measurements are then conducted on spring A and spring C. The rod is pressured at 30 bar, 40 bar, 50 bar and 70 bar with 20% Xe/He, 10% Xe/He and pure He. The pressure calculated by the model is compared to the reference pressure measured by the manometer with an acceptance interval of 15% defined by our industrial partner (Fig. 7). Signal processing is performed at $f_1 = 3.68$ MHz for measurement in 20% and 10% Xe/He and at $f_2 = 7.95$ MHz for pure He.

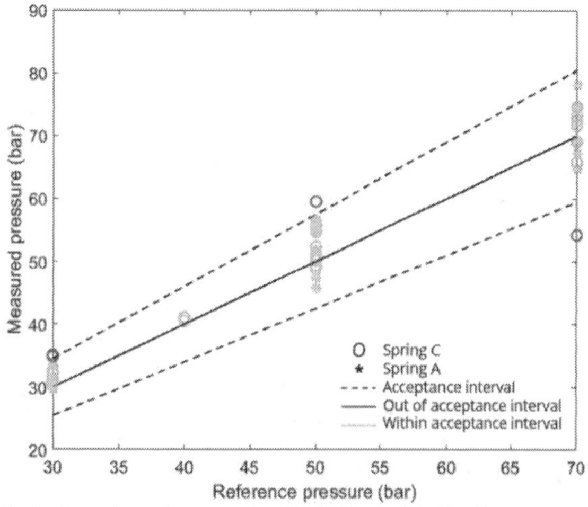

Fig. 7. Comparison of pressure calculated by the model with reference pressure measured by the manometer (black line) with an acceptance interval of $\pm$15% (dashed line). Measurements are performed at ambient room temperature (20 °C). Measurements are conducted in two zones containing spring A (O) and spring C (*). Calculated pressures lying out of the acceptance interval are red and calculated pressures within the acceptance interval are green.

Fig. 7 shows 50 measurements conducted in different gas mixtures and at different pressures covering a large range values of the fuel cycle. For an acceptance interval of $\pm$15% over the reference pressure, the model is capable of calculating pressures with a success ratio of 90%.

V. Conclusion

A non-destructive ultrasound-based technique for measuring the internal pressure and composition of PWR fuel rods has been developed. This work is part of a long-term collaboration between EDF and IES. The pressure is related to the amplitude of the acoustic signal. A theoretical model considering the effects of the pressure, gas composition and the presence of a spring in the measurement zone has been developed. This model is based on explicit equations describing the losses obtained through empirical work. The calibration process is now less time-consuming and exclusively related to the sensor, rather than varying spring configurations. This model has been successfully used to calculate the pressure from measurements obtained with different spring dimensions, achieving a success rate of 90%.

Acknowledgements

The authors would like to warmly thank G. Lévèque, former professor at IES, who mainly developed the theoretical aspects of the model presented in this paper.

They would also like to thank SONAXIS for manufacturing and supplying the sensors.

References

[1] D.R. Olander, Fundamental Aspects of Nuclear Fuel Elements, TID-26711-P1, US Department of Energy, 1976 (Chapter 15).
[2] J. Siefken, SCRAP/RELAP5/MOD3.3 Code Mannual, NUREG/CR-6150, vol.2, Rev. 2, INEL-96/0422, 2001.
[3] D. Baron, P. Thevenin, R. Largenton, R. Masson, S. Pujet, R. Arnaud, Cyrano 3 the EDF fuel performance code especially designed for engineering applications, in: Water Reactor Fuel Performance Meeting Proceeding, Seoul, 2008.
[4] J.Y. Ferrandis, E. Rosenkrantz, G. Leveque, D. Baron, J.C. Segura, G. Cecilia, O. Provitina, Full-scale Hot Cell Test of an Acoustic Sensor Dedicated to Measurement of the Internal Gas Pressure and Composition of a LWR Nuclear Fuel Rod, IEEE Transactions on Nuclear Science, 60(4), 2894-2897, 2013, https://doi.org/10.1109/TNS.2013.2264111
[5] G. Leveque, E. Rosenkrantz and D. Baron, "Acoustic Injection into a Pressurized Gas Through a Metallic Wall", Acta Acustica united with Acustica, vol. 93, pp. 722-729, 2007.
[6] E. Rosenkrantz, J.Y. Ferrandis, G. Leveque, D. Baron, "Ultrasonic measurement of gas pressure and composition for nuclear fuel rods", Nuclear Instruments and Methods in Physics Research Section A: Accelerators, Spectrometers, Detectors and Associated Equipment, Volume 603, Issue 3, 21 May 2009, Pages 504-509, https://doi.org/10.1016/j.nima.2009.02.028
[7] E. Rosenkrantz, J.Y. Ferrandis, G. Leveque, D. Baron and P. Thevenin, "Ultrasonic method for nuclear fuel rods pressure and gas composition released measurement," 2009 1st International Conference on Advancements in Nuclear Instrumentation, Measurement Methods and their Applications, Marseille, France, 2009, pp. 1-5 https://doi.org/10.1109/ANIMMA.2009.5503773
[8] "Caractérisation d'un gaz confiné à l'aide d'un capteur acoustique - Application aux crayons combustibles nucléaires", Ref. in 113, 2010 (in French), https://doi.org/10.51257/a-v1-in113
[9] R. J. White, S. B. Fisher, P. M. A. Cook, R. Stratton, C. T. Walker, and I. D. Palmer, "Measurement and analysis of fission gas release from BNFL's SBR MOX fuel," Journal of Nuclear Materials, vol. 288, no. 1, pp. 43-56 (2001).
[10] J.J. Hurly, J.W. Schmidt, S.J. Boyes, and M.R. Moldover, "Virial equation of state of helium, xenon, and helium-xenon mixtures from speed-of-sound and Burnett PρT measurements," International Journal of Thermophysics, vol. 18, no. 3, pp. 579, 579-634 (1997).
[11] P. Vasseur, J.M. Barthoulot, G. Leveque and J.Y. Ferrandis, (2016). *Acoustic sensors devoted to the detection of leaking irradiated fuel rod in spent fuel pool.* American Nuclear Society-ANS, 555 North Kensington Avenue, La Grange Park, IL 60526 (United States).

$CeBr_3$ gamma-ray logging probe qualification for uranium mining applications

Thomas Marchais[1,*], Bertrand Pérot[1], Hervé Toubon[2], Youcef Bensedik[2], Romain Mieszkalski[2], Sebastien Hocquet[2], Dragomir Savov[2], Matthew Schubert [3], Bruno Legros[3], Patrick Van Eyll[3], Michaël Pernette[3]

[1]CEA, DES, IRESNE Institute, DTN, SMTA, Nuclear Measurement Laboratory, France
[2]Orano Mining, France
[3]Advanced Logic Technology, Luxembourg
* Corresponding author: Thomas.MARCHAIS@cea.fr

Abstract— The development and optimization of gamma-ray spectrometry tools are crucial for ORANO Mining and geophysical exploration. This study presents a comprehensive study on the ALT QL40-SGR-2G spectrometric probe, equipped with a cerium bromide (CeBr3) detector, aimed at its detailed characterization and the validation of its numerical model through various experimental setups and MCNP (Monte Carlo N-Particle) simulations. The study described energy calibration, resolution and detection efficiency measurements at a 15 cm distance, crystal length characterization, probe rotation effects on performance, counting loss estimation, and response profile determination in a sand silo setup. These experiments aimed to validate the MCNP model of the probe and understand factors affecting its performance. Utilizing standard gamma sources, energy resolution and detection efficiency of the probe were evaluated, revealing a close match between experimental data and MCNP calculations, with a 2.4% average discrepancy across the energy spectrum. This validation highlights the reliability of the numerical model and the robustness of the probe in diverse measurement scenarios. Further investigations on the effective length of crystal confirmed the manufacturer's specifications, while tests on the probe's rotational impact showed minimal effect on spectral response (< 4%). The use of the 2 sources method allowed for accurate dead time estimation, which is crucial for correcting counting losses in high-radiation environments. In addition, this study involved the spatial response of the probe using a ^{137}Cs source within a sand silo. This experiment demonstrated the potential of this probe in simulating complex geological formations and validated the MCNP model accuracy in predicting spatial response, which is crucial for borehole modelling. The successful correlation between experimental findings and MCNP simulations enhances confidence in using the QL40-SGR-2G probe for accurate gamma-ray detection and analysis in varied settings.

Index Terms— $CeBr_3(Ce)$ probe for uranium borehole logging, MCNP simulation, experimental calibration, dead time

I. INTRODUCTION

The ALT QL40-SGR-2G probe, which is used for detecting gamma emissions within a wellbore, employs either a CeBr3 or BGO. For this study, we utilized a $CeBr_3$ detector measuring 20 mm x 96 mm. Fig. 1 illustrates the probe's design, which is 1089 mm long.

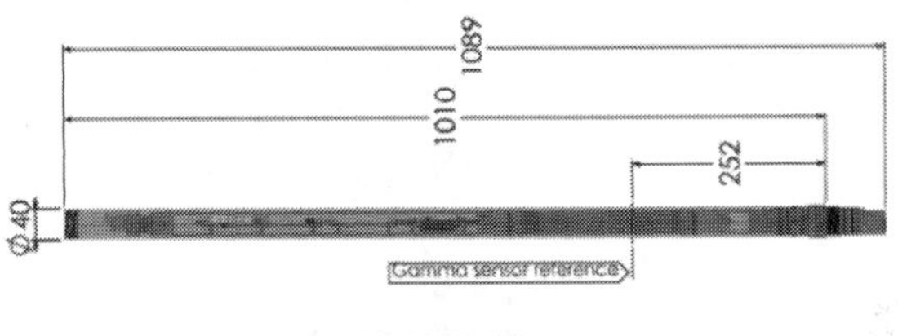

Fig. 1. QL40-SGR-2G CeBr3 probe.

This spectrometric probe was partially qualified in a previous study [1] on a standard block in Bessines, France [2], [3]. A significant discrepancy was observed in the total counting rate at high uranium content between MCNP [4] numerical simulations and experimental results. Consequently, it was decided to further characterize the probe in the Nuclear Measurement Laboratory at CEA Cadarache using standard sources. This process involved several steps:

- energy calibration, estimation of the probe's resolution and its detection efficiency at a distance of 15 cm, that permitted to compare experimental data with calculations performed using MCNP transport code,
- measurement of the crystal's length using a motorized table,
- assessment of the effect of the probe rotation on its performance,
- estimation of counting losses using the two-source method,
- determination of the response profile in a sand silo.

The results of the previous highlighted points are presented in the subsequent sections, allowing for the qualification of the MCNP model of the probe and the estimation of various influential factors intrinsic to the probe.

II. LABORATORY QUALIFICATION

A. *Energy resolution and detection efficiency*

To characterise the energy calibration of the probe, its resolution and its detection efficiency at 15 cm, we used six standard sources available in the laboratory: ^{60}Co, ^{88}Y, ^{133}Ba, ^{137}Cs, ^{152}Eu, and ^{241}Am. Table 1 shows the characteristics emission of each standard sources. It should be noted that the gamma emissions have been selected to have no parasitic lines.

TABLE I
MAIN GAMMA EMISSION OF SIX STANDARDS SOURCES [5]

Source	Activity (Bq)	Gamma energy (keV)	Gamma intensity (%)
^{60}Co	28.2 x 10^3	1173.23	99.85
		1332.49	99.98
^{88}Y	10.1 x 10^3	898.04	93.7
		1836.07	99.35
^{133}Ba	29.9 x 10^2	356.01	62.05
^{137}Cs	33.5 x 10^3	661.66	85.01
^{152}Eu	40.1 x 10^3	121.78	28.41
		244.70	7.55
		344.28	26.59
		964.08	14.50
		1408.01	20.85
^{241}Am	60.2 x 10^3	59.54	35.92

Fig. 2 shows the experimental configuration and corresponding measured spectra, with 2048 channels, for the six standards sources. These six sources are positioned 15 cm from the centre of the crystal.

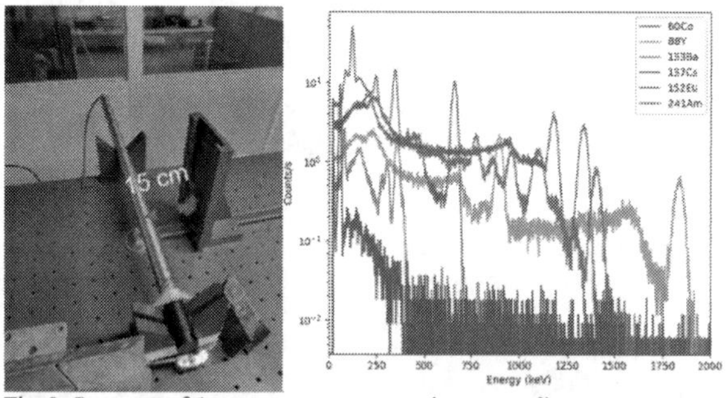

Fig. 2. Geometry of 6 sources measurment and corresponding spectra.

The previous spectra were calibrated thanks to the well-known emission lines of the 6 standard sources, see Table I. Thanks to this calibration, a 2-h background measurement without radioactive sources was carried out, see Fig. 3. This measurement shows that no contribution of background noise is to be noted in the selected lines of this study. In addition, we can clearly see the characteristic peaks of a background noise, 1460 keV from ^{40}K and 609 keV from ^{214}Bi.

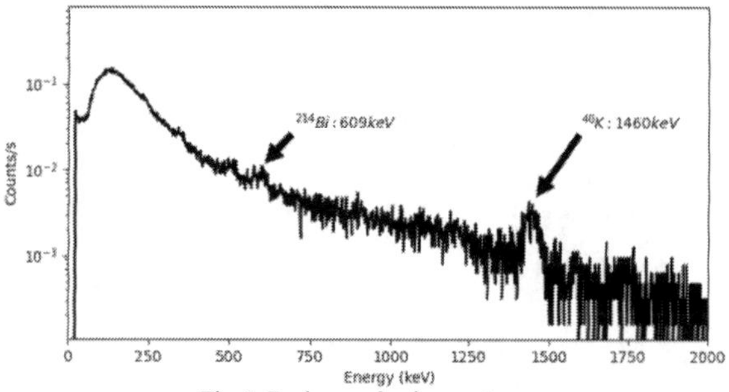

Fig. 3. Background noise spectrum

The energy resolution, denoted as FWHM (Full Width at Half Maximum), is estimated for each Gaussian peak and can be expressed by the following formula:

$$\mathrm{FWHM(E)} = A + B \times \sqrt{E + C \times E^2} \quad (1)$$

With:

- E: the energy of the Gaussian peak in keV,
- A, B, C: three coefficients respectively in keV, $keV^{1/2}$ and keV^{-1}.

The following figure shows the FWHM estimated thanks to the 6 sources measurements and formula (1). In addition, an example of Gaussian curve fitting is provided for the characteristic gamma emission at 662 keV from ^{137}Cs.

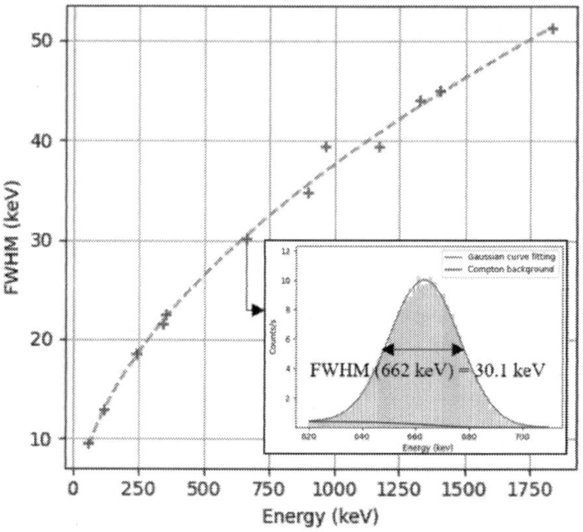

Fig. 4. Energy resolution of the probe as function of the energy and example of gaussian curve fitting for the characteristic gamma emission at 662 keV from ^{137}Cs.

We notice an energy resolution of 4.5% (30.1 keV) at 662 keV from ^{137}Cs source. This resolution corresponds to the classic resolution of a CeBr3 detector [6]. For equation (1), we found A = 0.22 keV, B = 1.17 $keV^{1/2}$ and C = 2.70 x 10^{-5} keV^{-1}.

The detection efficiency, denoted $Eff(15\ cm)$, is estimated for each peak according:

$$Eff(15\ cm) = \frac{Sn(E)}{Act(S) * I(E)} \quad (2)$$

With:

- $Sn(E)$: net area of the Gaussian peak
- $Act(S)$: activity of the source, see Table 1,
- $I(E)$: gamma intensity at energy E, see Table 1.

In addition, and in order to compare with simulated values, we decided to simulate the geometry of measurement with MCNP6 transport code [4], see Fig. 5. The objective is to validate the numerical model of the $CeBr_3$ probe. This modelled geometry is based on manufacturer characteristics and has already been used to model concrete blocks of increasing uranium content [1]. We subsequently modelled the measurement geometry of the 6 point sources described in part A.

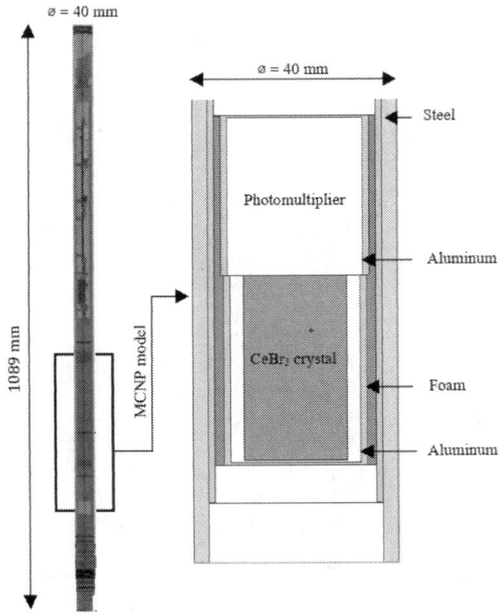

Fig. 5. QL40-SGR-2G probe from Advanced Logic Technology (ALT) and its associated MCNP model.

The following figure shows the detection efficiency of the probe at 15 cm, estimated through the measurements from six sources and formula (2), with the experimentation shown in red and MCNP calculations depicted in blue. The uncertainty plotted for the experimental values includes those due to statistical uncertainty and the activity of standard sources (3%). The uncertainty on the MCNP values is not visible on the figure because of very low values (<1%).

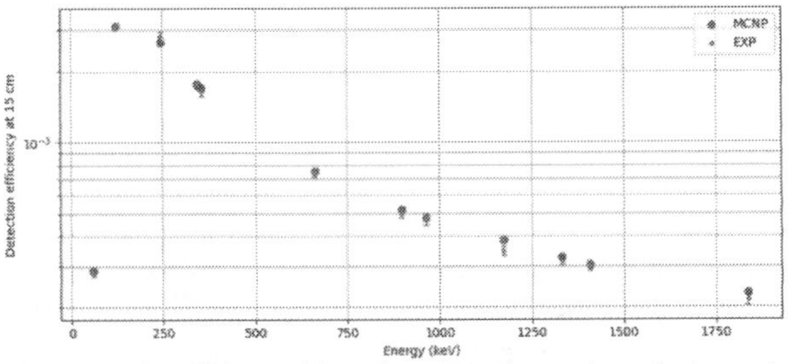

Fig. 6. Detection efficiency of the probe at 15 cm for experimental values and MCNP calculations.

The previous figure shows a very good agreement between the experiment and the MCNP calculation with only a 2.4% average discrepancy across the entire energy range.

B. Detector length estimation

To ascertain the length of the crystal but also its uniformity of response, we utilised a 1 mm collimated slit of lead, a ^{137}Cs source, and a motorised table which, through 42 measurements, enabled us to estimate the crystal's length with a step of 1 mm. The uncertainty in the measured crystal dimension is 2 mm, attributable to the distance between the probe and the source, as well as the collimated slit.

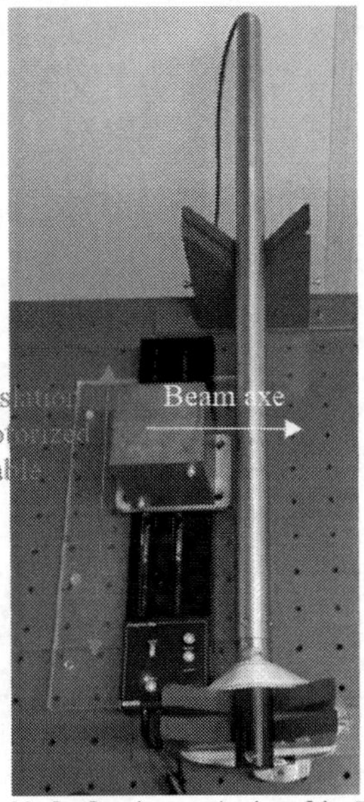

Fig. 7. Motorized table for fine characterization of the CeBr3 probe crystal.

The following figure illustrates, for the 42 measurements, the net area of the 662 keV peak of ^{137}Cs as a function of the position of the source.

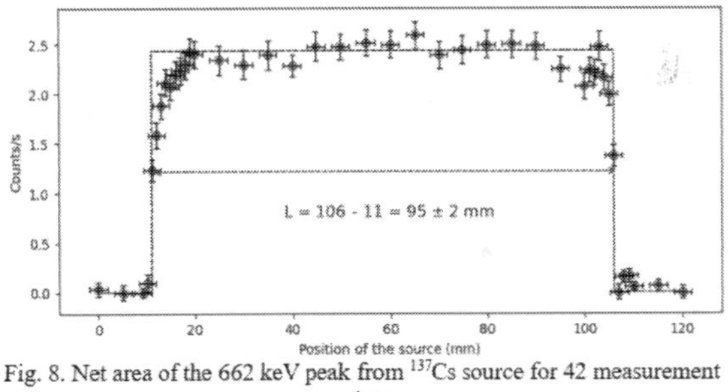

Fig. 8. Net area of the 662 keV peak from ^{137}Cs source for 42 measurement points.

A crystal length of 95 ± 2 mm was determined, which is in good agreement with the manufacturer's data (96 mm). The uniformity of the probe's response over the length of its crystal is relatively good. The fluctuations in the net area observed when we are in the crystal, between 11 mm and 106 mm, are attributable to the presence of absorbing elements close to the crystal or to a poor collection of charges inside it.

C. Probe rotation effect

In order to examine the uniformity of the spectral response of the $CeBr_3$ probe whilst it was rotated within the borehole, we opted to rotate the probe about its axis at a pitch of 45°.

Throughout these measurements, sources of ^{137}Cs, ^{60}Co, and ^{88}Y were employed at the same time to characterise the angular response across an energy spectrum ranging from 662 keV (^{137}Cs) to 1836 keV (^{88}Y). The subsequent figure illustrates the deviation for each net area estimated relative to the mean value. Moreover, the total count rate is also presented in the case of using the probe in total count rate mode. The uncertainty plotted for the experimental values includes those due to statistical uncertainty (≈ 1%).

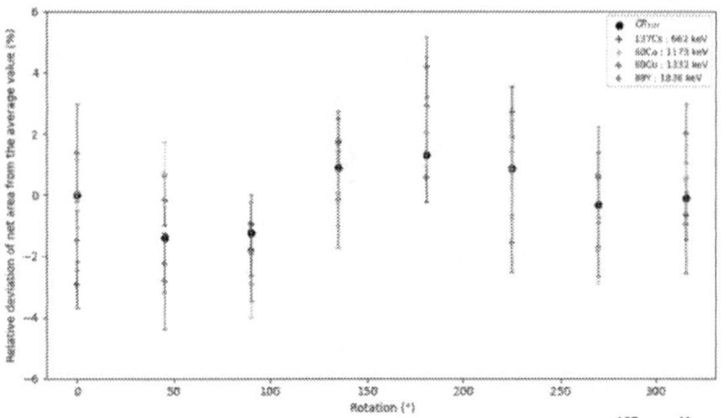

Fig. 9. Probe rotation effect of the signal of 3 standards sources (^{137}Cs, ^{60}Co, and ^{88}Y) and total count rate.

We can observed a very low effect of rotation (< 4%) for each points of measurement. The effect of rotation is lower than 1% for the total count rate.

D. *Counting loss and dead time*

In order to verify the dead time correction carried out by the $CeBr_3$ probe, it was decided to quantify the counting losses using the "2 sources" method [7]:

- a reference source of ^{137}Cs is placed at a fixed position near the $CeBr_3$ probe, which emits gamma radiation at 662 keV not coinciding with any other emission from the second source used (^{60}Co). The net area at 662 keV on the reference spectrum is S_{Ref} = 210 counts.s^{-1}, for a total counting rate of 1073 counts.s^{-1} and a dead time of 0.16% when no disturbing source is not used.

- a disturbing source of ^{60}Co is then approached as the detector progresses in order to increase the total counting rate and the dead time. Six positions of the disturbing source, so six total count rates, have been measured. At each step, we measure the net area at 662 keV due to the reference source. Upon contact with the probe, the disturbing source induces a counting rate of 167,200 counts.s^{-1} and a dead time of 16.35%.

The following figure represents the counting losses in the net area at 662 keV as a function of the total counting rate on the spectra. Each measurement is carried out for an active time of 120 seconds.

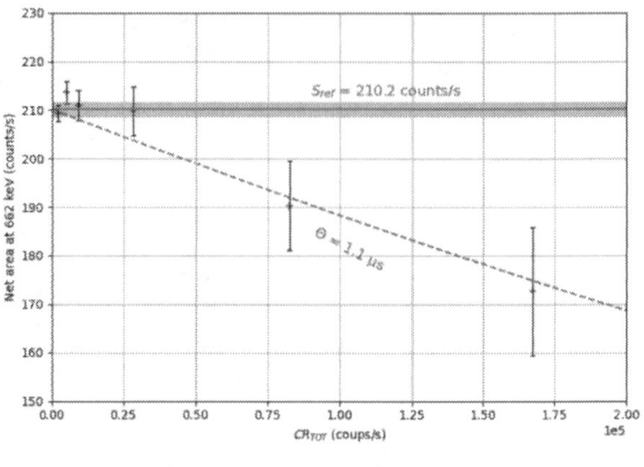

Fig. 10. Counting losses in the net area at 662 keV as a function of the total counting rate. In red the value of the net area at 662 keV when the ^{137}Cs source is alone. In green, the adjustment made to the data to estimate the dead time.

Thanks to these measurements, we are able to accurately estimate the dead time θ, of this probe (green curve in Fig. 10). The paralyzable model [7] can be directly used to estimate the dead time.

$$Sn(662\ keV) = S_{ref} \times e^{-CR_{TOT} \times \theta} \quad (3)$$

With:

- $Sn(662\ keV)$: net area at 662 keV, in counts.s^{-1},
- S_{ref}: net area at 662 keV on the reference spectrum without disturbing source, S_{ref} = 210 counts.s^{-1},
- CR_{TOT}: total count rate, in counts.s^{-1},
- θ: dead time, in second.

We found a θ = 1.1 µs with this experimentation (see green curve in Fig. 10). This value is in relative good agreement with the manufacturer's data, θ = 0.8 µs.

E. *Spatial response of the probe in a sand silo*

Finally, to complete the characterization of the $CeBr_3$ probe, we decided to study its spatial response to a mono-energetic source of ^{137}Cs in a sand silo. This source was selected because of this gamma emission at 662 keV, near to the characteristic peak of a uranium decay chain at 609 keV of ^{214}Bi. For this, we used a Fontainebleau sand silo, already used in previous studies for measurements using a neutron probe [8]. This silo, shown in Fig. 11, with a weight of 1.5 tonnes, measures 1.50 metres high and has an internal diameter of 100 cm. The density of the matrix is equal to 1.6. Four smaller stainless steel tubes, each 6 cm in diameter, are positioned at different radial distances (n°1: 13.2 cm, n°2: 22.7 cm, n°3: 32.7 cm, and n°4: 47 cm) from the centre to introduce the ^{137}Cs source.

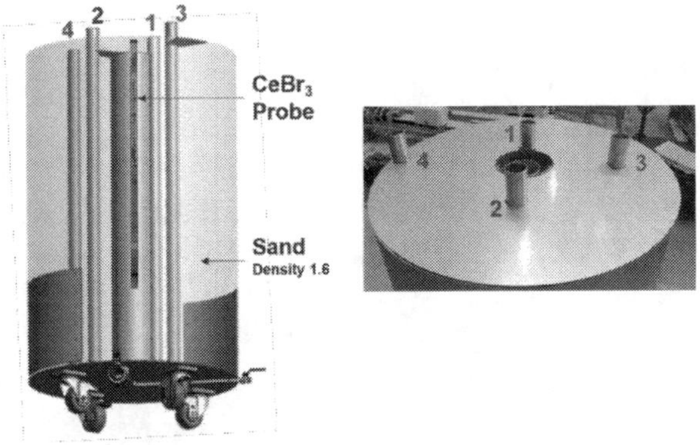

Fig. 11. On the left, the silo as well as the position of the probe and the 4 instrumentation tubes. On the right, a photograph of the top of the silo in order to observe the position of the 4 tubes used to insert the Cs source

At the same time, an MCNP model of the sand silo was developed using the $CeBr_3$ probe model presented previously in order to validate the simulated response of the probe in a real environment, see Fig. 12.

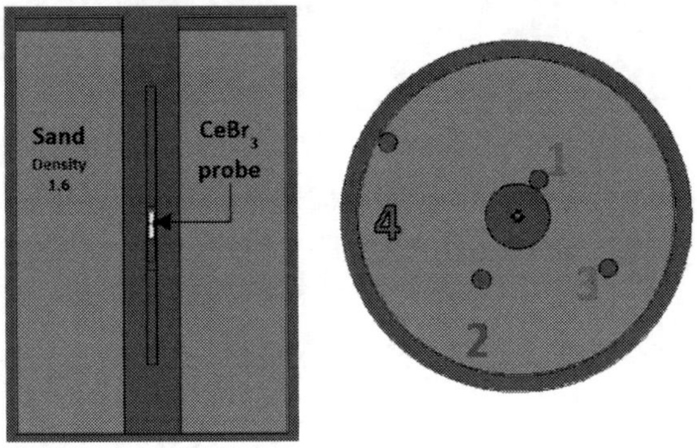

Fig. 12 : MCNP model of the sand silo

The Fig. 13, compares, for the 4 positions of the instrumentation tube (n°1, n°2, n°3, and n°4) and various depths of the source within the silo, the MCNP predictions and the measurements. All values were normalized to the value of 110 cm, corresponding to the position where the centre of the crystal is directly in front of the source, to facilitate comparison of each profile within the same figure.

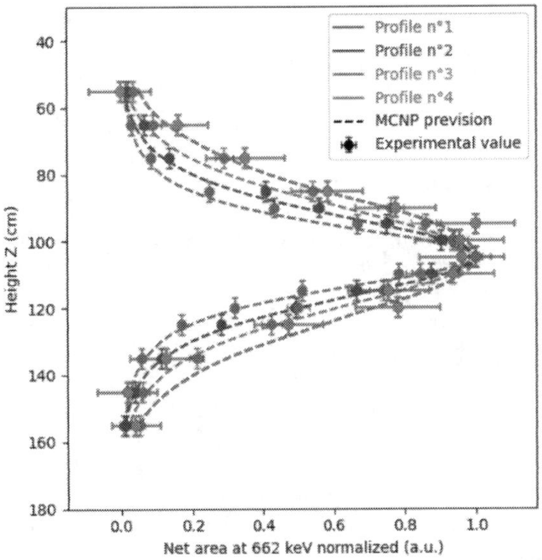

Fig. 13. Axial response of the $CeBr_3$ probe

The axial response of the probe is modelled with high accuracy by MCNP. The Table 2 displays the full width at half maximum (FWHM) of each profile, detailing both the experimental values, denoted as $FWHM_{EXP}$, and the MCNP estimations, referred to $FWHM_{MCNP}$.

TABLE II
Comparison of axial response (FWHM) between MCNP calculation and experimental value

Profile	$FWHM_{EXP}$ (cm)	$FWHM_{MCNP}$ (cm)	$\frac{FWHM_{MCNP} - FWHM_{EXP}}{FWHM_{EXP}}$
n°1	25.8	25.8	0.0%
n°2	33.5	32.6	-2.7%
n°3	41.4	38.8	-6.3%
n°4	43.0	46.0	7.0%

We notice that the full width at half maximum of each profile response increases following the distance from the probe. For example, the $FWHM_{EXP}$ = 25.8 cm for the profile closest to the probe while $FWHM_{EXP}$ = 46 cm for the profile farthest from the centre of the silo.

III. CONCLUSIONS

A comprehensive series of laboratory tests and simulations were used in this study to characterization the ALT QL40-SGR-2G spectrometric probe, equipped with a CeBr3 detector. The investigation explored crucial points such as energy resolution, detection efficiency, crystal length, probe rotation effects, counting losses, and spatial response in a sand silo context. The experimental results obtained demonstrated the probe's capabilities and reliability in various conditions, but also validated its MCNP numerical model.

The numerical model and probe design are robust due to their close agreement with experimental data and MCNP simulations, particularly in detection efficiency (2.4%) and spatial response characterization. Thanks to this study, we can now use the probe in a wider range of geophysical application and environmental monitoring tasks, with enhanced confidence in the accuracy of its measurements and the interpretations derived from them. The validated MCNP model can be further utilized to explore more complex geological scenarios, including heterogeneous formations and those with varying levels of radioactivity. The use of advanced modelling can lead to the creation of more sophisticated algorithms for the identification and quantification of materials in the subsurface.

References

[1] T. Marchais *et al.*, "Experimental validation of a CeBr3 gamma-ray logging probe MCNP model," *EPJ Web Conf.*, vol. 288, p. 05003, 2023, doi: 10.1051/epjconf/202328805003.

[2] "Detailed MCNP Simulations of Gamma-Ray Spectroscopy Measurements With Calibration Blocks for Uranium Mining Applications | IEEE Journals & Magazine | IEEE Xplore." Accessed: Feb. 08, 2024. [Online]. Available: https://ieeexplore.ieee.org/document/8268552

[3] B. Pérot *et al.*, "Development of Gamma Spectroscopic Tools for Uranium Ore Samples and Borehole Exploration," in *2022 IEEE Nuclear Science Symposium and Medical Imaging Conference (NSS/MIC)*, Italy: IEEE, Nov. 2022, pp. 1–5. doi: 10.1109/NSS/MIC44845.2022.10398898.

[4] Los Alamos National Laboratory, "MCNP6, User's manual – Version 1.0 - LA-CP-13-00634, Rev. 0 – May 2013 – Denise B. Pelowitz, editor." 2013.

[5] M. M. Be, Ch. Dulieu, and V. Chiste, "NUCLEIDE-LARA, a library for alpha, X and gamma emissions sorted by increasing energy," France, 2008.

[6] F. G. A. Quarati *et al.*, "Scintillation and detection characteristics of high-sensitivity CeBr3 gamma-ray spectrometers," *Nucl. Instrum. Methods Phys. Res. Sect. Accel. Spectrometers Detect. Assoc. Equip.*, vol. 729, pp. 596–604, Nov. 2013, doi: 10.1016/j.nima.2013.08.005.

[7] G. F. Knoll, *Radiation detection and measurement*, 3rd ed. New York: Wiley, 2000.

[8] V. Fondement *et al.*, "Development of a neutron probe to perform a combined measurement of uranium concentration and hydrogen porosity for mining applications," *Nucl. Instrum. Methods Phys. Res. Sect. Accel. Spectrometers Detect. Assoc. Equip.*, vol. 1059, p. 168888, Feb. 2024, doi: 10.1016/j.nima.2023.168888.

Effect of turbulence on ultrasound propagation for vibration measurements of PWR assemblies in the HERMES P loop of the POSEIDON platform.

A. Paveyranne[1,2], F. Baudry[2], P. Latil[1], E. Rosenkrantz[2], F. Augereau[2], G. Ricciardi[1], JY. Ferrandis[2,*]
[1]CEA, DES/IRESNE/DTN/STCP/LETH, Cadarache F-13108, Saint-Paul-Lez-Durance (France)
[2]IES, University of Montpellier, CNRS, Montpellier (France)
(*) jean-yves.ferrandis@umontpellier.fr

***Abstract*— This study focuses on the propagation of ultrasonic waves through a turbulent water flow representative of the hydraulic environment near a mixing grid in a Pressurized Water Reactor (PWR). Ultrasonic measurement, which relies on accurate knowledge of the acoustic velocity, is an effective non-intrusive technique for detecting displacements under harsh conditions. However, in turbulent flows, local velocity fluctuations induced by vortices can disrupt wave propagation, leading to measurement dispersion.**

An experimental setup was implemented to quantify these effects by comparing ultrasonic time-of-flight fluctuations upstream and downstream of a vortex-generating obstacle. The 1σ standard deviation from multiple acquisitions was used to characterize the variability of the measurements. The results show that dispersion increases with vortex density, but remains below 15 µm even in the most turbulent regions. In contrast, the Reynolds number alone does not adequately reflect this variability, which depends more strongly on the local vortex density.

These results confirm the feasibility of ultrasonic techniques in turbulent flows, provided that the influence of local flow structures is taken into account—particularly for vibration monitoring of mixing grids in PWR cores.

***Keywords* — acoustic, non-destructive testing, assembly, vibration, turbulent flow, pressurized water reactors.**

I. Introduction

This work presents a study of ultrasonic wave propagation through a turbulent flow representative of the flow near a mixing grid in a Pressurized Water Reactor (PWR) fuel assembly.

Acoustic measurement is a non-destructive testing technique that enables accurate determination of a distance or displacement, provided that the acoustic velocity of the media traversed is known [1]. Conversely, it can also be used to determine the acoustic velocity if the distance is known. This method is particularly well suited to harsh environments, as it does not require visual access, unlike techniques such as laser interferometry.

It is therefore relevant to investigate how an ultrasonic wave propagates through a known fluid medium—such as water—when the fluid is in motion, and in particular to assess whether the flow alters the local acoustic velocity [2].

In laminar flow conditions, studies have shown that the acoustic wave is simply advected by the flow in the direction of propagation. This principle is exploited in ultrasonic flowmeter technologies [3]. However, in the case of turbulent flow, the wave propagation is disrupted by local velocity fluctuations within the fluid. These fluctuations locally modify the apparent acoustic velocity [4].

The real acoustic velocity in water can then be expressed as:

$$v_{real} = v_{theoretical} + \delta v \tag{1}$$

δv denotes the perturbation induced by turbulence.

Quantifying this deviation δv is essential in order to apply a correction to the measurement or simply consider it as an uncertainty. If this correction is neglected and the acoustic velocity is assumed to remain constant and equal to its theoretical value, an error will be introduced in the estimation of the distance or displacement.

To gain a better understanding of the influence of turbulent flow on ultrasonic wave propagation, it is relevant to first investigate the isolated impact of a single vortex. This can be achieved by generating Von Karman instabilities [5] and analyzing the behavior of an ultrasonic wave propagating through these vortical structures. The acoustic setup is positioned perpendicular to the direction of the fluid flow. It is important to note that a vortex exhibits non-zero transverse velocity components relative to the main flow direction. Experimental results reveal a sinusoidal fluctuation in the time of flight of the ultrasonic wave as it passes through the vortex. This variation is directly attributed to the transverse velocities induced by the vortex field, which locally accelerate or decelerate the propagation of the ultrasonic wave.

It is then essential to quantify the measurement fluctuations induced by a turbulent flow, which is characterized by the presence of multiple vortices of varying scales.

In this paper, the motivation and relevance of this analysis will be firstly presented. Then, the experimental setup implemented to conduct the study will be described in detail.

Finally, results obtained on ultrasonic wave propagation within turbulent flow will be analyzed and discussed.

II. Interest of the study

Pressurized Water Reactors (PWRs) represent the dominant technology in the French nuclear fleet. The core of a PWR consists of fuel assemblies composed of metallic grids that hold 17 by 17 fuel rods. Each rod is a metallic tube filled with uranium pellets where nuclear fission occurs. The rods are not rigidly fixed to the grids but are held in place by springs and dimples at the mixing grids. This configuration allows for thermal expansion but also introduces mechanical contact points that can lead to vibration phenomena.

Under such conditions, vibrations may occur between the mixing grid and the fuel rods. Two main types of vibration have been identified:

- Fretting, a form of mechanical wear resulting from vibrations induced by the high-velocity coolant flow (serving as both moderator and heat transfer fluid), with axial flow rates around 500 m³/h and transverse components reaching up to 75 m³/h. This generates vibrations typically around 70 Hz with amplitudes of approximately 10 µm [6].
- Seismic-induced vibrations, which may also excite the fuel assemblies, but with significantly lower frequencies (< 10 Hz) and larger amplitudes (> 500 µm) [7].

These vibration phenomena can lead to mechanical degradation of the fuel rods, potentially resulting in cladding failure and the release of radioactive material into the primary circuit, thus compromising reactor safety and performance.

The present study aims to evaluate the feasibility of measuring such vibrations, particularly those induced by seismic activity, under realistic conditions. It is therefore assumed that the grid displacements can reach up to 500 µm, and the goal is to assess whether these vibrations can still be detected accurately in the presence of turbulent flow.

To better understand and ultimately model these phenomena, experimental facilities such as HERMES P, located on the POSEIDON platform at CEA Cadarache, have been developed. These experimental loops replicate full-scale (1:1) thermal-hydraulic conditions of a PWR core, excluding the irradiation aspect, with water temperatures approaching 320 °C, pressures around 155 bar, and flow rates on the order of 500 m³/h.

To measure the vibrations without altering the geometry of the test section, non-destructive testing (NDT) techniques must be employed. Ultrasonic measurement is a promising approach, as it allows for displacement tracking without physical contact or external disturbance. However, to ensure the reliability of such measurements in PWR-like conditions, it is essential to thoroughly understand the behavior of ultrasound propagation in turbulent flow regimes [8], which forms the central focus of this study.

III. Experimental setup

The following section provides a description of the test section as well as the implemented acoustic setup, as detailed in Figure 1.

A. Test section

The test section consists of a closed-loop water circuit with a square flow channel measuring 5 cm × 5 cm in cross-section and 1.20 m in height, made of PMMA [9]. The flow is driven by a pump with a capacity ranging from 0 to 75 m³/h. Although this flow rate is significantly lower than that observed in a Pressurized Water Reactor (PWR) or in the HERMES P test loop, the flow regime remains dynamically comparable due to a maximum Reynolds number of 4.2×10^5, compared to approximately 5×10^5 for a PWR.

An obstacle is placed at the center of the test section to generate vortex shedding phenomena similar to those occurring around a fuel assembly. This configuration makes it possible to distinguish two study regions: an upstream zone characterized by flow without vortex shedding, and a downstream zone with a high density of vortical structures. Such a geometry allows the analysis of the effects of turbulence and vortex density on ultrasonic wave propagation under conditions representative of a PWR environment.

B. Ultrasonic device

The experimental setup employs two 10 MHz ultrasonic transducers positioned face-to-face on either side of the channel, enabling through-transmission measurements. The transducers are mounted using lateral flanges attached to the channel walls; these flanges can be moved along the vertical axis, allowing acoustic measurements to be taken either upstream or downstream of the central obstacle.

It is important to note that the ultrasonic waves propagate perpendicularly to the main flow direction. As a result, the waves are not advected by the flow itself, but their propagation path can be locally disturbed by the presence of vortical structures within the turbulent flow.

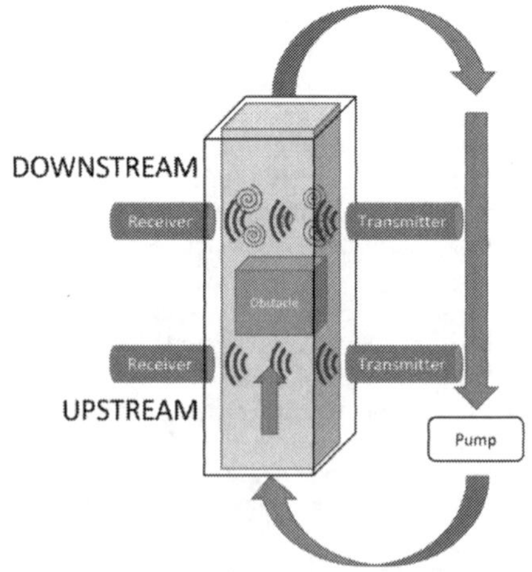

Figure 1: Diagram showing the test section with the position of the acoustic device upstream or downstream of the obstacle.

IV. Signal acquisition and processing

The ultrasonic path between the two sensors passes successively through three media: the first PMMA wall of the channel, then the flowing water, and finally a second PMMA wall before reaching the receiving sensor. This configuration involves three distinct layers—PMMA, water, and PMMA—for which the respective acoustic velocities must be known in order to accurately interpret the measured time of flight. Since the thicknesses of these media remain constant throughout the experiments, and the acoustic velocity in PMMA does not vary over time, only the acoustic velocity in water may fluctuate. The measured time of flight corresponds to the first detected signal, that is, the fastest path between the emitter and the receiver, meaning a path without any reflections within the media, and is given by the following expression:

$$TOF = \frac{d_{PMMA1}}{v_{PMMA1}} + \frac{d_{water}}{v_{real}} + \frac{d_{PMMA2}}{v_{PMMA2}} \quad (2)$$

with *TOF* representing the time of flight, *d* the distances, and v_{PMMA} the theoretical acoustic velocity of PMMA. The term v_{real} defined in Equation (1), represents the theoretical acoustic velocity of sound in water, with an additional perturbation due to turbulence δv. In Equation (2), all terms are constant except for v_{real}. By correcting Equation (2) to remove the constant PMMA terms, the following Equation (3) can be obtained:

$$TOF = \frac{d_{water}}{v_{real}} + K \quad (3)$$

where K is a constant term.

Thus, a fluctuation in the corrected time of flight originates only from a variation in the real acoustic velocity in water, since the terms are not independent and the water distance remains constant. The goal is then to quantify these fluctuations.

Given that it is difficult to accurately measure the real acoustic velocity in water in real time to correct for fluctuations, we assume that the real velocity is constant and equal to the theoretical velocity. Under this assumption, and according to Equation (3), the time-of-flight fluctuations can be interpreted as measurement fluctuations on the water distance.

The error in the distance measurement is calculated using Equation (4), under the assumption that the acoustic velocity of sound in water remains constant.

$$\Delta_{distance} = \Delta_{TOF} \times v_{water\ theoretical} \quad (4)$$

To evaluate the fluctuations in the time of flight, ten acquisitions of the first signal (as described by Equation 2) are performed for each configuration, while keeping the sensor positions and the flow rate/Reynolds number constant. These sets of measurements are repeated at different flow rates, both upstream and downstream of the obstacle.

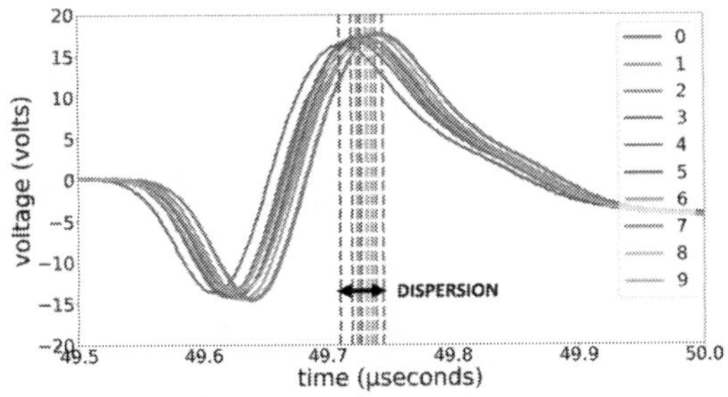

Figure 2: Example of time-of-flight dispersion measured with peak detection (maximum) for a batch of 10 measurements with the same flow rate and sensor position

For each set, the time of flight is determined by identifying the peak of the received signal, as shown in Figure 2. The mean value is calculated and fluctuation (dispersion) is then quantified by calculating the standard deviation at 1 σ of the time-of-flight values. The standard deviation serves as a relevant indicator of measurement variability around the mean value. It follows that a larger standard deviation reflects greater fluctuations in the acoustic velocity of the water, induced by turbulence.

By referring back to Equation (4), it is logical to substitute the uncertainties with the standard deviation measured at 1σ, leading to Equation (5).

$$\sigma_{distance} = \sigma_{TOF} \times v_{water\ theoretical} \quad (5)$$

In this way, TOF uncertainties caused by turbulence-induced speed perturbations are measured by the standard deviation, and can be translated as an error or uncertainty in the distance measurement.

V. Results and discussion

The results represent the standard deviation in distance as a function of the Reynolds number (flow rate) in Figure 3, for sensor positions located upstream or downstream of the obstacle.

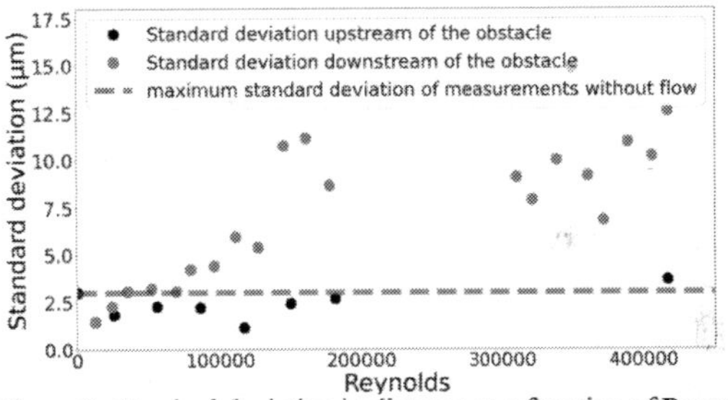

Figure 3: Standard deviation in distance as a function of Reynolds number for sensors placed upstream and downstream of the obstacle.

It is observed that when the sensor is positioned downstream of the obstacle—i.e., in the region with a high density of vortical structures—the standard deviation increases with the flow rate, following an approximately linear trend up to a Reynolds number of 1.5×10^5. However, it is important to highlight that the standard deviation never exceeds 15 µm for a distance of 5 cm of water, even under the most turbulent conditions.

Conversely, when the sensor is located upstream of the obstacle—where the flow exhibits low vorticity—the standard deviation remains nearly constant regardless of the Reynolds number. Even at a Reynolds number equal to 4.2×10^5, the standard deviation remains below 4 µm, for a distance of 5 cm of water.

It should also be noted that a nonzero standard deviation is recorded even in the absence of flow. This is attributed to the inherent limitations of the measurement system. Nonetheless, repeatability and reproducibility tests have demonstrated that, without flow, the standard deviation consistently remains below 3 µm, which is considered the baseline measurement uncertainty for this experimental setup.

It is important to emphasize that the Reynolds number is not a reliable indicator for assessing the dispersion of acoustic measurements caused by turbulence. Instead, the variability is more strongly correlated with the local density of vortical structures. Therefore, the observed fluctuations in time-of-flight measurements can be more accurately attributed to the presence of vortices in the measurement region.

VI. Conclusion

This study has assessed the impact of turbulent flow on ultrasonic wave propagation within a channel representative of conditions in a Pressurized Water Reactor (PWR). The results demonstrate that turbulence, particularly the local density of vortical structures, induces measurable fluctuations in the time of flight of ultrasonic waves. These fluctuations, quantified via the standard deviation, reflect a dispersion in the measurements that can reach up to 15 μm in regions with higher vortex density. This means that the mean value is determined with an uncertainty of ±15 μm but also that the accuracy of the measurement cannot be less than 15 μm for a distance of 5 cm of water. It is also shown that the Reynolds number alone is not a reliable indicator of this dispersion. The primary factor influencing measurement variability is the local density of vortices.

These knowledge of the accuracy of distance measurements is of great importance for the development of acoustic monitoring methods under realistic PWR conditions, especially for the detection of mechanical vibrations such as those induced by seismic excitations. The results confirm the feasibility of acoustic measurements for displacements on the order of 500 μm (seismic events), as measurements can be made with an accuracy of 15 μm even in the presence of significant turbulence, if the water distance is equal to 5 cm.

However, displacement measurements, such as mechanical vibration detection, are typically performed using reflected ultrasonic signals. It would therefore be of interest to investigate the impact of vortical structures on measurements performed in reflection, rather than in transmission as done in the present study.

References

[1] Pierce, Allan D., "Acoustics : an introduction to its physical principles and applications" in Springer , 2019.

[2] Mazue, G., Viennet, R., Hihn, J. Y., Bonnet, D., Barthes, M., Bailly, Y., & Albaïna, I. (2015). Influence of a perpendicular liquid flow on a cleaning process using 20 kHz ultrasound: Characterization of the agitation at vicinity of the surface opposite to the transducer. The Canadian Journal of Chemical Engineering, 93(2), 201-205.

[3] Mattingly, G. E., Mattingly, G. E., & Yeh, T. T. (2000). NIST's ultrasonic technology assessment program to improve flow measurements. US Department of Commerce, National Institute of Standards and Technology.

[4] Barthes, M., Mazue, G., Bonnet, D., Viennet, R., Hihn, J. Y., & Bailly, Y. (2015). Characterization of the activity of ultrasound emitted in a perpendicular liquid flow using Particle Image Velocimetry (PIV) and electrochemical mass transfer measurements. Ultrasonics, 59, 72-78.

[5] Massacret, N., Moysan, J., Ploix, M. A., Jeannot, J. P., & Chaouch, N. (2016, June). Study of ultrasonic propagation through vortices for acoustic monitoring of high-temperature and turbulent fluid. In WCNDT 2016.

[6] Rubiolo, P. R., & Young, M. Y. (2009). On the factors affecting the fretting-wear risk of PWR fuel assemblies. Nuclear Engineering and Design, 239(1), 68-79.

[7] Ricciardi, G. (2016). Fluid–structure interaction modelling of a PWR fuel assembly subjected to axial flow. Journal of Fluids and Structures, 62, 156-171.

[8] Bieder, U., Falk, F., & Fauchet, G. (2014). LES analysis of the flow in a simplified PWR assembly with mixing grid. Progress in Nuclear Energy, 75, 15-24.

[9] Aït Oucheggou, O. (2020). Ecoulement de particules autour d'un obstacle: application au colmatage dans un générateur de vapeur (Doctoral dissertation, Aix-Marseille).

Detector Noise in X-ray Computed Tomography Examinations of Irradiated Tristructural Isotropic (TRISO) Fuels

William Chuirazzi[1,*], Brian Gross[1], Joshua Kane[1], Swapnil Morankar[1], and Rahul Reddy Kancharla[1]
[1]Idaho National Laboratory, USA
(*) william.chuirazzi@inl.gov

Abstract—**Tristructural isotopic (TRISO) fuel is a promising next-generation fuel form for advanced nuclear reactors. TRISO fuel consists of sub-millimeter diameter uranium-bearing fuel kernels encapsulated in multiple layers of carbon and ceramic materials. As part of the assessment of TRISO fuels, post irradiation examination (PIE) is performed to provide information on changes in microstructure, mechanical properties, and potential degradation. X-ray computed tomography (XCT) is a nondestructive technique that collects a series of two-dimensional radiographs as a function of sample rotation and mathematically reconstructs them into a three-dimensional dataset that provides volumetric information on the sample's internal features. This work focuses on PIE of irradiated TRISO fuels using XCT with a focus on examining the impact of highly radioactive fuels on the images generated with the instrument's CsI detector. Despite examining samples with over 2.00 Sv/hr on contact, techniques such as frame-averaging as well as a lack of interaction of high-energy gamma-rays with the detector ensure there is no noticeable impact on the quality of the CT images due to detector noise.**

Keywords —**X-ray computed tomography, TRISO fuel, post-irradiation examination, nondestructive examination.**

I. Introduction

Tristructural isotropic (TRISO) nuclear fuel is a next-generation fuel form consisting of a fuel kernel (usually a mixture of uranium carbide and oxide, referred to as UCO) surrounded by different carbon based layers. These layers include the porous buffer layer, the inner pyrolytic carbon layer, the SiC layer, and the outer pyrolytic carbon layer [1]. Several thousand individual particles are loaded into a graphite matrix creating a fuel form known as a compact in the United State's Department of Energy's Advanced Gas Reactor (AGR) program. Originally conceived in the 1950's as part of the Dragon Project [2], interest has been renewed in TRISO fuel as it is a fuel type for Hight Temperature Gas Reactors (HTGRs). Some advantages of the HGTR design with TRISO fuel include high peak operating temperatures (up to ~1200 °C and above), effective fission product retention, and enhanced fuel performance [3]. TRISO fuel is also under consideration for a variety of microreactor designs, further motivating the need for further study of TRISO-based fuel. While there are slight variations in layer thickness between different irradiation campaigns, a typical depiction of an individual TRISO particle form the AGR-5/6/7 campaign is shown in *Fig. 1(a)* while *Fig. 1(b)* displays a general TRISO compact from the AGR program.

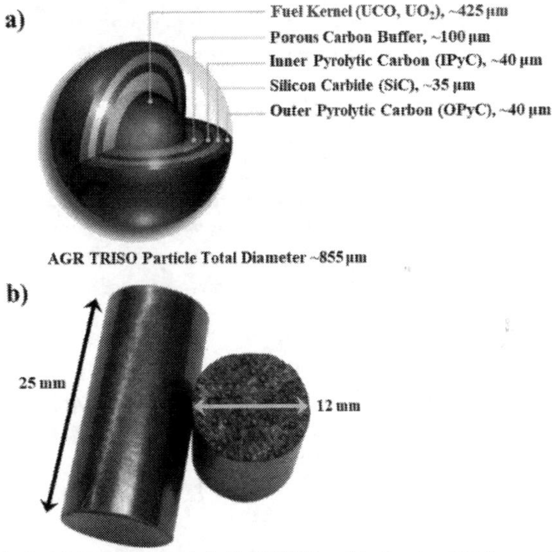

Fig. 1. (a) Depiction of an individual TRISO particle showing all its layers with associated layer thicknesses for AGR TRISO particles. (b) A photo of an AGR compact which consists of several thousand TRISO particles embedded in a graphite matrix. Images courtesy of the U.S. Department of Energy's Advanced Gas Reactor program.

Post-irradiation examination (PIE), the study of material after it has undergone in-core irradiation, is useful for providing a more comprehensive understanding of fuel behavior, including mechanisms of degradation, and the materials' performance under the harsh environmental conditions of a nuclear reactor. PIE helps improve fuel performance, enhance reactor safety, and informs the development of next-generation nuclear fuels by defining a safe range of conditions the fuel can operate within. Traditional destructive PIE exams, such as metallography and electron microscopy, can provide detailed information about microstructural changes and fuel integrity

[4].

X-ray computed tomography (XCT) is a nondestructive examination technique that has recently been incorporated into the PIE process to examine individual particles as well as fuel compacts. X-ray CT is a measurement of X-ray attenuation through a sample. A series of two-dimensional (2D) radiographs are captured as a function of sample rotation. These 2D projections are then reconstructed to produce a three-dimensional (3D) volume of the sample [5]. These data can then inform traditional destructive examination techniques by targeting regions of interest and can also be meshed to create direct input into modeling efforts.

As XCT has been used to examine highly radioactive nuclear fuel, the question arises whether the sample's radioactivity is directly affecting the measurement itself by inducing a signal on the equipment used to detect the X-rays transmitting through the sample. While the ability to examine highly radioactive material directly reduces the timespan from testing to information and ultimately expedites the implementation of new nuclear concepts, it is imperative that the high levels of radioactivity of these samples does not impact the measurement in a meaningful way, as this could compromise the data collected. The goal of this work is therefore to investigate what effect, if any, radioactive samples have on XCT measurements.

II. THEORY

In the AGR program, TRISO compacts were irradiated in the Advanced Test Reactor to determine the performance envelope of the fuel [6,7]. During this irradiation the uranium kernels absorbed neutrons to produce fission as well as activation. Once the compacts were removed from the reactor, the activated fuel continued to undergo radioactive decay, releasing a variety of ionizing radiation, including gamma-rays, alpha- and beta-particles. The source term that was recorded in the as-run physics analysis report for the AGR-5/6/7 Compact 5-3-1 [8] was updated to account for radioactive decay from the time of reactor discharge to the anticipated date of XCT examination using ORIGEN 2.2 [9]. The source term showed some of the most radioactive isotopes to be Cs-137, Pu-241, Sr-90, Pm-147, Ce-144, and Ru-106. A total of 96 emitting isotopes were identified.

The X-ray microscope utilized in this work uses a complementary metal-oxide semiconductor (CMOS) based detector. In this detector type, a semiconductor photodiode is used to create an electrical charge when exposed to incident radiation (light or otherwise). The voltage is proportional to the strength of the initial signal. Transistors are then used to amplify the voltage values and convert them to a digital signal. An analog-to-digital (ADC) converter then transforms these voltage values into a digital number that can be used to create an image. Note that a key feature of CMOS technology is that this entire process occurs directly on the sensor itself [10].

One concern with using CMOS detectors for imaging radioactive samples is that the ionizing radiation emitted from the samples themselves can induce charge directly on the detector. This can add noise to images and may introduce additional imaging artifacts. One process for mitigating this concern is that for each radiograph several images are taken (either 5 or 20 images per projection depending on detector binning) and then a median image is calculated from the set of images. It is unlikely for ionizing radiation emitting from a radioactive sample to impinge on the same pixel of the detector repeatedly due to the stochastic nature of radioactive decay. The median image therefore reduces the impact of the random radiation from the sample on the image produced by the detector.

III. EXPERIMENTAL

All measurements were performed at Idaho National Laboratory's (INL) Irradiated Materials Characterization Laboratory (IMCL).

A. X-ray Computed Tomography Equipment

The X-ray computed tomography measurements were performed using a ZEISS Xradia 620 Versa X-ray Microscope (XRM) [11]. In this work the flat panel detector (Varex 2315, Varex Imaging, Salt Lake City, UT, USA) was used to image the complete volume of the compact. This detector is a CMOS detector utilizing a high efficiency microcolumnar CsI scintillator. *Fig. 2* shows the internals of the ZEISS Xradia 620 Versa as staged for imaging AGR TRISO compacts.

Fig. 2. Labeled photograph showing the setup in the ZEISS Xradia 620 Versa at IMCL for imaging irradiated AGR TRISO compacts.

B. Compacts

During the PIE of irradiated AGR TRISO compacts a total of five individual compacts have been examined- two from the AGR-3/4 irradiation campaign [6] and three from the AGR-5/6/7 irradiation campaign [7]. A graded approach was taken to examining radioactive compacts- first XCT was performed on unirradiated compacts with natural uranium, followed by unirradiated compacts with fresh fuel, before AGR-3/4 compacts and finally AGR-5/6/7 compacts were examined. The radiological dose rates of each compact on their approximate date of examination are shown in Table I.

TABLE I
RADIOLOGICAL DOSE RATES FOR AGR COMPACTS EXAMINED WITH XCT

		Compact ID				
		AGR-3/4 12-4	AGR-3/4 7-1	AGR-5/6/7 5-3-1	AGR-5/6/7 3-6-1	AGR-5/6/7 2-1-3
Dose Rate on Contact (Sv/hr)	βγ	0.24	>0.5	13.18	>2.00	>0.50
	γ	0.03	0.12	1.20	1.40	1.99 (at ~12.7 cm)
Dose Rate @ 30 cm (Sv/hr)	βγ	0.03	0.04	0.10	0.11	0.10
	γ	0.00	0.01	0.02	0.02	0.03

Custom shielding of the compacts was necessary to protect laboratory personnel as well as equipment from the high radiation fields produced by the compacts [12]. The compact was placed in a radiologically clean aluminum container which was then mounted inside a tungsten "push-pop" shield. Operators transferred the shielded configuration into the X-ray microscope using long-reach tools. The shielded compact was placed inside instrument onto a stage which was coupled directly to the vertical stage inside the X-ray microscope. Once the compact was secure inside the X-ray microscope and the lid was removed, the compact could be remotely raised out of the shielding apparatus and into the X-ray beam path. A photograph of a compact and associated shielding is shown in *Fig.* 3. For more details on the design and implementation of the "push-pop" shielding apparatus, the interested reader is referred to [12].

Fig. 3. A photograph displaying a compact mounted with the "push-pop" shielding configuration within the X-ray microscope.

C. *Measurements*

During initial XCT scans of AGR-3/4 Compact 12-4 [6] as well as AGR-5/6/7 Compact 2-1-3 [7], a series of measurements were collected to attempt to quantify the impact the radiation from the compacts had on the XCT images themselves. First a dark image, an image acquired with the X-ray source off, was taken without any sample in the cabinet of the X-ray microscope. Once the compact and push-pop shielding were loaded into the X-ray microscope's cabinet, a dark image was taken of the compact shielded within the push-pop. The compact was then raised out of the shielding and a final dark image was then acquired.

A 100-pixel thick line profile was taken across the center of each image. The mean grayscale values across the line profile were calculated as were the associated error. These values were then compared across the different images to quantify the impact the radioactive compacts had on the resulting images.

Several complete CT acquisitions using different X-ray energies and imaging conditions were also taken over the course of ~3 days on AGR-5/6/7 Compact 3-6-1. Acquisition settings were modified to collect an open beam image (X-rays on but the compact removed from the field-of-view) every 500 radiographs. This evolution first consisted of a scan lasting approximately 18 hours and 45 minutes that utilized a 160 kvP Bremsstrahlung spectrum of X-rays. A total of 4801 radiographs were collected with each radiograph having an acquisition time of six seconds. Next, a scan taking approximately 35 hours and 30 minutes was collected using the same X-ray spectrum, but the count time of the 4801 radiographs was increased to 12 seconds. Finally, a scan using a Bremsstrahlung spectrum of X-rays with a kVp of 60 keV taking a total of approximately 27 hours was conducted. Each radiograph in this scan had an acquisition time of nine seconds and again a total of 4801 radiographs were captured. The different scan parameters were selected to try to get information on different components of the compact (by using different X-ray kVp) and for method develop for reducing the heavy metal artifacts from imaging nuclear fuel (longer exposure time).

A 102-pixel×102-pixel box was taken near the top-left corner of each open beam image as well as near the center of the image. The locations remained the same across each image and the median grayscale value within the box was measured. The goal of this measurement was to ascertain how the compact's radiation dose on the detector was impacting the detector's measurement over time.

IV. RESULTS

The results of dark images taken with the radioactive compacts in different shielding configurations (removed outside the instrument, inside the instrument but within the push-pop shielding, and inside the instrument completely exposed to the detector) are shown in *Fig. 4*. For AGR-3/4 Compact 12-4, the mean difference between the dark image of the compact removed from the X-ray microscope's cabinet and that of the dark image taken of the compact exposed directly to the detector with no shielding was 32.14 ± 0.22 (where the error is the standard deviation from the mean). While the mean value of the dark image when the compact was removed from the X-ray microscope was only 69.95 ± 0.68, it may seem like this ~46% increase from the dark image without a sample is significant. However, the detector has a 16-bit dynamic range, so the addition of ~32 counts to the baseline noise measurement from the compact's radiation represents an increased signal of less than 0.05% across the detector's entire dynamic range. This slight increase does not substantially impact the images produced of the compact. Likewise, the much more radioactive AGR-5/6/7 Compact 2-1-3 saw an increase in the dark image of 29.05 ± 1.21 grayscale counts when the radioactive sample was directly exposed to the detector. However, the initial noise in the detector with the compact removed was 2522.64 ± 22.93. The dark noise increase from the radioactive sample is therefore only ~1% of the initial background reading and represents a change of less than 0.05% across the detector's entire dynamic range. These increased signals are not substantial enough to meaningfully impact the images.

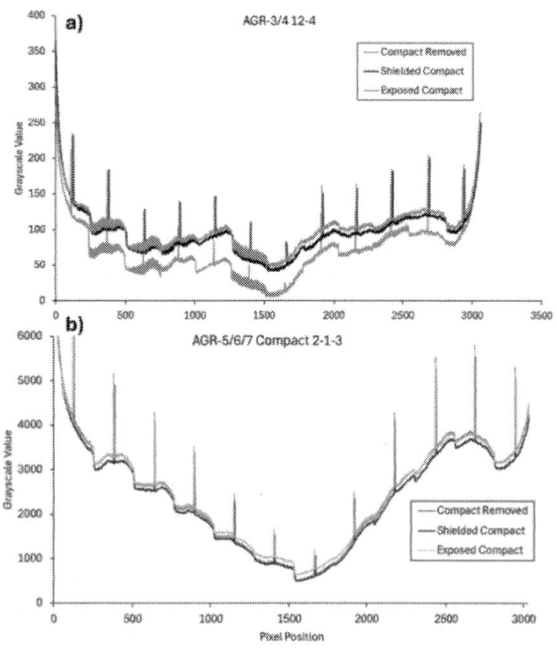

Fig. 4. Charts displaying the median value of a 100-pixel wide line profile across open beam images for (a) AGR-3/4 Compact 12-4 and (b) AGR-5/6/7 Compact 2-1-3.

Median grayscale values on different regions of the detector throughout the duration of the ~72-hour data acquisition for AGR-5/6/7 Compact 3-6-1 are shown in *Fig. 5*. The grayscale values in each region of the open-beam images within each acquisition are relatively consistent, pending minor fluctuations. Note that these fluctuations may be caused by X-ray source drift in the instrument and not by any radiation-induced degradation in the detector itself. One interesting observation is the increase in the signal in the middle of the detector compared to the edge of the detector for the lower energy (60 kVp scan). An explanation for this may be that the lower X-ray energies may get more attenuated in air when traversing the longer path length to the edge of the detector compared to the middle of the detector. More investigation is necessary to provide a more comprehensive explanation for this observation. Regardless, the incident radiation from the compact itself does not seem to significantly impact the open-beam images collected by the detector, implying that the detector performance is not degraded with its prolonged exposure to the radiation field of the compact.

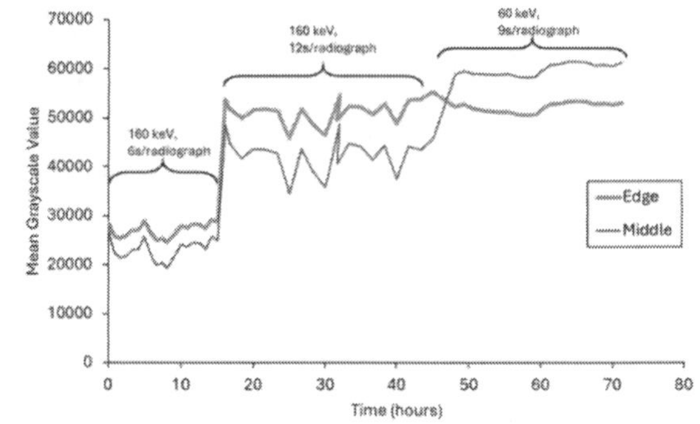

Fig. 5. Plot showing the mean grayscale value of a 102-pixel×102-pixel box at two locations in a series of open beam images as a function of time. The relative grayscale intensity across each exposure does not significantly change, indicating the detector is not experiencing significant effects from the radiation of the sample.

V. Conclusions

In this work, the impact of radiation fields from AGR TRISO compacts on a solid-state flat panel detector for X-ray computed tomography measurement is evaluated. Even when examining a compact with very high radiation dose (over 2.00 Sv/hr from just γ-rays on contact), these measurements suggest no appreciable level of noise is added to the collected data. The increase in signal on the detector from the radioactive compacts compared to no sample in the instrument was found to be minimal (less than 0.05% of the dynamic range of the detector). Likewise, there was not a noticeable change in the open-beam images generated over time when the detector was subjected to a sustained radiation field.

Frame-averaging multiple radiographs into a single median radiograph for each projection (as was done in this work) helps reduce the impact of incident radiation from the radioactive sample. Additionally, it is suspected that a large portion of the gamma-rays emitted from the samples have higher energy than the detector is optimized through, allowing them to pass through the detector without depositing significant energy into the detector and minimizing their impact of the images. Further simulation studies are underway to explore this hypothesis. In the meantime, this work shows that it is feasible to perform X-ray computed tomography on highly radioactive nuclear fuel specimens without fear of severe radiation induced damage to the detector or degradation of the images. Ultimately, the XCT examination of highly radioactive samples can help expedite the lifecycle from a novel nuclear concept's inception to its study and then commercial implementation.

ACKNOWLEDGMENT

This work was supported by the U.S. Department of Energy's Advanced Gas Reactor (AGR) program. This manuscript has been authored by Battelle Energy Alliance, LLC, under Contract No. DE-AC07–05ID14517 with the U.S. Department of Energy. The U.S. Government retains and the publisher, by accepting the article for publication, acknowledges that the U.S. Government retains a nonexclusive, paid-up, irrevocable, worldwide license to publish or reproduce the published form of this manuscript, or allow others to do so, for U.S. Government purposes.

REFERENCES

[1] Demkowicz, Paul A., Bing Liu, and John D. Hunn. "Coated particle fuel: Historical perspectives and current progress." *Journal of Nuclear Materials* 515 (2019): 434-450.
[2] Price, M. S. T. "The Dragon Project origins, achievements and legacies." *Nuclear engineering and design* 251 (2012): 60-68.
[3] Nabielek, Heinz, et al. "The performance of high-temperature reactor fuel particles at extreme temperatures." *Nuclear Technology* 84.1 (1989): 62-81.
[4] Degueldre, Claude, Johannes Bertsch, and Matthias Martin. "Post irradiation examination of nuclear fuel: Toward a complete analysis." *Progress in Nuclear Energy* 92 (2016): 242-253.
[5] Withers, Philip J., et al. "X-ray computed tomography." *Nature Reviews Methods Primers* 1.1 (2021): 18.
[6] Collin, Blaise P. *AGR 3/4 irradiation test final as run report*. No. INL/EXT-15-35550. Idaho National Lab. (INL), Idaho Falls, ID (United States), 2015.
[7] T Pham, Cam Binh, et al. *AGR-5/6/7 irradiation test final as-run report*. No. INL/EXT-21-64221-Rev000. Idaho National Lab. (INL), Idaho Falls, ID (United States), 2021.
[8] Werner, Christopher John, et al. *MCNP version 6.2 release notes*. No. LA-UR-18-20808. Los Alamos National Laboratory (LANL), Los Alamos, NM (United States), 2018.
[9] Croff, A. G. *ORIGEN2: a revised and updated version of the Oak ridge isotope generation and depletion code*. No. ORNL-5621. Oak Ridge National Lab.(ORNL), Oak Ridge, TN (United States), 1980.
[10] El Gamal, Abbas, and Helmy Eltoukhy. "CMOS image sensors." *IEEE Circuits and Devices Magazine* 21.3 (2005): 6-20.
[11] Chuirazzi, William, Rahul Reddy Kancharla, and Swapnil Morankar. "Laboratory-Based Micro-X-ray Computed Tomography of Energy Materials at Idaho National Laboratory." *JOM* 77.3 (2025): 1337-1353.
[12] Chuirazzi, William, et al. "Strategy to safely enable X-ray computed tomography examination of highly radioactive tristructural isotropic nuclear fuel." *Nuclear Engineering and Technology* (2025): 103726.

Fast neutron imaging using D-D generator and plastic scintillator – GEANT4 simulations

Filip Révai[1,*], Branislav Vrban[1], Jakub Lüley[1], Štefan Čerba[1], Vendula Vrtalová[1], Otto Glavo[1], Nikita Saito[1] and Vladimír Nečas[1]
[1]Slovak University of Technology, Faculty of Electrical Engineering and Information Technology, Slovakia
(*) filip.revai@stuba.sk

***Abstract*— This paper examines the potential of using a fast neutron technique called defectoscopy with a compact and efficient deuterium-deuterium (D-D) neutron generator and plastic scintillator. The paper describes the design of the apparatus and the methodology for obtaining images. In order to correctly interpret the results of the experiments, it is necessary to simulate the expected outcomes. For this purpose, simulations were performed using the Monte Carlo method in the GEANT4 code. A mockup of a printed circuit heat exchanger was modeled, and simulations were performed using multiple structural defects. The initial findings of our research are presented: GEANT4-simulated images and their processing for defects identification in combination with unidirectional surface sources. The results demonstrate the functionality and potential of our setup for fast neutron defectoscopy using a D-D generator.**

***Keywords* —D-D neutron generator, plastic scintillator, neutron imaging, Monte Carlo, GEANT4 simulations.**

I. Introduction

The fast neutron defectoscopy is a promising non-destructive testing method that uses a beam of neutrons to visualize the internal structure of materials and detect potential defects. The main advantage of this technique is that it allows us to see defects that are not visible with other imaging methods, such as X-ray imaging. This is due to the good penetration of fast neutrons into the sample material. In combination with a sensitive organic scintillator detector that converts neutron energy into a visible signal, this is an optimal setup for defectoscopy of small- and large-scale components. Our research focuses on using this imaging technique with a deuterium-deuterium (D-D) neutron generator and a plastic scintillator to capture digital images. The mockup of the Printed circuit hear exchanger [1] was modeled with defects that could occur during the device's lifetime. Then the simulations were performed using the GEANT4 program to obtain images suitable for comparison with experimental results. The results offer initial insights into our imaging device currently being developed and the images expected from the experiments. This paper presents the first Monte Carlo simulations performed using a surface neutron source and a plastic scintillator in GEANT4.

II. Methodology

A. Neutron generator

To the possible use of a neutron generator in a laboratory setting poses requirements to its compactness and simple operation. Neutron generators are often distinguished by the construction and energy of the neutrons they produce. One type of generator is the D-D generator, which is based on the deuterium-deuterium fusion reaction. Deuterium ions are produced by an ion source, often a microwave-driven electron cyclotron. These ions are then accelerated toward a target that is typically loaded or coated with either deuterium or titanium. Once on the target surface, the ions become implanted and subsequently interact with incoming ions. This leads to a D-D fusion reaction that emits neutrons with an average energy of 2.45 MeV. These neutron generators usually have a sealed design, which makes them more compact and easier to maintain. Their smaller size and simpler construction make them suitable for portable and field applications. The generator's design also allows samples to be placed closer to the source, enabling better, more precise measurements [2]. For our simulations, we postulated an "ideal" neutron generator furthermore denoted as a surface generator. A surface D-D generator is a type of neutron generator in which the reactions that produce neutrons take place at or near the sample material. This generator focuses the beam size on the target surface, thereby improving the image resolution. The surface generator is ideal because it eliminates the beam dispersion seen in real neutron generators and focuses the beam into a homogeneous line. It was designed for one purpose: to produce simulation results that validate the functionality of the imaging setup and yield comparable results to those of a commercial neutron generator for future measurements.

B. Organic scintillator

Organic scintillators are compact and effective solutions for fast neutron imaging, converting neutron energy into a visible light signal. The basic types of organic scintillators are plastic, liquid, and crystalline. Their high hydrogen content enables efficient detection through recoil protons created in elastic neutron scattering on hydrogen atoms. Their favorable pulse shape discrimination allows for clear separation between neutron and gamma radiation, improving measurement accuracy [3]. For this setup, a plastic scintillator was chosen.

These scintillators are widely used in fast neutron imaging setups, due to their fast response times, reasonable light yield, mechanical stability, and low-cost fabrication. The combination of a plastic scintillator and a D-D generator is commonly used for fast neutron imaging. This combination allows images to be captured with good spatial resolution and real-time imaging capability. Due to the large size of the scintillator, practical resolution and contrast can be achieved in hydrogen-rich materials [4].

C. *GEANT4*

GEANT4 is a powerful, versatile, and widely used Monte Carlo simulation toolkit for modeling particle interactions with matter, including neutrons. In the context of fast neutron imaging with a plastic scintillator, GEANT4 allows for the detailed modeling of various scenarios. One scenario is to model the interactions of neutrons with scintillator materials, the generation and transport of recoil protons from elastic scattering, and the transport and detection of scintillation photons. GEANT4 enables the construction of complex detectors and sample geometries and the specification of neutron source characteristics using several specialized physics models. It includes optical photon physics to model light creation and transport in the scintillator. When simulating a plastic scintillator with a D-D neutron surface source, GEANT4 optimizes the detector's size, geometry, material composition, and photodetector placement to maximize neutron detection efficiency and spatial resolution. Essentially, GEANT4 simulations allow for the accurate prediction and optimization of fast neutron imaging system performance [5].

III. Simulation Experiment

Four geometries of the mockup heat exchanger were modeled in GEANT4 for the simulation experiment. The geometries were created using stainless steel tubes filled with demineralized water arranged in a 38x38 grid. The scintillator detector was modeled as a 40x40x5 cm polystyrene prism.

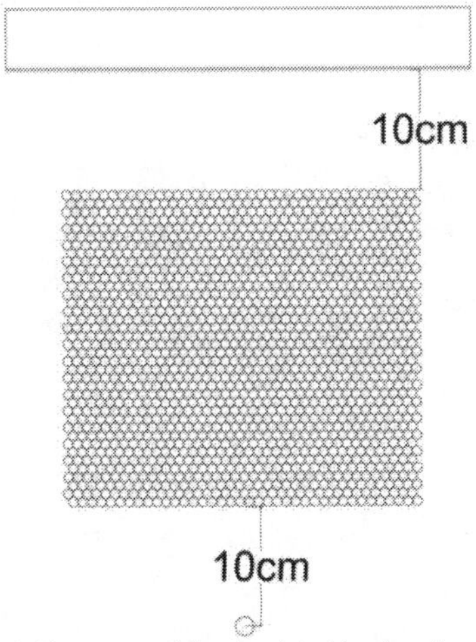

Fig. 1. Geometry of the simulated heat exchange mockup.

In front of the scintillator is an aluminum cover with a thickness of 0.2 cm. This cover is part of a stand that holds and protects the scintillator from outside effects. The mockup was placed 10 cm from the surface of the detector. The source was modeled as a surface D-D generator 10 cm from the edge of the mockup. Figure 1 shows the entire geometry setup for the simulations.

To simulate operational defects in the heat exchanger, three mockup geometries with setup defects were modeled, as shown in Figures 2, 3, and 4.

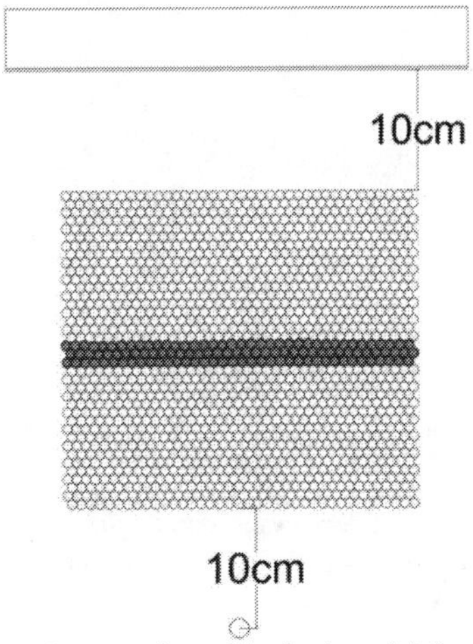

Fig. 2. Mockup with a horizontal defect.

The first defect, shown in Figure 2, is a "horizontal" defect. This defect is simulated as a tube without water. This defect can occur over a long period of time during which many small defects appear, causing the tubes to be plugged in and rendering them unable to operate.

The second defect, shown in Figure 3, is similar to the previous one; however, it is on the vertical side of the mockup.

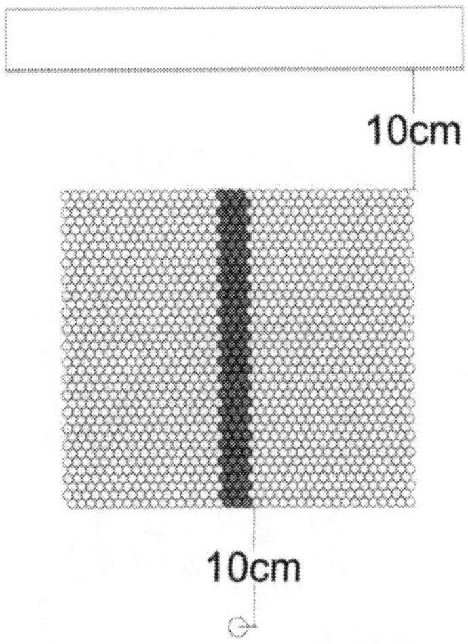

Fig. 3. Mockup with a vertical defect.

The third and final simulated defect is the "burst" defect. In this defect, a small area of the heat exchanger experiences a tube burst, which damages the adjacent tubes. Figure 4 shows this burst effect.

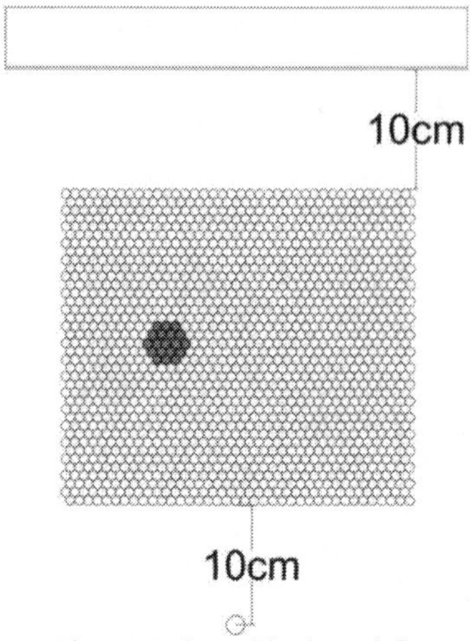

Fig. 4. Mockup with a burst defect.

These defects were chosen for simulation based on previous studies of heat exchanger defects during nuclear power plant operations [6].

IV. RESULTS

First, the main parameters of the simulations were defined. The simulations were carried out using the FTFP_BERT library in GEANT4, and the results were evaluated using the GEANT4 Scoring Tool. Separate simulations were performed for each defect geometry. The simulations were performed at 30-minute intervals, with an intensity of 10^6 n/s for the D-D surface source. The results of the simulation were then evaluated based on the stored energy in the scintillator and rendered as a heat map image. After normalizing the values in the interval [0;1], the final heat map image was obtained.

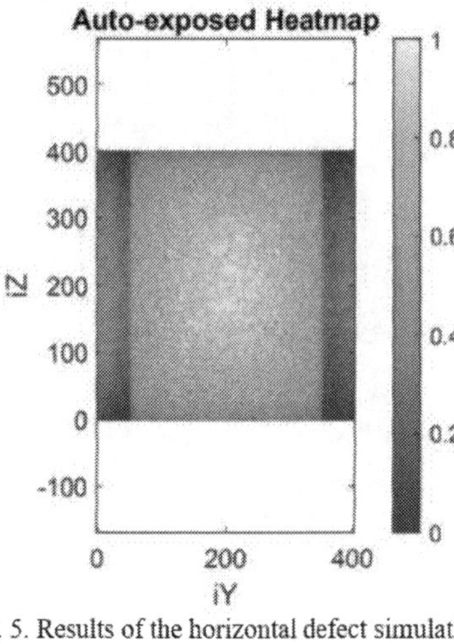

Fig. 5. Results of the horizontal defect simulation.

As can be clearly seen from Figure 5, this horizontal defect is difficult to detect using a surface D-D neutron generator. The heat map shows the light concentration in the middle of the image without any signs of the defect. This could be due to the density of the mockup. The density of the tubes remains the same as before and after the defect. The neutrons interact with the stainless-steel tubes filled with water, which makes it difficult to distinguish the defect. This problem can be partially solved by rotating the imaging setup. It is evident in the second geometry with a vertical defect, as well as in the results shown in Figure 6.

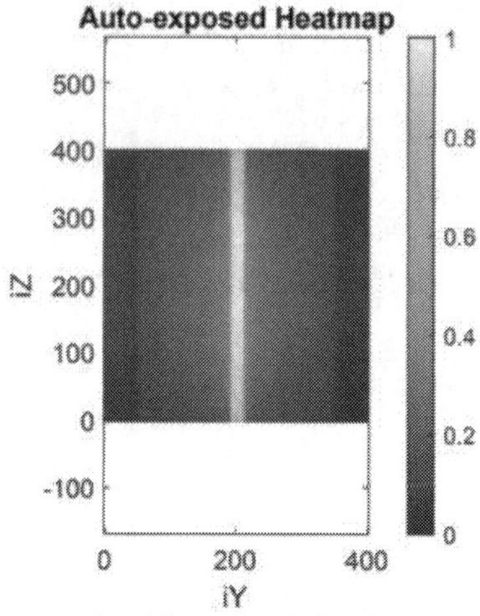

Fig. 6. Results of the vertical defect simulation.

The simulation of the vertical defect clearly shows the area where the defect was created. Since neutrons are not slowed down by any of the mockup's materials, the image is much brighter in the middle than on the sides.

The last simulation was about a burst defect, in which only a certain area of the mockup was damaged. Based on the horizontal simulation, it can be assumed that this defect will also not be clearly visible.

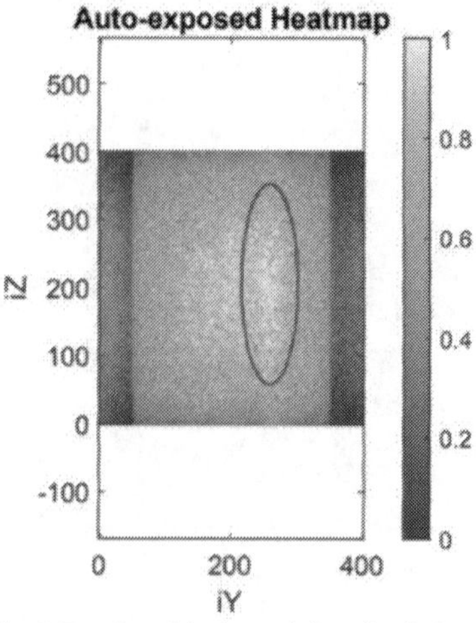

Fig. 7. Results of the burst defect simulation.

As Figure 7 shows, the assumption was partially correct because most of the scintillator brightened similarly to the first simulation. However, when we compare the images from these two simulations (Fig.5. and Fig.7.), we can see the defect on the right side of the image in Figure 7 (red circle). Although the simulated result seems reversed compared to geometry, it's actually in the same position because we are simulating an image based on the light signal coming out of the back of the scintillator.

V. CONCLUSIONS

This paper demonstrates the capability of using the GEANT4 model to simulate neutron interactions in a plastic scintillator, as well as the resulting light output signal and image. The initial simulations provided valuable insight into the expected results of using the surface D-D neutron generator with a plastic scintillator for practical use in the field of fast neutron imaging. However, this setup also revealed several shortcomings when imaging large industrial components with high material density. Because of the density of the mockup material, it was impossible to distinguish large defects from the image when they were covered by the mockup tubes. On the other hand, the defect was clearly visible when the device was rotated correctly. Even the burst defect was partially visible because of the difference in scintillator contrast. Another factor that can affect the visibility of a defect is the method used to process the final image. If we just take the raw image, the defects could be hard to spot. However, if we use a subtraction method, where we subtract the reference image (in our case, the image without the defect) from the image with the defect, the defect could be shown more clearly. This creates opportunities to optimize how parameters, such as the distance between the source and the detector, affect measurements and image resolution. Future studies will recreate this mockup geometry with setup defects using different neutron sources. One example is a radioisotope point source or a D-D neutron generator in the shape of a disk. This would allow us to compare the results and choose the most suitable source for this imaging setup.

ACKNOWLEDGMENT

This study has partially been financially supported by the Scientific Grant Agency of the Ministry of Education of Slovak Republic No. VEGA 1/0518/25 and the Slovak Research Development Agency APVV 22-0304.

REFERENCES

[1] B. Vrban, Š. Čerba, J. Lüley, V. Filová and V. Nečas "Printed circuit heat exchangers and fast neutron radiography". Eur. Phys. J. Special Topics, 2022.

[2] Tak Pui Lou "Compact D-D/D-T Neutron Generators and Their Applications", Dissertation thesis. University of California, Berkeley. Spring 2003

[3] J.T. Cremer, D.L. Williams, C.K. Gary, M.A. Piestrup, D.R. Faber, M.J. Fuller, J.H. Vainionpaa, M. Apodaca, R.H. Pantell, J. Feinstein "Large area imaging of hydrogenous materials using fast neutrons from a DD fusion generator", Nuclear Instruments and Methods, Vol. 675, 51-55, May 2012.

[4] Robert Adams, Robert Zboray, Horst-Michael Prasser "A novel fast-neutron tomography system based on a plastic scintillator array and compact D-D neutron generator", Applied Radiation and Isotopes, Vol. 107, 1-7, January 2016.

[5] J. Park, F.C.E. Teh, M.B. Tsang, K.W. Brown, Z. Chajecki, B. Hong, T. Lokotko, W.G. Lynch, J. Wieske, K. Zhu "neuSIM4: A comprehensive GEANT4 based neutron simulation code", Nuclear Instruments and Methods, Vol. 1065, August 2024.

[6] NEA "Collection and Analysis of Common-cause Failures of Heat Exchangers". ICDE Project Report. April 2013.

Detection of 2.223 MeV γ-rays for Water Identification in Spent Nuclear Fuel Assemblies

Damilola E. Folley[1]*, Brendan Green[1], Kartikey Mathur[1], Stephen Croft[1], Andrew Kennedy[1], and Malcolm J. Joyce[1,2]
[1]Lancaster University, United Kingdom
[2]Hybrid Instruments Ltd., United Kingdom
(*) d.folley@lancaster.ac.uk

***Abstract*— The detection of 2.223 MeV γ-ray emissions resulting from neutron capture on hydrogen, to monitor water presence in spent nuclear fuel (SNF) assemblies is described. Water detection is crucial because its ingress into dry storage systems or other containment environments poses safety risks and potential criticality concerns. This study thus aims to explore the use of the characteristic 2.223 MeV γ-rays for reliable detection of water in SNF assemblies, with a focus on how neutron interactions, such as scattering and absorption, impact γ-ray production and detection in different fuel configurations. This approach will provide a non-invasive and accurate approach for nuclear monitoring and maintaining nuclear safety. To achieve this objective, a Mirion Technologies' GR5021 high-purity germanium (HPGe) detector has been used to measure γ-ray emissions from neutron capture events under controlled conditions. Californium-252 (Cf-252) neutron sources were used to provide a well-characterised neutron field representative of those in SNF environments. Various test configurations were created using high-density polyethylene (HDPE) slabs and 3D-printed metal analogues. The HDPE slabs serve as analogues to simulate the neutron moderation characteristics of water, whereas the metal slabs serve as analogues for fuel-containing materials. Results showed the consistent detection of 2.223 MeV events, with γ-ray counts responding consistently with expectations as material thickness was increased. Overall, the results suggest that the 2.223 MeV γ-ray has potential as a reliable and sensitive indicator for detecting water in SNF assemblies, with significant implications for nuclear safety and fusion reactor environments.**

***Keywords*— neutron capture, hydrogen, water, γ ray, spent nuclear fuel.**

I. Introduction

THE long-term management of spent nuclear fuel (SNF) remains a central priority in the nuclear industry's commitment to safety and environmental responsibility. A key aspect of both wet and dry storage systems is to ensure that containment integrity is maintained, where the unintended presence or absence of water might introduce significant safety anomalies. In dry casks, water ingress poses risks such as corrosion, radiolysis, and most importantly, a decrease in the reactivity margin due to the neutron moderation properties of hydrogen. Conversely, in pool storage systems, confirming the continued presence of water is essential for adequate shielding and effective decay heat removal [1, 2].

Among the various methods for monitoring water in spent nuclear fuel (SNF) environments, passive γ-ray spectroscopy is a non-invasive and highly-sensitive possibility. A particularly distinctive signature is the 2.223 MeV γ-ray emitted when thermal neutrons are captured by hydrogen nuclei [3, 4]. Since neutrons are emitted from SNF via spontaneous fission and (α,n) reactions, this γ-ray line has potential as a direct indicator of hydrogenous materials, such as water. Previous research [3,5,6] has demonstrated the feasibility of this approach for safeguards and verification applications, with studies showing its effectiveness in detecting water signatures in storage pools and intact fuel assemblies.

Despite these advances, much of the existing research has been limited to simulations or homogeneous configurations, with there being relatively little experimental validation in heterogeneous fuel-like systems. This gap is significant because realistic SNF consignments can comprise complex geometries and diverse material compositions, influencing neutron scattering, absorption, and subsequent γ-ray production, significantly [2]. Addressing this challenge is essential for both fission and, potentially, fusion safety applications. Progress in the development of neutron detection techniques for various applications, as well as the methods of γ-ray measurement of hydrogen are well documented [1,7].

In this study, we investigate the detection of 2.223 MeV γ-rays experimentally, as a proxy for water in SNF. Using a high-resolution HPGe detector and a ^{252}Cf neutron source, we explore how different test materials, including high-density polyethylene (HDPE) and 3D-printed metal analogues, impact neutron capture rates and γ-ray emission under controlled conditions.

II. Experimental Method

Experiments were conducted with a ^{252}Cf source stored in 1 m × 1 m × 1 m water. Hydrogenous and metal analogues were used to simulate water dispersed in fuel-containing materials. High-density polyethylene (HDPE), with a density of 0.96 g/cm^3, served as a standard moderator material. The hydrogen atomic number density in the HDPE was 4.13×10^{22} atoms/cm^3. For the metal-based analogues, two different zinc-based materials were studied: zinc resin slabs measuring 15 cm × 21 cm × 6 cm, with a density of 3.60 g/cm^3, comprising 88.2% zinc and 11.8% resin, produced through advanced manufacturing processes. Additionally, commercially available natural zinc sheets (Metal Sheets Ltd, UK), made of 99.5% pure zinc with the remaining 0.5% consisting of Cu, Ti, Al, and other trace elements such as Fe, Cd, Sn, Mn and Mg, were investigated as a target sample. The zinc sheet dimensions were 300 mm × 200 mm × 6 mm.

To detect the capture γ-rays emitted from the sample materials described above, a portable GR5021 high-resolution reverse-electrode germanium (REGe) detector (Mirion Technologies, CANBERRA, Olen) with a crystal size of 90 mm ∅ × 165 mm and a CP5-PLUS-F-RDC-4 cryostat was used. The data acquisition was controlled by Genie 4.0 Software (Mirion Technologies). To ensure accurate energy assignments, the detector was calibrated using a standard range of point sources comprising ^{60}Co, ^{137}Cs, ^{241}Am, ^{22}Na, and ^{152}Eu. The energy resolution of the detector was evaluated as 1.91 keV (FWHM) at 1.3 MeV.

To reduce background radiation, the REGe detector, albeit resistant to neutron damage, was shielded with lead blocks and positioned perpendicular to a water-filled tank housing a ^{252}Cf neutron source with an activity of 3.7 MBq and a neutron emission rate of approximately 4.3×10^5 n/s. For each test configuration, the target samples were placed 20 cm from the detector and positioned at 45° from the source tank to optimise the detector's efficiency in capturing the γ-rays. All measurements were taken for a live time of 54000 s.

(a)

(b)

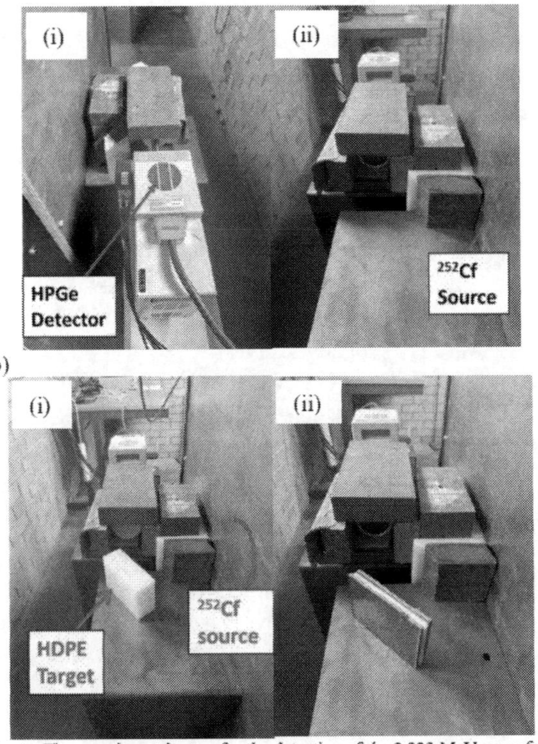

Fig. 1. The experimental setup for the detection of the 2.223 MeV γ-ray from (a) background radiation when the ^{252}Cf source is exposed, (i) with the detector shielded in a lead cave arrangement, (ii) with the end-cap view and source position from the tank, and (b) from the target sample placed at 45° from the source tank to optimise the detector's efficiency in capturing the γ-rays. The HDPE target in (i) is replaced with a zinc resin (ii) sandwiched with a 10 mm-thick HDPE. The acquisition time for each measurement is 15 hours.

III. Results

Fig. 2 presents the γ-ray energy spectra obtained from the experimental measurements. Fig. 2a shows the background energy spectrum from the (n, γ) reaction when the source is exposed. A 2.223 MeV γ-ray peak due to the water in the ^{252}Cf source tank is detected arising from the $^{1}H(n,\gamma)^{2}H$ reaction. Isotopes of ^{56}Fe and ^{57}Fe are observed in the spectrum and are thought to be due to the material composition of the source tank and the steel stand for the target in agreement with previous reports [8]. Other isotopes, e.g., ^{40}K, ^{208}Tl are evident from background radiation in the environment. Fig. 2(b) shows the spectrum measured from a 5 cm-thick HDPE slab when placed in front of the neutron source. A 2.223 MeV γ-ray peak is observed resulting in a peak area and intensity approximately twice that recorded in Fig. 2a without a target material.

(a)

(b)

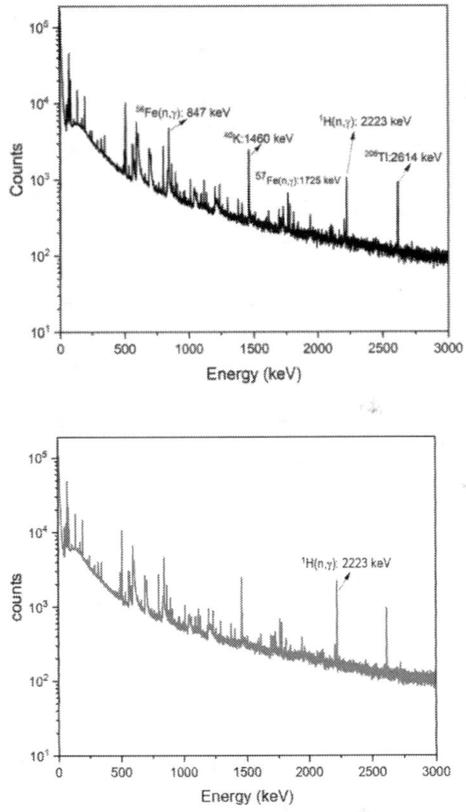

Fig. 2. γ-ray spectra of 2.223 MeV from (a) a background measurement (b) a 5-cm thick, 14 cm X 10 cm HDPE target, following 15 hours neutron exposure.

Fig. 3 shows the spectra produced when the 3-D analogues, that is, Zn resin and Zn sheets, were exposed to the neutron source. Fig. 3a shows the spectrum for the Zn resin which is consistent with that of Fig. 2. However, a reduction in counts is observed for the 2.223 MeV γ-ray peak in comparison with the HDPE spectra. This is as a result of the material composition and geometry of the target. Fig. 3b gives the spectra of the analogues at varying thicknesses. Each of the analogues was moderated with 10 mm of HDPE.

(a)

(b)

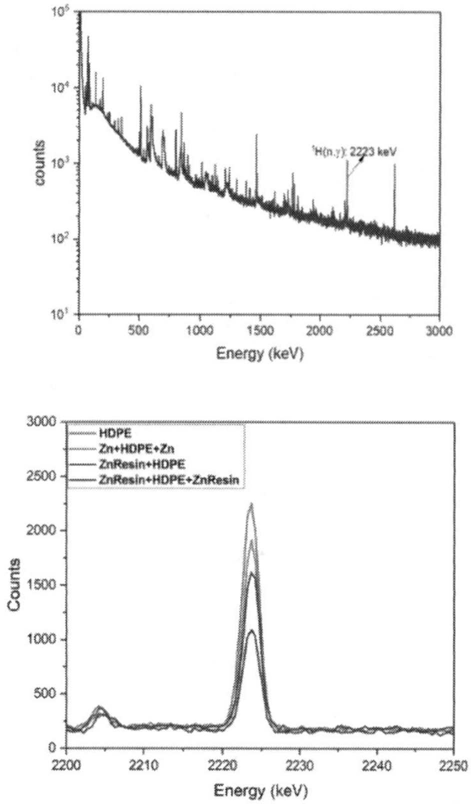

Fig. 3. (a) γ-ray spectra of 2.223 MeV corresponding to 10 mm-thick of Zn resin (105 mm × 150 mm) sandwiched with 10 mm of HDPE (b) 2.223 MeV γ-ray peak of fuel analogues at varying thicknesses. The dimensions of the Zn sheet are 300 mm × 200 mm × 6 mm.

Fig. 4 presents the measured energy spectra of HDPE at various thicknesses. The observed increase in the 2.223 MeV γ-ray peak arises from neutron moderation in hydrogen-rich HDPE, where elastic scattering slows neutrons from the ^{252}Cf source into the thermal range, leading to capture and emission of the 2.223 MeV line. At very large thicknesses, the peak intensity is expected to approach saturation due to self-shielding and neutron leakage.

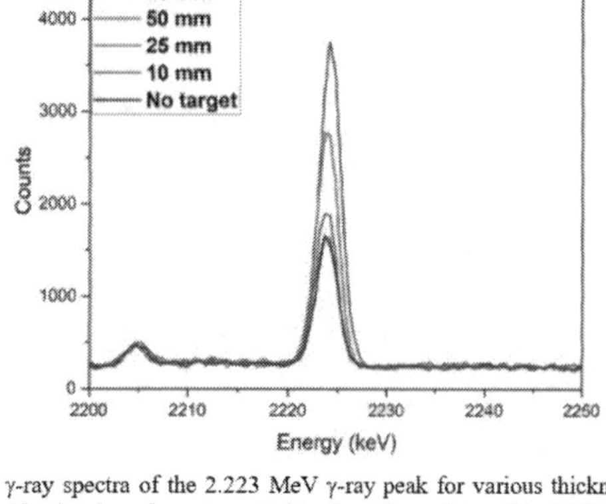

Fig. 4. γ-ray spectra of the 2.223 MeV γ-ray peak for various thicknesses of HDPE. The data have been truncated to show the regions of the 2.223 MeV γ-ray peak of hydrogen.

IV. Conclusions

The detection of the characteristic 2.223 MeV γ-ray from neutron capture of hydrogen in material analogues under consideration for SNF has been demonstrated. Consistent trends were observed across the measurements, confirming the reliability of the experimental approach. The variations in the γ-ray peak intensity with material thickness highlights the sensitivity of the technique to material composition and geometry. These results validate the use of hydrogenous and metal analogues for investigating neutron moderation and γ-ray production processes relevant to SNF scenarios.

Future research will focus on extending this study through simulations of SNF assemblies with varying water content, representing more realistic storage and accident conditions. In addition, systematic investigations will quantify the effect of foam cell density on neutron moderation and γ-ray emission. Experimental benchmarking using analogues representative of SNF will provide validation for computational results. Geometry-specific measurements using a number of bare ^{252}Cf neutron point sources will be conducted to further refine test configurations.

V. Acknowledgment

The support of the Engineering and Physical Sciences Research Council (EPSRC), via the grant EP/X038327/1, and that of the Nuclear Decommissioning Authority is acknowledged.

References

[1] B. Green et al. "Prompt γ-ray techniques based on neutron capture in hydrogen for the detection of water in nuclear legacy environments: a review", *Progress in Nuclear Energy* **189**, 2025: 105905.
[2] R.C. Ewing, "Long-term storage of spent nuclear fuel", *Nature Mater.*, **14**, 2015: 252–257.
[3] C. Binnersley, S. F. Ashley, P. Chard, A. Lansdell, G. O'Brien, P. Shaughnessy, M. J. Joyce, "Quantifying water in spent fuel assemblies with neutrons: A simulation-based approach", *Progress in Nuclear Energy*, **145**, 2022: 104110.
[4] Y. Nauchi et al., "Measurement of spent nuclear fuel burn-up using a new H (n, γ) method", *Nuclear Instruments and Methods in Physics Research Section A: Accelerators, Spectrometers, Detectors and Associated Equipment* **1050**, 2023: 168109.
[5] D. Chernikova, S.F. Naeem, K. Axell, N. Trnjanin, A. Nordlund, "A potential alternative/complement to the traditional thermal neutron based counting in Nuclear Safeguards and Security", *Nuclear Instruments and Methods in Physics Research Section A: Accelerators, Spectrometers, Detectors and Associated Equipment* **810**, 2016: 164-171.
[6] R. F. M. Abou Alo, A. Abdelhady, M. K. Shaat, "Neutronic Characterization for a Pressurized Water Reactor Spent Fuel Assembly", *Nuclear Science and Engineering*, **198**, (5) 2024: 1122-1130.
[7] A. Pietropaolo et al. "Neutron detection techniques from μeV to GeV. *Physics Reports* **875**, 2020: 1-65.
[8] D. E. Folley et al., "Comparative analysis of scintillators for detecting hydrogen capture γ rays", 2024 IEEE Nuclear Science Symposium (NSS), Medical Imaging Conference (MIC) and Room Temperature Semiconductor Detector Conference (RTSD), Tampa, FL, USA, 2024, pp. 1-1, doi: 10.1109/NSS/MIC/RTSD57108.2024.10654994.

Combined peak searching and machine learning algorithm for identification of radionuclides

Aliaksei Hutouski[1,*]
[1]Polimaster Europe, Lithuania
(*) hutouski@polimaster.com

***Abstract*—There are two main kinds of methods of radionuclides identification: peak searching (peak fitting) and machine learning (processing spectrum at once). In this work we present an identification method that combines peak searching and machine learning. As a peak searching routine convolution with the second derivative of Gauss function is used. For the machine learning part different algorithms were tested and MLPClassifier algorithm was selected. This identification method was trained using 540 real spectra and tested with 580 real spectra. As a result, all identifications were correct.**

***Keywords* —radionuclides identification, peaks fitting, machine learning.**

I. Introduction

Identification of radionuclides is an essential problem especially for border security, CBRN defense and field measurements. Since accurately identified natural occurring radioactive material (NORM) or medical radionuclide may prevent a lot of false alarms. And from the other side accurately identified shielded or masked special nuclear material (SNM) or industrial isotope can stop illegal radionuclide trafficking.

Two identification methods are in widespread use today. They are classical methods that process separate peaks (so called "peak-by-peak" methods) and newer methods that include different types of machine learning (ML) or spectrum convolution ("fit-at-once" methods).

Traditional "peak-by-peak" methods consist of several steps:

-peak searching,

-peak fitting,

-comparing acquired peaks to library,

-peak filtering to eliminate pile-up peaks, Compton edges, backscattering peaks and so on,

-and some kind of decision-making sub-algorithm that finalizes the result [1].

"Fit-at-once" methods work with the spectrum as the whole undividable thing using machine learning, neural networks or spectrum convolution with pre-training of the model [2].

Every type of algorithm has its own disadvantages. "Fit-at-once" methods require a very large training dataset of spectra and tend to show instability in identification when spectra are distorted by factors such as shielding or scattering. As for "peak-by-peak": large sensitivity to calibration deviations and necessity of complex decision-making algorithm that has ability to filter false peaks [3]. Detectors with medium resolution also face problems with overlapping peaks, which hinder the algorithm's ability to accurately estimate peak positions. Typical γ-spectrum of Cs^{137} acquired from CsI(Tl) detector presented on Fig. 1.

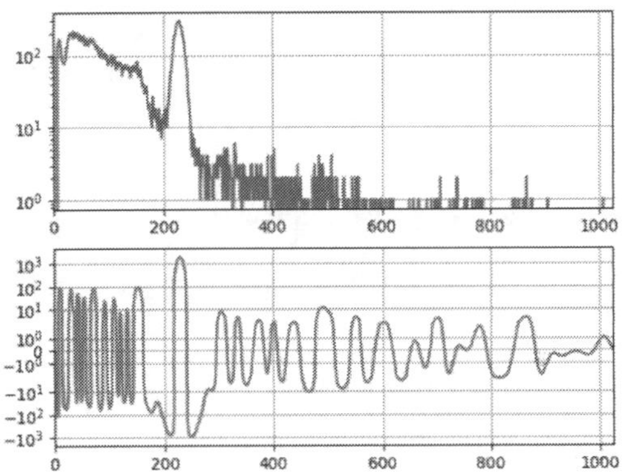

Fig. 1. Spectrum of Cs^{137} and its convolution with second Gauss function derivative

II. Methods

In our work we present the algorithm that is a combination of machine learning and peak searching. The first step of a new algorithm is a peak searching routine, but without peak fitting and filtering steps.

Peak searching was performed using convolution with the second derivative of Gauss function [4]. Then peak positions are compared with the gamma lines library and a peak presence matrix is formed. This matrix is composed of ones and zeros—where a zero indicates the absence of a peak in the spectrum corresponding to the gamma line library, and a one indicates its presence. The library must include every significant gamma line that can be detected and distinguished from others, based on the energy range and resolution of the chosen detector.

And then we use a pretrained ML algorithm as a decision-making tool, giving us an identified radionuclide as the result. Flowchart of the algorithm presented on Fig. 2.

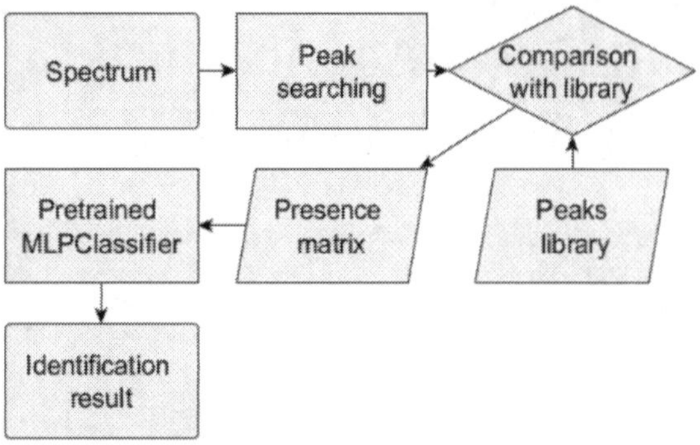

Fig. 2. Algorithm flowchart

Using peak searching as a first step for the ML algorithm can give us more accurate data than the whole unprocessed spectrum, more predictable and controlled behavior of the algorithm and overcome challenges with distorted spectra.

Peak searching routine, spectra and library input/output and peak presence matrix calculation created using Python 3. Machine learning algorithms used were implemented in Python library scikit-learn version 1.5.0 [5].

A comparison of different ML methods was performed with search of the optimal parameters for methods using GridSearchCV. Comparison was made between k-Nearest neighbors (k-NN), decision tree (DT), support vector classification (SVC), logistic regression classifier (LogRegCV) and neural network (MLPClassifier).

For model training and testing, 540 and 580 real spectra were used, respectively. These spectra were acquired using the Polimaster PM1401-K3 radioisotope identification device, equipped with a CsI(Tl) scintillation crystal measuring 2.5 cm in height and 2.5 cm in diameter, coupled with a SiPM. These spectra include NORM, SNM, medical and industrial radionuclides. A list of radionuclides and their categories is presented in Table I. SNM spectra include spectra from bare and shielded materials (10 mm steel or 70 mm of PMMA). Industrial spectra include bare, shielded (5 mm of steel) materials and high scattering, which was created using lead blocks. Dose equivalent rate during spectra acquisition is equal to 0.5 μSv/h and the acquisition time is 2 minutes according to ANSI N42.34 standard [6].

TABLE I
CATEGORIES OF SOURCES USED

NORM	SNM	Medical	Industrial
K^{40}	Pu^{239}	I^{131}	Am^{241}
U^{238}	U^{235}	Tc^{99m}	Ba^{133}
Ra^{226}		Tl^{201}	Co^{57}
Th^{228}			Co^{60}
			Eu^{152}
			Na^{22}

III. RESULTS

At the first step, all spectra were processed using a peak-searching algorithm to generate a peak presence matrix. Next, the spectra were divided into training and testing sets. Spectra of industrial radionuclides with shielding and scattering were included only in the testing set, while spectra of special nuclear materials (SNM) with shielding were included in both the training and testing sets. We performed an estimation of ML models by searching for the best model training parameters using the scikit-learn subroutine GridSearchCV.

The following parameters were found to be optimal:

- k-NN: algorithm='brute', leaf_size=10, n_neighbors=4, p=1.0, weights='distance'
- DT: criterion='gini', splitter='best', min_samples_split='2'
- SVC: degree=2, kernel='linear'
- LogRegCv: Cs=5, dual=True, n_jobs=-1, solver='liblinear'
- MLPClassifier: hidden_layer_sizes=100, activation='relu', solver='adam', alpha=0.0001.

In order to choose the best model, we compared them with the following rules: NORM identified as SNM, medical or industrial (false positive result) is better than SNM, medical or industrial identified as NORM (false negative result). If false identification has occurred between the radionuclides of the one group (for example Co^{57} predicted while Eu^{152} is true) we also consider it as a false positive result. The comparison of ML models presented in Table II.

TABLE II
ML MODELS COMPARISON

Model	False identifications	False positive	False negative
k-NN	6	6	0
DT	4	4	0
SVC	6	6	0
LogRegCV	5	5	0
MLPClassifier	0	0	0

Confusion matrix for the model with the best results (MLPClassifier) is presented on Fig.3. Confusion matrices for different models k-NN, DT, SVC and LogRegCV are presented on Fig. 4.

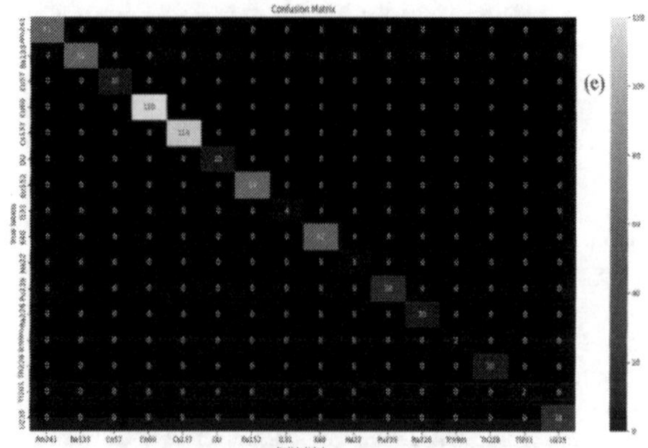

Fig. 3. Confusion matrix for MLPClassifier model with zero false results

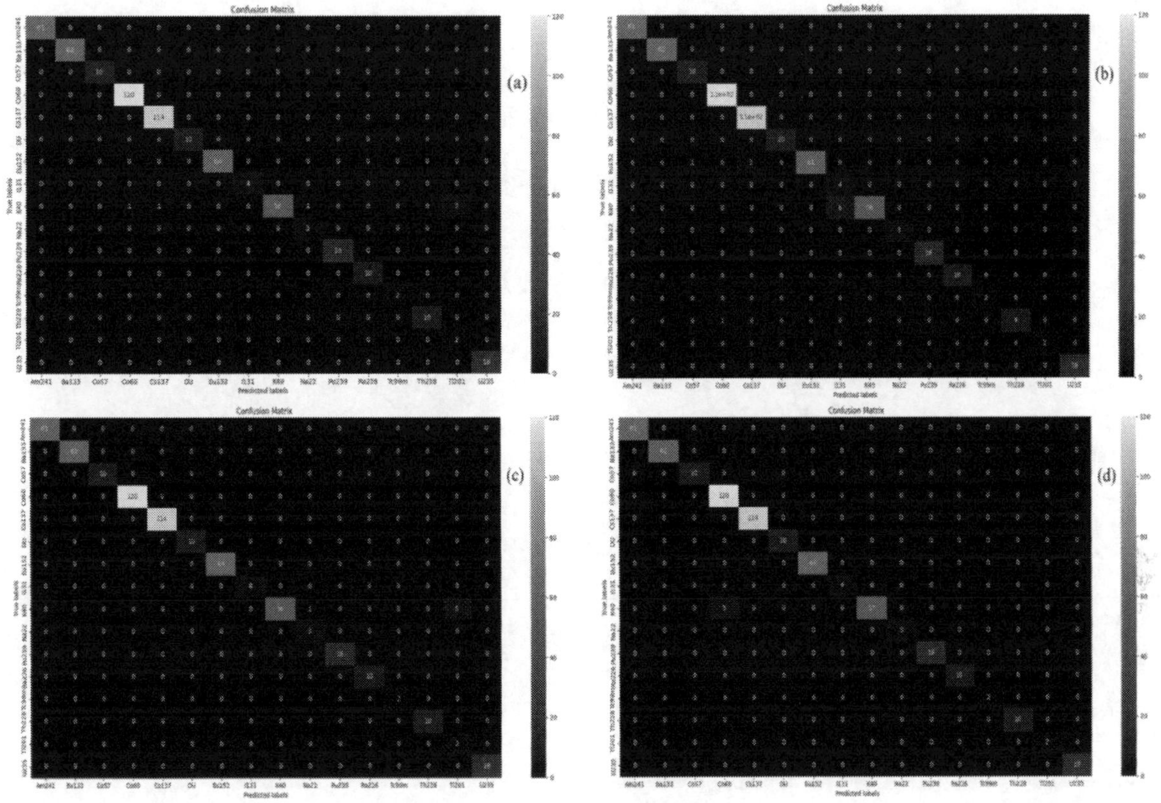

Fig. 4. Confusion matrices for different models. (a) k-NN, (b) DT, (c) SVC and (d) LogRegCV

IV. Conclusions

The combined algorithm demonstrated great performance on different real spectra from CsI(Tl) scintillation detector acquired in circumstances according to ANSI standard N42.34. Main advantages of our method are:

- following algorithm is stable in cases of distorted spectra (spectra with shielding of large scattering),
- we don't need to create complicated "if-else"-based decision-making system
- there is no need to subtract natural background before identification.

For the machine learning component, the MLPClassifier was selected, as it demonstrated zero false identifications on the testing spectra set.

However, it is still unable to identify a mixture of radionuclides because of the nature of ML method. To solve this problem our next step will be to reform the training spectra set by adding all possible combinations of radionuclides.

References

[1] G. R. Gilmore, "Computer analysis of gamma-ray spectra" in Practical gamma-ray spectrometry, 2nd ed., UK: John Wiley & Sons, 2008, pp. 183-203.

[2] M. Gomez-Fernandez, "Isotope identification using deep learning: An explanation," Nuclear Inst. And Methods in Physics Research, A, vol. 988, pp. 1-7, Dec. 2021, Art no. 164925, doi: https://doi.org/10.1016/j.nima.2020.164925.

[3] T. Burr, "Radio-Isotope Identification Algorithms for NaI γ Spectra", Algorithms, vol. 2, pp. 339-360, 2009, doi:10.3390/a2010339.

[4] A. Likar, "A peak-search method based on spectrum convolution", J. Phys. D: Appl. Phys., vol. 36, pp. 1903-1909, 2003, doi: 10.1088/0022-3727/36/15/323.

[5] J. Hao, "Machine Learning Made Easy: A Review of Scikit-learn Package in Python Programming Language", Journal of Educational and Behavioral Statistics, vol. 44, pp. 348-361, 2019, doi: 10.3102/1076998619832248.

[6] American National Standard Performance Criteria for Handheld Instruments for the Detection and Identification of Radionuclides, ANSI Standard N42.34, 2021.

Comparative analysis of automatic radionuclide identification in γ-ray spectrometry: machine learning versus statistical approaches

Dinh Triem Phan[1,*], Jérôme Bobin[2], Cheick Thiam[1] and Christophe Bobin[1]
[1]Université Paris-Saclay, List, CEA, Palaiseau, France
[2]Université Paris-Saclay, Irfu, CEA, Gif-sur-Yvette, France
(*) dinh-triem.phan@cea.fr

***Abstract*—Gamma-ray spectrometry is a widely used technique for identifying and quantifying γ-emitting radionuclides in many nuclear applications. Currently, there is a growing trend to address the problem of automatic identification of radionuclides by implementing machine learning (ML) approaches such as multilayer perceptrons (MLP) or convolutional neural networks (CNN). Alongside these ML methods, the statistical method based on full-spectrum analysis with Poisson likelihood yields reliable results. However, due to the lack of a common benchmark, a comparison of these methods has not yet been conducted. In this work, we introduce a benchmark to evaluate and compare these methods under three scenarios: (1) known spectral signatures, (2) spectra deformed by physical phenomena and (3) gain shift. A large dataset of 200000 simulated spectra was generated for each scenario, spanning a broad range of radionuclide combinations and mixing ratios. This work utilizes a dictionary of nine radionuclide spectral signatures generated with the Monte Carlo code Geant4 for a 3"×3" NaI(Tl) detector, combined with an experimental natural background. For the ML method, we trained a dedicated CNN model for each scenario, carefully optimized its hyperparameters, and adjusted its classification threshold. The results show that the statistical method consistently outperforms the ML approach across all scenarios, particularly under low-statistics conditions. Furthermore, the statistical method maintains a false positive rate close to the predefined level.**

***Keywords*—Gamma-ray spectrometry, Automatic identification, Machine Learning, Spectral unmixing, Convolutional neural networks**

I. Introduction

Gamma-ray spectrometry is a widely used technique for identifying and quantifying γ-emitting radionuclides, with widespread applications such as nuclear security, environmental monitoring and nuclear industry. Certain specialized tasks, for example, rapidly detecting illicit nuclear materials or performing *in situ* analysis after a nuclear incident, require methods that can identify radionuclides quickly and accurately while remaining accessible to non-experts, even under difficult conditions like low-count spectra. Traditional peak-fitting methods [1], which typically assume Gaussian noise, generally require significant user expertise and often struggle in these difficult scenarios, especially at low-statistics scenarios and spectra containing multiple radionuclides.

To overcome these challenges, machine learning (ML) approaches have recently gained attention as promising tools to provide automated, end-to-end solutions. Deep learning techniques such as convolutional neural networks (CNN) and multilayer perceptrons (MLP) [2]–[10] leverage large datasets of synthetic radionuclide mixtures to train models that can automatically identify radionuclides from γ-ray spectra. Once trained, these models have demonstrated strong performance across a range of detector types and measurement scenarios, delivering accurate and rapid radionuclide detection.

An alternative, more classical approach involves statistical methods based on full-spectrum Poisson-likelihood modeling. One such technique is full-spectrum unmixing, which treats an observed spectrum $y \in \mathbb{R}^M$ as a Poisson distribution of the theoretical spectrum Xa:

$$y \sim \mathcal{P}(Xa), \tag{1}$$

where $X = [X_{Bkg}, X_2, .., X_N] \in \mathbb{R}^{M \times N}$ contains the natural background (Bkg) and the normalized spectral signatures representing the detector's characteristic response to γ-photon emissions from $N-1$ radionuclides, and the vector $a \in \mathbb{R}^N$ represents the counting associated with each radionuclide. By applying the maximum likelihood estimation (MLE), this method estimates the counting of all radionuclides. When applied to NaI(Tl) and HPGe detectors, this approach delivers more accurate activity estimates and achieves lower decision thresholds compared to traditional peak-fitting techniques [11]–[14].

This work compares ML methods and statistical full-spectrum unmixing for radionuclide identification across three different test scenarios: (1) ideal conditions with known spectral signatures, (2) spectral deformation due to effects such as Compton scattering and attenuation, and (3) gain shift resulting from factors such as temperature variations. The evaluation is conducted on a simulated dataset of 200000 spectra for each scenario, containing nine radionuclides and Bkg. To ensure a fair comparison, we optimize the CNN architecture and hyperparameters and calibrate its classification thresholds.

The rest of this article is organized as follows:

- Section 2 presents the simulated dataset and the metrics used to evaluate identification performance.
- Section 3 describes the end-to-end machine learning and full-spectrum statistical unmixing approaches.

- Section 4 presents the results of these two methods across the three test scenarios.

II. Datasets and evaluation metrics

A. Datasets

Radionuclide identification can be performed with various detectors, including CdTe, LaBr3, NaI(Tl), and HPGe. Among these, the NaI(Tl) detectors are widely adopted because of their affordability, high detection efficiency which allows fast measurement capability. Therefore, this work uses a 3"×3" NaI(Tl) detector to evaluate ML and statistical unmixing techniques. These approaches can also be adapted for other detector types with different energy resolutions.

Developing an ML model requires a large dataset that reflects the main characteristics of the data. For radionuclide identification, this means creating diverse γ-ray spectra that include various radionuclides and their different contributions. Since acquiring such data through experiments is often slow and challenging, simulations using Monte Carlo codes like Geant4 [15] or MCNP [16] are commonly used. Synthetic spectra are produced by sampling from a Poisson distribution applied to weighted sums of normalized spectral signatures, with randomized choices for the number of radionuclides and their contributions.

Three datasets were created to address different scenarios: (1) ideal conditions with known spectral signatures, (2) spectral deformation due to physical phenomena and (3) gain shift.

a) **Scenario 1: Known spectral signatures:** This scenario represents the ideal case where the spectral signatures are known. The radionuclide library consists of nine radionuclides covering an energy range from 20 keV to 2 MeV: ^{57}Co, ^{60}Co, ^{99m}Tc, ^{123}I, ^{131}I, ^{133}Ba, ^{137}Cs, ^{152}Eu and ^{241}Am. These radionuclides are well-known γ-emitters in spectroscopy-based portal monitors (e.g., ANSI N42.38-2015 [17]) and are frequently studied in the literature. Spectral signatures were simulated using Geant4 [15], with the point source placed 20 cm from a 3"×3" NaI(Tl) detector. Each spectral signature contains 1024 channels with 2 keV per channel binning and a low-energy cutoff at 20 keV.

b) **Scenario 2: Deformed spectral signatures:** This scenario involves a source enclosed within a steel sphere of varying thickness, which deforms the spectral signatures through physical effects like attenuation and Compton scattering. The dataset, based on simulations from [18], covers sphere thicknesses from 0.001 mm up to 30 mm. For each radionuclide, 96 spectral signatures were generated by Geant4 for the NaI(Tl) detector. Due to the added complexity of spectral variability, the radionuclide library was shortened from nine to eight by eliminating ^{131}I. Fig. 1 displays the variation in ^{137}Cs spectral signatures with sphere thickness.

c) **Scenario 3: Gain shift:** This phenomenon, which can be caused by temperature variation in scintillation detectors, is another factor that can modify spectral signatures. The gain shift has been examined in recent research for its impact on automated radionuclide detection [2], [6], [7], [10]. For this scenario, the same set of radionuclides and spectral signatures from the known-signature case were used. The gain shift is modeled by adjusting the energies with the formula: $e_s = e_r \times (1 - \alpha)$ where e_s is the shifted energy, e_r is the reference energy and α represents the gain shift factor. An example illustrating the changes in the ^{137}Cs spectrum for various gain shift values is provided in Fig. 1.

For each of the three scenarios, synthetic datasets comprising 200000 simulated spectra were generated. These spectra were produced by applying a Poisson distribution to a linear combination of spectral signatures. Total counting values were drawn from a log-uniform distribution between 200 and 100000. The number of radionuclides present in each spectrum was randomly chosen from zero to four, while Bkg was always included, contributing at least 10% of the total counting. The relative contributions of radionuclides, expressed as mixing weights $z = a/\sum(a)$, were randomly assigned. The complete dataset was divided into training, validation, and testing subsets in proportions of 64%, 16% and 20%, respectively. In Scenario 2, steel sphere thicknesses were randomly chosen from 96 values spanning 0.001 mm to 30 mm, while in Scenario 3, the shift factor was uniformly sampled between -10% and 10%.

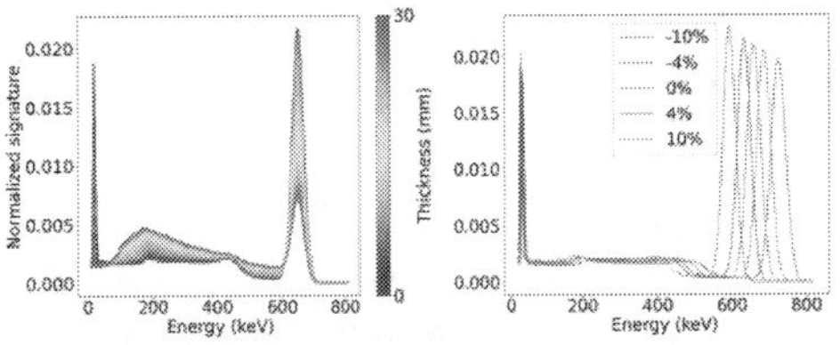

Fig. 1: These figures display ^{137}Cs spectral signatures: the left (Scenario 2) shows their change with steel thickness, and the right (Scenario 3) illustrates their variation with the gain shift.

B. Evaluation metrics

As radionuclide identification is a multi-label task (since multiple radionuclides can appear together in a single γ-ray spectrum), we assess the identification performance using these key metrics:

- False positive rate (FPR) across all radionuclides: the percentage of spectra where one or more absent radionuclides are predicted to be present.
- False negative rate (FNR) across all radionuclides: the percentage of spectra where one or more radionuclides present are not detected.
- Perfect prediction rate (PPR): the percentage of spectra where the model makes completely accurate identifications, with neither false positives nor false negatives.
- Recall per radionuclide (probability of detection): the percentage of present radionuclides that are correctly identified.

III. Automatic identification methods in γ-ray spectrometry

A. Machine learning method

Radionuclide identification in γ-ray spectrometry can be formulated as a multi-label classification since more than one radionuclide may contribute to a given spectrum. This differs from a multi-class classification, which restricts each sample to a single label. A standard approach in γ-ray spectrometry is to train the model with the average binary cross-entropy (BCE) across all radionuclides as the loss, using a sigmoid activation in the output layer. This design enables the network to consider all radionuclides at once and to provide a probability value between 0 and 1 for each of them, which can then be used to decide whether or not a particular radionuclide is present. This configuration is frequently applied in recent γ-ray spectrometry research [5]–[7].

In previous studies in γ-ray spectrometry, traditional ML techniques have often been applied. More recently, deep learning techniques have gained considerable interest. Methods such as multilayer perceptrons (MLP) [4], [9], [10] and convolutional neural networks (CNN) [2], [3], [5]–[8] are now commonly applied. Among these, CNN has demonstrated superior accuracy for radionuclide identification, which is why a CNN-based approach is adopted in this study.

The CNN model consists of multiple convolutional layers followed by fully connected (dense) layers. This structure is standard for classification tasks and is frequently used in γ-ray spectrometry [3], [5]–[8]. The overall design is shown in Fig. 2. Each convolutional block contains a 1D convolutional layer to identify key features, a max-pooling layer to decrease the spatial size, batch normalization to improve training speed and stability, and a ReLU activation. After the convolutional layers, the output is flattened and passed through multiple fully connected layers. Dropout is added after each dense layer to reduce overfitting.

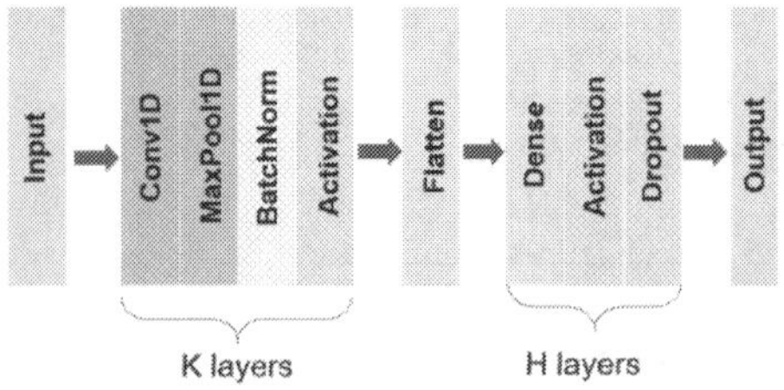

Fig. 2: The CNN architecture comprises multiple convolutional layers, followed by several fully connected layers.

The performance of CNN model depends strongly on their hyperparameters, including the number of convolutional and dense layers, filter sizes, learning rate and more. Selecting these parameters carefully is crucial for obtaining the optimal results. In ML, this optimization, referred to as hyperparameter fine-tuning, generally requires the exploration of several combinations in order to find the most efficient configuration. To ensure optimal performance and a fair comparison with the statistical method, hyperparameters were fine-tuned using Ray Tune [19] on the validation set described in Section II-A. Further implementation and training details can be found in the GammaBench GitHub repository.

The threshold value applied to the neural network's output plays a crucial role in the performance of the ML models. In most prior studies, this threshold is set to 0.5 [2]–[4], [6], [7], meaning that the output of every radionuclide above this level is predicted as present. To ensure a fair comparison with the statistical method introduced in the next section—which is based on a specified false positive rate—we adjust this threshold using the validation set to closely match the desired FPR.

B. Spectral unmixing method

In γ-ray spectrometry, the spectral unmixing method seeks to decompose a measured spectrum y into contributions from different radionuclides and Bkg. Throughout this work, Bkg is assumed to be known. As shown in Eq.1, the measured spectrum y is modeled as a Poisson distribution of Xa. Accordingly, the cost function is derived as the negative log-likelihood of this Poisson model:

$$L(y, X, a) = \sum_{m=1}^{M} ((Xa)_m - y_m log((Xa)_m)) \tag{2}$$

When the spectral signatures are not fixed (as in Scenarios 2 and 3), the matrix X can be expressed as a nonlinear function $f(\theta)$, where the parameter θ corresponds to physical factors such as the steel thickness in Scenario 2 or the shift factor in Scenario 3. Further details are provided in [18].

This method aims to estimate the counting vector a from the observed spectrum y by solving the optimization problem:

$$\hat{a}, \hat{\theta} = \underset{a \geq 0, \theta}{\operatorname{argmin}} L(y, f(\theta), a) \tag{3}$$

which can be addressed using algorithms such as NNPU [20] or SEMSUN [18].

The resulting solution can be used for radionuclide identification by setting a threshold on the estimated counting. However, directly minimizing the cost function in Eq.2 often leads to overfitting, which can cause the estimated counting for all radionuclides to be underestimated and may result in false positive identifications.

To address these challenges, additional regularization is required. A widely used strategy is to impose sparsity on the counting vector a. Within this spectral unmixing framework, the identification problem can then be reformulated as the following optimization task:

$$\hat{a}, \hat{\theta} = \underset{a \geq 0, \theta}{\operatorname{argmin}} L(y, f(\theta), a) \text{ subject to } ||a||_0 = K \tag{4}$$

where $||a||_0$ denotes the number of non-zero entries in a, which constrains the total number of identified radionuclides to no more than K.

To tackle this optimization, dedicated algorithms have been proposed. Specifically, P-OMP [20] addresses the fixed-signature case, while MoSeVa [21] deals with scenarios involving spectral variability.

These algorithms implicitly take sparsity into account by employing a greedy model selection procedure that tests models of progressively higher dimensionality. The process is controlled by a statistical criterion that determines when to stop. Specifically, using the likelihood ratio test (LRT) [22], this approach can correctly identify radionuclides while controlling the false positive rate (see [20] for more details).

IV. Results

A. Scenario 1: Known spectral signatures

Table I summarizes the results obtained with the CNN model and the spectral unmixing approach. Both methods produce FPR close to the expected 1%, and the unmixing method outperforms the CNN model on the other metrics, yielding a lower FNR and a higher PPR. Fig. 3 provides a more detailed view of FNR through the recall (1-FNR) for each radionuclide as a function of its total counting. Both methods achieve accurate identification when the number of counts is sufficiently high (above 1000 per radionuclide). However, under low-count conditions, the spectral unmixing method achieves better identification performance thanks to its incorporation of underlying physical properties.

The FPR reported above is computed across the entire test set. Although both methods achieve FPR close to the expected value on average, these rates can vary depending on the specific combination of radionuclides present. For example, Fig. 4 shows the variation of the FPR with respect to total counting and the number of radionuclides present. The ML method tends to keep the average FPR close to the expected level overall, but deviations may appear when the total counting or the number of radionuclides changes. In contrast, the spectral unmixing approach, based on statistical hypothesis testing, reliably keeps the FPR within the expected range.

TABLE I: Performance comparison between the ML method (CNN) and the spectral unmixing approach (P-OMP) on the test dataset for Scenario 1.

	PPR	FPR	FNR
CNN	95.53	0.94	3.60
Unmixing	97.19	0.98	1.88

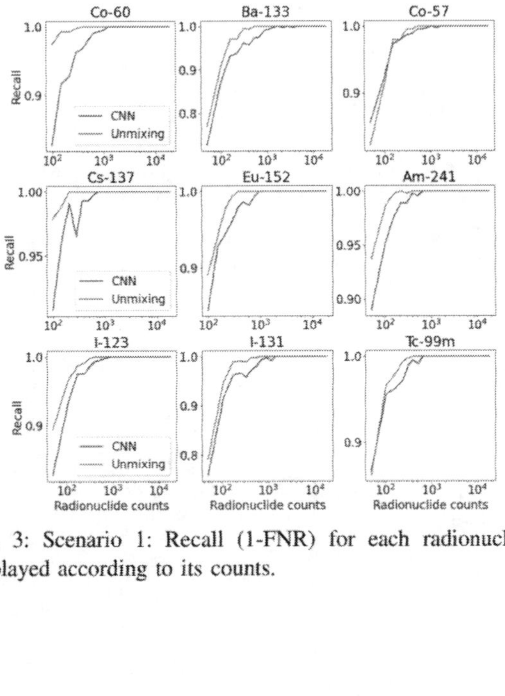

Fig. 3: Scenario 1: Recall (1-FNR) for each radionuclide displayed according to its counts.

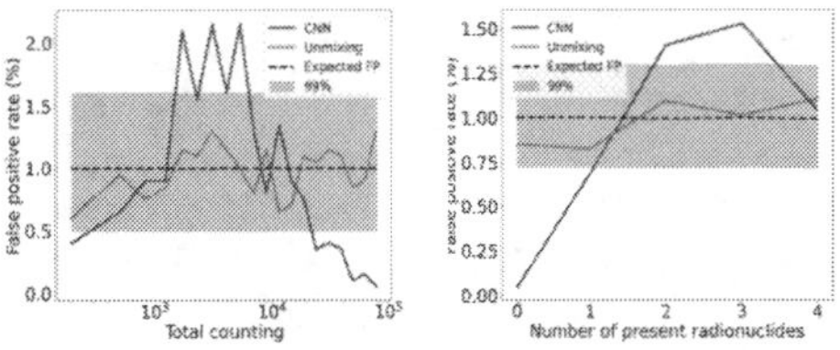

Fig. 4: Scenario 1: FPR displayed according to total counting (left) and number of present radionuclides (right).

B. Scenario 2: Spectral deformation

The first numerical experiment assesses the robustness of ML and spectral unmixing under spectral deformation by applying known (but inaccurate) spectral signatures from Scenario 1 to data with spectral deformation caused by varying steel thickness (from 0.001 mm to 1 mm). As shown in Table II, the performance of both methods decreases due to signature mismatch. The spectral unmixing method is more affected, often wrongly attributing parts of the spectrum to absent radionuclides, resulting in a higher FPR. Conversely, the ML method, which focuses more on peak positions, is less affected by inaccuracies in spectral signatures.

TABLE II: Scenario 2: Identification results under spectral deformation using known but inaccurate spectral signatures, with steel thickness varying between 0.001 mm and 1 mm. The CNN model used here is the same as in Scenario 1, while the spectral unmixing algorithm applied is P-OMP.

	PPR	FPR	FNR
CNN	85.04	4.53	11.02
Unmixing	82.82	12.22	5.86

To improve the robustness of radionuclide identification, it is essential to account for spectral deformation. Accordingly, new CNN models were trained on the dataset detailed in Section II-A, which includes spectral deformation. In the spectral unmixing approach, the MoSeVa algorithm replaces P-OMP to better address this spectral variability. Results shown in Table III indicate that the spectral unmixing approach outperforms the ML models, achieving lower FNR and higher PPR. Consistent with previous observations, the unmixing method demonstrates stronger detection capabilities, especially under low-count conditions (Fig. 5), while keeping the FPR close to the expected value (Fig. 6).

TABLE III: Scenario 2: Identification performance in the case of spectral variability with unknown spectral signatures. Steel thickness values range from 0.001 mm to 30 mm. The CNN model used was newly trained on this dataset, and the spectral unmixing method applied is MoSeVa.

	PPR	FPR	FNR
CNN	90.68	0.92	8.62
Unmixing	93.55	0.89	5.70

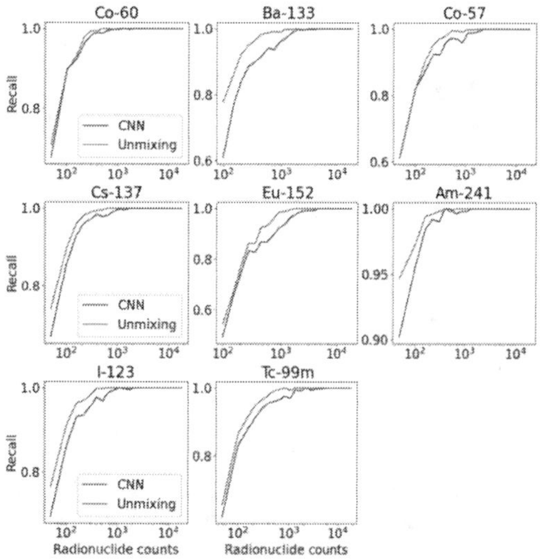

Fig. 5: Scenario 2: Recall (1-FNR) for each radionuclide displayed according to its counts.

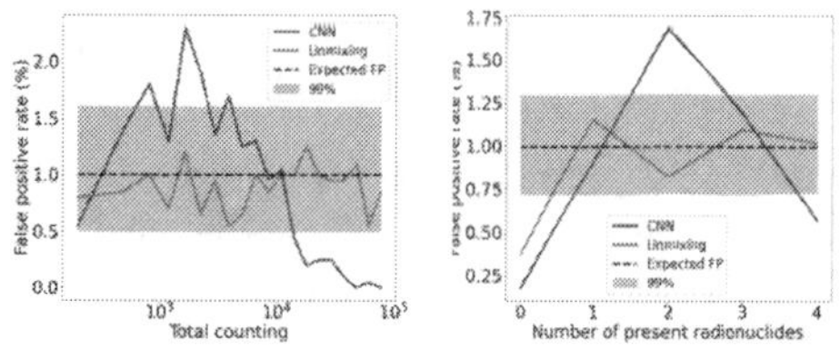

Fig. 6: Scenario 2: FPR displayed according to total counting (left) and number of present radionuclides (right).

C. Scenario 3: Gain shift

Similar to Scenario 2, both methods were first evaluated using known (but inaccurate) spectral signatures on the gain shift dataset. As shown in Table IV, the performance of all methods drops significantly, with the spectral unmixing method being more affected—consistent with observations from the previous scenario.

TABLE IV: Scenario 3: Identification performance in the case of gain shift with known incorrect spectral signatures. The shift factor is between -2% and 2%. The CNN model used here is the same as in Scenario 1, while the spectral unmixing algorithm applied is P-OMP.

	PPR	FPR	FNR
CNN	85.52	10.64	4.46
Unmixing	67.40	31.10	2.22

Subsequently, the gain shift effect is taken into account for both methods, following the same approach as in Scenario 2. The identification results, summarized in Table V, confirm the consistent finding across all scenarios: the spectral unmixing method outperforms the ML approach in identification.

TABLE V: Scenario 3: Identification performance in the case of gain shift with unknown spectral signatures. The shift factor is between -10% and 10%.

	PPR	FPR	FNR
CNN	93.75	1.11	5.35
Unmixing	96.59	0.90	2.60

V. Conclusion

This study presents a comparison of automatic identification between end-to-end machine learning and statistical unmixing methods in γ-ray spectrometry. The comparison covers three scenarios: (1) known spectral signatures, (2) spectral deformation due to physical effects and (3) gain shift. The analysis uses a library of spectral signatures simulated with Geant4 for nine radionuclides and an experimental natural background.

The results demonstrate that the statistical method consistently achieves better performance for all comparison metrics than the ML approach in all three scenarios while keeping the false-positive rate close to the expected value. Nevertheless, the effectiveness of the statistical approach can decline considerably if the spectral signatures are not accurately modeled. Thus, this full-spectrum method is best suited to situations where spectral signatures are known or can be accurately modeled. In contrast, when measurement conditions are uncertain or too complex to model, end-to-end machine learning provides a more appropriate alternative.

References

[1] G. Gilmore, *Practical gamma-ray spectroscopy*. John Wiley & Sons, 2008.

[2] D. Liang, P. Gong, X. Tang, P. Wang, L. Gao, Z. Wang, and R. Zhang, "Rapid nuclide identification algorithm based on convolutional neural network," *Annals of Nuclear Energy*, vol. 133, pp. 483–490, 2019.

[3] G. Daniel, F. Ceraudo, O. Limousin, D. Maier, and A. Meuris, "Automatic and real-time identification of radionuclides in gamma-ray spectra: a new method based on convolutional neural network trained with synthetic data set," *IEEE Transactions on Nuclear Science*, vol. 67, no. 4, pp. 644–653, 2020.

[4] J. Kim, K. Park, and G. Cho, "Multi-radioisotope identification algorithm using an artificial neural network for plastic gamma spectra," *Applied Radiation and Isotopes*, vol. 147, pp. 83–90, 2019.

[5] B. T. Koo, H. C. Lee, K. Bae, Y. Kim, J. Jung, C. S. Park, H.-S. Kim, and C. H. Min, "Development of a radionuclide identification algorithm based on a convolutional neural network for radiation portal monitoring system," *Radiation Physics and Chemistry*, vol. 180, p. 109300, 2021.

[6] A. N. Turner, C. Wheldon, T. K. Wheldon, M. R. Gilbert, L. W. Packer, J. Burns, and M. Freer, "Convolutional neural networks for challenges in automated nuclide identification," *Sensors*, vol. 21, no. 15, p. 5238, 2021.

[7] W. Kim, K. Ko, J. Park, S. Lee, H. Yun, and G. Cho, "Deep learning-based gamma spectroscopic analysis considering multiple variables for in situ applications," *Radiation Physics and Chemistry*, vol. 226, p. 112261, 2025.

[8] M. Kamuda, J. Zhao, and K. Huff, "A comparison of machine learning methods for automated gamma-ray spectroscopy," *Nuclear Instruments and Methods in Physics Research Section A: Accelerators, Spectrometers, Detectors and Associated Equipment*, vol. 954, p. 161385, 2020.

[9] M. Kamuda, J. Stinnett, and C. Sullivan, "Automated isotope identification algorithm using artificial neural networks," *IEEE Transactions on Nuclear Science*, vol. 64, no. 7, pp. 1858–1864, 2017.

[10] M. Kamuda and C. J. Sullivan, "An automated isotope identification and quantification algorithm for isotope mixtures in low-resolution gamma-ray spectra," *Radiation Physics and Chemistry*, vol. 155, pp. 281–286, 2019.

[11] J. Xu, J. Bobin, A. de Vismes Ott, and C. Bobin, "Sparse spectral unmixing for activity estimation in γ-ray spectrometry applied to environmental measurements," *Applied Radiation and Isotopes*, vol. 156, p. 108903, 2020.
[12] P. Malfrait, J. Bobin, and A. de Vismes Ott, "Spectral unmixing of multi-temporal data in gamma-ray spectrometry," *Nuclear Instruments and Methods in Physics Research Section A: Accelerators, Spectrometers, Detectors and Associated Equipment*, vol. 1045, p. 167547, 2023.
[13] C. Mano, C. Chapelle, A. D. M. Kabakian, P. Gross, H. Paradis, O. Delaune, and L. Patryl, "Algorithm development for low level radioxenon 2D spectra analysis: A first case of study using spectral unmixing for a β-γ detector," *Applied Radiation and Isotopes*, vol. 203, p. 111079, 2024.
[14] P. Malfrait, J. Bobin, and A. de Vismes Ott, "Online spectral unmixing in gamma-ray spectrometry," *Applied Radiation and Isotopes*, vol. 201, p. 111011, 2023.
[15] S. Agostinelli, J. Allison, K. a. Amako, J. Apostolakis, H. Araujo, P. Arce, M. Asai, D. Axen, S. Banerjee, G. Barrand *et al.*, "GEANT4—a simulation toolkit," *Nuclear instruments and methods in physics research section A: Accelerators, Spectrometers, Detectors and Associated Equipment*, vol. 506, no. 3, pp. 250–303, 2003.
[16] R. Forster and T. Godfrey, "MCNP-a general Monte Carlo code for neutron and photon transport," in *Monte-Carlo Methods and Applications in Neutronics, Photonics and Statistical Physics: Proceedings of the Joint Los Alamos National Laboratory-Commissariat à l'Energie Atomique Meeting Held at Cadarache Castle, Provence, France April 22–26, 1985*. Springer, 2006, pp. 33–55.
[17] "American national standard for performance criteria for spectroscopy-based portal monitors used in homeland security," *ANSI N42.38-2015 (Revision of ANSI N42.38-2006)*, pp. 1–59, 2016.
[18] D. T. Phan, J. Bobin, C. Thiam, and C. Bobin, "A hybrid machine learning unmixing method for automatic analysis of γ-spectra with spectral variability," *Nuclear Instruments and Methods in Physics Research Section A: Accelerators, Spectrometers, Detectors and Associated Equipment*, vol. 1060, p. 169028, 2024.
[19] R. Liaw, E. Liang, R. Nishihara, P. Moritz, J. E. Gonzalez, and I. Stoica, "Tune: A research platform for distributed model selection and training," *arXiv preprint arXiv:1807.05118*, 2018.
[20] R. André, C. Bobin, J. Bobin, J. Xu, and A. de Vismes Ott, "Metrological approach of γ-emitting radionuclides identification at low statistics: application of sparse spectral unmixing to scintillation detectors," *Metrologia*, vol. 58, no. 1, p. 015011, 2021.
[21] D. T. Phan, J. Bobin, C. Thiam, and C. Bobin, "Automatic identification and quantification of γ-emitting radionuclides with spectral variability using a hybrid machine learning unmixing method," *Radiation Physics and Chemistry*, vol. 232, p. 112654, 2025.
[22] S. S. Wilks, "The large-sample distribution of the likelihood ratio for testing composite hypotheses," *The annals of mathematical statistics*, vol. 9, no. 1, pp. 60–62, 1938.

Augmented X-rays: new functionalities for High energy X-rays Cargo scanner

Serge Maitrejean[1], Thi Xiu Le[1, *], Khanh Hung Tran[1], Najib Gadi[1], Abdellatif El-Jaafari[1], Benoit Beaurepaire[1]
[1] Smiths Detection France, 36 Rue Charles Heller, 94405 Vitry sur Seine, France
(*) thi-xiu.le@smithsdetection.com

***Abstract* — X-ray radioscopy at high energies (MeV) is a widely used technique for non-destructive inspection of cargo containers and trucks at ports and borders. Dual-energy imaging is an established method for cargo material discrimination, that works by estimating the average effective atomic number of penetrated objects along each X-ray path, based on the energy dependent mass attenuation coefficient of materials. However, due to 2D nature of the radioscopy image, it is currently difficult to understand how the objects are positioned in the cargo depth and material discrimination interest is limited when objects of different natures are superimposed along X-ray paths. In this article, we describe four totally new imaging functionalities which have been developed and validated by end-users during field tests. These functionalities, under a generic name Augmented X-rays (AXR), are based on the use of multi-column detectors (called Matrix detector) and new algorithms for the material discrimination. They are comprised of in-depth information, image stabilization, wall removal (removing of metallic objects) and de-overlapping of material discrimination. These functions help in pushing the current 2D cargo inspection close to pseudo 3D by localizing on which depth planes are the objects (either close to the accelerator or the detector) and extends the usefulness of material discrimination by discriminating objects one-by-one, helping ease the identification of low atomic number objects (explosives, drugs, cigarettes…).**

***Keywords* — Augmented X-rays, cargo inspection, depth information, advanced material discrimination.**

I. Introduction

Customs since the 90s, have used high energies X-ray scanners as a routine inspection tool for detecting smuggling or tax fraud in trucks at terrestrial borders or in containers at ports. Dual-energy systems allow for the discrimination of materials within cargo by estimating equivalent atomic number and material density based on the exponential attenuation law of X-rays, relating the transmitted intensity to the material attenuation coefficients. Different material discrimination algorithms exist such as α-curve, R-curve, H-L curve methods [2-4]. Recently, some studies tried to address the material discrimination for overlapped objects in cargo via virtual peeling method [5-6] but some prior knowledges of material objects inside the cargo were needed.

Until now, the current used technique is still 2D X-ray radioscopy. In this work, by using several-column detectors and a reconstruction algorithm, a kind of pseudo 3D image (2D X-ray image combined with depth information on object contours) can be obtained (in-depth information). In addition, X-ray image deformation due to unwanted motions can be stabilized by correcting "wave" artefacts. Furthermore, a new material discrimination method has been proposed so that we can not only have an instant view of objects made of organic and low atomic number in the cargo (wall removal) but also do the material discrimination for objects one by one even when screened by other materials (de-overlapping).

II. Methods

A. Methods based on the use of Matrix detector

Matrix detector is a type of detector comprising of several side-by-side columns of detection (Fig. 1 a). Fig. 1 b shows an example of image made of n pulses where each pulse containing 5 columns of data with a slight difference of scanning angle between these 5 columns.

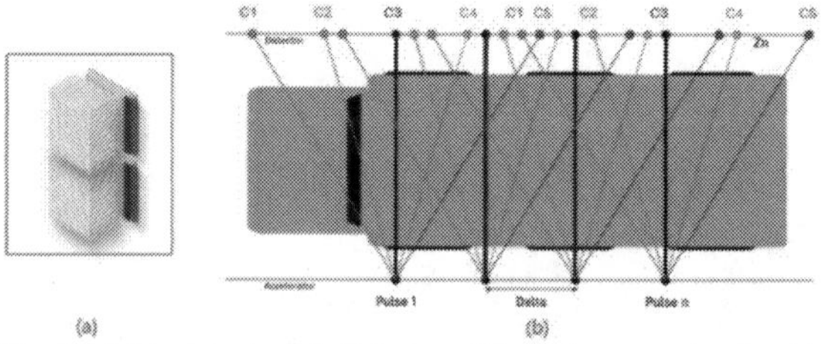

Fig. 1. Matrix detector (a); Principe of X-ray image made of multi-column detectors (b) where Delta is the distance between 2 consecutive pulses and C 1-5 are 5 detection columns.

1) Image reconstruction method and in-depth information

When data from different detection columns are simply put side by side, some artefacts can be observed on the image at the zones around the contours of objects due to the viewing angle difference. This happens when the displacement between two pulses is lower than the detector's width: an object can be seen several times by different detection columns at different pulses (Fig. 1 b). This effect, like stereoscopy at very low angles, is dependent on the depth of the objects. Therefore, to correct for these artefacts, the depth at which the objects are located must be estimated. This is done via a total variation minimization among all images reconstructed at different cargo depth planes. It provides us with a final image free of artefacts as well with a map of object depth positions within cargo.

2) Image stabilization

In general, two types of scanning mode are possible: mobile mode, where the scanner moves alongside stationary cargo, and passthrough mode, where cargo moves alongside the stationary scanner. The oscillations of the boom containing the detectors during the mobile scan, or uneven scanning ground used for the passthrough scans, spawn waves on the image. Based on the redundancy of the information given by different detection columns (Eq. 1) and using an energy minimization technique, the amplitude of the displacement can be estimated for each pixel in the image and therefore distortion effects can be corrected. The energy to be minimized is of the type:

$$E_T\left((\delta_x)_{ij},(\delta_y)_{ij}\right) = \sum_{\substack{n,p \\ p>n}} E_C\left(n,p,(\delta_x)_{ij},(\delta_y)_{ij}\right) + \alpha\left(\sum_{ij}(\nabla^2(\delta_x)_{ij} + \sum_{ij}(\nabla^2(\delta_y)_{ij}\right), \quad (1)$$

With $E_C\left(n,p,(\delta_x)_{ij},(\delta_y)_{ij}\right) = \sum_{ij}(I_p(i-(p-n)(\delta_x)_{ij}, j-(p-n)(\delta_y)_{ij}) - I_n(i,j))^2$

where (n, p) are 2 different detection columns among all detection columns (e.g. 5 in Fig. 1) and $((\delta_x)_{ij},(\delta_y)_{ij})$ are displacements between two adjacent pixels. The optimization constraint in Eq. 1 is since these displacements are regularly quite small.

B. New material discrimination method: the equivalent mass

The material discrimination is an estimation of atomic number based on energy dependency of mass attenuation coefficient. Practically, we use 2 energy spectrums which are emitted alternatively from pulse to pulse: High Mode (HM) & Low Mode (LM), for example 6 and 4 MeV. For each pixel of an X-Ray image, equivalent atomic number is calculated using different kinds of curves measured during a calibration phase as $-\ln T(E_{HM})$ vs $-\ln T(E_{HM}) + \ln T(E_{LM})$ (α curve: [2]); $-\ln T(E_{HM})$ vs $\ln T(E_{HM})/\ln T(E_{LM})$ (R curve: [3]) and $T(E_{HM})$ vs $T(E_{LM})$ (H-L curve: [4]) where T and lnT are transparency and logarithms of transparency. Despite the calibration phase, these curves stay highly non-linear for each material which makes them difficult to handle for more than estimating the average atomic number.

In this study, to obtain a quasi-linear representation of the material discrimination to facilitate the material decomposition of overlapped objects, we introduce a new material discrimination method called the equivalent mass. The main principle (Eq. 2) is to calculate for a chosen reference material (ref), an equivalent mass so we can achieve the same value of transparency by using the same energy spectrum (E).

$$T_r(M,Z,I) = \int_0^{E_{max}} I(E)e^{-\frac{\mu}{\rho}(E,Z\ \ )M}dE = \int_0^{E_{max}} I(E)e^{-\frac{\mu}{\rho}(E,Z_{ref})EqM}dE, \quad (2)$$

With M, Z, I are respectively mass and average atomic number crossed by X-rays and energy spectrum.

Like other material discrimination methods, a calibration phase of different materials (e.g. PMMA, Al, Fe, Pb) is needed. One of the calibrated materials is chosen as a reference material, and we plot the curves $(EqM_{HM} + EqM_{LM})/2$ vs $(EqM_{HM} - EqM_{LM})$ for all calibrated materials. Fig. 2 shows the calculation concept of the equivalent mass where for the reference material, i.e., steel $(EqM_{HM} - EqM_{LM})$ is zero and its curve lies on X axis. Note that experimental curves are quasi-linear.

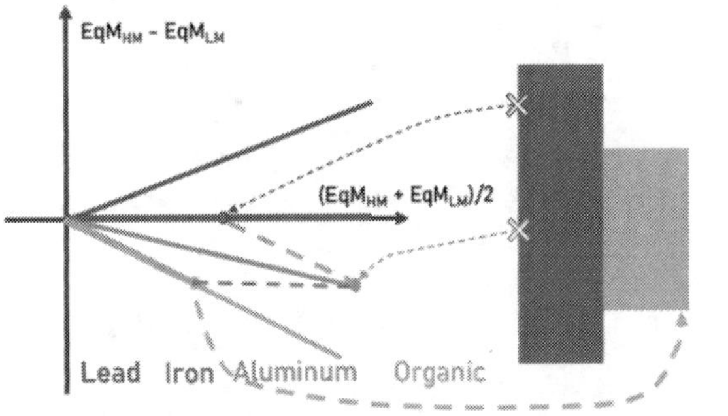

Fig. 2. The equivalent mass's schematic curves and the principle of atomic number's estimation of superimposed zones based on the quasi-linearity of the equivalent mass.

1) Wall removal

An image free of the reference material can be displayed by calculating $(EqM_{HM} - EqM_{LM})$ for each pixel. This is like the clinical dual-energy application where X-ray radioscopy image of human body was discomposed into bone and soft-tissue images [7-8]. By thresholding this image, it is possible to have a direct view of low atomic number objects (when the reference material is Aluminium or Iron). This possibility is named wall removal, as, in the case of containers for example, it removes digitally the container wall.

2) De-overlapping

As current material discrimination gives an average result on the whole thicknesses crossed by X-Rays, it doesn't directly discriminate objects, but only the superimposition of objects: e.g., a superimposition of an organic with steel objects can appear as aluminium. Based on the additive/quasi linear property of the equivalent mass (Fig. 2, the vector corresponding to the superimposition of two objects - the green one, is very close to sum of the two vectors representing the superimposed objects – the blue and the yellow ones) and an advanced segmentation process, it becomes possible to subtract, for each individual object in a cargo, all other objects which are superimposed, and to estimate its atomic number.

III. Results

A. In-depth information

As aforementioned, the objective of the in-depth function is to give the operators an estimation of how different objects are positioned in the cargo's depth. Fig. 3 shows an example of a cargo image where around the object's contours, different colours are added depending on their depth position in the cargo: red when objects are close to the accelerator, yellow-green when objects are in the middle of the cargo and blue when objects are close to the detector. From this in-depth information, the users can easily understand how the objects are positioned inside the container, helping to find the suspicious objects if needed. This is instrumental in shortening time required for a manual check of the cargo and fostering a better understanding of the cargo content.

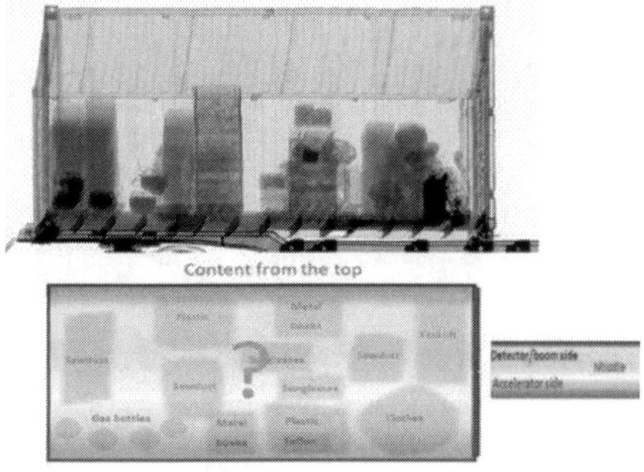

Fig. 3. Result of In-depth information. On top: the transmission image with the depth information displayed as the colour scale and Bottom left: the interpreted cargo content view from the top.

B. *Image stabilization*

Fig. 4 shows an example of a cargo image before (Fig. 4 a) and after stabilization (Fig. 4 b). In the corrected image, straight edges of objects appear as straight lines and no more waves can be seen on the top of the cargo, making the image more readable for the operators.

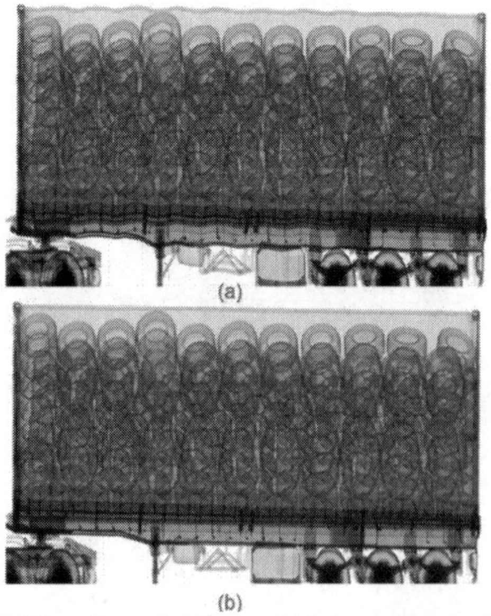

Fig. 4. Effect of Image Stabilization, (a) before and (b) after image stabilization.

C. *Wall removal*

Fig. 5 a shows an example image of a cargo filled of different objects made of different materials. This image was obtained during a European project - ENTRANCE's testing campaign. The corresponding wall removal image with Iron as reference material (Fig. 5 b) highlights cargo zones containing only objects with low atomic number (organic or intermediate materials), making the detection of drugs or explosives smuggling easier. The intensity of the wall removal image reflects the quantity (thickness) of low atomic number objects.

Fig. 5. Transmission image (a) and Wall Removal image (b) of the same truck.

D. *De-overlapping*

Fig. 6 a shows an image of the experimental setup where three bags of 1-kg of sugar (organic materials) were placed in different positions behind some centimetres of steel. By looking at its normal high energy discrimination image (Fig. 6 b), we can see that the background zone is made of steel (coloured as blue) while for the superimposed zones of steel screen and sugar bags, they are recognized as intermediate materials (coloured as green), corresponding to the definition of the equivalent atomic number of the whole thickness crossed by X-rays. By doing de-overlapping, as shown in Fig. 6 c & d, the superimposed zones are decomposed into 2 correct pure materials (organics – sugar bags coloured as orange and steel – the screen coloured as blue).

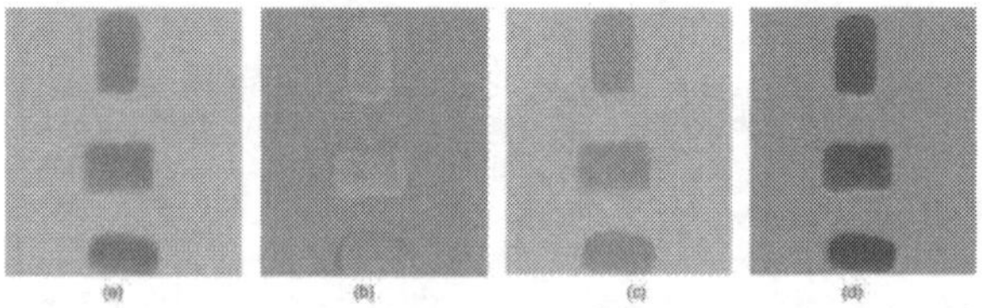

Fig. 6. De-overlapping of 1-kg sugar bags behind a steel screen with: Transmission image (a); results of Material Discrimination (b) and De-overlapping results (objects – sugar bags and the steel screen respectively: c, d).

E. *Combination of different AXR functions*

The four different functions of AXR (in-depth information, image stabilization, wall removal, de-overlapping) can be combined to fulfil the ultimate objective of the operators to check quickly and correctly the passing cargos. Fig. 7 shows a combination of in-depth and wall removal results. By looking at this image, the operators can have a quick idea of whether organic materials (suspicious materials) are placed in the cargo and where are they in the cargo's depth.

IV. Conclusions

In this article, we present four new imaging functionalities collectively referred to as AXR that can become a tool for customs organization to improve the efficiency of X-ray scanning practice and to reduce the time of manual checking:

- In-depth information is instrumental for a better understanding of cargo contents and necessary to know on what side of cargo are the threats;
- Image stabilization corrects unwanted deformation on X-ray images;
- Wall Removal gives an instant view of objects made of organics and low atomic number, therefore proving Backscatter-like images;
- De-overlap helps to discriminate objects one by one and not their superimposition, therefore, to extend the material discrimination usefulness.

Furthermore, the operators can combine different functions of AXR to refine the understanding of cargo contents.

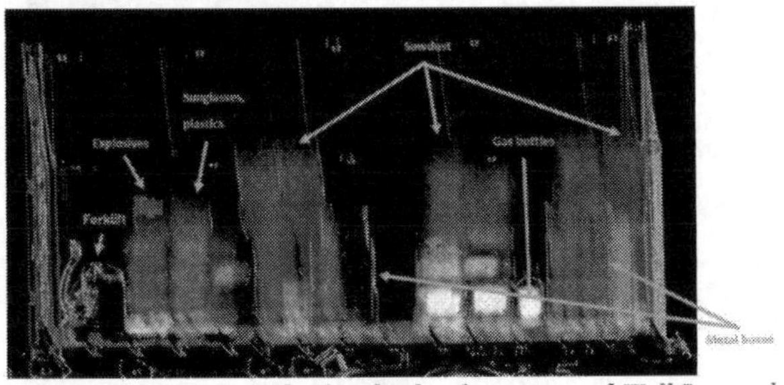

Fig. 7. Combination result of In-depth information and Wall Removal of the same container as Fig. 5.

Appendix

These works have been patented, the readers can find more information from [9-11].

Acknowledgment

The authors would like to thank the European commission for partly financing these works via two projects: CBORD (H2020 GA 653323) and ENTRANCE (H2020 GA 883424).

References

[1] https://www.smithsdetection.com/
[2] Novikov, V. L., S. A. Ogorodnikov, and V. I. Petrunin. "Dual energy method of material recognition in high energy introscopy systems." Вопросы атомной науки и техники, 1999.
[3] Ogorodnikov, S., and V. Petrunin. "Processing of interlaced images in 4–10 MeV dual energy customs system for material recognition." Physical Review Special Topics—Accelerators and Beams 5.10, 2002.
[4] Zhang, Guowei, Li Zhang, and Zhiqiang Chen. "An HL curve method for material discrimination of dual energy X-ray inspection systems." IEEE Nuclear Science Symposium Conference Record, 2005. Vol. 1. IEEE, 2005.
[5] Chen, Gongyin, et al. "Linatron mi6, the x-ray source for cargo inspection." Physics Procedia 66, 2015, pp. 68-74.
[6] Li, Liang, et al. "A dynamic material discrimination algorithm for dual MV energy X-ray digital radiography." Applied Radiation and Isotopes 114, 2016, pp. 188-195.
[7] Cardinal, H. Neale, and Aaron Fenster. "An accurate method for direct dual-energy calibration and decomposition." Medical physics 17.3, 1990, pp. 327-341.
[8] Cardinal, H. Neale, and Aaron Fenster. "Analytic approximation of the log-signal and log-variance functions of x-ray imaging systems, with application to dual-energy imaging." Medical physics 18.5, 1991, pp. 867-879.
[9] Determination of material, US2021356410A1, 18.11, 2021.
[10] Correction of images and depth information for detection with matrix, US2023012491A1, 19.01, 2023.
[11] Correction of images, US11983859B2, 14.05, 2024.

3D-Printed Modular Radiation Sources for Testing Radiation Detectors and Advancing Radioisotopes Identification Algorithms

Oluwadara Afolabi[1], Tanvir Hassan[1], Anil Prasad[1],
Nikolaos Kotsios[1], Liqian Li[1,*], and Ghaouti Bentoumi[1]
[1]Canadian Nuclear Laboratories, Canada
(*) liqian.li@cnl.ca

Abstract— **The international radiological and nuclear (RN) community recognizes improvised nuclear devices (INDs) as a significant security threat. To maintain border security, efficient and reliable detection solutions at points of entry are critical. Screening of containerized cargo for INDs and RN materials is primarily done using drive-through radiation portal monitors (RPMs). Globally, advanced computing algorithms, including machine learning and data analytics, are being developed and enhanced to improve the consistency and accuracy of RN material identification. However, these data analytics algorithms require augmentation with true positive scan data covering the full threat space, including scenarios involving INDs of varying intensities and shielding configurations. Due to the strict controls on large quantities of special nuclear materials in diverse geometries and isotopic compositions, many RPMs and portable detectors have been deployed without adequate testing against realistic IND threats. This has led to high false alarm rates, requiring time-consuming secondary screenings, and may also increase the probability of false negatives, allowing real threats to go undetected. Conversely, training algorithms using full-mass INDs introduce nuclear criticality risks. In this paper we present a 3D-printed radiation source designed to mimic high-mass solid INDs by distributing the radioactive material along a thin, hollow shell. Thanks to the self-shielding effect, the 3D-printed radiation source achieves radiological performance comparable to that of a solid high-mass source with less material. This approach offers a solution to challenges related to source availability and nuclear criticality in training environments. A scaled-down prototype was fabricated, and its radiological performance was experimentally evaluated.**

Keywords —**3D printing, Radiation source, Gamma detector, Isotope identification.**

I. Introduction

As nuclear technologies continue to advance, more countries are pursuing nuclear capabilities. Concurrently, improvised nuclear devices (INDs) have emerged as increasingly realistic security threats. Screening cargo for INDs at border crossings remains a critical first line of defense. The growing peaceful use of radioisotopes, e.g., for medical and industrial applications, makes accurate identification of INDs essential for the effective operation of border screening systems, particularly those using radiation portal monitors (RPMs) [1] . On the one hand, it is crucial to intercept all the INDs attempting to cross the border; on the other, false alarms triggered by medical or industrial sources must be minimized. Developing reliable algorithms for IND detection remains an active area of research among scientists and RPM manufacturers. Realistic radioactive sources are ideal for developing, testing, and benchmarking these algorithms and the RPM systems that implement them. While medical and industrial radioactive sources are relatively accessible, the use of special nuclear materials (SNMs) is tightly regulated. Even when SNMs are available, their deployment is often costly due to stringent safeguards and security requirements. Additionally, using large quantities of SNMs raises concerns about nuclear criticality. To address these challenges, researchers have proposed surrogate sources that use non-SNMs to radiologically mimic SNMs [2] . Another approach involves training sources that use small amounts of SNMs to simulate larger quantities [3] .

Canadian Nuclear Laboratories is investigating the use of additive manufacturing to produce modular radiation sources. By leveraging a shell-structured design, these sources can replicate the radiological performance of bulk SNMs using significantly smaller quantities, benefiting from the self-shielding properties of dense materials. This paper presents an initial prototype fabricated with natural uranium, focusing on the evaluation of the 1,001 keV gamma peak associated with the U-238 content.

II. Structure and Fabrication

This work employed a layered spherical shell design consisting of three concentric layers: an inner and outer shell made of plastic, and a thin intermediate layer containing the active material. In essence, the approach embeds a radiologically active layer between protective plastic layers using 3D printing. This configuration facilitates safe handling by encapsulating the radioactive material. Moreover, using a thin distributed layer of active material mitigates the need for bulk quantities, thereby drastically reducing the risk of nuclear criticality.

3D printing can create complex structures by depositing

materials layer by layer, typically in a vertical direction. This upward growth is straightforward, but when a design extends sideways, print quality often degrades. If the overhang angle exceeds a certain threshold, typically around 45 degrees, internal scaffolds become necessary to complete the print. In the initial trial, a spherical shell was 3D-printed in eight "orange slice" segments, as illustrated in Fig. 1. Each slice began with vertical growth, but as the geometry extended outward the overhang angle increased beyond the printable limit, requiring support structures (shown at the right side of Fig. 1). The addition of supports meant that the print job required 200% more time and 84% more material to print. This also raised the risk of compromising the containment of active material.

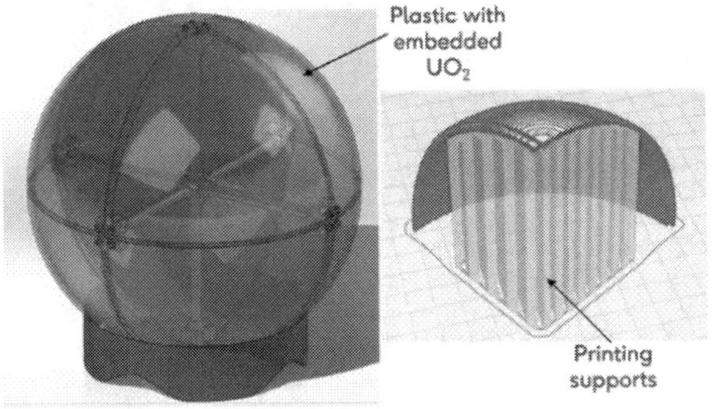

Fig. 1. Orange slice design (left, plastic shell with embedded UO_2) and one slice to print (right).

The second trial employed a "tennis ball" design, as shown in Fig. 2. This approach used two identical halves that could be 3D-printed without any support structures, as all overhang angles remained within the printable limit of 45 degrees. Compared to the orange-slice design, the tennis ball configuration resulted in significant savings in both print time and material consumption.

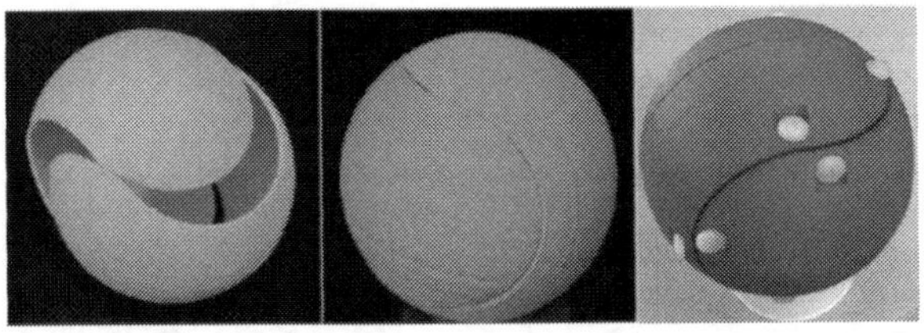

Fig. 2. Tennis ball design with a fully assembled demo.

To achieve a modular and adaptable design, the tennis ball structure was divided into eight nearly identical modules, each 3D-printed using UO_2-loaded filament. This modular approach allowed for the customization of individual segments, enabling the simulation of various radiological signatures by adjusting module properties as needed. Minor variations among the modules were introduced to enable secure mounting onto a 3D-printed internal frame. Fig. 3 shows the UO_2-loaded modules, the internal frame, and a partially assembled 1/8-scale prototype. As shown, the surface quality of the UO_2-loaded modules was suboptimal, partly due to the application of a rubber resin coating after printing, which was necessary to secure the active material within each module.

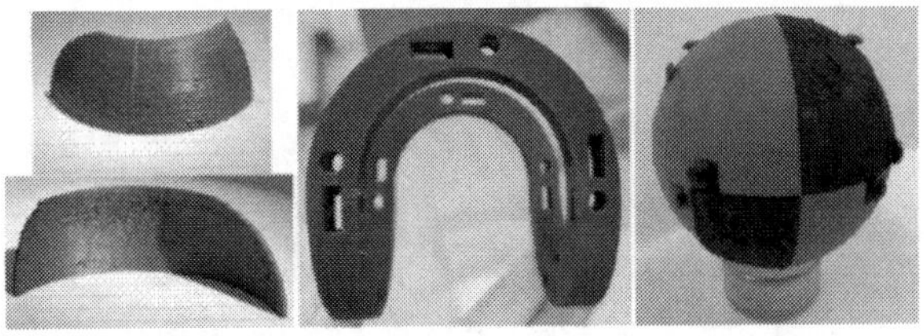

Fig. 3. UO_2-loaded modules (left), the internal frame (middle), and scaled-down assembly with only two active modules loaded (right).

To address the previously identified problem and enhance the structural integrity of the UO_2 shell, a new fabrication method based on mold injection was developed. In this approach, hollow tennis ball-style segments were first 3D-printed using high-temperature resin. Heated UO_2-loaded filament was then injected into the internal cavity of each segment, forming a fully sealed UO_2 layer encased within the protective resin shell. Notably, this hollow design retained the advantage of requiring no printing supports. Fig. 4 shows a cross-section of a mold-injected piece and the assembly of the mold-injected segments, representing a 1/3-scale prototype.

Fig. 4. Cross-section of a mold-injected piece (left), and assembly of the 3D-printed radiation source (right).

III. Gamma-Ray Assessment

As noted earlier, by focusing on the 1,001 keV gamma peak, depleted uranium (DU) and natural uranium (NU) can be used in place of highly enriched uranium (HEU) for testing purposes. DU and NU are significantly easier to obtain and handle, from both nuclear security and criticality safety standpoints. While this study used DU, the same principles apply when using HEU, where the focus would instead shift to the 185.7 keV gamma line. The primary objectives of the gamma assessment were twofold: (1) to evaluate whether the 3D-printed source emitted gamma rays isotopically, that is, with similar intensity when measured from different angles, and (2) to estimate the equivalent U-238 mass in a solid structure mimicked radiologically by the printed source. Measurements were carried out using a 76 mm × 76 mm NaI detector positioned 50 cm away from the source (center to center). The net count rate for the 1,001 keV peak was obtained using InterSpec software [4] , with background subtraction applied. Table 1 presents the measured count rates from three randomly selected angles, along with comparison data from a solid 9 cm cube of DU placed at the same location.

TABLE I
NET COUNT RATE OF THE 1,001 KEV PEAK

Sample	U-238 mass (g)	Count rate (cps)	Sigma (cps)
3D-source-angle-1	52.2	2.325	0.078
3D-source-angle-2		2.408	0.093
3D-source-angle-3		2.589	0.109
3D-source-average		2.441	0.054
DU-cube-9cm	13,887	69.303	0.420

The results in Table I show that the 3D-printed source produced comparable count rates across the three angles. Variations were within expected statistical uncertainties for 40-minute measurement durations. The slight variation observed is attributed to unfilled regions intentionally left in the prototype design to accommodate external mounting points.

To evaluate the efficiency of the printed design in mimicking greater mass using less material, a mass amplification factor, *M*, was defined. This factor represents the ratio of the gamma yield per unit mass of U-238 from the printed source to that from a solid structure. Using the average count rate from the printed source and comparing it with that from the solid DU cube, *M* can be determined. The gamma yield per gram of U-238 for the printed source, γ_{printed}, was calculated as

$$\gamma_{\text{printed}} = \frac{2.441\text{cps}}{52.2\text{ g}} = 0.04676\text{ cps/g}.$$

For the DU cube, the relative yield was

$$\gamma_{\text{DUcube}} = \frac{69.303\text{ cps}}{13887\text{ g}} = 0.00499\text{ cps/g}.$$

Thus, the mass amplification factor is

$$M = \frac{\gamma_{\text{printed}}}{\gamma_{\text{DUcube}}} = \frac{0.04676}{0.00499} \approx 9.37.$$

This indicates that, in terms of gamma emission at 1,001 keV, each gram of U-238 in the printed source was radiologically equivalent to approximately 9.37 grams of U-238 in this solid cube form. It is expected that mass amplification factor will change with the shape and the size of the solid target.

As the prototype scales up, the surface coverage of UO_2 is expected to increase, further improving uniformity and potentially increasing the mass amplification factor. Additionally, when the focus shifts from 1,001 keV of DU to 185.7 keV of HEU, the greater self-shielding effect of solid structures at lower energies will further enhance the amplification effect. Adding tungsten shielding with a thickness of 6.3 cm allowed for the study of the shielding effect on important analysis peaks, revealing that the reduction of the 185.7 keV peak was about 400 times greater than the reduction of the 1,001 keV peak. Preliminary estimates suggest that the mass amplification factor for U-235 could reach values on the order of 100 for HEU samples in the form of solid sphere with uranium mass in the order of 15 kg, with detailed characterization to be conducted through Monte Carlo simulations.

IV. CONCLUSIONS

A novel method for fabricating uranium-bearing radiation sources using 3D printing was developed and demonstrated. The approach involved mold-injecting UO_2-loaded material into a hollow, resin-printed shell, resulting in a fully sealed structure that is safer to handle and that does not require internal support scaffolding during print. The design also allows for future adaptation to include complete infilling, thanks to internally mounted features in the improved version. Gamma-ray measurements using the 1,001 keV peak of U-238 showed that the printed source emitted radiation uniformly across different angles, with variability consistent with statistical uncertainty. This confirms the viability of the design for isotropic radiation emission. When compared to a solid 9 cm DU cube, the prototype demonstrated a mass amplification factor of approximately 9.37, meaning that each gram of U-238 in the printed shell radiologically mimics nearly ten grams of solid U-238 at the reference energy. This factor is expected to increase further with larger source sizes due to improved surface filling ratios. Moreover, for scenarios involving HEU, where the 185.7 keV peak is dominant and self-shielding in solids is more pronounced, the mass amplification factor could be significantly greater, on the order of 100. These predictions will be validated through future Monte Carlo simulations.

The ability to customize and fabricate 3D-printed radiation sources emulating the gamma signatures of solid IND components provides a promising path forward for the development, testing, and benchmarking of radiation detection algorithms and systems. This approach offers a safe, flexible, and scalable solution to the long-standing challenge of training source availability.

REFERENCES

[1] International Atomic Energy Agency (IAEA), "Technical and Functional Specification for Border Monitoring Equipment," IAEA Nuclear Security Series No. 1 T, Vienna, Austria, 2006.
[2] B. A. Pohl and D. E. Archer, "133Ba as a gamma-ray surrogate source for 1kg HEU and 10g 239Pu and 252Cf as a neutron surrogate for Pu," Lawrence Livermore National Laboratory, UCRL-TR-202871, 2004.
[3] C. S. Blessinger, C. A. Pickett, J. H. Leckey, A. Demint, J. Gooch, and T. Hawk, "Design of HEU test objects," in 50th Institute of Nuclear Materials Management Conference, Tucson, AZ, USA, 2009.
[4] E. Chan, W. Johnson, D. Lee, and C. Morte, InterSpec Software, Sandia National Laboratories, 2025.

HV effect on neutron coincidence counting

Monika Risse*, Sebastian Chmel, Theo Köble, Olaf Schumann and Thorsten Teuteberg
Fraunhofer-Institut für Naturwissenschaftlich-Technische Trendanalysen, Euskirchen, Germany
(*) monika.risse@int.fraunhofer.de

***Abstract*— Neutron detection is important in the field of Nuclear Safeguards, Homeland Security, and CBRN (Chemical, Biological, Radiological, and Nuclear) defense. In these areas, precise as well as efficient detection and analysis of neutron emissions are necessary for identifying nuclear material or monitoring radioactive sources. On the one hand, neutrons are very difficult to shield and, considering the very low intrinsic background for neutrons, even weak sources can be detected. On the other hand, neutrons only interact weekly with the detector material and in most cases, they need to be moderated, making the detection a veritable challenge. Therefore, specialized materials and technologies, such as neutron-sensitive gases, are employed to facilitate detection.**

Normally, a count rate measurement is sufficient to determine that a neutron source is present, but other techniques must be used for more specific identification. In neutron coincidence counting, the temporal correlation of neutrons is considered to distinguish between the prompt emission of a bunch of neutrons, which often indicates the presence of nuclear material, and a random emission of neutrons from industrial sources like Am/Be. Therefore, this approach is particularly valuable in nuclear security or safeguards, as it enhances the detection capabilities for spontaneous fission and nuclear chain reactions. Neutron Multiplicity Counting (NMC) goes further by analyzing the frequency of single, double, triple, or higher-order coincidences, which provides even more detailed information concerning the fissile material.

Although many alternative materials have been developed in recent years due to Helium-3 (^{3}He) shortages, ^{3}He detectors still remain the standard. The ^{3}He isotope is highly effective for neutron detection due to its high cross-section for neutron capture, which allows these detectors to operate efficiently even at low neutron flux levels. A configuration of multiple ^{3}He detectors within a detection array can improve overall detection efficiency, and allows for more accurate coincidence analysis. This technique has found widespread application in nuclear monitoring and control, particularly in identifying and quantifying fissile materials, assessing radioactive sources, and ensuring compliance with international regulatory standards.

Measurements with a commercial neutron multiplicity counter (Fission Meter) using a set of 30 ^{3}He tubes were performed. Depending on the applied high voltage (HV), the count rate, Feynman variance (Y_{2F}), and Rossi-α values are examined. With increasing HV, the measured count rate rises, making a higher HV desirable for an increased efficiency of the detector. However, this increase also affects the Feynman variance and Rossi-α value in such a way that above a given HV value the measurement no longer qualifies for accurate analysis. In this case, the instrument can only be used for count rate measurements, and successful multiplicity measurements are no longer possible.

To distinguish nuclear, and in particular fissile, material from industrial neutron sources, the detector must be accurately calibrated and appropriately set to ensure reliable and valid results.

***Keywords* — Multiplicity counting, Feynman variance, Rossi-α, correlated and uncorrelated neutron sources.**

I. Introduction

The paper deals with the effect of different high voltage settings on neutron coincidence counting. First, an introduction to the measurement device used in this experiment, as well as a general concept of neutron multiplicity counting, is given. As part of that, two important parameters are explained: the Feynman variance Y_{2F} and the Rossi-α method. The impact of the high voltage, which is applied to the neutron detector's ^{3}He tubes, is assessed. Namely, the influence on the overall count rate, on the Feynman variance, and on the Rossi-α value is presented.

These measurements were initially performed to investigate the cause for a very high background-like signal in the Feynman variance of the Fission Meter of Fraunhofer INT [1].

II. Measurement Device and Setup

A. Source

For the measurements an Am/Li source with $5.3 \cdot 10^4$ n/s was used. This type of source does not emit neutrons in a correlated fashion and the neutron energy is quite small with a mean of 0.5 MeV. Reactions like the (n,2n) reaction are therefore suppressed or impossible. The expected Feynman variance of such a source is zero.

B. Measurement Device

The ORTEC Fission Meter is a portable neutron detection system designed for the rapid assessment of fissile material. It is equipped with 30 ^{3}He proportional counters. Fig. 1 shows the device and Table I gives the key technical specifications. It has two panels with 15 tubes each (a) and moderator material Polyethylene (PE) on the outside (b). The Fission Meter includes built-in electronics for signal processing, data acquisition, and real-time analysis. Both panels have separate high voltage power supplies. Each panel houses 8 preamplifiers, 7 of them are shared by two adjacent tubes. The Fission Meter collects the neutron count rate, the histogram distribution for 512 different gate lengths, and the Rossi-α distribution. From the histogram distributions, the Feynman Variance can be calculated for the different gate length [2], [3].

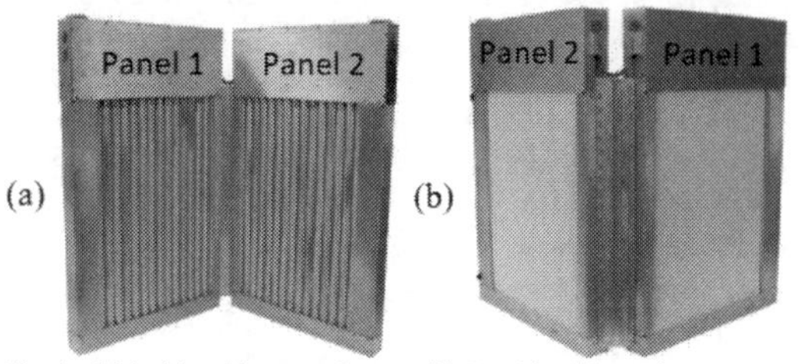

Fig. 1. (a) Inside without moderator. (b) Outside with moderator.

TABLE I
KEY TECHNICAL SPECIFICATIONS FISSION METER FROM ORTEC

Component	Specification
Detector Type	30 ^{3}He Proportional Counters (RS-P4-0819-106)
Tube Length	48.3 cm (19 in)
Tube Diameter	25.4 mm (1 in)
Fill Gas	^{3}He + CO_2 (quench gas), 7.5 bar pressure
Moderator Material	Polyethylene (25 mm thick on one side only)
Electronics	HV supply separate per panel, shared preamp/discriminator, FPGA

C. *HV Variation*

For high voltage adjustment, the Fission Meter was opened to provide access to the internal electronics. Each panel features its own independent high voltage supply. For the measurements with panel 2, the voltage was manually adjusted by turning the screw on the potentiometer which is located within the white-circled area. The actual high voltage level was measured using a probe placed at the designated test point which is indicated by the yellow circle. This setup allows for direct control and verification of the supply voltage during operation or initial setup. All measurements shown in this paper are done with panel 2.

Fig. 2. High Voltage adjustment of panel 2 at the open device, the change is done by turning the screw on the potentiometer in the white circle. The voltage measurement is done with a probe at the test point in the yellow circle.

III. NEUTRON MULTIPLICITY COUNTING

Neutron Multiplicity Counting is a passive, non-destructive technique used to characterize neutron-emitting materials by analyzing the statistical properties of emitted neutrons. The technique utilizes the difference in the temporal emission structure of correlated and uncorrelated sources. Correlated sources, such as the fissile isotopes ^{235}U, ^{238}U, and ^{252}Cf, emit time-correlated bursts of multiple neutrons as a result of spontaneous fission events. In contrast, uncorrelated neutron sources (e.g. industrial Am/Li or Am/Be sources) emit neutrons independently and randomly, and the emission follows a Poisson distribution, which corresponds to purely random emission statistics. In contrast, the correlated sources produce a deviation from the Poisson statistics.

Two important parameters are the Feynman variance Y_{2F} and Rossi-α, which are explained in the following subsections. Both parameters enable the reliable distinction between different neutron sources. The parameters are common in different communities.

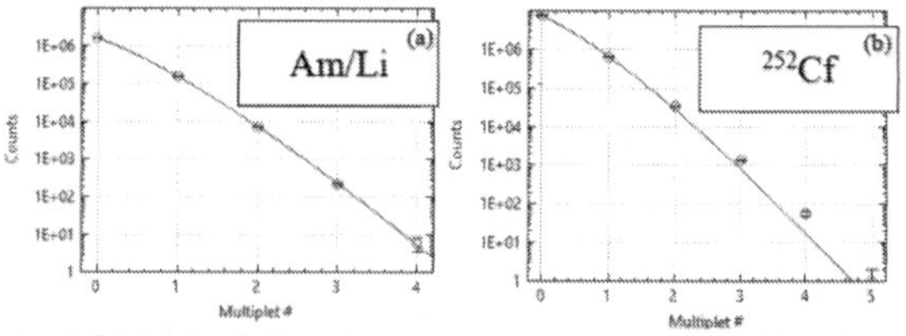

Fig. 3. Multiplet distribution: the measurement data are indicated by the red circles, the Poisson distribution is indicated with the blue line. (a) An uncorrelated source (Am/Li): the measurement data are on the Poisson line. (b) A correlated source (^{252}Cf): the measurement data do not follow the Poisson deviation for higher multiplicities.

A. *Feynman Variance – Y_{2F}*

The deviation of the measured neutron multiplicity histogram from a Poisson distribution is quantified by the Feynman variance (denoted Y_{2F}). The idea is that a Poisson distribution is characterized by having a variance that is equal to its mean value. By calculating variance and mean of a measured histogram and subtracting the two values, one obtains a metric that is zero for a pure Poisson distribution and positive otherwise. The Feynman variance is further normalized by the count rate (e.g. the mean). For uncorrelated neutron sources Y_{2F} should be zero, for correlated sources Y_{2F} is positive.

 https://doi.org/10.1051/epjconf/202533806005

Formula (1) gives the Feynman variance in dependence of the asymptotic value of the Feynman variance R_{2F}, and the neutron decay constant $\lambda=1/T_d$, wherein T_d is the neutron lifetime:

$$Y_{2F} = R_{2F} \cdot \left(1 - \frac{1-e^{-\lambda \cdot T}}{\lambda \cdot T}\right). \quad (1)$$

T is the gate width used for the multiplicity counting, it normally varies between 1 µs and 512 µs for a Fission Meter measurement. Due to the dependence of the neutron lifetime, Y_{2F} gives an indication of the material surrounding the neutron source. Fig. 4 shows the general behavior of the Feynman variance for the gate width or the applied high voltage. For uncorrelated neutron sources like Am/Li, the Feynman variance is expected to be zero (a). For correlated sources like ^{235}U, ^{238}U, other fissile materials, or ^{252}Cf, the value is greater than zero.

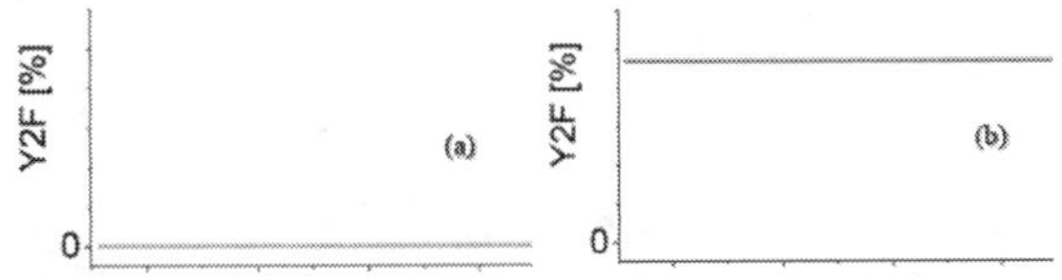

Fig. 4. Schematic diagrams of the general behavior of the Feynman variance Y_{2F} versus gate width or applied high voltage. (a) Uncorrelated source. (b) Correlated source.

B. Rossi-α

The Rossi-α technique is a time correlation measurement, that accumulates a histogram of time intervals between neutron detection events. For the Rossi-α the time difference to all following events (up to a fixed maximal time) are recorded. For an uncorrelated neutron source, such as Am/Li, the distribution is expected to be constant with respect to time, reflecting the random and independent nature of neutron emissions. In contrast, a correlated neutron source, such as ^{252}Cf or other fissile materials, produces neutrons in bursts from fission events. This results in a non-random time structure, and the Rossi-α distribution overlays a characteristic exponential decay, starting from the shortest time intervals. The implementation of the Rossi-α measurement in the Fission Meter deviates from this simple procedure due to the use of a data acquisition cycle of 512 bins, that result in a somewhat different shape of these Rossi-a curves.

Fig. 5 shows the Rossi-α measurements for the uncorrelated neutron source Am/Li and the correlated neutron source ^{252}Cf. The Am/Li source produces a straight line with a linear decline. The ^{252}Cf source yields an exponential drop in the first few channels, followed by a straight line with a decrease at longer times.

Due to the device-specific algorithm of the Fission Meter, time intervals are recorded only between 1 µs and 512 µs. As the Rossi-α histogram is built from start–stop event pairs, the maximum time interval of 512 µs can only occur if the start neutron is detected in the first time channel and the stop neutron in the last. All other combinations result in shorter intervals. Consequently, longer time intervals are statistically less likely to be recorded, leading to reduced counts in the highest channels and no flat line.

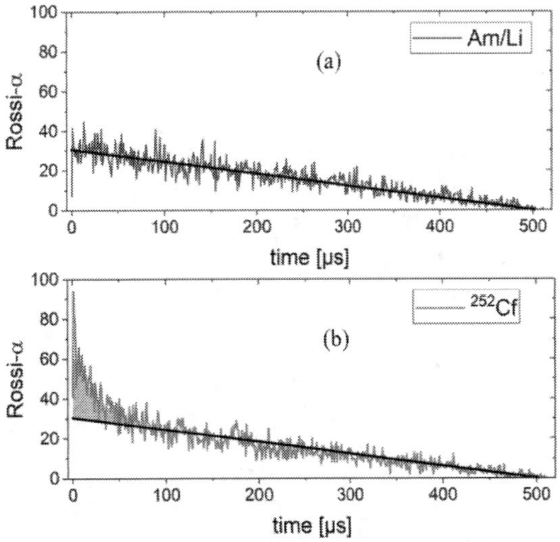

Fig. 5. (a) Rossi-α for an uncorrelated source (Am/Li). (b) Rossi-α for a correlated source (^{252}Cf). The black lines show the trend for the uncorrelated source; the deviation for the correlated source is indicated by the red shaded area. Due to the device specific algorithm the black line is not constant, but decreases.

IV. MEASUREMENTS

A. Count rate versus HV

In Fig. 6 the count rate versus the applied high voltage is displayed. A higher count rate is synonymous with a higher efficiency and therefore desirable. The diagram indicates that high voltages above 1800 V would be preferred as seems to lead to a plateau.

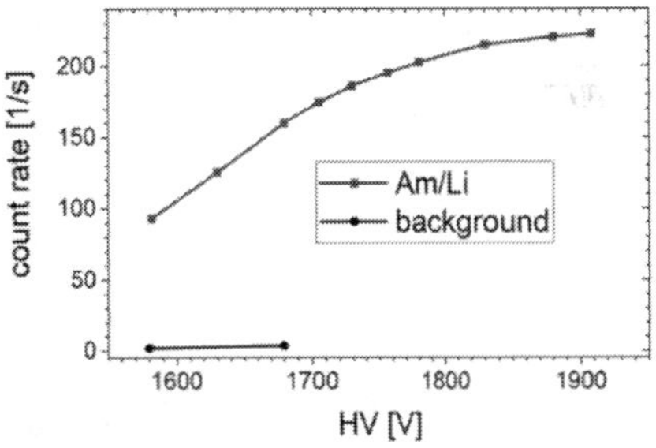

Fig. 6. Measurement with the Am/Li neutron source. Count rate measured in dependence of high voltage.

B. Feynman Variance versus HV

The Feynman variance Y_{2F} is found to be significantly influenced by the applied high voltage. Fig. 7 displays the Y_{2F} values measured as a function of high voltage using an Am/Li source. Since this source emits uncorrelated neutrons, Y_{2F} values are expected to remain around zero. However, for high voltage values of 1756 V and above, Y_{2F} clearly becomes non-zero.

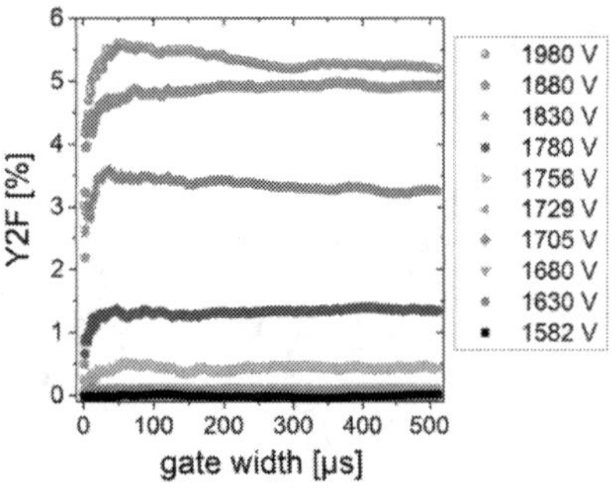

Fig. 7. Feynman variance Y_{2F} versus the gate width measured at different high voltages as indicated.

The Y_{2F} value is plotted against the HV for a gate length of 512 µs in Fig. 8.

At lower voltages the measured Y_{2F} values behave as expected and stay near zero. However, starting at approximately 1680 V, a marked increase in Y_{2F} is observed, with values rising well above zero.

This effect is critical when evaluating unknown neutron sources: if the high voltage is set too high, elevated Y_{2F} values may incorrectly suggest the presence of correlated neutrons. Such a misinterpretation could lead to the false identification of fissile material where none is present.

Moreover, excessive high voltage can distort neutron lifetime measurements, potentially resulting in incorrect conclusions regarding the moderating or absorbing properties of surrounding materials.

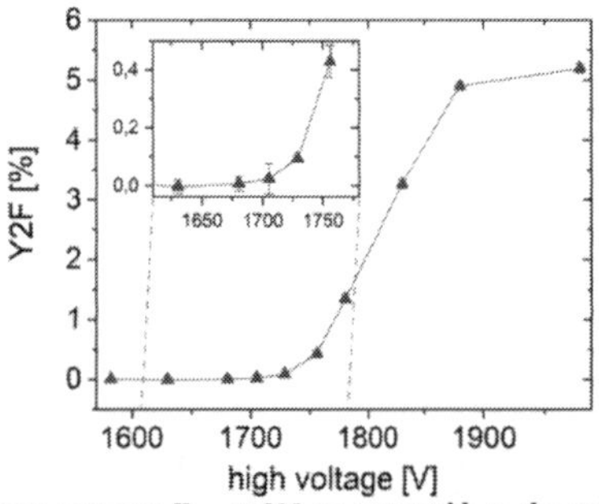

Fig. 8. *Feynman variance Y_{2F}, at 512 µs gate width in dependence of high voltage.*

C. *Rossi-α versus HV*

Fig. 9 shows the Rossi-α value as a function of time for the Am/Li source. Part (a) of Fig. 9 gives the results for lower high voltage (HV) values of 1582 V and 1630 V. As expected, the Rossi-α curves exhibit a slightly decreasing linear trend.
In contrast, Part (b) of Fig. 9 shows the behavior at higher HV values of 1830 V, 1880 V, and 1908 V. Here, the Rossi-α evolution deviates from the expected linear trend and shows a different behavior.

The effect in the Rossi-α measurements is very clear. In the second bin with a time distance of 1 µs a very strong peak is visible for larger high voltages.

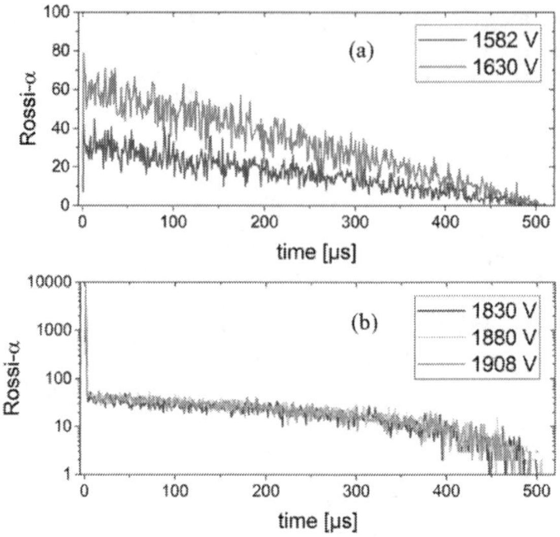

Fig. 9. *Rossi-α for an Am/Li source in dependence of the time for different high voltages as indicated. (a) Rossi-α with straight line. (b) Rossi-α with very strong peak in Bin 2.*

Fig. 10 presents the Rossi-α value from bin 2 plotted against the applied HV. At low HV, the Rossi-α remains small as expected, but from approximately 1680 V onwards, a clear increase is seen. This corresponds approximately to the point at which the effect also occurs within the Feynman variance, although the effect for Rossi-α is more significant.

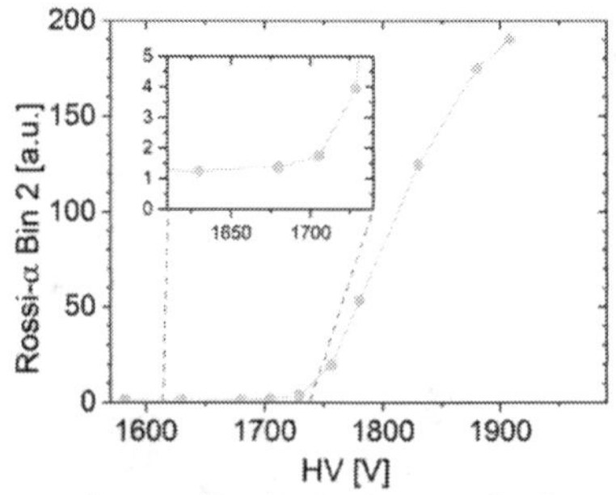

Fig. 10. *Rossi-α in the second bin (Δt=1 µs) measured with an Am/Li source in dependence of high voltage.*

The effect on the Rossi-α is within 1 µs, which is the time resolution and artificial electronic noise will count in. It seems there is the effect of double pulsing in which one neutron leads to a signal in the preamplifier which is counted in a fraction of the cases as two. This is in dependence of the present electronics and cannot be generalized in detail, but the effect in principle has to be considered for other detectors.

V. Conclusions

Neutron measurements require a good understanding of the measurement instrument and its operation. Proper device settings are critical, including knowing exactly what the displayed values mean and how to interpret them for reliable data analysis. For the ^3He proportional counter a parameter which has to be set correctly is the high voltage. It has a large impact on the Feynman variance Y_{2F} and Rossi-α. A too large high voltage causes Y_{2F} values that are always significantly above zero, indicating a correlated neutron source if an uncorrelated neutron source is measured. This could lead to misinterpretation of measurement results and a false identification of a neutron source as a correlated one, which could be a nuclear or fissile source. The Rossi-α value has too high values in the first bins and does not enable a correct analysis of the measurement data. The high voltage must be set appropriately for correct results: it must not be set too high and a lower efficiency has to be accepted.

REFERENCES

[1] Schumann et al. (2018). Simulation of a neutron multiplicity counter and comparison to validation experiments. Presented at ESARDA Annual Meeting. [Online]. Available: Simulation of a neutron multiplicity counter and comparison to validation experiments (fraunhofer.de). Accessed on: August 13, 2025.

[2] M.S. Rowland and R.A. Alvarez, "Real-time multiplicity counter", U.S. Patent 0 205 580 A1, Aug. 28, 2008.

[3] M.S. Rowland and N.J. Snyderman, "Fission meter and neutron detection using poisson distribution comparison", U.S. Patent 8 891 720 B2, Nov. 18, 2014.

Inter-laboratory comparison within the RESORAD Italian network on the natural occurring radioactive materials (NORMs)

Sonia Fontani[1(*)], Valeria Innocenzi[1], Giuseppe Menna[1]

Pierino De Felice[2], Marco Capogni[2], Mauro Capone[2], Pierluigi Carconi[2], Michele Croia[3], Aldo Fazio[2], Andrea Petrucci[2]

[1]**ISIN** - Ispettorato nazionale per la sicurezza nucleare e la radioprotezione, Via Capitan Bavastro, 116 – 00154 Roma; Italia

[2]**INMRI** (Istituto Nazionale di Metrologia delle Radiazioni Ionizzanti) ENEA – Casaccia R. C., Roma, Italia

[3]**N**uclear **R**esearch **R**eactors Laboratory ENEA – Casaccia R.C., Roma, Italia

(*)sonia.fontani@isinucleare.it

***Abstract* — Environmental radioactivity monitoring is of fundamental importance not only in routine situations but also in the event of accidental occurrences, such as the Fukushima incident in 2011, and in extraordinary situations, such as those currently experienced in war-torn countries with nuclear facilities. In Italy, the National Network for Environmental Radioactivity Surveillance (RESORAD), coordinated by ISIN, oversees the reliability of environmental radioactivity measurements. ENEA-INMRI develops, maintains, and disseminates measurement standards and units in the field of radioactivity through inter-laboratory comparisons (ILCs), calibration campaigns, and performance evaluation tests. This work presents the results of an ILC conducted for RESORAD laboratories, called NORM-2021, organized within a scientific collaboration between ISIN and ENEA-INMRI. The initiative focused on the measurement of naturally occurring radioactive materials (NORMs). Specifically, the aim of the ILC was to identify and quantify NORMs in natural volcanic sand. To perform the ILC, ENEA-INMRI prepared a solid reference material (MRS) by collecting sand samples from seven distinct volcanic lake sites in the Lazio region. An artificial imbalance in the uranium radioactive series was introduced into the MRS to assess the analytical capabilities of the participating laboratories. The results were analysed using indicators commonly employed in performance evaluations and inter-laboratory comparisons [ISO, 2010; ISO, 2016].**

***Keywords* —Environmental radioactivity, MRS. ILC.**

I. INTRODUCTION

Environmental radioactivity is a natural component of background radiation, originating from radioactive materials in our surroundings. Some radionuclides, like potassium-40 (^{40}K), occur naturally, while others such as strontium-90 (^{90}Sr) and technetium-99 (^{99}Tc) are mainly introduced through human activities. Certain isotopes, including tritium (^{3}H), have both natural and anthropogenic sources. Human actions, like nuclear weapons testing, can influence the levels and distribution of natural isotopes such as uranium-238 (^{238}U). Since Earth's formation, radioactivity has been pervasive, with soil containing isotopes like ^{40}K, radium-226 (^{226}Ra), ^{238}U, and thorium-232 (^{232}Th) with typical average activity concentrations. For instance, ^{40}K is commonly found in soil at levels around 370 Bq/kg, with values common ranging from 100 to 700 Bq/kg; some other radioisotopes often exhibit activity concentration in the range of 10-50 Bq/kg. Monitoring environmental radioactivity is essential during routine operations, accidents (e.g., Chernobyl and Fukushima), and conflicts involving nuclear facilities to assess potential risks.

In Italy, the National Network for Environmental Radioactivity Surveillance (RESORAD), coordinated by ISIN [1], is constituted by the laboratories of 21 Regional and Autonomous Provinces Agencies for Environmental Protection (ARPA/APPA) and 3 Experimental zoo-prophylactic Institutes. High data quality and robust measurement efficiency are essential to ISIN's role in ensuring radiation protection and nuclear safety. Under RESORAD, ISIN promotes national measurement reliability programs. ENEA-INMRI [2] is responsible for establishing standards and conducting inter-laboratory comparisons (ILCs). From 2020 to 2022, ISIN and ENEA collaborated on an ILC called NORM-2021, focused on measuring naturally occurring radioactive materials (NORMs) in volcanic sand, chosen for its high NORM content and logistical convenience within the Italian Lazio region. The objectives, methods and achieved results of this ILC are presented in this paper.

II. OBJECTIVES OF THE ILC NORM-2021

The specific objective of ILC NORM-2021 was defined as follows: "*Identification and quantification of gamma emitters present in a Solid Reference Material (MRS) composed of an environmental matrix of sand.*"

The MRS of ILC was prepared at the Laboratory of Radioactivity of ENEA-INMRI by collecting sand samples from seven distinct volcanic lake sites located in the Lazio region (Italy). In particular, each samples, with a volume of 1 L, was measured through gamma spectrometry at ENEA-INMRI, focusing on the gamma emissions of ^{40}K and the radioactive series of uranium and thorium. The sand from the site showing the highest concentration of NORMs, particularly from the thorium series, was chosen. The MRS was then produced using sand from the selected lake site, sampled near the shoreline, where the grain size is more uniform.

In order to introduce an element of novelty and make the progression of the ILC more interesting for testing the measurement capabilities of the participating laboratories, the organizers agreed to introduce an artificial imbalance into the uranium radioactive series by spiking the sampled sand with a known quantity of ^{226}Ra at the ENEA-INMRI laboratory. The added ^{226}Ra was 400 Bq/kg, increasing the original activity of ^{226}Ra-226 by a factor of 3 (from about 200 Bq/kg to about 600 Bq/kg).

Each participating laboratory was required to identify and quantity, through gamma spectrometry systems based on germanium detectors, the activity concentrations of gamma-emitting radionuclides present in a MRS with a sand matrix. This quantity represents the main Measured Parameter (MP) for the ILC. For the purpose of the ILC, the radionuclides (both natural and artificial) to be considered were gamma photon emitters with energies between 40 keV and 2 MeV.

In the MRS to be analysed, secular equilibrium conditions are not guaranteed for all radionuclides in which such a condition might potentially be established. Data analysis focused on radionuclides whose activity concentration does not depend on this condition. However, each participant laboratory could implement its own procedures to achieve radioactive equilibrium conditions and that allow them to extend and enhance their radiometric analysis. Measurement methods and experimental details (such as the volume of MRS used, measurement geometry, how to establish secular equilibrium, and the related methodology, etc.) were left to the discretion of the participant.

A further objective of the ILC was the determination, under measurement conditions applied to the sand sample, of the Minimum Detectable Activity (MDA) for some radionuclides not necessarily present in the sample matrix. This objective gave a support to the evaluation of potential sources of measurement error.

III. PREPARATION OF THE MRS

The preparation of the MRS for the ILC began in April 2021 with the collection, carried out by personnel from ISIN and INMRI, of approximately 80 liters of sandy material, also supported by logistical assistance from ARPA Lazio.

The sandy material, transported to ENEA-INMRI laboratories, underwent the following mostly physical preparatory steps:

- removal of coarse interfering materials (stones, leaves, twigs, pine needles, etc.);
- spreading on a large plastic sheet exposed to the sun outdoors until nearly completely dried (to remove the large amount of water present, given that the sample was collected near the shoreline);
- sieving the material through a 10-mesh sieve (2000 μm).

A portion, approximately 50 liters of the resulting material, was selected for subsequent treatment. Due to the impossibility of directly tracking the entire quantity of pretreated material (over 60 kg), a tracking approach using ^{226}Ra of a sub-aliquot, followed by dilution with identical untracked material, was adopted. To this end, five aliquots of 1 kg each were collected

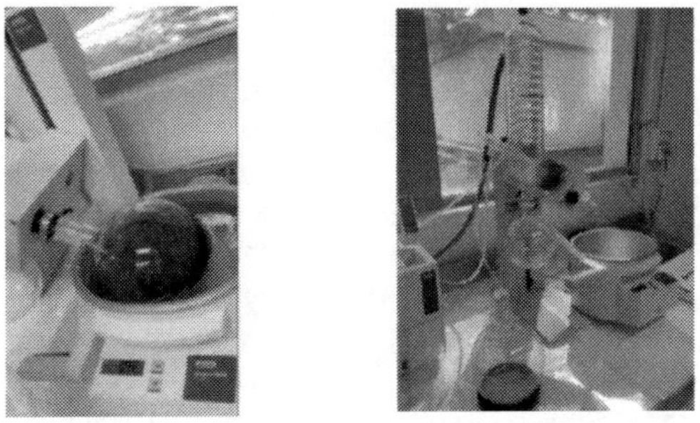

Fig. 1. Rotating evaporation system used by ENEA-INMRI for tracking MRS-2438 with ^{226}Ra using the acetone bath method

and individually spiked with a high-purity radionuclide solution of ^{226}Ra. The spiking of the sand matrix was performed using the "acetone bath" technique [3, 4, 5, 6]. This method allows for homogeneous and quantitative deposition of radionuclides onto a fine-grained matrix by covering the matrix with acetone, adding the tracer radionuclide, evaporating the acetone, and subsequently adsorbing the tracer onto the matrix itself. Controlled evaporation was carried out with a rotary evaporator (Figure 1).

Each 1 kg aliquot of dried, carefully weighed matrix was placed into the instrument's glass flask and then covered with pure acetone. The system was rotated for approximately 1 hour at the temperature of a thermostatic bath set to 40°C, allowing the solvent (acetone) to fully imbibe the matrix. Next, the specified amount of ^{226}Ra solution was added, and the system was kept in rotation for 2 hours to ensure homogeneous distribution of ^{226}Ra within the acetone.

Subsequently, the rotary evaporator's vacuum system was activated to slowly distil the solvent, permitting the radionuclide—concentrating during this phase—to adsorb evenly onto the active surface of the matrix. Once drying was complete, the tracked matrix was removed from the flask and subjected to radiometric measurement via gamma spectrometry to determine the activity concentration achieved.

The distilled acetone was also measured radiometrically to verify and to confirm the absence of radionuclide migration into the solvent.

The five prepared aliquots were finally added to the remaining inert sandy matrix and homogenized using a system for mixing granular solid materials.

The resulting MRS was assigned the code ENEA-MRS-2438. It was divided into 40 aliquots, each with a mass of 1.4 kg and a volume of 1 L, placed in polyethylene containers with screw caps. Each sample was uniquely identified with the code ENEA-MRS-2438 #An (with the number varying from 1 to 40 to indicate the specific aliquot used).

IV. MRS CHARACTERIZATION

The ENEA-MRS-2438 prepared at ENEA-INMRI was characterized for both homogeneity and radioactivity content, before sending to each participant laboratory a sample of it, identified by the #An code and prepared as above.

Homogeneity test was conducted successfully at the Radioactivity Laboratory of ENEA-INMRI by subjecting 19 of the 40 samples—packaged and ready for shipment to the participants—to spectrometric counting, treating the count rates, recorded at the 186 keV energy of ^{226}Ra, with the ANOVA variance test. The results of this analysis did not reveal any heterogeneity among the analysed samples. Therefore, any component due to non homogeneity among the prepared samples has been included in the uncertainty budget for the determination of the ILC Reference Values.

Reference values for the activity concentration of radionuclides present in the ENEA-MRS-2438 were determined by gamma ray spectrometry using the ENEA-INMRI calibrated p-type HPGe detector with a relative efficiency of 40% . The detector is housed inside a low-background well, constructed with 10 cm thick lead bricks, internally lined with copper and cadmium sheets. The setup is calibrated with ENEA-INMRI standard gamma-emitting radionuclide sources within the considered energy range. Additionally, for ^{226}Ra, a supplementary calibration was performed using the "internal standard" method, utilizing activity values of ^{226}Ra added during matrix tracing.

Measurements were carried out in three different measurement geometries:

a) Aluminium Marinelli beaker of 1 L volume;

b) Polypropylene (PP) container of 0.1 L volume;

c) Glass container of 0.1 L volume (two replicates on different samples).

The acquired spectra were analysed to determine the net peak area, applying corrections for measurement geometry, counting efficiency, self-absorption, sum effect, background, and decay. Based on the spectrometric data obtained and considerations related to radioactive equilibria, reference values for ENEA-MRS-2438 were assigned for each radionuclides of the three natural series of uranium, thorium and actinium and for ^{40}K and ^{137}Cs, the only artificial radionuclide detected in the MRS.

All the measurements were performed under standard conditions and the obtained reference values are therefore referred to the national radionuclides activity standards. sample.

For the uranium series, the values are provided for ^{210}Pb and for the radionuclides ^{234}Th, ^{234m}Pa, ^{226}Ra, ^{214}Pb, and ^{214}Bi assuming secular equilibrium within their respective decay series. The reference value for ^{238}U was derived based on the measured ^{235}U value knowing the natural isotopic composition of uranium. Only for two cases (^{234}U and ^{230}Th), despite results provided by some participant laboratories, measurements conducted at ENEA-INMRI did not allow for the production of a reference value. Therefore, it was determined as the average and median of the values provided by participants, in agreement with the UNI CEI EN ISO/IEC 17043:2010 standards.

Similar considerations were applied to the thorium and actinium series respectively, as well as for ^{40}K. It should be noted that ENEA-INMRI measurement systems are not dedicated to low-activity measurements; therefore, the detection of certain radionuclides by participants (e.g ^{234}U, ^{230}Th, ^{232}Th) does not necessarily imply false-positive results.

V. ILC'S ORGANIZATION

The participant laboratories were provided with instructions for performing measurements with their HPGe spectrometers and a sheet to record the results of their analyses. Each participant (with an assigned code) had to identify the radionuclides present in the sand matrix, quantify the activity concentration (by measurement), and determine the minimum detectable activity concentration (M.D.A.) under the applied measurement conditions, including for some radionuclides not necessarily present in the received sample.

The ENEA-MRS-2438 samples were distributed to the participants in sealed vials (Figure 2) containing the volume of material requested by the participant itself. They were shipped from ENEA-INMRI using standard express couriers, considering their activity level below the exemption limits. The samples were delivered as a definitive transfer for scientific collaboration. For full traceability, each participant was required to fill out a Takeover Declaration, sent by ENEA-INMRI, and return it to ENEA-INMRI duly signed. At the conclusion of the measurement campaign, the received sample can be retained by the participant as Certified Reference Material (CRM) accompanied by a Calibration Certificate of the MRS-2438 material used in the ILC, issued by ENEA-INMRI.

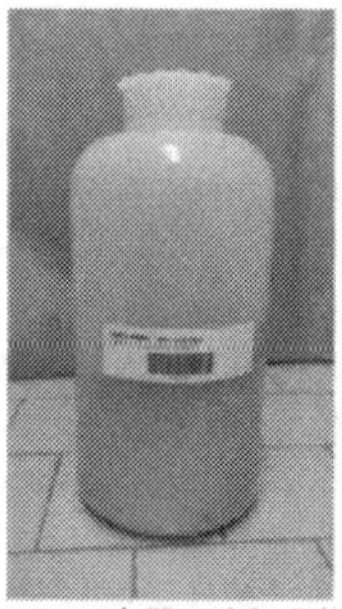

Fig. 2: Bottle of ENEA-MRS-2438 prepared for ILC NORM-2021.

VI. DATA ANALYSIS AND RESULTS

The analysis of the results provided by each participant was carried out in terms of indicators commonly used in performance evaluations and inter-laboratory comparisons [7, 8], including:

a) the percentage deviation $R(\%) = 100 * (M_{rv} - M)/M_{rv}$ between the measured value, M, provided by the participant and the reference value, M_{rv}, provided by ENEA-INMRI;

b) the normalized error or compatibility index $E_n = (M_{rv} - M)/\sqrt{U_{rv}^2 + U^2 - 2U_{cp}^2}$ defined, in terms of extended uncertainty (coverage factor k=2), U, associated to M, the extended uncertainty U_{rv} associated to M_{rv} and the uncertainty component, U_{cp}, in common to both the participant and the reference value provided by ENEA-INRMI. According to the ISO cited above, E_n value should be between ±1 for acceptable results.

All participants performed measurements with source-detector distances less than 5 cm, due to low activity values of the natural radionuclides contained in the analysed sample and in order to obtain enough count rate to minimize the statistical uncertainty. Consequently, a correction for the summation effect was applied both during calibration and measurement.

Some laboratories reported radionuclides that are probably not present in ENEA-MRS-2438. The erroneously detected radionuclides were labeled as "suspicious" false positives. Conversely, some other laboratories did not report radionuclides that were certainly present in ENEA-MRS-2438, and these were classified as "false negatives".

In Fig. 3 the distribution of the percentage deviation *R(%)* for each measured radionuclide is reported.

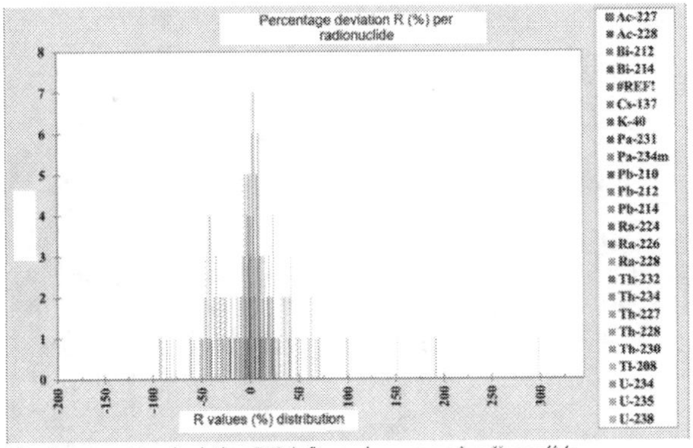

Fig. 3. percentage deviation R(%) for each measured radionuclide.

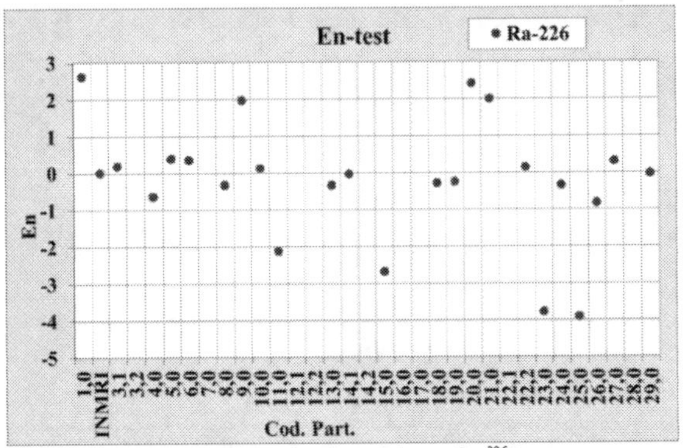

Fig. 4: E_n values of each participants for ^{226}Ra

The values of the compatibility index, E_n, are shown in Figures 4 and 5 for two radionuclides: ^{226}Ra and ^{235}U, respectively.

From these figures, it can be observed that:

a) the majority of the participant laboratories show an E_n value within the ±1, indicating the reliability of the RESORAD network for this type of measurement.;

b) the percentage of results with an E_n value outside the ±1 is greater for ^{226}Ra than ^{238}U, this mainly due to the artificial imbalance of ^{226}Ra introduced into the ENEA-MRS-2438.

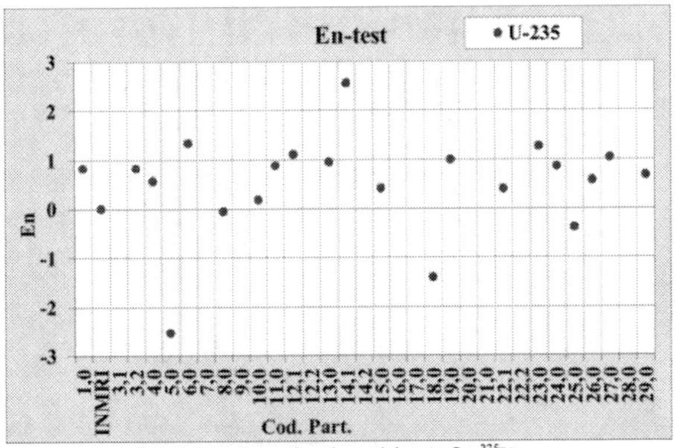

Fig. 5: E_n values of each participants for ^{235}U

VII. CONCLUSIONS

The Inter-Laboratory Comparison "NORM-2021" involved the participation of 28 laboratories from the Italian RESORAD network, reporting a total of 286 concentration values of activity, along with their associated uncertainties. The results show a good level of measurement accuracy.

Moreover, the ILC provided a valuable opportunity for in-depth discussions aimed at identifying and addressing critical issues observed in the measurement results. These insights were gained through bilateral interviews between the organizers and the participating laboratory, fostering a collaborative atmosphere that supported the continuous improvement of measurement practices within the network.

REFERENCES

[1] https://www.isinucleare.it/en
[2] https://www.inmri.enea.it/en/
[3] Dersch R., De Felice P., Jerome S.M., Pona C., De Sanoit J., Woods M.J. "Development of a Range of Gamma-Ray Emitting Radioactive Spiked Reference Materials for Environmental Monitoring - Part I", Nucl. Instr. and Meth. in Physics Research A 339 (1994) 55- 60.
[4] Biagini R., Dersch R., De Felice P., Jerome S.M., Perkin E.M.E., Pona C., De Sanoit J., Woods M.J. "Homogeneity Testing of Spiked Reference Materials", The Science of the Total Environment 173/174, 267-274, (1995).
[5] De Felice P., Iskandar D., Fazio A., Biagini R. "The Quality Control in the Production of Spiked Reference Materials for Gamma-Ray Spectrometry", Conference of the International Committee for Radionuclide Metrology, NIST (USA), 19-23 may 1997, Applied Radiation and Isotopes, 49 No 9-11, 1421-1427, (1998). BIAGINI
[6] De Felice P. "I materiali di riferimento dell'ENEA-INMRI", Tutto Misure, Anno IV n. 03, 253-258 (2002). ISO.
[7] International Organization for Standardization and International Electrotechnical Commission, ISO/IEC 17043:2010, "Conformity assessment - General requirements for proficiency testing", Online: https://www.iso.org/standard/29366.html
[8] International Organization for Standardization, ISO 13528:2015, "Statistical methods for use in proficiency testing by interlaboratory comparison", 2016.

Directional Detection and Fissile Material Identification with the nFacet 3D detector

Kaciel Béraud[1,*], Grégory Lehaut[1], and Antonin Vacheret[1]
[1]Laboratoire de Physique Corpusculure de Caen, France
*beraud@lpccaen.in2p3.fr

Abstract

Source localisation is a critical point in nuclear safeguard and security, but it is currently an overlooked concept. Source search using counters is the standard method but is very time-consuming when the environment is complex or unknown. We introduce the nFacet 3D system, a portable gamma and neutron radiameter with directional capabilities, to fill the gap of directional detectors. We present here the results of the measurement campaign done at the STUK facility in Finland to assess the performance of the system in the field. Infield resolution below 5° for both ^{137}Cs and ^{60}Co at 5 and 10 meters away was demonstrated using machine learning techniques. High sensitivity to neutrons was shown, allowing long-range detection of a source at least 50 meters away, with possible detection up to 80 or 100 meters away. Gamma source separation capabilities were shown using energy window selection techniques, and good stability across variable environments in terms of count rates was demonstrated. The high angular resolution, with high detection range, gamma separation capabilities, and stability across the environment enables source localisation and detection on a large scale, making nFacet suitable for a wide range or security-related applications.

Keywords **— neutron, gamma, direction, source localisation, machine learning, multimodal.**

I. Introduction

Source localisation in an unknown or critical environment, such as nuclear facilities, is a staple of nuclear safeguard and security. Checkpoints are set where flux control is essential, such as border crossings or restricted areas. Passive Radiation Portal Monitors (RPMs) are the historical technologies, prone to false alarms due to Naturally Occurring Radioactive Materials (NORMs) or tampering with high-activity sources nearby. After a trigger, localising the source is done by using handheld counters or droness but this procedure is time-consuming. Directional systems such as gamma cameras can be used to reduce the localisation time [2], but they have low efficiencies, especially for high energy gammas, limited field of view (FOV), and limited range. Due to the very nature of neutrons, very few systems like gamma cameras exist and suffer from the same limitations [6]. Some systems combine both radiation detection, achieving an integrated system [4]. Machine learning techniques are now used to increase the performance of reconstruction algorithms, improving the localisation capabilities of the system [7].

II. nFacet 3D

The detector technology introduced here, called nFacet3D, is a portable directional radiation monitor and dosimeter, capable of discriminating gamma rays, thermal and fast neutrons at the same time. The detector is composed of 4 modular detection units, each one composed of 16 active cells. Each detection cell is composed of a 5 cm PVT cube, for gamma detection and neutron moderation, and of a sheet of ^{6}LiF:ZnS(Ag) to capture thermal neutrons. Each detection unit reads the scintillation light through 8 silicon photomultipliers (SiPMs) with active temperature correction. The specifications of the system can be found in Table I, a mechanical drawing of the detector is shown in Fig. 1.

Table I
Specification of the nFacet3D system

nFacet 3D	Specifications
Modules	4
Channels per module	8 SiPMs
Scintillators per module	16 cubes
Gamma scintillator	PVT
Neutron scintillator	^{6}LiF:ZnS(Ag)
Field of View	4π
GARRn	1.019 at 10 mR/h
Autonomy	8 H
Weight	15 kg
Digitization rates	65 MHz
Acquisition mode	Waveform, Integration
Trigger	Amplitude, Time over threshold, Coincidence, Periodic
Saturation rates	10 kHz
Power draw	7 W
Connection	Ethernet
UI	Touchscreen

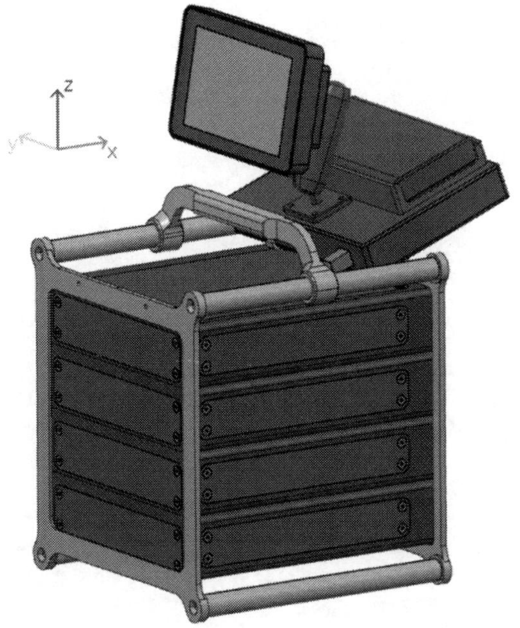

Fig. 1. nFacet 3D system CAD view.

The segmentation of the system enables the reconstruction of the direction information during a measurement in all 4π angles. Fig. 2 shows the count rates in different cells of the detector summed along the Z axis with a ^{60}Co source located in the direction $\varphi = 225°$.

To infer the direction of the source, a fully connected neural net is used for its simplicity, lightweight design, and short inference time. The input of the model is the fraction of counts per cube and the output is the Cartesian direction vector. The task is a regression, using the AdamW [5] optimizer and the PReLU with a single parameter for the activation function.

The model was trained on a selection of 11 sources covering the energy range from 1 keV to 15 MeV. The isotopes ranged from ^{241}Am, ^{9}Be-based neutron sources to ^{252}Cf's fission gamma rays. The nuclear data were loaded from the International Atomic Energy Agency (IAEA) Livechart Data API [3]. The data includes both the gamma rays and the X-rays of the isotopes.

The positions of each source for the training set were homogeneously randomized in space, occupying the volume of a hollow sphere, with an internal radius of 1 meter and an external radius of 8 meters.

The position for the test set were generated at a distance of 5 meters and linearly disposed on ϑ and φ angles, with a step of 5°.

The performance of the pre-trained model is shown in Fig. 3 for a ^{137}Cs source. Each pixel represents the direction of a test source, and the color is the direct angle between the predicted direction and the true direction.

The patterns visible on the figure are from the planar symmetry of the cubic formation of the system. The worst-case scenario is perpendicular to a face, resulting in a 6°

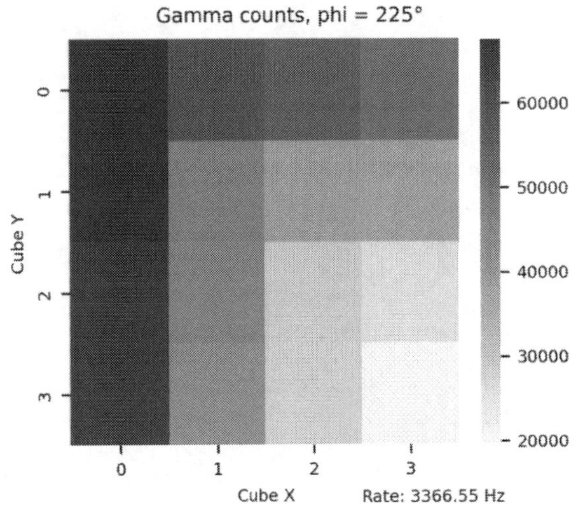

Fig. 2. Top view of the counts per cube, summed along the Z axis. A ^{60}Co source was placed in the direction of the detector's upper left corner of the figure.

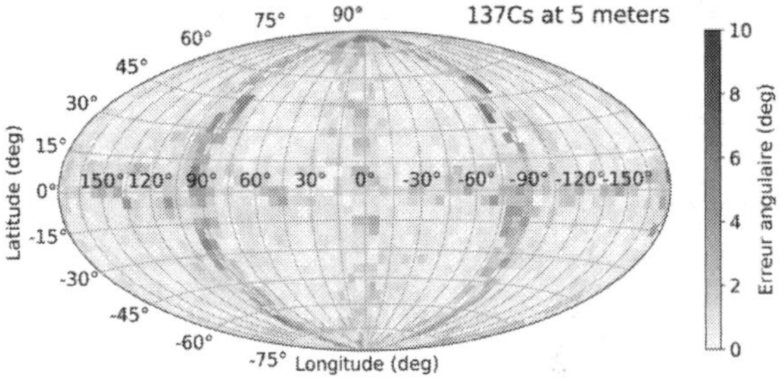

Fig. 3. Performance of a ^{137}Cs simulated source. The pattern comes from the symmetry planes of the cubic geometry of the nFacet system.

error. Along any X, Y, or Z plane, the performance is a little better but around the same 5° error value. The best-case scenario is around any corner of the system, which result in an error below 1° theoretically.

The model is then fine tuned with real data to improve the model's performance and learn the little variation of the detector real response.

III. Measurement

We participated in the SamLoc campaign held at the STUK facility, Finland. The objective of this campaign was to assess the capabilities of different technologies to localise a gamma source. Multiples measurements were made, with both neutrons and gamma rays, to assess different system capabilities, such as the angular resolution and the maximal detectable distance.

A. Angular resolution

For this, a 167.7 MBq ^{137}Cs and 77 MBq ^{60}Co gamma sources at 5 and 10 meters respectively were used in the

STUK's irradiation halls to determine the angular resolution of the system. The detector was manually turned around with a 22.5° step and the detector was vertically aligned with the source, fixing ϑ to 90°.

B. Maximal detectable distance

To determine the maximal detectable distance of a neutron source and assess the stability of the directional information over the distance, a 57 MBq ^{252}Cf, nominal activity of 500 MBq with a reference date of November 15 2016, was used outdoor on the STUK facility's parking lot. The distance was set from 3 to 50 meters away with a fixed approximate direction of $\vartheta = 90°$ and $\varphi = 180°$. The source was vertically aligned with the detector and the distance was measured with a ribbon tape of 50 meters length.

C. Spatial discrimination

During the outdoor measurement of the ^{252}Cf source, an unexpected ^{137}Cs source was inside the facility, waiting to be used, interfering with the gamma ray measurement of the neutron source. Exploiting that error, source separation in function of the energy and the relative intensity of each can be attempted. The ^{137}Cs source emit monoenergetic gamma rays of 662 keV, and the ^{252}Cf emits a continuous gamma ray spectrum from spontaneous fission, ranging from at least 100 keV to more than 20 MeV [8]. Using the source gamma signature, an energy selection can be done to isolate the higher source component.

D. Environmental stability

Finally, following the procedure in [1], done at the same facility, the effects of the environment on the count rate of nFacet were assessed in comparison to other technologies.

The objective was to assert the effect of the environment on the count rates for multiple technologies of passive neutron counters used in the nuclear weapon disarmament monitoring and verification process proposed by the International Partnership for Nuclear Disarmament Verification (IPNDV). The count rate as a function of the distance is modeled by (1).

$$R = \frac{A}{r^x} + B \tag{1}$$

Where R is the count rate, r the detector-source distance, and A, B, and x are the fitting parameters. Without scattering, the neutron count rates should follow an inverse square law, this model tries to take into account the neutron scattering from the environment, effectively reducing the power law of the equation. It's important to note that this model is non-physical, neutron scattering is nearly impossible to model and is one way to represent the scattering. The comparison will not be done on the value of the parameters but on the relative variation in different configurations.

Two indoor configurations and one outdoor were assessed. For the indoor configurations, one where the source is placed in the middle of the irradiation hall, limiting the backscattering and one where the source is placed 1 meter away from the back wall, increasing the scattering. The distance from the detector to the source was from 3 meters to 7 meters. The outdoor configuration result in little scattering, mostly from the air and the soil, the distance was set from 3 meters to 50 meters and reuse the same data acquired for the maximal detectable distance.

IV. Discussion and analysis

A. Angular resolution

The neural net model is used to predict the direction of the source for both ^{137}Cs and ^{60}Co, the background is scaled down to match the exact measurement time and removed from the counts of every cube. The latitude angle ϑ was every time very close to the correct value, 90°, the difference between the true φ angle and the predicted φ was computed.

The value and the error range are evaluated by resampling each cube count over a Poisson's law 100 times and computing the mean value and the dispersion of the φ angles of every prediction.

The result is shown in Fig. 4, in both cases, the uncertainty is below 5° with a confidence interval of 1, which is comparable or even better than typical gamma cameras at greater distances and with a 4π field of view [9].

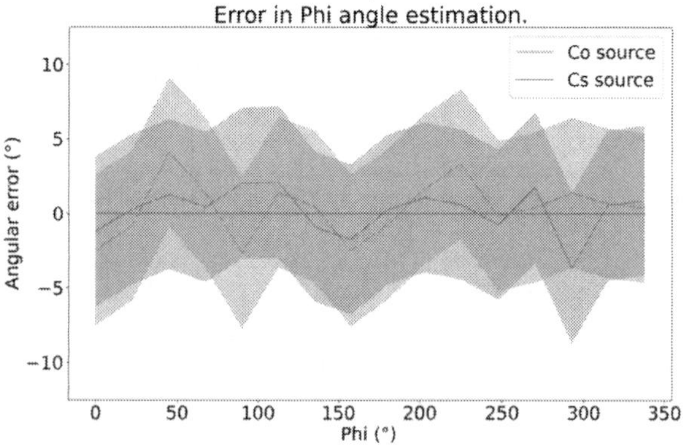

Fig. 4. Angular resolution for both ^{137}Cs and ^{60}Co sources.

B. Maximal detectable distance

The Signal over Background (SoB) is computed as (2), for no signals, the result will be 0. Following a similar equation as (1), (3) is used to fit the SoB over the distance, where A is the amplitude factor, r the distance, and x the power law. B is fixed to 0, forcing the equation to be null at infinity.

$$SoB = \frac{R_{measured} - R_{background}}{R_{background}} \tag{2}$$

$$SoB(r) = \frac{A}{r^x} \tag{3}$$

The SoB over the distance is shown in Fig. 5. At 50 meters, the SoB is still at 7, fitting with (3) and setting an

SoB threshold of 3, it appears the maximal detectable distance is 80 meters. Setting the SoB threshold to 2 increases the distance to 100 meters.

That high detectable distance can be explained by the skyshine effect, most of the neutrons are scattered in the air or the soil back to the detector, increasing the count rates. One other effect that can affect the count rates is the STUK's building. The detector was placed in the middle of the parking lot of the building at 11 and 16 meters from the walls, the configuration might have the effect of a satellite dish for neutrons, scattering back to the detector and thus increasing the count rates.

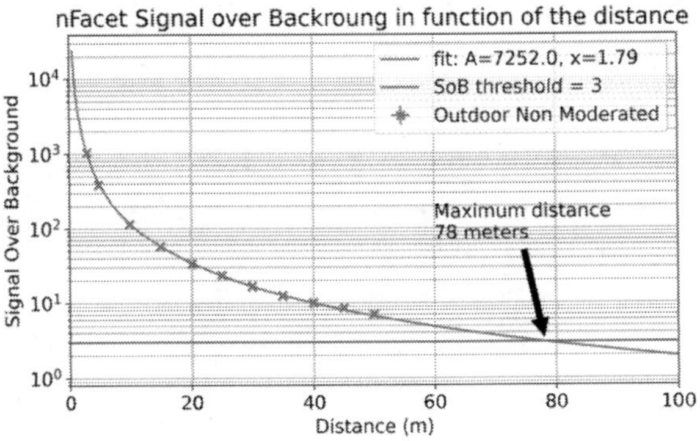

Fig. 5. Signal over background.

C. Spatial discrimination

Using the same data but looking at the gamma information instead of the neutron, the direction of the source was plotted as a function of the distance, in orange in Fig. 6. One should expect the prediction to be constant or worsen with the distance.

A shift of the direction from 180°, the correct value, to 320° can be observed after 10 meters. This shift represents the point the dominant gamma source swapped from the ^{252}Cf to the ^{137}Cs inside the facility.

A selection window in energy is applied to only keep the gamma rays that deposited more than 2 MeV in the detector. This selection has been done to select only the fission's gamma rays emitted by the ^{252}Cf source.

In Fig. 6, the green line shows the prediction of the direction for the fission's gamma rays only, the direction is close to the expected 180°, showing the effectiveness of the method. The variations and errors are due to the lower statistics of the measure, the intensity of these high-energy gamma rays decreases drastically the higher the energy is.

D. Environmental stability

The fits are done using (1) and the results are plotted in Fig. 7.

A few observations can be made across the 5 different configurations. The moderation of the source decreases the count rates, even if nFacet has a higher efficiency for thermal neutrons. This can be explained by the fact that

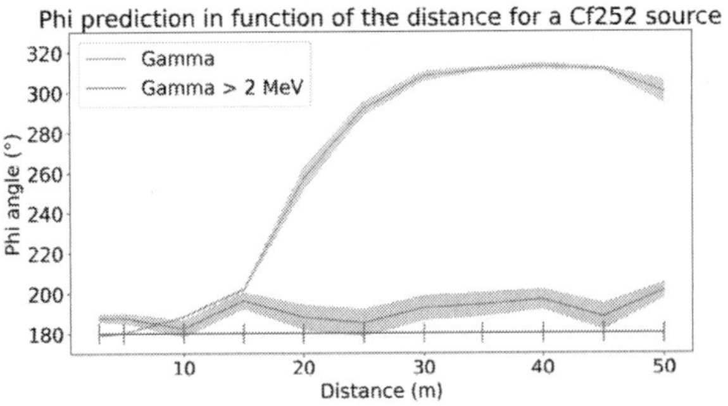

Fig. 6. Source direction in function of the distance. When applying an energy cut to only keep the fission's gamma from the ^{252}Cf source, a discrimination between the ^{137}Cs and the fission source is achieved.

thermal neutrons scattered are captured before reaching nFacet.

When the source is close to the wall, it increases the count rates. This is explained by the higher surface for the neutron to scatter back to nFacet, effectively increasing the neutron flux.

When in an outdoor configuration that approaches a low scattering factor, the count rate are at least a factor of 2 below the indoor configurations, which shows the impact of the environment when detecting neutrons.

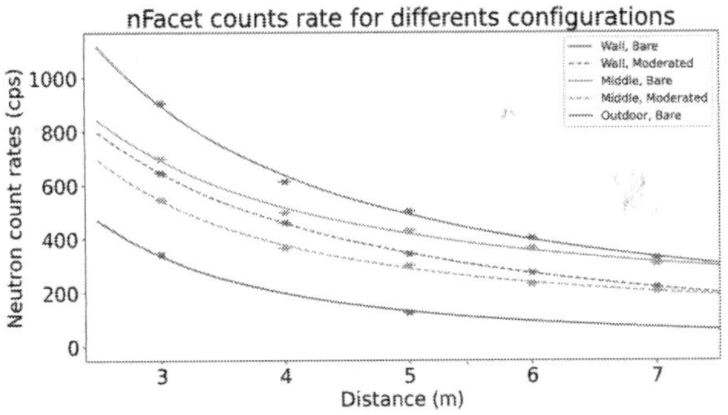

Fig. 7. Count rates as a function of the distance for all the configurations.

Looking at the fitting parameters, the variability can be compared to the ^{3}He counter, the reference. The A parameter, shown in Fig. 8a, representing the amplitude over a powered distance, is among the highest yet is the most stable in that category. The most stable is the reference ^{3}He but has a drastically lower efficiency. That's explained by the fact that nFacet is very sensitive to both thermal and fast neutrons.

In Fig. 8c, the x factor, the power law on the distance that represents the amount of neutron scattering, the closest to a value of 2 represents a detector little to not affected by neutrons backscattering. nFacet is the second best in stability, just behind the ^{3}He. It's notable to see that the x factor, in the case of the outdoor measurement, is very close to 2. It was unexpected to get such a high value in this configuration.

Finally, the B factor in Fig. 8b should be compatible with 0 if the model was realistic, since it's not the case it should be added to get better results.

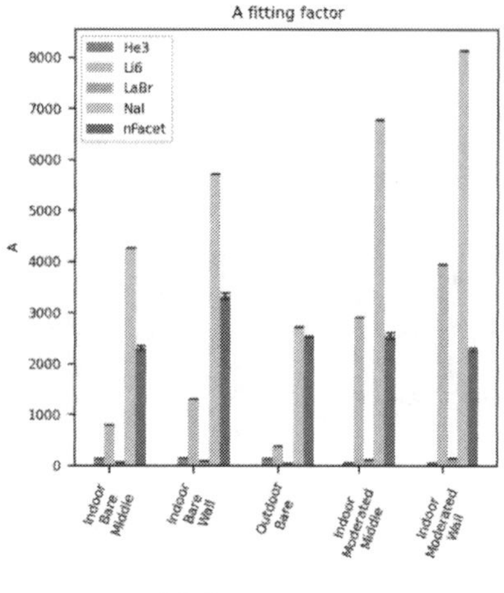

(a) A parameter.

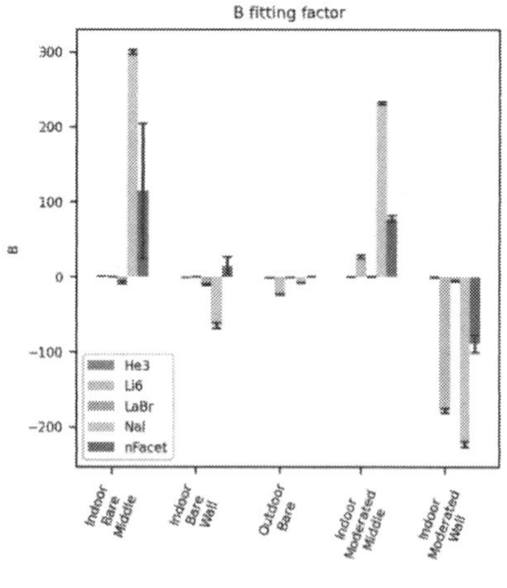

(b) B parameter.

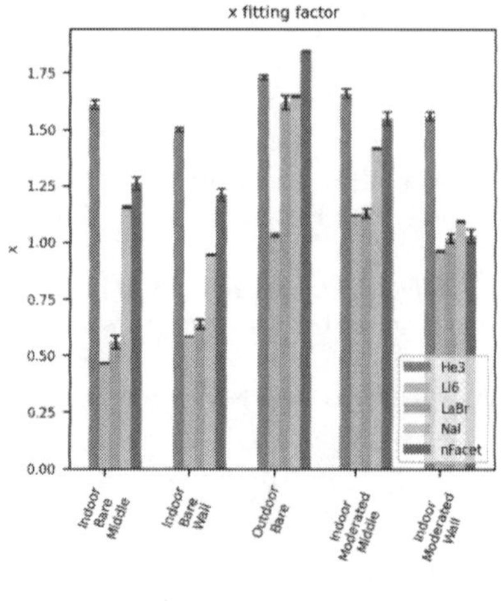

(c) x parameter.

Fig. 8. Fitting parameters for different neutron counter technologies.

V. Conclusion

The in-field performances of the nFacet 3D system were shown across a wide range of parameters. Using the segmentation of the system with an IA model, the direction of a source can be accurately estimated, with an angular resolution below 5° for two standard gamma sources at greater distances than typical gamma cameras due to the system's high efficiency. Gamma source separation has been achieved using the energy window technique, separating a low-energy gamma ray source and a high-energy source. A great sensitivity for neutrons has been shown, detecting a neutron source at least 50 meters away and possibly reaching 80 or 100 meters. The stability of the system has been compared to reference ^{3}He proportional counter in the context of nuclear disarmament monitoring and verification processes, showing good stability across multiple configurations. All these parameter are needed for area monitoring and security, making the nFacet 3D system perfectly adapted to this field of applications.

VI. Acknowledgements

We would like to thank Dr. Sakari Ihantola of nGate for the invitation. We would like to thank Dr. Jussi Huikari for his help during the measurement campaign and even after.

References

[1] V. Bogdanoff and K. Peräjärvi, "Improving the Confidence Associated with Passive Total Neutron Counting in the Nuclear Weapon Disarmament Verification Process," ESARDA Bulletin, vol. 65, pp. 2–9, Dec. 2023, DOI 10.3011/ESARDA.IJNSNP.2023.1.

[2] M. Gmar, M. Agelou, F. Carrel, and V. Schoepff, "GAMPIX: A new generation of gamma camera," Nucl. Instrum. Methods Phys. Res. A, vol. 652, pp. 638–640, Oct. 2011, DOI 10.1016/j.nima.2010.09.003.

[3] IAEA. *Livechart Data Download API.* [Online]. Available: https://www-nds.iaea.org/relnsd/vcharthtml/api_v0_guide.html Accessed on: Aug. 1, 2025.

[4] J. Lerendegui-Marco, V. Babiano-Suárez, J. Balibrea-Correa, L. Caballero, D. Calvo, I. Ladarescu, et al., "Simultaneous Gamma-Neutron Vision device: a portable and versatile tool for nuclear inspections," Eur. Phys. J. TI, vol. 11, no. 2, Feb. 2024, DOI 10.1140/epjti/s40485-024-00108-w.

[5] I. Loshchilov and F. Hutter. "Decoupled Weight Decay Regularization". *ICLR-2019.* 2019. [Online]. Available: https://arxiv.org/abs/1711.05101.

[6] N. Mascarenhas, J. Brennan, K. Krenz, P. Marleau, and S. Mrowka, "Results With the Neutron Scatter Camera," IEEE Trans. Nucl. Sci. vol. 56, no. 3, pp. 1269–1273, 2009, DOI 10.1109/TNS.2009.2016659.

[7] S. Pedemonte, L. Pierce, and K. Van Leemput, "A machine learning method for fast and accurate characterization of depth-of-interaction gamma cameras," Phys. Med. Biol. vol. 62, pp. 8376, Oct. 2017, DOI 10.1088/1361-6560/aa6ee5.

[8] S. Simakov and International Atomic Energy Agency International Nuclear Data Committee Vienna (Austria). *Evaluation of the Prompt Gamma-ray Spectrum from Spontaneous Fission of 252Cf.* Feb. 2024. DOI 10.61092/iaea.x6kd-w5qa.

[9] C. G. Wahl, W. Kaye, W. Wang, F. Zhang, J. Jaworski, Y. A. Boucher, et al., "Polaris-H Measurements and Performance," NSS/MIC, 2014 IEEE, Nov. 2014, DOI 10.1109/NSSMIC.2014.7431109.

Development of a radiation detection system for the prevention of illicit trafficking of special nuclear materials and radiation sources

Luís Marques[1,*], Francisca Moita[2], Alberto Vale[3] and Pedro Vaz[4]
[1]Centro de Investigação da Academia da Força Aérea, Instituto Universitário Militar, Portugal
[2]Instituto Superior Técnico, Universidade de Lisboa, Portugal
[3]Instituto de Plasmas e Fusão Nuclear, Instituto Superior Técnico, Universidade de Lisboa, Portugal
[4]Centro de Ciências e Tecnologias Nucleares, Instituto Superior Técnico, Universidade de Lisboa, Portugal
(*) lumarques@academiafa.edu.pt

***Abstract*—A comprehensive analysis of a novel dual-mode radiation detection system developed for preventing illicit trafficking of special nuclear materials and radioactive sources is presented. The system combines EJ-200 and EJ-426HD plastic scintillators with silicon photomultipliers to simultaneously detect gamma/beta radiation and neutrons, designed for integration with unmanned aerial vehicles and handheld operation. Through extensive experimental validation and comparative analysis against established detection methods, the system demonstrates superior performance across multiple evaluation criteria including localization accuracy, detection time, and cost-effectiveness.**

***Keywords* —Radiation detection, unmanned aerial vehicle, special nuclear material, informative path planning.**

I. Introduction

THE reinforcement of detection and control measures for Special Nuclear Materials (SNM) and other radioactive substances is a critical priority for global security and for ensuring the safe and sustainable use of radiation sources across sectors such as medicine, industry, energy, agriculture, space, and research. SNM—including plutonium and highly enriched uranium—pose substantial risks if diverted for non-peaceful applications, including the construction of improvised nuclear devices (INDs). Moreover, the illicit trafficking of other radioactive materials or sources raises concerns about their potential malicious use in radiological dispersion devices (RDDs), such as "dirty bombs," and radiological exposure devices (REDs), both of which can result in severe social, economic, and psychological disruption [1].

Therefore, effective control of radioactive sources throughout their lifecycle—"from cradle to grave"—is essential. Special attention is needed at critical points such as borders, seaports, and airports. While Radiation Portal Monitors (RPMs) are essential for detecting gamma and, in some cases, neutron emissions, they are expensive (due to high acquisition, installation and maintenance costs), fixed infrastructures that scan only one container or vehicle at a time. RPMs also rely on secondary inspections whenever alarms are triggered, requiring manual intervention and leading to long inspection times and increased operational costs (over 20 minutes for a single maritime container) [2-4]. Moreover, the use of handheld detectors requires proximity to the cargo, exposing operators to potential radiation hazards. If inconclusive, a full physical inspection is initiated—typically involving five personnel and requiring up to three hours to examine a standard 40-foot container thoroughly [4].

To overcome these limitations, literature proposes several complementary detection strategies, including:

- Spatially distributed detector networks: Compared to a single detection unit, it enhances coverage and confidence in identifying moving radioactive sources. However, it is impractical for large-scale, dynamic settings like container terminals [5];
- Portable RPMs: While intended to be transportable between sites, they are not genuinely mobile during operation due to their large size, high power consumption, and unsuitability for inspections in confined or elevated spaces [6];
- Vehicle-mounted systems: Effective for broad coverage but hindered by limited accessibility, terrain restrictions, and operational complexity [7].

Against this backdrop, recent developments have focused on compact and mobile systems capable of detecting and localizing radioactive sources in real-time. A key enabler is the integration of gamma source search algorithms—notably Maximum Likelihood Estimation (MLE) and Bayesian inference—which process sensor data to estimate the 3D location of a source. Among these, MLE has become particularly relevant due to its efficiency, accuracy, and compatibility with mobile sensor platforms [5,8].

However, a significant limitation of many existing approaches is their reliance on predefined scanning paths–typically grid or linear sweep trajectories–that fail to adapt to real-time variations in the radiation field. While these paths offer systematic coverage, they are often inefficient in terms of time and energy consumption, especially in environments with non-uniform container layouts, variable background radiation, or shielded sources that require adaptive sensing strategies. For instance, Sanada et al. [9] and Carrel et al. [10] implemented grid-based Unmanned Aerial Vehicle (UAV) surveys for post-accident mapping, which required extended flight times without real-time adaptation.

In contrast, the use of informative path planning—where the inspection path is dynamically updated based on real-time sensor data—has been shown to significantly improve detection performance and operational efficiency [11,12]. Marques et al. [12-14] implemented informative path planning strategies using profit functions to evaluate whether adjusting the detection system's path during container inspection yields sufficient benefit relative to its cost, guiding decisions to optimize inspection efficiency. These methods enabled a UAV to adjust its trajectory during flight to maximize information gain and reduce search time, achieving sub-meter localization of a 4 MBq ^{137}Cs source.

Building on these advances, this article summarizes the development, integration, and evaluation of a compact, dual-mode radiation detection system capable of detecting both gamma and neutron emissions. Designed for deployment on multirotor UAVs and in handheld mode, the system features low-cost plastic scintillators with Silicon Photomultiplier (SiPM) readout, a modular neutron moderator, and a lightweight architecture suitable for mobile platforms. It incorporates real-time MLE-based localization and informative path planning. The radiation detection can be integrated in small mobile platforms such as an UAV increasing its versatility and flexibility for Security scenarios. A comparison with existing radiation detection methods, such as the use of fixed RPM, van-mounted detector, and spatially distributed detector network, is also performed.

II. System Architecture and Design

A. Dual-Mode Detection System

The radiation detection system employs two complementary plastic scintillators optimized for different radiation types:

- Neutron Detection System: Features an EJ-426HD scintillator with ^{6}Li content, coupled with modular High-Density Polyethylene (HDPE) moderator plates. Experimental tests using an ^{241}Am-Be source demonstrated detection efficiency improvements of up to 79% by changing the moderator plates position relative to the source. The M40R40 configuration (shown in Fig. 1)—consisting of two 20 mm-thick plates on each side of the detector (with "M" indicating moderator thickness on the source side and "R" the reflector side)—offered the best compromise between weight and neutron detection efficiency to both moderated and unmoderated ^{241}Am-Be sources. Moreover, the detection system's low gamma sensitivity makes it well-suited for security applications by reducing false positives [15];
- Gamma/Beta Detection System: Utilizes a 110 mm diameter × 30 mm height EJ-200 plastic scintillator with a 32 μm titanium window for beta particle detection (Fig. 2). Thanks to its pancake shape, the EJ-200 achieves higher detection efficiency than a weight-equivalent CsI(Tl) scintillator (51 mm diameter × 51 mm height) for ^{137}Cs source distances between 1 and 5 meters, with a signal-to-noise ratio nearly three times greater than that of CsI(Tl) [16].

Both detectors use SiPM readout technology, enabling low-voltage operation (5V) suitable for battery-powered mobile platforms while maintaining excellent timing characteristics and radiation tolerance.

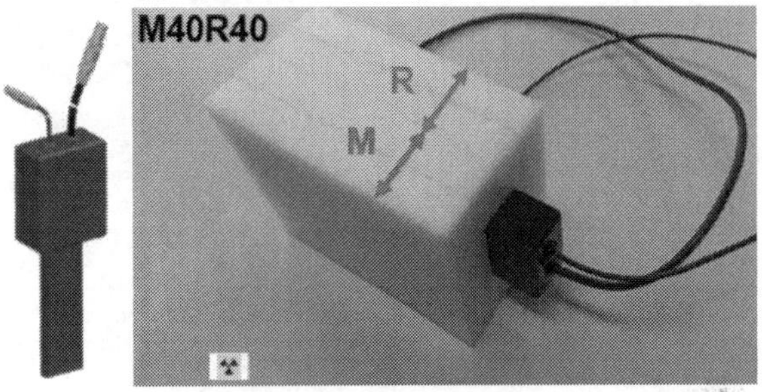

Fig. 1. EJ-426HD scintillator (left image) within the HDPE moderator using M40R40 configuration (right image).

Fig. 2. EJ-200 scintillator with Ti window, and detail of the 3D printed support.

B. Mobile Platform Integration

The system was designed for dual-mode operation either UAV platform or as handheld equipment. UAV integration utilizes a DJI Matrice 600 Pro, a hexacopter with 6 kg payload capacity, accommodating the 2.8 kg detection system through a custom carbon fiber mounting plate for optimal weight distribution (as shown in

Fig. *3*). Flight endurance ranges from 17-22 minutes, sufficient for typical container inspections [16,17].

Fig. 3. Radiation detection system integrated into an UAV (hexacopter).

C. *Real-Time Localization and Path Planning*

The system implements an MLE algorithm for three-dimensional source localization hidden inside a standard maritime container (20-foot long), operating at 1 Hz data acquisition rate synchronized with a Global Navigation Satellite System (GNSS) receiver and a Light Detection and Ranging (LiDAR) used for altitude measurements. An IPP algorithm based on profit functions was employed for dynamic trajectory optimization during inspection. The algorithm allowed to adapt the detection system path according to estimated source coordinates [12] or detected count rates [13,14], thereby improving localization accuracy and significantly reducing inspection time compared to predefined grid patterns.

III. Methodology and Evaluation Criteria

A. *Comparative Analysis Framework*

The performance evaluation employed a comprehensive set of criteria designed to assess operational effectiveness and technical improvement across multiple dimensions. The evaluation framework encompasses both quantitative performance metrics and qualitative operational characteristics essential for practical deployment in security scenarios. The primary evaluation criteria are:

- Source localization accuracy: Measured as Euclidean distance between estimated and true source position, this criterion serves as the primary figure-of-merit for system effectiveness. Sub-meter accuracy is critical for effective intervention and source isolation in security applications;
- Source detection/localization time: Essential for operational efficiency and throughput, measured from initial detection to confident source localization. Faster detection enables higher inspection rates and reduced operational costs;
- Search volume/area: Defines the region considered for the source position estimation;
- Platform versatility: Evaluates system adaptability across different deployment scenarios, including mobile vs. fixed installations and multi-platform compatibility;
- Detection capability: Assesses dual-mode gamma/neutron detection essential for SNM identification and security applications;
- Cost: High per-unit costs limits scalability, technology adoption and operational flexibility.

In security and defense scenarios, it is also crucial to select detection systems that demonstrate a high true positive rate—closely tied to system efficiency (ε_γ, ε_n) and sensitivity—to ensure reliable detection of weak radiation sources, such as SNM or shielded sources. At the same time, achieving a low false positive rate is equally important to minimize false alarms and operational disruptions, thereby preserving normal cargo flow. This performance balance enables effective discrimination between illicit or orphan sources and benign materials, including naturally occurring radioactive materials and medical radioisotopes, optimizing both threat identification and operational efficiency.

IV. Results and Comparative Analysis

A. *Source detection and localization methods comparison*

Experimental validation by Marques et al. [14,16,17], using 4 MBq ^{137}Cs and 1.45 GBq ^{241}Am-Be sources within maritime container mockups, highlighted the advantages of a compact mobile radiation detection system with high maneuverability when deployed on a UAV. This system outperformed other approaches, such as fixed RPM and spatially distributed detector networks, across the key evaluation criteria. The proposed system achieved localization accuracy better than 0.5 meters within 60 seconds, significantly surpassing established methods as shown in Table I.

TABLE I
COMPARISON OF GAMMA SOURCE DETECTION AND LOCALIZATION METHODS

Feature	Authors work [14,16,17]	RPMs [3,4]	Cook et al. [18]	Sharma et al. [5]
^{137}Cs activity	4 MBq	592 kBq [a]	281 kBq	189 kBq
Localization accuracy and algorithm	Within 0.5 m (MLE) [b]	- [c]	~ 1.5 m (Particle Filter)	Within 0.6 m (MLE)
Detection/ localization time	< 60 s	Few seconds [d]	300 s	120–180 s
Search volume/area	33 m^3	Up to 200 m^3	100 m^2	125 m^3
Platform versatility	UAV / handheld	Fixed structure	Spatially distributed detectors	Spatially distributed detectors
Gamma detection system	Compact EJ-200 scintillator	Large plastic scintillators [e]	22 NaI(Tl) scintillators	5 NaI(Tl) scintillators
Neutron detection system	EJ-426HD scintillator	Typically ^{3}He proportional tubes	No	No
Cost	+	+++	++	+

[a] Using Informative path planning.
[b] Minimum detectable activity following ANSI N42.35.
[c] Most RPMs can resolve the detection to the "vehicle" or "container" level.
[d] Detection only; Secondary inspection may take approximately 20 min.
[e] Polyvinyl toluene (PVT) plastic-based scintillators.

Fixed detector networks and traditional RPMs, often prove impractical for large, complex environments like seaports. While fixed cylindrical 51 mm × 51 mm NaI(Tl) scintillators can localize sources with moderate to high accuracy, their lack of mobility and real-time adaptability restricts operational flexibility [3,5,18]. Furthermore, RPMs tend to be heavy, costly, and reliant on scarce ^{3}He detectors (due to worldwide gas shortage), limiting scalability for widespread deployment [19]. In contrast, mobile systems such as the proposed EJ-200 plastic scintillator integrated with UAV platforms offer a lightweight, adaptable, and cost-effective solution better suited to the dynamic and demanding inspection needs of seaport environments.

B. Technological improvement

The proposed system's advantages stem from several key innovations:

- The use of IPP algorithms based on profit functions enable real-time path optimization, reducing inspection time compared to predefined grid patterns while improving localization accuracy [12-14,16];
- Dual-mode detection: Unique neutron detection capability using EJ-426HD with optimized HDPE moderators provides essential SNM identification capability;
- Cost-effectiveness: The use of EJ-200, a commercially available, low cost and compact plastic scintillator, mounted in a highly maneuverable UAV allowed to reduce acquisition and operational costs compared to fixed RPMs and spatially distributed detector networks;
- Platform versatility: The possible integration into an UAV or other small platform provides unprecedented operational flexibility for diverse security scenarios, allowing to access confined spaces or to perform inspections at heights (e.g. stacked containers).

V. Criteria Justification and Reasoning

A. Localization Accuracy Criterion

Sub-meter localization accuracy represents the fundamental requirement for effective source intervention and isolation in security scenarios. The criterion directly impacts operational success probability and response time effectiveness. The proposed system's source localization accuracy, better than 0.5 m, provides sufficient accuracy for targeted source removal while minimizing unnecessary cargo handling.

B. Detection Time Optimization

Rapid detection capability directly correlates with operational throughput and economic viability in high-volume inspection scenarios. The 60-second detection threshold enables inspection of multiple containers within standard operational windows while maintaining statistical confidence in source localization.

C. Neutron Detection Imperative

Dual-mode gamma/neutron detection capability represents an essential requirement for SNM identification, as many special nuclear materials exhibit characteristic neutron emission signatures. The unique neutron detection capability of the proposed system is a important tool for complex scenarios where weak or shielded sources can be presented.

D. Cost-Effectiveness Analysis

Economic viability determines system scalability and widespread deployment potential. The significative cost reduction compared to traditional radiation detection systems offers a valid alternative of complement equipment while maintaining detection performance enabling broader security coverage and enhanced threat detection probability.

VI. Future Development Areas

Future development areas that require further research include:

- Advanced signal processing: Integration of machine learning algorithms for pattern recognition and anomaly detection could enhance detection sensitivity and reduce false alarm rates. Deep learning approaches may enable automated threat classification and priority ranking for inspection resources;
- Increase light yield of plastic scintillators: Plastic scintillators generally feature low light yield (~10 photons/keV) and poor energy resolution, limiting their use in gamma spectroscopy [20]. Recent advances, such as incorporating iridium-based fluorescent complexes into high-Z-doped plastics, have improved both light output and resolution [21]. Additionally, optimizing scintillator geometry by using smaller volumes with higher-density SiPM coverage enhances light collection efficiency and signal strength, boosting detection of low-energy events while reducing internal attenuation and enabling better coupling with readout electronics [22];
- Spectroscopic enhancement: Development of hybrid detection systems combining plastic scintillators for gross counting with semiconductor detectors for spectroscopic identification could provide comprehensive characterization capability while maintaining mobility advantages;
- Emerging semiconductor materials such as TlBr and perovskite halide crystals have shown promising performance for room-temperature spectroscopic detection, offering a potential low-cost alternative with high efficiency [23];
- Energy windowing for background discrimination: By tailoring energy discrimination thresholds, it is possible to suppress background noise and isolate events within expected energy bands of interest. This approach does not enable full radionuclide identification using plastic scintillators but provides useful gross count differentiation, enhancing the system's operational capability when deployed in mixed radiation fields or cluttered environments [24,25];
- Multi-platform coordination: Implementation of cooperative detection strategies using unmanned vehicles could enable simultaneous multi-area surveillance and cross-verification of detection results, enhancing overall system reliability and coverage;
- Environmental robustness: Enhanced weather resistance and GNSS-denied navigation capabilities using LiDAR data and simultaneous localization and mapping algorithms could extend operational envelope to challenging environmental conditions and confined spaces;
- Autonomous operation: Development of fully autonomous inspection protocols with minimal human oversight could reduce operational costs and enable continuous monitoring capability in high-risk areas.

VII. Conclusion and Future Work

The developed radiation detection system represents a paradigm shift in mobile nuclear security technology, demonstrating improvement across all primary evaluation criteria while achieving unprecedented cost-effectiveness. The comprehensive comparative analysis reveals significant advantages over established detection methods, particularly in source localization accuracy, detection and localization time, and operational versatility.

The integration of IPP algorithms with dual-mode detection capability creates a synergistic effect that enhances efficiency and effectiveness.

The unique combination of sub-meter localization accuracy, rapid detection capability, neutron sensitivity, and extreme cost-effectiveness positions this system as a transformative technology for nuclear security applications. The demonstrated performance advantages justify adoption for maritime container inspection, border security, and emergency response scenarios where traditional fixed infrastructure proves inadequate or insufficient.

Future development efforts should focus on spectroscopic enhancement, multi-platform coordination, and autonomous operation capability to further extend the operational envelope and detection effectiveness. The modular, scalable design provides a foundation for continued technological advancement while maintaining the core advantages of mobility, cost-effectiveness, and operational flexibility that distinguish this approach from conventional detection methods. Other envisaged fields include applications in nuclear forensics, the inspection of nuclear facilities—both fission and fusion—and particle accelerator environments.

References

[1] International Atomic Energy Agency, "Combating illicit trafficking in nuclear and other radioactive material", International Atomic Energy Agency, Vienna, Tech. Rep. 6, 2008.

[2] ORTEC (Ametek), *High-Resolution Detection Systems for Interdiction of Nuclear Material Trafficking*, Technical Report, 2 April 2007.

[3] E. L. Connolly and P. G. Martin, "Current and Prospective Radiation Detection Systems, Screening Infrastructure and Interpretive Algorithms for the Non-Intrusive Screening of Shipping Container Cargo: A Review," *J. Nucl. Eng.*, vol. 2, no. 3, pp. 246–280, 2021. DOI. 10.3390/jne2030023.

[4] R. Downes, C. Hobbs, and D. Salisbury, "Combating nuclear smuggling? Exploring drivers and challenges to detecting nuclear and radiological materials at maritime facilities", Nonproliferation Review, vol. 26, pp. 83–104, 1-2 2019-01. DOI. 10.1080/10736700.2019.1610256.

[5] A. Sharma, M. Alsharif, and P. H. Rogers, "Fast Three-Dimensional Localization of a Gamma Source Using a Small Network of NaI Detectors," Sensors, vol. 19, no. 14, p. 3062, Jul. 2019. DOI. 10.3390/s19143062.

[6] U.S. Department of Homeland Security, Radiation Portal Monitors (RPMs): Market Survey Report, Systems Assessment and Validation for Emergency Responders (SAVER), Report No. SAVER-RPM-15, Mar. 2015.

[7] L. Marques, A. Vale, and P. Vaz, "State-of-the-art mobile radiation detection systems for different scenarios", *Sensors*, vol. 21, no. 4, 2021, DOI. 10.3390/s21041051.

[8] J.-C. Chin et al., "Identification of Low-Level Point Radioactive Sources Using a Sensor Network," ACM Trans. Sensor Netw., vol. 7, no. 3, Sep. 2010. DOI. 10.1145/1824766.1824769.

[9] Y. Sanada, M. Torii, and K. Yoshida, "Development of a Portable Radiation Mapping System Using a Small UAV for the Fukushima Nuclear Accident," IEEE Trans. Nucl. Sci., vol. 61, no. 5, pp. 2202–2208, Oct. 2014. DOI. 10.1109/TNS.2014.2339591.

[10] A. Carrel, P. M. D. Arsac, B. Perot, M. Amiel, and G. Pochon, "Gamma Source Localization With a Fleet of Mini-UAVs Using Particle Filtering," IEEE Trans. Nucl. Sci., vol. 59, no. 6, pp. 3114–3120, Dec. 2012. DOI. 10.1109/TNS.2012.2225630.

[11] Y. Brouwer, A. Vale, and R. Ventura, "Informative path planner with exploration–exploitation trade-off for radiological surveys in non-convex scenarios", Robotics and Autonomous Systems, vol. 136, p. 103 691, 2021. DOI. 10.1016/j.robot.2020.103691.

[12] L. Marques, R. Coito, S. Fernandes, T. Costa, A. Vale, and P. Vaz, "A mobile radiation detection system for security inspection and monitoring", *EPJ Web Conf.*, vol. 288, p. 06 001, 2023, DOI. 10.1051/epjconf/202328806001.

[13] L. Marques, R. Coito, T. Costa, S. Fernandes, A. Vale, and P. Vaz, "Path optimization of a mobile radiation detection system in security inspection", in *Developments and Advances in Defense and Security*, Á. Rocha, C. H. Fajardo-Toro, and J. M. Riola, Eds., ser. Smart Innovation, Systems and Technologies, vol. 423, Singapore: Springer, 2025, pp. 293–305, DOI. 10.1007/978-981-96-0235-3_24.

[14] L. Marques, R. Coito, T. Costa, S. Fernandes, A. Vale, and P. Vaz, "Radioactive source localization using a mobile radiation detection system featuring informed path-based decisions", *IEEE Transactions on Nuclear Science*, vol. 71, no. 5, pp. 1064–1071, 2024. DOI. 10.1109/TNS.2024.3354461.

[15] L. Marques, A. Vale, and P. Vaz, "Development of a portable neutron detection system for security and defense applications", in *Developments and Advances in Defense and Security*, Á. Rocha, C. H. Fajardo-Toro, and J. M. Riola, Eds., ser. Smart Innovation, Systems and Technologies, vol. 328, Singapore: Springer, 2023, pp. 283–293, DOI. 10.1007/978-981-19-7689-6_24.

[16] L. M. C. Marques, *Development of a Radiation Detection System Coupled to an Unmanned Aerial Vehicle for Security and Defence Applications*, Ph.D. dissertation, Dept. Technol. Phys. Eng., Instituto Superior Tecnico, Univ. Lisboa, Lisbon, Portugal, 2025.

[17] L. Marques, L. Félix, G. Cruz, *et al.*, "Neutron and gamma-ray detection system coupled to a multirotor for screening of shipping container cargo", *Sensors*, vol. 23, no. 1, 2023, DOI. 10.3390/s23010329.

[18] J. Cook, C. R. Smith, C. Ramirez, and N. S. V. Rao, "Particle filtering convergence results for radiation source detection," arXiv:2004.08953, 2020.

[19] R. T. Kouzes, J. H. Ely, L. E. Erikson, *et al.*, "Neutron detection alternatives to ^{3}He for national security applications", *Nuclear Instruments and Methods in Physics Research Section A: Accelerators, Spectrometers, Detectors and Associated Equipment*, vol. 623, no. 3, pp. 1035–1045, 2010. DOI. 10.1016/j.nima.2010.08.021.

[20] E. Technology, General purpose EJ-200, EJ-204, EJ-208, EJ-212, https://eljentechnology.com/products/plastic-scintillators/ej-200-ej-204-ej-208-ej-212 (accessed: June 4, 2025).

[21] T. J. Hajagos, C. Liu, N. J. Cherepy, and Q. Pei, High-Z Sensitized Plastic Scintillators: A Review, 2018-07. DOI. 10.1002/adma.201706956.

[22] Arktis, Mobile radiation monitoring system, https://www.arktis-detectors.com/de/products/mobile-radiation-monitoring-system/ (accessed June 12, 2025).

[23] A. Datta, P. Becla, and S. Motakef, "Novel Electrodes and Engineered Interfaces for Halide-Semiconductor Radiation Detectors", Scientific Reports, vol. 9, p. 9933, 2019. DOI. 10.1038/s41598-019-46360-z.

[24] J. Ely, R. Kouzes, J. Schweppe, E. Siciliano, D. Strachan, and D. Weier, "The use of energy windowing to discriminate SNM from NORM in radiation portal monitors", Nuclear Instruments and Methods in Physics Research Section A: Accelerators, Spectrometers, Detectors and Associated Equipment, vol. 560, no. 2, pp. 373–387, 2006. DOI. 10.1016/j.nima.2006.01.053.

[25] R. Hevener, M.-S. Yim, and K. Baird, "Investigation of energy windowing algorithms for effective cargo screening with radiation portal monitors", Radiation Measurements, vol. 58, pp. 113–120, 2013. DOI. 10.1016/j.radmeas.2013.08.004.

Portable Compton Camera for Enhanced Radioactive Waste Management

Víctor Babiano-Suárez[1,*], Javier Balibrea-Correa[1], Ion Ladarescu[1], Jorge Lerendegui-Marco[1] and César Domingo-Pardo[1]
[1]Instituto de Física Corpuscular, CSIC-UV, Spain
(*) victor.babiano@ific.uv.es

***Abstract*—The accurate characterization and management of radioactive waste are critical for the safe operation and decommissioning of nuclear facilities. This proceeding summarizes the development, validation, and performance evaluation of a novel Compton-imaging system enhanced with computer-vision techniques. Designed to be portable and cost-effective, this system aims to improve the detection and visualization of low- and medium-level radioactive waste. The system utilizes a state-of-the-art Compton camera integrated with advanced image reconstruction algorithms and real-time computer vision for enhanced performance. Experimental validation at the radioactive-waste disposal plant of El Cabril (Spain) demonstrated the capability of the system to accurately detect and characterize sealed drums of radioactive waste, achieving a technical readiness level of TRL 7. The results highlight the system's potential for creating detailed 3D tomographic reconstructions and its feasibility as a promising innovation for the nuclear industry, particularly for improving nuclear waste management and decommissioning processes.**

***Keywords* —Computer vision, Radiation detector, Compton Imaging, Monte Carlo technique, Machine Learning.**

I. Introduction

THE primary challenge of nuclear power is the generation and subsequent storage of radioactive waste. Accurate characterization of waste, including the identification of isotopes and their spatial distribution within storage containers, is a prerequisite for proper classification, treatment, and long-term disposal. Traditional methods for waste characterization often involve a combination of non-destructive assay techniques and destructive analysis, which can be time-consuming, costly, and pose potential radiation exposure risks to personnel. Although research groups around the world are actively working on developing new detection devices and techniques to address these challenges [1, 2, 3], a notable limitation of many current systems is their suboptimal detection efficiency and lengthy characterization time.

Compton imaging, a non-invasive technique, has emerged as a powerful tool for nuclear material detection and visualization due to its potential large field of view (FOV), high detection sensitivity, and portability. This work presents the results of field measurements using a novel Compton-imaging system that integrates computer-vision techniques enabling hybrid gamma-ray and visible-image reconstruction. This proceeding provides a summary of the system's design, experimental validation, and its potential impact on the nuclear industry, drawing from the detailed work presented in a recent publication [4].

II. Experimental Setup

The Compton imager used in this work was originally developed for nuclear astrophysics research [5] and is notable for its exceptional detection efficiency. This high efficiency is a result of its innovative design [6], which features an absorber (rear) detector with four times the area of the scatter (front) detector. This configuration significantly increases the number of scatter-absorber coincidence events [7], thereby boosting detection efficiency compared to conventional Compton cameras. Furthermore, the imager also includes a remotely controlled positioning drive that allows the adjustment of the distance between the scatter and absorber planes (d_f), enabling a strategic trade-off between detection efficiency and image resolution [8]. For this specific application, d_f was optimized to 35 mm to maintain high efficiency, which is especially beneficial for applications requiring short measurement times or real-time monitoring.

To implement the Compton imaging technique, the gamma-ray imager utilizes Position Sensitive Detectors (PSDs) for both scatter and absorber layers. The scatter detector consists of a single PSD, whereas the absorber is an array of four PSD units. Each PSD features a monolithic $LaCl_3$(Ce) crystal, with dimensions of 50×50 mm^2 face, by 10 mm and 25 mm thickness for the scatterer and absorber, respectively. Silicon Photomultipliers (SiPMs) (SensL ArrayJ-60035-65P-PCB), segmented into an 8×8 pixel array covering 50×50 mm^2, read out the scintillation photons. Application-Specific Integrated Circuit (ASIC)-based (TOFPET2) readout electronics from PETsys digitize the analog SiPM signals [9, 10].

To mitigate the impact of temperature variations on ASIC performance, 20×20 mm^2 Peltier cells (FPH1-7106NC) were thermally coupled to each ASIC [8]. The temperature near each ASIC was also logged at short intervals, enabling a numerical gain correction during offline data analysis. Under these conditions, the system achieved an average energy resolution of 7% FWHM at 662 keV. Advanced positioning algorithms were used to determine the 3D coordinates of gamma-ray interactions within each crystal, yielding spatial resolutions of approximately 2 mm FWHM in the transversal (XY) plane and 5 mm FWHM in the orthogonal Z-coordinate (depth of interaction) [11].

A key innovation of this work is the integration of a computer-vision system. This feature was implemented using a simple and cost-effective solution, consisting of an off-the-shelf RGB camera (Logitech HD C270) mounted at the front of the imager. This camera was connected to a laptop (ASUS GL552V), which also served as the primary data acquisition unit. The entire system was mounted on a compact trolley, which provides the mobility necessary for operators to precisely position the device for measurements.

Fig. 1. Measurement system placed in front of five radioactive bulks.

The entire system was tested at the field measurements conducted at the disposal facility of El Cabril, located in Córdoba, Spain [12], and managed by the national Waste Management Organization (WMO) for the management of radioactive waste, ENRESA [13]. Fig. 1 shows a picture of the experimental setup, with the measurements system positioned in front of an arrangement of five radioactive drums. Each steel barrel, with a radius of 28 cm, a height of 85 cm, and a maximum capacity of 700 kg, was filled with radioactive material. The latter generated during the decommissioning and regular operation of the power plant of Santa María de Garoña (GR), a BWR with 460 MW, and Ascó (AS), a pressurized-water reactor (PWR) of 1032 MW. The barrels labeled AS are filled with pressed and compacted cellulose, and their content is expected to be heterogeneous. In contrast, drums marked GR contain chemicals and other liquids that have been mixed with cement and solidified. Consequently, they are expected to be less heterogeneous than the AS barrels.

Most of the activity in the drums originates from the decay of ^{60}Co ($t_{1/2} = 5.27$ y), which is artificially produced via neutron activation of steel structures in the reactor. The activity of the five drums was previously characterized by the producer and validated by ENRESA. For the barrels labeled AS, several measurements are taken around the drum, with the average of those (performed at half-height) typically used as the final quoted contact dose rate (CDR). This procedure, however, is only accurate for barrels with a relatively homogeneous activity distribution. In contrast, for GR barrels, small but representative samples are taken before solidification for gamma spectrometry. The resulting activity is then extrapolated based on the drum's waste mass, assuming a homogeneous activity distribution. Both methods apply conversion factors that account for isotopic content and barrel composition to calculate the final activity presented in Table I. This value includes only the ^{60}Co activity that can be effectively measured on the surface of the bulk. For AS barrels, this quantity is considered the same as that obtained from the traditional method, while an attenuation factor is applied to GR barrels to account for their homogeneous radioactive content mixed with cement.

TABLE I
ESTIMATED ACTIVITY FOR DRUMS

Bulk ID	External activity
AS27954	284
AS28298	386
AS29448	38
GR15748	109
GR36256	112

Activity (MBq) due to ^{60}Co of the five barrels analysed in this work.

The five radioactive bulks listed in Table I were arranged in two different configurations for measurement. This experimental design aimed to demonstrate the large field of view and a sufficiently large image resolving power of the hybrid imaging system for detecting and monitoring variations in the nuclear inspection of a complex scenario. Fig. 2 shows the schematic view of one of these barrel configurations, along with the eight measurement positions where the gamma-ray imager was placed. For conciseness, only data from this configuration are presented herein. Nevertheless, comprehensive results from this experimental campaign are available in reference [4].

III. METHODOLOGY

The system described in the previous Section II is intended to provide a 2D image or heat map, using a color scale calibrated in activity units (Bq), which depicts the spatial distribution of the gamma-radiation overlaid onto a picture captured by the RGB camera. To this aim, the different techniques described in the following sections are employed.

A. Maximum-Likelihood based algorithm

In a first approach, the image of a gamma-ray source can be obtained using a back-projection algorithm [14]. This method involves superimposing the Compton cones onto a plane placed at the source position, oriented in parallel to the detection planes. However, the performance of this algorithm has been further improved by applying an iterative method for List Mode Likelihood Expectation Maximization (LMLEM) [15] for Compton Cameras [16, 17]. For this work, a far-field-of-view (FOV) approximation was employed. This approach assumes that the distance between the detector and the image plane is sufficiently large, allowing the system's efficiency to be treated as a constant. In our experimental setup, this condition was satisfied by positioning the gamma-ray imager at distances of around 2-3 m from the radioactive bulk arrangement. This specific configuration allowed the arrangement to fit within the central region of the FOV. In this region, the detection efficiency varies 10% according to Monte Carlo (MC) simulations.

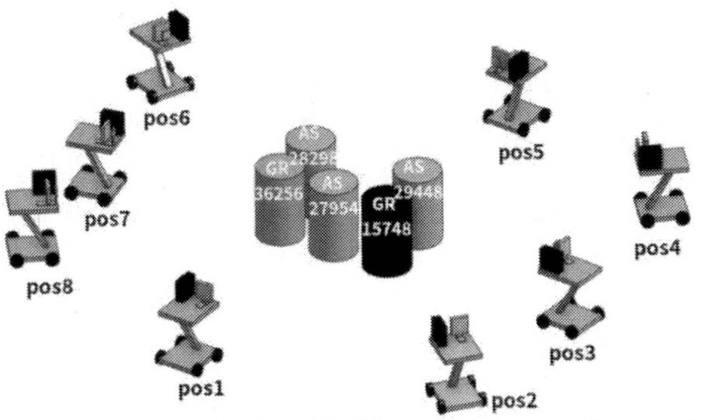

Fig. 2. CAD representation of the eight different measurement positions around the barrel arrangement.

B. *Computer vision*

In order to overlay the gamma-ray distribution image to the picture captured by the RGB camera, computer vision techniques are employed. On the one hand, the detection of fiducial markers allows the system to compute the distance to the barrel arrangement [1, 18]. Additionally, the markers' transversal plane (XY) size in pixels facilitated proper scaling of both Compton and RGB images through a transformation factor converting meters to pixels. To achieve this, an algorithm utilizing the Aruco C++ library [19, 20], was developed.

On the other hand, an alternative methodology based on the use of Artificial Intelligence (AI) was also implemented, enabling the individual identification of each barrel and yielding their X and Y coordinates in the image. This spatial information not only aids in resolving the Compton image within the region occupied by the drums but also offers an alternative method for getting the Z distance without the need for fiducial markers in the future. In order to accomplish this task, we employ a custom-trained model based on Tensorflow [21], which is an open-source library developed by Google for machine learning (ML) tasks, such as object detection.

C. *Real-time determination of external activity*

Calibrating the heat map in activity units requires estimating the activity of the bulk ensemble. In this work, we employed a method based on MC simulation, as outlined in [22]. A realistic model of the Compton camera was simulated using the C++ based GEANT4 toolkit [23]. The data from these MC simulations were then used to determine the apparent activity measured by each PSD in real-time. Although the accuracy of this method is approximately two to three times its stated value [22], it provides a fast, preliminary estimate that is highly useful for initial classification.

After generating the heat map of the gamma-radiation and overlaying it onto the visual image, the average activity is distributed over a flat surface according to the intensity of the Compton surface and providing the final 2D hybrid gamma-visual image.

D. *3D hybrid-image reconstruction*

Finally, the measurement of the barrel ensemble from different poses around it (see Fig. 2), enables the tomographic reconstruction of the 3D gamma-radiation distribution. To achieve this, the Compton cones are back-projected through the three-dimensional space that has been discretized into voxels. An effective attenuation coefficient of 1 m^{-1} is applied to take into account not only the attenuation of gamma-rays travelling through the drums but also a correction for the overlapping effect induced by the Compton arcs in the back-projection method. At the end, this process yields a complete 3D distribution of the registered gamma radiation, achieving a spatial resolution of 10 cm on each axis, as demonstrated by MC simulations [4].

IV. RESULTS

This section shows the results of applying the methodology outlined in the previous Section III to the data measured using the Compton imager presented in Section II and the barrel ensemble depicted in Fig. 2. First, the apparent activity estimated for the drum configurations. Here one is taking advantage of the large efficiency for single gamma-ray detection with the five large uncollimated $LaCl_3(Ce)$ crystals embedded in the Compton camera. Fig. 3 shows the overall apparent activity determined for each pose and compared to the total activity given by ENRESA (Table I). The very first remarkable aspect are the strong differences (factor 2-3) appreciated for some specific poses, despite the fact that the total activity remains constant. That is due to the rather different radiation absorption coefficient for GR barrels in comparison to AS, which is a factor 2 higher for the former. Indeed, for each pose GR barrels are expected to induce a large shielding effect for the radiation coming from the barrel(s) behind them, and a corresponding decrease in the measured apparent activity. On the contrary, AS-type barrels allow a much higher portion of high-energy gamma-rays to pass through them. Furthermore, it is worth noting that such a factor of 2-3 is comparable to the uncertainty expected for the implemented activity assessment methodology [22].

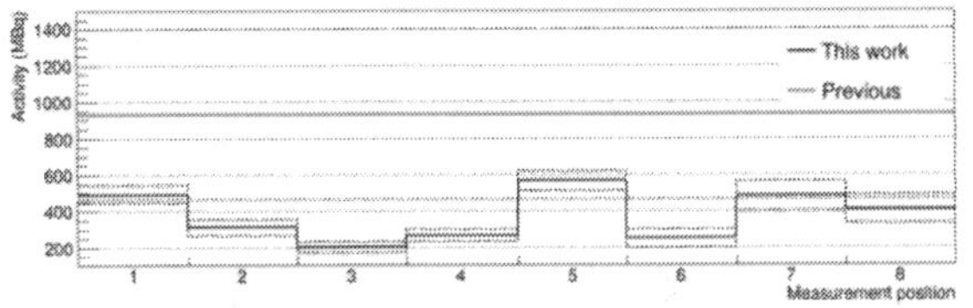

Fig. 3. Activity estimated for the drum ensemble at the eight measurement positions, compared to the previously available value.

Next, a brief discussion based on the 2D and 3D reconstructed images will be made. Fig. 4 displays the 2D hybrid gamma-visible images for the drum ensemble measured from positions 4, 5 and 6. As it can be seen, drums are enclosed in pink rectangles as a result of their detection by the AI model.

Moreover, the heat map calibrated in activity units shows the distribution of the gamma-radiation at each pose. Results are consistent with each other and, at each position, the hot-spot points to the most active barrels labelled as AS28298.

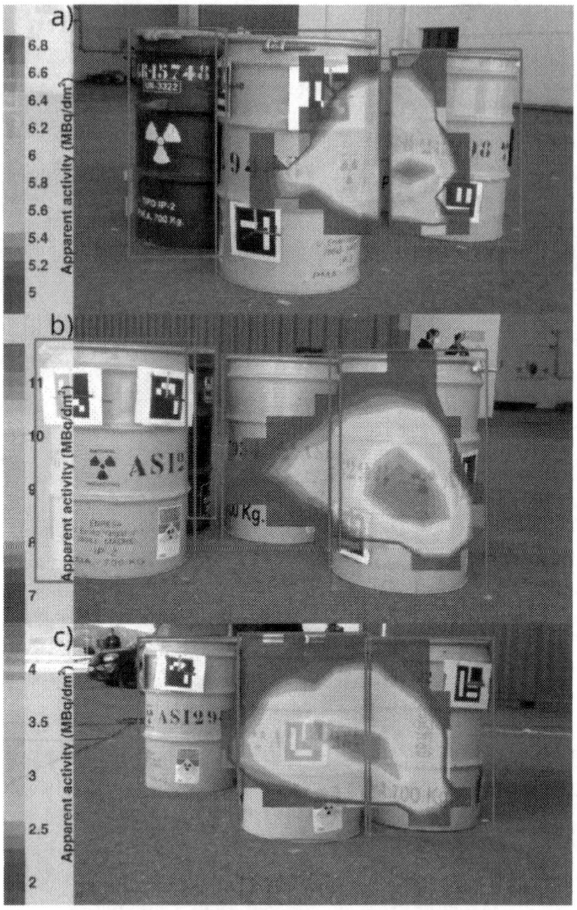

Fig. 4. 2D hybrid-images (gamma-visual) measured positioning the gamma imager at positions 4 (a), 5 (b) and 6 (c) around the drum ensemble.

Finally, the 3D gamma-distribution is shown in Fig. 5. This result is consistent with the previous interpretation that barrel AS28298 is the one showing largest overall apparent activity. This can be better appreciated in top view of the barrel ensemble shown in Fig. 5-b.

V. CONCLUSIONS AND OUTLOOK

In this work, we have demonstrated the effectiveness of a portable, high-efficiency gamma-ray imager combined with computer vision and image-segmentation techniques for the advanced analysis of radioactive waste. This system was successfully tested at the El Cabril disposal plant, managed by the national company ENRESA.

Our methodology allows for a real-time assessment of the overall apparent activity level of a drum or object of interest, a critical first step in any nuclear material inspection. The surface-activity estimated by the system was found to be in agreement with conventional dose-rate measurements within a factor of approximately two.

A key advantage of this system is its ability to produce a hybrid gamma-ray and visual image in real-time, providing a 2D map of the apparent surface-activity distribution. The high-efficiency design of the Compton camera enables rapid measurements—as short as two minutes for typical barrels—which is crucial for enhancing classification efficiency and reducing operator exposure.

In addition, we showed that by combining multiple measurements from different positions around an object, the system can provide a quantitative and visual assessment of activity levels and distributions. Both 2D hybrid images and 3D tomographic reconstructions were successfully created from a series of measurements, providing a comprehensive characterization of the different scenarios and barrels examined at the disposal plant.

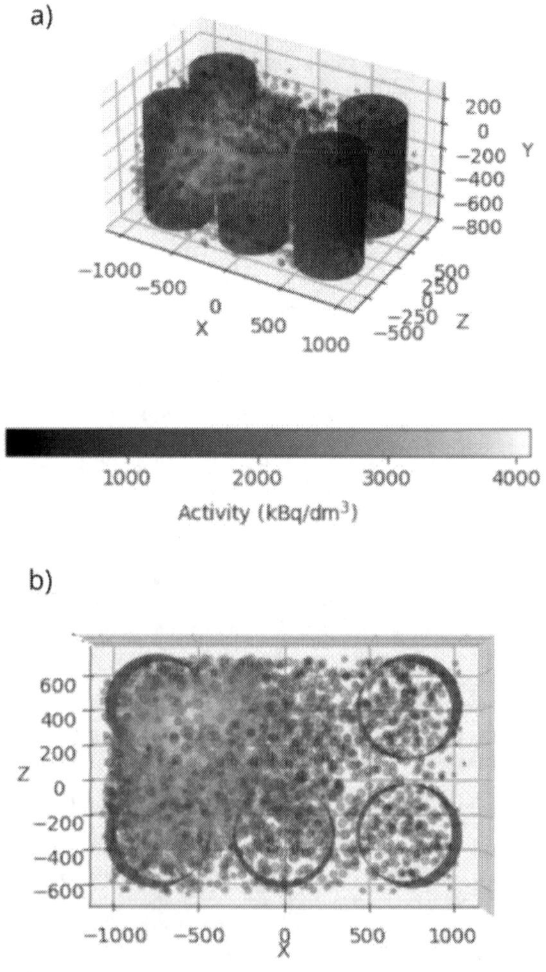

Fig. 5. Isometric (a) and elevation (b) views of the 3D distribution of gamma-radiation.

Ongoing work is focused on the analysis of data from a new

experimental campaign at the Santa María de Garoña nuclear power plant [24], which is currently undergoing decommissioning. This campaign tested several key improvements to the measurement system, designed to handle the large volume of radioactive bulk materials produced during this process. The updated system, shown in Fig. 6-a, incorporates three main innovations: a liquid cooling solution to mitigate temperature variations, a depth camera to accurately compute the distance to radioactive bulks without needing fiducial markers, and the addition of a second gamma-ray imager to reduce measurement time and enhance visual data. The results from this campaign are currently being analyzed.

Fig. 6. Pictures of the experimental campaign carried out at the Nuclear Power plant in Garoña.

Finally, we have also explored the applicability of the recently developed gamma-neutron vision system, GNVision [25, 26, 27] as a potential enhancement towards a more complete and reliable inspection of radioactive waste. This novel device, is an upgraded version of the Compton imager utilized in this work, which is able to simultaneously provide information about the spatial origin of gamma-ray and neutron sources. Therefore, it could help towards an enhanced management of radioactive residues that emit gamma-rays and neutrons, such as reactor spent fuel, by determining the presence of actinides, the amount of both types of radiation and their spatial distribution inside the bulk. These situations are displayed in Fig. 6-b and Fig. 6-c, which are pictures corresponding to the last experimental campaign carried out in Garoña [24].

Acknowledgment

We acknowledge our collaborator at ENRESA Dr. José Luis Leganés, whose insight and expertise were crucial for the interpretation of the results. He also provided assistance during the field measurements at the disposal plant of El Cabril and contributed with important comments that greatly improved the manuscript.

CDP and JLL acknowledge funding from Convenio CSIC-ENRESA "Desarrollo de un dispositivo para la identificación, cuantificación y distribución espacial de isótopos emisores gamma", 2014–2018. This work was also supported by the Spanish Ministerio de Ciencia e Innovación under grants PID2022-138297NB-C21, PID2019-104714GB-C21, FPA2017-83946-C2-1-P and CSIC for funding PIE-201750I26. Part of the hardware and software utilized in this work has been developed in the framework of the preceding HYMNS ERC-CoG Grant (Agreement Nr. 681740) led by CDP. The authors acknowledge the support provided by postdoctoral grants FJC2020-044688-I and ICJ220-045122-I, funded by MCIN/AEI/10.13039/501100011033.

References

[1] Caballero, L., Albiol Colomer, F., Corbi Bellot, A., Domingo-Pardo, C., Leganés, Olleros-Rodríguez, P. and Pérez Magán, D.L., "Gamma-ray imaging system for real-time measurements in nuclear waste characterisation," *Journal of Instrumentation,* vol. 13, no. 3, p. P03016, 2018.

[2] Tortajada, S., Albiol, F., Caballero, L., Albiol, A. and Leganés-Nieto, J.L, "A portable geometry-independent tomographic system for gamma-ray, a next generation of nuclear waste characterization," *Scientific Reports,* vol. 13, no. 12284, 2023.

[3] Fichet, R. et al., "A non-destructive and on-site digital autoradiography-based tool to identify contaminating radionuclide in nuclear wastes and facilities to be dismantled," *Journal of Radioanalytical and Nuclear Chemistry,* vol. 309, pp. 551-561, 2015.

[4] Babiano-Suárez, V., Balibrea-Correa, J., Ladarescu, I., Lerendegui-Marco, J. and Domingo-Pardo, C., "A computer-vision aided Compton-imaging system for radioactive waste characterization and decommissioning of nuclear power plants," *Nuclear Instruments and Methods in Physics Research Section A,* vol. 1076, p. 170449, 2025.

[5] High-sensitivitY Measurements of key stellar Nucleo-Synthesis reactions (HYMNS), *ERC-Consolidator Grant Agreement 681740, PI: C. Domingo Pardo..*

[6] Domingo-Pardo, C., "i-TED: A novel concept for high-sensitivity (n,g) cross-section measurements," *Nuclear*

Instruments and Methods in Physics Research A, vol. 825, pp. 78-86, 2016.

[7] Babiano-Suárez, V., Caballero, L., Calvo, D., Ladarescu, I., Olleros, P. and Domingo-Pardo, C., "Gamma-ray position reconstruction in large monolithic LaCl3 crystals with SiPM readout," *Nuclear Instruments and Methods in Physics Research A,* vol. 931, pp. 1-22, 2019.

[8] Babiano-Suárez, V., Balibrea-Correa, J., Caballero, L., Calvo, D., Ladarescu, I., Lerendegui-Marco, J., Mira-Prats, S. and Domingo-Pardo, C., "First i-TED demonstrator: A Compton imager with Dynamic Electronic Collimation," *Nuclear Instruments and Methods in Physics Research A,* vol. 953, p. 163228, 2020.

[9] Rolo, M. D. et al., "TOFPET ASIC for PET applications," *Journal of Instrumentation,* vol. 8, no. 2, p. C02050, 2013.

[10] Di Francesco, A. et al., "TOFPET 2: A high-performance circuit for PET time-of-flight," *Nuclear Instruments and Methods in Physics Research A,* vol. 824, pp. 194-195, 2016.

[11] Balibrea-Correa, J., Lerendegui-Marco, J., Babiano-Suarez, V., Caballero, L., Calvo, D., Ladarescu, I., Olleros-Rodriguez, P. and Domingo-Pardo, C. et al., "Machine Learning aided 3D-position reconstruction in large LaCl3 crystals," *Nuclear Instruments and Methods in Physics Research Section A,* vol. 1001, p. 165249, 2021.

[12] "El Cabril website," [Online]. Available: https://www.enresa.es/eng/index/activities-and-projects/el-cabril.

[13] "ENRESA website," [Online]. Available: https://www.enresa.es/eng/.

[14] Wilderman, S. J., Rogers, W. L., Knoll, G. F., and Engdahl, J. C., "Fast algorithm for list mode back-projection of Compton scatter camera data," *IEEE Transactions on Nuclear Science,* vol. 45, no. 3, pp. 957-962, 1998.

[15] Barrett, H.H., White, T. and Parra L.C., "List-mode likelihood," *Journal of the Optical Society of America,* vol. 14, no. 11, pp. 2914-2923, 1997.

[16] Wilderman, S.J., Clinthorne, N.H., Fessler, J.A. and Rogers, W.L., "List-mode maximum likelihood reconstruction of Compton scatter camera images in nuclear medicine," *1998 IEEE Nuclear Science Symposium and Medical Imaging Conference,* vol. 3, pp. 1716-1720, 1998.

[17] Wilderman, S.J., Fessler, J.A., Clinthorne, N.H., LeBlanc, J.W. and Rogers, W.L., "Improved modeling of system response in list mode EM reconstruction of Compton scatter camera images," *IEEE Transactions on Nuclear Science,* vol. 48, no. 1, pp. 111-116, 2001.

[18] Tortajada, S., Albiol, F., Caballero, L. and Leganés-Nieto, J.L., "A portable geometry-independent tomographic system for gamma-ray, a next generation of nuclear waste characterization," *Scientific Reports,* vol. 13, p. 12284, 2023.

[19] Garrido-Jurado, S., Muñoz-Salinas, R., Madrid-Cuevas, F. and Medina-Carnicer, R., "Generation of fiducial marker dictionaries using Mixed Integer Linear Programming," *Pattern Recognition,* vol. 51, 2015.

[20] Romero-Ramirez, F., Muñoz-Salinas, R. and Medina-Carnicer, R., "Speeded Up Detection of Squared Fiducial Markers," *Image and Vision Computing,* vol. 76, 2018.

[21] Abadi, M. et al., *TensorFlow: Large-Scale Machine Learning on Heterogeneous System,* https://www.tensorflow.org/, 2015.

[22] Ridikas, D., Feray, S., Cometto, M. and Damoy, F., "Non-destructive method of characterisation of radioactive waste containers using gamma spectroscopy and Monte Carlo techniques," *Radiation Protection Dosimetry,* vol. 115, no. 1-4, pp. 113-116, 2005.

[23] Allison, J. et al., "Recent developments in GEANT4," *Nuclear Instruments and Methods in Physics Research A,* vol. 835, pp. 186-225, 2016.

[24] Research and Development Funding by the Consejo de Seguridad Nuclear (CSN), Call 2024, Grant ID: SUBV-3-2024-CSIC PI: C. Domingo-Pardo., *Metodología y equipamiento Avanzado de Clasificación de residuos radiactivos mediante Imagen hibrida Neutrón-Gamma con amplio campo de visión (MACING).*

[25] Lerendegui-Marco, J., Cisterna, G., Hallam, J., Babiano-Suarez, V., Balibrea-Correa, J., Calvo, D., Ladarescu, I., de la Fuente, G., Gameiro, B., Sanchis-Molto, A., Torres-Sanchez, P. and Domingo-Pardo, C., "Imaging neutrons with a position-sensitive monolithic CLYC detector," *Nuclear Instruments and Methods in Physics Research Section A,* vol. 1079, p. 170594, 2025.

[26] Lerendegui-Marco, J., Babiano-Suárez, V., Balibrea-Correa, J., Caballero, L., Calvo, D., Ladarescu, I. and Domingo-Pardo, C., "Simultaneous Gamma-Neutron Vision device: a portable and versatile tool for nuclear inspections," *EPJ Techn Instrum,* vol. 11, no. 1, p. 2, 2024.

[27] J. Lerendegui-Marco, J. Balibrea-Correa, C. Domingo-Pardo, L. Caballero, V. Babiano-Suárez and I. Ladarescu, "Device for the simultaneous Detection, Identification, Quantification and/or location of gamma radiation and neutron sources". International Patent Patent WO2021229132A1, 2021.

[28] IAEA, Climate Change and Nuclear Power 2024, Geneva, 2024.

[29] IAEA, Classification of radioactive waste: safety guide, Vienna, 2009.

[30] Leganés, J. L., Uso de detectores de la radiación para el desarrollo de técnicas de reconstrucción espacial en residuos radiactivos, Valencia: Universidad de Valencia, 2019.

[31] Olleros, P., Caballero, L., Domingo-Pardo, C., Babiano, V., Ladarescu, I., Calvo, D., Gramage, P., Nacher, E.,

Tain J.L. and Tolosa, A., "On the performance of large monolithic LaCl3(Ce) crystals coupled to pixelated silicon photomultipliers," *Journal of Instrumentation,* vol. 13, no. 3, p. P03014, 2018.

[32] Babiano-Suárez, V., Lerendegui-Marco, J., Balibrea-Correa, J., Caballero, L., Ladarescu, I., Real, D., Domingo-Pardo, C. et al., "Imaging neutron capture cross sections: i-TED proof-of-concept and future prospects based on Machine-Learning techniques," *European Physical Journal A,* vol. 57, no. 6, p. 197, 2021.

[33] Celik, I. C., "Determination of Gamma-Ray Abundance in Photopeaks with Analytic and Numeric Approaches via CERN's ROOT Analysis Framework," *Avrupa Bilim ve Teknoloji Dergisi,* no. 36, pp. 160-167, 2022.

Active Fast Neutron Multiplicity Counting with DT Neutron Generator

Xiaobo LIU*, Jiansheng LI, Jian TANG, Yu JIN
Institute of Nuclear Physics and Chemistry, CAEP, China
(*) 13881190590@163.com

***Abstract*—The active fast neutron multiplicity counting (AFNMC) apparatus was consisted of an array with 36 pieces of liquid scintillator detectors in three rings cylindrical layout, multi-modules 500M14bit digitizers based on PXIe platform, the DT neutron generator, shielding cone and mechanical structure. Its center cavity was 35 cm in radius and 50 cm in height for assaying the standard container of plutonium item. The DT neutron generator was the type of alpha-associated to make the neutron injecting into the assay area being tagged and located on the bottom of detector array. The shielding cone was fixed between the DT generator and the lower ring detector with a hole to let the tagged neutron injecting to the sample assay area. The alpha signal was used as gating signal when doing neutron mutiplicity analysis. The experiments were performed with iron, lead, HEU metal in standard container for technical verification. The SDT values from both conventional and gating FNMC were obtained for comparison. The experimental results indicated this method dramatically reduced the random or non-correlated neutron signal noise and substantially promoted the D and T values sensitivity when the alpha signal gating was enabled.**

***Keywords* —active fast neutron multiplicity counting, AFNMC, alpha-associated DT neutron, generator liquid scintillator, digitizers, HEU, nuclear material assay**

I. INTRODUCTION

THE challenges has been existed for several decades on the HEU attribute assay in the fields of arms control and nuclear material assay and verification. Some detection options for HEU monitor were reviewed [1] and measurement approach suites for Uranium Mass Attribute System (UMAS) were evaluated [2]. The active neutron multiplicity counting (ANMC), originated from passive neutron multiplicity counting [3] is one of the most important method for U mass assay and accountability.

The ANMC has two branches according to the detected neutron spectrum, one is active thermal neutron multiplicity counting such as AWCC [4], [5], [6] and another is the active fast neutron coincidence [7]and multiplicity counting (AFNMC) [8], [9], [10]. Meanwhile the are mainly three categories of neutron sources used for active interrogation, one is the (alpha, n) neutron source such as Am-Li [11], Am-Be source, the other is the spontaneous self-fission neutron source such as Cf-252, and the last is the DD, DT neutron generator.

NMIS/NWIS was an innovative technology developed by J. T. Mihalczo [12], [13], [14]. in Oak Ridge National Laboratory for nuclear materials/weapon identification and multi-attributes measurements [15]. The interrogation source used were Cf-252 fission timing chamber [12, 14] earlier and upgraded with alpha associated DT generator [16], [17] later. Based on the measured time corelated neutron and gamma signature, the approaches for resolving the U attributes of mass [18], [19], enrichment [20], [21], shape and shielding or cladding material [22] were investigated.

In 2010s, we developed a thermal multiplicity counter (named NPL-NMC) for assaying the high mass plutonium metal items deposited in standard container [23], [24], [25], And a FNMC apparatus was constructed and verified by experiments of assaying plutonium items [26]. The alpha-associated DT neutron generator was integrated into this FNMC apparatus, making it be suitable for active interrogation (AFNMC). The conventional and alpha gating FNMC method was implemented for technical verification. The constitution of the apparatus, DT neutron generator, the calibration, the experiments and results were introduced in this paper.

II. AFNMC APPARATUS

The AFNMC apparatus was consisted of an array with 36 pieces of liquid scintillator detectors in three rings cylindrical layout, multi-modules 500M14bit digitizers based on PXIe platform, the DT neutron generator, shielding cone and mechanical structure, HV device, and software for control and data analysis. The generator was the type of alpha-associated to make the neutron injecting into the assay area being tagged and located on the bottom of detector array. The apparatus was assembled in the experimental hall. One PC located in measuring room was connecting the digitizers through Ethernet cable to control the digitizer remotely. The schematic picture was shown in Fig. 1.

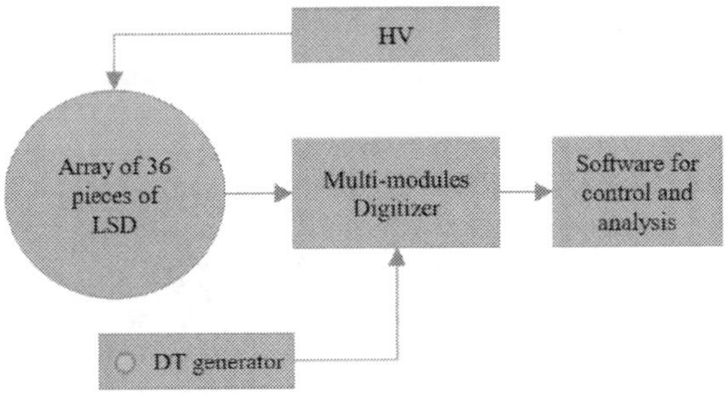

Fig. 1. Schematic of the AFNMC apparatus

A. DT Neutron Generator

The DT neutron generator was self-developed alpha-associated product [27] with compact dimension for portable application. The maximums strength for DT neutron emission was 2E7 s^{-1}. The schematic diagram and the photo for this generator were shown in Fig. 2 and Fig. 3.

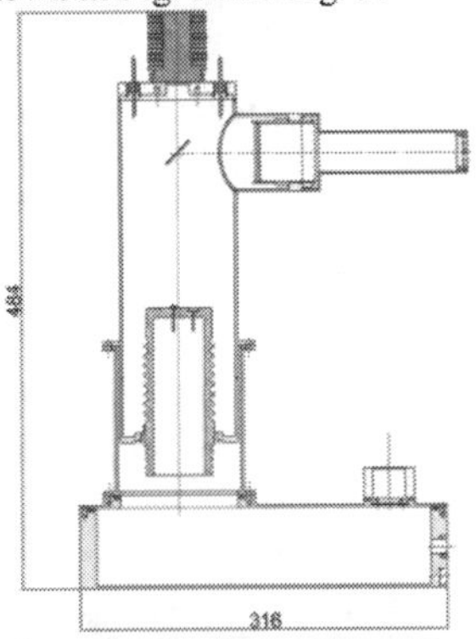

Fig. 2. Schematic diagram of the DT generator

Fig. 3. Photo for the DT generator

DT neutron generator was consisted of two parts: the neutron generator tube, including the alpha-tagging detector, with dimensions 150 mm × 330 mm × 490 mm and weight less than 13 kg; and the power supply and control unit - the black box as shown in Fig. 4. with dimensions 370 mm×250 mm×89 mm and weight less than 2 kg.

The ZnO: Ga scintillator fused on quartz substrate [28] was sealed in the generator tube for alpha particle detection. The effective area was 22 mm in diameter, and the distance to the tritium target was 66 mm. An ET 9125B photomultiplier tube was employed to convert the fluorescence burst to fast electronical pulse with 4ns in rising time and about 100 mV in amplitude, as shown in Fig. 4.

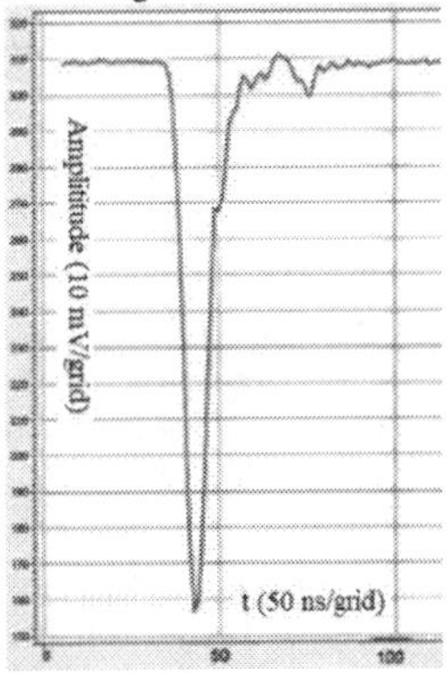

Fig. 4. Waveform of alpha signal

The DT neutron generator was installed on the bottom of detector array with 664 mm in distance from the tritium target to the assay cavity center. The generator was tilted by 5.5 degree to compensate the angle of deflection of tagged neutron.

B. Shielding Cone

A shielding cone made of brass was fixed between the DT generator and the lower ring detectors with a hole to let the tagged neutron injecting to the sample assay area. The shadow from the shielding cone covered all the liquid scintillator detectors to decrease the background counts induced by the untagged neutrons from generator. As shown in Fig. 5, the container (in grey), the shielding cone (in brown) and the generator (in green) were assembled from up to down.

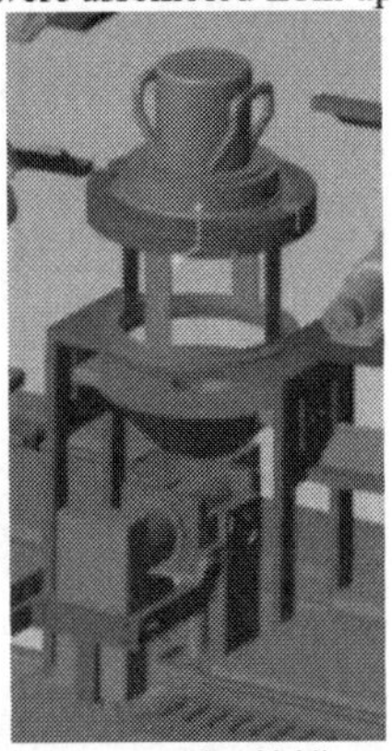

Fig. 5. Diagram for the generator, the shielding cone and the container

C. Array of 36 LSDs and Digitizers

The 36 pieces of liquid scintillator detectors (LSDs) were arranged in three rings cylindrical layout, with 16 prices in center plane and 10 pieced respectively in upper and lower plane. The array was installed on the mechanical structure through the aluminum circle belt and poles and divided two halves. Each half can move on the rails to open for sample loading or unloading and close for assaying.

The digitizers were comprised of 5 digitizer modules, one synchronization module, one PXIe imbedded controller and one 9-slot PXIe Chassis. There are 8 channels for each digitizer module, and totally 40 channels by 5 modules. The first 36 channels were used for signal form LSDs and the 37th channel was used for signal from alpha detector. The digitizers implemented detector's signal triggering, acquisition and real-time digital data processing for characteristics.

Please refer to the reference [26] for more details.

D. Software for control and processing

It has four main functions:

1) to control and configure the digitizers for synchronizing, signal acquisition and data processing;
2) to control and configure the DT neutron generator;
3) to do neutron and gamma discrimination and alpha signal discrimination, coincidence measurements, conventional and gating multiplicity counting analysis and uranium mass resolving with preset parameter and algorithm;
4) to receive and save the measuring data.

III. TEST AND CALIBRATIONS

A. Test of the DT Neutron Generator

The alpha signal was sampled with digitizers and processed for pulse amplitude analysis in preheated without neutron emitting status and in neutron emitting status. As shown in Fig. 6, it indicated that the maximum amplitude of alpha signal was about 100 mV, and the effective signal can be easily discriminated by amplitude with 35 mV thresholds.

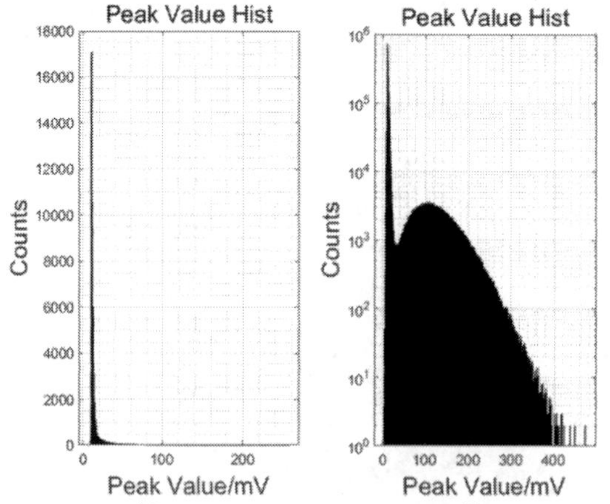

(a) (b)

Fig. 6. Amplitude distribution in a) preheated and b) emitting status

The cone area boundary for tagged neutron at the distance of 664 mm was measured with a miniature liquid detector with a diameter of 10mm and a thickness of 10mm. The mini detector was fixed in the center of the assay cavity through an arm attached to a moving slide rail. The rail can move in the horizontal plane precisely along X and Y directions step by step. The gated coincidence measurement was carried out in each step and the coincident counts distribution were obtained. As shown in Fig. 7, the boundary for tagged neutron was 100 mm, which covered the maximums radius for the sample to be assayed.

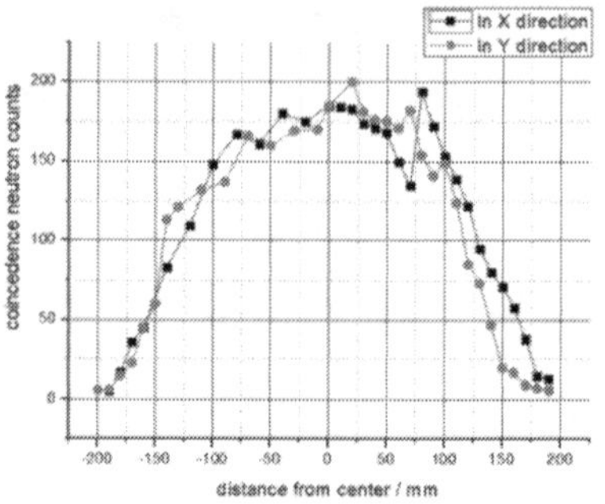

Fig. 7. Tagged neutron distribution along X and Y directions

B. Calibration for the Array of LSDs

The light output of each detector was calibrated with Cs-137 source. The nearly same light output for each detector were made by adjusting the voltage of HV device.

The Cf-252 source was used for PSD discrimination parameter calibration for each detector. The measured PSD data was binned by measured light output Qlong. The optimization of the discrimination value was determined and the FOM for the PSD performance was computed for each of "slice" of data.

The gated coincidence measurements were carried out for extracting the timing resolution and the time offset for each detector with Cf-252 fission chamber.

The passive multiplicity counting analysis for the Cf-252 source and the background was carried out.

Please refer to the reference [26] for more details.

IV. EXPERIMENTS AND RESULTS

The iron, lead and HEU metal samples respectively deposited in standard container were used for experiments. The background experiments for the empty and the container without sample in the assay cavity were also carried out. The methods for the experiments were TOF spectrum, conventional multiplicity counting and gating multiplicity counting. The measuring time for each experiment was 300s.

A. ToF from Alpha Signal

The TOF spectrum from one detector in the center ring were shown in Fig. 8. In the figures, the line in green refers to the gamma signal, the red or blue line refers to the neutron signal, and the black line refers to all the signal.

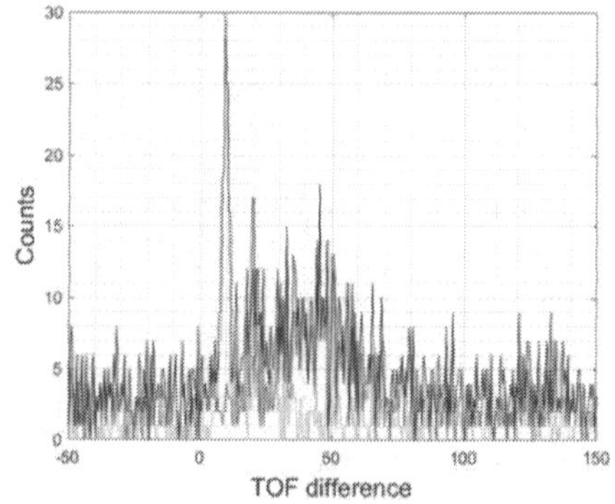

a) empty

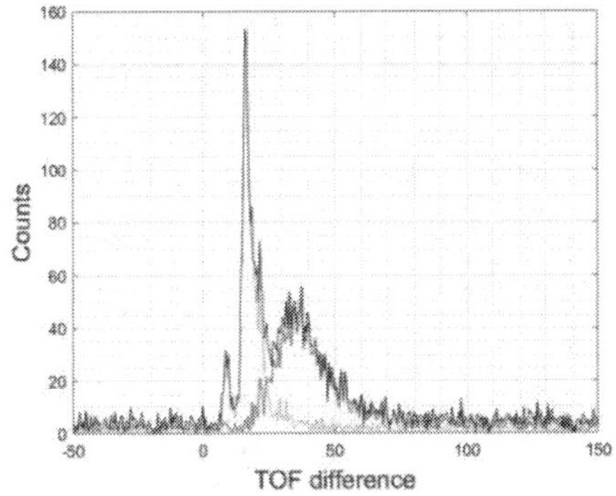

b) container without sample

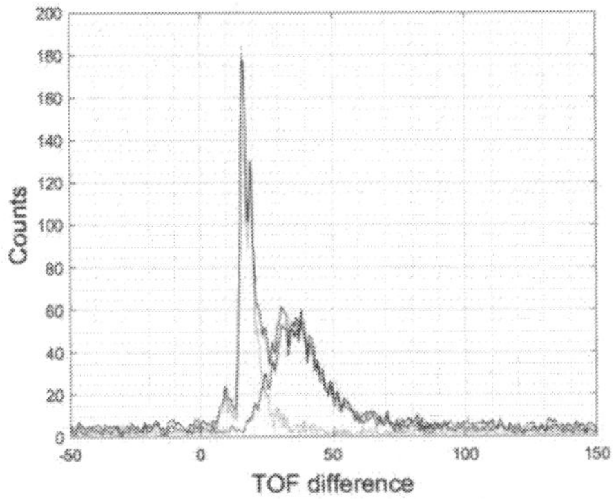

c) container with Fe

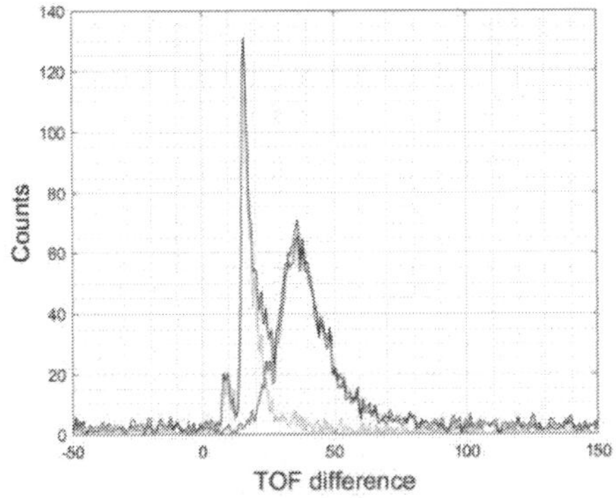

d) container with Pb

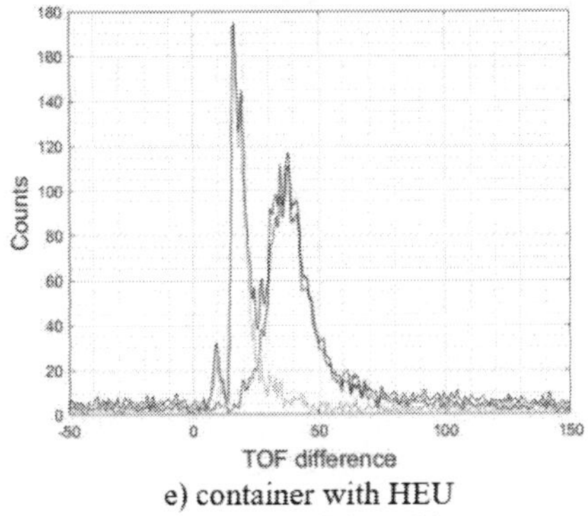

e) container with HEU

Fig. 8. Assaying results and biases.

There were three peaks (two gamma peaks and one neutron peak). The first peak was from inelastic gammas in generator and almost stable. The second peak was from inelastic gammas in the reaction of tagged DT neutron with the container and the sample. The third peak was from the scattered neutron in both inelastic and elastic reaction of tagged DT neutron with the container and the sample, and the fission neutrons if the sample was fissile nuclear material. There was a little increment in neutron counts for Pb sample compared Fe because the scatter cross section for Pb larger than Fe. There was a significant increment in neutron counts for HEU sample because the introduced fission contribution.

B. *AFNMC Experiments*

The gate window for multiplicity analysis was from 20 ns to 120 ns. The measured SDT (Single, double and triple) from conventional multiplicity counting method were shown in Table I. Because the yields of the generator were unstable in about several percent fluctuation, all the results were normalized to 35k cps based on the measured effective alpha count rates when doing active interrogation.

TABLE I
Measured SDT with conventional multiplicity analysis

	S/CPS	D/cps	T/cps
empty	1.415E4	108.6	0.506
container	1.575E4	120.9	0.708
Fe	1.490E4	114.9	0.608
Pb	1.612E4	133.0	0.722
HEU	1.630E4	206.8	9.523

The measured SDT (Single, double and triple) from gating active multiplicity counting method were shown in Table II. The S values

TABLE II
Measured SDT with gating active multiplicity analysis

	S/CPS	D/cps	T/cps
empty	109.9	0.70	0
container	307.9	2.19	0.010
Fe	331.4	2.36	0.017
Pb	413.6	3.53	0.019
HEU	696.1	39.44	3.841

From Table I and Table II, it indicated that the S rates were suppressed by about 25 ~50 times. The ratio of D/S and T/S for HEU sample were increased by about 4.5 times and 9.5 times respectively. In other words, the alpha gating FNMC method made the random or non-correlated neutron signal noise from the generator source totally cleared, and substantially promoted the D and T values sensitivity for the HEU mass attributes.

V. CONCLUSIONS

The active fast neutron multiplicity counting (AFNMC) apparatus was constructed. It consisted of an array with 36 pieces of liquid scintillator detectors, multi-modules 500M14bit digitizers, the DT neutron generator, shielding cone and mechanical structure. The DT neutron generator was the type of alpha-associated and the alpha signal was used as gating signal when doing AFNMC assay.

The experiments were performed with iron, lead, HEU metal in standard container for technical verification of this method.

The SDT values from both conventional and gating AFNMC were obtained for comparison. The experimental results indicated this method dramatically reduced the random or non-correlated neutron signal noise and substantially promoted the D and T values sensitivity when the alpha signal gating was enabled.

This research demonstrated a promising application for HEU mass attributes assaying in the fields of nuclear material assay and verification. Further work will be taking on such as the algorithm for HEU mass resolution, cross-talk correction, multiplication correction in the future.

REFERENCES

[1] Daren Norman, Seth McConchie, Mark E. Abhold, Martin Morgan-Reading, Highly Enriched Uranium (HEU) Detection Options, Los Alamos Nat. Lab., Los Alamos, NM, USA, Tech. Rep. LA-UR-14-21979, 2014.
[2] J. Knowles, E. Brubaker, S. McConchie, M. Williamson, Uranium Mass Attribute System (UMAS) Study: Evaluation of Options Against the Arms Control Evaluation Criteria, Pantex Plant and Y-12 National Security Complex, USA, Tech. Rep. Y/PM-20-005, 2020, DOI:10.2172/1608124.
[3] D. G. Langner, J. E. Stewart, M. M. Pickrell, M. S. Krick, N. Ensslin, and W. C. Harker, "Application guide to neutron multiplicity counting," Los Alamos Nat. Lab., Los Alamos, NM, USA, Tech. Rep. LA-13422-M, 1998, doi: 10.2172/1679.
[4] Menlove H O, "Description and operation manual for the active well coincidence counter", Los Alamos Nat. Lab., Los Alamos, NM, USA, Tech. Rep. LA-7823-M, 1979.
[5] Elhakim, El Gammalw. I. Zidane. "A proposed semi-empirical method for 235U mass calibration of the active-well neutron coincidence counter." Nuclear Instruments and Methods in Physics Research A, 565 (2006) 731–741.
[6] Ferrari, Francesca , and P. Peerani. "Performance of an active well coincidence counter for HEU samples." Radiation Measurements, 45.9(2010): 1034-1043.
[7] W.H. Geist, L.A. Carrillo, N. Ensslin, K. Ianakiev, D.R. Mayo, M.C. Miller , "Evaluation of a fast neutron coincidence counter for the measurements of uranium samples", Nuclear Instruments and Methods in Physics Research A, 470 (2001): 590–599.
[8] Hao Zhou, Hongtao Lin, Guorong Liu, et al. "A neutron multiplicity analysis method for uranium samples with liquid scintillators", Nuclear Nuclear Instruments and Methods in Physics Research A, 797(2015): 70–76
[9] J.L. Dolan, M.J.Marcath, M.Flaska, et al. "Active-interrogation measurements of fast neutrons from induced fission in low-enriched uranium." Nuclear Instruments & Methods in Physics Research A, 738.feb.(2014): 99-105.
[10] Di, Fulvio A. , et al. "Fast-neutron multiplicity counter for active measurements of uranium oxide certified material." Nuclear Instruments and Methods in Physics Research A, 907(2018): 248-257.
[11] Goddard, B. , et al. "Evaluation of Am-Li Neutron Spectra Data for Active Well Type Neutron Multiplicity Measurements of Uranium." Nuclear Instruments & Methods in Physics Research A, 830(2016): 256-264.
[12] Mihalczo J, Pare V, Valentine T, "Nuclear weapons identification system", IEEE Nuclear Science Symposium & Medical Imaging Conference (1996), 1: 44-45, DOI: 10.1109/nssmic.1996.590888
[13] Mattingly, J. K. , et al. "Physical and mathematical description of Nuclear Weapons Identification System (NWIS) signatures." Oak Ridge Y-12 Plant, Oak Ridge, Tennessee, Tech. Rep. Y/LB-15946-R1, 1997.
[14] Mihalczo J , Mullens J A , Mattingly J ,et al. "Physical description of nuclear materials identification system (NMIS) signatures." Nuclear Instruments & Methods in Physics Research A, 450(2000): 531-555.
[15] Mihalczo J T , Mattingly J K , Neal J S ,et al. "NMIS plus gamma spectroscopy for attributes of HEU, PU and HE detection." Nuclear Inst & Methods in Physics Research B, 213.Jan(2004):378-384.
[16] Hausladen P A , Bingham P R , Neal J S ,et al. "Portable fast-neutron radiography with the nuclear materials identification system for fissile material transfers." Nuclear Instruments & Methods in Physics Research B, 261.1-2(2007):387-390.
[17] Hausladen, P. A. , J. S. Neal , and J. T. Mihalczo . "An alpha particle detector for a portable neutron generator for the Nuclear Materials Identification System (NMIS)." Nuclear Inst. and Methods in Physics Research B, 241.1-4(2005): 835-838.
[18] Canion, Bonnie , S. Mcconchie , and S. Landsberger . "Point Kinetics Framework for Characterizing Prompt Neutron and Photon Signatures From Tagged Neutron Interrogation." IEEE Transactions on Nuclear Science, 64.7(2017): 1761-1768.
[19] Tweardy, Matthew C. , S. Mcconchie , and J. P. Hayward . "A Point Kinetics Model for Estimating Neutron Multiplication of Bare Uranium Metal in Tagged Neutron Measurements." IEEE Transactions on Nuclear Science, PP.7(2017):1-1.
[20] Mattingly J , Valentine T , Mihalczo J ,et al. "ENRICHMENT AND URANIUM MASS FROM NMIS FOR HEU METAL." (2000).
[21] Crye, Jason Michael, "ENRICHMENT DETERMINATION OF URANIUM METAL IN SHIELDED CONFIGURATIONS WITHOUT CALIBRATION STANDARDS. " PhD diss., University of Tennessee, 2013.
[22] T. A. Wellington, B. A. Palles, J. A. Mullens, et al. "Recent Fast Neutron Imaging Measurements with the Fieldable Nuclear Materials Identification System." Physics Procedia, 66(2015): 432-438.
[23] Ligao CHEN, Xiaobo LIU, Jian GONG, et al. Neutron multiplicity counter design and calibration[J]. Journal of Tsinghua University (Science and Technology), 2014, 54(2): 159-163
[24] Li-Gao, Chen , et al. "Variance analysis for passive neutron multiplicity counting." Nuclear Science and Techniques, 26.2(2015):54-58.
[25] Xiaobo Liu, Ligao Chen . "A two-parameter iterative approach of multiplication correction for assaying plutonium metal parts with neutron multiplicity counting." Nuclear Instruments and Methods in Physics Research A, 1016(2021):165779
[26] Xiaobo Liu, Jiansheng Li, Yu Jin, "FAST NEUTRON MULTIPLICITY COUNTING FOR HIGH MASS PLUTONIUM ASSAY", The 9th International Conference on Advancements in Nuclear Instrumentation Measurement Methods and their Applications, ANIMMA, Valencia, France, 2025.
[27] Jian Tang, Pu Zheng, Meng Liu, *et al.*, Zhang. The design and sealed operation of a portable alpha-tagging neutron generator[J]. Journal of Instrumentation, 2025, 20(07): 18.
[28] J.-X. Chen et al., "Development of the ZnO:Ga nanorod arrays as an alpha particle scintillation screen for the associated particle neutron generator", Appl. Phys. Lett. 120 (2022), 193502.

Estimation of the time of exposure from the differential fading of luminescent detectors

Gianpaolo Roina[1,*], Debora Siqueira Nascimento[1,2], Riccardo Ciolini[1], Mariagrazia Quattrocchi[3], Francesco d'Errico[1,4]
[1] School of Engineering, University of Pisa, Italy
[2] National Institute of Geophysics and Volcanology, Italy
[3] S.C. Fisica Sanitaria Azienda Toscana Nord Ovest - Lucca, Italy
[4] School of Public Health, Yale University, and Yale Center for Emergency Preparedness and Disaster Response, USA
(*) gianpaolo.roina@phd.unipi.it

Abstract— **Containment and surveillance measures for unattended or remotely monitored items and areas serve to detect, timestamp, and thereby deter the illicit movement and trafficking of radioactive material in the interval between two successive inspections. This study explores the feasibility of a passive solution that does not rely on complex electronic components, thus mitigating common vulnerabilities of active instruments, including risks of tampering and unauthorised data access. The signal loss in luminescent materials – dependent on both time and temperature and previously investigated as a means of estimating the time of exposure – proved applicable in this safeguards-relevant context as well. The differential fading of peaks II and III of the glow curve of LiF:Mg,Cu,P (GR-200A), compared to the stable dosimetric peak IV, provided a means to timestamp the moment of irradiation with reasonable accuracy in the short- and medium-term after exposure. While the analysis primarily focused on the behaviour of individual crystals, the averaging of multiple responses was also explored for comparison, to assess its potential for reducing estimation error.**

Keywords — **differential fading; tamper-proof Chain of Custody verification; thermoluminescence; nuclear safeguards.**

I. Introduction

To verify that States are upholding their obligations under non-proliferation agreements, the International Atomic Energy Agency (IAEA) implements a range of safeguard measures, including scheduled on-site inspections. These inspections aim to validate the accuracy and completeness of declarations submitted by facility operators. A central concern in this context is ensuring the Continuity of Knowledge (CoK) on safeguarded items and areas between successive inspector visits. To address this need, the IAEA deploys dedicated optical surveillance systems and tamper-indicating seals that register access events, unauthorised movements of nuclear material, and any tampering with containment structures or Agency equipment, samples, and data.

Despite their effectiveness, these monitoring systems typically incorporate sophisticated electronic architectures, which can present potential vulnerabilities to manipulation or covert interference. The present study explores the feasibility of a passive solution designed to detect the unauthorised removal of radioactive material from storage while also enabling retrospective estimation of the time at which such an event may have occurred. Additionally, application of the proposed approach to accidental radiation exposure scenarios might be possible, particularly in occupational settings, where accurate time reconstruction is essential for incident investigation and response.

The fading behaviour of luminescent materials under stimulation offers a promising method for estimating the time since exposure. While fading is typically regarded as a limitation in standard dosimetric practice, in this context it becomes an exploitable feature, offering a potential timestamping mechanism for radiation exposure events, particularly in scenarios relevant to nuclear safeguards and traceability. Early work in this direction – albeit aimed at dosimetric applications – was proposed by Moscovitch [1], who developed an automated method for estimating the elapsed time between irradiation and readout in LiF:Mg,Ti by analysing the fading behaviour of individual glow peaks. More recently, d'Errico et al. [2] extended this concept to safeguards-related scenarios, demonstrating the feasibility of estimating the time of exposure by comparing the fading signals of various luminescent materials. Building on this approach, our study advanced the concept by relying on a single material – LiF:Mg,Cu,P (GR-200A) – to obtain similar information, correlating the time elapsed since exposure with the fading of specific peaks in its thermoluminescent glow curve. The level of timestamping accuracy achieved with this method is discussed later in the paper.

The fading at room temperature of these peaks has been previously studied by Budzanowski et al. [3], who confirmed their exponential behaviour and the potential to estimate the time of exposure by examining the peak II-to-IV and III-to-IV area ratios. This methodology has been successfully applied in personal and environmental dosimetry to reconstruct the timing of high-dose-rate accidental exposures, with results that closely matched actual exposure times [4], validating the practicality of the approach.

While a more detailed discussion of the theoretical aspects of this research will be reported in a future publication, the present work offers an overview of the principal results.

II. Materials and Methods

This work used LiF:Mg,Cu,P-based thermoluminescent detectors (GR-200A), selected for their high sensitivity and suitability for low-dose applications. They are manufactured in disc form with a diameter of 4.5 mm and a thickness of 0.8 mm. The luminescence spectrum closely resembles that of LiF:Mg,Ti (TLD-100) [5], and the linearity of the dose response is maintained up to roughly 1 Gy [6], making it appropriate for the dose range under investigation.

LiF:Mg,Cu,P typically shows six glow peaks between 90 °C and 350 °C, but the short half-life of peak I limits its practical use, while peaks V and VI require reading temperatures above 240 °C, which can potentially affect crystal structure. Therefore, only peaks II, III, and IV were considered, with the latter used as a reference due to its relative insensitivity to post-irradiation fading. In contrast, peaks II and III undergo progressive signal loss over time, with markedly different decay rates – approximately one day and three months, respectively [7]. These differing stability characteristics enable time-resolved analysis of the luminescent response.

As a preliminary step, an initial set of 146 crystals was examined to identify the largest possible group of highly homogeneous crystals presenting the same response (±5%) to a given dose of 1 mGy. They were irradiated at the Nuclear Measurements Laboratory of the University of Pisa (Italy) – School of Engineering, using a 2.22 MBq Ra-226 radioactive source placed on top of a specially designed Plexiglas apparatus. The mass of each crystal was also accurately measured with an ABT 100-5NM KERN© analytical balance (readability of 0.01 mg), to relate the total and peak-specific generated charges to the unit mass of the phosphor, as mass ultimately influences the thermoluminescent light emitted for a given absorbed dose.

The 88 identified homogenous crystals were irradiated using a 6 MV photon beam from a medical linear accelerator at San Luca Hospital in Lucca (Italy), delivering a dose of 100 mGy (10 monitor units) over 6 seconds at the lowest available dose rate. Notably, the short duration of the irradiation resembles real-world scenarios in which a perpetrator carrying a stolen source would move swiftly past the detection system, resulting in only brief exposure times. A 1.5 cm solid water build-up layer was placed above the samples, with an additional 15 cm beneath them to ensure full backscatter conditions. To reconstruct the fading curves over the first week and the two months after exposure, numerous irradiation–readout cycles had to be performed, and permutation of the crystal readout order at each cycle ensured adequate temporal sampling of the individual elements during both intervals.

A Harshaw 4500 Manual TLD Reader operated via the WinREMS interface acquired the thermoluminescence signal under nitrogen flow using a resistive heating system that delivered a linear temperature profile from 40 °C to 240 °C over 200 seconds. Indeed, the adoption of the lowest available heating rate enhanced the resolution between peak III and the low-temperature tail of peak IV. Each readout produced a 200-point glow curve with peak maxima at approximately 110, 160, and 210 °C. A 10-second annealing phase at 240 °C followed the data acquisition, resulting in a residual signal below 2% of the measured value.

The software GlowFit v1.3 [8] was used to decompose the experimental glow curves into individual peaks using a first-order kinetic model with a linear heating profile. Initial estimates of the three kinetic parameters E, T_m and I_m (trap depth, and temperature and intensity of the maximum) associated with each peak are iteratively adjusted via the Levenberg–Marquardt algorithm to minimise the χ^2 error between the experimental and fitted curves.

Finally, a custom MATLAB script was used to process the glow-curve experimental data and the associated kinetic parameters to analyse the specifics of the three peaks, particularly the ratios between the areas underneath. The results are discussed in the following section.

III. Results and Discussion

The temporal variation of the peak II-to-IV and III-to-IV area ratios generated the fading curves required for the intended safeguards-related application. In the envisioned scenario, an inspector suspecting the illicit removal of nuclear material would be able to estimate the time t_{est} potentially elapsed since the removal by comparing the peak area ratios calculated from the crystal readouts with those reported in Fig. 1.

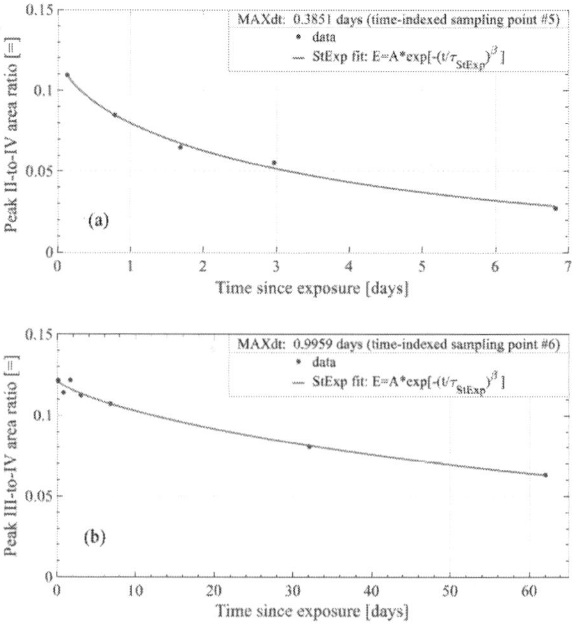

Fig. 1 Variation of the peak II-to-IV (a) and III-to-IV (b) area ratios with time after exposure for one of the crystals used in the study (illustrative example), showing experimental data and stretched exponential fit.

In mathematical terms, we introduced the fitting function

$$PAR^{X/IV} = f(t) \quad (1)$$

https://doi.org/10.1051/epjconf/202533806011

that approximates the experimental data points

$$\left(t_i, \mathrm{PAR}_i^{\frac{X}{IV}}\right) \qquad \forall i \tag{2}$$

for the generic peak X-to-IV area ratio (X = II or III). Its inverse, $t = f^{-1}(\mathrm{PAR}^{X/IV})$, allows estimation of t_{est} from the measured ratio:

$$t_{est} = f^{-1}\left(\mathrm{PAR}_{meas}^{\frac{X}{IV}}\right). \tag{3}$$

The overall data pattern exhibited exponential-like behaviour, as expected. However, a "stretched exponential relaxation" model [9] was preferred to the simpler one-term exponential decay due to its ability to capture the long-time fading tail of the experimental data (Fig. 2). This is attributable to the characteristic $t^{\beta-1}$ decrease in the decay rate $1/\tau_{StExp}$, that causes the graph of $\ln(f_{StExp})$ to exhibit a characteristic stretch when $0 < \beta < 1$:

$$f_{StExp}(t) = A \cdot \exp\left[-\left(\frac{t}{\tau_{StExp}}\right)^{\beta}\right] \qquad (\beta > 0) \tag{4}$$

Such behaviour can be explained as a linear superposition of simple exponential decays (note that $f_{StExp}(t)$ becomes exponential for a unit value of β), and it was already used by Werner [10] to describe complex luminescence decays.

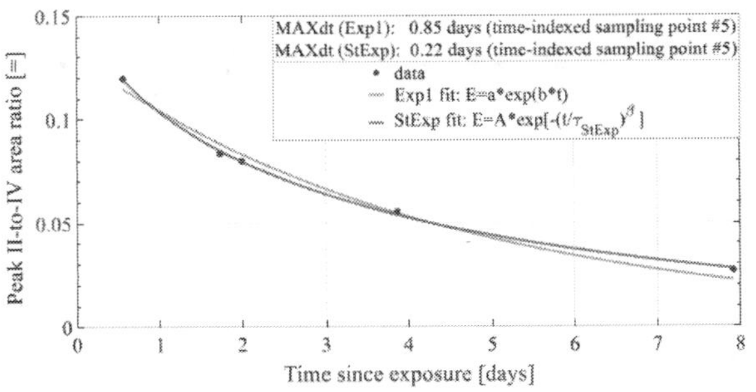

Fig. 2 Comparison of stretched exponential (red) and one-term exponential (green) fits to peak II-to-IV area ratios as a function of time since exposure in the single-crystal approach.

In order to compare various fitting models, as well as to quantify the accuracy of the time estimation, we defined the "maximum horizontal distance", MAXdt, as the largest of the absolute differences between the experimental fading times, t_i, and the values of the inverse function evaluated at the corresponding experimental peak area ratios $\mathrm{PAR}_i^{X/IV}$ (what is referred to as the i-th "horizontal distance", Δt_i):

$$\mathrm{MAXdt} = \max_i\{\Delta t_i\} = \max_i\left\{\left|t_i - f^{-1}\left(\mathrm{PAR}_i^{\frac{X}{IV}}\right)\right|\right\}. \tag{5}$$

Confirming prior considerations, for the vast majority (~90%) of the analysed crystals the value of MAXdt calculated using a stretched exponential relaxation was smaller than that corresponding to the one-term exponential model, indicating a better fit. Additional goodness-of-fit parameters were also evaluated (Table I) and further corroborated these results.

TABLE I
COMPARISON OF GOODNESS-OF-FIT PARAMETERS OF FITTING MODELS

GoF parameter	Ideal parameter value	Stretched exponential fit	One-term exponential fit
SSE	SSE → 0	1.987e-5	1.17e-4
R-square	$R^2 \to 1$	0.99602	0.9767
Adj. R-square	Adj $R^2 \to 1$	0.99205	0.96894
RMSE	RMSE → 0	0.0031513	0.0062277

SSE = Sum of Squares Due to Error; Adj. R-square = Degrees of Freedom Adjusted R-Square; RMSE = Root Mean Squared Error.

To extend the evaluation of our model performance, a statistical analysis of the estimation accuracy of the stretched exponential model was conducted for all crystals and both peak area ratios. The outcomes are presented in Figs. 3a and 4a as histograms illustrating the distribution of the maximum absolute differences between the experimental data and the fit. The corresponding cumulative percentage curves were superimposed to offer a comprehensive overview of the accuracy performance across the dataset.

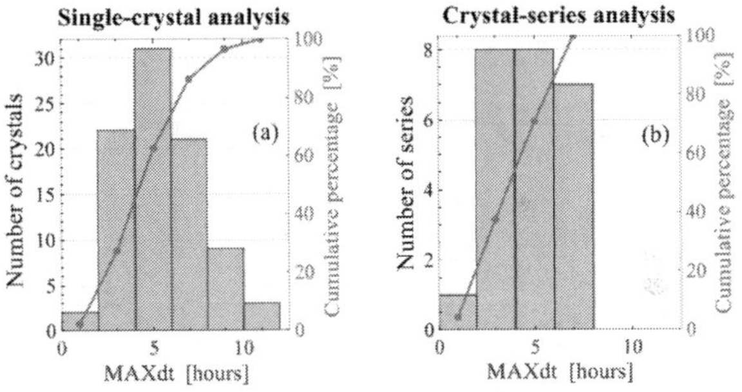

Fig. 3 Accuracy in estimating the time elapsed after exposure from peak II-to-IV area ratios for the single-crystal (a) and the crystal-series (b) approaches.

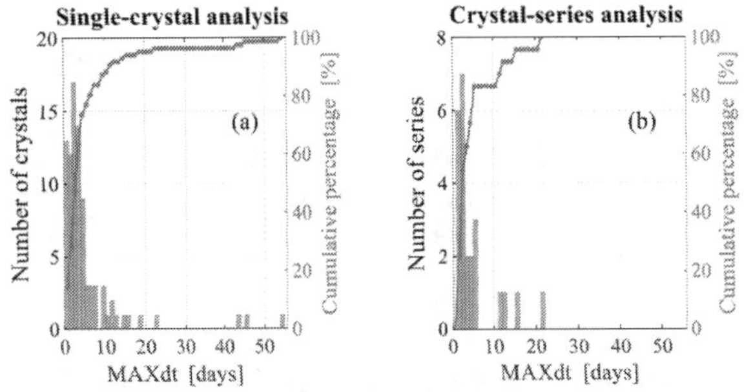

Fig. 4 Accuracy in estimating the time elapsed after exposure from peak III-to-IV area ratios for the single-crystal (a) and the crystal-series (b) approaches.

In the first week after exposure, the peak II-to-IV area ratio yielded an estimated exposure time of 8 hours or less from the actual event for 86% of the analysed crystals. A comparable result was observed over the first two months, with the peak III-to-IV area ratio achieving a deviation of less than 5 days in 74% of cases.

A similar analysis is shown in Figs. 5a and 6a, illustrating how the time-indexed sampling points with the lowest accuracy in the individual fading curves were spread across all crystals. As expected, for both peak area ratios accuracy dropped at longer fading times, as the decomposition software and the fitting model become less capable of capturing the disappearing low-temperature peaks. A comparable decline in accuracy was also observed for the peak II-to-IV area ratio at short fading times, where peak II shows the greatest variation.

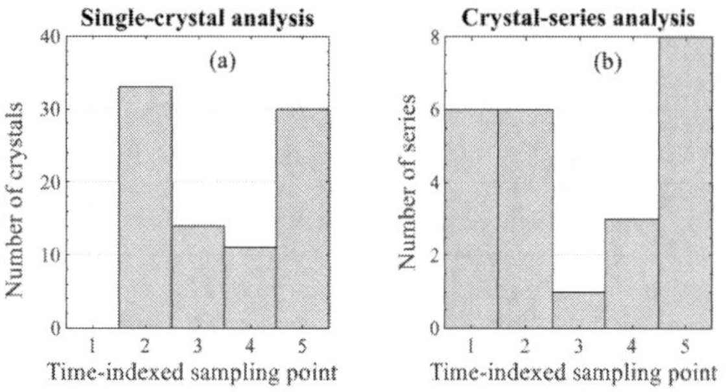

Fig. 5 Distribution of lowest-accuracy time-indexed sampling points for peak II-to-IV area ratios in the single-crystal (a) and the crystal-series (b) approaches.

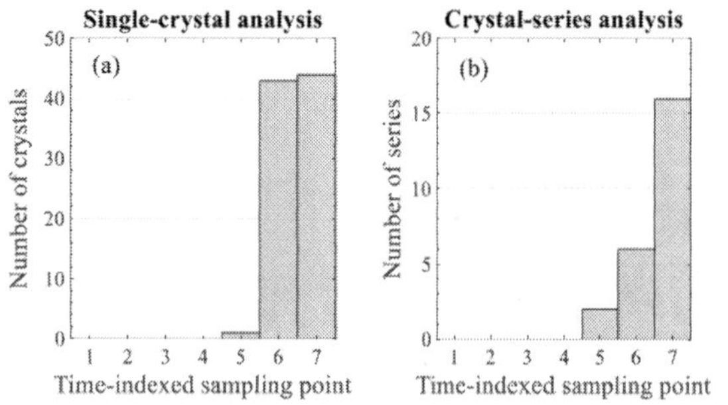

Fig. 6 Distribution of lowest-accuracy time-indexed sampling points for peak III-to-IV area ratios in the single-crystal (a) and the crystal-series (b) approaches.

While prioritising the analysis of individual crystal behaviours, we also examined the fading curves derived from series-averaged data. This complementary approach builds on the initial crystal selection process (see Materials and Methods), where groups of crystals with comparable responses were identified, and was explored solely for comparative purposes, to assess whether averaging could enhance the accuracy of the results. A total of 24 series were considered, each comprising a variable number of crystals. At each time sampling point, the measured peak area ratios of the crystals within a given series were averaged, and the resulting values were then fitted using the same stretched exponential model to produce the corresponding fading curve (Figs. 7-9).

To facilitate comparison of overall estimation accuracy across the full dataset, similar histograms with superimposed cumulative percentage curves were generated (Figs. 3b, 4b, 5b, 6b).

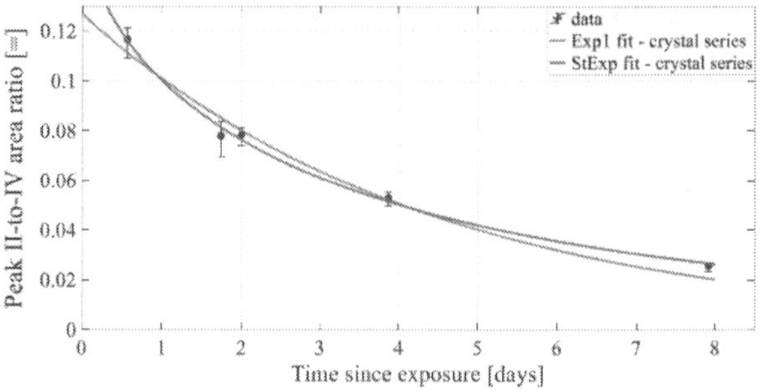

Fig. 7 Comparison of stretched exponential (red) and one-term exponential (green) fits to peak II-to-IV area ratios as a function of time since exposure in the crystal-series approach.

The series-based approach seems to produce slightly more accurate results in both intervals of interest, indicating that averaging may help reduce part of the inherent variability in the data. This is further supported by a numerical comparison of MAXdt values, which shows that, in just over half of the cases, the series-averaged method results in a lower estimation error. It is also worth mentioning that, in general, the maximum deviation for each series s tends to approximate the arithmetic mean of the MAXdt values of the N_s crystals composing it:

$$\mathrm{MAXdt}_s \approx \frac{1}{N_s}\sum_{i=1}^{N_s} \mathrm{MAXdt}_{i,s} \quad \forall s\,. \tag{6}$$

IV. Conclusions

This study evaluated the effectiveness of a passive method for detecting the unauthorised removal of radioactive material from storage, with the added capability of retrospectively estimating the time of the event. Building on previous research that connected time- and temperature-dependent signal loss in luminescent materials to post-exposure timing, the approach was expanded into a single-material framework (LiF:Mg,Cu,P) by relating the time factor to the differential fading of its low-temperature glow peaks. Fitting a stretched exponential model to the evolution of peak II-to-IV and III-to-IV area ratios in the individual fading behaviour of each analysed crystal has demonstrated the capacity to estimate the precise timing of the event within a margin of a few hours to a few days over the first week and two months after exposure. These estimates can be further refined, to some degree, by averaging signals from multiple crystals with similar responses. Since the half-life of the glow peaks influences the system's effectiveness, current research is investigating alternative luminescent materials with longer-lived peaks to broaden their applicability to applications featuring longer intervals between inspections.

References

[1] M. Moscovitch, "Automatic method for evaluating elapsed time between irradiation and readout in LiF-TLD", *Radiat. Prot. Dosim.*, vol. 17, no. 1-4, pp. 165-169, Dec. 1986, 10.1093/rpd/17.1-4.165.

[2] F. d'Errico, D. O. Junot, I. O. Polo, A. Chierici, R. Ciolini, D. N. Souza, et al., "Differential-fading optically stimulable materials for Nuclear Safeguards", in *Proc. INAC 2019*, Santos, SP, Brazil, 2019, pp. 3838-3843.

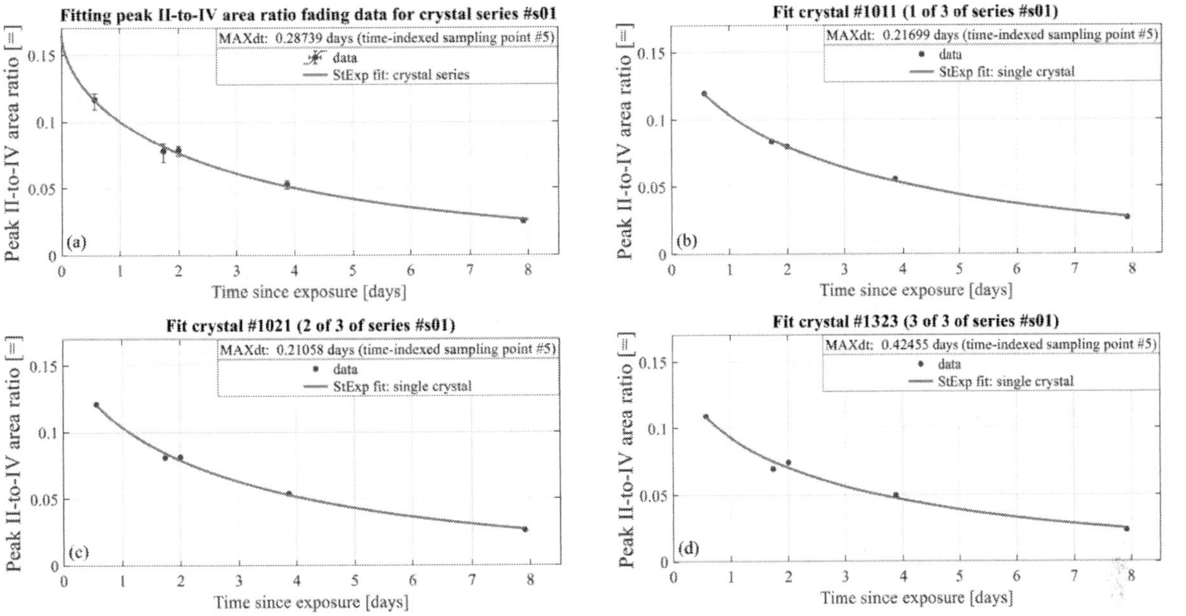

Fig. 8 Comparison of fading curves of peak II-to-IV area ratios versus time after exposure for a crystal series (a) and the individual crystals composing it (b–d). All data were fitted using a stretched exponential model.

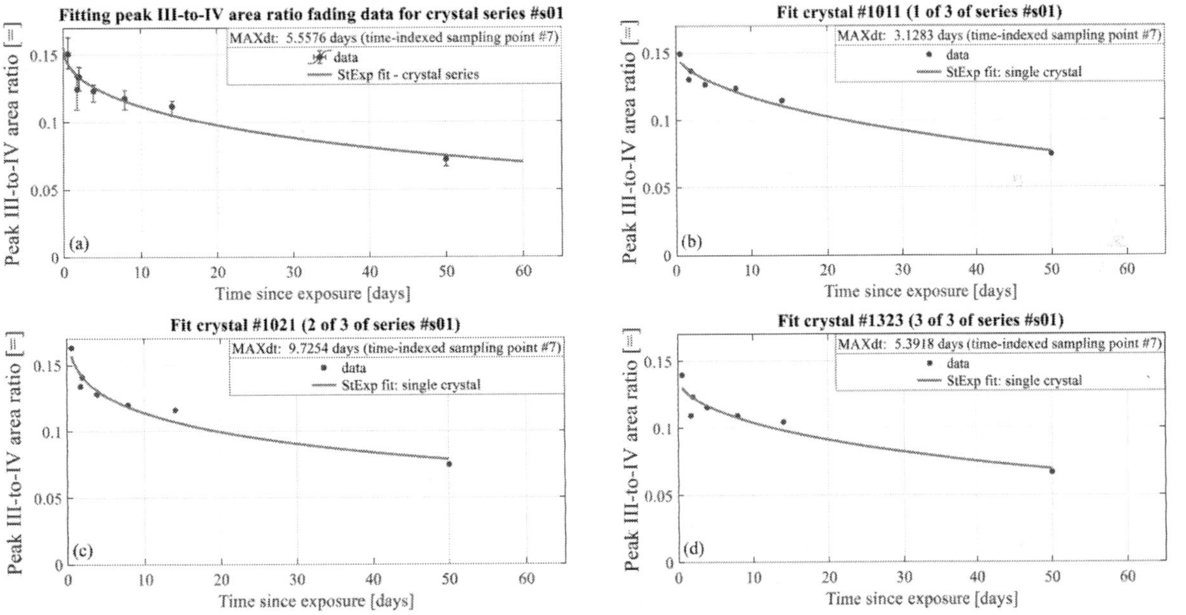

Fig. 9 Comparison of fading curves of peak III-to-IV area ratios versus time after exposure for a crystal series (a) and the individual crystals composing it (b–d). All data were fitted using a stretched exponential model.

[3] M. Budzanowski, J. C. Sáez-Vergara, J. M. Gómez-Ros, A. Romero, E. Ryba, "The fading of different peaks in LiF:Mg,Cu,P (MCP-N and GR-200A) TL detectors", *Radiat. Meas.*, vol. 29, no. 3–4, pp. 361–364, May 1998, 10.1016/S1350-4487(98)00028-6.

[4] M. Budzanowski, J. C. Sáez-Vergara, E. Ryba, P. Bilski, P. Olko, M. P. R. Waligórski, "Estimation of the time elapsed between exposure and readout using peak ratios of LiF:Mg,Cu,P (MCP-N, GR200A)", *Radiat. Prot. Dosim.*, vol. 85, no. 1, pp. 149–152, Sep. 1999, 10.1093/oxfordjournals.rpd.a032822.

[5] L. A. DeWerd, J. R. Cameron, W. Da-Ke, T. Papini, I. J. Das, "Characteristics of a new dosemeter material: LiF(Mg,Cu,P)", *Radiat. Prot. Dosim.*, vol. 6, no. 1–4, pp. 350–352, Dec. 1983, 10.1093/oxfordjournals.rpd.a082948.

[6] G. F. Knoll, "Miscellaneous detector types" in *Radiation Detection and Measurement*, 4th ed., Hoboken, NJ, USA: Wiley, 2010, ch. 19, sec. VII, pp. 755-757.
[7] E. Martella, "Dosimetric characteristics of LiF:Mg,Cu,P (GR-200A)", ASG–S.R.L., Italy, 1993. [Online] Available: https://www.mc2dna.com/wp-content/uploads/gr-200a.pdf.
[8] M. Puchalska, P. Bilski, "GlowFit—a new tool for thermoluminescence glow-curve deconvolution", *Radiat. Meas.*, vol. 41, no. 6, pp. 659–664, Jul. 2006, 10.1016/j.radmeas.2006.03.008.
[9] D. C. Elton, "Stretched exponential relaxation," *arXiv [cond-mat.dis-nn]*, Aug. 2018, 10.48550/arXiv.1808.00881.
[10] A. Werner, 1907, "Quantitative messungen der an- und abklingung getrennter phosphorescenzbanden", Ann. Phys., vol. 329, no. 11, pp. 164-190, Aug. 1907, 10.1002/andp.19073291111.

Fast Neutron Multiplicity Counting for High Mass Plutonium Assay

Xiaobo LIU*, Jiansheng LI, Yu JIN
Institute of Nuclear Physics and Chemistry, CAEP, China
(*) 13881190590@163.com

Abstract—**The fast neutron multiplicity counting (FNMC) apparatus was consisted of 36 pieces of liquid scintillator detectors in three rings cylindrical layout, multi-modules of 500M14bit digitizers based on PXIe platform, and mechanical structure. Its center cavity was 35 cm in radius and 50 cm in height for assaying the plutonium item in standard container. The digitizers implemented detector's signal acquisition, real-time digital data processing and data throughput. Each detector's light output was calibration with Cs-137 source for nearly equal response. The apparatus detection efficiency for fast neutron calibrated with Cf-252 source was about 3.2%. The fast neutron multiplicity counting analysis was verified with Cf-252 neutron source, then experiments were performed with a serial of different high mass plutonium metals. Two-parameter calibration procedure was used for mass resolving. The mass assay results were consistent with the sample nominal values within maximum bias of 6%. This research achievement demonstrated a promising application for nuclear material assay and verification.**

Keywords —**fast neutron multiplicity counting, FNMC, liquid scintillator, digitizers, plutonium, nuclear material assay**

I. INTRODUCTION

THE fast neutron multiplicity counting (FNMC) is originated from thermal neutron multiplicity counting[1]. As an innovative technology for nuclear material assay and verification, it does multiplicity analysis by detecting fast neutron directly in tens ns time scale rather than thermal neutron moderated by polyethylene medium in tens μs time scale.

The earliest report about fast coincidence neutron counter using liquid scintillator for plutonium bearing assay was from J. R. Wachter [2]. This prototype counter was setup with four detectors surrounding a 12.7cm-square sample chamber, and analogue electronics for pulse shape discrimination (PSD) and conventional fast timing modules with a resolving time of 30 ns was used for coincidence analysis.

In recently 20's years, the advances of fast ADC and high-speed FPGA technology paved the way for fast pulse digital acquisition and processing [3], later named the Digital Pulse Processing (DPP) technology [4], [5], [6]. The detector signal is processed directly by the fast digitizer in which basic information related to time and amplitude is computed on line by FPGA. This type of digitizers such as CAEN desktop DT5730 or VIM platform V1730 shows superior performance to the conventional analogue electronics [7],[8], and widely used in the research fields of nuclear physics, and nuclear safeguard.

There are different layout for fast multiplicity counting by combination of detector configurations, size and arrangement, introduce by Katherine Frame [9],.L. Nakae[10], D. L. Chichester[11], [12], Enqvist[13], Dolan[14], [15], [16], A. Di Fulvio[17] and Tony H. Shin[18], and Yuezhuang Liu [19], [20]. The experimental samples included plutonium metal plate, oxide plutonium, MOX pellet. The comparison between single-parameter calibration procedure and two-parameter calibration procedure was carried out for plutonium mass resolution [17, 18]. However, the fast neutron cross-talk between adjacent detectors also made significant contribution to the doubles and triples, and results [21], [22] indicated that the mass resolution from three parameter equation was not as good as expected even if the neutron cross-talk correction was considered.

In 2010s, we developed a thermal multiplicity counter (named NPL-NMC) for assaying the high mass plutonium metal items deposited in standard container [23], [24], [25]. By taking the advantages of FNMC for nuclear material assay and verification, a FNMC apparatus was constructed and experiments were carried out to assay these plutonium items. The constitution of the apparatus, the calibration, the experiments and results were introduced in this paper.

II. THE FNMC APPARATUS

The FNMC apparatus was consisted of an array of 36 pieces of liquid scintillator detectors (LSDs) in three rings cylindrical layout, mechanical structure, multi-modules of 500M14bit digitizers based on PXIe platform, HV device, and software for control and data analysis. The apparatus was assembled in the experimental hall. One PC located in measuring room was connecting to the digitizers through Ethernet cable to control the digitizer remotely. The schematic picture is shown in Fig. 1.

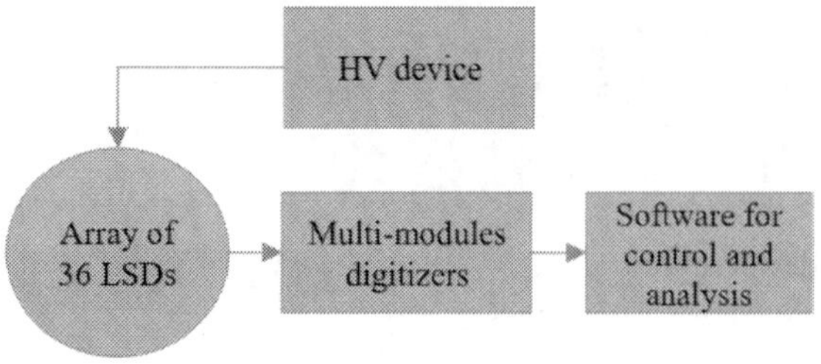

Fig. 1. Schematic diagram of the FNMC apparatus

A. Array of 36 LSDs

Each liquid detector was made up of a scintillator 3 inches in diameter by 2 inch in depth coupled with a 3 inch in diameter photomultiplier (PMT). A sleeve of 5 mm in thick Pb was used to wrap each detector's main body and the front end for shielding the low energy photos. This made to get a better performance between balancing the neutron gamma PSD discrimination and detection efficiency.

The 36 pieces of liquid scintillator detectors (LSDs) were arranged in three rings cylindrical layout, with 16 prices in center plane and 10 pieced respectively in upper and lower plane. All the detectors in the upper and lower plane were tilted in degree of 38.9 to make the axis line of the detector aiming to the array center. The distance from the chamber center to the detector front end surface is 35 cm.

The chamber cavity of array was 35 cm in radius and 50 cm in height, and suitable for the standard container.

As shown in Fig. 2, the array was installed on the mechanical structure through the aluminum circle belt and poles and divided two halves. Each half can move on the rails to open for sample loading or unloading and close for assaying.

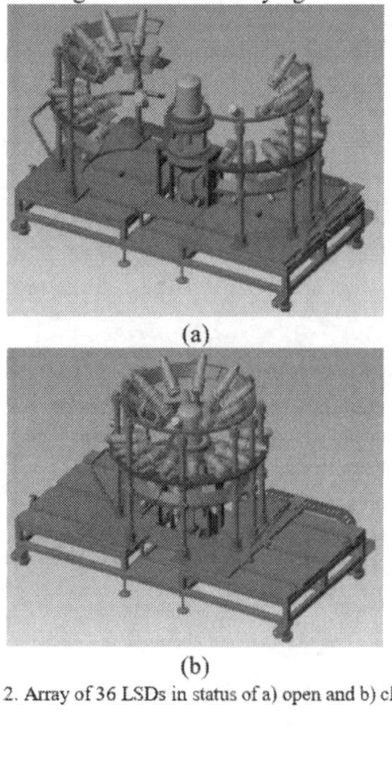

Fig. 2. Array of 36 LSDs in status of a) open and b) close

B. Multi-modules Digitizers

The digitizers were comprised of 5 digitizer modules, one synchronization module, one PXIe imbedded controller and one 9-slot PXIe Chassis, as shown in Fig. 3. Please note there are only 3 digitizer modules for illustrating.

Fig. 3. Multi-modules Digitizers

There are 8 channels for each digitizer module, and totally 40 channels by 5 modules but only 36 channels used. The specifications of the digitizer are 250 M Hz bandwidth, 500M Sample/s and 14-bit ADC resolution. The capability of pulse event rates is more than 50k cps for each channel.

The digitizers implemented detector's signal triggering and acquisition with ADC, and real-time digital data processing with FPGA for characteristics of signal such as triggering time, CFD timing stamp, pulse peak, Qlong and Qshort for PSD discrimination. It is almost the same as CAEN DPP [26], as shown in Fig. 4. The digitizers also manage the data throughput via PXIe bus of backplane with maximum 20 G b/s rate.

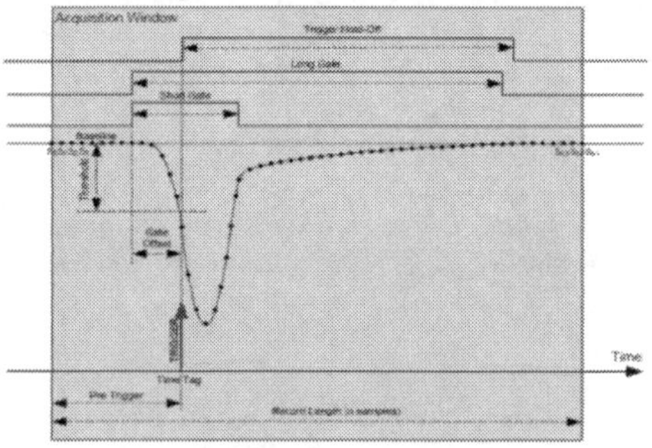

Fig. 4. Diagram summarizing the DPP-PSD parameters

C. Software for control and processing

It has three main functions:

1) to control and configure the digitizers for synchronizing, signal acquisition and data processing;
2) to do neutron and gamma discrimination, multiplicity counting analysis and plutonium mass resolving with preset parameters and algorithm;
3) to receive and save the measuring data.

III. CALIBRATIONS

A. Light output

The light output of each detector was calibrated with Cs-137 source. The nearly same light output for each detector were made by adjusting the voltage of HV device.

The histogram plots for the energy spectrum of Qlong, pulse peak, PSD scatter in 1-D and 2-D were shown in Fig. 5.

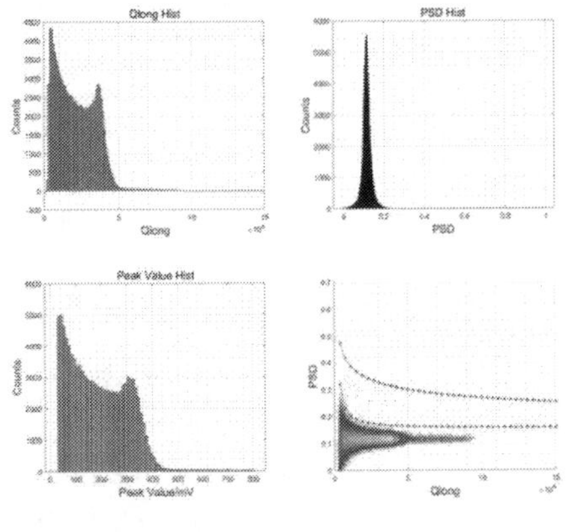

Fig. 5. The plots for Qlong, peak, PSD and 2-D scatter

B. PSD discrimination and detection efficiency

The Cf-252 source was used for PSD discrimination parameter calibration for each detector. The measured PSD data was binned by measured light output Qlong. The optimization of the discrimination value was determined and the FOM for the PSD performance was computed for each "slice" of data. An equation for the fit of discrimination point was used:

$$y = a \cdot x^b + c \quad (1)$$

The a, b and c are the coefficients. A line for upper limit neutron discrimination also was determined with (1) in different a, b and c. The typical PSD scatter in 2-D and FoM values for one detector was shown in Fig. 6.

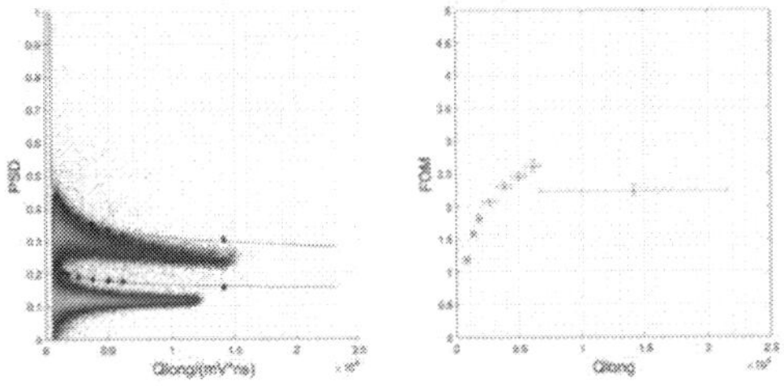

Fig. 6. The plots for PSD 2-D scatter and FoM

The neutron rates for each detector were obtained for three circumstances: in the standard container, naked, and in a 2-inch right cylinder chamber with 3mm in thickness of wall. The rates were shown in Fig. 7. It indicated that neutron rates for the detectors in the lower ring decreased by almost 40 % because the shielding from the bottom case of standard container.

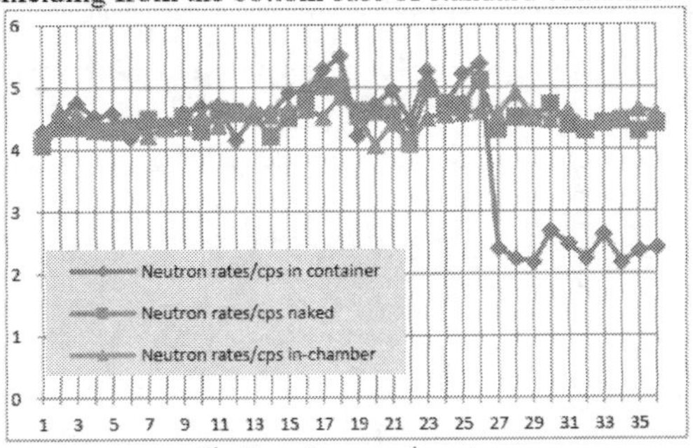

Fig. 7. neutron counting rates

The neutron source is 3mm in diameter by 6mm in height, and the neutron strength was 4.95 E3 s^{-1} at the experimental day. So, the detection efficiency for the FNMC apparatus was 2.95%, 3.29%, and 3.27% respectively. The standard container made the neutron detection efficiency of the apparatus decreased by 10.3%.

C. Time of flight (ToF)

The gated coincidence measurements were carried out for extract the timing resolution and the time offset for each detector with Cf-252 fission chamber [27], [28], [29]. It was fixed in the center of apparatus, and the fission timing signal was collected from the chamber and amplified for timing and gating. The typical ToF spectrum from one scintillator detector was shown in Fig. 8. The time offset was determined by tuning the gamma peak to be 1.50 ns. The FWHM of gamma peak was 1.30 ns, extracted by Gaussian fitting.

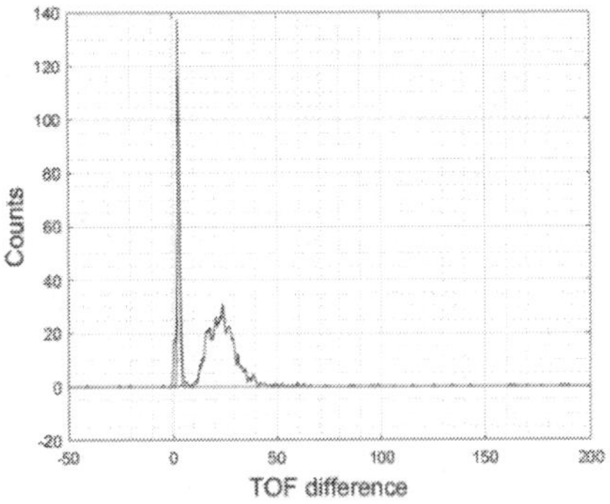

Fig. 8. The ToF spectrum

The average and standard deviation of gamma peak time for all the 36 detectors were 1.52 ns and 0.04 ns respectively. The average and standard deviation of gamma peak FWHM for all the 36 detectors were 1.27 ns and 0.07 ns respectively.

D. Test of multiplicity

The multiplicity counting analysis for the Cf-252 fission chamber and the background were carried out, and the measured SDT (Singles, Doubles and Triples) were shown in Table I.

TABLE I
Measured SDT for Cf-252 and background

	Measuring time/s	S/cps	D/cps	T/cps
Cf-252	300	2888 ±3.1	108.7 ±0.7	2.38 ±0.14
Cf-252	1000	2881 ±1.7	107.3 ±0.4	2.42 ±0.08
background	600	39.7±0.3	0.135±0.015	0

IV. EXPERIMENTS AND RESULTS

There were seven plutonium items used for FNMC assay. Each item was deposited in a standard container. The sketch diagram for the container was shown in Fig. 9.

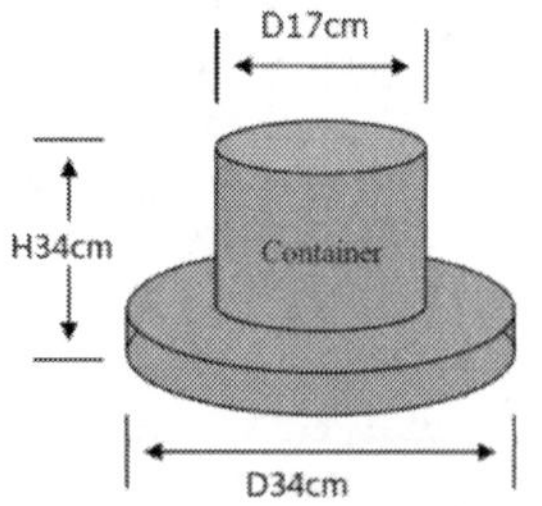

Fig. 9. Sketch diagram for the container

Each container was loaded by turn on the support base in the FNMC apparatus cavity, and the measuring time was 300s. The measured SDT were shown in Table II.

TABLE II
Measured SDT for plutonium items

	S/cps	D/cps	T/cps
1	385.5 ±1.1	9.38 ±0.19	0.257 ±0.049
2	416.6 ±1.2	10.07 ±0.20	0.277 ±0.046
3	898.0 ±1.7	42.35 ±0.45	2.827±0.192
4	830.7 ±1.7	20.68 ±0.28	0.498 ±0.067
5	968.2 ±1.8	24.31±0.31	0.690 ±0.077
6	2500.5 ±2.9	123.31 ±0.76	7.783 ±0.311
7	5086.5 ±4.1	553.18 ± 2.06	95.76 ±1.80

As the values of T for some items were less than 1 cps and the statistical uncertainty of counting were large rather than accepted, the two-parameter calibration procedure using S and D values was implemented for mass resolution. The equation was listed as following:

$$\frac{D}{m_{eff}} = A + B \cdot \left(\frac{D}{S}\right) + C \cdot \left(\frac{D}{S}\right)^2, \tag{2}$$

D refers to Doubles, in unit of cps, *S* refers to the Singles, in unit of cps, m_{eff} refers to the effective mass in items, in unit g, A, B and C refer to the coefficient from calibration, which were 0.082 ±0.022, 16.024 ±0.892 and 234.274 ±6.963 respectively.

The results of m_{eff} from assay were determined with (2), as shown in Fig. 10. The relative biases for these items were less than 6%.

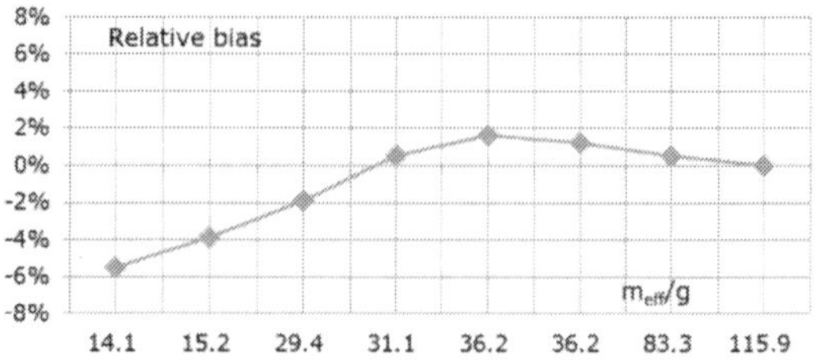

Fig. 10. Assaying results and biases.

V. CONCLUSIONS

The fast neutron multiplicity counting (FNMC) apparatus was constructed, calibrated and applied for assaying the high mass Pu items. The apparatus was consisted of an array of 36 pieces of liquid scintillator detectors in three rings cylindrical layout, multi-module of 500M14bit digitizers based on PXIe platform, and mechanical structure, and software for control and data analysis. The two-parameter calibration procedure using S and D values was implemented for mass resolution. The mass assay results were consistent with the sample nominal values within maximum bias of 6%.

This research achievements demonstrated a promising application for assaying high mass Pu items in the fields of nuclear material assay and verification. Further work will be taking such as the two parameter and three parameter equations for mass resolving, neutron cross-talk correction, multiplication correction and spectra analysis for oxidization in the future.

REFERENCES

[1] D. G. Langner, J. E. Stewart, M. M. Pickrell, M. S. Krick, N. Ensslin, and W. C. Harker, "Application guide to neutron multiplicity counting," Los Alamos Nat. Lab., Los Alamos, NM, USA, Tech. Rep. LA-13422-M, 1998, doi: 10.2172/1679.

[2] Wachter, J. R. , E. L. Adams , and N. Ensslin . "Prototype fast neutron counter for the assay of impure plutonium." Transactions of the American Nuclear Society (1987).

[3] Ny., Kornilov , et al. "Neutron spectroscopy with fast waveform digitizer." Nuclear Instruments & Methods in Physics Research A, 497.2-3(2003):467-478, doi: 10.1016/S0168-9002(02)01790-4.

[4] P. J. Sellin, G. Jaffar and S. D. Jastaniah, "Performance of digital algorithms for n/γ pulse shape discrimination using a liquid scintillation detector", 2003, IEEE Nuclear Science Symposium and Medical Imaging Conference Record.

[5] R. Aryaeinejad, J. K. Hartwell, D. F. Spencer, "Comparison between digital and analog pulse shape discrimination techniques for neutron and gamma ray separation", 2005 IEEE Nuclear Science Symposium and Medical Imaging Conference Record.

[6] Hennig, et al. "Development of 500 MHz Multi-Channel Readout Electronics for Fast Radiation Detectors. " IEEE Transactions on Nuclear Science 57.4(2010):2365-2370.

[7] D. Cester, M. Lunardon, G. Nebbia, L. Stevanato, G. Viesti, S. Petrucci, C. Tintori, "Pulse shape discrimination with fast digitizers," Nuclear Instruments and Methods in Physics Research A, Volume 748, 2014, 33-38, doi: 10.1016/j.nima.2014.02.032.

[8] C.S. Sosa, M. Flaska, S.A. Pozzi, "Comparison of analog and digital pulse-shape-discrimination systems," Nuclear Instruments and Methods in Physics Research A, 826, 2016, 72-79, doi: 10.1016/j.nima.2016.03.088.

[9] Katherine Frame, Willam Clay, Tim Elmont, Ernst Esch, Peter Karpius, Duncan MacArthur, Edward McKigney, Peter Santi, Morag Smith, Jonathan Thron, Richard Williams. "Development of a liquid scintillator neutron multiplicity counter (LSMC)" [J]. Nuclear Instruments and Methods in Physics Research A, 2007, 579(1): 192-195.

[10] Nakae, L.Recent. "Developments In Fast Neutron Detection And Multiplicity Counting With Verification With Liquid Scintillator" [R], Lawrence Livermore National Laboratory, USA, Tech. Rep. LLNL-CONF-502912-2011, 2011.

[11] D.L. Chichester; S.A. Pozzi; J.L. Dolan; M.T. Kinlaw; S.J. Thompson; A.C. Kaplan; M. Flaska; A. Enqvist; J.T. Johnson; S.M. Watson. "MPACT Fast Neutron Multiplicity System Prototype Development" [R], Idaho National Laboratory, USA, Tech. Rep. INL-EXT-13-30279, 2013.

[12] David L. Chichester;Scott J. Thompson;Mathew T. Kinlaw;James T. Johnson;Jennifer L. Dolan;Marek Flaska;Sara A. Pozzi. "Statistical estimation of the performance of a fast-neutron multiplicity system for nuclear material accountancy" [J]. Nuclear Instruments and Methods in Physics Research A, 2015, 784(1): 448-454.

[13] A, Andreas Enqvist , et al. "A combined neutron and gamma-ray multiplicity counter based on liquid scintillation detectors." Nuclear Instruments and Methods in Physics Research A, 652. 1(2011):48-51.

[14] Jennifer Lynn Dolan. "Safeguarding special nuclear material by detecting fast neutrons in liqiud scintillators" [D], University of Michigan, 2013.

[15] J. L. Dolan, M. Flaska, S. A. Pozzi, D. L. Chichester, "Passive measurements of mixed-oxide fuel for nuclear nonproliferation," Nuclear Instruments and Methods in Physics Section A, vol. 703, pp. 102-108, 2013.

[16] Jennifer L. Dolan, Marek Flaska, Alexis Poitrasson-Rivierea, Andreas Enqvistb, Paolo Peeranic, David L. Chichesterd, Sara A. Pozzia. "Plutonium measurements with a fast-neutron multiplicity counter for nuclear safeguards applications" [J]. Nuclear Instruments and Methods in Physics Research A, 2014, 763(Complete): 565-574.
[17] A. Di Fulvio, T.H. Shin, T. Jordan, C. Sosa, M.L. Ruch, S.D. Clarke, D.L. Chichester, S.A. Pozzia. "Passive assay of plutonium metal plates using a fast-neutron multiplicity counter" [J]. Nuclear Instruments and Methods in Physics Research A, 2017, 855(1): 92-101.
[18] Shin, Tony H.. "Fast-Neutron Multiplicity Counting Techniques for Nuclear Safeguards Applications" [D]. University of Michigan, 2019.
[19] Liu Y, Zhang S, Lin X, et al. "Research and Development of a Commercial Passive Fast Neutron Multiplicity Counting Device for Nuclear Material Accounting," in IEEE Transactions on Nuclear Science, vol. 72, no. 4, pp. 1631-1636, April 2025, doi: 10.1109/TNS.2025.3544881.
[20] Liu, Y., Zhang, S., Lin, X., et al. "Passive fast neutron multiplicity counting system based on a symmetric spherical design". Journal of Nuclear Science and Technology, 2025, 62(7), 631–638, doi: 10.1080/00223131.2025.2457597.
[21] Tony H. Shin; Matthew J. Marcath; Angela DiFulvio; Shaun D. Clarke; Sara A. Pozzi. "Neutron cross-talk characterization of liquid organic scintillators" [C].2015 IEEE Nuclear Science Symposium and Medical Imaging Conference (NSS/MIC). Nuclear Science, IEEE Symposium: IEEE Xplore Digital Library, 2015.
[22] Tony H. Shin; Michael Y. Hua; Matthew J. Marcath; David L. Chichester; Imre Pázsit; Angela Di Fulvio; Shaun D. Clarke; Sara A. Pozzi. "Neutron Multiplicity Counting Moments for Fissile Mass Estimation in Scatter-Based Neutron Detection Systems" [J]. Nuclear Science and Engineering, 2017, 188(3): 246-269.
[23] Ligao CHEN, Xiaobo LIU, Jian GONG, et al. "Neutron multiplicity counter design and calibration" [J]. Journal of Tsinghua University (Science and Technology), 2014, 54(2): 159-163
[24] Li-Gao, Chen , et al. "Variance analysis for passive neutron multiplicity counting." Nuclear Science and Techniques 26.2(2015):54-58.
[25] Liu, Xiaobo , and L. Chen . "A two-parameter iterative approach of multiplication correction for assaying plutonium metal parts with neutron multiplicity counting." Nuclear Instruments and Methods in Physics Research A, 1016(2021):165779
[26] User Manual UM2580 DPP-PSD Digital Pulse Processing for Pulse Shape Discrimination, CAEN S.p.A., ITALY, 2016.
[27] Hill, N. W. , et al. "Optimization of Nanosecond Fission Ion Chambers for Reactor Physics Applications." IEEE Transactions on Nuclear Science, 22.1(1975):686-690.
[28] LI Jian-sheng, ZHANG Yi, JIN Yu. *et al.* "Development of 252Cf fast fission chamber" [J]. Nuclear Electronics and Detection Technology, 2001, 21 (4), 264-267, DOI: CNKI:SUN:HERE.0.2001-04-007. In Chinese.
[29] GU Xian-bao, LI Guang-wu, ZHANG Wen-hui, ZHANG Gang, CHEN Yu. "Design and Test of a Thin Wall and Fast Pulse (252)Cf Ionization Chamber", Nuclear Physics Review, 2009,26(1): 37-40. in Chinese.

Toward neutron /gamma discrimination with proportional counter using artificial intelligence

Aya Kanj[1,*], Richard Babut[1], and Louis Roux[2]
[1]French nuclear safety and Radiation protection Authority (ASNR), PSE-HEALTH/SDOS/LMDN, F-13108, St-Paul-lez-Durance, France
[2]European organization for nuclear research (CERN), Geneva, Switzerland & with the University Claude Bernard Lyon1, CNRS, ILM UMR5306, Lyon, France

(*) aya.kanj@asnr.fr

***Abstract*— Extending the neutron detection capabilities of the ROSPEC SP2-1 proportional counter below 50 keV requires effective discrimination between neutron and gamma-ray signals at low energies. To address this challenge, we use a digital acquisition system, then apply signal processing and artificial intelligence techniques to analyze the data. The results showed that a CNN-based trained model can successfully distinguish noise from event signals in the recorded data. Key pulse features, such as rise time and amplitude, are extracted from the true signals to generate a two-dimensional plot of rise time versus amplitude, which facilitates the discrimination of neutron from gamma components. The application of the unsupervised clustering algorithm DBSCAN on this feature space shows limitations in accurately identifying low-amplitude gamma signals, while a measurement with only a gamma source confirms the presence of gamma events in the expected feature space. These results motivate the development of a supervised CNN-based approach to improve neutron/gamma discrimination.**

***Keywords* —ROSPEC, neutron/ gamma discrimination, proportional counter, CNN, low-energy neutrons.**

I. Introduction

THE Micro-irradiation, Neutron Metrology and Dosimetry Laboratory (i.e. the LMDN) from ASNR Cadarache in France, is responsible for the national references in the field of neutron dosimetry. For this purpose, LMDN produce reference neutron fields and has spectrometers to characterize them. Among these instruments is ROSPEC (ROtating SPECtrometer) [1], a multi-detector system comprising six spherical proportional counters designed for neutron detection and spectrometry across a wide energy range. Four of the counters, called SP2-1, SP2-4, SP2-10 and SP6, are designed to cover neutron energy ranges between 50 keV and 5 MeV. The fifth counter is a ^{3}He-filled detector designed to detect thermal neutrons (E < 25 meV), while the final counter, ^{3}He+^{10}B, has a boron-coated inner surface for detecting neutrons between 1 eV and 10 keV.

The aim of this work is to improve the ROSPEC performance by extending the energy range below 50 keV. This can be achieved by implementing neutron/ gamma discrimination in SP2-1 proportional counter, so that gamma event can be identified and removed.

In the SP2 proportional counter, neutrons interact with the hydrogen gas via elastic scattering. These interactions produce recoil protons, which can acquire energies up to that of the incident neutrons. The charged recoil protons produce primary ionization along its track until it is brought to rest by the gas.

Gamma rays interact within the chamber primarily through Compton scattering with the electrons of the gas atoms. The resulting recoil electron, which gains kinetic energy from the interaction, ionizes the gas along its path by producing ion-electron pairs. However, due to the limited stopping power of the gas, the electron can cross the entire chamber and deposits only a fraction of its total energy. As a result, there exists a maximum energy that an electron can deposit in the gas, independent of the initial gamma-ray energy. This limiting energy (approximately 50 keV), depending on the gas pressure, corresponds to the case where the electron's range is just equal to the chamber's diameter. Electrons with either lower or higher energy will ultimately deposit less energy in the gas.

In the SP2-1 proportional counter, with a diameter of 5.08 cm and operating at 0.748 atm gas pressure, electrons can deposit up to 18.2×10^{-3} MeV in the gas. This value was calculated using the ESTAR database [2], which provides electron stopping powers in various materials. To lower the minimum detectable neutron energy in the SP2-1 detector, currently set at 50 keV, which corresponds to the threshold defined by ROSPEC in analog acquisition mode, two main improvements will be implemented:

1. Switching from the default analog acquisition mode employed by ROSPEC to a digital acquisition mode, allowing access to raw data and enabling more flexible signal processing.
2. Applying neutron-gamma discrimination techniques for energies below the maximum gamma pulse energy 18.2 keV.

II. Method

A. Discrimination by Pulse Rise Time

The ranges in hydrogen are much lower for protons than for electrons of the same energy, in other words tracks will be much shorter for protons than electrons (indeed, electrons typically cross the entire chamber). The primary electrons generated along these tracks do not significantly contribute to the signal until they reach the narrow multiplication region near the central anode wire, where they generate secondary ionisation. If the particle's track is short, the primary electrons reach the anode in a brief time window, resulting in a short pulse rise time. Conversely, for longer tracks, the spread in arrival times is wider, leading to a longer rise time. As a result, neutron interactions produce pulses with short rise time (typically around one microsecond or less), while gamma interactions can lead to rise time as long as four microseconds.

B. Proposed Method for Neutron/Gamma Discrimination

To discriminate neutrons from gamma rays by pulse rise time, we will record each detected signal and calculate its rise time and amplitude. These parameters are then used to create a two-dimensional distribution, rise time as a function of amplitude. A clustering algorithm, specifically the DBSCAN (Density-Based Spatial Clustering of Applications with Noise) [3], will be applied to this distribution to separate neutron events from gamma ray events.

A numerical acquisition system is used by coupling the preamplifier of the SP2-1 counter with a CAEN digitizer DT5730 [4]. We used Wavedump [5] software that enables real-time signal monitoring and waveform recording.

The recorded signals pass through four steps of data processing:

1) Noise elimination

To detect low-energy signals (< 18.2 keV), we need to lower the detection threshold, which will consequently increase background signals. Therefore, the first step is to eliminate noise from the data. One solution is to use an AI-based method called Convolutional Neural Network (CNN) [6]. To be efficient, the algorithm must be trained with simulated signals (sigmoïde functions for neutrons and photons, and mixture of different operations on normal distribution for background signals). The CNN method applied to the measured data has proven to be effective, as it provides reliable probability estimates. These estimates are essential for setting a discrimination threshold (X) to classify gamma and neutron signals separately from noise. Specifically, a signal classified as a neutron or gamma with a probability ≥ X% is accepted as such, while signals below this threshold are considered noise. Fig.1 illustrates an example of CNN results obtained using the SP2-1 proportional counter irradiated with a mono-energetic neutron source of 42 keV.

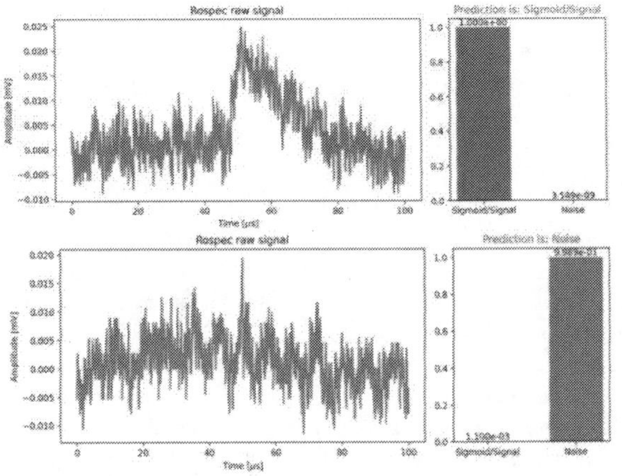

Fig.1. Discrimination between signal and noise by CNN method

2) Signal smoothing

Since the signals exhibit strong fluctuations due to noise caused by ROSPEC's electronics, and because we are detecting low-energy neutrons, these fluctuations become comparable to the signal amplitude. Therefore, it is necessary to smooth the signals to reduce these fluctuations and enable proper fitting. Fig. 2 shows an example of signal smoothing by applying a Savitzky-Golay filter.

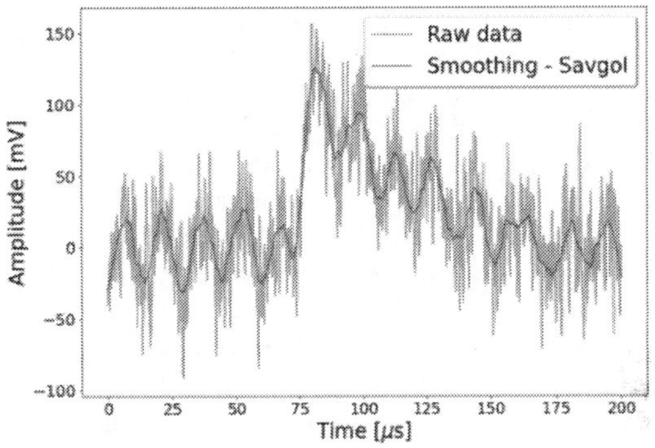

Fig. 2. Raw signal smoothing using Savitzky-Golay filter

3) Fitting the smoothed signal

The third step is to fit the smoothed signal. A typical output signal from the preamp associated to detection of a single event can be modeled by the mathematical expression:

$$f(t) = \begin{cases} b & si\, t > t_0 \\ A\left(e^{-\frac{t-t_0}{\tau_d}} - e^{-\frac{t-t_0}{\tau_r}}\right) + b & si\, t \leq t_0 \end{cases} \quad (1)$$

Where A is the amplitude related to energy deposited, t_0 is the pulse start time. τ_r and τ_d represent the rise and fall time, respectively. Fig. 3 shows the signal fitting using this model. This fit allows the extraction of pulse shape parameters such as amplitude and rise time, which will be used to generate the two- dimensional plot of rise time versus amplitude.

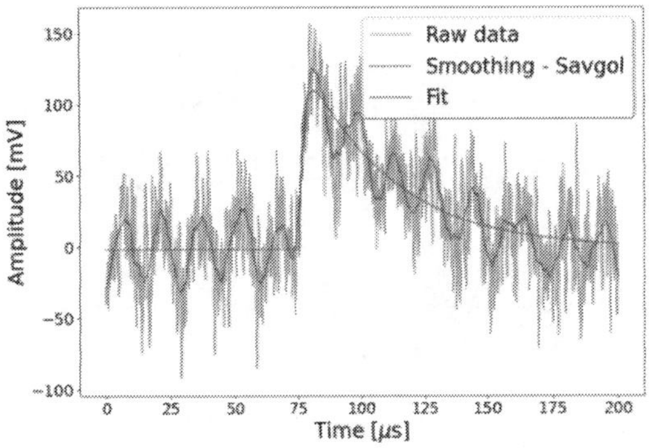

Fig. 3. Exponential fit of the smoothed signal

4) Applying DBSCAN method

The fourth step in the signal processing is applying the DBSCAN algorithm, implemented in Python, on the two-dimensional plot of rise time versus amplitude.

DBSCAN is a density-based clustering algorithm that groups together data points located in high-density regions, while labeling those in low-density areas as noise or outliers. It is particularly effective for detecting clusters of arbitrary shape without requiring the number of clusters in advance.

Consider a set of N data points $X \equiv \{x_n\}_{n=1}^{N}$. For each point x_n the algorithm defines its ε-neighborhood as follows:

$$N_\epsilon(x_n) = \{x \in X | \, d(x, x_n) < \epsilon\} \quad (2)$$

$N_\epsilon(x_n)$ are the data points that are at a distance smaller than ϵ from x_n. d(·,·) is the Euclidean distance metric.

The algorithm is initialized with the distance metric d(·,·), the neighborhood radius ϵ, and the minimum number of points minPts required to generate a cluster.

III. Results

In this section, we present the measurements conducted to discriminate between neutrons and gamma rays using an AmBe neutron-gamma source and a Bi-207 gamma source. Although our objective is neutron/gamma discrimination with the ROSPEC SP2-1 proportional counter, we start by applying the method on the ROSPEC SP6 detector. This choice is justified by the fact that the AmBe source emits neutrons with a main energy around 4 MeV, which lies within the energy detection range of the SP6 counter.

Fig. 4 shows measurement when SP6 is irradiated by both AmBe and Bi-207 sources. In this figure we can distinguish two components which are low amplitude and high rise time for gammas, and high amplitude and low rise time for neutrons. In order to confirm that high rise time low amplitude events can be identified as gammas, another measurement has been done by using Bi-207 gamma source alone. The results are shown in Fig. 5, where blue dots belong to gammas from Bi- 207 superimposed on the data taken previously when both AmBe and Bi-207 are present. The Bi-207 data occupy the same region of amplitude/rise time range as the high rise time locus in the previous measurement.

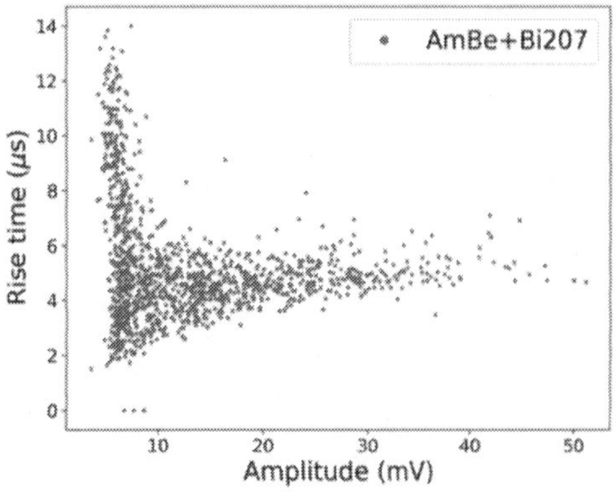

Fig. 4. Scatter plot showing the rise time and amplitude of pulses from SP6 exposed to both AmBe and Bi-207 sources.

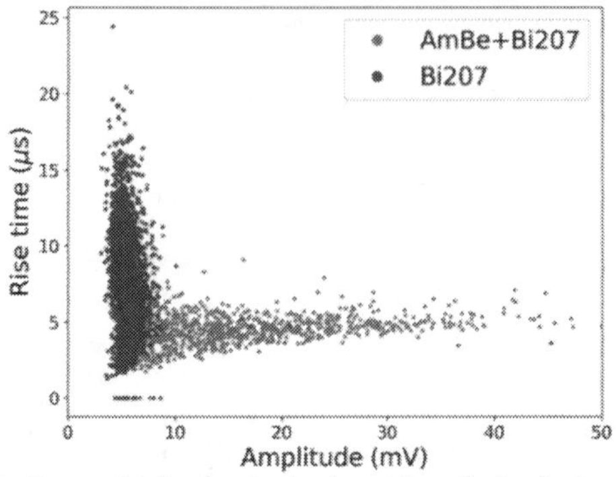

Fig. 5. Scatter plot showing the rise time and amplitude of pulses from SP6. The green dots correspond to data acquired with both AmBe and Bi-207 sources, while the blue dots represent a separate measurement with the Bi-207 source alone.

We applied the DBSCAN unsupervised machine learning clustering algorithm to distinguish between neutron and gamma clusters in the data shown in Fig. 4. The DBSCAN clustering result is presented in Fig. 6. All the points that are not part of a cluster (outliers) are labelled as noise. However, this method does not effectively identify gamma events in the region of low rise time and low amplitude. Nevertheless, when using the Bi- 207 gamma source alone (Fig. 5), gamma events in this region could be clearly observed.

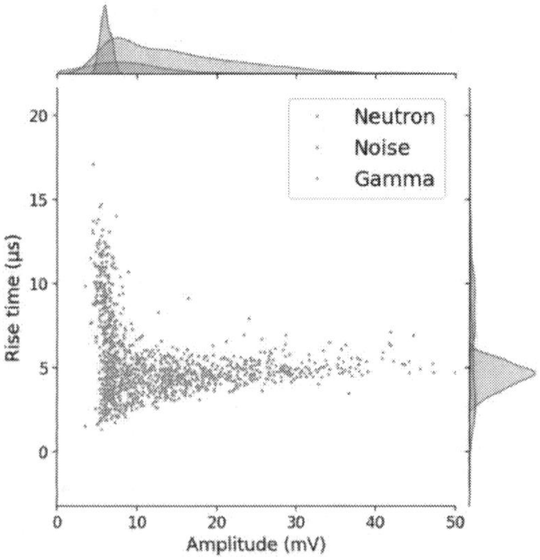

Fig. 6. DBSCAN clustering result.

IV. Conclusion

In this work, we proposed and implemented a method based on artificial intelligence to perform neutron/gamma discrimination in proportional counters, with the goal of extending the low-energy detection capabilities of the ROSPEC SP2-1 detector. By using digital acquisition, waveform processing, and a CNN-based method we were able to effectively separate noise from true signals and extract key pulse features such as rise time and amplitude. These parameters were then used in a clustering algorithm (DBSCAN) to distinguish between neutron and gamma-ray events.

DBSCAN clustering showed limitations in identifying low-amplitude gamma events, the use of a gamma-only source validated their presence in the expected region. Therefore, in future work, we plan to use a supervised machine learning method by extending the CNN beyond noise classification to perform neutron/gamma discrimination.

References

[1] H. Ing, T. Clifford, T. McLean, W. Webb, T. Cousins, and J. Dhermain, "ROSPEC _ A simple reliable high resolution neutron spectrometer," *Radiat. Prot. Dosimetry*, vol. 70, no. 1–4, pp. 273–278, 1997, [Online]. Available: https://doi.org/10.1093/oxfordjournals.rpd.a031959

[2] M. Berger, J. Coursey, and M. Zucker, "Estar, pstar, and astar : Computer programs for calculating stopping-power and range tables for electrons, protons, and helium ions," 1992, 1.21.

[3] M. Ester, H.-P. Kriegel, J. Sander, and X. Xu, "A Density-Based Algorithm for Discovering Clusters in Large Spatial Databases with Noise," in *KDD-96*, 1996, pp. 226–231. doi: 10.1016/B978-0-444-64165-6.03005-6.

[4] CAEN Tools for Discovery, "DT5730 / DT5730S." Accessed: Jul. 16, 2025. [Online]. Available: https://www.caen.it/products/dt5730/

[5] CAEN Tools for Discovery, "WaveDump." Accessed: Jul. 16, 2025. [Online]. Available: https://www.caen.it/products/caen-wavedump/

[6] Y. Lecun, Y. Bengio, and G. Hinton, "Deep learning," *Nature*, vol. 521, no. 7553, pp. 436–444, 2015, doi: 10.1038/nature14539.

Evaluating the Performance of a Photofission Prompt Neutron Detection System Based on Threshold Activation Scintillation Coupled with a 9 MeV Electron Accelerator

Luna Sobczak[1,*], Adrien Sari[1], and Frédérick Carrel[1]
[1]Université Paris-Saclay, CEA, List, 91120 Palaiseau, France
(*) luna.sobczak@cea.fr

Abstract—**Developments in photofission prompt neutron detection are crucial in nuclear instrumentation, whether for waste characterization and security applications. *Bremsstrahlung* photons, emitted by a linear electron accelerator (linac), can induce photofission reactions when interacting with actinides. These reactions then emit prompt and delayed neutrons and delayed gammas. However, the intense photon flash from the linac saturates detection systems, commonly He-3 proportional counters, making prompt neutron instantaneous detection challenging. This study evaluates a photofission prompt neutron detection system coupled with a 9 MeV linac to investigate the use of prompt neutrons as a third signature, alongside delayed neutrons and gammas. This to reduce uncertainties in detecting, differentiate and quantify actinide despite an attenuated interrogating flux or absorbed photofission particles by the environment. First, with Monte Carlo codes PHITS and MCNP6 and their respective nuclear data libraries (ENDF/B-VIII.0 and JENDL-5), we assessed the parameter of a photofission setup which consists in detecting the delayed 6.131 MeV gamma radiation emitted by N-16 decay, resulting from (n, α) activation of F-19. This approach avoids the impact of the linac's photon flash and minimizes the strong photoneutron background. Secondly, we conducted experimental tests with neutron sources and then with a 9 MeV electron accelerator, Linatron® M9A (Varex Imaging Corp.), at the SAPHIR platform (CEA Paris-Saclay, France), using machined polytetrafluoroethylene and a bismuth germanate scintillator to detect prompt neutrons from a depleted UMo sample. This work enhances prompt neutron detection developments and contributes to advancements in characterizing special nuclear materials (SNMs) by photofission for various applications.**

Keywords—**Linear electron accelerator (linac), Active Photon Interrogation (API), photofission, prompt neutrons, threshold activation detector (TAD), fluorine.**

I. Context and Challenges

A photon with enough energy (> 6 MeV) interacting with actinides, i.e., U-235, U-238, Pu isotopes, may generate a photofission reaction, emitting various particles at different emission time. The emission of prompt neutrons and prompt gammas occurs simultaneously with the fission fragments and the delayed particle are emitted later by the fission product [1, 2]. The photon beam is usually induced by a linear electron accelerator (linac), that converts electrons into high-energy *Bremsstrahlung* photons, by slowing them on heavy atomic nuclei target such as tungsten or tantalum for instance. Photofission is a reaction of great interest for research and development in the field of nuclear instrumentation and more specifically in non-destructive active nuclear measurement technique such as active photon interrogation (API) method. The latter finds many applications in the fields of quantification of fissile material for nuclear/radioactive waste package characterization, and storage [3–15], for homeland security or border control [16–19], to counter proliferation, and for fundamental nuclear physics [20].

The API method addresses passive emission detection difficulties and, furthermore, overcomes common attenuation of interrogating neutron beams by using photons which have an enhanced penetrating capability. In this context, prompt photofission neutron detection is particularly relevant to counter matrix effects in radioactive waste packages or shielding of SNM in cargo containers. That is to say, when measurements are complicated by self-absorption and the attenuation of photofission particles like gamma signatures and mainly delayed neutrons thermalization in hydrogenated materials. In spite of this, most applications regarding photofission particles emission concerns delayed neutrons and delayed gammas. This, because prompt neutrons are emitted instantaneously while the measurement system and electronics, commonly He-3 proportional counters, are saturated by the intense photon flash delivered by the linac and under a strong photoneutron background created in the conversion target and surrounding materials [21].

An indirect detection method based on threshold activation detectors (TAD) has been deployed since the earliest advancements in neutron detection for nuclear reactors and then applied to prompt neutron detection [22,23]. More recent progress has been led by developments with pentafluorostyrene plastic scintillator and CaF_2 TAD in nuclear waste package and homeland security applications [24–26]. It consists in delaying the measurement of prompt neutrons by detecting the delayed 6.131 MeV gamma radiation emitted by N-16 decay, resulting from (n, α) activation of F-19. These detectors enable firstly to bypass the gamma blinding by delaying the detection in time and secondly to avoid the background noise of photoneutrons produced in the tungsten target, collimator and surrounding materials during the photon beam production.

In this paper, we firstly present the photofission prompt neutron detection setup geometry designed based on the TAD indirect method and the Monte Carlo simulations we performed to assess its parameters. In a previous study, we evaluate with MCNP6 [27, 28] the photoneutron background produced in the tungsten of the accelerator considering the (γ, n), (γ, 2n) and (γ, *X*n) cross-sections and reaction threshold, also the photofission reaction rate in a uranium sample. Here, after reviewing these key quantities, we evaluate the (n, α) reaction rate on PTFE induced by the prompt neutron and photoneutron yield mentioned. Secondly, we report the results of two experimental campaigns conducted using machined polytetrafluoroethylene layers placed against a bismuth germanate scintillator. The first campaign involved tests with characterized AmBe and Cf-252 neutron sources from the LNHB (Laboratoire National Henri Becquerel). The second campaign utilized the 9 MeV linac of the SAPHIR platform at CEA Paris-Saclay to irradiate a depleted UMo sample.

II. Design of a detection system by simulation

To design the mentioned system, we modeled and simulated the following geometry with MCNP6 and PHITS [29, 30] Monte Carlo codes. The setup in Fig. 1 includes the accelerating portion of the linac in green, to wit, the pipe where the electron beam is accelerated. In addition, the conversion target on its copper stand, holding the *Bremsstrahlung* reactions and the collimator are both made of heavy nuclei, here tungsten. The sample is a depleted 7.5 cm-high uranium cylinder with a radius of 0.7 cm. The detector section is composed of a 7.62 cm × 7.62 cm bismuth germanate scintillator crystal ($Bi_4Ge_3O_{12}$), wrapped up with a fluoropolymer layer of polytetrafluoroethylene (PTFE, C_2F_4), known as Teflon®. The latter operates as the activated fluoride converter.

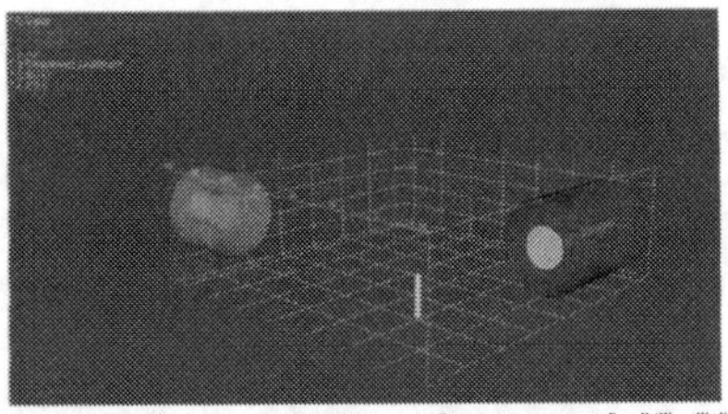

Fig. 1. MCNP6 photofission model visualization with 3D PHITS visualizer PHIG-3D

In a previous study we characterized the photofission rate and the corresponding background noise produced by a linac operating at 6, 9, 15 or 18 MeV. Given the 7 MeV energy threshold of the (γ, n) reaction on tungsten, the (γ, 2n) reactions involved for processes beyond 10 MeV, and the 6 MeV photofission barrier, prompt neutron detection frame was concluded more efficient with an operating energy of 9 MeV. Indeed, the photoneutron yield happens to be 90 times higher for a linac operated at 15 MeV compared to 9 MeV. Then calculation of the photofission reaction rate in a uranium sample leads to the conclusion that the signal-to-noise ratio is optimal for an accelerator operated at 9 MeV, i.e., for a photoneutron production about 24 times higher than the prompt neutron yield. In Table I, we registered different evaluated yield rates: both the prompt neutron yield with (γ, f) photofission reaction rate in uranium, the photoneutron yield with the (γ, *X*n) photonuclear reaction rate in tungsten target and collimator, and also the related signal to noise ratio.

Table I SIGNAL-TO-NOISE RATIO EVALUATION FOR A LINAC OPERATING AT 9 MEV

Electron energy (MeV)	(γ, f) ($n.s^{-1}$)	(γ, *X*n) ($n.s^{-1}$)	(γ, f)/(γ, *X*n) ratio
9	1.24×10^{11}	2.94×10^{12}	4.24×10^{-2}

Results obtained using the MCNP6 Monte Carlo code. A previous study employed also the PHITS code and both were run with their respective library (ENDF/B-VIII.0 and JENDL-5) and with the library of the other code to intercompare the impacts on the results.

The previous reaction rates indicate the prompt neutron (n_p) and photoneutron (n_γ) production per second and this gives the different neutron contributions in PTFE activation by (n, α) reactions. Given in Table II, the (n, α) reaction rate on PTFE induced by the two aforementioned neutron sources and the related ratio.

Table II PTFE ACTIVATION BY (n, α) REACTION

Electron energy (MeV)	(n_p, α) ($reac.s^{-1}$)	(n_γ, α) ($reac.s^{1}$)	(n_p, α)/(n_γ, α) ratio
9	4.57×10^{6}	1.26×10^{4}	3.63×10^{2}

Finally, a study on thickness optimization was conducted alongside the calculation of the signal-to-noise ratio given by the prompt neutron and photoneutron production, and the assessment of their contribution in the (n, α) reaction rate. This combined analysis led to the following conclusion regarding how the betas and gammas lose their energy in the materials:

→ 5 cm of PTFE (2.25 $g.cm^{-3}$) stops 25% of 6 MeV gammas and all beta particles.

→ 10 MeV β will deposit all its energy in 0.85 cm of BGO and 2.6 cm of PTFE.

From this, we conclude that we will be able to detect gamma particles but only the beta produced near the BGO in the very first centimetres of PTFE surrounding the crystal.

The evaluation of the reaction products' propagation led to efficiency and sensitivity calculation. The detection efficiency of the BGO is simulated with MCNP6 F8 tally, for the detection of beta and gamma particles produced in PTFE. In a first place, we assumed a homogeneous 6.1 MeV gamma source uniformly distributed within the PTFE volume and same procedure for beta particles with $^{16}_{7}N$ decay beta spectrum. We obtain the following efficiencies:

→ 5.17% for 6.1 MeV gamma

→ 0.68% for β spectrum of $^{16}_{7}N$ decay

The expected sensitivities are calculated for the following experimental configuration: a linac operated at 9 MeV and pulse frequency of 100 Hz, a 240 g depleted uranium sample, and an acquisition of 60 s after 60 s of irradiation. Given that the half-life of N-16 is 7.13 s, the equilibrium production–decay is achieved within approximately 36 s. Therefore, we optimize the N-16 production in PTFE and with a 60 s acquisition we nearly measure the full signal.

We obtain the following sensitivities, illustrating a favorable configuration for optimizing the signal-to-noise ratio:

→ $(1.01 \pm 0.01) \times 10^{-1}$ c.s^{-1} due to prompt neutrons

→ $(2.80 \pm 0.04) \times 10^{-4}$ c.s^{-1} due to photoneutrons

III. Evaluation of the performance in laboratory conditions

Ahead of the irradiation campaign at the SAPHIR platform, we moved the photofission prompt neutron detection system to be irradiated under laboratory conditions, that is to say, with metrologically characterized AmBe and Cf-252 neutron sources from the LNHB. This campaign allows to quantify and validate the discrepancy in the (n, α) reaction rate in Teflon as a function of the energy spectrum of the sources, whose uncertainty on the emission rates are very low. It also allows us to obtain an experimental reference value for the (n, α) reaction rate for a Cf-252 spectrum which is close to the prompt neutron spectrum.

The joint objective of the two following experiments is to activate with a neutron source the fluorine contained in a Teflon layer and identify the decay products from the reaction $^{19}_{9}F(\mathrm{n},\alpha)^{16}_{7}N$ with a scintillator. The experimental parameters of the setup in Fig. 2 consist in an AmBe or Cf-252 neutron source with a 9.51×10^{5} n.s^{-1} and 5.411×10^{6} n.s^{-1} neutron emission rate respectively, known to within less than two percent uncertainty, irradiating a 5 cm hollow cylindrical PTFE converter in which a 7.62 cm × 7.62 cm BGO scintillator is placed. The detector is powered by a +800 V high voltage supplied by a CAEN DT1470ET module and the data is acquired with a Dspec502 spectrometer from Ortec. The protocol is a simultaneous irradiation and acquisition of 2700 s after the energy calibration with the following gamma emitting sources: Co-60, Y-88, Na-22, AmBe.

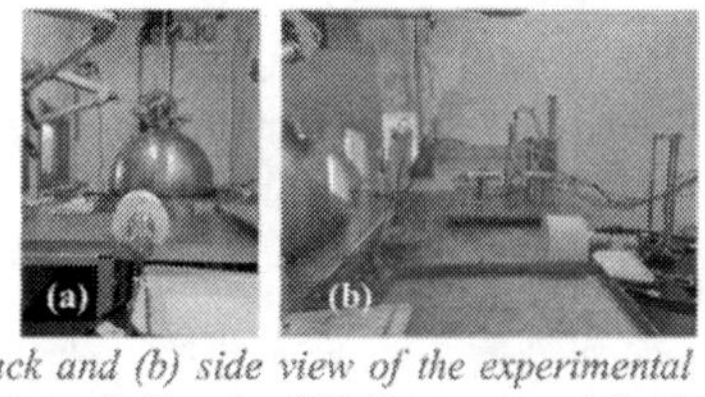

Fig. 2. (a) Back and (b) side view of the experimental setup at the LNHB casemate including the Cf-252 source and the BGO wrapped with the Teflon layer.

A. *Irradiation with an AmBe neutron source*

The gamma spectrum in Fig. 3 results from the activation of the PTFE layer by the AmBe neutron source (9.51×10^{5} n.s^{-1}) and includes the gamma emissions of the source itself, which has a photopeak at around 4.4 MeV. The high neutron energy spectrum of AmBe sources (around 4 MeV average energy) allowed us to identify the 6.131 MeV gamma product from the $^{19}_{9}F(\mathrm{n},\alpha)^{16}_{7}N$ reaction as the cross section reaches 0.1 mb for 2.5 MeV incident neutrons.

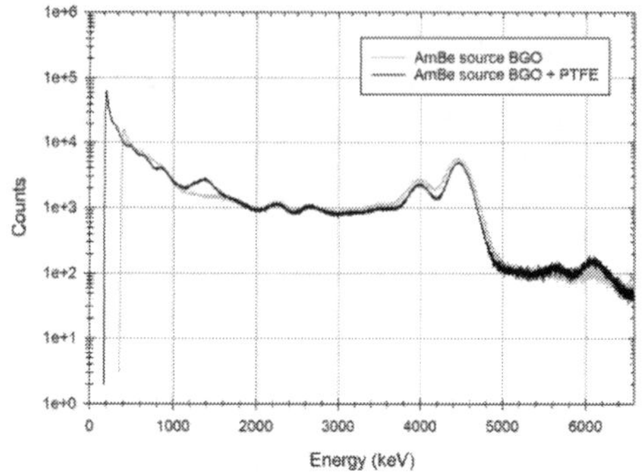

Fig. 3. Gamma spectra acquired after a 2700 s irradiation with an AmBe neutron source of a BGO scintillator (in blue) and a BGO wrapped in PTFE (in black).

Between 5.1 MeV and 7 MeV, the experimental peak net count rate obtained is of 11.06 ± 0.18 c.s^{-1}. Also, complementary simulations of (n, α) reaction rate on PTFE induced by the same AmBe neutron source will be modeled with MCNP6 to determine the (n, α) reaction rate and the corresponding sensitivity.

B. *Irradiation with a Cf-252 neutron source*

The gamma spectrum in Fig. 4 results from the activation of the PTFE layer by the Cf-252 neutron source (5.411×10^{6} n.s^{-1}). The energy spectrum of Cf-252 neutron source is close to a Watt fission spectrum, with an average energy of about 2.1 MeV.

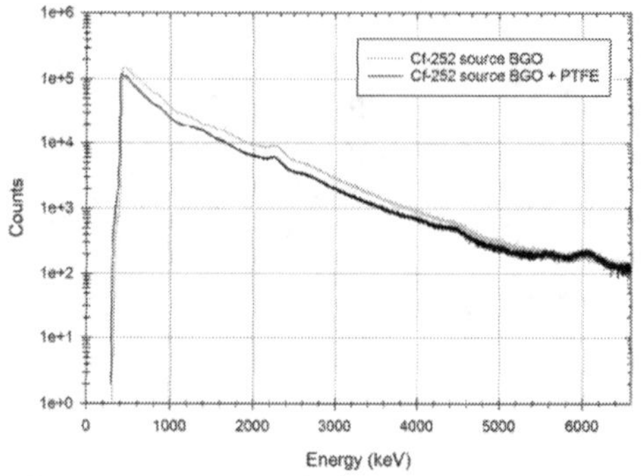

Fig. 4. Gamma spectra acquired after a 2700 s irradiation with a Cf-252 neutron source of a BGO scintillator (in blue) and a BGO wrapped in PTFE (in black).

 https://doi.org/10.1051/epjconf/202533806014

Once more, the neutron contribution on the gamma spectrum is calculated between 5.1 MeV and 7 MeV and the experimental peak net count rate obtained is 7.52 ± 0.21 c.s^{-1}. Complementary simulations of (n, α) reaction rate on PTFE induced by the Cf-252 neutron source will be modeled with MCNP6 to determine the (n, α) reaction rate and the corresponding sensitivity.

It should be noted that higher activation is observed with the AmBe source despite the Cf-252 source having a neutron emission rate nearly 6 times higher. This was expected and can be explained by the AmBe's harder energy spectrum.

IV. VALIDATION AT THE SAPHIR PLATFORM

Finally, we transferred the detection system at the SAPHIR platform. The source term is now a *Bremsstrahlung* photon beam produced by a linac operated at 9 MeV which irradiates the depleted uranium sample described earlier. All the other experimental parameters remain equivalent to the previous campaign at LNHB. The objective is to detect the decay product of the fluorine activation by the different neutron contributions and mainly the prompt neutrons from the photofission reactions in the sample. The protocol here consists in 60 s of acquisitions after 60 s of irradiation, following the energy calibration of the detector with Co-60, Cs-137 and AmBe gamma emitting sources.

With this method, we effectively highlight the signature of N-16. This demonstrates the feasibility of the detection system for a configuration based on a linac irradiation. The gamma spectrum in Fig. 5a shows an experimental peak net count rate of 2.94 ± 0.75 c.s^{-1} in the 5.1 MeV and 7 MeV region of interest. The agreement between experimental and simulation results requires further investigation, particularly to characterize the systematic errors in the experiment. In Fig. 5b, the spectrum is obtained after placing a lead shield between the uranium sample and the PTFE cylinder. This added layer is decreasing the background noise around our ROI, thereby changing the experimental net peak area to total area ratio in our ROI from 0.03% to 0.19% by stopping the high-energy delayed gammas emitted by the sample. This result demonstrates that further improvements and optimizations remain possible.

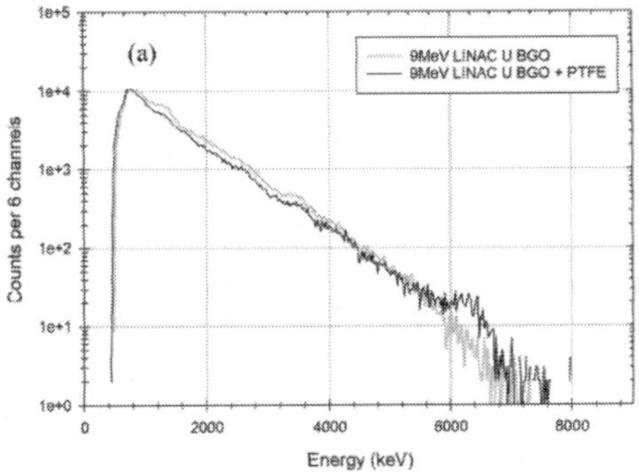

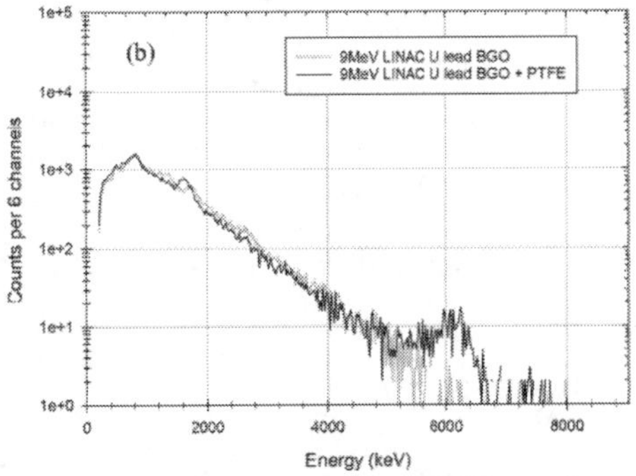

Fig. 5. Gamma spectrum acquired with (a) a BGO scintillator shielded with lead and (b) a BGO scintillator wrapped with PTFE and shielded with lead, after 60 s of irradiation and 60 s of acquisition with a 9 MeV linac.

V. CONCLUSIONS AND FUTURE DEVELOPMENTS

A photofission prompt neutron detection system was designed based on threshold activation detectors. After optimization through Monte Carlo simulations of the signal-to-noise ratio in a given configuration to enhance the detection sensitivity, we conduct several tests in laboratory conditions at the LNHB to validate the reliability of the system. Finally, we perform experiments at the SAPHIR platform using a 9 MeV linac to irradiate a uranium sample, produce photofission reactions and detect the gammas resulting from the fluorine activation from the prompt photofission neutrons.

Future work will include complementary acquisitions to characterize the signal resulting from the decay gamma emissions of N-16. Efforts will be made to optimize the reduction of photoneutron noise and improve the discrimination between prompt neutron contributions and high-energy delayed gammas. Simultaneous detection of delayed neutrons and prompt neutrons coupled with preliminary tests of inter-pulse acquisitions will also be performed.

REFERENCES

[1] N. Bohr and J. A. Wheeler, "The Mechanism of Nuclear Fission," Phys. Rev., vol. 56, no. 5, pp. 426–450, Sep. 1939, DOI. 10.1103/PhysRev.56.426.

[2] R. O. Haxby, W.E. Shoupp, W. E. Stephens, and W.H. Wells, "Photo-Fission of Uranium and Thorium," Phys. Rev., vol. 58, no. 1, p. 92, Jul. 1940, DOI. 10.1103/PhysRev.58.92.

[3] A. Lyoussi, J. Romeyer-Dherbey, F. Jallu, E. Payan, A. Buisson, G. Nurdin, *et al.*, "Transuranic waste detection by photon interrogation and on-line delayed neutron counting," Nuclear Instrum. and Methods in Phys. Res. B, vol. 160, no. 2, pp. 280–289, Feb. 2000, DOI. 10.1016/S0168-583X(99)00586-8.

[4] M. Gmar, J. Jeanneau, F. Lainé, H. Makil, B. Poumarède, and F. Tola, "Assessment of actinide mass embedded in large concrete waste packages by photon interrogation and photofission," Appl. Radiat. Isot., vol. 63, no. 5–6, pp. 613–619, Dec. 2005, DOI. 10.1016/j.apradiso.2005.05.010.

[5] F. Carrel, M. Agelou, M. Gmar, F. Lainé, B. Poumarède, and B. Rattoni, "Measurement of Plutonium in Large Concrete Radioactive Waste Packages by Photon Activation Analysis," IEEE Trans. Nucl. Sci., vol. 57, no. 6, pp. 3687–3693, Dec. 2010, DOI. 10.1109/TNS.2010.2087356.

[6] F. Carrel, M. Agelou, M. Gmar, and F. Lainé, "Detection of high-energy delayed gammas for nuclear waste packages characterization," Nuclear Instrum. and Methods in Phys. Res. A, vol. 652, no. 1, pp. 137–139, Oct. 2011, DOI. 10.1016/j.nima.2010.08.014.

[7] A. Sari, F. Carrel, F. Lainé, and A. Lyoussi, "Neutron interrogation of actinides with a 17 MeV electron accelerator and first results from photon and neutron interrogation non-simultaneous measurements combination," Nuclear Instrum. and Methods in Phys. Res. B, vol. 312, pp. 30–35, Oct. 2013, DOI. 10.1016/j.nimb.2013.06.020.

[8] F. Carrel, B. Charbonnier, R. Coulon, F. Lainé, S. Normand, and C. Salmon, "Characterization of Old Nuclear Waste Packages Coupling Photon Activation Analysis and Complementary Non-Destructive Techniques," IEEE Trans. Nucl. Sci., vol. 61, no. 4, pp. 2137–2143, Aug. 2014, DOI. 10.1109/TNS.2014.2304751.

[9] E. Simon, F. Jallu, B. Pérot, and S. Plumeri, "Feasibility study of fissile mass quantification by photofission delayed gamma rays in radioactive waste packages using MCNPX," Nuclear Instrum. and Methods in Phys. Res. A, vol. 840, pp. 28–35, Dec. 2016, DOI. 10.1016/j.nima.2016.09.047.

[10] I. Meleshenkovskii, T. Ogawa, A. Sari, F. Carrel, and K. Boudergui, "Optimization of a 9 MeV electron accelerator *Bremsstrahlung* flux for photofission-based assay techniques using PHITS and MCNP6 Monte Carlo codes," Nuclear Instrum. and Methods in Phys. Res. B, vol. 483, pp. 5–14, Nov. 2020, DOI. 10.1016/j.nimb.2020.10.002.

[11] M. Delarue, E. Simon, B. Pérot, P. G. Allinei, N. Estre, E. Payan, *et al.*, "Measurement of cumulative photofission yields of ^{235}U and ^{238}U with a 16 MeV *Bremsstrahlung* photon beam," Nuclear Instrum. and Methods in Phys. Res. A, vol. 1011, Art. no. 165598, Sep. 2021, DOI. 10.1016/j.nima.2021.165598.

[12] I. Meleshenkovskii, A. Elayeb, R. De Stefano, and A. Sari, "Feasibility study of the photofission technique for radiological characterization of 220-L concrete-lined nuclear waste drums using 7 or 9 MeV linacs," Nuclear Instrum. and Methods in Phys. Res. A, vol. 1029, Art. no. 166422, Apr. 2022, DOI. 10.1016/j.nima.2022.166422.

[13] M. Delarue, E. Simon, B. Pérot, P. G. Allinei, N. Estre, D. Eck, *et al.*, "New measurements of cumulative photofission yields of ^{239}Pu, ^{235}U and ^{238}U with a 17.5 MeV *Bremsstrahlung* photon beam and progress toward actinide differentiation," Nuclear Instrum. and Methods in Phys. Res. A, vol. 1040, Art. no. 167259, Oct. 2022, DOI. 10.1016/j.nima.2022.167259.

[14] A. Elayeb, A. Sari, I. Meleshenkovskii, R. De Stefano, A. Dabat-Blondeau, and Y. Moline, "Performance Assessment of a New Digital Readout Electronics for Nuclear Waste Drum Characterization by Photofission," in *2022 IEEE Nuclear Science Symposium and Medical Imaging Conference (NSS/MIC)*, Milan, Italy, Nov. 2022, pp. 1–2, DOI. 10.1109/NSS/MIC44845.2022.10399217.

[15] C. Carasco, D. Eck, B. Geslot, E. Payan, B. Pérot, J. Roult dit Rouaux, *et al.*, "Photofission delayed gamma-ray measurements in a large cemented radioactive waste drum during linac irradiation," Nuclear Instrum. and Methods in Phys. Res. A, vol. 1053, Art. no. 168360, Aug. 2023, DOI. 10.1016/j.nima.2023.168360.

[16] J. L. Jones, W.Y. Yoon, D. R. Norman, K. J. Haskell, J. M. Zabriskie, S.M. Watson, *et al.*, "Photonuclear-based, nuclear material detection system for cargo containers," Nuclear Instrum. and Methods in Phys. Res. B, vol. 241, no. 1–4, pp. 770–776, Dec. 2005, DOI. 10.1016/j.nimb.2005.07.242.

[17] J. L. Jones, D. R. Norman, K. J. Haskell, J. W. Sterbentz, W. Y. Yoon, S. M. Watson, *et al.*, "Detection of shielded nuclear material in a cargo container," Nuclear Instrum. and Methods in Phys. Res. A, vol. 562, no. 2, pp. 1085–1088, Jun. 2006, DOI. 10.1016/j.nima.2006.02.101.

[18] A. Sari, F. Carrel, A. Grabowski, F. Lainé, B. Espinosa, J. Poli, *et al.*, "Deployment of the First Photofission Measurement System Dedicated to SNM Detection in Europe: Outcomes and Future Prospects," in *2019 IEEE Nuclear Science Symposium and Medical Imaging Conference (NSS/MIC)*, Manchester, UK, 2019, pp. 1-2, DOI. 10.1109/NSS/MIC42101.2019.9059765.

[19] A. Sari, A. Elayeb, J. Piekar, R. De Stefano, G. Corre, Y. Moline, *et al.*, "Deployment of a 7 MeV Mobile Photofission Inspection System at the Rijeka Seaport for the Detection of SNM in Cargo Containers," in *2024 IEEE Nuclear Science Symposium (NSS), Medical Imaging Conference (MIC) and Room Temperature Semiconductor Detector Conference (RTSD)*, Tampa, FL, USA, 2024, pp. 1-2, DOI. 10.1109/NSS/MIC/RTSD57108.2024.10655794.

[20] A. Sari, A. Elayeb, R. De Stefano, I. Meleshenkovskii, M. Michel, and F. Carrel, "Evaluation of delayed neutron yields and time spectra from photofission of ^{238}U induced by *Bremsstrahlung* photons below 9 MeV," in *15th International Conference on Nuclear Data for Science and Technology (ND2022)*, virtual conference, May 2023, Art. No. 11002, DOI. 10.1051/epjconf/202328411002.

[21] A. Sari, "Characterization of photoneutron fluxes emitted by electron accelerators in the 4–20 MeV range using Monte Carlo codes: A critical review," Appl. Radiat. Isot., vol. 191, Art. no. 110506, Jan. 2023, DOI. 10.1016/j.apradiso.2022.110506.

[22] Z. Berant, A. Wolf, and R. Moreh, "Detection of fast neutron contamination in photon beams," Nuclear Instrum. and Methods, vol. 140, no. 1, pp. 109–112, Jan. 1977, DOI. 10.1016/0029-554X(77)90072-6.

[23] A. Wolf, and R. Moreh, "Utilization of teflon-covered Ge(Li) diodes for fast neutron detection," Nuclear Instrum. and Methods, vol. 148, no. 1, pp. 195–197, Jan. 1978, DOI. 10.1016/0029-554X(78)90353-1.

[24] T. Gozani, J. Stevenson, and M. J. King, "Neutron threshold activation detectors (TAD) for the detection of fissions," Nuclear Instrum. and Methods in Phys. Res. A, vol. 652, no. 1, pp. 334–337, Oct. 2011, DOI. 10.1016/j.nima.2011.01.029.

[25] M. Hamel, P. Sibczynski, P. Blanc, J. Iwanowska, F. Carrel, A. Syntfeld-Każuch, *et al.*, "A fluorocarbon plastic scintillator for neutron detection: Proof of concept," Nuclear Instrum. and Methods in Phys. Res. A, vol. 768, pp. 26–31, Dec. 2014, DOI. 10.1016/j.nima.2014.09.029.

[26] P. Sibczynski, J. Kownacki, M. Moszyński, J. Iwanowska-Hanke, A. Syntfeld-Każuch, A. Gójska, *et al.*, "Verification of threshold activation detection (TAD) technique in prompt fission neutron detection using scintillators containing ^{19}F," Journal of Instrumentation, vol. 10, Art. no. T09005, Sep. 2015, DOI. 10.1088/1748-0221/2015/09/T09005.

[27] J. A. Kulesza, T. R. Adams, J. C. Armstrong, S. R. Bolding, F. B. Brown, J. S. Bull, *et al.*, "MCNP® Code Version 6.3.0 Theory & User Manual," Los Alamos National Laboratory (LANL), Los Alamos, NM, USA, Rep. LA-UR-22-30006, Rev. 1, Sep. 2022, DOI. 10.2172/1889957.

[28] M. E. Rising, J. C. Armstrong, S. R. Bolding, F. B. Brown, J. S. Bull, T. P. Burke, *et al.*, "MCNP® Code Version 6.3.0 Release Notes," Los Alamos National Laboratory (LANL), Los Alamos, NM, USA, Rep. LA-UR-22-33103, Rev. 1, Jan. 2023, DOI. 10.2172/1909545.

[29] T. Sato, Y. Iwamoto, S. Hashimoto, T. Ogawa, T. Furuta, S. Abe, *et al.*, "Recent improvements of the particle and heavy ion transport code system – PHITS version 3.33," J. Nucl. Sci. Technol, vol. 61, no. 1, pp. 127–135, 2024, DOI. 10.1080/00223131.2023.2275736.

[30] Y. Iwamoto, S. Hashimotoa, T. Sato, N. Matsuda, S. Kuniedaa, Y. Çelikb, *et al.*, "Benchmark study of particle and heavy-ion transport code system using shield-ing integral benchmark archive and database for accelerator-shielding experiments," J. Nucl. Sci. Technol, vol. 59, no. 5, pp. 665–675, 2022, DOI. 10.1080/00223131.2021.1993372.

The photofission reaction against the illicit transport of nuclear material in cargo containers: A review on recent achievements and future challenges in active photon interrogation

Adrien Sari[1,*]
[1]Université Paris-Saclay, CEA, List, F-91120 Palaiseau, France
(*) adrien.sari@cea.fr

***Abstract*—X-ray scanning is widely used by customs for visual inspection of cargo containers at border controls. When Special Nuclear Material (SNM) such as uranium or plutonium is suspected, an active non-destructive nuclear measurement method serves as a second-line tool. The photofission reaction, central to the Active Photon Interrogation (API) technique, enables the direct detection of SNM. This paper reviews European advances in API over the past decade through three Horizon 2020 projects. In C-BORD (2015–2018), Europe's first photofission system was tested at the *Maasvlakte* terminal of the Port of Rotterdam (Netherlands), with experiments on depleted uranium and container mock-ups demonstrating integration into a stationary 9 MeV industrial X-ray scanning facility. The ENTRANCE project (2020–2023) evaluated a mobile 7 MeV X-ray scanner at the *Škrljevo* terminal of the Port of Rijeka (Croatia), investigating the 5–6 MeV actinide photofission threshold and assessing operational performance. The MULTISCAN 3D project (2021–2025) explored new approaches using high-energy photons from laser–plasma sources. Collectively, these developments have strengthened nuclear security and enhanced Europe's capacity to counter illicit SNM trafficking.**

***Keywords*—Border monitoring; Cargo containers; X-ray scanning; Active Photon Interrogation (API); Photofission; Special Nuclear Material (SNM).**

I. Introduction

THE concept of photofission—the induction of fission in heavy nuclei by high-energy photons—was first predicted [1] and later demonstrated experimentally [2] in early nuclear physics studies. The reaction occurs when photons with energies above ~5–6 MeV interact with fissile nuclei (^{235}U, ^{239}Pu, and others) or fertile nuclei (^{238}U, ^{240}Pu, and others), producing fission fragments along with prompt neutrons and gammas, while subsequent decay of the fission products emits delayed neutrons and gammas. A schematic of the photofission reaction is shown in Fig. 1.

Building on these foundations, Active Photon Interrogation (API) has been developed as a non-destructive nuclear measurement method [3]. API employs high-energy photons, typically generated by a linear electron accelerator (linac) via a *Bremsstrahlung* target, to induce photofission reactions on actinides. API overcomes the limitations of passive measurements and Active Neutron Interrogation (ANI), both constrained by matrix effects—passive methods relying on spontaneous SNM emissions, ANI limited in large volumes by an interrogative thermal neutron beam—using photofission signatures [4, 5].

API has been tested in realistic port environments with cargo containers, demonstrating the feasibility of both stationary [6] and mobile [7] systems, and confirming its potential for operational deployment at borders. Given the ongoing challenges in securing borders against SNM illicit trafficking and the rapid technological developments in photon sources and detection methods, a comprehensive review of European advances in API over the past decade is both timely and essential. This paper focuses on three Horizon 2020 projects: C-BORD, ENTRANCE, and MULTISCAN 3D. While the first two projects focused on linac-based systems, the MULTISCAN 3D project explored novel photon sources based on laser–plasma interactions, paving the way for next-generation photofission-based SNM detection.

The implementation of photofission-based inspection systems responds to the growing demand for robust nuclear security measures at European borders. These technologies not only enhance the detection of SNM but also support international trade by enabling efficient, non-intrusive inspections. Over the past decade, Europe has led in developing and testing these advanced systems, setting a benchmark for global security practices and demonstrating the feasibility of integrating API into operational port environments.

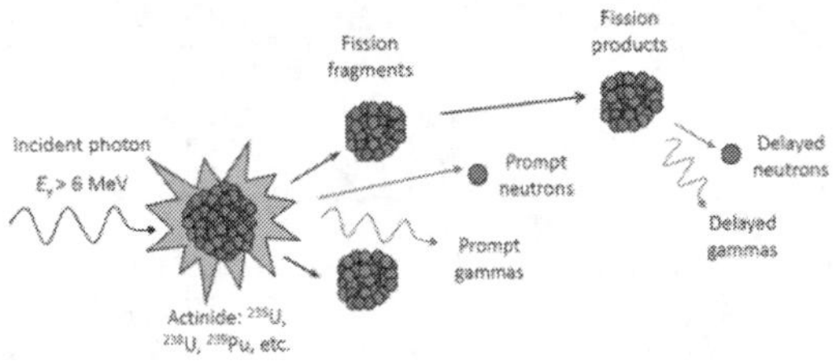

Fig. 1. Scheme of the photofission reaction. A high-energy photon interacts with a heavy nucleus (e.g., ^{235}U, ^{238}U, ^{239}Pu, etc.), inducing fission into two or three fragments and emitting prompt neutrons and gammas. Subsequent decay of the fission products produces delayed neutrons and gammas.

II. Pioneering photofission in Europe: The first inspection system

The C-BORD project (2015–2018) aimed to develop and demonstrate advanced technologies for non-intrusive inspection of cargo containers at European border controls. Within this framework, photofission (API) was deployed as a second-line, active non-destructive nuclear measurement method, complementing conventional X-ray scanning. A stationary photofission inspection system was installed in 2018 at the *Maasvlakte* terminal of the Port of Rotterdam (Netherlands) on an existing 9 MeV X-ray scanning facility operated by customs.

Several technical challenges had to be addressed during deployment. The geometry of the installation imposed a distance of more than 10 m between the linac and the container, while the photon beam was strongly collimated to a width of only a few millimeters. These constraints required careful alignment and optimization of the measurement setup.

Field experiments were conducted in 2018 using 10 ft mock-up containers loaded with depleted uranium (DU) samples and three representative cargo matrices: aluminum, fertilizer, and tobacco. In a realistic "blind test" configuration, the matrices and DU locations were prepared by customs, and their arrangement was unknown to the technical partners. The photofission subsystem was operated in a stationary "top linac" mode at 100 Hz, with active background acquisition followed by irradiation of the suspicious area. Measurements were performed at two positions, and results compared. A statistically significant increase of the neutron or gamma signal ($\geq 3\sigma$) triggered a red-light warning on the Graphical User Interface (GUI), while a green-light status indicated no anomaly. The setup included dedicated detectors for prompt neutrons, delayed neutrons, and delayed gammas (see Fig. 2).

These trials marked the first deployment of a photofission inspection system on an operational industrial site in Europe, demonstrating the feasibility of integrating active photon interrogation with existing high-energy X-ray scanning facilities.

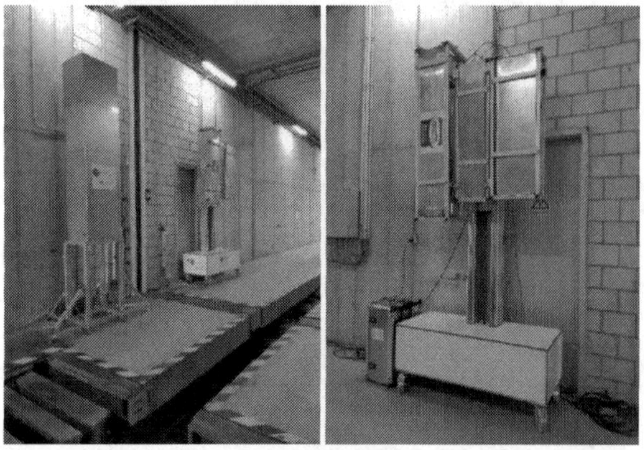

Fig. 2. Field trials of the C-BORD 9 MeV stationary photofission inspection system at the *Maasvlakte* terminal of the Port of Rotterdam (Netherlands, 2018) [6]. The setup included dedicated detectors for prompt neutrons, delayed neutrons, and delayed gammas, integrated with an X-ray scanner initially designed for visual inspection of cargo containers.

III. On the road to operational flexibility: Development of a mobile system

The ENTRANCE project (2020–2023) aimed to advance mobile photofission inspection systems for detecting Special Nuclear Material (SNM) at European ports. Photofission (API) was deployed as a second-line, non-destructive nuclear measurement method, complementing conventional X-ray scanning. The system was tested at the *Škrljevo* terminal of the Port of *Rijeka* (Croatia) in 2023, demonstrating operational capability under realistic conditions.

Deployment posed several technical challenges. Unlike the stationary C-BORD system, the ENTRANCE setup was fully mobile, installed on a truck, and relied on a 7 MeV linac, close to the 5–6 MeV photofission threshold. Accurate positioning relative to the container was critical for reliable measurements.

Field experiments were conducted using mock-up containers loaded with depleted uranium (DU) samples and various cargo matrices. The photofission subsystems provided delayed neutron and delayed gamma detection (see Fig. 3), with immediate feedback via a Graphical User Interface (GUI) showing green/red lights. Prior X-ray scanning allowed visual inspection and accurate localization of suspicious areas.

The ENTRANCE project successfully validated the feasibility of a 7 MeV mobile photofission inspection system, providing insights for flexible, rapid-response port operations.

Fig. 3. Field trials of the ENTRANCE 7 MeV mobile photofission inspection system at the *Škrljevo* terminal of the Port of Rijeka (Croatia, 2023) [7]. The setup included dedicated detectors for delayed neutrons and delayed gammas, integrated with an X-ray scanner initially designed for visual inspection of cargo containers.

IV. Toward a technological revolution: Exploring laser/plasma sources

The MULTISCAN 3D project (2021–2025) investigated next-generation photofission (API) inspection systems based on high-energy photons generated by laser–plasma sources. This pioneering approach aims to develop compact, high-brightness photon sources capable of inducing photofission in actinides. It enables flexible, rapid, and efficient SNM detection in cargo containers of various sizes and compositions. Laboratory-scale proof-of-concept experiments have been performed to demonstrate feasibility, leveraging the underlying laser–plasma interactions responsible for photon generation. These experiments highlight the potential for integrating advanced photon sources with existing inspection infrastructure. Detailed experimental results and a full technical characterization will be reported in a dedicated publication.

V. Conclusions and Future Challenges

A. Conclusions

Over the past decade, the European projects C-BORD, ENTRANCE, and MULTISCAN 3D have advanced photofission-based inspection technologies for cargo container security:

- **C-BORD (2015–2018)**
    - Stationary photofission inspection system deployed at the Port of Rotterdam (Netherlands).
    - Operated with a 9 MeV linac, integrated into an industrial X-ray scanning facility.
- **ENTRANCE (2020–2023)**
    - Mobile photofission inspection system deployed at the Port of Rijeka (Croatia).
    - Operated with a 7 MeV linac, demonstrating the feasibility of a mobile system embarked in a truck.
- **MULTISCAN 3D (2021–2025)**
    - Proof-of-concept of photofission experiment using a femtosecond laser-driven Inverse Compton Scattering (ICS) source.
    - First worldwide experiment at the intersection of laser, plasma, and nuclear physics.
    - Paves the way for next-generation inspection technologies.

B. Future challenges

- **Short-term outlook**
    - Development of a passive/active multi-threat mobile inspection system for detecting SNM, radioactive sources, narcotics, explosives.
    - Integration of multi-technology approaches to enhance detection efficiency.
- **Long-term outlook**
    - Expansion to emerging applications, including recycling and metallurgical industries.
    - Cross-technology synergy combining ANI, API, NAA, and PAA methods.
    - Strengthening interdisciplinary collaboration among nuclear physicists, laser/plasma experts, and border security practitioners.
- **Overall impact**
    - Highlights the growing potential of photofission technologies.
    - Sets the stage for the next generation of nuclear security tools, combining mobile deployment, high-precision photon sources, and advanced detection systems.
    - Strengthens border security and facilitates trade, while complementary nuclear techniques ensure robust detection of multiple threats.

Acknowledgment

The author gratefully acknowledges the active involvement and support of the project partners:

- **C-BORD (2015–2018)**
    - Smiths Detection, France
    - National Centre for Nuclear Research (NCBJ), Poland
    - Symetrica, UK
    - Dutch Customs Administration (DCA), Netherlands
    - CEA-List technical and coordination teams, France
- **ENTRANCE (2020–2023)**
    - Smiths Detection, France
    - Ruđer Bošković Institute (RBI), Croatia
    - The Ministry of Interior, the Ministry of Finance – Customs Administration, and the Port of Rijeka (*Luka Rijeka d.d.*), Croatia
    - CEA-List technical and coordination teams, France
- **MULTISCAN 3D (2021–2025)**
    - Ludwig-Maximilians-Universität München – Centre for Advanced Laser Applications (LMU-CALA), Germany
    - University of Padova (UNIPD), Italy
    - CEA-List technical and coordination teams, France

Funding

The C-BORD, ENTRANCE, and MULTISCAN 3D projects have received funding from the European Union's Horizon 2020 research and innovation programme under grant agreements No. 653323, 883424, and 101020100. The work presented in this presentation reflects only the authors' views and the Commission is not liable for any use that may be made of the information contained therein.

References

[1] N. Bohr and J. A. Wheeler, "The Mechanism of Nuclear Fission," Phys. Rev., vol. 56, no. 5, pp. 426–450, 1939, DOI. 10.1103/PhysRev.56.426.

[2] R. O. Haxby, W. E. Shoupp, W. E. Stephens, and W. H. Wells, "Photo-Fission of Uranium and Thorium," Phys. Rev., vol. 58, no. 1, p. 92, 1940, DOI. 10.1103/PhysRev.58.92.

[3] A. Lyoussi, *et al.*, "Transuranic waste detection by photon interrogation and on-line delayed neutron counting," Nuclear Instrum. and Methods in Phys. Res. B, vol. 160, no. 2, pp. 280–289, 2000, DOI. 10.1016/S0168-583X(99)00586-8.

[4] J. L. Jones, *et al.*, "Photonuclear-based, nuclear material detection system for cargo containers," Nuclear Instrum. and Methods in Phys. Res. B, vol. 241, no. 1–4, pp. 770–776, 2005, DOI. 10.1016/j.nimb.2005.07.242.

[5] J. L. Jones, *et al.*, "Detection of shielded nuclear material in a cargo container," Nuclear Instrum. and Methods in Phys. Res. A, vol. 562, no. 2, pp. 1085–1088, 2006, DOI. 10.1016/j.nima.2006.02.101.

[6] A. Sari, *et al.*, "Deployment of the First Photofission Measurement System Dedicated to SNM Detection in Europe: Outcomes and Future Prospects," in *2019 IEEE NSS/MIC/RTSD*, Manchester, UK, 2019, pp. 1-2, DOI. 10.1109/NSS/MIC42101.2019.9059765.

[7] A. Sari, *et al.*, "Deployment of a 7 MeV Mobile Photofission Inspection System at the Rijeka Seaport for the Detection of SNM in Cargo Containers," in *2024 IEEE NSS/MIC/RTSD*, Tampa, FL, USA, 2024, pp. 1-2, DOI. 10.1109/NSS/MIC/RTSD57108.2024.10655794.

Ultrasonic detection of radiolysis bubbles generated in bitumen for dose between 6 and 100 kGy

Didier Laux[1], Hanaa Houjeij[2], Céline Monsanglant[2], Georges Matta[2], Stéphane Poirier[2], François Millet[3] and Jean-Yves Ferrandis[1*]

[1]IES, Univ Montpellier, CNRS, Montpellier, France
[2]Autorité de Sûreté Nucléaire et de Radioprotection (ASNR), PSN-RES/SCA/LECEV, F-91400, Saclay, France
[3] Autorité de Sûreté Nucléaire et de Radioprotection (ASNR), PSE-ENV/SPDR/USDR, F-92260, Fontenay-aux-Roses, France

(*) jean-yves.ferrandis@umontpellier.fr

***Abstract*—The ARISE project, supported by the National French Research Center (CNRS) and the Nuclear Safety and Radiation Protection Authority (ASNR), began in 2023 to explore the feasibility of innovative ultrasonic methods to detect the appearance and growth of radiolysis bubbles in bituminized waste products (BWPs). More generally, the subject can cover the investigation of any opaque medium. In this communication, we present results on bitumen samples irradiated at the IRMA French facility, as part of the ASNR BOREE experimental platform. The samples were subjected to total absorbed doses of 6, 50 and 100 kGy with dose rates ranging from 155 to 1230 Gy/h. By using the velocity and attenuation of ultrasonic signals at around 500 kHz, combined with Foldy's model applied to polydisperse bubble clouds and X-Ray microtomography as validation technique, we demonstrate that relevant parameters can be estimated in a non-destructive manner in irradiated bitumen, namely: the void fraction, bubble radius and the polydispersity of the bubble size distribution. These results pave the way for non-destructive real time monitoring of bubble formation and growth under gamma irradiation in bitumen, as a part of the next phases of this extensive research program.**

***Keywords* —Ultrasounds, Bitumen, waste products, Bubble clouds, Gamma irradiation, Radiolysis, X-Ray Microtomography.**

I. Introduction

SINCE the 1960s, bitumen has been used by the French nuclear industry as an immobilization matrix for long-lived low-level (LL) and intermediate-level (IL) radioactive waste generated during spent fuel reprocessing. The process involves concentrating radionuclides by co-precipitating them with selected salts that exhibit strong chemical affinities with them. The resulting sludge is then filtered to increase its solid content, and the final residue—now containing the radionuclides—is embedded in a bituminous matrix through an extrusion process. The resulting bituminized waste products (BWPs) are subsequently conditioned and stored in metallic drums. Within the context of nuclear safety process, it is imperative to examine the thermal and under irradiation behavior of these BWPs. Radiolysis due to gamma self-irradiation occurs and leads to the formation of gas bubbles (essentially hydrogen) which can significantly elevate the fire and explosion risk when released [1]. Consequently, the nucleation and growth of these bubbles must be precisely determined. X-Ray microtomography, applied to pure bitumen and BWPs is commonly used to investigate such issues [2-5]. However, this technique is not suitable for direct in-line evaluation. In this framework, an extensive program (ARISE) has been initiated by ASNR (USDR and LECEV labs) and CNRS/Montpellier University (Acoustic Team of the Institute of Electronics and Systems, IES) in 2023. In our previous investigations we focused on the high frequency rheology of pure bitumen and on the influence of bubbles created by external irradiation on ultrasonic velocity [6,7]. We have demonstrated that the void fraction can be determined; however, further characteristics of the bubble cloud should be determined using ultrasonic waves attenuation in the medium, namely the median bubble radius and polydispersity of the bubble size distribution. In this communication, we focus on ultrasonic absorption by bubble clouds generated in pure bitumen through external gamma irradiation in the dose range of 6 to100 kGy. This study represents a first step prior to investigating BWPs. X-Ray microtomography will be used as a reference method. The irradiation procedure at IRMA facility, as well as X-Ray microtomography and ultrasonic techniques, will be presented. Then, Foldy's model for polydisperse bubble cloud in a viscoelastic medium will be applied to fit ultrasonic parameters (velocity and attenuation), in order to assess to what extent, the acoustic method can provide a complete characterization of the bubble cloud. Current limitations and future optimization of the method will be discussed at the end of this paper.

II. Measurement description

A. Samples and irradiation at the IRMA facility

Duran® glass bottles were filled with 180 g of a straight-run distillation bitumen with a penetration grade of 70/100, referred to as AZALT 70/100®, and supplied by TotalEnergies. These samples were then irradiated with gamma rays at the IRMA (IRradiation of Materials) facility, a ^{60}Co irradiation chamber, operated by ASNR and located at the CEA (French Alternative Energies and Atomic Energy Commission) Saclay research center. Four high activity sealed ^{60}Co sources, each measuring 45.2 cm in length and 1.1 cm in diameter, with a total activity of 453 TBq as of March 7, 2024 (the start date of the test), were used for the irradiation. The position of the sources can be adjusted to achieve the desired dose rates. Irradiation times, dose rates and total absorbed doses are listed in Table I.

TABLE I
IRRADIATION PARAMETERS OF BITUMEN SAMPLES

N°	Dose (kGy)	Dose rate (Gy/h)	Irradiation time (hours)
A	6.34 ± 0.64	155±16	~41
B	51.9 ± 5.2	1230±120	~42
C	50.8 ± 5.1	304±30	~167
D	101 ± 10	603±61	~167

B. X-Ray microtomography: data acquisition and processing

X-Ray microtomography tests were performed using a Bruker Skyscan 1173 micro-CT scanner with a voltage of 125 kV and an intensity of 61 µA. A full360° scan consisting of 1800 images required approximately 225 minutes. The acquired images were processed using a custom Python-based software developed in-house, to extract key bubble parameters such as volume fraction, median radius and the dispersion around the median radius. Image analysis involved six main steps: resizing, thresholding, removal of small particles, bubbles labelling and geometrical properties calculation, and data export to an Excel file. Further details on these processing steps are provided in [7].

C. Ultrasounds: experimental method and signal processing

In our previous studies [6], non-irradiated 70/100 bitumen was investigated with ultrasonic waves. So, its ultrasonic characteristics (velocity and attenuation) are well known. We first used a Duran bottle filled with non-irradiated bitumen as a reference. Two 500 kHz (Olympus - V101 RM VISIOSCAN) longitudinal transducers were coupled to the bottle with liquid honey. These sensors were excited with a JSR Ultrasonics DPR300 pulser/receiver generator. The acoustic signal transmitted through the bottle was transferred from an oscilloscope (LeCroy WaveJet 334) to a computer using a National Instruments (USB/GPIB HS+) acquisition card. The same procedure was subsequently repeated with a Duran bottle filled with irradiated bitumen samples. Then, an intercorrelation procedure was used to estimate the difference of time of flight between the two transmitted ultrasonic signals,

$$\Delta t = t_{irradiated} - t_{non-irradiated} \tag{1}$$

With Δt, the group velocity of the ultrasonic signal transmitted through the irradiated sample was estimated by the following relationship:

$$c_{irradiated} = \frac{e}{\Delta t + (e/c_{non-irradiated})} \tag{2}$$

In (2), e represents the propagation distance in bitumen and is equal to $55\,mm$. This velocity corresponds to the non-dispersive ultrasonic velocity in low frequency regime (See Part III.). Up to now, all ultrasonic measurements were performed at 10°C, as ultrasonic signals in bitumen are less attenuated at low temperatures due to rheological reasons, as reported in [6]. For this temperature, $c_{non-irradiated} = 1993\,m/s$ [6].
The Fast Fourier Transforms of the reference signal and the signal transmitted through irradiated bitumen were also calculated, and the ultrasonic attenuation caused by the bubble cloud was estimated using the following equation:

$$\alpha_{bubbles} = -\frac{1}{e} ln \frac{|FFT_{irradiated}|}{|FFT_{non-irradiated}|} \tag{3}$$

Signal acquisition and processing were performed using an in-house Python code. Diffraction effects caused by the finite size of the ultrasonic transducers were also taken into account, although the procedure is not described here. Further information concerning this aspect can be found in [8]. A photograph of a Duran bottle containing irradiated bitumen and the acquired ultrasonic signal are shown in Fig. 1.

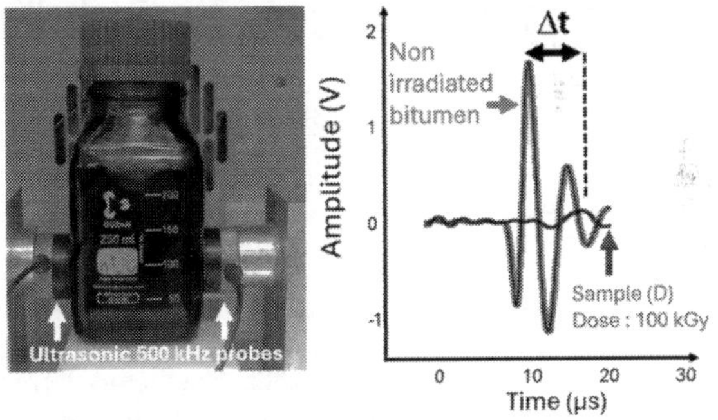

Fig. 1. Ultrasonic 500 kHz sensors, Duran bottle filled with irradiated bitumen and associated ultrasonic signal.

III. Ultrasonic waves interaction with bubble clouds

A bubble cloud excited by an acoustic wave can be seen as an oscillator assembly that may resonate at the Minnaert frequency F_M. For a monodisperse viscoelastic bubbly medium such as bitumen [9],

$$F_M \sim \frac{V_{To}}{\pi R_0}, \tag{4}$$

where V_{To} is the shear ultrasonic velocity in the non-porous matrix and R_0 the bubble radius. Far below F_M, the porous medium is seen as a homogeneous one, the acoustic attenuation is weak and the longitudinal acoustic velocity does not depend on frequency. For an acoustic frequency equal to F_M, resonance occurs and both acoustic velocity and attenuation increase a lot. At last, when frequency >>F_M, acoustic attenuation decreases, and acoustic velocity reaches longitudinal velocity V_0 in the non-porous matrix. This phenomenon has been extensively studied in the literature [9] and extended to polydisperse systems. Generally, the polydisperse bubble cloud (void fraction Φ, median radius R_0 and polydispersity factor ε) is described by a log-normal distribution

$$n(R) = \frac{n_0}{\varepsilon R\sqrt{2\pi}} \exp\left(-\left(\frac{(\ln(R/R_0))^2}{2\varepsilon^2}\right)\right), \quad (5)$$

where,

$$n_0 = \frac{3\Phi}{4\pi R_0^3 \exp\left(\frac{9\varepsilon^2}{2}\right)} \quad (6)$$

As bubbles lead to an attenuation of the acoustic signal, a complex wave number k* depending on pulsation ω can be defined so that,

$$c_L = \omega/Re(k^*) \quad (7)$$

$$\alpha = -Im(k^*) \quad (8)$$

k^* can be numerically calculated, with a Python code in our case, with the following relationship,

$$k^* = \frac{\omega^2}{c_0{}^2} + 4\pi\omega^2 \int_0^\infty \frac{Rn(R)dR}{\omega_0^2 - \omega^2 + 2ib\omega} \quad (9)$$

The factors b and ω_0^2 depend on the thermal and viscoelastic properties of the bubbles and the matrix and can be calculated using the relationships given in [10,11]. These details are not recalled here.

IV. Results

A. X-Ray microtomography

After rough data treatment, void fractions were obtained from X-Ray microtomography with an estimated accuracy of ±1.5%. Porosity histograms were then fitted using the log-normal law (5) for the median radius of bubbles estimation and polydispersity evaluation. All results are summarized in Table II.

TABLE II
PROPERTIES OF BUBBLE CLOUDS ESTIMATED BY X-RAY MICROTOMOGRAPHY

N°	Void fraction Φ (%) ± 1.5	Median radius (µm) ± 10 %	Polydispersity factor ε
A	0.7	380	35
B	7.4	500	19
C	6.7	540	25
D	11.4	580	25

B. Ultrasounds

In irradiated samples, longitudinal velocities ($V_{10^\circ C}$) measured at 10°C were respectively for samples A, B, C and D: 1944 m/s, 1800 m/s, 1813 m/s and 1660 m/s, with an accuracy ±40 m/s. In [7], our results suggest that this velocity corresponds to the velocity in low frequency domain. We also demonstrated that a first estimation of void fraction can be obtained with the following relationship:

$$\Phi = \frac{4}{3}\left[\frac{V_{To}}{V_o}\right]^2 \frac{(V_o^2 - V_{10^\circ C}^2)}{V_{10^\circ C}^2}, \quad (10)$$

where $V_o = 1993\, m/s$ and $V_{To} = 870\, m/s$ represent the longitudinal and shear velocities in bitumen with no bubbles, respectively.

This led to void fractions equal to 1.3% (A), 5.7% (B), 5.3% (C) and 11.2 %(D). Ultrasonic attenuation caused by the bubble clouds in the 100-200 kHz frequency range is shown in in Fig. 2. For sample (A), the bandwidth has been reduced because at low frequencies attenuation is too small. One can notice that even if 500 kHz sensors were used, the effective bandwidth after propagation in bitumen was modified because of sample attenuation equivalent to a filtering.

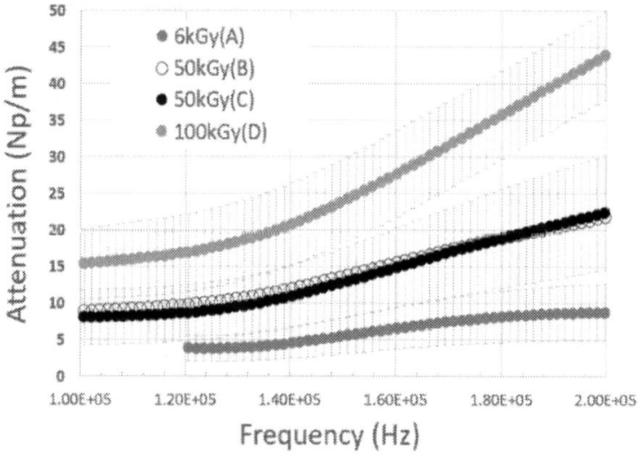

Fig. 2. Ultrasonic attenuation induced by bubble clouds versus frequency measured in irradiated samples

The model presented in the previous section was used to fit ultrasonic attenuation measured in order to deduce both the median radius and the polydispersity of the bubble cloud. Classical physical parameters required for the simulation (thermal conductivity, matrix density, surface tension, etc.) were extracted from literature. Matrix viscosity evolution versus frequency was described with a Carreau-Yasuda law, established in our previous works [6]. Regarding Fig. 2, for samples B, C and D at low frequencies, one can notice that the attenuation does not tend to 0 as predicted by the models. This is certainly due to an offset caused by the coupling between sensors and bottles, imperfect correction of sensors diffraction effects, edge effects in the Fast Fourier transform, low sensors efficiency at the considered frequency, etc. In order to adjust our data, the existing offset was corrected. The void fraction first estimated with (10), was slightly adjusted when necessary. The adjustment of experimental data is presented in Fig.3 for samples A, B and D. The adjustment for sample C is not presented for more clarity, as it led to results equivalent to those of sample B. The physical parameters of the bubble clouds, as estimated from the simulation, are listed in Table III.

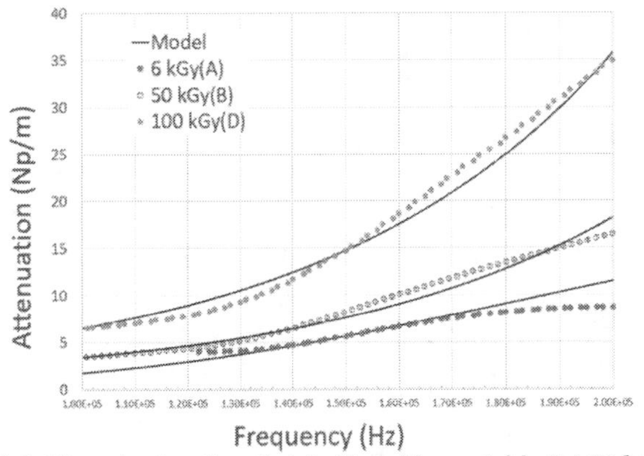

Fig.3. Ultrasonic attenuation adjusted with model presented in Part III for samples A, B and D

TABLE III
PROPERTIES OF BUBBLE CLOUDS ESTIMATED BY ULTRASOUND

N°	Void fraction Φ (%)± 1.5	Median radius (µm) ± 10 %	Polydispersity factor ε ± 10 %
A	1.5	530	40
B	5.7	500	17
C	5.3	500	25
D	11.4	540	18

Ultrasonic evaluation yields results that are consistent with X-ray microtomography, although the current level of uncertainty remains significant, especially for estimates of median radius and polydispersity.

V. Conclusions

Until now, only results at 10°C have been presented. Indeed, as previously explained in [6] ultrasonic attenuation increases up to 40-50°C, making measurements at room temperature more challenging due to the high thickness of the Duran flasks used in these experiments. We are currently developing smaller sample cells (thickness = 2 cm) made of aluminum holders. These new cells are expected to improve the contact between the sensors and the bitumen, allowing for a more accurate evaluation of ultrasonic attenuation versus frequency, reduce error bars, and measurements to be performed at room temperature. In addition, signal processing will also be adapted to assess the phase velocity of the ultrasonic signal over the sensor's bandwidth. Although the results presented in this study still require refinement, they are very encouraging and, in our opinion, clearly demonstrate that the ultrasonic approach constitutes a promising method for the non-destructive characterization of bubble clouds in bitumen used for nuclear waste immobilization.

Acknowledgment

The authors sincerely acknowledge Mejdi Neji (ASNR/PSE-ENV/SPDR/LETIS), Pascal Lemaitre and Soleiman Bourrous (ASNR/PSN-RES/SCA) for their precious help with the Python code dedicated to the 3D analysis of X-Ray microtomographic data. We also thank Gaetan Touzé from the LECBA laboratory at CEA for the 3D reconstructions performed using VgStudio, TotalEnergies for providing the AZALT 70/100 bitumen samples and Oluwaseyi Miracle Adebola for his help during his Master 2 internship. We are also very grateful to Valentin Leroy from Paris Cité University for his valuable insights into ultrasonic waves interaction with bubble clouds, which were particularly helpful for the simulation aspects. Finally, we sincerely acknowledge Lucie Millot for launching the ARISE project.

References

[1] L. Millot, H. Houjeij, G. Matta, J.Y. Ferrandis, D. Laux and C. Monsanglant- Louvet, "Radiolysis of Bituminous Radioactive Waste: A Comprehensive Review", EPJ Nuclear Sci. Technol., vol. 10, pp. 1-17, April 2024, Art no. 4, 10.1051/epjn/2024004

[2] M. Mouazen, A. Poulesquen and B. Vergnes, "Influence of Thermomechanical History on Chemical and Rheological Behavior of bitumen," Energy Fuels, vol. 25, no. 10, pp. 4614-4621, August 2011, 10.1021/ef201000j

[3] M. Mouazen, A. Poulesquen, F. Bart and B. Vergnes, "Effect of γ irradiation on nuclear bituminized waste products (BWP): X-ray microtomography and rheological characterization," J. Nucl. Mat., vol. 419, pp. 24-31, 2011, 10.1016/j.jnucmat.2011.07.025

[4] M. Mouazen, A. Poulesquen, F. Bart, J. Masson, M. Charlot and B. Vergnes, "Rheological, structural and chemical evolution of bitumen under gamma irradiation," Fuel Proc. Tech., vol. 114, pp. 144-153, 2013, 10.1016/j.fuproc.2013.03.039

[5] M. Mouazen, A. Poulesquen and B. Vergnes, "Correlation between thermal and rheological studies to characterize the behavior of bitumen," Rheol. Acta, vol. 50, pp. 169-178, 2011, 10.1007/s00397-011-0534-5

[6] D.Laux, K. Toulgoat, L.Millot and J-Y.Ferrandis, "Rheological investigation of bitumen, used for radioactive waste conditioning, with ultrasonic waves," EPJ Nuclear Sci. Technol., vol. 10, pp. 1-10, March 2024, Art no. 1, doi.org/10.1051/epjn/2024002

[7] D. Laux, G. Matta, H. Houjeij, C. Monsanglant, S. Poirier, F. Millet, V. Leroy, J-Y. Ferrandis. Study of radiolysis bubbles in bitumen used for nuclear waste management with ultrasounds and X-Ray microtomography. European Physical Journal – Nuclear Sciences &Technologies. EPJ-N. (2025). Accepted in June 2025. To be published.

[8] G. Lévêque, E. Rosenkrantz and D. Laux, "Correction of diffraction effects in sound velocity and absorption measurements," Meas. Sci. Technol., vol. 18, pp. 3458-3462, 2008, 10.1088/0957-0233/18/11/028

[9] V. Leroy, Y. Fan, A.L. Strybulevych, Bellido, J.H. Page and M.G. Scanlon, "Investigating the Bubble Size Distribution in Dough Using Ultrasound," in Bubbles in Food 2, G.M. Campbell, Ed: AACC International Press, 2008 pp. 51-60, 10.1016/B978-1-891127-59-5.50009-2K.

[10] W. Commander and A. Prosperetti, "Linear pressure waves in bubbly liquids: Comparison between theory and experiments," J. Acoust. Soc. Am., vol. 85, pp 732-746, 1989, 10.1121/1.397599

[11] A. Prosperetti, L. A. Crum and K. W. Commander, "Nonlinear bubble dynamics," J. Acoust. Soc. Am., vol. 83, pp. 502-514, 1988, 10.1121/1.396145

Bayesian Optimization for real-time γ-localization measurements with robotic platforms

Ioannis Tsitsimpelis[1*], Andrew West[2], Kartikey Mathur[1], Anže Jazbec[3], Luka Snoj[3], Shengshu Liu[2], Andrew Kennedy[1], Francis R. Livens[4], Barry Lennox[2], C. James Taylor[1], and Malcolm J. Joyce[1]

[1] School of Engineering, Lancaster University, United Kingdom
[2]School of Electrical and Electronic Engineering, University of Manchester, United Kingdom
[3]Research Centre for Radwaste Disposal, Williamson Research Centre and Dalton Nuclear Institute, University of Manchester, United Kingdom
[4]Reactor Physics Department, Jožef Stefan Institute, Slovenia

(*) i.tsitsimpelis3@lancaster.ac.uk

Abstract— **An adaptive approach driven by Bayesian Optimization is described for applications where remote radiation measurements made with robots are constrained by stringent upper thresholds on the mass and power payload of the necessary instrumentation, as well as by the time window within which measurements must be made, ultimately affecting their quality and maximum area coverage. The algorithm presented in this paper is applied to a gimbal assembly the comprises a collimated cerium bromide detector to perform γ-localization. A Gaussian Process models the angular distribution of radiation, and measurement locations are dynamically selected via a composite acquisition function, which combines the Expected improvement and Uncertainty Confidence Bounds functions. Convergence is driven by monitoring the rate of change of predictions and associated mean uncertainty. This approach enables accurate characterization of radiation fields while requiring up to 85% fewer measurements than conventional raster-type scanning. Its performance is evaluated in simulation, using previously obtained datasets used as measurement look-up tables, and validated in turn with hardware implementation, real-time scans.**

Keywords — **Remote measurements, Gamma localization, Bayesian Optimisation, Robotics.**

I. Introduction

Remote radiation measurements for monitoring and characterization are highly advantageous, as they mitigate the risks associated with manual assessments performed by radiation workers [1]. Additionally, these measurements can be conducted and augmented systematically, thereby enhancing the fidelity and confidence in the acquired knowledge of the target environment [2], [3]. However, their implementation can be particularly challenging due to stringent constraints related to accessibility and safety [4], [5]. These challenges are further constrained in applications where remote radiation measurements are conducted using robotic platforms, which are often limited by strict upper thresholds on the mass and power payload of the necessary instrumentation; and operational constraints such as the time window available for data acquisition due to e.g. the finite battery life of the robot [6]. Within these limitations, selective sampling strategies can be particularly advantageous, as they enable the prioritization of measurement locations that maximize area coverage while minimizing resource expenditure within the aforementioned constraints.

This research applies a well-established Bayesian Optimization (BO) technique in conjunction with a γ-radiation localization apparatus specifically designed for deployment on mobile robotic platforms. The apparatus is a gimbal-collimated detector assembly that can be mounted atop a robotic platform to enable the estimation of radiation distributions within its surroundings. The system identifies the directional distribution of γ-ray counts, providing the robot with situational awareness of radiation in all directions. To overcome the long dwell times associated with raster-scans (where the detector is rotated sequentially at discrete intervals), a dynamic sampling approach is implemented, whereby Gaussian Process Regression (GPR) together with a weighted exploration-exploitation acquisition function are used iteratively to efficiently sample and predict the radiation distribution of the angular input space. The evaluations presented hereinafter were carried out both in simulation and experimentally. The results suggest that reliable distributions can be derived at a fraction of the measurements that would be normally required by a raster-type scans; requiring up to 85% fewer measurements than conventional raster-type scans.

The rest of the paper is outlined as follows: Section II describes the instrumentation and BO algorithm. Section III presents results from simulated and real-time implementations. Section IV discusses advantages, limitations, and further work plans, and section V presents the concluding remarks.

II. Methods

A. Instrumentation

The radiation detection hardware comprises a miniature

cerium bromide ($CeBr_3$) detector with a 10 x 10 mm crystal coupled to a photomultiplier tube and a miniature High Voltage module; hermetically enclosed in a cylindrical aluminium body (110 mm long, 25 mm diameter, 60 g, Figure 1a). The detector's body is coupled to a novel lightweight collimator [7], made from a series of 3D printed porous rings coated with tungsten powder (Figure 1b), while its face is covered by two 3D printed tungsten filament slot discs (Figure 1c). The assembly is mounted to a gimbal, designed and 3D printed in-house (Figure 1d), to enable angular scans in azimuth and incline. The signal processing is carried out by a RedPitaya Field Programmable Gate Array (FPGA, model: STEMlab 125-14) board configured as a Multi-channel Analyzer (MCA) (Figure 1e). The instrumentation processes are integrated with Python (version 3.13) and the ubiquitous Robot Operating System (ROS2) middleware. Figure 1f shows a picture of the gimbal on top of the payload area of a Clearpath robotics Jackal platform.

For every angular position, a separate energy spectrum is recorded, thus allowing the derivation of net and energy resolved angular responses (accumulated counts per angle) to a radiation field. The resulting distribution depends on the collimation shape and material, as well as the activity and geometrical characteristics of the radiation field (e.g. dispersal); and the scan field of the collimated detector. For instance, the horizontal angular response of a vertical slot collimator to the x-ray and photopeak associated with a ^{137}Cs point source located at the centre of the scan resembles the shapes depicted in Figure 1g. In general, the shape depicted for the x-ray (in red circles) is similar for up to ~200 keV, whilst above that threshold it appears as per the angular distribution of the photopeak depicted in black circles.

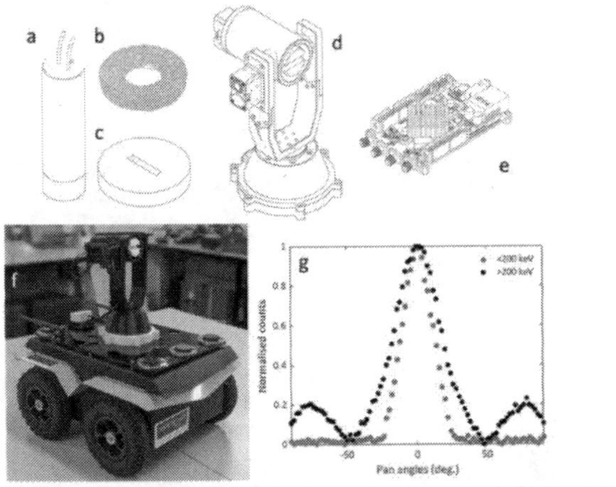

Figure 1: a) Drawing of the CeBr3 detector used in this research; b) Drawing of a porous 3D printed plastic ring coated with tungsten powder. Seven of these are used to cover the body of the detector; c) Drawing of a 3D-printed tungsten slot collimator. Two of these are placed in front of the detector's face; d) Drawing of the custom-made gimbal that houses the detector; e) Drawing of the Red Pitaya FPGA board used to record spectral data; f) A picture of the gimbal (housing a mock-up collimator) mounted on the payload bay of the Jackal robotic platform; g) A plot showing the typical horizontal angular responses observed at low and high gamma energies when using a vertical slot lead collimator. Here, the responses are based on a simple geometric scenario where a point source is located at the centre of the scan.

B. Bayesian Optimisation algorithm

The unknown radiation count distribution is modelled as a Gaussian Process (GP), with prior defined as:

$$f(\theta) \sim GP\big(\mu(\theta), k(\theta, \theta')\big) \quad 1$$

where $\mu(\theta) = \mathbb{E}[f(\theta)]$ is the mean function (estimated radiation distribution) against θ angles, and $k(\theta,\theta')$ is the covariance function defined as a 3/2 Matérn kernel. This is defined as:

$$k(\theta,\theta') = \sigma_f^2\left(1+\frac{\sqrt{3}\,|\theta-\theta'|}{l}\right)exp\left(-\frac{\sqrt{3}\,|\theta-\theta'|}{l}\right) \quad 2$$

where $\nu = 1.5$ is the smoothness control parameter, l is the length scale (in this case its value determines how quickly correlation decays at different angle distances), and σ_f^2 is an amplitude scaling factor.

The algorithm initializes by requesting a set of measurements at the ends of the input space and n equally spaced angular locations within the input space, specified by the user. The initial dataset thus contains n observations, consisting of angular positions θ_i and associated count measurements y_i. On the basis of this initial dataset, the covariance matrix $\boldsymbol{K}$ is calculated (with entries $\boldsymbol{K}_{i,j} = k(\theta_i, \theta_j)$) and subsequently used for the estimation of the radiation distribution across the whole input space. In particular, for each angular location with no recorded counts θ^*, the GP posterior has mean and variance computed as:

$$\boldsymbol{\mu}(\theta^*) = \boldsymbol{k}(\theta^*, \boldsymbol{\theta})[\boldsymbol{K} + \sigma_n^2 \boldsymbol{I}]^{-1}\boldsymbol{y} \quad 3$$

$$\boldsymbol{\sigma}^2(\theta^*) = k(\theta^*, \theta^*) - \boldsymbol{k}(\theta^*, \boldsymbol{\theta})[\boldsymbol{K} + \sigma_n^2 \boldsymbol{I}]^{-1}\boldsymbol{k}(\boldsymbol{\theta}, \theta^*) \quad 4$$

Where $k(\theta^*, \boldsymbol{\theta})$, being the cross-covariance function between the angle for which there is no associated count information, against all the angles for which this information exists; σ_n^2 is a noise variance hyper-parameter; $\boldsymbol{I}$ is the identity matrix; $\boldsymbol{y}$ is a vector of the current measurements associated with $\boldsymbol{\theta}$; and $k(\theta^*, \theta^*)$ is the auto-covariance function.

Once (3) and (4) are computed, they are in turn utilized by a composite acquisition function $\boldsymbol{w}(\boldsymbol{\theta})$, comprising a slightly modified Expected Improvement (EI) function and the Upper Confidence Bounds (UCB), and defined as:

$$\boldsymbol{w}(\boldsymbol{\theta}) = \alpha\, \boldsymbol{EI}(\boldsymbol{\theta}) + (1-\alpha)\, \boldsymbol{UCB}(\boldsymbol{\theta}) \quad 5$$

Where, α is a weight parameter that determines the contribution of each component. The modified EI function is defined as:

$$\boldsymbol{EI}(\theta) = (\boldsymbol{\mu}(\boldsymbol{\theta}) - f_{max} - \xi)\Phi(Z) + \boldsymbol{\sigma}(\boldsymbol{\theta})\phi(Z) + \beta\boldsymbol{\sigma}(\boldsymbol{\theta}) \quad 6$$

Where: ξ is a small constant to encourage exploration; f_{max}

is the maximum count value for the current locations for which associated counts exist; $\boldsymbol{\sigma(\theta)}$is the standard deviation; β is the modification the scales the contribution of the standard deviation to bias more exploration, $\Phi(Z)$ and $\phi(Z)$ are the cumulative distribution function (CDF) and probability density function (PDF) of the standard normal distribution, respectively; and Z is defined as:

$$Z = \frac{\boldsymbol{\mu(\theta)} - f_{max} - \xi}{\boldsymbol{\sigma(\theta)} + \epsilon} \quad 7$$

With ϵ a constant set to 1e-8 to avoid division by zero. The UCB function is designed to exploit areas of high activity and defined as:

$$\boldsymbol{UCB(\theta)} = \boldsymbol{\mu(\theta)} + \beta\boldsymbol{\sigma(\theta)} \quad 8$$

Upon computation of $w(\theta)$, the next sampling location (θ_{next}) is selected on the basis of

$$\theta_{next} = \operatorname{argmax}(\boldsymbol{w(\theta)}) \quad 9$$

In turn, a new measurement is carried out at the selected location and the model is updated. After a user-defined minimum number of iterations, convergence is evaluated on the basis of stability and uncertainty. For the former, the root mean square error is utilized, comparing the relative change of the current and previous predicted angular responses. For the latter, the mean estimated uncertainty is computed. These are defined as:

$$RMSE = \sqrt{\frac{1}{N}\sum_{i=1}^{N}\frac{\left(\boldsymbol{\mu}(\theta_i) - \boldsymbol{\mu}_{t-1}(\theta_i)\right)^2}{Var(\boldsymbol{\mu}_{t-1}(\boldsymbol{\theta})) + \epsilon}} \quad 10$$

and

$$\frac{1}{N}\sum_{i=1}^{N}\boldsymbol{\sigma}(\theta_i) \quad 11$$

When both metrics fall below user-defined thresholds, the algorithm terminates. The algorithm is implemented in Python using the Scikit-learn library.

III. Results

A. Simulation

In this simulation framework, previously obtained datasets function as look-up tables: every time the algorithm selects a measurement location, the corresponding recorded counts are retrieved from the dataset.

The black dots in Figure 2a collectively represent the gross spectral angular response of a vertically slotted, lead-collimated detector to a ^{152}Eu point source positioned at the centre of the scan. The raster scan spans 180° in the horizontal plane, with 2° angular intervals, totalling 91 measurements. The orange diamond markers represent the initial measurement locations requested during the initialization phase. The green triangles represent the measurements at locations subsequently selected by the algorithm. The dashed blue curve shows the final estimated response (predicted mean) across the input space, while the blue shaded area represents the associated uncertainty. In total, 14 measurement locations were "sampled" to approximate a radiation distribution that matches the raster scan ground truth very closely, requiring approximately 85% less measurements (14) than those carried out in the original raster scan (91).

Figure 2b presents a heatmap illustrating the progressive reduction of uncertainty along the input space as iterations increase. Similarly, Figure 2c displays a heatmap of the estimated radiation distribution at each iteration, with the top row showing the final estimate, also depicted in Figure 2a. Figure 2d visualizes the angle selection sequence through the combined acquisition heatmap: the bottom start of each blue vertical line represents the angle selected at each iteration, corresponding to the maximum value of the combined acquisition vector. Once selected, each angle is removed from the candidate set. It can be observed that the algorithm identifies and selects the central peak location early (prior to the second iteration of Gaussian Process Regression), and then alternates between sampling central and peripheral regions.

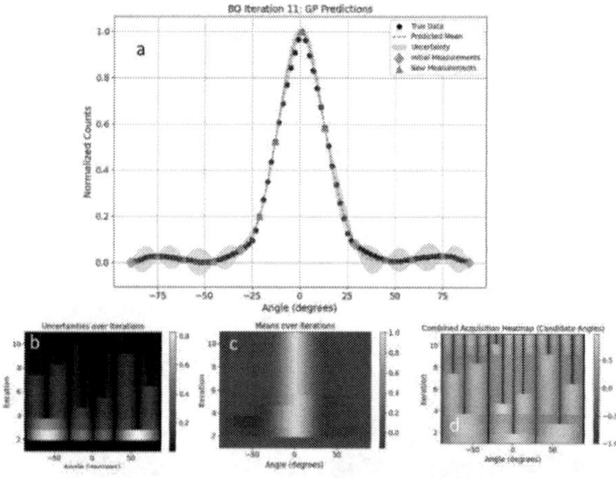

Figure 2: a) Gaussian Process (GP) predicted mean (dashed blue curve) and uncertainty (blue shaded area) after 11 BO iterations, compared against the true dataset (black dots). The orange diamonds represent initial measurements used for model initialization, while green triangles denote measurements acquired based on BO recommendations. The radiation source is centered at 0°, with the scan covering 180° horizontally at 2° intervals; b) Heatmap of GP predictive uncertainty evolution over successive iterations; c) Heatmap of the GP predicted mean radiation distribution over iterations, with the topmost row representing the final estimated distribution shown in (a); d) Combined acquisition function heatmap illustrating candidate angle selection at each iteration. Vertical blue lines indicate the angles selected by the algorithm, starting from the bottom.

Figure 3 presents the results of applying the BO algorithm using data that were obtained at the thermal irradiation port of the MARK II TRIGA research reactor (Jožef Stefan Institute, Slovenia). The experimental setup involved positioning the gimbal at a distance of 3.4 meters from the centre of the beam, with the scan covering a 180° angular range with a 6° angular resolution. Spectral measurements were acquired over a 60-

second integration time producing spectra across 32 angular positions. In Figure 3a, the algorithm is applied to the gross spectral counts, while in Figure 3b, it is applied on the angular response of the 511 keV γ -ray peak associated with electron-positron annihilation. In both cases, the adaptive sampling method demonstrates its capability to recover the true underlying radiation distribution while requiring only a 45 to 50% of the raster scan measurements.

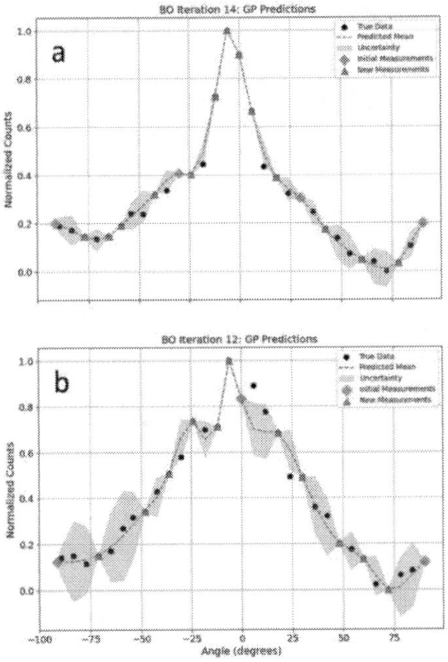

Figure 3: (a–b): In-situ tests conducted at the TRIGA research reactor's thermal irradiation port, and estimated radiation distributions produced by the BO algorithm on the gross spectra (a) and 511 keV peak (b). In each subplot, the black dots represent the true dataset obtained from full raster scans, the orange diamonds indicate initial measurements, and the green triangles mark the locations selected by the BO algorithm. The dashed blue curve shows the predicted mean distribution, and the shaded blue region represents the prediction uncertainty.

B. *Hardware implementation*

Figure 4 depicts results from the practical implementation of the algorithm using the gimbal-mounted collimated $CeBr_3$ detector with one and two ^{137}Cs sources. Each plot presents the BO-driven estimated distribution together with the sampled locations, and associated uncertainty. The distribution obtained from raster scans carried out independently is also depicted. The adaptive approach closely matches the detailed angular radiation distributions obtained separately via the exhaustive raster scans (carried out in 2° intervals) with only a quarter of measurements needed; thus, validating the results obtained in the previous subsection.

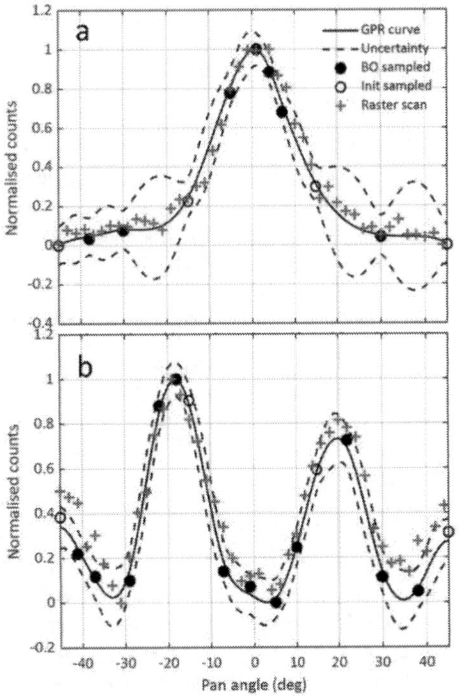

Figure 4: Experimental validation of Bayesian Optimization scans using the gimbal-mounted collimated $CeBr_3$ detector with (a) a single Cs-137 radiation source and (b) two Cs-137 radiation sources. The BO-sampled measurements (solid black circles) and initial sampling points (open black circles) are compared with reference raster scans (red crosses). The Gaussian Process Regression (GPR) predicted mean angular response (solid black line) and its uncertainty bounds (dashed black lines) suggest a near perfect match with the raster scan results.

IV. Discussion

The results demonstrate that substantial efficiency gains can be achieved using this adaptive sampling approach compared to sequential scanning. The algorithm consistently converges with significantly fewer measurements while yielding accurate angular radiation distributions. This reduction is critical in robotic radiation localization applications, where rapid and accurate assessments are of essence.

On the other hand, there are certain aspects to further address and refine in order to create a more robust implementation of the algorithm. Firstly, the choice of the kernel for the Gaussian Process, although adequate in the controlled scenarios presented here (point sources and beam port conditions), might be suboptimal or a poorly matched prediction tool for scenarios that involve dispersed contamination. Planned experiments involving such non-point source distributions will explore this hypothesis and the potential of implementing dynamic kernel selection.

Secondly, the current formulation of the acquisition function presents potential drawbacks: specifically, when the α parameter in equation 5 favours the influence of the Upper Confidence Bound (UCB) component, and as it becomes more dominant, the algorithm tends to over-exploit regions of high radiation intensity, resulting in more measurements than necessary. Therefore, a means of monitoring and dampening such behavior in real-time is essential, without hindering the exploration-exploitation balance. For instance, the geometric characteristics of the collimator, such as its field of view, can be utilized to impose larger angular intervals than 1° if the algorithm gets trapped in overexploitation.

Another aspect that needs addressing is that the aforementioned convergence criteria are satisfied using manually set thresholds. These are inherently heuristic and might not apply in all scenarios, such as the extreme contrast of low and high background between a laboratory equipped with point sources and an operating reactor. Implementing a real-time, automated setting of the thresholds based on observed measurement patterns could further optimize the sampling efficiency.

Finally, further research will focus on testing this algorithm with dispersed sources and subsequently testing its performance when integrated as the main radiation sensor for on-board high level decision-making algorithms, such as radiation-seeking or radiation-avoidance navigation

V. Conclusions

This paper proposed a Bayesian Optimization approach that offers a practical and efficient alternative to conventional scanning methods for γ-localization. The results suggest that it can maximize input space coverage without compromising accuracy, and hence it can be a useful instrumentation method for *in-situ* robotic characterization tasks. Future work will focus on enhancing its performance, such that it can be applied in broader and more complex sets of conditions.

Acknowledgments

We acknowledge the financial support of the Engineering and Physical Sciences Research Council (EPSRC) via grants no. EP/V026941/1 and EP/X038327/1. Malcolm Joyce acknowledges the receipt of a Wolfson Research Merit Award from the Royal Society.

References

[1] I. Tsitsimpelis, C. J. Taylor, B. Lennox, and M. J. Joyce, 'A review of ground-based robotic systems for the characterization of nuclear environments', *Progress in Nuclear Energy*, vol. 111, 2019, doi: 10.1016/j.pnucene.2018.10.023.

[2] A. West *et al.*, 'Use of Gaussian process regression for radiation mapping of a nuclear reactor with a mobile robot', *Sci Rep*, pp. 1–11, 2021, doi: 10.1038/s41598-021-93474-4.

[3] B. Nouri Rahmat Abadi *et al.*, 'CARMA II: A ground vehicle for autonomous surveying of alpha, beta and gamma radiation', *Front Robot AI*, vol. 10, Mar. 2023, doi: 10.3389/frobt.2023.1137750.

[4] H. Chapman *et al.*, 'Robots in nuclear decommissioning Safety Case Considerations for the Use of Robots in Nuclear Decommissioning', *VGB PowerTech*, vol. 5, no. 2020, 2020.

[5] H. Chapman, J.-P. Richardson, C. Fairbairn, A. Di Buono, and A. Gale, 'Best Practice Guidelines and Lessons Learned from Robotic System Deployment in Nuclear Decommissioning', *ATW International Journal for Nuclear Power*, vol. 68, pp. 53–59, 2023, [Online]. Available: https://orcid.org/0000-0001-7844-8890

[6] A. Chierici, R. Ciolini, A. Malizia, and F. d'Errico, 'Resource Constrained Electronics and Signal Processing for UAV Radiation Sensors', in *EPJ Web of Conferences*, A. Lyoussi, F. D'Errico, M. Carette, M. Joyce, R. Hodák, I. Jenčič, P. Le Dû, M. Morichi, S. Pospíšil, C. Reynard-Carette, L. Snoj, and L. Vermeeren, Eds., Nov. 2023, p. 10019. doi: 10.1051/epjconf/202328810019.

[7] I. Tsitsimpelis *et al.*, 'Low-density collimators for remote radiation characterisation tasks', *Nucl Instrum Methods Phys Res A*, vol. 1079, p. 170701, Oct. 2025, doi: 10.1016/j.nima.2025.170701.

Determination of Absolute Activities and Neutron Fluence Rates Using a Coincidence Method

Shokhrukh Mirzo Bakhodirov[1,*], Anja Seifert[1], Dirk Döhler[1], Pia Kahle[1] and Thomas Kormoll[1]
[1]AG Strahlungsphysik, Institute for Nuclear and Particle Physics, Technische Universität Dresden, Germany
(*) shokhrukh_mirzo.bakhodirov@mailbox.tu-dresden.de

***Abstract*— Activation studies are an important tool for nuclear decommissioning. By activating material samples in controlled conditions in a known neutron fluence rate, the expected activity in decommissioning can be estimated. This work focuses on determining the fluence rate of a moderated (α, n) neutron source, a key parameter for performing quantitative activity experiments. The fluence rate is determined by an absolute measurement of the accumulated activity of neutron-activated reference samples with a well-known composition using a βγ-coincidence setup. This method requires both single β- and γ- detection, as well as coincidence data. A main advantage of this technique is that the resulting measurement is mostly independent of individual detector efficiencies. The current setup uses a combination of two detectors, consisting of one γ- and one β-detector, paired with a multi-channel data acquisition system, simultaneously recording hits in the β- and γ-channels including their timestamps. Coincidences are extracted in the offline analysis from the stored data. Having singles and coincident data in one dataset reduces the impact of certain corrections, e.g. dead time. Additionally, Monte Carlo techniques were implemented to assess γ-interactions in the β-detector and to account for the fraction of fake coincidences. Samples, including aluminum (^{27}Al), gold (^{197}Au), Vanadium (^{51}V), sodium chloride (NaCl), and manganese (^{55}Mn) were activated. Calculated thermal fluence rates from weakly attenuating samples agree within uncertainties, with an average fluence rate of $\varphi_0 = (2.58 \pm 0.09) \times 10^5$ cm^{-2}s^{-1}.**

***Keywords* — nuclear decommissioning, activity, neutron fluence rate, βγ-coincidence method.**

I. Introduction

The decommissioning of nuclear facilities requires meticulous planning and assessment to ensure the safe management of radioactive materials and to minimize environmental impact. A key aspect of this process is the precise estimation of induced radioactivity in structural components, particularly in metals and concrete [1, 2]. These estimates inform waste classification, handling procedures, and disposal routes, which are crucial for planning cost-effective and regulatory-compliant dismantling operations [3].

Accurate activation assessments in decommissioning scenarios rely on reliable neutron fluence rate data, especially for thermal neutrons, which dominate in moderated neutron fields. In the framework of the EBENE, activity estimations for nuclear power plant structures are primarily carried out using detailed Monte Carlo (MC) simulations of neutron interactions within facility materials. To ensure confidence in these simulations, a dedicated validation campaign is performed using neutron activation analysis in a controlled moderated neutron field. Here, the neutron fluence rate of the source is simulated and experimentally validated through absolute activity measurements of irradiated samples using a βγ-coincidence method. This approach allows the accuracy of the simulations to be tested under realistic conditions. However, achieving sub-percent precision in activity measurements remains a technical challenge, particularly due to corrections for detector efficiency and geometry.

In this work, we present a methodology based on βγ-coincidence counting to determine the absolute activities of neutron activated samples, and consequently, the corresponding neutron fluence rates. While most existing literature on coincidence counting methods focuses on traditional 4πβγ-coincidence techniques [4–7], absolute activity determination can also be achieved using thin plastic scintillators as β-detectors. Although certain correction procedures from 4πβγ-coincidence counting, such as the use of energy selection in the γ-channel, dead-time and accidental coincidence corrections, can be adapted to this approach, the use of thin β-detectors introduces specific challenges. These include corrections for γ-sensitivity in the β-detectors and isotope-specific effects that are typically accounted for through efficiency extrapolation in 4π setups [7], but must be addressed differently here.

Our methodology implements dedicated MC simulations to account for relevant detector-specific and isotope-dependent correction factors.

II. Material and Methods

A. Experimental Setup

The experimental setup consists of two main components: an irradiation stand equipped with a moderated neutron source for activating prepared samples and a coincidence detection system for measuring the resulting radioactivity.

The irradiation stand is a one meter by one meter polyethylene (PE) cube with multiple (α, n) neutron sources placed at its center. These sources emit fast neutrons with energies up to 12 MeV, which are subsequently thermalized by the surrounding PE moderator. The highest thermal neutron fluence is achieved near the center of the box [8]. Activation samples are positioned close to this central region using a cylindrical PE rod, ensuring that both the sources and samples are fully embedded in a moderating environment (see Fig. 1). After activation, the samples have to be manually removed and placed in the detector setup.

For coincidence measurements, two separate detectors are typically used, i.e., one for detecting β-particles and one for γ-rays. The current setup employs a 2" × 2" $CeBr_3$ scintillator as a γ-detector, enclosed in aluminum shielding. The β-detector is a 1 mm thick cylindrical plastic (material: in-house development) scintillator covered by a 10 μm thick aluminum placed between two 30 μm thick plastic foils. The activated sample is placed between the β- and γ-detectors. For isotopes with short half-lives, the transfer from the irradiation position to the measurement setup is performed as quickly as possible (transfer time: ~ 10 s) to minimize decay losses before data acquisition. The geometry and arrangement of the detectors are shown schematically in Fig. 2.

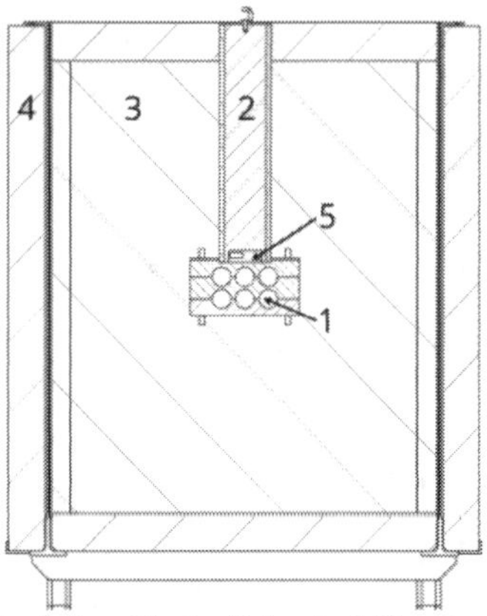

Fig. 1. Cross-sectional view of the irradiation stand. The setup consists of five main components: (1) (α, n) neutron sources, embedded in PE; (2) a cylindrical PE rod used to lower the samples; (3) the surrounding PE moderator volume; (4) borated PE walls for neutron shielding; and (5) a central cavity for sample placement during irradiation.

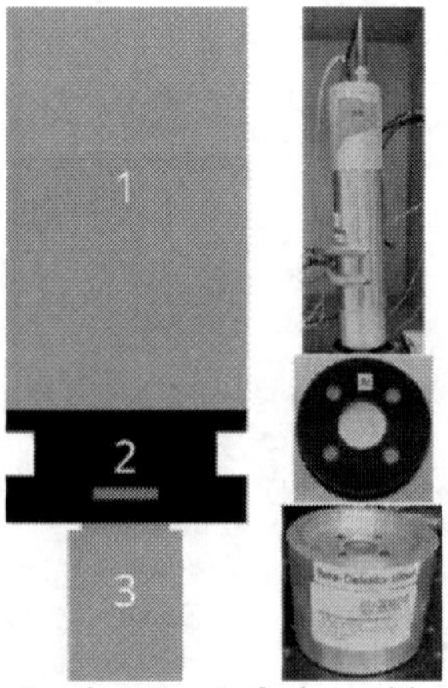

Fig. 2. Coincidence detection setup. Left: the spatial arrangement of the detectors, (1) γ-detector, (2) activated sample, and (3) β-detector. Right: photographs of the real components, (top) $CeBr_3$, (middle) sample holder with aluminum, and (bottom) light-tight plastic scintillator coupled to a PMT in an aluminum casing.

Light signals from both detectors are converted into electrical pulses and amplified using separate photomultiplier tubes (PMTs). The $CeBr_3$ detector is of type 51B51/2M-HV-CEBR-X by Scionix, the Netherlands [9]. The scintillator of the beta detector is coupled to a PDM04-9111-CN PMT by ET Enterprises, UK. Output of each PMT is fed into a data acquisition unit (DAQ), in-house development [10]. The analog signals are digitized by an analog-to-digital converter (ADC) operating at a sampling rate of 120 MHz. The digitized waveforms are then processed by a field-programmable gate array (FPGA), which calculates the signal integral. This integral is proportional to the total number of detected scintillation photons and thus serves as an estimate of the energy deposited by the primary particle in the scintillation crystal.

Each processed event is assigned a timestamp and stored together with its corresponding signal integral value in a list-mode format for offline analysis of singles and coincidence events.

In total, six samples with varying masses and sizes were used for activation. A summary of their properties is provided in Table I.

TABLE I. SAMPLES USED FOR ACTIVATION.

Sample	Mass (g)	Thickness (mm)	Thermal neutron capture cross section (b)	Half-life of activated isotope
Aluminum	3.426	2.20	0.231 ± 0.003	2.25 min
Vanadium	0.326 0.371	0.14 0.14	4.94 ± 0.04	3.74 min
Manganese	0.413	0.95	13.36 ± 0.05	2.579 h
Sodium chloride	0.418	2.25	0.525 ± 0.005	14.96 h
Gold	1.000	0.40	98.65 ± 0.09	2.69 d

For sodium chloride, the listed neutron capture cross section corresponds to sodium only. Two vanadium samples used are composed of two layers from the same sheet, effectively doubling the thickness and mass. Aluminum sample used is an Aluminum 7075 alloy, with a mass fraction of pure aluminum 87-91 %. The uncertainties in mass and thickness arise from the precision of the measuring instruments, which provide an accuracy of 0.5 mg and 0.005 mm, respectively. Neutron capture cross section values are taken from the ENDF/B-VIII.0 evaluated nuclear data library [11].

Sample thickness plays an important role when accounting for attenuation of the neutron fluence within the sample volume. This internal attenuation affects the estimation of the effective neutron fluence rate experienced by the sample. Corrections for self-shielding effects must therefore be considered in the analysis, particularly for materials with high neutron absorption cross section or significant thickness. In this case, such corrections are relevant for the gold, manganese, and sodium chloride samples. Although elemental sodium has a relatively low neutron capture cross section, the presence of chlorine, specifically ^{35}Cl with a much higher cross section of 44 b, results in significant overall attenuation within the sample.

B. Coincidence Analysis

As described in [4–7], the coincidence equations used to determine absolute activity vary depending on the isotope and the complexity of its decay scheme. In the ideal case, where each detector is sensitive only to its corresponding radiation type (i.e., the β-detector responds only to β-particles and the γ-detector only to γ-rays), and the βγ-emitter undergoes a simple decay (each decay emits exactly one β-particle and one γ-ray with 100 percent probability), the coincidence relation is given by the following set of equations:

$$N_\beta = N_{\text{Dec}}\varepsilon_\beta$$
$$N_\gamma = N_{\text{Dec}}\varepsilon_\gamma \quad (1)$$
$$N_c = N_{\text{Dec}}\overline{\varepsilon_\beta\varepsilon_\gamma},$$

where N_{Dec} is the total number of decays and N_β, N_γ and N_c are, respectively, the total number of detected β, γ and coincidences over the measurement time corrected for background; $\overline{\varepsilon_\beta\varepsilon_\gamma}$ is the volume-averaged product of the β- and γ-detector efficiencies, while ε_β and ε_γ are the volume-averaged efficiencies of the individual detectors. As shown in [4] that if at least one of the detectors has a uniform efficiency across the sample volume, the mean of the product equals the product of the means

$$\overline{\varepsilon_\beta\varepsilon_\gamma} = \varepsilon_\beta\varepsilon_\gamma. \quad (2)$$

It is clear that if (2) holds, then the solution to (1) is independent of the individual efficiencies. However, even for relatively simple decay schemes, such as that of ^{28}Al (shown in Fig. 3), the idealized coincidence relation (1) above is incomplete. When accounting for additional effects, the coincidence equations take the following form:

$$N_\beta = N_{\text{Dec}}I_\beta\big(\varepsilon_\beta + (1-\varepsilon_\beta)\varepsilon_{\beta\gamma}\big)$$
$$N_\gamma = N_{\text{Dec}}I_\gamma\varepsilon_\gamma \quad (3)$$
$$N_c = N_{\text{Dec}}I_\beta\big(\varepsilon_\beta\varepsilon_\gamma + (1-\varepsilon_\beta)\varepsilon_c\big),$$

where I_β and I_γ are the absolute intensities of β- and γ-emission, $\varepsilon_{\beta\gamma}$ is the γ-efficiency of the β-detector and ε_c is the probability of recording a coincidence when the associated β-particle is not detected, typically caused by the backscattered γ-rays. The second term in N_β accounts for cases where the β-particle is not detected, but the β-detector still registers a signal from the γ-ray. Similarly, the second term in N_c represents coincidences arising from such missed β-events, where the coincidence is caused by a backscattered γ-ray. It is assumed that the β-efficiency of the γ-detector is zero, which can accurately be achieved.

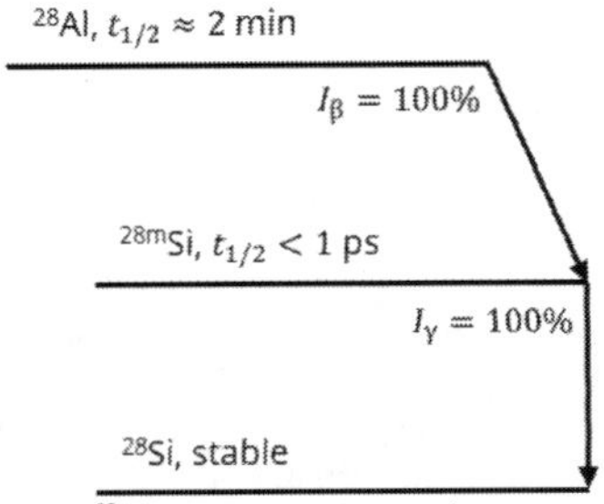

Fig. 3. Decay scheme of ^{28}Al. The indicated $t_{1/2}$ values correspond to the half-lives of the respective states. In this decay, ^{28}Al undergoes β^- decay followed by γ-emission as the nucleus de-excites from the metastable state ^{28m}Si.

In general, in order to compare correction magnitudes across all samples, (3) can be rewritten as:

$$N_{\text{Dec}} = \frac{N_\beta N_\gamma}{N_c}(1-\rho), \quad (4)$$

where ρ is the relative deviation from the idealized case described in (1). Finally, the saturation activity of the sample is calculated based on the corrected number of decays (4).

C. Simulation

To estimate the correction factors introduced in (4) for each of the isotopes listed in Table I, the measurement setup was reproduced in GEANT4 simulations [12], using the same detector and sample geometries as in the experiment. The simulations implement three built-in physics lists to account for the relevant physical processes: *G4EmPenelopePhysics*, *G4DecayPhysics*, and *G4RadioactiveDecayPhysics*. The production cut values were set to 0.1 mm.

For each sample, 5×10^6 decays were simulated using the corresponding detector and sample geometry. Based on the simulated detector counts (N_β, N_γ, N_c), the number of decays is calculated assuming the idealized coincidence case described in (1). The correction factor ρ is then derived from the ratio between the true and calculated number of decays as:

$$1-\rho = \frac{N_{\text{Dec}}^{\text{true}}}{N_{\text{Dec}}^{\text{calc}}}. \quad (5)$$

D. Neutron Field Characterization

To characterize the neutron field near the source, it is assumed to consist of two components: thermal and epithermal neutrons. The total spectrum f_E^{total} is expressed as

$$f_E^{\text{total}} = f_E^{\text{th}} = \frac{E}{E_0^2}e^{-\frac{E}{E_0}}, \text{for } E < E_t$$
$$f_E^{\text{total}} = f_E^{\text{th}} + f_E^{\text{epi}}, \text{for } E > E_t,$$

where f_E^{epi} is the epithermal spectrum with $1/E$ dependence; E_t is the transition energy at which the thermal and epithermal components are equal. It is set to be 0.197 eV based on simulations [8]; and E_0 corresponds to the characteristic energy at room temperature, equal to 25 meV. Fig. 4 shows the total spectrum f_E^{total} under the above assumed condition.

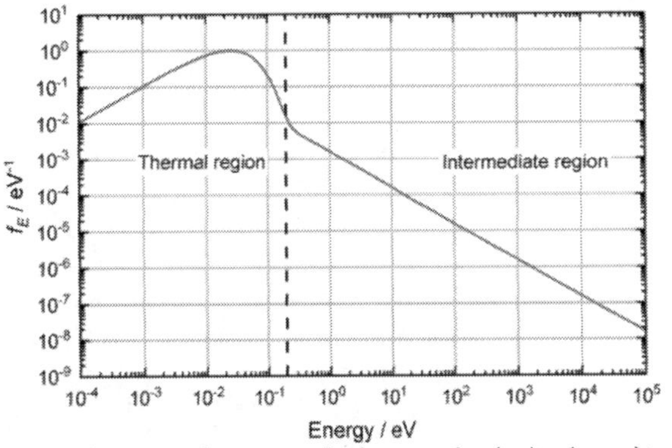

Fig. 4. Total spectrum of neutrons, under the assumption that it only consists of thermal and epithermal regions. Highlighted are the two energy regions.

However, due to the finite thickness of the samples, the neutron field inside the sample is attenuated. Consequently, the accumulation of activity A_0 within the sample is reduced because the sample effectively experiences the attenuated neutron fluence $\varphi_0 f_E^{\text{atten}}$, such that

$$A_0 = \int N_{\text{st}}\sigma(E)\varphi_0 f_E^{\text{atten}}\text{d}E, \quad (6)$$

where $\sigma(E)$ is the neutron activation cross section, N_{st} is the number of stable nuclei present in the sample and φ_0 is a normalization constant with units of cm^{-2}s^{-1}, representing the magnitude of the thermal neutron fluence. The parameter φ_0 is expressed in terms of the measured activity A_0 as

$$\varphi_0 = \frac{A_0}{\int N_{st}\sigma(E)\, f_E^{\text{atten}} \mathrm{d}E} = \frac{A_0}{r \int N_{st}\sigma(E)\, f_E^{\text{total}} \mathrm{d}E}, \quad (7)$$

where r is a measure of how much the field is attenuated inside the sample. The values for φ_0 are calculated under the assumption $r = 1$, which is only valid for very thin samples or isotopes with low neutron capture cross sections.

III. Results and Discussion

A. Coincidence Event Selection

Fig. 5 shows the time difference distribution between neighboring events detected in both channels for the NaCl measurement.

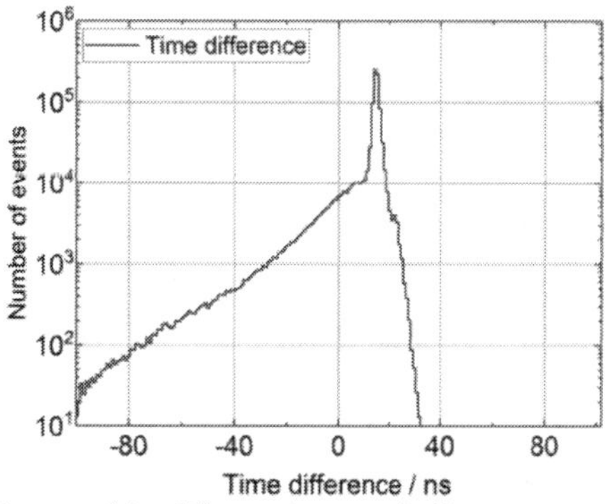

Fig. 5. Histogram of time differences between coincident events recorded from both channels for the NaCl measurement. Bin width: 1 ns.

The distribution features a sharp peak, shifted from zero, as well as broadened tails. The shift arises due to differences in signal propagation and processing between the β- and γ-detector channels. The broadened tails result from an effect known as time-walk, where signals of different amplitudes cross the discriminator threshold at different times. While the effect is not corrected in this analysis, it is minimized by the use of narrow coincidence time window and choosing a narrow energy window in the γ-detector, centered around the characteristic γ-ray energy of the activation product [13]. The full width of the sharp peak defines the resolving time of the setup, which in this case is approximately $\tau_R \approx 8$ ns.

Events used for the coincidence analysis are selected within the defined coincidence time window. These selected events form the energy spectra in each channel. Fig. 6 shows a two-dimensional histogram of the coincidence spectra from the NaCl activation measurement. The characteristic γ-ray emission lines of the ^{24}Na isotope at 1369 keV and 2754 keV are clearly visible. Since ^{24}Na emits two γ-rays in cascade per decay, the energy sum peak corresponding to the combined energies (1369 keV + 2754 keV) is also observed. Additionally, a noise peak appears in the low-energy region of the β-channel, primarily due to the low threshold set in that channel. This noise contribution is removed in the analysis by applying an energy cut in the β-channel to suppress non-physical events.

Fig. 7 shows the total and coincidence spectra recorded in the γ-detector from the activated NaCl sample. In the total spectrum, the characteristic γ-lines from ^{24}Na are visible alongside a background peak at 1461 keV originating from ^{40}K. This background line partially overlaps with the 1368 keV line from ^{24}Na. However, in the coincidence spectrum, this interference is no longer present. Since environmental background sources decay independently of the β-detector signal, they do not contribute to the coincidence events.

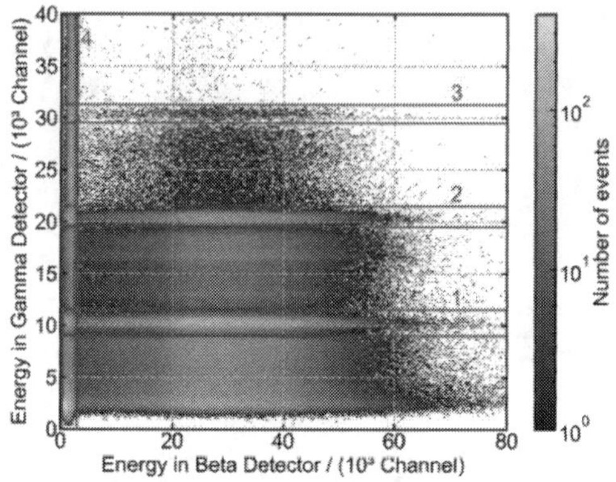

Fig. 6. Two-dimensional histogram of coincidence spectra from both channels. Four energy regions are highlighted: (1) 1369 keV and (2) 2754 keV γ-lines from ^{24}Na; (3) their sum peak at 4123 keV; (4) noise peak in the β-channel.

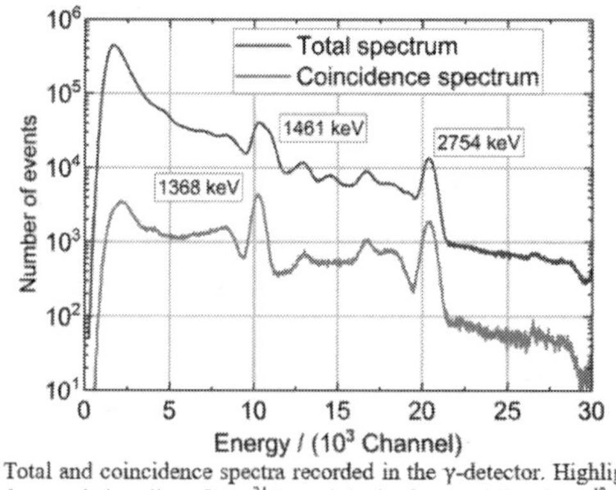

Fig. 7. Total and coincidence spectra recorded in the γ-detector. Highlighted are the characteristic γ-lines from ^{24}Na and the background line from ^{40}K. No interference with the background is observed in the coincidence spectrum.

As mentioned previously, a narrow energy window is chosen in the γ-detector, typically centered around the characteristic γ-emission peak of the corresponding isotope. The number of γ events N_γ and coincidence events N_c are then counted within the specified energy window to calculate the absolute activity of the activated sample. In contrast, the number of β-events N_β is counted over the full β-spectrum, with only the low-energy noise region excluded via an energy cut. This is because β-particles are emitted with a continuous energy distribution and the β-detector efficiency is mostly independent of β energy over the spectrum of interest, so applying an energy window would unnecessarily discard valid events and reduce counting statistics.

B. Correction Factors

As introduced in Section II, the total number of decays calculated for the idealized case (1) does not directly correspond to the true number of decays. Several effects contribute to this difference, including the sensitivity of the β-detector to γ-rays, false coincidences caused by backscattered γ-rays and possible deviations from the condition described in (2). Therefore, correction factors must be applied to the number of decays obtained for the idealized case.

Correction for the γ-sensitivity of the β-detector becomes particularly important for isotopes emitting low-energy γ-rays, such as ^{198}Au (412 keV). To assess this contribution for gold, two measurements were performed using the same activated sample. In the first, a 5 mm thick polymethyl methacrylate (PMMA) absorber was placed between the sample and the β-detector. In the second, the PMMA was removed, but the gold sample was positioned at the same height to maintain identical geometrical efficiency. The measurements were performed consecutively, each with a measuring time of 10 minutes. Fig. 8 shows the spectra obtained from both configurations. The γ contribution to the β-detector response is quantified as $N_\beta^{\text{with PMMA}}/N_\beta^{\text{without PMMA}} = (13.93 \pm 0.04)$ %.

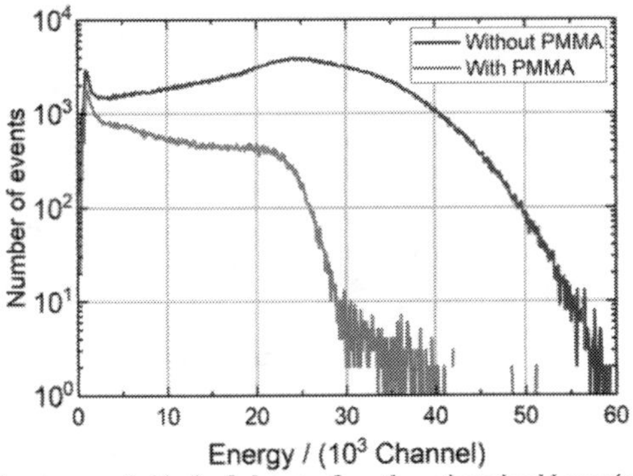

Fig. 8. Spectra recorded in the β-detector from the activated gold sample with and without a 5 mm PMMA absorber placed between the sample and the detector.

Correction factors for the rest of the samples were calculated by means of MC simulations as discussed above. Table II shows the estimated corrections factors, introduced in (5), for each sample. The correction values are less than 5 % across all samples.

TABLE II. ESTIMATED CORRECTION FACTORS.

Sample	$1-\rho$
Aluminum	0.980 ± 0.010
Vanadium	1.010 ± 0.006
Manganese	1.007 ± 0.006
Sodium chloride	0.956 ± 0.011
Gold	0.989 ± 0.015

The correction factor listed for gold in Table II corresponds to the value already adjusted for the γ contribution in the β detector, based on the PMMA measurement described above. The uncertainties listed for the correction factors are statistical. As the simulations were conducted under the assumption that the samples are located precisely at the detector center, potential systematic uncertainties arising from deviations in the actual sample positioning during measurements have not been accounted for.

C. Thermal Fluence Rate Calculation

Table III shows the corrected measured saturation activity $A_0(1-\rho)$ and the calculated thermal fluence rate φ_0.

TABLE III. CALCULATED THERMAL FLUENCE RATES FOR ALL SAMPLES.

Sample	$A_0(1-\rho)$ / kBq	φ_0 / 10^5 cm^{-2}s^{-1}
Aluminum	3.7 ± 0.1	2.676 ± 0.119
Vanadium	4.52 ± 0.06	2.550 ± 0.044
	5.03 ± 0.07	2.498 ± 0.043
Manganese	12.96 ± 0.08	2.293 ± 0.037
Sodium chloride	0.434 ± 0.005	2.288 ± 0.177
Gold	56.8 ± 0.9	1.513 ± 0.031

The uncertainty in the corrected activity is obtained by propagating the statistical uncertainties of the individual N_β, N_γ and N_c counts together with the uncertainty in the correction factor $1-\rho$. The uncertainty in the calculated thermal fluence rate φ_0 is obtained by propagating the uncertainties from both the corrected activity and cross section. The cross section uncertainties are taken from the nuclear data library [11]; if no value is provided, an uncertainty equal to 1 % of the cross section value at each energy point is assumed.

The calculated values of φ_0 for each sample do not all agree within uncertainties. Good agreement is obtained for samples with small neutron attenuation, while discrepancies arise for heavier isotopes with stronger self-shielding. These discrepancies arise because the calculation assumes no neutron attenuation inside the sample ($r = 1$). While this is a good approximation for thin samples or isotopes with small neutron capture cross sections, it is not valid for materials with stronger absorption. In those cases, the effective neutron fluence inside the sample is reduced, leading to systematically lower calculated values of φ_0.

IV. CONCLUSIONS AND OUTLOOK

A coincidence measurement setup was developed to determine the activities of activated samples and to calculate the thermal neutron fluence rate of a moderated neutron source from these activities. For each sample, correction factors were calculated to account for coincidence detection effects. The calculated fluence rates show agreement only for samples with small neutron field attenuation, while for samples with stronger attenuation a systematic discrepancy is observed. For the subset of consistent samples, the mean thermal fluence rate is $\varphi_0 = (2.58 \pm 0.09) \times 10^5$ cm^{-2}s^{-1}, corresponding to a relative uncertainty of about 3 % with a 95 % confidence level. This level of precision demonstrates the reliability of the method for thin or weakly attenuating samples and provides an experimental reference for validating the simulations.

Future work will focus on quantifying neutron self-shielding effects, for example by using MC simulations or analytic attenuation models, in order to improve the accuracy of the fluence determination for absorbing materials. Additionally, refinement of gamma sensitivity corrections for low-energy emitters will further reduce systematic uncertainties and strengthen the reliability of the method as a reference for validating simulations.

Acknowledgment

This work was funded by the German Federal Ministry of Education and Research (BMBF) under grant number 15S9447B.

References

[1] L. Schlömer, *et al.*, "Activation calculation for the dismantling and decommissioning of a light water reactor using MCNP^{TM} with ADVANTG and ORIGEN-S", EPJ Web of Conf., 153, 05020, 2017.
[2] J. Konheiser and A. Barkleit, "Development of a method for pre-activity and dose rate calculations of components in the reactor vicinity based on neutron fluence distributions", Hel.-Zen. Dre.-Ros. (HZDR), 133, Nov. 2024
[3] W. Alexander, *et al.*, "Decommissioning of nuclear power plants: Regulation, financing and production", DIW Data Doc., vol. 104, pp. 75–91, 2023.
[4] P. J. Campion, "The standardization of radioisotopes by the beta-gamma coincidence method using high efficiency detectors", Int. Jour. App. Rad. Iso., vol. 4, pp. 232–248, 1959.
[5] A. P. Baerg, "Measurement of radioactive disintegration rate by the coincidence method", Metrologia, vol. 2, pp. 23–32, 1966.
[6] R. Fitzgerald, *et al.*, "Uncertainties in $4\pi\beta - \gamma$ coincidence counting", Metrologia, vol. 52, pp. 86–96, 2015.
[7] A. P. Baerg, "The efficiency extrapolation method in coincidence counting", Nuc. Ins. Met., vol. 112, pp. 143–150, 1973.
[8] B. Keßler, *et al.*, "Accurate simulation of neutron fields using Tramo", presented at the 9th Int. Conf. ANIMMA, Valencia, Spain, Jun. 9-13, 2025.
[9] F. G. A. Quarati, *et al.*, "Scintillation and detection characteristics of high-sensitivity $CeBr_3$ gamma-ray spectrometers", Nuc. Ins. Met. Phys. Res. Sec. A, vol. 729, pp. 596–604, 2013.
[10] Y. Ma, *et al.*, "System noise of a digital pulse processing module for nuclear instrumentation", EPJ Web Conf., 225, 01012, 2020.
[11] D.A. Brown, *et al.*, "ENDF/B-VIII.0: The 8th major release of the nuclear reaction data library with CIELO-project cross sections, new standards and thermal scattering data", Nuc. Dat. Sheets, vol. 148, pp.1–142, Feb. 2018.
[12] S. Agostinelli, *et al.*, "GEANT4–a simulation toolkit", Nuc. Ins. Met. Phys. Res. Sec. A, vol. 506, pp. 250–303, Jul. 2003.
[13] J. Du, *et al.*, "A Time-Walk Correction Method for PET Detectors Based on Leading Edge Discriminators", IEEE Tran. Rad. Plas. Med. Sci., vol. 1, pp. 385-390, Sept. 2017.

Nuclear waste characterization using Bayesian methods

Achment Chalil[1,*], Rodolphe Antoni[1], Youness El Mahjouby[1], Cyrille Eleon[1], Mehdi Ben-Mosbah[1] and P.G. Allinei[1]
[1]CEA, DES, IRESNE, Nuclear Measurement Laboratory, 13108, Saint Paul-lez-Durance, France
(*) achment.chalil@cea.fr

***Abstract*— The relance of the nuclear industrial sector in France is an ambitious plan expected to meet the energy needs of the general public in the 21st century. Along with the construction of new nuclear power plants, advancements are taking place in the fields of the nuclear fuel cycle and nuclear waste management. In order to comply with the criticality limits, nuclear waste is stored in special drums and different types of nuclear measurements are employed in order to determine the amount of the nuclear matter (U, Pu) contained inside them. In this work, we present an experimental design for the measurement of waste drums using the SYMETRIC ^{3}He detector array located at CEA Cadarache. An experimental design using simulated data has been constructed for different parametrizations of waste drums in order to establish a model that has the potential of effectively predicting the calibration coefficient CP9, which is directly related to the amount of fissile matter inside the drum. Up until now, the models trained by the simulated data of the experimental design were exported by multiple linear regression. In an attempt reduce the statistical uncertainties, we have performed the analysis of the experimental design by applying two Bayesian methods: the Gaussian process and a Bayesian neural network. Both methods have resulted in a considerable improvement on several statistical metrics, showing promising performance for future applications.**

***Keywords* —Nuclear waste, Active neutron interrogation, Experimental design, MCNP, Bayesian methods**

I. Introduction

NUCLEAR waste characterization is a highly important procedure in the nuclear industry. Such waste may originate either from decommissioning operations of nuclear installations (reactors, accelerators) [1] , [2] or from fuel reprocessing procedures [3] . After a decommissioning operation or a fuel reprocessing operation, the waste is stored in specially fabricated waste drums, which have to be characterized in terms of various physical properties such as their activity or their radiological composition [4] . Such characterization is particularly challenging as there is limited a priori knowledge on the constituents of the drum, especially if legacy waste is considered, where quality control procedures were not yet implemented. Process fluctuations on homogeneous waste can also complicate the characterization [5] .

The Nuclear Measurement Laboratory at CEA Cadarache is specializing on the non-destructive nuclear measurements within the nuclear fuel cycle [[4] , [6] -[10]], including the characterization of waste drums. Its research activities are focused on the improvement of the relevant techniques as well as the application of new experimental or data analysis methods. Traditionally, the determination of the quantity of fissile material was done by using active neutron interrogation [11] . By using this method, the mass of the fissile material is proportional to the neutron counting rate. The constant of proportionality, is called calibration coefficient and is determined by a simulation-based experimental design [12] . Once the experimental design has been determined, classical statistical methods such multiple linear regression is used in order to determine a model for the calibration coefficient as a function of some measurable quantities, such as the signal of the internal monitor or the density of the matrix. The main problem is that classical methods such as OLS result in large uncertainties on the determination of the calibration coefficient and consequently for the mass of the fissile material. Such large uncertainties can pose difficulties in terms of radioprotection. It is thus important to try and minimize potential sources of uncertainties, either on the experimental part or on the statistical treatment of the experimental design.

The recent advancements in the field Machine Learning and Artificial Intelligence have permitted to tackle complex problems which were impossible to solve with classical statistical methods (such as image recognition). Along with the increase of computational power, Bayesian statistics has also become increasing popular in various fields. Because of their probabilistic nature, Bayesian methods provide an inherent way to quantify uncertainties, which is an important aspect on the determination of various physical quantities in nuclear measurements.

In the present work, two Bayesian methods will be applied in order to improve the performance and the uncertainty on the determination of the amount of fissile matter inside a waste drum via active neutron interrogation [11] , [12] . The first is the Gaussian process (GP) [13] , which is a non-parametric and probabilistic method which can solve regression problems while providing an inherent way to quantify the model uncertainty. The second method is the use of a Bayesian neural network (BNN) [14] . Within this approach on neural networks, prior distributions are given on the weights and biases of the network and in the end a posterior distribution is given on the output layers. This distribution can provide the uncertainty of the model.

The structure of the present article is as follows: Firstly, the simulation-based experimental design will be described along with the relevant physical quantities. Then, details will be given on the application of the Gaussian process and the Bayesian neutral network on the experimental design. Next, a comparison on the performance between the Bayesian methods and the classical OLS treatment will be given. Future perspectives on improving on these methods will be also given in the conclusions.

II. Experimental design

The main objective in order to estimate the mass of the fissile material inside a waste drum by using the Differential-Die Away

technique is the calculation of the calibration coefficient. As the fissile matter can consist of different isotopes ($^{233,\,235}$U, ^{239}Pu), it is usually convenient to express the amount of fissile material as the equivalent mass of ^{239}Pu, M(^{239}Pu). Then, the neutron counting rate, S, is proportional to this equivalent mass:

$$S = \mathrm{CP9} \times \mathrm{M}(^{239}\mathrm{Pu}) \quad (1)$$

The coefficient of proportionality, CP9, is called calibration coefficient. The determination of CP9 is quite complex, as it depends on the characteristics of the matrix and of the neutron detection setup. In the present work, it is determined by a simulation-based experimental design, using MCNP [15] to model matrices with different characteristics.

The experimental setup that is modeled in MCNP is the SYMETRIC detection array at CEA Cadarache (Fig. 1). It consists of a GENIE 16 neutron generator which emits neutrons at an energy of 14.1 MeV to the cavity where the waste drum is placed. These neutrons are then thermalized inside the cavity, made of graphite and polyethylene, and the matrix and induce fissions on the fissile matter. The prompt neutrons emitted from these fissions are then detected in the three neutron detection blocks, which consist of 33 ^{3}He detectors. An internal monitor is also used to monitor the neutron absorption inside the matrix [12].

By using the modeled experimental setup described above, an experimental design of different simulations can be run in order to establish a model for the CP9 as a function of measurable quantities. Such quantities can be the apparent density of the matrix (d), the internal monitor signal (IM) or the transmission signal (TS), among others. A total of 140 MCNP simulations were run and the different ranges of parameters varied are shown in Table I. Two materials have been used inside the matrix (PVC, CH_2) with varying density within a physically reasonable range, in order to explore the effect of non-homogeneous parts of the matrix on the CP9. It is important to note that the distribution of the density values of the 140 simulations of the experimental design is random, and within the ranges given in Table I.

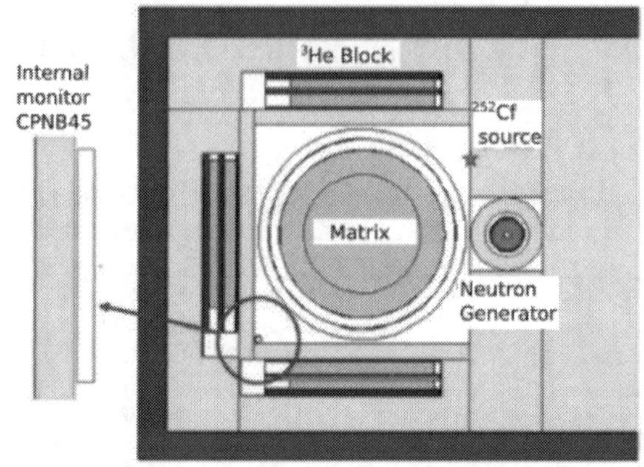

Fig. 1. MCNP model of the waste matrix inside SYMETRIC. The neutron generator, the internal monitor and the three ^{3}He detection blocks are shown. See text for details.

Table I
Units for Some Magnetic Properties

Material	Density
PVC	0.1-1
CH_2	0.1-1.65

In the end of 140 simulations, a dataset consisting of the calculated CP9 and all the other variables mentioned previously is created. Then a statistical method is used in order to establish a model, which will allow for the calculation of the CP9 as a function of different measurable quantities, such for example, the internal monitor signal or the density. The procedure is described in the following section.

III. Statistical models

Traditionally, the best model was chosen by using the OLS and a metric such as the residual standard error in order to evaluate the performance of each model. In this work, the performance of two other methods will be compared with the standard approach: Gaussian process regression and a Bayesian neural network.

A. Ordinary least-squares

In order to establish the optimal model for the experimental design, the stepwise regression method in conjunction with OLS is used [16] . The best model with the optimal predictor variables is determined by an iterative procedure based on the p-values of each term. The values of the coefficients of the model are determined by OLS.

The main advantage of this method is its simplicity and explicability. However, it can lead to diminished performance and relatively large uncertainties, as the model is not able to capture complex relationships between variables.

B. Gaussian process

Gaussian process regression is a non-parametric and probabilistic method which makes use of the multivariate gaussian distribution in order to fit a model to a dataset. It is a kernel-based method, meaning that it does not work with an explicit parametric model. The kernel expresses the covariance between two different points of the model and depends on some specific hyperparameters, depending on its structure.

The gaussian process can capture complex relationships between the different variables, allowing for the establishment of a much more performing model in comparison to OLS. Its probabilistic nature allows also for a direct way to calculate the uncertainty for each point of a model, providing thus a convenient means of uncertainty quantification. However, gaussian process is very prone to overfitting and special care must be taken in order to ensure the correct choice of the kernel. For the present work, the scikit-learn library implemented in Python has been used [17] .

C. Bayesian neural network

A Bayesian neural network is a neural network whose weights and biases are expressed by some prior distributions, following the Bayesian framework. Classical neural networks are not probabilistic methods, and their final results do not provide a measure of uncertainty. Usually, another method is used to quantify their uncertainty (such as bootstrapping) which is not usually straightforward. By introducing a Bayesian approach on neural networks, uncertainty quantification is facilitated.

The use of this type of networks has been proved particularly well-performing and have been used in various studies. However, as all types of neural networks, they are prone to overfitting and thus special attention needs to be taken on hyperparameter optimization, such as the number of layers/neurons. For the present work, the Pyro [18] library has been used, which provides an interface within PyTorch [19] for the implementation of Bayesian neural networks.

IV. RESULTS AND DISCUSSION

The best performing model using OLS is found to be:

$$CP9 = a_0 + a_1 MI + a_2 d + a_{12} MI \times d + a_{11} MI^2 + a_{22} d^2 \quad (2)$$

where the values of the coefficients are given in Table II. The performance of the model is shown on Fig. 2, where the CP9 calculated from Eq. 2 versus the CP9 calculated by MCNP is plotted. The root mean square error is calculated to be 2227, while the calculated 95% prediction intervals seem to cover the 96.4% of the data. As expected, the model is quite robust, and the performance seems to be stable over the whole range of the dataset.

In Fig. 3, the same dataset is fitted with a gaussian process model:

TABLE II
COEFFICIENTS OF THE MODEL OF EQ. 2

Coefficient	Value	Error
a_0	1.719×10^4	3.7×10^3
a_1	-0.65	0.26
a_2	-1.507×10^4	6.6×10^3
a_{12}	-0.432	0.17
a_{11}	3.085×10^{-5}	4.8×10^{-6}
a_{22}	1.037×10^4	3.9×10^3

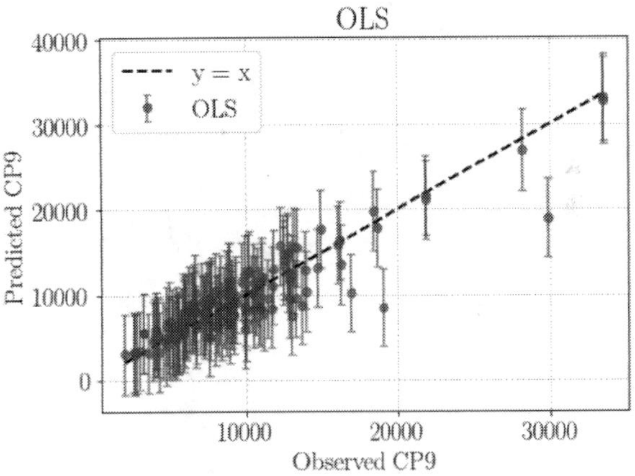

Fig. 2. Comparison of observed versus predicted values of the CP9 by using OLS. Error bars represent 95% prediction intervals.

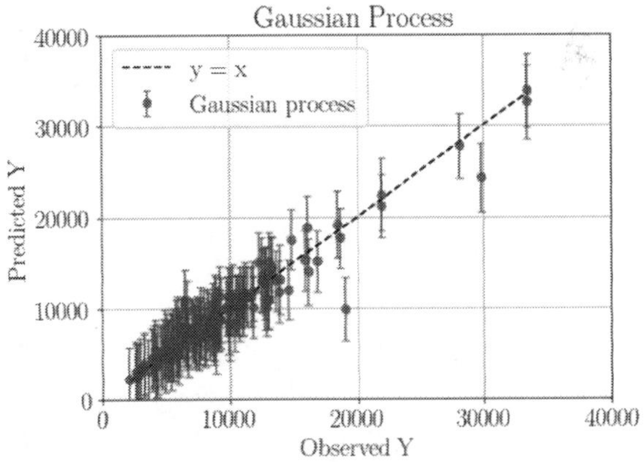

Fig. 3. Comparison of observed versus predicted values of the CP9 by using Gaussian process regression. Error bars represent 95% prediction intervals.

$$CP9 = \mathcal{GP}(d, MI, TS). \quad (3)$$

The kernel that is found to give the best results is of the form:

$$k = \text{Constant kernel} \times \text{RBF} + \text{White noise kernel} \quad (4)$$

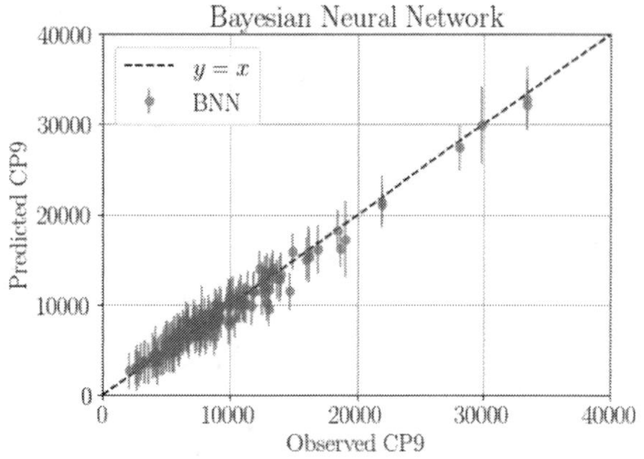

Fig. 4. Comparison of observed versus predicted values of the CP9 by using a Bayesian neural network. Error bars represent 95% prediction intervals.

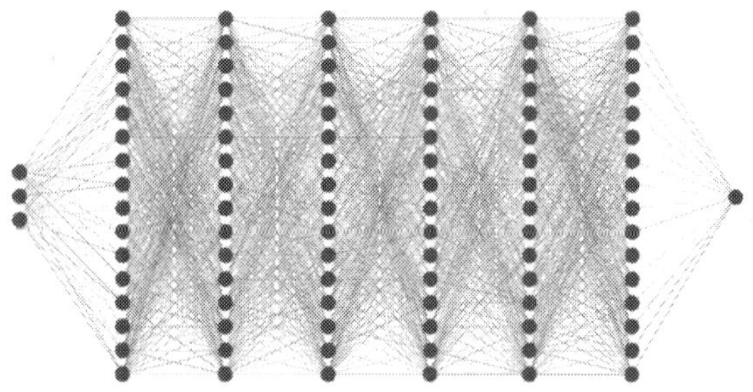

Fig. 5. Architecture of the Bayesian neural network used in this work. Three input variables are fed into a six-layer neural network, with each layer consisting of sixteen neurons. The output value gives the CP9.

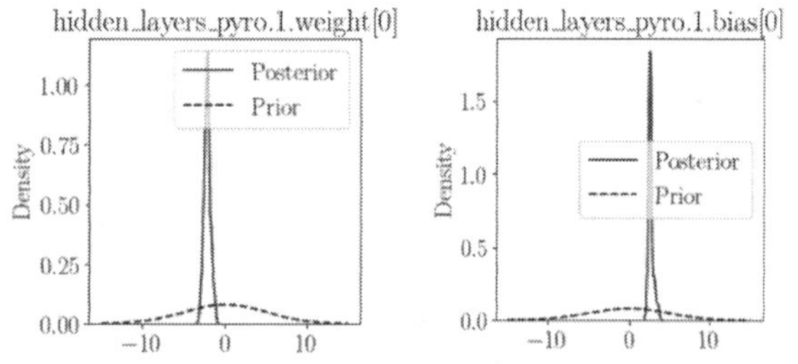

Fig. 6. Representative prior and posterior distributions for weights and biases in the Bayesian neural network used in this work.

where RBF is a Radial Basis Function kernel used to describe smooth functions and the white noise kernel describes the noise of the data, assuming that it follows a gaussian distribution. The performance over the whole dataset seems to be better than the case of OLS, as the root mean square error is now at 1470. The 95% coverage rate is calculated to be at 96.4%, at the same level as OLS. We observe also that the model is quite robust in terms of uncertainty evaluation throughout the whole dataset. This is because of the choice of the kernel, which well captures the behavior of the dataset. Other kernel combinations can easily result in overfitting and degraded performance in terms of

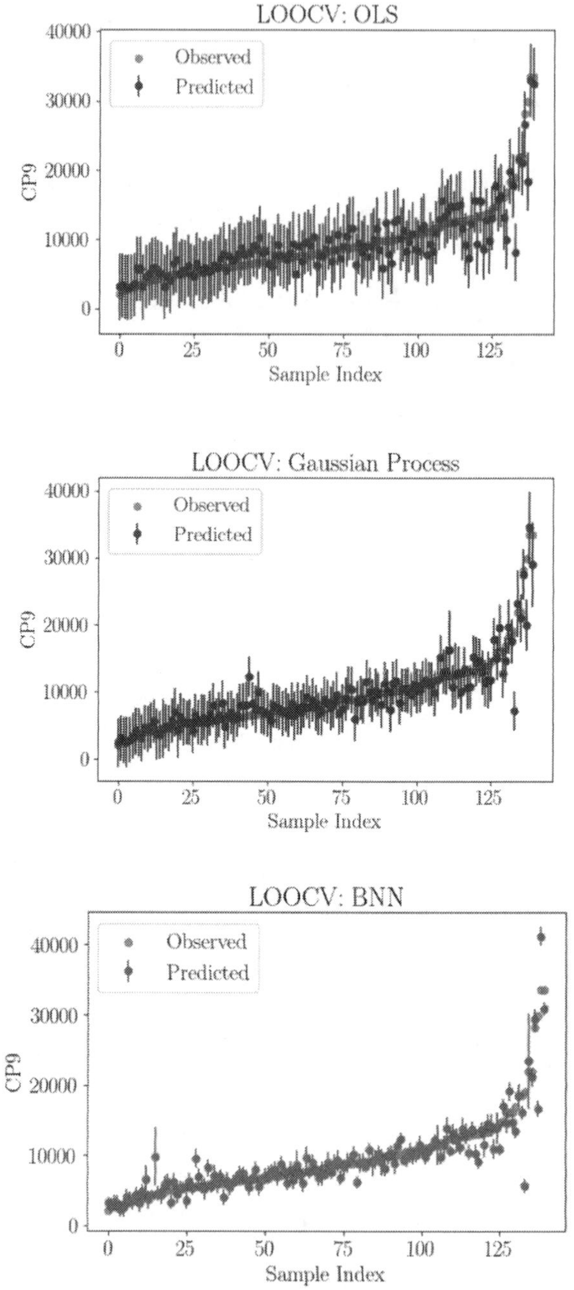

Fig. 7. LOOCV cross-validation summary plots for all three methods used in the present work. Every point was used as a test set in every iteration, while the model was trained in the remaining 139 data points. Then, the predicted value of the test point is compared to the observed (from MCNP) value and this is done for all points throughout the dataset. See text for details.

uncertainty.

In Fig. 4, the same dataset is fitted by using a Bayesian Neural network. The architecture of the network is shown in Fig. 5. It consists of 6 layers of 16 neurons and was chosen empirically. A selection of prior and posterior distributions for

TABLE III
METRICS FOR COMPARING THE THREE MODELS USED IN THIS WORK AFTER LEAVE-ONE-OUT CROSS VALIDATION.

Method	Metric	Value
OLS	RMSE.	2304
	95% Cov	96.4%
GP	RMSE	1944
	95% Cov	96.4%
BNN	RMSE	2198
	95% Cov	87.9%

several weights are shown on Fig. 6. All priors were chosen to follow a gaussian distribution with a sufficient large standard deviation. No significant effect on the posteriors was observed when increasing the standard deviation of the distribution. The performance of the network is sufficiently good, giving a mean square error of 936 when all data points are considered. However, it has to be noted that the hyperparameters of the network, that is, its architecture has not been optimized for the present work.

While the gaussian process and the Bayesian neural network seem to outperform the classical OLS method, the above results are not conclusive unless we compare them by means of cross validations methods. The method chosen for this work is the Leave-one-out cross validation (LOOCV) method, which is suitable for relatively small datasets as the one in the present work (140 data points). Within this iterative method, one data point is chosen each time as a test set and all the remaining points as the training set. The mean square error of the test set is then calculated for each regression method as a comparison metric. We repeat the same procedure until every data point is considered as a test point and in the end, the average mean square error is calculated. In Fig. 7, summary plots are shown that compare the predicted value of the test point with the observed value from MCNP. In Table III, the average mean square error and the 95% coverage rate are also shown.

After cross validation, it is evident that the Gaussian process is more performing on unseen data, followed by the Bayesian neural network and by OLS in the end. We also observe a more robust treatment of uncertainties by using the Gaussian process, compared to the Bayesian neural network, as there are points where the uncertainty can become very large when using the latter. As shown in Fig. 7, the BNN seems to not perform equally well in all regions. This could be solved by hyperparameter optimization. However, both methods seem to be performing better than OLS in terms of the mean square error.

V. CONCLUSIONS AND FUTURE DIRECTIONS

In this work, two Bayesian methods have been implemented in the model construction of the calibration coefficient CP9 and potentially lead to better characterization of the fissile matter inside the waste matrix. The results are deemed promising as there is a notable improvement compared to the traditional OLS method. After cross validation, both methods lead to a reduction (~ 15% for the GP and ~ 5% for the BNN) of the mean square error if every point is used iteratively as a test point, showing a promising alternative to traditional methods. The 95% coverage of the data seem to be robust for the Gaussian process but not for the case of the BNN, where the model seems to be overconfident (smaller uncertainties but they fail to cover close of 95% of the data). For this case, further investigations are required in order to arrive to a robust result.

It has to be noted that the potential of BNNs is not yet fully explored in this work as its hyperparameters were empirically chosen. Future work includes more exploration on such networks as well as the testing of other contemporary machine methods for model prediction (e.g. support vector machines, ridge kernel regression) and uncertainty quantification (e.g. conformal prediction).

REFERENCES

[1] International Atomic Energy Agency, "Decommissioning of nuclear power plants, research reactors and other nuclear fuel cycle facilities", IAEA Safety Standards Series No. SSG-47, Vienna, IAEA (2020)
[2] Y. Kim, C. Kim, W.D. Jung, "Sustainable decommissioning strategy for nuclear power plants: A systematic literature review", Sustainability, 14(10), 5947, (2022)
[3] C. Eleon et al., "Status of the nuclear measurement stations for the process control of spent fuel reprocessing at AREVA NC/La Hague," 4th International Conference on Advancements in Nuclear Instrumentation Measurement Methods and their Applications (ANIMMA), Lisbon, Portugal, 2015, pp. 1-5, (2015)
[4] Bertrand Pérot et al., The characterization of radioactive waste: a critical review of techniques implemented or under development at CEA, France, EPJ Nuclear Sci. Technol. 4 3 (2018)
[5] J. Pokorný et al., "Transient melt formation and its effect on conversion phenomena during nuclear waste vitrification–HT-ESEM analysis", Journal of the American Ceramic Society. Advance online publication (2024)
[6] V. Bottau et al., "High-Resolution Gamma Spectrometry of a Plutonium Bearing Waste Drum With High-Energy Reaction-Induced Gamma Rays", IEEE Transactions on Nuclear Science, vol. 67, no. 4, pp. 575-584, (2020).
[7] M. Ben Mosbah et al., "Boron-Coated Straws Imaging Panel Capability for Neutron Emission Computed Tomography for Source Localization Inside Radioactive Drums,", IEEE Transactions on Nuclear Science, vol. 69, no. 4, pp. 804-810 (2022)
[8] C. Eleon et al., "Boron-Coated Straws Imaging Panel Capability for Passive and Active Neutron Measurements of Radioactive Waste Drums", IEEE Transactions on Nuclear Science, vol. 67, no. 9, pp. 2096-2104 (2020)
[9] M. Delarue et al., "New measurements of cumulative photofission yields of 239Pu, 235U and 238U with a 17.5 MeV Bremsstrahlung photon beam and progress toward actinide differentiation", Nuclear Instruments and Methods in Physics Research Section A: Accelerators, Spectrometers, Detectors and Associated Equipment, Volume 1040, 2022, 167259, ISSN 0168-9002
[10] F. Salvador et al., "Low and high-resolution gamma-ray spectroscopy for the characterization of uranium contamination", Nuclear Instruments and Methods in Physics Research Section A: Accelerators, Spectrometers, Detectors and Associated Equipment, Volume 1065, 2024, 169549, ISSN 0168-9002

[11] J.T.Caldwell, "The Los Alamos second-generation system for passive and active neutron assays of drum-size containers", Tech. Rep., LA-441 10774-MS (1986)
[12] R. Antoni, C. Passard, B. Perot, F. Guillaumin, C. Mazy, M. Batifol, G. Grassi, Reduction of the uncertainty due to fissile clusters in radioactive waste characterization with the Differential Die-away Technique, Nuclear Instruments and Methods in Physics Research Section A: Accelerators, Spectrometers, Detectors and Associated Equipment, Volume 895, 2018,Pages 144-149, ISSN 0168-9002.
[13] T. Beckers, An Introduction to Gaussian Process models, arXiv:2102.05497v1 [eess.SY], 2021
[14] J. Arbel et al., A primer on Bayesian Neural Networks: Review and Debates, arXiv:2309.16314v1 [stat.ML] (2023)
[15] D. B. Pelowitz et al., "MCNP6™ User's Manual, Code Version 6.1.1 Beta", Los Alamos National Laboratory Tech. Rep. LA-CP-14-00745, Los Alamos, NM, USA, (2014)
[16] Draper, N. and Smith, H., Applied Regression Analysis, 2d Edition, New York: John Wiley & Sons, Inc (1981).
[17] F. Pedregosa et al., "Scikit-learn: Machine Learning in Python, Journal of Machine Learning Research", 12, 2825–2830 (2011)
[18] E. Bingham et al., "Pyro: Deep Universal Probabilistic Programming", Journal of Machine Learning Research, 20(28) 1–6 (2019)
[19] A. Paszke et al., "PyTorch: An Imperative Style, High-Performance Deep Learning Library", Advances in Neural Information Processing Systems (2019)

Simulation of Activation Experiments in an Irradiation Stand in the Context of Nuclear Decommissioning

Benjamin Keßler[1,2*], Shokhrukh Mirzo Bakhodirov[1], Pia Kahle[1], Steffen Taut[3], Jörg Konheiser[2], and Thomas Kormoll[1]

[1]Radiation Physics Group, Institute of Nuclear and Particle Physics, Technische Universität Dresden
[2]Department of Reactor Safety, Institute of Resource Ecology, Helmholtzzentrum Dresden Rossendorf
[3]Zentrales Radionuklidlabor, Technische Universität Dresden
(*) benjamin.kessler2@tu-dresden.de

Abstract

Monte Carlo simulations are a valuable tool to determine activated areas in nuclear power plants. This paper describes the simulation of an neutron irradiation stand containing 8 (α,n) sources and polyethylene shielding. The source spectrum of beryllium (α,n) sources is calculated using conservation of energy and momentum as well as the relevant cross-sections. Monte Carlo simulations are performed using a variety of codes with and without thermal scattering data. Simulation and experimental results are within the expected range. The thermal neutron fluence is about $2.2 \times 10^5\,\mathrm{cm^{-2}\,s^{-1}}$. The results provide a reliable basis to simulate the activation of nuclear power plant material samples.

I. INTRODUCTION

The decommissioning of nuclear power plants requires careful identification of activated materials and structures. Being able to calculate the activations enables a faster and more efficient decommissioning. The EBENE (Experimentally supported calculations of neutron fields and the resulting activities in spaces far from the reactor) project aims to determine the activation of concrete far from the reactor core. To facilitate this, it is necessary to be able to accurately simulate the spectral fluence rate of neutrons as well as to account for the composition and activation properties of the concrete. To test the accuracy of the simulation it needs to be verified in a known environment.

In this work the source spectrum of α-Be sources is estimated. The resulting spectrum serves as an input for Monte Carlo calculations in TRAMO [1] and Geant4 [2] to simulate the spectral neutron fluence in an irradiation stand. The simulation results are validated with experimental measurements of the saturation activity of well known samples (Au, Mn, V) in the irradiation stand.

II. DESCRIPTION

A. Setup

The geometry of the irradiation stand is modeled based on construction plans as shown in fig. 1. Sources are mounted to polyethylene (PE) rods and inserted from the front and back side. The sample is securely placed on to a PE column and lowered in from the top. Around the sources and sample is an approximately 50 cm thick layer of PE to thermalize the neutrons and to provide adequate shielding of the surroundings. The steel frame only covers the edges of the inner PE block and outside of the frame there is an additional 8 cm thick layer of PE with boric acid added in to further improve shielding. The simulations have shown that, by the time the neutrons reach the borated PE, the fluence is already reduced by four orders of magnitude. Therefore, the effect of the borated PE and the steel frame are negligible in the simulation of the neutron fluence at the sample position.

In the irradiation stand 8 $^9\mathrm{Be}(\alpha, \mathrm{n})$ sources with a total source strength of $1.1 \times 10^8\,\mathrm{s^{-1}}$ are used. Source-type and dimensions are only given for some sources and have to

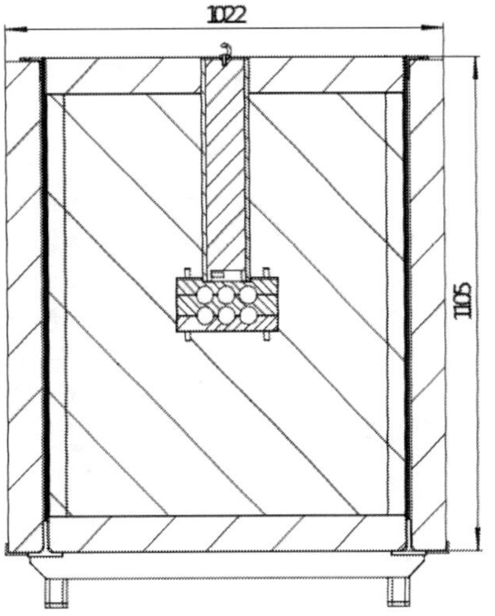

Figure 1: Schematic of the irradiation stand viewed from the front.

be assumed for the remaining ones.

B. Source Spectrum

Neutrons in Be(α, n)-sources are mainly produced by two reactions [3]. There are the $^9Be(\alpha, n)^{12}C$-reactions (1) to (3) that produce three distinct neutron groups depending on the final excitation state of the ^{12}C. Additionally the breakup-reaction $^9Be(\alpha, \alpha n)^8Be$ emits a neutron while the incident α-particle is not captured by the nucleus. This reaction mainly produces neutrons below 1 MeV. The remaining ^8Be also decays into two α-particles which do not have sufficient energy to produce additional neutrons. In general, the neutron spectrum of these sources greatly depends on the energy of the α-particles interacting with the ^9Be which in turn depends on the inner geometry and composition.

$$^9\mathrm{Be} + \alpha \longrightarrow {}^{13*}\mathrm{C} \longrightarrow {}^{12}\mathrm{C} + \mathrm{n}_0 \tag{1}$$
$$^9\mathrm{Be} + \alpha \longrightarrow {}^{13*}\mathrm{C} \longrightarrow {}^{12*}\mathrm{C} + \mathrm{n}_1 \tag{2}$$
$$^9\mathrm{Be} + \alpha \longrightarrow {}^{13*}\mathrm{C} \longrightarrow {}^{12**}\mathrm{C} + \mathrm{n}_2 \tag{3}$$
$$^9\mathrm{Be} + \alpha \longrightarrow {}^{8}\mathrm{Be} + \alpha + \mathrm{n}_b \tag{4}$$

To estimate the source spectrum from the $^9Be(\alpha, n)$ reaction a homogeneous distribution of Be inside the source was assumed. By using conservation of energy and momentum as well as the neutron production cross sections for the three different neutron groups the source spectrum in fig. 2 was computed. Since the $^{13*}C$ is moving relative to the lab the neutron spectrum needs to be Doppler broadened. This was done under the assumption that the neutron is emitted isotropically in the reference frame of the $^{13*}C$.

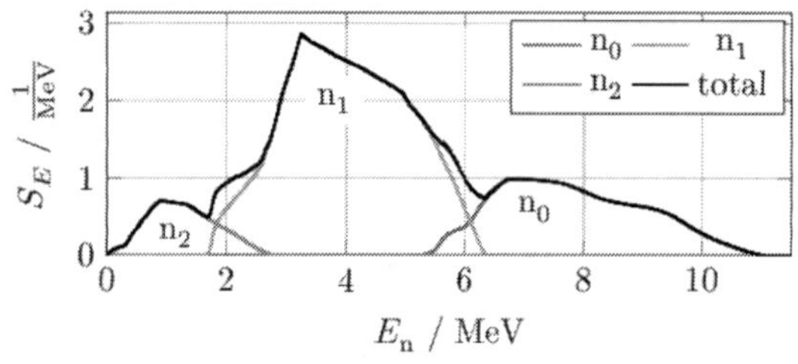

Figure 2: Contribution from the three neutron groups in the source spectrum after doppler broadening.

To determine the spectrum of neutrons produced in the breakup-reaction, conservation of energy and momentum is not sufficient since there are 3 particles left after the reaction. However, it is possible to calculate the number of breakup neutrons using the cross-section. Assuming the shape of the breakup neutron spectrum to fall off linearly until an upper energy limit produces results that are comparable to reference spectra. For ^{241}AmBe-sources it can be calculated that 10.8 % of the neutrons come from the breakup reaction.

The ISO-8529 [4] spectrum for ^{241}AmBe sources and a spectrum measured by the PTB [5], [6] can serve as a reference. Figure 3 shows both references together with our calculated spectrum. It is clear that our spectrum roughly follows the shape of the reference spectra while missing some of the structure. This additional structure is caused by an anisotropy in the emission of the the neutron from $^{13*}C$.

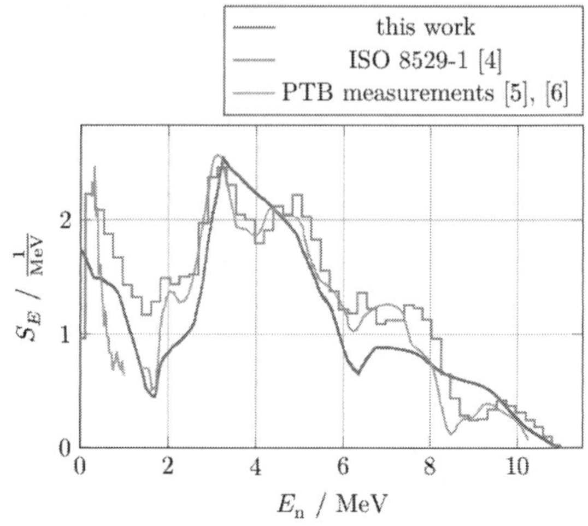

Figure 3: Our calculated ^{241}AmBe spectrum with breakup neutrons compared to the ISO 8529-1 reference and a measurement by the PTB.

C. Simulation

For the Monte Carlo simulation TRAMO and Geant4 version 11.03 with the physics list Shielding as well as NeutronHP were used. Shielding uses the free gas approximation for all isotopes while NeutronHP uses molecule specific thermal scattering data (TSL) for hydrogen in polyethylene [7]. TRAMO needs the cross-sections to be prepared using NJOY [8] and converted to a specific format. TRAMO provides reliable results when using the free-gas approximation, but its accuracy has not been thoroughly validated for cross-sections derived from TSL data.

Figure 5 shows the spatial distribution of the thermal fluence rate simulated by TRAMO on a plane through the center of the irradiation stand. The sample holder is marked in white. At the sample holder a fluence rate of $4.1 \times 10^5\,\mathrm{cm^{-2}\,s^{-1}}$ is calculated, with $2.2 \times 10^5\,\mathrm{cm^{-2}\,s^{-1}}$ in the thermal region below 200 meV.

D. Activation of Samples

For activation experiments samples that undergo β-decay quickly followed by the emission of a photon have been selected. This allows the use of a β- as well as a γ-detector in coincidence mode to measure the absolute activity of the sample without the need for an efficiency calibration. [9]

The simulations are done by placing a sample volume with the approximate shape of the sample and correct material in the Geant4 simulation. Geant4 allows to both record the spectral neutron fluence as well as the actual activation events in the simulation. While the recorded activation events are useful, using the spectral fluence output together with the activation cross-section reduces the uncertainty since all neutrons that enter the sample

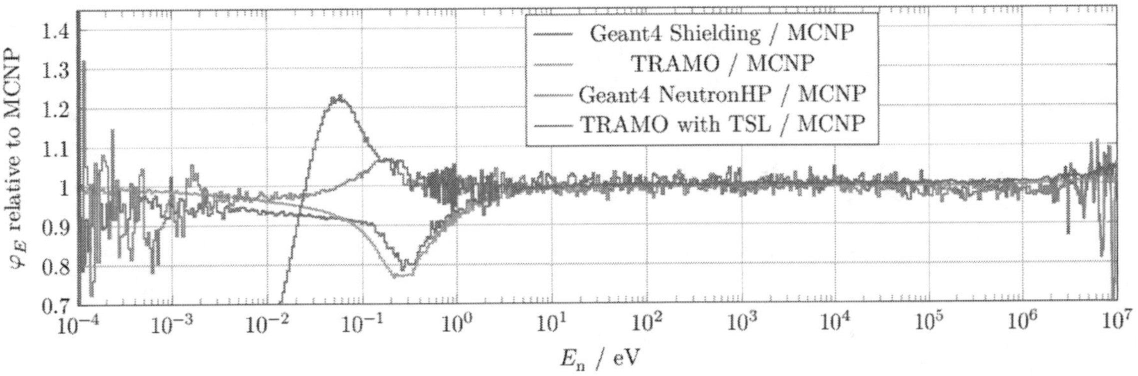

Figure 4: Spectral neutron fluence rate at the sample holder in a simplified geometry with polyethylene shielding compared to a calculation in MCNP with TSL data.

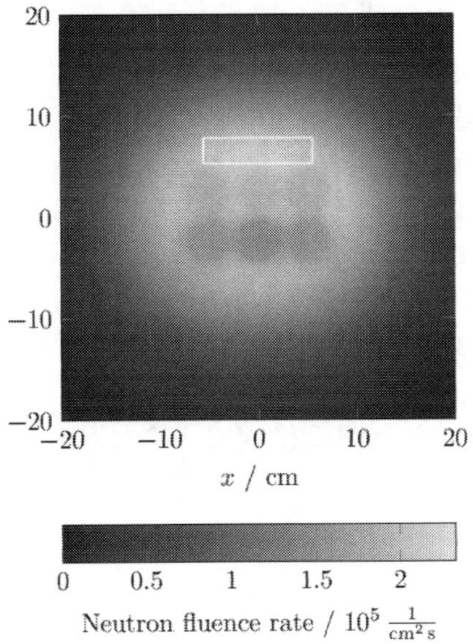

Figure 5: Map of the thermal neutron fluence rate below 200 meV inside the irradiation stand, simulated with TRAMO. The white outline marks the location of the sample holder.

volume contribute to the statistics. The activation rate A is then calculated using the number density n, volume V and the activation cross-section σ in (5). In saturation the activity of the sample is equal to the activation rate.

$$A = n \int \mathrm{d}E\,\sigma(E) \iiint_V \mathrm{d}^3\vec{r}\,\varphi_E(\vec{r}, E) \tag{5}$$

Self shielding effects reduce the mean neutron fluence in the sample and are therefore already accounted for by using the correct sample material in the simulation. However, since the neutron fluence is divided into energy groups it is important that the groups are fine enough to adequately represent the structure created by self shielding, otherwise the estimated activation rate will be too high. Table 1 shows the measured and simulated saturation activities of the samples.

Table 1: Experimental and simulated saturation activity with and without cadmium shielding

	Saturation Activity / kBq	
Sample	Experiment	Simulation
Au	56.8 ± 0.9	48.03 ± 0.25
Au with Cd	12.60 ± 0.20	4.00 ± 0.07
Mn	4.15 ± 0.05	7.95 ± 0.06
Mn with Cd	0.081 ± 0.001	0.236 ± 0.005
V	5.03 ± 0.07	5.357 ± 0.024
V with Cd	0.0195 ± 0.0006	0.1084 ± 0.0011

III. DISCUSSION

Even with the minimal approach of using conservation of energy and momentum and cross-sections it is still possible to calculate a satisfactory spectrum for ^{9}Be(α, n)-sources. Breakup neutrons have to be treated separately from this calculation. Because the scattering of neutrons in the irradiation stand tends to wash out the finer structure of the spectrum this simple approach appears to be adequate.

The discrepancy in the fluence rate between Geant4 and TRAMO on the poster could be traced back to a coding error on our side. The corrected Geant4 Shielding simulation shows much better agreement with TRAMO and MCNP in fig. 4. TRAMO with TSL data produces a similar behavior to Geant4 NeutronHP, however below 150 meV there are major discrepancies. The comparison with MCNP on a simplified model showed that improvements to TRAMO are necessary.

By comparing the activation measurement results with the simulation in table 1 it can be seen that without cadmium shielding the ratio between simulation and experiment is between 80 % and 200 %. Since it is hard to estimate the systematic uncertainties caused by differences between the construction plans and the actual irradiation stand the table only include statistical uncertainties, therefore it is to be expected that the differences can not be fully explained by the uncertainty. With cadmium the differences between simulation and experiment increases. Because cadmium has a very high capture cross-section for thermal neutrons it prevents almost all thermal neutrons from reaching the sample. Therefore the additional differences could be caused by a gradient in the neutron fluence rate at intermediate energies or uncertainties with the spectrum of the breakup neutrons.

IV. CONCLUSION AND OUTLOOK

Both Geant4 and TRAMO are well fitted for the simulation of intermediary neutrons and are well in agreement with MCNP. For the thermal region Geant4 produces good results while TRAMO still appears to have problems when using TSL data and needs more work. In the context of the EBENE project the neutron fluence far from the reactor core will mostly consist of thermal neutrons. Therefore both TRAMO without TSL data as well as Geant4 should be sufficient for the simulation.

The next steps in the project will be to simulate the activation of concrete samples in the irradiation stand to have a reference for the activation measurements of the concrete.

This work was funded by the Federal Ministry of Education and Research (BMBF) under contract number 15S9447A.

References

[1] H. Barz and J. Konheiser, "The Monte Carlo code TRAMO - capabilities and instructions for application," Dec. 1998.

[2] J. Allison et al., "Recent developments in geant4," *Nuclear Instruments and Methods in Physics Research Section A: Accelerators, Spectrometers, Detectors and Associated Equipment*, vol. 835, pp. 186–225, 2016, ISSN: 0168-9002. DOI: https://doi.org/10.1016/j.nima.2016.06.125. [Online]. Available: https://www.sciencedirect.com/science/article/pii/S0168900216306957.

[3] S. Croft, "The use of neutron intensity calibrated ^{9}Be(α, n)-sources as 4438 keV gamma-ray reference standards," 1989.

[4] I. O. for Standardization, *ISO 8529-1 – Reference Neutron Radiations – Part 1: Characteristics and Methods of Production.* Mar. 2000.

[5] H. Zill, "Das neutronenspektrum radioaktiver Be(α, n)-quellen im energiebereich unterhalb 1 MeV," *Zeitschrift Für Naturforschung Section A*, vol. 24, Jun. 1969. DOI: 10.1515/zna-1969-0822.

[6] H. Kluge, "Das neutronenspektrum radioaktiver Be(α, n)-quellen im energiebereich oberhalb 1 MeV," *Zeitschrift Für Naturforschung Section A*, vol. 24, Jun. 1969. DOI: 10.1515/zna-1969-0823.

[7] L. Thulliez, C. Jouanne, and E. Dumonteil, "Improvement of geant4 neutron-hp package: From methodology to evaluated nuclear data library," *Nuclear Instruments and Methods in Physics Research Section A: Accelerators, Spectrometers, Detectors and Associated Equipment*, vol. 1027, p. 166 187, 2022, ISSN: 0168-9002. DOI: https://doi.org/10.1016/j.nima.2021.166187. [Online]. Available: https://www.sciencedirect.com/science/article/pii/S016890022101055X.

[8] J. Conlin, A. Kahler, A. McCartney, and D. Rehn, "Njoy21: Next generation nuclear data processing capabilities," *EPJ Web of Conferences*, vol. 146, p. 09 040, Jan. 2017. DOI: 10.1051/epjconf/201714609040.

[9] S. M. Bakhodirov, A. Seifert, D. Döhler, P. Kahle, and T. Kormoll, "Determination of absolute activities and neutron fluence rates using a coincidence method," *EPJ Web Conf.*, 2025.

Towards Clinical Integration of a SPECT Detection Module for BNCT Dose Monitoring

F. Ghisio[1,2,*], T. Ferri[1,2], A. Caracciolo[1,2], S. Agosteo[2,3], S. Altieri[5,6], C. Fiorini[1,2], M. Carminati[1,2], L. Grisoni[2,3], G. Iaselli[7,8], Y. Ichikawa[4], D. Mazzucconi[2,3], N. Protti[5,6], L. Santini[1,2], K. Tsuchida[4], and G. Borghi[1,2]

***Abstract*—Boron Neutron Capture Therapy (BNCT) is a targeted radiotherapy that requires precise dose monitoring to ensure effective treatment. We initially report experimental measurements performed with the BeNEdiCTE detector, designed for SPECT-based dose monitoring in BNCT. The detector is based on a co-doped LaBr$_3$(Ce, Sr) monolithic crystal read out by Silicon Photomultipliers and compact electronics. Tomographic measurements were performed at the Laboratorio di Energia Nucleare Applicata in Pavia (Italy). Then, measurements with neutron conditions similar to those expected in clinical treatments were performed at the Nagoya University Accelerator-driven Neutron Source (Japan). The detector demonstrated the ability to detect the 478 keV gamma rays emitted during boron neutron capture reactions. Additionally, the detector successfully reconstructed 2D and 3D images of the boron distribution with sub-centimeter resolution. The measurements highlighted key areas for improvement to make the detector suitable for clinical use with patients, which will be presented in detail in this work. The boron present in the FR4 material used in the electronics boards was identified as a significant source of 478-keV gamma rays uncorrelated with the imaged object. To mitigate this, a new substrate material, RO3003, was tested, showing a substantial reduction in gamma rays emitted at 478 keV. Furthermore, the detector's electronics were optimized to improve count-rate capability. This involved reducing the dead time from 5.4 µs to 2.1 µs and refining an energy thresholding mechanism to process only events around 478 keV. These improvements will enhance the detector's performance, bringing it closer to clinical use with patients.**

***Index Terms*—boron neutron capture therapy, dose monitoring, spect detection module**

I. Introduction

BORON Neutron Capture Therapy (BNCT) is a targeted radiotherapy that exploits the selective accumulation of boron-10 compounds in tumor cells. Upon irradiation with an epithermal neutron beam, the ^{10}B atoms undergo a neutron capture reaction that produces high-linear-energy-transfer (high-LET) particles capable of destroying the tumor cells [1]. Thanks to the short range of these particles, BNCT enables micrometric precision treatment while sparing the surrounding healthy tissue [2].

[1] Dipartimento di Elettronica, Informazione and Bioingegneria, Politecnico di Milano, Milano, 20133, Italy.
[2] Istituto Nazionale di Fisica Nucleare, Sezione di Milano, Italy.
[3] Dipartimento di Energia, Politecnico di Milano, Milano, 20156, Italy.
[4] Graduate School of Engineering, Nagoya University, Furo-cho, Chikusa-ku, Nagoya 464-8603, Japan.
[5] Dipartimento di Fisica, Università di Pavia, Pavia, 27100, Italy.
[6] Istituto Nazionale di Fisica Nucleare, Sezione di Pavia, Italy.
[7] Dipartimento di Fisica, Politecnico di Bari, 70126, Bari, Italy.
[8] Istituto Nazionale di Fisica Nucleare, Sezione di Bari, Italy.
* Corresponding author: filippo.ghisio@polimi.it

The dose delivered to the patients depends on the local boron concentration and the neutron flux reaching the tumor. These quantities are currently estimated only before or after treatment. However, to exploit the selectivity of BNCT, real-time monitoring of the distribution of boron-neutron capture reactions is essential. A promising non-invasive solution is a SPECT imaging system detecting the 478 keV gamma rays emitted during the capture reactions [3]. To this purpose, the BeNEdiCTE detector was developed [4], [5].

In this work, we report some of the experimental measurements performed with the BeNEdiCTE detector in two facilities in Italy and Japan. We then describe in detail significant improvements performed on the BeNEdiCTE detector toward its clinical integration. These include the selection and test of boron-free materials for the electronics to minimize background, together with the development of faster electronics and a refined energy discrimination circuit to handle high rates.

II. System Architecture

The BeNEdiCTE detector features a co-doped lanthanum bromide monolithic scintillator crystal, i.e. LaBr$_3$ (Ce^{3+} + Sr^{2+}), provided by Saint Gobain. The crystal measures $5 \times 5 \times 2\ cm^3$ and it is optically coupled to an 8×8 matrix of Near Ultraviolet High-Density (NUV-HD) silicon photomultipliers (SiPMs) [6]. The SiPMs are from Fondazione Bruno Kessler (FBK), Trento, Italy and have a $30\ \mu m \times 30\ \mu m$ cell size. The SiPM signals are processed by four 16-channel GAMMA ASICs [7], developed at Politecnico di Milano, and subsequently digitized by four analog-to-digital converters (ADCs). The detector achieves a high energy resolution of approximately 2.9% at $662\ keV$ and 4% at $478\ keV$. This energy resolution is sufficient to distinguish the BNCT photopeak from the nearby $511\ keV$ photopeak of annihilation photons [8]. The $2\ cm$ thickness of the crystal ensures a good detection efficiency, reaching about 60% at $478\ keV$.

The system also includes a custom-designed channel-edge pinhole collimator, which ensures that the incoming radiation is effectively shaped and directed toward the BeNEdiCTE module to obtain an image. The collimator is made of lead and has dimensions of $9.4 \times 9.4 \times 10\ cm^3$, with a hole diameter of $0.5\ cm$ and an aperture angle of $9.52°$. Finally, the system features a neural network designed to reconstruct the interaction positions of the gamma rays on the LaBr$_3$ crystal [9]. The network architecture consists of an input layer with 64 neurons followed by two hidden layers with 25 neurons each, and an output layer with 2 neurons to predict the x and y coordinates.

III. Experimental Measurements

We performed experimental measurements with the BeNEdiCTE detector in two facilities in Italy and Japan. Tomographic measurements were performed at the Laboratorio di Energia Nucleare Applicata (LENA) in Pavia (Italy). Then, measurements with neutron beams similar to those expected in clinical treatments were performed at the Nagoya University Accelerator-driven Neutron Source (NUANS) facility (Japan).

At the LENA facility, a TRIGA MARK II reactor was used to irradiate two cylindrical vials with a height of 26.5 mm and an internal diameter of 6 mm positioned 1.4 cm apart. The vials were aligned with the neutron flux, while the detector and collimator were rotated around the vials at 0°, 60°, 120°, and 180° with respect to the beam axis. Both the sample-collimator and collimator-detector distances were set at 15 cm. In order to avoid activation of the readout electronics and scintillator crystal from thermal neutrons, a cadmium shielding case was employed around the detection module. The reactor power was set at 250 kW, corresponding to a neutron flux of approximately 3.14×10^6 $n/cm^2/s$, and the vials were irradiated for 10 minutes. During the measurements, two images were acquired for each of the four detector positions: an image of the vials containing the boron solution and an image obtained by replacing the two boron vials with identical samples filled only with distilled water.

The system demonstrates strong capabilities both in detecting the targeted radiation and in reconstructing 2D planar images of boron-containing vials. Using the Software for Tomographic Image Reconstruction (STIR), we successfully obtained a 3D reconstruction of two vials containing 7371 ppm of ^{10}B. Figure 1 shows the top view of the 3D reconstruction. This was accomplished using the four partial projections and 10 iterations of the Maximum Likelihood Expectation Maximization (MLEM) reconstruction algorithm.

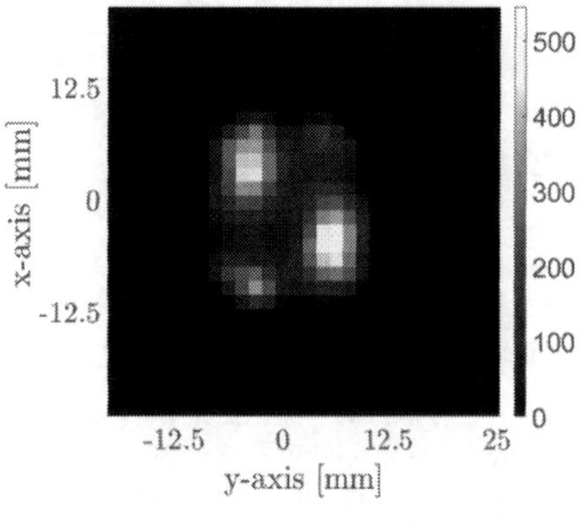

Figure 1. Top view of the 3D reconstruction of the two vials filled with 7371 ppm of ^{10}B, obtained with 10 iterations of the MLEM algorithm.

At the Nagoya University Accelerator-driven Neutron Source in Nagoya, Japan, the neutron beam was generated using a proton accelerator. The accelerator was set at 200 μA, resulting in a neutron flux at the sample position of approximately 1×10^7 $n/cm^2/s$, almost an order of magnitude higher than the one reached at LENA. Moreover, unlike the neutron beam at LENA, the neutrons used in this case are epithermal. Thus, we used a $20 \times 20 \times 20$ cm^3 acrylic box filled with distilled water as a phantom, and a cylindrical vial with a height of 26.5 mm and an internal diameter of 6 mm was inserted into the phantom. Different solutions with ^{10}B dissolved in distilled water were tested, down to a concentration of 1843 ppm. Acquisitions were also performed using a vial filled with distilled water only. The detector was positioned perpendicularly to the direction of the beam, at a distance of 50 cm from the vial, and the collimator was placed between the detector and the phantom. To minimize background around the energy of interest of 478 keV, shielding was designed to address emissions from neutron interactions with both the detector and the surrounding environment. The shielding was composed of three successive layers of polyethylene, cadmium, and lead surrounding the detector and the collimator.

The irradiation times vary among vials with different boron concentrations mainly because of accelerator instabilities, with the irradiation times chosen to minimize fluctuations and ensure stable conditions. The initial spectra obtained by irradiating a vial filled with distilled water only revealed a high gamma-ray background, also at 478 keV. This 478-keV background was due to the activation of the boron contained in the material used for printed circuit boards (PCBs) and in the borated polyethylene panels installed on the walls for radiation-protection purposes. An improvement strategy to address this issue will be discussed in Section IV.

Despite this issue, the system successfully detected an increase in the number of gamma rays at 478 keV when the borated vials were introduced. This has been achieved by reconstructing the energy spectrum using only the events identified by the neural network as originating within the collimator's FOV. Figure 2 shows the increase in the number of gamma rays emitted at 478 keV by the borated vial with a concentration of 3686 ppm, irradiated for 32 minutes, with respect to the vial filled with distilled water only.

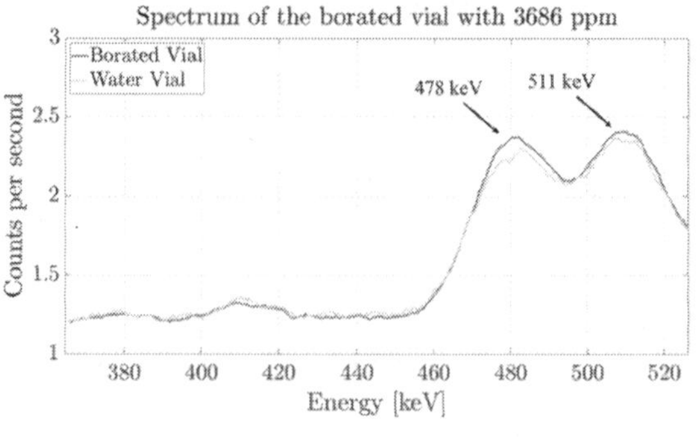

Figure 2. Energy spectrum showing the increase in 478 keV gamma rays for the vial with 3686 ppm compared to the vial filled with water only.

Furthermore, the system successfully imaged borated samples with different concentrations of ^{10}B. The image of the borated vial with a concentration of 3686 ppm is shown in Figure 3. This image was obtained by selecting the events under the 478 kev peak and feeding them to the neural network. Background subtraction was performed by removing the image of the water-filled vial from the boron source image. A Gaussian filter was then applied to reduce noise, and all events outside the 4.16 cm circular FOV were set to zero to highlight the signal of interest.

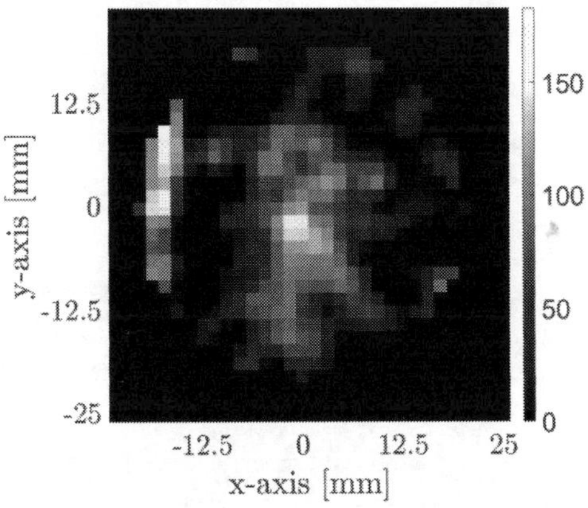

Figure 3. Reconstructed images of the borated vial with a boron concentrations of 3686 ppm irradiated for 32 minutes.

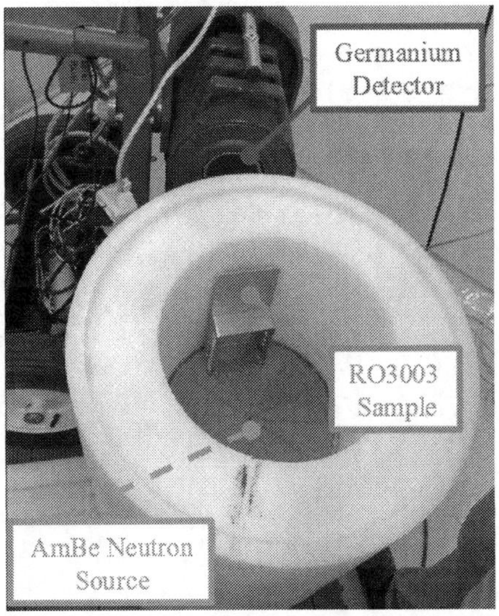

Figure 4. Experimental setup for the irradiation of the FR4 and RO3003 samples with an AmBe source. The germanium detector that captures the emitted gamma rays is also shown.

IV. Design Optimization Toward Clinical Integration

The measurements described in Section III have provided promising results, confirming the detector's potential for dose monitoring during a BNCT treatment. However, they also highlighted some limitations that need to be addressed before testing the detector in a real clinical facility in conditions equivalent to those present during patient treatments.

One key limitation is background interference at 478 keV. Monte Carlo simulations revealed that a significant source of this background comes from the boron in the Flame Retardant 4 (FR4) substrate material used in the detector's electronics. Indeed, FR4 is made of a combination of glass fiber and epoxy resin and contains boron trioxide (B_2O_3), with a boron content typically ranging from 5% to 10% by weight [10].

To address this issue, a new boron-free substrate for the PCBs has been identified, which is the RO3003 Laminate, produced by Rogers Corporation. This is an unreinforced material that does not contain glass fibers and lacks boron trioxide. Experimental measurements were conducted to compare boron content in FR4 and in this new material. Samples of both materials were exposed to a neutron field generated by an Americium Beryllium source and the resulting activation spectra were measured using a Germanium detector. Figure 4 shows the experimental setup.

The two samples measured 10×10 cm^2 and were irradiated for 10 minutes each. The resulting spectra obtained with the germanium detector are shown in Figure 5. A clearly evident peak at 478 keV can be observed when the FR4 sample was irradiated. In contrast, no peak at 478 keV was detected with the RO3003 sample. The broad shape of the 478 keV peak that can be seen in the figure is due to the fact that the gamma photon is emitted by a ^{7}Li nucleus in motion, immediately after the neutron capture reaction. Because the nucleus is recoiling at the moment of emission, the detected gamma energy is slightly shifted depending on the emission direction, resulting in Doppler broadening that is clearly visible thanks to the high resolution of the germanium detector.

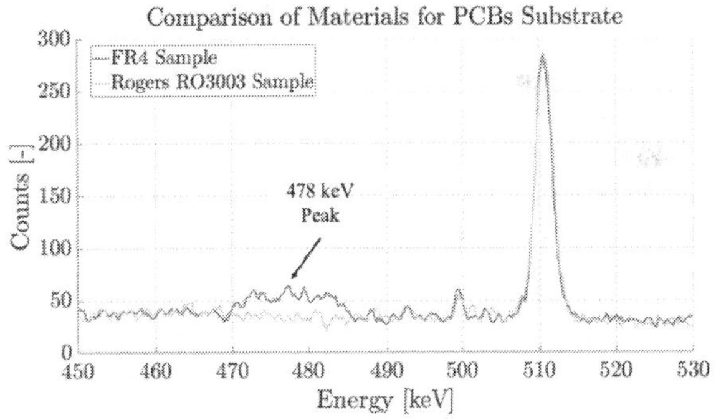

Figure 5. Comparison of the spectra obtained with a germanium detector when irradiating the FR4 and RO3003 samples. The 478 keV peak is broadened due to the Doppler effect.

Another limitation is the maximum count rate the detector can handle. At the Nagoya facility, the detector reached its limit at a neutron flux of 1×10^7 $n/cm^2/s$, about two orders of magnitude lower than the flux expected during patient treatment. Therefore, the electronics must be significantly improved to handle the higher rates expected in clinical scenarios.

To this purpose, a key focus will be the refinement of an energy discrimination mechanism that would limit the events to be processed only around the peak of interest. This mechanism employs an electronic circuit that monitors the total current from the SiPMs and compares it to two configurable thresholds, enabling signal integration and event processing only if the detected signal falls within a predefined energy window. A first version of this system was tested and provided important results, as highlighted in Figure 6. However, some areas for improvement have been identified to narrow the energy window and reduce the number of valid events, with the aim of processing only those corresponding to the 478 keV peak. By a sharp windowing of the signals around 478 keV, we expect to process less than 10% of the total events in the spectrum, significantly increasing the count rate capability.

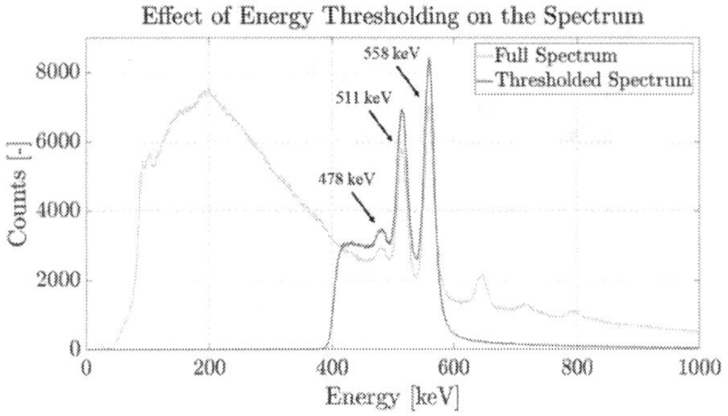

Figure 6. Comparison of spectra obtained with and without the use of the energy thresholding mechanism during a measurement at LENA.

To further improve the count rate, it is crucial to reduce the detector's dead time, which is the minimum time required between two events to register them as separate. To this purpose, the four ADCs used in the signal processing chain were replaced with faster models. Also, the integration time of the ASICs that process the SiPM signals was optimized. These improvements reduced the dead time from 5.4 μs to 2.1 μs.

The effect of the dead time can be seen by comparing the true interaction rate, also known as input count rate (ICR), with the recorded count rate, also known as output count rate (OCR). To have a precise estimation of this effect, we conducted experimental measurements using an event simulator from Nuclear Instruments to generate random pulses with Poissonian distribution that simulate those created by SiPMs during gamma-ray detection. The pulses were delivered to the BeNEdiCTE detector at a range of known input ICRs, recording the corresponding OCRs for each of them. This experiment has been repeated both with the configuration with a dead time of 5.4 μs and with the one with a dead time of 2.1 μs.

The results of the experiment are shown in Figure 7. The violet curve, representing the configuration with shorter dead time, is closer to the ideal behavior of a detector with no dead time. Additionally, the gap between the curves representing the two configurations increases at higher count rates, indicating that the improvement becomes more significant as the count rate rises. According to [11], measurement conditions where dead time losses exceed 30 − 40% should be avoided, as the errors associated with correcting these losses can be significant. The new configuration improves this limit, increasing the threshold where 30% dead time losses occur from 80 kHz to 200 kHz.

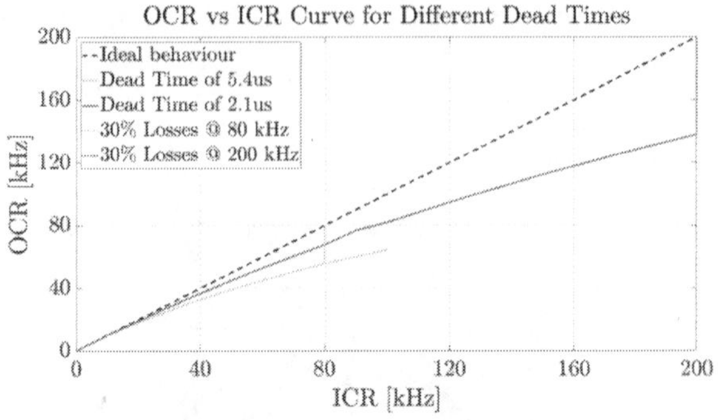

Figure 7. Comparison between the configuration with a dead time of 5.4 μs and the one with a dead time of 2.1 μs. The curves display the recorded output count rates versus the true input count rate. The values of ICR where dead time losses reach 30% is also highlighted for the two configurations.

We finally conducted experimental measurements in our laboratory at Politecnico di Milano using ^{133}Ba and ^{22}Na sources to further assess the improvements brought by the new electronics. Two measurements were performed by keeping the detector and the radioactive sources in the same position and changing the electronics boards to test both the configuration with a 5.4 μs dead time and the one with a 2.1 μs dead time. As shown in Figure 8, a higher number of events was recorded with the 2.1 μs dead time configuration, indicating that many events were missed when the detector had a higher dead time. The increase in the detected count rate, i.e. the OCR, which passes from 27.1 kHz to 30.3 kHz for this measurement, is consistent with theoretical expectations and further confirms the improvements introduced by the new electronics.

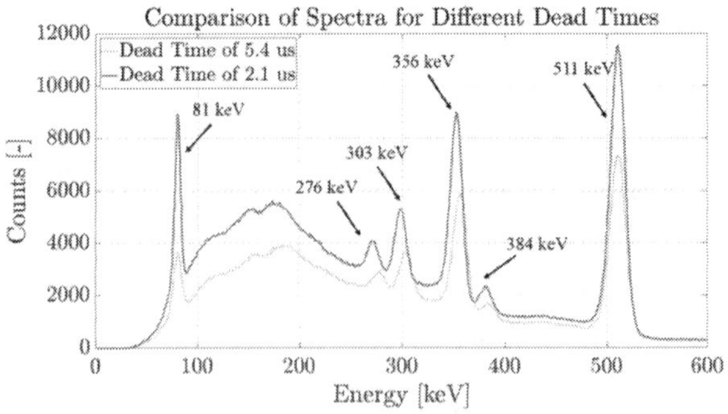

Figure 8. Spectra obtained irradiating the BeNEdiCTE detector with ^{133}Ba and ^{22}Na for different dead time configurations.

V. Conclusions

The BeNEdiCTE detector has shown promising results for real-time dose monitoring in BNCT, a critical requirement for maximizing treatment precision and safety. Through experimental measurements conducted in two facilities in Italy and Japan, the system has shown the capability of detecting the 478 keV gamma rays associated with boron neutron capture reactions. Furthermore, the system has been able to image borated vials with sub-centimeter resolution, providing also a 3D reconstruction of the samples.

The experimental measurements also served to highlight critical areas requiring optimization for clinical integration. Notably, the boron content in the commonly used FR4 substrate for PCBs was identified as a major source of 478-keV gamma rays uncorrelated with the samples. This insight guided the selection and successful testing of the boron-free RO3003 substrate, which exhibited no detectable activation at the energy of interest. Furthermore, key electronic upgrades further enhanced the system's performance. Reducing the dead time from $5.4\ \mu s$ to $2.1\ \mu s$, along with implementing an energy thresholding mechanism focused on the $478\ keV$ photopeak, enabled the system to manage higher event rates.

The improvements introduced have shown promising results but they have been validated only under controlled laboratory conditions. Therefore, future work will focus on testing these enhancements in a dedicated facility to verify their effectiveness in more realistic and clinically relevant environments. In any case, these advancements move the detector closer to clinical integration, enabling more precise and effective BNCT treatments.

VI. Acknowledgment

Part of this work has been financed by INFN (CSN5), Italy (SPOC Project), and by the PRIN PNRR 2022 funding program of the Ministero Dell'Università e della Ricerca (MUR), Italy (BNCT-SPECT project).

References

[1] W. Sauerwein, A. Wittig, R. Moss, and Y. Nakagawa, *Neutron Capture Therapy: Principles and Applications*. Springer Berlin Heidelberg, 2012.

[2] M. Suzuki, "Boron neutron capture therapy (BNCT): A unique role in radiotherapy with a view to entering the accelerator-based BNCT era," *International Journal of Clinical Oncology*, vol. 25, no. 1, 2020.

[3] T. Kobayashi, Y. Sakurai, and M. Ishikawa, "A noninvasive dose estimation system for clinical BNCT based on PG-SPECT," *Medical Physics*, vol. 27, no. 9, 2000.

[4] A. Caracciolo, T. Ferri, G. Borghi, *et al.*, "A compact scintillator-based detector with collimator and shielding for dose monitoring in boron neutron capture therapy," *Physics and Imaging in Radiation Oncology*, vol. 29, 2024.

[5] T. Ferri, A. Caracciolo, G. Borghi, *et al.*, "Imaging study and first measurements of a labr3 gamma detector for bnct applications," *IEEE Transactions on Nuclear Science*, vol. 71, no. 5, 2024.

[6] A. Gola, F. Acerbi, M. Capasso, *et al.*, "NUV-Sensitive Silicon Photomultiplier Technologies Developed at Fondazione Bruno Kessler," *Sensors*, vol. 19, no. 2, 2019.

[7] L. Buonanno, D. D. Vita, M. Carminati, and C. Fiorini, "Gamma: A 16-channel spectroscopic asic for sipms readout with 84-db dynamic range," *IEEE Transactions on Nuclear Science*, vol. 68, no. 10, 2021.

[8] A. Caracciolo, T. Ferri, G. Borghi, *et al.*, "A compact scintillator-based detector with collimator and shielding for dose monitoring in boron neutron capture therapy," *Physics and Imaging in Radiation Oncology*, vol. 29, 2024.

[9] T. Ferri, F. Rosellini, A. Caracciolo, *et al.*, "Gamma-Ray Position-of-Interaction Estimation in a Thick Monolithic $LaBr_3$ Detector Using Artificial Neural Networks," *IEEE Transactions on Radiation and Plasma Medical Sciences*, vol. 9, no. 3, 2025.

[10] G. Colombo, A. Caracciolo, D. Mazzucconi, *et al.*, "Study of the thermal neutron activation of a gamma-ray detector for BNCT dose monitoring," *Journal of Instrumentation*, vol. 19, no. 05, 2024.

[11] G. F. Knoll, *Radiation Detection and Measurement*, eng, 4th ed. Hoboken, NJ: Wiley, 2010.

Gamma ray monitors for MultiMod'Air, a mobile environmental monitoring platform

O. Limousin[1,*], C. Colin[2], G. Daniel[3], A. Deschamps[4], C. Fisher[5], S. Foucambert[1], V. Gros[6], G. Guthertz[4], T. Hedde[7], M. Mayne[8], S. Morilhat[7], L Mugherli[8], G. Trannoy[5], J. Trincal[7], A. Vanel[1]

[1] Université Paris-Saclay, Université Paris Cité, CEA, CNRS, AIM, F-91191 Gif-sur-Yvette, France
[2] CEA, DES, IRESNE, DEC, Cadarache F-13108 Saint-Paul-Lez-Durance, France
[3]Université Paris-Saclay, CEA, SGLS, F-91191, Gif-sur-Yvette, France
[4] Airparif, The air quality observatory for the Paris region, Paris, France
[5]Université Paris-Saclay, CEA, List, F-91120, Palaiseau, France
[6]Laboratoire des Sciences du Climat et de l'Environnement (LSCE), IPSL, CEA-CNRS-UVSQ, Université Paris-Saclay, Orme des merisiers, Gif sur Yvette, France
[7]CEA, DES, IRESNE, DTN, Cadarache F-13108 Saint-Paul-Lez-Durance, France
[8]Université Paris-Saclay, CEA, NIMBE, F-91191, Gif-sur-Yvette, France

(*) olimousin@cea.fr

Abstract— **Multimod'Air is a demonstrator of a mobile and energy self-sufficient platform for environmental measurement in any place of interest from cities, rural or industrial outdoor areas. The platform comprises a dozen of detectors including gaseous pollutant and fine particle sensors for pollution monitoring, as well as a complete weather station (rainfall intensity, wind speed, atmospheric pressure, relative humidity, temperature, solar irradiance and more) and gamma-ray monitors. The paper is specifically devoted to the gamma ray detectors for radiation monitoring with a focus on their architecture, real time and offline data analysis process and performance. Two different and complementary gamma ray monitors have been designed and integrated. The first prototype operates a Kromek GR1+ detectors while the second relies on the advanced multichannel Caliste-O detector technology initially designed for astrophysics applications and recently used for nuclear monitoring ground applications. Both detectors' electronics are coupled to a local ODROID XU4 microcomputer for data acquisition, management and communication. While GR1+ detector sends an integrated spectrum every minute, Caliste-O both sends aggregated spectra over all channels every minute and photon event list on demand. The two autonomous prototypes have been successfully tested outdoors, and the results of the measurements are presented.**

Keywords **—Gamma-ray spectrometry, anomaly detection, natural radioactivity, radon.**

I. Introduction

Multimod'Air is a CEA internal cross-disciplinary project aiming at developing a demonstrator of a mobile and energy self-sufficient platform for environmental measurements in any place of interest from cities, rural or industrial outdoor areas. Environmental real time monitoring, especially dedicated to outdoor air quality, is crucial for health as pollutants and particles are known to cause diseases and cancers [1-4]. Normative gases such as nitrogen oxides (NO_x), sulfur dioxide (SO_2), ozone (O_3), and fine particles (PM2.5, PM5) are conventionally monitored and mapped in fixed stations, mainly in cities, by different agencies responsible for monitoring air quality and providing regulation information as operated by AirParif for the greater Paris region. However, some areas are not monitored and could benefit of a temporary or permanent real-time diagnostic by means of a transportable platform for expertise such as industrial areas or rural places and fields during periods of fertilizer spreading for instance [5]. Moreover, some emerging pollutants are not monitored such as NH_3 even though they are known for their potential risk [6-8]. The reason is due to the difficulty to measure continuously and quantitatively such pollutants. Indeed, getting accurately low concentration values of such reactive pollutants through real-time measurements on site is still a challenge. Therefore, Multimod'Air project has been designed to meet this challenge, providing the community a light-weight, flexible, transportable and accurate environmental platform for gases and particles diagnostics. The platform relies on low-cost sensors and on advanced sensors, packed together into a single autonomous and multimodal system. The low-cost sensors are calibrated with respect to reference sensors and trained to correct their data using AI correction algorithms in real time. The advanced sensors bring new insights on environmental risks. Ultimately, the platform can be duplicated, customized and easily positioned in strategic places to map areas.

In addition to the gases and particle sensors, MultiMod'Air hosts a comprehensive deployable weather station as well as a natural radioactivity background monitor. The latter demonstrate the versatility and multimodal capability of the platform in terms of architecture. The platform is engineered to

 https://doi.org/10.1051/epjconf/202533809002

receive any type of additional sensor which can be hot plugged to its central computer ("central node") and operated independently and autonomously. Any additional technique can be added to the system in case of specific needs in given areas, such as water analysis, seismic monitors, magnetometers, ultrasonic sensors, brightening detectors, fire sensors, etc. All data processed in real time and stored in a cloud database can be corrected, calibrated, combined, and cross-analyzed, either online or offline.

The paper is specifically devoted to the gamma ray detectors for radiation monitoring. Their architecture, real time and offline data analysis process and performance are reported with examples.

Currently the detectors are deployed outdoor on two monitoring stations for evaluation in real conditions in Paris (AirParif station) and at CEA LSCE SIRTA station in a suburban area [9, 10]. Some observations after several months of continuous data acquisition are illustrated in the following sections.

II. MultiMod'Air Architecture

A. *Architecture*

MultiMod'Air platform prototypes currently comprise a dozen of detectors including gaseous pollutant sensors (NO_2, O_3, NH_3, N_3, Cl_2, O_x, H_2S, SO_2) and fine particle sensors (PM1, PM2.5, PM10) for pollution monitoring, as well as a complete weather station and gamma-ray monitors. All sensors are connected via RS485, UART, Analog, Ethernet, WiFi or Lora protocols to a central node through a microcomputer detailed further.

Cable-connected detectors can be deployed up to 30 meters away from their central node offering flexibility to fit any local installation constraints. Wireless systems can be deployed further.

B. *Central node computer*

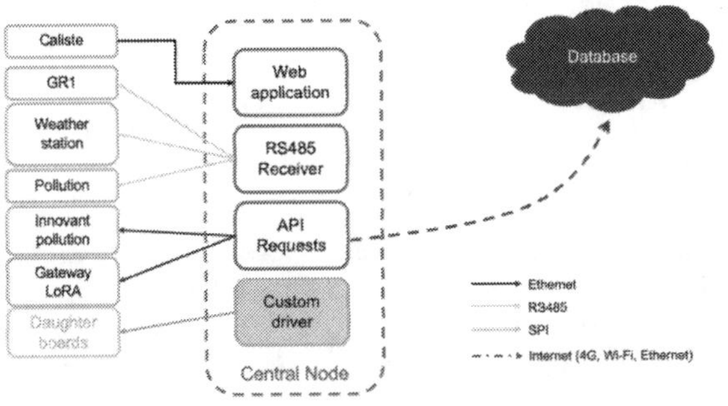

Fig. 1. MultiMod'Air architecture. The sensor boxes and applications (green frame) communicate to the central node with a variety of implemented protocols. The central node not only manage the data acquisition of the detectors but also performs real time calculation on the fly and preprocesses the data when needed (calibration, anomaly detection, time synchronization, …). The data or reduced data are pushed to a database into a secured cloud.

The "central node" computer is equipped with a full custom board based on NXP i.MX8 Cortex A-53 processor units with dedicated AI accelerators (NPU), that processes and combines the data in real time, involving conventional and/or AI algorithms. The raw data, the preprocessed data and the alerts are sent to a cloud server database accessed from anywhere by authorized users by means of a dedicated Graphical User Interface (GUI). This is relevant for online monitoring and for further offline analysis, verification and validation of data, as well as for training AI models. Their aim is to correct and predict data (see. Fig. 1).

The ORECHESTRA mother board is a full-custom design used in several internally developed acquisition systems [11, 12]. It is scalable and supports a range of on-board computer modules, from low-power processors to high-performance computing units, depending on the required computational load. Importantly, the board provides precise data timestamps via an integrated GNSS module.

C. *Data Base and monitoring tools*

Once the data are pushed into the remote database, daily monitoring and preliminary analysis can be performed using a dedicated GUI, specifically designed for MultiMod'Air. The *PyQt* interface can be deployed to any user easily. All aggregated data can be visualized, plotted and dumped for further analysis. In Fig. 2, a use-case is illustrated for the gamma-ray monitoring. Aggregated spectrum at a given time is displayed as well as timeseries where gamma-ray counts rates are monitored in specific lines of interest such as the Radon daughter product lines. An anomaly detection report is provided automatically. In case of an unexpected rate, a tentative analysis is performed to identify the radioisotopes in excess and an alert as well as a confidence level is released.

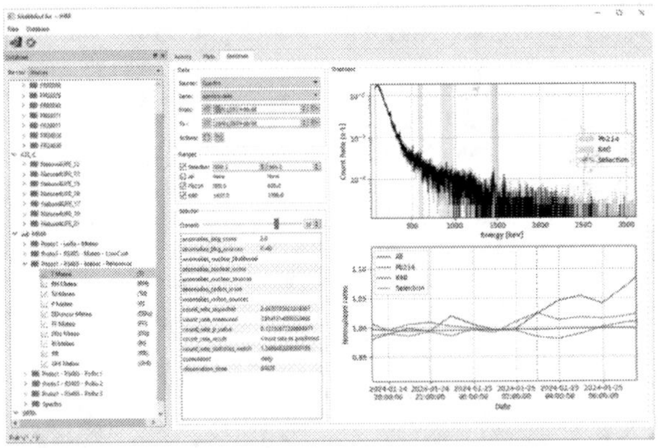

Fig. 2. MultiMod'Air graphical interface – example of a gamma-ray monitoring visualization including a spectrum, a gamma-ray line timeseries and anomaly detection report. On the left hand-side, the list of all sensors, in all deployed stations are easily accessible for the user.

III. Gamma-ray monitors

Two different gamma ray monitors have been developed in the project by a collaborative team involving CEA DRF/IRFU, DRT/LIST and DES/IRESNE.

A. *Gamma-ray monitors detection chains*

One gamma ray monitor is based on a Kromek GR1+ (1 cm^3 CdZnTe, coplanar grid single channel) coupled to a local ODROID XU4 microcomputer for data acquisition, management and communication towards the central node. A full custom data acquisition software has been designed in *C++* using the commercial GR1+ driver under Linux. A *Python* script manages the detector configuration and operations, collects the spectra and push the data to the central node via a USB/RS485 link.

The second sensor, Caliste-MMA, is based on Caliste-O detectors [13] (0.4 cm^3 CdTe Schottky diode, 256 independent spectrometry channels) initially developed at CEA/IRFU with 3D PLUS for space science [14] and used for gamma ray cameras [15]. Caliste-MMA is readout by a full-custom data acquisition system involving a Zynq 7020 System-on-Chip (SoC). The SoC embeds a FPGA and microprocessor to interface the Caliste-MMA sensors (ASIC configuration, High-Voltage management, Data Acquisition). It is routed via an Ethernet link to an ODROID XU4 computer as well, for data preprocessing such as energy calibration, histogramming, and data management. The onboard software is entirely written in *Python*. Even though the SoC could perform all operations without an ODROID, the later has been used to ease both the development and remote-control software updates through the central node via Ethernet.

Both detectors are fully autonomous. They can be configured, operated and reset remotely – depending on the environmental conditions, detector settings can be adjusted such as the low-energy threshold, high voltage, analog filters. Any hot channel can be disabled. The different detector geometries are fit into specific 3D printed carrier attached to a common waterproof and lighttight housing enabling a standard detector packaging definition. The latter is identical for any sensor in MultiMod'Air but the weather instrumentation.

B. Data acquisition scheme

While GR1+ detector sends an integrated spectrum every minute, Caliste-MMA both sends accumulated spectra over all channels every minute and photon event list on demand. Each event is timestamped with an accuracy of 10 ns and comprises the pixel number and the recorded pulse height. From the event list, accurate light curves can be produced and advanced spectral analysis can be performed filtering events from their geometry, energy range and time intervals.

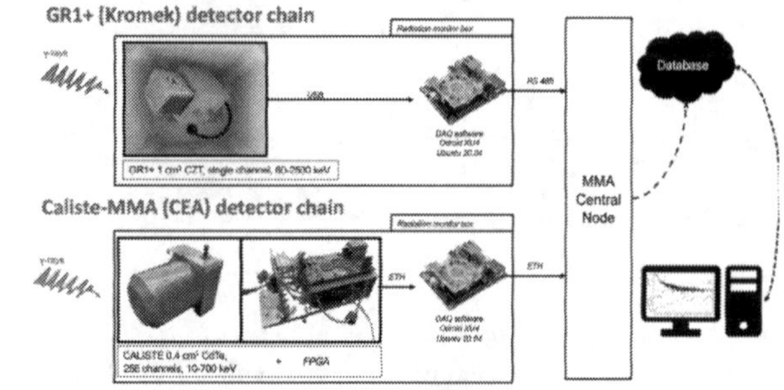

Fig. 3. Top panel: GR1+ detector chain. Bottom panel: Calste-MMA detector chain. Right panel: interface to Central node and database.

GR1+ and Caliste-MMA are complementary and may work together or independently. GR1+ is well suited for measurement from 60 to 2000 keV while Caliste-MMA shows high spectral response from 10 to 900 keV. Both detectors nicely cover the energy range where most of natural radioactivity lines are expected such as Uranium decay chain, Thorium decay chain and Potassium 40.

C. Anomaly detection pipeline

The central node computer collects the gamma-ray raw data and runs real time algorithm for data reduction and analysis on the fly: 1-minute spectra are aggregated to 10-minutes, 1-hour and 1-day accumulated spectra. Those are processed every minute, 10-minutes and hours respectively, to search for possible spectral anomalies with respect to a prerecorded gamma-ray background reference spectrum. Those spectra are continuously analyzed by an "anomaly detection algorithm" described hereafter. The algorithm is developed in Python and can be run on any operating system, online or offline.

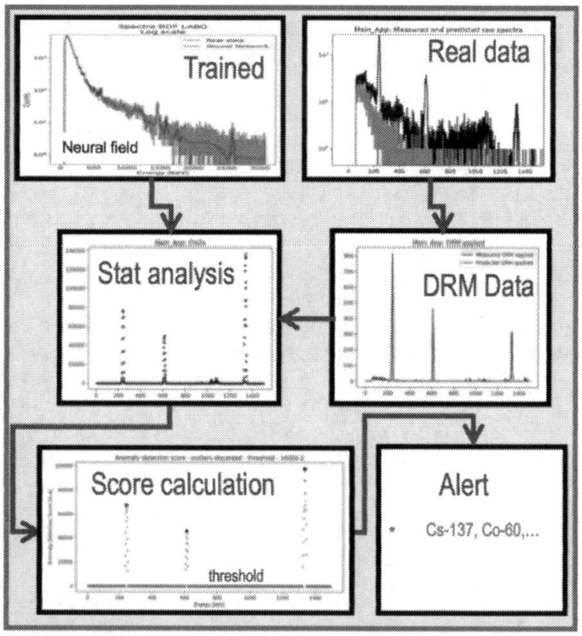

Fig. 4. Gamma-ray monitors "Anomaly Detection Algorithm". Based on a pre-recorded reference background measured locally, a background model is computed. The real data are processed and unfolded. Count rate excess is identified together with a confidence level score and an isotope identification is attempted. A confirmed alert if passed to the database.

In addition to a simple overall-count-rate monitoring, in case of an anomaly detection, the algorithm attempts an isotope identification and reports an alert. An anomaly detection confidence level is provided as well. A false alert might be released. The persistence of an alert at different time scales is a way to minimize false detection as well as the confidence score. The architecture and principle of the anomaly detection algorithm are illustrated in Fig. 4. A prerecorded background spectrum is recorded locally. It is adjusted using a neural field AI model to interpolate and transform the spectrum into a probability density function. It consists of training a Multilayer Perceptron that takes as input the energy and outputs the corresponding intensity, minimizing the Mean Squared Error loss function. Elsewhere, a detector response is previously calculated using a Monte Carlo simulation. The real data spectra and the randomly drawn reference spectrum corresponding to the detector live time are then subtracted and unfolded with a Least-Square inversion using Tikhonov regularization. The significance of count excess in the resulting spectrum are analyzed and identified, and a score is calculated to determine the alert level if needed.

The algorithm has been extensively tested with low radioactivity sources in the lab and detection threshold have been calibrated on real data.

IV. Real Environment Measurement Campaigns

Two complete MultiMod'Air platforms have been deployed in different real environments:

- One is deployed in Paris city, nearby the peripheric boulevard in an urban environment inside the AirParif "Porte Dorée" station [9]. This platform hosts a GR1+ gamma-ray monitor. MultiMod'Air stands few meters above the ground. The environment is essentially surrounded with buildings, roads and concrete or tarmac. The platform has been installed on site in October 2024 and has recorded data continuously.
- One is deployed in Saclay, nearby a rural/suburban area into the LSCE station, in-situ SIRTA platform, a national facility of the European Research Infrastructure ACTRIS [10]. This platform hosts a Caliste-MMA gamma-ray monitor. MultiMod'Air stands on a ~20-meter hill. The gamma ray detector is installed 2 m above the ground. The environment is essentially surrounded by agricultural cultivation fields and roads, as well as research center buildings. The platform itself is seating on a concrete slab. The platform has been completed with the Caliste-MMA detector in April 2025 and has recorded data continuously.

Both platforms host high-performance instrumentation for gas and particle monitoring. None of the reference sites hosts another reference gamma-ray monitor. Photograph of Fig. 5 shows the platforms as installed in Paris on AirParif site.

Fig. 5. MultiMod'Air sensor boxes are recognizable with their common shape. Each box is about 15×30×20 cm^3. The gamma ray monitor is marked by red square.

Hopefully, no anomaly has been reported during the operation. However, thanks to this instrumentation, natural radioactivity is monitored continuously. In the following, timeseries and spectra have been extensively analyzed against weather conditions. Some results are reported as examples and analyzed. Radon daily fluctuations measurements, natural radioactivity fluctuations with rainfalls, atmospheric muons rate dependency with atmospheric pressure are illustrated.

A. Kromek GR1+ chain deployed in Urban conditions

Fig 6 shows the very typical spectrum of the low-level natural radioactive background recorded in Paris between October 15, 2024 and June 3, 2025. The main gamma-ray lines of the natural background are labelled and easily recognizable.

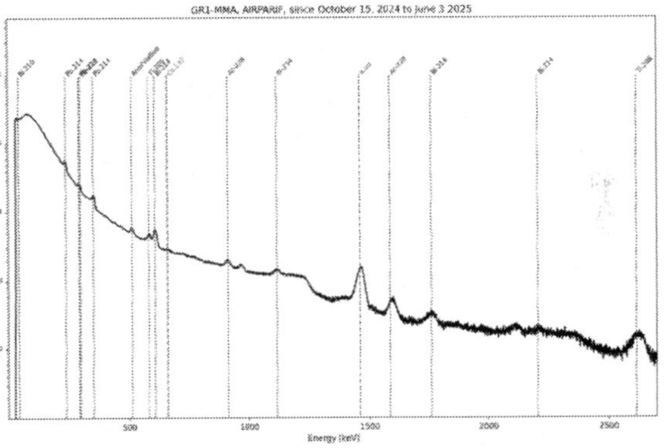

Fig. 6. MultiMod'Air GR1+ gamma-ray monitor in Paris. The spectrum is accumulated over 8 months of continuous data acquisition. The main background gamma-ray lines are identified.

It is worth noting that the detector is fairly stable over time and does not suffer greatly from temperature changes, although a temperature correction would help to achieve a slightly better energy resolution. The uncorrected spectrum is shown intentionally. The energy resolution is found to be ~15 keV FWHM at 609 keV. In fact, the counting statistics are pretty low, with a typical count rate of ~2-4 counts/sec/cm^3, and daily corrections are difficult to achieve automatically. A correction would improve the energy resolution by ~10%, by correcting for the detector offset variations with temperature.

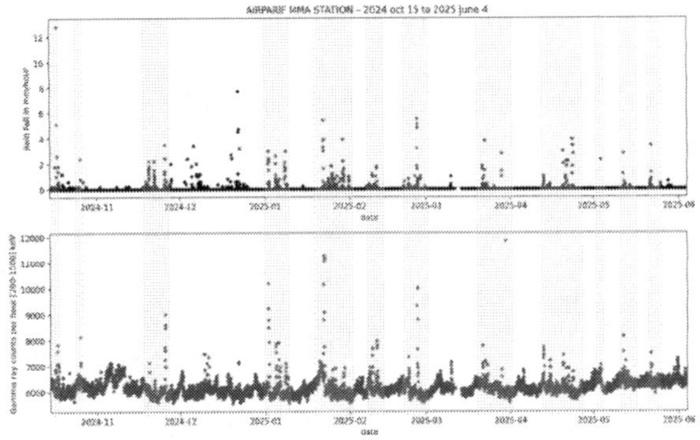

Fig. 7. Top panel: Rainfall intensity in Paris in mm/h. Bottom panel: Gamma-ray counts in the range [200-1500] keV in counts per hour.

During the measurement campaign, Paris experienced very rainy weather. Combining the gamma ray count timeseries and the rainfall intensity data from the same period enables the increase of the natural radioactivity to be correlated with the intensity of the rain, as seen in Fig. 7. As explained in [16, 17], rainfalls tend to wash out the aerosols in which radon daughter products are trapped, resulting in much higher radioactivity levels at ground level when it rains. Surprisingly, the count rate can double or more in the event of heavy rainfall, such as on January 22 or March 30 2025.

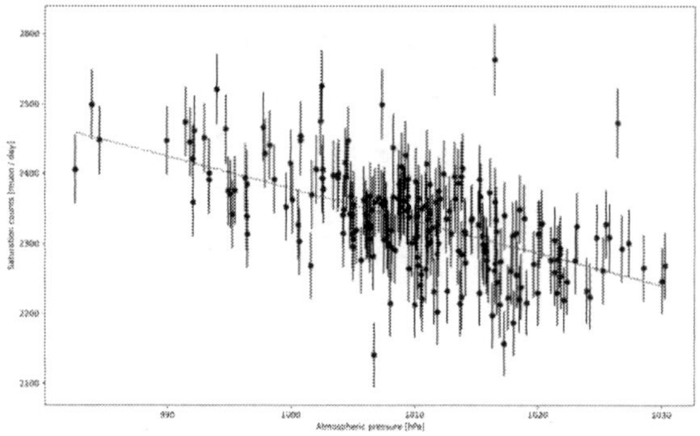

Fig. 8. Saturation rate in GR1+ attributed to cosmic ray muons against the local atmospheric pressure. Each point corresponds to 1-hour measurements. The anti-correlation is pretty clear.

On the other hand, the saturation channel of the GR1+ is essentially populated by cosmic ray muons at ground level. Muons saturate the 1cm^3 CdZnTe detector chain when they pass through. From the data, it is possible to observe a clear anti-correlation of the muons rates against the local atmospheric pressure as shown in figure 8. When the pressure is lower, the mean free path of cosmic ray muons in the atmosphere increases slightly, hence the saturation rate increases.

Rebinning the timeseries per day is sufficient for characterizing this effect using an 8-month dataset. This effect could be useful for anticipating thunderstorms from gamma-ray data. Indeed, when a thunderstorm occurs, the pressure drops suddenly and it is expected that the muon rate will increase accordingly. The field of view being extremely large, atmospheric pressure drops could be sensed distantly from the local weather station detecting cosmic ray muons excess, hopefully with a directivity measurement. This measurement is limited by statistics, but could be relevant for weather events lasting more than an hour, or with a larger detector volume to increase sensitivity.

B. Caliste-MMA chain deployed in suburban environment

The MultiMod'Air platform in Paris is located in an urban environment without any natural soil nearby. In contrast, the system installed at the LSCE in Saclay is located near an area of natural soil from which radon gas may escape.

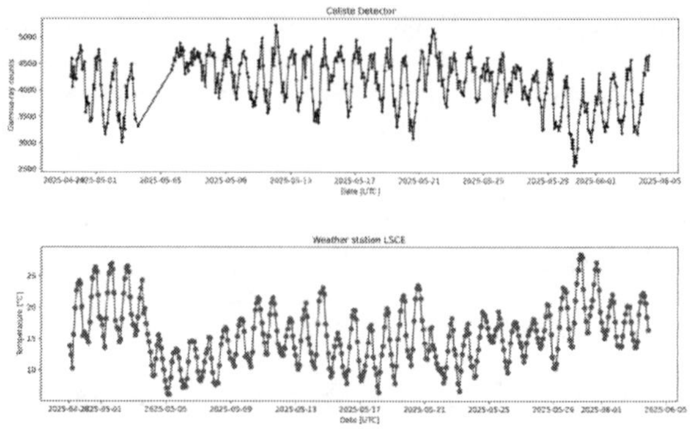

Fig. 9. Top panel: Caliste-MMA count rate in the range [20-700] keV. Bottom panel: Temperature in °C. Each point is about 90 minutes bin. The anti-correlation of the count rate and the temperature is very clear.

The Caliste-MMA detector was able to measure gamma-rays emitted by the Radon daughter products (Bi-214 and Pb-214 mainly). The count rate associated to those emitters is known to follow a daily cycle [18] when the rainfalls are not intense. Indeed, the cooler night air tends to let radon and radon daughter products stagnate near the ground, whereas during the day, the warmer air tends to disperse the products. Consequently, it is expected to see a lower rate in counting during the day than during the night, specifically when the diurnal temperature range is important (>10 to 15 °C or more).

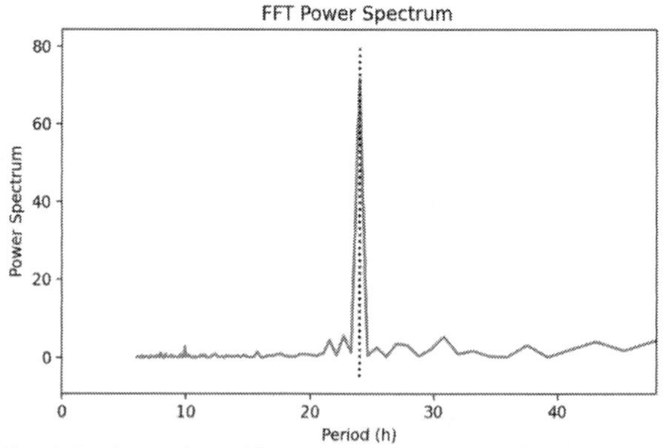

Fig. 10. Fourier transform of the gamma-ray counts times series of Fig. 9-top. A resonance peak is very clear at 24h period.

Those conditions were met during the period from April to June 2025 in Saclay. Fig 9 illustrates the daily fluctuations of gamma-ray counts correlated to temperature. Performing a Fourier transform on the count rate signal from the Caliste-MMA detector time series in the range from 20 to 700 keV makes it very clear that a resonance is detected at 24 hours (cf. Fig. 10). This is interpreted as the daily radon fluctuation.

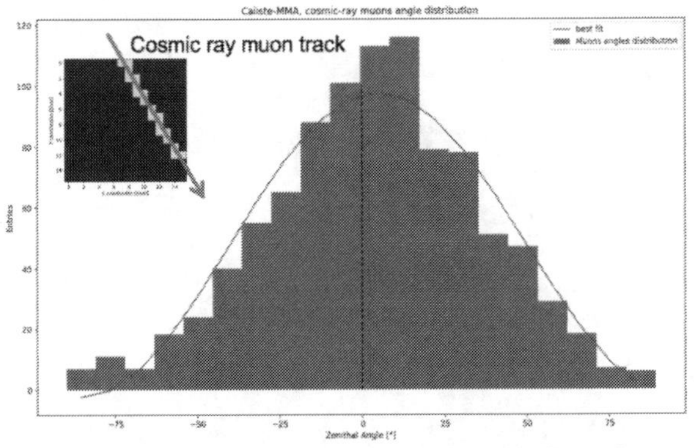

Fig. 11. *Distribution of the cosmic ray muons zenithal angles from Caliste-MMA recorded tracks. The distribution matches the expected distribution in red following a* $cos^2(\theta)$ *law. The offset of the centroid is due to poor alignment of the detector vertical axis.*

Finally, Caliste-MMA is a pixelated detector (800 µm pitch, 16×16 pixels). In contrast to the GR1+, passing cosmic-ray muons will not saturate the channels. The detector, which is installed edge-on, allows the trajectories of passing particles to be reconstructed. Monitoring the muon tracks and applying a Hough transform enables the zenithal angle distribution to be reconstructed and confirmed as matching the expected angular distribution (cf. Fig. 11).

V. Conclusions

MultiMod'Air is a multi-modal instrumented environmental monitoring full-custom platform. It hosts a dozen of gases and particle sensors as well as a comprehensive weather station and a gamma ray background monitor. The sensors are managed by a central node computer. The raw data as well as pre-processed data are pushed to a database in the cloud. The GUI helps monitor all the sensors and dump data for further analysis.

Two prototypes of the platform have been completed and deployed at two different sites for a demonstration: an urban site and a suburban site. After several months of successful, continuous operations, the gamma ray detectors have demonstrated stable performance, remote control and full autonomy capabilities. Data from the gamma-ray detectors was dumped and analyzed alongside weather sensor data. Both detectors produced high-quality data, and the gamma-ray background sensitivity to rainfall, daily Radon and daughter product fluctuations, and cosmic-ray muon dependency on local atmospheric pressure were successfully demonstrated with a small sensor.

In the next phase of the developments, the platform will be upgraded to a transportable version within a low weight vehicle, enabling environmental diagnostics anywhere. On the gamma ray detector side, a new version of Caliste-MMA shall be released with a higher quality crystal and a thicker sensor for higher sensitivity. Regarding the central node, real time analysis presently relies essentially on convectional algorithm for anomaly detection while this is a typical case for AI approaches. Moreover, AI is particularly important for real-time data correction, such as energy calibration dependent on temperature, reference background corrections dependent on rainfall and atmospheric pressure, and modelling daily radon fluctuations, all of which enhance the performance of anomaly detection.

VI. Acknowledgment

The authors are very thankful to AirParif for allowing the installation of the platform on their measurement site at Porte Dorée in Paris and for their technical assistance during the installation. The development of the MultiMod'Air platform and the associated study, including data acquisition and treatment, were entirely supported by a CEA financial grant in the framework of the CEA internal "Mission Numérique".

References

[1]- A. Juginović et al., "Health impacts of air pollution exposure from 1990 to 2019 in 43 European countries", Scientific Reports 11:22516, 2021.
[2] – K. Vohra et al., "Global mortality from outdoor fine particle pollution generated by fossil fuel combustion: Results from GEOS-Chem", Environmental Research, Volume 195, 110754, April 2021.
[3] – F Chen et al., "Breathing in danger: Understanding the multifaceted impact of air pollution on health impacts", Ecotoxicology and Environmental Safety, 280, 116532, 2024.
[4] – "Outdoor Air Pollution", IARC Monographs on the evaluation of carcinogenic risks to humans, Vol 109, 2016.
[5] – F. Borghi et al., "Monitoring of air pollution in agricultural settings: a systematic review of the literature", Research Square, 2023.
[6] – M. A. Sutton, S. Reis and S. M. H. Bakers, "Atmospheric ammonia, Detecting changes and environmental impacts", Springer Science & Business Media, 2009
[7] – A. Akhand and W-Y. Wu, "Current research on gaseous ammonia detecting and capture technologies", Current Opinion in Environmental Science & Health, Volume 36, 1005155, 2023.
[8] – J. Li, W. Xu, B. You and Y. Sun, "Dynamic variations of ammonia in various life spaces and seasons and the influences of human activities", Building and Environment 212, 108820, 2022.
[9] – https://www.airparif.fr
[10] – https://www.actris.eu
[11] – C. Fisher, A. Recoquillay, B. Chapuis, and P. Calmon, "Embedded Artificial Intelligence in Guided Wave SHM system: Signal processing, and data analysis" e-Journal of Nondestructive Testing, Volume 30, no. 06, Jun. 2025.
[12] – A. Recoquillay et al., "Optical fiber sensors for acoustic emission monitoring," e-Journal of Nondestructive Testing, Volume 29, no. 10, Oct. 2024.
[13] – D. Maier et al., "Second generation of portable gamma camera based on Caliste CdTe hybrid technology", Nuclear Instruments and Methods in Physics Research Section A: Accelerators, Spectrometers, Detectors and Associated Equipment, Volume 912, Pages 338-342, 2018.
[14] – A. Meuris et al., "Caliste HD: A new fine pitch Cd(Zn)Te imaging spectrometer from 2 keV up to 1 MeV," 2011 IEEE Nuclear Science Symposium Conference Record, Valencia, Spain, 2011, pp. 4485-4488, doi: 10.1109/NSSMIC.2011.6154695.
[15] – R. Le Breton et al., "The Spid-X gamma camera: A miniature gamma ray integral field spectrometer for nuclear industry applications", Nuclear Instruments and Methods in Physics Research Section A: Accelerators, Spectrometers, Detectors and Associated Equipment, Volume 1047, 167722, 2023.
[16] –C. Bottardi et al., "Rain rate and radon daughters' activity", Atmospheric Environment, 238, 117728, 2020.
[17] – J.-F. Mercier, "Increased environmental gamma-ray dose rate during precipitation: a strong correlation with contributing air mass", Journal of Environmental Radioactivity, Volume 100, Issue 7, 2009.
[18] – Winkler, R., et al., "Diurnal and seasonal variation of the equilibrium state between short-lived radon decay products and radon gas in ground-level air", Radiat Environ Biophys 40, 115–123, 2001.

A high-volume environmental radioactivity monitoring station integrating prompt and high-sensitivity γ-measuring systems

Massimo Altavilla[1,*], Monica Gattinoni[2], and Giorgio Angeli[2]
[1]National Inspectorate for Nuclear Safety and Radiation Protection (ISIN), Italy
[2]TNE - Technology Nuclear Electronics, Italy
(*) massimo.altavilla@isinucleare.it

***Abstract*—The automatic network of environmental radioactivity, REMRAD, managed by the National Inspectorate for Nuclear Safety and Radiation Protection (ISIN), consists of highly sensitive radiological monitoring stations through automatic sampling and spectrometric analysis systems of atmospheric particulate matter, deposited on a large filter splitted into 15 separate circular areas where the same particulate matter is collected. The extremely high detection sensitivity of the REMRAD stations is guaranteed both by the high sampling rate, up to 500 m^3/h, and by the measurement configuration adopted. The stations are equipped with both an "early alarm" system, based on a $LaBr_3(Ce)$ spectrometric detection system placed in an orthogonal position with respect to the sampled filter, patented by ISIN, and which guarantees measurement sensitivity below 10 mBq/m^3 on Cs-137, and a BeGe "Broad Energy Germanium" detector type, which operates in delayed mode on the same filter, guaranteeing sensitivity below 10 µBq/m^3, always with reference to the Cs-137 radionuclide.**

A customized management software integrates this powerful algorithm and allows multiple authorized recipients to receive prompt SMS and email in case of alarm or malfunctions.

Indeed, these stations can operate without any operator intervention for up to a month thanks to a software that manages all the hardware devices automatically performing data acquisitions, quantitative analysis and saving measurements in a dedicate database. The energy calibration of the "On line" detector is constantly checked and updated while Quality Assurance is automatically performed every filter change on the BeGe detector by measuring a source installed on the robot.

***Keywords* —REMRAD, radiological monitoring stations, "On line"**

I. Introduction

This work concerns a method for generating the absolute efficiency curve of a $LaBr_3(Ce)$ scintillation radiation detector with respect to a large glass fiber filter, divided into 15 circular active areas, contained within a multi-section structure called a "cassette". The absolute efficiency curve is calculated using the Monte Carlo code MCNP6.1, used to model the $LaBr_3(Ce)$ scintillation radiation detector and the "cassette" structure. The absolute efficiency curve is functional with respect to the $LaBr_3(Ce)$ scintillation radiation detector, a component of the high-volume airborne particulate matter sampling system, as it allows for the "quantification" of radionuclides, in terms of radiological activity (Becquerels [Bq]) deposited in the 15 active areas of the glass fiber filter, and the subsequent determination of the radioactive concentration in the air (Becquerels/m^3 [Bq/m^3]). The absolute efficiency curve is calculated with respect to the particulate matter deposited in the 15 circular active areas of the filter system. Each of the 15 active areas of the glass fiber filter contributes, with its own specific probability of particulate deposition, to the computation of the absolute efficiency curve. The same $LaBr_3(Ce)$ scintillation radiation detector also allows the energy discrimination of the radionuclides deposited in the 15 active areas of the glass fiber filter, with a high energy resolution, lower than a germanium (HpGe) type radiation detector but higher than other scintillation detectors such as sodium iodide (NaI(Tl)). At the end of a particulate collection cycle on the fiberglass filter and the subsequent decay phase, thanks to the use of a robotic system, the 15 particulate storage areas are cut to form the same number of sampling area filters.

The robotic system then collects the same filters and places them directly into a plastic container for subsequent measurement by the high-sensitivity system consisting of a Broad Energy Germanium detector (BEGe).

As is well known, automated high-volume airborne particulate matter sampling stations equipped with NaI(Tl) scintillation detectors are capable of continuously monitoring particulate matter deposited on a glass fiber filter. NaI(Tl) scintillation detectors are not calibrated in absolute efficiency with respect to the 15 active areas of the glass fiber filter, but they are capable of raising an alarm if a variation in the total count rate is detected. Thanks to the spectrometric capabilities of the NaI(Tl) scintillation detector itself, it is also capable of highlighting any variations in the total count rate due to radionuclides, both natural and artificial. The use of NaI(Tl) detectors does not allow the discrimination of multiple radionuclides that are close to each other in energy due to poor resolution and, furthermore, the lack of an absolute efficiency calibration with respect to the 15 active areas of the glass fiber filter does not allow the quantitative analysis of natural or artificial radionuclides present in the sampled particulate matter; it is only possible to make an "estimate" of the possible presence of natural or artificial radionuclides with respect to a predefined alarm threshold, calculated, as previously mentioned, on the basis of the rate of total counts.

An adequate early warning system must ensure: 1) greater discrimination of any radionuclides that may be present in atmospheric particulate matter, through the adoption of radiation detectors with higher energy resolution, such as $LaBr_3(Ce)$ scintillation radiation detectors, and, in particular, through modeling them with computer codes; 2) maximizing the absolute detection efficiency of the above radiation detector model, compared to the large-sized glass fiber filter, minimizing the "screening" effects caused by the various components that make up the high-volume airborne particulate matter sampling system and, finally, positioning the detector itself away from the high air flows present in the central part of the glass fiber filter, due to the high volumes of air aspirated by the sampling system; 3) the generation of an absolute detection efficiency curve, as a function of the probability of particulate deposition in the various areas of the large glass fiber filter, in such a way as to make the calculation of the radiological activity deposited on the filter itself as accurate as possible and, consequently, calculate more precisely the presence of any contamination in the air, due to radioactive particulate matter, to which the general public could be exposed.

II. Monitoring Station Description

The 460 mm × 285 mm fiber glass filter, which represents the main "part" of the monitoring station, is divided into 15 circular active areas with a diameter of 76 mm and it is contained within a multi-section structure called a "cassette" (Fig. 1).

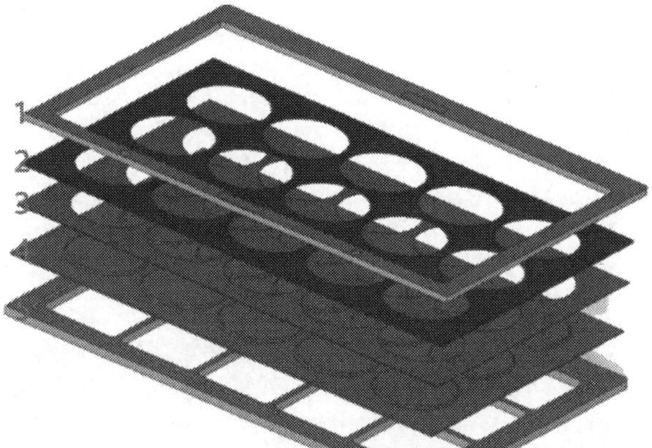

Fig. 1. "cassette" structure: (1) aluminum upper frame, (2) EPDM, (3) steel support layer, (4) steel mesh layer, (5) aluminum support "cassette body".

The "cassette" structure consists, from top to bottom (Fig. 1), of the following sections: Fig. 1 (1) an aluminum upper frame measuring 252 mm × 426 mm and a thickness of 6 mm; Fig. 1 (2) an EPDM (Ethylene Propylene Diene Monomer) rubber layer; Fig. 1 (3) a steel support layer for the EPDM rubber layer; Fig. 1 (4) a layer consisting of a steel mesh with square meshes of side 1.6 mm × 1.6 mm; Fig. 1 (5) An aluminum support called the "cassette body" with rectangular holes. The fiberglass filter is placed between point 1, the "aluminum frame," and point 2, the "EPDM rubber layer." The EPDM rubber layer (point 2) and the steel support layer (point 3) have 15 circular holes, 76 mm in diameter, inside them. The high-volume airborne particulate matter sampling system can guarantee air flow rates of 500 m^3/h. To obtain the efficiency curves, as a function of energy, it was necessary to reproduce the geometry of the radiological monitoring system in every detail, taking into account only those parts that contribute or could contribute to the definition of the efficiency curves. For this purpose, the "cassette" system (Fig. 1) was completely reproduced in every part using the MCNP6.1 code, as well as the steel support structures, the aluminum support base and the rubber "sealing" system.

The radiation detector used for continuous measurements on the filter is a $LaBr_3(Ce)$, which was chosen for its energy resolution and measurement characteristics typical of a scintillation detector. Furthermore, as an intrinsic characteristic of the detector itself, it operates at room temperature without the cooling requirements typical of an HpGe detector. The $LaBr_3(Ce)$ detector to be simulated with the Monte Carlo calculation code MCNP6.1 has physical dimensions of 3.81 cm × 3.81 cm and an energy resolution, corresponding to the Cs-137 energy line (662 keV), equal to almost 20 keV. In the case of gamma spectrometry, absolute efficiency calibration is particularly complex when large calibration standards or those with specific geometries, such as those represented by the "cassette" system, are not available. For this reason, special procedures are developed that combine the detector characteristics, the physical data describing the metallic material, and assumptions about the distribution of radionuclides in/on the material itself.

The absolute detection efficiency is determined based on the gamma emission energy of the radionuclide in question, the distribution of the radionuclide(s) in the material, the absorption properties of the material, the detection system used and the distance from the emission source. The efficiency functions are determined either with a procedure that combines the efficiency calibration results with the calculation model, or with a calibration procedure based exclusively on a numerical calculation model. The main advantage of numerical calibration is that it does not require specific radioactive calibration sources, as it is based on the characteristics of the detector, the geometric model of the object or surface being measured, and, finally, the detection efficiency calculation algorithm. Detector characterization is based on a mathematical expression that describes the detector's efficiency with respect to point sources; therefore, the characterization will depend on the reference gamma emission energies, the distance from the detector crystal, and the solid angle. The mathematical expression of the efficiency curve is determined through specific Monte Carlo simulations, modulating the simulation parameters so that the simulations coincide with actual radiometric measurements of point sources or surface sources. The above detector characterization method, defined for more complex systems such as hyperpure germanium (HpGe) detectors, is described in the scientific publication [1].

III. Prompt spectrometry system

The filter measurement is performed in two different phases. The first phase is performed using a 1.5" x 1.5" $LaBr_3(Ce)$ spectrometric detector. The detector head is centered on one of the lateral circular sections of particulate deposition. This measurement consists of twelve sequential acquisitions of 2 hours each and, since it is performed concurrently with the sampling phase, is defined as “online” or “prompt”.

Several simulations performed with the Monte Carlo code MCNP6.1 allowed the $LaBr_3(Ce)$ detector to be positioned vertically with respect to the cassette system, inside a PVC pipe placed immediately below the cassette structure, near the aluminum support called the “cassette body” (Fig. 1 (5)). This choice, in addition to "protecting" the detector from the air flow, allows for "maximizing" the absolute detection efficiency.

Fig. 2, however, presents a 3D representation of the same system, highlighting the various components and sections of the cassette system with the $LaBr_3(Ce)$ detector positioned inside the PVC pipe.

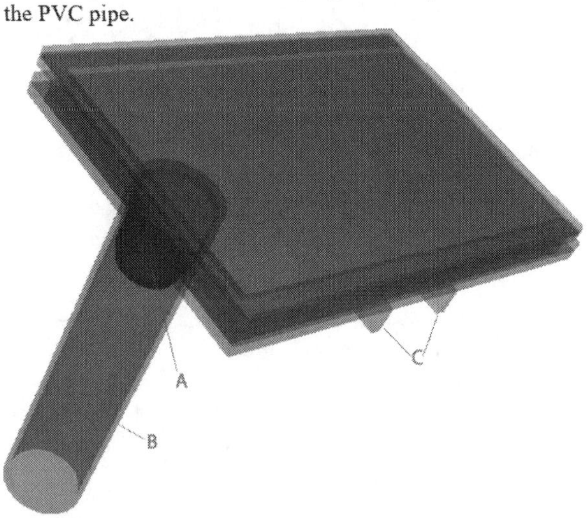

Fig. 2. 3D representation of the prompt spectrometry system with (A) defining the $LaBr_3(Ce)$ detector, (B) the PVC pipe and (C) the frame supporting the cassette system.

The result of this modeling led to the generation of absolute efficiency curves, relative to the position of the detector, with respect to the cassette system, described previously.

The prompt monitoring system configuration allowed to obtain values of minimum detectable concentration (MDC), considering an air flow on the filter equal to 500 m^3/h for 12 hours and a measurement time of 2 h, equal to 3.40 mBq/m^3 for the Cs-137 radioisotope and 2.36 mBq/m^3 for the I-131 radioisotope.

The prompt monitoring system has been validated using sealed multipeak sources, which allowed us to define the maximum percentage deviation between the certified activity of the multipeak source and the activity measured by the system under static conditions. The maximum percentage deviation was estimated to be between 10% and 20% depending on the reference energy of the generic radioisotope present in the multipeak source.

IV. High sensitivity spectrometry system

The second filter measurement phase is performed using a hyperpure germanium detector, type BeGe "Broad Energy Germanium," electrically cooled, installed inside a 10 cm thick cylindrical lead shield and internally lined with 1 mm of copper and 2 mm of tin. The upper part of the shield is made up of a sliding system with two panels. This measurement consists of a single 24-hour measurement with cumulative acquisitions saved every 2 hours. Between the first and second phases, the filter is automatically divided into fifteen sections of equal area corresponding to the fifteen particulate deposition areas. These are stacked inside a beaker that represents the measurement geometry of the second phase and left in a "rest" state for 24 hours, in order to reduce the contribution of short-lived, naturally occurring radionuclides present in the background and consequently reduce MDC of the measurement station. For this reason, the second measurement phase is defined as offline. In case of emergency, the cyclical online-rest-offline measurement procedure, lasting a total of 72 hours, can be modified by the operator to reduce analysis and response times.

The off-line monitoring system configuration allowed to obtain values of minimum detectable concentration (MDC), considering an air flow on the filter equal to 500 m^3/h for 24 hours sampling, 24 hours decay and 24 hours measurement, equal to 1.47 $\mu Bq/m^3$ for the Cs-137 radioisotope and 1.59 $\mu Bq/m^3$ for the I-131 radioisotope.

V. Monitoring station operation

Similar stations already exist around the world, but they are not fully automatic and do not provide a prompt alarm. One limitation is the quantitative analysis of gamma spectra, which requires expert review of calibrations and setup algorithms.

On the contrary REMRAD stations do not require to be manned: there is no need of operator intervention for up to a month and locally only the loading of new filters is required.

Experienced technicians can conveniently manage each station remotely, display spectra being acquired and latest isotopes quantities in Bq/m^3, check system status and weather information.

The accuracy of the results is guaranteed on the one hand by using the powerful algorithm described above and on the other by automatic energy calibration. Online detector is always perfectly calibrated in energy even in the event of temperature changes on the aspirated air by using natural isotopes peaks. Quality assurance is automatically performed every filter change on offline detector by measuring a source installed on the “arm” of a robot.

If alarm thresholds are exceeded, which can be set for each individual radionuclide, notifications can be sent via SMS and email in real time.

The alarm condition can set (either automatically or by operator input) the system in emergency mode: the filter is immediately measured by the germanium detector without waiting for decay.

In addition to any alarm signal, it is possible to enable the function of sending a text message daily which will warn of the correct operation of the system.

All spectra and measurement results are archived in monthly databases and can be queried and filtered by the "Measurement History" function.

VI. Conclusions

The current state of the art does not provide for the calculation of absolute detection efficiency relative to the probability of particulate deposition on the filter and, consequently, for the accurate measurement of natural and/or anthropogenic radionuclides present in aspirated particulate matter. The optimization of the measurement systems in both structural and analytical terms has allowed to achieve MDC values in the order of 5 mBq/m^3 for the on-line system and in the order of 2 $\mu Bq/m^3$ for the off-line system, compared to the reference radionuclide Cs-137.

The integration of all devices into a customized software allows authorized users to always have all the data at hand in real time and to receive notifications via SMS or email in the event of malfunctions or alarms being exceeded.

Radiological/nuclear early warning may be due to the release of radioactive particulate matter into the atmosphere from a nuclear accident occurring, for example, near national borders; the release of radioactivity into the atmosphere from nuclear tests conducted in areas even far from Italian territory; or, finally, the detection of radioactive particulate matter in the air may be associated with the improper use of nuclear materials within Italy, in the vicinity of the automatic radiological monitoring systems themselves.

References

[1] M. Altavilla, R. Remetti, "Development of a stochastic detection efficiency calibration procedure for studying collimation effects on a broad energy germanium detector," Nucl. Instr. and Meth. in Phys. Res. vol. A 712, pp. 157-161, 2013.

Discerning ^{90}Y in-growth from ^{90}Sr for in-situ groundwater monitoring with the ABACUS logging probe

Malcolm J. Joyce[1,2*], Jeremy Andrew[1,3], Manuel Bandala[1], Emma Budge[3], Frank Cave[2], Emma Coghill[3], Soraia Elísio[1], John MacGregor[3], Doug McAlister[3], Andrea O'Kane[3] and Jennifer Spence[3]
[1]Lancaster University, United Kingdom
[2]Hybrid Instruments Ltd., United Kingdom
[3]Nuclear Restoration Services, Dounreay, Scotland
(*) m.joyce@lancaster.ac.uk

***Abstract*— Measurements showing the potential to discern the in-growth of yttrium-90 (^{90}Y) from its parent strontium-90 (^{90}Sr) in aqueous media, based on direct β⁻ particle detection, are described. These have been made using an unshielded cerium bromide ($CeBr_3$) detector derived from a dedicated depth monitoring probe, hereafter referred to as 'ABACUS' in media stripped of yttrium. Whilst similar radiologically, ^{90}Y and ^{90}Sr differ in terms of their aqueous solubilities and associated tendencies to migrate as elemental species. Consequently, they may not always be in secular equilibrium due to their relative proportions in solution changing with time especially where groundwater flow or disturbances act to resuspend or wash out yttrium precipitates. A temporal count profile has been obtained of aqueous samples stripped of yttrium indicating the in-growth of ^{90}Y in this initially, Sr-only-containing groundwater sample. A plateau is observed 21 days consistent with secular equilibrium having been achieved (~ 8 half-lives of ^{90}Y) based on a combined scintillation response indicating not only characteristic features in the X-/γ-ray response but also the potential for the direct detection of ^{90}Y β⁻ particles in $CeBr_3$, evidenced by spectra consistent with the anticipated ^{90}Y β⁻ particle response. ABACUS is a groundwater X-/γ-ray prototype logging instrument designed for in-situ assessment of radioactivity in underground, water-logged environments, particularly in boreholes and sumps comprising a commercially available, 10 × 9.5 mm $CeBr_3$ scintillation crystal attached to a small, full-featured MCA Topaz-SiPM module, contained in a Ø 7 cm & 28 cm long cylindrical aluminium case. A first study presented at ANIMMA 2023 focused on the mathematical algorithm used to analyse the X-/γ-ray photon spectra from ABACUS. Here, the focus is on the implementation of a qualitative approach to discern the relative contribution of ^{137}Cs and $^{90}Sr/^{90}Y$ in solution with an unshielded ABACUS probe, based on comparison of X- and γ-ray photon spectra arising from the bremsstrahlung yield from $^{90}Sr/^{90}Y$, the ^{137}Cs γ-ray full-energy peak response and the direct β⁻ response, using spectral shape analysis. This not only enables spectral shapes linked to different source distributions and isotopes but also, given the very high-energy, β⁻ particle stimulus from ^{90}Y, the isolation of characteristic responses consistent with continuous β⁻ particle spectra and their semi-characteristic end-point energies, especially where detectors might be fabricated to enhance this response, i.e., to incorporate a thin window. Direct assessment of ^{90}Sr and ^{90}Y is known to be particularly challenging due to the short range of their β⁻ emissions in matter, relative to X-rays and γ rays, and their lack of characteristic γ rays in contrast to ^{137}Cs but not impossible especially given the high energies of said β⁻ particles, as this research will illustrate.**

***Keywords*— beta, gamma, spectroscopy, groundwater, logging.**

I. Introduction

In-situ characterisation of groundwater contaminated with strontium-90 (^{90}Sr) and its daughter yttrium-90 (^{90}Y) is complicated because they have few if any characteristic γ rays, only short-range β⁻ emissions and the latter have a relatively low bremsstrahlung yield [1].

Strontium-90 is a high-yield product (cf., 4.51%) of ^{235}U fission which decays ($t_{1/2}$=28.8 years, E_{Max}=0.546 MeV) to ^{90}Y entirely via β⁻ decay. The latter also decays by β⁻ emission, ($t_{1/2}$=2.67 days, E_{Max}=2.28 MeV) to zirconium-90, which is stable. Characterisation of strontium and yttrium is important because the radiotoxicity of both ^{90}Sr and ^{90}Y is a concern: ^{90}Sr and its daughter ^{90}Y are seeded directly to where strontium is sequestered in bone by the body, like calcium. Both ^{90}Sr and ^{90}Y are also prominent components of radioactive waste that are heat-generating, with facilities that store such waste that are party to ingress by water can yield water-borne contamination and hence significant, land-quality clean-up challenges.

The decay of ^{90}Sr to ^{90}Y constitutes a transformation of an alkaline earth (i.e., highly reactive and mobile with water) to a trivalent rare earth (insoluble in this case as the oxide yttria). Thus, strontium tends to remain in solution due to its high ionic potential (16 nm^{-1}) and hence changes in its groundwater abundance can be indicative of changes at its source. In contrast, yttrium forms insoluble carbonate complexes in neutral and alkaline aqueous environments which, being prone to precipitation, can inhibit ^{90}Y mobility arising instead as colloids or suspended particles [2]. Greater mobility of rare earth elements is likely in acidic environments [3] with co-precipitation in low-pH waters less likely with the dispersion of yttrium restricted by sorption on hydrous iron oxides and clays.

Given the above, measurements that require interventions to retrieve samples from challenging, high-dose situations, risk resuspension of the insoluble ^{90}Y-active component, thereby changing the sampling environment and augmenting the radiological risk. Detection systems that can discern ^{90}Y from ^{90}Sr efficiently might target such dynamic land quality issues, in-situ, thus potentially helping to better protect people and the environment. In this research, the use of an instrument (UK Patent No. P370180GB / CAB, 2024) is reported that appeals to this requirement.

II. Experimental method

A groundwater sample was collected from Pond D1204, location 3, at NRS Dounreay in 2020, diluted by a factor of 100, with 0.5 mL withdrawn in April 2023 to identify and quantify radioactivity present with of total α, total β, tritium and ^{137}Cs concentration listed in Table I.

TABLE I
TOTAL α, TOTAL β, TRITIUM AND ^{137}CS ACTIVITY CONCENTRATION FOR THE SAMPLES STUDIED IN THIS RESEARCH, AND RADIOMETRIC DATA FROM THE CALIBRATION SOLUTION FOR ^{241}AM, ^{137}CS AND ^{60}CO

Analysis	Activity / Bq ml^{-1}	γ-ray energy / keV and emission probability / %
α	<17.08	-
β	9178±1.88%	-
^{137}Cs	4810±4.94%	-
Tritium	69.2±1.33%	-
^{241}Am	40.168	59.54, 35.9
^{137}Cs	120.24	661.6, 85.1
^{60}Co	39.992	1333 (99.98%), 1172 (99.99%)

A strontium extraction technique was used to isolate ^{90}Sr in the sample from ^{90}Y using a chromatographic column containing a selective resin, with strontium thus removed from the column and then the ^{90}Y allowed to grow back [4]. The β^- activity of the ^{90}Y growth was corroborated by Cerenkov counting in a liquid scintillation counter. Strontium recovery was determined using ^{85}Sr to ^{85}Rb decay as a tracer, via the associated 514 keV γ-ray photon, with corrections made for grow-in and decay times [4].

The 'ABACUS' probe [1], comprising a cerium bromide ($CeBr_3$) detector, an integrated silicon photomultiplier (Scionix, Netherlands) and digital multichannel analyser (JCS Ltd., UK), was used in a cylindrical, aluminium housing to record photon spectra (primarily X- and γ rays and, it was anticipated, β^- particles with the housing removed).

A 105 ml undiluted fraction of the sample was used for the measurements and ABACUS used with and without the housing and held vertically, as per Fig. 1. X-/ γ-ray spectra were acquired for energy calibration with a 125-ml container containing triple γ-ray standards with the activity concentrations listed in Table 1. ABACUS was positioned to touch this container, with the latter arranged to lie on its side to optimise the surface area available for detection, and the ABACUS 8 mm from the top of the liquid and 40 mm from the benchtop, as per Fig. 1. A background measurement of an empty and decontaminated sample tin was also taken.

The 105 mL sample, stripped of ^{90}Y, was poured into a decontaminated metallic can, as shown in Fig. 1 and assessed with the $CeBr_3$ detector both with and without the ABACUS housing, and covered in a plastic sleeve to minimise contamination. Responses were counted successively for 3-hour long periods from the 5th April 2024 up to the 17th May 2024 (42 days). Additional radiometric analyses of the pond sample were conducted after 17-18 weeks to allow for ^{90}Y in-growth with the ABACUS aluminium probe housing in place to explore its effect on the recorded response of potentially attenuating the β^- response. For these measurements, the probe was positioned above the sample for a single, 3-hour measurement, both with and without the housing, as per Figs. 1a and b, respectively.

a)

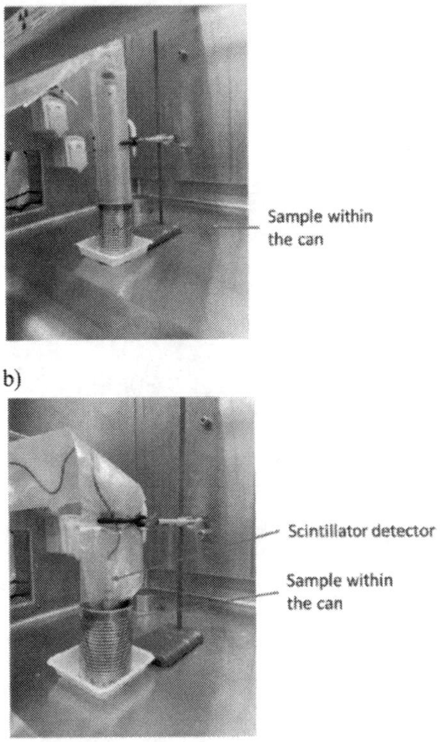

Fig. 1. The experimental setup: a) for assessment of the groundwater sample with the ABACUS probe housing in place to shield against β particles, and b) to measure ^{90}Y in-growth without the housing to optimise the β response, acquisition time of 3 hours per measurement.

III. Results

Energy spectra for the E_{Max} range 0-2500 keV are given in Figs. 2a and 2b, with and without the Al probe housing, respectively, for ^{90}Sr and ^{90}Y in secular equilibrium, prior to ^{90}Y stripping. Fig. 3a shows spectra for ^{90}Y fully stripped, and Fig. 3b for 0-2500 keV over time as dated. Note: most counts fall consistently within the energy window that extends to the average energy for each β emitter, i.e., 195.7 keV and 932.4 keV for ^{90}Sr and ^{90}Y, respectively.

IV. Conclusions

A ^{90}Sr-contaminated groundwater sample from a nuclear decommissioning environment has been stripped of ^{90}Y and the subsequent grow-in measured directly with a $CeBr_3$ detector. Whilst both β and bremsstrahlung responses are feasible sequelae for this process; the spectra evidence a continuum response consistent with the former whereas the yield of the latter is anticipated to be low. It is concluded that radiological dynamics in groundwater might be monitored directly and in-situ via β measurement with a small, inexpensive probe.

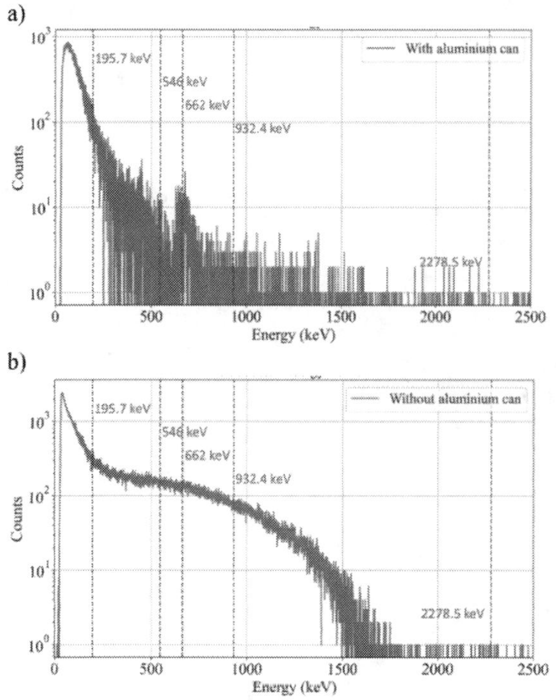

Fig. 2. Log_{10}/lin. energy spectra taken with $CeBr_3$ of contaminated groundwater with ^{90}Sr and ^{90}Y in secular equilibrium, with (a) and without (b) the Al probe housing. Vertical dashed red lines indicate the average and E_{Max} energies of ^{90}Sr and ^{90}Y and the γ-ray energy of ^{137}Cs at 662 keV.

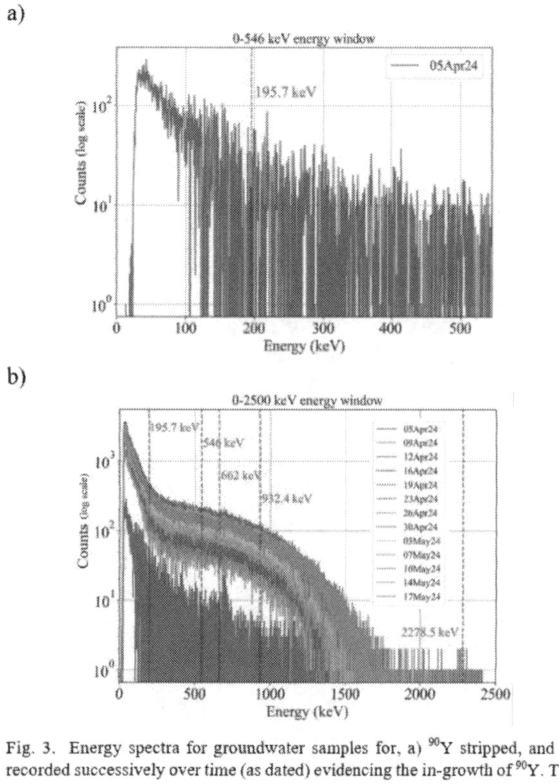

Fig. 3. Energy spectra for groundwater samples for, a) ^{90}Y stripped, and b) recorded successively over time (as dated) evidencing the in-growth of ^{90}Y. The vertical dashed red lines as for Fig. 2.

V. Acknowledgement

The support of the Engineering and Physical Sciences Research Council (EPSRC), via the Lancaster Impact Accelerator Account and the GREEN Centre for Doctoral Training, and the Royal Society, and the insight of Pete Burgess of Radiation Metrology Ltd. are acknowledged.

References

[1] S. C. Elísio, A. Bala, M. Bandala, J. Graham, A. Grievson and M. J. Joyce, "Point-spread Analysis of γ-ray/depth Spectra for Borehole Monitoring Applications". IEEE Trans. Nucl. Sci., vol. 70, no. 11, pp. 2506-2514, September 2023, 10.1109/TNS.2023.3319540.

[2] Yu. A. Balashov, A. B. Ronov, A. A. Migdisov, and N. V. Turanskaya, "The effect of climate and facies environment in the fractionation of the rare earths during sedimentation", Geochem. Int., vol. 10, pp. 951---969, 1964.

[3] D. Kaczor-Kurzawa, I. Wysocka and M. Chuchro, "The behavior of the rare earth elements and yttrium in groundwaters of the Holy Cross Mountains, SE Poland", J. Geochem. Explor. vol. 263, 107493, August 2024, 10.1016/j.gexplo.2024.107493.

[4] J. Andrew, Personal communication, 2024.

Toward quantitative bremsstrahlung medical imaging by custom gamma camera device

Marco Capogni[1], Evaristo Cisbani[2,*], Valerio Cosmi[3], Michele Croia[4], Fausto Giuliani[2], Giulia Limiti[5], Maurizio Lucentini[2], Silvio Murri[6], Paolo Musico[7], Maurizio Pilade[8], Fabio Santavenere[2], Alessandro Spurio[2]
[1]Istituto Nazionale di Metrologia delle Radiazioni Ionizzanti (INMRI), ENEA Casaccia R.C., (Rome), Italy
[2]Istituto Superiore di Sanità, Rome, Italy
[3]Technische Universiteit Delft, Delft, The Netherlands
[4] Nuclear Research Reactors Laboratory, ENEA Casaccia R.C., (Rome), Italy
[5]STMicroelectronics, Cornaredo, Milan, Italy
[6]Azienda Ospedaliera-Universitaria Sant'Andrea, UOD Fisica Sanitaria, Rome, Italy
[7]Istituto Nazionale di Fisica Nucleare, Sezione di Genova, Genoa, Italy
[8]Sapienza University of Rome, Rome, Italy
(*) evaristo.cisbani@iss.it

Abstract — **Bremsstrahlung radiation imaging may play an important role in the quantitative evaluation of the spatial-temporal distribution of the β^- radioactive emitters in order to optimize and personalize the methodology and dosage of radiopharmaceuticals in radiometabolic therapy.**

The present work is an attempt to investigates quantitative bremsstrahlung imaging aspects, using a configurable phantom based experimental apparatus with corresponding simulated model, to highlight critical issues and pitfalls, and to identify, whenever possible, directions to overcome, mitigate or likely take advantage of some of them. Procedure for precise phantom activity determination by portable counter, Monte Carlo fine tuning, definition of figures of merit for the choice of optimal image reconstruction parameters, and correction factors for the activity estimation are some of the main explored details to end up with a preliminary quantitative imaging obtained by simulation-measurement comparison.

Keywords —**Radiometabolic therapy, Quantitative Imaging.**

I. INTRODUCTION

THE specificity of radiometabolic therapy and the potential of a personalized approach make this treatment one of the most promising therapeutic protocol in clinical precision medicine. In fact, radiometabolic therapy is a clinical treatment that involves the administration of specific radiopharmaceutical to kill and prevent further development of cancer cells by targeting them at molecular level [1]. Its effectiveness depends on maximizing radiation damage to the tumor region while minimizing its potential toxicity to healthy organs and tissues. This is achieved byenhancing the uptake of the radiopharmaceutical by tumor cells and therefore selecting the most appropriate pharmaceutical ligands combined with radio isotopes emitting short-range, high energy transfer particles which provide highly biological effectiveness in the tumor region.

The quantitative determination of the spatial and temporal bio-distributions of the administered therapeutic agent, which posses patient specific characteristics, is a crucial aspects of personalized optimization of the radioisotope dosage; this is essential both for assessing therapy effectiveness and for verifying the actual administered dose during treatment. The necessarily information can be obtained non-invasively, possibly in combination with blood tests or other clinical analyses, by imaging the radiation emitted by the therapeutic radioactive substance, which can escape from the patient's body. Quantitative analysis of these images permits the estimation of distribution of the radiopharmaceutical isotope activities in each region of interest. The time dependent activity can then be related to relevant radiobiological quantities, *in primis* the absorbed dose.

Performing quantitative imaging in radiometabolic therapy is generally challenging due to different factors, including the peculiarities of involved radionuclides. Moreover, defining and quantifying the volumes of interest are also critical: images in nuclear medicine are functional-based, so the volumes identified from these images do not necessarily correspond to anatomical volumes, which are instead reconstructed in Computed Tomography (CT) or Magnetic Resonance Imaging (MRI).

In this context, the Bremsstrahlung radiation imaging technique may play an important role for radiopharmaceuticals with (almost) pure $\beta-$ (and $\beta+$) emitters, such as yttrium-90 (^{90}Y), used in radiometabolic therapy. While electrons perform their therapeutic task by releasing most of their energy to cells close to the point of radionuclide decay, the bremsstrahlung gamma radiation emitted from the same electrons can escape from the patient and be detected. However, this indirect radiation makes imaging quite complex, especially when administered radiopharmaceutical distribution shall be quantified. In fact, despite initial attempts at quantitative bremsstrahlung imaging dating back to 1980s and improvements on image reconstruction are progressing, clinical exploitation is still at the research level and not yet routinely adopted.

Therefore, the main aim of the proposed work has been the setup of an experimental apparatus and methodology to investigate quantitative bremsstrahlung imaging aspects, in order to highlight critical issues and pitfalls. Compared to most of the previous works in this field, we are considering relatively simplified, but as much as possible controlled experimental conditions: well-known emitter activities (^{90}Y), custom gamma camera including readout electronics and reconstruction algorithms; we also minimized attenuation and scattering effects by adopting, in the initial phase, simple, small phantoms. This approach should help to better disentangle physics and reconstruction effects that contribute to the bremsstrahlung quantitative imaging, allowing for a more straightforward analysis of potential sources of image degradation and identifying potential improvements. The data acquired on real measurements have been complemented by detailed Monte Carlo (MC) simulations and corresponding data analysis, which are then used to produce quantitative assessment but also better evaluate the overall imaging process.

II. Experimental Approach

The initial experimental apparatus used to study the quantitative bremsstrahlung radiation imaging (QBSI) aspects is characterized by the following main components:

- Radionuclide: weighted solution of yttrium-90 Dichloride ($^{90}YCl_2$)
- Portable TDCR (Triple-to-Double Coincidence Ratio) counter [2] for direct and precise on site activity measurement of the radioactive solution and related samples.
- Phantoms: two simple polymethyl methacrylate (PMMA) cylinders, 40 and 60 mm diameters, 13.9 and 19.7 mm heights, with 3 cylindrical cavities each filled by different ^{90}Y solutions, in the bottom side and a PMMA lid on top.
- Detector consists of a compact, custom gamma camera based on two detection heads with complementary collimators and sizes. The small head includes a pin hole tungsten collimator with 1.2 mm hole diameter, 2 mm thickness, an NaI(Tl) pixelated scintillator with 2 mm pitch, 50×50 mm^2 active area and 5 mm thickness, optically coupled to a single Hamamatsu H8500 photomultiplier. The other larger detection head is equipped by a 20 mm thick, 150×200 mm^2 active area defined by a parallel hexagonal holes lead collimator with 2.1 mm pitch and 0.3 mm septa, a pixellated NaI(Tl) scintillator 1.5 mm pitch, 6 mm thickness coupled to a 3×4 matrix of H8500 photomultipliers. At the time of measurement, this detector, originally designed for diagnostic scintigraphy research [3], was characterized and calibrated but not fully optimized for QBSI. Figure 1 shows a representative experimental setup with detectors and phantoms. The 832 photomultiplier pixels are readout by modular electronics based on the MAROC3 front-end ASIC, which allow single channel signal shaping, with configurable self-triggering capabilities [4].
- Monte Carlo simulator is based on OpenGate/Geant4 toolkit [5] used to identify optimal reconstruction methods and provide normalization factors for quantitative imaging of the activities evaluation. Figure 2 show an example of simulated model. For the simulation of the physics processes involved into the detector as a consequence of the ^{90}Y decay, the Penelope physics library (G4EmPenelopePhysics) has been adopted which should be preferred in the low energy region, and up to several MeV, for the simulation of the bremsstrahlung processes, compared to Livermore (G4EmLivermorePhysics) and Standard Physics (G4EmStandardPhysics) [6].

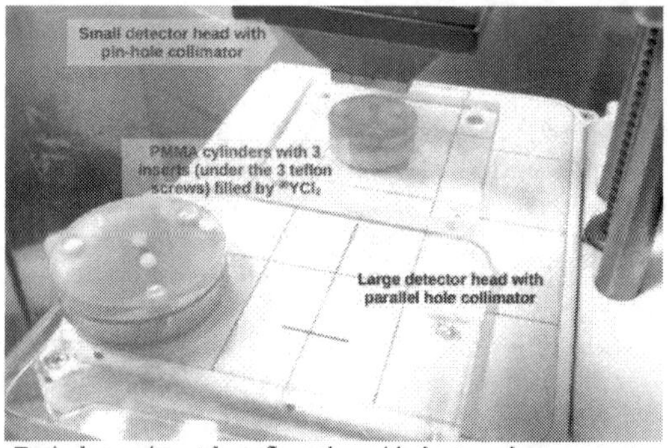

Fig. 1. Typical experimental configuration with the two phantoms on top of the large detector entrance window. $^{90}YCl_2$ liquid solution is deposited in the cavities under the three small Teflon screws in each PMMA phantom disk; on the upper side visible the pin-hole collimator of the small head.

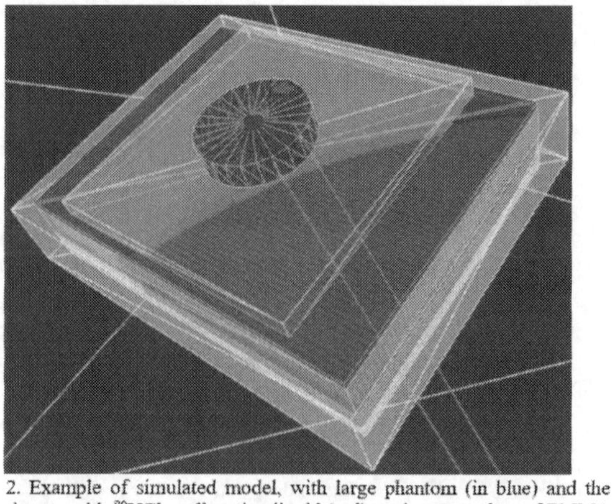

Fig. 2. Example of simulated model, with large phantom (in blue) and the three inserts with $^{90}YCl_2$ radioactive liquid (red) resting on a plate of PMMA (green) above the detector; in yellow the detector Pb external case. Note: inserts appear larger than they are simulated for a rendering of the graphics engine used.

Acquisitions of the two phantoms radiation with different amounts of $^{90}YCl_2$ on the inserts have been carried on using the dual head detectors and the corresponding setup modeled by the MC. The quantity of $^{90}YCl_2$ liquids in each insert (from

https://doi.org/10.1051/epjconf/202533809005

about 1.0 to 2.1 MBq) has been estimated combining the portable TDCR activity measurement (accuracy of 1 kBq) of the liquid before filling the inserts and the weights (0.1 mg accuracy) of the phantom after each insert filling.

III. DATA PROCESSING AND ANALYSIS

The image reconstruction from MC data offers the possibility to evaluate the effects of energy and hit cluster cuts, insert and background mask regions, and eventually determine the optimal trade-off.

Three indices have been defined to quantify the quality of the MC image reconstruction and therefore fine tune the energy window for the real data: the positional residue, the signal intensity and the noise level. The positional residue, a measure of spatial resolution, is the distance between the reconstructed hit position and the position of the ^{90}Y that generated the interacting particle (ground truth) projected on the sensor plane and normalized by the number of hits in the event. The signal intensity is the sum of intensities of those hits that have a positional residue below a certain threshold, which in the present analysis has been set to 2.1 mm, that is 3 times the expected resolution of the detector (σ=0.7 mm). The noise level is the complement of the signal intensity, that is the sum of intensities of those hits that have a positional residual larger than the 2.1 mm threshold.

Ideally the optimal setting corresponds to the maximum signal intensity and minimum positional residue and noise level; however these conditions do not occur for the same setting and therefore the overall optimization function shall be a weighted combination of the three conditions; the choice of such function is still under investigation and it somehow depends, in a clinical environment, on the final task of the image utilization. In the present preliminarily analysis, the signal intensity over the noise level (signal to noise ratio) has been maximized ending up with the 30 to 266 keV energy interval, corresponding to the bottom image in fig 3 where different energy cuts are applied to a typical simulation of the large phantom detected by the parallel-hole sensor head; as expected, the energy cuts have the larger impact; the choice of these cuts are mainly driven by the adopted collimator, that in the present work consists in the medium energy, diagnostic optimized, parallel hole described above.

The responses of the more than 700 anodes that make up the light sensors coupled to the scintillators should be, ideally, statistically identical when the detector is subjected to a spatially uniform irradiation (flood). For realizing this condition, the acquisition electronics allows to assign to each anode channel a gain (from 0 to 4, with 256 steps), which helps to equalize the response of the anodes during acquisition, which is critical for effective self-trigger logic (AND of at least 3 channels) during acquisition.

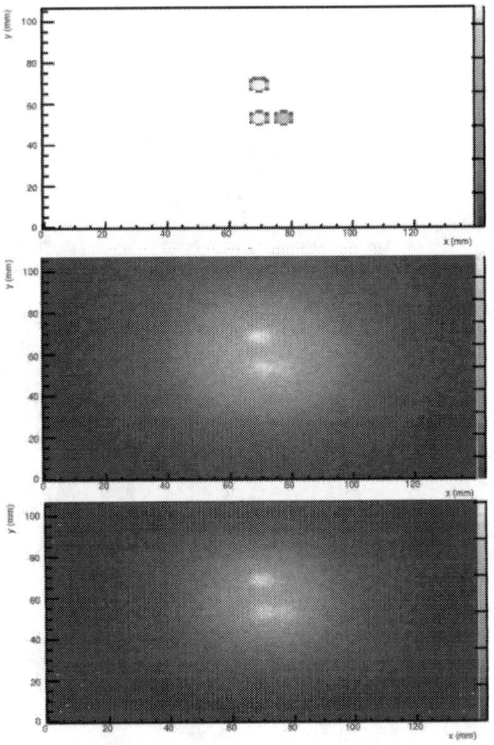

Fig. 3. Effects of hit energy and other cuts on reconstructed MC data; from top to bottom: Ground Truth phantom radioactive inserts; reconstructed image without cuts; 30 to 266 keV energy cut only and topological Compton mitigation dropping out events that present two distinct clusters of interaction in the scintillator.

A further equalization, generally finer, is performed on the acquired data, thanks to the fact that the charge collected by each anode is acquired separately and made available for image analysis and reconstruction.

Taking into account the MC simulation analysis for fine tuning of the optimal cuts in energy and cluster detection (for Compton scattering mitigation), Fig. 4 reports the reconstructed image of one of the acquired run of the two phantoms, where proper smearing is not applied to enhance the pixels structure of the detector system.

The simulations, in the aim of this work, is also used to estimate the activity of the sources starting from the insert counts extracted from the corresponding regions in the reconstructed images (as in Fig. 4): the counts on each region is the sum of the counts in the pixels which are totally or partially inside the circle of the region; counts in pixels shared by two regions are assigned to each region proportionally to the covered pixel area. The counts assigned to the insert are the counts of the corresponding region subtracted by the counts in the background region, scaled by their respective areas. The MC provides the fractions of photons emitted from one insert and detected in another region of the image; these fractions, reported in Table 1 for the same run of Fig. 4, are used to compute the expected activities in each insert from the measured raw counts.

https://doi.org/10.1051/epjconf/202533809005

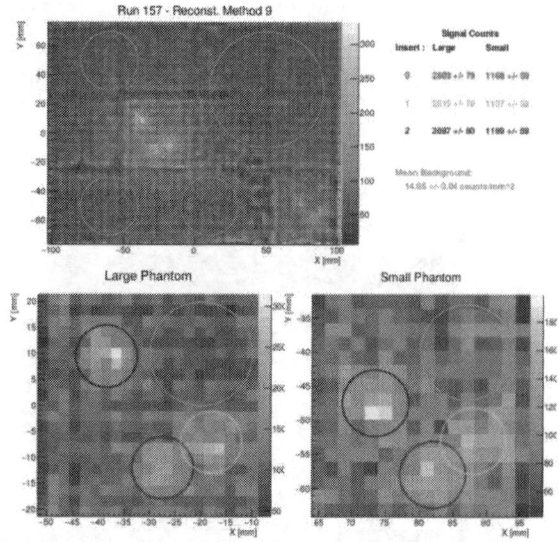

Fig. 4. Reconstructed image from real data (top); on the bottom part the zoom on the large (bottom-left) and small (bottom-right) phantoms with the 3 inserts filled with the beta- emitting $^{90}YCl_3$. The circles (masks) on the full image identify the regions used for the background evaluation, while in the zoomed images the smaller circles select the insert regions while the larger one the background. These areas are used to estimate the signal counts reported on top-right.

Fig. 5 shows the distribution of the preliminarily reconstructed activities (taking into account 4 ms dead time of the electronics) of the insert divided by the reference values extracted by the TDCR. These estimated activities, which are one of the main goal of the quantitative imaging, deviate on average from the "ground truth" by 3% (15% RMS) and 20% (6% RMS) for large and small inserts respectively.

TABLE I
FRACTIONS OF GAMMA MIGRATIONS

Destination image region	Insert origin					
	Large phantom			Small phantom		
	0	1	2	0	1	2
0	64.48	21.63	13.39	61.84	24.56	15.34
1	22.08	66.28	11.02	22.53	61.55	13.75
2	11.85	10.47	73.83	14.56	12.9	69.76
Background	1.59	1.62	1.76	1.07	0.99	1.15

The fraction of photons emitted by each insert (columns) and reconstructed in the image regions (inserts and background, on rows) for the large and small phantoms.

IV. CONCLUSIONS

The above outlined approach, which tries to investigate and possibly mitigate the critical aspects of the quantitative bremsstrahlung radiation imaging, is still work in progress; the preliminarily experimental procedures using sub-optimized detectors for bremsstrahlung gammas and supported by Monte Carlo simulations analyses provide an estimated activity distribution of ^{90}Y filled inserts within 30% from the reference values. In the current work, the choice of the gamma energy window combined to the assignment of the image pixels to the different sources of radiation (insert and background), has represented the most critical, and still not fully solved question.

Analysis of the pin hole collimator data, better integration of MC and real data reconstruction, identification of the most effective function for figures of merit optimization, are underway and likely applied also to new data from ^{166}Ho where some of the detector components have been optimized for gamma energy above 400 keV.

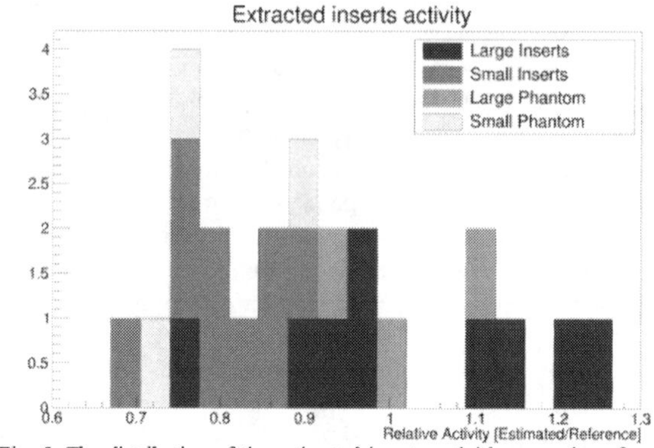

Fig. 5. The distribution of the estimated inserts activities over the reference values measured by the TDCR for three runs with different experimental configurations. Reported are also the total activity of each phantom respect to the reference.

REFERENCES

[1] S. St James, et al., "Current Status of Radiopharmaceutical Therapy". Int J Radiat Oncol Biol Phys. 2021 Mar 15;109(4):891-901, 10.1016/j.ijrobp.2020.08.035.
[2] M. Capogni, P. De Felice, "A prototype of a portable TDCR system at ENEA", Appl. Radiat. Isot. 93 (2014) 45-51, DOI: 10.1016/j.apradiso.2014.03.021
[3] F. Garibaldi, et al. (2010), "A novel high resolution and high efficiency dual head detector for molecular breast imaging: New results from clinical trials.", Nuclear Instruments and Methods in Physics Research Section A: Accelerators, Spectrometers, Detectors and Associated Equipment, 617(1):227–229. 11th Pisa Meeting on Advanced Detectors.
[4] A. G. Argentieri et al., A Multichannel Compact Readout System for Single Photon Detection: Design and Performances Nucl. Instr. Meth. Phys. Res., 2010
[5] OpenGate web page: https://opengate.readthedocs.io/en/latest/ (last visited 04/08/2025)
[6] Pandola, L., Andenna, C., and Caccia, B. (2015). Validation of the geant4 simulation of bremsstrahlung from thick targets below 3 mev. Nuclear Instruments and Methods in Physics Research Section B: Beam Interactions with Materials and Atoms, 350:41–48.

The ENEA TRIGA RC-1 facility inside the SECURE project: production of medical isotopes by neutron activation

Lucrezia Spagnuolo[1], Luigi Lepore [1,*], Letizia Cozzella [1], Tiziana Guarcini[1], Francesca Limosani[1], Angela Pagano[2], Simone Placidi[1], Luca Falconi[2], Valentina Fabrizio[2], Davide Formenton[2], Andrea Roberti[2], Marco Capogni[3]
[1] ENEA, NUC-IRAD-CRGR, Nuclear Material Characterization Laboratory and Nuclear Waste Management
[2]ENEA, NUC-IRAD-RNR, Research Nuclear Reactor Laboratory
[3]ENEA, NUC-INMRI, National Institute of Ionizing Radiation Metrology

ENEA Casaccia Research Centre, Via Anguillarese 301, 00123 Rome, Italy
(*) luigi.lepore@enea.it

***Abstract*— TRIGA (Training, Research, and Isotope production General Atomics) facilities are research reactors extensively used for neutron activation analyses and various neutron-related experiments and applications.**
Neutron fluxes in TRIGA reactors can be employed for radionuclide production through techniques such as Neutron Activation Analysis (NAA), enabling the generation of radioactive sources for technical applications and the production of medical isotopes. In this process, a non-radioactive material is exposed to neutrons, resulting in the generation of radioactive nuclei.
The ENEA TRIGA RC-1 reactor is involved in the EU-funded SECURE Project (HORIZON-EURATOM-2021-NRT-01 call, Strengthening the European Chain of sUpply for next generation medical RadionuclidEs, October 2022-September 2025), which aims to explore the feasibility of local radionuclide production for medical applications in Europe, including both currently established and innovative ways. Specifically, the ENEA team investigated the potential to produce terbium-161 (^{161}Tb), an isotope that shows promise for targeted radionuclide therapy and may offer advantages over lutetium-177 (^{177}Lu), which is currently used in cancer treatment. ^{161}Tb production exploits the reaction channel ^{160}Gd(n,γ)^{161}Gd(β$^-$)^{161}Tb, involving the neutron activation of a gadolinium (Gd) target that is highly enriched (more than 98%) in gadolinium-160 (^{160}Gd). After the appropriate chemical processes, both ^{161}Tb (the raw radionuclide precursor for radiopharmaceuticals used in cancer therapy) and ^{160}Gd oxide (which can be reused for future irradiation cycles) are extracted and purified.

***Keywords* — TRIGA reactor, terbium-161, neutron activation technique, medical isotopes.**

I. Introduction

In recent years, the decommissioning of major nuclear facilities dedicated medical radionuclide production has posed significant challenges to the fields of diagnostic imaging and radiotherapy. This situation has prompted extensive research and development efforts to identify new sources of supply and alternative production methods for these radionuclides. Such initiatives are essential for ensuring the continuity of medical applications, enhancing national self-sufficiency, and minimizing the risks of supply disruptions, ultimately contributing to more effective and sustainable healthcare.

With this purpose the ENEA TRIGA RC-1 joined the EU-funded SECURE Project (Strengthening the European Chain of sUpply for next generation medical RadionuclidEs). The project aimed to make a major contribution to the sustainability of medical radioisotope production and supply for its stable application in diagnostics and therapy within the European context. It focused on promising developments in innovative radionuclides, design of new irradiation targets and novel production routes for both currently used and newly emerging radioisotopes for medical applications.

One of the radioactive nuclides exploited is ^{161}Tb: this is an interesting radionuclide in cancer treatment, showing similar decay characteristics and chemical behaviour to the clinically employed ^{177}Lu. The therapeutic effect of ^{161}Tb may be enhanced due to the co-emission of a larger number of conversion and Auger electrons, which could be more effective in the treatment of the smallest metastases, as well as single cancer cells [1,2].

The ENEA Casaccia Research Centre contributed to the project by focusing on the production of the radionuclide ^{161}Tb. This is obtained by neutron activation of a gadolinium target, highly enriched in ^{160}Gd, with the reaction channel ^{160}Gd(n,γ)^{161}Gd(β$^-$)^{161}Tb at the TRIGA RC-1 Reactor (Fig.1).

Fig. 1. The ENEA TRIGA RC-1 at the ENEA Casaccia Research Center.

Upon completing the irradiation cycle, the target is removed and undergoes chemical processing to produce the final ^{161}Tb as a terbium chloride solution (serving as the precursor for the radiopharmaceutical product) and the following recovery of the raw ^{160}Gd material for the fabrication of new targets. Recovering the gadolinium raw material enables the initiation of subsequent irradiation cycles, thereby optimizing the utilization of the expensive ^{160}Gd target, minimizing overall costs.

This paper outlines the efforts undertaken by the ENEA staff, detailing the most significant irradiation tests, the related Monte Carlo simulations and deterministic calculations, as well as the final results obtained.

II. Methodology

The principal aim of ENEA's contribution to the project was the assessment of the producible activity within the ENEA TRIGA RC-1 reactor through neutron activation techniques.

Although neutron activation is used as a metrological method in analytical chemistry for elemental traces quantification, this does not preclude its use as a method for producing a radionuclide. Clearly, in the case of a metrological approach, the aim is to activate the sample just enough to perform the quantitative determination. On the other hand, in radionuclide production, the goal is to maximize the production of the radionuclide of interest.

From a technical standpoint, the reaction channel ^{160}Gd(n,γ)^{161}Gd(β$^-$)^{161}Tb does not exhibit a particularly favourable microscopic cross section. Consequently, irradiating a given amount of ^{160}Gd in a high-flux neutron source (e.g. $>1\cdot10^{14}$[cm^{-2} s^{-1}]) would be significantly more effective than in lower-flux facilities (e.g. $1\cdot10^{13}$[cm^{-2} s^{-1}]), such as a TRIGA reactor, usually. Furthermore, the irradiation time should be chosen to correspond to at least 1 to 2 mean-life of ^{161}Tb (i.e., approximately 10 to 20 days of continuous irradiation) to ensure adequate production. Under ideal conditions, the saturation activity — the maximum achievable activity for a fixed amount of target material — would be reached after roughly 30 days of continuous irradiation. However, achieving such extended irradiation durations is not feasible at the ENEA TRIGA RC-1 reactor for two main reasons:

a) the reactor core may lack the necessary excess reactivity to sustain such a prolonged irradiation cycle;

b) there are operational constraints, including a maximum irradiation time of 6 hours per day, which results in suboptimal utilization of the enriched gadolinium target material.

The expected irradiation schedule studied by the TRIGA RC-1 reactor staff, considering the previous mentioned limits, involves the placement of the sample at the Central Thimble position for 12 days, with 6-7 hours of irradiation per day, excluding Saturdays and Sundays. In total, the sample could be irradiated for approximately 72-84 hours.

The activity value (A) obtained at the end of irradiation could be estimated by the equation:

$$A = R\left(1 - e^{-\lambda t_{irr}}\right) \tag{1}$$

Where t_{irr} is the irradiation time (s), λ is the decay constant (s^{-1}), and R is the reaction rate, calculated as following:

$$R = N_t \int \sigma(E)\varphi(E)dE \tag{2}$$

where:

- N_t is the number of atoms within the sample;
- $\sigma(E)$ is the microscopic cross-section associated with the specific reactions;
- $\varphi(E)$ is the neutron flux.

In September 2024, an irradiation experiment was performed on a sample of 8.7 mg of Gd_2O_3 (containing Gd 98.2% enriched in ^{160}Gd, supplied by ILL, a partner of the SECURE project), the mass composition of the sample is reported in the table below.

TABLE I
MASS COMPOSITION OF THE SAMPLE TO BE IRRADIATED.

Isotope	Mass [mg]
^{152}Gd	0.0003
^{154}Gd	0.0037
^{155}Gd	0.0255
^{156}Gd	0.0355
^{157}Gd	0.0273
^{158}Gd	0.0437
^{160}Gd	7.4283
^{16}O	1.1325
^{17}O	0.0005
^{18}O	0.0026

The irradiation was conducted at the TRIGA RC-1 reactor, at the nominal maximum thermal power of 1 MW with the sample positioned at the bottom of the Central Thimble (the maximum available irradiation position) for approximately 77 hours.

The activity that can be produced from this amount of raw material under this irradiation condition was estimated using multiple methods and confirmed by experimental measurements of the sample after irradiation.

Below the description of the different approaches used to determine the activity values.

A. *FISPACT-II Code*

The FISPACT-II code is a computational tool capable of modelling activation, transmutation, and depletion processes induced by neutrons, protons, and other particles such as deuterons. It was developed by the United Kingdom Atomic Energy Authority at Culham.

This software is used to evaluate the final activity and activity concentration (applicable primarily to samples unaffected by self-shielding, such as those used in the September experiment), identify potential interfering radionuclides produced during irradiation, and determine the resulting inventory after several days of irradiation.

In this work FISPACT-II was initially used to estimate the producible activity of 1 gram of gadolinium oxide (Gd_2O_3) highly enriched in ^{160}Gd (98.2%), subjected to a comprehensive irradiation of 72 hours. The calculation was performed at the reactor's maximum power, i.e. 1 MW, using the highest neutron fluence rate available at the ENEA TRIGA RC-1 reactor, approximately $5 \cdot 10^{13}$[cm^{-2} s^{-1}] (measured at the bottom of the Central Thimble, the reactor core centre, through the ASTM E262 standard, see Fig.2).

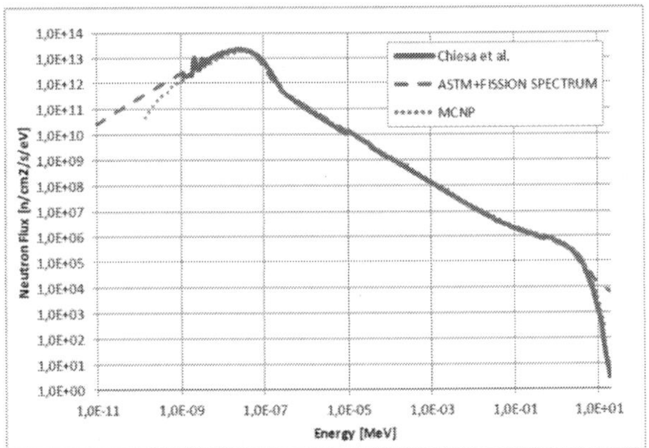

Fig. 2. Neutron Spectra at ENEA TRIGA RC-1 Central Thimble centre-point calculated with Chiesa et.al.[3], ASTM E262 methodology with fission spectrum added, and the MCNP code at 100 kW.

Applying the irradiation schedule, the estimated final activity concentration of ^{161}Tb in the sample is approximately 15 GBq/$g_{Gd\text{-}160}$ at end of irradiation (EOI).

The calculation does not account for neutron self-shielding within the sample; therefore, the resulting activity concentration represents the maximum achievable value. In cases where neutron self-shielding in real irradiated samples cannot be neglected, the results obtained from FISPACT calculation tend to be overestimated.

B. *MCNP Code*

MCNP (Monte Carlo N-Particle) is a code developed and maintained by Los Alamos National Laboratory, widely recognized for simulating neutron and gamma-ray transport. Reactor analysis codes based on Monte Carlo methods model neutron interactions—which occur stochastically due to their uncharged nature—with high precision. Simulation results can be highly accurate and reliable, provided that the user accurately reproduces the physical details of the real-world scenario within the virtual simulation environment.

The ENEA TRIGA RC-1 MCNPX model is based on detailed material composition retrieved from plant documentation and schemes [4,5] see Fig. 3.

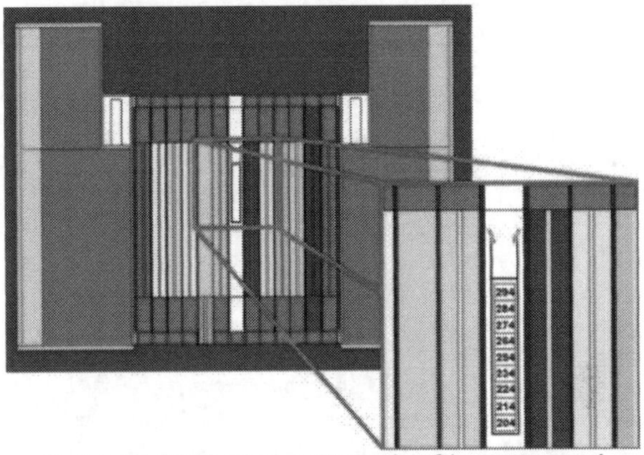

Fig. 3. ENEA TRIGA RC-1 in MCNP: section of the reactor core and particular of the Central Thimble irradiation position.

The MCNP input available at the ENEA Research Nuclear Reactor Laboratory can be employed for reactor calculations to determine neutron flux, reaction rates, and neutron self-shielding correction factors—parameters that are essential when irradiating samples containing large amounts of raw material. Simulating neutron self-shielding factors is also useful for understanding the sample's behaviour within the reactor configuration as the target mass is progressively increased. This approach is particularly important at low-power reactor facilities, where increasing the target mass is a key strategy to enhance the final activity of the desired radionuclide.

In this work the neutron self-shielding coefficient (G_{th} and G_{res} explained in the following) were determined in order to calculate the activity with the ASTM E262 standard.

The reaction rate is calculated by the code just associating a tally flux (f4) to the effective number of atoms inside the irradiated sample, returning the final activity of ^{161}Tb EOI.

MCNP has been used in reactivity calculations also, to verify perturbation induced by irradiation of samples to the reactor core assembly to be compliant with regulatory prescriptions.

C. *ASTM E262-97/08 Standard*

The ASTM method is a standardized test procedure for determining thermal neutron reactions and fluence rates using neutron activation techniques [5]. Its primary objective is to establish a general approach for quantifying unknown thermal neutron fluence rates. In this context, the method was employed both to characterize the neutron flux—by distinguishing its thermal and epithermal components—and to estimate the potentially producible activity of the target radionuclide.

The experimental neutron flux characterization of the channel has been performed using the general equation:

$$\varphi(E) = \frac{2}{\sqrt{\pi}}\,\varphi_0\sqrt{\frac{T_n}{T_0}}\,\frac{E}{(k\,T_n)^2}\,e^{-\frac{E}{kT_n}} + \varphi_e\,\Phi(E-\mu kT_n)\,\frac{1}{E^{1+\alpha}} \quad (3)$$

Where:

- φ_0 is the thermal component of the neutron flux;
- φ_e is the epithermal component of the neutron flux;
- α is a coefficient that accounts for the deviation of the epithermal flux from the 1/E trend;
- T_0 is the reference temperature;
- T_n is the neutron temperature;
- k is the Boltzmann constant;
- $\Phi(E)$ is the Heaviside step function which is equal to zero for $E<5kT_n$ and equal to one for $E\geq 5kT_n$.

The epithermal and thermal neutron flux components were obtained by Au-foils irradiated in bare and cadmium-covered configurations, according to the standard ASTM E262 procedure. Finally, the values of the components of the characterized neutron spectrum (for the experience of September 2024) are equal to $\varphi_0 = 1.19\cdot 10^{13} \pm 16\%$ [cm^{-2} s^{-1}] and $\varphi_e = 1.110\cdot 10^{12} \pm 3.5\%$ [cm^{-2} s^{-1}] at 1 MW reactor thermal power. The energy integrated value of the neutron flux experimentally measured and given by equation (3) is $\varphi(E) = 3.60\cdot 10^{13} \pm 9\%$ [cm^{-2} s^{-1}]. All stated uncertainties in previous values are given with a coverage factor k=1.

The final activity is calculated with the equation (1). In case of thermal reactors, it can be expressed as:

$$R = N_t[\varphi_0 G_{th}\sigma_0 g + \varphi_e(G_{res}I_0 + f)] \quad (4)$$

The various terms in the equation (4) are corrections factors. In particular:

- $G_{thermal}$ (G_{th}) takes into account that the sample introduced into the reactor will cause a local depression of the thermal neutron flux (0.025 eV – 0.55 eV);
- $G_{resonances}$ (G_{res}) takes into account that the sample introduced within the reactor will cause a local depression of the epithermal neutron flux (energies higher than the cadmium cutoff > 0.55 eV);
- g is the Westcott factor that takes into account the deviation from 1/v behaviour of the cross-section in the thermal energy range. This factor depends on the neutron temperature:

$$g = \frac{1}{v_0\sigma_0}\int_0^{\infty}\frac{4}{\sqrt{\pi}}\left(\frac{v}{v_0}\right)^3\left(\frac{T_0}{T}\right)^{\frac{3}{2}} e^{\left[-\left(\frac{v}{v_0}\right)^2\left(\frac{T_0}{T}\right)\right]}\sigma(v)dv \quad (5)$$

- I_0 is the resonance integral:

$$\int_{E_{cd}}^{1MeV}\frac{\sigma(E)}{E}dE \quad (6)$$

- f takes into account the presence of an epithermal component below the cadmium cut-off:

$$\int_{5kT}^{E_{cd}}\frac{\sigma(E)}{E}dE \quad (7)$$

The neutron self-shielding coefficients (G_{th} and G_{res}) were determined by means of MCNP calculations, while the other coefficients were determined using the Python code.

D. *HPGe measurement*

The HPGe (High Purity Germanium) spectrometry is a valuable tool when is necessary to detect, identify and quantify complex gamma spectra, e.g. from irradiated samples.

In this work, two different HPGe detector by Mirion Technologies were used, rated for measurements in the energy range of 60 keV – 3 MeV, energy and total efficiency calibrated with certified ^{133}Ba and ^{152}Eu standard sources. The measurement distance HPGe surface-irradiated sample is 200 mm. Software programming allows for sequential recording of spectra during natural decay of the sample.

In this work, the irradiated Gd_2O_3 sample described before, after a cooling period of 48 hours, was characterized. Up to 280 spectra have been acquired, the most significant in the first three days after the irradiation.

III. RESULTS

The results obtained with the above-mentioned methods are summarized in Table II [7].

TABLE II
RESULTING ACTIVITY OF THE SAMPLE IRRADIATED IN SEPTEMBER 2024.

Expected activation results from calculation			Experimental determination
FISPACT-II Activity [MBq]	MCNP Activity [MBq]	ASTM Activity [MBq]	HPGe Measured Activity [MBq]
106 (±78%)	110 (±78%)	103 (±78%)	110 (±6%)

As shown in Table II, the simulated and calculated results are in good agreement with the measured values.

The table also reports the errors calculated using the error propagation linear formula with a coverage factor of two (k=2). The measurement error, related to the HPGe experimental determination, amounts to 3% (k=1), averaging several measurements and considering uncertainties in count rates and detection efficiency, mainly. Regarding the errors associated with the calculation codes, the error for the MCNP code, the FISPACT-II code and the ASTM methodology, they are namely 39% (k=1). This extremely high value is because all parameters involved in its formulation (equation 4) are influenced by uncertainties in nuclear data (taken from [8], having considered ENDF/B-VIII, JEFF-3.3, and TENDL-2019 libraries). While the microscopic cross-section errors for gold are minimal (due to the widespread use of ^{197}Au(n,γ)^{198}Au reaction in neutron flux measurements), the error related to the gadolinium $^{160}Gd(n,\gamma)^{161}Gd$ spectral cross-section varies between 17% and 25% depending on the nuclear data library used. The combination of such values with the uncertainties of the experimentally determined neutron flux reported before give rise to final error in 39% (k=1). Although the uncertainty associated with this calculation methods appears relatively high, it is observed that the predicted mean values closely match the mean value of the experimental determination, demonstrating to be still useful for predictive purposes of the irradiated sample final activity.

About the activity concentration, given the mass of ^{160}Gd within the sample in 7.43 mg (see Table I) and the experimentally determined value in 110 MBq of ^{161}Tb, the final activity EOI per unit mass of gadolinium is approximately 15 $GBq_{Tb\text{-}161}$ per gram of ^{160}Gd.

IV. Conclusion

This work has demonstrated that the maximum production capacity of the ENEA TRIGA RC-1 reactor for ^{161}Tb to be 15 GBq per gram of ^{160}Gd (reaching the 15% of the saturation activity with a 77-hours irradiation schedule), as performed in the activation experiment in September 2024.

It is important to underline that the activity concentration obtained refers to the specific irradiation setup used in September 2024 (8.7 mg of Gd_2O_3, enriched to 98.2% in ^{160}Gd), in which the neutron self-shielding coefficient can be considered negligible. When irradiating larger amounts of material to maximize the overall final ^{161}Tb activity produced, the activity concentration is expected to be lower than the values obtained in this study.

Several calculations have been performed: irradiating approximately 125 g of Gd_2O_3 (enriched to 98.2% in ^{160}Gd) using the same irradiation protocol would result in a final activity of about 300 GBq, i.e. about 3 $GBq_{Tb\text{-}161}$ per gram of ^{160}Gd. Considering that the extraction and the chemical processing of the irradiated targets takes about one week (about one half-life of ^{161}Tb), the useful activity available for medical purposes is considered to be approximately 150 GBq. Given that a typical lutetium therapy involves six injections of 7 GBq each, administered at six-week intervals, the ^{161}Tb activity level reached is sufficient to treat around 20 patients per month.

In conclusion, ENEA's production capacity is sufficient to meet local demand for this radiopharmaceutical. However, ENEA could potentially contribute to production efforts aimed at supporting experimental studies or early-phase clinical trials.

Acknowledgment

The publication was created within the project SECURE funded by the European Union under grant agreement No. 101061230.

References

[1] N. Gracheva, C. Müller, Z. Talip, *et al.*, "Production and characterization of no-carrieradded ^{161}Tb as an alternative to theclinically-applied ^{177}Lu for radionuclidetherapy" *EJNMMI Radiopharmacy and Chemistry*, vol. 4, July 2019, Art. no. 12.

[2] S. McNeil, M. Van de Voord, C. Zhang, *et al.*, "«A simple and automated method for 161Tb purification and ICP-MS analysis of 161Tb" *EJNMMI Radiopharmacy and Chemistry*, vol. 7, Dec. 2022, Art. no. 31.

[3] D. Chiesa, M. Carta, V. Fabrizio, L. Falconi, A. Grossi and M. Nastasi, "Characterization of TRIGA RC-1 neutron irradiation facilities for radiation damage testing," The European Physical Journal Plus, vol. 135, March 2020, Art no. 349.

[4] J. F. Briesmeister, Ed., "MCNP: A General Monte Carlo N-Particle Transport Code," 2000.

[5] L. Falconi, N. Burgio, M. Palomba, E. Santoro e M. Carta, "A MCNPX TRIGA RC1 Experimental channels model for the design of a new neutronic diffraction facility" presented at RRFM Available: https://www.igorr.com/_media/proceedings:rrfm2016-transactions.pdf, Accessed on March 2016.

[6] American Society for Testing and Materials, "Standard test method for determining thermal neutron reaction and fluence rates by radioactivation techniques.", West Conshohocken, PA (United States), Tech. Rep. ASTM-E-262-97/XAB; TRN: 91440952, Aug. 1998.

[7] L. Lepore, L. Spagnuolo, L. Cozzella, T. Guarcini, F. Limosani, S. Placidi, A. Pagano, L. Falconi, V. Fabrizio, D. Formenton, M. Lammardo, A. Roberti, M. Capogni "ENEA TRIGA RC-1 reactor to medical radionuclides production within the EU Secure project, presented at RRFM Available: https://www.euronuclear.org/download/rrfm-2025/, Accessed on April 2025.

[8] N. Soppera, M. Bossant, E. Dupont, "JANIS 4: An Improved Version of the NEA Java-based Nuclear Data Information System", Nuclear Data Sheets, Volume 120, June 2014, Pages 294-296. Available: http://dx.doi.org/10.1016/j.nds.2014.07.071.

Hybrid Compton-PET imaging for Ion-Range Monitoring in Hadron Therapy

J. Balibrea-Correa[1],*, V. Babiano-Suarez[1], J. Lerendegui-Marco[1],P. Torres-Sánchez[1] , I. Ladarescu[1], M. Pallàs[2], B. Brusasco[2], J. Wulff[3], C. Bäumer[3], Severine Rossomme[4], Alexis Warnier[4],A. Tarifeño-Saldivia[1], C. Domingo-Pardo[1]

[1]Instituto de Física Corpuscular (CSIC-University of Valencia), Paterna, Spain
[2]Universitat Politècnica de Catalunya, Barcelona, Spain
[3]West German Proton Therapy Centre, Essen, Germany.
[4]Ion Beam Applications (IBA) , Belgium.
(*) javier.balibrea@ific.uv.es

Abstract— **Hadron therapy is a rapidly growing cancer treatment, offering high precision in targeting tumors while sparing healthy tissue. However, its full potential is limited by uncertainties in the range of therapeutic ions, largely due to current reliance on X-ray CT-based planning. Real-time, accurate range verification methods are needed to further improve treatment accuracy. Two main gamma-ray imaging techniques have been studied for in-beam range verification: Positron Emission Tomography and Prompt-Gamma Imaging. The first is sensitive to β^+ isotopes produced during irradiation but faces challenges like biological wash-out and low signal-to-background ratios. The latter uses prompt gamma-rays for real-time monitoring but is technically challenging due to high backgrounds and detection difficulties.**

Combining both into a hybrid PGI-PET approach leverages the strengths of both methods, offering high spatial and temporal resolution. In this study, the first experimental results using a hybrid imaging system in a cross configuration are presented. The experiment was carried out at the Western Proton Therapy Center in Germany with 100 and 125 MeV proton beams. Polyethylene phantoms were irradiated and imaged using both techniques during distinct beam on and beam off periods. Results demonstrate successful 2D reconstruction of both imaging techniques. The PGI images show the Bragg peak position as a maximum in the heat map, while the PET images offer high spatial resolution of β^+ byproducts, allowing indirect inference of the Bragg peak. These findings illustrate the potential of hybrid PGI-PET imaging for accurate, real-time verification of proton range in hadron therapy.

Keywords — **Hadron therapy, cancer, ion range monitoring, PGI, PET, hybrid system.**

I. Introduction

Over the past decades, hadron therapy has become one of the most promising treatments for certain types of cancer worldwide. According to the Particle Therapy Co-Operative Group [1], there are more than 100 facilities in operation, with over 350,000 patients already treated. Moreover, exponential growth in the use of this technique is expected worldwide, with more than 50 new facilities currently under construction or in the planning stages. A particular case is Spain with 10 facilities planned for the next 3 years.

The main advantage of hadron therapy, compared to conventional radiotherapy, is the ability to maximize the dose delivered to the target area while minimizing the dose to the surrounding healthy tissues. This reduces the probability of secondary cancers and other complications associated with conventional radiotherapy [2]. However, the full benefits of therapeutic hadron beams are still limited by uncertainties in ion range, or penetration depth, within tissue. In addition, current proton treatment plans rely on X-ray computed tomography techniques, where ion stopping power is estimated using semi-empirical photo-attenuation coefficients leading to range uncertainties of 1–3%, which, in turn, require conservative safety margins of up to 3.5% + 3 mm in this type of radiotherapy [3,4,5]. As a result, a significant volume of healthy tissue is irradiated to ensure the entire tumor is treated holding back the full benefits.

A real-time, sub-millimeter ion-range verification method could help improve this situation by enhancing the spatial accuracy of dose delivery [2,4,5]. This would increase the therapeutic benefits of hadron therapy and could potentially expand its application to other diseases [7–9].

Presently, there are two main γ-ray imaging techniques that have been thoroughly researched in clinical conditions for in-beam ion-range verification: Positron Emission Tomography (PET); and Prompt-Gamma Imaging (PGI).

PET was the first online methodology for ion range verification [10] and it is one of the most extensively researched in the field [11,12]. It is commonly based on ^{11}C, ^{13}N and ^{15}O

unstable β^+ isotopes produced as a consequence of nuclear reactions induced by primary ion-beam along the irradiated tissue [13-15].

The challenges of the PET technique strongly depend on the time lapse between irradiation and measurement. In this sense, the main challenges are related to biological wash-out effects and relatively low signal to background ratio [2,16,17]. This requires the use of innovative non-standard PET systems which are currently in development [12,18] .

PGI, on the other hand, relies on the secondary γ-rays emitted from the nuclear reactions occurring in the ion path along the patient tissues [19]. The technique was proposed in the early 2000s [20], and demonstrated by Min et al. in 2006 [21] by means of a mechanically collimated gamma-camera. Prompt γ-rays, with energies covering from a few keV up to 7-8 MeV, are specially well suited for real-time monitoring because of the high spatial and time correlation with the primary proton range (Bragg peak). However, PGI is extremely challenging because of the harsh experimental conditions during the irradiations, including large backgrounds and instantaneous high count rates [2,21-23].

A clinical slit-camera has been developed for ion range verification with very satisfactory results [24]. However, slit-cameras have a rather low detection efficiency which severely constrains their precision in the 1D-profile [25].

Both imaging techniques, PET and PGI, could be simultaneously combined in the so-called (PGI-PET) hybrid scheme as originally proposed by K. Parodi in 2016 [26]. Such imaging system would enable the possibility to combine PET inherent tomographic and high resolution features with the PGI high-yield prompt γ-rays closely linked in position and time to the Bragg peak. This idea could be implemented based on the combination of multiple Compton cameras in such geometric configuration that both imaging techniques can be applied simultaneously [27].

In previous works, we explored the feasibility of applying the hybrid PGI-PET technique to in-beam ion-range verification using two-planes Compton imagers [28–30]. In fact, PGI-PET can be fully exploited by utilizing pulsed beam structures with pauses between ion spills delivery. In this scheme, PGI is employed during the beam delivery periods, while PET is used during the waiting periods between ion spills. These time intervals are referred to in this work as in-spill and off-spill time periods, respectively. It is worth mentioning that depending of the in- off- spills time duration, the sensitivity of both imitating techniques could be adjusted aiming, for instance, to short- or long-lived β^+ isotopes.

In the present work we describe the first PGI-PET imaging results from an experimental campaign carried out in 2024 at WPE-Germany with the IBA Proteus-Plus, an Isochronous Normal Conducting Cyclotron.

II. Experimental setup

The experiment was carried out at the Western Proton Therapy Center (WPE) in Germany using an Isochronous Normal Conducting Cyclotron (*IBA-Proteus Plus*) in pencil mode with a size of about 3 mm. The proton beam, operated with energies of 100 and 125 MeV, was used in clinical operation but regulated to the minimal current possible (0.3 nA). An ad-hoc time structure was used during the whole experiment to fully exploit the PGI-PET technique with duty cycles of 400 ms in-spill and 1 to 2 s off-spill. A total of 40×10^9 protons were used for the individual irradiations in order to ensure sufficient statistics for all the configurations measured.

Figure 1: Picture of the experimental setup aligned with the beam isocenter.

The hybrid PGI-PET experimental setup consisted of four two-planes Compton Imagers in a cross configuration aligned with the proton beam isocenter as shown in Fig.1. This geometric configuration was preferred to the one used in previous works in order to enhance the PGI efficiency in exchange of restricting the PET FoV [30]. The distance between front faces of Compton imagers was set to 22.3 cm. The individual Compton imagers used in this experimental setup were originally optimized for (n,γ) cross section experiments [31-33], where the γ-ray signal strongly depend on the isotope under study and neutron induced backgrounds are a major issue.

Therefore, the setup is optimized for a broad γ-rays energy range that spans from a few hundreds of keV up to several MeV while minimizing the sensitivity to neutron interactions with the detectors. For further details the reader is referred to the aforementioned references.

The individual detectors of the Compton imagers are read by silicon photomultipliers (SensL ArrayJ-60035-65P-P) and readout using an acquisition system based on PETsys Front-End Board D version 2 (FEB/D-1024) [34]. A liquid cooling system ensured that the temperature of all the electronic components was maintained through the entire campaign, minimizing possible electronic gain drifts.

The data from all the detectors were registered into a high performance computer using a data-acquisition board with a maximum event rate of 100 MEvents/s thanks to its optical fiber SFP+ high-speed links, thus minimizing any loss due to the high count rate [34]. Dedicated GPU-assisted Machine Learning analysis routines were used to reconstruct the Compton and PET events [35,36]. The latter was fed into the Compton and PET imaging GPU-based reconstruction algorithms developed within our research group aiming for real time reconstruction [28-30,36-38].

In the current geometric configuration the PET Field of View is constrained by the size of the detectors in the absorber plane, yielding to a maximum of 120x120mm2 along the beam axis. On the other hand, the PGI Field of View extends at least twice the size of the PET. However, angular resolution and efficiency of Compton imaging decrease rapidly with angles larger than 45° respect to the sight line of the Compton image [28].

Several polyethylene phantoms of size 10×10×7.5 cm3 were irradiated during the experiment. The phantoms were placed along the center of the experimental setup as shown in Fig. 1.

III. Experimental procedure and first experimental results

The experiment was designed to minimize systematics and facilitate the comparison with experiments carried out in other facilities. Thus, the experimental procedure was as follows: For a fixed proton energy, either 100 or 125 MeV, a polyethylene phantom was placed in a well defined position along the beam axis. The phantom was irradiated using the ad-hoc pulsed beam registering all events in- and off-spill time periods as explained before. After the proton irradiation, the phantom was exchanged with a fresh one, thus removing any long-lived β^+ residual activity that could mask the results. The procedure was repeated at different positions covering a maximum of ±40 mm along the beam axis with respect to the center of the experimental setup.

In this work, for demonstration purposes. we present the first 2D reconstructed PGI and PET images results during in- and off-spill time periods, respectively, for 100 MeV protons. In both cases, the phantom frontal face position corresponds to the position +4 mm in the direction of the beam.

The PGI result for a broad FoV of 400 × 400 mm² is displayed in Fig. 2. The image was obtained by combining the individual Compton-reconstructed distributions at a distance of 11,3 mm from the frontal face of the detectors corresponding with the beam axis. All registered Compton events during the in-spill periods with energies above 500 keV were used for the Compton image reconstruction. The reconstructed heat map in Fig.2 is overlaid with the phantom position, shown as a light blue box, and the beam direction is indicated by a red arrow. The maximum in the heat map corresponds to the expected Bragg position within the phantom since, as explained previously, the Bragg peak position is closely related to the expected prompt γ-ray maximum emission location in the phantom. This is the first, but encouraging, result we have obtained for 100 MeV protons using the full statistics of 40×10^9 protons.

From the position of the maximum in the heat map, we can experimentally determine the location of the Bragg peak [36-38], thus enabling us to evaluate the methodology across a broad range of positions along the beam axis—in particular, over a range of 80 mm.

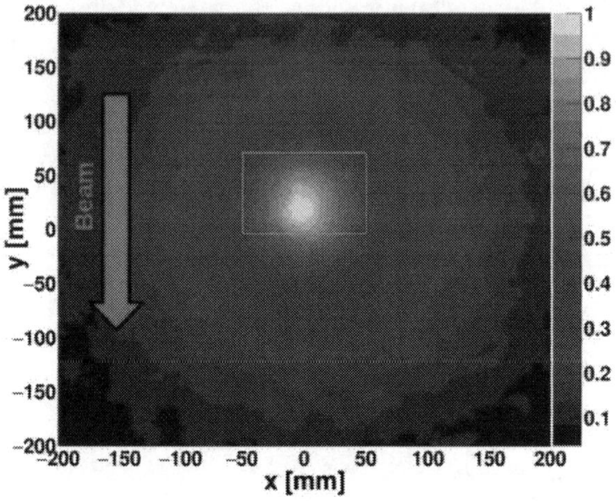

Figure 2: In-spill PGI 2D diagram reconstructed from one of the proton irradiations.

Fig. 3 presents the reconstructed 2D PET distribution for off-spill periods during the same proton irradiation, with a FoV of 120 × 120 mm². Full statistics from the irradiation were considered for the calculation as in the case of PGI. The image was obtained by projecting the 3D reconstruction, which combines information from all detectors simultaneously. Only time-coincident events with individual energies of 511 keV were considered for image calculation. As in Fig.2, the position of the phantom is overlaid, along with the beam direction.

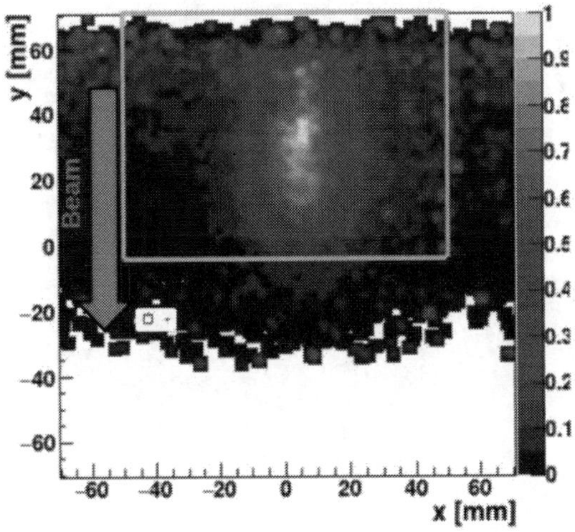

Figure 3: Off-spill PET 2D diagram reconstructed from one of the proton irradiations.

The restricted PET FoV is compensated by its unparalleled position resolution. Compared to the PGI distribution in Fig. 2, the PET distribution is much narrower, with some structure visible from the β^+ byproducts within the phantom. In this case, the Bragg peak is not directly measured but can be inferred from the 2D distribution, specifically from the deepest edge as previously described [36–38]. Repeating this methodology across all available positions will enable us to evaluate ion range verification using the PET technique.

IV. Conclusions and Outlook

In this study we report on the development and experimental evaluation of an hybrid PGI-PET system for proton range verification based on four high-efficiency Compton imagers in cross configuration. The experiments were conducted at the Western Proton Therapy Center (WPE) in Germany, utilizing an Isochronous Normal Conducting Cyclotron (IBA-ProteusPlus) operating in pencil beam mode with energies of 100 and 125 MeV. The hybrid imaging system is , optimized for both prompt gamma and β^+ emission detection. Polyethylene phantoms were irradiated at well-controlled positions along the beam axis, and an optimized pulsed beam structure was employed to separate in-spill (PGI) and off-spill (PET) data acquisition. Advanced GPU-assisted machine learning routines enabled efficient, high-throughput reconstruction of both PGI and PET images. The system allowed for the experimental determination of the Bragg peak position over a range of 80 mm along the beam axis.

The results demonstrate that the hybrid PGI-PET setup can successfully acquire and reconstruct 2D images for both prompt gamma and positron emission events during and after proton irradiation. The chosen cross configuration of Compton imagers significantly enhanced PGI detection efficiency with respect to previous setups. The PGI method enabled direct localization of the Bragg peak, observed as the maximum in the reconstructed heat map, while PET imaging provided a high spatial resolution distribution of β^+ byproducts, from which the Bragg peak could be inferred based on the distal edge of the activity profile. The experimental results show the feasibility of this hybrid approach for in situ ion range verification, offering complementary spatial and temporal information from both imaging modalities.

Results presented in this study correspond to one phantom position (+4 mm) and 100 MeV proton beam (Fig.2-3). This work is being extended to all measured positions and energies (125 MeV). This will allow one to quantify the accuracy of both techniques. A sensitivity study for different statistics will be also conducted in order to estimate the capability of performing ion-range verification in quasi real-time. From the hardware and software perspective, optimization of the detector configuration and imaging algorithms is anticipated to improve both efficiency and spatial resolution. We aim at adapting this methodology to more complex and clinically relevant scenarios, including anthropomorphic phantoms and, ultimately, patient studies. Continued development and integration of advanced machine learning algorithms are expected to further enhance real-time image reconstruction and quantitative analysis capabilities. Overall, this hybrid PGI-PET study represents a step ahead toward clinical implementation of real-time, accurate range verification in proton therapy.

Acknowledgment

This work derives from fundamental research carried out with funding from the European Research Council (ERC) under the European Union's Horizon 2020 research and innovation programme (ERC Consolidator Grant project HYMNS, with grant agreement No.~681740) and part of this work has been funded under ERC Proof-of-Concept Grant GNVISION (Grant Agreement 101113330). The authors acknowledge support from the Spanish Ministerio de Ciencia e Innovación under grants PDC2021-121536-C21, PID2019-104714GB-C21 and RTI2018-098117-B-C21. We acknowledge support from the Severo Ochoa Grant CEX2023-001292-S funded by MCIU/AEI. J. Balibrea-Correa is supported by grant ICJ220-045122-I and J. Lerendegui-Marco is supported by grant FJC2020-044688-I funded by MCIN/AEI/10.13039/501100011033 and CIAPOS/2022/020 funded by MICIN/AEI and Generalitat Valenciana. Ion Ladarescu was funded by the R&D grant PDC2022-133382-I00, funded by the Spanish Ministry of Science (MCIN/AEI/10.13039/501100011033).

A. References

[1] https://www.ptcog.site/
[2] A. Knopf *et al.* Physics in Medicine and Biology **58** 13 (2013)
[3] M. Moyers *et al.* Medical Dosimetry **35** 179-194 (2010)
[4] H. Paganetti *et al.* Physics in Medicine and Biology **57** 11 (2012)
[5] M. Yang *et al.* Physics in Medicine and Biology **57** 13 (2012)
[6] A. Kraan Frontiers in Oncology **5** (2015)

[7] M. Durante & H. Paganetti Reports on Progress in Physics **9** 096702 (2016)

[8] M. Durante et al. Nature Reviews Clinical Oncology **14** 483-495 (2017)

[9] M. Durante British Journal of Cancer 8 777-778 (2019)

[10] L. Llacer Nucl Sci Appl, 3 (1988), pp. 111-131

[11] W. Enghardt et al. NIM-A **525** (2004) pp. 284-288

[12] M. Bisogni et al. Journal of Medical Imaging 4 (2016) pp. 011005

[13] A. Kraan Particle Radiosurgery Conference 30 (2014) pp. 559-569

[14] M. Moteabbed et al. Physics in Medicine and Biology 56 (2011) pp. 1063

[15] K. Parodi et al. Recent Developments in Nuclear Medicine Imaging and Therapy 33 (2023) 22-34

[16] K. Parodi et al. International Journal of Radiation Oncology Biology Physics 68 (2007) pp. 920-934

[17] G. Shakirin et al. Physics in Medicine and Biology 56 (2011) pp 1281

[18] V. Ferrero et al. Scientific Reports 8 (2018) pp. 4100

[19] V. Bom et al. Physics in Medicine and Biology 57 (2011) pp. 297

[20] F. Stichelbaut PTCOG23 Conference Proceedings

[21] C. Min et al. Applied Physics Letters 89 (2006) pp. 183517

[22] E. Testa et al. Applied Physics Letters 93 (2008) pp. 093506

[23] J. Verburg and J. Seco Physics in Medicine and Biology 59 (2014) pp. 7089-7106

[24] J. Smeets et al. Physics in Medicine and Biology 57 (2012) pp. 3371-3405

[25] J. Krimer et al. Radiation Imaging Techniques and Applications 878 (2018) pp. 58-73

[26] K. Parodi NIM-A 809 (2016) pp. 113-119

[27] C. Lang et al. Journal of Instrumentation 9 (2014) pp. 01008

[28] J. Balibrea-Correa et al. EPJ WoC 261, (2022) pp. 05002

[29] J. Balibrea-Correa et al. Eur. Phys. J. Plus **137** (2022) pp. 1258

[30] J. Balibrea-Correa et al. arXiv:2504.11273 (2025)

[31] C. Domingo-Pardo NIM-A 825 (2016) 78-86

[32] V. Babiano et al. NIM-A 953 (2020) pp. 0168-9002

[33] C. Domingo-Pardo et al. 284 (2023) pp. 01018

[34] A. Di Francesco NIM-A 824 (2016) pp. 194-195

[35] J. Balibrea-Correa et al. NIM-A 1001 (2021) pp 165249

[36] J. Balibrea-Correa et al. IEEE NSS/MIC (2021) pp: 1-7

[37] J. Lerendegui et al. Scientific Reports 12 (2022) pp. 2735

[38] J. Balibrea et al Arxiv, 2504.11273

EPJ Web of Conferences **338**, 09008 (2025) https://doi.org/10.1051/epjconf/202533809008
ANIMMA 2025

Feasibility study of a 2D Amorphous Silicon-based Detector for Proton Therapy

Khalil El Achi[1]*, Alexis Warnier[2], Victor de Beco[2], Severine Rossomme[2], Eduardo Cortina Gil[1]
[1]*Université Catholique de Louvain*, Louvain-la-Neuve, Belgium
[2]*IBA Dosimetry*, Louvain-la-Neuve, Belgium
*khalil.elachi@uclouvain.be

***Abstract*—New particle therapy modalities, including ultra-high dose rate (UHDR) proton therapy, present various challenges for particle detectors used in patient-specific quality assurance (PSQA). Amorphous Silicon (a-Si) Flat Panel (FP) detectors are a powerful tool used to create a 2D image of an administered dose map. They are known for high radiation hardness and image resolution. To investigate and characterize the behavior of a-Si pixels under particle therapy irradiation, specifically proton irradiation, a 2.4 x 2.4 mm pixel array of a-Si pixels, photodiodes coupled with thin-film-transistors (TFTs), was acquired and tested. Before irradiation, electronic probing was performed to generate the I-V curves and to qualify the stability of the pixel current for different bias and gate voltages to understand the proper voltage regime for optimal pixel functionality. In addition, proton irradiation was performed using pencil beam scanning (PBS) in a medical facility for charge linearity verification through direct detection. The test was done in conjunction with an IBA Razor^{TM} Detector to compare each detector's performance. Finally, irradiation with a high-power green laser was performed to examine the performance of the individual components of the pixels under high-intensity irradiation.**

***Index Terms*—Amorphous Silicon, Flat Panel Detectors, Medical Dosimetry, Proton Therapy, and Patient-Specific Quality Assurance,**

I. Introduction to Particle Therapy

ROBUST cancer treatments often include radiation therapy modalities that involve delivering high-energy particles to targeted areas [1]. This results in depositing lethal doses only to tumor regions, thus minimizing the damage to healthy cells. Many particles are employed for particle therapy. Figure 1 showcases the energy deposition spectra for common particle therapies, which were generated in Geant4, a Monte Carlo (MC)-based simulation toolkit [2]. Several radiation therapy modalities utilize Gamma beams with energies ranging from $1 \sim 20$ MeV, which deliver the majority of the dose near the entry point and significant energy deposition to nearby healthy cells. In contrast, protons and heavy ions have the unique capability to deliver the majority of their energies at the end of their trajectory near their stopping point. A phenomenon resulting in the Bragg Peak. First observed by Robert Wilson [3], and has since been followed with immense progress in the field of radiation physics.

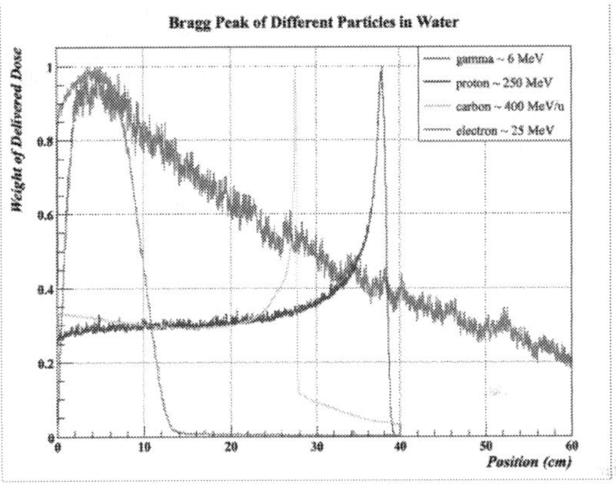

Fig. 1: Spectrum of deposited dose in water of different particles used in radiation therapy treatments.

A. Proton Therapy

A particular therapy modality that utilizes a high-energy proton beam with energies ranging from up to ~ 250 MeV. The beam is capable of penetrating healthy tissue, delivering minor non-lethal doses along its path, only delivering the lethal dose at the end of its trajectory to the targeted region. At present, the treatment harnesses a narrow and magnified proton beam via a set of scanning magnets that guide the beam to scan a desired region [4]. This can be visualized in Figure 2. Ultra-high-dose-rate (UHDR) proton flash therapy, a new mechanism in proton therapy treatment, is of great interest to several clinical, research, and commercial groups. The treatment is able to deliver extremely high doses in a few milliseconds with dose rates above 40 Gy/s. This culminates in reducing normal cell toxicity and error margins whilst treating mobile cells [5]. In comparison, intensity-modulated proton therapy (IMPT) only delivers doses with a dose rate of a couple of Gy/min [6].

*Khalil El Achi is at the Centre for Cosmology, Particle Physics and Phenomenology at the Université Catholique de Louvain, Louvain-la-Neuve, 1348, Belgium (email: khalil.elachi@uclouvain.be).
Eduardo Cortina Gil is at the Centre for Cosmology, Particle Physics and Phenomenology at the Université Catholique de Louvain, Louvain-la-Neuve, 1348, Belgium (email: eduardo.cortinagil@uclouvain.be)
Alexis Warnier is at IBA Dosimetry, Louvain-la-Neuve, 1348, Belgium (email: Alexis.Warnier@iba-group.com)
Victor de Beco is at IBA Dosimetry, Louvain-la-Neuve, 1348, Belgium (email: Victor.DeBeco@iba-group.com)
Severine Rossomme is at IBA Dosimetry, Louvain-la-Neuve, 1348, Belgium (email: Severine.Rossomme@iba-group.com)

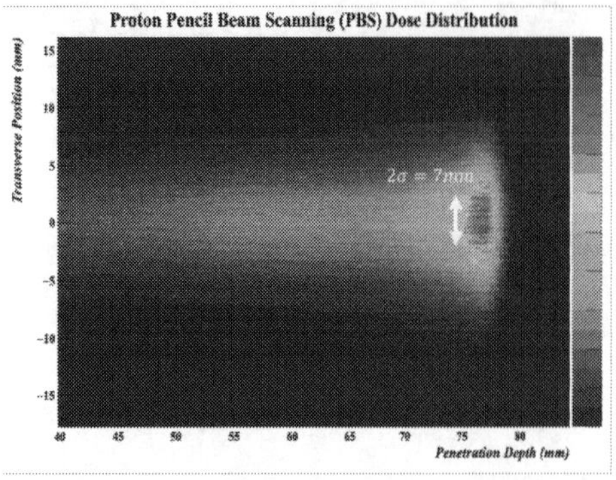

Fig. 2: Distribution of dose delivered via PBS utilizing a narrow-focused beam.

II. Semiconductor-based Detectors

Semiconductor-based detectors function through a prowess similar to that of gas-based detectors. The active region consists of a semiconducting junction, also known as the space-charge region. Generally, a reverse bias voltage is applied to the junction, creating an electric field. When exposed to different radiation sources, electron/hole pairs are created and thus drift via the created electric field. This motion induces a current on the electrode that is then analyzed via the readout electronics [7]. The current is then integrated to get the total induced charge that will then be converted to the deposited dose in the active region of the detector.

A. Amorphous Silicon (a-Si) in Particle Detectors

Amongst semiconducting materials that are employed in particle detectors, hydrogenated a-Si (a-Si:H) is a good candidate for investigating its usage because of its high radiation hardness and availability via commercial groups [8]. Radiation hardness is acquired as a result of random arrangements of atoms throughout the structure. The interlocking silicon bonds are not saturated, leading to dangling bonds and intermediate energy states in between the valence and conduction bands [9]. Furthermore, the addition of hydrogen passivation leads to the stabilization of these bonds and protects the materials from different defects triggered by radiation sources. Due to its capacity to resist heavy irradiation, a-Si detectors were of interest to several groups that have tested such materials specifically under heavy irradiation to assess their performance over time, such as [10] and [11].

B. PD and TFT Layout

An individual a-Si pixel is comprised of a photodiode (PD) coupled with a thin-film-transistor (TFT). The PIN diode formation is engineered by an intrinsic a-Si layer that is sandwiched by a highly doped p (acceptor dopants) and n (donor dopants) layers on either side. To reverse-bias the PIN diode, the p-layer must be connected to a negative bias voltage. Whilst the n-layer is connected to a base electrode, where the current is induced due to radiation exposure. Subsequently, the induced current will be stored in the source electrode of the TFT, which acts as a switcher. The gate is kept closed with a negative voltage that forbids the passage of the charges. Once the gate is set to open with a positive voltage, the charges pass to the drain electrode and are subsequently collected in the data acquisition system [12]. The process is illustrated in Figure 3.

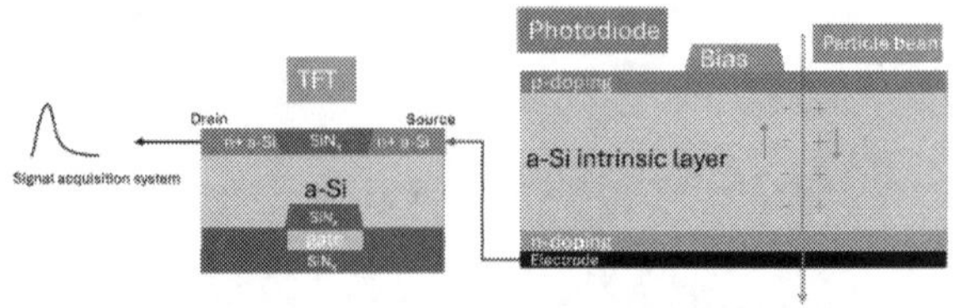

Fig. 3: Cross-section of the photodiode coupled with a TFT forming the a-Si pixel geometry

C. A-Si FP Detector Technology

From the standpoint of mass production, large-area 2D detectors for medical imaging are normally built using a-Si (PD + TFT) coupling. Figure 4 shows a simplified top-view of such an FP system.

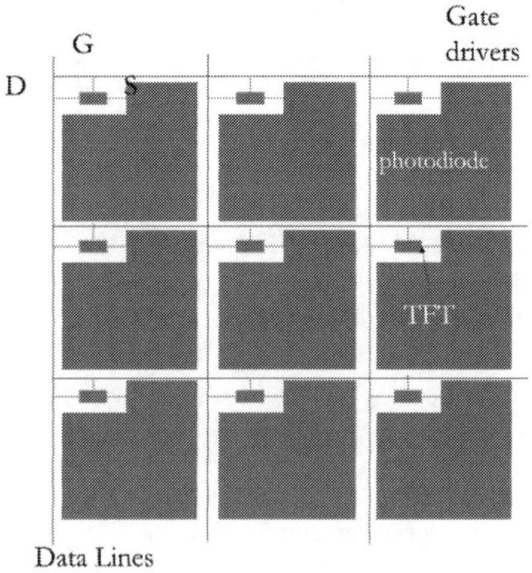

Fig. 4: Array of a-Si pixels comprised of a photodiode + TFT powered by gate and data lines

The pixels are fabricated on glass or plastic substrates and controlled via gate drivers that successively scan the detector row by row by introducing a positive voltage, which drains the induced current into the data lines and via the readout integrated circuits that integrate the measured current to get a designated charge value for each pixel [13]. The working mechanism of the TFTs can be visualized in Figure 5. By introducing a small drain voltage above the threshold voltage, a current should be measured on the drain electrode. Sweeping from negative to positive gate voltages shows almost a null value of the current measured when the gate voltage is negative, suggesting a "closed" state. When the gate voltage becomes positive, a large jump in the drain current of almost 8 order of magnitude from the lowest to the highest current measured is brought on, which switches on the readout of the pixel.

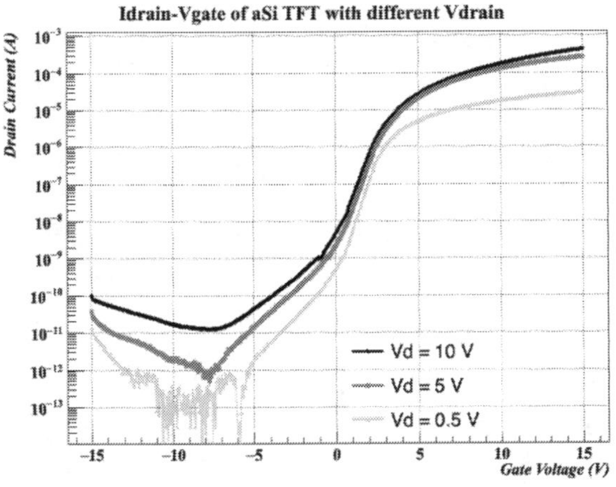

Fig. 5: The drain current vs. gate voltage of a-Si TFTs for different drain voltages.

Considering the advantage of radiation hardness and large-scale manufacturing capacity, a-Si FP detectors provide high image resolution, especially for quality assurance (QA). To put this into perspective, Figure 6 shows a pair of three PBS dose patterns measured by two flat panel detectors. The first set of dose patterns in Figure 6a is measured by an ionization chamber-based 2D detector. The IBA MatriXX Resolution^{TM} is used for dose distribution measurements, with an image resolution of 6.5 mm [14]. Figure 6b shows the second set of dose patterns measured by an a-Si FP detector. The IBA Sphinx Compact^{TM} is primarily used for PBS machine QA with an image resolution of 0.2 mm [15].

III. Project's Motivations and Objectives

This project endeavors to develop a semiconductor-based 2D FP detector capable of performing dosimetry measures in a wide range of clinical settings, which includes low-fluence X-ray radiation therapy and UHDR proton therapy. To proceed, a test key of an array of a-Si pixels with a 2.4 x 2.4 mm active area was acquired and thoroughly tested. Initially, the test key is probed to characterize leakage current and current stability. After wire-binding it, a high-power green laser is used to further characterize charge sensitivity, and linearity was tested. Such a-Si Pixels are usually used via indirect detection as X-ray imaging tools. A thick plastic scintillator is placed in front of the pixel board that produces visible light in the green spectrum when irradiated with X-rays [12]. But these types of detectors have shown their capacity in direct detection of high-energy proton beams to generate a dose map of the treatment plan [16]. Thus, it is also vital to perform irradiation measured with a proton PBS beam within conventional ranges of proton therapy to observe charge linearity compared to a calibrated detector.

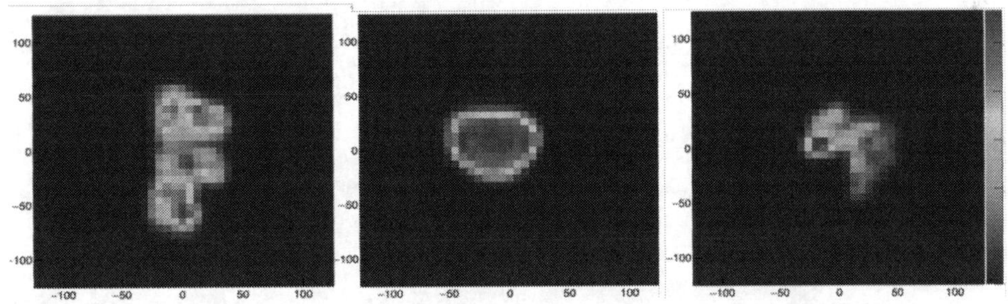

(a) PBS dose patterns measured by the IBA MatriXX Resolution^{TM}

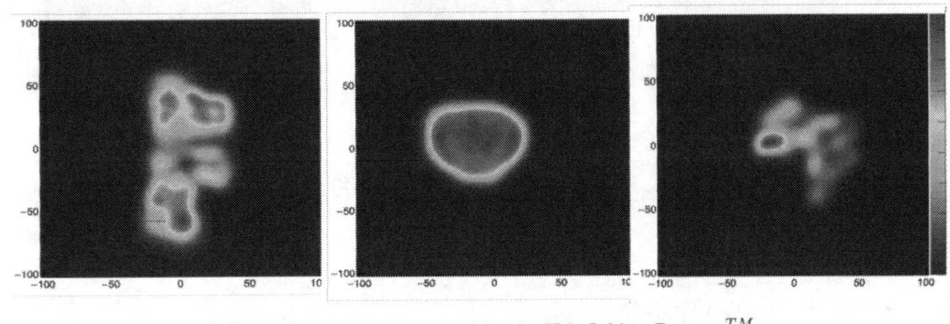

(b) PBS dose patterns measured by the IBA Sphinx Compact^{TM}

Fig. 6: Set of three PBS dose patterns measured by separate detectors showing the increased image-resolution brought by a-Si detectors

IV. Outcome & Results

A. Characterization of a-Si Pixels with PM8PS Semiconductor Probing Device

Before performing irradiation testing, the behavior of the pixels at different bias voltages was thoroughly examined using the PM8PS Semiconductor Probing Device. The devices consist of three probes that come into contact with the electrode pads, supplying different voltages and measuring the current flowing through each electrode.

The setup consists of placing the wafer on a light-isolated vacuum stage and manually placing the three probes on the three necessary electrodes: the gate, drain, and bias pads. The probes are guided via an optical microscope that exposes pixels to visible light via different intensities. The gate voltage is set at a constant voltage ($V_{gate} \geq 5V$) and the drain voltage remains floating at $0V$. Furthermore, the voltage is swept on the bias pad from $-14V \rightarrow 0V$, to remain under the reverse bias measuring condition. The setup is illustrated in Figure 7.

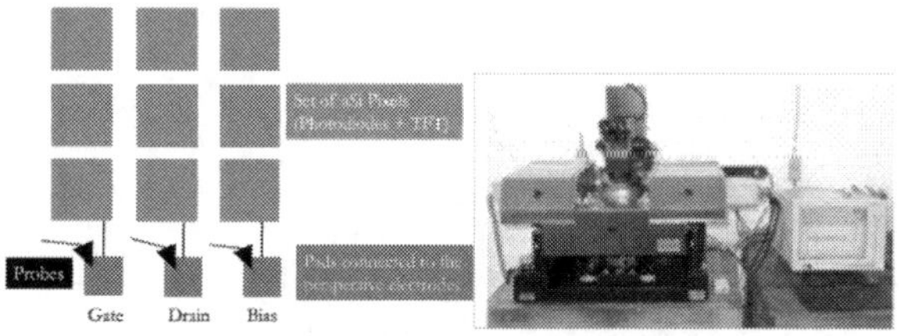

Fig. 7: Illustrating how the probing measurements were taken using the PM8PS Semiconductor Probing Device

The leakage current of both the a-Si PDs and a-Si pixels (PD +TFT) was measured, as seen in Figure 8a. This bias sweep is performed under full dark conditions. PDs exhibit a leakage current of -2 to $-3pA/mm^2$, while pixels exhibit a leakage current ranging between -14 to $-12pA/mm^2$. Note that both the photodiode and the TFT contribute to the measured leakage current.

The TFT introduces a significant contribution to the leakage current. With that being said, it is still orders of magnitude lower than the current value that is expected with different irradiation measurements. This will be discussed in more detail in Section IV-C.

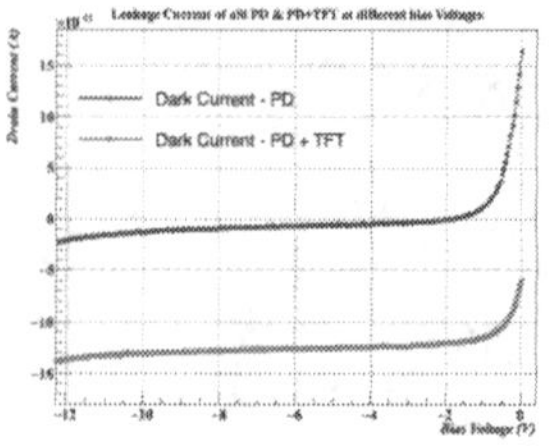

(a) The leakage current $/mm^2$ measured in dark conditions.

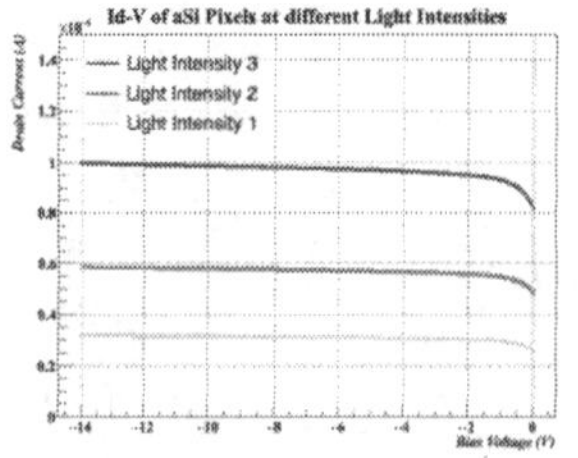

(b) Drain Current measured by the pixels at different light intensities.

Fig. 8: Drain Current measured in dark and light conditions to assess charge stability at different reverse bias voltages.

The measurements were repeated by exposing the pixels to visible light from the microscope at different intensities, Figure 8b. The I-V curve shows a linear current jump from one intensity to the other and adequate current stability. Thus, the region between $-12V$ to $-5V$ shows a slight increase in current output, but is fairly stable if the bias voltage fluctuates. Considering that the PDs are made of a-Si P-I-N diodes, the active region is restricted by the thickness of the intrinsic layer. Once the diode is fully depleted via the reverse bias voltage, the sensitivity of the pixels should not change much. With that, the charge collection efficiency starts to plateau at about $-3V$, above that voltage, the current stabilizes.

Note that only the response from the pixels is shown, since the behavior of PD and pixels is similar.

B. Charge and Current Acquisition with Laser Irradiation

The test board was bonded onto a circuit board and transformed into a functioning detector. Using a power supply with two outlets, the bias and gate voltages were supplied. The drain current was continuously read through the IBA DOSE-X^{TM} electrometer [17].

To test functionality and proper performance of the circuit board, it was placed inside a black box, which was used to shield off ambient light, with a high-power continuous green laser, as seen in Figure 9. The test board was irradiated by applying (1) bias voltages to test for the detector sensitivity and charge linearity at each voltage state, and (2) different gate voltages to verify the functionality of the gate switchers of the TFTs.

Fig. 9: The setup showing the a-Si Pixels and a high power laser in a black box

The pixels were irradiated for a set interval of time, and the integrated charge was read via the electrometer. Taking into account the results obtained in section IV-A, three discrete voltages were chosen $-12V$, $-9V$, and $-6V$ in the plateau region of the pixels' current output seen in Figure 8.

The integrated charges for each voltage are plotted in Figure 10a. The pixels exhibit pristine charge linearity at all bias voltages. The sensitivity of the pixels remains very similar at each bias voltage with an average increase of about 2.5% for the chosen voltages. This is in correlation to what was seen in section IV-A, where the current increased slightly within -6 and $-12V$. Thus, keeping the diode within a specified reverse bias voltage is imperative to maintain constant sensitivity during different irradiation measurements. Further tests will be conducted on some samples to determine the maximum reverse bias voltage that can be applied to the diodes before current breakdown occurs.

The same test was repeated, where the pixels were irradiated for a set time for the three selected bias voltages with a gate voltage sweep, as seen in Figure 10b. As predicted, when the gate voltage is more negative than the bias voltage, the pixel is in an "off" state and the charge is almost null. Whereas, when the gate voltage is $\geq 5V$ the pixel is "on" and the charge integrates equally in all voltages. This mirrors what was seen in Figure 5. The drain current of the TFTs was low at negative voltages and then rose to high and constant values at positive gate voltages.

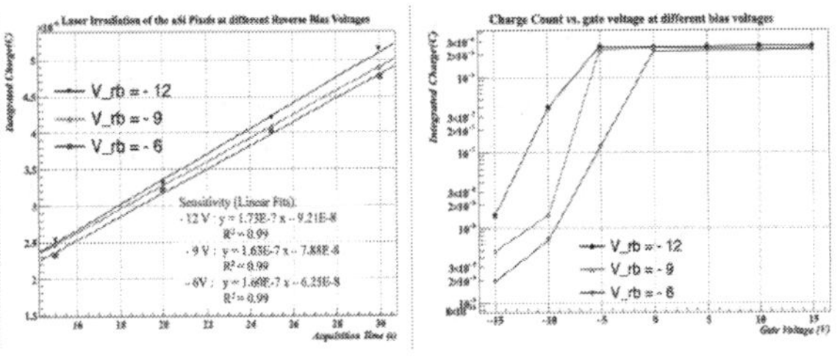

(a) Integrated charge vs. time at different bias voltages. (b) Integrated charge vs. gate voltage at different bias voltages.

Fig. 10: Irradiation of a-Si pixels with green laser at 1.5 A intensity under different conditions.

C. Irradiation Measurement with Proton Pencil Beam (PBS)

Irradiation measurements with a quasi-continuous PBS proton beam were performed in the Proton Therapy Center in Prague, Czech Republic. All measurements were conducted by delivering a 10 x 10 cm^2 uniform field. The beam was delivered with a constant current ($\sim$ $200nA$) at the exit of the cyclotron, whilst varying the delivered dose in Monitor Units (MU). The a-Si test board was placed in the middle of the field, along with the IBA RAZOR^{TM} Detector [18], as seen in Figure 11. The latter is composed of a p-doped c-Si diode used for small-field dosimetry. The silicon chip has a diameter of $0.6mm$ and a thickness of $20\mu m$.

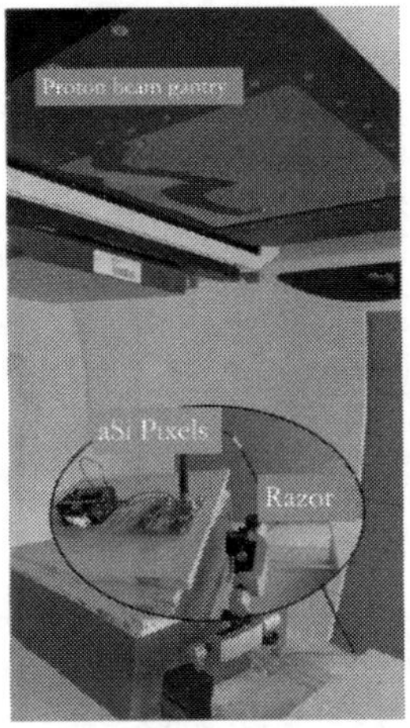

Fig. 11: Measurement setup of both the a-Si pixels and RAZOR^{TM} detectors in the PBS flat field region.

The RAZOR^{TM} detector and a-Si are placed side by side to compare the charge linearity due to proton irradiation of the test board with a calibrated diode, ensuring predictable and proper performance.

Both detectors performed properly under proton irradiation with charge linearity and proper current acquisition, with ambient light shielding in the beam room, as seen in Figure 12. The test board has proven its efficacy and capacity to be used as a dosimetry tool via direct detection of proton beams in clinical dose ranges.

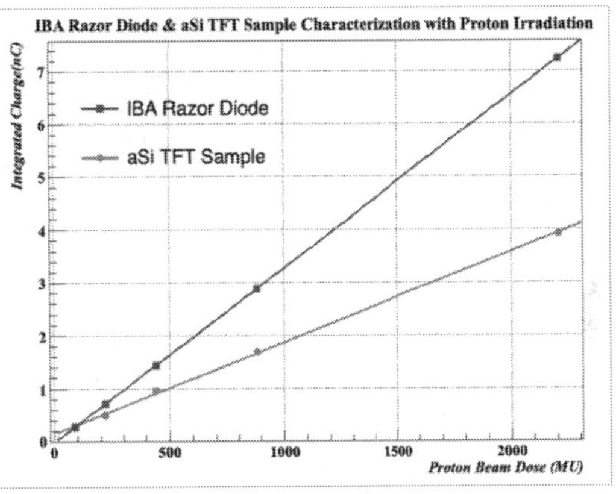

Fig. 12: a-Si pixels (red) and RAZOR^{TM} (blue) Detectors' charge acquisition vs. dose (MU)

V. Conclusion and Further Remarks

Large a-Si TFT FPs offer high radiation hardness and image resolution. In comparison with flat panel ionization chambers, a-Si FP offers about $30x$ more image resolution. The presented probing results show that a-Si Pixels are characterized by low leakage current and stable current with reverse bias voltage at several light intensities in the plateau region of the pixels' current output. The working principles of the a-Si pixels were characterized using a high-power green laser to confirm the functionality of the PDs in light detection and the TFTs as gate switchers. Furthermore, measurements with proton PBS beams show a good charge linearity for both detectors, without any saturation in conventional treatment doses. These tests verify the amicable position of a-Si TFT pixels as a particle detector for medical dosimetry in proton therapy QA. The diodes will be further tested with proton flash dose rates to assess issues with saturation and image lagging with improved circuitry to accommodate higher current throughput.

Acknowledgment

This work is a collaboration between the Université Catholique de Louvain and IBA Dosimetry and other partners, which is funded by Region Wallonne under Grant n9810. Our research endeavors are supported by IBA Dosimetry through access to their facilities, the use of their equipment, and the collaboration of their members, along with several departments in UCLouvain, such as the WELCOME platform for access to the semiconductor device probing station. All mentioned co-authors have contributed significantly to the success of this work through their assistance with equipment management, experimental methods, and academic literature.

Abbreviations and Acronyms

Abbreviation	Full Term
a-Si	Amorphous Silicon
c-Si	Crystalline Silicon
PD	Photodiode
TFT	Thin-Film-Transistor
PBS	Pencil Beam Scanning
UHDR	Ultra-high Dose Rate
FP	Flat Panel
IMPT	Intensity Modulated Proton Therapy
PSQA	Patient-Specific Quality Assurance
MU	Monitor Unit

TABLE I: List of abbreviations used throughout the paper

References

[1] M. Durante, R. Orecchia, and J. S. Loeffler, "Charged-particle therapy in cancer: Clinical uses and future perspectives," *Nature Reviews Clinical Oncology*, vol. 14, no. 8, pp. 483–495, 2017.

[2] A. S. et al., "Geant4 - a simulation toolkit," *Nuclear Intrumentation Methods*, no. 506, pp. 250–303, 2003.

[3] R. R. Wilson, "Radiological use of fast protons," *Radiology*, vol. 47, no. 5, pp. 487–491, 1946.

[4] E. C. Halperin, "Particle therapy and treatment of cancer," *The lancet oncology*, vol. 7, no. 8, pp. 676–685, 2006.

[5] B. Lin, D. Huang, F. Gao, *et al.*, "Mechanisms of flash effect," *Frontiers in oncology*, vol. 12, p. 995 612, 2022.

[6] M. Rovituso, C. F. Groenendijk, E. Van der Wal, *et al.*, "Characterisation of the hollandptc r&d proton beamline for physics and radiobiology studies," *Physica Medica*, vol. 130, p. 104 883, 2025.

[7] S. M. Sze, Y. Li, and K. K. Ng, *Physics of semiconductor devices*. John wiley & sons, 2021.

[8] N. Kishimoto, H. Amekura, K. Kono, and C. Lee, "Radiation resistance of amorphous silicon in optoelectric properties under proton bombardment," *Journal of nuclear materials*, vol. 258, pp. 1908–1913, 1998.

[9] M. Menichelli, M. Bizzarri, M. Boscardin, *et al.*, "Fabrication of a hydrogenated amorphous silicon detector in 3-d geometry and preliminary test on planar prototypes," *Instruments*, vol. 5, no. 4, p. 32, 2021.

[10] J. R. Woodyard and G. A. Landis, "Radiation resistance of thin-film solar cells for space photovoltaic power," *Solar Cells*, vol. 31, no. 4, pp. 297–329, 1991.

[11] L. Eglseer, S. Horvat, and H. Kroha, "Study of the long-term behavior of the sensitivity of amorphous silicon photo detectors under illumination," *Nuclear Instruments and Methods in Physics Research Section A: Accelerators, Spectrometers, Detectors and Associated Equipment*, vol. 568, no. 1, pp. 18–21, 2006.

[12] R. L. Weisfield, "Amorphous silicon tft x-ray image sensors," *International Electron Devices Meeting 1998. Technical Digest (Cat. No. 98CH36217)*, pp. 21–24, 1998.

[13] P. R. Granfors and R. Aufrichtig, "Performance of a 41× 41-cm amorphous silicon flat panel x-ray detector for radiographic imaging applications," *Medical physics*, vol. 27, no. 6, pp. 1324–1331, 2000.

[14] IBA Dosimetry, *Matrixx resolution™: Highest-resolution ionization chamber array for imrt & vmat quality assurance*, Product page on IBA Dosimetry website, Includes specs (6.5mm chamber pitch, 1,521 chambers over 25.3×25.3cm²), wireless operation, and integration with myQA® software, 2025. [Online]. Available: https://www.iba-dosimetry.com/product/matrixx-resolution.

[15] IBA Dosimetry, *Sphinx compact: All-in-one daily qa solution for pbs proton therapy*, Product page on IBA Dosimetry website, Compact phantom + detector system for daily PBS machine QA (¡10min), TG-224 compliant, integrates with Lynx detector and myQA® software, 2025. [Online]. Available: https://www.iba-dosimetry.com/product/sphinx-compact.

[16] M. Martišíková, B. Hesse, O. Nairz, and O. Jäkel, "Test of an amorphous silicon detector in medical proton beams," *Nuclear Instruments and Methods in Physics Research Section A: Accelerators, Spectrometers, Detectors and Associated Equipment*, vol. 633, S259–S261, 2011.

[17] IBA Dosimetry, *Dose-x: Next-generation reference class electrometer*, Product page on IBA Dosimetry website, Includes specifications, supported correction factors, API, and interface features, 2025. [Online]. Available: https://www.iba-dosimetry.com/product/dose-x.

[18] IBA Dosimetry, *Razor™ detector: High-performance diode detector for small field dosimetry*, Product page on IBA Dosimetry website, Technical overview, specifications (chip size, active area, effective point), and application in SRS/SBRT small-field dosimetry, 2025. [Online]. Available: https://www.iba-dosimetry.com/product/razor-detector.

The POSiCS handheld gamma-ray camera for radio-guided surgery

Domenico della Volpe[1,*], Alberto Gola[2], Aramis Raiola[1], Cyril Alispach[1], Fabio Acerbi[2], Habib Zaidi[3], and Hossein Arabi[3]

[1] Dept. of Nuclear and Corpuscular Physics (DPNC), Université de Genève, Switzerland
[2] Center for Sensors and Devices (SD), Fondazione Bruno Kessler (FBK), Italy
[3] Division of Nuclear Medicine and Molecular Imaging, Geneva University Hospital, Switzerland

(*) Corresponding author: domenico.dellavolpe@unige.ch

***Abstract*—Intraoperative imaging techniques such as Sentinel Lymph Node Biopsy (SLNB) and Radio Occult Lesion Localization (ROLL) are adopted as a standard in breast cancer and melanoma surgery. Traditional 1D gamma probes, while widely used, lack spatial imaging capability, limiting their effectiveness near injection sites. To address this, we present POSiCS: a compact, wireless gamma camera designed for real-time intraoperative imaging. POSiCS employs a high-density scintillator and novel position-sensitive SiPMs developed by FBK, enabling sub-millimeter spatial resolution with only 8 readout channels. The device supports interchangeable collimators for adjustable sensitivity and resolution, and is optimized for detection of ^{99m}Tc. Performance tests demonstrate a spatial resolution of 1.4 mm and a sensitivity of up to 481 cps/MBq, as well as reliable energy discrimination at 140 keV. Phantom studies confirm the system's ability to resolve closely spaced sources with low activity levels, reinforcing its suitability for SLNB and ROLL. With a patent filed and promising initial results, POSiCS offers a portable, high-performance solution that enhances surgical precision and reduces procedure time.**

***Index Terms*—Intraoperative imaging, Gamma camera, Sentinel lymph node biopsy, Silicon photomultiplier, Spatial resolution, Surgical guidance**

I. The POSiCS-2 camera

A. Introduction

RADIO-Guided Surgery (RGS) is a well-established surgical approach, which is performed in oncological procedures such as Sentinel Lymph Node Biopsy (SLNB) and Radio Occult Lesion Localization (ROLL). In these procedures, the injection of a radiotracer, which is absorbed by metabolic hot spots in the body, is used to guide surgeons to the lesion to be resected during surgical procedures. RGS typically relies on 1D gamma probes, which offer limited directional information, as these devices are only capable of measuring the decay rate of the radiotracer in the direction they are pointed. To address this, 2D Intraoperative Gamma Cameras (IGCs) have emerged as a promising alternative, enabling real-time image visualization of the radioactive source, enhanced localization near injection sites, and reduced operating times [1]–[9].

We present here POSiCS-2: a novel, wireless handheld gamma camera tailored for SLNB and ROLL procedures [10], [11]. The system utilizes a high-density scintillator, coupled with position-sensitive Linearly-Graded Silicon Photomultipliers (LG-SiPMs) developed by FBK, which enables high spatial resolution with a minimal number of readout channels [12]–[14]. Designed for compatibility with Tc-99m radiotracers, the camera also includes two custom-designed collimators to optimize for either sensitivity or resolution.

A European patent (EP 24183916.6) has been filed for the POSiCS camera. In this contribution, we report on its prototype design and present a comprehensive evaluation of its performance, including sensitivity, spatial and energy resolution, and phantom imaging results.

B. The POSICS-2 Concept: A New Paradigm for RGS

The POSICS-2 project introduces a novel approach to intraoperative imaging through a compact, fully wireless gamma camera. Unlike conventional probes, this system offers real-time visual imaging of radiolabeled tissues during surgery. Within the landscape of Intraoperative Gamma Cameras (IGCs), POSiCS-2 stands out by integrating position-sensitive silicon photomultipliers (LG-SiPMs) with a pixelated scintillator and a collimator system. This results in a lightweight, fully wireless device that represents a significant innovation compared to current state-of-the-art cameras. [4], [5], [8], [15].

The unique selling proposition of POSICS-2 lies in its ease of use, portability, and performance. The device is battery-powered, weighs less than 400 grams, and features an external display for viewing the acquired images. These images can also be displayed on monitors already available in surgical rooms, making the device ideally suited for use in the operating theatre. POSiCS produces high-resolution images in real-time, enhancing the surgeon's ability to localize and excise target tissues precisely. This approach is expected to significantly reduce surgical invasiveness, decrease postoperative complications such as lymphedema, and improve recovery outcomes. Additionally, faster decision-making enabled by visual imaging may help shorten surgery times.

The POSiCS device offers a field of view (FOV) of 30.8×30.8 mm^2, optimized for imaging small metabolically active areas (such as sentinel lymph nodes) and for providing clear definition of their margins, for large lesions.

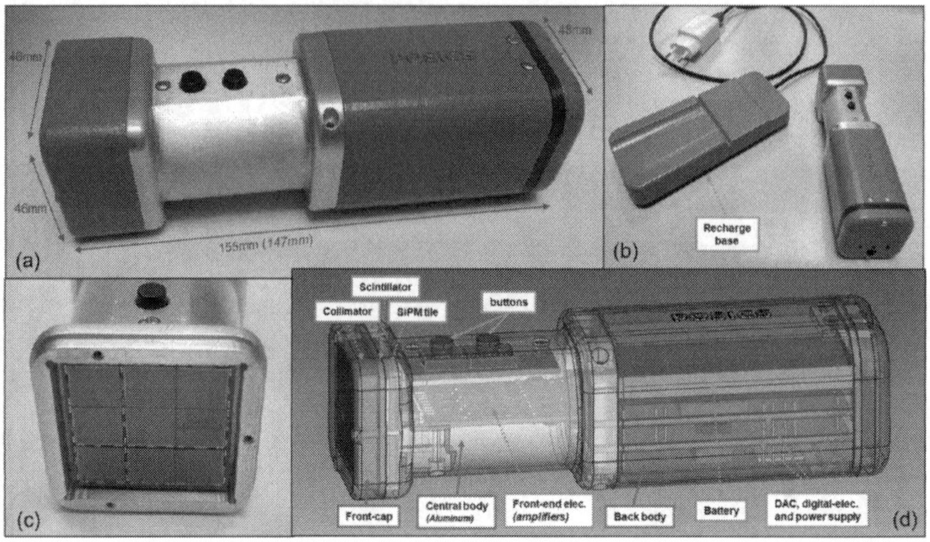

Fig. 1. (a) 1st prototype module of the POSiCS camera with LEHS collimator. (b) POSiCS camera module with its charging wireless base. (c) Front view of the LG-SiPMs array used for the optical readout of the scintillator. (d) Mechanical drawing of the POSiCS camera, showing also the battery housing and the PCBs configuration. Figure from [11].

Its compactness is achieved by a matrix of LG-SiPMs which allow to have a simple and low power electronics These sensors allow accurate localization of gamma interactions with sub-millimeter precision using only a few analog channels, significantly simplifying readout electronics [10]–[14].

The device is equipped with a thin LYSO:Ce scintillator, chosen for its high stopping power and excellent light yield. It includes two interchangeable tungsten collimators to accommodate different clinical requirements: the Low Energy High Sensitivity (LEHS) collimator maximizes count rates, while the Low Energy High Resolution (LEHR) collimator provides greater spatial accuracy at the cost of sensitivity. This modular approach allows clinicians to adapt the imaging system to the specific demands of each surgical procedure.

Its fully wireless operation, ergonomic design, and lightweight build allow surgeons to handle the camera with ease during interventions. A standout feature of the system is its fast data acquisition and real-time image reconstruction, ensuring optimal performance in dynamic surgical environments.

II. Performance Evaluation

A. Spatial Resolution

The spatial resolution of the POSiCS system was characterized using a ^{57}Co point source (122 keV), which closely mimics Tc-99m emission energy. Images were acquired at multiple distances in air and through PMMA blocks to simulate tissue scattering. Gaussian fits were used to determine the Full Width at Half Maximum (FWHM) of the reconstructed point sources.

At zero source-to-detector distance, resolutions of 1.9 ± 0.1 mm for LEHS and 1.4 ± 0.1 mm for LEHR were recorded. As expected from the known behavior of parallel-hole collimators, spatial resolution decreases with increasing distance from the detector. However, this effect is less pronounced with the LEHR collimator. At clinically relevant depths (around 3 cm), the spatial resolution is reduced to 2.7 ± 0.1 mm for the LEHR and 5.5 ± 0.1 mm for the LEHS collimator when imaging through PMMA. With this level of resolution, we expect to reliably distinguish lymph nodes or tumor margins within tissue with millimetric precision, even at greater depths within the body.

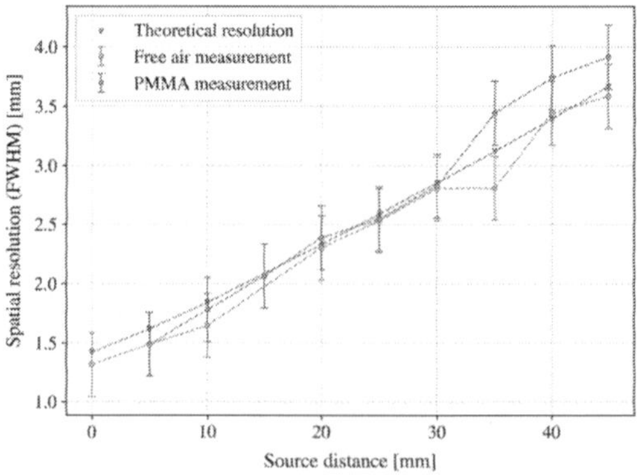

Fig. 2. Spatial resolution as a function of distance for the LEHR collimator. The degradation in resolution is only due to the collimator, while the presence of the material (PMMA) has a negligible impact.

B. Sensitivity Measurements

To evaluate detection efficiency, a uniform Tc-99m source was placed in front of the detector, and the registered count rate was normalized to the known source activity (measured with a calibrated probe). At a working distance of 2 cm, the

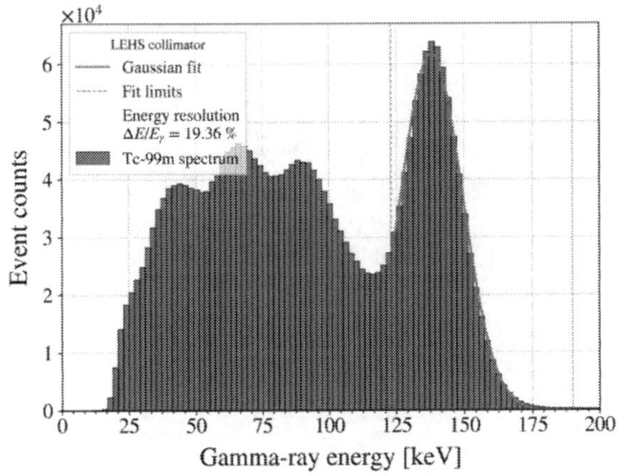

Fig. 3. Energy spectrum of the POSiCS camera with the LEHS collimator after energy calibration. The photopeak is clearly distinguishable from the Compton scattering spectral region and the background noise.

LEHS collimator achieved a sensitivity of 481 ± 14 cps/MBq, while the LEHR reached 134 ± 8 cps/MBq.

These results are consistent with the design goals and compare favorably with existing intraoperative gamma cameras in terms of sensitivity [4], [5].

C. Energy Resolution

The energy response of the camera was assessed using a uniform illumination with a flat Tc-99m source. A pixel-by-pixel calibration was performed to correct charge variations across the FOV. The measured energy resolution at 140 keV is approximately 20% for both collimator types, which is sufficient to discriminate Tc-99m emissions from background noise and scatter-induced events

III. Phantom Study and Living Tissues Tests

A. Phantom Studies

To evaluate the imaging capabilities, we carried out phantom tests that simulated clinical scenarios. A mouse phantom filled with low-activity Tc-99m sources was imaged using the LEHS collimator. Each cavity contained 3 MBq of activity, diluted in saline solution. The POSiCS camera showed a strong ability to resolve closely spaced cavities that represent key anatomical structures like lymph nodes and tumors. Figure 4 shows four scans of the phantom mouse, each consisting of a 1-minute data acquisition. The white borders indicate the theoretical sizes of the imaged structures, which align well with the actual cavities. Interestingly, the cavity representing the mouse liver appears fainter than the surrounding reconstructed sources. This is due to its lower radiotracer concentration, which is expected given the liver's larger volume compared to the other cavities.

These results demonstrate the system's capability to localize small, closely spaced sources in real time, even with minimal injected activity.

B. First animal tests

The POSiCS camera was recently evaluated for in vivo imaging capabilities in a preclinical setting. A series of tests were conducted on murine models bearing subcutaneously implanted tumors, with an average growth period of approximately two weeks. To assess the distribution and uptake dynamics, an initial intratumoral radiotracer injection was performed.

Imaging was carried out using an energy window of 120–160 keV, and the resulting data are presented in Figure 5. Following tracer administration, a 30-minute uptake period was allowed, during which no activity monitoring was performed, prior to imaging. Upon initiating image acquisition, the tumor was immediately localized. Notably, the approximate position of the lesion could be identified in under one second of acquisition time. Here, acquisition time is defined as the sum of the detector exposure time and the wireless data transmission delay to the host system.

After less than 5 seconds of acquisition, the spatial margins of the tumor were already clearly discernible, and further acquisition did not yield substantial improvements in image resolution or lesion delineation. Additionally, an elevated background signal was observed in the region adjacent to the lesion, likely due to hepatic uptake of the tracer. This is consistent with the anatomical positioning of the liver posterior to the tumor implantation site.

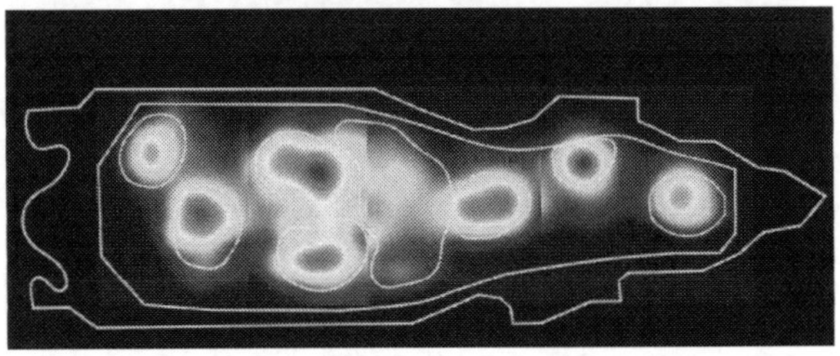

Fig. 4. Imaging of a plastic mouse phantom (Bioemtech Mice Phantom [16]) with the LEHS collimator, illustrating source separation and anatomical localization.

IV. Conclusions

We presented the development and first performance evaluation of the POSiCS handheld gamma camera, a compact and wireless device designed for intraoperative imaging in SLNB and ROLL procedures. The system leverages a pixelated scintillator coupled with position-sensitive LG-SiPMs, offering high spatial resolution (1.4–1.9 mm at contact) and sensitivity (up to 481 cps/MBq), depending on the collimator used.

The camera's design emphasizes portability, ease of use, and real-time wireless imaging, enabling efficient lesion localization even in low-activity scenarios. Phantom and animal studies confirmed the device's ability to produce accurate images in under 1 second, facilitating rapid surgical decision-making.

These results establish POSiCS as a promising tool for enhancing surgical precision and reducing operative times. Ongoing developments include system integration for clinical

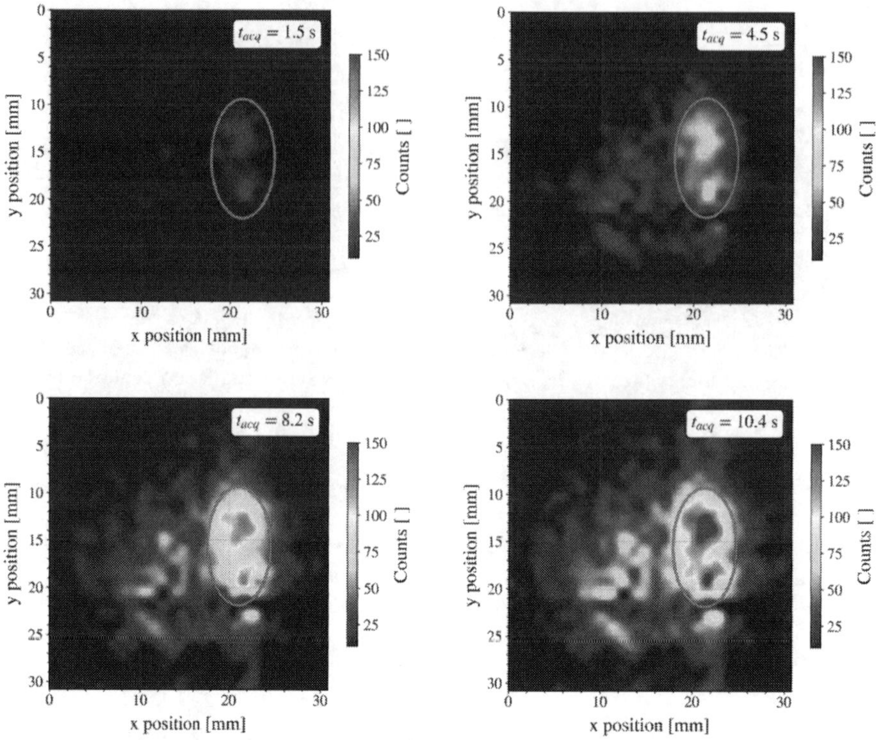

Fig. 5. Time evolution of in-vivo tumor acquisition from the mice tests. Image reconstruction takes place in a few seconds, reconstructing the correct size of the lesion (estimated through SPECT/CT scan and drawn as a red ellipse in the image). It was also possible to clearly distinguish the tumor from the background uptake of the liver.

trials and the extension of imaging capabilities to additional isotopes, such as Lu-177, for theranostic applications.

References

[1] G. Mariani *et al.*, "Radioguided sentinel lymph node biopsy in malignant cutaneous melanoma," *J. Nucl. Med.*, vol. 43, no. 6, pp. 811–827, Jun. 2002. [Online]. Available: https://jnm.snmjournals.org/content/43/6/811

[2] G. Mariani *et al.*, "Radioguided sentinel lymph node biopsy in breast cancer surgery," *J. Nucl. Med.*, vol. 42, no. 8, pp. 1198–1215, Aug. 2001. [Online]. Available: https://jnm.snmjournals.org/content/42/8/1198

[3] G. Marin *et al.*, "Accuracy and precision assessment for activity quantification in individualized dosimetry of 177Lu-dotatate therapy," *EJNMMI Phys.*, vol. 4, Dec. 2017, Art. no. 15. doi:10.1186/s40658-017-0174-7

[4] A. L. Farnworth and S. L. Bugby, "Intraoperative gamma cameras: A review of development in the last decade and future outlook," *J. Imaging*, vol. 9, May 2023. doi:10.3390/jimaging9050102

[5] M. Tsuchimochi and K. Hayama, "Intraoperative gamma cameras for radioguided surgery: Technical characteristics, performance parameters, and clinical applications," *Phys. Med.*, vol. 29, no. 2, pp. 126–138, 2013. doi:10.1016/j.ejmp.2012.05.002

[6] D. Hellingman and S. Vidal-Sicart, "The Use of Intraoperative Small and Large Field of View Gamma Cameras for Radioguided Surgery," in *Radioguided Surgery*, Cham, Switzerland: Springer, 2016, pp. 35–56. doi:10.1007/978-3-319-26051-8_3

[7] A. Soluri *et al.*, "Small field of view, high-resolution, portable gamma-camera for axillary sentinel node detection," *Nucl. Instrum. Methods Phys. Res., Sect. A*, vol. 569, no. 2, pp. 273–276, 2006. doi:10.1016/j.nima.2006.08.027

[8] P. Paredes *et al.*, "Radioguided occult lesion localisation in breast cancer using an intraoperative portable gamma camera: First results," *Eur. J. Nucl. Med. Mol. Imaging*, vol. 35, no. 2, pp. 230–235, Feb. 2008. doi:10.1007/s00259-007-0640-x

[9] H. Arabi, A. S. Saberi Manesh, and H. Zaidi, "Innovations in dedicated PET instrumentation: from the operating room to specimen imaging," *Phys. Med. Biol.*, vol. 69, no. 11, 11TR03, 2024. doi:10.1088/1361-6560/ad4b92

[10] F. Acerbi *et al.*, "POsition-sensitive SiPMs compact and scalable beta-camera (POSiCS)," Public Deliverable, ATTRACT Final Conference, May 2019. [Online]. Available: https://phase1.attract-eu.com/wp-content/uploads/2019/05/POSiCS.pdf

[11] F. Acerbi *et al.*, "Compact and handheld SiPM-based gamma camera for radio-guided surgery and medical imaging," *Instruments*, vol. 9, no. 2, 2025. doi:10.3390/instruments9020014

[12] A. Gola *et al.*, "A novel approach to position-sensitive silicon photomultipliers: First results," in *Proc. IEEE Nucl. Sci. Symp. Med. Imaging Conf. (NSS/MIC)*, 2013, pp. 1–4. [Online]. Available: https://api.semanticscholar.org/CorpusID:24489596

[13] F. Acerbi, S. Merzi, and A. Gola, "Large area tiles of position-sensitive silicon photomultipliers," in *Proc. 13th Int. Conf. Position Sensitive Detectors (PSD13)*, Oxford, U.K., Sep. 2023. [Online].

[14] F. Acerbi, S. Merzi, and A. Gola, "Position-sensitive silicon photomultiplier arrays with large-area and sub-millimeter resolution," *Sensors*, vol. 24, Jul. 2024. doi:10.3390/s24144507

[15] A. Raiola *et al.*, "Quantitative determination of spatial resolution and linearity of position-sensitive LG-SiPMs at sub-millimeter scale via Ricean distribution fitting," *Nucl. Instrum. Methods Phys. Res. A*, vol. 1080, 170774, Nov. 2025. doi:10.1016/j.nima.2025.170774

[16] Biometech AB, "Fillable mouse phantom for preclinical imaging," 2024. [Online]. Available: https://bioemtech.com/bioemtech-fillable-mouse-phantom-in-the-journal-of-nuclear-medicine-jnm/

Microdosimetry of kidney cells exposed to ^{243}Am and X-ray irradiation

Pia Kahle[1,*], Christian Senwitz[2,3], Anne Heller[2,3], Anja Seifert[1], Steffen Taut[3] and Thomas Kormoll[1]
[1] *Institute of Nuclear and Particle Physics, Technische Universität Dresden, 01069 Dresden,* Germany
[2] *Professorship of Radiochemistry and Radioecology, Technische Universität Dresden, 01069 Dresden*, Germany
[3] *Central Radionuclide Laboratory, Radiation Protection Office, Technische Universität Dresden, 01069 Dresden,* Germany
(*) pia.kahle@tu-dresden.de

***Abstract—In vitro* experiments with rat kidney cells exposed to ^{243}Am in cell culture medium show a decreasing cell viability with increasing americium concentration. Both the chemical cytotoxicity of americium and the dose caused by ionizing radiation have to be considered as possible causes. To determine the dose rate, a model which was reduced to the dosimetric relevant circumstances was set up. Since the size of the cells is in the micrometer range α-radiation makes the biggest contribution to the dose rate. Using the characteristic behaviour of α-particles the dose rate can be calculated from fluence and stopping power. Biological aspects like uptake of Americium into the cells as well as inaccuracies arising from inhomogeneity and dynamic behaviour in biological systems were taken into account. The dose rate was considered pointwise in the cell center and averaged over the whole cell. Calculations show that dose rates up to the Gy/h range were achieved at the highest americium concentrations of 8·10^{-4} M. For an experimental comparison of the effect in dependence of the dose reference samples of untreated kidney cells were exposed to analog doses in an X-ray field. Taking relative biological effectiveness into account the cell viability curves depending on the dose show good agreement.**

***Keywords* —Microdosimetry, Relative Biological Effectiveness**

I. Introduction

In the fuel cycle of nuclear power plants various heavy nuclides are produced and can be released into the environment. Therefore the impact on living systems is of particular interest for research. This work focuses on experiments with rat kidney cells exposed to ^{243}Am. With increasing ^{243}Am concentration the cells show, among others, decreasing viability [1]. Since ^{243}Am is radioactive, the following analysis examines if dose caused by ionizing radiation has to be considered as an influencing parameter on the cell viability.

^{243}Am decays by emitting α-particles with a range of 40 µm in water. The size of the cells is also in the µm range, therefore α-particles deposit a significant amount of their energy in the cells. α-decay transfers about 5 MeV to the α-particles, which is considerably more than usually emitted by γ- and β-conversion. Furthermore the range of beta- and low energy gamma-radiation is in the range of millimeters, which means the transferred proportion of the total energy to the cells is much lower than for α-particles.

Therefore it is expected, that the dose caused by β-, and γ-radiation is negligible compared to the dose caused by α-particles. But the dose caused by α-particles may have impact on the cell viability. In the following the dose caused by a ^{243}Am solution covering the cells will be determined analytically.

II. Methods

A. Contribution from α-particles

Kidney cells from rats (NRK-52E) were covered with a solution of ^{243}Am for 24 h (see Fig. 1). Therefore, the activity is distributed homogenously in the volume above the cells. The experiments were carried out with different concentrations of ^{243}Am (10^{-6} – $8 \cdot 10^{-4}$ M). Calculations were done with the highest used concentration of $8{\cdot}10^{-4}$ M.

Additionally the cells internalize some amount of ^{243}Am. Inside the cells, the concentration of ^{243}Am and consequently the activity per volume amounts 0.5 % to 1 % of the applied concentration respectively the activity outside the cells. The percentage of the uptake depends on the concentration of ^{243}Am outside the cells. [1]

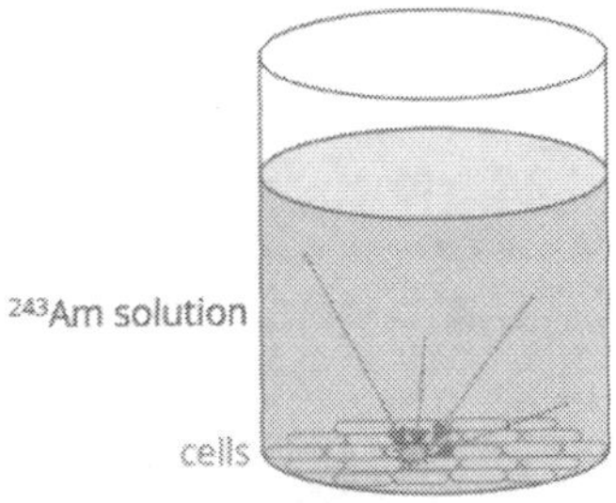

Fig. 1. Experimental setup.

^{243}Am decays to ^{239}Np by emitting α-particles with energies between 5.2 MeV and 5.3 MeV. Since α-particles move nearly in a straight line and every point of their path is clearly assigned to a stopping power value, the mean imparted energy per mass (dose) can be calculated for a certain point in the cells. For this purpose, a model was created taking only the dosimetric relevant aspects into account. A two-dimensional view of this model is given in Fig. 2.

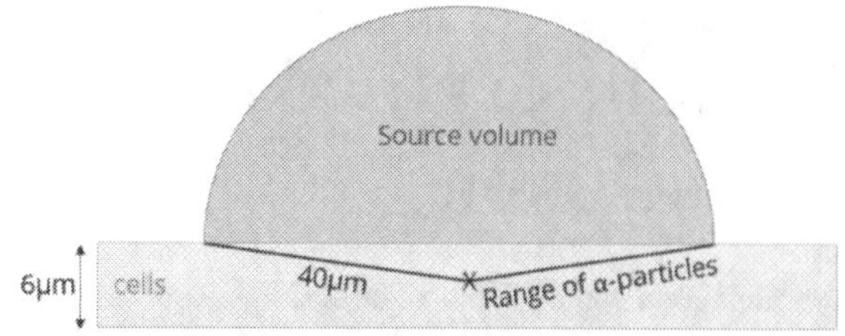

Fig. 2. Microdosimetric model for in vitro experiments.

In the experiment, the cells form a monolayer (see Fig. 4a). Because of that, from the dosimetric point of view, they are a continuous layer and the width of the cells is not relevant. ^{243}Am and, thus, also the activity is homogenously distributed in a solution above the cells. Since the dose is considered for one point inside the cells and α-particles have a clearly defined range, only those α-particles which originate within a fixed radius around this point are relevant. Consequently, a hemisphere above the point of interest is considered.
The dose caused by α-particles can be calculated by the following formula:

$$D(r) = \frac{1}{V_{\text{source}}} \int \Phi(r) \cdot \frac{S}{\rho}(E(r)) \mathrm{d}V_{\text{source}}$$

with dose D, particlefluence Φ, mass collision stopping power $\frac{S}{\rho}$ and source volume V_{source}. Dose is the mean energy $\mathrm{d}E$ imparted by ionizing radiation to matter of mass $\mathrm{d}m$:

$$D = \frac{\mathrm{d}E}{\mathrm{d}m}$$

Fluence is the number of particles $\mathrm{d}N$ incident on a sphere of cross-sectional area $\mathrm{d}A$:

$$\Phi = \frac{\mathrm{d}N}{\mathrm{d}A}$$

Collision stopping power describes the average energy loss $\mathrm{d}E$ of charged particles due to inelastic collisions per path length $\mathrm{d}x$:

$$\frac{S}{\rho} = \frac{1}{\rho} \frac{\mathrm{d}E}{\mathrm{d}x}$$

Fluence can be calculated from the concentration of ^{243}Am in the solution and its half-life.
Stopping power in dependence on the distance of the particles from the origin were taken from the NIST astar database [2]. It is assumed that the stopping power in the cells is equal to the stopping power in water. This approximation is very close to reality and simplifies the model considerably.
The α-particles originate both from the cell culture medium above the cells (extracellular) and from the uptake of ^{243}Am inside the cells (intracellular).
In the following, the calculation of the dose will be described starting with the dose from extracellular α-radiation.
Since the activity is homogenously distributed along the radius of the source volume, the stopping power is integrated over the path of the α-particles. The lower integration limit corresponds to the inner border of the extracellular source volume.

Consequently, the lower integration limit depends on the polar angle. The maximum range of the α-particles defines the upper integration limit.
Next, integration is carried out over the polar angle. The integration limits depend on the height of the cells.
To cover all dimensions of the source volume, integration has to be carried out over the azimuthal angle from 0 to 2π as well.
Expressed in spheric coordinates and with integration limits the formula is:

$$\dot{D} = \frac{1}{V_{\text{source}}} \int_0^{2\pi} \int_0^{\theta} \int_{R_0(\theta)}^{R} \frac{A f_\alpha}{4\pi r^2} \frac{S}{\rho}(E(r))\, r^2 \sin\theta \, \mathrm{d}r \, \mathrm{d}\theta \, \mathrm{d}\varphi$$

The calculation of the intracellular dose is analog to the calculation of the extracellular dose. However the integration limits in the radial dimension have to be changed.
The upper integration limit corresponds to the lower integration limit from the extracellular calculation and the lower integration limit is set to 0.
Moreover, for the dose from intracellular α-radiation, the particles which originate from the sides of the point under consideration, must be taken into account. In this area, the stopping power is integrated radially over the entire range of the α-particles, because they can originate from every point within the maximum range.

B. *Contribution from photons and electrons*

Due to the small scale of the model (3 mm filling height of the cell culture well), γ-radiation (44 keV - 334 keV) is not expected to make a single interaction, because the mean free path is between 4 cm and 8 cm.
X-rays are emitted with 12 keV – 23 keV in this decay chain. Their mean free path amounts 3 mm to 1.7 cm. Consequently an interaction with the cells can happen. Nevertheless the kinetic energy of the X-rays is less than 0.5 % of the α-energies.
Electrons are emitted by the daughter nucleus ^{239}Np with 341 keV respectively 438 keV maximum β-energy. The corresponding maximum range is 1.4 mm in water. Therefore the stopping power of electrons is much lower than the stopping power of α-particles.
For these reasons β- and γ-radiation is neglected in this consideration.

III. Results

A. *Dose rate*

The results of the calculations are given in Tab. 1.

Tab. 1: Radiation contributions to the dose rate at the cell center

Origin of radiation	**Dose rate in Gy/h**
extracellular α-radiation ($8 \cdot 10^{-4}$ M ^{243}Am)	1.70
intracellular α-radiation (estimated uptake: $3.5 \cdot 10^{-6}$ M ^{243}Am)	0.003

Values were not only calculated for the center of the cells (results in Tab. 1), but also for different points distributed over the cell thickness. The average of them is 1.75 Gy/h at the highest ^{243}Am concentration of $8 \cdot 10^{-4}$ M.

The contribution of intracellular α-radiation to the total dose is very small compared to extracellular α-radiation. On the one hand only 0.5 % to 1 % of the ^{243}Am applied is accumulated inside the cells. On the other hand, within the range of the α-particles, the volume inside the cells is very small compared to the hemispheric volume above the cells.

B. *Inaccuracy from varying size of the cells*

A crucial parameter for the dose rate at the center of the cells is the height of the cells. Fig. 3 shows the dependence of the dose rate to the height of the cells. With increasing height of the cells the extracellular contribution decreases while the intracellular contribution increases. That is because with increasing cell volume there is less extracellular volume from where α-particles can reach the cells. But concurrently, there is more intracellular volume from where α-particles can originate.
Before irradiation, the cells have a height of 6 μm. Because of the cells being stressed upon exposure to ^{243}Am, they swell to approximately twice their size (see Fig.4). In consequence the total dose rate decreases during exposure. Since the exact time course of the swelling is not known, inaccuracies arise in the analysis. To estimate the impact of swelling on the dose rate, it was assumed, that the height of the cells increases linear with time. Under this assumption the mean dose rate of 24 h exposure time is 1.54 Gy/h at the highest ^{243}Am concentration of $8 \cdot 10^{-4}$ M.

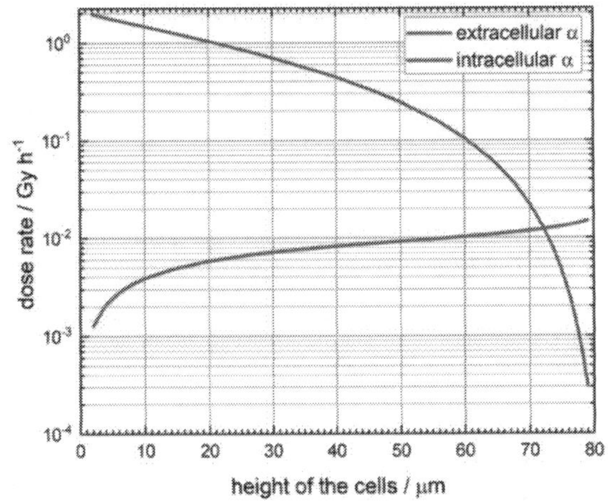

Fig. 3. Influence of the cell height on the dose rate.

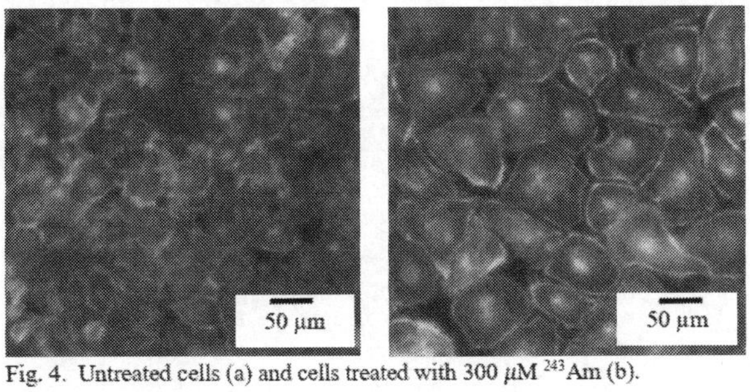

Fig. 4. Untreated cells (a) and cells treated with 300 μM ^{243}Am (b).

C. *Reference irradiations with X-ray*

To compare if the effect fits to the determined dose, reference samples of untreated kidney cells were exposed to analog doses in an X-ray operated with 50 kV and 50 mA. Dose in the X-ray field was measured by tissue equivalent dosimeters, which use optical stimulated luminescence of Berylliumoxide.
For the comparison of damage caused by different kinds of radiation Relative Biological Effectiveness (RBE) has to be taken into account. The RBE model says that different kinds of radiation need a different amount of dose to cause the same effect. The difference is described by the RBE-factor.

$$D_{X-\mathrm{ray}} = D_{\alpha} \cdot \mathrm{RBE} \rightarrow \text{same effect}$$

Fig. 5 shows the results of the irradiation experiments. For both treatments cell viability decreases with increasing dose.
By applying two x-axis, one for each kind of radiation, the data points can be moved to the same effect to read out the RBE factor.
The viability measurements of the cells treated with α-radiation are shown in red, as well as the corresponding x-axis (dose) at the top of the diagram. In blue, the results and the corresponding x-axis of the cells treated with X-rays are shown. Within the uncertainties the measurement points show the same effect at a RBE-factor of about 2.3.
The RBE depends on various parameters like the endpoint of the measurement, the cell line or the Linear Energy Transfer. Consequently, experiments with mixed α-energies, like in this case, are not comparable to experiments in a helium beam.

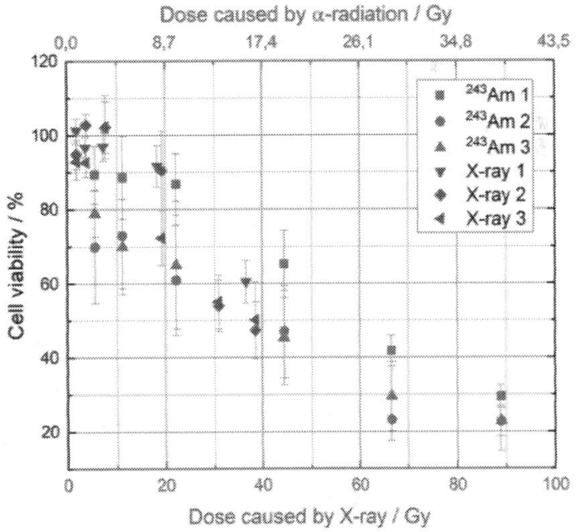

Fig. 5. Viability of cells treated with α-radiation (red) and X-rays (blue) in dependence of the applied dose.

IV. CONCLUSIONS

The complex system of cells in a radioactive solution was simplified to the dosimetrically relevant circumstances.
Using this model, it is possible to calculate the dose rate caused by α-particles at certain points by looking at fluence and stopping power. This analytical method is faster and easier to handle compared to a detailed simulation.

The result of the analysis is 1.70 Gy/h in the center of the cell. Alpha-radiation from inside the cells contribute 0.003 Gy/h. Averaged over the whole cell the dose rate amounts 1.75 Gy/h. The only dosimetric relevant parameter with a non-negligible inaccuracy is the height of the cells. Since the cells are swelling to approximately twice their size while irradiation, the average dose rate is about 1.5 Gy/h.

After 24 h exposure time, dose rates in the Gy/h range are likely to cause a decrease in cell viability, as it was observed in the experiments.

To compare the calculation to an experiment with known dose rate, the cells were irradiated with X-ray. In comparison to the irradiation with ^{243}Am, the results are in good agreement with the RBE model. With X-rays, about 2.3 times the dose was necessary to cause the same effect as with the ^{243}Am-solution. It has to be mentioned that RBE values are difficult to compare because they depend on many different parameters in the experiment. Furthermore, the inaccuracy of the viability measurements influence the accuracy of the determined RBE value.

The measurements with X-rays demonstrated that the calculated dose rate is in the right order of magnitude. To proof the accuracy of the calculation, further experiments with exact known dose rates under the same circumstances as in the corresponding model would be necessary. The calculation would also be applicable to other orders of magnitude.

Acknowledgment

This work was partly funded by the German Federal Ministry of Education and Research (BMBF) under grant numbers 15S9447B and HE 7054/2-1.

References

[1] A. Heller, C. Senwitz et al., "Am/Cm(III) and DTPA/LIHOPO interactions with kidney cells *in vitro* studied by bioassays, luminescence spectroscopy, and microdosimetry," submitted to Ecotoxicology and Environment Safety

[2] Berger, M.J., Coursey, J.S., Zucker, M.A., and Chang, J. (2005), *ESTAR, PSTAR, and ASTAR: Computer Programs for Calculating Stopping-Power and Range Tables for Electrons, Protons, and Helium Ions* (version 1.2.3). [Online] Available: http://physics.nist.gov/Star [*2025, August 31*]. National Institute of Standards and Technology, Gaithersburg, MD.

Compton imaging for dosimetry and real time monitoring in boron neutron capture therapy

Pablo Torres-Sánchez[1,*], Sebastián Valladares[1], Javier Balibrea-Correa[1], Jorge Lerendegui-Marco[1], Víctor Babiano-Suarez[1], Bernardo. Gameiro[1], Ion Ladarescu[1], Caterina Michelagnoli[2], Jean-Michel Daugas[2], Ulli Koester[2], Patricia Álvarez-Rodríguez[2,3], Cristina Méndez-Malagón[3,4], Maribel Porras-Quesada[3], María Pedrosa-Rivera[5], M José Ruiz-Magaña[6], Carmen Ruiz-Ruiz[4], Ignacio Porras[5], César Domingo Pardo[1]

[1] Instituto de Fisica Corpuscular (CSIC-UV), Spain
[2] Institut Laue-Langevin, France
[3] Instituto de Biopatología y Medicina Regenerativa, Centro de Investigación Biomédica, Universidad de Granada, Spain
[4] Departamento de Bioquímica y Biología Molecular III e Inmunología, Universidad de Granada, Spain
[5] Departamento de Física Atómica, Molecular y Nuclear, Universidad de Granada, Spain
[6] Departamento de Biología Celular, Universidad de Granada, Spain

(*) Pablo.Torres@ific.uv.es

Abstract—**Boron Neutron Capture Therapy (BNCT) is an emerging radiotherapy technique that exploits the high neutron capture cross-section of ^{10}B to deliver targeted high-LET radiation to tumor cells while sparing surrounding healthy tissue. Effective clinical implementation requires accurate in vivo determination of boron concentration during treatment to enable precise dose monitoring. The i-TED Compton Camera Array, originally developed for high count-rate neutron-induced reaction measurements at CERN, offers promising capabilities for BNCT dose monitoring due to its low neutron sensitivity, high efficiency, and large field-of-view. We report on a series of experiments at the FIPPS beamline (ILL, France) to evaluate i-TED performance under progressively challenging background and count rate conditions, including borated water and hydrogen-rich environments mimicking clinical scenarios. Results show that while the system can identify the 478 keV gamma-rays from boron under realistic boron-to-hydrogen ratios, high count rates induce gain shifts that degrade low-energy gamma detection. Strategies to mitigate these effects, including optimized scintillator thicknesses and angular event selection, are proposed. These findings provide a clear path toward adapting i-TED for real-time BNCT dose monitoring in clinical environments.**

Keywords — **Boron Neutron Capture Therapy, Compton Imaging, Dosimetry, Gamma-ray Imaging**

I. INTRODUCTION

Boron Neutron Capture Therapy (BNCT) is an experimental form of radiotherapy of growing interest in the last decade. It is based on the administration of a non-toxic, non-radioactive drug labelled with ^{10}B, which is selectively uptaken by cancerous cells. ^{10}B has a very large neutron capture cross-section, releasing an alpha particle and a ^{7}Li nucleus with high LET [1]. Therefore, a neutron irradiation of the patient after they have been injected with the boron compound is expected to damage or kill cancerous cells while sparing healthy tissues nearby. For this reason, BNCT has been proposed for amorphous and infiltrating tumors in critical organs, such as glioblastomas or head and neck cancers [2].

BNCT has experienced an expansion in recent years due to the development of accelerator-based neutron sources of enough intensity to provide treatment in clinical settings, which are now available in Japan as part of the National Health System, with other facilities under clinical trials in China, Korea and Finland. This has come together with the discovery of boron carriers of easy manufacturing and administration, low levels of toxicity and sufficient tumor-to-normal cell selectivity, such as BPA.

One of the major challenges still to be addressed in BNCT is the boron concentration determination during a treatment [3], which would enable an accurate dose monitorization and report, providing a strong support to the efficacy and safety of the treatments by allowing to reduce current margins applied to avoid damaging healthy tissues. Current methodologies are prone to inaccuracies and are based on several assumptions that propagate to large uncertainty levels. Several proposals have been suggested so far, particularly taking profit from the 478 keV gamma ray emitted in 94% of the boron neutron capture reactions. Those proposals are based on the SPECT [4, 5, 6] and the Compton imaging techniques [7, 8, 9]. In general, SPECT proposes devices with higher intrinsic resolution. However, the bulky designs with large collimators near the patient irradiation area generate secondary dose to the patient and increase significantly the background rate in the detectors, thereby reducing the sensitivity to 478 keV gamma rays and thus to boron concentrations much higher than those applied in clinical use [10]. Compton cameras, on the other hand, offer the advantage of a reduced volume, allowing to close the patient-to-detector distance, which in turn improves spatial resolution. This comes with the drawback of very large count rates, which could easily overwhelm the acquisition systems of the detectors when operating under nominal clinical conditions.

The i-TED Compton Camera Array, developed under the

HYMNS ERC project for neutron-induced reaction cross-section measurements of astrophysical interest at n_TOF (CERN) [11], is an excellent starting point for the development of a dose monitoring device for BNCT. This has also been proven by its application towards other medical applications such as in range verification in proton- and heavy-ion therapy [12, 13]. Due to its original design devoted to operating in high-count rate conditions in a very intense neutron radiation environment, which resembles that of a BNCT treatment, its low neutron sensitivity, high efficiency, and large field-of-view provide an outstanding candidate for this application [14].

In a recent campaign, we performed the first pilot tests of the use of i-TED for BNCT, by measuring small amounts of boron and generating images of the location of the boron neutron capture reaction in the FIPPS beamline at ILL (Grenoble, France). Those tests were performed in vacuum conditions and under a very low background environment, and were also used to measure the boron uptake of several cell lines of interest. [15]

The successful results from that pilot campaign have then been followed by a subsequent campaign which we have devoted to the identification of major challenges and elements for the improvement of this device towards its clinical application in BNCT, under the AMA ERC project.

II. MATERIALS AND METHODS

The second experiment at ILL was focused on progressively exposing the i-TED to increasingly higher background and count rate conditions, in order to test its robustness and extract valuable input on key aspects to develop the next version of the camera, fully oriented and optimized for its use in BNCT.

The experiment was performed at the FIPPS beamline, in this case under free-air conditions, a neutron flux of 10^8 n/cm^{2}s in a pencil beam of 1.5 cm in diameter. The full array of four i-TED modules in cross-configuration was used to maximize detection efficiency, thus enabling 3D imaging reconstruction.

A varied set of different configurations was used, including, from teflon bags with a boron deposit (lowest background), and successively, petri dishes with a boron deposit, petri dishes filled with borated water, a polyethylene block with a petri dish inside; and finally, a water container, filled with clean and borated water, with the petri dish inside (highest background). Fig. 1. Shows some of those configurations.

Fig. 1. Pictures from different setups used during the experiment at ILL used to analyze the performance of i-TED under progressively more challenging conditions in terms of background radiation. The *top* picture uses a 8 ml borated water dish, the *center* picture shows a polyethylene block, with one borated water dish inside, and the *bottom* picture shows a water container, also with a borated water dish inside.

Apart from that experimental campaign, another recent development is the implementation of a rotatory gantry, which enables the use of a single i-TED module for 3D image reconstruction with tomographic capabilities, and therefore minimizing the impact of the device in the treatment, both as a volume located near the patient that could reduce the positioning options, and also reducing the secondary neutron radiation redirected to the patient through scattering from the detector. This rotatory gantry, together with the List-Mode Maximum Likelihood Expectation Maximization (LM-MLEM) image reconstruction algorithm, developed by our group [14] have been recently validated by means of point-like gamma radiation sources, as shown in Fig. 2. An additional effort has been taken to optimize the performance of such algorithms by the use of parallelized algorithms, capable of being executed both in CPU and GPU architectures [16], leading to close to real-time data processing in the steps from data filtering to 3D image generation in the case of GPU.

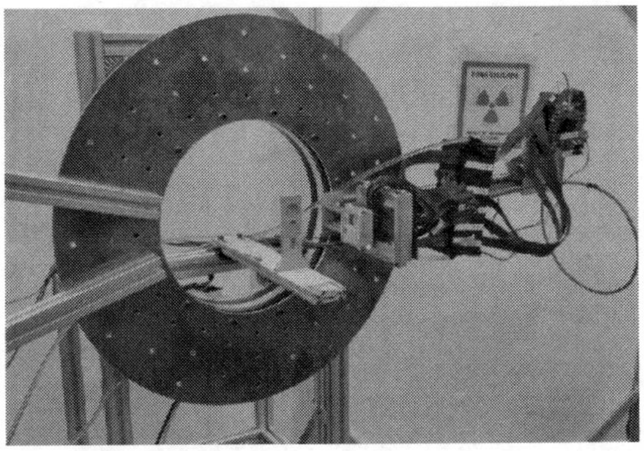

Fig. 2. Rotatory system used to validate the 3D imaging system with tomographic capabilities. One i-TED module and its electronic data acquisition system is coupled to the rotary gantry, and a counter-weight is added in the opposite side of the setup. A radioactive sample is located at the center of the field of view, along the rotation axis.

III. Results and Discussion

The campaign at ILL provided a set of measurements with very valuable information regarding the challenges that a BCNT treatment poses to dose monitoring devices.

Those measurements were performed both with and without boron to test the sensitivity and background levels, and also check the impact of boron-related events in the overall count rate. Fig. 3. shows the add-back coincidence energy deposited spectra for those sets of measurements of increasing complexity. For the simplest cases (teflon bag and dry dish with extra shielding around the detectors, magenta and violet), count rates are low enough (<70 kHz) that pileup and dead time effects do not affect the acquisition. In general, the spectra are dominated by the events produced by boron (478 keV), present both in the environment (due to boron-based neutron absorbers in the walls of the experimental hall and the beam dump), and particularly the 2.224 keV gamma-rays generated by neutron captures in hydrogen. This contribution dominates the spectra once any hydrogen-rich material is introduced to the beam, namely water or polyethylene present inside or around the dishes. In the experiment, the amounts of hydrogen relative to boron were kept as similar as possible as in an actual treatment, in order to keep a reliable signal to background ratio. For that reason, boron concentrations of 65 and 100 ppm of ^{10}B were used, which are typical low and high boron concentrations inside the tumors.

Once the additional shielding was removed (cyan), the overall count rates go over 100 kHz and pileup starts to be noticeable as a soft increase in the high energy tail of the spectrum.

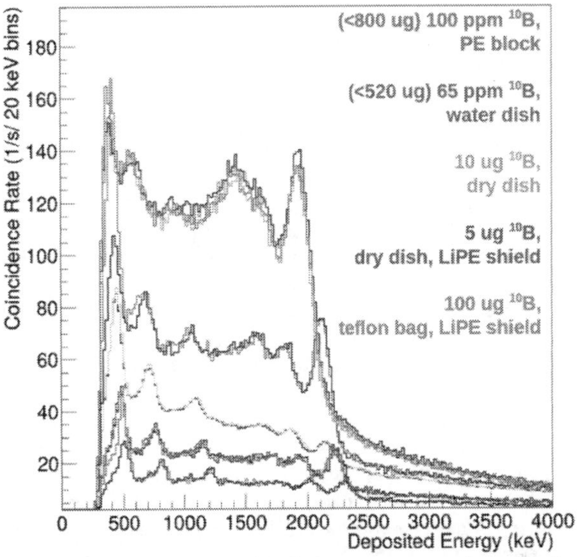

Fig. 3. Add-back coincidence deposited energy spectra obtained under increasingly higher background and count rate conditions, from a small teflon bag holding 100 μg of ^{10}B, to a dry dishes with small amounts of 10B deposited on it, with and without an extra neutron shielding; with a borated water dish and with an additional polyethylene block around it. Black lines corresponding to each of the colour ones, refer to the equivalent measurement without boron in the sample, thus acting as dummy or reference measurement for each configuration.

Moreover, a shift in the estimated energy deposited starts to appear evident, which further worsens as the conditions become harsher. A linear relation between the observed count rate and the energy shift of the peaks is clearly observed, with a similar trend for all deposited energies, as can be noticed both for the 478 and 2224 keV lines.

That fact requires special attention for boron imaging, since that gain shift hinders the detection of low energy gamma rays, in particular if they fall below the detection threshold. This is crucial in the case of the signal detected in the scatterer crystal, where typically the energy deposited is lower. In our case, once we reach 250 kHz, that gain shift is so large that the boron peak is not observed anymore. This occurred, with our detection settings, still far from the maximum expected acquisition count rate achievable by the PETsys electronics, around 500-600 kHz.

On the plus side, at the highest count rate for which 478 keV gamma-rays are still observed, our detection system is clearly capable of identifying the 478 keV gamma rays coming from the boron sample over the background (green line). Those conditions, though not to the level of beam intensity and count-rates, do more or less match the expected boron to hydrogen concentration ratios in the tumor and nearby tissues, which ensures an adequate sensitivity level for our detector, once the count rate and related issues are removed.

These observations summarize a strong influence of high count rates on the performance of the detectors, and therefore a series of strategies need to be applied to upgrade i-TED towards BNCT use in clinical environments.

The use of thinner scintillator crystals is a key aspect to reduce the count rates and their related effects. In particular, by changing the nominal 15 mm and 25 mm thicknesses of scatterer and absorber crystals to 6 mm and 10 mm, respectively, the total count rate can be reduced by a factor of 3, while maintaining or even improving the coincidence to overall count rate ratio. This means that the detector is more efficient at identifying coincidence events, even if the total count rate reduces. A second aspect of this thickness reduction is particularly relevant to the scatterer. The total energy deposited in the scatterer is usually smaller than to the absorber, and often, for small scattering angles, this deposited energy falls below the detection threshold. Thinner crystals allow to reduce the value of that detection threshold, as our previous studies with scintillators of various thicknesses have shown. Additionally, this reduction of the scatterer detection angle enables yet another signal enhancement technique, related to back-scattering events (those in which the original gamma ray scatters in the absorber detector and then is absorbed at the scatterer, which cannot be distinguished by timing due to the short distance between the two events). For 478 keV gamma rays, the back-scattering events usually are mixed with events coinciding with forward scatterings of 60-70°. Therefore, if low energy deposited events can be detected, and therefore many low angle scattering coincidences can be used for imaging, then by rejecting all events above 60° and only keeping those with lower reconstructed angles, the signal-to-background ratio greatly improves, by removing artifacts usually observed in tangential sides of the reconstructed images generated from misidentified back-scattering events.

IV. CONCLUSIONS

The experimental campaigns at ILL have tested the feasibility of using the i-TED Compton Camera for detecting the 478 keV gamma rays from ^{10}B neutron capture under conditions approaching those of a clinical BNCT treatment. The device exhibits adequate sensitivity to clinically relevant boron concentrations even in hydrogen-rich environments, provided that count rate–induced gain shifts are mitigated. Our observations highlight the critical impact of high count rates on spectral stability and low-energy gamma detection, which must be addressed to ensure robust imaging performance. Reducing scintillator crystal thicknesses offers a practical means to lower total count rates, improve coincidence efficiency, and enable enhanced signal-to-background discrimination through optimized angular filtering. Together with the integration of a rotatory gantry and real-time image reconstruction algorithms, these upgrades position i-TED as a strong candidate for BNCT dose monitoring systems. Future work will focus on implementing these design changes and validating their performance under full-intensity, clinically relevant neutron irradiation conditions.

ACKNOWLEDGMENTS

This work builds upon research conducted under the ERC Consolidator Grant project HYMNS (grant agreement No. 681740) and has been supported by the ERC Proof-of-Concept Grant project AMA (grant agreement No. 101137646). We also acknowledge funding from the Spanish Ministerio de Ciencia e Innovación under grant PID2022-138297NB-C21.

REFERENCES

[1] W. Sauerwein *et al.* "Neutron Capture Therapy: Principles and Applications", Springer Nature (2025) https://link.springer.com/book/10.1007/978-3-642-31334-9

[2] T. D. Malouff, et al. "Boron Neutron Capture Therapy: A Review of Clinical Applications" in Frontiers in Oncology (2021), vol. 11, 601820, https://doi.org/10.3389/fonc.2021.601820

[3] IAEA, "Advances in Boron Neutron Capture Therapy", Non-serial publications, IAEA, Vienna (2023). https://www.iaea.org/publications/15339/advances-in-boron-neutron-capture-therapy

[4] M. Manabe *et al.* "Study on measuring device arrangement of array-type CdTe detector for BNCT-SPECT", Rep. Pract. Oncol. Radiather. (2016), 21:2, pp. 102-107, https://doi.org/10.1016/j.rpor.2015.04.002

[5] B. Hales *et al.* "Predicted performance of a PG-SPECT system using CZT primary detectors and secondary Compton-suppression anti-coincidence detectors under near-clinical settings for boron neutron capture therapy" Nucl. Instr. Meth. A. (2017), 875, pp. 51-56, https://doi.org/10.1016/j.nima.2017.09.009

[6] I. Murata *et al.* "Design of SPECT for BNCT to measure local boron dose with GAGG scintillator", Appl. Radiat. Isot. (2022), 181, 110056, https://doi.org/10.1016/j.apradiso.2021.110056

[7] T. Lee *et al.* "Monitoring the distribution of prompt gamma rays in boron neutron capture therapy using a multiple-scattering Compton camera: A Monte Carlo simulation study", Nucl. Instr. Meth. A. (2015), 798, pp. 135-139, https://doi.org/10.1016/j.nima.2015.07.038

[8] C. Gong *et al.* "Optimization of the Compton camera for measuring prompt gamma rays in boron neutron capture therapy", Appl. Radiat. Isot. (2017), 124, pp. 62-67. https://doi.org/10.1016/j.apradiso.2017.03.014

[9] Z. Hou *et al.* "Boron concentration prediction from Compton camera image for boron neutron capture therapy based on generative adversarial network", Appl. Radiat. Isot. (2022), 186, 110302, https://doi.org/10.1016/j.apradiso.2022.110302

[10] T. Ferri *et al.* "Design and Validation of a SPECT Prototype for Treatment Monitoring in BNCT and First Experimental Tomographic Results", IEEE. Trans. Radiat. Plasma. Med. Science. (2025), https://doi.org/10.1109/TRPMS.2025.3562079

[11] C. Domingo-Pardo, "i-TED: A novel concept for high-sensitivity (n,γ) cross-section measurements", Nucl. Instr. Meth. A (2016), vol. 825, pp. 78-86 https://doi.org/10.1016/j.nima.2016.04.002

[12] J. Lerendegui-Marco *et al.* "Towards machine learning aided real-time range imaging in proton therapy", Sci. Rep. (2022), 12, 2735 https://doi.org/10.1038/s41598-022-06126-6

[13] J. Balibrea-Correa *et al.* "Hybrid Compton-PET Imaging for ion-range verification:A Preclinical Study for Proton-, Helium-, and Carbon-Therapy at HIT", Submitted to Eur. Phys. Jour. Plus (2025) https://arxiv.org/abs/2504.11273

[14] P. Torres-Sánchez *et al.* "The potential of the i-TED Compton camera array for real-time boron imaging and determination during treatments in Boron Neutron Capture Therapy", Appl. Radiat. Isot. (2025), 217, 111649 https://doi.org/10.1016/j.apradiso.2024.111649

[15] J. Lerendegui-Marco *et al.* "First pilot tests of Compton imaging and boron concentration measurements in BNCT using i-TED", Appl. Radiat. Isot. (2025), 225, 112009 https://doi.org/10.1016/j.apradiso.2025.112009

[16] B. Gameiro *et al.* "Towards Real Time Compton Imaging in Demanding Conditions", Lecture Notes in Computer Science, vol 15386, (2025). Springer, Cham. https://doi.org/10.1007/978-3-031-90203-1_47

Dosimetry with a phoswich detector

Nicolás Ávila[1,*], Dirk Döhler[1], Pia Kahle[1], Anja Seifert[1] and Thomas Kormoll[1]
[1]Institute of Nuclear and Particle Physics, Technische Universität Dresden, Germany
(*) nicolas_alejandro.avila_perez@mailbox.tu-dresden.de

***Abstract*—A phoswich detector concept based on the combination of a high-Z element and a plastic element with a free running analogue-to-digital converter is presented in this work for the simultaneous measurement of operational dosimetric quantities $H^*(10)$ and $H'(0.07)$ in photon radiation fields. By means of a Pulse Shape Discrimination method and down to an energy of 60 keV $\left(^{241}\mathrm{Am}\right)$ in continuous photon radiation fields, the signal arriving at the common light sensor was separated into components coming from each element of the detector. In this context, crosstalk between the sensitive volumes represented an obstacle and was solved with a proper covering over one of the elements. Additionally, an algorithm for the calculation of the intended quantities based on the measured deposited doses was developed and tested using Monte Carlo methods with monoenergetic and X-ray radiation fields. Simulations set a validity region for the algorithm to the energy range between 20 keV and 3 MeV in the case of monoenergetic radiation and a lower energy cut of 15 keV in the case of X-ray fields. Furthermore, the algorithm was tested experimentally using ^{241}Am (60 keV) and ^{137}Cs (662 keV) sources against reference values from an OD-01 ionization chamber. Results showed a good sensitivity to the incoming photon energy, reflected in the variation of the ratio between the measured doses with the considered energies. At the same time, discrepancies between the measured and reference values of the operational quantities highlighted the need for a better calibration of the algorithm.**

***Keywords* —radiation protection, phoswich detector, pulse shape discrimination methods, dose measurements in photon radiation fields**

I. Introduction

Determination of ICRU operational dosimetric quantities such as ambient dose equivalent $H^*(10)$ and directional dose equivalent $H'(0.07)$ in photon radiation fields is of great importance in radiation protection monitoring. In particular, the simultaneous estimation of more than one of these quantities provides a better picture of the radiation field and the exposure due to it. Significant advances in this scope are described in [1] for the simultaneous measurement of $H^*(10)$ and $H'(0.07)$ using scintillation detectors in combination with a digital signal processing unit, with special relevance in low energy photon fields, such as those encountered in medical applications including mammography and other X-ray techniques. These types of setups typically involve an arrangement of several detectors, each of them calibrated to measure only one of the intended quantities and coupled to an individual light sensor. In order to overcome the latter aspect, and enable the development of more cost effective devices, the proposal in this work is to use a phoswich detector.

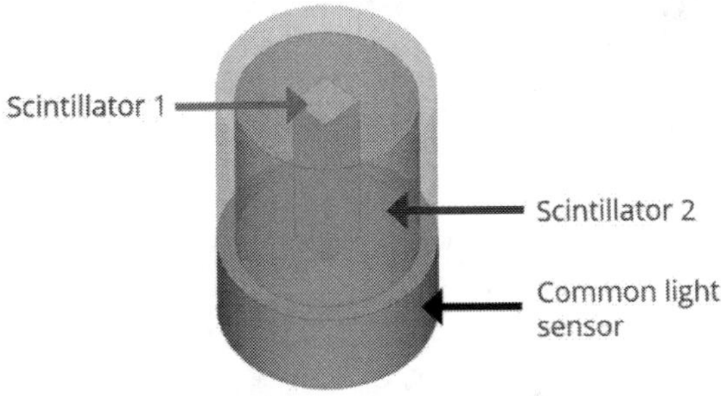

Fig. 1. Sketch of a phoswich detector. In this example, one scintillation material is embedded in another. The compound detector is optically coupled to a common photomultiplier tube.

A phoswich detector is a combination of scintillation materials optically coupled to a common light sensor. An example is depicted in Fig. 1. Such a detector exploits differences in the decay times of different scintillation materials and therefore in the shapes of the pulses arriving at the photomultiplier tube. These differences can, based on a Pulse Shape Discrimination (PSD) analysis, enhance the coincidence and anti-coincidence detection ability of a single photomultiplier tube, as described in [2] and illustrated in Fig. 2.

For the scope of this study, it is important not only to choose the sensitive volumes for the phoswich detector such as the difference in the decay times is as large as possible but also such as a proper sensitivity in the energy range of interest is achieved. The latter is motivated mainly by the behavior of the operational quantities in the low energy regime, relevant for the medical applications mentioned in this section. This behavior is characterized by the fluence to dose conversion coefficients introduced by ICRU [3] and shown in Fig. 3. A good sensitivity to low energies is important for estimations of $H'(0.07)$, whose contribution is significant in that region. A high Z material such as Gadolinium Aluminium Gallium Garnet (GAGG) offers a high cross section for photon full absorption at low energies. On the other hand, due to its tissue equivalence and therefore to mimic $H^*(10)$ dose deposition, a plastic scintillator was chosen as the second element. This combination of materials is also convenient due to its difference in the decay times: about 100 ns for GAGG and typically in the range between 2 ns and 3 ns for plastic.

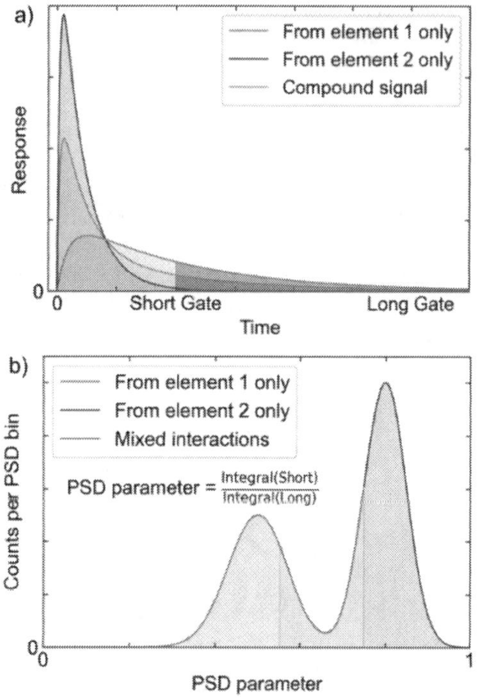

Fig. 2. Scheme of a PSD analysis on the signals from a phoswich detector. In (a), voltage pulses coming from the photomultiplier tube, where different pulse shapes can be appreciated depending on whether the interaction occurred in one of the elements or in both. In (b), the histogram for the PSD parameter, here defined as the ratio between integrals over the time gates of (a), and used for event discrimination.

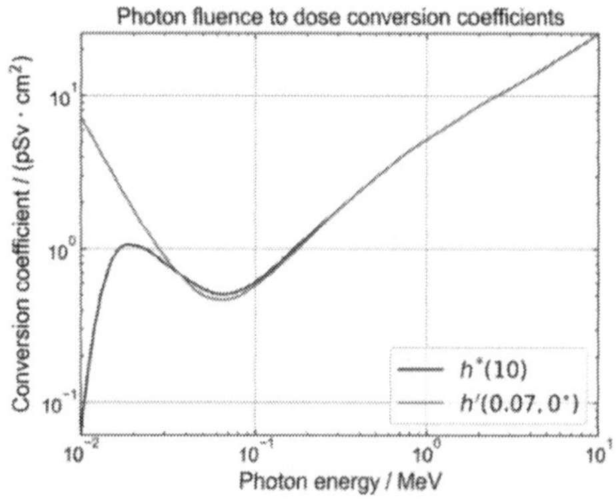

Fig. 3. Photon fluence to dose conversion coefficients for $H^*(10)$ and $H'(0.07)$ provided by ICRU [3] and here labeled as $h^*(10)$ and $h'(0.07, 0°)$, respectively.

It is therefore feasible to use a detection system with one single light sensor, and use the extra information now available by means of a proper PSD analysis to provide simultaneously the value of a second operational quantity.

In the following sections, the PSD capability of different phoswich detector prototypes using the proposed material choice will be explored in measurements with ^{137}Cs (662 keV) and ^{241}Am (60 keV) sources at the Institute of Nuclear and Particle Physics (IKTP) of Technische Universität Dresden, Germany. The performance of such a detector for dosimetry measurements of $H^*(10)$ and $H'(0.07)$ will be tested against reference values in experiments with the same sources. For that purpose, a suitable algorithm for the calculation of the intended quantities based on the measured deposited doses will be formulated from simulation results.

II. Materials and Methods

Different arrangements of the two detectors have been tested and are shown in Fig. 4. The dimensions for the arrangements in Fig. 4 (b), (c) and (d) are 21 mm diameter × 20 mm height for the cylindrical (or semicylindrical) plastic element and 7.5 mm × 7.5 mm × 20 mm for the GAGG crystal. Dimensions for the arrangement in Fig. 4 (a) are similar, except for the height, which is 21 mm. An additional component consisting of an aluminium foil wrapping the GAGG crystal is included in prototypes shown in Fig. 4 (b) and (c).

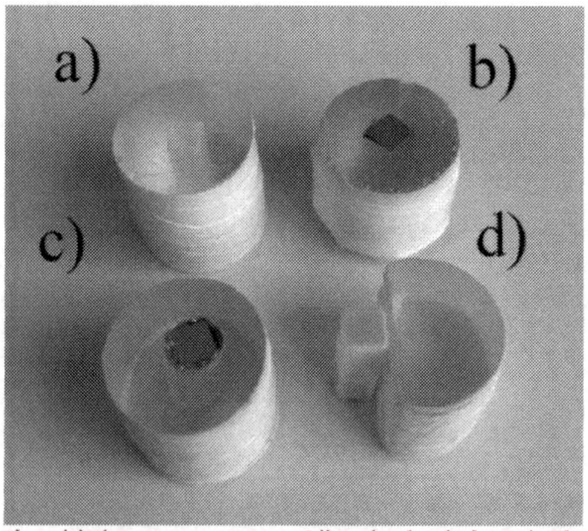

Fig. 4. Phoswich detector arrangements follow the sketch shown in Fig. 1 with a 7.5 mm × 7.5 mm × 20 mm GAGG crystal accompanied by a cylindrical plastic scintillator with diameter 21 mm and height 20 mm, except for prototype (a), which has a height of 21 mm.

During measurements, the phoswich detector is coupled to a photomultiplier tube (model PDM 04-9111-CN) by ETEnterprises, which is in turn connected to a free running 14 bit analogue-to-digital converter with a sampling rate of 125 MHz (DAQ125) from Serious Dynamics, Dresden, Germany [4].

A. Experimental Setups

Experimental work involved two phases: measurements in a) for energy calibration and PSD analysis; and in b) for dosimetry.

a) Energy calibration and PSD: Photon radiation fields were characterized using ^{137}Cs and ^{241}Am sources at the IKTP. The setup, seen in Fig. 5, shows the phoswich detector coupled to the photomultiplier. The scintillator is covered with a foil made of plastic and aluminium to shield it from light. The thickness of the foil is 100 μm, thin enough to allow low energy radiation measurements.

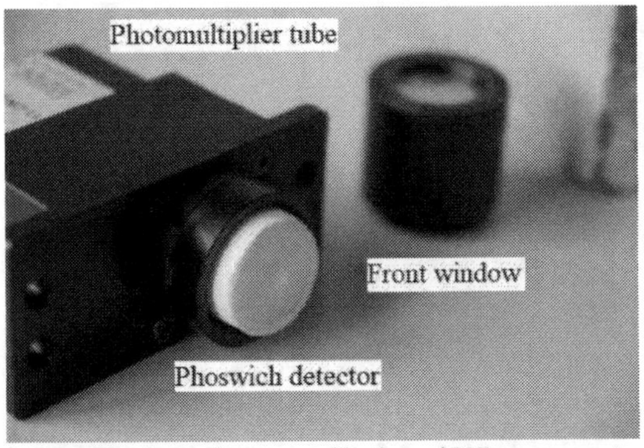

Fig. 5. Experimental setup for energy calibration and PSD measurements.

b) Dosimetry measurements: Values of $H'(0.07)$ and $H^*(10)$ were measured for a collimated source using the setup in Fig. 6. Values obtained with the phoswich detector were compared with reference values given by an ionization chamber OD-01 by STEP Sensortechnik.

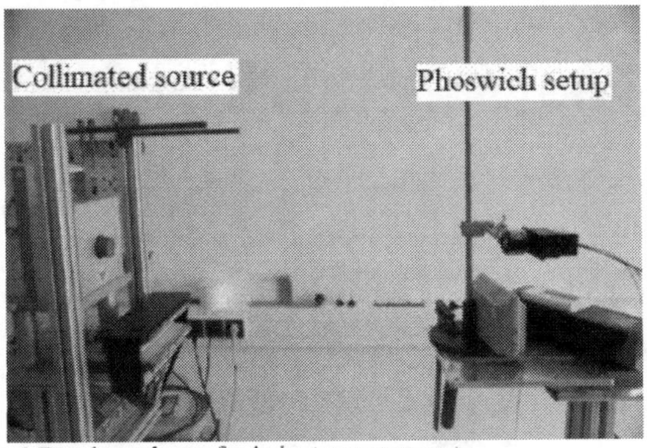

Fig. 6. Experimental setup for dosimetry measurements.

B. Simulation

The goal of the simulations was to develop an algorithm capable of estimating $H'(0.07)$ and $H^*(10)$ based on the measured doses on each sensitive volume of the detector. In this context, a phoswich configuration obeying the sketch in Fig. 1 was replicated in GEANT4 [5] simulations.

Dimensions for such a representation were chosen following the experimental setup: 7.5 mm × 7.5 mm × 20 mm for the GAGG crystal, 20 mm diameter and 20 mm height for the organic scintillator. The physics list used for the relevant processes was *G4EmPenelopePhysics*. The material for the plastic scintillator was set to ICRU 4-element tissue defined in [3]. This is a convenient choice to mimic $H^*(10)$ dose deposition in tissue, as suggested in Section I., and therefore useful to obtain $H^*(10)$ directly from the dose deposition in plastic.

For energy values between 10 keV and 10 MeV, 10^7 events consisting of monoenergetic photons of the given energy where simulated parallel to the cylinder axis and homogeneously distributed over the detector face. Doses on each sensitive volume were recorded and values for the operational quantities were estimated using the simulated fluence and the conversion coefficients shown in Fig. 3.

III. Results and Discussion

A. Simulation

Fig. 7 shows the ratio between the scored dose in the organic scintillator and the value of $H^*(10)$, over the range of simulated energies. As the first step of the algorithm, the ratio is taken as approximately constant over the energy region between 20 keV and 3 MeV. This suggests that $H^*(10)$ can be estimated directly from the measured dose in the plastic element by means of a multiplicative factor, labeled in the figure as α, as

$$\frac{\text{Dose}_{\text{Pl}}}{H^*(10)} \simeq \text{constant} := \alpha\,,$$

$$\rightarrow H^*(10) = \frac{1}{\alpha}\cdot \text{Dose}_{\text{Pl}}, \qquad (1)$$

with $\alpha \simeq 0.726$ Gy/Sv. Other organic materials were explored in the simulations; however, the quantity $\frac{\text{Dose}_{\text{Pl}}}{H^*(10)}$ showed the least variations for a meaningful energy range when tissue is selected.

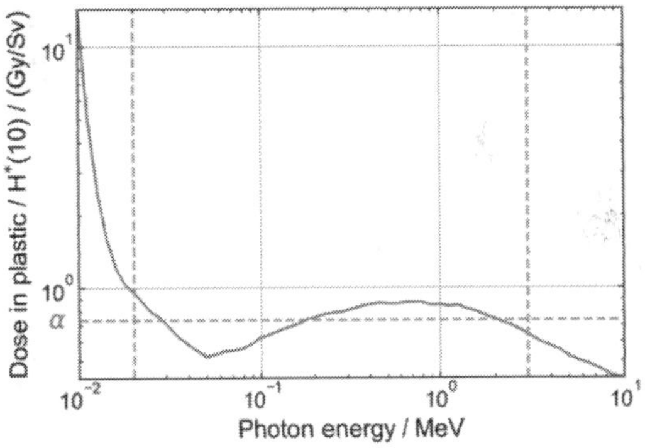

Fig. 7. The ratio between dose in the organic scintillator and $H^*(10)$ is taken as constant over a wide energy range as the first step of the algorithm. This approximation makes sense only for energies within the range defined by the red-dashed lines. The green-dashed line represents the constant value taken for that region and estimated to be $\alpha \simeq 0.726$ Gy/Sv.

On the other hand, Fig. 8 illustrates how, given a measured ratio between the dose in GAGG and the dose in plastic, it is possible to calculate the ratio between $H'(0.07)$ and $H^*(10)$, labeled in the figure as β.

This completes the algorithm as follows:

1. Measure the absorbed doses in the sensitive volumes,
2. using (1) and the dose in the plastic scintillator element, compute $H^*(10)$,

3. with the ratio between the measured doses as starting point, follow the curve in Fig. 8 to calculate the ratio between the operational quantities, labeled as $\beta\left(\frac{\mathrm{Dose_{GAGG}}}{\mathrm{Dose_{Pl}}}\right)$,

4. compute $H'(0.07)$ as

$$H'(0.07) = \beta\left(\frac{\mathrm{Dose_{GAGG}}}{\mathrm{Dose_{Pl}}}\right) \cdot H^{*}(10). \qquad (2)$$

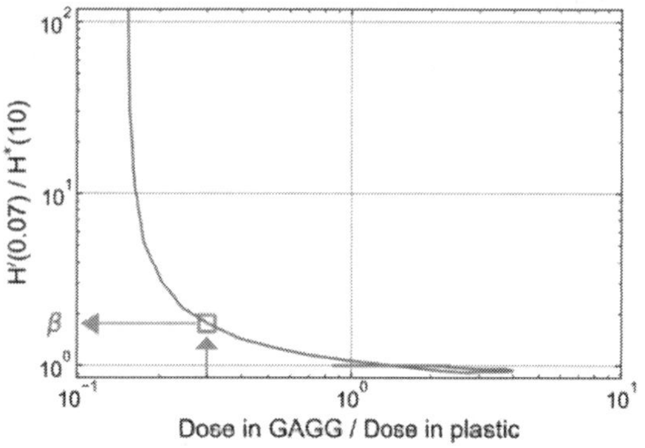

Fig. 8. The red arrows on the graph indicate how to go from a measured doses ratio to the corresponding ratio between the operation quantities.

Simulations to test the algorithm were performed for monoenergetic photon beams, showing good results within the energy range defined in Fig. 7; and for X-ray spectra, with best results when the low energy range of the spectrum was not included (see Fig. 9).

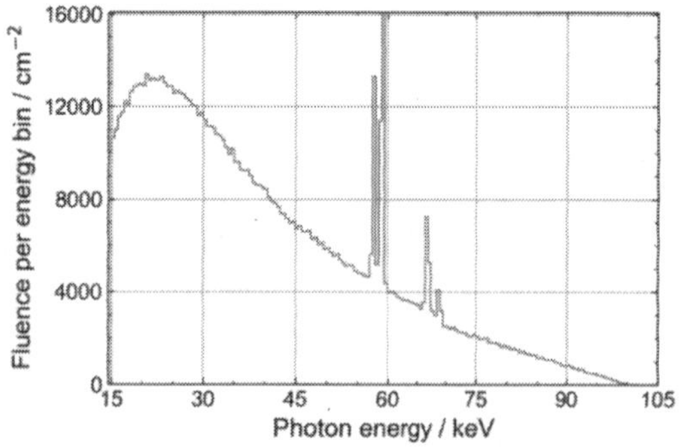

Fig. 9. X-ray spectrum taken from the catalogue of Physikalisch-Technische Bundesanstalt (PTB) [6]. It was used in the simulations with an energy cut for energies below 15 keV.

B. Energy calibration and PSD

Fig. 10 (a) shows a preliminary measurement carried out with the detector shown in Fig. 4 (a) using a ^{137}Cs source. Crosstalk effects were evident, as only events from the GAGG crystal appeared as a well-defined branch in the two dimensional histogram. This is explained by the overlap between the emission spectrum of the plastic element and the absorption spectrum of GAGG, resulting in light emitted from plastic stimulating scintillation in GAGG and not being recorded by the photomultiplier. The problem was addressed by wrapping the GAGG crystal in aluminium foil (see Fig. 4 (b) and (c)). The same type of measurement using the detector in Fig. 4 (b) is shown in Fig. 10 (b), where events from plastic are now visible.

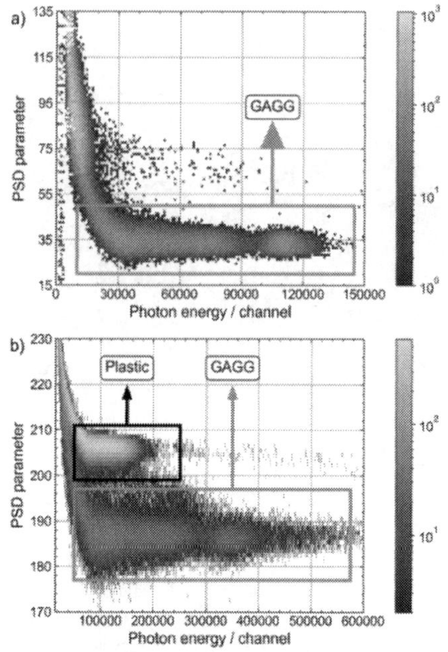

Fig. 10. Wrapping the GAGG crystal in aluminium foil shows a significant suppression of the crosstalk between the sensitive volumes. The measurement was done without the foil in (a) and with the foil in (b).

Alternatively, the detector shown in Fig. 4 (d) offers a more convenient way to dicouple the sensitive volumes and does not evidence a significant detriment to the branches due to crosstalk (see Fig. 11). Detector in Fig. 4 (d) will be used for the subsequent analysis.

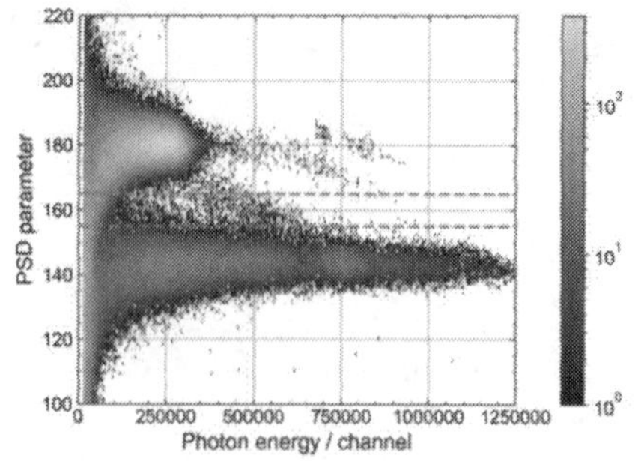

Fig. 11. Measurement with ^{137}Cs using the detector in Fig. 4 (d). Red-dashed lines for signal separation are drawn.

The discrimination lines drawn in Fig. 11 are used as follows: events above the upper line are taken as events purely from plastic, while events below the lower line are taken as events from GAGG. Results from this analysis are shown in Fig. 12.

https://doi.org/10.1051/epjconf/202533809012

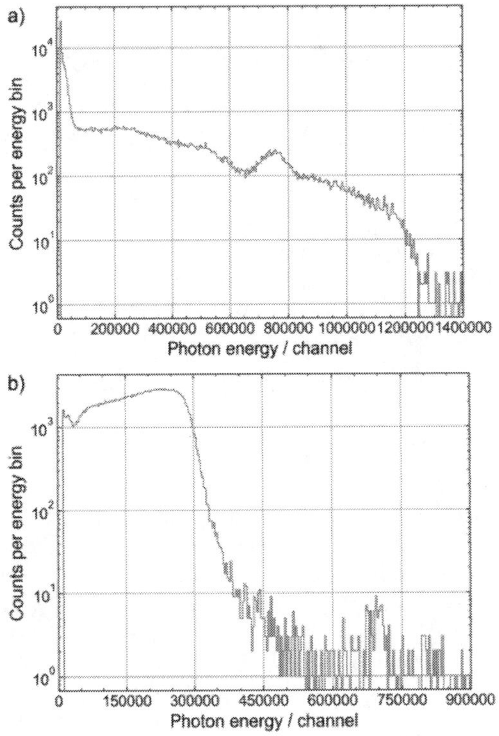

Fig. 12. Signal separation following Fig. 11 exhibits (a) the full absorption peak in GAGG events and (b) the Compton region in events from plastic.

Analysis using ^{241}Am produced similar results, showing the 60 keV peak in both elements. The current lower energy limit for a meaningful signal separation is set to 60 keV. The GAGG element was calibrated using the full absorption peaks, while plastic was calibrated using either also this peak (when visible) or the Compton edge.

C. Dosimetry measurements

After energy calibration, the algorithm presented in Section III. A. was tested experimentally. Table I summarizes the results.

TABLE I
DOSIMETRY MEASUREMENTS

		$H^*(10)$ / μSv		$H'(0.07)$ / μSv	
Source	$\frac{\text{Dose}_{\text{GAGG}}}{\text{Dose}_{\text{Pl}}}$	Reference	Measured	Reference	Measured
^{241}Am (60 keV)	2.11	0.38	0.46	0.35	0.42
^{137}Cs (662 keV)	0.48	3.41	6.27	3.33	7.65

Discrepancies between the reference and measured values can be attributed to the difference between the simulated geometry and the geometry of the phoswich detector used experimentally. A highlight from the results is that a good responsiveness with the photon energy, evidenced in the variation of the quantity $\frac{\text{Dose}_{\text{GAGG}}}{\text{Dose}_{\text{Pl}}}$ with the two considered energies, was achieved.

IV. CONCLUSIONS AND OUTLOOK

From the conducted experiments, the discrimination ability of a phoswich detector was studied and shown to be possible with the current methods down to an energy of 60 keV in continuous photon radiation fields. Crosstalk between the sensitive volumes was identified as a challenge in certain configurations and was addressed with a proper covering over one of the elements. Different signal filter techniques will be studied to improve the signal discrimination. The detector performance under X-ray radiation fields is also of great interest and will be investigated.

Simulations were carried out to develop an algorithm for the measuring system to evaluate the operational quantities based on measured values of the deposited doses in the sensitive volumes. Dosimetry measurements with the proposed algorithm showed a good sensitivity to the photon energy and were compared with reference values. As suggested by the discrepancies, the algorithm needs to be adjusted for a better matching to the reference values. The proposed approaches here include to refine the geometry in the simulations and to calibrate the algorithm experimentally.

REFERENCES

[1] T. Werner, *et al.*, "Dose Rate Measurements in Pulsed Radiation Fields by Means of an Organic Scintillator", in *Proc. EPJ Web Conf.*, Prague, Czech Republic, 253, 09002, 2021.

[2] D. H. Wilkinson, "The Phoswich-A Multiple Phosphor", in *Review of Scientific Instruments*, 23 (8), 414–417, 1952.

[3] International Commission on Radiation Units and Measurements, "ICRU Report 57. Conversion Coefficients for Use in Radiological Protection against External Radiation", in *Journal of the ICRU*, vol. os29, no. 2, Aug. 1998.

[4] Y. Ma, *et al.*, "System noise of a digital pulse processing module for nuclear instrumentation", in *Proc. EPJ Web Conf.*, Portorož, Slovenia, 225, 01012, 2020.

[5] S. Agostinelli, *et al.*, "GEANT4–a simulation toolkit", in *Nuclear Instruments and Methods in Physics Research Section A*, vol. 506, pp. 250–303, Jul. 2003.

[6] U. Ankerhold, "Catalogue of X-ray spectra and their characteristic data: ISO and DIN radiation qualities, therapy and diagnostic radiation qualities, unfiltered X-ray spectra", Physikalisch-Technische Bundesanstalt (PTB), 2000.

Experimental proof-of-concept and first field tests of the dual gamma-neutron imager GN-Vision

A. Sanchis-Moltó[1], J. Lerendegui-Marco[1,*], G. Cisterna[1], P. Álvarez-Rodríguez[1], J. Balibrea-Correa[1], V. Babiano-Suárez[1], J-M. Daugas[1], G. de la Fuente[1], B. Gameiro[1], I. Ladarescu[1], C. Méndez-Malagón[1], C. Michelagnoli[1], I. Porras[1], A. Navarro-Mocholi[1], M. Porras-Quesada[1], C. Ruiz-Ruiz[1], P. Torres-Sánchez[1], S. Valladarés[1] and C. Domingo-Pardo[1]

[1]Instituto de Física Corpuscular, CSIC-Universitat de València, Spain

(*) Jorge.lerendegui@ific.uv.es

***Abstract*—GN-Vision is a compact dual-modality imaging system designed to simultaneously localize the spatial origin of γ-ray and slow neutron sources, with promising applications in hadron therapy, particularly Boron Neutron Capture Therapy (BNCT). This contribution presents results from the first experimental neutron imaging campaign using an early prototype of GN-Vision, conducted at the Institut Laue-Langevin (ILL) with a thermal neutron beam. The primary objective was to validate the system's neutron imaging capabilities. Additionally, we investigated thermal neutron absorption in water phantoms containing ^{10}B, with experimental results benchmarked against GEANT4 simulations. The second part of the work reports on recent advancements towards the development of a full GN-Vision demonstrator. In particular, we present the characterization of the spatial and energy response of a CLLBC scintillator to both neutrons and γ -rays using the compact and scalable PETsys electronics. The promising results highlight the potential of CLLBC to enable integrated neutron and gamma imaging in the first complete GN-Vision prototype.**

***Keywords* — BNCT, CLYC detector, CLLBC scintillation crystal, dual neutron-gamma imaging.**

I. MOTIVATION

Hadron therapy is currently experiencing an increasing interest due to the physical and radiobiological characteristics of the different particles employed (protons, neutrons, heavy ions) that offer many advantages in comparison with conventional radiotherapy [1]. Particle therapy offers more localized treatments focused on tumour cells resulting in less damage in the surrounding healthy tissues compared to conventional radiotherapy.

Particularly, Boron Neutron Capture Therapy (BNCT) takes advantage of the ^{10}B high cross section with thermal neutrons [2] to produce the ^{10}B(n,α)^{7}Li reaction in order to kill the tumour cells. In this reaction, an alpha particle is released and, in 94% of the cases, an excited state of ^{7}Li which decays to the ground state emitting a γ-ray with a fixed energy of 478 keV.

Thus, the effectiveness of this therapy relies on the release of a large amount of energy through heavy particles with high Linear Energy Transfer (LET) by means of the irradiation of ^{10}B-loaded tumour cells with thermal neutrons. Moreover, ranges of both emitted particles, alphas and lithium atoms, are approximately 5 μm and 9 μm for their respective energies, thus, the energy deposition takes place inside the cell. Therefore, unlike other particle therapies such as Proton Therapy (PT) in which the dose is spatially focused, BNCT displays selectivity at the cellular level, making it a promising therapy for scattered tumours with difficult treatment such as glioblastoma multiforme, meningioma, head and neck cancers, lung cancers and breast cancers [3].

In order to carry the ^{10}B to the tumour cells, the latter is introduced in the molecule 4-borono-L-phenylalanine (BPA) which has a high affinity for this kind of cells [4]. Once the radioisotope is injected in the patient, its concentration is expected to be higher in the tumour region than in healthy tissues [5], typical concentrations are 65 and 20 ppm respectively [6]. Nevertheless, the measurement of these ^{10}B concentrations is still one of the greatest challenges to overcome in BNCT [4]. Therefore, treatment planning in BNCT is rather unconventional and based only on estimates of the boron tumour uptake, that can be made via a previous PET scan of the patient, from which the tumour dose is estimated, and an estimate of the dose at the organs of risk based on analytics of the boron in blood.

In this context, a system capable of simultaneously detecting and imaging γ-rays and slow neutrons (thermal and epithermal up to 1 keV) would be very convenient. As for BNCT, this device would allow to perform real-time dosimetry through the detection and imaging of the 478 keV γ-rays produced in the 94% of the ^{10}B reactions with thermal neutrons [6, 7]. It should be noted that, in BNCT, a significant fraction of the secondary dose comes from neutrons. Therefore, imaging and detection of these particles would provide really valuable information not only on the imparted dose in the tumour region but also on other peripheral regions in which the risk of developing a secondary cancer may become significant.

In this work, the GN-Vision concept based on the previous i-

TED Compton imager is introduced in Sec. II. Since the γ-imaging capability of GN-Vision has been developed in previous works, exploring the neutron imaging capability of the detector has become the focus of the current experimental work. Thus, Sec. III describes the first prototype designed for neutron imaging based on CLYC-6 scintillation crystal with a resistor-network based readout electronics used in the experimental campaign at Institut Laue-Langevin (ILL). This campaign was carried out with the aim of proving the imaging capabilities [8] and exploring the potential applicability of GN-Vision to BNCT. The experimental results obtained are shown and discussed in Sec. III.B. On the other hand, in the pursuit of integrating γ-ray and neutron imaging, another experimental setup is explored and described in Sec. IV. This setup includes the CLLBC scintillation crystal with a PETsys ASIC-based readout electronics, the same electronics used for the i-TED detector. Finally, Sec. V provides a summary of the results and perspectives of the project.

II. THE GN-VISION DEVICE

GN-Vision is a compact and portable device capable of spatially locating a thermal and epithermal neutron or γ-emitting source. It has been built on the foundations of the i-TED imager [9], an array of Compton cameras that has been developed in recent years [10, 11, 12, 13]. These Compton imagers have been successfully utilized in nuclear astrophysics experiments at CERN n_TOF [14, 15, 16, 17, 18] and afterwards adapted for medical and nuclear security applications [6, 7, 19, 20, 21, 22].

Thus, γ-ray imaging is expected to be achieved in GN-Vision using the Compton imaging technique, following the technical implementation of the previous i-TED detector. The latter is structured in two planes, in which position sensitive detectors are used, in particular, $LaCl_3$ monolithic scintillation crystals. The first plane, known as *scatterer*, contains a single 50 mm × 50 mm × 15 mm $LaCl_3$ crystal in which the incident γ-rays are expected to undergo Compton scattering and subsequently become detected in the larger second plane. The latter, known as *absorber* plane, is composed of an array of four 50 mm × 50 mm × 25 mm $LaCl_3$ crystals, thicker than those in the scatterer plane, in order to absorb the scattered γ-rays. To perform imaging, the position sensitive $LaCl_3$ crystals are coupled to the Onsemi 8 × 8 SiPM array (ArrayJ-60035-64P-PCB) and connected to the PETsys readout electronics. Table I displays a comparison between both devices and their features.

In order to perform simultaneous neutron and γ-ray imaging with the GN-Vision concept, two conditions must be fulfilled: the active detection material employed for the first detection plane must be sensitive to both kind of particles and, at the same time, be able to discriminate between them. This is the reason why a CLYC-6 scintillation crystal (afterwards CLLBC is also considered) is introduced in the first detection plane of GN-Vision, instead of the single $LaCl_3$ crystal used in i-TED. The inclusion of this new scintillation crystal implies that the latter not only works as a Compton scatterer, thus enabling γ-rays imaging with same working principle as i-TED, but also works as a slow neutron detector. Neutron detection in the so-called CLYC-6 and CLLBC scintillation crystals, whose compositions are Cs_2LiYCl_6:Ce and $Cs_2LiLaBr_{4.8}Cl_{1.2}$:Ce respectively, lies on their ^{6}Li content. This isotope presents a high absorption cross section for thermal neutrons, ~940 b [23], producing the $^{6}Li(n,\alpha)^{3}H$ reaction in which an α particle and a tritium with a total energy of ~4.78 MeV are released.

Once the neutrons are detected and discriminated from γ-rays, neutron imaging is performed through a passive collimation technique. To achieve this, in the first prototype of GN-Vision a collimator is placed in front of the scintillator crystal in the scatterer plane, allowing GN-Vision to work as a pinhole camera [24]. As for the current work, a pinhole collimator made of 95% ^{6}Li-enriched polyethylene ($^{6}LiPE$) is used to increase neutron absorption, thus improving the resolution and reducing the number of background counts. Fig. 1 shows a scheme of the whole detector.

Table I. Comparison of the main components and features of the i-TED and GN-Vision detectors.

	i-TED	GN-Vision
First plane	$LaCl_3$50 mm × 50 mm × 15 mm	CLYC-6 / CLLBC
Second plane	4$LaCl_3$50 mm × 50 mm × 25 mm	4$LaCl_3$50 mm × 50 mm × 25 mm
γ-ray imaging	Compton (active)	Compton (active)
Neutron imaging	-	Pinhole camera (passive)

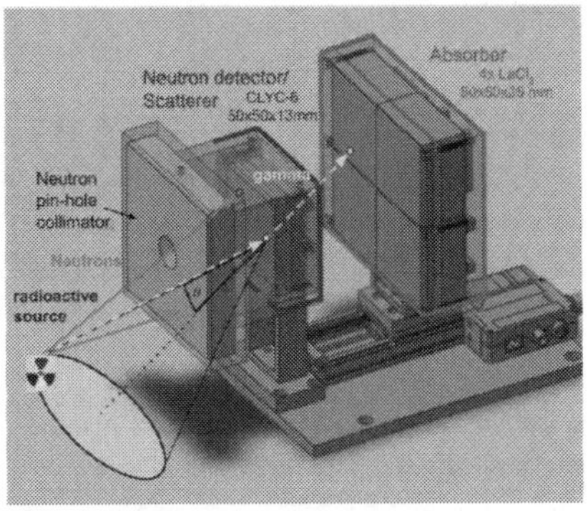

Fig. 1. Scheme of the GN-Vision system in which the most relevant components are highlighted.

III. FIRST TESTS OF NEUTRON IMAGING FOR BNCT

A. Experiment at ILL

The first experimental campaign involving GN-Vision took part in the Fission Product Prompt Gamma-Ray Spectrometer (FIPPS) beamline at ILL. The existing reactor produces the most intense continuous neutron flux in the world in the moderator region, 1.5 x 10^{15} n/s/cm^2 [24]. The thermal neutron flux reaching the FIPPS station is around 5 x 10^{7} n/s/cm^2 in a spot of ~1.5 cm.

The experimental setup and disposition of the detector with respect to the thermal neutron beam are shown in Fig. 2. As seen in the sketch, the neutron beam traverses a distance of ~3 m in air before reaching the target, which will act as a neutron scatterer and, therefore, as the neutron source for our neutron detector, enabling to perform imaging of the target. In order to explore the potential application of GN-Vision to BNCT, a series of phantoms with different ^{10}B concentrations were used as targets. Finally, a neutron dump consisting of a concrete wall covered with borated gum is placed 20 cm away. While the beam profile at FIPPS is well defined, the spread in the position of our setup is uncertain due to the presence of a thick aluminum window followed by ~3 m of air.

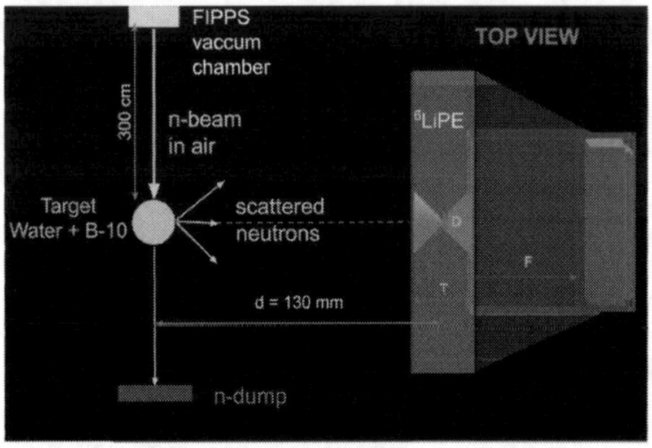

Fig. 2. Scheme of the experimental setup at ILL.

For this campaign, a reduced version of the GN-Vision system has been used, including only the module involved in neutron imaging. The latter includes the ^{6}LiPE (95% ^{6}Li-enriched polyethylene) pinhole collimator and the first position-sensitive detection plane comprising a CLYC-6 scintillator crystal (dimensions 50 mm × 50 mm × 13 mm) coupled to the Onsemi 8 × 8 SiPM array, with a pixel size of ~6 mm (ArrayJ-60035-64P-PCB), the same model used in the i-TED detector. Regarding the readout electronics, the AiT 4-Channel Active Base Readout instrumentation from AiT (model: AB4T-ARRAY64P [25]) was employed. The details related to the development of this GN-Vision module together with characterization of the CLYC crystal and the AiT electronics have been reported in [8].

B. First neutron images: experiment and simulations

Previous works have presented the first experimental results aiming to proof the imaging capabilities and resolving power of this first neutron imaging prototype [8]. In this work, the first results related to the potential applicability of GN-Vision to BNCT are presented. In this line, small PMMA dishes of ~1.75 cm radius and 1.2 cm thickness filled with water and different ^{10}B concentrations have been employed as targets. As previously mentioned, the targets used will act not only as neutron scatterers but also as thermal neutron sources from the detector's point of view. It should be noted that the thickness of the dish walls is completely negligible, thus, the relevant interaction occurs between thermal neutrons and water.

Fig. 3. PMMA dish used as a neutron target at the experimental campaign in ILL.

Different water phantoms were used in this campaign. Nevertheless, the present work aims to present the first experimental results based on a single target and configuration, in which the dishes are placed sideways from the detector's point of view, as it can be appreciated in Fig. 3. The dishes were filled with water with three different ^{10}B concentrations: 0, 20 and 100 ppm (parts per million). These ^{10}B concentrations are chosen to be as realistic as possible since ~65 ppm is the standard concentration used for simulations of BNCT in brain tumours with BPA as the carrier molecule [27, 28, 29]. Additionally, several measurements were performed: a first 60-second short measurement and a longer one of 300-600 seconds, to gather more statistics.

Given the complex geometry and materials of the dishes, GEANT4 simulations of the detector and the experimental setup have been performed. These simulations allow one to establish a comparison between simulated and experimental results, hence achieving a deeper understanding of the experimental results. For these simulations, the geometry of the detector and targets have been accurately designed to closely resemble the real experiment. Moreover, the simulation setup is in vacuum and has no walls or surroundings. This is a key discrepancy with the experiment which will cause the simulation images to have less background counts than the experimental ones. Apart from this, a pencil-beam of thermal neutrons (~25 meV) is simulated. The pencil-beam configuration is taken as an approximation since the real size and beam profile are not known in that position. Nevertheless, the neutron beam in the experiment is known to have a finite diameter as described in Sec. III.A. More details on the GEANT4 application employed can be found in [24].

The processes of position reconstruction in the detector and thermal neutron event selection are different for the experimental images and those obtained from GEANT4 simulations. On the one hand, as for the experimental case, thermal neutron events are selected introducing a cut in the deposited energy, in order to obtain only events from ^{6}Li(n,α)^{3}H reaction and pulse-shape discrimination techniques are employed to discriminate between neutron and γ events [8, 30]. On the other hand, event selection in GEANT4 builds upon the so-called *hits*. Each *hit* represents the interaction of a particle with the active volume of the detector, and the deposited energy, momentum and secondary particles are recorded. In order to select the thermal neutron events, two restrictions are applied:

- Since the interaction of thermal neutrons with ^{6}Li nuclei produces an alpha particle and a triton, the restriction simply consists of requiring the energy deposition by any of those particles.
- A cut in energy in the range [4.5, 5] MeV is applied in order to select the thermal neutron events, in which a total of 4.78 MeV are deposited in the crystal.

After the selection of thermal neutron events in the detector, image reconstruction is performed both in the experiment and simulations following the imaging working principle of a pinhole camera. Thus, to obtain the image, two operations are performed: inversion and rescale. The latter is carried out by using a scale factor determined by the distance from the pinhole to the object and the distance between the pinhole and the detector (so-called focal distance). This is explained in more detail in [8, 24].

Both experimental and simulation images are displayed in the top and bottoms panels in Fig. 4 respectively. It should be noted that for the simulation images, the target considered is sketched in red for clarity. As for the experimental ones, because of the systematic errors in target positioning in the experiment, this shape is not sketched.

As it can be appreciated in the experimental images (top panel, Fig. 4), as the boron concentration increases, the region with higher count cumulation (i.e. the area that reflects the extension of the target) gets smaller. On the other hand, in the simulation images (bottom panel, Fig. 4), we can observe that the region where most of the neutrons are scattered (yellow region) is more condensed in the centre of the target compared to its distribution in the experimental images. A possible cause behind this difference could be related to the beam conformation in the simulation and experiment. In the first, an ideal pencil-beam was simulated whereas for the experiment, the beam is known to have a finite diameter.

Aiming at studying the potential of neutron imaging with GN-Vision applied to BNCT, it is interesting to quantify the ^{10}B thermal neutron absorption, a measure of the delivered dose via the $^{10}B(n,\alpha)^{7}Li$ reaction, from the obtained neutron imaging. In order to establish a truthful comparison between the images experimentally obtained for the phantoms with different ^{10}B concentrations, counts have been normalised to the live measurement time. This simply stands for the fact that the colour scale numbers are directly comparable to those from the experimental images since the units are counts/second. Despite this normalisation, a visual comparison does not result satisfactory enough to extract proper conclusions, since the maximum value of the 2D image is not a reliable parameter to quantify the ^{10}B neutron absorption by itself.

Hence, to more accurately quantify this neutron absorption, different Regions of Interest (ROIs) are defined in which neutron counts are integrated and then compared to the results of the water dish (with no ^{10}B concentration), used as a baseline:

$$\% \, absorption = \frac{n_{water} - n_{B-10}}{n_{water}}, \qquad (1)$$

where n_{water} and n_{B-10} stand for the neutron counts in the water dish image and the image with the corresponding ^{10}B concentration for each considered ROI respectively. The absorption analysis has been performed for the experimental images and the ones reconstructed from the GEANT4 simulations. Table II and Table III show the computed percentages of ^{10}B neutron absorption for the experimental and simulated images respectively.

Table II. Percentage of thermal neutron B-10 absorption evaluated for the experimental images obtained for the PMMA dishes shown in Fig. 3 for different ROIs.

Percentage of neutron absorption in ^{10}B			
ROI area	Water Dish	20ppm ^{10}B	100ppm ^{10}B
13×20	-	-26.0 ± 0.4	-32.3 ± 0.4
13×22	-	-26.0 ± 0.4	-34.4 ± 0.3

Table III. Percentage of thermal neutron B-10 absorption evaluated for the simulated images obtained for the PMMA dishes shown in Fig. 3 for different ROIs.

Percentage of neutron absorption in ^{10}B			
ROI area	Water Dish	20ppm ^{10}B	100ppm ^{10}B
20×20	-	-0.3 ± 0.7	-5.2 ± 0.7
10×10	-	1.6 ± 1.1	-3.5 ± 1.0
14×20	-	-0.5 ± 0.7	-5.2 ± 0.7
20×32	-	-1.1 ± 0.6	-5.8 ± 0.6

As it can be appreciated, there is a significant difference between the neutron ^{10}B absorption for the resulting images obtained from the experiment and simulations. Among the possible hypothesis to explain these discrepancies between the experimental and simulation results we find: the beam conformation, since the simulated pencil-beam does not necessarily correspond to the actual beam in the experiment; the absence of the G4NeutronHP thermal scattering libraries in the GEANT4 simulation and, finally, a possible experimental deviation from the assumed ^{10}B concentration. Further analysis should be conducted in order to properly assess the impact of these factors on the discrepancies observed in ^{10}B thermal neutron absorption.

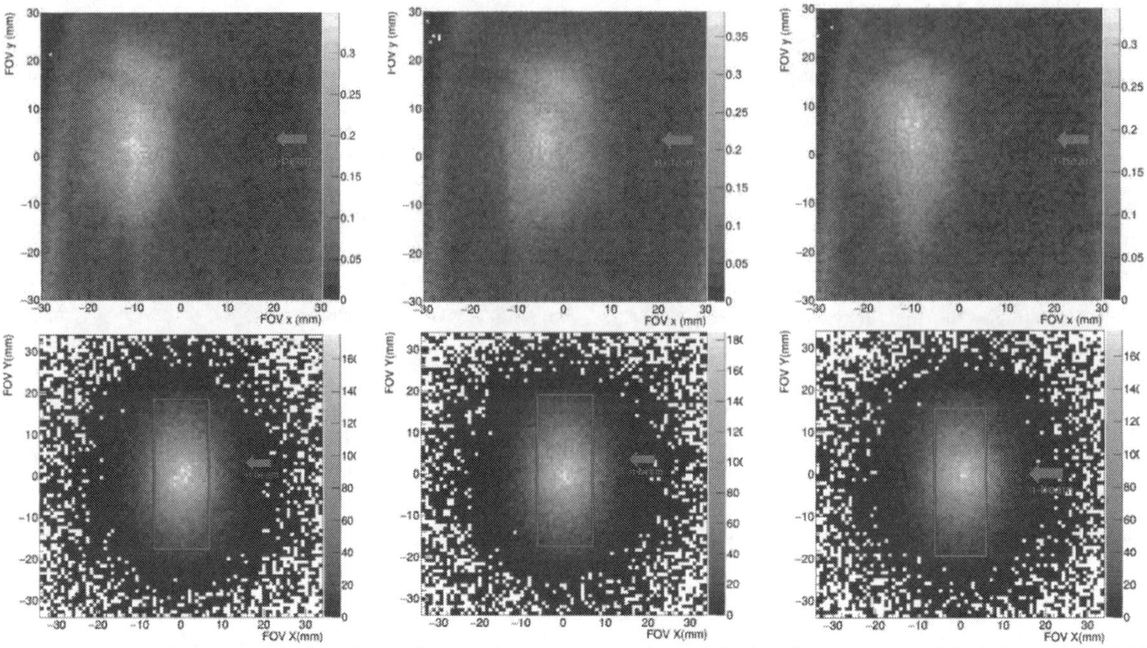

Fig. 4. Experimental (upper row) and simulation (lower row) neutron images obtained for the sideways position of the dish containing 0, 20 and 100 ppm B-10 concentration in water. The color scale units are counts/second (upper row) and counts (lower row)

IV. ADVANCES TOWARDS A COMPLETE GN-VISION DEMONSTRATOR

A. Challenges and status

In the second part of this work, we sought to make further progress in the integration of the neutron and γ-ray imaging capabilities in a single device.

The integration of the CLYC crystal with PETsys electronics used in the predecessor i-TED is not possible because of its moderate light yield (20000 photons/MeV) and slower scintillation decay components in comparison with $LaCl_3$ crystal (45000 photons/MeV). This has led to the design of the first prototype of neutron imaging detector through the combination of CLYC scintillation crystal and AiT readout electronics, as explained in more detail in [24, 31]. More recently, tests with an alternative scintillation crystal with gamma-neutron discrimination capabilities and higher light yield than the CLYC, $Cs_2LiLaBr_{4.8}Cl_{1.2}$:Ce abbreviated simply as CLLBC, have been carried out, obtaining promising results towards the integration of neutron and γ-ray imaging in a device. In this work we have proven for the first time that the CLLBC readout with PETsys TOFPET2 ASIC electronics [32] works correctly and unlocks the possibility of integrating neutron and γ-ray imaging in the GN-Vision device with a single readout system. The first steps in the characterization of the CLLBC crystal are presented hereafter.

B. CLLBC crystal characterization

B.1. CLLBC: Neutron and γ-ray response

In order to run the first tests of the CLLBC scintillator crystal with PETsys electronics, the experimental setup shown in Fig. 5 was used. Two different radioactive sources were used: an encapsulated source of ^{137}Cs, emitting a 662 keV γ-ray following its β-decay, and ^{252}Cf, emitting neutrons by means of spontaneous fission. The latter is introduced in a polyethylene shielding to partially moderate the fast neutron spectrum of ^{252}Cf [24]. In the setup, a CLLBC crystal (cylindrical 50 mm diameter x 10 mm thick) is coupled to the same Onsemi 8 x 8 SiPM array (ArrayJ-60035-64P-PCB) and connected to the PETsys readout electronics.

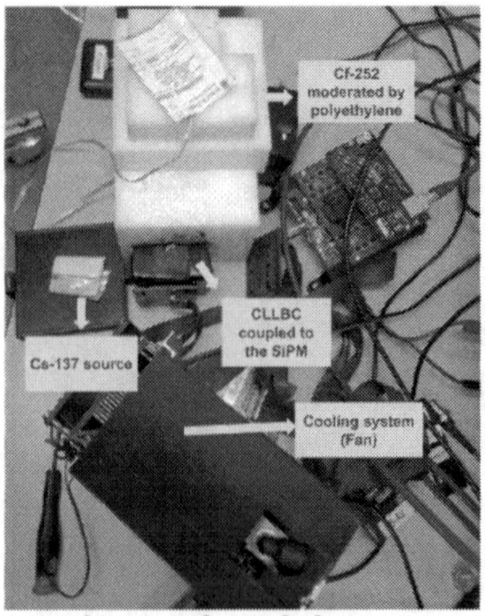

Fig. 5. Experimental setup used to study the spectroscopic and spatial response of the CLLBC with PETsys electronics to neutrons and gamma sources.

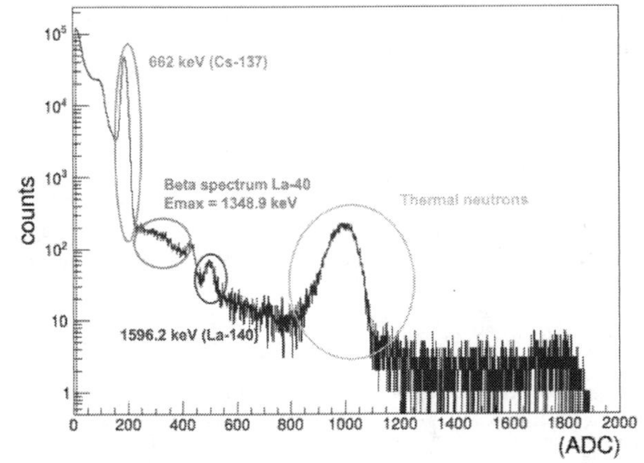

Fig. 6. Deposited energy spectrum (in arbitrary ADC units) measured with the CLLBC scintillation crystal when exposed to Cs-137 and Cf-252 sources.

The energy spectrum obtained is displayed in Fig. 6. In the latter, the 662 keV γ-peak from ^{137}Cs manifests clearly, preceded by the Compton continuum. Due to the natural presence of ^{140}La in the crystal, the β spectrum of 1348.9 keV endpoint and γ-rays of 1596.2 keV are registered. Furthermore, a clear high-energy peak (several MeV) can be observed, and corresponds to thermal neutrons detection since it is not observed in the background measurement (using only the ^{137}Cs).

The spatial distribution of scintillation light collected in the SiPM array coupled to the CLLBC allows one to reconstruct the position of interaction of neutrons and γ-rays. From the measured distribution, the 3D position of the interaction inside the crystal is reconstructed using the same analytic reconstruction algorithms implemented in the processing codes for the i-TED detector [11, 13]. Thus, Fig. 7 illustrates the reconstructed positions of interaction in the transverse plane (XY) of the detector for all the events (Fig. 7a) and for a particular energy selection which corresponds to the thermal neutron peak (Fig. 7b). The upper image presents a uniform distribution of counts with a high concentration on the crystal's left side. This is consistent with the positions of the ^{252}Cf and ^{137}Cs sources with respect to the crystal in the experimental setup. The ^{252}Cf is placed right in front of the detector, thus, a uniform distribution of counts among the crystal's surface is expected. As for the ^{137}Cs source, it is placed next to the detector, on its left side (see Fig. 5). Additionally, both for γ-ray and neutron events, the cylindrical shape of the CLLBC crystal is correctly reconstructed. However, a clear compression effect related to light reflection near the edges of the crystal is observed [33].

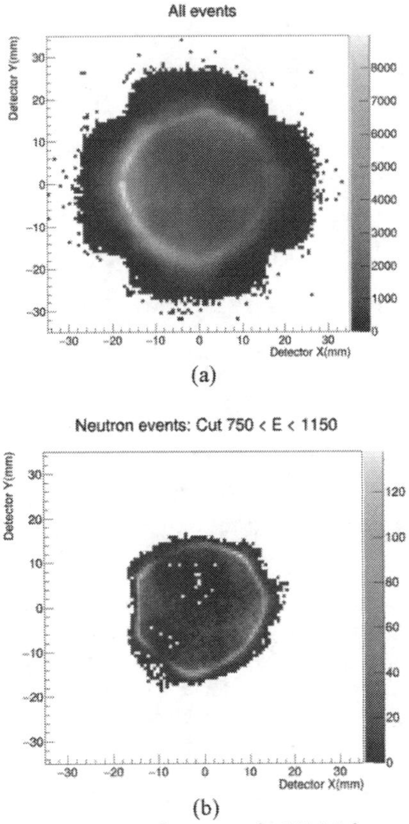

Fig. 7. Position reconstructed images in the CLLBC detector plane for (a) all the reconstructed events and (b) thermal neutron events.

B.2. Energy resolution using a ^{137}Cs source

In order to perform the first measurements using the combination CLLBC integrated with PETsys electronics readout, an optimisation of the PETsys electronics parameters is required. This process involves the modification of parameters such as the charge integration window or the SiPM threshold (minimum charge value required to trigger the acquisition on individual SIPMs) to optimize the energy resolution. To assess this, the 662 keV γ-ray in ^{137}Cs source was used, as this is the reference energy typically employed to evaluate the energy resolution.

The results obtained are displayed in Fig. 8, which shows the variation of the energy resolution as a function of the charge integration window, which is the parameter with the largest impact in the energy resolution. As seen in the graph, the minimum resolution is achieved with the ~500 ns integration window. According to this, we can conclude that an energy resolution of 6.5% is obtained for the CLLBC when the electronics parameters are optimised.

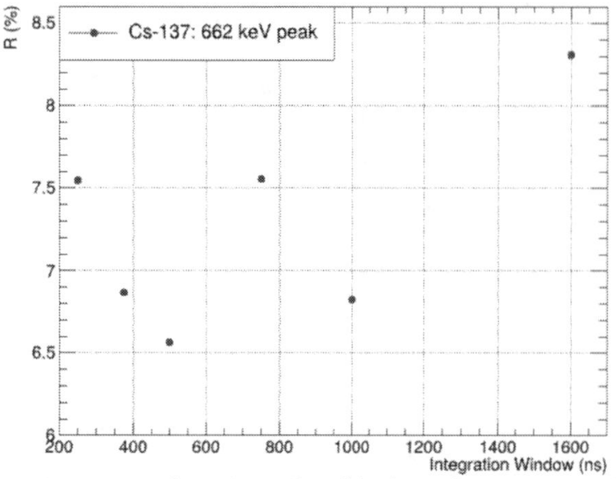

Fig. 8. Energy resolution (%) evaluated for the 662 keV gamma-ray of Cs-137 as a function of the integration window (ns).

V. SUMMARY AND OUTLOOK

Hadron Therapy is experiencing a great booming over the last years. Novel therapies such as Boron Neutron Capture Therapy (BNCT) offer different and appealing approaches to conventional radiotherapy and are clear candidates to transform and improve the prospects of many patients with resistant and difficult to treat cancers. Nevertheless, BNCT is currently facing many challenges, such as the necessity for real-time dose monitoring.

In this context, a Gamma-Neutron Vision detector (GN-Vision) capable of performing simultaneously slow neutron and γ-ray imaging offers possible solutions to overcome the current limitations of these therapies. GN-Vision builds upon the basis of the previous i-TED detector, which works as a Compton camera. The first prototype of the neutron imaging module has been developed using the CLYC-6 scintillation crystal, capable of discriminating between γ-rays and neutrons, integrated with AiT electronics readout and a pinhole ^{6}LiPE collimator.

In line with the above, a twofold objective is pursued for this work. On the one hand, the first experimental neutron imaging campaign with GN-Vision has been carried out at ILL. ^{10}B-containing water dishes were studied as targets and a study of thermal neutron absorption by this isotope has been performed, establishing a comparison between the experimental images and the ones obtained through the GEANT4 simulations. There is a significant difference between the percentage of absorption experimentally measured (~30%) and that obtained from the simulation (~5%) for 100 ppm ^{10}B concentration in water. Further research is required to accurately determine the main factors of this discrepancy.

The second objective of this work is the characterization of the CLLBC crystal, in the context of the GN-Vision detector development. The latter has proven to be the first crystal able to discriminate between neutrons and γ-rays compatible with PETsys electronics, unlocking the possibility of eventually performing neutron and γ-ray imaging in a single device. The energy resolution in CLLBC using ^{137}Cs has been studied, obtaining 6.5%, which improves the energy resolution of CLYC-6 with AiT electronics (8.9%). Moreover, neutron and γ-ray response in CLLBC has been evaluated using ^{252}Cf and ^{137}Cs sources. A thermal neutron event selection is performed introducing a cut in the energy deposited, thus selecting those events where the thermal neutron interacts through the ^{6}Li(n,α)^{3}H reaction. These successful results have motivated the deployment of the first full demonstrator of GN-Vision using the CLLBC crystal in the first detection plane. The first proof-of-concept experiment of the dual imaging capability will start soon.

ACKNOWLEDGMENT

This work builds upon research conducted under the ERC Consolidator Grant project HYMNS (grant agreement No. 681740) and has been supported by the ERC Proof-of-Concept Grant projects GNVISION (Grant Agreement 101113330) and AMA (Grant Agreement 101137646). We acknowledge funding from the Universitat de València through the Valoritza i Transfereix Programme, under grant UV-INV_PROVAL21-1924580. We also acknowledge funding from the Spanish Ministerio de Ciencia e Innovación under grant PID2022-138297NB-C21. We acknowledge support from the Severo Ochoa Grant CEX2023-001292-S funded by MCIU/AEI. Additionally, ASM acknowledges the JAE Intro ICU grant from CSIC; JLM and JBC thank the support provided by postdoctoral grants FJC2020-044688-I and ICJ220-045122-I, funded by MCIN/AEI/10.13039/501100011033 and the European Union NextGenerationEU/PRTR; JLM also acknowledges the postdoctoral grant CIAPOS/2022/020 funded by the Generalitat Valenciana and the European Social Fund, and BG acknowledges a PhD grant (PRE2023_IFIC_141). Financial support from the Institute Laue Langevin during the experimental campaign is also gratefully acknowledged.

REFERENCES

[1] R Orecchia, A Zurlo, A Loasses, M Krengli, G Tosi, S Zurrida, *et al.*, «Particle beam therapy (hadrontherapy): basis for interest and clinical experience,» *European Journal of Cancer, Volume 34, Issue 4, p459-468,* 1998.

[2] R. M. Sealock and J. C. Overley, «10-B(*n*,*α*)7-Li, 7-Li* differential cross section measurements between 0.2 and 1.25 MeV,» *Phys. Rev. C,* vol. 13, pp. 2149--2158, 1976.

[3] T.D. Malouff, D.S. Seneviratne, D.K. Ebner, W.C. Stross, M.R. Waddle, D.M. Trifiletti, *et al.*, «Boron Neutron Capture Therapy: A Review of Clinical Applications,» *Front Oncol.*, p. 11:601820, 2021.

[4] Advances in Boron Neutron Capture Therapy, Vienna: INTERNATIONAL ATOMIC ENERGY AGENCY, 2023. [Online]. Available: https://www.iaea.org/publications/15339/advances-in-boron-neutron-capture-therapy

[5] R.F. Barth, P. Mi, and W Yang, «Boron delivery agents for neutron capture therapy of cancer,» *Cancer Communications,* vol. 38, nº 1, p. 35, 2018.

[6] P.Torres-Sánchez, J. Lerendegui-Marco, J. Balibrea-Correa, V. Babiano-Suárez, B. Gameiro, I. Ladarescu, *et al.*, «The potential of the i-TED Compton camera array for real-time boron imaging and determination during treatments in Boron Neutron Capture Therapy,» *Applied Radiation and Isotopes,* vol. 217, p. 111649, 2025.

[7] J. Lerendegui-Marco, J. Balibrea-Correa, P. Álvarez-Rodríguez, V. Babiano-Suárez, B. Gameiro, I. Ladarescu, *et al.*, «First pilot tests of Compton imaging and boron concentration measurements in BNCT using i-TED,» *Applied Radiation and Isotopes,* vol. 225, p. 112009, 2025.

[8] J. Lerendegui-Marco, G. Cisterna, J. Hallam, V. Babiano-Suárez, J. Balibrea-Correa, D. Calvo, *et al.*, «Imaging neutrons with a position-sensitive monolithic CLYC detector,» *Nuclear Instruments and Methods in Physics Research Section A,* vol. 1079, p. 170594, 2025.

[9] C. Domingo-Pardo, «i-TED: A novel concept for high-sensitivity (n,gamma}) cross-section measurements,» *Nuclear Instruments and Methods in Physics Research A,* vol. 825, pp. 78-86, 2016.

[10] P. Olleros, L. Caballero, C. Domingo-Pardo, V. Babiano, I. Ladarescu, D. Calvo, *et al.*, «On the performance of large monolithic {LaCl}3(Ce) crystals coupled to pixelated silicon photosensors,» *Journal of Instrumentation,* vol. 13, n° 03, pp. P03014--P03014, 2018.

[11] V. Babiano, L. Caballero, D. Calvo, I. Ladarescu, P. Olleros and C. Domingo-Pardo, «γ-Ray position reconstruction in large monolithic LaCl3(Ce) crystals with SiPM readout,,» *Nuclear Instruments and Methods in Physics Research Section A,* vol. 931, pp. 1-22, 2019.

[12] V. Babiano, J. Balibrea, L. Caballero, D. Calvo, I. Ladarescu, J. Lerendegui, *et al.*, «First i-TED demonstrator: A Compton imager with Dynamic Electronic Collimation,» *Nuclear Instruments and Methods in Physics Research Section A,* vol. 953, p. 163228, 2020.

[13] J. Balibrea-Correa, J. Lerendegui-Marco,V. Babiano-Suárez, L. Caballero, D. Calvo, I. Ladarescu,*et al.*, «Machine Learning aided 3D-position reconstruction in large LaCl3 crystals,» *Nuclear Instruments and Methods in Physics Research Section A,* vol. 1001, p. 165249, 2021.

[14] V. Babiano-Suárez, J. Lerendegui-Marco, J. Balibrea-Correa, L. Caballero, D. Calvo, I. Ladarescu, *et al.*, «Imaging neutron capture cross sections: i-TED proof-of-concept and future prospects based on Machine-Learning techniques,» *Eur. Phys. J. A,* vol. 57, p. 197, 2021.

[15] J. Lerendegui-Marco, V. Babiano-Suárez, J. Balibrea-Correa, V. Babiano-Suárez, L. Caballero, D. Calvo,*et al.*, «Compton Imaging and Machine-Learning techniques for an enhanced sensitivity in key stellar (n,γ) measurements,» *EPJ Web Conf.,* vol. 260, p. 10002, 2022.

[16] C. Domingo-Pardo, V. Babiano-Suarez, J. Balibrea-Correa, L. Caballero, I. Ladarescu, J. Lerendegui-Marco, *et al.*, «Compton imaging for enhanced sensitivity (n,γ) cross section TOF experiments: Status and prospects,» *EPJ Web of Conf.,* vol. 284, p. 01018, 2023.

[17] J. Lerendegui-Marco, V. Babiano-Suárez, J. Balibrea-Correa, C. Domingo-Pardo, I. Ladarescu, A. Tarifeño-Saldivia, *et al.*, «New detection systems for an enhanced sensitivity in key stellar (n,γ) measurements,» *EPJ Web Conf.,* vol. 279, p. 13001, 2023.

[18] C. Domingo-Pardo, V. Babiano-Suarez, J. Balibrea-Correa, L. Caballero, I. Ladarescu, J. Lerendegui-Marco, *et al.*, «Advances and new ideas for neutron-capture astrophysics experiments at CERN n_TOF,» *The European Physical Journal A,* vol. 59, n° 1, p. 8, 2023.

[19] J. Lerendegui-Marco, J. Balibrea-Correa, V. Babiano-Suárez, I. Ladarescu and C. Domingo-Pardo , «Towards machine learning aided real-time range imaging in proton therapy,» *Scientific Reports,* vol. 12, n° 1, p. 2735, 2022.

[20] J. Balibrea-Correa, J. Lerendegui-Marco, I. Ladarescu, C. Guerrero, T. Rodríguez-González, M. C. Jiménez-Ramos,*et al.*, «Hybrid in-beam PET- and Compton prompt-gamma imaging aimed at enhanced proton-range verification,» *Eur. Phys. J. Plus,* vol. 137, n° 11, p. 1258, 2022.

[21] V. Babiano-Suárez, J. Balibrea-Correa, I. Ladarescu, J. Lerendegui-Marco and César Domingo-Pardo, «A computer-vision aided Compton-imaging system for radioactive waste characterization and decommissioning of nuclear power plants,» *Nuclear Instruments and Methods in Physics Research Section A,* vol. 1076, p. 170449, 2025.

[22] J. Balibrea-Correa, J. Lerendegui-Marco, I. Ladarescu, S. Morell, C. Guerrero, T. Rodríguez-González,*et al.*, «Hybrid Compton-PET Imaging for ion-range verification:A Preclinical Study for Proton-, Helium-, and Carbon-Therapy at HIT,» *Eur. Phys. J. Plus,* vol. 140, n° 870, 2025.

[23] Z. Youxiang, J. Yongli and W. Jimin, «The evaluation of cross sections for n + 6Li reaction,» vol. 1, n° 6, pp. p. 43-45, 2002.

[24] J. Lerendegui-Marco, V. Babiano-Suárez, J. Balibrea-Correa, L. Caballero, D. Calvo, I. Ladarescu, *et al.*, «Simultaneous Gamma-Neutron Vision device: a portable and versatile tool for nuclear inspections,» *EPJ Techniques and Instrumentation,* vol. 11, n° 1, p. 2, 2024.

[25] FIPPS (Fission Product Prompt gamma–ray Spectrometer) at Institut Laue-Langevin, [Online]. Available: https://www.ill.eu/for-ill-users/instruments/instruments-list/fipps/description/instrument-layout

[26] AiT, «AB4T-ARRAY64P, 4-Channel Active Base for the SensL ArrayC/J-60035-64P-PCB,» 2010. [Online]. Available: https://www.ait-instruments.com/AB4T_ARRAY64P_p/ab4tarray64p.htm

[27] O.E Kononov, V.N Kononov, M.V Bokhovko, V.V Korobeynikov, A.N Soloviev, A.S Sysoev , *et al.*, «Optimization of an accelerator-based epithermal neutron source for neutron capture therapy,» *Applied Radiation and Isotopes,* vol. 61, pp. 1009-1013, 2004.

[28] A. Verdera, P. Torres-Sánchez, J. Praena and I. Porras, «Study of the out-of-field dose from an accelerator-based neutron source for boron neutron capture therapy,» *Applied Radiation and Isotopes,* vol. 212, p. 111458, 2024.

[29] P. Torres-Sánchez, I. Porras, N. Ramos-Chernenko, F. Arias de Saavedra and J. Praena , «Optimized beam shaping assembly for a 2.1-MeV proton-accelerator-based neutron source for boron neutron capture therapy,» *Sci. Rep.,* vol. 11, n° 7576, pp. 1--12, 2021.

[30] A. Giaz, L. Pellegri, F. Camera, N. Blasi, S. Brambilla, S. Ceruti, *et al.*, «The CLYC-6 and CLYC-7 response to γ-rays, fast and thermal neutrons,» *Nuclear Instruments and Methods in Physics Research Section A,* vol. 810, pp. 132-139, 2016.

[31] J. Lerendegui-Marco, J. Hallam, G. Cisterna, A. Sanchis-Moltó, J. Balibrea-Correa, V. Babiano-Suárez ,*et al.*, «First experimental results and optimization study of the portable neutron-gamma imager GN-Vision,» *Applied Radiation and Isotopes,* vol. 224, p. 111826, 2025.

[32] A. Francesco, R. Bugalho, L. Oliveira, A. Rivetti, M. Rolo, J.C. Silva, *et al.*, «TOFPET 2: A high-performance circuit for PET time-of-flight,» *Nuclear Instruments and Methods in Physics Research A,* vol. 824, pp. 194-195, 2016.

[33] J. Balibrea-Correa, J. Lerendegui-Marco, D. Calvo, L. Caballero, V. Babiano, I. Ladarescu, *et al.*, «A first prototype of C6D6 total-energy detector with SiPM readout for neutron capture time-of-flight experiments,» *Nuclear Instruments and Methods in Physics Research A,* vol. 985, p. 164709, 2021.

ScintiPulses: A Python Package to Simulate Signal from Scintillation Detectors

Romain Coulon[1,*]
[1] Bureau International des Poids et Mesures, Pavillon de Breteuil, Sèvres, Cedex, F-92312, France,
(*) romain.coulon@bipm.org

***Abstract*— The scintiPulses Python package provides a versatile and comprehensive toolkit for simulating pulse signals from scintillation detectors. It accurately models scintillation pulse dynamics while incorporating various noise sources, including shot noise generated by the inhomogeneous statistical process of fluorescence. The package offers extensive customization through a wide range of parameters, making it adaptable to different scintillation detectors and signal processing applications. With the increasing adoption of digital electronics enabling offline processing of scintillation signals, it addresses the critical need for high-quality training datasets in the development of AI-driven signal processing algorithms. This open-source package is distributed at https://pypi.org/project/scintiPulses/ and available for collaborative development on GitHub https://github.com/RomainCoulon/scintiPulses. This paper presents an overview of the package's features.**

***Keywords* —Scintillation detectors, Digital signal processing, Signal emulator, Artificial intelligence, Machine learning.**

I. Introduction

Scintillation detectors are essential tools in various fields, including nuclear medicine, particle physics, environmental monitoring and nuclear security. These detectors convert energy from incident particles into light, which is subsequently transformed into electrical signals. With the advent of fast digitizers, the digitized representation of scintillation signals has become standard, enabling advanced digital signal processing techniques. Unlike analog electronics, which are limited to linear operations, digital processing unlocks new possibilities for more sophisticated signal analysis.

Accurate simulation of scintillation signals is essential for the development and validation of signal processing algorithms. In recent years, various pulse generators have been introduced, including the commercially available CAEN DT5810 module, a simulator implemented in Wolfram Mathematica® [1], and another developed using LabView [2]. These tools serve as digital twins of physical scintillation systems, enabling users to optimize parameter settings under simulated conditions that are difficult or impractical to replicate experimentally.

While these pulse generators are valuable for many applications, they fall short in one critical aspect: the generation of high-fidelity, tagged simulated data suitable for training advanced AI algorithms. Specifically, they do not adequately model the stochastic nature of scintillation light generation. Although some forms of noise are considered, these simulators typically overlook the shot noise that arises from the discrete nature of photon emissions during fluorescence.

To effectively train AI-based models for particle detection, it is imperative to have access to simulated signals that closely replicate the true statistical properties of scintillation events. This includes accounting for inherent randomness and quantum effects that traditional simulators tend to neglect. The scintiPulses package addresses this need by simulating realistic scintillation pulse signals. It incorporates shot noise from scintillation charges and other noise sources to produce high-fidelity data. This capability is particularly useful for generating large datasets to train machine learning and deep learning models.

II. Description of the package

A. Initialization

The simulation begins by generating a sequence of voltage signals based on the sampling frequency f_s of the analog-to-digital conversion (ADC) and the sequence duration t_N. A time vector $t = (t_1, t_2, \cdots, t_i, \cdots, t_N)$ is created, where the interval between consecutive time points satisfies $t_i - t_{i-1} = 1/f_s$.

B. Nuclear decay process

The nuclear decay process is simulated given the input count rate λ, modelled as a Poisson process. The decay source is assumed to have a constant activity within the sequence's time scale. The time intervals between pulses, denoted Δ, is a random variable drawn from an exponential distribution:

$$P(\Delta = \delta_k) = \text{Exponential}\left(-\frac{1}{\lambda}\right). \qquad (1)$$

The arrival times θ_k of events are generated by iteratively sampling the inter-arrival time Δ, as described in Algorithm I. For each event k, the deposited energy E_k in the scintillator is obtained from a user-defined vector of energies $Y = (E_1, E_2, \dots, E_k, \dots)$, typically produced by Monte Carlo simulation of particle transport and interactions. When the number of events N exceeds the dimension of Y, a bootstrapping procedure is applied to extend the dataset. Each index j_k in the discrete time vector t corresponds to the interaction time θ_k, thereby emulating temporal quantization.

The resulting vector $y^{(0)} = \left(y_1^{(0)}, y_2^{(0)}, \cdots, y_i^{(0)}, \cdots, y_N^{(0)}\right)$ stores the deposited energies E_k at their respective time indices j_k, representing the energy of each interaction event.

Algorithm I
Simulation of the nuclear decay process

Initialize $k = 0$, $y^{(0)} = (0, 0, \dots)$.

While $\theta_k < t_N$,

- Sample $\delta_k \sim \text{Exponential}\left(-\frac{1}{\lambda}\right)$.
- Compute $\theta_{k+1} = \theta_k + \delta_k$.
- Determine the time index
  - $j_{k+1} = \underset{1 \le i \le N}{\operatorname{argmin}}(t_i \ge \theta_{k+1})$.
- Sample E_k from $Y(E)$.
- Update $y_{j_{k+1}}^{(0)} = E_k$.
- Increment $k := k + 1$.

In many applications, the signal does not follow a homogeneous Poisson process and may contain hidden structures or correlations. As an alternative to Algorithm I, *scintiPulses* allows users to provide a custom timestamp vector θ.

C. Theoretical illumination function

Ionizing particles deposit energy in scintillators by exciting and ionizing atoms or molecules, thereby generating excited states in proportion to the deposited energy. These states decay radiatively, producing scintillation photons. A scintillator's performance is commonly characterized by its light yield L, defined as the number of scintillation photons produced per unit of deposited energy. For efficient detection, the emitted photons must have wavelengths compatible with transmission through the scintillator matrix (quantified by the optical efficiency η_o) and effective conversion to photoelectrons at the photodetector (described by its quantum efficiency η_q).

To maximize detection efficiency and energy resolution, scintillators are engineered to optimize the production of photoelectrons. The expected number of charges $\langle n_{e,k} \rangle$ produced during an event depositing energy E_k is given by:

$$\langle n_{e,k} \rangle = L\eta_o\eta_q E_k, \qquad (2)$$

The temporal profile of the scintillation light output, referred to as the illumination function (as introduced by Birks in *Theory and Practice of Scintillation Counting* [4]), typically follows an exponential decay. In practice, the scintillation signal is well described by a bi-exponential decay function $v^{(0)}(t)$, incorporating both prompt and delayed components. These are characterized by decay constants τ_1 and τ_2, and a delayed fraction p_{delayed}:

$$v^{(0)}(t) = \left(1 - p_{\text{delayed}}\right)\frac{n_e}{\tau_1}\exp\left(-\frac{t}{\tau_1}\right) + p_{\text{delayed}}\frac{n_e}{\tau_2}\exp\left(-\frac{t}{\tau_2}\right), \qquad t \ge 0, \qquad (3)$$

In organic scintillators, the prompt component arises from singlet excited states S_1 decaying radiatively to the ground state S_0 (i.e., $S_1 \rightarrow S_0$). This process competes with delayed mechanisms involving triplet states T_1, which may form via intersystem crossing. Delayed emission occurs through triplet-triplet annihilation ($T_1 + T_1 \rightarrow S_1 \rightarrow S_0$) or phosphorescence ($T_1 \rightarrow S_0$). The delayed fluorescence, governed by τ_2, is more pronounced for charged particles compared to electrons, enabling pulse shape discrimination—useful, for example, in identifying recoil protons from neutron interactions or alpha particles [3].

In inorganic scintillators, energy deposition primarily generates electron–hole pairs and excitons within the crystal lattice. These excitations may become trapped at luminescent centers (e.g., activator ions), leading to prompt luminescence via radiative transitions. Delayed emission in these materials originates from carrier trapping and subsequent thermal de-trapping within the bandgap, contributing to afterglow and slow scintillation tails.

For each event k, the digitized illumination function is represented in the discrete time for both types of scintillators as:

$$v_{j_k:N}^{(0)} := v_{j_k:N}^{(0)} + \left(1 - p_{\text{delayed}}\right)\frac{\langle n_{e,k} \rangle}{\tau_1 f_s}\exp\left(-\frac{t_{j_k:N}}{\tau_1}\right) + p_{\text{delayed}}\frac{\langle n_{e,k} \rangle}{\tau_2 f_s}\exp\left(-\frac{t_{j_k:N}}{\tau_2}\right), \qquad (4)$$

where $t_{j_k:N}$ represents the time steps from the interaction time t_{j_k} to the end of the sequence, and ": =" denotes updating the charge signal to account for pulse pileup. The discrete form of the illumination function (Eq. 4) corresponds to a histogram of the mean number of charges per time step, as illustrated by the blue signal in Fig. 1 (blue signal).

D. Quantum illumination model (Shot noise)

The *scintiPulses* package incorporates a quantum illumination model, as originally described by [5], to model the discrete nature of charge production in scintillation detectors. This functionality is essential for generating realistic scintillation pulses, as it captures the impact of shot noise: a type of noise arising from the quantized and stochastic nature of photon emission and charge generation. By doing so, the model introduces fluctuations into the ideal exponential decay of Eq. 4, particularly visible as small charge clusters and pulse trains in the trailing edge of the signal. These features become increasingly prominent under conditions such as low energy deposition, poor scintillator light yield, or substantial delayed fluorescence. In these scenarios, shot noise and resultant after-pulses can introduce significant metrological perturbations, potentially affecting the accuracy of charge quantification and signal interpretation [6].

For each energy deposition event k, the number of exited states leading to prompt photons $n_{sk}^{(\text{promt})}$ and delayed radiative transitions $n_{sk}^{(\text{delayed})}$ are modeled as random variables with expected values: $\langle n_{sk}^{(\text{promt})} \rangle = \left(1 - p_{\text{delayed}}\right)LE_k$ and $\langle n_{sk}^{(\text{delayed})} \rangle = p_{\text{delayed}}LE_k$. In organic scintillators, these quantities follow Poisson statistics, while in inorganic scintillators, due to excitation correlations, a truncated Normal distribution (TN) is used instead, with a variance scaled by the Fano Factor F related to the scintillant crystal [7].

During each sampling interval $[t_{i-1}, t_i]$, the number of emitted photons $(n_{pi}^{(\text{promt})}, n_{pi}^{(\text{delayed})})$ from prompt and delayed

https://doi.org/10.1051/epjconf/202533810001

populations $(n_{si}^{(\text{promt})}, n_{si}^{(\text{delayed})})$ follows a Binomial distribution. The transition probabilities η_{tp} and η_{td} from prompt and delayed emissions are derived from first-order decay kinetics: $\eta_{\text{tp}} = 1 - \exp(-(f_s\tau_1)^{-1})$ and $\eta_{\text{td}} = 1 - \exp(-(f_s\tau_2)^{-1})$.

The final number of photoelectrons is determined by two successive binomial processes: the first accounts for the optical collection efficiency η_o, and the second for the photodetector's quantum efficiency η_q. By the law of total probability, the total number of collected charges during the interval is drawn from a single binomial distribution with parameters, $n_{pi}^{(.)}$ and $\eta_o\eta_q$.

The full implementation of this inhomogeneous Compound Binomial-Binomial process is described in Algorithm II, which models the hierarchical stochastic process of the scintillation signal.

ALGORITHM II
THE QUANTUM ILLUMINATION MODEL

Initialize: $v_{1:N}^{(1)} = (0, 0, \ldots)$, $\eta_{\text{tp}} = 1 - \exp(-(f_s\tau_1)^{-1})$, $\eta_{\text{td}} = 1 - \exp(-(f_s\tau_2)^{-1})$.

For all k:

- $\langle n_{sk}^{(\text{promt})}\rangle = (1 - p_{\text{delayed}})LE_k$
- $\langle n_{sk}^{(\text{delayed})}\rangle = p_{\text{delayed}}LE_k$
- If $F = 1$:
  - $n_s^{(\text{promt})} \sim \text{Poisson}\left(\langle n_{sk}^{(\text{promt})}\rangle\right)$
  - $n_s^{(\text{delayed})} \sim \text{Poisson}(\langle n_{sk}^{(\text{delayed})}\rangle)$
- Else:
  - $n_s^{(\text{promt})} \sim \text{TN}\left(\langle n_{sk}^{(\text{promt})}\rangle, F\langle n_{sk}^{(\text{promt})}\rangle\right)$
  - $n_s^{(\text{delayed})} \sim \text{TN}(\langle n_{sk}^{(\text{delayed})}\rangle, F\langle n_{sk}^{(\text{delayed})}\rangle)$
- $i = 0$
- While $n_s^{(\text{promt})} \neq 0$ and $j_k + i < N$:
  - $n_{pi}^{(\text{promt})} \sim \text{Binomial}(n_s^{(\text{promt})}, \eta_{\text{tp}})$
  - $n_{ei}^{(\text{promt})} \sim \text{Binomial}\left(n_{pi}^{(\text{promt})}, \eta_o\eta_q\right)$
  - $n_s^{(\text{promt})} := n_s^{(\text{promt})} - n_{pi}^{(\text{promt})}$
  - $i := i + 1$
- $l = 0$
- While $n_s^{(\text{delayed})} \neq 0$ and $j_k + l < N$:
  - $n_{pi}^{(\text{delayed})} \sim \text{Binomial}(n_s^{(\text{delayed})}, \eta_{\text{td}})$
  - $n_{ei}^{(\text{delayed})} \sim \text{Binomial}\left(n_{pi}^{(\text{delayed})}, \eta_o\eta_q\right)$
  - $n_s^{(\text{delayed})} := n_s^{(\text{delayed})} - n_{pi}^{(\text{delayed})}$
  - $l := l + 1$
- $v_{j_k:N}^{(1)} := v_{j_k:N}^{(1)} + n_{e,j_k:j_k+i}^{(\text{promt})} + n_{e,j_k:j_k+l}^{(\text{delayed})}$

Fig. 1 illustrates the simulated response for a 10 keV energy deposition. The comparison between the theoretical illumination (blue) and the quantum illumination signal (orange) reveals the characteristic shot noise-induced fluctuations, especially pronounced at low energies due to limited photon statistics.

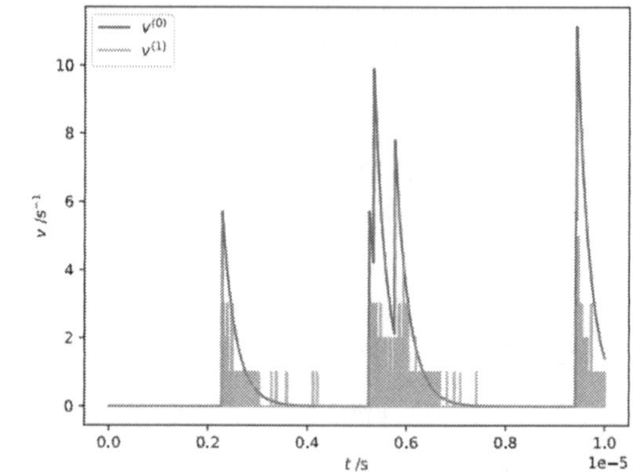

Fig. 1. Simulated signal with $f_s = 5 \times 10^8$ Hz, $t_N = 10$ µs, $\lambda = 10^6$ s^{-1}, $L = 10$ keV^{-1}, $\eta_o = 1, \eta_q = 1$, $E = 10$ keV.

E. *After-pulses*

After-pulses, including delayed charges originating from the scintillation process, pose significant challenges for counting measurements with scintillators. These after-pulses often necessitate the implementation of specific strategies, such as the extended dead-time approach [8], [9], to mitigate their effects.

In addition to delayed charges from the scintillation shot noise, another mechanism contributes to after-pulses in photomultiplier tubes (PMTs), with residual gas ionization being a significant factor. When accelerated photoelectrons collide with residual gas molecules inside the PMT, they ionize these molecules, creating positive ions. Under the influence of the PMT's electric field, these ions are drawn back toward the photocathode. Upon impacting the photocathode surface, they release secondary electrons, resulting in after-pulses. This phenomenon has been extensively studied and documented [10], [11], [12].

This type of after-pulses is observed to occur randomly after a primary pulse, but no comprehensive model currently exists to describe their behavior. To address this limitation, *scintiPulses* implements a surrogate model to simulate after-pulses.

The model defines after-pulses with three key parameters:

- p_A: the probability that a primary charge contributes to an interaction with a molecule of residual gas during its multiplication process;
- τ_A: mean delay time of the after-pulse; and
- σ_A: time spread of the delay.

The delay time θ of an after-pulse is seen as a random variable $P(\theta = \delta_{Ai})$, given in Eq. 5, drawn from a truncated normal distribution, where τ_A is the mean delay, σ_A is the standard deviation, and the truncation ensures the delay remains within the bounds of the signal duration $[0, N - i]$.

$$P(\theta = \delta_{Ai}) = \text{TruncatedNormal}(\tau_A, \sigma_A^2, 0, N - i). \quad (5)$$

The number of charges contributing to the after-pulse is modeled as a Binomial random variable given in Eq. 6, parameterized by the number of charges from the primary pulse

$v_i^{(1)}$ and the probability p_{A}, where j is the delayed time index corresponding to the occurrence of the after-pulse.

$$v_j^{(2)} \sim \mathrm{Binomial}\left(v_i^{(1)}, p_{\mathrm{A}}\right). \tag{6}$$

The parameters p_{A}, τ_{A}, and σ_{A} can be experimentally determined for a specific photodetector setup. The simulation of after-pulses is detailed in Algorithm III.

ALGORITHM III
SIMULATION OF AFTER-PULSES

For all $i \in [\![1, N]\!]$ and if $v_i^{(1)} > 0$:

- Sample the delay:
  - $\delta_{\mathrm{A}i} \sim \mathrm{TruncatedNormal}(\tau_{\mathrm{A}}, \sigma_{\mathrm{A}}^2, 0, N - i)$,
- Compute the delayed index:
  - $j = i + \lfloor \delta_{\mathrm{A}i} f_{\mathrm{s}} \rfloor$,
- Sample the after-pulse charges:
  - $v_j^{(2)} \sim \mathrm{Binomial}\left(v_i^{(1)}, p_{\mathrm{A}}\right)$.

Fig. 2 presents an example of a signal containing after-pulses, which occur randomly following the primary pulses. If left untreated, these after-pulses can significantly affect counting estimations that assume a pure Poisson process, due to their correlation with the primary pulse.

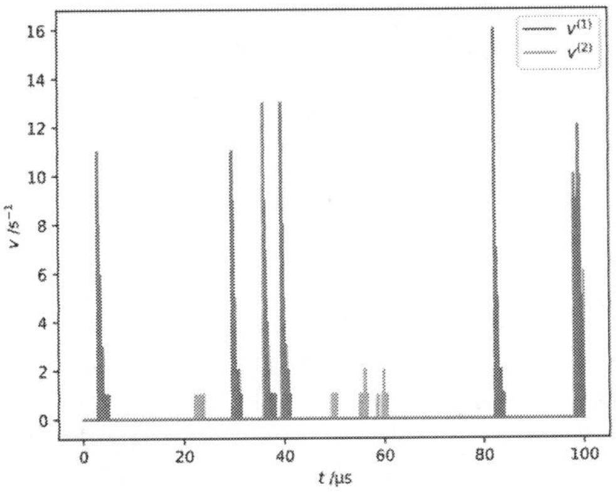

Fig. 2. Simulated signal with $f_s = 5 \times 10^8$ Hz, $t_N = 100$ µs, $\lambda = 5 \cdot 10^4$ s^{-1}, $L = 10$ keV^{-1}, $\eta_{\mathrm{o}} = 1, \eta_{\mathrm{q}} = 1$, $E = 10$ keV, $\tau_{\mathrm{A}} = 20$ µs, $p_{\mathrm{A}} = 0.01$, $\sigma_{\mathrm{A}} = 0.5$ µs. The after-pulses are visible in the ranges [20,25] µs and [50,60] µs. The brown colour represents the superimposition of the two signals $v^{(1)}$ and $v^{(2)}$.

F. Dark noise

Photodetectors operating at room temperature produce noise even in the absence of incident photons, primarily due to the thermionic emission of electrons from the active region of the photodetector (*e.g.*, the photocathode). This noise, referred to as dark noise, manifests as unexpected single charges whose arrival times follow a Poisson process characterized by an average appearance frequency f_{D}.

The time between two consecutive dark noise charges, d_k, is a random variable D governed by an exponential distribution:

$$P(D = d_k) = \mathrm{Exponential}\left(-\frac{1}{f_{\mathrm{D}}}\right). \tag{7}$$

The arrival times T_k of dark noise events are sampled from the random variable D using the procedure described in Algorithm IV.

ALGORITHM IV
SIMULATION OF DARK NOISE

Initialize: $k = 0$, $v^{(3)} := v^{(2)}$

While $T_k < t_N$,

- If $k > 0$, update the signal:
  - $v_{l_k}^{(3)} := v_{l_k}^{(3)} + 1$.
- Sample the time interval for the next dark noise charge:
  - $d_k \sim \mathrm{Exponential}\left(-\frac{1}{f_{\mathrm{D}}}\right)$.
- Update the arrival time:
  - $T_{k+1} = T_k + d_k$.
- Find the corresponding time index:
  - $l_{k+1} = \underset{1 \le i \le N}{\mathrm{arginf}}(T_{k+1} \geq t_i)$.
- Increment k: $k := k + 1$.

Fig. 3 illustrates the simulated signal with inactive after-pulses ($v^{(2)}$ in blue), and the simulated signal with after-pulses activated ($v^{(3)}$ in orange). These events appear as randomly occurring single charges and, unlike after-pulses, are not correlated with primary interaction events.

In counting measurements, dark noise generally does not impact results if the lower energy threshold is set sufficiently high (e.g., above the single photoelectron pulses). However, for low-energy applications, dark noise becomes significant and can degrade measurement accuracy. To mitigate its effects, filtering techniques such as coincidence counting are often employed [13], [14].

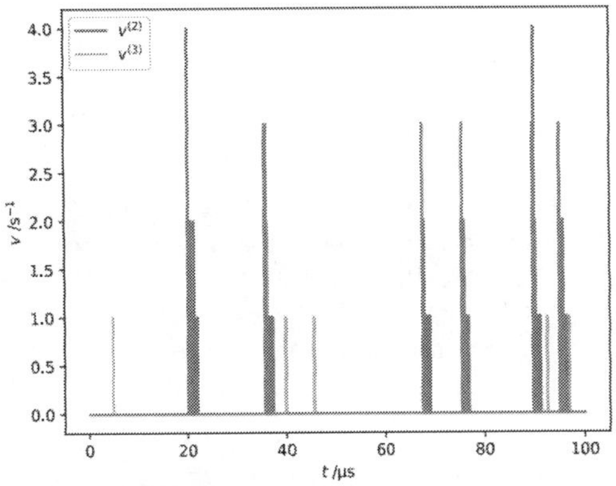

Fig. 3. Simulated signal with $f_s = 5 \times 10^8$ Hz, $t_N = 100$ µs, $\lambda = 5 \times 10^4$ s^{-1}, $L = 10$ keV^{-1}, $\eta_{\mathrm{o}} = 1, \eta_{\mathrm{q}} = 1, E = 10$ keV, $f_{\mathrm{D}} = 5 \cdot 10^4$ s^{-1}. The dark noise pulses are visible around 5 µs, 40 µs, 45 µs and 95 µs. The brown colour represents the superimposition of the two signals $v^{(2)}$ and $v^{(3)}$.

G. *Transimpedance*

The conversion of charges to voltage is performed by a conversion device, such as a photomultiplier tube (PMT). This process is characterized by:

- a capacitance C_1, which determines the average voltage produced when a scintillation photon is converted into a charge;
- an energy-to-charge spread parameter σ_C, which introduces fluctuations in the charge-to-voltage conversion;
- a time spread parameter τ_s, which defines the pulse shape of the charge-induced voltage, reflecting the transit spread of the electron bunch during the multiplication process.

The model assumes a Gaussian pulse shape for the smoothing operation, where the voltage signal $v^{(4)}$ is obtained by convolving the charge signal $v^{(3)}$ with a truncated Gaussian kernel g_j given in Eq. 8, where H is the half-width of the kernel. The resulting voltage is negative by convention and can be inverted using the parameter $I = -1$.

$$g_j = \exp\left(-\frac{t_j^2}{2\tau_s^2}\right), \qquad q = -H \dots H,$$
$$\text{with} \sum_{q=-H}^{H} \exp\left(-\frac{t_q^2}{2\tau_s^2}\right) = 1. \tag{8}$$

The procedure is summarized in Algorithm V.

ALGORITHM V
SIMULATION OF TRANSIMPEDANCE

For all $i \in [\![1, N]\!]$:

- Sample the charge-to-voltage conversion coefficient:
  - $k_C \sim \text{Normal}(1, \sigma_C^2)$.
- Perform the smoothing operation with the Gaussian kernel:
  - $v_i^{(4)} = -I \cdot k_C \cdot C_1^{-1} \sum_{j=-H}^{H} v_{i-j}^{(3)} g_j$.

Fig. 4 illustrates single charge pulses simulated with the transimpedance model. The impact of the parameters σ_C and τ_s on the resulting signal is clearly visible:

- σ_C causes fluctuations in the pulse height, reflecting the spread in energy-to-charge conversion,
- τ_s broadens the pulse shape, representing the temporal spread induced by electron transit and multiplication processes.

This transimpedance model provides a realistic representation of the charge-to-voltage conversion process, accounting for both amplitude variability and temporal smoothing effects.

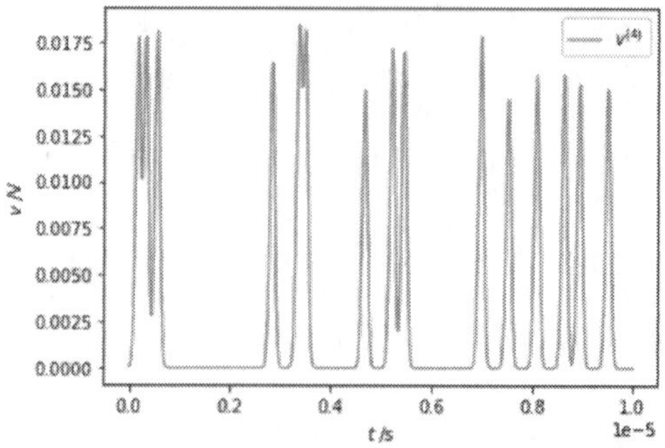

Fig. 4. Simulated signal of thermionic pulses with $f_s = 5 \times 10^8$ Hz, $t_N = 10$ µs, $f_D = 10^6$ s^{-1}, $C_1 = 1.6 \times 10^{-19}$ F, $\sigma_C = 0.1$ and $\tau_s = 50$ ns.

H. *Johnson-Nyquist noise*

Johnson-Nyquist noise arises from the thermal agitation of charge carriers in the passive electronic components of the photodetector. This thermal noise is modeled in Algorithm VI as Gaussian white noise and introduces random fluctuations into the voltage signal. The noisy voltage $v_i^{(5)}$ is obtained by adding a noise component $v_{\text{JN},i}$, sampled from a centered Normal distribution with a standard deviation equal to the root mean square (RMS) value V_{RMS} of the Johnson-Nyquist noise.

ALGORITHM VI
SIMULATION OF JOHNSON-NYQUIST NOISE

For all $i \in [\![1, N]\!]$:

- Sample the Johnson-Nyquist noise:
  - $v_{\text{JN},i} \sim \text{Normal}(0, V_{\text{RMS}}^2)$.
- Add the noise to the clean signal:
  - $v_i^{(5)} = v_i^{(4)} + v_{\text{JN},i}$.

Fig. 5 illustrates the noised signal $v_i^{(5)}$ alongside its clean counterpart $v_i^{(4)}$. The Johnson-Nyquist noise appears as a high-frequency component, which can be challenging to filter using linear low-pass filters without distorting the underlying charge signal. This issue is particularly pronounced for fast photodetectors characterized by a low time spread parameter τ_s, where signal bandwidth and noise overlap significantly.

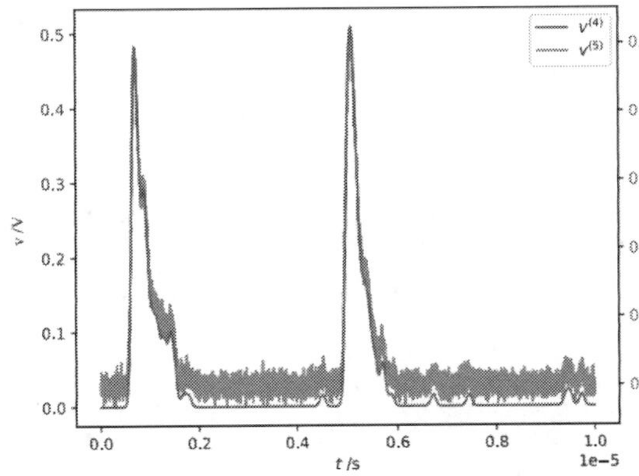

Fig. 5. Simulated signal with $f_s = 5 \times 10^8$ s^{-1}, $t_N = 10$ µs, $\lambda = 2 \times 10^5$ s^{-1}, $L = 10$ keV^{-1}, $\eta_o = 1, \eta_q = 1, E = 10$ keV, $V_{RMS} = 10$ mV.

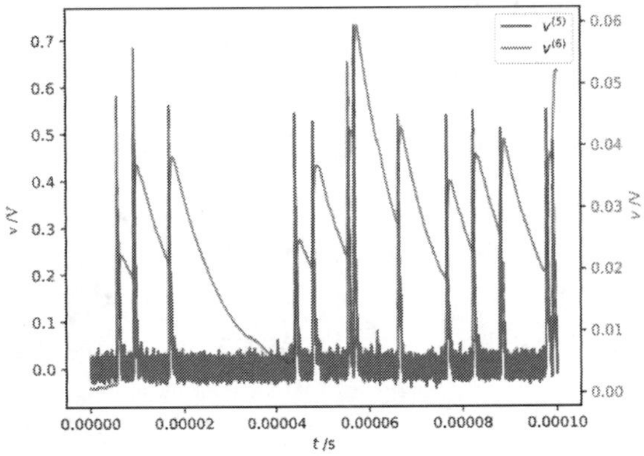

Fig. 6. Simulated signal from preamplifier with $f_s = 5 \times 10^8$ s^{-1}, $t_N = 100$ µs, $\lambda = 1 \times 10^5$ s^{-1}, $L = 10$ keV^{-1}, $\eta_o = 1, \eta_q = 1, E = 10$ keV, $G_1 = 1$, $\tau_{RC} = 10$ µs.

I. Preamplification stage

While *scintiPulses* is primarily designed to support digital signal processing, it also includes traditional linear filtering from preamplification and amplification stages.

It offers the option to simulate the voltage signal after a preamplification stage, where an RC filter is applied to the signal $v_i^{(5)}$. The filter is characterized by a time constant τ_{RC} and a gain G_1. The resulting filtered signal $v_i^{(6)}$ is computed iteratively as described in Algorithm VII.

ALGORITHM VII
SIMULATION OF THE PREAMPLIFICATION STAGE

For all $i \in [\![2, N]\!]$:

- Compute the filter coefficient:
  - $\alpha = (f_s\tau_{RC} + 1)^{-1}$.
- Apply the RC filter:
  - $v_i^{(6)} = \alpha v_i^{(5)} + (1 - \alpha)v_{i-1}^{(6)}$.
- Apply the gain:
  - $v_i^{(6)} := G_1 v_i^{(6)}$.

Fig. 6 presents an example of an RC-filtered signal with a time constant $\tau_{RC} = 10$ µs. The RC filter smooths the voltage signal, reducing high-frequency components while amplifying the overall signal by a factor G_1. The time constant τ_{RC} determines the extent of signal smoothing, with larger values leading to a slower response and stronger attenuation of high-frequency components. After this preamplification step, the energy is proportional to the pulse height.

J. Amplification stage

In nuclear instrumentation, it is common to include an amplification stage to enhance the preamplifier signal to accelerate the baseline restauration and optimize the encoding with a Multi-Channel Analyzer (MCA). This stage is often implemented as a CR^n filter, which processes the voltage $v_i^{(6)}$. The filter is characterized by a time constant τ_{CR}, a voltage gain G_2, and an order n_{CR} (typically equal to 1 or 2) [15], [16]. The simulation of the amplification stage is described in Algorithm VIII.

ALGORITHM VIII
SIMULATION OF THE AMPLIFICATION STAGE

Initialize: $v^{(CR^0)} = v^{(6)}$ for $\mu = 1, \ldots, n_{CR}$, and $\alpha = (f_s\tau_{CR} + 1)^{-1}$.

For all $\mu \in [\![1, n_{CR}]\!]$ and for all $i \in [\![2, N]\!]$:

- Update the filtered signal:
  - $v_i^{(CR^\mu)} = \alpha\left(v_{i-1}^{(CR^\mu)} + v_i^{(CR^{\mu-1})} - v_{i-1}^{(CR^{\mu-1})}\right)$.
- Update the intermediate signal for the next iteration:
  - $v^{(CR^{\mu-1})} = v^{(CR^\mu)}$.

Apply the voltage gain: $v^{(7)} = G_2 v^{(CR^{n_{CR}})}$

It can be observed the issue of baseline instability with the amplification stage due the poles and zeros of the transfer function which could be mitigate by tuning τ_{CR} or with higher filter order. Fig. 7 illustrates the amplified signal $v^{(7)}$. The CR filter serves as a fast amplifier that enhances signal rise times while simultaneously filtering low-frequency components. However, baseline instability can arise due to the poles and zeros of the filter's transfer function, which introduces signal drift. With the advent of fast digitizers, it is increasingly common to eliminate the preamplification and amplification stages traditionally required with analog electronics and MCAs, thereby reducing distortion and minimizing information loss [17], [18].

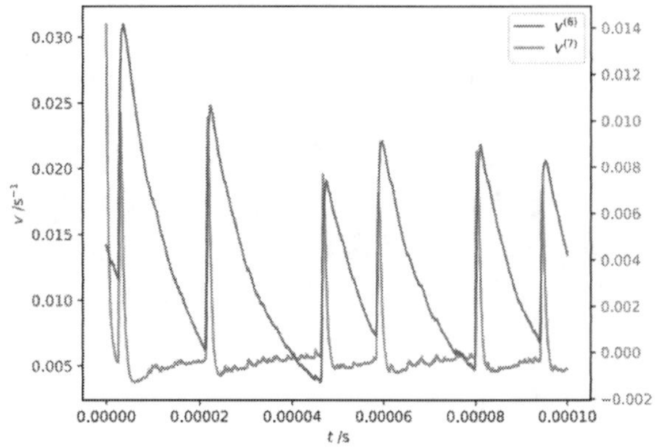

Fig. 7. Simulated signal from the fast amplifier with $f_s = 5 \times 10^8$ s^{-1}, $t_N = 100$ µs, $\lambda = 1 \times 10^5$ s^{-1}, $L = 10$ keV^{-1}, $\eta_o = 1$, $\eta_q = 1$, $E = 10$ keV, $G_1 = 1$, $\tau_{RC} = 10$ µs, $\tau_{CR} = 0.5$ µs, $n_{CR} = 1$.

K. *Digitization*

The anti-aliasing filter is a crucial frontend component in an ADC (Analog-to-Digital Converter) to prevent aliasing of the signal. This filter is implemented as a low-pass Butterworth filter of order n_B and cutoff frequency f_n, which must satisfy the condition $f_n \geq f_s/2$ (the Nyquist frequency f_n).

The filter is constructed using the Butter method and applied using the filtfilt function from the scipy.signal library. In practice, the cutoff frequency f_n is often set at $0.4f_s$, ensuring sufficient information retention while leaving a safety margin relative to the Nyquist frequency. A Butterworth filter of order $n_B = 4$ is typically chosen as the minimal low-pass filter applied to the acquired voltage signal $v^{(7)}$. After filtering, the ADC encodes the voltage signal within a defined dynamic range V_S and a resolution R. The simulation procedure is detailed in Algorithm IX.

ALGORITHM IX
SIMULATION OF THE DIGITIZATION

Define the voltage resolution: $\rho = 0.5V_S/2^R$.

Apply the Butterworth low-pass filter: $v^{(7*)} = \mathrm{Butterworth}(v^{(7)}, f_n, n_B)$.

For all $i \in [\![1, N]\!]$:

- If $|v_i^{(7*)}| < V_S$:
  - Quantize the signal:
    - $v_i^{(8)} = \mathrm{round}(v_i^{(7*)}/\rho) \times \rho$,
- Otherwise, clip the signal dynamic range limits:
  - $v_i^{(8)} = \mathrm{sgn}(v_i^{(7*)})V_S$

A synoptic diagram of the code is presented in Fig. 8.

- Input parameters are shown in white boxes with black borders.
- Output quantities are represented in white boxes with green borders.
- Functions systematically executed are shown in plain blue boxes.
- Optional functions appear in plain orange boxes.

The settings for the entire simulation are implemented as arguments of the main function in the package.

The code returns:
- Time-index vector J.
- Energy vector Y.
- Voltages at different stages: $v^{(0)}, v^{(1)}, \ldots, v^{(8)}$ suitable to build a large variety training datasets.

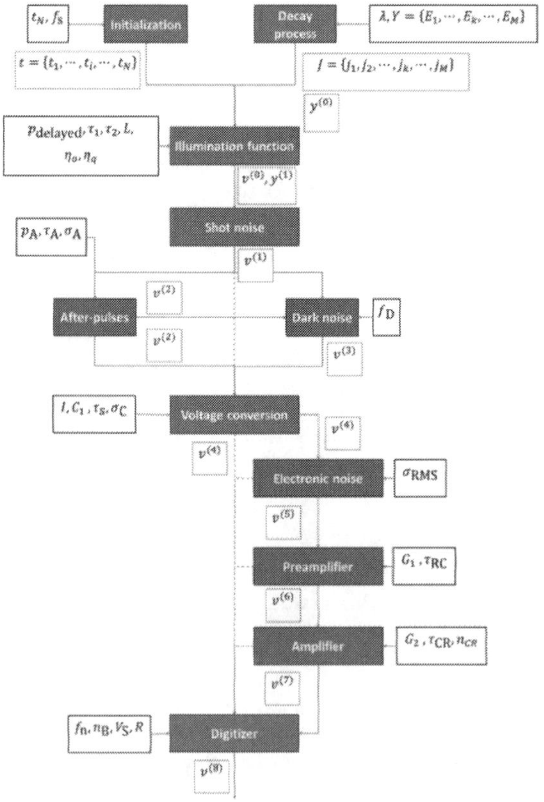

Fig. 8. Synoptic of the scintiPulses code.

III. CONCLUSIONS

The scintiPulses package is a robust and versatile tool for simulating pulse signals from scintillation detectors, providing a high degree of flexibility in defining parameters such as decay times, energy distributions, and various noise sources. Its implementation of quantum illumination, incorporating shot noise from the inhomogeneous stochastic processes, ensures realistic and physically accurate signal outputs. This makes the package particularly well-suited for advanced signal processing applications, including the training of AI models.

By offering rich and customizable datasets that capture the intricacies of scintillation signals, *scintiPulses* enables the development and evaluation of cutting-edge techniques in digital signal processing. It facilitates the design of AI algorithms capable of addressing challenges such as noise filtering, pattern recognition, and high-precision measurements in scintillation systems.

References

[1] J. Pechousek *et al.*, "Software emulator of nuclear pulse generation with different pulse shapes and pile-up," *Nucl Instrum Methods Phys Res A*, vol. 828, pp. 81–85, Aug. 2016, doi: 10.1016/j.nima.2016.05.032.

[2] H. I. Salih and A. A. Arafah, "Modeling and simulation of scintillation pulses for crystal identification applications," *Arab Journal of Nuclear Sciences and Applications*, vol. 49, no. 2, pp. 123–131, Apr. 2016.

[3] E. Montbarbon *et al.*, "*N* -(2-Ethylhexyl)carbazole: A New Fluorophore Highly Suitable as a Monomolecular Liquid Scintillator," *Chemistry - A European Journal*, vol. 22, no. 34, pp. 12074–12080, Aug. 2016, doi: 10.1002/chem.201601749.

[4] J. B. Birks, *The Theory and Practice of Scintillation Counting*, Pergamon P., vol. 27. Elsevier, 1964. doi: 10.1016/C2013-0-01791-4.

[5] R. Coulon, M. Becht, J. Dumazert, Q. Lecomte, and S. Garti, "Simulation of Scintillation Signal as a Help in Phoswich Systems Conception," in *2017 IEEE Nuclear Science Symposium and Medical Imaging Conference (NSS/MIC)*, IEEE, Oct. 2017, pp. 1–3. doi: 10.1109/NSSMIC.2017.8533083.

[6] C. Bobin, C. Thiam, B. Chauvenet, and J. Bouchard, "On the stochastic dependence between photomultipliers in the TDCR method," *Applied Radiation and Isotopes*, vol. 70, no. 4, pp. 770–780, 2012, doi: 10.1016/j.apradiso.2011.12.035.

[7] V. Bora *et al.*, "Estimation of Fano factor in inorganic scintillators," *Nucl Instrum Methods Phys Res A*, vol. 805, pp. 72–86, Jan. 2016, doi: 10.1016/j.nima.2015.07.009.

[8] A. P. Baerg, K. Munzenmayer, and G. C. Bowes, "Live-Timed Anti-Coincidence Counting with Extending Dead-Time Circuitry," *Metrologia*, vol. 12, no. 2, pp. 77–80, Apr. 1976, doi: 10.1088/0026-1394/12/2/006.

[9] J. Bouchard, "MTR2: a discriminator and dead-time module used in counting systems," *Applied Radiation and Isotopes*, vol. 52, no. 3, pp. 441–446, Mar. 2000, doi: 10.1016/S0969-8043(99)00192-X.

[10] V. A. Morozov, N. V. Morozova, and P. Budzyński, "Delayed electron emission in photomultiplier tubes," *Nucl Instrum Methods Phys Res A*, vol. 1053, p. 168323, Aug. 2023, doi: 10.1016/j.nima.2023.168323.

[11] N. Akchurin and H. Kim, "A study on ion initiated photomultiplier afterpulses," *Nucl Instrum Methods Phys Res A*, vol. 574, no. 1, pp. 121–126, Apr. 2007, doi: 10.1016/j.nima.2007.01.093.

[12] G. A. Morton, H. M. Smith, and R. Wasserman, "Afterpulses in Photomultipliers," *IEEE Trans Nucl Sci*, vol. 14, no. 1, pp. 443–448, 1967, doi: 10.1109/TNS.1967.4324452.

[13] R. Coulon, V. Kondrasovs, Q. Lecomte, and J. Dumazert, "Multilayer phoswich scintillators for neutron / gamma discrimination," *Radiat Meas*, vol. 117, pp. 57–62, Oct. 2018, doi: 10.1016/j.radmeas.2018.07.011.

[14] R. Coulon *et al.*, "The new international reference system for pure α- and pure β-emitting radionuclides and some electron capture decaying radionuclides by liquid scintillation counting," *J Radioanal Nucl Chem*, vol. 331, no. 8, pp. 3221–3230, Aug. 2022, doi: 10.1007/s10967-022-08337-7.

[15] Y. Liu, J. Zhang, L. Liu, S. Li, and R. Zhou, "Implementation of real-time digital CR-RC shaping filter on FPGA for gamma-ray spectroscopy," *Nucl Instrum Methods Phys Res A*, vol. 906, pp. 1–9, Oct. 2018, doi: 10.1016/j.nima.2018.05.020.

[16] H.-Q. Zhang, Z.-D. Li, B. Tang, and H.-X. Wu, "Optimal parameter choice of CR–RCm digital filter in nuclear pulse processing," *Nuclear Science and Techniques*, vol. 30, no. 7, p. 108, Jul. 2019, doi: 10.1007/s41365-019-0638-7.

[17] R. Coulon, V. Gressier, and C. Michotte, "Test of a digitizer to process the pulse signal from the 3 photomultiplier tubes of a TDCR liquid scintillation counter," *Applied Radiation and Isotopes*, vol. 192, p. 110598, Feb. 2023, doi: 10.1016/j.apradiso.2022.110598.

[18] R. Coulon, C. Michotte, M. Nonis, and S. Courte, "Comparative Analysis of the CAEN Digitizer DT5730SB and the MTR2 Module for SIRTI Applications," in *Proceedings of the ICRM 2025*, 2025.

A Neural Network Approach for Online Reconstruction of Bremsstrahlung Spectra Produced by Electron Accelerator

Lucas Tasinato[1, *], David Tisseur[2], Florent Kuntz[1], Nicolas Arbor[3], Abdallah Lyoussi[2]
[1] Aerial-CRT, 250 Rue Laurent Fries, 67400 Illkirch Graffenstaden, France
[2] CEA, DES, IRESNE, DER, Cadarache F-13108, Saint-Paul-Lez-Durance, 13108, France
[3] Université de Strasbourg, IPHC, 23 rue du Loess, 67037 Strasbourg, France
(*) l.tasinato@aerial-crt.com

Abstract— **The characterization of bremsstrahlung spectra generated by electron accelerators is becoming increasingly crucial, particularly in radiation processing applications such as sterilization of medical devices or food irradiation. The growing transition from isotopic to electric irradiators presents new challenges related to the control of beam properties. In this context, the technology resource center Aerial in collaboration with CEA/IRESNE and IPHC is looking to develop a tool and methodology enabling the online characterization of the bremsstrahlung spectra generated in its feerix[1] installation. This step is very important for their irradiation operations to ensure precise dose deposition in the sample and to precisely estimate the activation product when the photon energy exceeds the photonuclear reaction threshold. Information on the energy spectrum is also a key input for Monte Carlo simulations, which are increasingly used in radiation processing. However, conventional direct and indirect spectrometry methods are limited in meeting the challenges of Aerial's high energy and high-power irradiation platform. In this study, we propose a new theoretical approach based on neural networks to solve an ill-posed inverse problem, enabling the reconstruction of bremsstrahlung spectra from depth-dose measurements. This approach is motivated by the limitations of previously discussed regularization methods and existing neural network approaches. We focus here on an analytical approach for generating realistic training and validation datasets, consisting of Bremsstrahlung spectra and their corresponding dose distributions in any medium. This neural network approach will also be compared with other methods reported in the literature.**

***Keywords* — Electron accelerator, Bremsstrahlung spectra, Inverse method, Unfolding, Deep Learning, Neural Network.**

I. MONITORING THE BREMSSTRAHLUNG SPECTRUM PRODUCED BY ELECTRON ACCELERATOR IN RADIATION PROCESSING

This paper is contextualized by recent trends in radiation processing. During the last decade, a shift from isotopic irradiators to electron accelerator-based irradiators has been encouraged. The reasons behind this shift are multiple, including concerns about the malicious use of nuclear material as well as the management of nuclear waste.

However, it raises new concerns regarding the monitoring of the spectrum produced by such high-energy, high-power installations. Indeed, with isotopic irradiators, the energy and dose rate of the beam are precisely known, whereas the characteristics of beams produced by electron accelerators are subject to technical perturbations, disruptions, and component wear. Accurately characterizing the bremsstrahlung spectrum of an irradiation facility is crucial for several reasons. First, it ensures reliable and effective irradiation of samples. It also makes it possible to estimate activation products when photon energies surpass photonuclear reaction thresholds. Moreover, as Monte Carlo simulations are increasingly used in radiation processing, accurate knowledge of the input spectrum is once again essential.

However, characterizing the bremsstrahlung spectrum is at the same time very challenging due to the ultra-high dose rate and the long irradiation period in facilities used for radiation processing. For instance, in Aerial's state-of-the-art platform feerix, the dose rate reaches no more than 100 Gy/s at 40 cm of the conversion target. Conventional spectrometers saturate or are destroyed due to those extreme conditions. Alternative inverse methods already exist and were discuss in previous paper [2]. Those methods are efficient, but they are not suitable for Aerial's challenges as the associated experimental setup either too large to fit in feerix or require high maintenance. The Table I enumerates the advantages and disadvantages of the inverse methods found in literature.

TABLE I
A SUMMARY TABLE OF THE ADVANTAGES AND DISADVANTAGES OF INDIRECT SPECTRAL CHARACTERIZATION METHODS DEVELOPED TO DATE, IN THE CONTEXT OF AERIAL'S CHALLENGES. THIS IS DISCUSSED IN DETAIL IN [2].

Methods	Compact Device	Low maintenance	Online Measurement
Monte Carlo	+	+	−
Compton Scattering	−	+	+
Photo-activation	+	−	−
Transmission	−	+	−
Deposited Dose	−	+	−

In this paper we will discuss new insight about a method based on the resolution of the inverse problem allowing to unfold the spectra from depth dose measurement. This paper is following discussion made to develop a setup compatible with Aerial's challenges [2].

II. Unfolding Bremsstrahlung spectra from depth dose measurements: the limitation of existing reconstruction methods

A. Introducing the inverse problem

Before discussing the limitations of existing approaches, we can briefly review the principle behind this method. For a more detailed introduction, many papers have extensively covered this topic such as [3], [4]. If we discretize the incident spectrum in k energy groups, the depth dose measured by our device can be expressed as a contribution of the depth dose produced by each energy group in the incident spectrum, weighted by their intensity in the spectrum, as expressed in (1):

$$d = \sum_{i=1}^{k} M_1 s_1 + \dots + M_k s_k \qquad (1)$$

With d, the total depth dose measured in any medium, M_i, the depth dose associated to each energy group and s_i, the intensity of the i-th energy group in the spectrum. This representation can easily be converted in a matrix form as a linear problem (2):

$$d = Ms \qquad (2)$$

With $d = \begin{pmatrix} d_1 \\ \vdots \\ d_l \end{pmatrix}$ the measured depth-dose,
$M = (M_1 \quad \dots \quad M_k)$ the response matrix composed of k depth dose vectors M_k from each energy group and $s = \begin{pmatrix} s_1 \\ \vdots \\ s_k \end{pmatrix}$, the spectrum vector whose coefficients are to be evaluate.

Solving this inverse problem is made difficult by its ill-posed nature, where the noise on the measured depth dose is not explained by the response matrix which leads to unrealistic solutions if no constraints are applied during the resolution process. More robust methods have been applied to this problem such as regularization methods, Bayesian optimization methods or Deep Learning approaches. We previously selected two of them that appear compatible with online measurement due to their potential for fast computation.

B. Limitation of Regularization methods

In a previous paper [2], we discussed the optimization of a device compatible with online measurement through the lens of regularization methods. This method was already introduced in other papers such as [5], [6]. It basically consists of solving a least-squares problem by adding a side constraint on the shape of the solution. It can be formulated as (3):

$$J_\alpha = min_\alpha \left\{ ||Ms - d||^2 + \alpha ||\Gamma s||^2 \right\} \qquad (3)$$

Where α is the regularization parameter and Γ the regularization operator.

One of its advantages is the simplicity with which the solution s_α of the minimization problem can be calculated analytically as (4):

$$s_\alpha = (M^T M + \alpha \Gamma^T \Gamma) M^T d \qquad (4)$$

The regularization operator Γ depends strictly on the nature of the problem, i.e., on the characteristics of the solution we need to unfold. Many operators can be used such as the identity operator or the first and second derivative operator. In previous work, it was shown that the first derivative operator was more suited for the spectral reconstruction [2].

As shown in our earlier work, it is possible to retrieve precise spectra as long as the response matrix is representative of the problem. However, we faced a few limitations which constraint the use of regularization method in this context. First of all, even if the reconstruction is effective overall, it performs poorly in low-intensity regions. As a result, the maximum energy of the spectra, a parameter of critical importance for Aerial, cannot be reconstructed with this approach (see Fig 1).

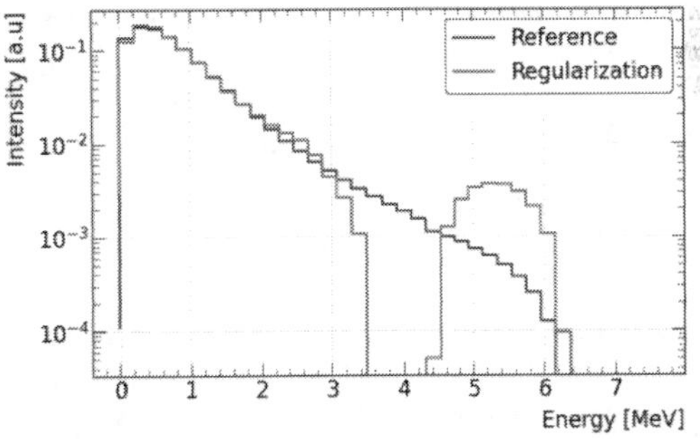

Fig 1. Illustration of the limitations of regularization methods, which fail to reconstruct the maximum of energy in low-intensities regions of the spectrum.

Another limitation of these methods is the repeatability of the measurement. To simulate the measurement of the N depth dose under the same conditions, we added random noise to the reference depth dose and evaluated the repeatability of the reconstruction.

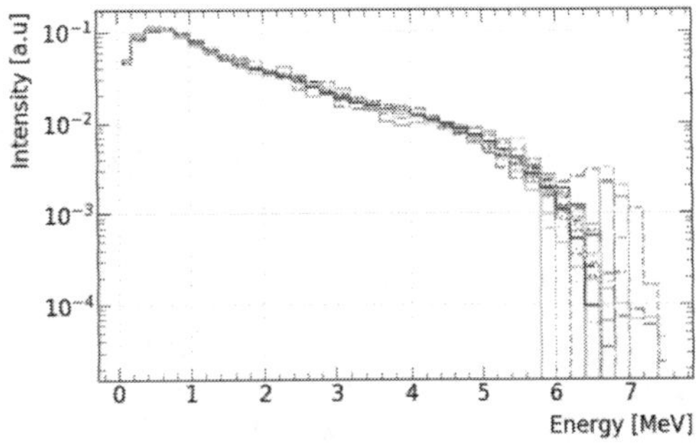

Fig 2. Impact of noise on the repeatability and stability of regularization-based reconstructions (for 100 reconstructions).

As shown in Fig 2, the reconstruction is very unstable in the last bin of the unfold spectra even if we optimize the choice of regularization parameter for each reconstruction, which is not instantaneous in practice, taking approximately 2 minutes per reconstruction. More importantly, it is obviously not possible to assess precisely the maximal energy of our spectrum. Thus, this method is not suitable for robust online reconstruction of the bremsstrahlung spectrum in the context of Aerial's challenges.

C. Limitation of existing neural network-based methods

Another promising approach was presented in [7]. In this paper the authors presented a method based on Deep Learning and especially on neural networks. The idea is to train a neural network to learn the relationship from the depth dose to the corresponding spectra. To do so, it is necessary to have a substantial dataset, composed of spectra and their associated dose deposits defined in the same conditions as the upcoming measurement. One of the main challenges is obtaining this type of dataset. Building it from measurements is impossible due to insufficient knowledge of the spectra from each accelerator, while simulating such a large amount of data is also problematic, as we aim to generate at least 10,000 pairs of spectra and doses.

In this paper, the authors proposed an approach to build an analytic set of data by generating spectra from five different functions, all of them, combination of exponentials. The associated dose deposits are then calculated with the simulated response matrix using (2).

With this dataset, the authors are then able to train a neural network taking the different depth doses as input and the spectra as output with a hidden layer of 2000 neurons. After training their neural network they performed their first reconstruction on experimental data and managed to approach the real spectrum from different 6 MeV LINAC as shown in Fig 3.

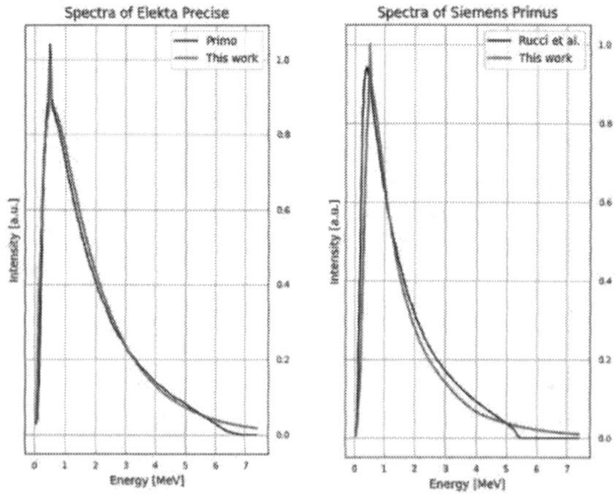

Fig 3. Illustration from Torres-Diaz et al. [6], showing experimental spectra reconstructions on two medical accelerators.

One notorious advantage of this method is the computational time to obtain the unfolded spectra once the training is done. Indeed, with such methods the unfolded spectrum can be calculated in less than a second which is compatible with our constraints of online measurement. However, the method suffers some limitations regarding our problem. Since the dataset is composed of combinations of exponential functions, none of the reference spectra display a maximum energy. It is reflected on the reconstructed spectra which cannot estimate this key parameter since they never learned it during their training, as shown in Fig 3.

Despite these limitations, a neural network approach seems promising, which is why we decided to explore a new approach allowing for more robust and accurate reconstruction of the incident spectra

III. A MORE ROBUST AND ACCURATE RECONSTRUCTION MODEL BASED ON NEURAL NETWORKS

A. Building a realistic dataset

Herein, the objective was to propose a new method for generating a realistic dataset that accurately accounts for the maximum energy of the spectrum. For this purpose, we attempted to find a parametric function that could fit realistic spectra from simulations. We simulated these spectra using Gate 9.3 [8], taking into consideration the electron spectra on feerix and the geometry of the conversion target, along with potential variations of these parameters.

From these simulations, we attempted to find a parametric function that could fit each of the spectra. The best candidate is based on a generalized beta prime function, completed by a linear adjustment in the last bin. It can be expressed as follows, using only seven parameters:

$$s(E, bin_0, \alpha, \beta, p, q, E_{cut}, E_{max}) = s = [s_0, s_{inf}, s_{sup}] \quad (5)$$

$$With \begin{cases} s_0 = bin_0 \\ s_{inf} = \dfrac{p\left(\frac{E}{q}\right)^{\alpha p-1}\left(1+\left(\frac{E}{q}\right)^{p}\right)^{-\alpha-\beta}}{qB(\alpha,\beta)} \\ s_{sup} = \dfrac{s_{inf}(E_{cut})}{E_{cut}-E_{max}}(E - E_{max}) \end{cases} \quad (6)$$

Where bin_0 being the first bin of the spectrum, α,β,p, and q are the parameters of the generalized beta prime function, $B(\alpha, \beta)$ is the beta function, E_{cut}, is the position of the cut between the generalized beta prime function and the linear function, and finally E_{max}, the maximal energy of the spectrum. Fig 4 illustrates the role of each parameter.

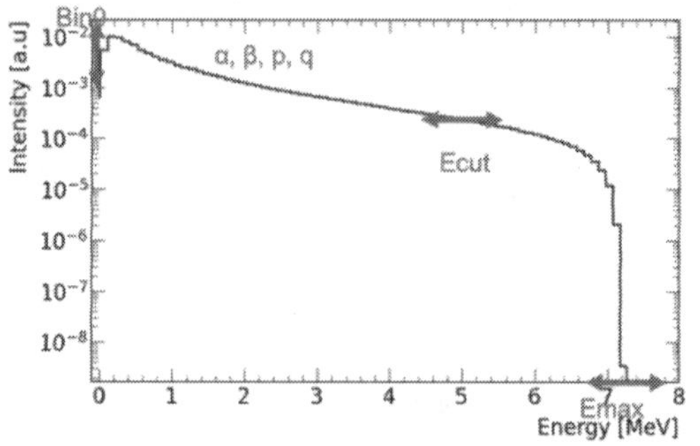

Fig 4. Illustration of the role of the 7 parameters in our parametric representation of the Bremsstrahlung spectra.

After fitting each function from the sensitivity analysis, we are able to retrieve ranges of variation for every parameter.

From these ranges, we can then randomly and uniformly select parameters to build spectra that are representative of our problem. From these spectra and with the appropriate simulated response matrix, we can build the associated dose deposits using (2). We then have a realistic dataset that represents key parameters such as the maximal energy of the spectra. Fig 5 and Fig 6 illustrate the variation range of 10 000 spectra and depth doses generated in ~2 seconds using this approach. The range of variation of each parameter used to build this dataset is given in Table II.

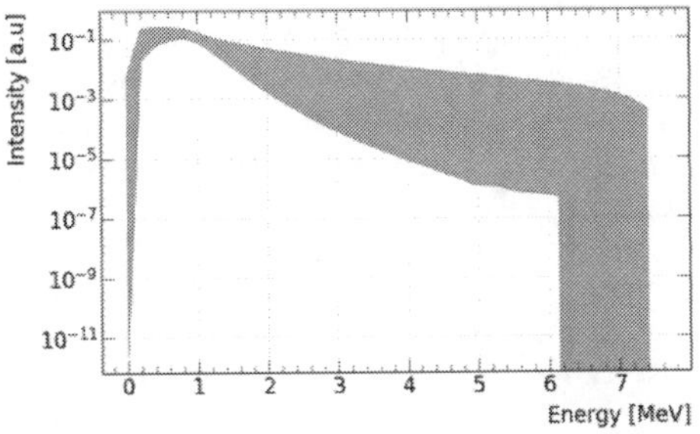

Fig 5. Variation range of the spectra produced for our dataset.

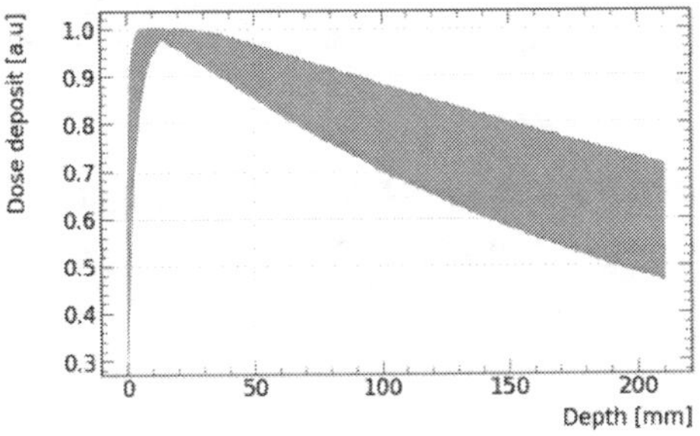

Fig 6. Variation range of the depth dose associated with each spectrum computed for the training dataset.

TABLE II
RANGE OF VARIATION OF EACH PARAMETER USED TO BUILD THE DATASET IN FIG 5 AND FIG 6.

bin_0	α	β	p
[0,02-0,03]	[0.35-0,65]	[2,87-5,38]	[1.96-3,64]
q	E_{cut}	E_{max}	
[0,53-0,99]	[3,9-7,4]	[5,6-8,5]	

B. *Training our neural network*

For the training of the neural network, we differ from previous methods by not learning the full spectra in the output, but rather their parametric representation (i.e., the 7 parameters as shown in Fig 7). As input, we used the 300 bins of the depth-dose profile, calculated via the response matrix.

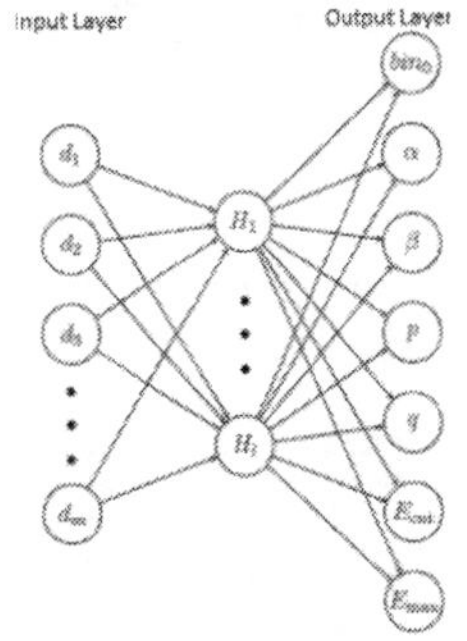

Fig 7. Architecture of our neural network, which takes the depth doses as input and outputs the 7 parameters of the parametric representation of the associated spectrum.

In terms of the architecture of our neural network, we added three hidden layers, a batch normalization layer, and a dropout layer as follows:

- Input layer (300 neurons)
- Hidden layer (256 neurons)
- Batch Normalization layer
- Hidden layer (128 neurons)
- Dropout layer (0.2 %)
- Hidden layer (64 neurons)
- Output layer (7 neurons)

As the activation function, we used the Rectified Linear Unit (ReLU). For the optimizer, we used the Adaptive Moment Estimation (Adam). Finally, we set the mean squared error as our cost function. To reinforce the stability of our training, we added L2 kernel regularization to each of our hidden layers. The neural network is then trained with a dataset of 10,000 spectra/dose pairs across 300 epochs.

C. Validating our neural network model

Once the model is trained, we proceed to its validation. To do so, we generated a validation dataset, independent of the training dataset, which we input into our trained model. For each depth dose, we reconstruct the seven parameters and then the associated spectra. An illustration of a reconstructed spectrum is presented in Fig 8.

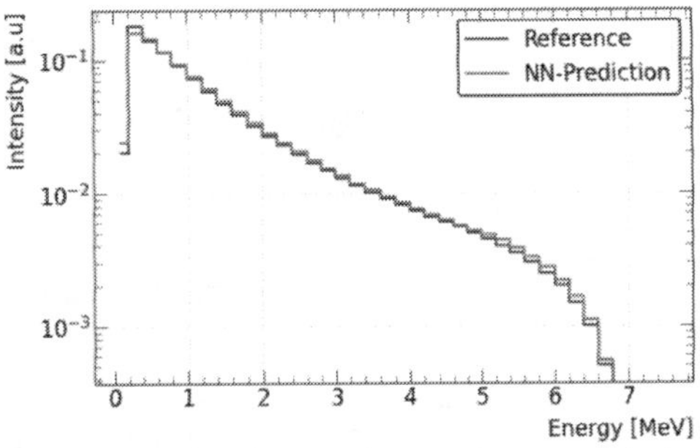

Fig 8. An example of reconstructed spectra using our model. The predicted parameters are $[E; bin_0; \alpha; \beta; p; q; E_{cut}; E_{max}]$ = [0.02; 0.17; 1.11; 5.12; 1.88; 6.63; 0.68] and the RMSE between the reference and predicted spectrum is RMSE = 0.003.

To assess the general efficiency of our model, we looked at the root mean squared errors (RMSE) between the reference and predicted spectra and plotted their distribution. An example is given in Fig 9, for a validation dataset of 1,000 spectra/depth dose pairs, to which we also compared the reconstruction efficiency of previously discussed methods ([2] and [7]) on the same validation dataset.

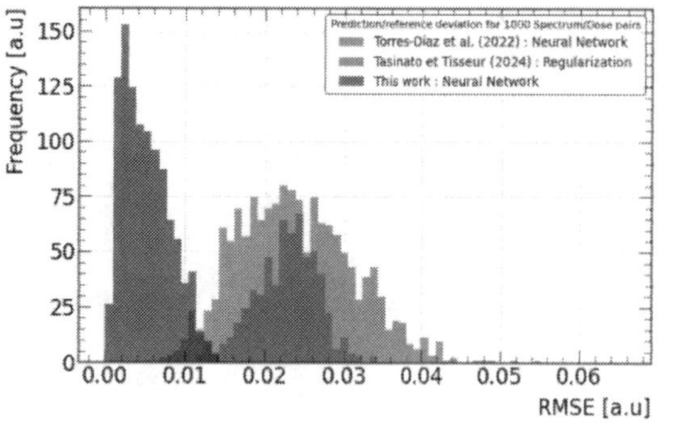

Fig 9. Comparison of the reconstruction performance of our model with that of other methods from the literature, evaluated on the same dataset. We computed the root mean square error between each predicted spectrum and its reference spectrum.

As shown in Fig 9, the overall RMSE between the reference and predicted spectra is lower with our new model than with the neural network [7] or with our previous regularization method. This illustrates a more accurate unfolding of the spectra, not only in the high-energy parts (which have low intensity and thus contribute less to the total RMSE), but mainly also in high-intensity areas of the spectrum. This improvement is most likely due to our more constrained method, where we do not reconstruct the full spectra themselves, but rather the parameters of a realistic representation of the Bremsstrahlung spectra.

As for the high-energy part of the spectra, as we discussed, an important parameter for Aerial is the estimation of the maximal energy in the spectrum. We estimate the ratio between the predicted and reference E_{max} in the same validation dataset. The resulting distribution is presented in Fig 10.

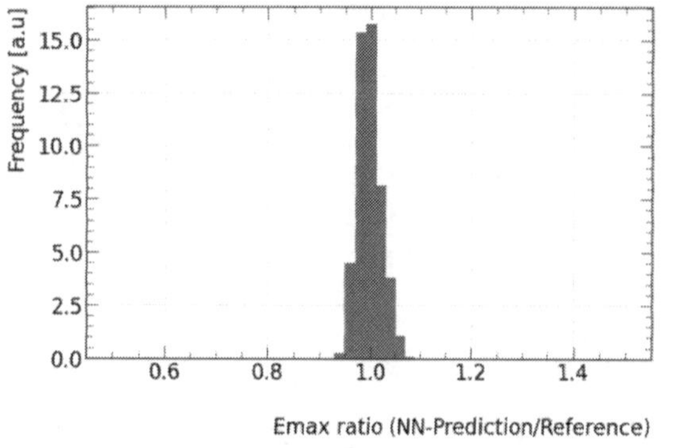

Fig 10. Ratio between the reference and predicted maximal energy E_{max} for 1000 spectra reconstructions.

Again, we observe good efficiency in the reconstruction of the maximal energy. We have a maximal deviation of ±2.5% at k=1, which is compatible with national standards and normative documents, such as NF EN ISO 11137-1 [9], which discuss the monitoring of the bremsstrahlung spectra produced by electron accelerators in the context of medical sterilization.

IV. Conclusions and perspectives

In this paper, we discussed a new approach for the reconstruction of Bremsstrahlung spectra produced by electron accelerators. This method has proven to be more efficient than previously developed methods (regularization and previous deep learning methods) in terms of overall reconstruction efficiency. The improvement is partly due to our constrained method, where we do not reconstruct the entire spectra themselves but rather their parametric representation. This method implies that the shape of the spectra can be described at any time by our parametric representation. This is a strong yet credible assessment for bremsstrahlung spectra produced on platforms similar to feerix, which can be endorsed by a sensitivity analysis of the spectra produced on the platform.

One advantage of this method is also its ability to retrieve key parameters of the spectra, such as their maximal energy. This parameter is important for Aerial to meet ISO standards recommendations, which require monitoring of the maximum emission of the bremsstrahlung spectra to be within <5%.

A multiparametric study as well as an experimental validation of the methods are being performed. These findings will be detailed in a future publication, which will also explore how to optimize the experimental device to improve reconstruction efficiency with our new neural network approach.

Acknowledgment

The authors would like to thank both the ANRT (Association Nationale de Recherche et Technologie) and the CEA (RBNEW project) for co-funding this research work.

References and Footnotes

A. References

[1] F. Kuntz, A. Nasreddine, N. Ludwig, et A. Strasser, « Feerix, a Novel Irradiation Platform for R&D, Education and Training », 2022.

[2] L. Tasinato et D. Tisseur, « Development of a method for the on-line spectral characterization of Bremsstrahlung photons from an electron accelerator », *Radiation Physics and Chemistry*, vol. 215, p. 111374, févr. 2024, doi: 10.1016/j.radphyschem.2023.111374.

[3] B. Armbruster, R. J. Hamilton, et A. K. Kuehl, « Spectrum reconstruction from dose measurements as a linear inverse problem », *Phys. Med. Biol.*, vol. 49, n° 22, p. 5087, oct. 2004, doi: 10.1088/0031-9155/49/22/005.

[4] B. Juste, R. Miró, G. Verdú, S. Díez, et J. M. Campayo, « Linac photon spectra reconstruction using a depth dose gradient TSVD methodology based on Monte Carlo simulation », in *2011 4th International Conference on Biomedical Engineering and Informatics (BMEI)*, oct. 2011, p. 786-790. doi: 10.1109/BMEI.2011.6098411.

[5] B. Juste, R. Miró, G. Verdú, S. Díez, et J. M. Campayo, « Bremsstrahlung Spectrum Reconstruction from Gradient Depth Dose Curves Obtained in a Water Phantom », *Nuclear Technology*, vol. 175, n° 1, p. 175-181, juill. 2011, doi: 10.13182/NT11-A12287.

[6] Z. Shafahi, S. Sina, et R. Faghihi, « Comparison of TSVD, MTSVD, and Tikhonov unfolding methods for reconstruction of X-ray spectra », *Radiation Physics and Chemistry*, vol. 166, p. 108437, janv. 2020, doi: 10.1016/j.radphyschem.2019.108437.

[7] J. Torres-Díaz, G. B. Grad, et E. V. Bonzi, « Measurement of linear accelerator spectra, reconstructed from percentage depth dose curves by neural networks », *Physica Medica*, vol. 96, p. 81-89, avr. 2022, doi: 10.1016/j.ejmp.2022.02.019.

[8] S. Jan *et al.*, « GATE: a simulation toolkit for PET and SPECT », *Phys Med Biol*, vol. 49, n° 19, p. 4543-4561, oct. 2004, doi: 10.1088/0031-9155/49/19/007.

[9] ISO 11137-1:2006. Sterilization of health care products — Radiation — Part 1: Requirements for development, validation and routine control of a sterilization process for medical devices. International Organization for Standardization, Geneva.

https://doi.org/10.1051/epjconf/202533810003

Investigating radiation effects on LSF0102 level shifters

Pierluigi Casolaro[1,2*], Vincenzo Izzo[2], Riccardo Vari[3], Claudio Principe[2], Antonio Vanzanella[2], Paolo Di Meo[2], Antonio Anastasio[2], Arbab Imtiaz[1,2,4], and Alberto Aloisio[1,2]
[1]Department of Physics "Ettore Pancini", Università degli Studi di Napoli Federico II, Complesso Univ. Monte S. Angelo, Napoli, I-80126, Italy
[2]INFN Sezione di Napoli, Complesso Univ. Monte S. Angelo, Napoli, I-80126, Italy
[3]INFN Sezione di Roma, Roma, Italy
[4]Dipartimento di Fisica e Astronomia G. Galilei, Università degli Studi di Padova, Padova, Italy
(*) pierluigi.casolaro@unina.it

Abstract— **We present the results of a Total Ionizing Dose (TID) test on Texas Instruments LSF0102 level shifters irradiated with ^{60}Co gamma rays up to a total dose of 20 kGy. Non-monotonic variations were observed in the supply current, high-level output voltage, and output characteristics as a function of dose. Worst-case scenario in terms of highest supply current and lowest high-level output voltage occurs at a specific critical dose. This behavior should be carefully considered when evaluating such devices for use in environments with high radiation levels.**

Keywords **—Total ionizing dose, gamma-ray effects, MOSFET circuits.**

I. Introduction

Level shifters enable communications between circuits powered with different voltage domains. In complex data acquisition systems presently used in high-energy physics experiments and space applications, level shifters are crucial devices deployed to match logic levels of different integrated circuits, including microprocessors, FPGAs, analog-to-digital converters and other peripherals. Bidirectional level shifters in MOS technologies employ pass transistors which are susceptible to threshold voltage V_t shifts impacting proper voltage translation. Understanding Total Ionizing Dose (TID)-induced degradation in level shifters is therefore essential to ensure the long-term reliability of such devices. In this work, we tested LSF0102 level shifters from Texas Instruments up to a dose of 20 kGy from a ^{60}Co gamma source. Differently from other architectures, LSF family is based on simple MOS transistors, allowing to pinpoint the degradation of the most important device parameters as a function of the dose.

II. Materials and methods

A. *Level shifters*

Fig. 1 shows a schematic layout of the TI LSF0102 under test. In this level shifter [1-5], T1 (MOSFET n-type enhancement) is configured as a MOS diode, with its gate and drain shorted. In this way, T1 sets the gate voltage to the other transistors in the device (all of them being MOSFET n-type enhancement), which translate digital signals across voltage domains. The gates are connected to a capacitance C_f tied to ground, to filter high frequency noise [1]. We studied the translation between 1.2 V and 3.3 V power domains. The input signal V_{in} is fed into the source of T2.

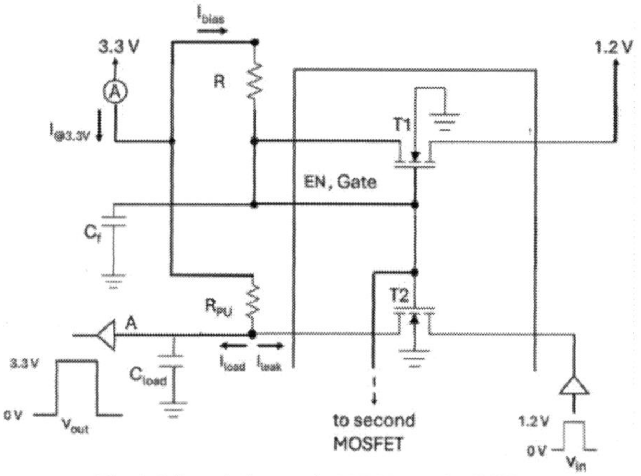

Fig. 1: Schematic layout of a TI LSF0102 level shifter.

When the difference between the gate voltage V_g with respect to ground and V_{in} exceeds V_t+V_{OV} [6] (with V_{OV} being the overdrive voltage, which for this technology is typically in the order of 500 mV):

$$V_g - V_{in} > V_t + V_{OV} \tag{1}$$

T2 turns on and transfers the voltage to the drain. Conversely, when

$$V_g - V_{in} \lesssim V_t + V_{OV} \tag{2}$$

T2 turns off and the voltage translation is ruled by the pull-up resistor R_{PU} biased to 3.3 V. The scheme in Fig. 1 shows also an ammeter used to measure the current in the 3.3 V domain, as discussed in the following.

B. *TID test*

TID irradiations were performed at the CERN CC60 facility equipped with a ~3 TBq ^{60}Co gamma source. Fig. 3 shows a picture (left) of the board housing ten DUTs exposed to the ^{60}Co gamma source.

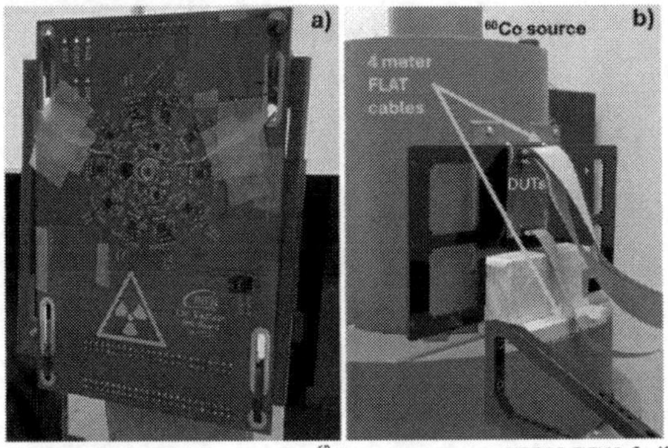

Fig. 2: DUT board (a) exposed to ^{60}Co gamma rays at CERN CC60 facility (b).

The board also hosts two thermocouples and a radiochromic film for both temperature and dose monitoring [7]. During the test, the temperature was constantly measured and found to be 23.5 degrees. As shown in Fig. 3 (right), the board was placed on a stage in front of the gamma source where the dose rate was of 327 Gy/h. We irradiated for about 60 hours yielding a total dose to the DUT of 20 kGy. Following the methodology of our previous irradiation test [8], the DUTs were powered and remotely controlled using a Raspberry Pi 3 Model B+. We used a Tektronix AFG31152 Function Generator, configured to transmit a serial stream at 1.2 V at a frequency of 200 kHz, a Tektronix MSO46 oscilloscope and an Agilent N6705B DC Power Analyzer. During the test, we monitored DUT current consumption, as well as variations in amplitude and rise time of the waveforms translated in the 3.3 V domain. Connections between the DUT board and the input signals and power supplies, as well as between the DUT board and the remote control board, were implemented using 4-meter-long flat cables. Before and after the test, we measured the output characteristics with a B1500A Semiconductor Device Parameter Analyzer.

III. Results

A. Characteristic curves

Fig. 3 shows the I_{bias}-V_{gs} curves of T1 before and after irradiation, taken in configuration shown in the inset (MOS diode).

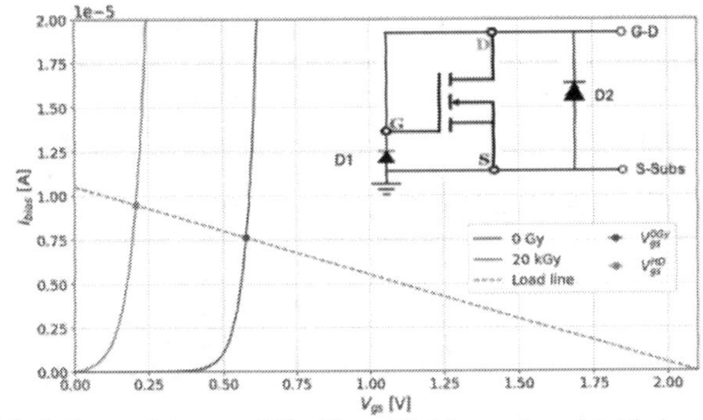

Fig. 3: Characteristic curves, T1 load line and their intersection points. The inset shows the configuration of the level shifter in this measurement.

Diode D1 protects the gate against potentially dangerous voltage surges. Diode D2 is the body diode, typical of MOS structures, between drain and substrate. For $V_{ds} < 0$ V, D1 and D2 are on, while the MOS diode is reverse-biased. For $V_{ds} > 0$ V, D1 and D2 are reverse-biased, while the MOS diode is on. With reference to Fig. 3, the intersection of the I_{ds}-V_{ds} curves with the load line:

$$I_{bias} = \frac{V_{cc}}{R_{PU}} - \frac{1}{R_{PU}} \cdot V_{gs} \quad (3)$$

yields the following operating points before and after irradiation:

$$V_{gs}^{0Gy} = (578 \pm 5)\ mV \text{ at } I_{bias}^{0Gy} = (7.61 \pm 0.03)\ \mu A \quad (4)$$

and

$$V_{gs}^{HD} = (210 \pm 5)\ mV \text{ at } I_{bias}^{HD} = (9.46 \pm 0.03)\ \mu A \quad (5)$$

where the superscripts "0Gy" and "HD" indicate doses of 0 Gy and 20 kGy respectively and $V_{cc} = 2.1\ V$ is the voltage difference between the power domains. V_g is obtained by adding the source voltage, which is fixed at 1.2 V, to V_{gs}. Hence:

$$V_g^{0Gy} = (1780 \pm 5)\ mV \quad (6)$$

and

$$V_g^{HD} = (1410\ \pm 5)\ mV \quad (7)$$

These variations in the operating point result from shifts in the V_t of T1, which are clearly visible in the output characteristics I_{ds}-V_{ds} at different V_{gs} values, as shown in Fig. 4.

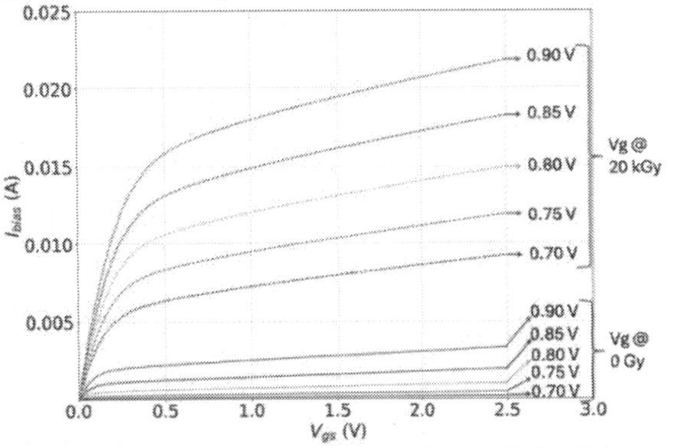

Fig. 4: Output characteristics of T1 MOSFET before and after irradiation.

Ionizing radiation induces charge trapping in the gate oxide and produces interface states, leading to threshold voltage V_t shifts and increased subthreshold leakage in MOSFETs [9]. It should be noted that the characteristics of all transistors are consistent with those presented in Fig. 4, both prior to and following irradiation.

B. Power supply current

The trend of the current $I_{3.3V}$ drawn in the 3.3 V domain by the ten DUTs altogether during the test is shown in Fig. 4.

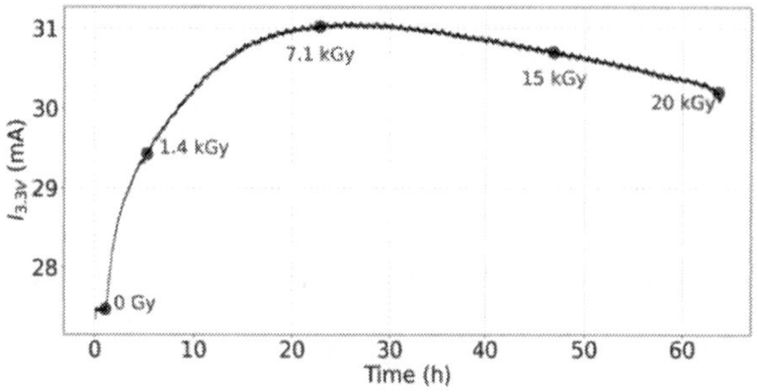

Fig. 5: Current measured by the power supply in the 3.3 V domain.

$I_{3.3V}$ increases, reaching a maximum at 7.1 kGy, and then decreases slightly toward the end of the irradiation. Despite relatively small percentage difference

$$\Delta = \frac{I_{3.3V}^{20kGy} - I_{3.3V}^{0Gy}}{I_{3.3V}^{20kGy}} \times 100 \sim 10\% \quad (8)$$

in current between the start and end of the irradiation, such a trend reflects the presence of V_t shift in TID damage. From Fig. 1, one can see that:

$$I_{3.3V} = I_{bias} + I_{load} + I_{leak} \quad (9)$$

From (4) and (5), the bias current for the ten DUTs contributes to less than 1% to $I_{3.3V}$ shown in Fig. 5. Since I_{load} is mostly unaffected by irradiation, any change in $I_{3.3V}$ should be attributed to V_t variations in T2, directly affecting I_{leak}.

C. Waveforms

Fig. 6 shows waveforms taken in A (see Fig. 1), at 0 Gy for three different values of the pull-up resistor, R_{PU}=33 kΩ, R_{PU}=68 kΩ and R_{PU}=100 kΩ.

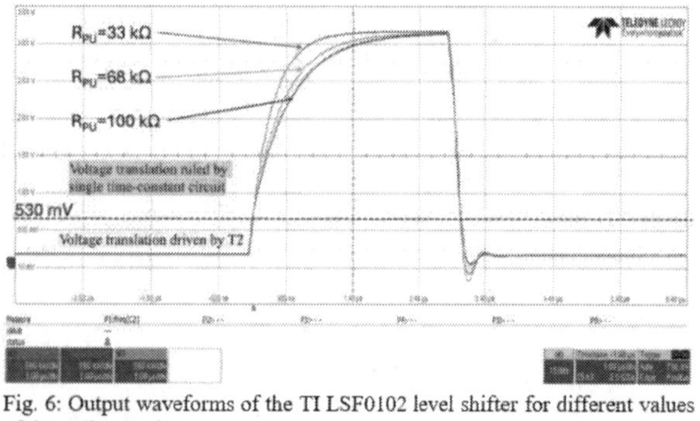

Fig. 6: Output waveforms of the TI LSF0102 level shifter for different values of the pull-up resistor.

Fig. 6 clearly shows the two-stage operation of the level shifter. From (6) and (1), V_t+V_{OV}=1250 mV and V_g=1780 mV, therefore for input voltages up to 530 mV, T2 is on and the voltage translation is actively driven. Beyond this value, T2 goes off and the translation is ruled by a single time-constant circuit with $\tau = R \cdot C_{load}$, where C_{load} is the sum of the receiver input and cable stray capacitances.

Fig. 7 shows the rising edge and high-level voltage V_{OH} reached by the output waveforms at different doses during the TID test.

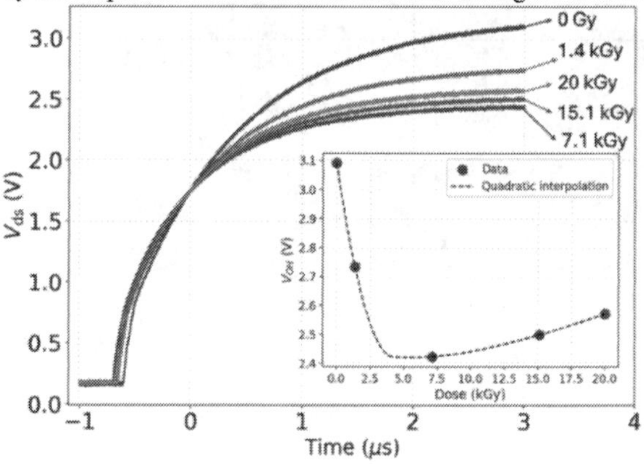

Fig. 7: Output waveforms of the TI LSF0102 level shifter for different doses.

The inset of Fig. 7 shows the non-monotonic trend of V_{OH} as a function of dose. To guide the eye, experimental data are interpolated with a quadratic spline (dashed curve) using the Scipy "interp1d" function. V_{OH} decreases from 3.09 V at 0 Gy to 2.42 V at 7.1 kGy, before partially recovering to 2.57 V at 20 kGy at the end of the irradiation. V_{OH} is given by

$$V_{OH} = Vcc - R \cdot (I_{load} + I_{leak}) \quad (10)$$

therefore the observed trend in Fig. 7 reflects the one followed by $I_{3.3V}$ in Fig. 5. Since I_{load} can be considered nearly constant, V_{OH} variations are proportional to I_{leak} changes. These findings reveal a critical dose threshold at approximately 7.1 kGy, characterized by a minimum in V_{OH} and a corresponding peak in the supply current $I_{3.3V}$.

IV. Conclusions

We investigated the TID response of TI LSF0102 level shifters exposed to ^{60}Co gamma irradiation up to 20 kGy at the CERN CC60 facility. Our results show that ionizing radiation induces changes in the output characteristics of T1 and T2, as well as in the operating point of T1 configured as MOS diode. The supply current increases as a function of dose up to a maximum at 7.1 kGy and then it slightly decreases toward the end of the irradiation. We found that variations in the supply current are mostly due to changes in T2 I_{leak}, which are in turn proportional to changes in V_{OH}. This behavior is consistent with a shift in V_t, typically observed in MOS technologies exposed to TID radiation. Our work shows that V_{OH} is non-monotonic with dose, with a critical dose threshold at approximately 7.1 kGy, characterized by a minimum in V_{OH} and a corresponding maximum in the supply current $I_{3.3V}$. This aspect should be taken carefully into consideration when evaluating the maximum dose tolerable by the devices.

ACKNOWLEDGMENT

The authors would like to thank the staff of the CERN CC60 facility. This work was supported by the Italian Research Infrastructure "PRP@CERIC - Pathogen Readiness Platform for CERIC ERIC upgrade" (IR0000028) and by the PRIN 2022 PNRR - OPTICS, both funded by EU - Next Generation EU.

REFERENCES

[1] Texas Instruments, LSF0102 2-Channel Auto-Bidirectional Multi-Voltage Level Translator for Open-Drain and Push-Pull Applications datasheet (Rev. B), Dallas, TX, USA, Apr. 2024. [Online]. Available: https://www.ti.com/lit/pdf/SDLS972

[2] J. Guan, "Top Questions About Auto Bi-Direction LSF Family Translators," Texas Instruments, Dallas, TX, USA, App. Note SCEA144, Dec. 2024. [Online]. Available: https://www.ti.com/lit/pdf/SCEA144

[3] R. Land, "Voltage-Level Translation With the LSF Family," Texas Instruments, Dallas, TX, USA, App. Rep. SLVA675B, Dec. 2014, Revised Mar. 2015. [Online]. Available: https://www.ti.com/lit/pdf/SLVA675

[4] S. Rao and A. Ozer, "Biasing Requirements for TXS, TXB, and LSF Auto-Bidirectional Translators," Texas Instruments, Dallas, TX, USA, App. Rep. SCEA060, Oct. 2017. [Online]. Available: https://www.ti.com/lit/pdf/SCEA060

[5] Texas Instruments, "Translate Voltages for I2C for PCA9306 LSF0102 TXS0102," Texas Instruments, Dallas, TX, USA, App. Note SCET011, Sept. 2022. [Online]. Available: https://www.ti.com/lit/pdf/SCET011

[6] A. S. Sedra and K. C. Smith, *Microelectronic Circuits*, 7th ed. New York, NY, USA: Oxford Univ. Press, 2015.

[7] P. Casolaro, L. Campajola, and F. Di Capua, "The physics of radiochromic process: one calibration equation for all film types," *J. Instrum.*, vol. 14, Aug. 2019, Art. no. P08006, doi: 10.1088/1748-0221/14/08/P08006

[8] P. Casolaro, V. Izzo, R. Vari, M. D'Angelantonio, A. Vanzanella, C. Principe, and A. Aloisio, "TID damage assessment on LVDS links for the ATLAS muon barrel spectrometer readout system," *J. Instrum.*, vol. 20, no. 01, Art. no. C01023, Jan. 2025, doi: 10.1088/1748-0221/20/01/C01023.

[9] J. R. Schwank *et al.*, "Radiation Effects in MOS Oxides," IEEE Trans. Nucl. Sci., vol. 55, no. 4, pp. 1833-1853, Aug. 2008, doi: 10.1109/TNS.2008.2001040

Real-time gamma-neutron discrimination with a trainable polynomial kernel

Matyáš Hlavinka[1,*], Zdeněk Matěj[1], Jiří Čulen[1], Jan Král[1], and Václav Přenosil[1]
[1]Faculty of Informatics, Masaryk University, Czech Republic
(*) hlavinka@mail.muni.cz

***Abstract*—This paper presents an implementation of gamma-neutron pulse shape discrimination by a support vector machine polynomial decision function in a field-programmable gate array. The training is carried out on a conventional computer using widespread Python libraries. The hardware architecture is designed to allow parameter changes on demand, enabling tuning of hyperparameters and kernel coefficients without requiring re-synthesis. A cubic kernel is compared against a linear kernel which was developed alongside it for non-biased comparison. Both are designed to be viable for real-time classification. The particularities of the designs are explored. The cubic kernel makes use of two stand-alone state machines to keep the sequential data pipelined without interference between the sampled pulses. The results show the trade-off between separation quality, numerical accuracy and physical on-board requirements of the implementations. The separation quality is demonstrated on two datasets, one with a noticeable overlap, to assess any benefits the cubic kernel may bring.**

***Keywords*—gamma-neutron discrimination, support vector machines, field-programmable gate array**

I. Introduction

GAMMA-NEUTRON discrimination methods are used in mixed radiation fields to differentiate neutrons from gamma rays, since neutrons cannot be detected independently. Scintillation materials are used to detect interactions of ionizing radiation. Fast neutrons interact with sensitive scintillators indirectly through a process known as elastic scattering [1]. The responses generated by a gamma ray and an elastic scattering of a fast neutron are similar, but exhibit slightly different characteristics, which can be exploited to differentiate them.

A highly effective method for real-time discrimination is charge comparison, which involves integrating the tail portion of the signal response, as neutron waveforms typically exhibit a longer decay time [2].

Support vector machines (SVM) have been used as offline classifiers for gamma-neutron discrimination already [3]–[5]. Exhibiting good separation results when trained for specific responses, the method can be used to mimic the conventional charge comparison method. The radial basis function (RBF) kernel is used in the cited works with good results. Although implementable in hardware [6], the footprint is very large and too demanding for scintillation pulses in real time. Furthermore, [7] points to the linear kernel being more suitable than RBF for many-featured datasets - which the response pulses definitely are, as they are vectors of considerable length.

A well-trained SVM can offer two key improvements over the charge comparison method. In an offline setting, the mapping of the vectors into a higher dimension can possibly enhance the discrimination of the pulses, especially when selecting just some features. The second use case is simpler real-time discrimination in hardware using just sample amplitudes. The core of the decision function is a simple dot product. The charge comparison method has to find the maximum amplitude, implement a more robust integral method such as Simpson's rule and finally calculate a ratio by division, which is more complex to synchronize together. A pipelined linear SVM n/γ classifier has already been shown to be implementable in a field-programmable gate array (FPGA) [8].

The aim of this paper is to find the possible benefits of the polynomial kernel against its cost. The constraint of real-time processing plays a big role in the design methods and compromises taken. A cubic kernel is compared with a linear kernel implementation developed alongside it to properly assess whether the much more complex polynomial architecture is worth exploring further.

The target device *xc7a100tcsg321-1* is used to test the proposed architecture. While the numerical results are obtained from a testbench version, all of the designs have to pass on-board implementation in a synthesis tool of choice to be truly viable. This includes correct clocking, timing requirements and board utilization. Any model which can not be physically implemented is discarded, regardless of its accuracy. Fixed-point arithmetic is used for its speed. The accuracy is compared to native Python functions.

II. Training

A. Support Vector Machines

Although we are mainly concerned about the particularities of gamma-neutron discrimination and the later hardware implementation, it is necessary to provide an insight to support vector machines themselves. In the context of this paper, we regard SVM as a supervised machine learning method for binary classification. The input vector is made up of sampled amplitudes for hardware simplicity. This tests the ability of the method to generalize based on inferred characteristics but requires samples from the whole energy range for both classes.

The SVM used works with a soft-margin classifier which introduces a penalty hyperparameter C used in misclassification. This parameter affects the strictness of the classifier. Smaller values of C lead to a more lenient penalization and thus a wider hyperplane margin. This allows some pulses to be misclassified in order to better separate the majority of data. This does not affect the decision functions.

Using definitions from [9], binary classification uses function $f : X \subseteq \mathbb{R}^n \to \mathbb{R}$ for the input $\mathbf{x} = (x_1, ..., x_n)'$, where the decision rule is driven by $sgn(f(\mathbf{x}))$. Function f is defined for SVM in [10] and shown in (1). Here, the intercept value is added as well for clarity.

$$f(x) = \sum_{i=1}^{l} \alpha_i y_i K(x, x_i) + b_0 \tag{1}$$

The intercept value b_0, the support vectors, and their coefficients are learned from data under specific constraints described in [10], [11] and previously mentioned papers working with SVM discrimination, such as [3]. The method finds such parameters, that the separating hyperplane it creates provides the widest margin between the classes. As this paper is mainly concerned with the hardware implementation of the decision functions, we mainly focus on them only.

The linear kernel is just a dot product, meaning the support vectors can be reduced into a single weight vector $\mathbf{w}$ by their linear combination, making the decision function into:

$$f(x) = w \cdot x + b_0 \tag{2}$$

The polynomial kernel can map the feature space into a higher dimension, where a linear separation may be found. The polynomial dependencies between the support vectors and the input are not as easily reducible. We are dealing with a high number of features, therefore the mapping cannot be done explicitly without an unreasonable number of terms. Calculating the kernel is much more efficient. The modified decision function from *scikit.learn* which introduces another normalizing parameter γ is:

$$f(x) = \sum_{i=1}^{n} \alpha_i y_i (\gamma(x \cdot x_i) + c)^d + b_0, \tag{3}$$

where $\alpha_i y_i$ are signed dual coefficients, d is the polynomial degree, γ and c are the scaling factor and bias term respectively, b_0 is a global bias term and $\mathbf{x_i}$ is a corresponding support vector.

B. Data

There are two main sets of data this paper works with. The first one is from a Cf-252 source, captured by a detector containing p-Terphenyl sampled by Keysight EXR104A at 16 GSa/s with a 10-bit resolution. This yields 1000 samples per pulse with the sampling speed resulting in a great, mostly smooth representation of the responses which can be used for checking SVM quality. The second set of data is from the side channel of the VR-1 training reactor. Captured by the Picoscope 6426E digitizer at a 12-bit resolution and 1.25 GSa/s. The detector used is the Ne-323 liquid scintillator with an area of 2.2 cm^2, coupled via EJ-550 to a probe at -1500 V. With 250 samples per pulse and 100 pre-trigger samples, these pulses are more representative of the usual capture. This data also introduces noise and overlap in the groups.

C. Pulse separation

The captured data is separated offline by the charge comparison method with gate zero being at the maximum amplitude. Two timed gates are employed and the separation parameter D is calculated as the ratio of the tail part of the pulse against the whole charge from gate zero as:

$$\begin{aligned} Q1 &= \int_{t_1}^{t_2} U(t)\, dt \\ Q2 &= \int_{t_0}^{t_2} U(t)\, dt \\ D &= \frac{Q1}{Q2} \end{aligned} \tag{4}$$

The gates are chosen in a ratio 1:16 with the shorter gate being 48 samples for the Keysight data and 8 for the latter. The resulting scatter plots are shown in Fig. 1. The vertical line at the right side of Fig. 1a is value saturation created by clipping pulses. They should be discarded in training and results alike.

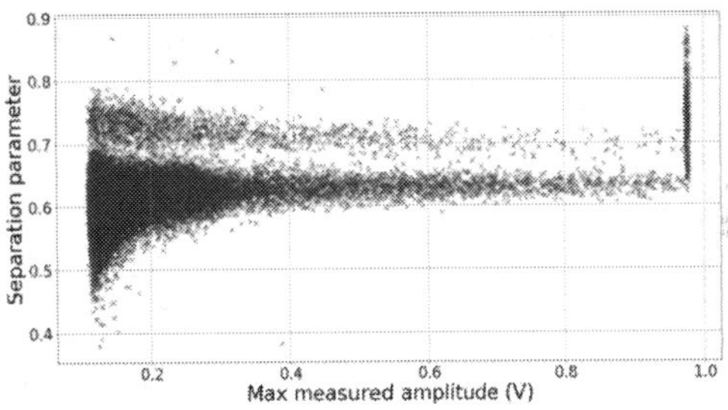

(a) Keysight separation

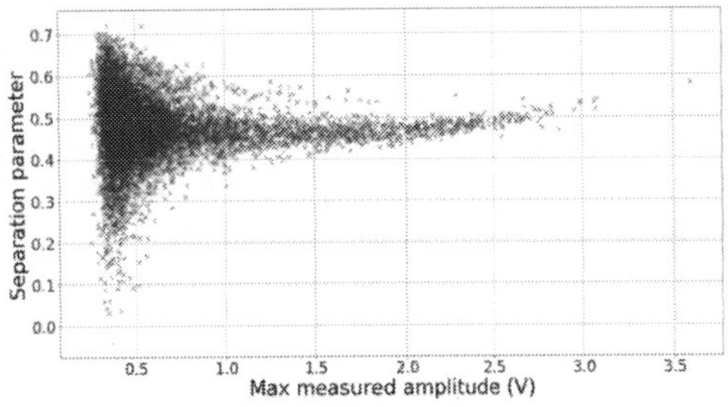

(b) Picoscope overlap

Fig. 1: Charge comparison method visualized.

The value at which the separation parameter divides the responses is chosen based on the assumed split between the two groups. Only the obviously separated pulses are chosen by filtering all of the overlapping parts around the split and trimming off high amplitudes for cleaner results. The training data is prepared by selecting two equally sized sets of gamma-rays and neutrons. The input for the classifier is the whole waveform of the pulse with a tag of its class based on the resulting parameter of the charge comparison method described earlier.

The data is fitted in Python into the SVC class of $sklearn.svm$ which is based upon the widely used $libsvm$ library. Using grid search, we find the optimal hyperparameters by using $cross\_val\_score$ to evaluate the fit based on multiple metrics such as accuracy, precision and recall. The polynomial kernel has an additional crucial requirement of its quality based on the count of support vectors. The FPGA is limited in resources so it cannot support a high number of vectors. For demonstration purposes the maximum amount was chosen to be 32. This can eliminate some more ideal fits but it is a trade-off for real-time calculation.

III. Hardware architecture

A. Introduction and particularities

It is crucial to have the option to change vectors without having to re-synthetize the whole design. This means no fixed values for wave length, vector number and the weights themselves. We have to hold these values in memory and access them dynamically. BRAM modules are used for this purpose. Limits of FPGA design still apply. The RAM is dual-port to disallow input data interfering with weight fetching. Re-writing the memory can be done by setting the appropriate flags. The writing has to be done sequentially but the design assumes no measurement is being done while inputting the configuration.

With each new input sample, the address is automatically incremented to fetch the correct weight feature. This means the whole process moves forward in sync with the input vector and no addressing mismatch can happen.

The main component which moulded the architecture around itself is the DSP48E1 slice used in AMD devices we are interested in. An important detail to bring up is that without daisy-chaining, the multiplication in this DSP slice is executed on *18×25-bit* registers, resulting in a 43-bit output extended to 48 bits [12]. This in turn limits the available widths of registers which can be used, because we do not want to use more clock cycles than necessary. Any register which has to be used as an input to the product has to abide by these widths. The digitizers operate with less precision regardless, therefore this should not be an issue. Using fixed-point arithmetic on pulses in this amplitude range allows for 14 out of 18 bits dedicated to the fractional part and one to the sign. This gives us a range of ± 8 with a small enough step of 2^{-14}. This is crucial to balance out in the cubic polynomial as the power can quickly cross this threshold when dealing with sums of this many samples.

B. Linear

Only the reduced weight vector **w** needs to be stored. The data input is sequential and made up of sample amplitudes. Synthesis can be limited to a single DSP slice by creating a separate multiplication module instead of calculating the product directly inside the state machine. By correctly changing the weights each clock cycle, the total DSP slice cost for a linear kernel is two. One for multiplication and another for a separate accumulation of the whole result. The footprint scales very favourably. It is only dictated by the maximum wave length we allow, independent of any SVM parameters other than vector length.

The pipeline is separated into two state machines. DSP slices output a result two clock cycles after the input is assigned and fetching the weight value from RAM takes one cycle. This means we are working with a constant delay of three cycles. When the machine detects an end of a pulse, a flag is set. This switches our logic to propagate the last three accumulated results into our output while also resetting the accumulator, because the multiplier module is already working with a new pulse.

C. Polynomial

Polynomial kernels cannot be as efficiently reduced as the linear one. We need to respect the contribution of each vector in the final equation. The inner product relies on each feature. A hard cap on the number of vectors must be set to a power of two to allow efficient operations. Important thing is the fact we need to hold $18 \times (samples + 1) \times vectors$ bits. A support vector is a waveform from the training data set with extra bits for its dual coefficients. The reasonable choice for a proof-of-concept implementation was 32 vectors which takes up 18 kB of memory for shorter waves and 72 kB for the Keysight data. If there are less vectors than the maximum, the rest of memory is padded by zeroes. These zero value vectors are used in the calculation as well. The normalization parameter $coef0$ would impact the result even with the dot product being zero, so their dual coefficient must be set to zero. The memory needs to be fully rewritten on a new setup.

The RAM module must fetch a feature from each kernel on every input. This was implemented by saving a transposed, concatenated value of features at the same position in their corresponding vectors. We mask the other values by just selecting specific bits in the bit string and pass it onto the stand-alone kernel modules. In order for the loading to be realizable, it is performed on a feature basis into an unpacked array and then flattened to a big register. This means the input can be an integer and we do not have to come up with special data types. The values of gamma, dual coefficients and coef0 need to be managed and kept in memory as well.

The main logic of the kernel follows the linear implementation as the dot product remains the same. The output state machine has to have more states to accommodate the polynomial exponentiation. The entire logic is shown in Fig. 2. The linear kernel can output the final sum directly after correcting for the clock delays.

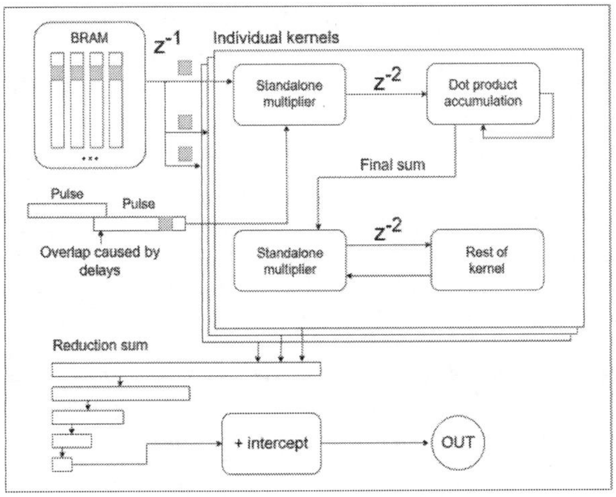

Fig. 2: Polynomial decision function architecture diagram.

The formula for the cubic kernel, which needs to be calculated for every support vector x_i in separate modules, is as follows:

$$(\gamma\langle x, x_i\rangle + c)^3 \tag{5}$$

Even while using the result registers directly from the multiplication module, the total output state count comes to 13 due to propagation delays. The dot product can be accumulated into an arbitrarily large register, but it is necessary for the result of every subsequent operation to fit into 18 bits.

The gamma value is calculated from the number of features and data variance, taken from [13]:

$$\frac{1}{n_{features} \cdot \mathrm{Var}(x_0, x_1, \ldots, x_n)} \tag{6}$$

This division means the sum is effectively trimmed from its 25-bit representation to 18 bits without losing any precision unnecessarily. This happens right before the power is calculated, where having two 25-bit values is an issue. One of the multiplicands is sign extended back to match the width requirement of one of our fixed DSP multiplier's input. This bitshift arithmetically works when the number of samples per pulse is larger than 2^7 which should be most of the time.

The resulting kernel outputs are passed into a large register. A parallel reduction sum is implemented to calculate the final result in logarithmic time. The polynomial approach is much more resource-heavy. It requires three DSP blocks per support vector, takes up more memory and has to produce the final sum in another module. The delay in this approach should ultimately not matter, as long as the output state and final sum take less cycles together than the number of samples in a full pulse, because the calculation is fully pipelined,

IV. Testing

A. *Python*

As the hardware aims to replicate the Python results, we are interested in the separation quality of the Python models. The models are numerically checked with five-fold cross-validation. Graphical methods are used to check the class designation quality. The visual check is performed by assigning the class predicted by the SVM model to the original charge comparison vectors, because the whole pulses are multidimensional. This 2D representation neatly visualizes the decision boundary formed by the model and can be directly compared to the original separation by a straight line which is shown in Fig. 3. The SVM classification in Fig. 4 shows a slight curvature in the low energy spectrum of both groups which corresponds to the assumed spread in the overlapping section of the original discrimination. Do note, that this overlap classification cannot be proven as correct, which can be seen later in the results.

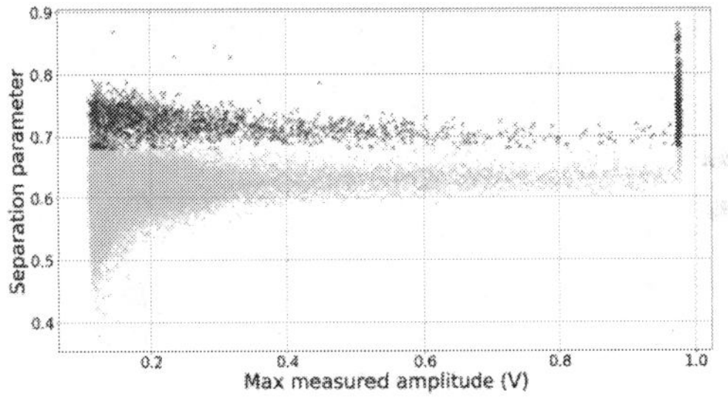

Fig. 3: Charge comparison method classification. Gamma-rays in orange, neutrons in blue. Notice the strict border between the classes.

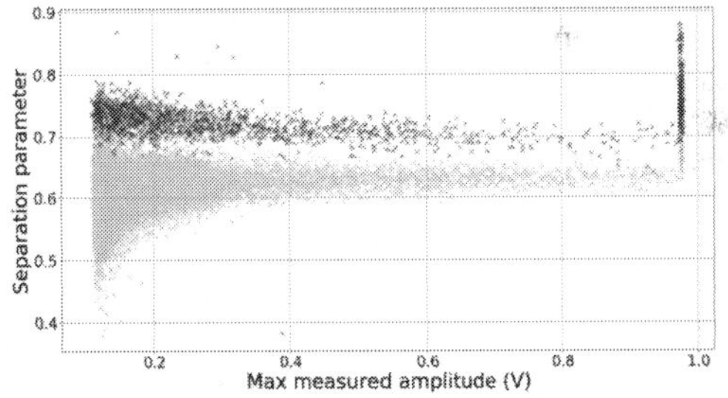

Fig. 4: Linear SVM classification. Gamma-rays in orange, neutrons in blue. The border is more relaxed and allows some expected overlap to emerge.

B. *cocotb*

cocotb is a Python verification environment used for the simulation of designs in hardware description languages. The designs were imported into *Vivado* to check the implementation and real-world constraints, but *cocotb* allows for efficient fuzzing and complex testbenches that would be difficult to do in native SystemVerilog. We utilized it being a Python library by directly importing sample values into the testbench of the SVM and comparing Verilog results to the native Python predictions on floats.

Having many kernels and further operations can create a compound error in the numeric values. The kernels must be able to correctly mimic float results. The numbers are manually converted from native Python floats into 18-bit fixed-point binary strings using two's complement representation. The sign is remembered at start to correctly process numbers rounding to zero.

The fixed-point precision can be moved arbitrarily, depending on the maximum range of our data. The selected number of precision bits from the fixed total should be tested in a standalone testbench to ensure that extreme values are not saturated when they shouldn't be. This is pure data representation and does not affect synthesis in any way as long as the results can be interpreted later.

The main testbench also checks correct timing and interactions between modules on subsequent pulses. Any missing, stolen, or overlapping features between multiple pulses can lead to a sign change and will impact the results. The most important metric is the class choice, but we can also measure the numerical error which helps with fine-tuning. This is important in the case of pulses found at the border of the hyperplane where a sign change can happen easily.

V. RESULTS

A. Numerical accuracy

The first metric of the hardware decision function is it's numerical accuracy. Any mismatch with Python is undesirable, because the classification accuracy is derived from the original model. Bad numerical results would mean the module is unusable even with perfect offline separation.

The implementations were tested in *cocotb* with 5000 random pulses for each dataset that was used in testing the offline SVM. We are interested in the percentage of decisions that match the Python model's class prediction and the average numerical deviation of the results respectively. The misclassification happens due to rounding errors. Results are shown in Table I.

TABLE I: Percentage of matches

Dataset	Linear		Polynomial	
	% match	*deviation*	*% match*	*deviation*
Picoscope 1	99.98	0.00372	99.90	0.00301
Picoscope 2	100	0.00269	100	0.00595
Picoscope 3	99.98	0.00321	99.94	0.00745
Picoscope 4	99.98	0.00286	99.90	0.00684
Keysight	99.99	0.00284	100	0.00286
Total	99.99	0.00306	99.95	0.00522

Both kernels managed to almost match the Python calculation completely. The SVM needs to be balanced between accuracy and the way parameters interact numerically. The hyperparameters impact the size of the coefficients and the fixed-point registers have limited capabilities so great care is needed. The polynomial kernel suffers much more from this effect. A scaling parameter should be introduced in the Python code to divide all of the dual coefficients by the same value. The intercept value should then be divided as well to conform to the new scale. This loses some precision but avoids overflows. The choice for this divider depends on the dataset and should be experimented with to obtain the best numeric precision.

B. Classification quality

The quality of the classification heavily depends on the underlying dataset and chosen training data. For our purposes, more strict handling of misclassification expectedly returned results more closer to the charge comparison method. This is desirable as the overlapping pulses cannot be reliably distinguished from each other and the wider margin amplifies this uncertainty. The hyperparameter C that turned out the best as a compromise between accuracy and vector number was 64 for both kernels. The cubic kernel then uses the standard $coef0$ of 1. The ranges tested were $\{2^{-4}, 2^{-3}, \ldots, 2^{10}\}$ for C and $\{0, 0.5, \ldots, 2\}$ for $coef0$. Fig. 5 shows the raw result of decision functions from the trained SVMs.

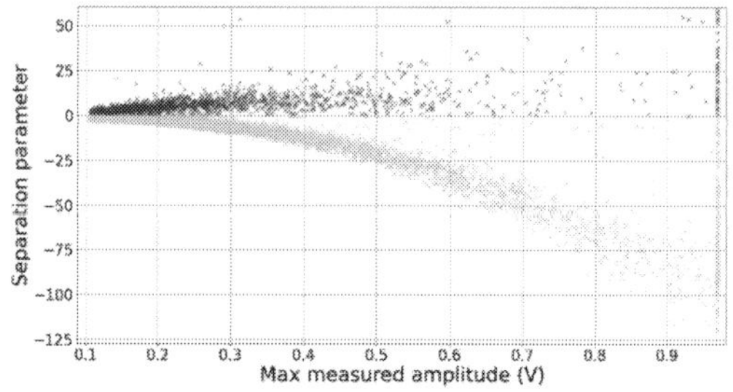

(a) Cubic kernel SVM

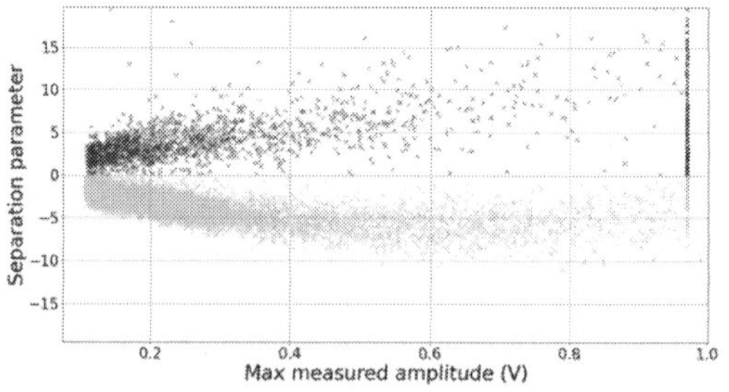

(b) Linear kernel SVM

Fig. 5: The decision output for Keysight data. Positive numbers (blue) represent neutrons.

The maximum measured amplitude is used the same way as in the charge comparison method. The separation parameter is the final sum output of the decision function before the sign is applied. This directly visualizes the split between classes as seen by the SVM. The cubic kernel displays more divergence in the higher amplitudes with the separation parameter also being more pronounced due to the power it applies. However, the lower amplitude values are more uncertain. The linear kernel has a more noticeable gap between the two groups. This is reflected in Fig. 6 where the cubic kernel has an overrepresentation of certain values, while the linear kernel

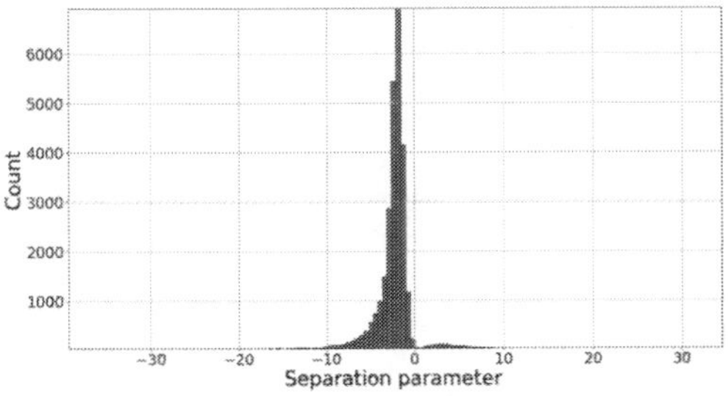

(a) Cubic kernel histogram

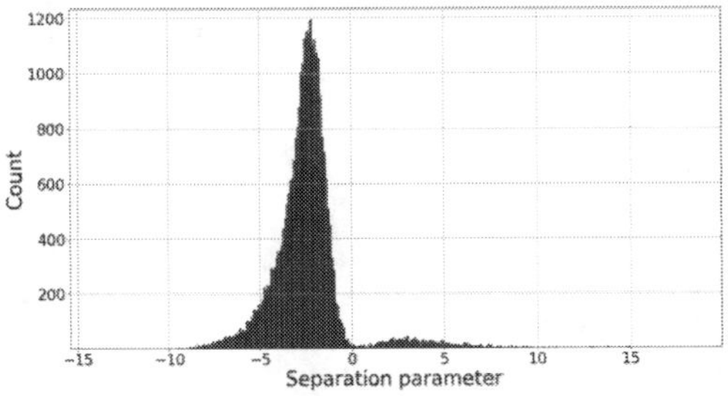

(b) Linear kernel histogram

Fig. 6: Separation histograms.

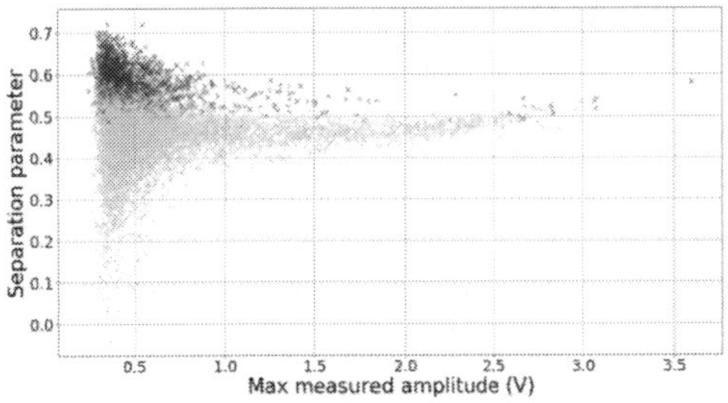

(a) Cubic kernel prediction

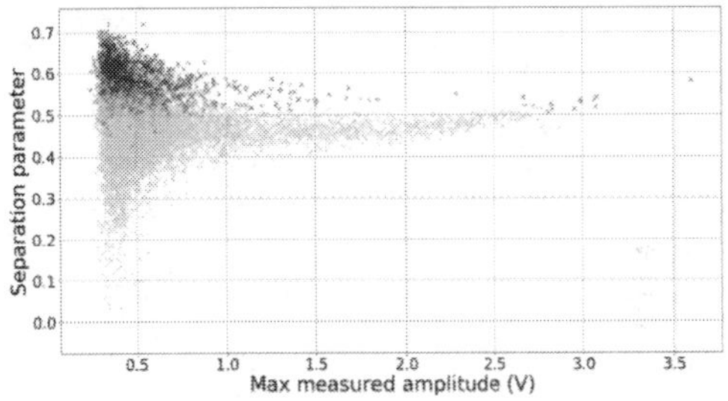

(b) Linear kernel prediction

Fig. 7: Classification of an overlap. Neutrons in blue.

is more evenly distributed. Both kernels have achieved a good split and separation, replicating the quality of the charge comparison method.

C. Results on an overlap

Both SVMs were retrained for the Picoscope datasets which exhibit a significant overlap when separated with the charge comparison method. The underlying data is noisy and most of it is overlayed on top of each other. Trying different gate lengths still yields very uncertain results. The aim of trying SVM on this dataset was to see whether this problem can be improved by the higher-dimensional mapping.

The dataset was split by a parabola to better represent the possible class distribution. The cubic kernel is more conservative and better follows the parabola, indicating a possible overfit. The linear kernel made some large guesses in the lower amplitudes even with the same penalty parameter C. Both kernels were able to classify a second measurement from the same source similarly as the training split, displaying generalization capability. The separation results for both of them are overlapped in the same way the charge comparison is and no solid improvement was made. The main advantage of the cubic kernel was better replication of the given border. Results are presented in Fig. 7.

D. Utilization and overhead

Timing was met for both designs in all categories. The polynomial kernel has an obvious flaw in not being reducible to a single weight vector. The linear kernel uses just 2 DSP slices, while the cubic one uses 3 per every support vector. The number of LUTs and flip-flops needed scaled up roughly by two times the number of vectors. This is due to larger memory operations, an additional module and special copying of features. This held true for the shorter, 250-sample versions of the SVMs with 32 maximum support vectors. RAMB36E1 was used as a general purpose memory block for the dual port RAM.

VI. Conclusion

This paper presented hardware architecture for linear and polynomial SVM decision functions along with testing their qualities. Support vector machines proved to be capable of discriminating the data on the same level as the charge comparison method while being possible to calculate in real time. However, without normalization, the method must be retrained for each source.

The polynomial kernel did not demonstrate sufficient benefits to outweigh the large hardware cost compared to the linear kernel. On the contrary, the polynomial kernel tended to overfit the many-featured training data and suffered from being harder to tune numerically. The architecture may be better for specific features or needs, but as of now, it is advisable not to explore it further in the case of real-time n/γ discrimination.

References

[1] G. F. Knoll, *Radiation Detection and Measurement*, 4th ed. Hoboken, NJ: John Wiley & Sons, 2010. [Online]. Available: https://indico-tdli.sjtu.edu.cn/event/171/contributions/2123/attachments/982/1592/Knoll4thEdition.pdf. Accessed on: Feb. 20, 2025.

[2] M. Amiri, *Neutron/gamma-ray measurement and discrimination* [online]. Doctoral dissertation, Masaryk University, Faculty of Informatics, Brno, 2014. Available: https://is.muni.cz/th/j2qsw/

[3] X. Yu, J. Zhu, S. Lin, L. Wang, H. Xing, C. Zhang, *et al.*, "Neutron–gamma discrimination based on the support vector machine method," in *Nucl. Instrum. Methods Phys. Res. A*, vol. 777, pp. 80–84, 2015. doi: 10.1016/j.nima.2014.12.087

[4] P. Krömer, Z. Matej, P. Musilek, V. Prenosil, and F. Cvachovec, "Neutron-gamma classification by evolutionary fuzzy rules and support vector machines," in *2015 IEEE International Conference on Systems, Man, and Cybernetics*, Hong Kong, China, 2015, pp. 2638–2642, doi: 10.1109/SMC.2015.461.

[5] H. Arahmane, A. Mahmoudi, E.-M. Hamzaoui, Y. Ben Maissa, and R. Cherkaoui El Moursli, "Neutron-gamma discrimination based on support vector machine combined to nonnegative matrix factorization and continuous wavelet transform," *Measurement*, vol. 149, p. 106958, 2020. doi: 10.1016/j.measurement.2019.106958.

[6] M. H. Yacoub, S. M. Ismail, L. A. Said, A. H. Madian, and A. G. Radwan, "Support vector machine reconfigurable hardware implementation on FPGA," *Franklin Open*, vol. 7, p. 100115, 2024. doi: 10.1016/j.fraope.2024.100115.

[7] C.-W. Hsu, C.-C. Chang, and C.-J. Lin, *A Practical Guide to Support Vector Classification [online]*, Department of Computer Science, National Taiwan University, 2003, updated 2016. [Online]. Available: https://www.csie.ntu.edu.tw/~cjlin/papers/guide/guide.pdf. Accessed on: Feb. 21, 2025.

[8] W. Zhang, T. Wu, B. Zheng, S. Li, Y. Zhang, and Z. Yin, "A real-time neutron-gamma discriminator based on the support vector machine method for the time-of-flight neutron spectrometer," *Plasma Sci. Technol.*, vol. 20, no. 4, p. 045601, Mar. 2018. doi: 10.1088/2058-6272/aaaaa9.

[9] N. Cristianini and J. Shawe-Taylor, *An Introduction to Support Vector Machines and Other Kernel-Based Learning Methods*. Cambridge, UK: Cambridge University Press, 2000.

[10] C. Cortes and V. N. Vapnik, "Support-vector networks," *Machine Learning*, vol. 20, no. 3, pp. 273–297, Sep. 1995. doi: 10.1007/BF00994018.

[11] V. N. Vapnik, *Estimation of Dependences Based on Empirical Data*, New York: Springer, 2006, Information Science and Statistics series. [Online] Available: https://link.springer.com/book/10.1007/0-387-34239-7. Accessed on: Apr. 13, 2025.

[12] AMD, *7 Series DSP48E1 Slice User Guide*, [Online] Available: https: //docs.amd.com/v/u/en-US/ug479_7Series_DSP48E1. Accessed on: Apr. 15, 2025.

[13] Scikit-learn developers, "SVC," scikit-learn.org, [Online], Available: https://scikit-learn.org/stable/modules/generated/sklearn.svm.SVC.html. Accessed on: May 20, 2025.

Applications of 4H Silicon Carbide Radiation Detectors in Fission and Fusion Reactor Environments

Frank H. Ruddy[1*], Abdallah Lyoussi[2], Christophe Destouches[2], Laurent Ottaviani[3], Christelle Reynard-Carette[3], and Olivier Llido[2]
[1]Ruddy Consulting, 2162 Country Manor Dr., Mt. Pleasant, South Carolina, 29466, USA
[2]CEA, DES, IRESNE, DER, Cadarache F-13108, Saint-Paul-Lez-Durance, 13101, FRANCE
[3]Aix Marseille Univ, Université de Toulon, CNRS, IM2NP, Marseille, FRANCE
(*) FrankHRuddy@gmail.com

***Abstract*—Semiconductor radiation detectors based on the 4H polytype of Silicon Carbide (4H-SiC) have many advantages for high-temperature, high-radiation and mixed-radiation applications. The wide band gap of 4H-SiC (3.27 eV) allows measurements at temperatures up to 800 °C and probably much higher. Conventional lower band-width semiconductors such as silicon or germanium are limited by thermally generated noise to measurements at lower temperatures and can require external cooling. 4H-SiC has a high thermal conductivity, high breakdown voltage, and high electron saturation velocity. Furthermore, 4H-SiC detectors have also been demonstrated to have outstanding radiation hardness. These properties make 4H-SiC Schottky barrier and p-n diode detectors compatible with radiation monitoring applications in both fission and and fusion reactors. For fission reactors, measurements of the neutron fluence rate and energy spectrum characteristics are required and monitoring environments with temperatures up to and exceeding 800 °C can be encountered. For fusion reactors, extremely high-temperature monitoring environments will also be encountered. In both cases, the detector service lifetime in the extreme radiation environment is a prime consideration. 4H-SiC Schottky barrier detectors have been demonstrated to be operational after ^{137}Cs gamma ray doses up to 22.7 MGy and fast-neutron (E> 1 MeV) fluences up to 1.7 x 10^{17} cm^{-2}. 4H-SiC detectors have been demonstrated to provide linear count-rate responses as a function of fluence-rate for thermal, epithermal and fast fission neutrons. In addition, linear fluence-rate responses have been demonstrated for deuterium-deuterium (D-D) and deuterium-tritium (D-T) fusion neutrons. In the latter case, the 14-MeV neutron energy exceeds the thresholds for $^{12}C(n,\alpha)^{9}Be$ and $^{28}Si(n,\alpha)^{25}Mg$ reactions and allows peaks to be observed from neutron reactions with the SiC detector. The energy widths of these reaction peaks can be used to deduce information on the plasma temperature as well as neutron yield for fusion devices. 4H-SiC detectors with lithium neutron convertor foils can be used to monitor $^{6}Li(n,\alpha)^{3}H$ and $^{7}Li(n,n'\alpha)^{3}H$ reaction rates to directly measure tritium breeding rates in fusion reactors. In this paper we will review relevant developments in 4H-SiC radiation detectors and discuss their potential and planned applications in fission and fusion reactors.**

***Keywords* —4H-silicon carbide (4H-SiC), Schottky barrier diode, wide bandgap, semiconductor, neutron detector, fission reactor, fusion reactor.**

I. Introduction

Silicon Carbide (SiC) radiation detectors were first demonstrated more than sixty years ago [1]. Their development was facilitated by the inception of hot-wall epitaxy manufacturing techniques leading to the first SiC radiation detectors based on high-quality epitaxial materials in 1997 [2]. In the past few decades, the advantages and versatility of SiC radiation detectors have been demonstrated in a variety of applications [3-7].

Among the advantages of SiC are low leakage current, a maximum breakdown field that is eight times that of silicon allowing higher applied biases and more efficient charge collection, a saturated drift velocity nearly twice that of silicon resulting in low charge trapping, high thermal conductivity, and a high band gap (3.27 eV at 300K for 4H-SiC). The higher band gap compared to conventional semiconductors such as silicon or germanium leads to lower thermally induced leakage currents and allows the use of SiC detectors at room temperature and elevated temperatures up to at least 800 °C [8].

SiC radiation detectors have been proposed for many applications where conventional semiconductor detectors are limited by operating environment conditions [9,10,11]. Among the more promising applications are power monitoring in fission reactors and characterization of the output of nuclear fusion devices.

II. Applications in Fission Reactor Environments

To ensure safe operation of nuclear power reactors as well as test and research reactors accurate power monitoring is required as part of the instrument and control (I&C) system. I&C systems infer power by monitoring neutron fluence rates in key reactor locations.

Historically, reactor neutron detectors have been limited to metallic designs such as gas-filled detectors (BF_3 or boron-lined proportional counters), fission chambers, or self-powered detectors. The use of neutron detectors based on semiconductors such as silicon is not practical, because narrow band gap semiconductors cannot operate reliably at temperatures higher than about 45 °C and are susceptible to radiation damage.

Neutron detectors based on SiC diodes have been demonstrated for thermal-neutron [12-15], epithermal-neutron [16,17], and fast-neutron [18] detection over the range of energies and fluence rates encountered in fission-neutron reactors. Because photoelectrons and Compton-scattered electrons only deposit a small amount of energy in the detectors resulting in low pulse-rate responses, gamma rays are separable from the neutron response based on pulse-height discrimination in all cases [12-18].

The thermal-neutron response of SiC detectors was found to be linear over seven orders of magnitude in neutron fluence rate, and their precision compared to a National Institute of Standards and Technology (NIST) double fission chamber was found to be better than 0.6% [13,14].

Epithermal neutron response-rate measurements were done at the Penn State Breazeale TRIGA reactor at low power following higher power gamma activation of the core in order to simulate high gamma-to-neutron environments such as spent fuel radiation environments. In the resultant radiation fields, 35,000-45,000 R/hr with an epi-cadmium neutron fluence rate of 600 cm^{-2}-$sec.^{-1}$, measurement precisions of better than 1.8% and 0.6% were demonstrated for neutrons and gamma rays, respectively [16,17].

Thermal- and fast-neutron response measurements were carried out at the Ohio State University Research Reactor using SiC Schottky diode detectors with and without a ^{6}LiF neutron-conversion layer [18]. A linear response rate was demonstrated over a factor of twenty in reactor power, and the precision relative to the reactor I&C instrumentation was better than ±0.6%. The relative precision of the SiC detector measurements was better than ±0.2%. Current-mode operation was also demonstrated [18].

SiC neutron detectors have been proposed [19] as an alternative to metallic detectors in order to take advantage of their rapid pulse-rate response and their compatibility with modern digital electronics. Prototype thermal-neutron detector arrays based on SiC Schottky diodes equipped with ^{6}LiF neutron-convertor layers were assembled for comparison measurements with conventional pressurized water reactor (PWR) I&C instruments. Irradiations were carried out using the Cornell University Ward Laboratory Triga research reactor. The absolute responses of the SiC neutron detectors were also measured relative to radiometric gold foil measurements [19].

In commercial PWRs, reactor power is characterized in three operating ranges: source startup range, intermediate range, and power range [20]. The neutron detectors are located in detector well locations in the reactor cavity located between the reactor pressure vessel and the biological shield. BF_3 gas-filled detectors operating in the pulse mode are used to monitor neutron fluence rates in the source startup range. Boron-lined proportional counters operating in the current mode are used in the intermediate range. Current-mode operation is used in the intermediate range, because individual pulses are not distinguishable as a result of high count rates. Since boron-lined proportional counters are sensitive to gamma rays, a second proportional counter with no boron lining is also used, and its response is subtracted from the boron-lined chamber response, a practice called gamma compensation. In the power range, mainly prompt gamma rays from fission reactions in the reactor core are detected, and the gamma-ray count rate is proportional to reactor power. Therefore, gamma compensation is not necessary in the power range, and single proportional chambers can be used rather than lined/unlined pairs [20].

Three prototype SiC neutron-detector arrays were assembled to simulate the sensitivities of the gas-filled detectors used in the three power ranges [19]. The count-rate response of each of these arrays as a function of thermal-neutron fluence rate as determined by gold foil activation is shown in Fig. 1.

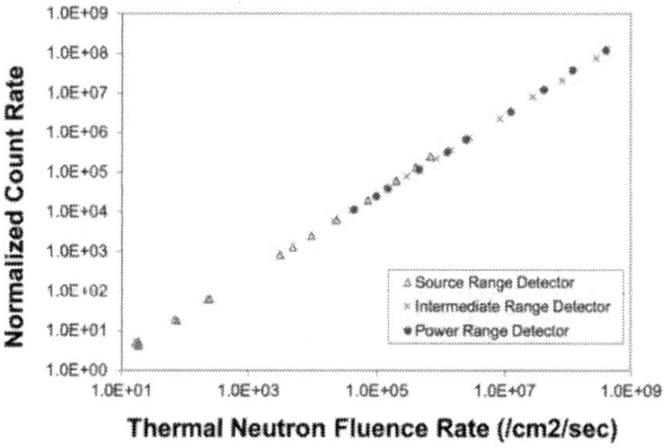

Fig. 1. Normalized detector count rate as a function of neutron-fluence rate for prototype SiC source, intermediate and power range neutron detector arrays. Figure adapted from reference [19].

The SiC prototype detector responses were linear with thermal-neutron fluence rate over the entire range of the measurements, and significant fluence-rate overlap of the prototype responses was observed. Note that the range of responses of the SiC source-range and power-range detector prototypes overlapped indicating that the intermediate-range prototype was not needed. In fact, the authors speculated that a single prototype could be designed to cover the entire fluence rate range [19].

SiC detectors offer the following advantages compared to conventional gas-filled detectors:

- The need for gamma compensation and Campbell-mode operation is eliminated, because gamma-rays can be discriminated against on the basis of pulse height.
- SiC detector arrays require much less space than conventional I&C neutron detectors.
- Two, or perhaps a single SiC detector array, is capable of covering the entire reactor power range from start-up to full power.
- Pulse-mode operation is possible over the entire reactor power range enabling better compatibility with modern digital instrumentation.
- Although current-mode operation has been demonstrated for SiC neutron detectors [18],

 https://doi.org/10.1051/epjconf/202533810005

pulse-mode operation avoids current drift that can be caused by radiation damage effects in SiC that interfere with charge collection efficiency.

SiC neutron detectors were shown to be compatible with the power monitoring needs of PWRs in the operating nuclear power reactor fleet, but their deployment was hindered by scheduling and logistics constraints. The emergence of next-generation and advanced reactors may provide an opportunity for their deployment.

New reactor designs under development range from microreactors and small modular reactors (SMRs) to high-power reactors based on liquid-alkaline metal cooling, molten-salt reactors and high-temperature gas reactors [21]. All of these reactors present unique challenges for the design of I&C for reactor power monitoring and safety requirements.

For example, the space constraints placed on the size of I&C neutron detectors are likely to be more demanding for microreactors and SMRs. Neutron detectors based on SiC diode arrays can be configured in a variety of shapes and generally require less space than gas-filled detectors, boron-lined proportional counters, fission chambers, or self-powered detectors.

Microreactors under development employ innovative TRISO fuels and high-temperature gas cooling or fluoride-salt cooling necessitating a high-temperature environment for the I&C neutron detectors. SMRs and higher-power designs based on molten-chloride fast-reactor technology also can present high-temperature environments. Alkaline-metal cooled reactors including sodium- and potassium-cooled designs involve operating environments up to 850 °C [10,11]. SiC neutron detectors have been shown to be operational at temperatures up to 800 °C [8] and are likely to be capable of operation at much higher temperatures. The sublimation temperature for 4H-SiC is 2857 °C.

The other advantages demonstrated for SiC neutron detectors in PWRs [19] will likely also prove useful for applications in emerging future-generation reactors.

III. Applications in Fusion-Reactor Environments

Fusion is produced in dense, high-temperature, magnetically confined plasmas or by inertial confinement. The primary energy-producing reactions are:

$$^{3}H + ^{2}H \rightarrow ^{4}He + n \quad (E_n = 14.1\text{ MeV}), \qquad (1)$$

$$^{2}H + ^{2}H \rightarrow ^{3}He + n \quad (E_n = 2.5\text{ MeV}) \qquad (2)$$

Although the first reaction, the D-T reaction, produces most of the energy, both reactions as well as other secondary reactions will take place.

Advanced diagnostic systems such as neutron cameras, neutron energy spectrometers, and neutron flux monitors are under development. At present, the primary wide-band gap semiconductor planned for fusion-monitoring applications is synthetic diamond [22,23]. Although SiC has been identified as a promising material for fusion-monitoring applications [10,11], it has not been installed at a Tokamak device [24].

4H-SiC Schottky barrier detectors have been demonstrated for detection of 14-MeV neutrons from a deuterium-triton (D-T) accelerator [25-30]. The fast-neutron response results from ionization produced in the detector active volume mainly by the products of the following reactions:

$$^{12}C\,(n,n')^{12}C \qquad (3)$$

$$^{28}Si(n,n')^{28}Si \qquad (4)$$

$$^{12}C(n,n')3\alpha \qquad (5)$$

$$^{12}C(n,\alpha)^{9}Be \qquad (6)$$

$$^{28}Si(n,\alpha)^{25}Mg \qquad (7)$$

$$^{12}C(n,p)^{12}B \qquad (8)$$

$$^{28}Si(n,p)^{28}Al \qquad (9)$$

Other multi-particle breakup reactions equivalent to the $^{12}C(n,n')3\alpha$ reactions are present as well as reactions with the less-abundant isotopes of carbon and silicon.

A typical SiC 14-Mev D-T neutron-response spectrum is shown in Fig.2 [27]. A Si detector neutron-response spectrum is shown for comparison. Most of the pulse-height responses

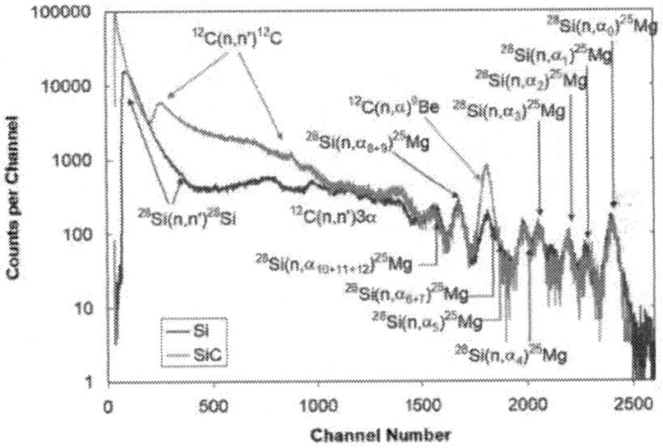

Fig. 2. Neutron-response spectrum for a 100-μm x 28.3-mm^2 Schottky barrier diode exposed tor 14-MeV neutrons. A Si detector response spectrum is shown for comparison. Figure adapted from reference [27].

are from elastic and inelastic neutron-scattering reactions on ^{12}C and ^{28}Si resulting in the continua shown in Fig. 2. However, for those reactions that result in only charged-particle products, such as the (n,α) reactions, peaks will be formed in the spectrum if the recoil-ion products of the reaction dissipate all of their energy within the detector active volume. The most prominent peak present in the SiC spectrum is from the $^{12}C(n,\alpha)^{9}Be$ reaction, where a total recoil energy of 8298.8 keV is deposited by the reaction products. In the case of the $^{28}Si(n,\alpha)^{25}Mg$ reactions, the total recoil energy deposited is 11,346.3 kV for reactions that lead to the ground state of ^{25}Mg. However, reaction branches to excited states of ^{25}Mg followed by gamma de-excitation are also present leading to a set of peaks. An enlarged view of the reaction peaks is shown in Fig. 3. The ground state and the first four excited states (designated as α_1, α_2, α_3, and α_4,) show clearly resolved peaks for the $^{28}Si(n,\alpha)^{25}Mg$ reactions. The fifth, sixth, and seventh excited-state peaks are obscured by the $^{12}C(n,\alpha)^{9}Be$ peak in the SiC spectrum, and peaks for the eighth and nineth excited states, which are separated by only 63 keV, are not resolved in either the SiC or Si response spectra.

All of the neutron-reaction peaks have Gaussian shapes, and the full width at half maximum (FWHM) is 2.18% for the

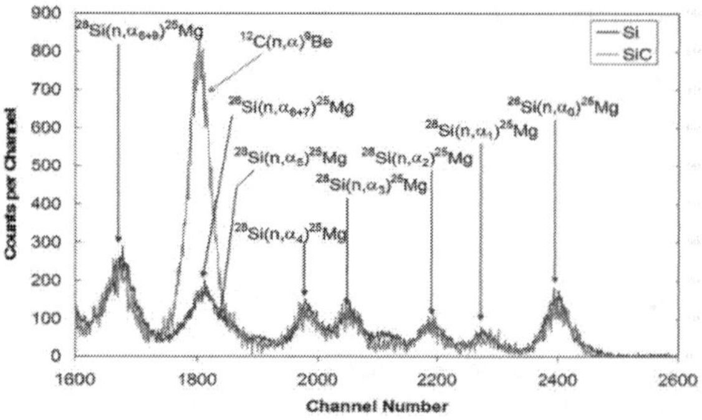

Fig. 3. Enlarged view of the neutron-induced reaction peaks shown in Fig.2. Figure adapted from reference [27].

^{12}C(n,α)^{9}Be peak after correcting for the interfering peaks from the ^{28}Si(n,α)^{25}Mg reaction. Since the total energy deposited by the ^{12}C(n,α)^{9}Be reaction is 8298.8 keV, the corresponding energy width is 181 keV. The FWHM for the ground-state ^{28}Si(n,α)^{25}Mg peak is 2.40% corresponding to an energy width of 272 keV. Note that these FWHM values are much higher than the 55.4-keV (0.66%) FWHM measured with an identical SiC detector for the 8375.9-keV alpha-decay peak of ^{213}Po [31]. The increased peak widths result from the more complex nature of the energy losses by the ^{9}Be and ^{25}Mg heavy ions where more energy goes into non-ionizing energy losses than for alpha particles. The 10% higher peak width for the ground-state ^{28}Si(n,α)^{25}Mg peak compared to the ^{12}C(n,α)^{9}Be peak results from higher non-ionizing energy losses for ^{25}Mg compared to ^{9}Be and the higher overall energy dissipated from the ^{28}Si reaction.

As a result of the nature of the energy losses by the ^{9}Be and ^{25}Mg heavy ions, deriving the energy scale for pulse-height response spectra is complex. A plot of the energies of the ^{28}Si(n,α)^{25}Mg reaction peaks as a function of observed peak-centroid channel number is shown in Fig. 4 [27,32]. A straight line is observed for the resolved ^{28}Si(n,α)^{25}Mg reaction peaks.

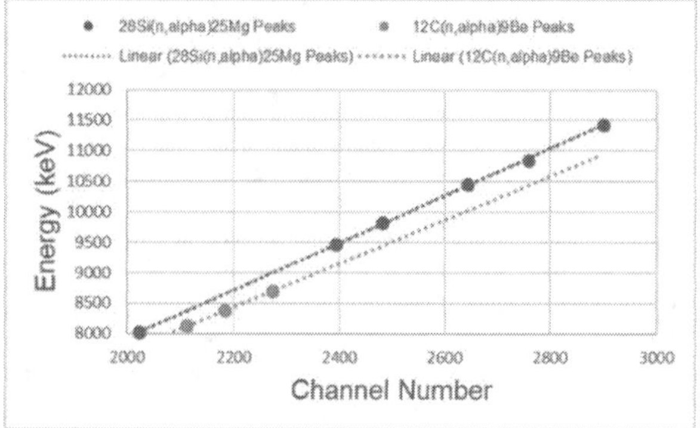

Fig. 4. Reaction energy for 14-MeV neutron-induced (n,α) reaction peaks as a function of observed channel number in a SiC detector response spectrum. Derived from data reported in references [27] and [32].

Although only a single reaction peak is observed for the ^{12}C(n,α)^{9}Be reaction, the energy scale can be derived by utilizing the dependence of the observed energy on the detector angle with respect to the beamline for a D-T generator. For a D-T generator operated at an acceleration potential of 100 kV, the resulting energies for angles of 45°, 90°, and 135° are 14.574 MeV, 14.008 MeV, and 13.619 MeV, respectively. The peak channels observed at these three angles are plotted as a function of corresponding neutron energy in Fig. 4 resulting in a second straight-line response [32].

It has been demonstrated that the total count rates for 4H-SiC detectors are linear as a function of D-T neutron fluence rate [33]. Therefore, SiC detectors are capable of monitoring both the intensity and energy characteristics of neutrons produced in fusion reactors.

The advantages of SiC detectors for monitoring fusion reactors include the following:

- SiC detectors can operate stably at very high temperatures.
- SiC detectors are resistant to radiation damage effects from gamma-ray, neutron, and charged-particle exposures.
- SiC neutron detectors provide high-resolution reaction peaks for 14-MeV fusion neutrons.
- SiC charged-particle detectors can be used to monitor tritium breeding.

ITER measurement requirements include diagnostic systems to monitor the following key parameters [34]:

- Fusion power.
- Power density.
- Neutron and alpha source profiles.
- The fuel ratio (η_T/η_D) from 14- and 2.5-MeV neutron measurements.
- Neutron fluence on the plasma-facing components
- Energy distribution of reacting particles (thermal and fast ions).

Fusion power and density measurements are planned [34] by using activation foils coupled with active measurements with ^{238}U and ^{235}U microfission chambers to determine fast- and thermal-neutron fluence rates and distributions [35-38]. ITER requires neutron source strength measurements from 10^7 to 3.2 x 10^{21} s^{-1} [35] with a temporal resolution of 1 ms and an accuracy of ±10 % [34]. It is proposed to cover this wide dynamic range using microfission chambers by using a combination of pulse-mode operation and current-mode operation including Campbell mode measurements for gamma-ray discrimination [37]. The present design requires operation in a 350 °C environment.

SiC thermal-neutron detectors using ^{6}LiF neutron-convertor layers are a viable alternative to microfission chambers. It is anticipated that the seven order of magnitude range can be covered using pulse-mode operation and gamma-ray discrimination on the basis of pulse height thereby eliminating the complexities of current-mode and Campbell-mode operation. The ^{6}LiF/SiC thermal-neutron detectors will easily meet the 350 °C operating requirement.

Vertical and radial neutron cameras are planned for ITER to evaluate neutron emission profiles from the neutron plasma source, determine η_T/η_D, and to aid the characterization of fusion power density, alpha source density, neutron and alpha profiles, and ion temperature profiles [34].

The vertical neutron camera employes ^{238}U microfission chambers to measure E > 1 MeV neutrons and CVD diamond detectors to measure 12-16 MeV neutrons. SiC fast-neutron detectors can replace both of these detectors. Fluence rates for E > 1 MeV neutrons can be inferred from the overall SiC neutron response spectrum, and 12-16 MeV fluence rates can be derived from the ^{12}C and ^{28}Si reaction peaks. Changes in thermal-neutron sensitivity caused by ^{239}Pu buildup ($^{238}U[n,\gamma]^{239}U \rightarrow ^{239}Pu$) in the microfission chambers are avoided. Furthermore, SiC detectors can be expected to be more resistant to neutron-induced radiation damage than diamond detectors [39]. In addition, recent studies have shown that 4H-SiC radiation detectors are much more stable at elevated temperatures than diamond detectors [40,41]. Diamond detectors were shown to suffer charge-collection efficiency losses up to 60% between 200 and 300 °C, whereas 4H-SiC detectors were stable up the maximum temperature studied, 500 °C [41].

The vertical neutron camera will view the plasma directly through a collimator and is anticipated to have neutron fluence rates from 10^5 – 10^{10} cm^{-2}s^{-1}. SiC fast-neutron detectors can easily cover this fluence-rate range using pulse-mode operation.

A high-resolution neutron spectrometer is planned for ITER mainly to measure η_T/η_D . The spectrometer uses proton-recoil spectrometry and time-of-flight measurements to deduce neutron energies [34]. A combination of fast organic scintillators and plastic scintillators are the primary detectors. A particle identifier array comprised of SiC pixel detectors might be a viable alternative to avoid the interactions of the anticipated magnetic fields on the photomultipliers needed for the scintillators.

A tangential neutron spectrometer is also planned using tangential lines of sight to study fast-ion characteristics. In the non-nuclear phase charge-exchange ions will be monitored, and D-T neutron spectra will be measured in the nuclear phase using CVD diamond detectors [34]. SiC detectors are capable of supplementing or potentially replacing the diamond detectors and supply information from ^{28}Si threshold reaction peaks in addition to the ^{12}C reaction-peak data supplied by both SiC and diamond detectors [9,10].

IV. Conclusions

SiC neutron and charged-particle detectors have many potential applications for both fission and fusion reactor monitoring.

In the case of fission reactors, 4H-SiC detectors have the following advantages compared to conventional I&C detectors:

- Pulse-mode operation is possible over a wide range of neutron fluence rates.
- Gamma ray pulses can be discriminated on the basis of pulse height – gamma-ray compensation and Campbell-mode operation are unnecessary.
- The thermal-neutron, fast-neutron, and gamma-ray responses are all linear.
- The shape of the fast-neutron scattering response continua can be used to deduce information on the incident fast-neutron energy spectrum.
- Losses in SiC charge collection efficiency and the resultant reduced pulse heights caused by accumulated radiation damage do not affect pulse-mode count rates and gamma-ray discrimination
- Operation is possible in high-temperature environments up to at least 800 °C.

In the case of fusion-reactor applications, 4H-SiC detectors have all of the above advantages plus the following:

- Monoenergetic D-T fusion reactions produce threshold reactions peaks in the detector response spectra – the mean pulse heights and widths of these peaks can be used to deduce information on the peak energies and widths of D-T fusion-neutrons.
- Energy and intensity information can be obtained for charged-particle products such as ^2H, ^3H, ^3He and ^4He produced in fusion environments.
- Tritium breeding can be monitored directly using 4H-SiC detectors coupled with ^6Li and ^7Li neutron convertor layers.

These advantages will lead to increasing applications of 4H-SiC neutron and charged-particle detectors in both fission and fusion reactors.

References

[1] R. Babcock and H. Chang, "Silicon Carbide Neutron Detectors for High-Temperature Operation," in Reactor Dosimetry, Vol. 1, Vienna, Austria, International Atomic Energy Agency, 1963, p 613.

[2] F. H. Ruddy, A. R. Dulloo, J. G. Seidel, S. Seshadri and L. B. Rowland, "Development of a Silicon Carbide Radiation Detector," *IEEE Trans. Nucl. Sci.*, vol. 45, pp 536-541, June 1998. http://dx.doi.org/10.1109/23.682444

[3] F. Nava, G. Bertuccio, A. Cavallini and E. Vittone, "Silicon Carbide and Its Use as a Radiation Detector Material," *Meas. Sci. Technol.*, vol. 19, pp 1-25, Aug. 2008. http://iopscience.iop.org/0957-0233/19/10/102001

[4] F. H. Ruddy, "Silicon Carbide Radiation Detectors: Progress, Limitations and Future Directions," *Materials Research Society Online Proceedings* vol. 5761, Nov. 2013. http://dx.doi.org/10.1557/opl.2013.1142

[5] K. C. Mandal, J. W. Kleppinger, and S. K. Chaudhuri, "Advances in High-Resolution Radiation Detection Using 4H-SiC Epitaxial Layer Devices," *Micromachines*, vol. 11, pp 254-281, Feb. 2020. http://dx.doi.org/10.3390/mi11030254

[6] I. Capan, "4H-SiC Schottky Barrier Diodes as Radiation Detectors: A Review" *Electronics* 11, no. 4: 532, Feb. 2022. https://doi.org/10.3390/electronics11040532.

[7] M. DeNapoli, "SiC detectors: A review on the use of silicon carbide as radiation detection material', *Front Phys.*, 10:898833., Oct. 2022. https://doi.org/10.3389/fphy.2022.898833

[8] P. G. Neudeck, D. J. Spry, L. Chen, N. F. Prokop and M. J. Krasowski, "Demonstration of 4H-SiC digital integrated circuits above 800 °C," *IEEE Electron Device Lett.*, vol. 38, no. 8, pp. 1082-1085, 2017. http://dx.doi.org/10.1109/LED.2017.2719280 .

[9] S. Metzger, H. Henschel, O. Kohn, W. Lennartz, "Silicon Carbide Radiation Detector for Harsh Environments," *IEEE Trans. Nucl.Sci.*, vol. 49, pp 1351-1355, June 2002. http://doi.org/10.1109/RADECS.2001.1159258

[10] F.H. Ruddy, L. Ottaviani, A. Lyoussi, C, Destouches, O. Palais and C. Reynard-Carette, "Performance and applications of silicon carbide neutron detectors in harsh nuclear environments", *EPJ Web Conf.* Volume 253 (2021), https://doi.org/10.1051/epjconf/202125311003

[11] F. H. Ruddy, L. Ottaviani, A. Lyoussi, C, Destouches, O. Palais and C. Reynard-Carette, "Silicon carbide neutron detectors for harsh nuclear

environments: a review of the state of the art", *IEEE Trans. Nucl. Sci.* 69 (4) (2022) 792–803. http://dx.doi.org/10.1109/TNS.2022.3144125

[12] S. Seshadri, A.R. Dulloo, F.H. Ruddy, J. G. Seidel, and L. B. Rowland, "Demonstration of a SiC neutron detector for high-radiation environments," *IEEE Trans. Electron Devices*, vol. 46, pp. 567-571, March 1999. http://doi.org/10.1109/16.748878

[13] A. R. Dulloo, F. H. Ruddy, J. G. Seidel, J. M. Adams, J. S. Nico, and D. M. Gilliam, "The Neutron Response of Miniature Silicon Carbide Semiconductor Detectors," *Nucl. Instr Meth. A*, vol. 422, pp 47-48, Feb. 1999. https://doi.org/10.1016/S0168-9002(98)01060-2

[14] A. R. Dulloo, F. H. Ruddy, J. G. Seidel, J. M. Adams, J. S. Nico, and D. M. Gilliam, "The Thermal Neutron Response of Miniature Silicon Carbide Semiconductor Detectors," *Nuclear Instruments and Methods A*, vol. 498, pp 415-423, Feb. 2003. https://doi.org/10.1016/S0168-9002(02)01987-3

[15] A. Lo Giudice, F. Fasolo, E. Durisi, C. Manfredotti, E. Vittone, F. Fizzotti, A. Zanini, and G. Rosi, "Performance of 4H-SiC Schottky Diodes as Neutron Detectors," *Nuclear Instruments and Methods A*, vol. 583, pp 177-180, Dec. 2007. http://dx.doi.org/10.1016/j.nima.2007.08.241

[16] A. R. Dulloo, F. H. Ruddy, J. G. Seidel, C. Davison, T. Flinchbaugh, and T. Daubenspeck, "Simultaneous Measurement of Neutron and Gamma-Ray Radiation Levels from a TRIGA Reactor Core Using Silicon Carbide Semiconductor Detectors," *IEEE Transactions on Nuclear Science*, vol. 46, pp 275-279, June1999. http://dx.doi.org/10.1109/23.775527

[17] A. R. Dulloo, F. H. Ruddy, J. G. Seidel, T. Flinchbaugh, C. Davison, and T. Daubenspeck, "Neutron and Gamma Ray Dosimetry in Spent-Fuel Radiation Environments Using Silicon Carbide Semiconductor Radiation Detectors," in *Reactor Dosimetry: Radiation Metrology and Assessment*, ASTM STP 1398, (John G. Williams, David W. Vehar, Frank H. Ruddy, and David M. Gilliam, [Eds.]), West Conshohoken, Pennsylvania, USA, American Society for Testing and Materials, 2001, pp 683-690. http://dx.doi.org/10.1520/STP13657S

[18] F. H. Ruddy, J. G. Seidel, T. E. Blue, and D. W. Miller, "Reactor Power Monitoring Using Silicon Carbide Fast Neutron Detectors," in PHYSOR-2006 – American Nuclear Society's Topical Meeting on Reactor Physics, LaGrange Park, Illinois, USA, American Nuclear Society, 2006.

[19] F. H. Ruddy, A. R. Dulloo, J. G. Seidel, F. W. Hantz and L. R. Grobmyer, "Nuclear Reactor Power Monitoring Using Silicon Carbide Semiconductor Radiation Detectors", *Nuclear Technology* vol.140, pp 198-208, November 2002. http://dx.doi.org/10.13182/NT02-A3333

[20] G. F. Knoll, Radiation Detection and Measurement, 4th Edition, New York, NY, USA, Wiley, 2019, pp 540-542.

[21] United States Department of Energy Advanced Reactor Demonstration Program, https://www.energy.gov/ne/advanced-reactor-demonstration-program .

[22] A. Muraro,L. Giacomelli, M. Nocente, M. Rebai, D. Rigamonti, F. Belli, P. Calvani, J. Figueiredo, M. Girolami, G. Gorini, G. Grosso, A. Murari, S. Popovichev, D. M. Trucchi, and M. Tardocchi,, "First neutron spectroscopy measurements with a pixelated diamond detector at JET", *Rev. Sci. Instr.* 87, 11D833, September 2016, https://doi.org/10.1063/1.4961557 .

[23] M. Pillon, M. Angelone, A. Krása, A. J. M. Plompen, P. Schillenbeeckx, and M. L. Sergi, "Experimental response functions of a single-crystal diamond detector for 5-20.5 MeV neutrons", *Nucl. Instr Meth. A*, vol. 640, No. 1, June 2011, pp 185-201, https://doi.org/10.1016/j.nima.2011.03.005.

[24] M. H. Kushoro, S. Colombi, G. Croci, A. Dal Molin, E. Gallo, M. Nocente, M. Osipenko, O. Putignano, M. Rebai, D. Rigamonti, M. Tardocchi, and G. Gorini, "SiC detector response to tokamak neutron spectra mock-up validated on experimental results", *Fusion Engineering and Design*, vol. 214, May 2025, 114875. http://dx.doi.org/10.1016/j.fusengdes.2025.114875

[25] F. H. Ruddy, A. R. Dulloo, B. Petrović, and J. G. Seidel, "Fast Neutron Spectrometry Using Silicon Carbide Detectors", *Reactor Dosimetry in the 21st Century*, 2003, pp. 347-355, https://doi.org/10.1142/9789812705563_0043 .

[26] F. H. Ruddy, A. R. Dulloo, J. G. Seidel, M. K. Das, S-H. Ryu, and A. K. Agarwal, "The Fast Neutron Response of 4H Silicon Carbide Semiconductor Radiation Detectors", *IEEE Trans. Nucl. Sci.*, vol. 53, No. 3, pp. 1666-1670, June 2006, http://dx.doi.org/10.1109/TNS.2006.875151

[27] F. H. Ruddy, J. G. Seidel, and F. Franceschini, "Measurements of the Recoil-Ion Response of Silicon Carbide Detectors to Fast Neutrons," in Reactor Dosimetry State of the Art 2008, (W. Voorbraak, L. Debarberis, P. D'hontd, and J. Wagemans, Eds.), London, England, World Scientific, 2009, pp 77-84. http://dx.doi.org/10.1142/9789814271110_0014

[28] O. Obraztsova, L. Ottaviani, A. Klix, T. Döring, O. Palais, A. Lyoussi, "Comparing the response of a SiC and a sCVD Diamond detectors to 14 MeV neutron radiation," *IEEE Transactions on Nuclear Science*, vol. 65, pp. 2380-2384, Sept. 2018. https://doi.org/10.1109/TNS.2018.2848469

[29] R. Ferone, F. Issa, D. Szalkai, A. Klix, L. Ottaviani, S. Biondo, V. Vervisch, L. Vermeeren, R. Saenger, A. Lyoussi, "SiC-based neutron detector in quasi-realistic working conditions: efficiency and stability at room temperature and high temperatures under fast neutron irradiations," in *2015 4th International Conference on Advancements in Nuclear Instrumentation Measurement Methods and their Applications (ANIMMA)*, Institute of Electrical and Electronics Engineers, pp. 1–4, 2015. http://dx.doi.org/10.1109/ANIMMA.2015.7465545

[30] D. Szalkai, R. Ferone, F. Issa, A. Klix, M. Lazar, A. Lyoussi, L. Ottaviani, P. Tutto, V. Vervisch, "Fast Neutron Detection with 4H-SiC Based Diode Detector up to 500 °C Ambient Temperature," *IEEE Transactions on Nuclear Science*, vol. 63, pp 1491-1498, June 2016. http://dx.doi.org/10.1109/TNS.2016.2522921

[31] F. H. Ruddy, J. G. Seidel, H. Chen, A. R. Dulloo, and S-H. Ryu, "High-Resolution Alpha-Particle Spectrometry Using 4H Silicon Carbide Detectors," *IEEE Trans. Nucl. Sci.*, vol. 53, pp. 1713-1718, 20063. http://dx.doi.org/10.1109/TNS.2006.875155

[32] F.H. Ruddy, J. G. Seidel and A. R. Dulloo, "Fast Neutron Dosimetry and Spectrometry Using Silicon Carbide Semiconductor Detectors", *Journal of ASTM International* Vol 3, Online Issue 3, March 2006. http://dx.doi.org/10.1520/STP45469S

[33] A. R. Dulloo, F. H. Ruddy, J. G. Seidel, and B. Petrović, "Monitoring of D-T Accelerator Neutron Output in a PGNAA System Using Silicon Carbide Detectors," *Applications of Accelerators in Research and Industry – Sixteenth International Conference*, American Institute of Physics Conference Proceedings, vol. 576, p 499, September 2001. http://dx.doi.org/10.1063/1.1395358

[34] L. Bertalot, V. Krasilnikov, L. Core, A. Saxena, N. Yukhnov, R. Barnsley, and M. Walsh., "Present Status of ITER Neutron Diagnostics Development", *J. Fusion Energy*, Vol. 38, pp. 283–290, July 2019. http://dx.doi.org/10.1007/s10894-019-00220-w

[35] T. Nishitani, S. Kasai, L. C. Johnson, K. Ebisawa, C. Walker, and T. Ando, "Neutron Monitor Using Microfission Chambers for the International Thermonuclear Experimental Reactor", *Review of Scientific Instruments*, Vol. 70, No. 1, January 1999. http://dx.doi.org/10.1063/1.1149297

[36] A. O. Kovalev, A. R. Polevoi, E. I. Polunovskiy, L. Bertalot, Y. A. Kashchuk, D. V. Portnov, A. Loarte, M. Loughlin, and S. D. Pinches, "Simulations of Fusion Power Measurements by Monitors of Neutron Flux in Evolving ITER Plasma", *Journal of Fusion Energy*, Vol. 39, No. 1, , April 2020, pp.40-52, http://dx.doi.org/10.1007/s10894-020-00232-x

[37] A. O. Kovalev, R. N. Rodionov, V. A. Vorobiev, D. V. Portnov, T. M. Kormilitsyn, Y. G. Vysokih, S. Y. Obudovsky, and Y. A. Kashchuk, "Measurements of ITER Fusion Power by Neutron Flux Monitors". *Physics of Atomic Nuclei*, Vol. 86, Suppl, 2, February 2024, pp.S187-S197 http://dx.doi.org/10.1134/S1063778823140077.

[38] Z. Wen, G. Yuan, Y. Li, Q. Yang, J. Li, L. Zhao, L. Wei and J. Zhang, "Neutronic Analysis and Measurement Performance Assessment of ITER Neutron Flux Monitor System", *Fusion Engineering and Design*, Vol. 187, 113364, January 2023. http://dx.doi.org/10.1016/j.fusengdes.2022.113364

[39] F. H. Ruddy, S. K. Chaudhuri, and K. C. Mandal. "A review of the effects of fast-neutron irradiation on the performance of 4H-SiC Schottky barrier detectors." *IEEE Transactions on Nuclear Science*, Vol.71, No. 5, pp. 1056-1063, May 2024. http://dx.doi.org/10.1109/TNS.2024.3355293

[40] Q. Potiron, "Advanced Solid State Neutron Detectors for Fast Neutron Measurement and Monitoring in Fusion and Fission Environments", Ph.D. Thesis, Aix Marseilles University, December 10, 2024, p 248.

[41] Q. Potiron, C. Destouches, M. Houry, O. Llido, A. Lyoussi, C. Reynard-Carette, L. Dubus, P. Malard, P. Legou and B. Cheymol, "Evaluating Solid-State Neutron Detectors for Measuring 14 MeV Neutrons at High Temperatures", *Fusion Engineering and Design* (to be published).

Fast Simulation of Gamma-Ray Logs for Uranium Exploration in Roll-Front Deposits

Arthur Pellet-Rostaing[1,*], Bertrand Pérot[1], Thomas Marchais[1], Nadia Pérot[2], Hervé Toubon[3], Youcef Bensedik[3], Sebastien Hocquet[3], Christophe Pouet[4], and Mitra Fouladirad[5]
[1]CEA, DES, IRESNE, DTN, SMTA, Nuclear Measurement Laboratory, France
[2]CEA, DES, IRESNE, DER, SESI, Laboratory for the Study and Modeling of Systems, France
[3]Orano Mining, France
[4]Aix-Marseille Université, CNRS, Centrale Méditerranée, I2M, France
[5]Aix-Marseille Université, CNRS, Centrale Méditerranée, M2P2, France
(*) arthur.pellet-rostaing@cea.fr

***Abstract*—Orano Mining is evaluating the potential of the CeBr₃ spectrometric gamma-ray logging probe developed by Advanced Logic Technology (ALT) for estimating uranium concentration in roll-front deposits, where decay chain disequilibrium disrupts its relationship with total gamma count rate. The Nuclear Measurement Laboratory of CEA IRESNE, in Cadarache, France, is working on predictive algorithms capable of fully exploiting the shape of the recorded gamma spectra. However, the current number of logged wells is insufficient to properly train such algorithms. This calls for the creation of a diverse database of simulated gamma-ray logs, using the Monte Carlo N-Particle (MCNP) transport code. Conventional analog simulations based on MCNP pulse-height (F8) tally provide accurate estimates but are too computationally expensive for this task. Achieving low statistical uncertainty in every energy bin necessitates simulating a large number of particle histories. Moreover, modeling borehole measurements involves simulating numerous probe locations. Designing a more efficient yet sufficiently accurate simulation procedure is therefore crucial. We propose a 2-step method, where the flux reaching the probe surface and the detector response to this flux are simulated separately. The bulk of runtime reduction is achieved in the first step, through the use of MCNP point detector (F5) tally, which acts as a variance reduction technique. However, treating the detector as a point may lead to significant bias. We conducted an experimental validation on Orano CIME calibration blocks (Bessines, France) and observed good agreement between measured and simulated spectra. These findings support the 2-step approach as a viable option for building our database.**

***Keywords* —Uranium mining, gamma logging, CeBr₃ probe, fast simulation.**

I. Introduction

ORANO Mining is partnering with the Nuclear Measurement Laboratory of CEA IRESNE to develop new techniques for reconstructing uranium grade vertical profiles in roll-front deposits from gamma-ray logs, which correspond to series of gamma measurements performed at near-adjacent depths. Since uranium ore bodies can be as thin as a few decimeters (see Fig. 1), Orano Mining acquires gamma-ray spectra at 10-cm intervals to enable fine-grained spatial estimations.

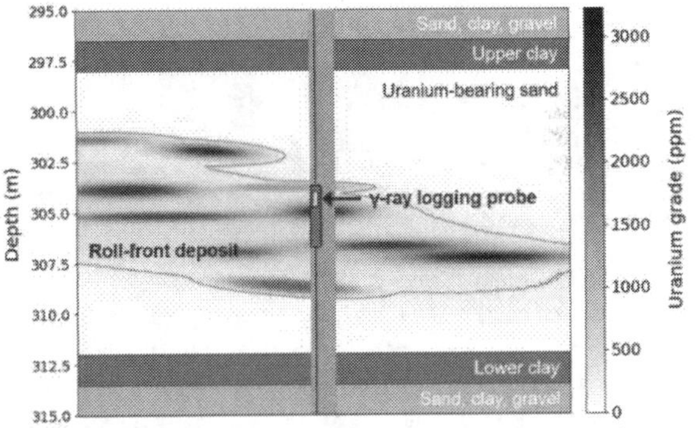

Fig. 1. Simplified 2D schematic of gamma borehole logging in a uranium roll-front deposit.

Until recently, gamma logging has been confined to profiling total count rates. At secular equilibrium, the total gamma count rate is directly proportional to uranium content. In this case, computing a detector-specific calibration coefficient is sufficient to provide an accurate estimate. Calibration coefficients accounting for borehole geometry, tubing effects, and uranium self-absorption have been determined experimentally for NaI detectors using calibration blocks [1]. However, the linear relationship between total count rate and uranium grade is disrupted in roll-front deposits due to the presence of radioactive disequilibrium in the ^{238}U decay chain (see Fig. 2). The imbalance between the activities of ^{238}U and those of its daughter nuclides stems first from the differential leaching of uranium and radium. Although uranium solubility in groundwater is influenced by the redox potential of the medium, the same does not apply to radium, one of its decay products. This results in the formation of radium-depleted and radium-enriched regions. Secular equilibrium may also be disrupted by radon evaporation during borehole drilling, radon being a volatile noble gas that tends to escape from the host rock through pores and fractures. Emissions from decay products of ^{222}Rn, ^{214}Pb, and ^{214}Bi are responsible for the majority of the recorded gamma signal. Relying solely on total count rate might therefore lead to inaccurate estimates.

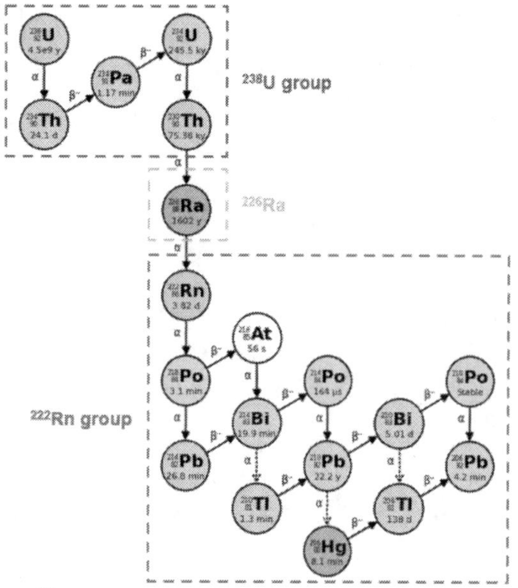

Fig. 2. ^{238}U decay chain, with radionuclide groups involved in radioactive disequilibrium identified by dotted rectangles. The ^{222}Rn group accounts for 95% of the total gamma count rate at secular equilibrium.

This has prompted a shift toward exploiting spectral information. The characterization of uranium content based on spectrometric measurements with high-purity germanium detectors (HPGe) has been carried out on ore samples [2, 3]. These high-resolution detectors are unfortunately ill-suited to *in situ* measurements because of their cost, their volume, and the need to cool the germanium crystal. Orano Mining has therefore turned to scintillator-based spectrometric probes for borehole logging. For example, the performance of a $LaBr_3(Ce)$ gamma probe in quantifying uranium content in ore samples despite U/Rn disequilibrium has been evaluated [4]. Orano Mining is now appraising the capabilities of the $CeBr_3$ probe developed by the Advanced Logic Technology (ALT) company [5, 6].

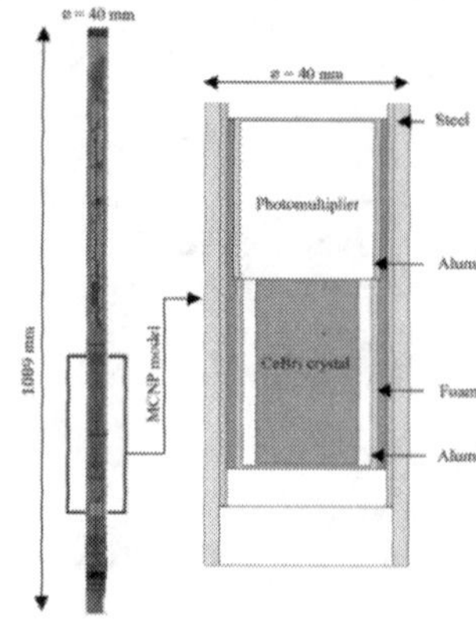

Fig. 3. ALT QL40-SGR-2G logging probe equipped with a $CeBr_3$ crystal, and its MCNP model [6].

Measurement conditions inside boreholes are particularly demanding. The low energy resolution of the $CeBr_3$ detector and the short acquisition times imposed by the high probe speed inside the borehole, the latter leading to increased measurement uncertainty, make usual estimation techniques, such as net peak area analysis [7], fail. This motivates the adoption of automated predictive algorithms. However, their performance hinges on the quality and the quantity of data they are trained on. Supervised methods need to be trained on a labeled dataset that comprises pairs of observations and corresponding target values. Acquiring accurate underground uranium profiles requires numerous core extractions, a process which is both expensive and time-consuming. To date, the number of logged boreholes for which laboratory-analyzed core samples are available remains scarce, which hinders the training of supervised algorithms. A suggested workaround involves building a synthetic dataset of gamma-ray logs using MCNP6.3 (Monte Carlo N-Particle) simulations [8].

So far, methods for estimating uranium grade from gamma-ray logs have been designed without taking geological stratification into account [9, 10]. This assumption needs to be revised when the goal is to reconstruct precise uranium depth profiles. This implies modeling fine-scale vertical fluctuations of uranium content (see Fig. 4a). Variations in other geological (bulk density) and drilling parameters (borehole diameter) also need to be introduced into the simulations to enhance realism. The number of simulations needed to cover the full range of geological configurations then becomes intractable.

II. 2-STEP (F5+F8) SIMULATION METHOD

Due to the low probability of gamma particles reaching the detector, the conventional analog simulation method relying on MCNP pulse-height (F8) tally requires sampling a significant number of particles (10^{10} per radionuclide group) to sufficiently reduce statistical uncertainty in each of the 292 energy channels of the gamma spectrum. As a result, the average runtime per simulation amounts to 580 h.

We propose an alternative approach which consists in dividing the simulation into two distinct parts. The first step involves computing the gamma flux incident on the probe surface with the efficient, semi-deterministic MCNP point detector tally (F5), as described below. In the second step, the simulated flux is projected onto the probe to estimate the energy deposition spectrum in the $CeBr_3$ crystal using the more time-consuming analog F8 tally (see Fig. 4c). This step requires only a small number of source particles.

The flux on the probe surface can be approximated very efficiently thanks to MCNP point detector (F5) tally (see Fig. 4b). The F5 tally is referred to as a next-event estimator, meaning that it evaluates during each particle history a contribution to the detector at every source emission or collision site. Each of these contributions is weighted by the probability of scattering towards the point detector along a direct path, accounting for geometric and attenuation effects. The number of contributions per particle is thus drastically increased. Acting as a variance reduction technique, the F5 tally achieves high statistical precision with far fewer particle histories. On average, F5 simulations complete in only 15 h. Moreover, vertically stacking F5 tallies (see Fig. 4b) enables flux estimation at multiple probe locations simultaneously, further decreasing simulation time.

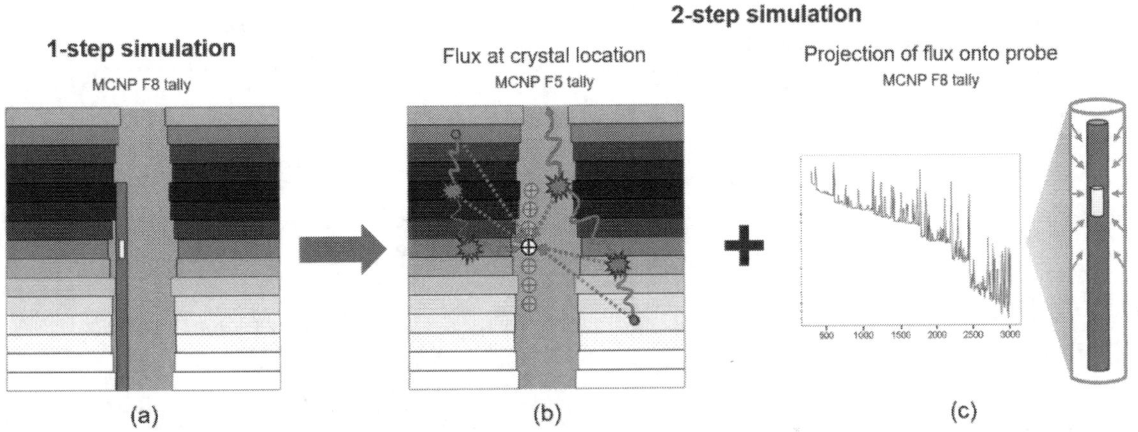

Fig. 4. Simulation of a gamma-ray spectrum acquired within a vertically stratified formation. (a) Conventional 1-step (F8) simulation including the probe geometry. (b) Estimation of the probe flux by MCNP point detector (F5) tally. Straight dotted arrows represent direct contributions computed at source and collision sites. (c) Projection of the simulated flux onto the probe to compute the energy deposition in the crystal using MCNP F8 tally.

Given the pointwise nature of the F5 tally, runtime reduction comes at the price of physical simplification. The tally only captures the energy distribution of the flux at a single point. Information about particle positions and angles of incidence is inevitably lost. This may introduce a bias when projecting the flux onto the probe. Automated algorithms designed to predict uranium grade profiles are likely to exploit not only total count rates, but also subtle shape-related features. If these characteristics were to be altered, prediction accuracy could be undermined. We therefore evaluate the extent of a potential bias by comparing 2-step simulated spectra with measurements taken in calibration blocks designed to replicate actual borehole conditions as closely as possible.

III. Experimental validation on Bessines standard blocks

The Orano CIME (Innovation Center for Extractive Metallurgy) calibration station in Bessines, France, includes seven cubic concrete blocks with 70-cm edges (see Fig. 5). Each block is built from a carefully controlled homogeneous mixture of fine sand, cement, and crushed uranium ores. A central air-filled hole of 8 cm diameter allows the insertion of the radiometric probes. Their uranium content, as well as their elemental composition, has been estimated through chemical analysis. Uranium concentrations range from 0 to 9700 ppm_U (milligrams of uranium per kilogram of ore). A gamma-ray spectrometry measurement campaign using an HPGe detector has also been carried out [2]. The estimated concentrations, reported in Table I, agree with those obtained from chemical analysis. Orano Mining has determined thorium and potassium mass fractions of 20 ppm and 3.7%, respectively. However, the absence of the 2 615 keV photopeak, characteristic of ^{208}Tl, a daughter nuclide of ^{232}Th, in the recorded gamma-ray spectra (see Fig. 6), indicates that no thorium is present in the concrete matrix. Emissions from the ^{232}Th decay chain are therefore not modeled in the simulations. All blocks are at secular equilibrium. The B1 block contains no uranium and is therefore not considered in the remainder of the study.

Table I. Uranium mass concentrations (in ppm_U) of Orano CIME calibration blocks estimated by chemical analysis and HPGe gamma-ray spectroscopy

Block ID	Chemical analysis (in ppm_U)	HPGe gamma-ray spectroscopy (in ppm_U)	Bulk density (g/cm^3)
B2	500	459 (± 3.7%)	1.92
B3	1000	958 (± 1.9%)	1.89
B4	1900	1960 (± 2.3%)	1.91
B5	2900	2800 (± 3.7%)	1.92
B6	4800	4550 (± 3.3%)	1.93
B7	9700	9530 (± 2.0%)	1.92

In prior work, MCNP 1-step (F8) simulations of $CeBr_3$ probe measurements inside each standard block were carried out using uranium concentrations determined by chemical analysis [6]. In the present study, the simulated uranium concentrations correspond to those derived from HPGe measurements. Fig. 5 shows the block geometry as implemented in MCNP. Acquisition times for the experimental spectra range from 612 s to 1170 s.

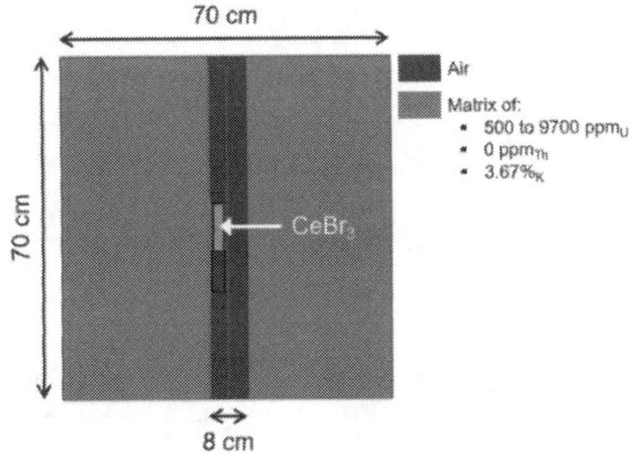

Fig. 5. MCNP simulation of ALT $CeBr_3$ logging probe positioned inside a Bessines standard block.

Table II reports the relative differences between the total count rate C_{TOT} measured above the electronic cutoff, set at an equivalent energy of 80 keV, and the 1-step (F8) or 2-step (F5+F8) simulations, for the six calibration blocks. We observe satisfactory experimental agreement for both approaches. The 2-step method tends to yield slightly higher count rates than the 1-step method. Relative to experimental values, the 2-step method more frequently overestimates the C_{TOT}.

TABLE II. RELATIVE DIFFERENCES IN TOTAL COUNT RATES ABOVE THE 80-KEV THRESHOLD BETWEEN MEASUREMENTS AND 1-STEP (F8) AND 2-STEP (F5+F8) SIMULATIONS

Block ID	$\frac{\text{1-step} - \text{EXP}}{\text{EXP}}$	$\frac{\text{2-step} - \text{EXP}}{\text{EXP}}$
B2	-2.6%	-0.8%
B3	0.9%	2.7%
B4	1.7%	3.6%
B5	0.7%	2.6%
B6	-1.3%	0.5%
B7	5.3%	7.2%

Next, we evaluated the accuracy of both methods in reconstructing the spectral shape. The experimental gamma spectrum recorded in block B3 (958 ppm_U), along with the simulated spectra, is displayed in Fig. 6. For this uranium concentration, which is typical of roll-front deposits, the overall shape of the experimental spectrum is well reproduced by both simulation methods. However, we observe a slight overestimation of the low-energy bump in both simulations, which is a bit larger with the 2-step approach.

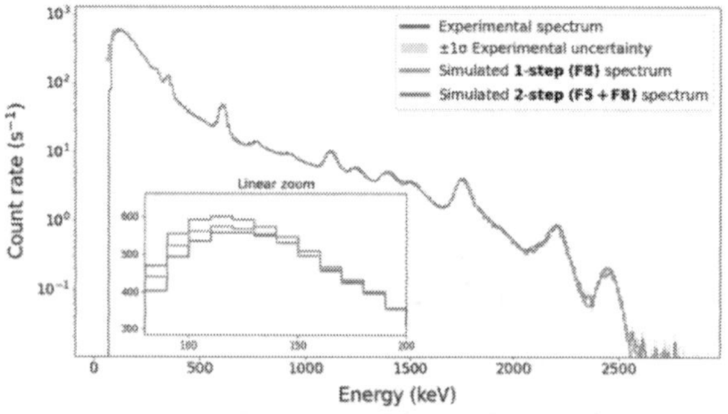

Fig. 6. Gamma-ray spectra measured with the $CeBr_3$ probe inserted in the B3 block (958 ppm_U), and corresponding 1-step and 2-step simulations

We then performed a comparison between the count rates in the two energy windows (120-170 and 960-1040 keV) specified by the K_{APC} (Peak-to-Compton) indicator, developed in earlier work for uranium grade estimation [4], for all calibration blocks. The count rate in the low-energy region is overestimated by MCNP 1-step simulation for all blocks, except B2 (see Fig. 7). Conversely, mid-energy count rates are all underestimated. This distortion of the spectrum becomes more pronounced as uranium grade increases. The same pattern is observed with the 2-step approach, and is slightly amplified.

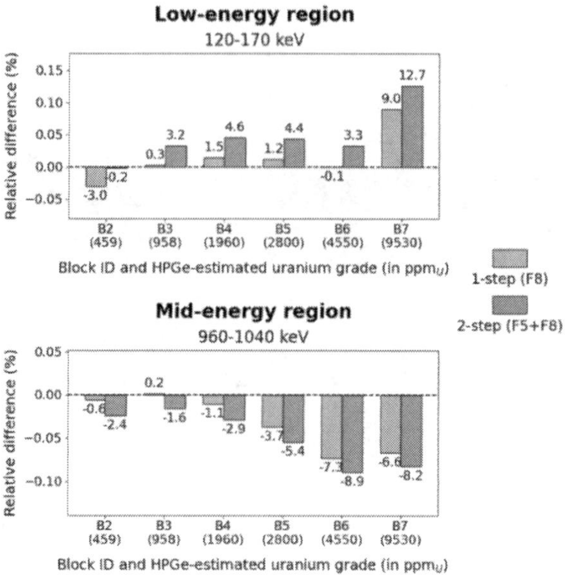

Fig. 7. Relative differences between experimental count rates and those simulated with 1-step and 2-step methods in low- and mid-energy regions (120-170 and 960-1040 keV, respectively), computed for the six standard blocks of Orano CIME calibration station (Bessines, France).

IV. CONCLUSIONS

We designed a 2-step MCNP simulation approach for fast gamma-ray log generation. By employing MCNP point-detector (F5) tally as a variance reduction technique, we achieved a substantial reduction in computation time. Experimental validation using Orano CIME calibration blocks at Bessines (France) shows satisfactory agreement between simulations and measurements, both in terms of total count rate and overall spectral shape. The physical simplification involved in the nonanalog flux estimation phase leads, however, to greater spectral distortion than the conventional 1-step method. Given the small magnitude of this bias, the 2-step approach proves to be a well-suited candidate for building a comprehensive database of gamma-ray logs to train predictive algorithms.

The calibration blocks, however, reproduce borehole geometry only imperfectly. In particular, logging operations in roll-front deposits take place within water-filled boreholes. This difference in the surrounding medium may affect the accuracy of the 2-step approach. Moreover, the potential impact of geological vertical heterogeneities on the experiment-simulation bias remains to be assessed, as the calibration blocks are homogeneous.

Another promising strategy for efficiently generating gamma-ray logs is to bypass MCNP altogether, instead of trying to accelerate the simulations. This could be achieved by developing a surrogate model, *i.e.*, an emulator based on a machine learning algorithm trained to predict gamma-ray spectra directly from geological and drilling input parameters. The surrogate model is trained on an experimental design built from MCNP 1-step simulations spanning various geological configurations. This design must be carefully constructed to ensure that the trained emulator can replicate MCNP outputs as faithfully as possible. We are currently exploring this approach and plan to validate it against both 1-step reference simulations and actual borehole logging data to assess its reliability.

REFERENCES

[1] C. Carasco, B. Pérot, J.-L. Ma, H. Toubon, and A. Dubille-Auchère, "Improving gross count gamma-ray logging in uranium mining with the NGRS probe," *IEEE Trans. Nucl. Sci.*, vol. 65, no. 3, pp. 919–923, Mar. 2018, doi: 10.1109/TNS.2018.2800909.

[2] T. Marchais, B. Pérot, C. Carasco, P.-G. Allinei, P. Chaussonnet, J.-L. Ma, *et al.*, "Detailed MCNP simulations of gamma-ray spectroscopy measurements with calibration blocks for uranium mining applications," *IEEE Trans. Nucl. Sci.*, vol. 65, no. 9, pp. 2533–2538, Sep. 2018, doi: 10.1109/TNS.2018.2797312.

[3] T. Marchais, B. Pérot, C. Carasco, J.-L. Ma, P.-G. Allinei, H. Toubon, *et al.*, "The use of self-induced X-ray fluorescence in gamma-ray spectroscopy of uranium ore samples," *EPJ Web Conf.*, vol. 225, Jan. 2020, Art. no. 05003, doi: 10.1051/epjconf/202022505003.

[4] B. Pérot, T. Marchais, P.-G. Allinei, H. Toubon, Y. Bensedik, R. Roupillou, *et al.*, "Development of gamma spectroscopic tools for uranium ore samples and borehole exploration," in *2022 IEEE Nucl. Sci. Symp. Med. Imag. Conf. (NSS/MIC)*, Milan, Italy, Nov. 2022, pp. 1–5, doi: 10.1109/NSS/MIC44845.2022.10398898.

[5] "Home," Advanced Logic Technology (ALT). [Online]. Available: https://www.alt.lu/. Accessed on: Jul. 28, 2025.

[6] T. Marchais, B. Pérot, P.-G. Allinei, H. Toubon, Y. Bensedik, R. Mieszkalski, *et al.*, "Experimental validation of a $CeBr_3$ gamma-ray logging probe MCNP model," *EPJ Web Conf.*, vol. 288, Nov. 2023, Art. no. 05003, doi: 10.1051/epjconf/202328805003.

[7] T. Marchais, B. Pérot, C. Carasco, J.-L. Ma, P.-G. Allinei, H. Toubon, *et al.*, "Characterization of uranium ore samples by HPGe gamma-ray spectroscopy," *IEEE Trans. Nucl. Sci.*, vol. 67, no. 4, pp. 654–661, Apr. 2020, doi: 10.1109/TNS.2020.2966824.

[8] J. A. Kulesza, T. R. Adams, J. C. Armstrong, S. R. Bolding, F. B. Brown, J. S. Bull, *et al.*, "MCNP Code Version 6.3.0 Theory & User Material". Los Alamos National Laboratory, Los Alamos, NM, USA, Tech. Rep. LA-UR-22-30006, Rev. 1, Sep. 2022. [Online]. Available: https://www.osti.gov/biblio/1889957. Accessed on: Jul. 28, 2025.

[9] P. G. Killeen, "Gamma ray spectrometric methods in uranium exploration–application and interpretation," in *Geophysics and Geochemistry in the Search for Metallic Ores*, P. J. Hood, Ed., Geol. Surv. Can., Econ. Geol. Rep. 31, pp. 163–229, 1979. doi: 10.4095/106049.

[10] Procédé d'évaluation de la teneur en uranium par spectrométrie gamma dans un puits de forage et dispositif associé, by T. Marchais, B. Pérot, C. Carasco, H. Toubon, and Y. Bensedik. (2021, Jun. 18), Patent FR3104739A1, [Online]. Available: https://patents.google.com/patent/FR3104739A1. Accessed on: Jul. 28, 2025.

Utilization Of Periodic Nanostructures To Improve Detection Efficiency And Time/Energy Resolution Of Inorganic Scintillators

Stuti Surani[1,*], Roman Samulyak[1], Federico Scurti[1], Douglas E. Wolfe[1,2], and Marek Flaska[1]
[1]Ken and Mary Alice Lindquist Department of Nuclear Engineering
[2]Department of Materials Science and Engineering
The Pennsylvania State University
(*) srs6376@psu.edu

***Abstract*—Inorganic scintillators are commonly used in various radiation detection applications due to their excellent energy resolution, reliable performance, relatively low cost, and high detection efficiency. However, many inorganic scintillators have high refractive indices and experience significant light losses at the collection surface caused by total internal reflection (TIR). This project employs optimized periodic nanostructures, known as photonic crystals (PHCs), to recover some of the light losses. A PHC layer creates an improved optical pathway between the scintillator and the photosensor for the trapped photons through constructive light interference. Enhancing the light extraction from an inorganic scintillator improves its energy resolution, time resolution, and the overall detection efficiency. In this work, the effects of an optimized PHC coupling with a LYSO scintillator are experimentally demonstrated. The PHC geometry is optimized through tailored simulations, manufactured with electron beam lithography, and characterized with radiation measurements to quantify improvements in light output and energy resolution. A 10 x 10 x 3 mm^3 LYSO sample shows an improvement of 28% in light output and 13% in energy resolution with an optimized 2D block structure Si_3N_4 PHC structure. Future work will focus on expanding the experimental framework to other inorganic scintillators such as NaI and $LaBr_3$, which are expected to have even greater light output improvements owing to their higher refractive index and total internal reflection.**

***Keywords* —Photonic crystals, nanostructures, light output, inorganic scintillators.**

I. Introduction

Inorganic scintillators are used in various radiation detection applications, such as nuclear security, medical imaging, fundamental physics, and industrial processes [1], [2]. Inorganic scintillators are preferred for their high light yield, versatility, and lower cost compared to semiconductors; however, they lack the high energy resolution of semiconductors.

While most inorganic scintillators have a relatively high light yield, their resulting light output is reduced due to various light loss mechanisms, such as absorption within the crystal, crystal defects, refraction through the sides of the crystal, and, most importantly, total internal reflection (TIR) at the scintillator-photosensor interface. Reducing the light loss and improving the light output from state-of-the-art scintillators would improve their energy resolution, time resolution, and detection efficiency, thereby broadening the application space for inorganic scintillators.

Photonic crystals (PHCs) are periodic nanometer-scale structures that can, when optimized and properly manufactured, improve the light output of inorganic scintillators by substantially reducing TIR light losses. These nanostructures, composed of a high refractive index material, function as periodically alternating dielectric media deposited directly on the scintillator's surface, as shown in Fig. 1. PHCs provide an optimized optical medium to transmit photons of specific wavelengths through constructive interference and diffraction [2], [3], [5]. These nanostructures can be optimized in terms of their material composition and size to enhance constructive and suppress destructive interference at the light collection surface of an inorganic scintillator, thereby improving the overall light output.

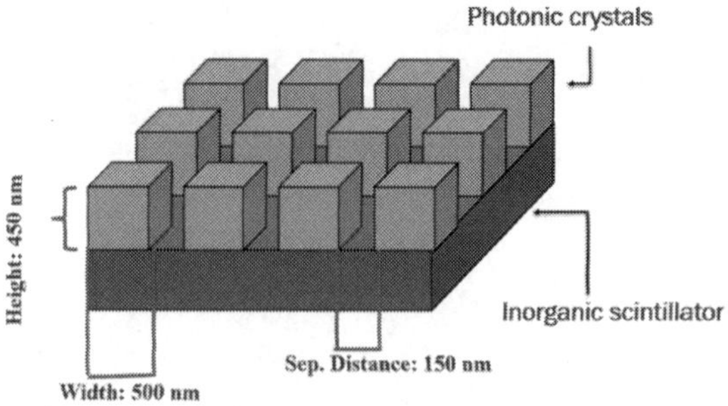

Fig. 1. Schematic for a simple 2D PHC structure deposited on an inorganic scintillator.

For this work, the LYSO – Si_3N_4 PHC combination was explored to assess improvements in light output. The LYSO scintillator was selected for testing due to its non-hygroscopic nature, lower cost compared to other inorganic scintillators, and its prevalence in the literature for PHC applications [6], [7]. Si_3N_4 was chosen as the PHC material due to its ease of nanostructure manufacturing and its refractive index being comparable to that of LYSO.

This paper presents the experimental characterization of PHC-deposited scintillators and demonstrates improvements in light output and energy resolution for LYSO samples.

II. Methodology

This research was broadly divided into three interlinked steps: PHC optimization simulations, PHC manufacturing, and radiation measurements, as shown in Fig. 2.

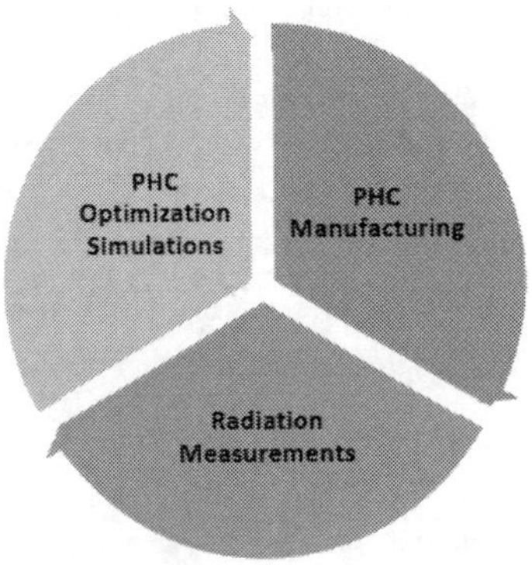

Fig. 2. Schematic of project flowchart.

The PHC optimization simulations combined two codes: a Monte Carlo code, Geant4 [8], for modeling the scintillation effect and light transport within the scintillator, and a deterministic code, OptiFDTD [9], for optimizing the PHC geometry and modeling the light transport through the PHCs. The PHC optimization geometry is shown in Fig. 3, for a given scintillator (10 x 10 x 3 mm^3 LYSO), with the simulated 2D PHC block structure geometry of the given PHC material (Si_3N_4).

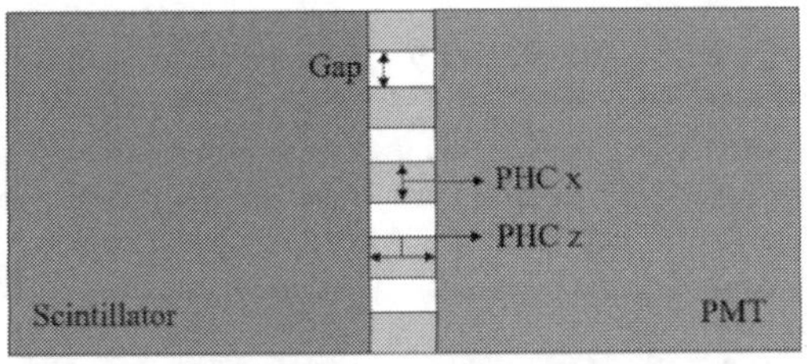

Fig. 3. Schematic of simulated PHC geometry for optimization.

In Fig. 3, the variables of interest were the thickness of the PHC (PHC z), the side length of the PHC (PHC x), and the spacing between the PHCs (Gap). These variables were individually iterated to optimize for the maximum light transmission to an ideal photosensor (PMT). Through this simulation process, the optimized PHC geometry was selected for manufacturing.

The optimized PHC geometry was then manufactured with electron beam lithography and reactive ion etching for a 10 x 10 x 3 mm^3 LYSO scintillator. The PHC deposition was performed for 10 LYSO samples, with a maximum PHC coverage of 90% and some imperfections. A PHC deposited LYSO sample is shown in Fig. 4, with both a visual inspection and an SEM image showing the PHC structure.

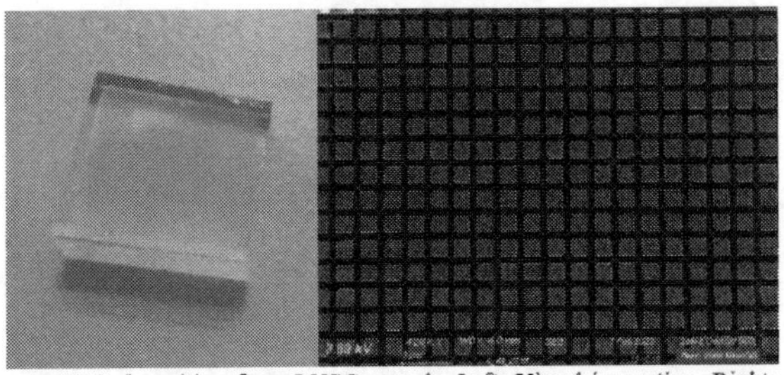

Fig. 4. PHC deposition for a LYSO sample. Left: Visual inspection. Right: SEM image.

The final step involved radiation measurements to assess the improvement in light output and energy resolution from PHC deposition. 10 LYSO samples were characterized without PHCs (called "Bare") and with PHCs (called "PHC"). The radiation measurements were performed with a Cs-137 gamma source with more than 100000 counts in the photopeak. The measurements were taken for all 10 samples without Teflon coating and for 7 samples with Teflon coating. All measurements were performed with a dry coupling between the scintillator and photosensor (Hamamatsu H10580 PMT), to ensure extra consistency during the measurements.

The radiation measurement setup is shown in Fig. 5. 3D-printed holders were constructed to ensure consistent geometry across the various measurements. The PMT was connected to a high voltage source set to -1200 V and a DT5730 digitizer for pulse processing. The resulting digital waveforms were separately analyzed to obtain the Cs-137 energy spectra.

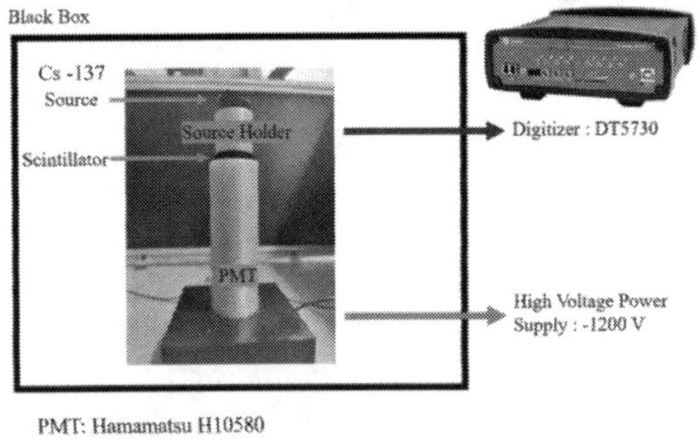

Fig. 5. Radiation measurement setup.

The measured spectra were used to calculate the change in energy resolution and light output for the Bare and PHC samples. Efforts were undertaken to understand and account for the uncertainty in the radiation measurements due to electronic noise and human error by repeating the measurement for five Bare LYSO samples, each measured 10 times, which resulted in a maximum error estimate of 10 channels for the photopeak and 0.3% for the energy resolution, as shown in Table I.

TABLE I
STATISTICAL DATA FOR TEN MEASUREMENTS PERFORMED FOR FIVE BARE SAMPLES TO MEASURE STANDARD DEVIATION ERROR.

	Photopeak Location		**Energy Resolution%**	
Sample	Average	Standard Deviation	Average	Standard Deviation
1	671.2	6.4	12.2	0.1
2	674.8	8.6	14.0	0.2
3	678.2	7.4	12.2	0.2
4	677.6	9.2	12.3	0.2
5	683.5	10.4	12.7	0.3

III. RESULTS

The PHC simulation and radiation measurement results for a LYSO scintillator coupled with Si_3N_4 PHC structures are presented in this section. The simulations provide an optimized geometry with ideal improvement in transmission, while the experimental results show realistic improvements for samples manufactured in the lab.

A. PHC Optimization Simulation Results

The PHC simulations varied the 2D block structure Si_3N_4 PHC geometry to maximize light output from a LYSO scintillator. The various PHC geometry iterations are shown in Fig. 6, which presents the percentage transmission for each iteration, corresponding to a specific Gap, PHC x, and PHC z value. The bare LYSO sample was calculated to have a transmission of 38.3%. The iterations occurred in nested loops, with the outermost loop controlling the Gap variable, which changed every 100 iterations. From the peaks in the transmission values from iteration #100 to iteration #500, it can be inferred that the optimal Gap value lies between 0.2 and 0.5 um.

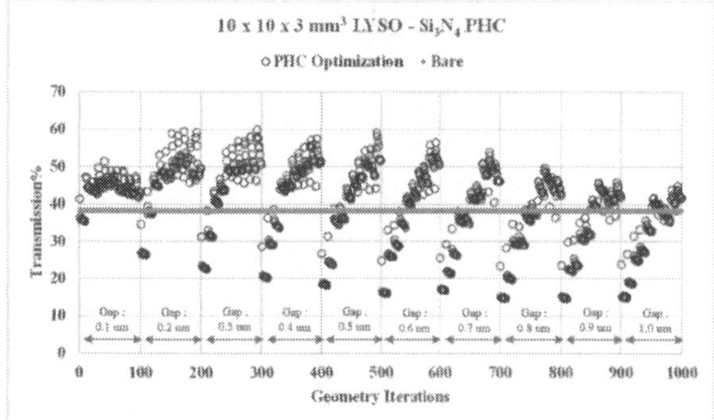

Fig. 6. Geometry iterations for optimization of PHC structure in a LYSO – Si_3N_4 PHC combination.

By keeping a constant PHC x (1 um) and PHC z (0.3 um) value, it was observed that the transmission was not significantly impacted by changes in Gap from 0.2 to 0.5 um, as shown in Table II.

TABLE II
PHC OPTIMIZATION RESULTS FOR THE LYSO – Si_3N_4 PHC COMBINATION.

Gap (um)	PHC x (um)	PHC z (um)	Transmission%	Improvement%
0.2	1	0.3	59.1	54.4
0.3	1	0.3	59.9	56.7
0.4	1	0.3	57.4	50.0
0.5	1	0.3	59.0	54.2

These simulations indicated that using Si_3N_4 PHCs could enhance light transmission from the LYSO scintillator by more than 50%. The refractive index of LYSO is close to that of Si_3N_4 over the wavelengths of interest, resulting in minimal light losses for this coupling. These simulations form the foundation for the PHC manufacturing efforts with the LYSO scintillator.

While the simulations can determine the optimized PHC geometry, they only model perfect PHC structures and coupling, which is not representative of real-world applications. An average light output improvement for the selected PHC geometry, with a Gap of 0.4 um to 0.45 um, PHC x of 0.9 um to 1 um, and PHC z of 0.3 um to 0.4 um, was simulated to be 43% for full coverage.

Since the maximum expected PHC coverage was 90%, the expected ideal simulation improvement was 39% for the given PHC geometry. The simulation toolkit with Geant4 and OptiFDTD can fully simulate an ideal PHC structure for any material and geometry. Future work will incorporate manufacturing defects in the simulations to make them more realistic.

B. Experimental Radiation Measurement Results

Ten LYSO samples were deposited with an optimized PHC geometry. However, many samples had manufacturing imperfections, which led to poorer quality, coverage, and suboptimal coupling. The results for one of the best samples are shown here.

The manufactured Si_3N_4 PHC LYSO sample had approximately 90% PHC coverage and PHC geometry with PHC x 0.95 um, PHC z 0.3 um, and Gap 0.42 um, as measured with SEM imaging. The uncalibrated energy spectra for the Bare and PHC measurements are shown without Teflon coating in Fig. 7, and with a Teflon coating in Fig. 8. Both configurations resulted in an improved light output calculated from the relative shift in the photopeak location, as well as in a better energy resolution, as shown in Table III.

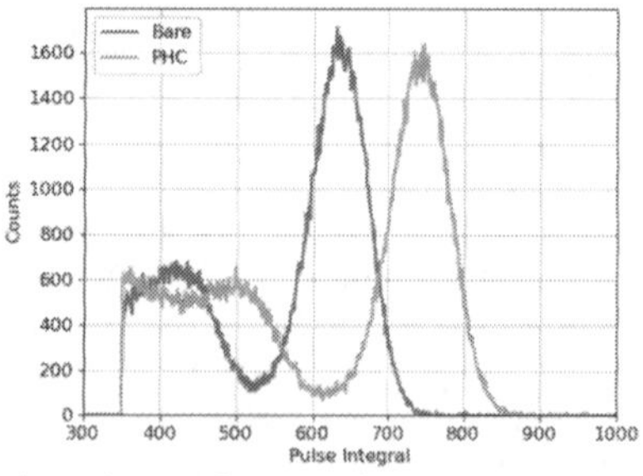

Fig. 7. Pulse integral spectra for Bare and PHC measurements without Teflon coating and with dry coupling for a LYSO scintillator coated with Si_3N_4 PHC.

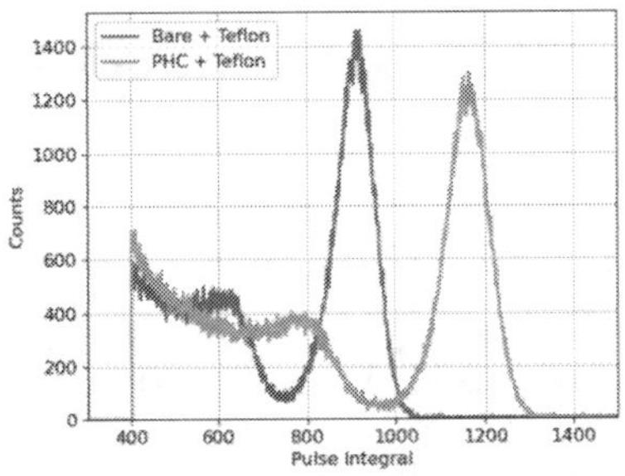

Fig. 8. Pulse integral spectra for Bare and PHC measurements with Teflon coating and dry coupling for a LYSO scintillator coated with Si_3N_4 PHC.

TABLE III
RELATIVE PERCENTAGE IMPROVEMENTS IN LIGHT OUTPUT AND ENERGY RESOLUTION FOR BARE AND PHC MEASUREMENTS, WITH AND WITHOUT TEFLON COATING.

	Photopeak Location		**% Energy Resolution**		**% Improvement**	
Measurement	Bare	PHC	Bare	PHC	Light Output	Energy Resolution
No Teflon Coating	634	741	14.4	12.8	**16.9**	**11.4**
With Teflon Coating	909	1162	11.8	10.3	**27.7**	**13.1**

In these tests, both the light output and energy resolution were improved, with a maximum improvement in light output of almost 28% and energy resolution of 13%, with Teflon coating. The results demonstrated the effectiveness of PHCs in improving the light output and energy resolution. However, the manufactured PHC results have yet to reach the potential light improvement of 39% as estimated by the simulations. This discrepancy was attributed to the imperfections in the manufactured PHC geometry and quality, as well as suboptimal optical coupling during the radiation measurements. Further testing with additional LYSO samples is needed to better align the experimental results with the simulated expectations. Additionally, future work will also focus on making the simulations more realistic by incorporating fabrication defects and imperfections.

IV. Conclusions

This work experimentally demonstrates the potential of PHCs in improving the light output and energy resolution for LYSO scintillators, with a specific sample exhibiting the light output improvement of 28% and with the energy resolution improvement of 13%. Enhancing scintillator light output and energy resolution without modifying the bulk material opens new avenues for detector optimization in nuclear security, medical imaging, space instrumentation, industrial applications, and beyond. For some applications, PHC-coated scintillators may begin to approach or even rival the performance of some room-temperature semiconductor detectors, possibly offering a potential lower-cost alternative. Future research will aim to minimize manufacturing uncertainties in PHC dimensions and periodicity, while improving the overall quality of manufactured PHCs. Simulations will incorporate manufacturing defects and explore different PHC shapes for optimized light output. Additionally, different scintillator materials such as $LaBr_3$ and NaI that are expected to have better improvement owing to their higher refractive indices will be explored for PHC applications.

Acknowledgment

This research is sponsored by the Defense Threat Reduction Agency (DTRA) as part of the Interaction of Ionizing Radiation with Matter University Research Alliance (IIRM-URA) under contract number HDTRA1-20-2-0002.

References

[1] C. Dujardin *et al.*, "Needs, Trends, and Advances in Inorganic Scintillators," *IEEE Trans Nucl Sci*, vol. 65, no. 8, pp. 1977–1997, 2018, doi: 10.1109/TNS.2018.2840160.

[2] S. E. Derenzo, M. J. Weber, E. Bourret-Courchesne, and M. K. Klintenberg, "The quest for the ideal inorganic scintillator," *Nucl. Instrum. Methods Phys. Res. Sect. Accel. Spectrometers Detect. Assoc. Equip.*, vol. 505, no. 1, pp. 111–117, Jun. 2003, doi: 10.1016/S0168-9002(03)01031-3.

[3] M. Kronberger, E. Auffray, and P. Lecoq, "Probing the Concepts of Photonic Crystals on Scintillating Materials," *IEEE Trans. Nucl. Sci.*, vol. 55, no. 3, Art. no. 3, Jun. 2008, doi: 10.1109/TNS.2008.922827.

[4] M. Salomoni, R. Pots, E. Auffray, and P. Lecoq, "Enhancing Light Extraction of Inorganic Scintillators Using Photonic Crystals," *Crystals*, vol. 8, no. 2, Art. no. 2, Feb. 2018, doi: 10.3390/cryst8020078.

[5] J. D. Joannopoulos, S. G. Johnson, J. N. Winn, and R. D. Meade, *Photonic crystals: Molding the flow of light*. Princeton University Press, 2011. Accessed: Dec. 08, 2023. [Online]. Available: https://collaborate.princeton.edu/en/publications/photonic-crystals-molding-the-flow-of-light

[6] F. Liu *et al.*, "Large energy resolution improvement of LYSO scintillator by electron beam lithography method," *AIP Adv.*, vol. 10, no. 2, p. 025101, Feb. 2020, doi: 10.1063/1.5136077.

[7] Z. Zhu *et al.*, "Improved light extraction of LYSO scintillator by the photonic structure from a layer of anodized aluminum oxide," *Nucl. Instrum. Methods Phys. Res. Sect. Accel. Spectrometers Detect. Assoc. Equip.*, vol. 786, pp. 1–4, Jun. 2015, doi: 10.1016/j.nima.2015.03.030.

[8] S. Agostinelli *et al.*, "Geant4—a simulation toolkit," *Nucl. Instrum. Methods Phys. Res. Sect. Accel. Spectrometers Detect. Assoc. Equip.*, vol. 506, no. 3, Art. no. 3, Jul. 2003, doi: 10.1016/S0168-9002(03)01368-8.

[9] "OptiFDTD Overview," Optiwave. Accessed: Dec. 08, 2023. [Online]. Available: https://optiwave.com/optifdtd-overview/

Fast neutron detector based on vertically aligned carbon nanotubes

Jean Nippert[1*], Ugo Forestier Colleoni[2], Mathieu Pinault[2], and David Tisseur[1**]
[1]CEA, IRESNE, 13115 Saint-Paul-lès-Durance, France
[2]Université Paris-Saclay, CEA, CNRS UMR 3685, NIMBE,91191 Gif-sur-Yvette, France
(*) jean.nippert@cea.fr
(**) david.tisseur@cea.fr

***Abstract*—The unique physical properties of carbon nanotubes have spurred numerous applications across various fields. In this study, we investigate their electrical properties to develop a detector for fast neutron flux measurement. Previous research has demonstrated their potential in dosimetry for thermal neutrons and gamma radiation, using random networks of carbon nanotubes. For the measurement of fast neutron flux, we utilize vertically aligned multi-walled carbon nanotubes and monitor the changes in their electrical resistance under neutron irradiation. The vacancies generated in the nanotube structure due to irradiation directly affect their electrical properties, allowing us to deduce the neutron flux from the variation in resistance. Future fusion reactors, such as ITER, are expected to generate intense neutron fluxes in high-temperature environments. These fluxes are challenging for traditional detectors, which tend to degrade quickly under such conditions. Carbon nanotube carpets, with their low density reduce the number of interactions caused by neutron irradiation, thereby extending the device's operational lifetime. Here the simulation shows that neutron irradiation induces local damages to the molecular structure of the carbon nanotubes. A first test at the CABRI research reactor (CEA Cadarache) has demonstrated the viability of the project.**

***Keywords* —Neutron flux detector, vertically aligned carbon nanotubes, fast neutron.**

I. Introduction

Carbon nanotubes (CNTs) exhibit unique electrical, thermal, and mechanical properties, making them promising candidates for advanced radiation detection. Unlike other materials like diamond or SiC, the CNTs can resists to intense flux due to its very low density, and also can be used in difficult environmental conditions such as high temperature, like those expected in the ITER Tokamak experiment. In the context of nuclear applications, a study by [1] used carbon nanotubes, single-wall (SWNT) and multi-wall (MWNT) for neutron and gamma radiation dosimetry. They demonstrated the feasibility of the technique with significant differences between SWNT and MWNT, the latter are less sensitive to gamma irradiation, which allows for the selective measurement of the neutron flux and results in more predictable behavior. This is a useful asset in a reactor environment where both neutrons and gamma rays are emitted by the nuclear reactions. More recently, a simulation study on vertically aligned carbon nanotubes used for fast neutron detection has been investigated [2]. It demonstrates that vertically aligned SWNTs behave like a network of parallel resistors. The damages generated on the molecular structure of carbon nanotubes by neutron irradiation have a strong impact on their electrical properties. This change in the resistance has a global effect on the carbon nanotube carpet and its resistance change can be quantitatively correlated to a neutron flux measurement.

This work will use carpets of multi-walled carbon nanotubes to measure fast neutron flux using the evolution of their electrical resistance through irradiation. However, other parameters such as the temperature can affect the electrical properties of carbon nanotubes [3], and must be taken into account. The need to measure fast neutron flux comes from the emergence of large-scale deuterium-tritium fusion reactors, for safety, plasma and in situ tritium generation monitoring.

II. Prototype

The first prototypes are in the form of a disk with a 12 mm diameter. They are available in two configurations of multi-walled carbon nanotube carpets: a high-density version and a low-density one. The high-density carpet produced on an thin aluminum substrate [4] is composed of 100 x 10^9 NT.cm^{-2} of carbon nanotubes with an average inner diameter of 6 nm and an outer diameter of 7.6 nm. The low-density carpet contains 10 times fewer nanotubes growing on a thin stainless-steel substrate with an average inner diameter of 9 nm and a larger outer diameter of 27.4 nm. Both configurations are produced with two lengths of nanotubes, 100 µm and 200 µm. The vertically aligned carbon nanotube (VACNT) carpets are produced at CEA Saclay in the Laboratory of nanometric structures (LEDNA) using a Catalytic Chemical Vapor Deposition (CCVD) technique [5].

To be used as a detection device, it required to measure the electrical resistance between the two faces of the nanotubes' carpets. To achieve it, VACNT are metallized with 2 µm of aluminum using a Physical Vapor Deposition (PVD) technique. The Fig. 1 shows a Scattering Electron Microscope (SEM) image of the top of a high-density carpet of vertically aligned carbon nanotube metallized with 2 µm of aluminum. It shows that the aluminum is uniformly deposited on the surface of the carpet without penetrating inside it. Each face is then connected to an ohmmeter using gold wires and BNC-type cables.

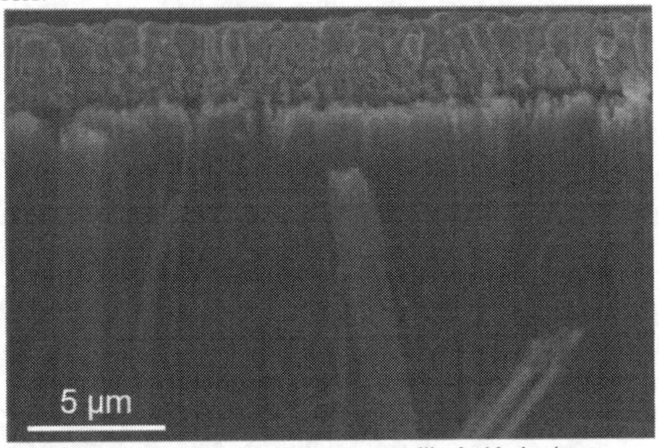

Fig. 1. SEM observation of a nanotubes carpet metallized with aluminum

III. Simulations

A simulation study was conducted using GEANT4 [6]. Several geometries were investigated for the different substrate configurations. In the simulation, each nanotube is modeled as a tube with realistic inner and outer diameters, filled with carbon at a density similar to that of graphite. The computational power demand is high for such a geometry, so it was simplified to reduce the calculation time. Due to the low overall density of the carpet, this simplification does not affect the simulation results. The diameter used for the simulation is 0.2 mm with a length of the nanotubes of 50 µm. The aim of the simulations is to understand how neutrons interact with the carbon from the molecular structure of the nanotubes and to point out the difference in interactions between the high and low-density carpets. Geant4 simulations were performed using the QGSP_BERT_HP physics list which is appropriate for neutron interactions below 20 MeV with high precision. The cross-sections used are extracted from the ENDF/B-VIII.0 database [7]. The reactions involved in the carbon nanotube irradiations are the elastic scattering (n,n) and the inelastic scattering (n,n'), (n,α) and (n,γ). The (n,n) reaction cross-section dominates at all energies but the (n,n') and (n,α) reactions contribution are not negligible, starting at 4.8 and 6.2 MeV, respectively. In the context of D-T fusion the neutron spectrum peak at 14 MeV where the main contribution comes from the (n,n) and (n,n') reactions as it can be seen in Fig. 2, where the reactions are represented in purple and grey, respectively. It shows that these two reactions represent almost 90% of the total cross-section (in red) at 14 MeV with the (n,α) reaction more than 10 times lower and the (n,γ) reaction completely negligible.

Another interaction is important for understanding the damages generated by neutrons on the molecular structure of the multi-walled carbon nanotubes. The multi-scattering interaction in Geant4 represents multiple scattering of the same particle on the atoms of the irradiated material. In this case the particles involved in multi-scattering events are the recoil nucleus bumping into the carbon atoms of the molecular structure of the nanotubes. Fig. 3 shows the energy deposition at each step of the multi-scattering event and it shows that most of the energy is above 1 keV, an energy sufficient enough to break the bonds and eject the atom from the molecular structure.

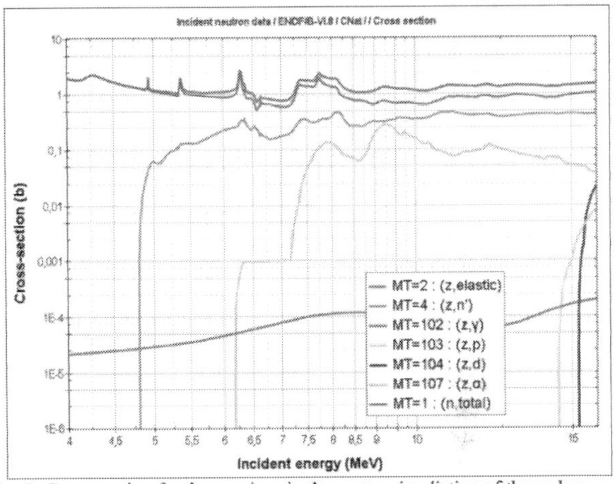

Fig. 2. Cross-section for the reactions in the neutron irradiation of the carbon nanotubes extracted with the Janis software from the ENDF/B-VIII.0 database [7].

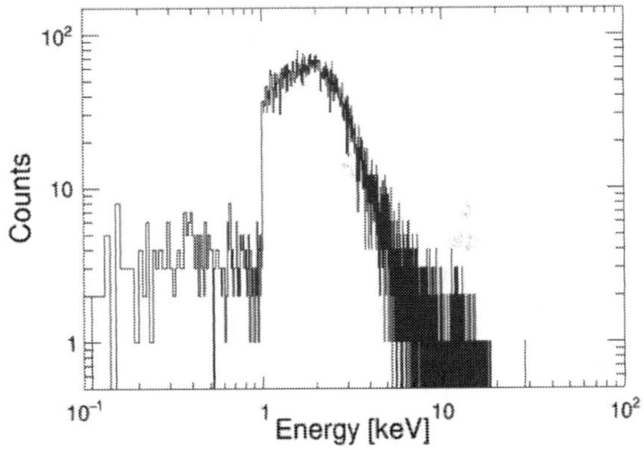

Fig. 3. Energy deposed by the recoil nucleus on an atom from the molecular structure of the carbon nanotube at each step.

The elastic and inelastic scattering involving the incoming neutron will now be called primary reactions, and the multi-scattering secondary reactions. The simulations show that for the secondary reactions, the distance between two interactions involving the same nucleus is very short. As shown in Fig. 4, most of the secondary interactions happen in a range lower than 2 nm. It means that when a neutron interacts with the carbon, the recoil nucleus generates localized damages in the nanotubes structure. The ratio $R_{sp} = N_s/N_p$ between the number of primary interaction N_p and the number of secondary interactions N_s can provide information on the mean size of the localized damages. The simulation shows that this number vary with the size of the nanotubes. For instance, a dense configuration on an aluminum substrate will have a ratio of $R_{sp} = 2.32$ while it goes to $R_{sp} = 3.40$ for the low-density configuration on a stainless-steel substrate. This ratio is mainly related to the thickness of the nanotubes of each configuration. However, the nanotubes for the low-density configuration are more than ten times thicker than those for the high-density, which means that the thinner nanotubes will be more impacted than the thicker ones.

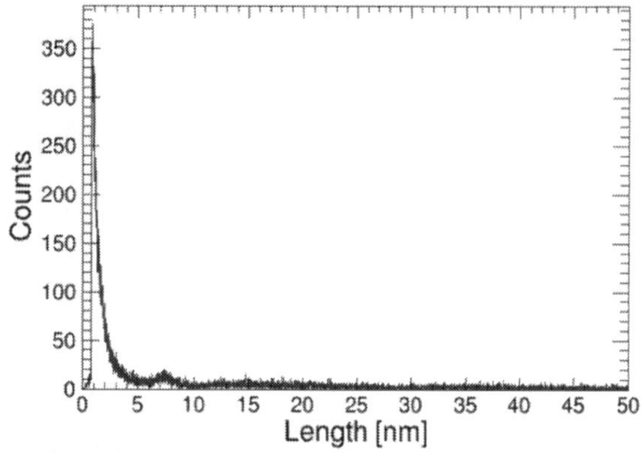

Fig. 4. Distance between two multi-scattering interactions involving the same recoil nucleus.

If we consider a simple model where the metallized nanotube carpet represents a circuit of parallel resistors and the damages generated by the incident neutron on the molecular structure of a nanotube vary its electrical resistance by a constant factor, the expected evolution of the total resistance R_{tot} will follow the equation:

$$R_{tot} = a \times \frac{R_{NT}}{a(N-X)+X} \quad (1)$$

With a the factor of variation of the resistance, R_{NT} the nanotube resistance, N the number of nanotubes in the carpet and X the number of damaged nanotubes. This indicates that the variation is not expected to be linear with the neutron flux and requires characterization.

IV. Irradiation test at CABRI facility

A first irradiation has been performed to test the viability of the project. The nanotube carpet was a high-density configuration metallized with 2 μm of aluminum. Before irradiation, it showed an electrical resistance of 1.6 +/- 0.1 Ω. The irradiation was performed at the CABRI research reactor in CEA Cadarache under a stationary regime of 100 kW. The neutron flux at the detector placement was around $7e^{10}$ $n.cm^{-2}.s^{-1}$ during 7 hours with a spectrum mostly composed of thermal neutrons. The resistance has been measured 45 days after the irradiation for radioprotection reasons, reaching 3.8 +/- 0.2 Ω. The electrical resistance increased by factor 2.385, which is significant enough to confirm the potential of vertically aligned carbon nanotubes as a sensing medium for neutron flux measurement. The irradiated carpet was characterized using a SEM analysis. The image can be seen in the bottom of the Fig. 5, the carbon nanotubes do not show signs of molecular structure changes with the SEM observation compared to an observation of a carpet from the same batch (on top of the Fig. 5), but it will be complemented with a Raman analysis to better conclude about it.

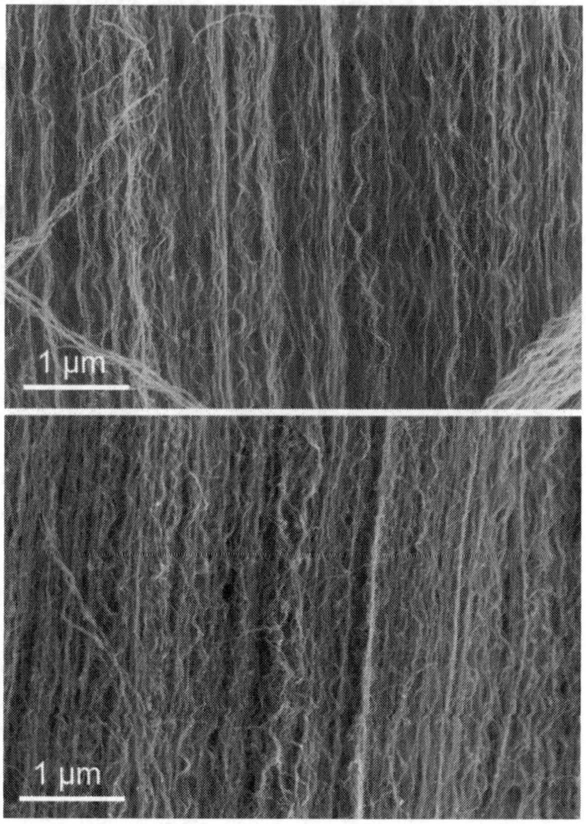

Fig. 5. SEM observation of the carbon nanotube carpets before (on top, observation of a carpet from the same batch) and after (at the bottom) irradiation at CABRI.

V. Conclusions

Vertically aligned carbon nanotube carpets have interesting electrical properties that are promising for neutron flux measurement. The physical process relies on neutrons interacting with the carbon nuclei, breaking the molecular bonds of the molecular structure. The simulations show localized damages mostly from the recoil nuclei. The detector's response is not expected to be linear and needs to be characterized. A first test at the CABRI facility proved that the electrical resistance of a nanotube carpet increases with thermal neutron irradiation. Tests for online measurement of fast neutron flux will be performed in the near future. The SEM analysis of the nanotube carpet has not yet provided evidence of morphology changes, but Raman analysis might provide more conclusive information on carbon structure evolution.

ACKNOWLEDGEMENTS

The authors would like to thank the INSNU program of the CEA and the PE-PTC VACAFON project for their financial support. They also thank the LEXIC team for their support during the irradiation at the CABRI reactor.

REFERENCES

[1] A. J. Nelson, "Carbon nanotube based dosimetry of neutron and gamma radiation," Virginia Tech, 2016.

[2] J. A. G. Gallardo, S. B. Pupo, M. Petaccia et J. Gervasoni, «Nuclear properties of carbon nanotubes for fast neutron detection,» *Journal of Energy and Power Engineering,* vol. 12, pp. 192-196, 2018.

[3] A. Zahab, L. Spina, P. Poncharal and C. Marlière, "Water-vapor effect on electrical conductivity of a single-walled carbon nanotube mat," *Phys. Rev. B,* vol. 62, pp. 10000 - 10003, 2000.

[4] F. Nassoy *et al.*, «Single-Step Synthesis of Vertically Aligned Carbon Nanotubes Forest on Aluminium Foils,» *Nanomaterials,* vol. 9, p. 1590, 2019.

[5] M. Pinault, V. Pichot, H. Khodja, P. Launois, C. Reynaud et M. Mayne-L'Hermite, «Evidence of sequential lift in growth of multiwsalled caron nanotubes multilayers,» *NanoLetters,* vol. 5, n° 112, pp. 2394-2398, 2005.

[6] J. Allison *et al.*, «Recent developments in Geant4,» *NIM-A,* vol. 835, pp. 186-225, 2016.

[7] D. C. M. Brown *et al.* «ENDF/B-VIII. 0: THe 8th major release of the nuclear reaction data library wit CIELO-projetct cross-section, new standards and thermal scattering data,» *Nucl. Data Sheets,* vol. 148, n° 11, 2018.

Comparison of 4H-SiC Radiation Detectors Based on PN Junctions and Schottky Contacts

Nikola Kurucová[1,*], Andrea Šagatová[1] , Bohumír Zaťko[2], Olivier Palais[3], Laurent Ottaviani[3], and Damien Barakel[3]
[1]Institute of Nuclear and Physical Engineering, Slovak University of Technology in Bratislava, Slovakia
[2]Institute of Electrical Engineering, Slovak Academy of Sciences, Bratislava, Slovakia
[3]Aix-Marseille University, University of Toulon, CNRS, IM2NP, Marseille, France
(*) nikola.kurucova@stuba.sk

***Abstract*—Silicon carbide (SiC) is a material that shows great promise in the fabrication of semiconductor radiation detectors intended for harsh environments, such as space missions, nuclear reactors, or direct neutron detection. This is primarily attributable to its outstanding radiation hardness and thermal stability. The present work proposes a comparative study of two types of SiC detectors based on different blocking type systems: PN junction and Schottky contact. All detectors were fabricated from 350 µm-thick 4H-polytype-SiC substrates with various thicknesses of epitaxial layers forming the active region. Detectors with three different thicknesses of the epitaxial layer were evaluated for each detector type: 20, 60, and 100 µm for PN junction detectors, and 25, 50, and 100 µm for Schottky detectors. An investigation was conducted to establish a comparison between the electrical properties and alpha spectrometric performance of the samples. Current-voltage measurements revealed a lower leakage current in Schottky-type detectors, resulting in reduced detector noise. Alpha spectra demonstrated better energy resolution for Schottky-contact detectors. Detectors with intermediate epitaxial layer thicknesses (50–60 µm) exhibited optimal performance, characterized by low leakage current, moderate depletion voltage, and efficient charge collection. Conversely, very thin layers resulted in incomplete charge collection, while the thickest layers required high bias voltages and exhibited increased leakage currents. The findings of this study underscore the technological benefits of Schottky contacts and the necessity of optimized epitaxial layer design.**

***Keywords* —Semiconductor radiation detectors, Silicon carbide, PN junction, Schottky contacts.**

I. Introduction

SILICONE CARBIDE (SiC) has emerged as a highly promising material for the development of semiconductor radiation detectors designed for harsh environments, such as space monitoring or nuclear reactor instrumentation. Of the various silicon carbide polytypes (3C, 6H and 4H), 4H-SiC is considered the most suitable for detector fabrication thanks to its high electron mobility and saturation velocity, as well as its availability in high-quality commercial wafers. The wide bandgap (3.23 eV at 300 K), high displacement energy (20 – 35 eV) and superior thermal conductivity (~ 280 $W \cdot m^{-1} \cdot K^{-1}$ at room temperature [1]) of 4H-SiC result in exceptional radiation and temperature resistance, enabling operation at temperatures up to 700 °C [2, 3]. These characteristics make 4H-SiC a strong alternative to conventional materials such as silicon (Si), particularly for applications involving high radiation fields or elevated temperatures. Additionally, the presence of low-Z carbon atoms makes 4H-SiC suitable for direct neutron detection [4, 5], in addition to the detection of alpha particles [6, 7], γ- and X-rays [8 - 10].

An important factor in the design and performance of semiconductor radiation detectors is the type of used blocking system. For 4H-SiC detectors, two main structures can be employed: Schottky barrier diodes and PN junction-based detectors. Apart from the structure configuration, the thickness of the epitaxial layer plays a crucial role in determining the device's depletion characteristics and electric field distribution. Consequently, it has a significant impact on the charge collection performance and energy resolution of the device. Therefore, the blocking structure type and thickness of the active epitaxial layer must both be carefully considered when designing high-performance 4H-SiC radiation detectors.

To address these considerations, a comparative study was conducted focusing on 4H-SiC radiation detectors fabricated with either a Schottky contact or a PN junction system. Three detectors with similar active layer thicknesses (approximately 25 µm, 50 µm and 100 µm) were tested and analyzed for each blocking structure type. The study aimed to evaluate the impact of blocking structure type and epitaxial layer thickness on the detectors' electrical behavior and spectrometric performance, particularly focusing on leakage current at full depletion voltage, energy resolution and charge collection efficiency.

II. Instrumentation and Measurements

A. Detector Structures

Investigated detector structures were based on nitrogen-doped 4H-SiC epitaxial layers, which were grown using liquid phase epitaxy (LPE) on n-type substrates that were

350 μm thick. Each wafer included a 0.5 μm undoped buffer layer to smooth the interface between the substrate and the epitaxial layer. These base materials, consisting of the substrate and epitaxial layer, were purchased from commercial suppliers and subsequently processed into functional detectors in research laboratories of the Institute of Electrical Engineering, Slovak Academy of Sciences (IEE SAS) in Bratislava, Slovakia (Schottky contact based detectors) and at the Institute of Materials, Microelectronics and Nanosciences of Provence (IM2NP) of Aix-Marseille University (AMU), France (PN junction based detectors).

A total of six detector groups were examined, comprising two complementary sets based on Schottky barrier or PN junction configurations. Three epitaxial thicknesses were selected for both sets to enable comparison of depletion layer behavior: 25 μm, 50 μm and 100 μm for the Schottky-based devices and 20 μm, 60 μm and 100 μm for the PN junction-based devices. These thicknesses were chosen to be comparable between the two types of detectors, which allows for a valid comparison of the behavior between Schottky barrier and PN junction devices.

The wafers for Schottky detectors were supplied by two manufacturers: LPE S.p.A. (Italy), which is now part of ASM, provided the 25 μm and 50 μm thick epitaxial layer, while the 100 μm samples were sourced from CREE (USA), which currently operates as Wolfspeed, Inc. According to the manufacturers, the nominal uncompensated donor concentration in the epitaxial layers used for Schottky detectors is below 1×10^{14} cm^{-3} for the 25 μm and 50 μm thick layers and below 2×10^{14} cm^{-3} for the 100 μm layer. The wafers for PN junction detectors were supplied by NOVASiC (France). For the PN junction structures, the specified donor concentration of epitaxial n-layers is below 1×10^{14} cm^{-3} (± 30%).

The fabrication of the Schottky barrier detectors, including surface preparation and metallisation, was carried out at IEE SAS in Bratislava, Slovakia. Following cleaning in boiling acetone and isopropyl alcohol, the samples were rinsed in deionised water and dried with nitrogen. A circular Schottky contact with a diameter of 3 mm was deposited on the epitaxial surface using high-vacuum electron-beam evaporation of a 15/25 nm Ni/Au bilayer. The backside of each wafer was then coated with a 15/30/80 nm Ti/Pt/Au multilayer to form an ohmic contact. Fig. 1 shows a schematic cross-section of this detector structure.

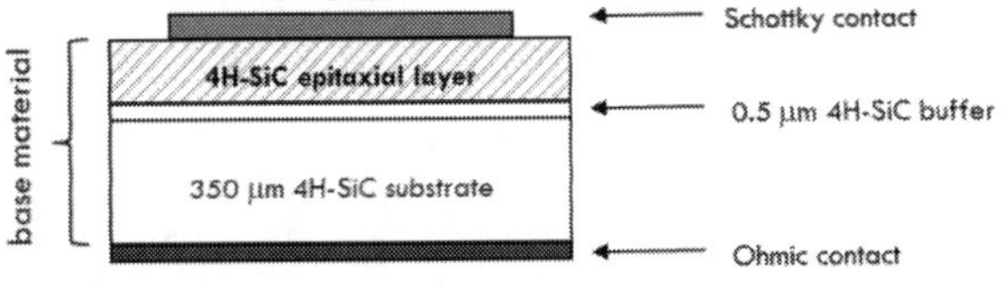

Fig. 1. A cross-sectional view of the 4H-SiC detector structure with Schottky contact.

The PN junction detectors were fabricated at the IM2NP laboratory of AMU, France. These devices are composed of p$^+$n epitaxial junctions, grown on a 350 μm-thick n$^+$-type 4H-SiC substrate from a wafer provided by NOVASiC (cf. Fig. 2). The n-type epitaxial layer exhibits a low doping level of approximately 2×10^{14} cm^{-3} with nitrogen. This doping corresponds to the concentration of donor atoms. The p$^+$-type epitaxial layer exhibits a thickness of 1.3 μm and a high aluminum doping level of approximately 10^{19} cm^{-3}. The concentration of acceptor ions is 1×10^{21} cm^{-3}, with an ionization coefficient of approximately 1%, since Al has a deep energy level in SiC, typically located around 0.2 eV above the valence band. Subsequently, a stack of Ni/Ti/Al/Ni layers, with a thickness of 100 nm, was deposited and annealed to establish the ohmic contact, which is designated as the anode. In order to reduce contact resistance, a 1 μm-thick layer of Al was then deposited using the electron beam evaporation technique. Finally, a 555 nm-thick Au layer was deposited on the surface of the detector. The cathode was fabricated by depositing a stack of 100-nm-thick Ti/Ni layers on the opposite side of the 4H-SiC substrate. The resulting device structure is illustrated in Fig. 2.

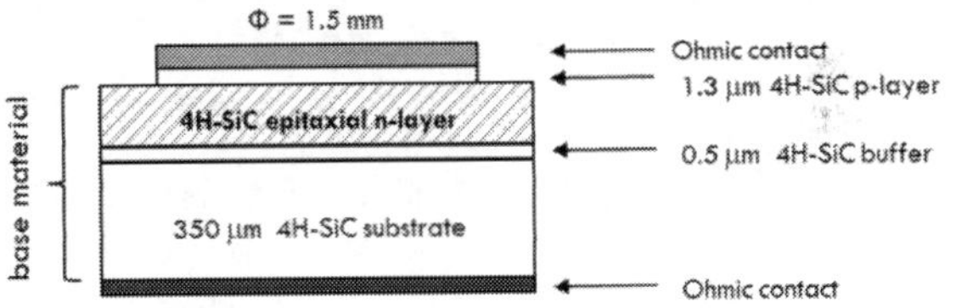

Fig. 2. A cross-sectional view of the 4H-SiC detector structure with PN junction.

These unified designs allow for direct comparisons of electrical and spectrometric characteristics between the two contact technologies while maintaining equivalent material properties and depletion thicknesses.

B. *Electrical and Spectrometric Characterization*

Electrical measurements were performed to evaluate the leakage current behavior of the fabricated detectors. A custom-made picoammeter with an integrated voltage source (up to 1000 V) was used to record the current-voltage (I-V) characteristics. To minimize external noise and ensure measurement stability, all detectors were placed in a shielded metal enclosure and kept in the dark during the experiments. The measurements were carried out at room temperature.

Alpha spectrometric measurements were performed to evaluate the charge collection efficiency and energy resolution of the fabricated detectors. Selected detectors were connected to a standard spectrometric chain comprising a charge-sensitive preamplifier CAEN A1422 and digital amplifier CAEN HEXAGON with a high-voltage source for detector power supplying. All measurements were controlled by CAEN mc^2 Analyzer software. Fig. 3 shows a schematic of the experimental setup used for alpha spectrometry. A triple-isotope alpha source ^{239}Pu^{238}Pu^{244}Cm was used to irradiate the detectors. All measurements were carried out at room temperature and in a vacuum to minimize alpha particle energy loss in air.

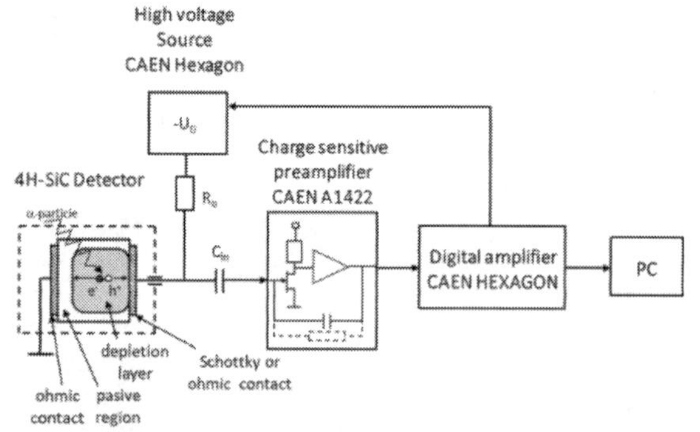

Fig. 3. Schematic diagram of the experimental setup used for alpha spectrometry with a triple-isotope alpha source $^{239}Pu^{238}Pu^{244}Cm$ emitting particles with energies of 5156.6 keV, 5499.0 keV, and 5804.8 keV.

III. Results and Discussion

A. Current-voltage measurement

The electrical performance of the fabricated detectors was evaluated by measuring their reverse I-V characteristics under dark conditions. Particular attention was given to leakage current levels and breakdown voltages, since these parameters directly affect noise levels and determine the detectors' safe operational range. As signal formation relies on the efficient collection of radiation-induced charge carriers within the depleted region, low leakage currents and the full depletion of the epitaxial layer at moderate bias voltages are essential for the reliable operation of the detectors.

The reverse I–V characteristics of six sets of 4H-SiC detectors were measured. These sets comprised devices with Schottky contacts and PN junctions, each of which was fabricated with three different thicknesses of SiC epitaxial layer. The Schottky-type detectors featured active layers measuring 25 µm, 50 µm and 100 µm, while the PN junction detectors were prepared with layers measuring 20 µm, 60 µm and 100 µm in thickness. Representative I–V curves for each group are presented in Fig. 4.

No clear saturation currents or breakdown voltages were observed within the applied voltage range up to 1000 V. Therefore, leakage current values were compared at bias voltages corresponding to full depletion, as previously established from capacitance–voltage (C–V) measurements for the Schottky-type detectors [11]. For the Schottky samples, full depletion of the 25 µm detector was observed at approximately 30 V with a leakage current of 6.5 pA; full depletion of the 50 µm detector was observed at approximately 130 V with a leakage current of 6.2 pA; and current saturation of the 100 µm detector was observed to occur near 400 V with a leakage current of approximately 10^{-9} A. However, full depletion was not reached up to the measurement limit of 600 V.

For the PN junction detectors, leakage currents of 6.1 pA, 8.6 nA and approximately 10^{-5} A were measured at depletion voltages of 30 V, 130 V and around 400 V for the 20 µm, 60 µm and 100 µm thick samples, respectively.

Comparing the leakage current values of Schottky and PN junction detectors reveals significant differences in their electrical performance. For the thinnest samples (25 µm for Schottky and 20 µm for PN), the leakage currents were of the same order of magnitude, both being in the range of several picoamperes. However, for detectors with thicker epitaxial layers, the PN junction devices exhibited significantly higher leakage currents — up to three orders of magnitude greater at intermediate thicknesses (50–60 µm) and up to four orders of magnitude higher for the 100 µm samples.

These results clearly demonstrate that Schottky-type detectors exhibit lower leakage currents, resulting in reduced detector noise. When comparing detectors with different thicknesses of epitaxial layers, a consistent trend was observed for both types of devices: an increase in thickness required a higher reverse bias voltage to achieve full depletion, accompanied by a corresponding increase in leakage current. This directly impacts the overall noise performance, as thicker active layers lead to higher detector noise.

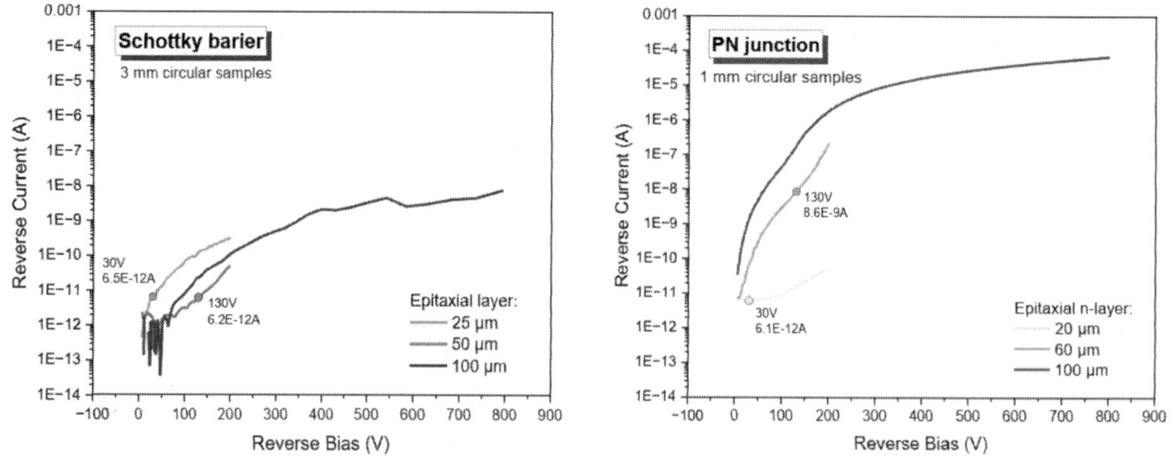

Fig. 4. Representative I-V characteristics of 4H-SiC detectors based on Schottky barrier and PN junction with varying epitaxial layer thicknesses. In cases where the detector thickness was below 60 µm, the applied bias was limited to 200 V due to current instabilities observed at higher voltages.

B. Alpha spectrometry

Alpha spectrometric measurements were carried out to evaluate the energy resolution and charge collection efficiency of the fabricated detectors when irradiated with a triple-isotope alpha source. This source emitted alpha particles with energies of 5156.6 keV, 5499.0 keV and 5804.8 keV from its ^{239}Pu, ^{238}Pu and ^{244}Cm isotopes, respectively. The detectors were biased at voltages ranging from 20 V to 200 V for thin and medium-thickness samples and up to 250 V for those with a 100 µm thick epitaxial layer. These voltage ranges were selected to approach or achieve full depletion of the active region, based on C–V measurements. All measurements were performed in a vacuum to minimize the loss of energy by alpha particles in air. The representative alpha spectra obtained with Schottky-type and PN junction detectors are presented in Fig. 5.

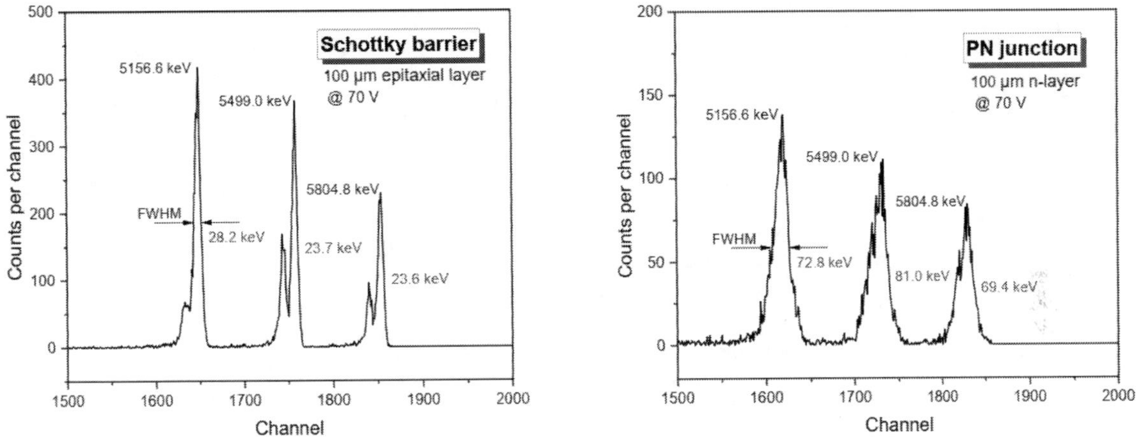

Fig. 5. Representative alpha spectra measured at 70 V for 4H-SiC detectors with 100 µm epitaxial layers based on Schottky barrier and PN junction, acquired using a triple-isotope $^{239}Pu/^{238}Pu/^{244}Cm$ alpha source in vacuum.

The recorded spectra were analyzed to determine the full width at half maximum (FWHM) of each photopeak, which represents the energy resolution of the detectors. The results obtained indicate that both detector types have the capability to resolve all three alpha peaks, though there are notable differences in resolution and peak structure. As Fig. 5 shows, spectra measured at 70 V for detectors with a 100 µm thick epitaxial layer demonstrate that the Schottky-type detector exhibits sharper peaks and improved energy resolution. In addition to the three main photopeaks corresponding to ^{239}Pu, ^{238}Pu and ^{244}Cm, secondary peaks are observed on the low-energy side of each main peak. The measured FWHM values for the Schottky-type detector were 28.2 keV, 23.7 keV and 23.6 keV, whereas the PN junction detector showed broader peaks with FWHM values of 72.8 keV, 81.0 keV and 69.4 keV, respectively. These differences highlight the superior spectrometric performance of the Schottky-type detector in terms of energy resolution under identical biasing conditions. In the case of PN junction detectors, the significantly thicker top contact compared to Schottky detectors acts as a dead layer, which degrades alpha particle energy resolution. Table 1 provides a summary of the achieved FWHM values for all Schottky and PN junction detector types with three different epitaxial layer thicknesses. The results obtained across all sample types and the entire voltage range (20 - 250 V) demonstrate that the Schottky contact detectors consistently achieve better energy resolution, particularly for medium and thick detectors (50 - 100 µm).

TABLE I
REPRESENTATIVE FULL WIDTH AT HALF MAXIMUM (FWHM) VALUES OBTAINED FOR SCHOTTKY AND PN JUNCTION 4H-SIC DETECTORS WITH DIFFERENT EPITAXIAL LAYER THICKNESSES ACROSS THE FULL BIAS VOLTAGE RANGE (20 - 250 V).

Thickness of epi layer (µm) Sch./PN	FWHM (keV)					
	Schottky detectors			**PN junction detectors**		
@E (keV)	**5156.6**	**5499.0**	**5804.8**	**5156.6**	**5499.0**	**5804.8**
25/20	74.9	87.3	83.0	72.8	84.8	134.1
50/60	27.5	21.6	19.9	83.7	91.0	82.2
100/100	26.4	21.5	21.0	61.0	71.3	65.4

To estimate the charge collection efficiency (CCE) of the 4H-SiC detectors, a Si detector with a known CCE of 100% was used as a reference under identical irradiation conditions. CCE was calculated using the following formula (1):

$$CCE_{SiC} = \frac{\varepsilon_{Si}}{\varepsilon_{SiC}} \cdot \frac{S_{SiC}}{S_{Si}}, \quad (1)$$

where ε_{Si} = 3.6 eV and ε_{SiC} = 7.8 eV are the mean energies required to generate one electron–hole pair in Si and 4H-SiC, respectively. S_{SiC} and S_{Si} are the slopes of the linear calibration curves (channels per keV) obtained from the three alpha peaks for the 4H-SiC and Si detectors. As illustrated in Fig. 6, the calculated CCE values for both Schottky and PN junction detectors demonstrate that charge collection efficiency approaches 100% for all samples at reverse bias voltages above

50 V. However, full charge collection was not achieved for the thinnest detectors - 25 μm for Schottky-type and 20 μm for PN junction devices - with CCE reaching only about 50% and 90%, respectively. This incomplete charge collection is consistent with the simulated alpha particle ranges in 4H-SiC: approximately 16.3 μm for ^{238}Pu, 17.9 μm for ^{239}Pu and 19.4 μm for ^{244}Cm. In such thin detectors, a significant proportion of charge carriers are generated near or beyond the limits of the depleted volume, especially at lower bias voltages. The reduced CCE in the 20 μm PN junction sample can be attributed to insufficient depletion depth relative to the alpha particle range. However, the 25 μm Schottky detector should, in theory, allow full charge collection under sufficient bias conditions. Consequently, the significantly lower CCE measured in 25 μm Schottky samples is presumed to result from material or structural defects in the detectors. This results in recombination losses and reduced effective CCE.

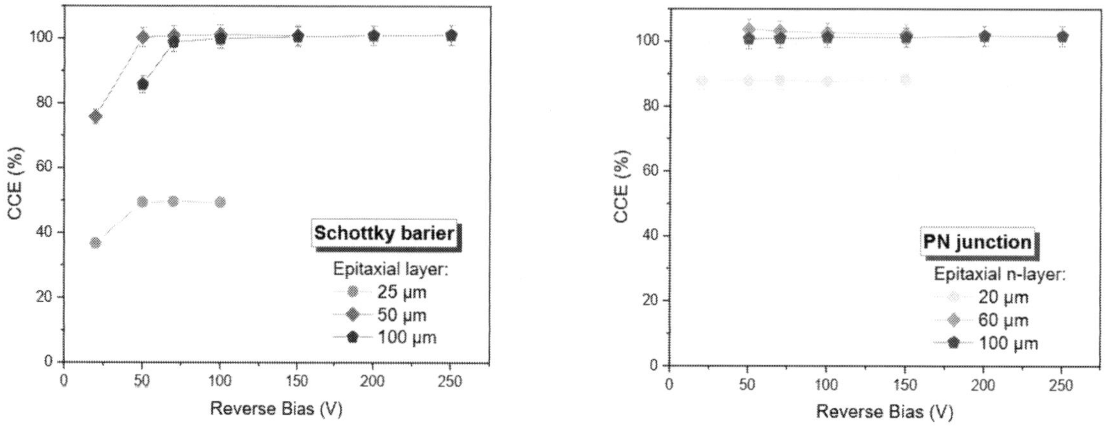

Fig. 6. Calculated CCE of 4H-SiC detectors with Schottky barrier and PN junction at various reverse bias voltages.

IV. Conclusion

This study focused on a comparative investigation of 4H-SiC detectors using two different contact systems: Schottky barrier and PN junction. Additionally, the impact of varying thickness of the epitaxial layer on detector performance was examined. Two sets of samples were fabricated and analyzed: Schottky-contact detectors with epitaxial layer thicknesses of 25 μm, 50 μm and 100 μm; and PN-junction detectors with layer thicknesses of 20 μm, 60 μm and 100 μm. The performance of the detectors was evaluated from electrical (I-V measurements) and spectrometric (alpha particle irradiation) perspectives.

Electrical characterization revealed that, at the required bias voltages for full depletion, the leakage currents were similar for the thinnest samples of both detector types. However, for intermediate and thicker epitaxial layers (50 - 60 μm and 100 μm), the PN junction detectors exhibited leakage currents up to three to four orders of magnitude higher than the Schottky detectors. Furthermore, the thickest detectors (100 μm) required significantly higher reverse bias voltages to reach full depletion, which contributed to an increased leakage current. In particular, the leakage currents were 6.5 pA (25 μm), 6.2 pA (50 μm) and approximately 10^{-9} A (100 μm) for the Schottky detectors, and 6.1 pA (20 μm), 8.6 nA (60 μm) and around 10^{-5} A (100 μm) for the PN junction detectors. These results suggest that Schottky-type detectors offer superior electrical performance, particularly with regard to lower leakage currents and consequently lower detector noise. Their favorable electrical behavior makes them better suited to applications requiring low-noise, high-sensitivity detection.

The superior performance of Schottky-type detectors was also confirmed by alpha spectrometry results. While both detector types could resolve the three alpha peaks from the mixed-isotope source, the Schottky detectors demonstrated significantly better energy resolution. Specifically, FWHM values of 28.2 keV, 23.7 keV and 23.6 keV were achieved with the Schottky detectors, compared to 72.8 keV, 81.0 keV and 69.4 keV with the PN junction detectors. This is primarily due to the thicker contacts of PN detectors resulting in a larger dead layer, which degrades the energy resolution. Furthermore, the CCE of the detectors reached nearly 100% for all the medium and thickest detectors. By contrast, incomplete charge collection was observed in the thinnest detectors, with CCE values of approximately 50% and 90% for the 25 μm Schottky and 20 μm PN junction devices, respectively.

Overall, the results demonstrate that detectors with Schottky contacts exhibit superior electrical and spectrometric properties compared to those based on PN junctions.

Acknowledgment

This work was partially supported by grants of the Slovak Research and Development Agency Nos. APVV-22-0382, DS-FR-22-0012 and DS-FR-24-0020. The author acknowledges using OpenAI's ChatGPT for language correction and stylistic editing, the scientific content and interpretations remain the sole responsibility of the author.

References

[1] R. Wei *et al.*, "Thermal conductivity of 4H-SiC single crystals," *J. Appl. Phys.*, vol. 113, no. 5, 053503, Feb. 2013, doi: 10.1063/1.4790134.

[2] N. Gál, L. Hrubčín, A. Šagátová, G. Vanko, E. Kováčová, and B. Zaťko, "High-resolution alpha-particle detector based on Schottky barrier 4H-SiC detector operated at elevated temperatures up to 500 °C," *Appl. Surf. Sci.*, vol. 635, p. 157708, Oct. 2023, doi: 10.1016/j.apsusc.2023.157708.

[3] F. H. Ruddy, L. Ottaviani, A. Lyoussi, C. Destouches, O. Palais, and C. Reynard-Carette, "Silicon Carbide Neutron Detectors for Harsh Nuclear Environments: A Review of the State of the Art," *IEEE Trans. Nucl. Sci.*, vol. 69, no. 4, pp. 792–803, Apr. 2022, doi: 10.1109/TNS.2022.3144125.

[4] B. Zaťko et al., "Radiation detector based on 4H-SiC used for thermal neutron detection," *J. Instrument.*, 11, C11022, Nov. 2016, doi: 10.1088/1748-0221/11/11/C11022.

[5] V. Valero *et al.*, "In-core thermal and fast neutron measurements with 4H-SiC P+N junction diodes in the JSI TRIGA Mark II research reactor," *EPJ Nucl. Sci. Technol.*, vol. 11, July 2025, doi: 10.1051/epjn/2025024.

[6] B. Zaťko *et al.*, "Schottky barrier detectors based on high quality 4H-SiC semiconductor: Electrical and detection properties," *Appl. Surf. Sci.*, vol. 461, pp. 276–280, Dec. 2018, doi: 10.1016/j.apsusc.2018.08.082.

[7] B. Zaťko *et al.*, "Study of Schottky barrier detectors based on a high quality 4H-SiC epitaxial layer with different thickness," *Appl. Surf. Sci.*, vol. 536, 147801, Jan. 2021, doi: 10.1016/j.apsusc.2020.147801.

[8] G. Bertuccio, R. Casiraghi, and F. Nava, "Epitaxial silicon carbide for X-ray detection," *IEEE Trans. Nucl. Sci.*, vol. 48, no. 2, pp. 232–233, Apr. 2001, doi: 10.1109/23.915369.

[9] G. Bertuccio, R. Casiraghi, A. Cetronio, C. Lanzieri, and F. Nava, "Silicon carbide for high resolution X-ray detectors operating up to 100°C," *Nucl. Instrum. Methods Phys. Res., Sect. A*, vol. 522, no. 3, pp. 413–419, Apr. 2004, doi: 10.1016/j.nima.2003.11.413.

[10] G. Bertuccio, S. Caccia, D. Puglisi, and D. Macera, "Advances in silicon carbide X-ray detectors," *Nucl. Instrum. Methods Phys. Res., Sect. A*, vol. 652, no. 1, pp. 193–196, Oct. 2011, doi: 10.1016/j.nima.2010.08.046.

[11] N. Kurucová, A. Šagátová, E. Kováčová, and B. Zaťko, "Effect of Epitaxial Layer Thickness on the Electrical Properties of SiC Detectors," to be published in *AIP Conf. Proc.*, 2025.

SiC-based detectors for fast neutron measurements in a multi-energy field

Quentin Potiron[1*], Christophe Destouches[2], Michael Houry[3], Olivier Llido[2], Abdallah Lyoussi[2], and Christelle Reynard-Carette[1]

[1]Aix Marseille Univ, Université de Toulon, CNRS, IM2NP, Marseille, France.
[2]CEA, DES, IRESNE, DER, F-13108 Saint-Paul-lez-Durance, France.
[3]CEA, IRFM, F-13108 Saint-Paul-lez-Durance, France.
(*) quentin.potiron@iter.org

***Abstract*—Wide-bandgap semiconductor detectors like silicon carbide (SiC) are promising for fusion and fission environments due to their radiation hardness and thermal/mechanical stability. The 4H-SiC diodes studied, with a 60 µm "p-n" junction, collect charge carriers generated by neutron reaction products and gammas. The resulting signal provides information on fast neutron energy. Previous studies in mono-energetic fields showed promising results for estimating fast neutron flux, but characterizing the sensor's response in multi-energetic fields remains challenging. In this work, the calibration of 4H-SiC response functions was performed using Time-of-Flight (ToF) measurements at the Neutron For Science (NFS) facility of GANIL, which allowed the construction of a response matrix. The presented results are promising, providing a response matrix with enough energy channels that can be used for neutron spectrum unfolding purposes.**

***Keywords*—Fast neutrons, Semiconductor Detectors, Silicon Carbide, Radiation Detection.**

I. Introduction

WIDE-bandgap semiconductors-based detectors such as Silicon Carbide (SiC) or CDV Diamond had been efficiently used as fast neutron detectors and have shown interesting performances for many applications related to radiation detection in constrained environments [1]–[3] Indeed, neutron detection mechanisms with detectors based on semiconductor materials rely on indirect methods in which neutrons interact with specific atomic nuclei and produce ionizing charged particles. These detectors have the advantage of having threshold nuclear reactions with fast neutrons as seen in TABLE I, while being insensitive to thermal neutrons. This characteristic gives them interesting spectrometric properties [4], [5] and they are therefore widely studied in the literature such as for the topic of cross section evaluation [6], [7].

TABLE I
THE THRESHOLD NEUTRON REACTIONS

Reaction	Reaction threshold energy [MeV]
$^{28}Si(n,n')^{28}Si$	1.8
$^{12}C(n,n')^{12}C$	4.8
$^{28}Si(n,\alpha)^{25}Mg$	2.8
$^{28}Si(n,p)^{28}Al$	4.0
$^{12}C(n,\alpha_0)^{9}Be^{(g.s)}$	6.2

II. Methods

The experimental campaign was carried out at the GANIL facility in Caen, France. The SPIRAL-2 project enhances this facility with a superconducting LINAC supporting intense proton, deuteron, and heavy ion beams (up to 5mA, 40MeV for deuterons, 14MeV/A for heavy ions). Equipped with two ion sources, a Radio-Frequency Quadrupole (RFQ), and 26 superconducting cavities [8], it powers the Neutron For Science (NFS) facility, located 9.5m underground, featuring a neutron production area called converter room, shown in Fig. 2. and a time-of-flight chamber.

Two sensors were positioned as shown in Fig. 1. Due to proximity to the beam dump and magnets, conditions were suboptimal, with magnetic fields and scattered neutrons affecting measurements. The thinner sensor (S60) was mounted on the tube guide, while the thicker sensor was placed on a chassis beneath the magnet.

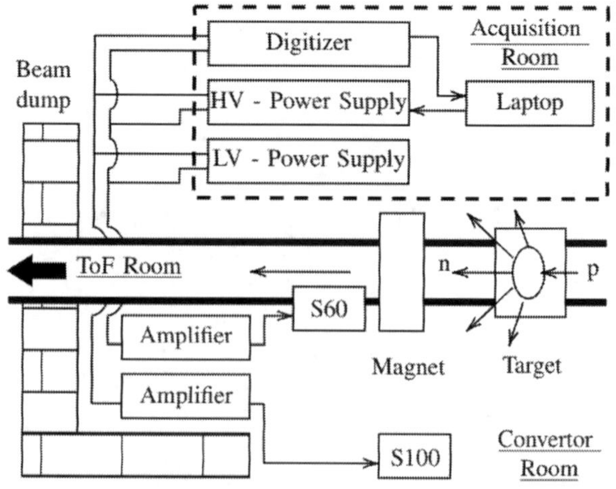

Fig. 1. Diagram of the NFS setup. The dashed rectangle indicates the remote acquisition room. For simplicity, only one output cable is depicted for each amplifier device. S60 and S100 represent SiC sensors with thicknesses of 60 m and 100 m, respectively.

The acquisition system used for the calibration of sensor responses was valided in a previous work using a neutron generator producing 14 MeV neutrons [9]. It consists of the following components:

Fig. 2. Convertor room of the NFS facility at GANIL. The red star indicates the position of the neutron production target, the green and blu stars respectively indicate the position of the SiC detectors of 60 and 100m thickness.

- A high-voltage Source (HV) provided by CAEN, delivering the voltage required to operate the detector. It is controlled via a laptop.
- A current preamplifier (PA), the C2 model, manufactured by Cividec Instrumentation, ensureing optimal coupling between the detector output and the rest of the counting system. The C2 broadband amplifier features a 2 GHz analog bandwidth and 40 dB gain, minimizing noise sources that could distort the signal.
- A digitizer, the Acqiris SA230E, which samples the amplified signal This 14-bit digitizer operates at a sampling rate of 4 GS/s, capturing current pulse shapes up to 2 GHz. It includes a long acquisition memory of up to 8 GB and a Xilinx Kintex UltraScale FPGA for real-time signal processing. A dedicated peak listing firmware was also integrated providing a range of programmable parameters for peak detection, enabling real-time identification and characterization of peaks. Detected peaks are stored in DDR4 SDRAM8 memory as peak descriptors. For each new trigger event, a 256-bit trigger descriptor is generated including metadata such as peak position, amplitude, and width.

The Time-of-Flight (ToF) method enables the measurement of neutron energy by calculating the time a neutron takes to travel a known distance from the source to the detector. The creation time, t_0, is inferred relative to other events, occurring when deuterons hit the target as discribed in Fig. 3. Then, gamma spallation products travel a given distance at the speed of light. Subsequent signals correspond therefore to neutron-induced reactions, and are obtained with time-of-flight measurement such that:

$$t_{ToF} = t_n - t_\gamma + \frac{d}{c} \tag{1}$$

With:
c the speed of light,
d the distance between the target and the detector,
t_n the time of the neutron-induced reaction,
t_γ the time of the gamma flash,
t_{ToF} the time-of-flight of the neutron from the target to the detector.

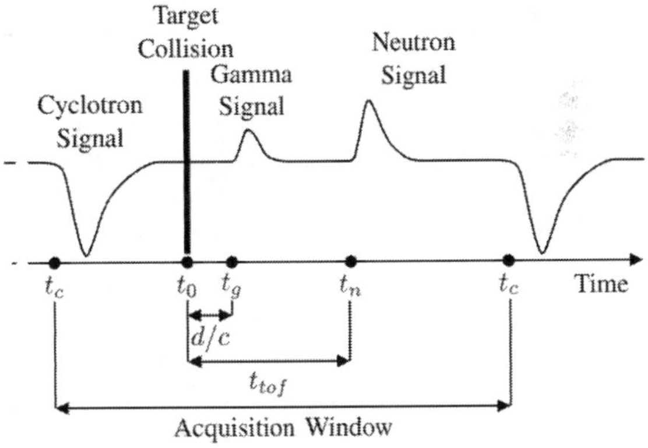

Fig. 3. Principle of the Time-of-Flight (TOF) technique. The time difference between the emission of neutrons and their detection is used to determine their energy.

Using a current amplifier with digitizer firmware configured to record waveform of the signals, current pulses are recorded. The later are characterized by a peak, corresponding to the neutron-induced or gamma-induced reaction. The current pulses are defined by their amplitude, width, and shape, which influence the transport and collection of charge carriers.

III. Results

To calculate the neutron time-of-flight necessary for determining neutron energy, it is essential to distinguish the gamma-ray flash generated by spallation reactions. The gamma flash peak is then identified as the first peak in the pulse timestamp distribution, analyzed by the peak listing firmware, as shown in Fig. 4.

The contributions near the first peak, indicated by the dashed line, are attributed to delayed gamma rays. In this study, the FWHM of the prompt gamma-flash peak is measured to be approximately 2 ns based on experimental results, demonstrating the timing precision of the detectors used. This visualization provides initial insights into the pulse distribution within the NFS neutron field. Indeed, the plot of event timestamps highlights the various reaction channels shaping the ToF

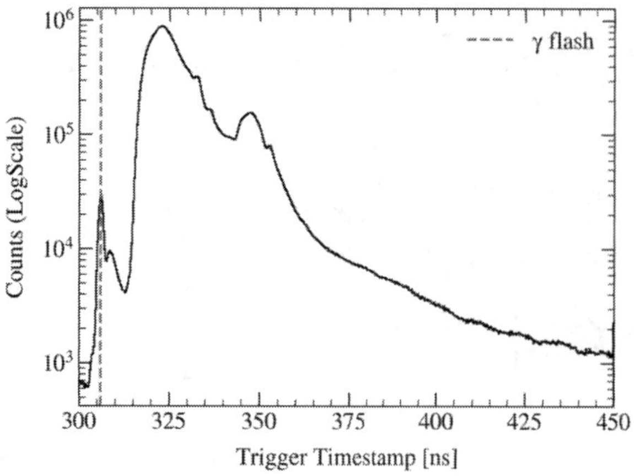

Fig. 4. Pulses timestamp distribution. The dashed lined guides the eyes to the prompt gamma flash peak.

spectrum. Pulses with timestamps around 320 ns correspond to the highest-energy neutrons, once corrected for the distance and gamma flash time-of-flight, while lower-energy neutrons are detected at later times.

The conversion of neutron time-of-flight into neutron energy is derived using the relativistic expression for total energy, as follows:

$$E_n = m_n c^2 \cdot \left(\frac{1}{\sqrt{1-\beta^2}} - 1 \right) \tag{2}$$

With:
E_n the neutron energy,
m_n the neutron mass at rest,
c the speed of light,
$\beta = \frac{v}{c}$ the ratio of neutron velocity to the speed of light.

From the time of fligh distribution of the recorded current pulses, and using the relativistic expression for total energy, the neutron energy distribution is computed, as shown in Fig. 5.

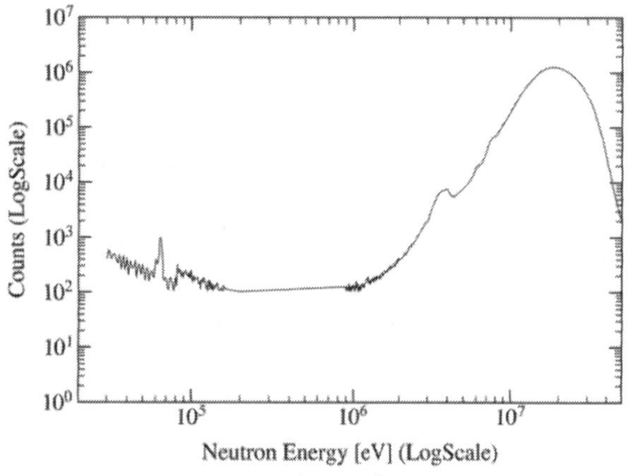

Fig. 5. Neutron energy distribution computed using the time of fligh distribution of Fig. 4 and equation (2).

It can be observed that few events are recorded at low energies, with one (n,γ) reaction on ^{28}Si, with a resonance at 68 keV, observed with a peak in Fig. 5. In comparison, for higher energies, the numerous threshold reactions listed in TABLE I, create highly energetic products from neutron captures.

These findings confirm the initial observation that the sensor response is strongly influenced by neutron kinetic energy, as expected from the nuclear reactions occurring across the broad energy range of the NFS neutron field.

In parallel, the deposited energy in the SiC sensor shown in Fig. 6, is computed from the recorded current pulses such that:

$$E_{dep} = \frac{\int v(t)dt}{q} \cdot E_W \tag{3}$$

With:
E_{dep} the deposited energy in the sensor,
$v(t)$ the voltage recorded by the digitizer,
q the elementary charge of an electron,
E_W the energy required to create an electron-hole pair in 4H-SiC, which is approximately 7.78 eV.

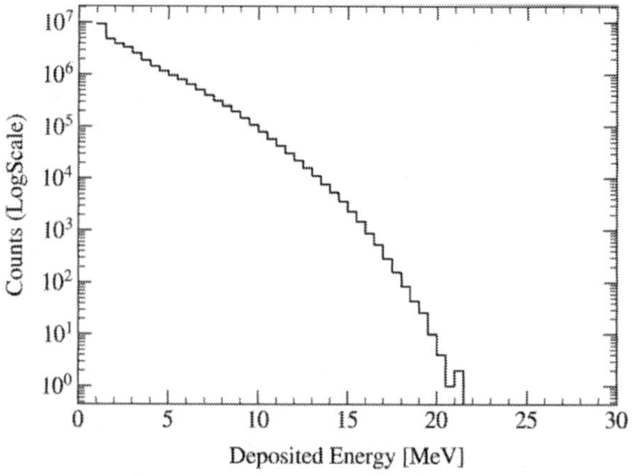

Fig. 6. Deposited energy distribution obtained with 4H-SiC sensor at NFS.

As anticipated in a continuous neutron field with neutron energies reaching up to 40 MeV as for the NFS neutron field, the deposited energy distribution is broad and lacks distinct features, such as peaks corresponding to full energy deposition, signatures of specific reactions, which are typically observed in mono-energetic neutron fields.

Considering the constraints of sensor proximity to the target and the resolution limitations of the digitizer and sampling rate, the matrix is constructed with a neutron energy bin width of 2 MeV. This width ensures full coverage of the NFS neutron field's energy range using equally spaced bins. For deposited energy distributions, a finer bin width of 0.1 MeV is used to provide a detailed representation of the relationship between deposited energy and neutron energy. Consequently, the sensor response matrix, comprising 20 neutron energy bins and 300 deposited energy bins, is illustrated in Fig. 7.

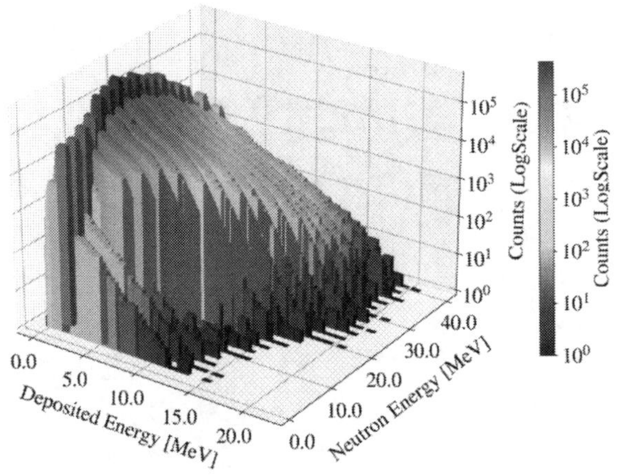

Fig. 7. Experimental response matrix of 4H-SiC sensor. The data are computed using both the neutron energy distribution Fig. 5 and the deposited energy distribution Fig. 6.

Each bin in the two-dimensional matrix corresponds then to a coefficient (R_{ij}), representing the sensor's response to the neutron flux at a given energy (E_i) within energy bin (j). This relationship involves the convolution of the neutron spectrum with the sensor response.

IV. Conclusion

The study focused on the use of Silicon Carbide (4H-SiC) sensors for neutron detection. SiC sensors are chosen for their robustness in harsh environments, such as nuclear facilities. The research aims to construct a sensor response matrix to relate neutron energy to energy deposition during interactions. The experimental campaign was conducted at the GANIL facility in Caen, France, the Neutron For Science (NFS) facility hosting a time-of-flight room. Two SiC sensors were positioned in convertor room, at less than 2 meters from the trget. Data acquisition involved fast current amplifiers, digitizers with a high sampling rate (4 GS/s), and a dedicated peak listing firmware that enables triggering events at the accelerator's high frequency (80 MHz).

The sensor response matrix is constructed using experimental data collected. Its primary purpose is to establish the relationship between neutron energy and the energy deposited during each interaction. For a specific incident neutron energy that induces nuclear reactions and generates charged particles depositing energy within the sensor, the matrix records the number of corresponding events. For neutron energies determined via time-of-flight measurements, the collected charge from current pulses is used to populate the bins of the sensor response matrix.

The response functions, derived from the neutron energy bin division are now available. The next step involves applying these sensor response functions with the iterative inversion method for neutron spectrum unfolding purposes in a multi-energetic neutron field.

Acknowledgment

This work has been made possible thanks to the NFS/GANIL facility staff, in particular Sébastien Gautrot, Jean-Claude Foy and X. Ledoux. The project leading to this publication has received funding from the Excellence Initiative of Aix-Marseille University - A*Midex, a French "Investissements d'Avenir" program AMX-19-IET-013.

References

[1] M. De Napoli, "SiC detectors: A review on the use of silicon carbide as radiation detection material," *Frontiers in Physics*, vol. 10, Oct. 2022.

[2] F. H. Ruddy, L. Ottaviani, A. Lyoussi, C. Destouches, O. Palais, and C. Reynard-Carette, Christelle "Silicon Carbide Neutron Detectors for Harsh Nuclear Environments: A Review of the State of the Art," *IEEE Transactions on Nuclear Science*, vol. 69, no. 4, pp. 792-803 Apr. 2022.

[3] Q. Potiron, C. Destouches, L. Dubus, M. Houry, O. Llido, A. Lyoussi, C. Reynard-Carette, P. Legou, P. Malard, and B. Cheymol, Evaluating Solid-State Neutron Detectors for Measuring 14 MeV Neutrons at High Temperatures Fusion Engineering and Design, in-press, 2025.

[4] F. H. Ruddy, "The fast neutron response of 4H silicon carbide semiconductor radiation detectors, *IEEE Transactions on Nuclear Science*, vol. 53, pp. 1666-1670, 2006.

[5] R. W. Flammang, J. Seidel and F. H. Ruddy, "Fast neutron detection with silicon carbide semiconductor radiation detectors," *Nuclear Instruments and Methods in Physics Research Section A: Accelerators, Spectrometers, Detectors and Associated Equipment*, vol. 579, no. 1, pp. 177-179 Aug. 2007.

[6] E. Griesmayer, P. Kavrigin and C. Weiss, "Neutron Cross Section Measurements with Diamond Detectors," *Engineering of Scintillation Materials and Radiation Technologies*, pp. 187-193, Sep. 2019.

[7] C. Weiss, "A CVD Diamond Detector for (n,α) Cross-Section Measurements," Ph.D. dissertation, Atomic Institute of the Austrian Universities, Vienna University of Technology., Geneva, June 13, 2014.

[8] X. Ledoux, "The Neutrons for Science Facility at SPIRAL-2", *Nuclear Data Sheets*, vol. 119, pp. 353-356, 2014.

[9] Q. Potiron, C. Destouches, L. Dubus, M. Houry, O. Llido, A. Lyoussi, L. Ottaviani, C. Reynard-Carette, and F.H. Ruddy, Estimation of the Fast-Neutron Fluence Rate with a 4H-SiC Based Detector in a 14.1 MeV Mono-Energetic Field. IEEE Transactions on Nuclear Science, 1-1. doi:10.1109/TNS.2023.3348147, 2024.

Development of a transportable neutron spectrometer for operational dosimetry within the DONEUT project

M. Labalme[1]*, S. Bozec[2], Y. Brelet[1], G. de Bouteiller[1,2], A. Hemmer[1,2], E. Moncanis[1,2], A. Roussel[1,2], C.A. Simonetti[1,2], J.L. Trolet[2]
[1]LPC Caen, Normandie Univ, ENSICAEN, UNICAEN, CNRS/IN2P3, Caen, France
[2]Marine Nationale, EAMEA, Cherbourg-en-Cotentin, France
(*) labalme@lpccaen.in2p3.fr

Abstract—**Accurate neutron dose rate estimation demands precise measurement of both kinetic energy and neutron flux. This study introduces a multi-instrumented transportable neutron spectrometer/dose rate meter, developed as part of the DONEUT project. The prototype, optimized for 0–20 MeV neutron dose rates, is based on a polyethylene moderator with thermal neutron detectors at different depths to reconstruct energy spectra via unfolding algorithms. The initial DONEUT V1 design consists of a 25 cm diameter polyethylene cylinder with 32 embedded ZnS:LiF scintillators, coupled to a SiPM matrix via light guides and wavelength-shifting fibers. GEANT4 Monte-Carlo simulations guided detector configuration and design optimization. Unfolding methods (MAXED and GRAVEL) were applied to various moderator volumes and neutron energy spectra, including Am/Be and ^{252}Cf sources. This led to a "dome" prototype, DONEUT V1+, which improves vertical angles response uniformity and reduces weight. A sixth detection depth was added to enhance deconvolution robustness, particularly in the epithermal range, while the number of detectors was reduced from 32 to 24. A custom Python implementation of the GRAVEL algorithm was developed, replacing traditional codes, increasing flexibility and processing speed for energy spectra reconstruction. Experimental tests with moderated and unmoderated Am/Be sources were performed, benchmarked against NNS measurements, demonstrated dose equivalent rate (H*(10)) accuracy within 15%, even at low dose rates (~1 µSv/h) over a 10-minute measurement. The prototype reliably assesses dose rates from 1 to 100 µSv/h. A new version, DONEUTV2, insensitive to gamma background, is under development to extend the range to several mSv/h**

Keywords —**neutron, dose rate meter, unfolding algorithm, ZnS: LiF.**

I. Introduction

Accurate measurement of neutron dose equivalent rates is a critical requirement in nuclear safety and radiation protection, particularly in demanding environments such as nuclear-powered submarines. These operational settings necessitate compact, reliable, and fast-response instruments capable of characterizing neutron fields with high precision. The DONEUT (DOsimétrie NEUTrons) project was initiated to address this need, aiming to develop a transportable neutron spectrometer able to provide real-time or near-real-time spectral and dose rate information.

Fig. 1. Pictures of the three version of developed prototypes, DONEUT V1 (2021) in the left, DONEUT V1+ (2024) in the center and DONEUT V2 (2025) in the right.

The project, launched in 2019, conducted at the Laboratoire de Physique Corpusculaire de Caen (LPC) in partnership with the French Navy's Nuclear Physics Laboratory (LPN) at EAMEA, has evolved through several design iterations of spectrometer prototypes, as presented in Fig. 1. The primary objective was to create a system capable of measuring neutron fluence and energy distributions across a broad energy range from thermal energies up to 20 MeV, and from these, derive ambient and individual dose equivalents (H*(10) and Hp(10)) thanks to unfolding algorithms. It consists on designing a transportable neutron detector prototype able to measure dose equivalents in compliance with a stringent set of specifications derived from both civilian and military nuclear requirements:

- Ambient dose equivalent detection thresholds ranging from 1 µSv/h to 10 mSv/h;
- Maximum measurement time of 10 minutes;
- Neutron energy spectrum coverage from 0.025 eV to 20 MeV;
- Dose equivalent calculation accuracy of 20% or better;
- Gamma/neutron discrimination to ensure environmental insensitivity;

- Isotropic response for omnidirectional source measurements;
- Portability: < 15 kg and < 60 cm in diameter.

In line with these specifications, several prototypes were developed and evaluated under laboratory conditions: DONEUT V1 between 2021 and 2023, DONEUTV1+ in 2024 and DONEUT V2 from 2025. A synthesis of the key findings is presented in the following sections.

II. INSTRUMENT DESIGN AND OPTIMIZATION

The DONEUT spectrometer is based on the principle of a multi-instrumented moderator volume, similar in concept to Bonner spheres [1] but adapted for compactness and rapid measurement. The core of the system is a polyethylene (PE) moderator, 25 cm diameter cylinder or sphere, within which several thermal neutron detectors are embedded at various depths. This configuration allows the system to be sensitive to different neutron energy ranges in only one measurement, enabling neutron energy spectral reconstruction through unfolding techniques.

To optimize the spectrometer's geometry and detectors configuration, extensive simulations using the GEANT4 Monte Carlo toolkit were conducted. These simulations evaluated various moderator shapes and detector placements, using neutron sources such as Am/Be and ^{252}Cf, both moderated and unmoderated. The unfolding of simulated detector responses was performed using the MAXED and GRAVEL algorithms from the UMG 3.3 software suite [2].

The first prototype, referred to as DONEUT V1 developed from 2019 [3], consisted of a 25 cm diameter PE cylinder housing 32 ZnS:LiF (0.3mm x 10mm x 20mm) scintillation detectors arranged in concentric layers at five different depths (Fig. 1.). Each detector is composed of two layers of ZnS:LiF glued to a 3 mm thick plexiglass light guide, coupled to a Kuraray Y11(200) wavelength-shifting fiber that transmitted scintillation light to a 64-channels silicon photomultiplier (SiPM) matrix. The readout electronics is based on the CAEN DT5202 module, which integrates two Citiroc-1A chips and provides 64 channels with temperature compensation. Simulation results exhibit good energy spectra reconstructions and dose rates evaluations in agreement with the specification, ±20%, regardless of the energy spectrum.

III. EXPERIMENTAL VALIDATION

Experimental campaigns were carried out at LPC Caen and LPN EAMEA Cherbourg to validate the performance of the DONEUT prototypes. Measurements were conducted using Am/Be neutron sources, both bare and moderated with a 25 cm PE sphere, as well as ^{137}Cs gamma sources to assess gamma sensitivity.

The first prototype, DONEUT V1, was extensively tested with Am/Be sources for neutron dose rates ranging from 1 μSv/h to 100 μSv/h. Results obtained for both energy spectra and dose rates evaluations showed good agreement between DONEUT (using MAXED or GRAVEL unfolding algorithms) and the reference NNS spectrometer (Detec) [4], with deviations within ±20% for various source configurations and dose rates as illustrated by the example presented in Fig. 2. The agreement is of the order of 5%. These tests validated the concept. Nevertheless, this prototype showed limitations versus the vertical angle and a high sensitivity to the inner detectors.

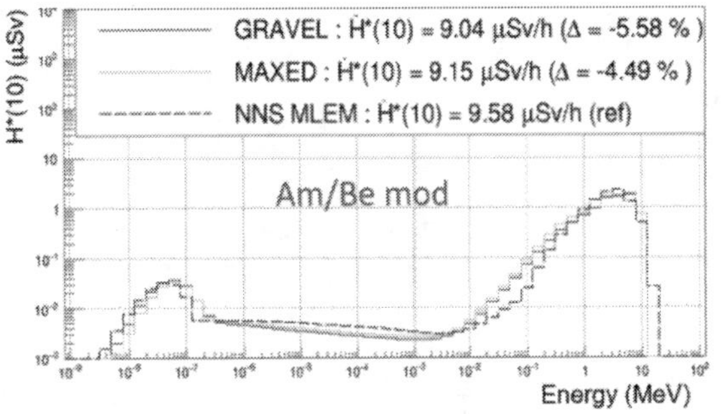

Fig. 2. Dose rate vs neutron energy measured at LPN Cherbourg by DONEUT V1, unfolded with GRAVEL and MAXED and compared to the NNS results used as reference. Neutron source: Am/Be moderated in a 25cm diameter polyethylene sphere. The dose rate integrated on the all-energy spectra are calculated.

New simulations were performed in 2024 whom results led to the development of an improved prototype, DONEUT V1+, featuring a dome-shaped moderator. This new geometry improved vertical angle response uniformity and reduced the overall weight of the system. Additionally, the number of detectors was reduced from 32 to 24, and a sixth depth layer was introduced to improve sensitivity in the epithermal range. This new prototype was extensively tested with Am/Be sources, maintaining measurement accuracy while improving portability and robustness.

IV. DEVELOPMENT OF A PYTHON BASED UNFOLDING SOFTWARE

One of the key innovations of the DONEUT project was the development of a custom Python-based implementation of the GRAVEL unfolding algorithm. This new tool, referred to as DSP (Déconvolution Sous Python), was inspired by the work of T. Doležal [5] and offers significant improvements in processing speed and flexibility compared to the UMG3.3 software. While UMG required up to 20 minutes for data processing, the DSP algorithm reduced this time to under 30 seconds.

DSP algorithm was first validated through GEANT4 simulations. An example of the results obtained for an unmoderated Am/Be source is presented in Fig. 3. Better agreements have been obtained for DONEUT V1+ and DSP compared to the previous prototype for both moderated and unmoderated neutron energy spectra.

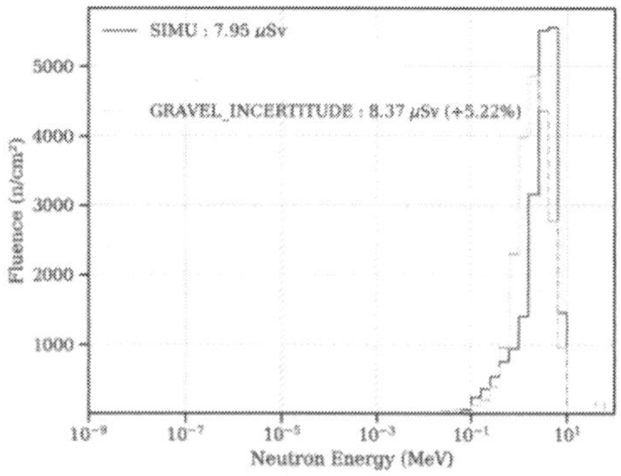

Fig. *3*. DSP unfolding algorithm (GRAVEL INCERTITUDE) results obtained with DONEUT V1+ in GEANT4 simulations with an unmoderated Am/Be source.

The new prototype using DSP was then extensively tested in laboratory, and results were compared to the one obtained using the NNS. An example of results obtained with a moderated Am/Be source at LPN is presented in Fig. 4. DSP produced energy spectra and dose rate estimates that were consistent with those obtained using the commercial NNS spectrometer, while avoiding the thermal peak overestimation as it has been observed with UMG. A good agreement on dose rates is obtained, within ±10% The improved performance obtained thanks to the DSP algorithm was particularly evident in the epithermal energy range, where it provided more accurate spectral reconstructions.

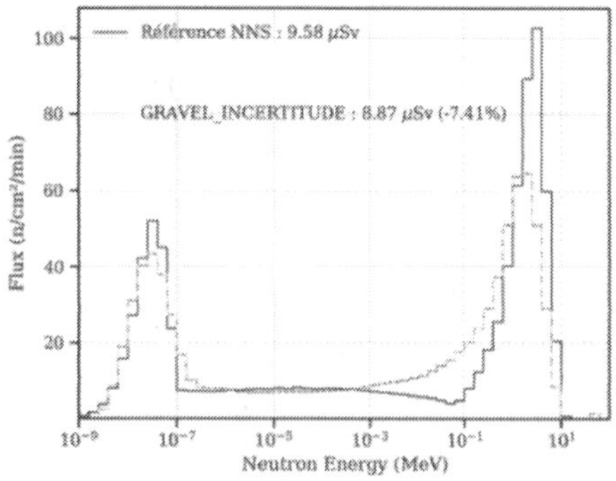

Fig. *4*. DSP unfolding algorithm (GRAVEL INCERTITUDE) results obtained with DONEUT V1+ in tests with a moderated Am/Be source and compared to results obtained with the NNS.

This new algorithm is the one used since 2024 in simulations and measurements. The results demonstrated that the DONEUT V1 and V1+ prototypes could achieve dose rate accuracies within 15–20% over a wide range of neutron fluxes, from 1 μSv/h to 100 μSv/h. The system exhibited good angular response uniformity, with deviations less than ±10% up to 45° from the vertical axis. Furthermore, the spectrometer showed low sensitivity to gamma background, with a neutron-to-gamma dose equivalent ratio (H*(10)γ / H*(10)n) of approximately 2.

V. Conclusions and perspectives

The DONEUT project successfully delivered a compact, accurate, and fast-response neutron spectrometer for operational dosimetry in complex radiation environments achieving a 20% accuracy on neutron dose rates compared to the NNS reference.

Future developments will focus on further miniaturization, integration of neural network-based unfolding algorithms, and extensive field testing. The transition to DONEUT V2 in 2025 marked a significant leap in hardware design. The new prototype featured a more compact and spherical PE moderator weighing less than 10 kg. Detector modules were redesigned to integrate the scintillator, the fiber, and one SiPM into a sealed, modular unit, improving mechanical robustness and ease of deployment. The readout electronics was upgraded with new home-made front-end electronics and to a "trigger-less" acquisition system based on the LPC-made FASTER digitizer in QDC mode (125 MHz, 12 Bits) [6]. This system supports long charge integration windows, which are better suited to the slow light emission of ZnS scintillators than the one proposed by the CAEN module. Preliminary tests of individual detectors demonstrated the ability to operate at high dose rates (>500 μSv/h) while maintaining good energy resolution and a very good neutron/gamma discrimination even for dose rates ratio H*(10)γ / H*(10)n ~50.

Acknowledgment

Authors thanks the region Normandie for its financial support via a RIN projet emergent.

References

[1] T.W. Bonner, Richard L. Bramblett , Ronald I. Ewing. « A new type of neutron spectrometer». Nuclear Instruments and Methods 9.1 (1960), p. 1-12.
[2] M. Reginato, B. Wiegel, A. Zimbal and F. Langner, "UMG 3.3, Analysis of data measured with spectrometers using unfolding techniques," Energy Agency of the OECD (NEA), 2004.
[3] C.A. Simonetti, M. Labalme, J.L. Trolet and P. Mary, «Deconvolution methods used for the development of a neutron spectrometer", EPJ Web of Conferences 253, 09003 (2021)
[4] J. Dubeau et al., "A neutron spectrometer using nested moderators," Radiation Protection Dosimetry, vol. 150, no. 2, pp. 217, 2012.
[5] T. Doležal, "Unfolding the AmBe Neutron Spectra using GRAVEL and MLEM," Department of Engineering Physics, Air Force Institute of Technology, 2021. https://github.com/tylerdolezal/Neutron-Unfolding
[6] https://faster.in2p3.fr/

https://doi.org/10.1051/epjconf/202533810012

Optimization of a VCO-Based Pixelated Particle Detector for particle discrimination and tracking

Karine Coulié[1*], Wenceslas Rahajandraibe[1], Valentin Valero[1] and Laurent Ottaviani[1]
[1]Aix Marseille Université, CNRS, Université de Toulon, IM2NP UMR 7334, Marseille, France
(*) karine.coulie@im2np.fr

Abstract— **The motivation of this work is to provide a system available for particle tracking and discrimination. The chosen methodology consists in using an indirect detection to avoid noise consideration linked to the use of a wide band amplifier. The detector is dimensioned by electrical simulations (Cadence) for the signal conditioning part and a device simulator (Synopsis) for detection matrix. Then, we demonstrate the ability to track different types of particles and to establish a clear relationship between input current characteristics and the corresponding output signal, thereby enabling particle identification**

Keywords— **Sensor, detection, VCO, radiation**

I. Introduction

Semiconductor radiation detectors, also known as solid-state detectors, are categorized based on their applications and the types of particles they detect. These detectors work by collecting and amplifying charge within a semiconductor volume, such as a p-n junction [1-2]. They have several advantages over other types of detectors, such as gas-filled counters. For example, they generally offer faster charge-collection times, allowing them to handle higher counting rates. They are also more compact, which enables the measurement of intensity variations over small distances, thereby improving resolution. For high-sensitivity applications, a pixel-based design with multiple sensors can be used [2-8], which is the focus of this work.

Detecting and identifying particles requires conditioning the signal generated by the sensor. Most existing conditioning systems directly read the current produced when an ionizing particle passes through the material. However, for embedded systems detecting low-energy particles, this method may not be the most suitable as the current signature can be a narrow pulse lasting only a few nanoseconds. Therefore, the readout chain must have a high enough resolution (*i.e.*, optimized Signal-to-Noise-Ratio) for accurate post-processing. A fast-response amplifier is also needed, which increases both current consumption and circuit complexity. Designing high-precision, low-power, wideband amplifiers is a significant challenge. Consequently, most practical detectors use Charge Sensitive Amplifiers (CSA) [9] for direct current reading. While effective, this approach provides limited information about the actual current shape and no information about the electrical signature of the particles themselves. In applications like gas prospecting or medical therapy, knowing the electrical signature of the particles is crucial. Additionally, having information about the generated current shape at the detector's output simplifies post-processing.

To address these challenges, a new approach was developed using a ring oscillator as conditioning method [10]. The system is based on the reading of information related to indirect output parameters of the detection chain, specifically voltage variation, instead of directly measuring the sensor's current. This method avoids many of the design problems mentioned earlier. The information is then extracted by correlating the system's initial oscillating signal with the signal after a particle has passed through the detector. Then the requirement for particle recognition is to link the output information (the voltage variation of the oscillating signal) to the input information (the current stimuli).

The full detection system (VCO + matrix) will be described in section II. Section III addresses the particle tracking methodology. Section IV focuses on the characterization of the VCO response. The last section concludes the paper.

II. Presentation of the System

The proposed detection system is composed of a detection matrix together with a readout circuit based on a high frequency voltage-controlled oscillator. The operating is synthesized in this section.

A. The VCO for signal conditioning

The detection chain has been designed and implemented on 130nm CMOS (Complementary Metal-Oxide-Semiconductor) SOI (Silicon On Insulator) technology, then simulated at circuit level using "Spectre" simulator (SPICE-based) under Cadence *Virtuoso* © CAO tool.

The concept was presented in [10-13]. The VCO chain is composed of three parts: a CMOS based pixel detector, a shaping circuit based on a Voltage Controlled ring Oscillator (VCO) and a system for frequency and amplitude detection (Fig. 1). This system allows the evaluation of the circuit sensitivity to radiation by measuring the oscillator responses. In the very first versions of the system, the operating frequency was 1MHz and the VCO was based on a 0,35µm Bulk Silicon substrate CMOS. A solution based on another CMOS process (130nm SOI), allowed to reach 4.8GHz [14]. This increase of the operating frequency, through different releases of the VCO, made it possible the development of new methods for signal

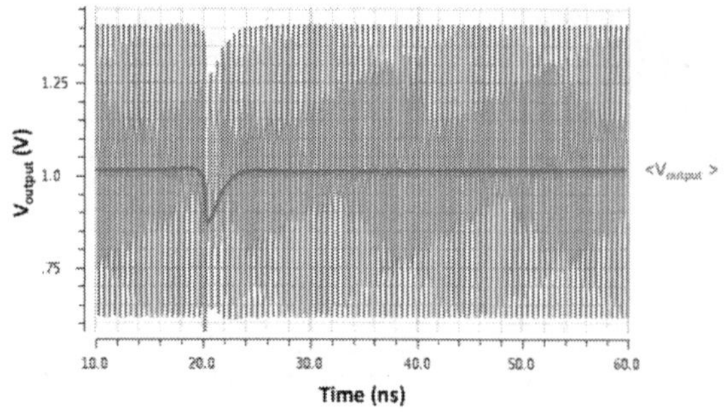

Fig.1. Variation of the output voltage versus time for the indirect detection chain (VCO based).

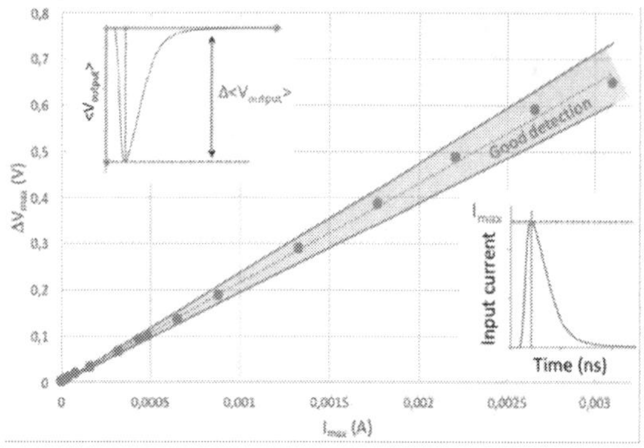

Fig.2. Variation of the average output signal (ΔV_{max}) versus the maximum of the input current (I_{max}).

recognition. In this paper, we adopt the operating frequency of 4.3 GHz. At this frequency, the shape of the signal is directly reproduced at the output of the VCO. The advantage of this solution is a direct recognition of the signal shape as an amplitude modulation of the VCO output frequency.

B. Calibration of the detection system

The key point of the system lies in its ability to extract information about the input current, generated by a particle traversing the detection matrix, through the analysis of the VCO chain output signal. This can be achieved through the analysis of various parameters as, for instance, the variation of the average output signal (ΔVmax) versus the maximum of the input current (Imax) (Fig. 2). A calibration curve can be provided by analyzing the response of the system when simplified analytical currents are injected. Through this calibration curve, the output parameters can be linked to the input currents, which could allow the incoming particle identification. To get reliable results, the resolution of the system has to be checked. Then a "good detection" zone is defined. This zone corresponds to a linear response of the system in the range of ±10 %. Outside this range, the system resolution is considered degraded [15].

C. The detection matrix

In order to improve the efficiency of the detection and the tracking ability, a *3x5* matrix is modeled and analyzed (Fig.3). The input current is no longer the simplified current used for the VCO characterization. Now, the current source comes from

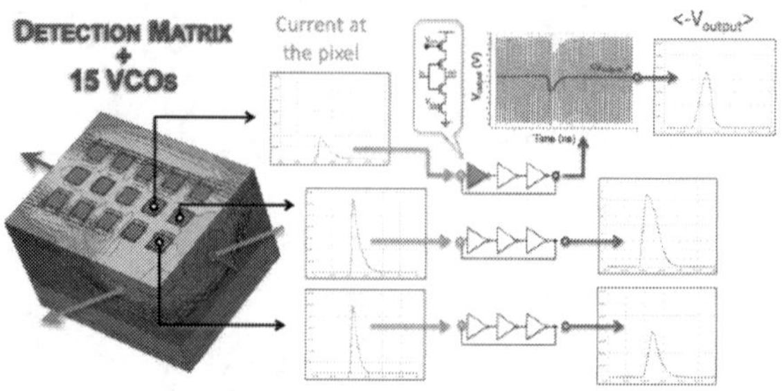

Fig.3. Left: Simulation of the 3x5 matrix; middle: currents generated by the particle at three pixels; right: corresponding signal at the output of the VCO.

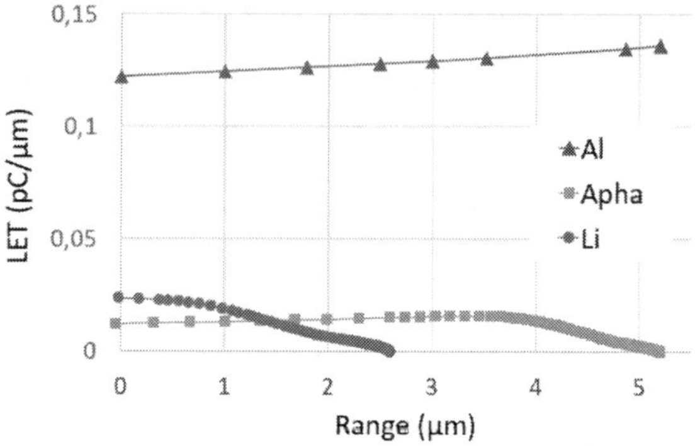

Fig.4. LET variation versus range for the three studied ions: Alpha, Lithium and Aluminum.

the realistic simulation of the matrix using TCAD device simulation tools (Synopsis ©). The effect of the ion strike is simulated using the *Heavy Ion* module of Synopsis [16], considering an electron-hole pair column centered on the ion track axis. The Linear Energy Transfer (LET) is defined as the energy lost by the particle, by unit of length and varies along the track depending on the initial energy of the particles [17]. An actual variation of the LET was integrated in our simulations, based on the value given by SRIM tables [18]. For many years, neutron – boron 10 interaction is a concern for microelectronics reliability [19]. Then we decided to study the particles generated by the main interaction of thermal neutron with boron-10, *i.e.* one Li with an initial energy of 0.9 MeV and one Alpha with an initial energy of 1.5 MeV (Fig. 4). The LET variations of these particles along the track are shown in Fig. 4. The case of a more energetic particle is also be considered. This is a 50 MeV Aluminum which could be produced by the interaction of fast protons with silicon, for instance (Fig. 4) [20]. The particles are generated at 50 ps.

An example of the generation of a particle in the whole structure is shown in Fig. 3. This figure presents the currents generated by the particle at three pixels and the corresponding signal at the output of the VCO. If a 3x5 matrix is simulated, then fifteen currents will be collected and injected at the input of the VCO. So, fifteen voltage variations will be analyzed at the VCO output.

III. Tracking: proof of concept

A. Aluminum tracking: horizontal case in the middle of the matrix

The firs case examines the crossing of the Aluminum ion in

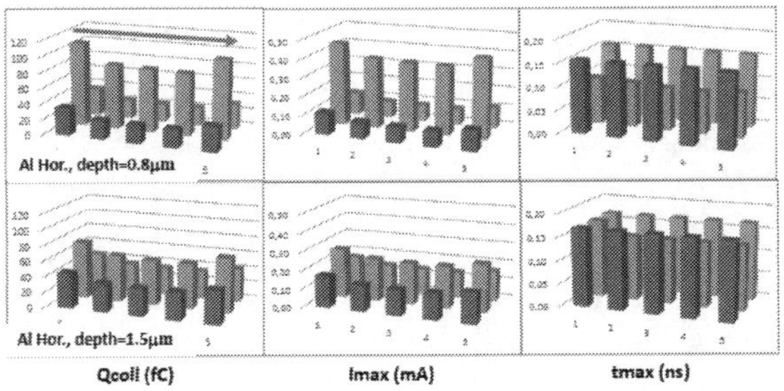

Fig. 5. Total collected charge (Qcoll), current peak (Imax) and time corresponding to this current peak (tmax) for each pixel of the detection matrix, for two different depths: 0.8 µm or 1.5 µm.

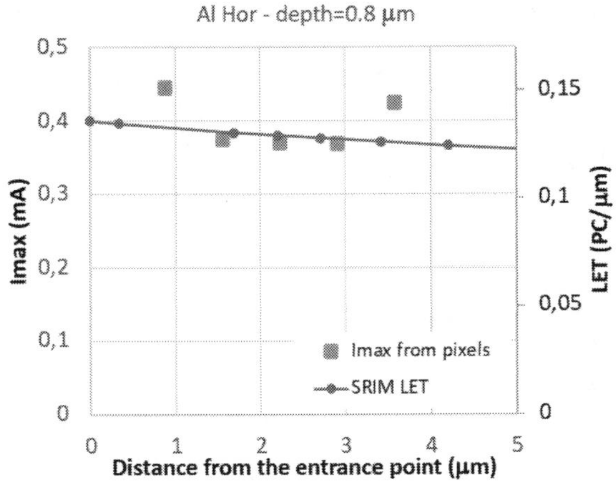

Fig. 6. Current peak (squares) and LET (circles) versus the distance from the striking point of the ion in the matrix – Aluminum ion case – depth=0.8µm.

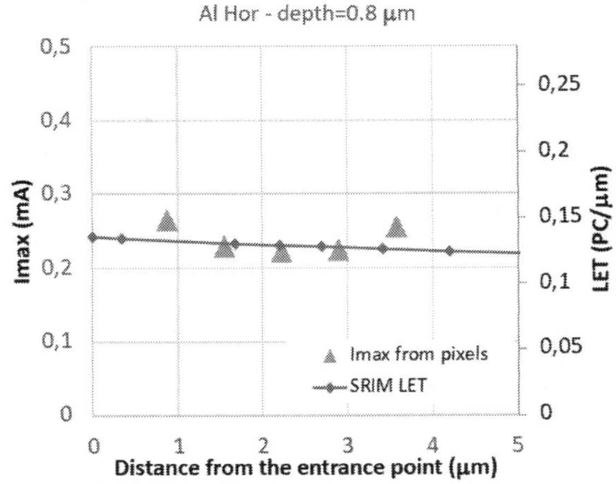

Fig. 7. Current peak (squares) and LET (circles) versus the distance from the striking point of the ion in the matrix – Aluminum ion case – depth=1.5µm.

the 3x5 detection matrix. In Fig. 5, three metrics are analyzed: the total collected charge at a pixel (Qcoll), the current peak (Imax) and the time corresponding to this peak current (tmax). The particle is simulated parallel to the z-plane at two different depths: 0.8µm or 1.5µm. The direction is (100). The starting point of the generation is on the middle of the y-direction (red arrow). In a previous study [21], it was shown that the highest current at a pixel corresponds to the highest electron density and the lowest current corresponds to the lowest charge density. Thus, the current variation can provide information about the charge density in the detection cell. It was also observed that the current shape can give information about the distance between the charge generation and the contact. Indeed, the contact located just above the ion corresponds to the shortest current delay.

As expected, the signals at the left and the right of the ion axis provide the same response symmetrically with respect to the generation axis. The highest currents and collected charges and the shortest time are obtained along the ion axis. If the ion is located deeper into the matrix, the amplitude between the lowest and highest values is reduced because the charges are shared more evenly among the fifteen pixels. The variation of the time is similar along the axis because the charge generation is almost instantaneous. On the contrary, for both depths, the Imax and Qcoll values varies along the track. This can be correlated to the LET variation used as input into the generation function of the device simulator (Fig. 6 and 7).

In Fig. 6 and 7, the current peak variation along the ion axis (symbols) is directly compared the LET profile of the Aluminum (symbols + line). The correlation is quite good excepted concerning the two pixels on the boundary. Which provided higher-than-expected values. It can be explained by charge sharing at the contacts. Indeed, as studied in previous papers [21], contacts influence each other. Thus, for the same generated ion density, the charges collected at a contact at the edge of a structure will not be shared the same way as charges collected at a contact surrounded by other contacts. The charge under the edge contact will be higher than the one under a surrounded contact. which is consistent with the behaviors observed in Fig. 6 and 7. However, the correlation is better for depth=1.5µm. Indeed, as observed in Fig. 5, the amplitude between the lowest and highest values of the current peak is reduced at the depth what can smooth charge sharing effect.

B. Secondary particles tracking: neutron-boron 10 interaction

The second study deals with the particles generated by the primary interaction between thermal neutron with boron-10, which produces one lithium nucleus (Li) with an initial energy of 0.9 MeV and one Alpha particle (He) with an initial energy of 1.5 MeV. These two secondary particles are simulated as moving parallel to the z-plane at depth=0.8 µm. Two configurations are investigated: in the first (Fig. 8), the Alpha (He) particles propagates along the (110) crystallographic direction (Diago.), while and lithium (Li) moves along in the

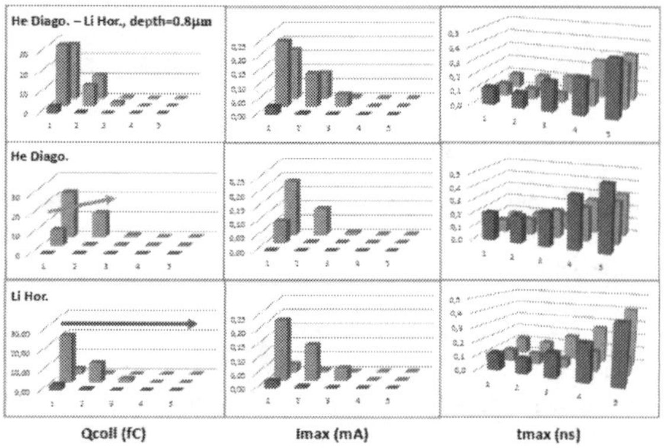

Fig. 8. Total collected charge (Qcoll), current peak (Imax) and time corresponding to this current peak (tmax) for each pixel of the detection matrix. Three configurations generated at depth=0.8µm: simultaneous generation of He Diago. and Li Hor., He Diago. alone and Li Hor. alone.

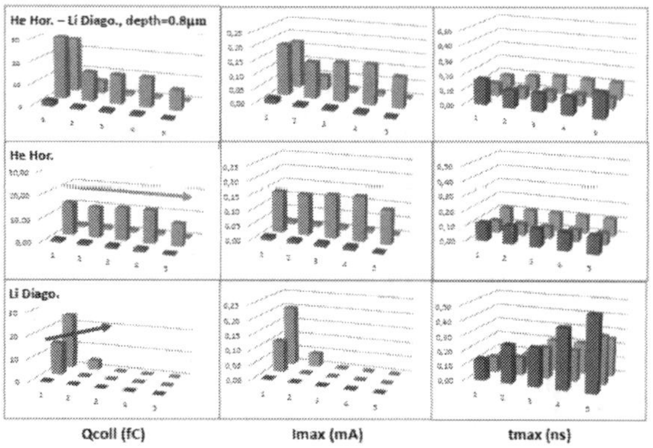

Fig. 9. Total collected charge (Qcoll), current peak (Imax) and time corresponding to this current peak (tmax) for each pixel of the detection matrix. Three configurations generated at depth=0.8µm: simultaneous generation of He Hor. and Li Diago., He Hor. alone and Li Diago. alone.

(100) direction (Hor.). In the second configuration (Fig. 9), the directions are reversed: the Alpha (He) travels along the (100) direction (Hor.) and lithium ion (Li) along the (110) direction (Diago.). Fig. 8 presents three scenarios: simultaneous generation of He Diago. and Li Hor., He Diago. alone and Li Hor. alone. In Fig. 9, three configurations are also considered: simultaneous generation of He Hor. and Li Diago., He Hor. alone and Li Diago. alone. As for the Aluminum, three metrics are analyzed: the total collected charge at a pixel (Qcoll), the current peak (Imax) and the time corresponding to this current peak (tmax). The starting point of the generation is on the middle of the *y*- direction (green or red arrow).

The highest currents and collected charges values and the shortest time give information about the charge density and the distance between the contact and the track. As expected, for Li. Hor. (Fig. 8) and He Hor. (Fig. 9), the signals at the left and the right of the ion axis provide the same response symmetrically with respect to the generation axis. The current peak for these cases can be correlated to the LET variation as with the Aluminum ion case (Fig. 10 and 11). The better correlation is obtained for the Lithium (Fig. 10). The correlation for the alpha is also very good but, as for the Aluminum, a variation is observed for the two contacts at the edge of the matrix. This is not observed for the Lithium because the charge at the end is zero.

For the He. Diago (Fig. 8) and Li Diago. (Fig. 9), the peak current and collected charge are the highest, and the time the lowest for the contact located on the top left corner of the figure. This is consistent because the generation is just under this contact. The two other significant contacts on the left also exhibit values consistent with the expected charge variation.

The study of He or Li alone are consistent with results obtained for the simultaneous generation of both particles. This is visible on Fig. 8 and 9.

C. Case study analysis

The cases of Lithium and Alpha particles are further analyzed in this section to reconstruct the trajectory within the detection matrix.

First case concerns the Lithium ion (Fig. 12). The time corresponding to the peak current (tmax) for the three highest current values is very similar, with a maximum difference of approximatively 20 ps between the first the third peaks. This

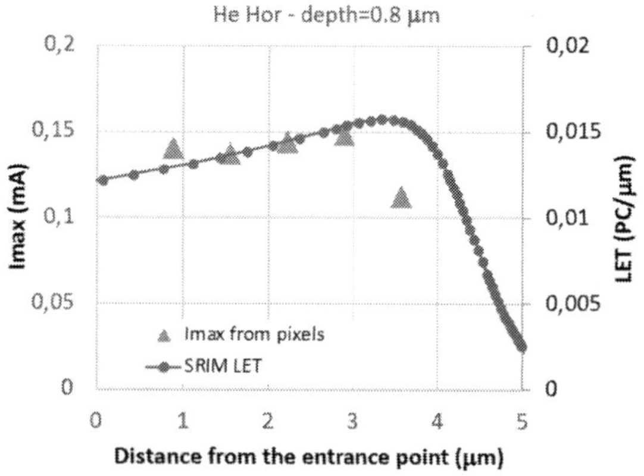

Fig. 10. Current peak (triangles) and LET (circles) versus the distance from the striking point of the ion in the matrix – He Hor. case – depth=0.8µm.

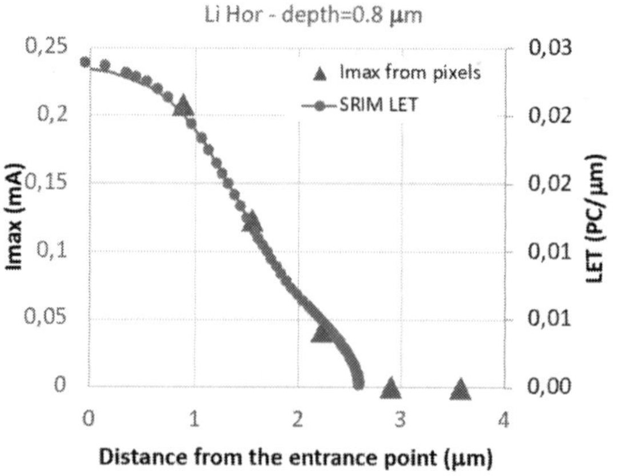

Fig. 11 Current peak (triangles) and LET (circles) versus the distance from the striking point of the ion in the matrix – Li Hor. case – depth=0.8µm.

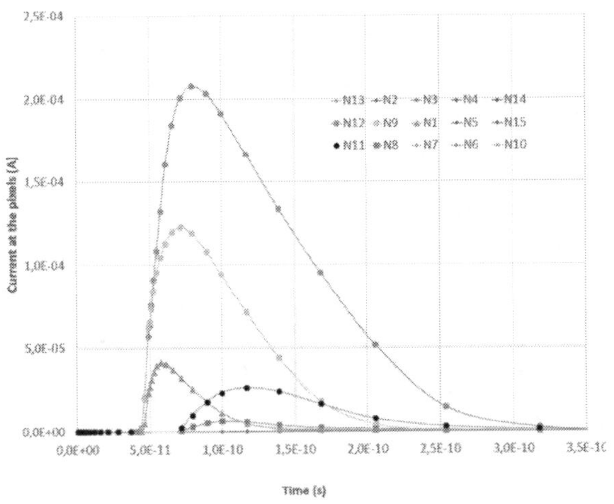

Fig. 12. Lithium generation in the 15-pixels matrix at depth=0.8μm.

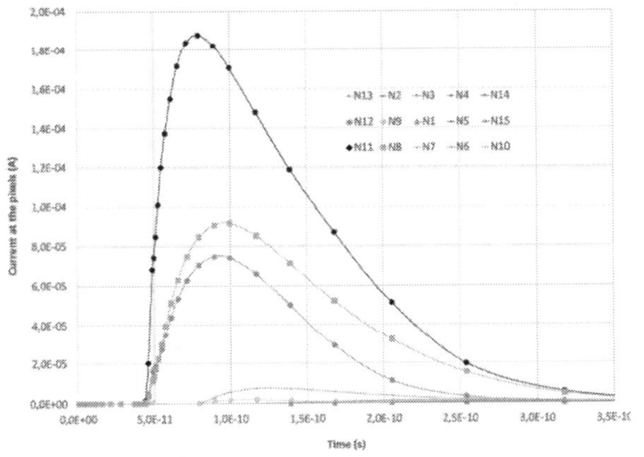

Fig. 13. Alpha generation in the 15-pixels matrix at depth=0.8μm.

parameter may vary slightly depending on the charge density.

Indeed, for a same location, the current peak will be reached later if the charge density is very high. That what we observe in Fig. 12. The currents measured on the left side of the matrix (N13, N2, N3, N4, N14) and on the right side (N11, N8, N7, N6, N10) exhibit the same response symmetrically with respect to what should be the generation axis. The corresponding times for these contacts are at least twice those of the of the middle contacts. This suggests that the particle path is probably located closer to the central region of the matrix than to its edges. Accounting for Fig. 12, we can assume that the ion has been generated under the middle contacts (N12, N9, N1, N5, N15) and that the charge density decreases progressively from contact N12 to contact N1. This observation is consistent with the Li. Hor. configuration presented in Fig. 8.

Second case concerns the Alpha particle (Fig. 13). In this configuration, the signal responses are not symmetric. The highest current exhibit the shorter tmax, suggesting that the maximum of the charge density is probably located under contact N11. The other closest contacts are probably contacts N8 and N12. The highest charge density is for contact N8. The current peak for N9 and N1 is lowest what is consistent with the assumption of a track diagonally generated under N11 what would correspond to the He. Diago. configuration of Fig. 8.

IV. Back to the VCO analysis

As demonstrated in a previous work, two particles were sequentially injected into the VCO to evaluate the tracking capability based on the VCO response. The average output voltage corresponding to the highest currents peaks for both particles is shown in Fig. 14. The associated current and voltage metrics are summarized in Fig. 15. All the currents are accurately reproduced, excepted for the N10 current associated with the Aluminum ion. Although this current is sufficiently large to be detected by the VCO, its peak amplitude exceeds the dynamic range of the circuit, resulting in saturation. This saturation phenomenon was previously analyzed in [22]. Nevertheless, tracking remains feasible for both high-energy and low-energy particles, despite the occurrence of saturation.

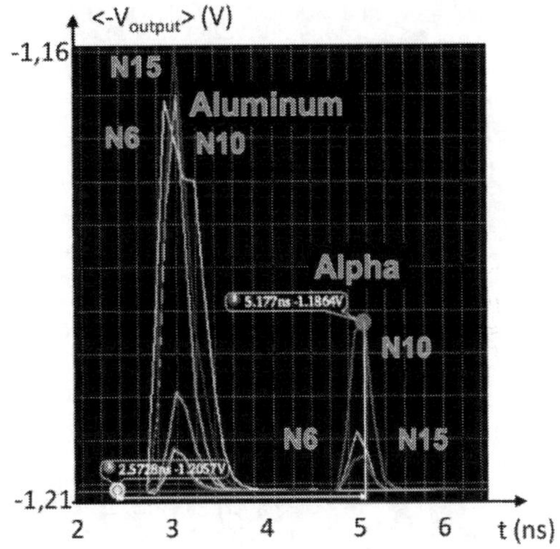

Fig. 14. Mean voltage for the highest currents of Aluminum – Diago. (left), and Alpha – Diago. (right).

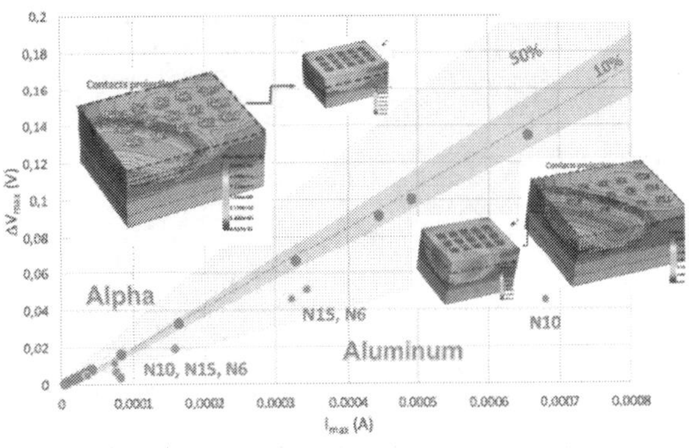

Fig. 15. Linearity curve and detected currents concerning the Aluminum ion or the Alpha particle crossing in a fifteen contacts matrix.

V. Conclusion

The detection chain presented in this work is based on an indirect particle detection through a voltage-controlled oscillator (VCO). This paper primarily focuses on the detection matrix and investigates the feasibility of ion tracking under various configurations. The analysis of the signals extracted from the matrix showed the possibility to recognize different particle responses by observing three key metrics : the total collected charge, the current peak, and time corresponding to this peak.

Then the maximum signal generation time gives information about the distance from the contact to the particle and the current peak can give information about the local charge density. The total collected charge correlates with the energy deposited by the particles.

Finally, with a VCO oscillating at 4.3GHz we have demonstrated the ability to track different types of particles and to establish a clear relationship between input current characteristics and the corresponding output signal, thereby enabling particle identification.

Future studies will focus on spectrometric analysis using oscillators operating at various frequencies to enable the discrimination of particles with different characteristics (possibility to address various environments). Addressing this objective will require further investigation into power consumption and address associated technological challenges related to material implementation.

References

[1] C. Fiorini, A. Pullia, E. Gatti, A. Longoni, and W. Buttler, "A monolitic implementation of a switch-current 'wheel' amplifier for multi-channel silicon drift detectors, *IEEE Trans.* Nucl. Sci., vol. 46 n°3 p. 161, 1999.

[2] D. Hussona,, A. Bozierb, S. Higueretc, T.D. Lec, A. Nourreddinea, "AlphaRad, a new integrated CMOS System-on-Chip for high efficiency alpha particles counting" , *Nuclear Instruments and Methods in Physics Research A 569.*, pp. 845-852 161, 2006.

[3] I. Perić, C. Kreidl, P. Fischer, "Particle pixel detectors in high-voltage CMOS technology–New achievements", Nuclear Instruments and Methods in Physics Research A, Vol. 650, Issue 1, p. 158-162, 2011.

[4] M. Trocmé, S. Higueret, D. Husson, A. Nourreddine, T.D. Lê, "A new compact device for efficient neutron counting using a CMOS active pixel sensor", Radiation Measurements, pp. 1100-1103, 2008.

[5] D. Passeri, P. Placidi, L. Verducci, P. Ciampolini, G. Matrella, A. Marras, G/M. Bilei, "High-resolution CMOS particle detectors: design and test issues", 331-334 Vol.1, *2003* IEEE Nuclear Science *Symposium Conference Record.*

[6] M. Campbell, V. Havranek, E. HEIjne, T. Holy, J. Idarraga, J. Jakubek, C. Lebel, C. Leray, X. Llopart, J. Novotny, S. Pospisil, L. Tlustos, Z. Vykydal, "Charge collection from proton and alpha particle tracks in silicon pixel detector devices", 2007 IEEE Nuclear Science *Symposium Conference Record.*

[7] M. J. Baudot ; G. Bertolone ; A. Brogna ; G. Claus ; C. Colledani et al., "First test results Of MIMOSA-26, a fast CMOS sensor with integrated zero suppression and digitized output", 2009 IEEE Nuclear Science *Symposium Conference Record, 2009.*

[8] Tomas Baca ; Martin Jilek ; Petr Manek ; Petr Stibinger ; Vladimir Linhart ; Jan Jakubek ; Martin Saska, "Timepix Radiation Detector for Autonomous Radiation Localization and Mapping by Micro Unmanned Vehicles", 2019 IEEE/RSJ International Conference on Intelligent Robots and Systems (IROS), 3-8 Nov. 2019.

[9] Ying Zhang, Christine Hu-Guo, Daniel Husson, The-Duc Le , Stéphane Higueret, Yann Hu, "A High-Sensitivity Low-Power CMOS Sensor for a Future Neutron Personal Dosimeter", *in IEEE Trans. On Nucl. Sc.*, vol.59, no.4, pp. 1465-1471, August 2012.

[10] Rahajandraibe W., Aziza H., Coulié-Castellani K., Micolau, "Device and Method for Detecting Radiation Particles"
G. Patent AMU/CNRS/UAPV – Septembre 17, 2015 WO2015136220, https://patentscope.wipo.int/search/en/detail.jsf?docId=WO2015136220

[11] K. Castellani-Coulié, H. Aziza, W. Rahajandraibe, G. Micolau, J.-M. Portal, "Development of a CMOS Oscillator Concept for Particle Detection and Tracking", *IEEE Trans Nucl. Sci.*, Vol. 60 , Issue 4 , pp 2450 - 2455 , Aug. 2013. DOI: 10.1109/TNS.2013.2254723

[12] K. Coulié-Castellani, W. Rahajandraibe, G. Micolau, H. Aziza, J.-M. Portal, "Improvement of a detection chain based on a VCO concept for microelectronic reliability under natural radiative environment", *IEEE LATW Conference*, 2015.

[13] S. Ben Krit, K. Coulié-Castellani, W. Rahajandraibe, L. Ottaviani, G. Micolau, H. Aziza, J.-M. Portal, "Comparison of a Readout Chain Dedicated to the Signal Conditioning of a Particle Detector and an Innovative Chain Based on a VCO Concept", *RADECS 2015*, Moscou, Russia, September 2015. DOI: 10.1109/RADECS.2015.7365592

[14] K. Coulié, W. Rahajandraibe and L. Ottaviani, "Particles discrimination and tracking through a matrix of detection coupled with a voltage-controlled oscillator", *ANIMMA 2023, EPJ Web Conf.*, 288 (2023) 10017 DOI: https://doi.org/10.1051/epjconf/202328810017

[15] K. Coulié, W. Rahajandraibe, H. Aziza, G. Micolau and R. Vauché, "Detection limit of a VCO based detection chain dedicated to particles recognition and tracking," in ANIMMA 2017 – Advancements in Nuclear Instrumentation Measurement Methods and their Applications, Volume 170, 2018.

[16] TCAD Sentaurus user's manual (Synopsys).

[17] A. Akkerman, J. Barak, D. Emfietzoglou, "Ion and electron track-structure and its effects in silicon: model and calculation", *Nuclear Instruments and Methods in Physics Research*, pp. 319-336, 2005.

[18] www.srim.org

[19] Vasiliy V. Markelov and Michael G. Tverskoy, "Evaluation of LET Spectra Produced by High Energy Protons in Si", *RADECS 2015*, Moscou, Russia, September 2015. DOI: 10.1109/RADECS.2015.7365592

[20] M. Gedion, F. Wrobel, F. Saigné, R. D. Schrimpf and J. Mekki, "Monte Carlo Simulations to Evaluate the Contribution of Si Bulk, Interconnects, and Packaging to Alpha-Soft Error Rates in Advanced Technologies," in IEEE Transactions on Nuclear Science, vol. 57, no. 6, pp. 3121-3126, Dec. 2010, doi: 10.1109/TNS.2010.2085446.

[21] K. Coulié, W. Rahajandraibe and L. Ottaviani, "On the Improvement of VCO Based Matrix Particle Detector for High Resolution Particles Recognition and Tracking," 2020 20th European Conference on Radiation and Its Effects on Components and Systems (RADECS), Toulouse, France, 2020, pp. 1-6, doi: 10.1109/RADECS50773.2020.9857723

[22] K. Coulié, W. Rahajandraibe and L. Ottaviani, "Particle Identification and Tracking by the Use of a Pixel-Based Semiconductor Radiation Detector Coupled with Voltage Controlled Oscillators", ANIMMA 2021, EPJ Web Conf., 253 (2021) 09007
DOI: https://doi.org/10.1051/epjconf/202125309007

A Novel Measurement Setup for the Data Acquisition and Evaluation for Hydrogen-Filled Proportional Detectors and Stilbene Scintillation Detectors

Vincent Melzer[1,*], Sascha Weichel[1], Robert Reisdorf[1,ǂ], Marco Viebach[1], Carsten Lange[1], and Antonio Hurtado[1]
[1]Chair of Hydrogen and Nuclear Energy, Dresden University of Technology, Germany
(*) vincent.melzer@tu-dresden.de

***Abstract*—Within the framework of the NAUTILUS research project, the fast neutron spectrum of the Training Reactor AKR-2 at the Dresden University of Technology is being determined with proton recoil detectors. Hereby, two different detector technologies are utilized for the evaluation of different energy ranges of fast neutrons: Hydrogen-filled proportional detectors for energies from a few to roughly 1500 keV, and stilbene scintillation detectors for energies above 1 MeV. This publication presents the data acquisition system *neXmess* for both detector types currently in development and its implementation in the measurement chain at the AKR-2. Therefore, recent experimental results – functionality test results of the data acquisition system, exemplary pulses, pulse shape discrimination histograms and detector characterization results for hydrogen-filled proportional detectors – are being presented.**

***Keywords* —neutron spectrum, fast neutrons, data acquisition systems, neutron detectors, hydrogen-filled proportional detectors, stilbene scintillation detectors**

I. Introduction

THE experimental determination of fast neutron spectra is a crucial aspect in the characterization of research reactors. Therefore, a main goal of the NAUTILUS research project [1] is the determination of the fast neutron spectrum of the Training Reactor AKR-2 at the Dresden University of Technology in Germany. This is especially performed by means of proton recoil counters whereas two complementary types of detectors are used: Hydrogen-filled proportional detectors (HPDs) and stilbene scintillation detectors.

Both types of detectors have been used in neutron spectrometry for decades [2] [3] [4] whereas each detector type covers a distinct and overlapping portion of the neutron spectrum. HPDs are used for low to intermediate fast neutron energies, ranging from a few to roughly 1500 keV. Stilbene detectors are suited for higher fast neutron energies, effectively covering the range above 1 MeV. As illustrated by Tagziria and Hansen [5], and Klein [6], a combined use of both detector types can enable an assessment of the complete fast neutron spectrum.

However, the pulse shapes generated by both detector types differ significantly which poses specific challenges for the associated data acquisition system. For HPDs, pulses are produced with a rise time of several tens of microseconds, followed by a plateau and a slow decay, resulting in total pulse durations of several hundred microseconds up to the millisecond range. In contrast, stilbene detectors generate short pulses with durations of only a few tens of nanoseconds.

These differences are not only in duration but also in how deposited energy and incident particle type are determined from the signals. For HPDs, the deposited energy is typically derived from the pulse height while pulse shape discrimination methods rely on the rise time in combination with the pulse amplitude (as for example demonstrated in Hawkes [7]). In contrast, for stilbene detectors, the deposited energy is extracted via pulse integration, and pulse-shape discrimination is performed using the ratio of a short integral to the total pulse integral, i.e. the pulse integral fraction (see for example Ruch et al. [8]).

Due to these fundamentally different signal characteristics and analysis requirements, the novel data acquisition system *neXmess* which can measure pulses of both detector types is being developed in the NAUTILUS research project.

This publication presents the current state of the *neXmess* system. The measurement chain at the AKR-2 is described, including all its components. Experimental results from the *neXmess* data acquisition system are shown, covering signal-generator test measurements that verify the functionality of the trigger logic and the baseline restoration filter of the system. Pulse shapes and discrimination capabilities are analyzed for HPDs and stilbene detectors using measured pulses and the according 2D histograms of the pulse-shape discrimination quantities. Characterization results for HPDs, including energy calibration, resolution, and relative gain, are also presented.

II. Materials

The current measurement chain at the AKR-2 including the *neXmess* data acquisition system is illustrated in the structure diagram depicted in Fig. 1. In the following sections, the individual components of the measurement chain are discussed.

ǂ R. Reisdorf, now at Helmholtz-Zentrum Dresden-Rossendorf, 01328 Dresden, Germany

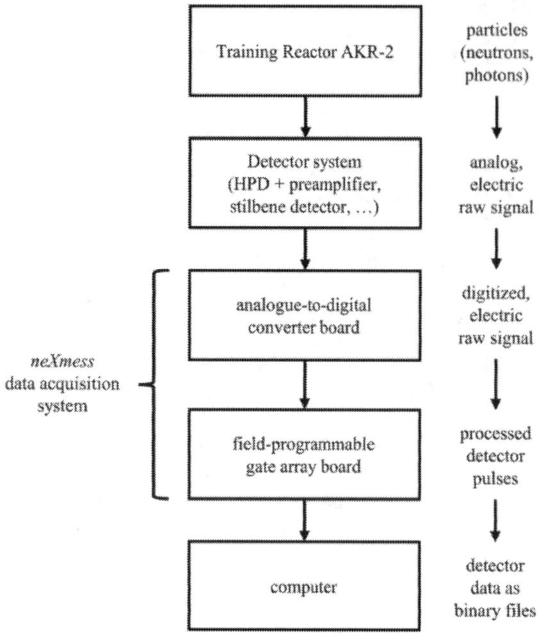

Fig. 1. Structure diagram of the measurement chain at the AKR-2

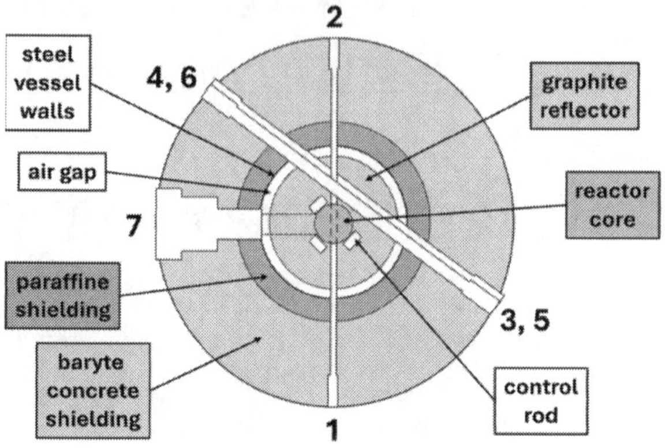

Fig. 2. Cross-sectional diagram of the AKR-2

Fig. 3. Photo of a HPD ("NOK" model)

A. Training Reactor AKR-2

The Training Reactor AKR-2 (German: **A**usbildungs**k**ern**r**eaktor) is described in Lange and Bernt [9]. It is a homogeneous, thermal zero-power reactor with a nominal thermal power of 2 W and is located at the Chair of Hydrogen and Nuclear Energy at the Dresden University of Technology in Germany. Since March 22nd, 2005, it has been used for research, education and public relations.

The reactor core is of cylindrical shape with a diameter of 250 mm and a critical height of 275 mm. It is constructed from two separate halves which are made of layered fuel plates consisting of a mixture of uranium dioxide (enrichment 19.8 % uranium-235) and polyethylene. To achieve criticality, the lower core is cojoined with the upper core half. The reactor core furthermore is enclosed by a high-purity graphite reflector, and layers of paraffine and baryte concrete as shielding. For controlling and for reactor safety three control rods containing cadmium are used. Multiple fission chambers and a gamma-compensated ionization chamber are permanently mounted inside the reactor and operated for monitoring the neutron flux level. The AKR-2 furthermore offers five experimental channels with their openings labeled from 1 to 7. A cross section of the AKR-2 is depicted in Fig. 2.

B. Detectors

A selection of the available HPDs and stilbene detectors is presented in the following sections. They are summarized in Tab. I.

1) Hydrogen filled proportional detectors

Five HPDs used at the AKR-2 were produced under the trademark name "NOK" at the former Instytut Badań Jądrowych (IBJ) in Świerk, Poland. A photograph is shown in Fig. 3.

The construction of the detectors is thoroughly described in Jánský et al. [10]. Mainly, they consist of a spherical stainless steel cathode shell and an inside anode tungsten wire. By means

TABLE I
PROPORTIES OF THE AVAILABLE DETECTORS AT THE AKR-2

NOK detectors						Stilbene detectors	
Detector	#	Pressure of H_2	Pressure of CH_4	Calibration additive	Operating voltage	Serial number	Operating voltage
NOK 0.543	7/75	0.5 atm	0.05 atm	^{3}He, ca. 0.01 Tr	+900 V – +1660 V	S2AC1498	-730 V
NOK 145	1/75	1.0 atm	0.1 atm	^{3}He, ca. 0.01 Tr	+1200 V – +1960 V	S2AC1499	-540 V
NOK 160	1/72	1.0 atm	–	adsorbed BF_3	+1500 V – +1800 V		
NOK 343	5/75	3.0 atm	0.3 atm	^{3}He, ca. 0.01 Tr	+1540 V – +2720 V		
NOK 443	2/75	4.0 atm	0.2 atm	^{3}He, ca. 1 Tr	+2300 V – +2500 V		

https://doi.org/10.1051/epjconf/202533810013

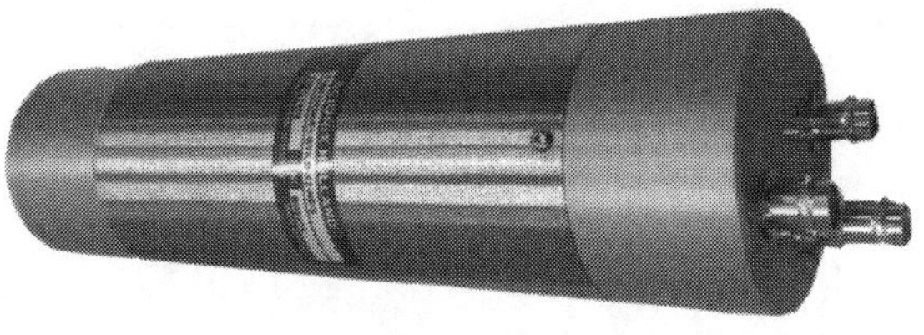

Fig. 4. Photo of a stilbene detector ("S2AC1499" model by Scionix)

Fig. 5. Photo of the *neXmess* data acquisition system

of a C-to-SHV connector, the bias voltage is applied, and the detector signal is transmitted to a preamplifier. Inside the measurement chain at the AKR-2, a "NHQ 206L" high-voltage source by iseg and a "Model 2006E" preamplifier by Canberra is used. The active detector volume inside the spherical shell consists of hydrogen gas. Additional methane is used as quenching additive. Traces of helium-3 and adsorbed BF_3 are utilized for characterization purposes.

In contrast to stilbene detectors, HPDs currently are scarcely used. However, research on the use of NOK detectors in neutron spectroscopy is performed by the ÚJV Řež in the Czech Republic [11], [12], [13].

2) Stilbene detectors

Two stilbene detectors at the AKR-2 were produced by Scionix with the model's name "51B51/2M-E1-STILB-X2-NEG". A photo of one of the stilbene detectors is shown in Fig. 4.

The stilbene crystal has a diameter of 51 mm and a height of 51 mm. It is encased in a body of aluminum and coupled to a "ET 9266" photomultiplier tube by ET Enterprises. By means of a SHV connector the bias voltage is applied. Hereby, a "3106 D" high-voltage source by Canberra is used. The detector signal is transmitted via a BNC connector.

C. Data acquisition system

The data acquisition system *neXmess* consists of an analog-to-digital converter (ADC) board which is connected to a field-programmable gate array (FPGA) board. Furthermore, it is enclosed by a casing. A photo of the *neXmess* data acquisition system is shown in Fig. 5.

As ADC board, a modified version of the "EVAL-FMCDAQ3-EBZ" by Analog Devices is used. It utilizes a "AD9680" ADC which digitizes electronic signals with a configurable sampling rate. In the *neXmess* data acquisition system, a sampling rate of 500 MHz was set.

As FPGA board, the "Zynq 7000 SoC ZC706 Evaluation Kit" by Advanced Micro Devices is used. It utilizes a "Zynq-7000 XC7Z045-2FFG900C SoC" which features a "Kintex-7" FPGA and "dual-core ARM Cortex-A9 MPCore" processor.

On the FPGA, custom created logic modules are used for signal filtering (signal inversion, offset correction, FIR filter, baseline restoration), pulse triggering and timestamp generation.

Custom C++ server software on the FPGA board is used to configure, start and stop measurements, and to transmit the measurement data. An additional custom Python software "Coincer" is used to for the control of the server software. For the evaluation of the detector pulse data, individual Python scripts are used.

III. Methods and Measurements

A. Test of the data acquisition system

Core components of the *neXmess* data acquisition system include a trigger logic unit and a baseline restoration filter. To verify their functionality, test measurements were carried out using a signal generator configured to produce rectangular pulses superimposed on a slow sinusoidal signal. Hereby, one measurement M1.1 with and one measurement M1.2 without activation of the baseline restoration filter was performed. For both measurements, the following parameters were configured: A measurement time of 1 ms, a trigger window size (hereafter referred to as "long-gate") of 100 samples, a number of samples before the triggering point (hereafter referred to as "pre-gate") of 20 samples and a trigger threshold of 500 ADC units.

B. Experiments with hydrogen-filled proportional detectors

A series of measurements was performed using the HPD "NOK343" to evaluate the change of its pulse shapes under varying conditions. The detector was operated at multiple bias voltages and exposed to two different types of radiation sources: the mixed neutron and photon field of the AKR-2 and a set of gamma calibration sources. The specific configurations are listed in Tab. II. For all measurements, a long-gate of 10000 samples and a pre-gate of 1000 samples was configured. Due to the general energy deposition of neutron-induced events in the detector being bigger than those of photon-induced events, a trigger threshold of 30 ADC units was set for measurements M2.1 to M2.3 while for measurement M2.4 a trigger threshold of 20 ADC units was applied.

Additional characterization measurements were performed to determine the energy calibration, energy resolution and relative gain of the NOK detectors. The specific measurement configurations are listed in Tab. III. For all measurements, a long-gate of 10000 samples was set, along with a pre-gate of 2000 samples and a trigger threshold 30 ADC units.

C. Experiments with stilbene detectors

A measurement M4 with stilbene detector "S2AC1499" was performed in the field of the AKR-2 to investigate the pulse shapes of the detector. The measurement time was set to 2 h 15 min. As a long-gate, 80 samples were configured, along with a pre-gate of 10 samples and a trigger threshold of 20 ADC units.

IV. Data Analysis

A. Test of the data acquisition system

The measured pulses with the *neXmess* data acquisition system of measurements M1.1 and M1.2 are depicted Fig. 6.

B. Pulse shapes of both detector types

Exemplary pulses of measurements M2.1 and M4 are depicted in Fig. 7 and Fig. 8. For the HPD, the pulse height and risetime were annotated. As for the stilbene detector, the integration boundaries for the long and short integral are depicted.

C. Pulse-shape discrimination histograms

For measurements M2.1 to M2.4, the HPD pulses were further smoothed in post-processing. Then their pulse heights as well as their risetimes between 30 % and 80 % of the pulse's heights were determined and sorted into a histogram. The results are depicted in Fig. 9.

For measurement M4, short and long window sizes of 16 samples and 40 samples were used for the calculation of the individual pulse integrals. The pulse integral fraction was determined by division of the short window pulse integral by the long window pulse integral. Additionally, for a better assessment of the starting point of integration, a constant fraction discrimination algorithm was applied beforehand. All determined pulse integrals and pulse integral fractions were sorted in a histogram (see Fig. 10).

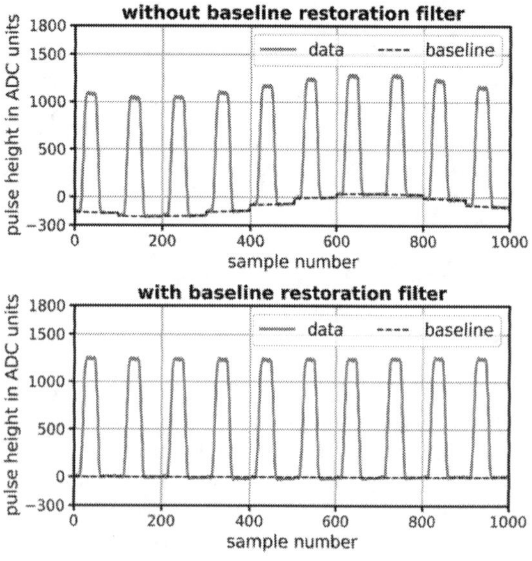

Fig. 6. Test pulses measured by the *neXmess* data acquisition system without (measurement M1.1, top) and with (measurement M1.2, bottom) baseline restoration filter

TABLE II
PULSE SHAPE EVALUATION MEASUREMENTS WITH DETECTOR "NOK343"

Index	Radiation source	Voltage	Measurement time
M2.1	AKR-2	+2700 V	63 min
M2.2	AKR-2	+2500 V	76 min
M2.3	AKR-2	+2300 V	67 min
M2.4	gamma calibration sources	+2700 V	20 min

TABLE III
CHARACTERIZATION MEASUREMENTS FOR HYDROGEN-FILLED PROPORTIONAL DETECTORS

Index	Detector	Voltage	Measurement time	Calibration additive
M3.1	NOK0.543	+1300 V	1 h 1 min	^{3}He
M3.2	NOK0.543	+1400 V	1 h 5 min	^{3}He
M3.3	NOK0.543	+1500 V	1 h 7 min	^{3}He
M3.4	NOK0.543	+1600 V	1 h 19 min	^{3}He
M3.5	NOK145_75	+1200 V	1 h	^{3}He
M3.6	NOK145_75	+1300 V	1 h 29 min	^{3}He
M3.7	NOK145_75	+1400 V	1 h 6 min	^{3}He
M3.8	NOK145_75	+1500 V	1 h 9 min	^{3}He
M3.9	NOK145_75	+1600 V	1 h	^{3}He
M3.10	NOK145_75	+1700 V	1 h 18 min	^{3}He
M3.11	NOK160	+1500 V	1 h 57 min	adsorbed BF_3
M3.12	NOK160	+1525 V	1 h 42 min	adsorbed BF_3
M3.13	NOK160	+1550 V	1 h 19 min	adsorbed BF_3
M3.14	NOK160	+1575 V	50 min	adsorbed BF_3
M3.15	NOK160	+1600 V	1 h	adsorbed BF_3
M3.16	NOK160	+1625 V	1 h	adsorbed BF_3
M3.17	NOK160	+1650 V	1 h	adsorbed BF_3
M3.18	NOK160	+1675 V	1 h	adsorbed BF_3
M3.19	NOK160	+1700 V	1 h	adsorbed BF_3
M3.20	NOK160	+1725 V	1 h	adsorbed BF_3
M3.21	NOK160	+1750 V	1 h	adsorbed BF_3
M3.22	NOK160	+1775 V	1 h	adsorbed BF_3
M3.23	NOK160	+1800 V	1 h	adsorbed BF_3
M3.24	NOK343	+1900 V	1 h 31 min	^{3}He
M3.25	NOK343	+2000 V	1 h 26 min	^{3}He
M3.26	NOK343	+2100 V	1 h 52 min	^{3}He
M3.27	NOK343	+2200 V	1 h 5 min	^{3}He
M3.28	NOK343	+2300 V	1 h 6 min	^{3}He
M3.29	NOK343	+2400 V	1 h 3 min	^{3}He
M3.30	NOK343	+2500 V	1 h 15 min	^{3}He
M3.31	NOK343	+2600 V	1 h	^{3}He
M3.32	NOK343	+2700 V	1 h 2 min	^{3}He
M3.33	NOK443	+2300 V	1 h	^{3}He
M3.34	NOK443	+2325 V	1 h	^{3}He
M3.35	NOK443	+2350 V	1 h	^{3}He
M3.36	NOK443	+2375 V	1 h	^{3}He
M3.37	NOK443	+2400 V	1 h	^{3}He
M3.38	NOK443	+2425 V	1 h	^{3}He
M3.39	NOK443	+2450 V	1 h	^{3}He
M3.40	NOK443	+2475 V	1 h	^{3}He
M3.41	NOK443	+2500 V	1 h	^{3}He

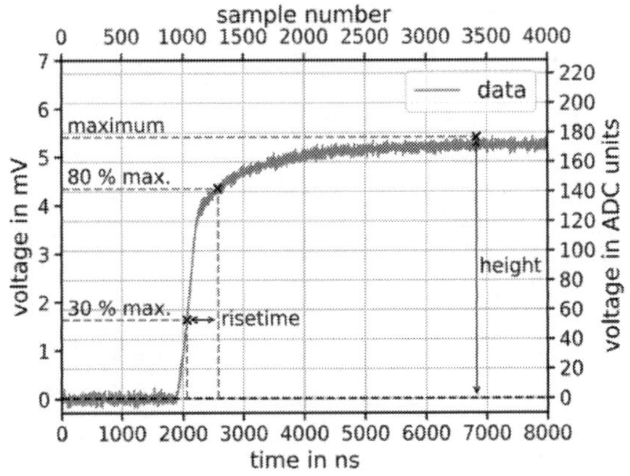

Fig. 7. Exemplary HPD pulse of measurement M2.1, zoomed into samples 0 to 4000 and annotated with its pulse height and risetime

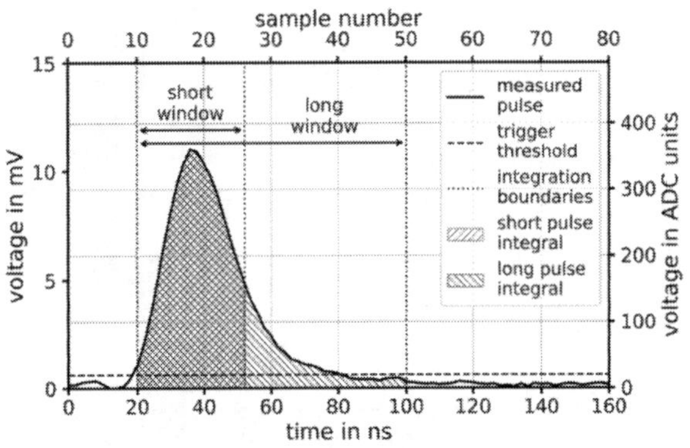

Fig. 8. Exemplary stilbene detector pulse of measurement M4, annotated with the integration boundaries for the determination' of their respective pulse integrals

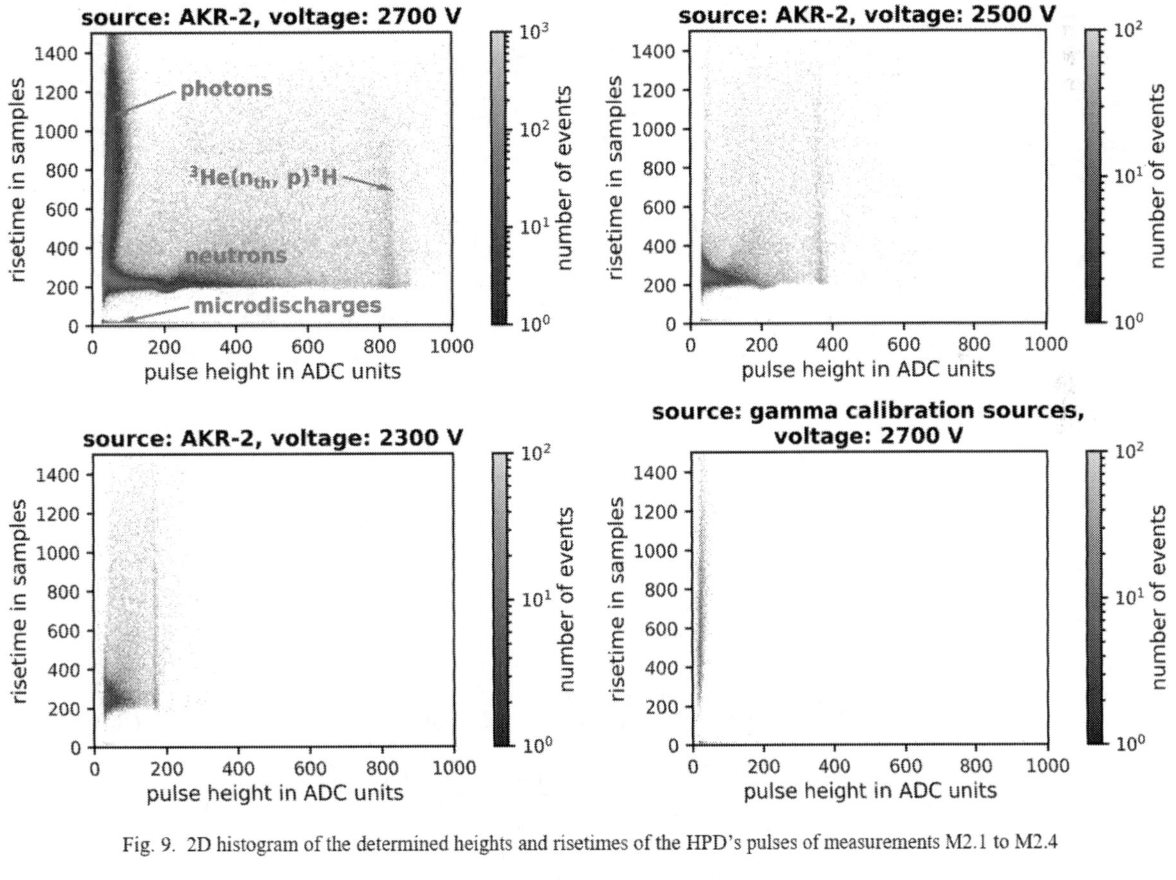

Fig. 9. 2D histogram of the determined heights and risetimes of the HPD's pulses of measurements M2.1 to M2.4

D. Characterization of the hydrogen-filled proportional detectors: Energy calibration, energy resolution and relative gas gain

For measurements M3.1 to M3.41, the HPD's pulses were further smoothed in post-processing. Then, their pulse heights as well as their risetimes between 30 % and 80 % of the pulses' heights were determined. For a filtering of the neutron-induced detector pulses, only the pulse heights above 10 ADC units were sorted into histograms. Finally, the characteristic peaks of the detectors' additives which develop due to thermal neutron capture were evaluated. Summarizing information on the calibration peaks is listed in Tab. IV.

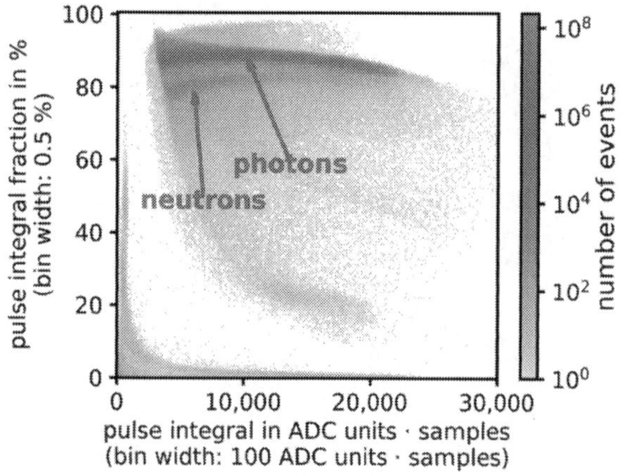

Fig. 10. 2D histogram of the determined pulse integrals and pulse integral fractions for the stilbene detector pulses of measurement M4

TABLE IV
CHARACTERISTIC THERMAL NEUTRON CAPTURE PEAKS OF THE HPDS' CALIBRATION ADDITIVES

Additive	Reaction	Peak	Energy
^{3}He	$^{3}He + n_{th} \rightarrow p + {}^{3}H$	Full energy	0.764 MeV
adsorbed BF_3	$^{10}B + n_{th} \rightarrow \alpha + {}^{7}Li$	^{7}Li edge	1.777 MeV
adsorbed BF_3	$^{10}B + n_{th} \rightarrow \alpha + {}^{7}Li$	α edge	1.015 MeV
adsorbed BF_3	$^{10}B + n_{th} \rightarrow \alpha + {}^{7}Li + \gamma$	^{7}Li edge	1.470 MeV
adsorbed BF_3	$^{10}B + n_{th} \rightarrow \alpha + {}^{7}Li + \gamma$	α edge	0.840 MeV

As a function for approximating the peak, a normal distribution on a linear background was used:

$$N(h) = a \cdot \exp\left\{-\frac{1}{2}\left(\frac{h-\mu}{\sigma}\right)^2\right\} + m \cdot h + n$$

Here, $N(h)$ denotes the number of events for the pulse height h. The normal distribution is described by a normalization factor a, the mean value μ, and the standard deviation σ. For the assumed linear background, m denotes the slope and n the offset for the number of events.

The energy calibration was performed by assigning the known energy value of the respective characteristic peak to the peaks position. A linear calibration function was determined by regression:

$$E(h) = m_{\mathrm{cal}} \cdot h + n_{\mathrm{cal}}$$

Energy calibration curves for the detectors NOK160 and NOK443 are depicted in Fig. 11 and Fig. 12.

For the energy resolution, the full width at half maximum (FWHM) of the peaks was divided by the peaks position. The energy resolutions for the detectors NOK145_75, NOK160, NOK343 and NOK443 are depicted in Fig. 13. As the energy resolution of detector NOK0.543 was poorer than the other detectors, it is depicted separately in Fig. 14.

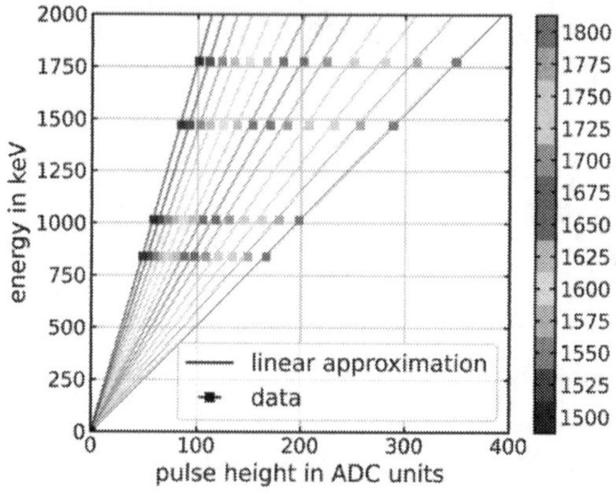

Fig. 11. Energy calibration of the NOK160 detector

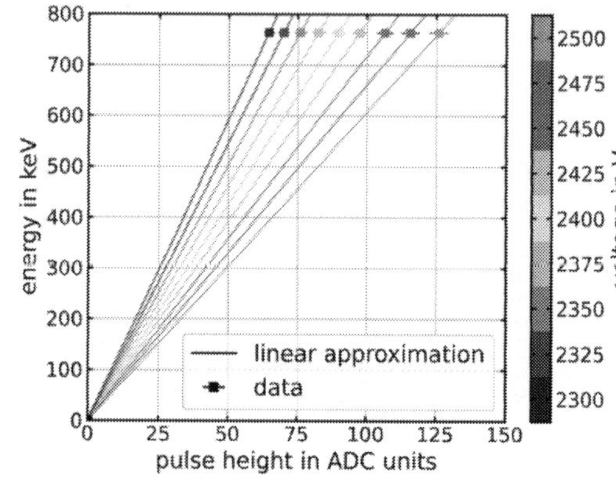

Fig. 12. Energy calibration of the NOK443 detector

The relative gains $M_{\mathrm{rel}}(U)$ for the voltages U of all detectors were all calculated with

$$M_{\mathrm{rel}}(U) = \frac{\mu(U)}{\mu(U_{\mathrm{rel}})}$$

where $\mu(U)$ denotes the characteristic peak's position for the voltage U. For the reference measurement, measurement M3.32 with detector NOK343 at voltage $U_{\mathrm{ref}} = +2700$ V was selected. Additionally, as there were four peaks for all measurements with the detector NOK160, an according weighting of M_{rel} was performed. The results were further evaluated as a Diethorn plot and are depicted in Fig. 15 [14].

Parallel to the approximation of the peaks in the pulse-height histograms, the fit parameters uncertainties were determined as well. By error propagation, the uncertainties of the energy calibrations, energy resolutions and relative gains were calculated.

V. DISCUSSION

A. Performance of the data acquisition system

As illustrated in Fig. 6, the *neXmess* data acquisition system reliantly triggers detector signals for the configured long-gates

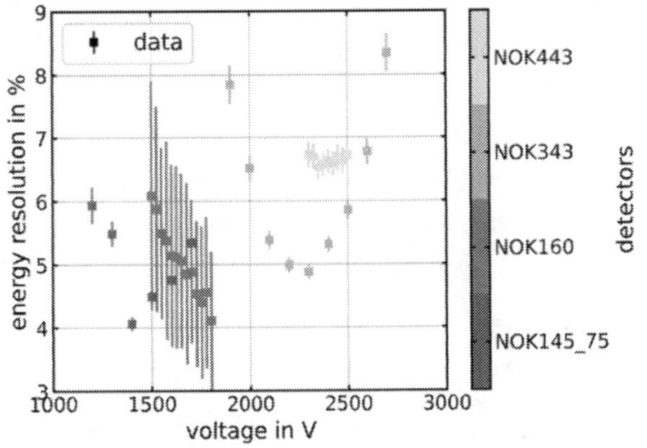

Fig. 13. Energy resolution of the detectors NOK145_75, NOK160, NOK343 and NOK443

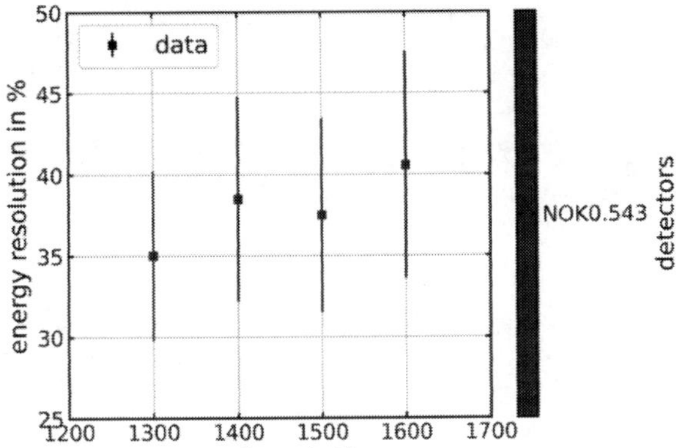

Fig. 14. Energy resolution of the detector NOK0.543

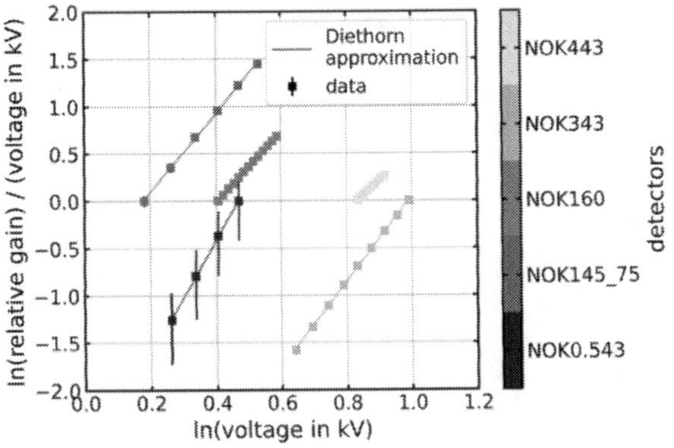

Fig. 15. Relative gain of the NOK detectors as Diethorn plot

and pre-gates. With the elimination of the added sinusoidal test perturbation as the baseline, the baseline restoration filter's functionality was demonstrated.

B. Pulse shapes of both detector types

For the HPDs, the measured pulses (see Fig. 7) develop within several hundred nanoseconds to a few microseconds. The steep incline is followed by a flat top and a slow decline with the latter being governed by the time constant of the preamplifier. As discussed, the heights and risetimes of the pulses can be used to distinguish between pulses of different particle origin. This is illustrated in Fig. 9 where detected events induced by fast and thermal neutrons, by photons as well as by microdischarges [15] in the detector are clearly distinguishable. As expected, the mean pulse height increases with higher applied voltage.

The pulses of the stilbene detectors (see Fig. 8) have a shorter length compared to the HPD's pulses of some ten nanoseconds. As discussed, pulse integrals and pulse integral fractions can be used to discriminate between neutron- and photon-induced pulses in stilbene detectors. This is depicted in Fig. 10. The upper branch represents photon-induced, the lower branch neutron-induced detector events.

C. Characterization of the hydrogen-filled proportional detectors

For all detectors, the energy calibration, energy resolution and relative gain at different voltages could be successfully determined.

The detectors NOK0.543, NOK145_75, NOK343 and NOK443 use ^{3}He as calibration additive. Therefore, the according peak of the neutron capture reaction at 764 keV in the pulse-height histogram was evaluated. As NOK160 utilizes adsorbed BF_3 as calibration additive, the four corresponding peaks of the secondary particle escapes of the neutron capture reaction in ^{10}B at and 840 keV, 1015 keV, 1470 keV and 1777 keV were evaluated.

The energy calibrations of all detectors exhibit a decreasing slope for increasing bias voltages. As depicted in Fig. 11, the energy calibration for detector NOK160 is justifying the assumption of linearity that is applied for the other detectors (see Fig. 12).

For the detectors NOK145_75, NOK160, NOK343 and NOK443 the energy resolutions lie in the range of 4 % to 8.5 % with varying uncertainties. The NOK0.543 detector's energy resolution ranges at about 40 %.

The NOK343 detector exhibits the highest total gain, yielded at 2700V operating voltage. For all the detectors, the determined relative gain agrees with the linearity of the Diethorn approximation derived for the electrical field of a cylindrical gas detector [14].

VI. Conclusion and Outlook

In this publication, the current state of the *neXmess* data acquisition system and its integration in the underlying measurement chain at the AKR-2 reactor at Dresden University of Technology was presented. Additionally, recent experimental results for measurements with HPDs and stilbene detectors were shown.

The functionality of the trigger logic as well as the baseline restoration filter was depicted by means of measurement data recorded with a signal generator. For HPDs and stilbene detectors, the corresponding pulse shapes and particle discrimination capabilities were discussed. Experimental characterization results of the HPDs in the form of determined energy calibrations, energy resolutions, and relative gains at different voltages were presented.

As upcoming steps, it is planned to fully integrate the pulse evaluation, especially neutron-gamma discrimination, for both detector types into the FPGA's hardware logic. With this, data rates to be handled by the downstream hardware (computer, LAN, etc.) can be substantially reduced which significantly shortens the processing time for data evaluation. Additionally, a middleware for the *neXmess* data acquisition system is currently being developed. It separates the backend (FPGA configuration, streaming of measurement data, etc.) from the frontend to enable an easier development of both ends.

Finally, the fast neutron spectrum of the AKR-2 will be assessed by means of deconvoluting proton recoil histograms that were measured by the *neXmess* system and the detectors presented here. For this purpose, corresponding response matrixes are currently simulated with FLUKA [16] [17] and Geant4 [18] [19] [20].

ACKNOWLEDGMENT

This work is part of the research projects NAUTILUS and DeepCore funded by the German Federal Ministry of Research, Technology, and Space (BMFTR) under the reference numbers 02NUK079 and 02NUK063, resp. The authors also want to thank Michal Košťál (Centrum výzkumu Řež, Řež, CZ) and Zdeněk Matěj (Masarykova univerzita, Brno, CZ), as well as the members of their respective working groups, especially in regards of HPD research. Additionally, the authors want to express their gratitude to the technicians Erik Findeisen and Tobias Langnickel, as well as to the student Ryan Wickremasinghe.

REFERENCES

[1] M. Viebach, A. Knospe, C. Lange, B. Merk, and A. Hurtado, "NAUTILUS: A Project for the Development of Experimental Methods for Investigating Innovative Approaches to Nuclear Waste Management and to Nuclear Safety," in Proc. PHYSOR 2024, San Francisco, CA, US, 2024, pp. 316-325.

[2] E. F. Bennett, "Fast Neutron Spectroscopy by Proton-Recoil Proportional Counting," NSE, vol. 27, no. 1, pp. 16-27, Jan. 1967, 10.13182/NSE67-A18038.

[3] P. W. Benjamin, C. D. Kemshall and J. Redfearn, "A High Resolution Spherical Proportional Counter," Nucl. Instrum. Methods, vol. 59, no. 1, pp. 77-85, Feb. 1968, 10.1016/0029-554X(68)90347-9.

[4] H. W. Broek, and C. E. Anderson, "The Stilbene Scintillation Crystal as a Spectrometer for Continuous Fast-Neutron Spectra," Rev. Sci. Instrum., vol. 31, no. 10, pp. 1063-1069, Oct. 1960, 10.1063/1.1716815.

[5] H. Tagziria, and W. Hansen, "Neutron spectrometry in mixed fields: proportional counter spectrometers," Radiat. Prot. Dosim, vol. 107, no. 1-3, pp. 73-93, Nov. 2003, 10.1093/oxfordjournals.rpd.a006389.

[6] H. Klein, "Neutron spectrometry in mixed fields: NE213/BC501A liquid scintillation spectrometers," Radiat. Prot. Dosim, vol. 107, no. 1-3, pp. 95-109, Nov. 2003, 10.1093/oxfordjournals.rpd.a006391.

[7] N. P. Hawkes, "Pulse shape discrimination in hydrogen-filled proportional counters by digital methods," Nucl. Instrum. Methods Phys. Res. A., vol. 574, no. 1, pp. 133-136, Apr. 2007, 10.1016/j.nima.2007.01.099.

[8] M. L. Ruch, M. Flaska, and S. A. Pozzi, "Pulse shape discrimination performance of stilbene coupled to low-noise silicon photomultipliers," Nucl. Instrum. Methods Phys. Res. A., vol. 793, no. 1, pp. 1-5, Sep. 2015, 10.1016/ 10.1016/j.nima.2015.04.053.

[9] C. Lange and N. Bernt, "The Education and Research Reactor AKR-2 and its Experimental Programme for Education," in Proc. PHYSOR, Cancún, MX, 2018, pp. 2705-2720.

[10] B. Jánský, Z. Turzík, J. Racek, E. Novák, and M. Karáč, "Analýza vlastností vodíkového sférického proporcionálního detektoru používaného ve spektrometrii neutron," ÚJV Řež, Řež, CZ, Rep. 11860-R, 2002.

[11] B. Jánský and E. Novák, "Neutron spectrometry with spherical hydrogen proportional detectors," Nucl. Instrum. Methods Phys. Res. A., vol. 735, pp. 390-398, Jan. 2014, 10.1016/j.nima.2013.08.034.

[12] B. Jánský, J. Rejchrt, E. Novák, and Anatoly Blokhin. Analysis of cross sections on iron and oxygen using Cf-252 neutron source. Presented at ND 2019 – International Conference on Nuclear Data for Science and Technology. [Online]. Available: https://www.epj-conferences.org/articles/epjconf/abs/2020/15/epjconf_nd2019_18005/epjconf_nd2019_18005.html. Accessed on: Apr. 25, 2023.

[13] A. Horna et al., "Innovative Methods of Signal Processing for Hydrogen-Filled Proportional Counters," IEEE Sens. J., vol. 25, no. 11, pp. 19159-19167, June 2025, 10.1109/JSEN.2025.3558688.

[14] W. Diethorn, "A Methane Proportional Counter System for Natural Radiocarbon Measurements," thesis, Dept. Chem., Carnegie Inst. of Tech., Pittsburgh, PA, US, 1956. pp. 69-76.

[15] K. M. Heeger, S. R. Elliott, R. G. H. Robertson, M. W. E. Smith, T. D. Steiger, and J. F. Wilkerson, "High-voltage microdischarge in ultra-low background 3He proportional counters," IEEE Trans. Nucl. Sci., vol. 47, no. 6, pp. 1829-1833, Dec. 2000, 10.1109/23.914454.

[16] T. T. Böhlen, F. Cerutti, M. P. W. Chin, A. Fassò, A. Ferrari, P. G. Ortega, et al., "The FLUKA Code: Developments and Challenges for High Energy and Medical Applications," NDS, vol. 120, pp. 211-214, June 2014, 10.1016/j.nds.2014.07.049.

[17] A. Ferrari, P.R. Sala, A. Fassò, and J. Ranft, "FLUKA: a multi-particle transport code," CERN, Geneva, CH, Rep. CERN-2005-10, INFN/TC_05/11, SLAC-R-773, 2005.

[18] J. Allison, K. Amako, J. Apostolakis, P. Arce, M. Asai, T. Aso, et al., "Recent developments in GEANT4," Nucl. Instrum. Methods Phys. Res. A, vol. 835, pp. 186-225, Nov. 2016, 10.1016/j.nima.2016.06.125.

[19] J. Allison, K. Amako, J. Apostolakis, H. Araujo, P. Arce Dubois, M. Asai, et al., "Geant4 developments and applications," IEEE Trans. Nucl. Sci., vol. 53, no. 1, pp. 270-278, Feb. 2006, 10.1109/TNS.2006.869826.

[20] S. Agostinelli, J. Allison, K. Amako, J. Apostolakis, H. Araujo, P. Arce, et al., "GEANT4—a simulation toolkit," Nucl. Instrum. Methods Phys. Res. A, vol. 506, no. 3, pp. 250-303, July 2003, 10.1016/S0168-9002(03)01368-8.

Assessment of Gamma-Ray Spectrum Transmission with Scintillator Shading Variations for Determining Minimal Transmission Diameter

Tadeas Zbozinek[1,2,*], Michal Jelinek[1], and Bretislav Mikel[1]
[1]Institute of Scientific Instruments of the CAS, v. v. i., The Czech Republic
[2]Department of Microelectronics FEEC BUT, The Czech Republic
(*) zbozinek@isibrno.cz

***Abstract*— Gamma-ray spectroscopy is essential for identifying radioactive sources, but direct coupling of scintillators to photodetectors risks exposing sensitive electronics to radiation damage in high-radiation environments. Optical fibres offer a safer alternative by transmitting scintillation light over long distances, although differences in diameter can reduce spectral resolution for larger crystals, making precise spectroscopy difficult. To address this issue, we introduce an experimental setup that uses an adjustable iris and aperture plate to simulate fibre cores of various diameters, paired with scintillators of different sizes. We systematically measured the transmitted light intensity and spectral peak resolution at each aperture setting to analyse the relationship between core diameter, scintillator size, and spectral resolution. By optimising the balance between scintillator size and effective aperture, we can retain the advantages of larger crystals, enabling spectral analysis with smaller detection diameters. This work offers practical guidelines for selecting optimal configurations of detector and scintillator sizes, which are essential for monitoring gamma radiation with an optimal balance of high resolution and sufficient sensitivity.**

***Keywords* — Scintillation, Aperture, ionising radiation, optical fibre,**

I. Introduction

THE utilisation of ionising radiation encompasses a wide range of applications. As radiation is hazardous, facilities, including nuclear plants, require a stable monitoring system to measure personnel dose continuously. Standard methods, like ionising chambers, semiconductors and scintillators, provide sufficiently precise measurements. Difficulties occur in sites with higher radiation dose rates, experimental sites and spatially limited areas. These areas face measurement challenges due to electronic interference, size and access limitations, and radiation-induced damage [1]. The effects that influence the evaluation, particularly the damage caused by radiation, can ultimately shorten the lifespan of the equipment [2,3]. An optical fibre can be introduced into the measurement system to overcome these restrictions, maintaining distance from radiation sources and enabling access to confined spaces.

Several methods utilising optical fibres are used in the field of radiation measurement. Absorption techniques take advantage of changes in optical fibre attenuation during radiation exposure [4,5]. The Bragg grating technique observes wavelength shifts in gratings incorporated into fibres [4,6]. However, from the perspective of this paper, the most interesting method lies in luminescence-based detection, specifically radiation-induced luminescence: scintillation.

Plastic fibres are often used within these techniques, as they provide an efficient, easier-to-manufacture, and more cost-effective way to measure radiation. To preserve radiation sensing, enhance resistance, and create a probe-type device, the scintillation material should not be doped into the fibre but used as a separate crystal mounted at the fibre's end [7,9]. This approach ensures good linear counting properties, flexibility in scintillator material and shape choice, radiation resistance, and the ability to measure over longer distances [9]. In this field, several probes were tested using plastic fibres, which are easily manufactured in larger diameters and allow larger scintillators to be used [10 – 15].

However, from a radiation resistance perspective, the optical properties of plastic fibres are less resistant to radiation damage than those of silica or fluoride-doped silica fibres [9,16,17,18]. The silica optical fibres can also be used for counting with theoretically better transportation and radiation resistance properties, but are commercially restricted to a diameter of 1 – 1.5 mm [9,19].

We focus on adding an appropriate scintillator material and adjusting its volume; this work can enhance the detection system's efficiency and sensitivity. However, an undesirable effect has been observed in our design: the signal from the scintillator is impacted by diameter mismatches at the fibre-scintillator interface, resulting in degraded optical fibre transmission. Therefore, the pulse shape is altered, quickly reducing energy resolution and making spectral analysis impossible.

Measurement is proposed to address issues related to the silica fibre's dimensional limitations and to clarify the relationship between scintillator size and fibre diameter. Replacing an optical fibre with a shading aperture allows us to adjust the proportion of light reaching the detector. Gradual shading enables us to observe changes in the gamma spectrum and identify transmission limits. This approach provides an approximated simulation of the varying light fraction from the scintillator necessary for transmitting a resolvable gamma spectrum, from which the appropriate optical fibre diameter can be determined. Consequently, the successful characterisation of the mentioned measurement allows us better to understand the behaviour of the optical fibre setup, guiding us in a more effective design of our fibre-based measurements.

II. Methodology

Our previously presented work's high-radiation-resistance real-time monitoring system combines optical fibres with a scintillator crystal [9]. We use large-diameter (1 and 1.5 mm) silica multimode, step-index polished fibres FP1500URT (supplied by Thorlabs) to maximise light collection at the scintillator side (Fig. 1). Yet these large multimode fibres exhibit high modal dispersion, distorting the scintillation pulse. Only a fraction of each pulse is collected, and transporting that small signal through the fibre further limits the pulse shape. Although a high numerical aperture (NA) improves collection efficiency, it also increases modal dispersion and broadens pulses. With a small collection diameter and significant modal dispersion, achieving the required level of gamma-spectrum resolution and pulse discrimination is difficult. Initially, we focused on the fraction of light collected by the optical fibre and pulse amplitude, as preserving amplitude is essential for gamma spectroscopy.

By replacing the optical fibre with an iris of similar diameter, we can simulate its effect on light collection and predict system behaviour. We measured relative light collection (i.e., the fraction of scintillator light captured) by varying the iris opening. The iris was replaced with sets of shading apertures to obtain more precise results and adapt the measurement for differently sized crystals. This adjustment allows for replicating the measurement and decreases the distance between the crystal and the photocathode of the photomultipliers, which was created due to the iris dimensions. It is worth noting that both shading methods simulate optical fibre only partially, as their numerical aperture approaches 1. In contrast, optical fibre is limited by material properties, achieving NA 0.5 with current fibres in use. Still, we can determine how many photons we need from the scintillators for the spectrum resolution.

III. Experimental measurement

To study the shading effect, we measured three NaI(Tl) crystals to ensure that proportionality holds across different sizes. A common 2″ in diameter NaI(Tl) was first tested with an iris; then smaller 1″x2″ and 1″x1″ cylinders were tested using a shading aperture, alongside the large one again for consistency. All measurements used a ^{137}Cs source to define the gamma spectrum clearly due to its single peak at 662 keV.

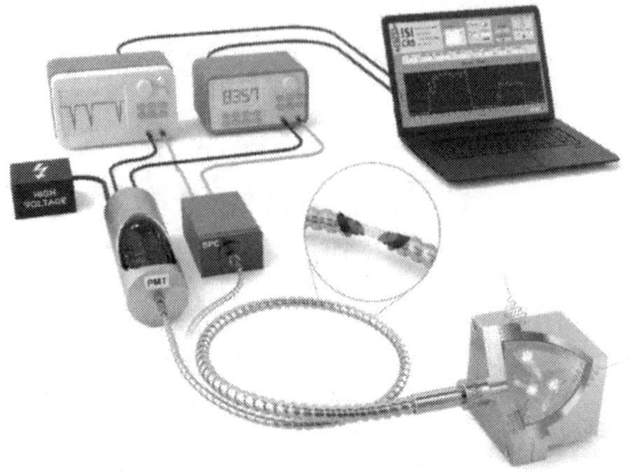

Fig. 1: Measurement of ionizing radiation activity with an optical fibre [9]

A. Experimental setup

The setup comprised Saint-Gobain NaI(Tl) scintillators, an iris or shading aperture, and an ET Enterprises 9266KB photomultiplier tube. The PMT output went directly into a CAEN DT5751 digitiser for pulse-amplitude recording and spectrum evaluation. Experiments with the iris suffered from an excessive distance between the scintillator and PMT, so a shading aperture was introduced to minimise the gap to 0.1 mm. The experiment was enclosed in a light-tight black box to eliminate ambient-light interference. Schematic measurement is illustrated in Figure 2.

B. Measurement

Each spectrum was recorded with a ten-minute counting interval to ensure adequate statistics. Initially, the 2" crystal was placed next to the iris adjacent to the PMT, then the iris diameter was gradually decreased, and measurements were taken at each setting. The diameter was reduced in 5 mm steps, except for the first step due to iris size limitations, resulting in the sequence (48, 45, 40, 35, 30, 25, 20, 15) mm. Next, to compensate for the spectrum shift caused by shading, the PMT gain was increased so that the photopeak is positioned on the same digitiser channel, facilitating direct comparison. Finally, the procedure was repeated for all three crystals using shading apertures whose hole diameters were scaled relative to each crystal's face area.

C. Evaluation

After all measurement series were completed, each series was evaluated separately. For the large 2″ NaI(Tl), we determined the smallest surface/diameter at which

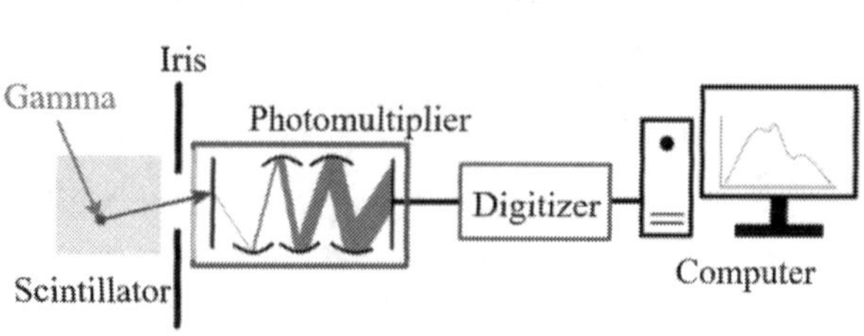

Fig. 2: Schematic illustration of the experiment

the photopeak remained recognisable and the individual parts of the gamma spectrum (the Compton continuum and Compton edge) were still separable. These results were then used to set the minimum aperture diameter for subsequent measurements and to calculate the resolution for each crystal size. Finally, all crystals were compared to illustrate the relationship between crystal size and required aperture diameter.

IV. Results discussion

In the first measurement, we used an iris with a decreasing diameter. As the aperture closed and the shaded area increased, the number of photons reaching the detector decreased, causing the photopeak channel in the gamma spectrum to shift lower (channel number is proportional to collected photon count). The resulting graph is shown in Fig. 3. With this setup, we could transmit the spectrum with a 35 mm diameter (47% of surface area). By adjusting the PMT gain, the minimum aperture diameter at which parts of the spectrum remained visible was 15 mm (9 % of scintillator surface), with a resolution of 19,6 %. With further reductions in aperture diameter, the spectrum becomes dominated by noise and barely readable, making it difficult to resolve even a single peak, let alone multiple ones. This is why we observe a minimal transmission diameter corresponding to 15 mm iris diameter (9% of the original surface area).

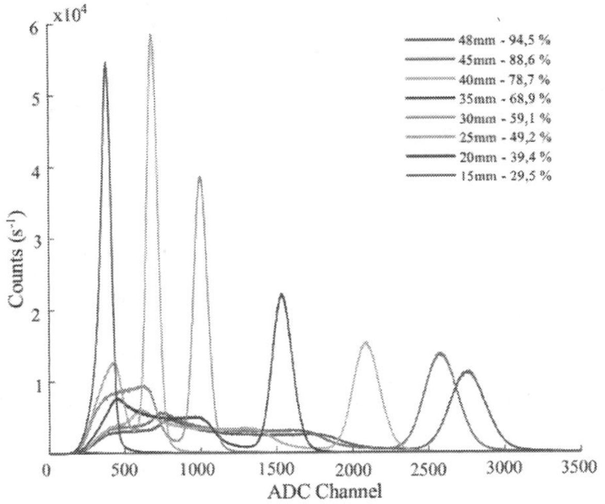

Fig. 3: The energy spectra of gamma radiation obtained by shading the scintillation detector NaI(Tl); each curve corresponds to the spectrum of a specific opening of the iris (Caesium 137) [20]

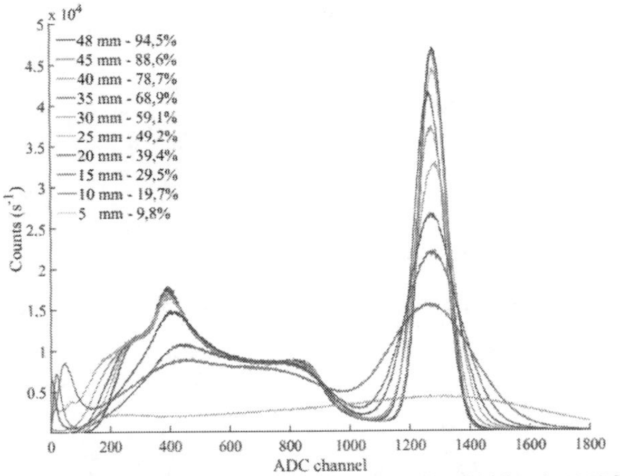

Fig. 4: Gamma-ray energy spectra (137Cs) obtained by shielding the NaI(Tl) scintillation detector - Gradual increase of the photomultiplier gain to align the peak positions [20]

After these observations, all three crystal sizes were measured under the same conditions, with resulting spectra plotted in Fig. 6, which shows the same trend for all sizes. The resolutions with corresponding diameters are summarised in Table 1 and shown graphically in Fig. 5. A direct comparison of all measured diameters is provided with the corresponding photopeak resolutions.

TABLE 1

Comparison of aperture diameter (exposed surface to PMT) and calculated resolution for different sizes of crystals

Scintillator size					
2"x 2"		1"x 2"		1"x 1"	
ø (mm)	Res. (%)	ø (mm)	Res. (%)	ø (mm)	Res. (%)
15.0	19.6	7.0	23.5	7.0	29.2
20.0	16.8	10.0	19.6	10.0	25.8
25.0	15.7	12.5	17.6	12.5	24.6
30.0	14.5	15.0	17.4	15.0	22.4
35.0	14.0	18.0	15.0	18.0	22.5
40.0	13.5	20.0	14.2	20.0	21.4
45.0	13.1	22.0	13.8	22.0	21.8
48.0	13.1	24.0	13.8	24.0	21.3

The minimum resolvable resolution for symmetrically shaped crystals at 9 % scintillator area is 19.6 %. Symmetrical scintillators maintain the same spectral shape and resolution; only the channel numbers differ (due to different sizes). For the elongated crystal (1″x2″), a resolution of 23.5 % at 9 %

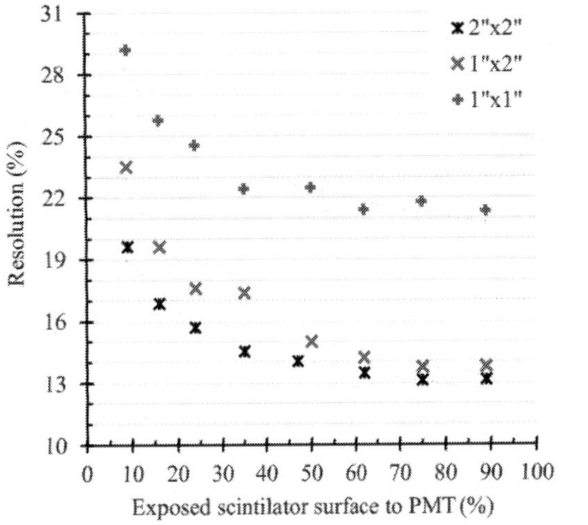

Fig. 5: Comparison of the variation in energy resolution during shading for scintillation crystals of three different sizes

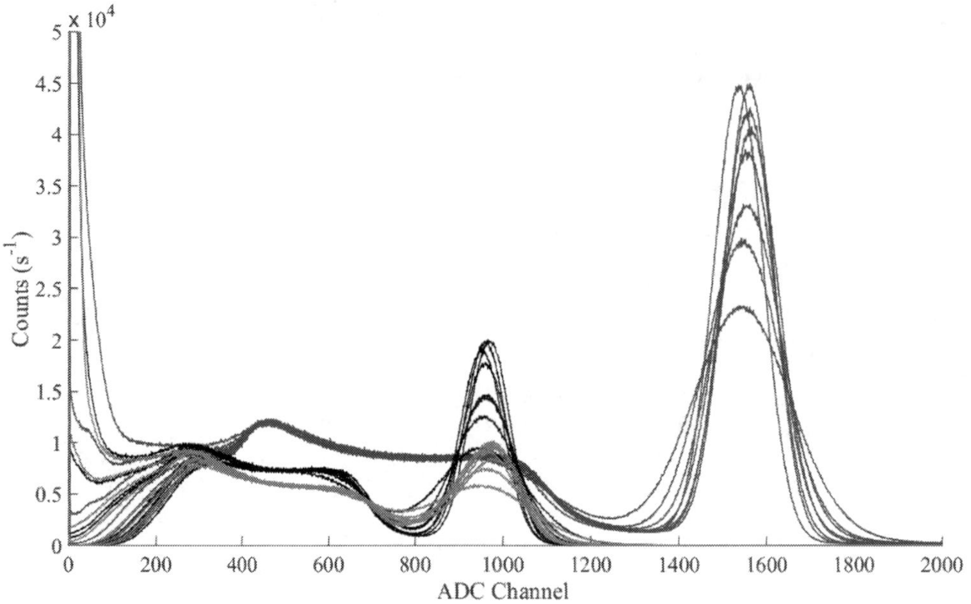

Fig. 6: Gamma-ray energy spectra (137Cs) obtained by shielding three differently sized NaI(Tl) scintillation crystals - Gradual increase of the photomultiplier gain to align the peak positions for each crystal. Sizes: 2"x 2"(blue), 1"x 2" (black) and 1"x 1"(red).

scintillator area was observed, likely because its geometry is better suited for capturing gamma photons along the crystal's long axis, in which the etalon was placed (in comparison to 1" x 1" scintillator, where resolution is 29.2 %). This approach preserves the advantages of a larger crystal while allowing gamma spectrum evaluation even with reduced resolution due to the smaller collector diameter, thus demonstrating that spectral transmission remains possible even when connected to much smaller optical fibres.

V. CONCLUSION

Three NaI(Tl) scintillators of different sizes were gradually shaded and the resulting gamma spectra analysed. The progressive reduction of the aperture diameter (the fraction of scintillation light captured) produced similar behaviour across all crystal sizes. Resolution remains relatively stable until more than 50% of the scintillator face is shaded, beyond which it increases rapidly; spectral features and resolution exhibit the same tendency. To transmit a recognisable spectrum from a cylindrically symmetric NaI(Tl) crystal, at least 9% of the crystal face must remain exposed. Resolutions achieved were 19.6% for the 2″ crystal, 23.5% for the 1″×2″ crystal, and 29.2% for the 1″×1″ crystal. Further reductions in the exposed area introduce excessive noise and render the spectrum unreadable.

These findings support a new approach to scintillator–fibre dosimetry with full gamma-spectrum resolution. By clarifying the relationship between fibre properties and scintillator geometry, crystal size can be optimised to maximise count rate and increase efficiency, enabling compact and flexible detector designs. The results suggest that, for fibre transfer, a small crystal with asymmetric dimensions is required to achieve spectral transmission. This opens the way for flexible and robust gamma-ray spectroscopy in high-dose, spatially constrained, and hard-to-access environments.

REFERENCES

[1] G. F. Knoll, *Radiation detection and measurement*, 4th ed. Hoboken: Wiley, c2010.

[2] E. Garutti and Y. Musienko, "Radiation damage of SiPMs", *Nuclear Instruments and Methods in Physics Research Section A: Accelerators, Spectrometers, Detectors and Associated Equipment*, vol. 926, 2019, doi: 10.1016/j.nima.2018.10.191.

[3] V. A. J. Van Lint, "The physics of radiation damage in particle detectors", *Nuclear Instruments and Methods in Physics Research Section A: Accelerators, Spectrometers, Detectors and Associated Equipment*, vol. 253, no. 3, 1987, doi: 10.1016/0168-9002(87)90532-8.

[4] S. O'Keeffe *et al.*, "A review of recent advances in optical fibre sensors foriin vivo/idosimetry during radiotherapy", *The British Journal of Radiology*, vol. 88, no. 1050, 2015, doi: 10.1259/bjr.20140702.

[5] J. Hou, Z. Feng, G. Ma, W. Zhang, Z. Meng, and Y. Li, "Research on the Intrinsic Sensing Performance of an Optical Fiber Dosimeter Based on Radiation-Induced Attenuation", *Sensors*, vol. 25, no. 12, Jun. 2025, doi: 10.3390/s25123716.

[6] K. Krebber, H. Henschel, and U. Weinand, "Fibre Bragg gratings as high dose radiation sensors?", *Measurement Science and Technology*, vol. 17, no. 5, Apr. 2006, doi: 10.1088/0957-0233/17/5/s26.

[7] E. Lewis *et al.*, "Terbium-doped gadolinium oxysulfide (Gd2O2S:Tb) scintillation-based polymer optical fibre sensor for real time monitoring of radiation dose in oncology", in *SPIE Proceedings*, SPIE, May 2014. doi: 10.1117/12.2058572.

[8] D. McCarthy *et al.*, "Radiation Dosimeter Using an Extrinsic Fiber Optic Sensor", *IEEE Sensors Journal*, vol. 14, no. 3, 2014, doi: 10.1109/jsen.2013.2284857.

[9] M. Jelinek, O. Cip, J. Lazar, and B. Mikel, "Design and Characterisation of an Optical Fibre Dosimeter Based on Silica Optical Fibre and Scintillation Crystal", *Sensors*, vol. 22, no. 19, Sep. 2022, doi: 10.3390/s22197312.

[10] D. Dohler, L. Alshut, L. Konig, T. Teichmann, T. Werner, and T. Kormoll, "Characterization of a Fiber Optic Radiation Sensor Prototype for Nuclear Dismantling", *IEEE Transactions on Nuclear Science*, vol. 69, no. 8, 2022, doi: 10.1109/tns.2022.3189787.

[11] S. Yamamoto, I. Aoki, and T. Higashi, "Optical fiber-based ZnS(Ag) detector for selectively detecting alpha particles", *Applied Radiation and Isotopes*, vol. 169, 2021, doi: 10.1016/j.apradiso.2020.109495.

[12] W. Yoo, S. Shin, D. Lee, K. Jang, S. Cho, and B. Lee, "Development of a Small-Sized, Flexible, and Insertable Fiber-Optic Radiation Sensor for Gamma-Ray Spectroscopy", *Sensors*, vol. 15, no. 9, Aug. 2015, doi: 10.3390/s150921265.

[13] K. Watanabe, "Applications of scintillators in optical-fiber-based detectors", *Japanese Journal of Applied Physics*, vol. 62, no. 1, Nov. 2022, doi: 10.35848/1347-4065/ac90a5.

[14] L. Ding, Q. Wu, Q. Wang, Y. Li, R. M. Perks, and L. Zhao, "Advances on inorganic scintillator-based optic fiber dosimeters", *EJNMMI Physics*, vol. 7, no. 1, Oct. 2020, doi: 10.1186/s40658-020-00327-6.

[15] J. H. Kim *et al.*, "Gamma-ray Spectroscopy Using Inorganic Scintillator Coated with Reduced Graphene Oxide in Fiber-Optic Radiation Sensor", *Photonics*, vol. 8, no. 12, Nov. 2021, doi: 10.3390/photonics8120543.

[16] J. Li *et al.*, "Radiation Damage Mechanisms and Research Status of Radiation-Resistant Optical Fibers: A Review", *Sensors*, vol. 24, no. 10, May 2024, doi: 10.3390/s24103235.

[17] S. Girard *et al.*, "Radiation Effects on Silica-Based Optical Fibers: Recent Advances and Future Challenges", *IEEE Transactions on Nuclear Science*, vol. 60, no. 3, 2013, doi: 10.1109/tns.2012.2235464.

[18] M. Kovacevic *et al.*, "Measurement of 60CO gamma radiation induced attenuation in multimode step-index POF at 530 nm", *Nuclear Technology and Radiation Protection*, vol. 28, no. 2, 2013, doi: 10.2298/ntrp1302158k.

[19] S. Kodama *et al.*, "Fiber-read radiation monitoring system using an optical fiber and red-emitting scintillator for ultra-high-dose conditions", *Applied Physics Express*, vol. 13, no. 4, Mar. 2020, doi: 10.35848/1882-0786/ab77f7.

[20] T. Zbožínek, M. Jelínek, and B. Mikel, "Evaluation of the gamma-ray spectrum transmitted upon alteration in scintillator shading", in *Proceedings II of the 30st Conference STUDENT EEICT 2024: Selected papers*, Brno: Brno University of Technology, Faculty of Electrical Engineering and Communication, 2024, pp. 197-201. doi: 10.13164/eeict.2024.197.

Neutron field spatial distribution from the DD neutron source determined by the KOSTKA experimental device

Vendula Vrtalová[1*], Branislav Vrban[1], Jakub Lüley[1], Štefan Čerba[1], Matúš Sedlák[2], Filip Révai[1], Otto Glavo[1] and Vladimír Nečas[1]
[1]Slovak University of Technology in Bratislava, Faculty of Electrical Engineering and Information Technology, Institute of Nuclear and Physical Engineering, Ilkovičova 3, 841 04 Bratislava, Slovakia
[2]Slovak Academy of Sciences, Institute of Physics, Dúbravská cesta 9, 845 11 Bratislava, Slovakia
(*) vendula.vrtalova@stuba.sk

Abstract—**The spatial distribution of neutrons plays a crucial role in characterizing the neutron source. In this experiment, solid-state nuclear track detectors made of poly-allyl diglycol carbonate (PADC) were selected for this purpose. These detectors have already proven effective in fast neutron measurements, and their dimensions, together with their flexibility, enabled simultaneous measurements in a 3D geometry. For this task, the KOSTKA device in combination with the KOSTKACode was employed, allowing the attachment of additional materials depending on the experimental requirements. The experimental data obtained was then compared with simulations carried out using DROSG-2000.**

***Keywords* —D-D, neutrons, PADC, SSNTD**

I. Introduction

BOTH the research and practical applications of the neutron sources require the determination of the spatial distribution of the emitted neutrons. Such characterization is essential for the assessment of the radiation protection, the calculations of neutron transport, and optimization of the experimental setups. The traditional method often involves Bonner spheres spectrometers [1], providing information on the angular distribution of both yield and energy of the produced neutrons. Such approach is very time demanding, the time-consumption increases with the required increment of the measurement step. This work introduces the method based on the passive Solid-State Nuclear Track Detectors (SSNTDs) made of Poly-Allyl Diglycol Carbonate (PADC), also known under the trademark CR-39. These detectors are often used for their sensitivity to the neutrons in the energy range between 0.1 MeV and 10 MeV [2], however, their sensitivity can be improved by incorporating the convenient converter layer. In general, SSNTDs detect ions through the creation of the latent track in the material of the detector. The latent track is usually enlarged by etching to be visualized using the optical microscope. Neutron detection is based on neutron interactions with the constituents of the PADC polymer, which has a high content of hydrogen atoms. These allow for elastic scattering, resulting in the recoil protons' registration within the PADC detector.

Since the unit cost of the PADCs is low, it is convenient to apply these detectors in the spatial distribution measurement, enabling the simultaneous measurement at multiple positions. For this purpose, the 3D-printed KOSTKA device was designed, together with the Python-based software tool KOSTKAcode, to determine the spatial distribution of the measured value.

The applicability of the conjunction of the PADCs, KOSTKA device and KOSTKAcode was verified in the experiment with the Tandem electrostatic accelerator of TANDETRON type, where the Deuterium-Deuterium (D-D) fusion reaction was initiated to create the fast neutron field. The results of the measurement were compared to the simulations in DROSG-2000 [3].

II. Methodology

The presented work involved several metrics and methods, which will be described in the following paragraphs.

A. PADC detectors

The TASLImage system [4] is used to process the measurement using PADCs at the Slovak Technical University in Bratislava (STU) since 2021 [4], together with the PADC detectors of TASTRAK type, manufactured to be used in the combination with the TASLImage system. The etching procedure is based on the NaOH solution of the molarity of 6.25 M and temperature of 85 °C. The etching lasts for 2 hours and 50 minutes and is followed by the neutralization of the detectors in the demineralized water, and the glacial acetic acid. The neutralized detectors are usually dried in dryer before the analysis.

The analysis using TASLImage system is based on the 20x focus optical microscope with the CCD camera, controlled with the associated software. The analysis is performed automatically, up to 49 detectors can be analyzed during one session. For each detector, the system takes 80 photos of the etched surface of the detector, corresponding to the total area of 1 cm^2. The system counts the etched tracks corresponding to the recoiled protons, the etched tracks density is then proportional to the measured dose. The system calculates the dose according to its internal calibration; however, it is possible to calibrate the data manually. Since the incident neutrons interact withing the whole volume of the PADC detector, both the front side (facing the neutron source) and the back side of the detectors were analyzed.

B. Experimental setup

The experiment was performed at the Tandem electrostatic accelerator of TANDETRON type with the terminal voltage up to 2 MV. The accelerator was designed to deliver protons, deuterons and alpha particles of energies between 100 keV and 4 MeV (in the case of the alpha particles, the maximal energy is 6 MeV). It can be operated with both water-cooled solid targets and gaseous targets.

In the presented experiment, the gaseous target was used, with the target chamber made of stainless steel. The chamber had a cylindrical shape with a height of 37 mm and the radius of 12 mm. The thickness of the stainless-steel wall was approx. 1 mm. The pressure of the deuterium gas inside the target chamber was 1 bar, and the chamber was separated from the beamline with a 12 μm Ni foil window. The deuterium ions were accelerated to the energy of 2.20 MeV towards the target, with the current of 800 nA. The energy loss of the deuterium ions in the Ni window was determined to be 1.21 MeV, according to SRIM simulation [6]. Thus, the energy of the deuteron ions inside the target chamber was approximately. 0.99 MeV.

The PADC detectors were mounted in the KOSTKA device (see Fig. 1). KOSTKA is a cubic 3D-printed modular device, enabling the installation of up to 25 PADC detectors per side (with an exception for the access to the beamline – the corresponding values were assessed artificially). The device was designed to surround the neutron source placed in the middle of KOSTKA. The modularity of KOSTKA also allows for the installation of additional materials, such as aluminum (in Fig. 1), if the measurement could be affected by other products of the reactions within the target, for example, in the case of the D-D reaction, the secondary protons could reach energy up to 17.5 MeV. The aluminum degrader was not used during the presented experiment. The KOSTKA device was mounted on the aluminum construction to avoid the additional background signal. In the presented experiment, the PADC detectors were exposed to the fast neutrons from the D-D reaction for 60 minutes. The measurement time was set to obtain sufficient etched track densities for all PADC detectors involved (in the order of hundreds etched tracks per 1 cm^2).

Fig. 1. The rear view of the KOSTKA device with aluminum plates and PADC detectors, installed around the target chamber.

C. KOSTKAcode

The Python-based code called KOSTKAcode was developed to determine the spatial distribution of the measured value, based on the data obtained from the measurement in the cubic geometry. It was designed to be used with KOSTKA device, but it is scalable, enabling the processing of similar measurements regardless of the size of the cube.

The KOSTKAcode is based on the Monte Carlo generator of pseudo-random numbers, defined in Eq. 1 based on [7] as

$$X_{k+1} = a \cdot X_k \cdot \mathrm{mod}\, c, \qquad (1)$$

where X_k is the current pseudo-random number, X_{k+1} is the following pseudo-random number, and a and c are constant, corresponding to $a = 5e19$ and $c = 2e48$. The KOSTKAcode generates pairs of pseudo-random, defining the direction of the given projection. The measured value, corresponding to the given direction within the cube is then projected onto a surface of the sphere. The KOSTKAcode provides the information of the measured value (in the case of this study, the etched track density) corresponding to the given solid angle. Also, it provides the visualization of the data (see Fig. 2).

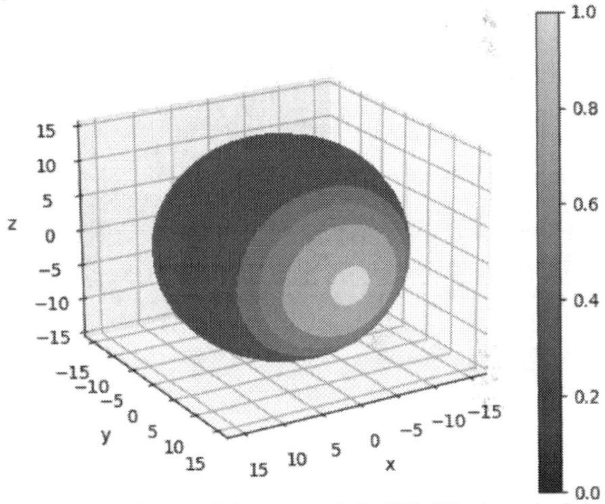

Fig. 2. *Visualization of the output of the KOSTKAcode, with slicing corresponding to individual elements of the solid angle. The results are normalized to the element with the maximal projected value.*

Since the KOSTKA device does not allow measurement in the whole 4π geometry (due to the necessity of the access of the beamline to the target), the values corresponding to the rear side detectors were assessed artificially, from the values obtained from the detectors installed on the rear side, weighted according to the values measured on the front side of the KOSTKA device.

D. DROSG-2000

The DROSG-2000 code was developed to simulate accelerator-based two-body neutron source reactions. It consists of three executables NEUYIE, WHIYIE, and TIMREV, among which the NEUYIE was used to determine the spatial distribution of the yield and the neutron energy.

The simulation for the verification of the experimental results was defined for the gaseous target, and the current of the deuterium ions of 800 nA, to reflect the experimental conditions. The simulation was performed for deuteron ions of initial energy of 0.5, 1.0, 1.5., 2.0, and 2.5 MeV.

III. RESULTS

The results of the simulation in DROSG-2000 are presented in Fig. 3. It shows that the relative neutron yield (related to the sum of the yields for the individual soli angles) from D-D reaction prefers the forward neutron direction, and that the neutron spatial distribution is dependent on the energy of the incident ion.

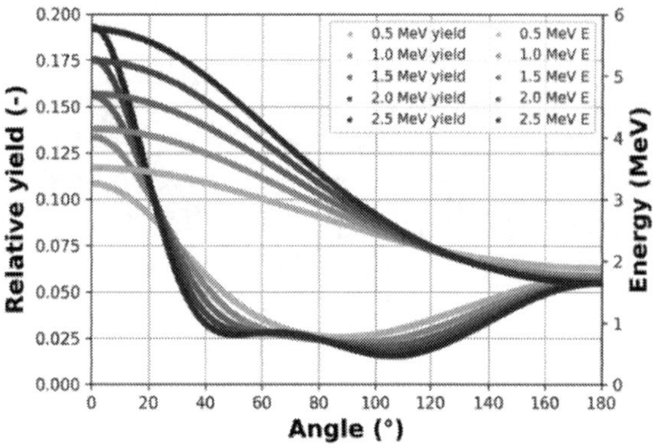

Fig. 3. The spatial distribution of the relative neutron yield and the neutron energy, according to the simulation in DROSG-2000 four various incident deuteron energies.

For the comparison of the experimental results to the results of the simulation in DROSG-2000, the relative yield output of the KOSTKAcode was calculated as the relative yield from DROSG-2000.

Fig. 4 shows the comparison of the experimental results for the front sides of the detectors to the simulated results corresponding to the deuteron ion energy of 1.0 MeV.

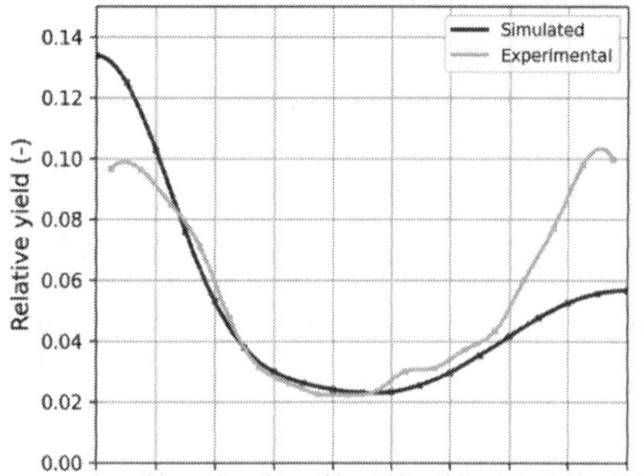

Fig. 4. The comparison of the relative yields obtained from the simulation in DROSG-2000 and from the experiment for the front sides of the detectors.

The comparison of the experimental and simulated results shows that the measurement overestimated the yield on the rear side of the KOSTKA device, corresponding to the angle above 140°. This is most probably caused by the method of assessing values on the rear side, since the PADCs could not be placed to the whole 4π space. Since Fig. 4shows the comparison of the relative yields, this is probably also the reason of the underestimation of the experimental results for the angles below 40°. Another cause of the deviation might be also the change of the response of PADCs to fast neutrons with the change of the neutron energy.

The same comparison for the back sides of the detectors is shown in Fig. 5.

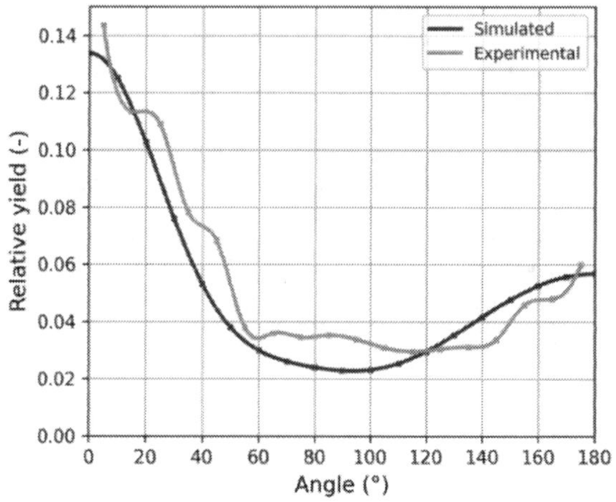

Fig. 5. The comparison of the relative yields obtained from the simulation in DROSG-2000 and from the experiment for the back sides of the detectors.

The experimental results for the back sides of the detectors agree with the simulated results. However, the effect of the change of energy response should be determined. The increase of the experimentally determined relative yield for angles between 60° and 120° is most probably caused by the neutron scattering from the floor and ceiling of the laboratory. This effect might be visible also for the results of the front sides, is the effect of the calculation of the values for the rear side would be suppressed.

IV. CONCLUSIONS

The presented work aimed to characterize the spatial distribution of the neutrons produced in D-D fusion reaction at TANDETRON accelerator. SSNTDs, specifically PADC detectors, were used for fast neutron detection. The new 3D-printed KOSTKA experimental device was designed for this purpose, together with the development of the Python-based KOSTKACode software tool for the processing of measured data. The experimental results were compared to the theoretical values obtained from DROSG-2000. The results obtained for the back sides of the PADC detectors were in agreement with the simulated results. In the case of the front sides, the discrepancy was the most probably caused by the method of determination of the values of the etched tracks density for the rear side of the KOSTKA device. Another important source of discrepancy could be the change in the energy response of

PADC detectors, regarding the significant angular dependency of the produced neutron energies.

Acknowledgment

This work was partially supported by the Slovak Research and Development Agency under the project No. APVV-22-0304 and by the EU NextGenerationEU through the Recovery and Resilience Plan for Slovakia under the project No. 09I03-03-V04-00132.

References

[1] Varela, A., R. Policroniades, F. Jiménez, and J. Calvillo. "The Use of a Bonner Sphere Spectrometer for Determining the Spatial Distribution of Neutron Fields." *Nuclear Instruments and Methods in Physics Research Section A: Accelerators, Spectrometers, Detectors and Associated Equipment* vol. 428 iss. 2–3, pp. 439–45, 1999, 10.1016/S0168-9002(99)00017-0.

[2] Domingo, C., M. De-San-Pedro, M. J. García-Fusté, M. T. Romero, K. Amgarou, and F. Fernández. "Estimation of the Response Function of a PADC Based Neutron Dosimeter in Terms of Fluence and Hp(10)." *Radiation Measurements*, vol. 50, pp. 82–86, Mar 2013, 10.1016/J.RADMEAS.2012.02.016.

[3] M. Drosg, "DROSG-2000: Neutron Source Reactions. Data files with computer codes for 60 accelerator-based two-body neutron source reactions," International Atomic Energy Agency, Nuclear Data Section, Vienna, Austria, IAEA-NDS-87, Rev. 10, Mar. 2017. [Online]. Available: https://www-nds.iaea.org/public/libraries/drosg2000/iaeands-v12.01.pdf. Accessed on: Aug. 28, 2025.

[4] Track Analysis Systems Ltd. (TASL), "TASTRAK PADC (CR-39) solid state nuclear track detector," Bristol, U.K. [Online]. Available: https://www.tasl.co.uk/. Accessed on: Aug. 28, 2025.

[5] B. Vrban, J. Lüley, Š. Čerba, and V. Nečas, "Response of plastic track detector of the TASTRAK type to neutrons of Pu–Be source," *Radiat. Prot. Dosim.*, vol. 198, no. 9–11, pp. 687–692, Aug. 2022, 10.1093/rpd/ncac120.

[6] J. F. Ziegler, M. D. Ziegler, and J. P. Biersack, "SRIM—The stopping and range of ions in matter (2010)," *Nucl. Instrum. Methods Phys. Res., Sect. B*, vol. 268, nos. 11–12, pp. 1818–1823, Jun. 2010, doi: 10.1016/j.nimb.2010.02.091.

[7] D. H. Lehmer, "Mathematical methods in large-scale computing units," in *Proc. of a Second Symposium on Large-Scale Digital Calculating Machinery, Annals of the Computation Laboratory of Harvard University*, vol. 26. Cambridge, MA, USA: Harvard Univ. Press, 1951, pp. 141–146.

Experimental analysis of Schottky-barrier 4H-SiC-detector-response to fast neutrons: comparison with Si and diamond detectors

Andrea Šagátová[1*], Tomáš Slavíček[2], Radu Emanuel Mihai[2,3], Rudolf Sýkora[2], Abdallah Lyoussi[4], Christophe Destouches[4], Oliver Llido[4], Mehdi Ben Mosbah[5], Benedikt Bergmann[2], Stanislav Pospíšil[2], Matej Balušík[1], Eva Kováčová[6] and Bohumír Zaťko[6]

[1]Institute of Nuclear and Physical Engineering, Faculty of Electrical Engineering and Information Technology, Slovak University of Technology, SK-84104 Bratislava, Slovak Republic
[2]Institute of Experimental and Applied Physics, Czech Technical University in Prague, Husova 240/5, CZ 110 00 Prague 1, Czech Republic
[3]Horia Hulubei National Institute for R&D in Physics and Nuclear Engineering, No. 30 Reactorului Street, Măgurele, Ilfov, Romania
[4]CEA, DES, IRESNE, DER, Cadarache F-13108, Saint-Paul-Lez-Durance, 13108, France
[5]CEA, DES, IRESNE, DTN, Cadarache F-13108, Saint-Paul-Lez-Durance, 13108, France
[6]Institute of Electrical Engineering, Slovak Academy of Sciences, Dúbravská cesta 9, SK-841 04 Bratislava, Slovak Republic
*andrea.sagatova@stuba.sk

***Abstract*— The response of 4H-SiC detectors to 17 MeV neutrons was experimentally analysed using detectors composed solely of carbon or silicon. The detectors were tested using three different neutron energies (14.5 MeV, 15 MeV and 17 MeV) at two different facilities with T(d,n) neutron generators, the Van de Graaff laboratory at IEAP CTU in Prague and at the DANAIDES facility at CEA research center in Cadarache. The calibration via interactions on ^{28}Si nuclei, the (n, α) nuclear reaction to ground and excited states of ^{25}Mg, was made. The standard deviation of experimental values measured by Si detector from prepared calibration curve, was of 0.073 MeV, which represents the uncertainty of less than 1% in this energy range (≈ 10 MeV). The experiment showed that the pulse height defect affects the measured spectra the way that the peak energies from the C interactions are shifted by about 0.5 MeV to higher values when the calibration from the Si interactions is used.**

***Keywords* — 4H-SiC detector, Si detector, diamond detector, fast-neutron spectroscopy.**

I. Introduction

THE 4H-polytype SiC remains promising semiconductor material for detector preparation. Its wide bandgap of 3.23 eV at room temperature [1] determines the operability of the prepared detectors even at relatively high temperatures up to several hundred degrees Celsius [2]. The high breakdown voltage of 4 MV/cm enables high drift velocity of radiation-generated charge carriers, ensuring fast detector response. Moreover, 4H-SiC stands out in a relatively high atomic displacement energy (20–35 eV) compared with other semiconductor materials, making it very promising in terms of resistance against radiation damage. The exceptional thermal and radiation resistance of SiC as a detector material implies its application for today's nuclear reactors, or in other radiation and thermally harsh technologies. Nuclear reactors are characterized by neutron radiation fields, for the registration of which the 4H- SiC detectors could be suitable. The 4H-SiC detectors were studied for registration of fast neutrons in the energy range of 1 to 18.3 MeV [3-7]. In [3,6] the 4H-SiC detector response to 14 MeV neutrons was also compared to diamond detectors. In this paper we have prepared single pad Schottky-type detectors based on 4H-SiC epitaxial material dedicated to registration of fast neutrons. Their response to 17 MeV neutrons was compared and analysed with both the silicon and the diamond detectors.

II. Experiment

The SiC detector structures, characterized in [7], were fabricated from an 80 μm thick nitrogen-doped 4H-SiC epitaxial layer (with a donor doping level ≈ 7×10^{13} cm^{-3}) on 4-inch 4H-SiC wafer. Detector was created by Schottky electrode of 3 mm in diameter (Table I) and opposite full area ohmic contact. For comparison, the silicon and carbon detectors were prepared at the Institute of Electrical Engineering SAS in Bratislava as well (Table I). The Schottky barrier type Si detector was prepared from 380 μm thick substrate by a 5×5 mm^2 square electrode as it is described in [8] and the carbon detector was manufactured from 500 μm thick high quality monocrystalline diamond substrate as used for detector preparation in [9] but this time with Au circle electrode of 3.5 mm diameter evaporated on both sides of the substrate. The three main types of detectors are illustrated in Fig. 1.

TABLE I
DETECTOR PARAMETERS

Material and Type	Area	Active Thickness
4H-SiC Schottky	Ø 3 mm	80 µm
C – monocrystalline diamond	Ø 3.5 mm	500 µm
Si Schottky	5×5 mm^2	380 µm

Fig. 1. Photographs of manufactured detectors from left to right: Si, SiC and diamond as bare detectors in the top row and encapsulated with BNC connection in bottom row.

Each tested detector was connected to spectrometric chain consisting of a charge sensitive preamplifier and a digital multi-channel analyser CAEN Hexagon with a high voltage source for detector, connected to the computer. Three detectors, one of each type, were placed around the neutron source in such geometry that each detector was exposed to neutrons of the same energy and flux. As a source of fast neutrons, the T(d,n) generators at two different facilities were used. The 14.5 MeV neutrons impinged detectors at DANAIDES facility at CEA Cadarache and the 15 and 17 MeV neutrons were obtained at Van de Graaff (VdG) laboratory of IEAP CTU in Prague. The Fig. 2 and Fig. 3 show photographs of the arrangement of detectors connected to preamplifiers during experiments at DANAIDES and VdG facilities.

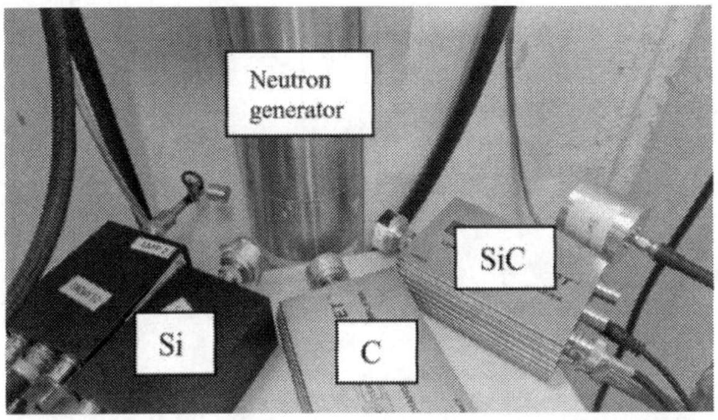

Fig. 2. Photograph of detector arrangement around neutron source at DANAIDES facility of CEA in Cadarache. Three evaluated detectors are labelled through their preamplifiers. The others were part of another experiment.

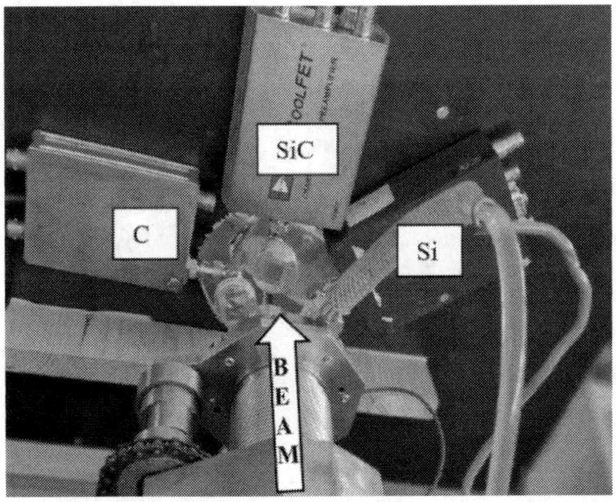

Fig. 3. Photograph of detector arrangement around the neutron source at VdG facility of IEAP CTU in Prague.

III. DISCUSSION

The SiC detectors register fast neutrons through scattering and nuclear reactions on Si and C nuclei, where it serves as a target and detection medium in one. The most relevant interactions of fast neutrons together with their Q-values from ENDF/B-VIII.1 [10] are listed in Table II. In the case of 17 MeV neutrons the energy released by interaction products, which detector registers in the spectrum, is shown in the right column of the table. In case of elastic scattering, the maximum energy transferred to the scattered nucleus is at a head on collision defined by:

$$T_{A,max} = \left\{1 - \left(\frac{A-1}{A+1}\right)^2\right\} \times T_n \,, \qquad (1)$$

with A as the mass number of the nucleus and T_n the energy of impinging neutron. Then, as a result of neutron scattering on Si or C nuclei, the nuclei gain at most, the following fraction of neutron's kinetic energy:

$$T_{Si,max} = 0.133\, T_n \,, \qquad (2)$$

$$T_{C,max} = 0.284\, T_n \,, \qquad (3)$$

which will be 2.26 and 4.83 MeV, respectively, for 17 MeV neutrons.

TABLE II
THE MOST PROBABLE INTERACTIONS OF NEUTRONS WITH SIC NUCLEI IN GROUND STATE WITH THEIR PRODUCTS' ENERGIES FOR 17 MEV INTERACTING NEUTRONS

Interaction	Q-value (MeV)	Products' energy (MeV)
^{28}Si(n,α)^{25}Mg	-2.65	14.35
^{12}C(n, α)^{9}Be	-5.702	11.298
^{12}C(n, n')3α edge	-7.367	9.633
Scattering edge for C	0	4.83
Scattering edge for Si	0	2.26

According to values in Table II all listed interactions of 17 MeV neutrons should be recognizable in measured spectra except for the last one, scattering on Si, most of which cannot be distinguished from detector noise due to its low energy.

The aim of this paper is to experimentally break down the 17 MeV-neutron response in 4H-SiC detectors by resolving monoatomic spectral contributions using detectors composed solely of carbon or silicon. The (n, α) nuclear reaction on ^{28}Si has the highest energy yield for reaction products, so the peak will be the most right in the measured spectrum. Moreover, in this experimental case, the ^{25}Mg nucleus can be produced in one of 15 excited states, listed in Table III, which would be consistent with another 15 peaks towards left in the spectrum. This is why we analysed the spectra measured by Si detector first. Neutrons of chosen energies, 14.5, 15 and 17 MeV, irradiated the detectors separately. In measured spectra we assigned the corresponding energies (Table III) to the peaks from ^{28}Si(n, α) reaction to prepare the calibration curve.

TABLE III
REACTION ^{28}SI(N, α)^{25}MG PRODUCTS' ENERGIES FOR 14.5, 15 AND 17 MEV NEUTRONS ENTERING THE REACTION. THE Q-VALUES ARE FROM ENDF/B-VIII.1 [10]

Label of ^{25}Mg energy level	Q-value (MeV)	Products' energy (MeV) for 14.5 MeV	Products' energy (MeV) for 15 MeV	Products' energy (MeV) for 17 MeV
0	2.65	11.85	12.35	14.35
1	3.235	11.265	11.765	13.765
2	3.625	10.875	11.375	13.375
3	4.262	10.238	10.738	12.738
4	4.615	9.885	10.385	12.385
5	5.214	9.286	9.786	11.786
6	5.388	9.112	9.612	11.612
7	5.451	9.049	9.549	11.549
8	6.055	8.445	8.945	10.945
9	6.064	8.436	8.936	10.936
10	6.558	7.942	8.442	10.442
11	6.621	7.879	8.379	10.379
12	6.71	7.79	8.29	10.29
13	6.927	7.573	8.073	10.073
14	7.009	7.491	7.991	9.991
15	7.361	7.139	7.639	9.639

The raw spectra measured by Si detector irradiated with fast neutrons are shown in Fig. 4. The spectra were smoothed for better peak distinguishing by Savitzky-Golay method in OriginPro2022 program [11]. The result can be seen in Fig. 5 with counts for 17 MeV neutrons scaled 50 times to prevent the spectra overlapping. The numbers marked with an asterisk (*) represent the energy level of ^{25}Mg corresponding to a given peak. Some of the peak energies are so close together that the detector energy resolution is insufficient for distinguishing them. In this case, the overlapping peaks are labelled with a slash (/) as a divider. Labelled peaks in Fig. 5 were used to determine the calibration curve shown in Fig. 6. The chosen colour of each experimental point agrees with the colour of spectrum of its origin in Fig. 5. The R-Square of 0.99844 of the line fitted through experimental points confirms the correctness of the assignment of peaks to given energy states.

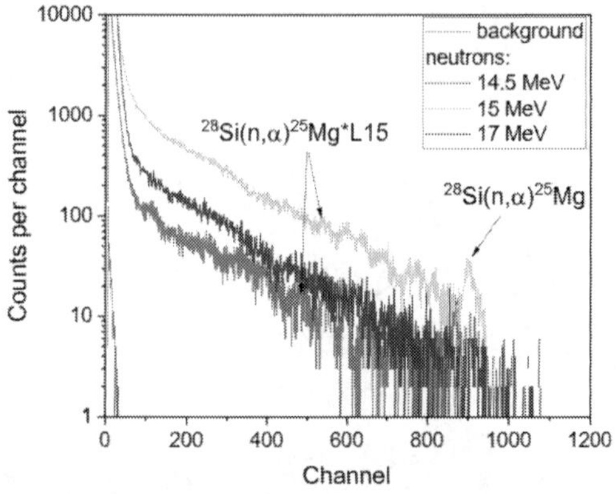

Fig. 4. The raw spectra measured by Si detector irradiated with 14.5, 15 and 17 MeV neutrons.

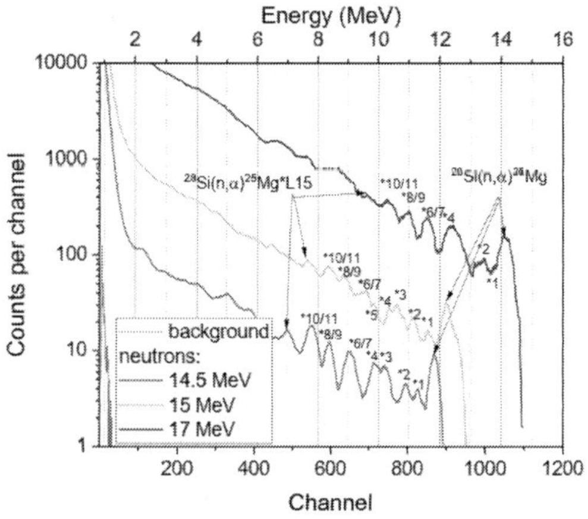

Fig. 5. The smoothed spectra measured by Si detector irradiated with 14.5, 15 and 17 MeV neutrons.

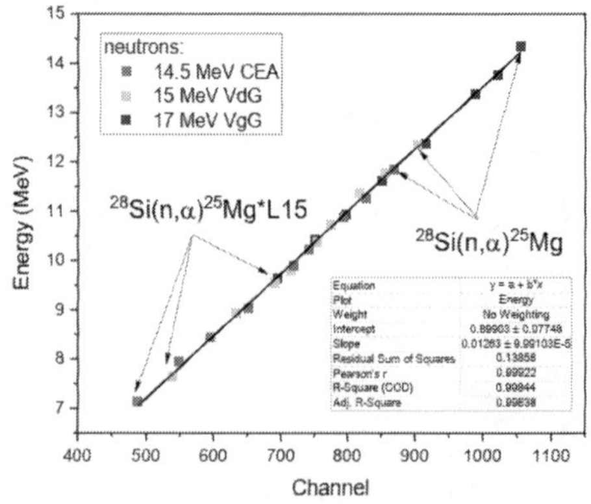

Fig. 6. The calibration curve for the Si detector.

Moreover, the dispersion of deviations of experimental values from the calibration curve (see Fig. 7) confirms their statistical origin and excludes systematic measurement error. The standard deviation of these values is of 0.073 MeV, which at measured energies around 11 MeV represents the uncertainty below 1%.

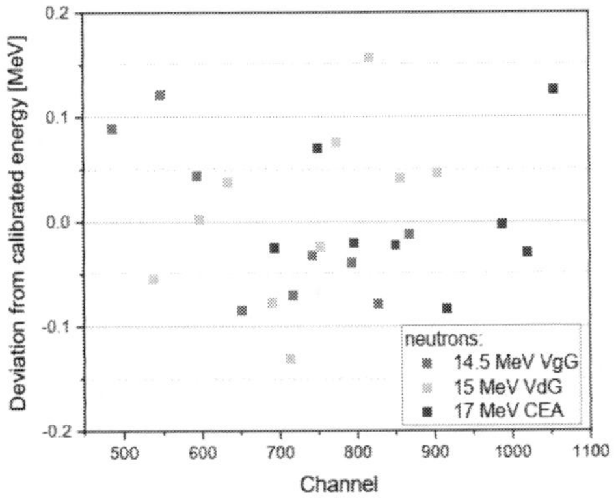

Fig. 7. The deviations of experimental values from the calibration curve measured by Si detector in Fig.6.

Fig. 8 represents the response of three different detectors with SiC, Si and diamond sensors, to 17 MeV neutrons. The calibration curve from Fig. 6 was used to obtain the red spectrum with the energy x-axis here and allowed to arrange the top energy axes in Fig. 5.

First the Si detector spectrum was displayed, then other two spectra were overlaid, first fitting the known peaks from (n, α) reaction on ^{28}Si (the SiC detector) and then the diamond detector spectrum according to (n,α) reaction on ^{12}C (Table II). Besides mentioned (n, α) reactions, the detectors containing carbon (blue and green spectrum) show the ^{12}C(n, n')3α nuclear reaction with calculated edge at 9.63 MeV. The calculated values of mentioned peaks are written above them in Fig.8. However, one can observe, that the position of two peaks from carbon interaction is placed at higher energies on x-axis than the calculated value determines. This shift can be explained by the pulse height defect [12] of ions produced during nuclear reactions of neutrons with carbon and silicon nuclei in the detector. The pulse height defect, the charge deficit during charge collection, is more pronounced with ^{25}Mg ions from neutron interaction with ^{28}Si than in the case of ^{9}Be or ^{4}He ions from neutron interaction with ^{12}C. For detector calibration the peaks from neutron interaction with ^{28}Si were used, so it can be expected that the peaks from neutron interaction with ^{12}C will be overestimated in this calibration. We can see that the ^{12}C(n, α) peak is at 11.8 MeV position, which is about 0.5 MeV more than calculated position (11.3 MeV). Also, the edge from ^{12}C(n, n')3α reaction seems to appear higher than at 9.63 MeV. On the other hand, the peaks from ^{28}Si interactions fit exactly to their energy on x-axis. Similar effect was analysed in [3] with SiC and diamond detectors irradiated by 14 MeV neutrons.

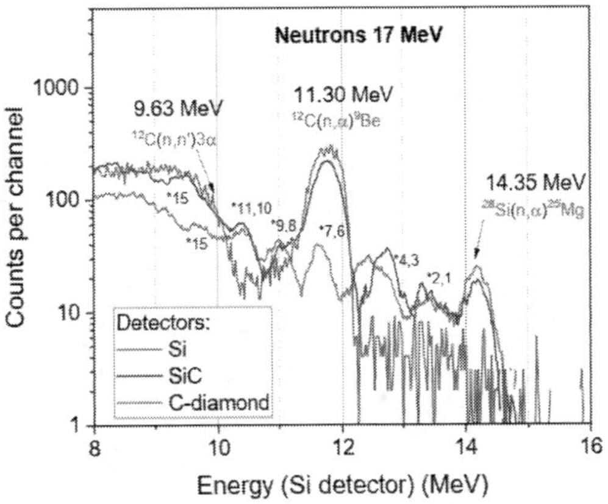

Fig. 8. The spectra measured using 4H-SiC (blue), Si (red) and diamond (green) detector when irradiated by 17 MeV neutrons.

Experimental comparison of Si-, SiC- and diamond-detector response to fast neutrons shows that when using the energy calibration from the interaction on Si, the difference in the peak positions from the interactions on carbon will be approximately 0.5 MeV, so the pulse height defect should be considered in this energy range in this height. Detailed peak energy-position determination including the pulse height defect effect can help in neutron spectrometry by semiconductor detectors, where via machine learning the neutron spectra can be reconstructed from the SiC detector response to neutrons [13].

IV. Conclusions

The 4H-SiC epitaxial Schottky barrier detector was studied as a detector of fast neutrons. The response of the SiC detector to 17 MeV neutrons was experimentally compared to the response of detectors based on solely silicon (Si detector) or solely carbon (chemical vapour deposited monocrystalline diamond detector). The Si detector was calibrated by three different neutron energies 14.5, 15 and 17 MeV provided by two different facilities. The results proved the correct determination of peaks from neutron interactions on Si and C nuclei at studied detectors and revealed the pulse height defect effect on measured spectra. The peaks from neutron interactions on carbon nuclei were shifted to higher energies by about 0.5 MeV with respect to the calibration from Si interactions. The study of this phenomenon in detail is the subject of our future research.

Acknowledgment

This work was partially supported by grants of the Slovak Research and Development Agency No. APVV-22-0382, DS-FR-24-0020APVV-SK-CZ-21-0116 and of the Scientific Grant Agency of the Ministry of Education of the Slovak Republic and the Slovak Academy of Sciences No. 2/0084/20.

The work profited from funding from the Ministry of Education, Youth and Sports under grant agreement No. 8X23035 in the frame of the 'Timepix SiC detector' international project under the DANUBE2022 region strategy programme. Measurements were funded by the EU NextGenerationEU through the Recovery and Resilience Plan for Slovakia under the project No. 09I05-03-V02-00073 and the LM2018108 grant of MEYS, Czech Republic.

References

[1] Silicon Carbide, Basic Parameters. Available: http://www.ioffe.ru/SVA/NSM/Semicond/SiC/bandstr.html#Basic

[2] N. Gál, L. Hrubčín, A. Šagátová, G. Vanko, E. Kováčová, and B. Zaťko, "High-resolution alpha-particle detector based on Schottky barrier 4H-SiC detector operated at elevated temperatures up to 500 °C," *Appl. Surf. Sci.*, vol. 635, Art. 157708, Oct. 2023, DOI: 10.1016/j.apsusc.2023.157708.

[3] F. H. Ruddy, A. R. Dulloo, J. G. Seidel, M. K. Das, S. H. Ryu, and A. K. Agarwal, "The fast neutron response of 4H silicon carbide semiconductor radiation detectors," *IEEE Trans. Nucl. Sci.*, vol. 53, no. 3, pp. 1666–1670, Jun. 2006, DOI 10.1109/TNS.2006.875151.

[4] J. H. Ha, S. M. Kang, H. S. Kim, S. H. Park, N. H. Lee, T. Y. Song, J. H. Lee, H. Park, and J. Kim, "4H SiC PIN type semiconductor detector for fast neutron detection," *Prog. Nucl. Sci. Technol.*, vol. 1, no. 1, pp. 237–239, Feb. 2011, DOI 10.15669/pnst.1.237.

[5] A. F. Flammang, "Fast-neutron detection with 4H-SiC Schottky diodes: effects of polyethylene converters," *Nucl. Instrum. Methods Phys. Res.* A, vol. 579, no. 1, pp. 177-179, Aug. 2007, DOI 10.1016/j.nima.2007.04.034.

[6] H. Rebai et al., "Comparison of thick 4H-SiC pad response to 14 MeV neutrons versus diamond," *Nucl. Instrum. Methods Phys. Res.* A, vol. 946, no. 1, pp. 1-7, Dec. 2019, DOI 10.1016/j.nima.2019.162637.

[7] B. Zaťko, " Detection of fast neutrons using 4H-SiC radiation detectors," *EPJ Web Conf.,* vol. 288, no. 1, Art. 03004, Nov. 2023. DOI 10.1051/epjconf/202328803004.

[8] L. Hrubčín, B. Zaťko, and E. Kováčová, "Silicon radiation detectors with rectifier junction prepared by different technological procedures," *AIP Conf. Proc.*, vol. 3251, Art. 080005, Oct. 2024, DOI 10.1063/5.0239936.

[9] B. Zaťko, M. Varga, E. Kováčová, and A. Šagátová, "A study of particle detectors based on single crystal diamond substrate," *J. Instrum.*, vol. 19, no. 11, Art. C11016, Nov. 2024, DOI: 10.1088/1748-0221/19/11/C11016.

[10] D. A. Brown et al. "The 8th Major Release of the Nuclear Reaction Data Library with CIELO-project Cross Sections, New Standards and Thermal Scattering Data ENDF/B-VIII.1, " *Nuclear Data Sheets*, vol. 148, pp.1-142, Feb. 2018. Available: https://www.nndc.bnl.gov/endf/

[11] OriginLab, Available: https://www.originlab.com/

[12] B. Zaťko, L. Hrubčín, A. Šagátová, P. Boháček, O.M. IvanovM. Sekáčová, E. Kováčová, Y.B. Gurov and V.A. Skuratov, " Study of the pulse height defect of 4H-SiC Schottky barrier detectors in heavy ion detection, ," *AIP Conf. Proc.*, vol. 2411, Art. 070007, Nov. 2021. DOI: 10.1063/5.0067353.

[13] E. Belfiore et al., "Reconstruction of Neutron Spectra Using Silicon Carbide Detectors in Monoenergetic Fields with Machine Learning Approach," *EPJ Web Conf.,* to be published.

Reconstruction of Neutron Spectra Using Silicon Carbide Detectors in Monoenergetic Fields with Machine Learning Approach

Enrica Belfiore[1,*], Achment Chalil[1], Mehdi Ben Mosbah[1], Rodolphe Antoni[1], Olivier Llido[2], Abdallah Lyoussi[2], Christophe Destouches[2], Tomas Slavicek[3], Andrea Šagátová[4], and Jean-Emmanuel Groetz[5]

[1]CEA, DES, IRESNE, DTN, Cadarache F-13108, Saint-Paul-Lez-Durance, 13108, France
[2]CEA, DES, IRESNE, DER, Cadarache F-13108, Saint-Paul-Lez-Durance, 13108, France
[3]Institute of Experimental and Applied Physics, Czech Technical University in Prague, Husova 240/5,110 00 Prague, Czech Republic
[4]Institute of Nuclear and Physical Engineering, Faculty of Electrical Engineering and Information Technology, Slovak University of Technology, SK-841 04 Bratislava, Slovak Republic
[5]Université Marie et Louis Pasteur, CNRS, Chrono-environnement (UMR 6249), F-25000 Besançon, France

(*) enrica.belfiore@cea.fr

Abstract— **Recent advancements in machine learning have shown significant promise in nuclear applications, particularly in optimizing reactor operations, core design, and improving neutron spectroscopy techniques. This study introduces a novel approach that leverages a non-parametric and probabilistic Bayesian learning method to predict neutron spectra based on data obtained from solid-state detectors, with a specific focus on the performance of 4H-polytype Silicon Carbide detectors under various monoenergetic neutron fields. The experimental measurements were conducted at the Van de Graaff accelerator facility in Prague and at the TOTEM facility in Cadarache. In the first installation, the Silicon Carbide detector was exposed to multiple monoenergetic neutron beams generated via p+T, D+T, and D+D reactions. In the second infrastructure, the detector was subjected to a monoenergetic beam generated by GENIE16 D-T neutron generator. The measurements were post-processed to collect the deposited energy in the detector which is used as input for the machine learning algorithm in order to obtain the reconstruction of neutron spectra. Preliminary findings demonstrate that the machine learning-based approach can successfully reconstruct monoenergetic normalized neutron spectra with good accuracy, showcasing its potential for improving neutron spectroscopy with semiconductor sensors.**

Keywords **—Semiconductor sensors, neutron spectrum reconstruction, machine learning algorithm.**

I. Introduction

Silicon carbide (SiC) is a promising semiconductor material for solid-state radiation detectors intended for deployment in harsh environments characterized by high temperatures and intense radiation fields, such as those encountered within nuclear reactors, radioactive waste management facilities, and high-radiation dismantling and decommissioning operations. The suitability of SiC for fast neutron spectroscopy arises from its advantageous properties, including a wide bandgap, high radiation resistance, rapid response times, and robust thermal stability. [1]

However, a persistent challenge in fast neutron spectroscopy—regardless of detector type—is the reconstruction of neutron energy spectra, which typically requires complex signal processing and unfolding techniques. In the case of conventional systems such as organic scintillators, activation detectors, or Bonner spheres, spectral reconstruction often relies on unfolding methods involving linear inversion of matrix, which are prone to numerical instabilities due to the ill-conditioned nature of the response matrices. Semiconductor detectors like SiC offer a different paradigm: thanks to their sensitivity to charged particles from specific neutron-induced reactions (e.g., (n,α), (n,p)), they allow a direct link between the incident neutron energy and the measurement, and potentially enable alternative data-driven approaches that bypass traditional unfolding. In this study, we explore the use of machine learning algorithms to reconstruct neutron spectra from the response of SiC detectors, leveraging their unique combination of material properties and reaction mechanisms. [2] Specifically, the feasibility of using 4H-Silicon Carbide detectors in the reconstruction of neutron spectra was investigated under various monoenergetic neutron fields. The experimental measurements were conducted in two facilities: the Van de Graaff accelerator in Prague, where the SiC detector was exposed to quasi-monoenergetic neutron beams generated via p+T, D+T, and D+D reactions, and the TOTEM facility at

CEA Cadarache, where the detector was subjected to a quasi-monoenergetic beam produced by the GENIE16 D-T neutron generator. The SiC detector used in these experiments measured 2 mm × 3 mm × 60 μm.

The machine learning-based approach developed in this study employs a training database constructed from simulated detector responses to synthetic spectra generated as combinations of various mathematical functions. Testing was performed using simulations of detector responses to experimental spectra from the IAEA Compendium database. [3] All simulations were carried out using the PHITS code, [4] with a Gaussian Process model employed as the machine learning algorithm due to its suitability for handling complex, nonlinear relationships and its inherent ability to provide a measure of uncertainty of predictions.

Preliminary findings indicate that this machine learning approach can accurately reconstruct the normalized monoenergetic neutron spectra, offering a promising alternative to conventional unfolding techniques that often suffer from numerical instability. [5] The ongoing research at the IRESNE Institute of CEA Cadarache, particularly within the Department of Nuclear Techniques and the Department of Reactor Study, aims to further assess the feasibility and reliability of this technique in complex mixed-field environments such as reactor cores. [6, 7] As a first step in validating this methodology, the present work focuses on the reconstruction of monoenergetic neutron spectra, serving as a proof of concept for the proposed machine learning approach.

This paper is structured as follows: Section 2 details the experimental setup and measurements performed in the two facilities. Section 3 describes the post-processing and data analysis techniques, including the implementation of the machine learning algorithm used for this analysis. Section 4 presents the results collected and section 5 outlines the conclusions and future research directions.

II. Experimental campaigns

The detection of fast neutrons in SiC detectors is achieved through interactions between incoming neutrons and Si or C nuclei via nuclear reactions and scattering that generate secondary ionizing particles. As these particles traverse the semiconductor material, they produce electron–hole pairs. The resulting charge carriers are then drifted by an electric field established by an externally applied bias voltage, inducing a measurable pulse (signal). [8]

In this study, the investigated sensor is a 4H-polytype SiC detector with dimensions of 2 mm × 3 mm and a sensitive layer thickness of 60 μm. As described in the introduction, this detector was exposed to various monoenergetic neutron fields at two distinct facilities: the Van de Graaff accelerator facility in Prague [9] and the TOTEM facility at CEA Cadarache. In this section the two facilities are described, including the setup configuration adopted during the measurements.

The experimental setup employed at both facilities was consistent and included the SiC detector connected to an homemade current preamplifier with a gain of 55 dB and a bandwidth of 75 MHz, a CAEN DT5533EM module high voltage power supply and an Acqiris SA230E digitizer 14-bit ADC Module, 4 GS/s.

A. Measurements at Van de Graaff facility, Prague

The Van de Graaff accelerator laboratory, part of the Institute of Experimental and Applied Physics (UTEF) at the Czech Technical University (CTU) in Prague, is dedicated to fundamental and applied research in experimental nuclear physics. Its applications encompass testing and calibration of radiation detectors, as well as the development of analytical techniques relevant to various domains, including material and surface analysis, high-energy physics, and space-related research. [10] The facility functions primarily as a tunable source of monoenergetic neutrons over a broad energy range (100 keV to 19 MeV), generated through reactions involving charged particles such as protons or deuterons. Neutron production is achieved via D-T, D-D, and p-T reactions, making the facility particularly suitable for investigating detector responses under well-characterized neutron fields.

In the present study, the SiC detector was exposed to several monoenergetic neutron beams produced at this facility. Specifically, the results reported in this paper correspond to neutron energies of 15.3 MeV, 0.78 MeV and 4.180 MeV.

B. Measurements at TOTEM facility, Cadarache

The TOTEM platform at CEA Cadarache is a versatile experimental facility dedicated to the development, testing, and maintenance of various devices and instrumentation utilized across different research and operational activities. Within this platform, the measurements reported in this study were conducted in the DANAIDES irradiation casemate, a specialized environment equipped to support neutron irradiation experiments. The primary neutron source used in the DANAIDES casemate is the GENIE-16 D-T neutron generator. [11] This device produces high-intensity, monoenergetic neutron beams via the deuterium-tritium (D-T) fusion reaction, resulting in a neutron energy of 14.1 MeV, accompanied by a minor contribution from scattered neutrons.

III. Data processing and Machine Learning algorithm

This section details the data post-processing used to obtain the spectrum of the energy deposited in the sensor and the machine learning algorithm used for this analysis.

A. Deposited energy in SiC detector

The pulses collected during the SiC detector measurements were processed using a Python routine designed to extract the deposited energy within the detector. The Python routine identifies the peak—either maximum or minimum—from the acquired waveform and defines a time window based on the first zero crossings on both the left and right sides of the peak. Any offset present in the waveform is subtracted before further analysis. To ensure the accuracy of the pulse data, some pulses are discarded based on an analysis of the Fisher coefficient. This coefficient is used to assess the asymmetry of the pulse distribution. Specifically, the Fisher's Skewness Coefficient

[12] quantifies the degree of skewness in the pulse shape, with values greater than 0.5 indicating an asymmetric distribution. In such cases, the pulse is retained for further analysis, as this typically suggests a valid signal. In contrast, pulses with a Fisher coefficient less than 0.5 exhibit symmetry, which is indicative of noise that tends to follow a Gaussian distribution. These pulses are excluded from the analysis.

Subsequently, for the pulses considered valid, the deposited energy is proportional to the surface area of each pulse, as the pulse area reflects the amount of charge (Q) induced in the detector. This relationship can be described by the following equation:

$$Q = \frac{1}{AZ} \int x(t)\, dt\ , \qquad (1)$$

where A represents the internal gain and Z the impedance of the pre-amplifier. The collected charge Q is directly proportional to the number of electron-hole pairs N generated in the detector, divided by the elementary charge q. To estimate the deposited energy, the first ionization energy E_w is taken into account, leading to the following equation:

$$E_{dep} = N\, E_w = \frac{Q}{q}\, E_w\ . \qquad (2)$$

Here, E_{dep} represents the total energy deposited in the detector. [13] The calculated deposited energy was then used as input for the machine learning algorithm to reconstruct the neutron spectra.

B. Gaussian Process algorithm

To generate the datasets required for the machine learning approach, extensive simulations were performed using the PHITS (Particle and Heavy Ion Transport code System) code. The SiC detector was modeled as a parallelepiped comprising only the depleted layer, deliberately excluding electrodes and ohmic contacts from the model. This modeling choice is based on the fact that, although neutron interactions can occur throughout the detector volume, only those interactions taking place within the depleted region contribute to the measurable signal, as charge collection is limited to this active zone. All relevant nuclear data were taken from the JENDL-4 nuclear data library. [14] The primary output of these simulations was the energy deposited within the detector, which served as the input for constructing the training and testing datasets utilized in the machine learning analysis. The training dataset (base-learning set) was generated by simulating the detector's response to a wide range of synthetic neutron spectra. These spectra were not derived from physical measurements but were constructed as linear combinations of mathematical functions (e.g., Gaussians, exponentials) chosen to span a representative variety of spectral shapes encountered in real-world neutron fields. While not physically measured, these synthetic spectra were designed to reflect plausible distributions in energy and shape, allowing the machine learning model to generalize across a broad range of scenarios. A total of 544 simulations were conducted to comprehensively capture a wide range of possible spectral shapes and intensities. The detailed construction of this dataset is provided in [3]. The testing dataset (base-test set) was developed by simulating the response of the detector to 120 experimentally measured spectra obtained from the IAEA Compendium database. This database serves as a comprehensive repository of neutron spectra, primarily intended for radiation protection applications, providing a valuable benchmark for assessing the theoretical robustness of the proposed machine learning approach under realistic conditions.

In contrast to the theoretical analysis previously employed in [3], this study adopts a probabilistic approach based on Gaussian Process Regression (GPR) [15] to explore the feasibility of the machine learning reconstruction methodology with the real experimental data collected. A Gaussian Process is a non-parametric, probabilistic model that defines a distribution over functions rather than just predicting discrete outputs. It operates under the assumption that any finite set of points follows a joint Gaussian distribution. Given training data, GPR infers a continuous function that best fits the data points while providing a prediction interval for each point. The principal advantage of GPR lies in its ability to provide not only predictions but also a probabilistic measure of confidence associated with each prediction. Unlike deterministic approaches, GPR offers flexibility in modeling complex functions without requiring predefined functional forms, making it highly adaptable to various underlying data distributions. [16] Moreover, it effectively manages noise within the dataset by producing smooth approximations through covariance functions, which define the similarity between data points. Despite these advantages, GPR presents certain drawbacks, primarily related to its high computational complexity. The training process involves inverting a covariance matrix of size nxn, where n is the number of training points. This results in a computational cost of $O(n^3)$, [17] making GPR impractical for very large datasets. However, in this study, the size of the training set (544 spectra) remains sufficiently small to keep the computational load manageable. This holds true regardless of whether the spectra are monoenergetic or complex: what matters is the number of training points, not the spectral structure itself. Therefore, GPR remains computationally efficient for the current dataset and could be applied to multi-energetic spectra as well, provided the dataset size remains within a similar range. Additionally, the computational time required for GPR is comparable to that of other widely used methods, such as Kernel Ridge Regression, while providing superior flexibility and probabilistic interpretation of the results.

Further work may involve expanding the dataset and exploring more scalable models, such as Bayesian Neural Networks, but such developments lie beyond the scope of the present analysis. In this study, the GPR model utilizes a composite kernel composed of a Matern kernel, an RBF (Radial Basis Function) kernel, and a White kernel to address different aspects of the data. The Matern kernel, configured with a smoothness parameter ν = 1.5, is particularly suited to capturing irregular patterns and moderate smoothness, making it well-adapted to experimental data, which often exhibit natural fluctuations. The RBF kernel, characterized by a length scale l = 100, models gradual variations and contributes to the overall smoothness of the prediction. Noise is explicitly incorporated through a White kernel with an initial noise level of 1, enabling the model to manage observational uncertainties and avoid overfitting. The kernel structure is formulated as a combination of these components:

$$ConstantKernel \ x \ Matern + ConstantKernel \ x \ RBF + WhiteKernel$$

Hyperparameter optimization of each kernel is performed by maximizing the log-marginal likelihood. The algorithm uses using the L-BFGS-B algorithm with 100 independent restarts in order to avoid conversion to possible local minima and to ensure thus robust convergence. The model is trained using the *GaussianProcessRegressor* class from Scikit-Learn python package.

IV. Results and discussion

In this section, the results of the energy deposited within the SiC sensor are presented in the form of histograms, alongside the reconstruction of the neutron spectrum obtained through the machine learning algorithm. The data analysis is conducted by applying the two-body kinematic equations detailed below, considering the interactions between silicon and carbon atoms and incident neutrons. These interactions can lead to various nuclear reactions or scattering, which are summarized in Table 1 with their corresponding threshold energies. [18]

Table I
Si-n and C-n threshold reactions

Reaction	Threshold energy [MeV]
$^{28}Si(n,n)^{28}Si$	-
$^{28}Si(n,n')^{28}Si^{*}$	1.8
$^{28}Si(n,\alpha)^{25}Mg$	2.76
$^{28}Si(n,p)^{28}Al$	4
$^{12}C(n,n)^{12}C$	-
$^{12}C(n,n')^{12}C^{*}$	4.8
$^{12}C(n,\alpha_0)\ ^{9}Be$	6.42
$^{12}C(n,\alpha_1)\ ^{9}Be$	8.09
$^{12}C(n,n')3\alpha$	8.43

It is important to note that the cross-section for the direct production of 3α particles is notably low. This reaction predominantly occurs as a secondary process rather than a primary interaction. Specifically, it can manifest through various channels, including the de-excitation of carbon nuclei from excited states (second or third excitation level) following inelastic scattering. Additionally, it may arise from the de-excitation of beryllium nuclei from their first or second excited states. The results of neutron spectrum reconstruction are presented considering the equally log-spaced energy grid employed for both the training and testing datasets. This approach was adopted to ensure consistency with the data structure provided by the IAEA Compendium database, which was initially divided into 55 energy bands spanning from $2 \ x \ 10^{-8}$ MeV to 44 MeV. However, due to the limited efficiency of the SiC detector at energies below $5 \ x \ 10^{-2}$ MeV and the presence of numerous empty bands, the number of energy intervals was reduced to 25, covering the range from $5 \ x \ 10^{-2}$ MeV to 15.5 MeV. The deposited energy results are presented as histograms normalized to their integral, corresponding to the exact input used for the machine learning model. The number of counts under each peak is estimated, and the corresponding statistical uncertainty is evaluated as √N, where N is the number of counts. These uncertainties are reported as error bars in the following figures.

A. *Results collected at Van de Graaff facility, Prague*

The SiC sensor was exposed at the Van de Graaff facility emitting 15.3 MeV neutrons, generated by a DT reaction. This neutron energy exceeds the threshold energy reactions listed in Table 1, indicating that all the reactions are energetically possible. The energy deposited within the sensor is shown in Fig. 1a. A total of 6117 pulses were collected, with an uncertainty between 2 and 45%.

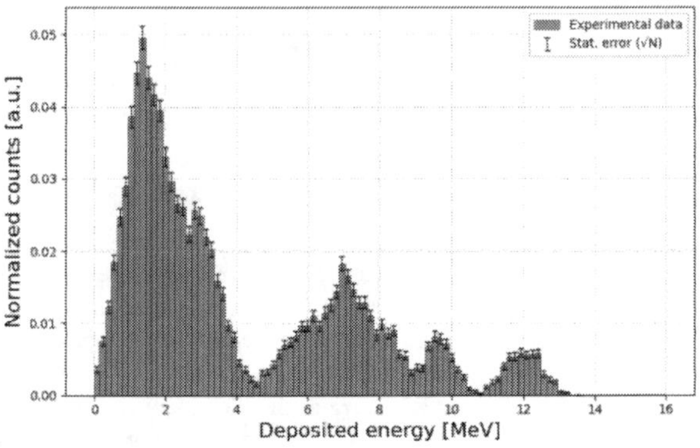

(a)

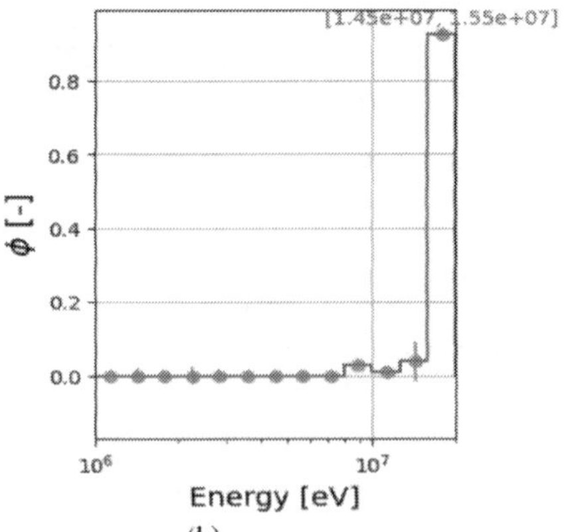

(b)

Fig. 1. (a) Deposited energy spectrum obtained from the SiC detector exposed to the 15.3 MeV (b) Reconstructed 15.3 MeV neutron spectrum using the Gaussian Process algorithm.

The reaction heat associated with each interaction can be expressed as:

$$Q = \left(\sum \Delta M_{initial} - \sum \Delta M_{final}\right)c^2 \quad (3)$$

where c is the speed of light and $\Delta M = Z \cdot m_p + N \cdot m_n - m_A$, with Z the number of protons of mass m_p, N the number of neutrons of mass m_n and m_A the nucleus mass. According to the data from National Nuclear Data Center (NNDC), [19] the reaction heat for the $^{12}C(n,\alpha\ )\,^9Be$ process is approximately -5.7 MeV. By applying the law of energy conservation and the two-body kinematic equations, the reaction heat for this interaction can be further expressed as:

$$Q - E^*_{exc} = \left(m_n + m_{12_C} - m_\alpha - m_{9_{Be}}\right)c^2 - E^*_{exc} = E_\alpha + E_{k_{9_{Be}}} - E_{k_n} \quad (4)$$

with $E^*_{exc_{9_{Be}}}$ the excitation energy in which the ^9Be particle is released and E_k the kinetic energy.

From this, the total kinetic energy deposited in the sensor is given by:

$$E_{k_{tot}} = E_{\alpha_0} + E_{k_{9_{Be(gs)}}} - E^*_{exc} = Q + E_{k_n} - E^*_{exc} \quad (5)$$

Finally, if the ^9Be nucleus is emitted in its ground state, i.e. $E^*_{exc} = 0$, then, based on Equation 5 and considering an incident neutron energy $E_{k_n} = 15.3$ MeV, the kinetic energy deposited within the sensor is:

$$E_{k_{tot}} = Q + E_{k_n} = -5.7\ MeV + 15.3 MeV = 9.6\ MeV \quad (6)$$

A similar consideration applies to the other excited states from which the ^9Be particle can be emitted. These states correspond to excitation energies of 1.68, 2.4, and 3 MeV, resulting in deposited energies ranging between 6.6 and 7.92 MeV. This energy range also coincides with the production of 3α particles. The production of alpha particles is expected to be the most significant reaction within this energy range because, at 15.3 MeV, the probability of alpha emission is the highest among the interactions between carbon atoms and neutrons. This probability even surpasses the cross-section associated with elastic scattering, indicating that alpha production is the dominant process at these energy levels. Finally, for interactions involving silicon, the most distinguishable reaction observed in Figure 1a is the $^{28}Si(n,\alpha)^{25}Mg$ process, where the magnesium particle is released in its ground state. This reaction is associated with a reaction heat of approximately -2.56 MeV. Consequently, the total kinetic energy resulting from this reaction is given by:

$$E_{k_{tot}} = E_{\alpha_0} + E_{k_{25_{Mg(gs)}}} - E^*_{exc} = Q + E_{k_n} - E^*_{exc} =$$
$$= -2{,}65\ MeV + 15{,}3\ MeV = 12{,}6\ MeV \quad (7)$$

which corresponds to the peak located at the highest energy in the deposited energy spectrum.

The consistency between the deposited energy spectrum and the expected data derived from cross-section analysis using the JENDL-4 nuclear data library supports the reliability of the experimental results.

In this case, the machine learning algorithm accurately identifying the energy bin into which the monoenergetic neutron energy falls, as shown in Figure 1b. The low uncertainty provided by the GP model in identifying the correct bin — 2% for the selected bin and 7% for the preceding one — suggests that the input data (deposited energy vector) are highly informative, the signal is clear and distinct, and the corresponding region is well represented in the training set.

The other monoenergetic neutron beam considered in this study has an energy of 0.78 MeV coming from a pT reaction. At this energy, the incident neutrons do not exceed the threshold energies required for any of the reactions listed in Table 1. Therefore, only elastic scattering interactions with carbon and silicon atoms are expected to occur. Assuming the sensor atomic composition is approximately 50% carbon and 50% silicon, the most significant interaction is expected to be neutron elastic scattering with silicon atoms. This is due to the fact that the elastic scattering cross-section for silicon is approximately an order of magnitude higher than that for carbon, as reported in the JENDL-4 nuclear data library. Moreover, because silicon atoms are heavier than carbon atoms, the energy transfer during elastic scattering is on average lower. As a consequence, the deposited energy spectrum is expected to be shifted toward lower energies, with the maximum approximately at $E^{Si}_{max} = 0.10\ MeV$ for Si- neutron interactions and $E^{C}_{max} = 0.21\ MeV$ for C- neutron interactions, as shown in Figure 2a. In this case, 5716 pulses were collected, with an associated uncertainty between 12% and 62%.

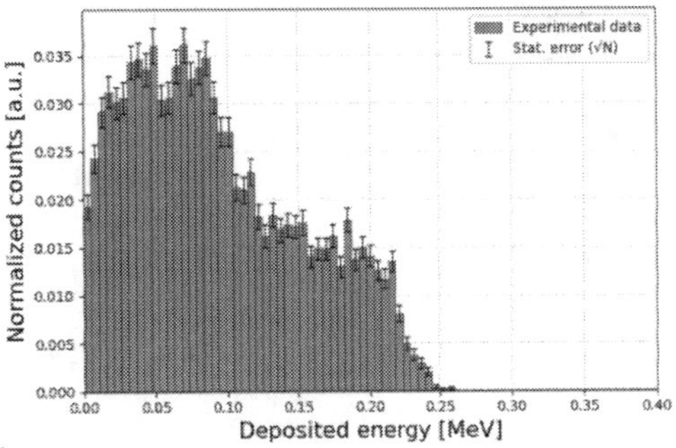

(a)

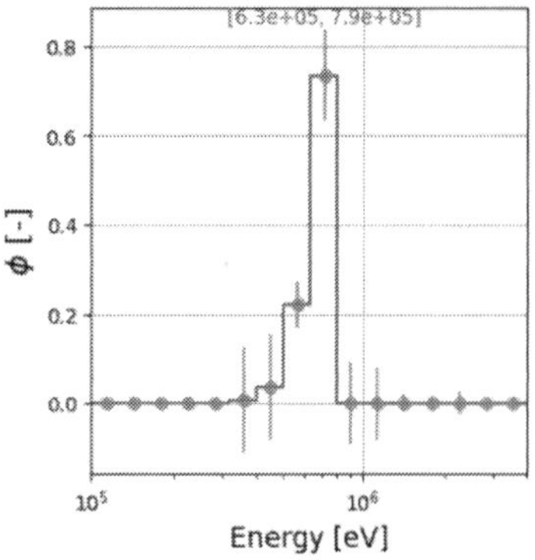

(b)

Fig. 2. (a) Deposited energy spectrum obtained from the SiC detector exposed to the 0.78 MeV (b) Reconstructed 0.78 MeV neutron spectrum using the Gaussian Process algorithm.

The machine learning algorithm, when applied to this dataset as input, successfully identifies the low-energy region, corresponding to the expected neutron energy and interaction type. This outcome is primarily attributed to the fact that neutron scattering interactions with these nuclei are well-characterized and show consistent agreement across various nuclear data libraries, as reported in [20]. Consequently, the reconstruction algorithm is able to accurately identify the corresponding energy bin in Figure 2b. However, the statistical quality of this dataset is relatively low. and due to the limited number of pulses, the prediction interval provided by the Gaussian Process model is broader compared to the previous case, ranging from 23% to 36%. This increased uncertainty arises from the probabilistic nature of the algorithm, which relies on the availability of sufficient training data to achieve accurate predictions. In scenarios where data scarcity is present, the Gaussian Process tends to produce wider prediction intervals to account for the increased uncertainty in the reconstruction process.

The final neutron energy analyzed in this paper for the Van de Graaff facility is 4.18 MeV from a DD reaction, with the corresponding energy deposition in the SiC detector presented in Figure 3a. Approximately 4821 pulses were collected, with uncertainties ranging from 1% to 45%.

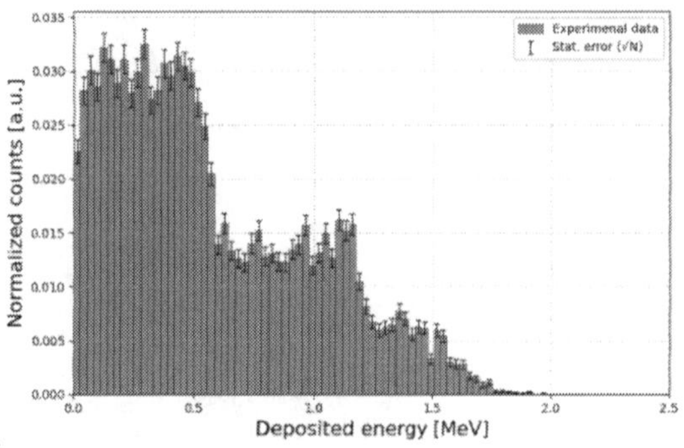

(a)

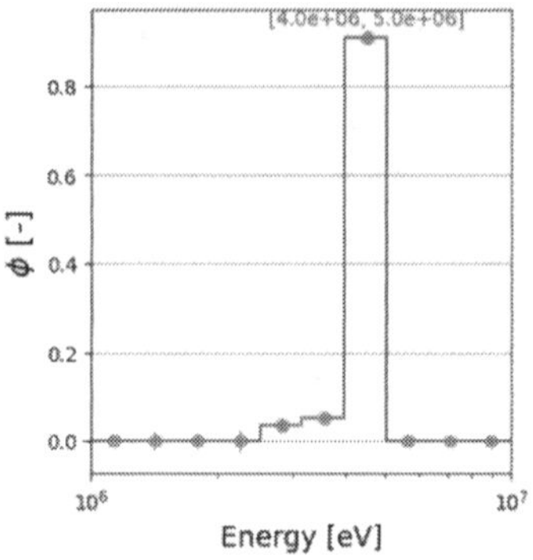

(b)

Fig. 3. (a) Deposited energy spectrum obtained from the SiC detector exposed to the 4.18 MeV (b) Reconstructed 4.18 MeV neutron spectrum using the Gaussian Process algorithm.

In this case, the incident neutrons on the SiC sensor possess an energy higher than the threshold energy for Si-n interactions but remain below the threshold required to initiate C-n reactions, outside of elastic reactions. According to the cross-section data, the predominant interaction within this energy range is elastic scattering with Si atoms. This is reflected in the spectrum by a peak shifted towards $E_{max}^{Si} = 0.54\, MeV$, accompanied by a high-energy tail primarily resulting from elastic scattering events involving carbon atoms in smaller quantities, as well as Mg/alpha production. The experimental results enable the learning algorithm to accurately reconstruct the energy bin of interest. The excellent statistical quality of the experimental deposited energy spectrum results in very low uncertainty in the prediction, with prediction intervals below 2% across all bins. The only noticeable deviation is a slight broadening of the prediction interval — up to 5% — for energy bins just below 4 MeV, suggesting the presence of lower-energy neutrons, likely due to backscattering effects.

B. Results collected at TOTEM facility, Cadarache

At this facility, the SiC sensor was exposed to a 14.1 MeV neutron beam produced by the GENIE-16 DT neutron generator. The deposited energy spectrum corresponding to this energy is well-documented in the literature. [21] The spectrum obtained from our measurements, shown in Figure 4a, is consistent with previously reported data, with the exception of a slightly higher-than-expected scattering component. This discrepancy is likely attributed to neutron backscattering, as the generator is positioned near a concrete wall within the DANAIDES irradiation casemate of the TOTEM facility. In this configuration, extended measurement periods yielded a total of 82579 pulses, with associated uncertainties varying between 1% and 36%.

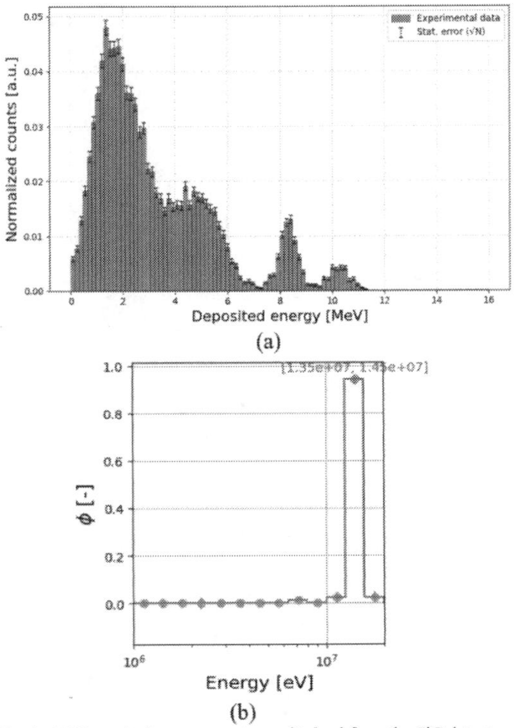

Fig. 4. (a) Deposited energy spectrum obtained from the SiC detector exposed to the 14.1 MeV (b) Reconstructed 14.1 MeV neutron spectrum using the Gaussian Process algorithm.

The measurements clearly reveal the expected characteristic peaks, particularly the prominent peak at 8.3 MeV associated with alpha/Be production in the ground state. This peak is commonly used for assessing resolution in such sensors. Higher-energy peaks are attributed to Mg/alpha or Al/proton production, where the Mg and Al particles can be released in various excited states. Additionally, the bump observed between 2 and 6 MeV is primarily due to the production of 3 alpha particles. The accurate representation of these peaks enables the machine learning algorithm to successfully identify the energy bin corresponding to 14 MeV, as shown in Figure 4b, with an uncertainty always below 3%. In addition, dedicated experimental campaigns at GANIL (TOF) and AMANDE facilities can be planned in order to achieve improved metrology and validation of the proposed approach. A comparison with reference measurements obtained using Bonner sphere spectrometry will also be carried out to further benchmark the methodology.

V. Conclusions and Perspectives

This paper presents a machine learning algorithm based on the Gaussian Process model for reconstructing neutron spectra in solid-state detectors. To validate the reconstruction capability with experimental data, two measurement campaigns were conducted at the Van de Graaff and TOTEM facilities, where a 60 μm thick SiC sensor was exposed to various monoenergetic neutron beams. The experimental results of the deposited energy in the sensor are consistent with both simulations and reference values reported in the literature, displaying the characteristic peaks that were thoroughly analyzed in Section 4.

The accurate representation of these peaks in the deposited energy allows the learning algorithm to reliably identify the correct interval in which the neutron beam to which the sensor was exposed falls. This outcome provides a proof of concept for the feasibility of normalized neutron spectrum reconstruction through this approach in monoenergetic condition, serving as a potential alternative or complementary technique to conventional unfolding methods.

Further steps toward validating this methodology include a more detailed investigation of the energy grid discretization, as well as the analysis of more complex and diverse spectra, such as those originating from isotopic sources or nuclear reactors, which are currently under study. Future work should also involve comparisons with other machine learning models to improve both optimization and robustness. Moreover, the possibility of extending the energy range of interest through the use of Bayesian Neural Networks will be explored.

Acknowledgment

This work is part of an international collaboration within the 'Timepix SiC detector' project, supported by the DANUBE Region Strategy program: by the Slovak Research and Development Agency under the Contract no. DS-FR-22-0012, by the Ministry of Education, Youth and Sports of Czech Republic by the Contract no. 8X23035 and 49925ZL.

References

[1] F. H. Ruddy, et al., "Performance and Applications of Silicon Carbide Neutron Detectors in Harsh Nuclear Environments", in EPJ Web of Conferences, vol. 253, 2021.

[2] E. Belfiore, et al., "Theoretical analysis of neutron spectra measurement with SiC detectors using a machine learning technique", in Journal of Instrumentation, vol. 19, 2024.

[3] R.V. Griffith, J. Palfalvi and U. Madhvanath, Compendium of neutron spectra and detector responses for radiation protection purposes, Tech. Rep. STI/DOC/010/318, Technical Reports Series No. 318, IAEA (1990).

[4] T. Sato, et al., "Features of particle and heavy ion transport code system (PHITS) version 3.02", J. Nucl. Sci. Technol., 2017.

[5] E. Belfiore, et al., "Proof of Concept of the Prediction of Neutron Spectra using Solid Detectors via Machine Learning Approach", in 2024 IEEE Nuclear Science Symposium (NSS), Medical Imaging Conference (MIC) and Room Temperature Semiconductor Detector Conference (RTSD), Tampa, FL, USA, 2024, pp. 1-2, doi: 10.1109/NSS/MIC/RTSD57108.2024.10658457.

[6] E. Belfiore, et al., "SiC Detector Thickness Optimization for Enhanced Response Variability", in EPJ Web of Conference, vol. 302, 2024.

[7] E. Belfiore, et al., "Comparative assessment of feature extraction for fast neutron spectra prediction with machine learning algorithms using a CVD diamond detector", in NIM-A, 2025, 10.1016/j.nima.2025.170951

[8] D. Munteanu, et al. "Basic Mechanisms of Single-Event Occurrence in Silicon Carbide Semiconductor under Terrestrial Atmospheric Neutron Irradiation", in Electronics, vol. 12, 2023, pp.4468.

[9] Z. Kohout, et al., "Electrostatic separator as a source of low energy protons (32 keV – 400 keV) on the Van de Graaff accelerator", in Journal of Instrumentation, vol. 14, 2019.

[10] Laboratory of the VdG accelerator, https://aladdin.utef.cvut.cz/projekty/vdg/

[11] V. Fondement, et al., "Development of a neutron probe to perform a combined measurement of uranium concentration and hydrogen porosity for mining applications", in NIM-A, vol. 1059, 2024.
[12] Shorna, Ummay Salma and Hossain, Md. Forhad, "A Robust Approach to Determine the Coefficient of Skewness and an Alternative Form of Boxplot", Available at SSRN: https://ssrn.com/abstract=4436052 or .
[13] Q. Potiron, et al., "Modelling of a SiC Based Detector for the Interpretation of 14.1 MeV Neutrons Measurements", in EPJ Web of Conferences, vol. 288, 2023.
[14] K. Shibata, et al., "JENDL-4.0: A New Library for Nuclear Science and Engineering", in Journal of Nuclear Science and Technology, vol. 48, 2011.
[15] E. Schulz, et al., "A tutorial on Gaussian process regression: Modelling, exploring, and exploiting functions", in Journal of Mathematical Psychology, vol. 85, 2018.
[16] A. Gelman, et al. "Bridges between deterministic and probabilistic models for binary data", in Statistical Methodology, vol. 7, 2010.
[17] M. Belyaev, et al., "Computationally efficient algorithm for Gaussian Process regression in case of structured samples", in Comput. Math. and Math. Phys., vol. 56, 2016.
[18] Q. Potiron, et al., "Estimation of the Fast-Neutron Fluence Rate With a 4H-SiC-Based Detector in a 14.1-MeV Mono-Energetic Field", in IEEE Transactions on Nuclear Science, vol. 71, 2024.
[19] B. Shu, et al., "Ongoing Updates to NNDC Web Services", in EPJ Web of Conferences, vol. 284, 2023.
[20] E. Belfiore, et al., "Cross Section Discrepancies in Neutron-Carbon Interactions at 14 MeV: a Sensitivity and Uncertainty Analysis", in M&C Conference Proceeding, 2025.
[21] M. Osipenko, et al., "Response of a diamond detector sandwich to 14 MeV neutrons" in NIM-A, vol. 10, 2016.

EducTUM - An Interactive Platform for Education and Training in the non-destructive analysis of radioactive materials

Thomas Bücherl*, Thomas Narr, Anton Kastenmüller, Christoph Lierse von Gostomski
Technische Universität München - ZTWB Radiochemie München, Germany
(*) Thomas.buecherl@tum.de

Abstract— **The decommissioning of nuclear facilities poses major challenges for the declaration of the resulting waste. This will be further exacerbated by changes to existing regulations and the introduction of new ones, also with regard to existing (legacy) waste. As dismantling will continue for decades to come, this requires the long-term maintenance of the relevant specific specialist skills. The EducTUM project (https:\\educTUM.de) aims to provide basic and advanced training, preserve skills and impart general information in the fields of radioactive waste characterization and product control. The combination of online measurements with real measuring equipment and real radioactive samples using (commercial) programs sets it apart from a "normal" Wikipedia webpage. It is aimed at a wide range of people, from those with a general interest in the subject to users and technical experts. The way in which knowledge is imparted therefore depends on the level of knowledge of the individual groups of people.**

Keywords —**education, training, radioactive material, non-destructive analysis decommissioning, dismantling.**

I. Introduction

THE decommissioning of nuclear facilities poses major challenges for the declaration of the resulting waste. This will be further exacerbated by changes to existing regulations and the introduction of new ones, also with regard to existing (legacy) waste. As dismantling will continue for decades to come, this requires the long-term preservation of the relevant specific specialist skills.

This can only be achieved through continuous further and advanced training of the people already involved, while at the same time training the next generation of technical and scientific experts.

In all these cases, demographic trends must be taken into account, which predict a dramatic age-related decline in existing knowledge due to retirement (Fig. 1).

For this reason, urgent measures are needed to preserve the knowledge that currently still exists and to pass it on.

Such a process should also specifically target various groups. From the general public, which, due to various accidents and political decisions in the (recent) past, often has a negative, but frequently unfounded, opinion on the subject of radioactivity, radioactive waste and nuclear technology, to authorities and their experts, personnel working in the field of nuclear technology, scientists, and manufacturers of measuring systems, all should be addressed and able to access or learn information in a suitable form.

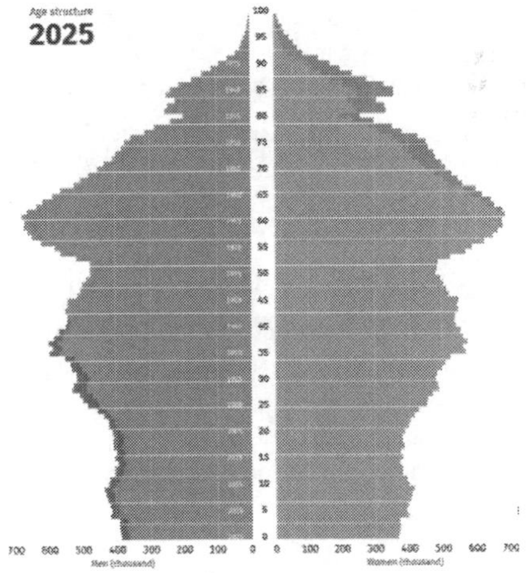

Fig. 1. Coordinated Population Projection for Germany. Variant 1: Moderate development in fertility and life expectancy with low net migration [1].

For conservation and provision knowledge in the field of non-destructive analysis of radioactive materials from decommissioning and dismantling by education, qualification, sustainability, training of the next generation of technical and scientific staff, the knowledge platform EducTUM (https:\\eductum.de) is actually under development and is made available online as development progresses. The corresponding hardware and software are being developed based on a teaching concept created for this purpose. EducTUM is supplemented by the provision of various (commercial) software, some of which can also be tested online at real measurement stations or downloaded for personal use.

II. Conception

A. Areas of Application

In its current phase, the EducTUM project focuses on the non-destructive characterization of radioactive materials and containers enclosing them. To this end, the measurement methods Gamma spectrometry (GS), segmented Gamma scanning (SGS), Gamma radiography and Gamma tomography will initially be considered [1]. In a future expansion, it is planned to also consider the areas of radiation protection, quality control and others.

B. *Teaching Concept*

An important point is the selection and implementation of a suitable curriculum. This defines the teaching content and objectives and must be determined separately for each area. Based on the defined content and objectives, a suitable method of teaching the material must be determined.

It is important to consider the target audience when deciding on the appropriate teaching method. The information to be conveyed must be presented in a completely different way to the "common man" with no prior knowledge than to a scientist or technician who has been working in the field for years. This led to a division of the target audience into the groups "beginners", "advanced learners", "users", and "experts", each with individual teaching concepts.

For practical reasons, which suggest independence from time and place in the transfer of knowledge, some form of distance learning must be made possible. Traditionally, textbooks, journals, periodicals, etc. have been used for this purpose. However, this form of knowledge transfer is now considered "old-fashioned." It also makes it difficult to take new findings into account, correct errors or unclear descriptions, etc., and is therefore now mostly seen as a supporting measure.

With the global availability of the Internet and the web-based presences that can be created on it, which provide not only pure text-based but also audiovisual information for a wide variety of devices, the local and temporal dependence of information retrieval has been largely eliminated on the one hand, while the amount of available information has increased enormously on the other, which in turn raises the problem of selecting the information that is actually relevant. More recently, the question of the credibility of the information found has also become increasingly important.

Social media represent a further development of general web-based information dissemination and is a widely used source – especially among the younger generation, that prefers to convey information in short videos.

Based on this brief (and certainly incomplete) overview of media for knowledge transfer, social media should really be the obvious choice. However, recent studies (e.g.[8][9]) also highlight possible long-term problems associated with their use.

This led to the decision to create an interactive website as the basis for knowledge transfer. The respective teaching content is presented differently for each of the four target groups mentioned above. The results of a study [2] were incorporated here, which investigated whether a teaching concept should preferably follow the pattern of problem-solving followed by instruction (PS-I) or instruction followed by problem-solving (I-PS).

The results of the study showed a preference for the PS-I method, primarily because the productive failures that occur in this method generate a positive learning effect, which also has a positive impact on the long-term development of learning effectiveness. However, the study, which was conducted with undergraduates and students at higher levels of education, also indicates that undergraduates showed a slight preference for the I-PS method. Consequently, the teaching concept for the "beginners" and "advanced learners" was developed based on the I-PS method. After the introductory explanations on a topic, varying questions are asked to check learning progress and, if necessary, provide further and additional assistance for better understanding. For the other two groups ("users", "experts"), elements of the PS-I method are used. Here, too, the individual topics are explained in detail according to the respective target group and, where possible, illustrated with practical examples. However, users have the option of first settling for a general description (and applying this knowledge in practice) before accessing more detailed and comprehensive information if required.

In addition, these two groups have direct access—as do the other two groups indirectly—to further tools for calculations and evaluations, as well as access to (real) online measurements and evaluations using various (commercial) programs and measurement systems.

III. Implementation

A. *Basic Hardware Concept*

The basic hardware concept consists of a server hosted by a commercial provider, which can be accessed via the Internet using a desktop PC, laptop, tablet, smartphone, etc. The information provided on the server can be used via a browser without access restrictions. This requires a responsive website design.

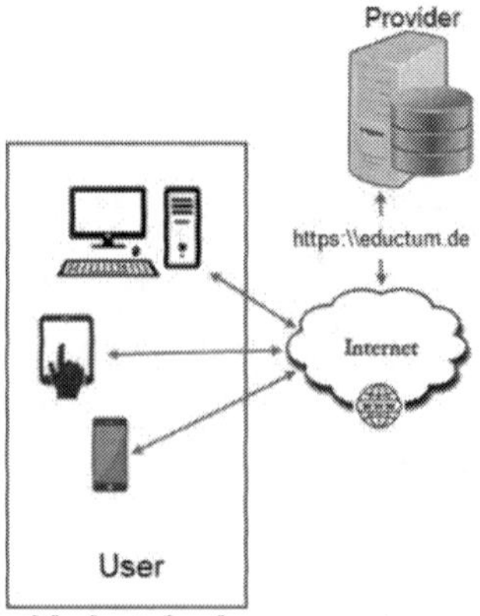

Fig. 2. Design of the basic hardware concept.

B. *Basic Software Concept*

The basic software concept is implemented using the content management system Joomla! and a YOOtheme template. The landing page is shown in Fig. 3.

Currently, all interfaces and content are only available in German. Until the planned English version is completed, most of the content can be displayed in the desired language using

https://doi.org/10.1051/epjconf/202533811001

Fig. 3 Landing page of EducTUM (https:\\eductum.de)

the translation options available in modern browsers.

The content is currently divided into three main categories: measurement methods, evaluation and analysis, and information (see large icons in right column in Fig. 4). Each category has subcategories assigned to it via a mega menu. For example, the measurement methods category includes segmented gamma scanning, gamma spectrometry, radiography, and tomography (see Fig. 4 below).

Fig. 4. Mega menu with the main topics segmented gamma-scanning, gamma-spectrometry, radiography and tomography (from left to right). The topic segmented gamma-scanning is subdivided in the chapters introduction (Übersicht), beginner (Einsteiger), advanced (Fortgeschrittene), user (Anwender), and expert (Experten), respectively.

For each subcategory, there is a menu item with general information (introduction) and four additional menu items for the different levels ("beginner", "advanced", "user", "expert") (Fig. 4). The latter provides access to individual information in the form of text, audiovisual components, questions, tools, links, etc. During implementation, attention was paid not only to user-friendliness (e.g., access to information with as few mouse clicks as possible) but also to appropriate linguistic presentation. In the "beginner" and "advanced" sections, a more colloquial, and in some areas sometimes even casual, style is used, and technical terms and foreign words are avoided as far as possible. In the "users" and "experts" sections, attention is paid to a technically correct presentation of the facts, which also uses technical terms and foreign words. Where necessary, cross-references to corresponding explanations of technical terms are included.

The concept presented above and its implementation are no different from numerous other implementations. But what sets it apart from the rest?

C. *Advanced Hardware and Software Concept*

EducTUM's unique selling point is based on an expanded hardware concept (Fig. 5). Registered EducTUM users have access to a protected server installed at the RCM (Fig 6). This server provides access to additional hardware and software. The hardware currently consists of a gamma spectrometry measuring station with a NaI, a $LaBr_3$, and an HPGe detector. It is to be expanded with additional typical measuring devices, such as dose rate probes. A wide variety of radioactive sources are placed in front of the detectors as measurement objects and are exchanged at irregular intervals.

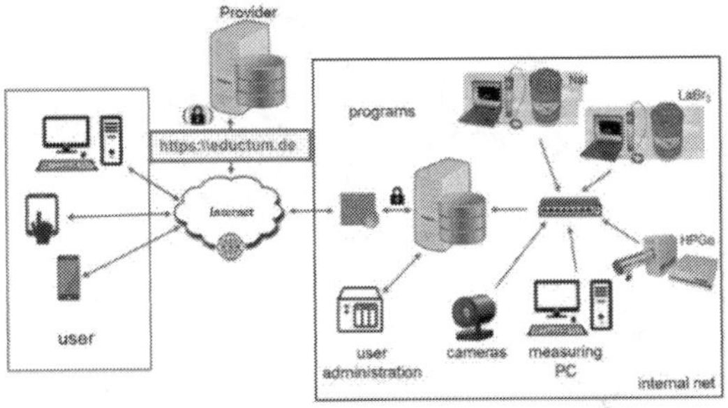

Fig. 6. Layout of the advanced implementation with access to the online gamma-spectrometry measuring system.

The commercially available gamma spectrometry program LVis [10][11] is currently available for use at the gamma spectrometry measurement station. With this program, every user has access to every detector at any time and can perform measurements on the radioactive samples provided. A large part of the specific settings of the measurement parameters, such as the number of channels, gain, etc., can be individually adjusted.

This is achieved through the following measures: At RCM, the detectors with typical measurement parameters are integrated into conventional gamma spectrometry systems that continuously record measurement data. This data is forwarded to a software-based transmitter, which calculates incremental spectra from this data, i.e., the data recorded by the respective detector during a short period of time (e.g., 1 second). This data is transferred via a network to another computer, where it is recorded by a software-based broker and forwarded to the respective instance of the gamma spectrometry program LVis. Each

Fig. 5 Login-webpage for accessing the programs

user can use their own LVis instance completely independently of other users, record the measurement data, analyze it completely, and save it in a user-specific file folder so that it can be reused for later analysis.

The gamma spectrometry program LVis can be used with virtually no restrictions. Only the uploading and downloading of your own data is prohibited for security and copyright reasons. Numerous video tutorials support the practical application of the program.

In the near future, we plan to make additional (commercially) available software available. This includes a program for performing and analyzing segmented gamma spectrometry (SGS) measurements, but without the integration of real gamma scanners, which is not feasible for obvious reasons. But measured data sets will be provided for the analysis of SGS measurement data.

We hope that a preliminary agreement already in place with another manufacturer will enable us to offer another commercial evaluation program for gamma spectrometry measurements in the near future.

Additional non-commercial software for analyzing gamma spectrometry measurements, specifically of samples containing uranium and plutonium, will be available for download.

D. Example

One example of the use of EducTUM and its practical application is the gamma spectrometry internship conducted with students from the Technical University of Munich at the end of 2024. The aim of this practical experiment was to test the extended hardware and software concept in terms of its suitability for the direct implementation of acquired knowledge in practical applications using current (commercial) measurement and analysis software, some of which is already in use at many measurement stations worldwide.

Based on detailed practical instructions implemented in EducTUM in the gamma spectrometry section, which included numerous references to relevant topics and detailed step-by-step instructions for performing online measurements and their evaluations, as well as references to the corresponding video tutorials on the gamma spectrometry program, the practical course was successfully completed with very positive feedback. Nevertheless, comments from students led to a partial revision of the instructions and individual software modules for using the hardware.

IV. Conclusions

The EducTUM project aims to provide basic and advanced training, maintain skills and impart general information in the fields of radioactive waste characterization and product control. It is aimed at a wide range of people, from those with a general interest in the subject to users and technical experts. The way in which knowledge is imparted therefore depends on the level of Source and LVis based coincidence summing correction in HPGe

gamma-ray spectrometry knowledge of the individual groups of people.

The project's strength lies in combining knowledge acquisition via conventional websites, supported by videos, animations, etc., with the practical application of this knowledge using (commercial) programs in online measurements with real detectors and real radioactive samples, as well as the use of other programs.

The EducTUM project is currently still in the development phase and is available in German. The starting point of the EducTUM website is https:\\educTUM.de. In the first phase of its development, which is currently underway, the focus is on non-destructive metrological investigations of radioactive waste packages and, in particular, on measurement methods that use gamma and X-rays. Other subject areas are to be included in the course of further development.

Content is being added on an ongoing basis. Plans are in place to make it available in other languages (initially English).

Acknowledgment

The authors thank the Federal Ministry of Research, Technology and Space for funding the project under support code 15S9443 and the Bavarian State Ministry of Environment and Consumer Protection for funding most of the measuring equipment.

References

[1] T. Buecherl, C. Lierse, E. Kaciniel, T. Neukel, K. Krebs, "Determination of gamma emitters in radioactive waste containers by advanced gamma scanning". American Society of Mechanical Engineers., 1995, S. p. 903–907.

[2] G. F. Knoll, Radiation detection and measurement, 3rd ed. New York: Wiley, 2000.

[3] C. Lierse, H. Goebel, E. Kaciniel, T. Buecherl, und K. Krebs, „Application of digital radiography for the non-destructive characterization of radioactive waste packages", in *Fifth international conference on radioactive waste management and environmental remediation – ICEM '95: Proceedings. Volume 2: Management of low-level waste and remediation of contaminated sites and facilities*, 1995. [Online]. Available at: https://inis.iaea.org/search/search.aspx?orig_q=rn:27048654

[4] R. J. Estep, T. H. Prettyman, und G. A. Sheppard, „Tomographic Gamma Scanning to Assay Heterogeneous Radioactive Waste", Nuclear Science and Engineering, Bd. 118, Nr. 3, S. 145–152, Jan. 1994, doi: 10.13182/NSE94-A19380

[5] B. Pérot u. a., „The characterization of radioactive waste: a critical review of techniques implemented or under development at CEA, France", EPJ Nuclear Sci. Technol., Bd. 4, S. 3, Jan. 2018, doi: 10.1051/epjn/2017033.

[6] Moderate development in fertility and life expectancy with low net migration (from: Deutsches Statistisches Bundesamt, *https://service.destatis.de/bevoelkerungspyramide/index.html#!y=2025&l=en*

[7] *T. Sinha, M. Kapur, "When Problem Solving Folloed by Instruction Works: Evidence for Productive Failure," Review of Educational Research,* vol. 91, no. 5, Oct. 2021, pp. 761-798.

[8] *Anthes L. „Was machen TikTok & Co mit unserem Gehirn?" 18. Apr. 2023. https://scilogs.spektrum.de/hirn-und-weg/was-machen-tiktok-co-mit-unserem-gehirn,* Accessed on: July 28, 2025.

[9] *J. Skowronek, A. Seifert, S. Lindberg, "The mere presence of a smartphone reduces basal attentional performance," Sci Rep 13, 9363 (2023). https://doi.org/10.1038/s41598-023-36256-4,* Accessed on: July 28, 2025.

[10] https://www.ortec-online.com/products/software/labvision, accessible end of August 2025.

[11] J. Lee, H. Kim, Y. U. Kye, D. Y. Lee, J. K. Kim, W. S. Jo, Y.-R. Kang, "Source and LVis based coincidence summing correction in HPGe gamma-ray spectrometry", Nuclear Engineering and Technology, Vol. 54, Issue 5, May 2022, Pages 1754-1759, https://doi.org/10.1016/j.net.2021.11.008

From real life to scale model (and back again): the life of CROCUS in the student's mind

Vincent Lamirand[1,2*]
[1]École polytechnique fédérale de Lausanne (EPFL), Switzerland
[2] Paul Scherrer Institute (PSI), Switzerland

(*) vincent.lamirand@epfl.ch

***Abstract*— The CROCUS nuclear reactor is primarily motivated by education, from public visits to advanced teaching at the Bachelor's, Master's, and doctoral levels, as well as professional training. It is a uranium-fueled, light-water moderated, zero power reactor with a maximum power of 100 W. As simple as an operational reactor can be, CROCUS effectively supports conceptual understanding through its tangible materiality, offering virtually no distance from the system, no significant risk, and negligible dose rates in its vicinity.**

Once the basic concepts and systems are grasped, however, a deeper effort is required to fully comprehend the machine and the underlying physics. Various tools are employed to accompany this learning process, including Monte Carlo modeling, operator licensing for PhD students, and tangible scale models – the focus of this contribution. Following the limited use of an earlier 3D-printed model, a sturdier and more manipulation-friendly version was built using LEGO® bricks. This model reproduces all key features with minimal approximations, helping to visualize otherwise hidden elements, such as the startup source and the expansion vases.

We present the educational objectives, the learning methods applied, and the benefits of using CROCUS and its models. The impact of the LEGO® model was evaluated in two teaching contexts: a large group of 2nd-year Physics Bachelor students during a single 4-hour experiment, and a smaller group of 1st-year Nuclear Engineering Master students over a semester-long practical course. Statistical analysis indicates improvement in the Master students' understanding of CROCUS and reactor functioning.

***Keywords* — zero-power nuclear reactor; reactor physics; education; STEM; active learning; trialogical learning approach; affordance; CROCUS; LEGO®.**

I. Introduction

THE importance of sufficient, high-quality education in the nuclear field has been recently highlighted by academia, networks, and international institutions, particularly with access to experimental facilities [1]–[6]. EPFL operates the CROCUS reactor for both teaching and research purposes [7], [8]. Whereas research activities span a broad range of topics, the educational mission of the laboratory, the team, and the facility is equally central [9], [10]. Educational activities range from introducing basic radiological and nuclear concepts to public visitors, to advanced education at the Bachelor's, Master's, and doctoral levels, as well as training programs for professionals.

CROCUS is a uranium-fueled, light-water moderated, zero power reactor with a maximum power of 100 W [11], [12]. As simple as an operational nuclear reactor can be, its apparent simplicity is one of its greatest strengths: visible and accessible fuel rods, reactor vessel and hydraulic systems, neutron detectors and power monitors, etc. This accessibility aligns with the pedagogical concept of *affordance*: CROCUS provides the tangible materiality that bridges the teacher's explanations with student's learning processes [13], [14]. In a *trialogical learning approach*, students can walk around the vessel, lean over it to view the inside and the fuel, and even, to an extent, "touch" the core [15]. With negligible dose rates and no safety risk in normal operation, there is virtually no distance between the learner and the object of the study. Nevertheless, CROCUS remains a complex system from an educational perspective. Beyond its apparent transparency lie physical, instrumental, and mechanical specificities that require deeper engagement to understand once basic concepts are understood. To support this, a variety of tools are employed, all rooted in *active learning* principles.

For advanced students, typically at the Master's level, Monte Carlo reactor models provide valuable insights into CROCUS's design and behavior. Whereas computational work complements hands-on experience, it remains static and intangible, offering no direct sense of the system's operation and inner workings. At the other end of the spectrum, PhD students with experimental projects in CROCUS complete the reactor operator license, acquiring in-depth operational knowledge through direct responsibility for the machine. They come to know CROCUS better than anyone for their specific experimental applications, having operated it for and by themselves. This learning path is highly effective but also time-intensive.

To bridge the gap for intermediate learners, and as a complement during hands-on learning courses, tangible models have been developed. The first one, a detailed 1:5 scale 3D-printed replica based on the CAD model of CROCUS, captures the reactor's appearance with high fidelity. However, it is delicate, costly, and requires careful handling. It is not well-suited for manipulation but rather serves as an excellent demonstration tool. Moreover, it cannot easily reveal hidden components such as the startup source or the water intake and expansions vases. Whereas it captures the reactor's appearance, this model's pedagogical affordance is limited by the same restrictions as the actual reactor, without having the same impressive appearance.

To address these limitations, a second model was built entirely from LEGO® bricks. Preserving the relative scale and key features of CROCUS with minimal approximation, the LEGO® model is robust, easy to manipulate, and allows for interactive exploration. The core can be "unloaded" similarly to the real one, and it can be disassembled without being damaged. It includes movable parts, such as the spillway and absorber rods, and the vessel can be opened to reveal the internals. Crucially, it exposes normally hidden elements, enabling clearer visualization and comprehension of the reactor's inner workings.

In this contribution, we present the educational objectives and challenges of communicating reactor physics and operation, and the role of both the real CROCUS reactor and its models in addressing them. Additionally, We also report on an evaluation of the LEGO® model's impact on classroom learning outcomes.

II. CROCUS FOR TEACHING

A. Hands-on courses in CROCUS

By internal regulation, CROCUS is primarily dedicated to teaching: research activities are systematically given second priority. It supports a wide range of hands-on experimental courses, starting with EPFL Physics Lab II and III (PHYS-204/211 and PHYS-319/320, 8 ECTS, in French) and Radiation and Reactor Experiments (PHYS-451, 6 ECTS, in English).

The Physics Lab II course is aimed at 2nd-year Bachelor students and consists of a series of weekly hands-on sessions throughout the academic year. In this framework, and among other experiments provided by other laboratories, each week a different group of eight students visit the reactor facility to perform an approach to critical. Given the specific nature of reactor physics in relation to the rest of the curriculum, the experiment is designed so that theory and practice are covered in a single four-hour session.

In Physics Lab III, 3rd-year Bachelor students select three experiments per semester from a panel offered in the physics facilities. Each experiment runs one day per week over four weeks, i.e. four full days. At CROCUS, two groups of 2-3 students are hosted simultaneously, with constant success, each pursuing a different objective: one group carries out neutron flux measurements, including reactor kinetics; the other performs gamma spectrometry experiments, including foil activation.

The Radiation and Reactor Experiments course is primarily offered to Nuclear Engineering Master's students. Due to increasing batch sizes in recent years, students from other fields have not been accommodated for some time. In this course, students perform 13 experiments, weekly over a semester, ranging from basic radiation detection to advanced reactor physics. Radiation detection experiments range from simple particle detection to complex settings, e.g. coincidence measurements, and radiation protection applications. Reactor-related experiments include: approach to critical, control rod worth, power calibration using activation, and reactor noise.

Recent research developments have purposefully enriched these sessions, for instance by integrating the SAFFRON array for noise measurements. New types of experiments are also under consideration, such as pile-oscillation and neutron imaging [16]–[22].

Beyond these courses, CROCUS also hosts sessions within various other frameworks, such as collaborations with other schools and universities, e.g. HEPIA (Geneva, Switzerland) and PHELMA (Grenoble, France), or professional training programs (e.g. PSI Reactor School). Hands-on sessions are frequently organized for dissemination in European projects (e.g. CORTEX) or as a central objective in education-oriented initiatives (e.g. GRE@T-PIONEeR) [10], [23].

B. Active learning methods

Evidence across STEM disciplines (Science, Technology, Engineering, and Mathematics) as well as nuclear engineering specifically, shows that *active learning* reduces failure rates, improve exam performance, and broadens the objectives of teaching [24]–[26]. By definition, hands-on experiments –and, to some extent, facility visits– constitute active learning sessions: they require participants to do more than passively listen to the educator. Typical features include:

- Purposiveness: acting, with a clear goal; e.g., estimating the next reactor state
- Reflectiveness: assessing outcomes; e.g., "does this value make sense?"
- Situation-driven constraints: operating within time, system, and facility limitations
- Complexity/adversity: addressing unexpected behavior, from systems or partners; e.g., a system not responding as predicted, requiring troubleshooting.

This level of engagement and consequent cognitive processing foster deeper understanding and longer knowledge retention.

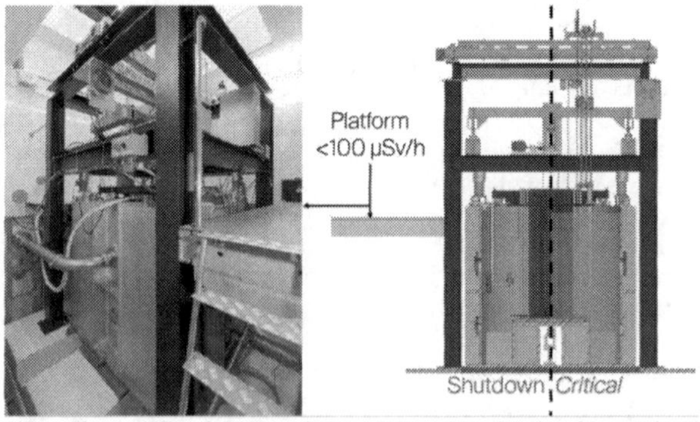

Fig. 1. Photograph and CAD rendering of CROCUS, showing the vessel and surrounding walk-around platform, as well as the vessel in both critical (filled with water) and shutdown (drained) states.

For CROCUS, the concept of *affordance* captures a major educational advantage of the facility [13], [14]. The reactor's design makes systems and components directly reachable and inspectable: vessel internals and fuel can be viewed and approached closely from a platform at vessel level (see Fig. 1); auxiliary systems (water cistern, piping, etc.) are exposed for

https://doi.org/10.1051/epjconf/202533811002

demonstration; and ambient dose rates are low enough to permit extended stays in the cavity by students and visitors. Particularly striking moments, such as the slow, supervised opening of the heavy cavity door (≈1.5 m thick), create a visceral impression that reinforces learning and motivates further inquiry. At shutdown, the absence of water in the vessel expose otherwise hidden internals, producing a strong demonstrative "wow" effect that anchors abstract concepts in tangible reality: "so, this a reactor core".

The facility's overdesign also enables students to work in operational areas, interact with operators in the control room, and observe typical reactor systems, including secondary and tertiary hydraulic circuits. This physical accessibility supports a *trialogical learning approach*, in which small teams interact with a shared artefact to build collective knowledge; an experience that differs qualitatively from purely individual (monological) or discussion-based (dialogical) learning [15].

III. Tangible Models of CROCUS

A. Developing a LEGO® model of CROCUS

CROCUS is a uniquely accessible reactor, yet many students still struggle to visualize its full system layout, even after hands-on experiments and extended visits. Some components and features remain hidden from direct view, and once the visit is over it becomes difficult to discuss systems without a physical reference. Affordance is not absolute: the educational value of the facility can still be enhanced through targeted improvements that increase visibility, manipulability, and alignment with pedagogical goals and proceedings.

The idea of creating a tangible model usable directly where experimental sessions take place, i.e., in the control room, emerged from this statement more than a decade ago. Building on a complete CAD model of CROCUS developed with a former student, a 3D-printed 1:5 scale replica was built by him with the support of the EPFL Additive Manufacturing Workshop (see Fig. 2). This model reproduces CROCUS in detail, down to its individual elements such as the spillway, the thermocouples, and the 500 fuel rods. Whereas the high fidelity is highly effective for discussing visible features in class, it offers little improvement for visualizing hidden components, and its size, fragility, and cost prohibit direct manipulation, limiting its pedagogical usefulness. Inappropriate manipulation of students results in fuel rods and configuration havoc on a regular basis, and even parts breaking oftentimes.

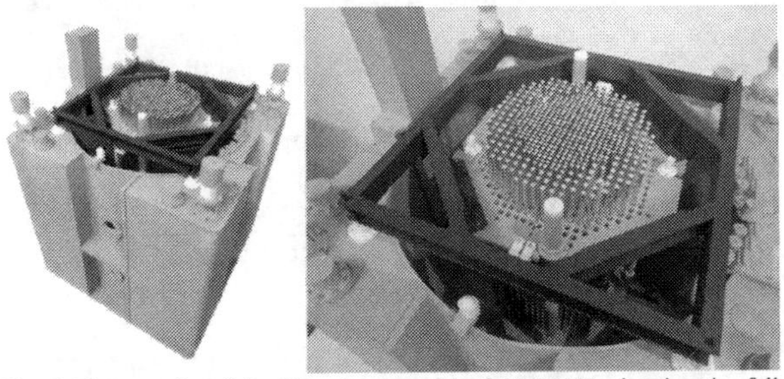

Fig. 2. Photographs of the 3D-printed model of CROCUS, showing the full assembly and close-up of the vessel top, with visible fuel rods, grid and supporting frame.

To overcome these limitations, the design of a LEGO® model was explored with the following objectives:

- Benefit from the availability, durability, and relatively low cost of LEGO® bricks
- Reproduce the overall layout of CROCUS
- Preserve all main features, with meaningful, functional interactions (e.g., moving parts)
- Maximize visualization possibilities (e.g., openable sections)
- Ensure robustness and ease of handling

The development was carried out on personal time, whereas EPFL financed the purchase of the bricks. Two iterations of CAD design were created using LEGO's official BrickLink Studio 2.0 software [27], with parts bought via the associated community platform. This allowed testing of several design options, such as the movable spillway mechanism. The first version proved too fragile, prompting structural reinforcements in the second iteration. The current model can be held and manipulated using one hand without damage, loose parts, or accidental disassembly. If a disassembly happens, building it back remains easy.

B. Features of the LEGO® model of CROCUS

Built from about 650 bricks, the LEGO® model is both durable and fully accessible, allowing users to see exactly how each component fits together and operates. The design strikes a balance between accuracy and functionality: scale, movable parts, and representations of the core, fuel rods, and control mechanisms were all chosen to capture the essence of CROCUS without compromising ease of handling or robustness (see Fig. 3 and Fig. 4). Key features include:

- Removable fuel-rod core with two distinct fuel zones
- Openable reactor vessel for full access to internals
- Sliding boron-carbide absorber rods
- Movable spillway
- Start-up neutron source with shielding and channel
- Exposed hydraulic and instrumentation systems
- Viewing window into the vessel interior

The model is built in modular sections, allowing inspection of individual sections, and operations to be "performed" as in the real reactor: for example, loading and unloading the core from the vessel.

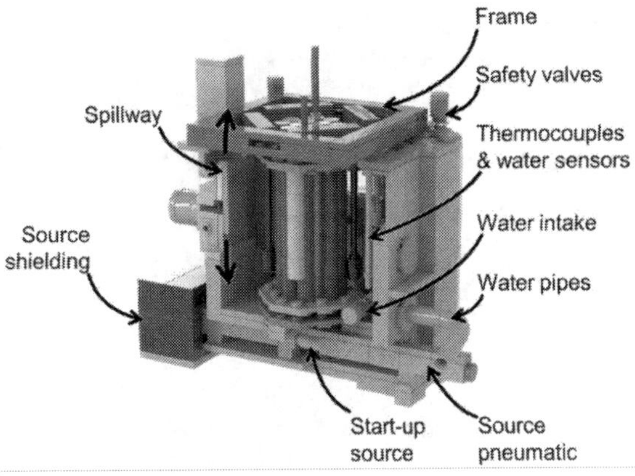

Fig. 3. Cross section of the complete LEGO® model of CROCUS.

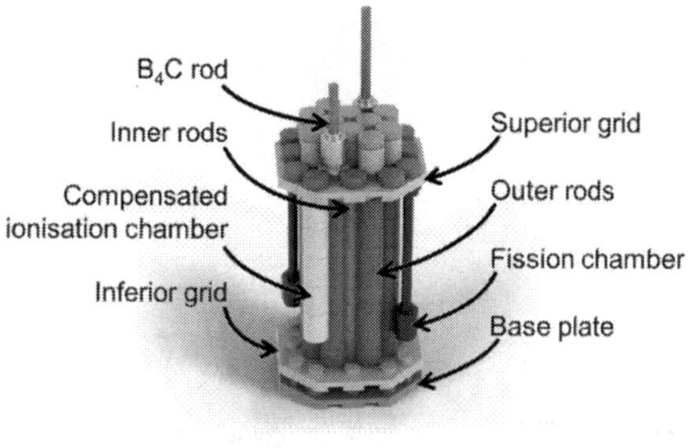

Fig. 4. Unloaded core of the LEGO® model.

IV. Testing the Educational Value of the Model

A. Materials and Methods

Courses and participants

The educational value of the LEGO® CROCUS model was assessed in two courses (see II.A):

- Physics Lab II (Spring 2025): 96 2nd-year BSc physics students taking a single 4-hours "approach to critical" experiment (~8 new students per week for 12 weeks)
- Radiation and Reactor Experiments (Autumn 2024): 28 1st-year MSc students in Nuclear Engineering completing a 13-week sequence of 4-hours experiments

For each courses, students were assigned either a Control group (no LEGO® model) or a Test group (exposed to the LEGO® model). For Physics Lab II, assignment was done by alternating weeks, yielding 51 in the Control condition, and 45 in the Test condition. For Radiation and Reactor Experiments, students were randomly assigned to the Tuesday or Wednesday session: the Tuesday session (no LEGO®) returned 14 answers, the Wednesday session (with LEGO®) returned 13 answers; one student did not participate. The 3D-printed model was available to all groups.

Questionnaire

Assessment used a short questionnaire (available under request only, to allow for more statistics in the coming years) comprising 10 multiple-choice items (i.e., questions) focused on reactor features and instrumentation, plus an open comment field for qualitative feedback on the perceived model's usefulness. The item topics were:

1. Fuel
2. Zero-power
3. Hydraulics
4. Criticality control
5. Start-up source (principle)
6. Start-up source (shielding)
7. Expansion vases
8. Safety blades
9. Power monitoring (principle)
10. Power monitors (instruments)

Each item offered four response options:

- Good: the correct answer
- Incorrect: a plausible-but-incorrect distractor
- "Terrible": a clearly incorrect option
- and "I'm not sure."

The questionnaire was not graded and students were informed that participation would not affect course marks, in order to avoid biases in their answers (cheating, etc.)

Administration and anonymization

Administration differed between courses:

- Physics Lab II: paper questionnaires were distributed and completed at the end of the 4-hour session. Students returned anonymous, hand-written forms (no names or student IDs). The teacher ensured individual and autonomous completion.
- Radiation and Reactor Experiments: the questionnaire was distributed by e-mail after the semester examination period and replies were collected electronically over a long period of time. Identifiers were removed prior to analysis to ensure anonymity. Individual completion was requested and assumed.

We note that the two administration modes (in-person paper vs. post-semester e-mail) may introduce mode and timing effects, and therefore represent two distinct evaluation contexts.

Scoring and analysis

Multiple-choice responses were scored as one of four categories: good, incorrect, "terrible", and unsure. For analysis, we computed i) for each student an overall score per Good, Terrible, and Unsure categories (sum, range 0-10) to assess understanding, misconception, and uncertainty, respectively; and ii) computed item-level percent correct by group. Group differences (Test vs. Control) were assessed using Student t-tests, per answer category and category-aggregated. Statistical significance was evaluated at $\alpha = 0.05$. In practice, this consists in comparing the mean of differences between groups over all items, with the standard deviation of the samples, and comparing the result with a tabulated value for a degree of freedom d_f.

$$t = \frac{\bar{d}}{s_d/\sqrt{n}}, \quad (1)$$

Where :

- n is the number of items, and $d_f = n - 1$
- $\bar{d} = \frac{\sum_i^n d_i}{n}$ is the mean of differences between control and test groups for all items
- $s_d = \sqrt{\left(\sum_i^n (d_i - \bar{d})^2\right)/(n-1)}$ is the standard deviation of differences.

For 10 items, i.e. a degree of freedom of 9, and $\alpha = 0.05$, the critical t-value t_{crit} is 1.833. If $t > t_{\text{crit}}$, the null hypothesis can be rejected, and we may conclude that an effect is present.

Open-ended comments were analyzed qualitatively by coding recurring themes (e.g., improved visualization, suggestions for model improvements, remaining confusions). Qualitative findings complement quantitative results.

Aggregated items and composite score

Because items 5/6 (start-up source principle and shielding) and 9/10 (power monitoring: principle and instruments) probe closely related concepts, we specified aggregated scores (5-6 and 9-10) computed as the mean of the two items. Aggregation reduces the number of independent tests and often increases robustness against single-item noise.

We also computed a composite item-level metric S_i per item defined in percentage-point difference of Good and Terrible values as:

$$S_i = G_i - T_i \quad (2)$$

This composite aims at observing convergent improvement: more correct and fewer wrong answers. The Uncertain values are not included, as their increase could be considered either positive or negative depending on the other categories results.

Results on aggregated items and on composite S-scores are indicated and discussed along other results.

Procedure and data integrity

Participation was voluntary, students gave informed consent. No student identifiers were retained in the analysis dataset. Completed paper forms were digitized immediately after collection; electronic responses were exported and anonymized prior to analysis. Incomplete answers were assigned "unsure".

Limitations

Key limitations include the differing administration modes between the two courses (which may affect response rates and candor), potential non-response bias in the post-semester e-mail survey, teacher variation, language variation, and the relatively small sample size in the Master's course. For Physics Lab II, each of the three teachers took care of an even number of weeks to balance teacher effects; for the Master's course, the Tuesday and Wednesday sessions were taught by two different teachers of equivalent background and interest. Ceiling/floor effects (sum of answers cannot be lower than 0 or higher than 10) are possible given the 0-10 score range and were considered when interpreting results. These limitations are acknowledged in the Results and Discussion hereafter.

B. *Results and Discussion*

All item-level outcomes are reported in Fig. 5, on the following page; per-student item counts by category are provided in Appendix A. For clarity, we remind the reader that for each student we scored the 10 multiple-choice items into three categories: Good (correct answer), Terrible (clearly incorrect option), and Unsure ("I'm not sure"). For each student we computed the per-category totals (integer 0-10). Group summary statistics (means and standard deviations) and group differences (Test − Control, in percentage points) are reported in Table I below. Wherever appropriate we report differences as percentage-point changes; positive values indicate either better (Good, Unsure) or worsened performance (Terrible) in the Test group (the group exposed to the LEGO® model). In the table, green is used for positive impact, orange for negative, "white" for negligible.

Three complementary levels of analysis were carried out:

- Descriptive, distribution-based of Control vs. Test for the three core categories Good/Terrible/Unsure, presented in Table I, to identify the general trends, and the statistical qualities.
- Descriptive, item-by-item comparisons (visualized in Fig. 5) to identify which concepts were most affected by exposure to the model.
- Statistical testing on item-level results, based on Student t-tests with corresponding assumptions, for all possible cases: aggregated or not, Good, Terrible, Uncertain, and Composite scores.
- Analysis of comments, rate of positive and negative feedbacks, and identification of themes and aspects mentioned.

Distribution analysis

Physics Lab II (BSc, n = 96).

Baseline performance is moderate: students averaged 5.0 correct answers out of 10 (50%), with a standard deviation (SD) of about 21 percentage points. The most notable model-related change is a reduction in the rate of Unsure responses: the Test group shows a drop of roughly 5 percentage points in unsure responses compared to Control, with a corresponding lowered standard deviation of −4%. Mean changes in Good and Terrible rates are small (on the order of ±0.3 percentage points) and the overall variability across students is substantial. In other words, at the Bachelor level the LEGO® model appears to reduce students' uncertainty about some reactor concepts without markedly increasing the absolute correct-answer rate.

Radiation & Reactor Experiments (MSc, n = 28).

Baseline performance is higher among Master students: mean correct is about 8.0/10 (80%), and SD about 13 percentage points. As compared to BSc, the Test group shows a more pronounced and consistent improvement: mean Good-answer rate increased by approximately +7 percentage points, accompanied by a similar reduction in Terrible answers, about −7 percentage points. Corresponding standard deviations decreased modestly (≈ 3-5 percentage points), indicating a somewhat tighter distribution of responses in the Test group. The rate of Uncertain answers is unaffected. Overall, the LEGO® model has a stronger and more positive effect at the Master level than at the Bachelor level.

TABLE I
CHARACTERISTICS OF THE RESULTS DISTRIBUTIONS AND CORRESPONDING DIFFERENCES

Category	Quantity	Physics Lab II: All students (/10; %)	Physics Lab II: Students without (/10; %)	Physics Lab II: Students with LEGO® (/10; %)	Physics Lab II: Difference (/10; %)	Radiation and Reactor Experiments: All students (/10; %)	Radiation and Reactor Experiments: Students without (/10; %)	Radiation and Reactor Experiments: Students with LEGO® (/10; %)	Radiation and Reactor Experiments: Difference (/10; %)
Good answers	Average	5.0	4.9	5.1	+0.3	8.0	7.6	8.4	+0.7
	Standard dev.	21%	21%	21%	0%	13%	14%	11%	−3%
Terrible answers	Average	1.5	1.3	1.6	+0.3	0.4	0.8	0.1	−0.7
	Standard dev.	11%	11%	11%	0%	7%	8%	3%	−5%
Uncertain answers	Average	1.6	1.8	1.3	−0.5	0.6	0.6	0.6	0
	Standard dev.	17%	19%	14%	−4%	7%	8%	7%	−1%

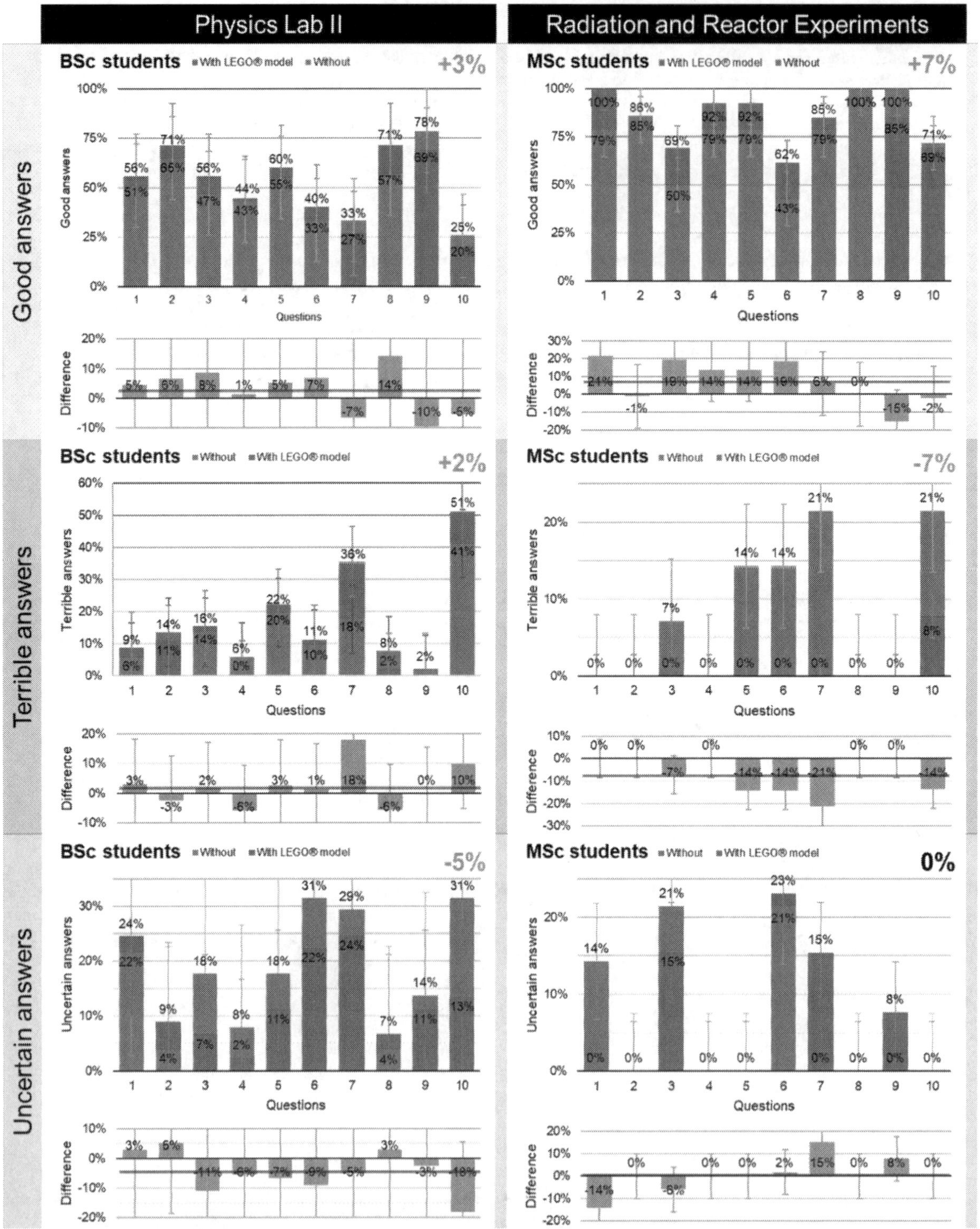

Fig. 5. Rates and differences of Good (top), Terrible (middle) and Uncertain (bottom) answers for each item, between students with (Test, red) and without (Control, gray) the LEGO® model, for the Physics Lab II (BSc students, left), and the Radiation and Reactor Experiments (MSc, right). Indicated error bars correspond to the sample standard deviation, thus assuming a normal distribution for all cases. The mean of differences is indicated as well in the top right corner of each case, with the coloring indicating benefit (green), negative impact (orange), negligible effect (black).

 https://doi.org/10.1051/epjconf/202533811002

Item-level analysis

Fig. 5 presents item-by-item rates (Test vs. Control) and the Test − Control differences for both courses.

Physics Lab II (BSc):

- The Test group shows mild improvements on most items; exceptions are items 7, 9 and 10 where either no improvement or a small adverse change is observed.
- Items 7 and 10, in particular, display increased rates of Terrible answers in the Test group (+18 and +10 percentage points respectively) and therefore deserve follow-up investigation (possible confusion introduced by the model or by the way those concepts were taught).
- The strongest and most consistent effect across items is the reduction in Unsure responses, suggesting improved confidence or clarity even where the correct-answer rate does not change strongly.

Radiation and Reactor Experiments (MSc):

- The Test group exhibits improved Good rates on nearly all items; notable exceptions with no improvement are items 2, 8 and 10, and item 9 shows a sensible negative change (−15%).
- The already low incidence of Terrible answers was substantially reduced for non-zero items (3, 5, 6, 7, 10) in the Test group, to the point that Terrible answers were effectively absent for all items but 10.
- Low Unsure rates were largely unchanged, except for a decrease on item 1 and an increase on item 7.

These item-level patterns suggest the LEGO® model supports conceptual consolidation for many topics (visible as increased Good answers and decreased Terrible answers among Master Students), while for some concepts it either had no effect or may have introduced confusion (notably a small set of items in both courses). More importantly, the topics most concerned by the improvements are the hidden and atypical features targeted by the model: spillway, hydraulics, source, and expansion vases (especially wrt. Terrible); and more surprisingly, the fuel (2-zone configuration, Good and Uncertain).

Statistical tests

The Student t-test was applied to all comparisons. Distribution checks indicate that the normality hypothesis is reasonable for the Good answers, but not demonstrated for the Terrible category (low statistics), and less convincing for the Uncertain one (skewed, peaked at low values). These would require further checks, e.g. Mann-Whitney U. Nevertheless, the t-test should still allow for the results presented hereafter.

Despite the expected negligible or positive impact of the model, we cannot exclude a priori an effect in either direction. Thus we report both one-sided (unipolar) and two-sided (bipolar) tests. The relevant critical t values at $\alpha = 0.05$ are:

- for 9 degrees of freedom: 1.833 (uni) and 2.262 (bi),
- for 7 degrees of freedom: 1.895 (uni) and 2.365 (bi).

Results for all comparisons and aggregated scores are summarized in Table II.

Physics Lab II (BSc):

For the Bachelor course, the only statistically significant change is a reduction in the Uncertain category, and this significance holds only for the one-sided test. Aggregating items does not strengthen the effect: there is no detectable impact on the Good or Terrible categories, and the composite score does not indicate a significant overall change.

Radiation and Reactor Experiments (MSc)

For the Master course, disregarding Uncertain (no clear effect), consistent improvements are observed. The one-sided tests show significant effects for both the Good and Terrible categories, which is coherent with the hypothesis that the model helps clarify concepts (increasing correct answers and reducing clear misunderstandings). Under the two-sided test, the effect for Good is not significant, while the effect for Terrible is substantial, near the two-sided critical value. The Composite score yields the strongest result: it is statistically significant under both one-sided and two-sided tests.

Comments

The analysis indicates that 60% and 92% of the students found the model useful, in Physics Lab II and Radiation and Reactor Experiments, respectively; 20% and 8% found it not useful (with an additional 20% of BSc students not answering).

The main positive feedback (see Table III) highlights how the model helps in visualizing the core (16/6 comments for BSc/Msc) thanks to its simplification (4/2), its ability to provide a clear overview (2/3) at reduced scale (1/1), and the possibility of opening it (2/1) to see inside and reveal hidden features (4/2) such as the expansion tanks (0/2), the neutron source (0/1), or the neutron detectors (2/0). Other points raised include its engaging format (1/4), the advantage of being able to use during experiments in class and in interaction (1/3), and, explicitly, its manipulability (2/3). Finally, only among MSc students (who answered later), 2 noted that the model helped in memorizing the features, thus supporting knowledge retention.

Negative feedback was limited. The model was judged unnecessary because the rest of the material was considered sufficient (2/1), or due to lack of time to use it (1/0). A final comment mentioned the lack of colors as drawback (1/0).

TABLE II
T-VALUES AND STATISTICAL SIGNIFICANCE FOR ALL CATEGORIES, SCORES, AND HYPOTHESES (t_{crit}: 1.833/2.262 (d_f=9); 1.895/2.365 (d_f=7))

	Physics Lab II						Radiation and Reactor Experiments					
Category	t(9)	unipolar	bipolar	$t_{agg}(7)$	unipolar	bipolar	t(9)	unipolar	bipolar	$t_{agg}(7)$	unipolar	bipolar
Good	1.11	No	No	1.36	No	No	2.07	Yes	No	2.31	Yes	No
Terrible	1.06	No	No	0.78	No	No	2.88	Yes	Yes	2.34	Yes	*No*
Uncertain	2.23	Yes	No	1.74	No	No	0.19	No	No	0.01	No	No
Composite	0.06	No	No	0.30	No	No	**3.07**	**Yes**	**Yes**	**3.19**	**Yes**	**Yes**

TABLE III
ANALYSIS OF STUDENTS' COMMENTS

Theme	BSc comments	MSc comments
Visualize core / internal structure	16	6
Manipulable / interactive	3	4
Simplifies / highlights essentials	4	2
Engaging	1	4
Helps memorization	0	2
Not useful: material sufficient	2	1
Not useful: no time	1	0
Criticism: lack of color	1	0

V. CONCLUSIONS

CROCUS is an exceptional tool for active learning thanks to its unique affordance. Still, several of its features remain difficult for students to internalize, due to system complexity, and partial concealment of components. To address this, we designed and built a sturdy and manipulable LEGO® model to complement direct facility access and hands-on experiments, enabling further exploration in line with active learning principles and use, especially for the nuclear engineering education.

We evaluated the model in two hands-on courses with different structures and student levels. In the BSc course, observed effects were limited, likely due to a large variability in that population (SD ≈ 20%). In the MSc course, however, item-level comparisons and statistical testing indicate a clear pedagogical benefit: the rate of correct answers increased while the incidence of clearly incorrect answers decreased. The model increased the composite score by +14.6% ($t(7) = 3.19$, 95% confidence interval: +3 to +26%; $p = 0.015$, two-sided).

Additionally, the students' feedback in the open-ended comments supports the model's pedagogical value for conveying internal structure and facilitating interaction during experiments, as well as enhancing memorization, while suggesting minor usability improvements. Consequently, we recommend the use of suitable tangible models, including in hands-on experimental contexts.

Looking ahead, if time and team allow, the study will be continued over coming academic years to improve statistical confidence, better quantify the impact for MSc students, and test for effects in the BSc course. A power-based back-of-the envelope estimate suggests that, if the effect size observed for MSc students also applies to BSc level, a sample on the order of 200-300 students would be required to detect it reliably.

ACKNOWLEDGMENT

The author would like to thank Elisa Cattaruzza for her guidance on pedagogical theory and its application; LRS for financial support of this rather unusual project; Michael Chadwick for designing and building the 3D-printed model; the colleagues Thomas Guilbaud and Michel Saliba for using the model in class and collecting questionnaires; the Physics and Nuclear Engineering students for their contribution to the study and for promoting the model and the LEGO Ideas® project; colleagues for patiently enduring countless conversations on this overfocused topic; LEGO® for providing the author's early mechanical education; their two children for inspiring the work; and their wife for her understanding of the many hours spent, and countless pieces scattered around the house.

APPENDIX A – ANSWER DISTRIBUTIONS PER STUDENT

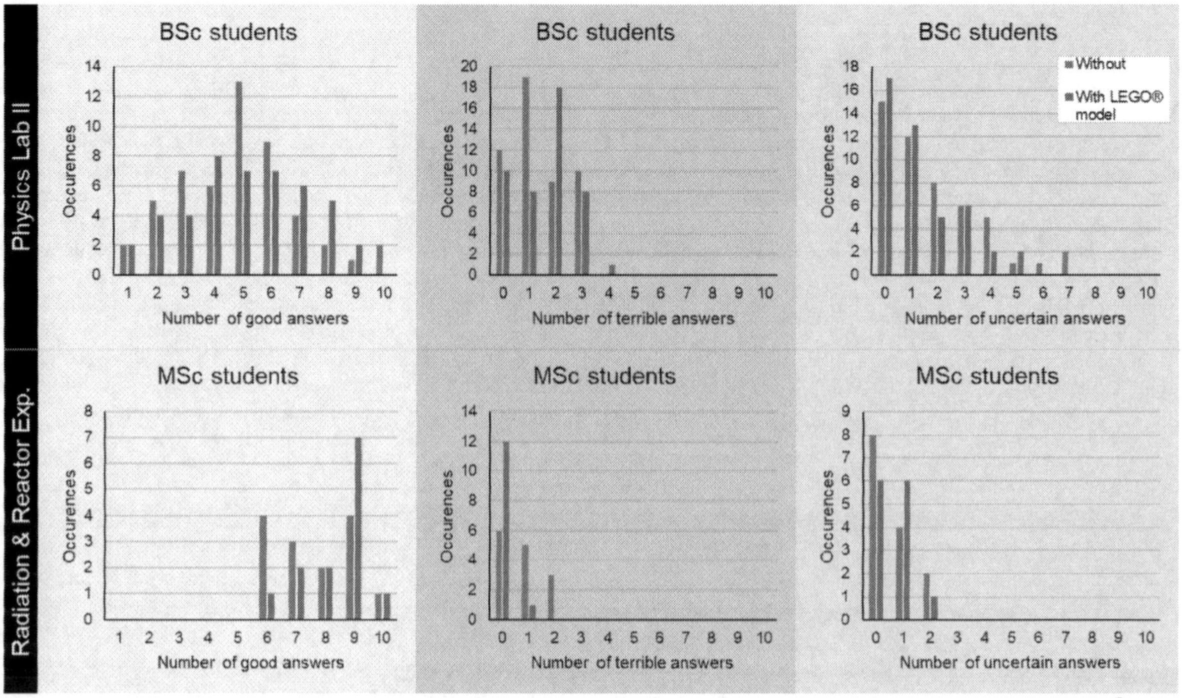

Fig. 6. Distributions of number of answers per category (green/Good, red/Terrible, blue/Uncertain), with (red/Test) and without (gray/Control) the LEGO® model.

REFERENCES

[1] R. Garbil, S. Ben Hadj Hassine, P. Blaise, C. Ferry, G. L. Pavel, J. Starflinger, C. Demazière, K. Simola, M. Němec, and Š. Čerba, "Education, training and mobility, knowledge management: towards a common effort to ensure a future workforce in Europe and abroad," *EPJ Nuclear Sciences and Technologies*, vol. 9. EDP Sciences, 2023.

[2] "Nuclear Educational Networks: Experience Gained and Lessons Learned," IAEA-TECDOC-2007, 2022.

[3] OECD/NEA and IAEA, "Measuring Employment Generated by the Nuclear Power Sector," 2018.

[4] Global Forum on Nuclear Education Science Technology and Policy, "Strengthening the future of the nuclear sector's workforce," *OECD/NEA*, 16-Mar-2023. [Online]. Available: https://oecd-nea.org/jcms/pl_79281/strengthening-the-future-of-the-nuclear-sector-s-workforce. [Accessed: 13-Aug-2025].

[5] OECD/NEA, *Nuclear Education and Training: From Concern to Capability*. Paris: OECD Publishing, 2012.

[6] OECD/NEA, "The demise of zero power reactors: From concern to action (workshop)," 22-Jun-2023. [Online]. Available: https://www.oecd-nea.org/jcms/pl_79783/the-demise-of-zero-power-reactors-from-concern-to-action. [Accessed: 13-Aug-2025].

[7] V. Lamirand, M. Hursin, G. Perret, P. Frajtag, O. Pakari, and A. Pautz, "Future experimental programmes in the CROCUS reactor," in *RRFM/IGORR 2016, Berlin (Germany) 13-17 March*, 2016, no. 02–2016, pp. 284–292.

[8] V. Lamirand, "Ten springs of experiments in CROCUS," *EPJ Web Conf*, vol. 288, p. 04026, 2023.

[9] V. Lamirand and A. Pautz, "A New Lecture on Radiation Detection at EPFL," in *ANIMMA 2019*, 2019, no. June, p. 2019.

[10] C. Demazière, C. Stöhr, Y. Zhang, O. Cabellos, S. Dulla, N. Garcia-Herranz, R. Miró, R. Macian, M. Szieberth, C. Lange, M. Hursin, and S. Strola, "Enhancing higher education through hybrid and flipped learning: Experiences from the GRE@T-PIONEeR project," *Nuclear Engineering and Design*, vol. 421, May 2024.

[11] J. M. Paratte, R. Früh, U. Kasemeyer, M. a. Kalugin, W. Timm, and R. Chawla, "A benchmark on the calculation of kinetic parameters based on reactivity effect experiments in the CROCUS reactor," *Ann Nucl Energy*, vol. 33, no. 8, pp. 739–748, May 2006.

[12] U. Kasemeyer, R. Früh, J. M. Paratte, and R. Chawla, "Benchmark on Kinetic Parameters in the CROCUS Reactor," in *International Reactor Physics Experiments Handbook (IRPhE)*, no. 4440, OECD, Ed. 2007, p. 94.

[13] P. C. Lippman and E. Matthews, "Affordance Theory in Relationship to the Learning Environment," in *Encyclopedia of Educational Innovation*, R. Peters Michael A. and Heraud, Ed. Singapore: Springer Singapore, 2020, pp. 1–6.

[14] G. Kalmpourtzis and M. Romero, "An affordance-based framework for the design and analysis of learning activities in playful educational robotics contexts," *Interactive Learning Environments*, vol. 32, no. 6, pp. 2744–2757, 2024.

[15] A. Iannaccone and E. Cattaruzza, *Expériences sociomatérielles - Objets, interactions, espaces*, Editions Alphil. Presses Universitaires Suisses, 2024.

[16] F. Vitullo, "Miniature and Minimalistic Neutron Detectors for Online High-Resolution Experiments in the Zero-Power Reactor CROCUS," PhD thesis, EPFL, Lausanne, 2022.

[17] F. Vitullo, V. Lamirand, K. Ambrožič, L. Braun, D. Godat, P. Frajtag, and A. Pautz, "Design of a 150-miniature detectors 3D core-mapping system for the CROCUS reactor," *EPJ Web Conf*, vol. 253, p. 04023, Nov. 2021.

[18] T. Ligonnet, N. Weiss, V. Lamirand, A. Laureau, and A. Pautz, "POLLEN: A Pile-Oscillator for the BLOOM Experimental Program," *EPJ Web Conf*, vol. 288, p. 04018, 2023.

[19] T. Ligonnet, A. Laureau, P. Simbsler, A. Pautz, and V. Lamirand, "Design of an Open-Loop Pile-Oscillation Program in the CROCUS Reactor," *IEEE Trans Nucl Sci*, vol. 71, no. 5, pp. 1001–1009, May 2024.

[20] T. Ligonnet, D. Clément, N. Weiss, G. Noguère, A. Laureau, O. Pakari, A. Dupont-Bembinoff, A. Pautz, and V. Lamirand, "Pile-oscillation results for BLOOM with local and 3D global measurements," in *ANIMMA-2025*, 2025.

[21] M. Saliba, V. Lamirand, W. Monange, F. Vitullo, M. Hursin, A. Pautz, and O. Pakari, "Measurement of the Prompt Neutron Decay Constant in a Zero-Power Reactor Using a Novel 3D Detector System," *Nuclear Science and Engineering*, pp. 1–9, Mar. 2025.

[22] M. Saliba, E. Lehmann, P. Trtik, V. Lamirand, A. Pautz, and O. Pakari, "New neutron imaging station at the CROCUS zero-power reactor," in *ANIMMA-2025*, 2025.

[23] C. Demazière, P. Vinai, M. Hursin, S. Kollias, and J. Herb, "Overview of the CORTEX project," in *Proceedings of PHYSOR 2018*, 2018, pp. 2971–2980.

[24] S. Freeman, S. L. Eddy, M. McDonough, M. K. Smith, N. Okoroafor, H. Jordt, and M. P. Wenderoth, "Active learning increases student performance in science, engineering, and mathematics," *Proc Natl Acad Sci U S A*, vol. 111, no. 23, pp. 8410–8415, Jun. 2014.

[25] C. Demazière, "Using active learning in hybrid learning environments," in *International Conference on Physics of Reactors: Transition to a Scalable Nuclear Future, PHYSOR 2020*, 2020, vol. 2020-March, pp. 2419–2429.

[26] S. U. Egarievwe, "Development of Active-Learning Units in Nuclear Engineering," *EPJ Web Conf*, vol. 253, p. 10003, 2021.

[27] The LEGO Group, "BrickLink Studio," *https://www.bricklink.com/v3/studio/download.page*. 2018.

AUTHOR INDEX